DYNAMICS OF STRUCTURES

DYNAMICS OF STRUCTURES
Theory and Applications to Earthquake Engineering

Anil K. Chopra

University of California at Berkeley

Fourth Edition

Prentice Hall

Boston Columbus Indianapolis New York San Francisco Upper Saddle River
Amsterdam Cape Town Dubai London Madrid Milan Munich Paris Montréal Toronto
Delhi Mexico City São Paulo Sydney Hong Kong Seoul Singapore Taipei Tokyo

Vice President and Editorial Director, ECS:
   Marcia J. Horton
Executive Editor: *Holly Stark*
Vice President, Production: *Vince O'Brien*
Senior Managing Editor: *Scott Disanno*

Art Director: *Jayne Conte*
Art Editor: *Greg Dulles*
Cover Design: *Bruce Kenselaar*
Manufacturing Buyer: *Lisa McDowell*
Executive Marketing Manager: *Tim Galligan*

Cover Photo: Transamerica Building, San Francisco, California. The motions shown are accelerations recorded during the Loma Prieta earthquake of October 17, 1989 at basement, twenty-ninth floor, and forty-ninth floor. Courtesy Transamerica Corporation.

Credits and acknowledgments for material from other sources and reproduced, with permission, in this textbook appear on appropriate page within text.

Many of the designations by manufacturers and seller to distinguish their products are claimed as trademarks. Where those designations appear in this book, and the publisher was aware of a trademark claim, the designations have been printed in initial caps or all caps.

The author and publisher of this book have used their best efforts in preparing this book. These efforts include the development, research, and testing of the theories and programs to determine their effectiveness. The author and publisher make no warranty of any kind, expressed or implied, with regard to these programs or the documentation contained in this book. The author and publisher shall not be liable in any event for the incidental or consequential damages in connection with, or arising out of, the furnishing, performance, or use of these programs.

Library of Congress Cataloging-in-Publication Data on File

6 7 8 9 10 11 12 13 14 V092 18 17 16 15 14

Prentice Hall
is an imprint of

www.pearsonhighered.com

ISBN 10:   0-13-285803-7
ISBN 13: 978-0-13-285803-8

Dedicated to Hamida and Nasreen with gratitude for suggesting the idea of working on a book and with appreciation for patiently enduring and sharing these years of preparation with me. Their presence and encouragement made this idea a reality.

Overview

Contents

Foreword

The need for a textbook on earthquake engineering was first pointed out by the eminent consulting engineer, John R. Freeman (1855–1932). Following the destructive Santa Barbara, California earthquake of 1925, he became interested in the subject and searched the Boston Public Library for relevant books. He found that not only was there no textbook on earthquake engineering, but the subject itself was not mentioned in any of the books on structural engineering. Looking back, we can see that in 1925 engineering education was in an undeveloped state, with computing done by slide rule and curricula that did not prepare the student for understanding structural dynamics. In fact, no instruments had been developed for recording strong ground motions, and society appeared to be unconcerned about earthquake hazards.

In recent years books on earthquake engineering and structural dynamics have been published, but the present book by Professor Anil K. Chopra fills a niche that exists between more elementary books and books for advanced graduate studies. The author is a well-known expert in earthquake engineering and structural dynamics, and his book will be valuable to students not only in earthquake-prone regions but also in other parts of the world, for a knowledge of structural dynamics is essential for modern engineering. The book presents material on vibrations and the dynamics of structures and demonstrates the application to structural motions caused by earthquake ground shaking. The material in the book is presented very clearly with numerous worked-out illustrative examples, so that even a student at a university where such a course is not given should be able to study the book on his or her own time. Readers who are now practicing engineering should have no difficulty in studying the subject by means of this book. An especially interesting feature of the book is the application of structural dynamics theory to important issues in the seismic response and design of multistory buildings. The information presented in this book will be of special value to those engineers who are engaged in actual seismic design and want to improve their understanding of the subject.

Although the material in the book leads to earthquake engineering, the information presented is also relevant to wind-induced vibrations of structures, as well as man-made motions such as those produced by drophammers or by heavy vehicular traffic. As a text-book on vibrations and structural dynamics, this book has no competitors and can be rec-ommended to the serious student. I believe that this is the book for which John R. Freeman was searching.

George W. Housner
California Institute of Technology

Preface

PHILOSOPHY AND OBJECTIVES

This book on dynamics of structures is conceived as a textbook for courses in civil engineering. It includes many topics in the theory of structural dynamics, and applications of this theory to earthquake analysis, response, design, and evaluation of structures. No prior knowledge of structural dynamics is assumed in order to make this book suitable for the reader learning the subject for the first time. The presentation is sufficiently detailed and carefully integrated by cross-referencing to make the book suitable for self-study. This feature of the book, combined with a practically motivated selection of topics, should interest professional engineers, especially those concerned with analysis and design of structures in earthquake country.

In developing this book, much emphasis has been placed on making structural dynamics easier to learn by students and professional engineers because many find this subject to be difficult. To achieve this goal, the presentation has been structured around several features: The mathematics is kept as simple as each topic will permit. Analytical procedures are summarized to emphasize the key steps and to facilitate their implementation by the reader. These procedures are illustrated by over 120 worked-out examples, including many comprehensive and realistic examples where the physical interpretation of results is stressed. Some 500 figures have been carefully designed and executed to be pedagogically effective; many of them involve extensive computer simulations of dynamic response of structures. Photographs of structures and structural motions recorded during earthquakes are included to relate the presentation to the real world.

The preparation of this book has been inspired by several objectives:

- Relate the structural idealizations studied to the properties of real structures.
- Present the theory of dynamic response of structures in a manner that emphasizes physical insight into the analytical procedures.
- Illustrate applications of the theory to solutions of problems motivated by practical applications.
- Interpret the theoretical results to understand the response of structures to various dynamic excitations, with emphasis on earthquake excitation.
- Apply structural dynamics theory to conduct parametric studies that bring out several fundamental issues in the earthquake response, design, and evaluation of multistory buildings.

This mode of presentation should help the reader to achieve a deeper understanding of the subject and to apply with confidence structural dynamics theory in tackling practical problems, especially in earthquake analysis, design, and evaluation of structures, thus narrowing the gap between theory and practice.

EVOLUTION OF THE BOOK

Since the book first appeared in 1995, it has been revised and expanded in several ways, resulting in the second edition (2001) and third edition (2007). Prompted by an increasing number of recordings of ground motions in the proximity of the causative fault, Chapter 6 was expanded to identify special features of near-fault ground motions and compare them with the usual far-fault ground motions. Because of the increasing interest in seismic performance of bridges, examples on dynamics of bridges and their earthquake response were added in several chapters. In response to the growing need for simplified dynamic analysis procedures suitable for performance-based earthquake engineering, Chapter 7 was expanded to provide a fuller discussion relating the earthquake-induced deformations of inelastic and elastic systems, and to demonstrate applications of the inelastic design spectrum to structural design for allowable ductility, displacement-based design, and seismic evaluation of existing structures. Chapter 19 (now Chapter 20) was rewritten completely to incorporate post-1990 advances in earthquake analysis and response of inelastic buildings. Originally limited to three building codes—United States, Canada, and Mexico—Chapter 21 (now Chapter 22) was expanded to include the Eurocode. The addition of Chapter 22 (now Chapter 23) was motivated by the adoption of performance-based guidelines for evaluating existing buildings by the structural engineering profession.

In response to reader requests, the frequency-domain method of dynamic analysis was included, but presented as an appendix instead of weaving it throughout the book. This decision was motivated by my goal to keep the mathematics as simple as each topic permits, thus making structural dynamics easily accessible to students and professional engineers.

WHAT'S NEW IN THIS EDITION

Dynamics of Structures has been well received in the 16 years since it was first published. It continues to be used as a textbook at universities in the United States and many other countries, and enjoys a wide professional readership as well. Translations in Japanese, Korean, Chinese, Greek, and Persian have been published. Preparation of the fourth edition provided me with an opportunity to improve, expand, and update the book.

    Chapter 14 has been added, requiring renumbering of Chapters 14 to 22 as 15 to 23 (the new numbering is reflected in the rest of the Preface); Chapters 5 and 16 underwent extensive revision; Chapters 12 and 13 have been expanded; and Chapters 22 and 23 have been updated. Specific changes include:

- Chapter 14 on nonclassically damped systems has been added. This addition has been motivated by growing interest in such systems that arise in several practical situations: for example, structures with supplemental energy-dissipating systems or on a base isolation system, soil–structure systems, and fluid-structure systems.

- Chapters 5 and 16 on numerical evaluation of dynamic response have been rewritten to conform with the ways these numerical methods are usually implemented in computer software, and to offer an integrated presentation of nonlinear static analysis—also known as pushover analysis—and nonlinear dynamic analysis.

- A section has been added at the end of Chapter 12 to present a general version of the mode acceleration superposition method for more complex excitations, such as wave forces on offshore drilling platforms.

- Chapter 13 has been extended to include two topics that so far have been confined to the research literature, but are of practical interest: (1) combining peak responses of a structure to individual translational components of ground motion to estimate its peak response to multicomponent excitation; and (2) response-spectrum-based equations to determine an envelope that bounds the joint response trajectory of all simultaneously acting forces that control the seismic design of a structural element.

- Chapters 22 and 23 have been updated to reflect the current editions of building codes for designing new buildings, and of performance-based guidelines and standards for evaluating existing buildings.

- The addition of Chapter 14 prompted minor revision of Chapters 2, 4, 6, 10, and 12.

- Several new figures, photographs, worked-out examples, and end-of-chapter problems have been added.

Using the book in my teaching and reflecting on it over the years suggested improvements. The text has been clarified and polished throughout, and a few sections have been reorganized to enhance the effectiveness of the presentation.

SUBJECTS COVERED

This book is organized into three parts: I. Single-Degree-of-Freedom Systems; II. Multi-Degree-of-Freedom Systems; and III. Earthquake Response, Design, and Evaluation of Multistory Buildings.

Part I includes eight chapters. In the opening chapter the structural dynamics problem is formulated for simple elastic and inelastic structures, which can be idealized as single-degree-of-freedom (SDF) systems, and four methods for solving the differential equation governing the motion of the structure are reviewed briefly. We then study the dynamic response of linearly elastic systems (1) in free vibration (Chapter 2), (2) to harmonic and periodic excitations (Chapter 3), and (3) to step and pulse excitations (Chapter 4). Included in Chapters 2 and 3 is the dynamics of SDF systems with Coulomb damping, a topic that is normally not included in civil engineering texts, but one that has become relevant to earthquake engineering, because energy-dissipating devices based on friction are being used in earthquake-resistant construction. After presenting numerical time-stepping methods for calculating the dynamic response of SDF systems (Chapter 5), the earthquake response of linearly elastic systems and of inelastic systems is studied in Chapters 6 and 7, respectively. Coverage of these topics is more comprehensive than in texts presently available; included are details on the construction of response and design spectra, effects of damping and yielding, and the distinction between response and design spectra. The analysis of complex systems treated as generalized SDF systems is the subject of Chapter 8.

Part II includes Chapters 9 through 18 on the dynamic analysis of multi-degree-of-freedom (MDF) systems. In the opening chapter of Part II the structural dynamics problem is formulated for structures idealized as systems with a finite number of degrees of freedom and illustrated by numerous examples; also included is an overview of methods for solving the differential equations governing the motion of the structure. Chapter 10 is concerned with free vibration of systems with classical damping and with the numerical calculation of natural vibration frequencies and modes of the structure. Chapter 11 addresses several issues that arise in defining the damping properties of structures, including experimental data—from forced vibration tests on structures and recorded motions of structures during earthquakes—that provide a basis for estimating modal damping ratios, and analytical procedures to construct the damping matrix, if necessary. Chapter 12 is concerned with the dynamics of linear systems, where the classical modal analysis procedure is emphasized. Part C of this chapter represents a "new" way of looking at modal analysis that facilitates understanding of how modal response contributions are influenced by the spatial distribution and the time variation of applied forces, leading to practical criteria on the number of modes to include in response calculation. In Chapter 13, modal analysis procedures for earthquake analysis of classically damped systems are developed; both response history analysis and response spectrum analysis procedures are presented in a form that provides physical interpretation; the latter procedure estimates the peak response of MDF systems directly from the earthquake response or design spectrum. The procedures are illustrated by numerous examples, including coupled lateral-torsional response of unsymmetric-plan buildings and torsional response of nominally symmetric buildings. The chapter ends with response spectrum-based procedures to consider all simultaneously acting forces that control the design of a structural element, and to estimate the peak response of a structure to multicomponent earthquake excitation. The modal analysis procedure is extended in Chapter 14 to response history analysis; of nonclassically damped systems subjected to earthquake excitation. For this purpose, we first revisit classically damped systems and re-cast the analysis procedures of Chapters 10 and 13 in a form that facilitates their extension to the more general case.

Chapter 15 is devoted to the practical computational issue of reducing the number of degrees of freedom in the structural idealization required for static analysis in order to recognize that the dynamic response of many structures can be well represented by their first few natural vibration modes. In Chapter 16 numerical time-stepping methods are presented for MDF systems not amenable to classical modal analysis: systems with nonclassical damping or systems responding into the range of nonlinear behavior. Chapter 17 is concerned with classical problems in the dynamics of distributed-mass systems; only one-dimensional systems are included. In Chapter 18 two methods are presented for discretizing one-dimensional distributed-mass systems: the Rayleigh–Ritz method and the finite element method. The consistent mass matrix concept is introduced, and the accuracy and convergence of the approximate natural frequencies of a cantilever beam, determined by the finite element method, are demonstrated.

Part III of the book contains five chapters concerned with earthquake response design, and evaluation of multistory buildings, a subject not normally included in structural dynamics texts. Several important and practical issues are addressed using analytical procedures developed in the preceding chapters. In Chapter 19 the earthquake response of linearly elastic multistory buildings is presented for a wide range of two key parameters: fundamental natural vibration period and beam-to-column stiffness ratio. Based on these results, we develop an understanding of how these parameters affect the earthquake response of buildings and, in particular, the relative response contributions of the various natural modes, leading to practical information on the number of higher modes to include in earthquake response calculations. Chapter 20 is concerned with the important subject of earthquake response of multistory buildings deforming into their inelastic range. Part A of the chapter presents rigorous nonlinear response history analysis; identifies the important influence of modeling assumptions, key structural parameters, and ground motion details on seismic demands; and determines the strength necessary to limit the story ductility demands in a multistory building. Recognizing that rigorous nonlinear response history analysis remains an onerous task, the modal pushover analysis (MPA) procedure—an approximate analysis procedure—is developed in Part B of the chapter. In this procedure, seismic demands are estimated by nonlinear static analyses of the structure subjected to modal inertia force distributions. Base isolation is the subject of Chapter 21. Our goal is to study the dynamic behavior of buildings supported on base isolation systems with the limited objective of understanding why and under what conditions isolation is effective in reducing the earthquake-induced forces in a structure. In Chapter 22 we present the seismic force provisions in four building codes—*International Building Code* (United States), *National Building Code of Canada, Eurocode,* and *Mexico Federal District Code*—together with their relationship to the theory of structural dynamics developed in Chapters 6, 7, 8, and 13. Subsequently, the code provisions are evaluated in light of the results of dynamic analysis of buildings presented in Chapters 19 and 20. Performance-based guidelines and standards for evaluating existing buildings consider inelastic behavior explicitly in estimating seismic demands at low performance levels, such as life safety and collapse prevention. In Chapter 23, selected aspects of the nonlinear dynamic procedure and of the nonlinear static procedure in these documents—ATC-40, FEMA 356, and ASCE 41-06—are presented and discussed in light of structural dynamics theory developed in Chapters 7 and 20.

A NOTE FOR INSTRUCTORS

This book is suitable for courses at the graduate level and at the senior undergraduate level. No previous knowledge of structural dynamics is assumed. The necessary background is available through the usual courses required of civil engineering undergraduates. These include:

- Static analysis of structures, including statically indeterminate structures and matrix formulation of analysis procedures (background needed primarily for Part II)
- Structural design
- Rigid-body dynamics
- Mathematics: ordinary differential equations (for Part I), linear algebra (for Part II), and partial differential equations (for Chapter 17 only)

By providing an elementary but thorough treatment of a large number of topics, this book permits unusual flexibility in selection of the course content at the discretion of the instructor. Several courses can be developed based on the material in this book. Here are a few examples.

Almost the entire book can be covered in a one-year course:

- *Title:* Dynamics of Structures I (1 semester)

  Syllabus: Chapter 1; Sections 1 and 2 of Chapter 2; Parts A and B of Chapter 3; Chapter 4; selected topics from Chapter 5; Sections 1 to 7 of Chapter 6; Sections 1 to 7 of Chapter 7; selected topics from Chapter 8; Sections 1 to 4 and 9 to 11 of Chapter 9; Parts A and B of Chapter 10; Sections 1 and 2 of Chapter 11; Parts A and B of Chapter 12; Sections 1, 2, 7, and 8 (excluding the CQC method) of Chapter 13; and selected topics from Part A of Chapter 22

- *Title:* Dynamics of Structures II (1 semester)

  Syllabus: Sections 5 to 7 of Chapter 9; Sections 3 to 5 of Chapter 11; Parts C and D of Chapter 12; Sections 3 to 11 of Chapter 13; selected parts of Chapters 14, 15, 17, 19 to 21, and 23; and Appendix A.

The selection of topics for the first course has been dictated in part by the need to provide comprehensive coverage, including dynamic and earthquake analysis of MDF systems, for students taking only one course.

Abbreviated versions of the outline above can be organized for two quarter courses. One possibility is as follows:

- *Title:* Dynamics of Structures I (1 quarter)

  Syllabus: Chapter 1; Sections 1 and 2 of Chapter 2; Sections 1 to 4 of Chapter 3; Sections 1 and 2 of Chapter 4; selected topics from Chapter 5; Sections 1 to 7 of Chapter 6; Sections 1 to 7 of Chapter 7; selected topics from Chapter 8; Sections 1 to 4 and 9 to 11 of Chapter 9; Parts A and B of Chapter 10; Part B of Chapter 12; Sections 1, 2, 7, and 8 (excluding the CQC method) of Chapter 13.

- *Title:* Dynamics of Structures II (1 quarter)

  Syllabus: Sections 5 to 7 of Chapter 9; Sections 3 to 9 of Chapter 13; and selected topics from Chapters 19 to 23

A one-semester course emphasizing earthquake engineering can be organized as follows:

- *Title:* Earthquake Dynamics of Structures

  Syllabus: Chapter 1; Sections 1 and 2 of Chapter 2; Sections 1 and 2 of Chapter 4; Chapters 6 and 7; selected topics from Chapter 8; Sections 1 to 4 and 9 to 11 of Chapter 9; Parts A and B of Chapter 10; Part A of Chapter 11; Sections 1 to 3 and 7 to 11 of Chapter 13; and selected topics from Chapters 19 to 23.

Solving problems is essential for students to learn structural dynamics. For this purpose the first 18 chapters include 373 problems. Chapters 19 through 23 do not include problems, for two reasons: (1) no new dynamic analysis procedures are introduced in these chapters; (2) this material does not lend itself to short, meaningful problems. However, the reader will find it instructive to work through the examples presented in Chapters 19 to 23 and to reproduce the results. The computer is essential for solving some of the problems, and these have been identified. In solving these problems, it is assumed that the student will have access to computer programs such as MATLAB or MATHCAD. Solutions to these problems are available to instructors as a download from the publisher.

In my lectures at Berkeley, I develop the theory on the blackboard and illustrate it by transparencies of the more complex figures in the book; enlarged versions of many of the figures, which are suitable for making transparencies for use in the classroom, are available to instructors as a download from the publisher. Despite requests for a complete set of PowerPoint slides, they have not been developed because I do not think this approach is the most effective strategy for teaching dynamics of structures.

A NOTE FOR PROFESSIONAL ENGINEERS

Many professional engineers encouraged me during 1980s to prepare a book more comprehensive than *Dynamics of Structures, A Primer,* a monograph published in 1981 by the Earthquake Engineering Research Institute. This need, I hope, is filled by the present book. Having been conceived as a textbook, it includes the formalism and detail necessary for students, but these features should not deter the professional from using the book, because its philosophy and style are aimed to facilitate learning the subject by self-study.

For professional engineers interested in earthquake analysis, response, design, and evaluation of structures, I suggest the following reading path through the book: Chapters 1 and 2; Chapters 6 to 9; Parts A and B of Chapter 10; Part A of Chapter 11; and Chapters 13 and 19 to 23.

REFERENCES

In this introductory text it is impractical to acknowledge sources for the information presented. References have been omitted to avoid distracting the reader. However, I have included occasional comments to add historical perspective and, at the end of almost every chapter, a brief list of publications suitable for further reading.

YOUR COMMENTS ARE INVITED

I request that instructors, students, and professional engineers write to me (chopra@ce.berkeley.edu) if they have suggestions for improvements or clarifications, or if they identify errors. I thank you in advance for taking the time and interest to do so.

Anil K. Chopra

Acknowledgments

I am grateful to the many people who helped in the preparation of this book.

- Professor Rakesh K. Goel, a partner from beginning to end, assisted in numerous ways and played an important role. His most significant contribution was to develop and execute the computer software necessary to generate the numerical results and create most of the figures.
- Professor Gregory L. Fenves read the first draft of the original manuscript, discussed it with me weekly, and provided substantive suggestions for improvement.
- Six reviewers—Professors Luis Esteva, William J. Hall, Rafael Riddell, C. C. Tung, and the late George W. Housner and Donald E. Hudson—examined a final draft of the original manuscript. They provided encouragement as well as perceptive suggestions for improvement.
- Professors Gregory L. Fenves and Filip C. Fillipou advised on revising Chapters 5 and 16, and commented on the final draft.
- Dr. Ian Aiken provided the resource material—including photographs—and advice on revising Sections 7.10.1 and 7.10.2 on supplemental energy dissipation devices.
- Dr. Charles Menun, whose research results were the basis for new Section 13.10, provided extensive advice on preparation of this section and reviewed several drafts.
- Professor Oscar Lopez and Dr. Charles Menun, whose research results were the basis for new Section 13.11, provided advice and reviewed a final draft.

- Several reviewers—Professors Michael C. Constantinou, Takeru Igusa, George C. Lee, Fai Ma, and Carlos E. Ventura—suggested improvements for the final draft of Chapter 14.

- Six experts advised on the interpretation of the updated versions of the four building codes in Chapter 22: Yousef Bozorgnia and Ronald O. Hamburger (*International Building Code*); Jagmohan L. Humar (*National Building Code of Canada*); Eduardo Miranda (*Mexico Federal District Code*); and Peter Fajfar and Michael N. Fardis (*Eurocode*).

- Several professors who have adopted the book in their courses have over the years suggested improvements. Some of the revisions and additions in this edition were prompted by recommendations from Professors Stavros A. Anagnostopoulos, Michael C. Constantinou, Kincho Law, and Jose M. Roesset.

- Many former students have over the years assisted in preparing solutions for the worked-out examples and end-of-chapter problems and helped in other ways: Ashraf Ayoub, Ushnish Basu, Shih-Po Chang, Juan Chavez, Chatpan Chintanapakdee, Juan Carlos De la llera, Rakesh K. Goel, Garrett Hall, Gabriel Hurtado, Petros Keshishian, Allen Kwan, Wen-Hsiung Lin, Charles Menun, and Tsung-Li Tai. Han-Chen Tan did the word processing and graphics for the original Solutions Manual for the 233 problems in the first edition.

- Several students, present and former, assisted in the preparation of new material in the fourth edition: Juan Carlos Reyes solved examples and end-of-chapter problems in Chapters 5, 14, and 16; and prepared figures. Yvonne Tsui generated the numerical results for Section 13.10 and prepared preliminary figures. Neal Simon Kwong solved the examples and prepared the figures in Sections 12.14 and 13.11, and finalized the figures for Section 13.10. Eric Keldrauk developed the results for Figure 11.4.3.

- Charles D. James, Information Systems Manager for NISEE at the University of California, Berkeley, helped in selecting and collecting the new photographs.

- Claire Johnson prepared the text for the revised and new parts of the manuscript, and also assembled and edited the Solutions Manual.

- Barbara Zeiders served as the copy editor for this edition, just as she did for the first three.

- Professor Joseph Penzien assumed my duties as Associate Editor of *Earthquake Engineering and Structural Dynamics* from June 1993 until August 1994 while I was working on the original book.

I also wish to express my deep appreciation to Professors Ray W. Clough, Jr., Joseph Penzien, Emilio Rosenblueth, and A. S. Veletsos for the influence they have had on my professional growth. In the early 1960s, Professors Clough, Penzien, and Rosenblueth exposed me to their enlightened views and their superbly organized courses on structural dynamics and earthquake engineering. Subsequently, Professor Veletsos, through his research, writing, and lectures, influenced my teaching and research philosophy. His work, in collaboration with the late Professor Nathan M. Newmark, defined the approach adopted for parts of Chapters 6 and 7; and that in collaboration with Professor Carlos E. Ventura defined the presentation style for Chapter 14.

This book has been influenced by my own research experience in collaboration with my students. Since 1969, several organizations have supported my research in earthquake engineering, including the National Science Foundation, U.S. Army Corps of Engineers, and California Strong Motion Instrumentation Program.

This book and its revised editions were prepared during sabbatical leaves, a privilege for which I am grateful to the University of California at Berkeley.

Anil K. Chopra

PART I

Single-Degree-of-Freedom Systems

Equations of Motion, Problem Statement, and Solution Methods

PREVIEW

In this opening chapter, the structural dynamics problem is formulated for simple structures that can be idealized as a system with a lumped mass and a massless supporting structure. Linearly elastic structures as well as inelastic structures subjected to applied dynamic force or earthquake-induced ground motion are considered. Then four methods for solving the differential equation governing the motion of the structure are reviewed briefly. The chapter ends with an overview of how our study of the dynamic response of single-degree-of-freedom systems is organized in the chapters to follow.

1.1 SIMPLE STRUCTURES

We begin our study of structural dynamics with *simple* structures, such as the pergola shown in Fig. 1.1.1 and the elevated water tank of Fig. 1.1.2. We are interested in understanding the vibration of these structures when subjected to a lateral (or horizontal) force at the top or horizontal ground motion due to an earthquake.

We call these structures *simple* because they can be idealized as a concentrated or lumped mass m supported by a massless structure with stiffness k in the lateral direction. Such an idealization is appropriate for this pergola with a heavy concrete roof supported by light-steel-pipe columns, which can be assumed as massless. The concrete roof is very stiff and the flexibility of the structure in lateral (or horizontal) motion is provided entirely by the columns. The idealized system is shown in Fig. 1.1.3a with a pair of columns supporting the tributary length of the concrete roof. This system has a lumped mass m

Figure 1.1.1   This pergola at the Macuto-Sheraton Hotel near Caracas, Venezuela, was damaged by earthquake on July 29, 1967. The Magnitude 6.5 event, which was centered about 15 miles from the hotel, overstrained the steel pipe columns, resulting in a permanent roof displacement of 9 in. (From the Steinbrugge Collection, National Information Service for Earthquake Engineering, University of California, Berkeley.)

equal to the mass of the roof shown, and its lateral stiffness k is equal to the sum of the stiffnesses of individual pipe columns. A similar idealization, shown in Fig. 1.1.3b, is appropriate for the tank when it is full of water. With sloshing of water not possible in a full tank, it is a lumped mass m supported by a relatively light tower that can be assumed as massless. The cantilever tower supporting the water tank provides lateral stiffness k to the structure. For the moment we will assume that the lateral motion of these structures is small in the sense that the supporting structures deform within their linear elastic limit.

We shall see later in this chapter that the differential equation governing the lateral displacement $u(t)$ of these idealized structures without any external excitation—applied force or ground motion—is

$$m\ddot{u} + ku = 0 \tag{1.1.1}$$

where an overdot denotes differentiation with respect to time; thus $\dot{u}$ denotes the velocity of the mass and $\ddot{u}$ its acceleration. The solution of this equation, presented in Chapter 2, will show that if the mass of the idealized systems of Fig. 1.1.3 is displaced through some initial displacement $u(0)$, then released and permitted to vibrate freely, the structure will oscillate or vibrate back and forth about its initial equilibrium position. As shown in Fig. 1.1.3c, the same maximum displacement occurs oscillation after oscillation; these oscillations continue forever and these idealized systems would never come to rest. This is unrealistic,

Figure 1.1.2   This reinforced-concrete tank on a 40-ft-tall single concrete column, located near the Valdivia Airport, was undamaged by the Chilean earthquakes of May 1960. When the tank is full of water, the structure can be analyzed as a single-degree-of freedom system. (From the Steinbrugge Collection, National Information Service for Earthquake Engineering, University of California, Berkeley.)

of course. Intuition suggests that if the roof of the pergola or the top of the water tank were pulled laterally by a rope and the rope were suddenly cut, the structure would oscillate with ever-decreasing amplitude and eventually come to rest. Such experiments were performed on laboratory models of one-story frames, and measured records of their free vibration

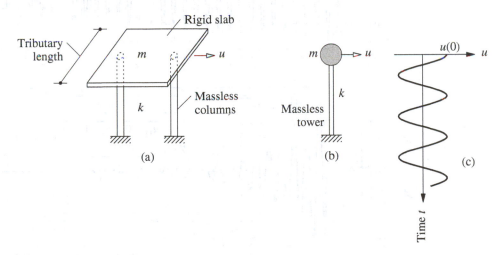

Figure 1.1.3   (a) Idealized pergola; (b) idealized water tank; (c) free vibration due to initial displacement.

response are presented in Fig. 1.1.4. As expected, the motion of these model structures decays with time. Observe that the motion of the plexiglass model decays more rapidly relative to the aluminum frame.

(a)

Figure 1.1.4   (a) Aluminum and plexiglass model frames mounted on a small shaking table used for classroom demonstration at the University of California at Berkeley (courtesy of T. Merport); (b) free vibration record of aluminum model; (c) free vibration record of plexiglass model.

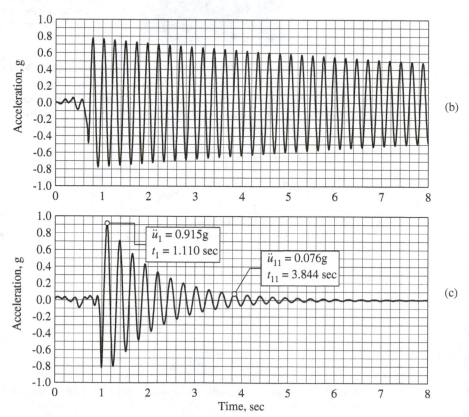

(b)

$\ddot{u}_1 = 0.915g$
$t_1 = 1.110 \text{ sec}$

$\ddot{u}_{11} = 0.076g$
$t_{11} = 3.844 \text{ sec}$

(c)

The process by which vibration steadily diminishes in amplitude is called *damping*. The kinetic energy and strain energy of the vibrating system are dissipated by various damping mechanisms that we shall mention later. For the moment, we simply recognize that an energy-dissipating mechanism should be included in the structural idealization in order to incorporate the feature of decaying motion observed during free vibration tests of a structure. The most commonly used damping element is the viscous damper, in part because it is the simplest to deal with mathematically. In Chapters 2 and 3 we introduce other energy-dissipating mechanisms.

1.2 SINGLE-DEGREE-OF-FREEDOM SYSTEM

The system considered is shown schematically in Fig. 1.2.1. It consists of a mass m concentrated at the roof level, a massless frame that provides stiffness to the system, and a viscous damper (also known as a dashpot) that dissipates vibrational energy of the system. The beam and columns are assumed to be inextensible axially.

This system may be considered as an idealization of a one-story structure. Each structural member (beam, column, wall, etc.) of the actual structure contributes to the inertial (mass), elastic (stiffness or flexibility), and energy dissipation (damping) properties of the structure. In the idealized system, however, each of these properties is concentrated in three separate, pure components: mass component, stiffness component, and damping component.

The number of independent displacements required to define the displaced positions of all the masses relative to their original position is called the number of *degrees of freedom* (DOFs) for dynamic analysis. More DOFs are typically necessary to define the stiffness properties of a structure compared to the DOFs necessary for representing inertial properties. Consider the one-story frame of Fig. 1.2.1, constrained to move only in the direction of the excitation. The static analysis problem has to be formulated with three DOFs—lateral displacement and two joint rotations—to determine the lateral stiffness of the frame (see Section 1.3). In contrast, the structure has only one DOF—lateral displacement—for dynamic analysis if it is idealized with mass concentrated at one location, typically the roof level. Thus we call this a *single-degree-of-freedom* (SDF) *system*.

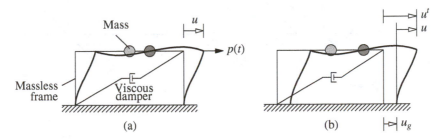

Figure 1.2.1   Single-degree-of-freedom system: (a) applied force $p(t)$; (b) earthquake-induced ground motion.

Two types of dynamic excitation will be considered: (1) external force $p(t)$ in the lateral direction (Fig. 1.2.1a), and (2) earthquake-induced ground motion $u_g(t)$ (Fig. 1.2.1b). In both cases u denotes the relative displacement between the mass and the base of the structure.

1.3 FORCE–DISPLACEMENT RELATION

Consider the system shown in Fig. 1.3.1a with no dynamic excitation subjected to an externally applied static force f_S along the DOF u as shown. The internal force resisting the displacement u is equal and opposite to the external force f_S (Fig. 1.3.1b). It is desired to determine the relationship between the force f_S and the relative displacement u associated with deformations in the structure during oscillatory motion. This force–displacement relation would be linear at small deformations but would become nonlinear at larger deformations (Fig. 1.3.1c); both nonlinear and linear relations are considered (Fig. 1.3.1c and d).

To determine the relationship between f_S and u is a standard problem in static structural analysis, and we assume that the reader is familiar with such analyses. Thus the presentation here is brief and limited to those aspects that are essential.

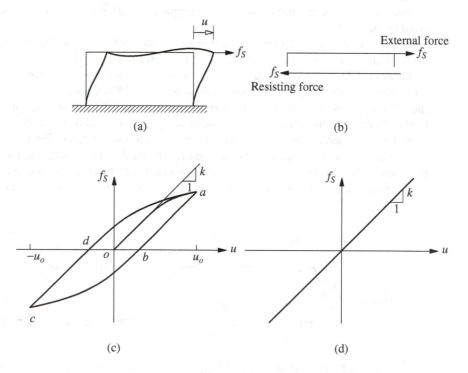

Figure 1.3.1

1.3.1 Linearly Elastic Systems

For a *linear system* the relationship between the lateral force f_S and resulting deformation u is linear, that is,

$$f_S = ku \tag{1.3.1}$$

where k is the lateral stiffness of the system; its units are force/length. Implicit in Eq. (1.3.1) is the assumption that the linear f_S–u relationship determined for small deformations of the structure is also valid for larger deformations. This linear relationship implies that f_S is a single-valued function of u (i.e., the loading and unloading curves are identical). Such a system is said to be *elastic*; hence we use the term *linearly elastic system* to emphasize both properties.

Consider the frame of Fig. 1.3.2a with bay width L, height h, elastic modulus E, and second moment of the cross-sectional area (or moment of inertia)[†] about the axis of bending $= I_b$ and I_c for the beam and columns, respectively; the columns are clamped (or fixed) at the base. The *lateral stiffness* of the frame can readily be determined for two extreme cases: If the beam is rigid [i.e., flexural rigidity $EI_b = \infty$ (Fig. 1.3.2b)],

$$k = \sum_{\text{columns}} \frac{12EI_c}{h^3} = 24\frac{EI_c}{h^3} \tag{1.3.2}$$

On the other hand, for a beam with no stiffness [i.e., $EI_b = 0$ (Fig. 1.3.2c)],

$$k = \sum_{\text{columns}} \frac{3EI_c}{h^3} = 6\frac{EI_c}{h^3} \tag{1.3.3}$$

Observe that for the two extreme values of beam stiffness, the lateral stiffness of the frame is independent of L, the beam length or bay width.

The lateral stiffness of the frame with an intermediate, realistic stiffness of the beam can be calculated by standard procedures of static structural analysis. The stiffness matrix of the frame is formulated with respect to three DOFs: the lateral displacement u

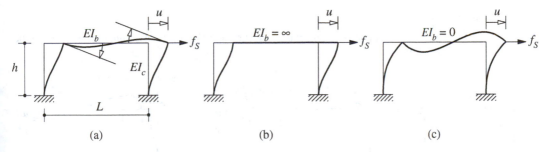

(a)   (b)   (c)

Figure 1.3.2

[†]In this book the preferred term for I is *second moment of area* instead of the commonly used *moment of inertia*; the latter will be reserved for defining inertial effects associated with the rotational motion of rigid bodies.

and the rotations of the two beam–column joints (Fig. 1.3.2a). By static condensation or elimination of the rotational DOFs, the lateral force–displacement relation of Eq. (1.3.1) is determined. Applying this procedure to a frame with $L = 2h$ and $EI_b = EI_c$, its lateral stiffness is obtained (see Example 1.1):

$$k = \frac{96}{7} \frac{EI_c}{h^3} \tag{1.3.4}$$

The lateral stiffness of the frame can be computed similarly for any values of I_b, I_c, L, and h using the stiffness coefficients for a uniform flexural element presented in Appendix 1. If shear deformations in elements are neglected, the result can be written in the form

$$k = \frac{24EI_c}{h^3} \frac{12\rho + 1}{12\rho + 4} \tag{1.3.5}$$

where $\rho = (EI_b/L) \div (2EI_c/h)$ is the *beam-to-column stiffness ratio* (to be elaborated in Section 18.1.1). For $\rho = 0$, ∞, and $\frac{1}{4}$, Eq. (1.3.5) reduces to the results of Eqs. (1.3.3), (1.3.2), and (1.3.4), respectively. The lateral stiffness is plotted as a function of ρ in Fig. 1.3.3; it increases by a factor of 4 as ρ increases from zero to infinity.

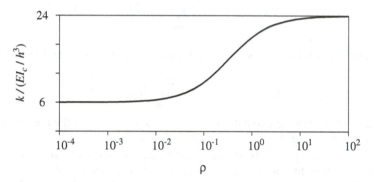

Figure 1.3.3 Variation of lateral stiffness, k, with beam-to-column stiffness ratio, ρ.

Example 1.1

Calculate the lateral stiffness for the frame shown in Fig. E1.1a, assuming the elements to be axially rigid.

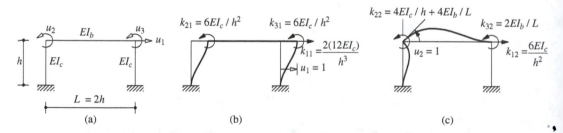

Figure E1.1

Solution   This structure can be analyzed by any of the standard methods, including moment distribution. Here we use the definition of stiffness influence coefficients to solve the problem.

The system has the three DOFs shown in Fig. E1.1a. To obtain the first column of the 3×3 stiffness matrix, we impose unit displacement in DOF u_1, with $u_2 = u_3 = 0$. The forces k_{i1} required to maintain this deflected shape are shown in Fig. E1.1b. These are determined using the stiffness coefficients for a uniform flexural element presented in Appendix 1. The elements k_{i2} in the second column of the stiffness matrix are determined by imposing $u_2 = 1$ with $u_1 = u_3 = 0$; see Fig. E1.1c. Similarly, the elements k_{i3} in the third column of the stiffness matrix can be determined by imposing displacements $u_3 = 1$ with $u_1 = u_2 = 0$. Thus the 3×3 stiffness matrix of the structure is known and the equilibrium equations can be written. For a frame with $I_b = I_c$ subjected to lateral force f_S, they are

$$\frac{EI_c}{h^3} \begin{bmatrix} 24 & 6h & 6h \\ 6h & 6h^2 & h^2 \\ 6h & h^2 & 6h^2 \end{bmatrix} \begin{Bmatrix} u_1 \\ u_2 \\ u_3 \end{Bmatrix} = \begin{Bmatrix} f_S \\ 0 \\ 0 \end{Bmatrix} \tag{a}$$

From the second and third equations, the joint rotations can be expressed in terms of lateral displacement as follows:

$$\begin{Bmatrix} u_2 \\ u_3 \end{Bmatrix} = -\begin{bmatrix} 6h^2 & h^2 \\ h^2 & 6h^2 \end{bmatrix}^{-1} \begin{bmatrix} 6h \\ 6h \end{bmatrix} u_1 = -\frac{6}{7h} \begin{bmatrix} 1 \\ 1 \end{bmatrix} u_1 \tag{b}$$

Substituting Eq. (b) into the first of three equations in Eq. (a) gives

$$f_S = \left(\frac{24EI_c}{h^3} - \frac{EI_c}{h^3} \frac{6}{7h} \langle 6h \quad 6h \rangle \begin{bmatrix} 1 \\ 1 \end{bmatrix} \right) u_1 = \frac{96}{7} \frac{EI_c}{h^3} u_1 \tag{c}$$

Thus the lateral stiffness of the frame is

$$k = \frac{96}{7} \frac{EI_c}{h^3} \tag{d}$$

This procedure to eliminate joint rotations, known as the *static condensation method,* is presented in textbooks on static analysis of structures. We return to this topic in Chapter 9.

1.3.2 Inelastic Systems

Determined by experiments, the force–deformation relation for a structural steel component undergoing cyclic deformations expected during earthquakes is shown in Fig. 1.3.4. The initial loading curve is nonlinear at the larger amplitudes of deformation, and the unloading and reloading curves differ from the initial loading branch; such a system is said to be *inelastic*. This implies that the force–deformation relation is path dependent, i.e., it depends on whether the deformation is increasing or decreasing. Thus the resisting force is an implicit function of deformation:

$$f_S = f_S(u) \tag{1.3.6}$$

The force–deformation relation for the idealized one-story frame (Fig. 1.3.1a) deforming into the inelastic range can be determined in one of two ways. One approach is to use methods of nonlinear static structural analysis. For example, in analyzing a steel structure with an assumed stress–strain law, the analysis keeps track of the initiation and spreading of yielding at critical locations and formation of plastic hinges to obtain the initial loading

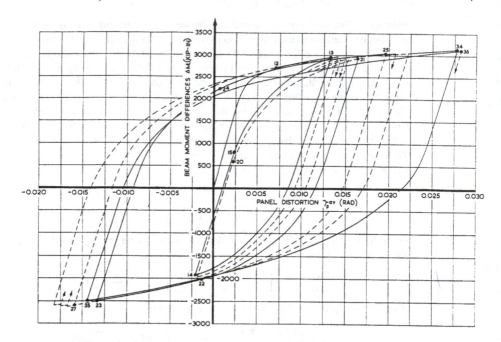

Figure 1.3.4 Force–deformation relation for a structural steel component. (From H. Krawinkler, V. V. Bertero, and E. P. Popov, "Inelastic Behavior of Steel Beam-to-Column Subassemblages," *Report No. EERC 71-7*, University of California, Berkeley, 1971.)

curve (*o–a*) shown in Fig. 1.3.1c. The unloading (*a–c*) and reloading (*c–a*) curves can be computed similarly or can be defined from the initial loading curve using existing hypotheses. Another approach is to define the inelastic force–deformation relation as an idealized version of the experimental data, such as in Fig. 1.3.4.

We are interested in studying the dynamic response of inelastic systems because many structures are designed with the expectation that they will undergo some cracking, yielding, and damage during intense ground shaking caused by earthquakes.

1.4 DAMPING FORCE

As mentioned earlier, the process by which free vibration steadily diminishes in amplitude is called *damping*. In damping, the energy of the vibrating system is dissipated by various mechanisms, and often more than one mechanism may be present at the same time. In simple "clean" systems such as the laboratory models of Fig. 1.1.4, most of the energy dissipation presumably arises from the thermal effect of repeated elastic straining of the material and from the internal friction when a solid is deformed. In actual structures, however, many other mechanisms also contribute to the energy dissipation. In a vibrating building these include friction at steel connections, opening and closing of microcracks in concrete, and friction between the structure itself and nonstructural elements such as partition walls. It seems impossible to identify or describe mathematically each of these energy-dissipating mechanisms in an actual building.

As a result, the damping in actual structures is usually represented in a highly idealized manner. For many purposes the actual damping in a SDF structure can be idealized satisfactorily by a linear viscous damper or dashpot. The damping coefficient is selected so that the vibrational energy it dissipates is equivalent to the energy dissipated in all the damping mechanisms, combined, present in the actual structure. This idealization is therefore called *equivalent viscous damping*, a concept developed further in Chapter 3.

Figure 1.4.1a shows a linear viscous damper subjected to a force f_D along the DOF u. The internal force in the damper is equal and opposite to the external force f_D (Fig. 1.4.1b). As shown in Fig. 1.4.1c, the damping force f_D is related to the velocity $\dot{u}$ across the linear viscous damper by

$$f_D = c\dot{u} \tag{1.4.1}$$

where the constant c is the *viscous damping coefficient*; it has units of force × time/length.

Unlike the stiffness of a structure, the damping coefficient cannot be calculated from the dimensions of the structure and the sizes of the structural elements. This should not be surprising because, as we noted earlier, it is not feasible to identify all the mechanisms that dissipate vibrational energy of actual structures. Thus vibration experiments on actual structures provide the data for evaluating the damping coefficient. These may be free vibration experiments that lead to data such as those shown in Fig. 1.1.4; the measured rate at which motion decays in free vibration will provide a basis for evaluating the damping coefficient, as we shall see in Chapter 2. The damping property may also be determined from forced vibration experiments, a topic that we study in Chapter 3.

The equivalent viscous damper is intended to model the energy dissipation at deformation amplitudes within the linear elastic limit of the overall structure. Over this range of deformations, the damping coefficient c determined from experiments may vary with the deformation amplitude. This nonlinearity of the damping property is usually not considered explicitly in dynamic analyses. It may be handled indirectly by selecting a value for the damping coefficient that is appropriate for the expected deformation amplitude, usually taken as the deformation associated with the linearly elastic limit of the structure.

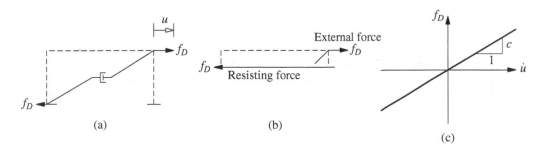

(a)   (b)   (c)

Figure 1.4.1

Additional energy is dissipated due to inelastic behavior of the structure at larger deformations. Under cyclic forces or deformations, this behavior implies formation of a force–deformation hysteresis loop (Fig. 1.3.1c). The damping energy dissipated during one deformation cycle between deformation limits $\pm u_o$ is given by the area within the hysteresis loop *abcda* (Fig. 1.3.1c). This energy dissipation is usually not modeled by a viscous damper, especially if the excitation is earthquake ground motion, for reasons we note in Chapter 7. Instead, the most common, direct, and accurate approach to account for the energy dissipation through inelastic behavior is to recognize the inelastic relationship between resisting force and deformation, such as shown in Figs. 1.3.1c and 1.3.4, in solving the equation of motion (Chapter 5). Such force–deformation relationships are obtained from experiments on structures or structural components at slow rates of deformation, thus excluding any energy dissipation arising from rate-dependent effects. The usual approach is to model this damping in the inelastic range of deformations by the same viscous damper that was defined earlier for smaller deformations within the linearly elastic range.

1.5 EQUATION OF MOTION: EXTERNAL FORCE

Figure 1.5.1a shows the idealized one-story frame introduced earlier subjected to an externally applied dynamic force $p(t)$ in the direction of the DOF u. This notation indicates that the force p varies with time t. The resulting displacement of the mass also varies with time; it is denoted by $u(t)$. In Sections 1.5.1 and 1.5.2 we derive the differential equation governing the displacement $u(t)$ by two methods using (1) Newton's second law of motion, and (2) dynamic equilibrium. An alternative point of view for the derivation is presented in Section 1.5.3.

1.5.1 Using Newton's Second Law of Motion

The forces acting on the mass at some instant of time are shown in Fig. 1.5.1b. These include the external force $p(t)$, the elastic (or inelastic) resisting force f_S (Fig. 1.3.1), and the damping resisting force f_D (Fig. 1.4.1). The external force is taken to be positive in the direction of the x-axis, and the displacement $u(t)$, velocity $\dot{u}(t)$, and acceleration $\ddot{u}(t)$ are also positive in the direction of the x-axis. The elastic and damping forces are shown

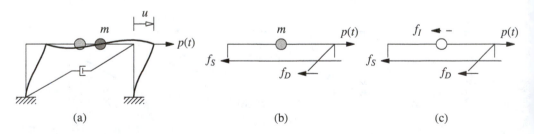

(a)   (b)   (c)

Figure 1.5.1

acting in the opposite direction because they are internal forces that resist the deformation and velocity, respectively.

The resultant force along the x-axis is $p - f_S - f_D$, and Newton's second law of motion gives

$$p - f_S - f_D = m\ddot{u} \quad \text{or} \quad m\ddot{u} + f_D + f_S = p(t) \tag{1.5.1}$$

This equation after substituting Eqs. (1.3.1) and (1.4.1) becomes

$$m\ddot{u} + c\dot{u} + ku = p(t) \tag{1.5.2}$$

This is the equation of motion governing the deformation or displacement $u(t)$ of the idealized structure of Fig. 1.5.1a, assumed to be linearly elastic, subjected to an external dynamic force $p(t)$. The units of mass are force/acceleration.

This derivation can readily be extended to inelastic systems. Equation (1.5.1) is still valid and all that needs to be done is to replace Eq. (1.3.1), restricted to linear systems, by Eq. (1.3.6), valid for inelastic systems. For such systems, therefore, the equation of motion is

$$m\ddot{u} + c\dot{u} + f_S(u) = p(t) \tag{1.5.3}$$

1.5.2 Dynamic Equilibrium

Having been trained to think in terms of equilibrium of forces, structural engineers may find D'Alembert's principle of *dynamic equilibrium* particularly appealing. This principle is based on the notion of a fictitious *inertia force*, a force equal to the product of mass times its acceleration and acting in a direction opposite to the acceleration. It states that with inertia forces included, a system is in equilibrium at each time instant. Thus a free-body diagram of a moving mass can be drawn, and principles of statics can be used to develop the equation of motion.

Figure 1.5.1c is the free-body diagram at time t with the mass replaced by its inertia force, which is shown by a dashed line to distinguish this fictitious force from the real forces. Setting the sum of all the forces equal to zero gives Eq. (1.5.1b),[†] which was derived earlier by using Newton's second law of motion.

1.5.3 Stiffness, Damping, and Mass Components

In this section the governing equation for the idealized one-story frame is formulated based on an alternative viewpoint. Under the action of external force $p(t)$, the state of the system is described by displacement $u(t)$, velocity $\dot{u}(t)$, and acceleration $\ddot{u}(t)$; see Fig. 1.5.2a. Now visualize the system as the combination of three pure components: (1) the stiffness component: the frame without damping or mass (Fig. 1.5.2b); (2) the damping component: the frame with its damping property but no stiffness or mass (Fig. 1.5.2c); and (3) the mass component: the roof mass without the stiffness or damping of the frame (Fig. 1.5.2d).

[†]Two or more equations in the same line with the same equation number will be referred to as equations a, b, c, etc., from left to right.

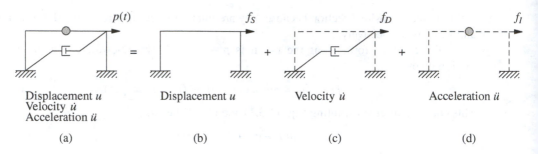

Figure 1.5.2   (a) System; (b) stiffness component; (c) damping component; (d) mass component.

The external force f_S on the stiffness component is related to the displacement u by Eq. (1.3.1) if the system is linearly elastic, the external force f_D on the damping component is related to the velocity $\dot{u}$ by Eq. (1.4.1), and the external force f_I on the mass component is related to the acceleration by $f_I = m\ddot{u}$. The external force $p(t)$ applied to the complete system may therefore be visualized as distributed among the three components of the structure, and $f_S + f_D + f_I$ must equal the applied force $p(t)$ leading to Eq. (1.5.1b). Although this alternative viewpoint may seem unnecessary for the simple system of Fig. 1.5.2a, it is useful for complex systems (Chapter 9).

Example 1.2

A small one-story industrial building, 20 by 30 ft in plan, is shown in Fig. E1.2 with moment frames in the north–south direction and braced frames in the east–west direction. The weight of the structure can be idealized as 30 lb/ft^2 lumped at the roof level. The horizontal cross bracing is at the bottom chord of the roof trusses. All columns are W8 × 24 sections; their second moments of cross-sectional area about the x and y axes are $I_x = 82.8$ in^4 and $I_y = 18.3$ in^4, respectively; for steel, $E = 29,000$ ksi. The vertical cross-bracings are made of 1-in.-diameter rods. Formulate the equation governing free vibration in (**a**) the north–south direction and (**b**) the east–west direction.

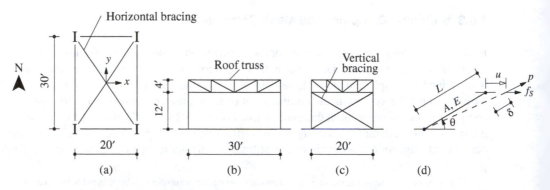

Figure E1.2   (a) Plan; (b) east and west elevations; (c) north and south elevations; (d) cross brace.

Solution   The mass lumped at the roof is

$$m = \frac{w}{g} = \frac{30 \times 30 \times 20}{386} = 46.63 \text{ lb-sec}^2/\text{in.} = 0.04663 \text{ kip-sec}^2/\text{in.}$$

Because of the horizontal cross-bracing, the roof can be treated as a rigid diaphragm.

(a) *North–south direction.*   Because of the roof truss, each column behaves as a clamped–clamped column and the lateral stiffness of the two moment frames (Fig. E1.2b) is

$$k_{\text{N-S}} = 4 \left(\frac{12 E I_x}{h^3} \right) = 4 \frac{12(29 \times 10^3)(82.8)}{(12 \times 12)^3} = 38.58 \text{ kips/in.}$$

and the equation of motion is

$$m\ddot{u} + (k_{\text{N-S}}) u = 0 \tag{a}$$

(b) *East–west direction.*   Braced frames, such as those shown in Fig. E1.2c, are usually designed as two superimposed systems: an ordinary rigid frame that supports vertical (dead and live) loads, plus a vertical bracing system, generally regarded as a pin-connected truss that resists lateral forces. Thus the lateral stiffness of a braced frame can be estimated as the sum of the lateral stiffnesses of individual braces. The stiffness of a brace (Fig. E1.2d) is $k_{\text{brace}} = (AE/L) \cos^2 \theta$. This can be derived as follows.

We start with the axial force–deformation relation for a brace:

$$p = \frac{AE}{L} \delta \tag{b}$$

By statics $f_S = p \cos \theta$, and by kinematics $u = \delta/\cos \theta$. Substituting $p = f_S/\cos \theta$ and $\delta = u \cos \theta$ in Eq. (b) gives

$$f_S = k_{\text{brace}} u \qquad k_{\text{brace}} = \frac{AE}{L} \cos^2 \theta \tag{c}$$

For the brace in Fig. E1.2c, $\cos \theta = 20/\sqrt{12^2 + 20^2} = 0.8575$, $A = 0.785 \text{ in}^2$, $L = 23.3 \text{ ft}$, and

$$k_{\text{brace}} = \frac{0.785(29 \times 10^3)}{23.3 \times 12} (0.8575)^2 = 59.8 \text{ kips/in.}$$

Although each frame has two cross-braces, only the one in tension will provide lateral resistance; the one in compression will buckle at small axial force and will contribute little to the lateral stiffness. Considering the two frames,

$$k_{\text{E-W}} = 2 \times 59.8 = 119.6 \text{ kips/in.}$$

and the equation of motion is

$$m\ddot{u} + (k_{\text{E-W}}) u = 0 \tag{d}$$

Observe that the error in neglecting the stiffness of columns is small: $k_{\text{col}} = 2 \times 12 E I_y/h^3 = 4.26 \text{ kips/in.}$ versus $k_{\text{brace}} = 59.8 \text{ kips/in.}$

Example 1.3

A 375-ft-long concrete, box-girder bridge on four supports—two abutments and two symmetrically located bents—is shown in Fig. E1.3. The cross-sectional area of the bridge deck

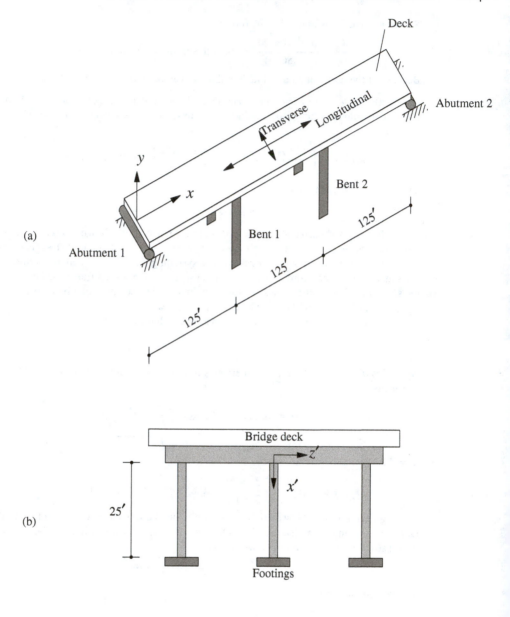

(a)

(b)

(c)

Figure E1.3

is 123 ft^2. The weight of the bridge is idealized as lumped at the deck level; the unit weight of concrete is 150 lb/ft^3. The weight of the bents may be neglected. Each bent consists of three 25-ft-tall columns of circular cross section with $I_{y'} = I_{z'} = 13$ ft^4 (Fig. E1.3b). Formulate the equation of motion governing free vibration in the longitudinal direction. The elastic modulus of concrete is $E = 3000$ ksi.

Solution   The weight per unit length lumped at the deck level is $(123 \times 1)150 = 18.45$ kips/ft. The total weight lumped at the deck level is

$$w = 18.45 \times 375 = 6919 \text{ kips}$$

and the corresponding mass is

$$m = \frac{w}{g} = \frac{6919}{32.2} = 214.9 \text{ kip-sec}^2/\text{ft}$$

The longitudinal stiffness of the bridge is computed assuming the bridge deck to displace rigidly as shown in Fig. E1.3c. Each column of a bent behaves as a clamped–clamped column. The longitudinal stiffness provided by each bent is

$$k_{\text{bent}} = 3 \left(\frac{12EI_{z'}}{h^3} \right) = 3 \left[\frac{12(3000 \times 144)13}{(25)^3} \right] = 12{,}940 \text{ kips/ft}$$

Two bents provide a total stiffness of

$$k = 2 \times k_{\text{bent}} = 2 \times 12{,}940 = 25{,}880 \text{ kips/ft}$$

The equation governing the longitudinal displacement u is

$$m\ddot{u} + ku = 0$$

1.6 MASS–SPRING–DAMPER SYSTEM

We have introduced the SDF system by idealizing a one-story structure (Fig. 1.5.1a), an approach that should appeal to structural engineering students. However, the classic SDF system is the mass–spring–damper system of Fig. 1.6.1a. The dynamics of this system is developed in textbooks on mechanical vibration and elementary physics. If we consider the spring and damper to be massless, the mass to be rigid, and all motion to be in the direction of the x-axis, we have an SDF system. Figure 1.6.1b shows the forces acting on the mass; these include the elastic resisting force, $f_S = ku$, exerted by a linear spring of stiffness k, and the damping resisting force, $f_D = c\dot{u}$, due to a linear viscous damper. Newton's second law of motion then gives Eq. (1.5.1b). Alternatively, the same equation is obtained using D'Alembert's principle and writing an equilibrium equation for forces in the free-body diagram, including the inertia force (Fig. 1.6.1c). It is clear that the equation of motion derived earlier for the idealized one-story frame of Fig. 1.5.1a is also valid for the mass–spring–damper system of Fig. 1.6.1a.

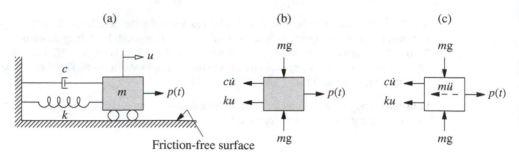

Figure 1.6.1   Mass–spring–damper system.

Example 1.4

Derive the equation of motion of the weight w suspended from a spring at the free end of the cantilever steel beam shown in Fig. E1.4a. For steel, $E = 29{,}000$ ksi. Neglect the mass of the beam and spring.

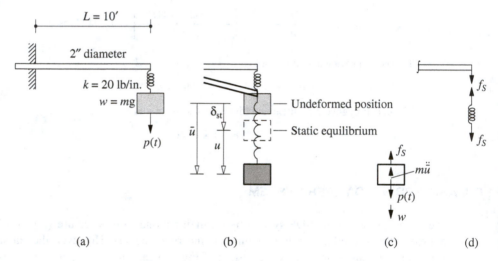

Figure E1.4   (a) System; (b) undeformed, deformed, and static equilibrium positions; (c) free-body diagram; (d) spring and beam forces.

Solution   Figure E1.4b shows the deformed position of the free end of the beam, spring, and mass. The displacement of the mass $\bar{u}$ is measured from its initial position with the beam and spring in their original undeformed configuration. Equilibrium of the forces of Fig. E1.4c gives

$$m\ddot{\bar{u}} + f_S = w + p(t) \tag{a}$$

where

$$f_S = k_e \bar{u} \tag{b}$$

and the effective stiffness k_e of the system remains to be determined. The equation of motion is

$$m\ddot{\bar{u}} + k_e \bar{u} = w + p(t) \tag{c}$$

The displacement $\bar{u}$ can be expressed as

$$\bar{u} = \delta_{st} + u \tag{d}$$

where δ_{st} is the static displacement due to weight w and u is measured from the position of static equilibrium. Substituting Eq. (d) in Eq. (a) and noting that (1) $\ddot{\bar{u}} = \ddot{u}$ because δ_{st} does not vary with time, and (2) $k_e \delta_{st} = w$ gives

$$m\ddot{u} + k_e u = p(t) \tag{e}$$

Observe that this is the same as Eq. (1.5.2) with $c = 0$ for a spring–mass system oriented in the horizontal direction (Fig. 1.6.1). Also note that the equation of motion (e) governing u, measured from the static equilibrium position, is unaffected by gravity forces.

For this reason we usually formulate a dynamic analysis problem for a linear system with its static equilibrium position as the reference position. The displacement $u(t)$ and associated internal forces in the system will represent the dynamic response of the system. The total displacements and forces are obtained by adding the corresponding static quantities to the dynamic response.

The effective stiffness k_e remains to be determined. It relates the static force f_S to the resulting displacement $\bar{u}$ by

$$f_S = k_e \bar{u} \tag{f}$$

where

$$\bar{u} = \bar{u}_{spring} + \bar{u}_{beam} \tag{g}$$

where $\bar{u}_{beam}$ is the deflection of the right end of the beam and $\bar{u}_{spring}$ is the deformation in the spring. With reference to Fig. E1.4d,

$$f_S = k\bar{u}_{spring} = k_{beam}\bar{u}_{beam} \tag{h}$$

In Eq. (g), substitute for $\bar{u}$ from Eq. (f) and the $\bar{u}_{spring}$ and $\bar{u}_{beam}$ from Eq. (h) to obtain

$$\frac{f_S}{k_e} = \frac{f_S}{k} + \frac{f_S}{k_{beam}} \quad \text{or} \quad k_e = \frac{kk_{beam}}{k + k_{beam}} \tag{i}$$

Now $k = 20$ lb/in. and

$$k_{beam} = \frac{3EI}{L^3} = \frac{3(29 \times 10^6)[\pi(1)^4/4]}{(10 \times 12)^3} = 39.54 \text{ lb/in.}$$

Substituting for k and k_{beam} in Eq. (i) gives

$$k_e = 13.39 \text{ lb/in.}$$

As mentioned earlier, the gravity forces can be omitted from the formulation of the governing equation for the system of Fig. E1.4 provided that the displacement u is measured from the static equilibrium position. However, the gravity loads must be considered if they act as either restoring forces (Example 1.5) or as destabilizing forces (Example 1.6).

Example 1.5

Derive the equation governing the free motion of a simple pendulum (Fig. E1.5a), which consists of a point mass m suspended by a light string of length L.

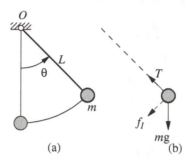

Figure E1.5   (a) Simple pendulum; (b) free-body diagram.

(a)   (b)

Solution   Figure E1.5a shows the displaced position of the pendulum defined by the angle θ measured from the vertical position, and Fig. E1.5b shows the free-body diagram of the mass. The forces acting are the weight mg, tension T in the string, and D'Alembert's fictitious inertia force $f_I = mL\ddot{\theta}$.
    Equilibrium of the moments of forces about O gives

$$mL^2\ddot{\theta} + mgL\sin\theta = 0 \tag{a}$$

This is a nonlinear differential equation governing θ.
    For small rotations, $\sin\theta \simeq \theta$ and the equation of motion [Eq. (a)] can be rewritten as

$$\ddot{\theta} + \frac{g}{L}\theta = 0 \tag{b}$$

Example 1.6

The system of Fig. E1.6 consists of a weight w attached to a rigid massless bar of length L joined to its support by a rotational spring of stiffness k. Derive the equation of motion. Neglect rotational inertia and assume small deflections. What is the buckling weight?

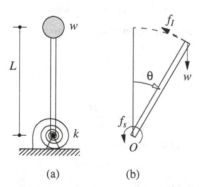

(a)   (b)   **Figure E1.6**

Solution   Figure E1.6b shows the displaced position of the system defined by the angle θ measured from the vertical position and the free-body diagram, which includes the weight w, the spring force (moment) $f_S = k\theta$, and D'Alembert's fictitious inertia force $f_I = (w/g)L\ddot{\theta}$.

Equilibrium of the moments about O gives

$$f_I L + f_S = mgL \sin \theta$$

or

$$\frac{w}{g} L^2 \ddot{\theta} + k\theta = wL \sin \theta \qquad (a)$$

For small rotations $\sin \theta \simeq \theta$ and Eq. (a) can be rewritten as

$$\frac{w}{g} L^2 \ddot{\theta} + (k - wL)\theta = 0 \qquad (b)$$

Observe that the gravity load reduces the effective stiffness of the system. If the weight $w = k/L$, the effective stiffness is zero and the system becomes unstable under its own weight. Thus, the buckling load (or weight) is

$$w_{\text{cr}} = \frac{k}{L} \qquad (c)$$

1.7 EQUATION OF MOTION: EARTHQUAKE EXCITATION

In earthquake-prone regions, the principal problem of structural dynamics that concerns structural engineers is the behavior of structures subjected to earthquake-induced motion of the base of the structure. The displacement of the ground is denoted by u_g, the total (or absolute) displacement of the mass by u^t, and the relative displacement between the mass and ground by u (Fig. 1.7.1). At each instant of time these displacements are related by

$$u^t(t) = u_g(t) + u(t) \qquad (1.7.1)$$

Both u^t and u_g refer to the same inertial frame of reference and their positive directions coincide.

The equation of motion for the idealized one-story system of Fig. 1.7.1a subjected to earthquake excitation can be derived by any one of the approaches introduced in Section 1.5. Here we choose to use the concept of dynamic equilibrium. From the free-body diagram including the inertia force f_I, shown in Fig. 1.7.1b, the equation of dynamic equilibrium is

$$f_I + f_D + f_S = 0 \qquad (1.7.2)$$

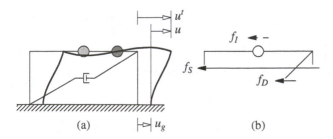

(a)   (b)   **Figure 1.7.1**

Only the relative motion u between the mass and the base due to structural deformation produces elastic and damping forces (i.e., the rigid-body component of the displacement of the structure produces no internal forces). Thus for a linear system, Eqs. (1.3.1) and (1.4.1) are still valid. The inertia force f_I is related to the acceleration $\ddot{u}^t$ of the mass by

$$f_I = m\ddot{u}^t \tag{1.7.3}$$

Substituting Eqs. (1.3.1), (1.4.1), and (1.7.3) in Eq. (1.7.2) and using Eq. (1.7.1) gives

$$m\ddot{u} + c\dot{u} + ku = -m\ddot{u}_g(t) \tag{1.7.4}$$

This is the equation of motion governing the relative displacement or deformation $u(t)$ of the linearly elastic structure of Fig. 1.7.1a subjected to ground acceleration $\ddot{u}_g(t)$.

For inelastic systems, Eq. (1.7.2) is valid, but Eq. (1.3.1) should be replaced by Eq. (1.3.6). The resulting equation of motion is

$$m\ddot{u} + c\dot{u} + f_S(u) = -m\ddot{u}_g(t) \tag{1.7.5}$$

Comparison of Eqs. (1.5.2) and (1.7.4), or of Eqs. (1.5.3) and (1.7.5), shows that the equations of motion for the structure subjected to two separate excitations—ground acceleration $\ddot{u}_g(t)$ and external force $= -m\ddot{u}_g(t)$—are one and the same. Thus the relative displacement or deformation $u(t)$ of the structure due to ground acceleration $\ddot{u}_g(t)$ will be identical to the displacement $u(t)$ of the structure if its base were stationary and if it were subjected to an external force $= -m\ddot{u}_g(t)$. As shown in Fig. 1.7.2, the ground motion can therefore be replaced by the *effective earthquake force* (indicated by the subscript "eff"):

$$p_{\text{eff}}(t) = -m\ddot{u}_g(t) \tag{1.7.6}$$

This force is equal to mass times the ground acceleration, acting opposite to the acceleration. It is important to recognize that the effective earthquake force is proportional to the mass of the structure. Thus the structural designer increases the effective earthquake force if the structural mass is increased.

Although the rotational components of ground motion are not measured during earthquakes, they can be estimated from the measured translational components and it is of interest to apply the preceding concepts to this excitation. For this purpose, consider the cantilever tower of Fig. 1.7.3a, which may be considered as an idealization of the water tank of Fig. 1.1.2, subjected to base rotation θ_g. The total displacement u^t of the mass is

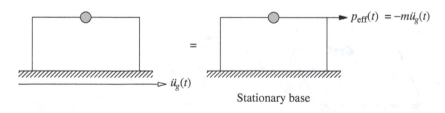

Figure 1.7.2 Effective earthquake force: horizontal ground motion.

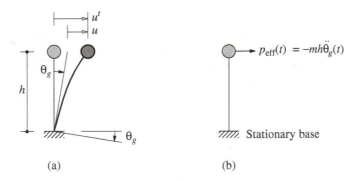

Figure 1.7.3   Effective earthquake force: rotational ground motion.

made up of two parts: u associated with structural deformation and a rigid-body component $h\theta_g$, where h is the height of the mass above the base. At each instant of time these
displacements are related by

$$u^t(t) = u(t) + h\theta_g(t) \tag{1.7.7}$$

Equations (1.7.2) and (1.7.3) are still valid, but the total acceleration $\ddot{u}^t(t)$ must now be
determined from Eq. (1.7.7). Putting all these equations together leads to

$$m\ddot{u} + c\dot{u} + ku = -mh\ddot{\theta}_g(t) \tag{1.7.8}$$

The effective earthquake force associated with ground rotation is

$$p_{\text{eff}}(t) = -mh\ddot{\theta}_g(t) \tag{1.7.9}$$

Example 1.7

A uniform rigid slab of total mass m is supported on four columns of height h rigidly connected to the top slab and to the foundation slab (Fig. E1.7a). Each column has a rectangular
cross section with second moments of area I_x and I_y for bending about the x and y axes,
respectively. Determine the equation of motion for this system subjected to rotation $u_{g\theta}$ of the
foundation about a vertical axis. Neglect the mass of the columns.

Solution   The elastic resisting torque or torsional moment f_S acting on the mass is shown in
Fig. E1.7b, and Newton's second law gives

$$-f_S = I_O\ddot{u}^t_\theta \tag{a}$$

where

$$u^t_\theta(t) = u_\theta(t) + u_{g\theta}(t) \tag{b}$$

Here u_θ is the rotation of the roof slab relative to the ground and $I_O = m(b^2 + d^2)/12$ is the
moment of inertia of the roof slab about the axis normal to the slab passing through its center
of mass O. The units of moment of inertia are force $\times$ (length)^2/acceleration. The torque f_S
and relative rotation u_θ are related by

$$f_S = k_\theta u_\theta \tag{c}$$

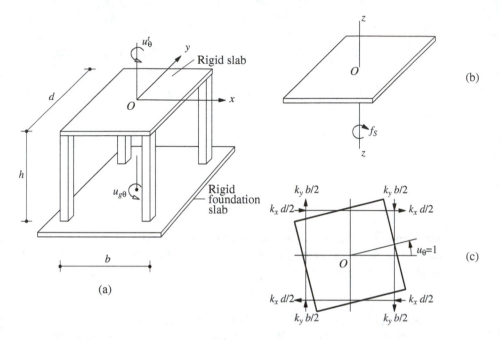

Figure E1.7

where k_θ is the torsional stiffness. To determine k_θ, we introduce a unit rotation, $u_\theta = 1$, and identify the resisting forces in each column (Fig. E1.7c). For a column with both ends clamped, $k_x = 12EI_y/h^3$ and $k_y = 12EI_x/h^3$. The torque required to equilibrate these resisting forces is

$$k_\theta = 4\left(k_x \frac{d}{2}\frac{d}{2}\right) + 4\left(k_y \frac{b}{2}\frac{b}{2}\right) = k_x d^2 + k_y b^2 \qquad \text{(d)}$$

Substituting Eqs. (c), (d), and (b) in (a) gives

$$I_O \ddot{u}_\theta + (k_x d^2 + k_y b^2)u_\theta = -I_O \ddot{u}_{g\theta} \qquad \text{(e)}$$

This is the equation governing the relative rotation u_θ of the roof slab due to rotational acceleration $\ddot{u}_{g\theta}$ of the foundation slab.

1.8 PROBLEM STATEMENT AND ELEMENT FORCES

1.8.1 Problem Statement

Given the mass m, the stiffness k of a linearly elastic system, or the force–deformation relation $f_S(u)$ for an inelastic system, the damping coefficient c, and the dynamic excitation—which may be an external force $p(t)$ or ground acceleration $\ddot{u}_g(t)$—a fundamental problem in structural dynamics is to determine the response of an SDF system: the idealized one-story system or the mass–spring–damper system. The term *response* is used in a general sense to include any response quantity, such as displacement, velocity, or acceleration of the mass; also, an internal force or internal stress in the structure. When the excitation is an external force, the response quantities of interest are the displacement or deformation $u(t)$, velocity $\dot{u}(t)$, and acceleration $\ddot{u}(t)$ of the mass. For earthquake excitation, both the total (or absolute) and the relative values of these quantities may be needed. The relative displacement $u(t)$ associated with deformations of the structure is the most important since the internal forces in the structure are directly related to $u(t)$.

1.8.2 Element Forces

Once the deformation response history $u(t)$ has been evaluated by dynamic analysis of the structure (i.e., by solving the equation of motion), the element forces—bending moments, shears, and axial forces—and stresses needed for structural design can be determined by static analysis of the structure at each instant in time (i.e., no additional dynamic analysis is necessary). This static analysis of a one-story linearly elastic frame can be visualized in two ways:

1. At each instant, the lateral displacement u is known to which joint rotations are related and hence they can be determined; see Eq. (b) of Example 1.1. From the known displacement and rotation of each end of a structural element (beam and column) the element forces (bending moments and shears) can be determined through the element stiffness properties (Appendix 1); and stresses can be obtained from element forces.

2. The second approach is to introduce the *equivalent static force*, a central concept in earthquake response of structures, as we shall see in Chapter 6. At any instant of time t this force f_S is the static (slowly applied) external force that will produce the deformation u determined by dynamic analysis. Thus

$$f_S(t) = ku(t) \tag{1.8.1}$$

where k is the lateral stiffness of the structure. Alternatively, f_S can be interpreted as the external force that will produce the same deformation u in the stiffness component of the structure [i.e., the system without mass or damping (Fig. 1.5.2b)] as that determined by dynamic analysis of the structure [i.e., the system with mass, stiffness, and damping (Fig. 1.5.2a)]. Element forces or stresses can be determined at each time instant by static analysis of the structure subjected to the force f_S determined from Eq. (1.8.1). It is unnecessary to introduce the equivalent static force concept for the mass–spring–damper system because the spring force, also given by Eq. (1.8.1), can readily be visualized.

For inelastic systems the element forces can be determined by appropriate modifications of these procedures to recognize that such systems are typically analyzed by time-stepping procedures with iteration within a time step (Chapter 5).

Why is external force in the second approach defined as $f_s(t)$ and not as $f_I(t)$? From Eq. (1.7.2), $-f_I(t) = f_S(t) + f_D(t) = ku(t) + c\dot{u}(t)$. It is inappropriate to include the velocity-dependent damping force because for structural design the computed element stresses are to be compared with allowable stresses that are specified based on static tests on materials (i.e., tests conducted at slow loading rates).

1.9 COMBINING STATIC AND DYNAMIC RESPONSES

In practical application we need to determine the total forces in a structure, including those existing before dynamic excitation of the structure and those resulting from the dynamic excitation. For a linear system the total forces can be determined by combining the results of two separate analyses: (1) static analysis of the structure due to dead and live loads, temperature changes, and so on; and (2) dynamic analysis of the structure subjected to the time-varying excitation. This direct superposition of the results of two analyses is valid only for linear systems.

The analysis of nonlinear systems cannot, however, be separated into two independent analyses. The dynamic analysis of such a system must recognize the forces and deformations already existing in the structure before the onset of dynamic excitation. This is necessary, in part, to establish the initial stiffness property of the structure required to start the dynamic analysis.

1.10 METHODS OF SOLUTION OF THE DIFFERENTIAL EQUATION

The equation of motion for a linear SDF system subjected to external force is the second-order differential equation derived earlier:

$$m\ddot{u} + c\dot{u} + ku = p(t) \qquad (1.10.1)$$

The initial displacement $u(0)$ and initial velocity $\dot{u}(0)$ at time zero must be specified to define the problem completely. Typically, the structure is at rest before the onset of dynamic excitation, so that the initial velocity and displacement are zero. A brief review of four methods of solution is given in the following sections.

1.10.1 Classical Solution

Complete solution of the linear differential equation of motion consists of the sum of the complementary solution $u_c(t)$ and the particular solution $u_p(t)$, that is, $u(t) = u_c(t) + u_p(t)$. Since the differential equation is of second order, two constants of integration are involved. They appear in the complementary solution and are evaluated from a knowledge of the initial conditions.

Example 1.8

Consider a step force: $p(t) = p_o, t \geq 0$. In this case, the differential equation of motion for a system without damping (i.e., $c = 0$) is

$$m\ddot{u} + ku = p_o \tag{a}$$

The particular solution for Eq. (a) is

$$u_p(t) = \frac{p_o}{k} \tag{b}$$

and the complementary solution is

$$u_c(t) = A \cos \omega_n t + B \sin \omega_n t \tag{c}$$

where A and B are constants of integration and $\omega_n = \sqrt{k/m}$.

The complete solution is given by the sum of Eqs. (b) and (c):

$$u(t) = A \cos \omega_n t + B \sin \omega_n t + \frac{p_o}{k} \tag{d}$$

If the system is initially at rest, $u(0) = 0$ and $\dot{u}(0) = 0$ at $t = 0$. For these initial conditions the constants A and B can be determined:

$$A = -\frac{p_o}{k} \qquad B = 0 \tag{e}$$

Substituting Eq. (e) in Eq. (d) gives

$$u(t) = \frac{p_o}{k}(1 - \cos \omega_n t) \tag{f}$$

The classical solution will be the principal method we will use in solving the differential equation for free vibration and for excitations that can be described analytically, such as harmonic, step, and pulse forces.

1.10.2 Duhamel's Integral

Another well-known approach to the solution of linear differential equations, such as the equation of motion of an SDF system, is based on representing the applied force as a sequence of infinitesimally short impulses. The response of the system to an applied force, $p(t)$, at time t is obtained by adding the responses to all impulses up to that time. We develop this method in Chapter 4, leading to the following result for an undamped SDF system:

$$u(t) = \frac{1}{m\omega_n} \int_0^t p(\tau) \sin[\omega_n(t - \tau)]\, d\tau \tag{1.10.2}$$

where $\omega_n = \sqrt{k/m}$. Implicit in this result are "at rest" initial conditions. Equation (1.10.2), known as *Duhamel's integral*, is a special form of the convolution integral found in textbooks on differential equations.

Example 1.9

Using Duhamel's integral, we determine the response of an SDF system, assumed to be initially at rest, to a step force, $p(t) = p_o, t \geq 0$. For this applied force, Eq. (1.10.2) specializes to

$$u(t) = \frac{p_o}{m\omega_n} \int_0^t \sin[\omega_n(t-\tau)]\,d\tau = \frac{p_o}{m\omega_n}\left[\frac{\cos\omega_n(t-\tau)}{\omega_n}\right]_{\tau=0}^{\tau=t} = \frac{p_o}{k}(1-\cos\omega_n t)$$

This result is the same as that obtained in Section 1.10.1 by the classical solution of the differential equation.

Duhamel's integral provides an alternative method to the classical solution if the applied force $p(t)$ is defined analytically by a simple function that permits analytical evaluation of the integral. For complex excitations that are defined only by numerical values of $p(t)$ at discrete time instants, Duhamel's integral can be evaluated by numerical methods. Such methods are not included in this book, however, because more efficient numerical procedures are available to determine dynamic response; some of these are presented in Chapter 5.

1.10.3 Frequency-Domain Method

The Laplace and Fourier transforms provide powerful tools for the solution of linear differential equations, in particular the equation of motion for a linear SDF system. Because the two transform methods are similar in concept, here we mention only the use of Fourier transform, which leads to the *frequency-domain method* of dynamic analysis.

The Fourier transform $P(\omega)$ of the excitation function $p(t)$ is defined by

$$P(\omega) = \mathcal{F}[p(t)] = \int_{-\infty}^{\infty} p(t)e^{-i\omega t}\,dt \tag{1.10.3}$$

The Fourier transform $U(\omega)$ of the solution $u(t)$ of the differential equation is then given by

$$U(\omega) = H(\omega)P(\omega) \tag{1.10.4}$$

where the complex frequency-response function $H(\omega)$ describes the response of the system to harmonic excitation. Finally, the desired solution $u(t)$ is given by the inverse Fourier transform of $U(\omega)$:

$$u(t) = \frac{1}{2\pi}\int_{-\infty}^{\infty} H(\omega)P(\omega)e^{i\omega t}\,d\omega \tag{1.10.5}$$

Straightforward integration can be used to evaluate the integral of Eq. (1.10.3) but contour integration in the complex plane is necessary for Eq. (1.10.5). Closed-form results can be obtained only if $p(t)$ is a simple function, and application of the Fourier transform method was restricted to such $p(t)$ until high-speed computers became available.

The Fourier transform method is now feasible for the dynamic analysis of linear systems to complicated excitations $p(t)$ or $\ddot{u}_g(t)$ that are described numerically. In such situations, the integrals of both Eqs. (1.10.3) and (1.10.5) are evaluated numerically by the *discrete Fourier transform method* using the *fast Fourier transform* algorithm developed in 1965. These concepts are developed in Appendix A.

Figure 1.10.1   These two reinforced-concrete dome-shaped containment structures house the nuclear reactors of the San Onofre power plant in California. For design purposes, their fundamental natural vibration period was computed to be 0.15 sec assuming the base as fixed, and 0.50 sec considering soil flexibility. This large difference in the period indicates the important effect of soil–structure interaction for these structures. (Courtesy of Southern California Edison.)

The frequency-domain method of dynamic analysis is symbolized by Eqs. (1.10.3) and (1.10.5). The first gives the amplitudes $P(\omega)$ of all the harmonic components that make up the excitation $p(t)$. The second equation can be interpreted as evaluating the harmonic response of the system to each component of the excitation and then superposing the harmonic responses to obtain the response $u(t)$. The frequency-domain method, which is an alternative to the *time-domain method* symbolized by Duhamel's integral, is especially useful and powerful for dynamic analysis of structures interacting with unbounded media. Examples are (1) the earthquake response analysis of a structure where the effects of interaction between the structure and the unbounded underlying soil are significant (Fig. 1.10.1), and (2) the earthquake response analysis of concrete dams interacting with the water impounded in the reservoir that extends to great distances in the upstream

Figure 1.10.2 Morrow Point Dam, a 465-ft-high arch dam, on the Gunnison River, Colorado. Determined by forced vibrations tests, the fundamental natural vibration period of the dam in antisymmetric vibration is 0.268 sec with the reservoir partially full and 0.303 sec with a full reservoir. (Courtesy of U.S Bureau of Reclamation.)

direction (Fig. 1.10.2). Because earthquake analysis of such complex structure–soil and structure–fluid systems is beyond the scope of this book, a comprehensive presentation of the frequency-domain method of dynamic analysis is not included. However, an introduction to the method is presented in Appendix A.

1.10.4 Numerical Methods

The preceding three dynamic analysis methods are restricted to linear systems and cannot consider the inelastic behavior of structures anticipated during earthquakes if the ground shaking is intense. The only practical approach for such systems involves numerical time-stepping methods, which are presented in Chapter 5. These methods are also useful for evaluating the response of linear systems to excitation—applied force $p(t)$ or ground motion $\ddot{u}_g(t)$—which is too complicated to be defined analytically and is described only numerically.

1.11 STUDY OF SDF SYSTEMS: ORGANIZATION

We will study the dynamic response of linearly elastic SDF systems in free vibration (Chapter 2), to harmonic and periodic excitations (Chapter 3), to step and pulse excitations (Chapter 4), and to earthquake ground motion (Chapter 6). Because most structures are designed with the expectation that they will deform beyond the linearly elastic limit during major, infrequent earthquakes, the inelastic response of SDF systems is studied in Chapter 7. The time variation of response $r(t)$ to these various excitations will be of interest. For structural design purposes, the maximum value (over time) of response r contains the crucial information, for it is related to the maximum forces and deformations that a structure must be able to withstand. We will be especially interested in the peak value of response, or for brevity, *peak response*, defined as the maximum of the absolute value of the response quantity:

$$r_o \equiv \max_t |r(t)| \qquad (1.11.1)$$

By definition the peak response is positive; the algebraic sign is dropped because it is usually irrelevant for design. Note that the subscript o attached to a response quantity denotes its peak value.

FURTHER READING

Clough, R. W., and Penzien, J., *Dynamics of Structures*, McGraw-Hill, New York, 1993, Sections 4-3, 6-2, 6-3, and 12-6.

Humar, J. L., *Dynamics of Structures*, 2nd ed., A. A. Balkema Publishers, Lisse, The Netherlands, 2002, Chapter 9 and Section 13.5.

APPENDIX 1: STIFFNESS COEFFICIENTS FOR A FLEXURAL ELEMENT

From the slope deflection equations, we can write the stiffness coefficients for a linearly elastic, prismatic (uniform) frame element. These are presented in Fig. A1.1 for an element of length L, second moment of area I, and elastic modulus E. The stiffness

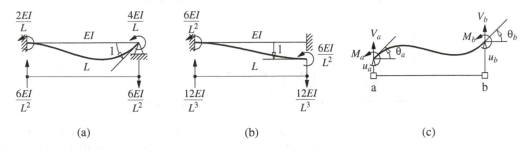

Figure A1.1

coefficients for joint rotation are shown in part (a) and those for joint translation in part (b) of the figure.

Now consider the element shown in Fig. A1.1c with its two nodes identified as a and b that is assumed to be axially inextensible. Its four degrees of freedom are the nodal translations u_a and u_b and nodal rotations θ_a and θ_b. The bending moments at the two nodes are related to the four DOFs as follows:

$$M_a = \frac{4EI}{L}\theta_a + \frac{2EI}{L}\theta_b + \frac{6EI}{L^2}u_a - \frac{6EI}{L^2}u_b \tag{A1.1}$$

$$M_b = \frac{2EI}{L}\theta_a + \frac{4EI}{L}\theta_b + \frac{6EI}{L^2}u_a - \frac{6EI}{L^2}u_b \tag{A1.2}$$

The shearing forces at the two nodes are related to the four DOFs as follows:

$$V_a = \frac{12EI}{L^3}u_a - \frac{12EI}{L^3}u_b + \frac{6EI}{L^2}\theta_a + \frac{6EI}{L^2}\theta_b \tag{A1.3}$$

$$V_b = -\frac{12EI}{L^3}u_a + \frac{12EI}{L^3}u_b - \frac{6EI}{L^2}\theta_a - \frac{6EI}{L^2}\theta_b \tag{A1.4}$$

At each instant of time, the nodal forces M_a, M_b, V_a, and V_b are calculated from u_a, u_b, θ_a, and θ_b. The bending moment and shear at any other location along the element are determined by statics applied to the element of Fig. A1.1c.

PROBLEMS

1.1–
1.3   Starting from the basic definition of stiffness, determine the effective stiffness of the combined spring and write the equation of motion for the spring–mass systems shown in Figs. P1.1 to P1.3.

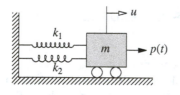

Figure P1.1

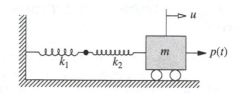

Figure P1.2

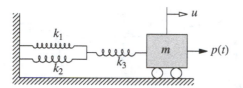

Figure P1.3

1.4   Derive the equation governing the free motion of a simple pendulum that consists of a rigid massless rod pivoted at point O with a mass m attached at the tip (Fig. P1.4). Linearize the equation, for small oscillations, and determine the natural frequency of oscillation.

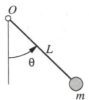

O

L

θ

m **Figure P1.4**

1.5   Consider the free motion in the xy plane of a compound pendulum that consists of a rigid rod suspended from a point (Fig. P1.5). The length of the rod is L and its mass m is uniformly distributed. The width of the uniform rod is b and the thickness is t. The angular displacement of the centerline of the pendulum measured from the y-axis is denoted by $\theta(t)$.
    (a) Derive the equation governing $\theta(t)$.
    (b) Linearize the equation for small θ.
    (c) Determine the natural frequency of small oscillations.

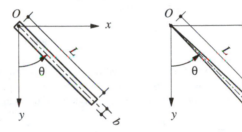

Figure P1.5   **Figure P1.6**

1.6   Repeat Problem 1.5 for the system shown in Fig. P1.6, which differs in only one sense: its width varies from zero at O to b at the free end.

1.7   Develop the equation governing the longitudinal motion of the system of Fig. P1.7. The rod is made of an elastic material with elastic modulus E; its cross-sectional area is A and its length is L. Ignore the mass of the rod and measure u from the static equilibrium position.

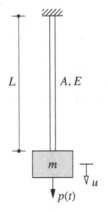

L A, E

m

u

$p(t)$ **Figure P1.7**

1.8 A rigid disk of mass m is mounted at the end of a flexible shaft (Fig. P1.8). Neglecting the weight of the shaft and neglecting damping, derive the equation of free torsional vibration of the disk. The shear modulus (of rigidity) of the shaft is G.

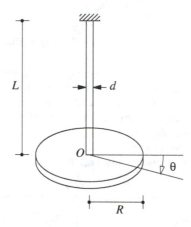

Figure P1.8

1.9– Write the equation governing the free vibration of the systems shown in Figs. P1.9 to P1.11.
1.11 Assuming the beam to be massless, each system has a single DOF defined as the vertical deflection under the weight w. The flexural rigidity of the beam is EI and the length is L.

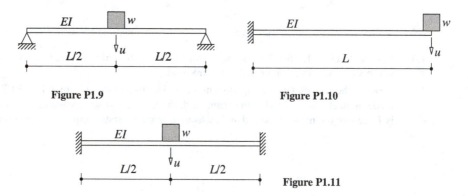

Figure P1.9   **Figure P1.10**

Figure P1.11

1.12 Determine the natural frequency of a weight w suspended from a spring at the midpoint of a simply supported beam (Fig. P1.12). The length of the beam is L, and its flexural rigidity is EI. The spring stiffness is k. Assume the beam to be massless.

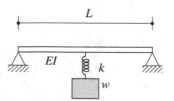

Figure P1.12

1.13   Derive the equation of motion for the frame shown in Fig. P1.13. The flexural rigidity of the beam and columns is as noted. The mass lumped at the beam is m; otherwise, assume the frame to be massless and neglect damping. By comparing the result with Eq. (1.3.2), comment on the effect of base fixity.

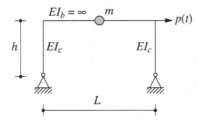

Figure P1.13

1.14   Write the equation of motion for the one-story, one-bay frame shown in Fig. P1.14. The flexural rigidity of the beam and columns is as noted. The mass lumped at the beam is m; otherwise, assume the frame to be massless and neglect damping. By comparing this equation of motion with the one for Example 1.1, comment on the effect of base fixity.

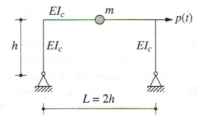

Figure P1.14

1.15–   Write the equation of motion of the one-story, one-bay frame shown in Figs. P1.15 and P1.16.
1.16   The flexural rigidity of the beam and columns is as noted. The mass lumped at the beam is m; otherwise, assume the frame to be massless and neglect damping. Check your result from Problem 1.15 against Eq. (1.3.5). Comment on the effect of base fixity by comparing the two equations of motion.

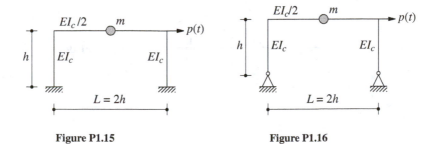

Figure P1.15   **Figure P1.16**

1.17   A heavy rigid platform of weight w is supported by four columns, hinged at the top and the bottom, and braced laterally in each side panel by two diagonal steel wires as shown in

Fig. P1.17. Each diagonal wire is pretensioned to a high stress; its cross-sectional area is A and elastic modulus is E. Neglecting the mass of the columns and wires, derive the equation of motion governing free vibration in (**a**) the x-direction, and (**b**) the y-direction. (*Hint:* Because of high pretension, all wires contribute to the structural stiffness, unlike Example 1.2, where the braces in compression do not provide stiffness.)

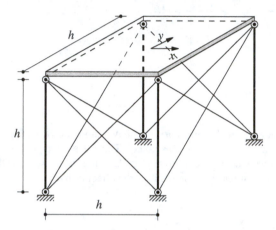

Figure P1.17

1.18   Derive the equation of motion governing the torsional vibration of the system of Fig. P1.17 about the vertical axis passing through the center of the platform.

1.19   An automobile is crudely idealized as a lumped mass m supported on a spring–damper system as shown in Fig. P1.19. The automobile travels at constant speed v over a road whose roughness is known as a function of position along the road. Derive the equation of motion.

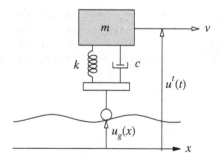

Figure P1.19

2

Free Vibration

PREVIEW

A structure is said to be undergoing *free vibration* when it is disturbed from its static equilibrium position and then allowed to vibrate without any external dynamic excitation. In this chapter we study free vibration leading to the notions of the natural vibration frequency and damping ratio for an SDF system. We will see that the rate at which the motion decays in free vibration is controlled by the damping ratio. Thus the analytical results describing free vibration provide a basis to determine the natural frequency and damping ratio of a structure from experimental data of the type shown in Fig. 1.1.4.

Although damping in actual structures is due to several energy-dissipating mechanisms acting simultaneously, a mathematically convenient approach is to idealize them by equivalent viscous damping. Consequently, this chapter deals primarily with viscously damped systems. However, free vibration of systems in the presence of Coulomb friction forces is analyzed toward the end of the chapter.

2.1 UNDAMPED FREE VIBRATION

The motion of linear SDF systems, visualized as an idealized one-story frame or a mass–spring–damper system, subjected to external force $p(t)$ is governed by Eq. (1.5.2). Setting $p(t) = 0$ gives the differential equation governing free vibration of the system, which for systems without damping ($c = 0$) specializes to

$$m\ddot{u} + ku = 0 \qquad (2.1.1)$$

Free vibration is initiated by disturbing the system from its static equilibrium position by imparting the mass some displacement $u(0)$ and velocity $\dot{u}(0)$ at time zero, defined as the instant the motion is initiated:

$$u = u(0) \qquad \dot{u} = \dot{u}(0) \tag{2.1.2}$$

Subject to these initial conditions, the solution to the homogeneous differential equation is obtained by standard methods (see Derivation 2.1):

$$u(t) = u(0) \cos \omega_n t + \frac{\dot{u}(0)}{\omega_n} \sin \omega_n t \tag{2.1.3}$$

where

$$\omega_n = \sqrt{\frac{k}{m}} \tag{2.1.4}$$

Equation (2.1.3) is plotted in Fig. 2.1.1. It shows that the system undergoes vibratory (or oscillatory) motion about its static equilibrium (or undeformed, $u = 0$) position; and that this motion repeats itself after every $2\pi/\omega_n$ seconds. In particular, the state (displacement and velocity) of the mass at two time instants, t_1 and $t_1 + 2\pi/\omega_n$, is identical: $u(t_1) = u(t_1 + 2\pi/\omega_n)$ and $\dot{u}(t_1) = \dot{u}(t_1 + 2\pi/\omega_n)$. These equalities can easily be proved, starting with Eq. (2.1.3). The motion described by Eq. (2.1.3) and shown in Fig. 2.1.1 is known as *simple harmonic motion*.

The portion a–b–c–d–e of the displacement–time curve describes one cycle of free vibration of the system. From its static equilibrium (or undeformed) position at a, the mass moves to the right, reaching its maximum positive displacement u_o at b, at which time the velocity is zero and the displacement begins to decrease and the mass returns back to its equilibrium position c, at which time the velocity is maximum and hence the

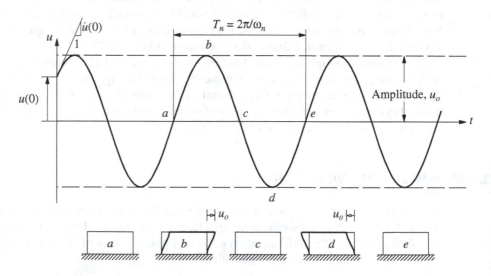

Figure 2.1.1   Free vibration of a system without damping.

mass continues moving to the left, reaching its minimum displacement $-u_o$ at d, at which time the velocity is again zero and the displacement begins to decrease again and the mass returns to its equilibrium position at e. At time instant e, $2\pi/\omega_n$ seconds after time instant a, the state (displacement and velocity) of the mass is the same as it was at time instant a, and the mass is ready to begin another cycle of vibration.

The time required for the undamped system to complete one cycle of free vibration is the *natural period of vibration* of the system, which we denote as T_n, in units of seconds. It is related to the *natural circular frequency of vibration*, ω_n, in units of radians per second:

$$T_n = \frac{2\pi}{\omega_n} \tag{2.1.5}$$

A system executes $1/T_n$ cycles in 1 sec. This *natural cyclic frequency of vibration* is denoted by

$$f_n = \frac{1}{T_n} \tag{2.1.6}$$

The units of f_n are hertz (Hz) [cycles per second (cps)]; f_n is obviously related to ω_n through

$$f_n = \frac{\omega_n}{2\pi} \tag{2.1.7}$$

The term *natural frequency of vibration* applies to both ω_n and f_n.

The natural vibration properties ω_n, T_n, and f_n depend only on the mass and stiffness of the structure; see Eqs. (2.1.4) to (2.1.6). The stiffer of two SDF systems having the same mass will have the higher natural frequency and the shorter natural period. Similarly, the heavier (more mass) of two structures having the same stiffness will have the lower natural frequency and the longer natural period. The qualifier *natural* is used in defining T_n, ω_n, and f_n to emphasize the fact that these are natural properties of the system when it is allowed to vibrate freely without any external excitation. Because the system is linear, these vibration properties are independent of the initial displacement and velocity. The natural frequency and period of the various types of structures of interest to us vary over a wide range, as shown in Figs. 1.10.1, 1.10.2, and 2.1.2a–f.

The natural circular frequency ω_n, natural cyclic frequency f_n, and natural period T_n defined by Eqs. (2.1.4) to (2.1.6) can be expressed in the alternative form

$$\omega_n = \sqrt{\frac{g}{\delta_{st}}} \qquad f_n = \frac{1}{2\pi}\sqrt{\frac{g}{\delta_{st}}} \qquad T_n = 2\pi\sqrt{\frac{\delta_{st}}{g}} \tag{2.1.8}$$

where $\delta_{st} = mg/k$, and where g is the acceleration due to gravity. This is the static deflection of the mass m suspended from a spring of stiffness k; it can be visualized as the system of Fig. 1.6.1 oriented in the vertical direction. In the context of the one-story frame of Fig. 1.2.1, δ_{st} is the lateral displacement of the mass due to lateral force mg.

Figure 2.1.2a Alcoa Building, San Francisco, California. The fundamental natural vibration periods of this 26-story steel building are 1.67 sec for north–south (longitudinal) vibration, 2.21 sec for east–west (transverse) vibration, and 1.12 sec for torsional vibration about a vertical axis. These vibration properties were determined by forced vibration tests. (Courtesy of International Structural Slides.)

Figure 2.1.2b Transamerica Building, San Francisco, California. The fundamental natural vibration periods of this 49-story steel building, tapered in elevation, are 2.90 sec for north–south vibration and also for east–west vibration. These vibration properties were determined by forced vibration tests. (Courtesy of International Structural Slides.)

Figure 2.1.2c   Medical Center Building, Richmond, California. The fundamental natural vibration periods of this three-story steel frame building are 0.63 sec for vibration in the long direction, 0.74 sec in the short direction, and 0.46 sec for torsional vibration about a vertical axis. These vibration properties were determined from motions of the building recorded during the 1989 Loma Prieta earthquake. (Courtesy of California Strong Motion Instrumentation Program.)

Figure 2.1.2d   Pine Flat Dam on the Kings River, near Fresno, California. The fundamental natural vibration period of this 400-ft-high concrete gravity dam was measured by forced vibration tests to be 0.288 sec and 0.306 sec with the reservoir depth at 310 ft and 345 ft, respectively.

Figure 2.1.2e   Golden Gate Bridge, San Francisco, California. The fundamental natural vibration periods of this suspension bridge with the main span of 4200 ft are 18.2 sec for transverse vibration, 10.9 sec for vertical vibration, 3.81 sec for longitudinal vibration, and 4.43 sec for torsional vibration. These vibration properties were determined from recorded motions of the bridge under ambient (wind, traffic, etc.) conditions. (Courtesy of International Structural Slides.)

Figure 2.1.2f   Reinforced-concrete chimney, located in Aramon, France. The fundamental natural vibration period of this 250-m-high chimney is 3.57 sec; it was determined from records of wind-induced vibration.

The undamped system oscillates back and forth between the maximum displacement u_o and minimum displacement $-u_o$. The magnitude u_o of these two displacement values is the same; it is called the *amplitude of motion* and is given by

$$u_o = \sqrt{[u(0)]^2 + \left[\frac{\dot{u}(0)}{\omega_n}\right]^2} \tag{2.1.9}$$

The amplitude u_o depends on the initial displacement and velocity. Cycle after cycle it remains the same; that is, the motion does not decay. We had mentioned in Section 1.1 this unrealistic behavior of a system if a damping mechanism to represent dissipation of energy is not included.

The natural frequency of the one-story frame of Fig. 1.3.2a with lumped mass m and columns clamped at the base is

$$\omega_n = \sqrt{\frac{k}{m}} \qquad k = \frac{24EI_c}{h^3} \frac{12\rho + 1}{12\rho + 4} \tag{2.1.10}$$

where the lateral stiffness comes from Eq. (1.3.5) and $\rho = (EI_b/L) \div (2EI_c/h)$. For the extreme cases of a rigid beam, $\rho = \infty$, and a beam with no stiffness, $\rho = 0$, the lateral stiffnesses are given by Eqs. (1.3.2) and (1.3.3) and the natural frequencies are

$$(\omega_n)_{\rho=\infty} = \sqrt{\frac{24EI_c}{mh^3}} \qquad (\omega_n)_{\rho=0} = \sqrt{\frac{6EI_c}{mh^3}} \tag{2.1.11}$$

The natural frequency is doubled as the beam-to-column stiffness ratio, ρ, increases from 0 to ∞; its variation with ρ is shown in Fig. 2.1.3.

The natural frequency is similarly affected by the boundary conditions at the base of the columns. If the columns are hinged at the base rather than clamped and the beam is rigid, $\omega_n = \sqrt{6EI_c/mh^3}$, which is one-half of the natural frequency of the frame with clamped-base columns.

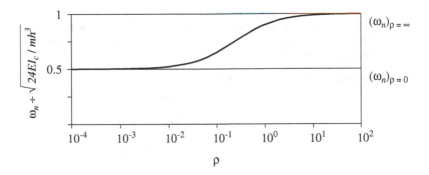

Figure 2.1.3   Variation of natural frequency, ω_n, with beam-to-column stiffness ratio, ρ.

Derivation 2.1

The solution of Eq. (2.1.1), a linear, homogeneous, second-order differential equation with constant coefficients, has the form

$$u = e^{\lambda t} \tag{a}$$

where the constant λ is unknown. Substitution into Eq. (2.1.1) gives

$$(m\lambda^2 + k)e^{\lambda t} = 0$$

The exponential term is never zero, so

$$(m\lambda^2 + k) = 0 \tag{b}$$

Known as the *characteristic equation*, Eq. (b) has two roots:

$$\lambda_{1,2} = \pm i\omega_n \tag{c}$$

where $i = \sqrt{-1}$. The general solution of Eq. (2.1.1) is

$$u(t) = a_1 e^{\lambda_1 t} + a_2 e^{\lambda_2 t}$$

which after substituting Eq. (c) becomes

$$u(t) = a_1 e^{i\omega_n t} + a_2 e^{-i\omega_n t} \tag{d}$$

where a_1 and a_2 are complex-valued constants yet undetermined. By using the Euler relations, $e^{ix} = \cos x + i \sin x$ and $e^{-ix} = \cos x - i \sin x$. Equation (d) can be rewritten as

$$u(t) = A \cos \omega_n t + B \sin \omega_n t \tag{e}$$

where A and B are real-valued constants yet undetermined. Equation (e) is differentiated to obtain

$$\dot{u}(t) = -\omega_n A \sin \omega_n t + \omega_n B \cos \omega_n t \tag{f}$$

Evaluating Eqs. (e) and (f) at time zero gives the constants A and B in terms of the initial displacement $u(0)$ and initial velocity $\dot{u}(0)$:

$$u(0) = A \qquad \dot{u}(0) = \omega_n B \tag{g}$$

Substituting for A and B from Eq. (g) into Eq. (e) leads to the solution given in Eq. (2.1.3).

Example 2.1

For the one-story industrial building of Example 1.2, determine the natural circular frequency, natural cyclic frequency, and natural period of vibration in **(a)** the north–south direction and **(b)** the east–west direction.

Solution   **(a)** *North–south direction:*

$$(\omega_n)_{\text{N–S}} = \sqrt{\frac{38.58}{0.04663}} = 28.73 \text{ rad/sec}$$

$$(T_n)_{\text{N–S}} = \frac{2\pi}{28.73} = 0.219 \text{ sec}$$

$$(f_n)_{\text{N–S}} = \frac{1}{0.219} = 4.57 \text{ Hz}$$

(b) *East–west direction:*

$$(\omega_n)_{\text{E–W}} = \sqrt{\frac{119.6}{0.04663}} = 50.64 \text{ rad/sec}$$

$$(T_n)_{\text{E–W}} = \frac{2\pi}{50.64} = 0.124 \text{ sec}$$

$$(f_n)_{\text{E–W}} = \frac{1}{0.124} = 8.06 \text{ Hz}$$

Observe that the natural frequency is much higher (and the natural period much shorter) in the east–west direction because the vertical bracing makes the system much stiffer, although the columns of the frame are bending about their weak axis; the vibrating mass is the same in both directions.

Example 2.2

For the three-span box girder bridge of Example 1.3, determine the natural circular frequency, natural cyclic frequency, and natural period of vibration for longitudinal motion.

Solution

$$\omega_n = \sqrt{\frac{k}{m}} = \sqrt{\frac{25,880}{214.9}} = 10.97 \text{ rad/sec}$$

$$T_n = \frac{2\pi}{10.97} = 0.573 \text{ sec}$$

$$f_n = \frac{1}{0.573} = 1.75 \text{ Hz}$$

Example 2.3

Determine the natural cyclic frequency and the natural period of vibration of a weight of 20 lb suspended as described in Example 1.4.

Solution

$$f_n = \frac{1}{2\pi}\sqrt{\frac{g}{\delta_{\text{st}}}} \qquad \delta_{\text{st}} = \frac{w}{k_e} = \frac{20}{13.39} = 1.494 \text{ in.}$$

$$f_n = \frac{1}{2\pi}\sqrt{\frac{386}{1.494}} = 2.56 \text{ Hz}$$

$$T_n = \frac{1}{f_n} = 0.391 \text{ sec}$$

Example 2.4

Consider the system described in Example 1.7 with $b = 30$ ft, $d = 20$ ft, $h = 12$ ft, slab weight $= 0.1$ kip/ft^2, and the lateral stiffness of each column in the x and y directions is $k_x = 1.5$ and $k_y = 1.0$, both in kips/in. Determine the natural frequency and period of torsional motion about the vertical axis.

Solution From Example 1.7, the torsional stiffness k_θ and the moment of inertia I_O are

$$k_\theta = k_x d^2 + k_y b^2 = 1.5(12)(20)^2 + 1.0(12)(30)^2 = 18{,}000 \text{ kip-ft/rad}$$

$$I_O = m\frac{b^2 + d^2}{12} = \frac{0.1(30 \times 20)}{(32.2)}\left[\frac{(30)^2 + (20)^2}{12}\right] = 201.86 \text{ kip-sec}^2\text{-ft}$$

$$\omega_n = \sqrt{\frac{k_\theta}{I_O}} = 9.44 \text{ rad/sec} \qquad f_n = 1.49 \text{ Hz} \qquad T_n = 0.67 \text{ sec}$$

2.2 VISCOUSLY DAMPED FREE VIBRATION

Setting $p(t) = 0$ in Eq. (1.5.2) gives the differential equation governing free vibration of SDF systems with damping:

$$m\ddot{u} + c\dot{u} + ku = 0 \qquad (2.2.1a)$$

Dividing by m gives

$$\ddot{u} + 2\zeta\omega_n\dot{u} + \omega_n^2 u = 0 \qquad (2.2.1b)$$

where $\omega_n = \sqrt{k/m}$ as defined earlier and

$$\zeta = \frac{c}{2m\omega_n} = \frac{c}{c_{\text{cr}}} \qquad (2.2.2)$$

We will refer to

$$c_{\text{cr}} = 2m\omega_n = 2\sqrt{km} = \frac{2k}{\omega_n} \qquad (2.2.3)$$

as the *critical damping coefficient*, for reasons that will appear shortly; and ζ is the *damping ratio* or *fraction of critical damping*. The damping constant c is a measure of the energy dissipated in a cycle of free vibration or in a cycle of forced harmonic vibration (Section 3.8). However, the damping ratio—a dimensionless measure of damping—is a property of the system that also depends on its mass and stiffness. The differential equation (2.2.1) can be solved by standard methods (similar to Derivation 2.1) for given initial displacement $u(0)$ and velocity $\dot{u}(0)$. Before writing any formal solution, however, we examine the solution qualitatively.

2.2.1 Types of Motion

Figure 2.2.1 shows a plot of the motion $u(t)$ due to initial displacement $u(0)$ for three values of ζ. If $c < c_{\text{cr}}$ or $\zeta < 1$, the system oscillates about its equilibrium position with a progressively decreasing amplitude. If $c = c_{\text{cr}}$ or $\zeta = 1$, the system returns to its equilibrium

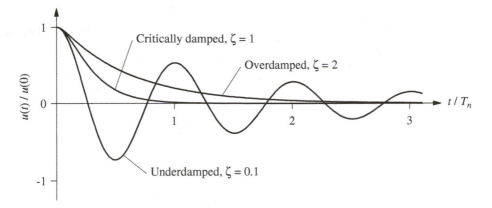

Figure 2.2.1 Free vibration of underdamped, critically damped, and overdamped systems.

position without oscillating. If $c > c_{cr}$ or $\zeta > 1$, again the system does not oscillate and returns to its equilibrium position, as in the $\zeta = 1$ case, but at a slower rate.

The damping coefficient c_{cr} is called the *critical damping coefficient* because it is the smallest value of c that inhibits oscillation completely. It represents the dividing line between oscillatory and nonoscillatory motion.

The rest of this presentation is restricted to *underdamped systems* ($c < c_{cr}$) because structures of interest—buildings, bridges, dams, nuclear power plants, offshore structures, etc.—all fall into this category, as typically, their damping ratio is less than 0.10. Therefore, we have little reason to study the dynamics of *critically damped systems* ($c = c_{cr}$) or *overdamped systems* ($c > c_{cr}$). Such systems do exist, however; for example, recoil mechanisms, such as the common automatic door closer, are overdamped; and instruments used to measure steady-state values, such as a scale measuring dead weight, are usually critically damped. Even for automobile shock absorber systems, however, damping is usually less than half of critical, $\zeta < 0.5$.

2.2.2 Underdamped Systems

The solution to Eq. (2.2.1) subject to the initial conditions of Eq. (2.1.2) for systems with $c < c_{cr}$ or $\zeta < 1$ is (see Derivation 2.2)

$$u(t) = e^{-\zeta \omega_n t} \left[u(0) \cos \omega_D t + \frac{\dot{u}(0) + \zeta \omega_n u(0)}{\omega_D} \sin \omega_D t \right] \qquad (2.2.4)$$

where

$$\omega_D = \omega_n \sqrt{1 - \zeta^2} \qquad (2.2.5)$$

Observe that Eq. (2.2.4) specialized for undamped systems ($\zeta = 0$) reduces to Eq. (2.1.3).

Equation (2.2.4) is plotted in Fig. 2.2.2, which shows the free vibration response of an SDF system with damping ratio $\zeta = 0.05$, or 5%. Included for comparison is the free

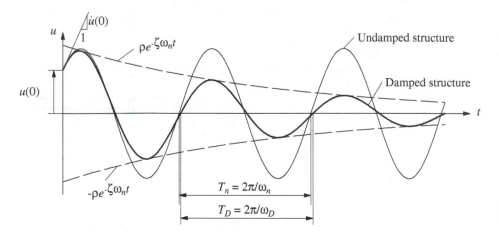

Figure 2.2.2 Effects of damping on free vibration.

vibration response of the same system but without damping, presented earlier in Fig. 2.1.1. Free vibration of both systems is initiated by the same initial displacement $u(0)$ and velocity $\dot{u}(0)$, and hence both displacement–time plots start at $t = 0$ with the same ordinate and slope. Equation (2.2.4) and Fig. 2.2.2 indicate that the *natural frequency of damped vibration* is ω_D, and it is related by Eq. (2.2.5) to the natural frequency ω_n of the system without damping. The *natural period of damped vibration*, $T_D = 2\pi/\omega_D$, is related to the natural period T_n without damping by

$$T_D = \frac{T_n}{\sqrt{1 - \zeta^2}} \tag{2.2.6}$$

The displacement amplitude of the undamped system is the same in all vibration cycles, but the damped system oscillates with amplitude decreasing with every cycle of vibration. Equation (2.2.4) indicates that the displacement amplitude decays exponentially with time, as shown in Fig. 2.2.2. The envelope curves $\pm \rho e^{-\zeta \omega_n t}$, where

$$\rho = \sqrt{[u(0)]^2 + \left[\frac{\dot{u}(0) + \zeta \omega_n u(0)}{\omega_D} \right]^2} \tag{2.2.7}$$

touch the displacement–time curve at points slightly to the right of its peak values.

Damping has the effect of lowering the natural frequency from ω_n to ω_D and lengthening the natural period from T_n to T_D. These effects are negligible for damping ratios below 20%, a range that includes most structures, as shown in Fig. 2.2.3, where the ratio $\omega_D/\omega_n = T_n/T_D$ is plotted against ζ. For most structures the damped properties ω_D and T_D are approximately equal to the undamped properties ω_n and T_n, respectively. For systems with critical damping, $\omega_D = 0$ and $T_D = \infty$. This is another way of saying that the system does not oscillate, as shown in Fig. 2.2.1.

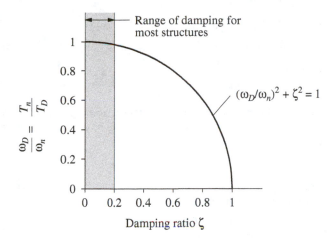

Figure 2.2.3 Effects of damping on the natural vibration frequency.

The more important effect of damping is on the rate at which free vibration decays. This is displayed in Fig. 2.2.4, where the free vibration due to initial displacement $u(0)$ is plotted for four systems having the same natural period T_n but differing damping ratios: $\zeta = 2, 5, 10,$ and 20%.

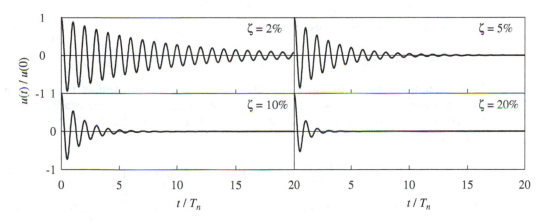

Figure 2.2.4 Free vibration of systems with four levels of damping: $\zeta = 2, 5, 10,$ and 20%.

Derivation 2.2

Substituting Eq. (a) of Derivation 2.1 into Eq. (2.2.1b) gives

$$\left(\lambda^2 + 2\zeta\omega_n\lambda + \omega_n^2\right)e^{\lambda t} = 0$$

which is satisfied for all values of t if

$$\lambda^2 + 2\zeta\omega_n\lambda + \omega_n^2 = 0 \tag{a}$$

Equation (a), which is known as the *characteristic equation*, has two roots:

$$\lambda_{1,2} = \omega_n\left(-\zeta \pm i\sqrt{1-\zeta^2}\right) \tag{b}$$

which are complex-valued for $\zeta < 1$. The general solution of Eq. (2.2.1b) is

$$u(t) = a_1 e^{\lambda_1 t} + a_2 e^{\lambda_2 t} \tag{c}$$

which after substituting Eq. (b) becomes

$$u(t) = e^{-\zeta \omega_n t}\left(a_1 e^{i\omega_D t} + a_2 e^{-i\omega_D t}\right) \tag{d}$$

where a_1 and a_2 are complex-valued constants as yet undetermined and ω_D is defined in Eq. (2.2.5). As in Derivation 2.1, the term in parentheses in Eq. (d) can be rewritten in terms of trigonometric functions to obtain

$$u(t) = e^{-\zeta \omega_n t}(A \cos \omega_D t + B \sin \omega_D t) \tag{e}$$

where A and B are real-valued constants yet undetermined. These can be expressed in terms of the initial conditions by proceeding along the lines of Derivation 2.1:

$$A = u(0) \qquad B = \frac{\dot{u}(0) + \zeta \omega_n u(0)}{\omega_D} \tag{f}$$

Substituting for A and B in Eq. (e) leads to the solution given in Eq. (2.2.4).

We now make two observations that will be useful later: (1) λ_1 and λ_2 in Eq. (b) are a complex conjugate pair, denoted by λ and $\bar{\lambda}$; and (2) a_1 and a_2 must also be a conjugate pair because $u(t)$ is real valued. Thus, Eq. (c) can be written as

$$u(t) = be^{\lambda t} + \bar{b}e^{\bar{\lambda} t} \tag{g}$$

where b is a complex-valued constant.

2.2.3 Decay of Motion

In this section a relation between the ratio of two successive peaks of damped free vibration and the damping ratio is presented. The ratio of the displacement at time t to its value a full vibration period T_D later is independent of t. Derived from Eq. (2.2.4), this ratio is given by the first equality in

$$\frac{u(t)}{u(t + T_D)} = \exp(\zeta \omega_n T_D) = \exp\left(\frac{2\pi \zeta}{\sqrt{1 - \zeta^2}}\right) \tag{2.2.8}$$

and the second equality is obtained by utilizing Eqs. (2.2.6) and (2.1.5). This result also gives the ratio u_i/u_{i+1} of successive peaks (maxima) shown in Fig. 2.2.5, because these peaks are separated by period T_D:

$$\frac{u_i}{u_{i+1}} = \exp\left(\frac{2\pi \zeta}{\sqrt{1 - \zeta^2}}\right) \tag{2.2.9}$$

The natural logarithm of this ratio, called the *logarithmic decrement*, we denote by δ:

$$\delta = \ln \frac{u_i}{u_{i+1}} = \frac{2\pi \zeta}{\sqrt{1 - \zeta^2}} \tag{2.2.10}$$

If ζ is small, $\sqrt{1 - \zeta^2} \simeq 1$ and this gives an approximate equation:

$$\delta \simeq 2\pi \zeta \tag{2.2.11}$$

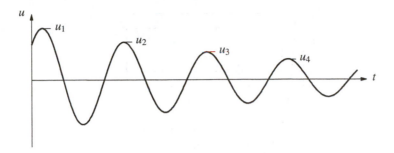

Figure 2.2.5

Figure 2.2.6 shows a plot of the exact and approximate relations between δ and ζ. It is clear that Eq. (2.2.11) is valid for $\zeta < 0.2$, which covers most practical structures.

If the decay of motion is slow, as is the case for lightly damped systems such as the aluminum model in Fig. 1.1.4, it is desirable to relate the ratio of two amplitudes several cycles apart, instead of successive amplitudes, to the damping ratio. Over j cycles the motion decreases from u_1 to u_{j+1}. This ratio is given by

$$\frac{u_1}{u_{j+1}} = \frac{u_1}{u_2}\frac{u_2}{u_3}\frac{u_3}{u_4} \cdots \frac{u_j}{u_{j+1}} = e^{j\delta}$$

Therefore,

$$\delta = (1/j) \ln (u_1/u_{j+1}) \simeq 2\pi\zeta \qquad (2.2.12)$$

To determine the number of cycles elapsed for a 50% reduction in displacement amplitude, we obtain the following relation from Eq. (2.2.12):

$$j_{50\%} \simeq 0.11/\zeta \qquad (2.2.13)$$

This equation is plotted in Fig. 2.2.7.

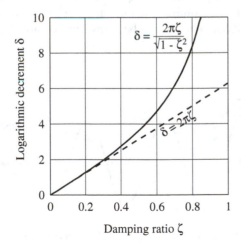

Figure 2.2.6 Exact and approximate relations between logarithmic decrement and damping ratio.

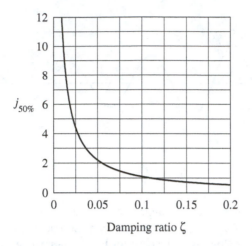

Figure 2.2.7 Number of cycles required to reduce the free vibration amplitude by 50%.

2.2.4 Free Vibration Tests

Because it is not possible to determine analytically the damping ratio ζ for practical structures, this elusive property should be determined experimentally. Free vibration experiments provide one means of determining the damping. Such experiments on two one-story models led to the free vibration records presented in Fig. 1.1.4; a part of such a record is shown in Fig. 2.2.8. For lightly damped systems the damping ratio can be determined from

$$\zeta = \frac{1}{2\pi j} \ln \frac{u_i}{u_{i+j}} \qquad \text{or} \qquad \zeta = \frac{1}{2\pi j} \ln \frac{\ddot{u}_i}{\ddot{u}_{i+j}} \qquad (2.2.14)$$

The first of these equations is equivalent to Eq. (2.2.12), which was derived from the equation for $u(t)$. The second is a similar equation in terms of accelerations, which are easier to measure than displacements. It can be shown to be valid for lightly damped systems.

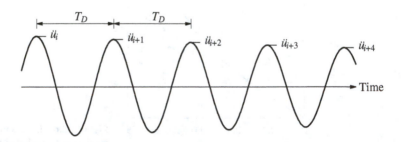

Figure 2.2.8 Acceleration record of a freely vibrating system.

The natural period T_D of the system can also be determined from the free vibration record by measuring the time required to complete one cycle of vibration. Comparing this with the natural period obtained from the calculated stiffness and mass of an idealized system tells us how accurately these properties were calculated and how well the idealization represents the actual structure.

Example 2.5

Determine the natural vibration period and damping ratio of the plexiglass frame model (Fig. 1.1.4a) from the acceleration record of its free vibration shown in Fig. 1.1.4c.

Solution   The peak values of acceleration and the time instants they occur can be read from the free vibration record or obtained from the corresponding data stored in a computer during the experiment. The latter provides the following data:

Peak	Time, t_i (sec)	Peak, $\ddot{u}_i$ (g)
1	1.110	0.915
11	3.844	0.076

$$T_D = \frac{3.844 - 1.110}{10} = 0.273 \text{ sec} \qquad \zeta = \frac{1}{2\pi(10)} \ln \frac{0.915\text{g}}{0.076\text{g}} = 0.0396 \text{ or } 3.96\%$$

Example 2.6

A free vibration test is conducted on an empty elevated water tank such as the one in Fig. 1.1.2. A cable attached to the tank applies a lateral (horizontal) force of 16.4 kips and pulls the tank horizontally by 2 in. The cable is suddenly cut and the resulting free vibration is recorded. At the end of four complete cycles, the time is 2.0 sec and the amplitude is 1 in. From these data compute the following: **(a)** damping ratio; **(b)** natural period of undamped vibration; **(c)** stiffness; **(d)** weight; **(e)** damping coefficient; and **(f)** number of cycles required for the displacement amplitude to decrease to 0.2 in.

Solution   **(a)** Substituting $u_i = 2$ in., $j = 4$, and $u_{i+j} = 1$ in. in Eq. (2.2.14a) gives

$$\zeta = \frac{1}{2\pi(4)} \ln \frac{2}{1} = 0.0276 = 2.76\%$$

Assumption of small damping implicit in Eq. (2.2.14a) is valid.

(b) $T_D = \dfrac{2.0}{4} = 0.5$ sec; $T_n \simeq T_D = 0.5$ sec.

(c) $k = \dfrac{16.4}{2} = 8.2$ kips/in.

(d) $\omega_n = \dfrac{2\pi}{T_n} = \dfrac{2\pi}{0.5} = 12.57$ rad/sec;

$m = \dfrac{k}{\omega_n^2} = \dfrac{8.2}{(12.57)^2} = 0.0519$ kip-sec^2/in.;

$w = (0.0519)386 = 20.03$ kips.

(e) $c = \zeta(2\sqrt{km}) = 0.0276\left[2\sqrt{8.2(0.0519)}\right] = 0.0360$ kip-sec/in.

(f) $\zeta \simeq \dfrac{1}{2\pi j} \ln \dfrac{u_1}{u_{1+j}}$; $j \simeq \dfrac{1}{2\pi (0.0276)} \ln \dfrac{2}{0.2} = 13.28$ cycles $\sim$ 13 cycles.

Example 2.7

The weight of water required to fill the tank of Example 2.6 is 80 kips. Determine the natural vibration period and damping ratio of the structure with the tank full.

Solution

$$w = 20.03 + 80 = 100.03 \text{ kips}$$

$$m = \frac{100.03}{386} = 0.2591 \text{ kip-sec}^2/\text{in.}$$

$$T_n = 2\pi \sqrt{\frac{m}{k}} = 2\pi \sqrt{\frac{0.2591}{8.2}} = 1.12 \text{ sec}$$

$$\zeta = \frac{c}{2\sqrt{km}} = \frac{0.0360}{2\sqrt{8.2(0.2591)}} = 0.0123 = 1.23\%$$

Observe that the damping ratio is now smaller (1.23% compared to 2.76% in Example 2.6) because the mass of the full tank is larger and hence the critical damping coefficient is larger.

2.3 ENERGY IN FREE VIBRATION

The energy input to an SDF system by imparting to it the initial displacement $u(0)$ and initial velocity $\dot{u}(0)$ is

$$E_I = \frac{1}{2}k[u(0)]^2 + \frac{1}{2}m[\dot{u}(0)]^2 \tag{2.3.1}$$

At any instant of time the total energy in a freely vibrating system is made up of two parts, kinetic energy E_K of the mass and potential energy equal to the strain energy E_S of deformation in the spring:

$$E_K(t) = \frac{1}{2}m[\dot{u}(t)]^2 \qquad E_S(t) = \frac{1}{2}k[u(t)]^2 \tag{2.3.2}$$

Substituting $u(t)$ from Eq. (2.1.3) for an undamped system leads to

$$E_K(t) = \frac{1}{2}m\omega_n^2 \left[-u(0) \sin \omega_n t + \frac{\dot{u}(0)}{\omega_n} \cos \omega_n t \right]^2 \tag{2.3.3}$$

$$E_S(t) = \frac{1}{2}k \left[u(0) \cos \omega_n t + \frac{\dot{u}(0)}{\omega_n} \sin \omega_n t \right]^2 \tag{2.3.4}$$

The total energy is

$$E_K(t) + E_S(t) = \frac{1}{2}k[u(0)]^2 + \frac{1}{2}m[\dot{u}(0)]^2 \tag{2.3.5}$$

wherein Eq. (2.1.4) has been utilized together with a well-known trigonometric identity.

Thus, the total energy is independent of time and equal to the input energy of Eq. (2.3.1), implying conservation of energy during free vibration of a system without damping.

For systems with viscous damping, the kinetic energy and potential energy could be determined by substituting $u(t)$ from Eq. (2.2.4) and its derivative $\dot{u}(t)$ into Eq. (2.3.2). The total energy will now be a decreasing function of time because of energy dissipated in viscous damping, which over the time duration 0 to t_1 is

$$E_D = \int f_D \, du = \int_0^{t_1} (c\dot{u})\dot{u} \, dt = \int_0^{t_1} c\dot{u}^2 \, dt \qquad (2.3.6)$$

All the input energy will eventually get dissipated in viscous damping; as t_1 goes to ∞, the dissipated energy, Eq. (2.3.6), tends to the input energy, Eq. (2.3.1).

2.4 COULOMB-DAMPED FREE VIBRATION

In Section 1.4 we mentioned that damping in actual structures is due to several energy-dissipating mechanisms acting simultaneously, and a mathematically convenient approach is to idealize them by equivalent viscous damping. Although this approach is sufficiently accurate for practical analysis of most structures, it may not be appropriate when special friction devices have been introduced in a building to reduce its vibrations during earth-quakes. Currently, there is much interest in such application and we return to them in Chapter 7. In this section the free vibration of systems under the presence of Coulomb friction forces is analyzed.

Coulomb damping results from friction against sliding of two dry surfaces. The friction force $F = \mu N$, where μ denotes the coefficients of static and kinetic friction, taken to be equal, and N the normal force between the sliding surfaces. The friction force is assumed to be independent of the velocity once the motion is initiated. The direction of the friction force opposes motion, and the sign of the friction force will change when the direction of motion changes. This necessitates formulation and solution of two differential equations, one valid for motion in one direction and the other valid when motion is reversed.

Figure 2.4.1 shows a mass–spring system with the mass sliding against a dry surface, and the free-body diagrams for the mass, including the inertia force, for two directions of

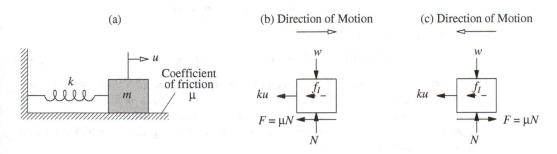

Figure 2.4.1

motion. The equation governing the motion of the mass from right to left is

$$m\ddot{u} + ku = F \tag{2.4.1}$$

for which the solution is

$$u(t) = A_1 \cos \omega_n t + B_1 \sin \omega_n t + u_F \tag{2.4.2}$$

where $u_F = F/k$. For motion of the mass from left to right, the governing equation is

$$m\ddot{u} + ku = -F \tag{2.4.3}$$

for which the solution is

$$u(t) = A_2 \cos \omega_n t + B_2 \sin \omega_n t - u_F \tag{2.4.4}$$

The constants A_1, B_1, A_2, and B_2 depend on the initial conditions of each successive half-cycle of motion; $\omega_n = \sqrt{k/m}$ and the constant u_F may be interpreted as the static deformation of the spring due to friction force F. Each of the two differential equations is linear, but the overall problem is nonlinear because the governing equation changes every half-cycle of motion.

Let us study the motion of the system of Fig. 2.4.1 starting with some given initial conditions and continuing until the motion ceases. At time $t = 0$, the mass is displaced a distance $u(0)$ to the right and released from rest such that $\dot{u}(0) = 0$. For the first half-cycle of motion, Eq. (2.4.2) applies with the constants A_1 and B_1 determined from the initial conditions at $t = 0$:

$$A_1 = u(0) - u_F \qquad B_1 = 0$$

Substituting these in Eq. (2.4.2) gives

$$u(t) = [u(0) - u_F] \cos \omega_n t + u_F \qquad 0 \le t \le \pi/\omega_n \tag{2.4.5}$$

This is plotted in Fig. 2.4.2; it is a cosine function with amplitude $= u(0) - u_F$ and shifted in the positive u direction by u_F. Equation (2.4.5) is valid until the velocity becomes zero again at $t = \pi/\omega_n = T_n/2$ (Fig. 2.4.2); at this instant $u = -u(0) + 2u_F$.

Starting from this extreme left position, the mass moves to the right with its motion described by Eq. (2.4.4). The constants A_2 and B_2 are determined from the conditions at the beginning of this half-cycle:

$$A_2 = u(0) - 3u_F \qquad B_2 = 0$$

Substituting these in Eq. (2.4.4) gives

$$u(t) = [u(0) - 3u_F] \cos \omega_n t - u_F \qquad \pi/\omega_n \le t \le 2\pi/\omega_n \tag{2.4.6}$$

This is plotted in Fig. 2.4.2; it is a cosine function with reduced amplitude $= u(0) - 3u_F$ and shifted in the negative u direction by u_F. Equation (2.4.6) is valid until the velocity becomes zero again at $t = 2\pi/\omega_n = T_n$ (Fig. 2.4.2); at this time instant $u = u(0) - 4u_F$.

At $t = 2\pi/\omega_n$ the motion reverses and is described by Eq. (2.4.2), which after evaluating the constants A_1 and B_1 becomes

$$u(t) = [u(0) - 5u_F] \cos \omega_n t + u_F \qquad 2\pi/\omega_n \le t \le 3\pi/\omega_n \tag{2.4.7}$$

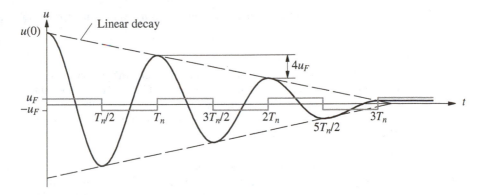

Figure 2.4.2   Free vibration of a system with Coulomb friction.

This is a cosine function with its amplitude reduced further to $u(0) - 5u_F$ and shifted, as before, in the positive u direction by u_F.

The time taken for each half-cycle is π/ω_n and the duration of a full cycle, the natural period of vibration, is

$$T_n = \frac{2\pi}{\omega_n} \tag{2.4.8}$$

Observe that the natural period of a system with Coulomb damping is the same as for the system without damping. In contrast, viscous damping had the effect of lengthening the natural period [Eq. (2.2.6)].

In each cycle of motion, the amplitude is reduced by $4u_F$; that is, the displacements u_i and u_{i+1} at successive maxima are related by

$$u_{i+1} = u_i - 4u_F \tag{2.4.9}$$

Thus the envelopes of the displacement–time curves are straight lines, as shown in Fig. 2.4.2, instead of the exponential functions for systems with viscous damping.

When does the free vibration of a system with Coulomb friction stop? In each cycle the amplitude is reduced by $4u_F$. Motion stops at the end of the half-cycle for which the amplitude is less than u_F. At that point the spring force acting on the mass is less than the friction force, $ku < F$, and motion ceases. In Fig. 2.4.2 this occurs at the end of the third cycle. The final rest position of the mass is displaced from its original equilibrium position and represents a permanent deformation in which the friction force and spring force are locked in. Shaking or tapping the system will usually jar it sufficiently to restore equilibrium.

Damping in real structures must be due partly to Coulomb friction, since only this mechanism can stop motion in free vibration. If the damping were purely viscous, motion theoretically continues forever, although at infinitesimally small amplitudes. This is an academic point, but it is basic to an understanding of damping mechanisms.

The various damping mechanisms that exist in real structures are rarely modeled individually. In particular, the Coulomb frictional forces that must exist are not considered explicitly unless frictional devices have been incorporated in the structure. Even with such devices it is possible to use equivalent viscous damping to obtain approximate results for dynamic response (Chapter 3).

Example 2.8

A small building consists of four steel frames, each with a friction device, supporting a reinforced-concrete slab, as shown schematically in Fig. E2.8a. The normal force across each of the spring-loaded friction pads is adjusted to equal 2.5% of the slab weight (Fig. E2.8c). A record of the building motion in free vibration along the x-axis is shown in Fig. E2.8d. Determine the effective coefficient of friction.

Solution

1. *Assumptions:* (a) the weight of the frame is negligible compared to the slab. (b) Energy dissipation due to mechanisms other than friction is negligible, a reasonable assumption because the amplitude of motion decays linearly with time (Fig. E2.8d).

2. *Determine T_n and u_F.*

$$T_n = \frac{4.5}{9} = 0.5 \text{ sec} \qquad \omega_n = \frac{2\pi}{0.5} = 4\pi$$

$$4u_F = \frac{5.5 - 0.1}{9} = 0.6 \text{ in.} \qquad u_F = 0.15 \text{ in.}$$

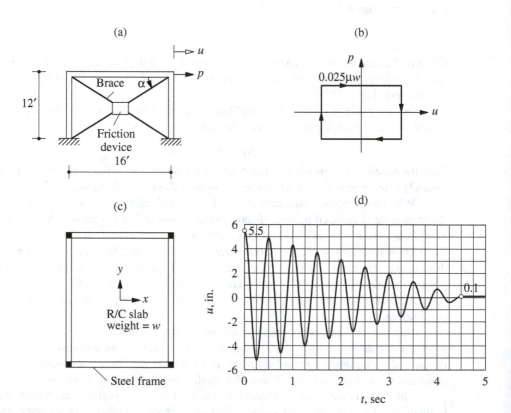

Figure E2.8

3. *Determine the coefficient of friction.* The friction force along each brace is $\mu(0.025w)$ and its component in the lateral direction is $(0.025\mu w)\cos\alpha$. The total friction force in the lateral direction due to the four braces, two in each of the two frames, is

$$F = 4(0.025\mu w)\cos\alpha = (0.1\mu w)\left(\frac{16}{20}\right) = 0.08\mu w$$

$$u_F = \frac{F}{k} = \frac{0.08\mu w}{k} = \frac{0.08\mu mg}{k} = \frac{0.08\mu g}{\omega_n^2}$$

$$\mu = \frac{u_F\omega_n^2}{0.08g} = \frac{0.15(4\pi)^2}{0.08g} = 0.767$$

PROBLEMS

2.1   A heavy table is supported by flat steel legs (Fig. P2.1). Its natural period in lateral vibration is 0.5 sec. When a 50-lb plate is clamped to its surface, the natural period in lateral vibration is lengthened to 0.75 sec. What are the weight and the lateral stiffness of the table?

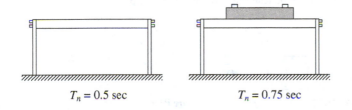

$T_n = 0.5$ sec $T_n = 0.75$ sec   **Figure P2.1**

2.2   An electromagnet weighing 400 lb and suspended by a spring having a stiffness of 100 lb/in. (Fig. P.2.2a) lifts 200 lb of iron scrap (Fig. P2.2b). Determine the equation describing the motion when the electric current is turned off and the scrap is dropped (Fig. P2.2c).

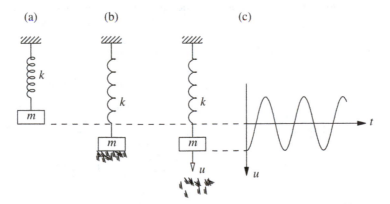

Figure P2.2

2.3   A mass m is at rest, partially supported by a spring and partially by stops (Fig. P2.3). In the position shown, the spring force is $mg/2$. At time $t = 0$ the stops are rotated, suddenly releasing the mass. Determine the motion of the mass.

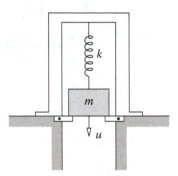

Figure P2.3

2.4   The weight of the wooden block shown in Fig. P2.4 is 10 lb and the spring stiffness is 100 lb/in. A bullet weighing 0.5 lb is fired at a speed of 60 ft/sec into the block and becomes embedded in the block. Determine the resulting motion $u(t)$ of the block.

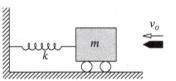

Figure P2.4

2.5   A mass m_1 hangs from a spring k and is in static equilibrium. A second mass m_2 drops through a height h and sticks to m_1 without rebound (Fig. P2.5). Determine the subsequent motion $u(t)$ measured from the static equilibrium position of m_1 and k.

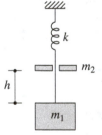

Figure P2.5

2.6   The packaging for an instrument can be modeled as shown in Fig. P2.6, in which the instrument of mass m is restrained by springs of total stiffness k inside a container; $m = 10$ lb/g and $k = 50$ lb/in. The container is accidentally dropped from a height of 3 ft above the ground.

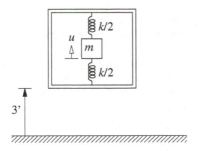

Figure P2.6

Assuming that it does not bounce on contact, determine the maximum deformation of the packaging within the box and the maximum acceleration of the instrument.

2.7 Consider a diver weighing 200 lb at the end of a diving board that cantilevers out 3 ft. The diver oscillates at a frequency of 2 Hz. What is the flexural rigidity EI of the diving board?

2.8 Show that the motion of a critically damped system due to initial displacement $u(0)$ and initial velocity $\dot{u}(0)$ is

$$u(t) = \{u(0) + [\dot{u}(0) + \omega_n u(0)]\,t\}\,e^{-\omega_n t}$$

2.9 Show that the motion of an overcritically damped system due to initial displacement $u(0)$ and initial velocity $\dot{u}(0)$ is

$$u(t) = e^{-\zeta \omega_n t}\left(A_1 e^{-\omega_D' t} + A_2 e^{\omega_D' t}\right)$$

where $\omega_D' = \omega_n \sqrt{\zeta^2 - 1}$ and

$$A_1 = \frac{-\dot{u}(0) + \left(-\zeta + \sqrt{\zeta^2 - 1}\right)\omega_n u(0)}{2\omega_D'}$$

$$A_2 = \frac{\dot{u}(0) + \left(\zeta + \sqrt{\zeta^2 - 1}\right)\omega_n u(0)}{2\omega_D'}$$

2.10 Derive the equation for the displacement response of a viscously damped SDF system due to initial velocity $\dot{u}(0)$ for three cases: **(a)** underdamped systems; **(b)** critically damped systems; and **(c)** overdamped systems. Plot $u(t) \div \dot{u}(0)/\omega_n$ against t/T_n for $\zeta = 0.1$, 1, and 2.

2.11 For a system with damping ratio ζ, determine the number of free vibration cycles required to reduce the displacement amplitude to 10% of the initial amplitude; the initial velocity is zero.

2.12 What is the ratio of successive amplitudes of vibration if the viscous damping ratio is known to be **(a)** $\zeta = 0.01$, **(b)** $\zeta = 0.05$, or **(c)** $\zeta = 0.25$?

2.13 The supporting system of the tank of Example 2.6 is enlarged with the objective of increasing its seismic resistance. The lateral stiffness of the modified system is double that of the original system. If the damping coefficient is unaffected (this may not be a realistic assumption), for the modified tank determine **(a)** the natural period of vibration T_n, and **(b)** the damping ratio ζ.

2.14 The vertical suspension system of an automobile is idealized as a viscously damped SDF system. Under the 3000-lb weight of the car the suspension system deflects 2 in. The suspension is designed to be critically damped.
(a) Calculate the damping and stiffness coefficients of the suspension.
(b) With four 160-lb passengers in the car, what is the effective damping ratio?
(c) Calculate the natural frequency of damped vibration for case (b).

2.15 The stiffness and damping properties of a mass–spring–damper system are to be determined by a free vibration test; the mass is given as $m = 0.1$ lb-sec^2/in. In this test the mass is displaced 1 in. by a hydraulic jack and then suddenly released. At the end of 20 complete cycles, the time is 3 sec and the amplitude is 0.2 in. Determine the stiffness and damping coefficients.

2.16 A machine weighing 250 lb is mounted on a supporting system consisting of four springs and four dampers. The vertical deflection of the supporting system under the weight of the machine is measured as 0.8 in. The dampers are designed to reduce the amplitude of vertical vibration to one-eighth of the initial amplitude after two complete cycles of free vibration. Find the following properties of the system: (a) undamped natural frequency, (b) damping ratio, and (c) damped natural frequency. Comment on the effect of damping on the natural frequency.

2.17 Determine the natural vibration period and damping ratio of the aluminum frame model (Fig. 1.1.4a) from the acceleration record of its free vibration shown in Fig. 1.1.4b.

2.18 Show that the natural vibration frequency of the system in Fig. E1.6a is $\omega_n' = \omega_n(1 - w/w_{cr})^{1/2}$, where ω_n is the natural vibration frequency computed neglecting the action of gravity, and w_{cr} is the buckling weight.

2.19 An impulsive force applied to the roof slab of the building of Example 2.8 gives it an initial velocity of 20 in./sec to the right. How far to the right will the slab move? What is the maximum displacement of the slab on its return swing to the left?

2.20 An SDF system consisting of a weight, spring, and friction device is shown in Fig. P2.20. This device slips at a force equal to 10% of the weight, and the natural vibration period of the system is 0.25 sec. If this system is given an initial displacement of 2 in. and released, what will be the displacement amplitude after six cycles? In how many cycles will the system come to rest?

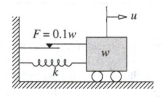

Figure P2.20

3

Response to Harmonic and Periodic Excitations

PREVIEW

The response of SDF systems to harmonic excitation is a classical topic in structural dynamics, not only because such excitations are encountered in engineering systems (e.g., force due to unbalanced rotating machinery), but also because understanding the response of structures to harmonic excitation provides insight into how the system will respond to other types of forces. Furthermore, the theory of forced harmonic vibration has several useful applications in earthquake engineering.

In Part A of this chapter the basic results for response of SDF systems to harmonic force are presented, including the concepts of *steady-state response, frequency-response curve,* and *resonance.* Applications of these results to experimental evaluation of the natural vibration frequency and damping ratio of a structure, to isolation of vibration, and to the design of vibration-measuring instruments is the subject of Part B; also included is the concept of *equivalent viscous damping.* This concept is used in Part C to obtain approximate solutions for the response of systems with rate-independent damping or Coulomb friction; these results are then shown to be good approximations to the "exact" solutions. A procedure to determine the response of SDF systems to periodic excitation is presented in Part D. A Fourier series representation of the excitation, combined with the results for response to harmonic excitations, provides the desired procedure.

PART A: VISCOUSLY DAMPED SYSTEMS: BASIC RESULTS

3.1 HARMONIC VIBRATION OF UNDAMPED SYSTEMS

A harmonic force is $p(t) = p_o \sin \omega t$ or $p_o \cos \omega t$, where p_o is the *amplitude* or maximum value of the force and its frequency ω is called the *exciting frequency* or *forcing frequency;* $T = 2\pi/\omega$ is the *exciting period* or *forcing period* (Fig. 3.1.1a). The response of SDF systems to a sinusoidal force will be presented in some detail, along with only brief comments on the response to a cosine force because the concepts involved are similar in the two cases.

Setting $p(t) = p_o \sin \omega t$ in Eq. (1.5.2) gives the differential equation governing the forced harmonic vibration of the system, which for systems without damping specializes to

$$m\ddot{u} + ku = p_o \sin \omega t \qquad (3.1.1)$$

This equation is to be solved for the displacement or deformation $u(t)$ subject to the initial conditions

$$u = u(0) \qquad \dot{u} = \dot{u}(0) \qquad (3.1.2)$$

where $u(0)$ and $\dot{u}(0)$ are the displacement and velocity at the time instant the force is applied. The particular solution to this differential equation is (see Derivation 3.1)

$$u_p(t) = \frac{p_o}{k} \frac{1}{1 - (\omega/\omega_n)^2} \sin \omega t \qquad \omega \neq \omega_n \qquad (3.1.3)$$

The complementary solution of Eq. (3.1.1) is the free vibration response determined in Eq. (e) of Derivation 2.1:

$$u_c(t) = A \cos \omega_n t + B \sin \omega_n t \qquad (3.1.4)$$

and the complete solution is the sum of the complementary and particular solutions:

$$u(t) = A \cos \omega_n t + B \sin \omega_n t + \frac{p_o}{k} \frac{1}{1 - (\omega/\omega_n)^2} \sin \omega t \qquad (3.1.5)$$

The constants A and B are determined by imposing the initial conditions, Eq. (3.1.2), to obtain the final result (see Derivation 3.1):

$$u(t) = \underbrace{u(0) \cos \omega_n t + \left[\frac{\dot{u}(0)}{\omega_n} - \frac{p_o}{k} \frac{\omega/\omega_n}{1 - (\omega/\omega_n)^2} \right] \sin \omega_n t}_{\text{transient}}$$

$$+ \underbrace{\frac{p_o}{k} \frac{1}{1 - (\omega/\omega_n)^2} \sin \omega t}_{\text{steady state}} \qquad (3.1.6a)$$

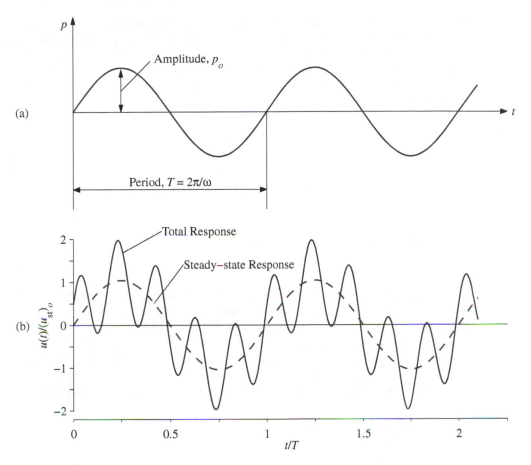

Figure 3.1.1   (a) Harmonic force; (b) response of undamped system to harmonic force; $\omega/\omega_n = 0.2$, $u(0) = 0.5 p_o/k$, and $\dot{u}(0) = \omega_n p_o/k$.

Equation (3.1.6a) has been plotted for $\omega/\omega_n = 0.2$, $u(0) = 0.5 p_o/k$, and $\dot{u}(0) = \omega_n p_o/k$ as the solid line in Fig. 3.1.1. The $\sin \omega t$ term in this equation is the particular solution of Eq. (3.1.3) and is shown by the dashed line.

Equation (3.1.6a) and Fig. 3.1.1 show that $u(t)$ contains two distinct vibration components: (1) the $\sin \omega t$ term, giving an oscillation at the forcing or exciting frequency; and (2) the $\sin \omega_n t$ and $\cos \omega_n t$ terms, giving an oscillation at the natural frequency of the system. The first of these is the *forced vibration* or *steady-state vibration,* for it is present because of the applied force no matter what the initial conditions. The latter is the *free vibration* or *transient vibration,* which depends on the initial displacement and velocity. It exists even if $u(0) = \dot{u}(0) = 0$, in which case Eq. (3.1.6a) specializes to

$$u(t) = \frac{p_o}{k} \frac{1}{1 - (\omega/\omega_n)^2} \left(\sin \omega t - \frac{\omega}{\omega_n} \sin \omega_n t \right) \qquad (3.1.6b)$$

The transient component is shown as the difference between the solid and dashed lines in Fig. 3.1.1, where it is seen to continue forever. This is only an academic point because the damping inevitably present in real systems makes the free vibration decay with time (Section 3.2). It is for this reason that this component is called *transient vibration*.

The steady-state dynamic response, a sinusoidal oscillation at the forcing frequency, may be expressed as

$$u(t) = (u_{\text{st}})_o \left[\frac{1}{1 - (\omega/\omega_n)^2} \right] \sin \omega t \tag{3.1.7}$$

Ignoring the dynamic effects signified by the acceleration term in Eq. (3.1.1) gives the static deformation (indicated by the subscript "st") at each instant:

$$u_{\text{st}}(t) = \frac{p_o}{k} \sin \omega t \tag{3.1.8}$$

The maximum value of the static deformation is

$$(u_{\text{st}})_o = \frac{p_o}{k} \tag{3.1.9}$$

which may be interpreted as the static deformation due to the amplitude p_o of the force; for brevity we will refer to $(u_{\text{st}})_o$ as the *static deformation*. The factor in brackets in Eq. (3.1.7) has been plotted in Fig. 3.1.2 against ω/ω_n, the ratio of the forcing frequency to the natural frequency. For $\omega/\omega_n < 1$ or $\omega < \omega_n$ this factor is positive, indicating that $u(t)$ and $p(t)$ have the same algebraic sign (i.e., when the force in Fig. 1.2.1a acts to the right, the system would also be displaced to the right). The displacement is said to be *in phase* with the applied force. For $\omega/\omega_n > 1$ or $\omega > \omega_n$ this factor is negative, indicating that $u(t)$ and

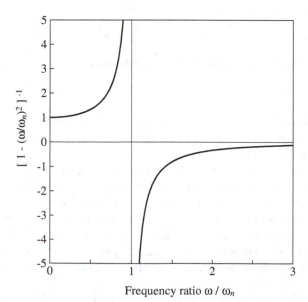

Frequency ratio ω / ω_n **Figure 3.1.2**

$p(t)$ have opposing algebraic signs (i.e., when the force acts to the right, the system would be displaced to the left). The displacement is said to be *out of phase* relative to the applied force.

To describe this notion of phase mathematically, Eq. (3.1.7) is rewritten in terms of the amplitude u_o of the vibratory displacement $u(t)$ and *phase angle* ϕ:

$$u(t) = u_o \sin(\omega t - \phi) = (u_{st})_o R_d \sin(\omega t - \phi) \tag{3.1.10}$$

where

$$R_d = \frac{u_o}{(u_{st})_o} = \frac{1}{|\,1 - (\omega/\omega_n)^2\,|} \quad \text{and} \quad \phi = \begin{cases} 0° & \omega < \omega_n \\ 180° & \omega > \omega_n \end{cases} \tag{3.1.11}$$

For $\omega < \omega_n$, $\phi = 0°$, implying that the displacement varies as $\sin \omega t$, in phase with the applied force. For $\omega > \omega_n$, $\phi = 180°$, indicating that the displacement varies as $-\sin \omega t$, out of phase relative to the force. This phase angle is shown in Fig. 3.1.3 as a function of the frequency ratio ω/ω_n.

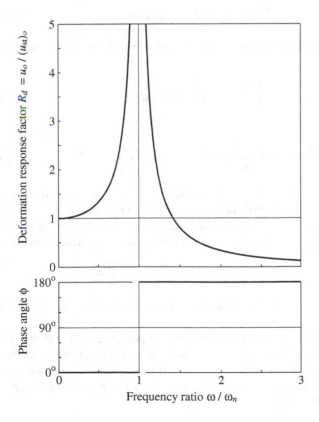

Figure 3.1.3   Deformation response factor and phase angle for an undamped system excited by harmonic force.

The *deformation* (or *displacement*) *response factor* R_d is the ratio of the amplitude u_o of the dynamic (or vibratory) deformation to the static deformation $(u_{st})_o$. Figure 3.1.3, which shows Eq. (3.1.11a) for R_d plotted as a function of the frequency ratio ω/ω_n, permits several observations: If ω/ω_n is small (i.e., the force is "slowly varying"), R_d is only slightly larger than 1 and the amplitude of the dynamic deformation is essentially the same as the static deformation. If $\omega/\omega_n > \sqrt{2}$ (i.e., ω is higher than $\omega_n\sqrt{2}$), $R_d < 1$ and the dynamic deformation amplitude is less than the static deformation. As ω/ω_n increases beyond $\sqrt{2}$, R_d becomes smaller and approaches zero as $\omega/\omega_n \rightarrow \infty$, implying that the vibratory deformation due to a "rapidly varying" force is very small. If ω/ω_n is close to 1 (i.e., ω is close to ω_n), R_d is many times larger than 1, implying that the deformation amplitude is much larger than the static deformation.

The *resonant frequency* is defined as the forcing frequency at which R_d is maximum. For an undamped system the resonant frequency is ω_n and R_d is unbounded at this frequency. The vibratory deformation does not become unbounded immediately, however, but gradually, as we demonstrate next.

If $\omega = \omega_n$, the solution given by Eq. (3.1.6b) is no longer valid. In this case the choice of the function $C \sin \omega t$ for a particular solution fails because it is also a part of the complementary solution. The particular solution now is

$$u_p(t) = -\frac{p_o}{2k}\omega_n t \cos \omega_n t \qquad \omega = \omega_n \qquad (3.1.12)$$

and the complete solution for at-rest initial conditions, $u(0) = \dot{u}(0) = 0$, is (see Derivation 3.2)

$$u(t) = -\frac{1}{2}\frac{p_o}{k}(\omega_n t \cos \omega_n t - \sin \omega_n t) \qquad (3.1.13a)$$

or

$$\frac{u(t)}{(u_{st})_o} = -\frac{1}{2}\left(\frac{2\pi t}{T_n}\cos\frac{2\pi t}{T_n} - \sin\frac{2\pi t}{T_n}\right) \qquad (3.1.13b)$$

This result is plotted in Fig. 3.1.4, which shows that the time taken to complete one cycle of vibration is T_n. The local maxima of $u(t)$, which occur at $t = (j - 1/2)T_n$, are $\pi(j - 1/2)(u_{st})_o$—$j = 1, 2, 3, \ldots$—and the local minima, which occur at $t = jT_n$, are $-\pi j(u_{st})_o$—$j = 1, 2, 3, \ldots$. In each cycle the deformation amplitude increases by

$$|u_{j+1}| - |u_j| = (u_{st})_o[\pi(j+1) - \pi j] = \frac{\pi p_o}{k}$$

The deformation amplitude grows indefinitely, but it becomes infinite only after an infinitely long time.

This is an academic result and should be interpreted appropriately for real structures. As the deformation continues to increase, at some point in time the system would fail if it is brittle. On the other hand, the system would yield if it is ductile, its stiffness would decrease, and its "natural frequency" would no longer be equal to the forcing frequency, and Eq. (3.1.13) or Fig. 3.1.4 would no longer be valid.

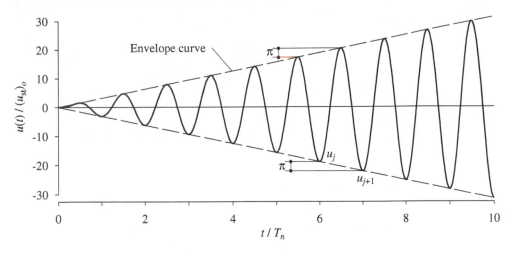

Figure 3.1.4   Response of undamped system to sinusoidal force of frequency $\omega = \omega_n$; $u(0) = \dot{u}(0) = 0$.

Derivation 3.1

The particular solution of Eq. (3.1.1), a linear second-order differential equation, is of the form

$$u_p(t) = C \sin \omega t \tag{a}$$

Differentiating this twice gives

$$\ddot{u}_p(t) = -\omega^2 C \sin \omega t \tag{b}$$

Substituting Eqs. (a) and (b) in the differential equation (3.1.1) leads to a solution for C:

$$C = \frac{p_o}{k} \frac{1}{1 - (\omega/\omega_n)^2} \tag{c}$$

which is combined with Eq. (a) to obtain the particular solution presented in Eq. (3.1.3).

To determine the constants A and B in Eq. (3.1.5), it is differentiated:

$$\dot{u}(t) = -\omega_n A \sin \omega_n t + \omega_n B \cos \omega_n t + \frac{p_o}{k} \frac{\omega}{1 - (\omega/\omega_n)^2} \cos \omega t \tag{d}$$

Evaluating Eqs. (3.1.5) and (d) at $t = 0$ gives

$$u(0) = A \qquad \dot{u}(0) = \omega_n B + \frac{p_o}{k} \frac{\omega}{1 - (\omega/\omega_n)^2} \tag{e}$$

These two equations give

$$A = u(0) \qquad B = \frac{\dot{u}(0)}{\omega_n} - \frac{p_o}{k} \frac{\omega/\omega_n}{1 - (\omega/\omega_n)^2} \tag{f}$$

which are substituted in Eq. (3.1.5) to obtain Eq. (3.1.6a).

Derivation 3.2

If $\omega = \omega_n$, the particular solution of Eq. (3.1.1) is of the form

$$u_p(t) = Ct \cos \omega_n t \tag{a}$$

Substituting Eq. (a) in Eq. (3.1.1) and solving for C yields

$$C = -\frac{p_o}{2k} \omega_n \tag{b}$$

which is combined with Eq. (a) to obtain the particular solution, Eq. (3.1.12).

Thus the complete solution is

$$u(t) = A \cos \omega_n t + B \sin \omega_n t - \frac{p_o}{2k} \omega_n t \cos \omega_n t \tag{c}$$

and the corresponding velocity is

$$\dot{u}(t) = -\omega_n A \sin \omega_n t + \omega_n B \cos \omega_n t - \frac{p_o}{2k} \omega_n \cos \omega_n t + \frac{p_o}{2k} \omega_n^2 t \sin \omega_n t \tag{d}$$

Evaluating Eqs. (c) and (d) at $t = 0$ and solving the resulting algebraic equations gives

$$A = u(0) \qquad B = \frac{\dot{u}(0)}{\omega_n} + \frac{p_o}{2k}$$

Specializing for at-rest initial conditions gives

$$A = 0 \qquad B = \frac{p_o}{2k}$$

which are substituted in Eq. (c) to obtain Eq. (3.1.13a).

3.2 HARMONIC VIBRATION WITH VISCOUS DAMPING

3.2.1 Steady-State and Transient Responses

Including viscous damping the differential equation governing the response of SDF systems to harmonic force is

$$m\ddot{u} + c\dot{u} + ku = p_o \sin \omega t \tag{3.2.1}$$

This equation is to be solved subject to the initial conditions

$$u = u(0) \qquad \dot{u} = \dot{u}(0) \tag{3.2.2}$$

The particular solution of this differential equation is (from Derivation 3.3)

$$u_p(t) = C \sin \omega t + D \cos \omega t \tag{3.2.3}$$

where

$$C = \frac{p_o}{k} \frac{1 - (\omega/\omega_n)^2}{[1 - (\omega/\omega_n)^2]^2 + [2\zeta(\omega/\omega_n)]^2}$$

$$D = \frac{p_o}{k} \frac{-2\zeta\omega/\omega_n}{[1 - (\omega/\omega_n)^2]^2 + [2\zeta(\omega/\omega_n)]^2} \tag{3.2.4}$$

The complementary solution of Eq. (3.2.1) is the free vibration response given by Eq. (e) of Derivation 2.2:

$$u_c(t) = e^{-\zeta\omega_n t}(A \cos \omega_D t + B \sin \omega_D t)$$

where $\omega_D = \omega_n \sqrt{1 - \zeta^2}$. The complete solution of Eq. (3.2.1) is

$$u(t) = \underbrace{e^{-\zeta \omega_n t}(A \cos \omega_D t + B \sin \omega_D t)}_{\text{transient}} + \underbrace{C \sin \omega t + D \cos \omega t}_{\text{steady state}} \qquad (3.2.5)$$

where the constants A and B can be determined by standard procedures (e.g., see Derivation 3.1) in terms of the initial displacement $u(0)$ and initial velocity $\dot{u}(0)$. As noted in Section 3.1, $u(t)$ contains two distinct vibration components: *forced vibration* (excitation frequency ω terms) and *free vibration* (natural frequency ω_n terms).

Equation (3.2.5) is plotted in Fig. 3.2.1 for $\omega/\omega_n = 0.2$, $\zeta = 0.05$, $u(0) = 0.5 p_o/k$, and $\dot{u}(0) = \omega_n p_o/k$; the total response is shown by the solid line and the forced response by the dashed line. The difference between the two is the free response, which decays exponentially with time at a rate depending on ω/ω_n and ζ; eventually, the free response becomes negligible, hence we call it *transient response*; compare this with no decay for undamped systems in Fig. 3.1.1. After awhile, essentially the forced response remains, and we therefore call it *steady-state response* and focus on it for the rest of this chapter (after Section 3.2.2). It should be recognized, however, that the largest deformation peak may occur before the system has reached steady state; see Fig. 3.2.1.

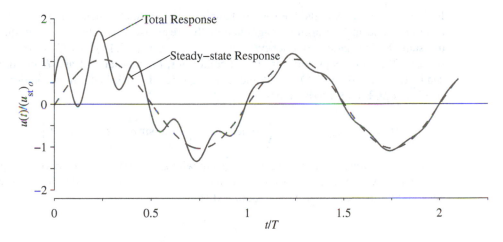

Figure 3.2.1   Response of damped system to harmonic force; $\omega/\omega_n = 0.2$, $\zeta = 0.05$, $u(0) = 0.5 p_o/k$, and $\dot{u}(0) = \omega_n p_o/k$.

Derivation 3.3

Dividing Eq. (3.2.1) by m gives

$$\ddot{u} + 2\zeta \omega_n \dot{u} + \omega_n^2 u = \frac{p_o}{m} \sin \omega t \qquad (a)$$

The particular solution of Eq. (a) is of the form

$$u_p(t) = C \sin \omega t + D \cos \omega t \qquad (b)$$

Substituting Eq. (b) and its first and second derivatives in Eq. (a) gives

$$[(\omega_n^2 - \omega^2)C - 2\zeta\omega_n\omega D]\sin\omega t + [2\zeta\omega_n\omega C + (\omega_n^2 - \omega^2)D]\cos\omega t = \frac{p_o}{m}\sin\omega t \qquad \text{(c)}$$

For Eq. (c) to be valid for all t, the coefficients of the sine and cosine terms on the two sides of the equation must be equal. This requirement gives two equations in C and D which, after dividing by ω_n^2 and using the relation $k = \omega_n^2 m$, become

$$\left[1 - \left(\frac{\omega}{\omega_n}\right)^2\right]C - \left(2\zeta\frac{\omega}{\omega_n}\right)D = \frac{p_o}{k} \qquad \text{(d)}$$

$$\left(2\zeta\frac{\omega}{\omega_n}\right)C + \left[1 - \left(\frac{\omega}{\omega_n}\right)^2\right]D = 0 \qquad \text{(e)}$$

Solving the two algebraic equations (d) and (e) leads to Eq. (3.2.4).

3.2.2 Response for $\omega = \omega_n$

In this section we examine the role of damping in the rate at which steady-state response is attained and in limiting the magnitude of this response when the forcing frequency is the same as the natural frequency. For $\omega = \omega_n$, Eq. (3.2.4) gives $C = 0$ and $D = -(u_{st})_o/2\zeta$; for $\omega = \omega_n$ and zero initial conditions, the constants A and B in Eq. (3.2.5) can be determined: $A = (u_{st})_o/2\zeta$ and $B = (u_{st})_o/2\sqrt{1 - \zeta^2}$. With these solutions for A, B, C, and D, Eq. (3.2.5) becomes

$$u(t) = (u_{st})_o\frac{1}{2\zeta}\left[e^{-\zeta\omega_n t}\left(\cos\omega_D t + \frac{\zeta}{\sqrt{1 - \zeta^2}}\sin\omega_D t\right) - \cos\omega_n t\right] \qquad \text{(3.2.6)}$$

This result is plotted in Fig. 3.2.2 for a system with $\zeta = 0.05$. A comparison of Fig. 3.2.2 for damped systems and Fig. 3.1.4 for undamped systems shows that damping lowers each peak and limits the response to the bounded value:

$$u_o = \frac{(u_{st})_o}{2\zeta} \qquad \text{(3.2.7)}$$

For lightly damped systems the sinusoidal term in Eq. (3.2.6) is small and $\omega_D \simeq \omega_n$; thus

$$u(t) \simeq (u_{st})_o\underbrace{\frac{1}{2\zeta}(e^{-\zeta\omega_n t} - 1)\cos\omega_n t}_{\text{envelope function}} \qquad \text{(3.2.8)}$$

The deformation varies with time as a cosine function, with its amplitude increasing with time according to the envelope function shown by dashed lines in Fig. 3.2.2.

The amplitude of the steady-state deformation of a system to a harmonic force with $\omega = \omega_n$ and the rate at which steady state is attained is strongly influenced by damping.

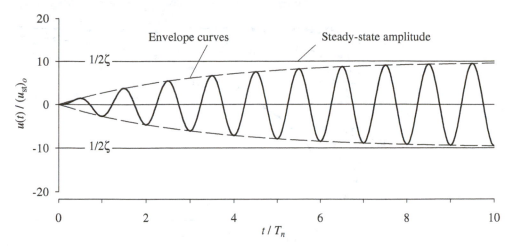

Figure 3.2.2   Response of damped system with $\zeta = 0.05$ to sinusoidal force of frequency $\omega = \omega_n$; $u(0) = \dot{u}(0) = 0$.

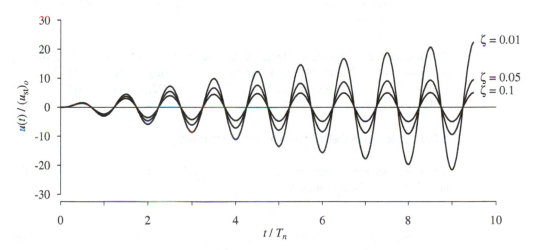

Figure 3.2.3   Response of three systems—$\zeta = 0.01$, 0.05, and 0.1—to sinusoidal force of frequency $\omega = \omega_n$; $u(0) = \dot{u}(0) = 0$.

The important influence of the damping ratio on the amplitude is seen in Fig. 3.2.3, where Eq. (3.2.6) is plotted for three damping ratios: $\zeta = 0.01$, 0.05, and 0.1. To study how the response builds up to steady state, we examine the peak u_j after j cycles of vibration. A relation between u_j and j can be written by substituting $t = jT_n$ in Eq. (3.2.8), setting $\cos \omega_n t = 1$, and using Eq. (3.2.7) to obtain

$$\frac{|u_j|}{u_o} = 1 - e^{-2\pi \zeta j} \qquad (3.2.9)$$

This relation is plotted in Fig. 3.2.4 for $\zeta = 0.01$, 0.02, 0.05, 0.10, and 0.20. The discrete points are joined by curves to identify trends, but only integer values of j are meaningful.

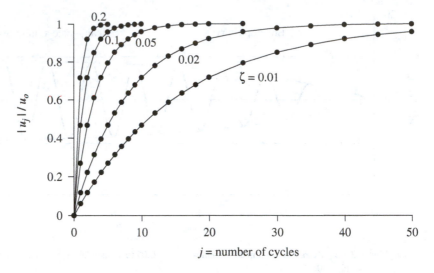

Figure 3.2.4   Variation of response amplitude with number of cycles of harmonic force with frequency $\omega = \omega_n$.

The lighter the damping, the larger is the number of cycles required to reach a certain percentage of u_o, the steady-state amplitude. For example, the number of cycles required to reach 95% of u_o is 48 for $\zeta = 0.01$, 24 for $\zeta = 0.02$, 10 for $\zeta = 0.05$, 5 for $\zeta = 0.10$, and 2 for $\zeta = 0.20$.

3.2.3 Maximum Deformation and Phase Lag

The steady-state deformation of the system due to harmonic force, described by Eqs. (3.2.3) and (3.2.4), can be rewritten as

$$u(t) = u_o \sin(\omega t - \phi) = (u_{st})_o R_d \sin(\omega t - \phi) \qquad (3.2.10)$$

where the response amplitude $u_o = \sqrt{C^2 + D^2}$ and $\phi = \tan^{-1}(-D/C)$. Substituting for C and D gives the *deformation response factor*:

$$R_d = \frac{u_o}{(u_{st})_o} = \frac{1}{\sqrt{[1 - (\omega/\omega_n)^2]^2 + [2\zeta(\omega/\omega_n)]^2}} \qquad (3.2.11)$$

$$\phi = \tan^{-1} \frac{2\zeta(\omega/\omega_n)}{1 - (\omega/\omega_n)^2} \qquad (3.2.12)$$

Equation (3.2.10) is plotted in Fig. 3.2.5 for three values of ω/ω_n and a fixed value of $\zeta = 0.20$. The values of R_d and ϕ computed from Eqs. (3.2.11) and (3.2.12) are identified. Also shown by dashed lines is the static deformation [Eq. (3.1.8)] due to $p(t)$, which varies with time just as does the applied force, except for the constant k. The steady-state motion is seen to occur at the forcing period $T = 2\pi/\omega$, but with a time lag $= \phi/2\pi$; ϕ is called the *phase angle* or *phase lag*.

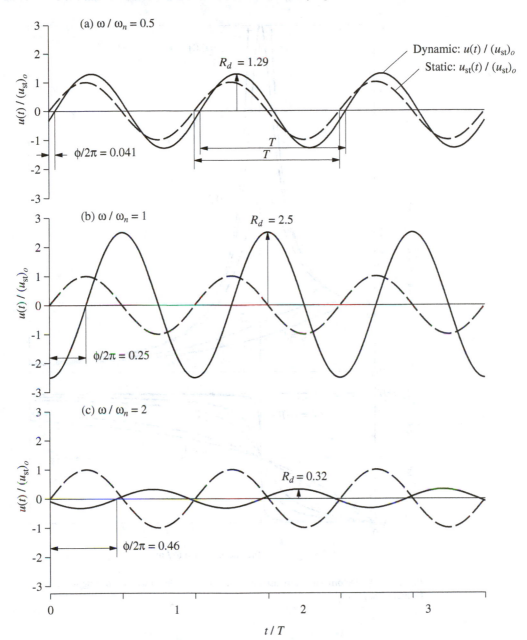

Figure 3.2.5 Steady-state response of damped systems ($\zeta = 0.2$) to sinusoidal force for three values of the frequency ratio: (a) $\omega/\omega_n = 0.5$, (b) $\omega/\omega_n = 1$, (c) $\omega/\omega_n = 2$.

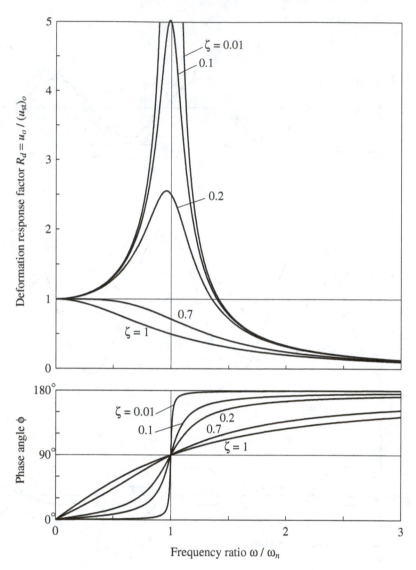

Figure 3.2.6 Deformation response factor and phase angle for a damped system excited by harmonic force.

    A plot of the amplitude of a response quantity against the excitation frequency is called a *frequency-response curve*. Such a plot for deformation u is given by Fig. 3.2.6, wherein the deformation response factor R_d [from Eq. (3.2.11)] is plotted as a function of ω/ω_n for a few values of ζ; all the curves are below the $\zeta = 0$ curve in Fig. 3.1.3. Damping reduces R_d and hence the deformation amplitude at all excitation frequencies.

The magnitude of this reduction is strongly dependent on the excitation frequency and is examined next for three regions of the excitation-frequency scale:

1. If the frequency ratio $\omega/\omega_n \ll 1$ (i.e., $T \gg T_n$, implying that the force is "slowly varying"), R_d is only slightly larger than 1 and is essentially independent of damping. Thus

$$u_o \simeq (u_{st})_o = \frac{p_o}{k} \qquad (3.2.13)$$

This result implies that the amplitude of dynamic response is essentially the same as the static deformation and is controlled by the stiffness of the system.

2. If $\omega/\omega_n \gg 1$ (i.e., $T \ll Tn$, implying that the force is "rapidly varying"), R_d tends to zero as ω/ω_n increases and is essentially unaffected by damping. For large values of ω/ω_n, the $(\omega/\omega_n)^4$ term is dominant in Eq. (3.2.11), which can be approximated by

$$u_o \simeq (u_{st})_o \frac{\omega_n^2}{\omega^2} = \frac{p_o}{m\omega^2} \qquad (3.2.14)$$

This result implies that the response is controlled by the mass of the system.

3. If $\omega/\omega_n \simeq 1$ (i.e., the forcing frequency is close to the natural frequency of the system), R_d is very sensitive to damping and, for the smaller damping values, R_d can be several times larger than 1, implying that the amplitude of dynamic response can be much larger than the static deformation. If $\omega = \omega_n$, Eq. (3.2.11) gives

$$u_o = \frac{(u_{st})_o}{2\zeta} = \frac{p_o}{c\omega_n} \qquad (3.2.15)$$

This result implies that the response is controlled by the damping of the system.

The phase angle ϕ, which defines the time by which the response lags behind the force, varies with ω/ω_n as shown in Fig. 3.2.6. It is examined next for the same three regions of the excitation-frequency scale:

1. If $\omega/\omega_n \ll 1$ (i.e., the force is "slowly varying"), ϕ is close to $0°$ and the displacement is essentially in phase with the applied force, as in Fig. 3.2.5a. When the force in Fig. 1.2.1a acts to the right, the system would also be displaced to the right.

2. If $\omega/\omega_n \gg 1$ (i.e., the force is "rapidly varying"), ϕ is close to $180°$ and the displacement is essentially of opposite phase relative to the applied force, as in Fig. 3.2.5c. When the force acts to the right, the system would be displaced to the left.

3. If $\omega/\omega_n = 1$ (i.e., the forcing frequency is equal to the natural frequency), $\phi = 90°$ for all values of ζ, and the displacement attains its peaks when the force passes through zeros, as in Fig. 3.2.5b.

Example 3.1

The displacement amplitude u_o of an SDF system due to harmonic force is known for two excitation frequencies. At $\omega = \omega_n$, $u_o = 5$ in.; at $\omega = 5\omega_n$, $u_o = 0.02$ in. Estimate the damping ratio of the system.

Solution   At $\omega = \omega_n$, from Eq. (3.2.15),

$$u_o = (u_{st})_o \frac{1}{2\zeta} = 5 \tag{a}$$

At $\omega = 5\omega_n$, from Eq. (3.2.14),

$$u_o \simeq (u_{st})_o \frac{1}{(\omega/\omega_n)^2} = \frac{(u_{st})_o}{25} = 0.02 \tag{b}$$

From Eq. (b), $(u_{st})_o = 0.5$ in. Substituting in Eq. (a) gives $\zeta = 0.05$.

3.2.4 Dynamic Response Factors

In this section we introduce deformation (or displacement), velocity, and acceleration response factors that are dimensionless and define the amplitude of these three response quantities. The steady-state displacement of Eq. (3.2.10) is repeated for convenience:

$$\frac{u(t)}{p_o/k} = R_d \sin(\omega t - \phi) \tag{3.2.16}$$

where the *deformation response factor* R_d is the ratio of the amplitude u_o of the dynamic (or vibratory) deformation to the static deformation $(u_{st})_o$; see Eq. (3.2.11).

Differentiating Eq. (3.2.16) gives an equation for the velocity response:

$$\frac{\dot{u}(t)}{p_o/\sqrt{km}} = R_v \cos(\omega t - \phi) \tag{3.2.17}$$

where the *velocity response factor* R_v is related to R_d by

$$R_v = \frac{\omega}{\omega_n} R_d \tag{3.2.18}$$

Differentiating Eq. (3.2.17) gives an equation for the acceleration response:

$$\frac{\ddot{u}(t)}{p_o/m} = -R_a \sin(\omega t - \phi) \tag{3.2.19}$$

where the *acceleration response factor* R_a is related to R_d by

$$R_a = \left(\frac{\omega}{\omega_n}\right)^2 R_d \tag{3.2.20}$$

Observe from Eq. (3.2.19) that R_a is the ratio of the amplitude of the vibratory acceleration to the acceleration due to force p_o acting on the mass.

The dynamic response factors R_d, R_v, and R_a are plotted as functions of ω/ω_n in Fig. 3.2.7. The plots of R_v and R_a are new, but the one for R_d is the same as that in Fig. 3.2.6. The deformation response factor R_d is unity at $\omega/\omega_n = 0$, peaks at $\omega/\omega_n < 1$, and approaches zero as $\omega/\omega_n \to \infty$. The velocity response factor R_v is zero at $\omega/\omega_n = 0$, peaks at $\omega/\omega_n = 1$, and approaches zero as $\omega/\omega_n \to \infty$. The acceleration response factor R_a is zero at $\omega/\omega_n = 0$, peaks at $\omega/\omega_n > 1$, and approaches unity as $\omega/\omega_n \to \infty$. For $\zeta > 1/\sqrt{2}$ no peak occurs for R_d and R_a.

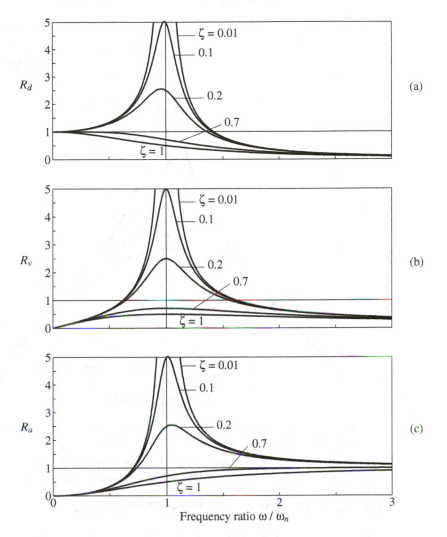

Figure 3.2.7 Deformation, velocity, and acceleration response factors for a damped system excited by harmonic force.

The simple relations among the dynamic response factors

$$\frac{R_a}{\omega/\omega_n} = R_v = \frac{\omega}{\omega_n} R_d \qquad (3.2.21)$$

make it possible to present all three factors in a single graph. The R_v–ω/ω_n data in the linear plot of Fig. 3.2.7b are replotted as shown in Fig. 3.2.8 on four-way logarithmic graph paper. The R_d and R_a values can be read from the diagonally oriented logarithmic scales that are different from the vertical scale for R_v. This compact presentation makes it

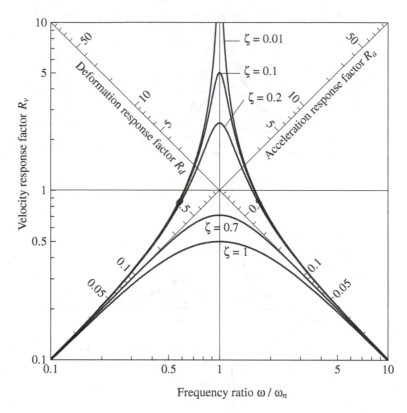

Figure 3.2.8 Four-way logarithmic plot of deformation, velocity, and acceleration response factors for a damped system excited by harmonic force.

possible to replace the three linear plots of Fig. 3.2.7 by a single plot. The concepts underlying construction of this four-way logarithmic graph paper are presented in Appendix 3.

3.2.5 Resonant Frequencies and Resonant Responses

A *resonant frequency* is defined as the forcing frequency at which the largest response amplitude occurs. Figure 3.2.7 shows that the peaks in the frequency-response curves for displacement, velocity, and acceleration occur at slightly different frequencies. These resonant frequencies can be determined by setting to zero the first derivative of R_d, R_v, and R_a with respect to ω/ω_n; for $\zeta < 1/\sqrt{2}$ they are:

Displacement resonant frequency: $\quad \omega_n\sqrt{1-2\zeta^2}$

Velocity resonant frequency: $\quad \omega_n$

Acceleration resonant frequency: $\quad \omega_n \div \sqrt{1-2\zeta^2}$

For an undamped system the three resonant frequencies are identical and equal to the natural frequency ω_n of the system. Intuition might suggest that the resonant frequencies for a damped system should be at its natural frequency $\omega_D = \omega_n\sqrt{1 - \zeta^2}$, but this does not happen; the difference is small, however. For the degree of damping usually embodied in structures, typically well below 20%, the differences among the three resonant frequencies and the natural frequency are small.

The three dynamic response factors at their respective resonant frequencies are

$$R_d = \frac{1}{2\zeta\sqrt{1 - \zeta^2}} \qquad R_v = \frac{1}{2\zeta} \qquad R_a = \frac{1}{2\zeta\sqrt{1 - \zeta^2}} \qquad (3.2.22)$$

3.2.6 Half-Power Bandwidth

An important property of the frequency response curve for R_d is shown in Fig. 3.2.9, where the *half-power bandwidth* is defined. If ω_a and ω_b are the forcing frequencies on either side

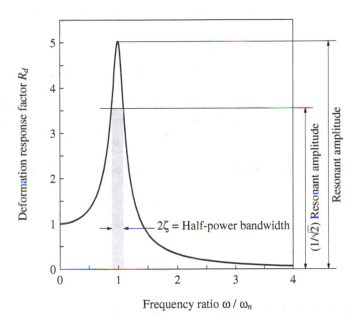

Figure 3.2.9   Definition of half-power bandwidth.

of the resonant frequency at which the amplitude u_o is $1/\sqrt{2}$ times the resonant amplitude, then for small ζ

$$\frac{\omega_b - \omega_a}{\omega_n} = 2\zeta \qquad (3.2.23)$$

This result, derived in Derivation 3.4, can be rewritten as

$$\zeta = \frac{\omega_b - \omega_a}{2\omega_n} \qquad \text{or} \qquad \zeta = \frac{f_b - f_a}{2f_n} \qquad (3.2.24)$$

where $f = \omega/2\pi$ is the cyclic frequency. This important result enables evaluation of damping from forced vibration tests without knowing the applied force (Section 3.4).

Derivation 3.4

Equating R_d from Eq. (3.2.11) and $1/\sqrt{2}$ times the resonant amplitude of R_d given by Eq. (3.2.22), by definition, the forcing frequencies ω_a and ω_b satisfy the condition

$$\frac{1}{\sqrt{[1 - (\omega/\omega_n)^2]^2 + [2\zeta(\omega/\omega_n)]^2}} = \frac{1}{\sqrt{2}} \frac{1}{2\zeta\sqrt{1 - \zeta^2}} \tag{a}$$

Inverting both sides, squaring them, and rearranging terms gives

$$\left(\frac{\omega}{\omega_n}\right)^4 - 2(1 - 2\zeta^2)\left(\frac{\omega}{\omega_n}\right)^2 + 1 - 8\zeta^2(1 - \zeta^2) = 0 \tag{b}$$

Equation (b) is a quadratic equation in $(\omega/\omega_n)^2$, the roots of which are

$$\left(\frac{\omega}{\omega_n}\right)^2 = (1 - 2\zeta^2) \pm 2\zeta\sqrt{1 - \zeta^2} \tag{c}$$

where the positive sign gives the larger root ω_b and the negative sign corresponds to the smaller root ω_a.

For the small damping ratios representative of practical structures, the two terms containing ζ^2 can be dropped and

$$\frac{\omega}{\omega_n} \simeq (1 \pm 2\zeta)^{1/2} \tag{d}$$

Taking only the first term in the Taylor series expansion of the right side gives

$$\frac{\omega}{\omega_n} \simeq 1 \pm \zeta \tag{e}$$

Subtracting the smaller root from the larger one gives

$$\frac{\omega_b - \omega_a}{\omega_n} \simeq 2\zeta \tag{f}$$

3.2.7 Steady-State Response to Cosine Force

The differential equation to be solved is

$$m\ddot{u} + c\dot{u} + ku = p_o \cos \omega t \tag{3.2.25}$$

The particular solution given by Eq. (3.2.3) still applies, but in this case the constants C and D are

$$C = \frac{p_o}{k} \frac{2\zeta(\omega/\omega_n)}{[1 - (\omega/\omega_n)^2]^2 + [2\zeta(\omega/\omega_n)]^2}$$

$$D = \frac{p_o}{k} \frac{1 - (\omega/\omega_n)^2}{[1 - (\omega/\omega_n)^2]^2 + [2\zeta(\omega/\omega_n)]^2} \tag{3.2.26}$$

These are determined by the procedure of Derivation 3.3. The steady-state response given by Eqs. (3.2.3) and (3.2.26) can be expressed as

$$u(t) = u_o \cos(\omega t - \phi) = (u_{\text{st}})_o R_d \cos(\omega t - \phi) \tag{3.2.27}$$

where the amplitude u_o, the deformation response factor R_d, and the phase angle ϕ are the same as those derived in Section 3.2.3 for a sinusoidal force. This similarity in the steady-state responses to the two harmonic forces is not surprising since the two excitations are the same except for a time shift.

PART B: VISCOUSLY DAMPED SYSTEMS: APPLICATIONS

3.3 RESPONSE TO VIBRATION GENERATOR

Vibration generators (or shaking machines) were developed to provide a source of harmonic excitation appropriate for testing full-scale structures. In this section theoretical results for the steady-state response of an SDF system to a harmonic force caused by a vibration generator are presented. These results provide a basis for evaluating the natural frequency and damping of a structure from experimental data (Section 3.4).

3.3.1 Vibration Generator

Figure 3.3.1 shows a vibration generator having the form of two flat baskets rotating in opposite directions about a vertical axis. By placing various numbers of lead weights in the baskets, the magnitudes of the rotating weights can be altered. The two counterrotating masses, $m_e/2$, are shown schematically in Fig. 3.3.2 as lumped masses with eccentricity $= e$; their locations at $t = 0$ are shown in (a) and at some time t in (b). The x-components of the inertia forces of the rotating masses cancel out, and the y-components combine to produce a force

$$p(t) = (m_e e \omega^2) \sin \omega t \tag{3.3.1}$$

By bolting the vibration generator to the structure to be excited, this force can be transmitted to the structure. The amplitude of this harmonic force is proportional to the square of the excitation frequency ω. Therefore, it is difficult to generate force at low frequencies and impractical to obtain the static response of a structure.

3.3.2 Structural Response

Assuming that the eccentric mass m_e is small compared to the mass m of the structure, the equation governing the motion of an SDF system excited by a vibration generator is

$$m\ddot{u} + c\dot{u} + ku = (m_e e \omega^2) \sin \omega t \tag{3.3.2}$$

Figure 3.3.1 Counterrotating eccentric weight vibration generator.

The amplitudes of steady-state displacement and of steady-state acceleration of the SDF system are given by the maximum values of Eqs. (3.2.16) and (3.2.19) with $p_o = m_e e \omega^2$. Thus

$$u_o = \frac{m_e e}{k} \omega^2 R_d = \frac{m_e e}{m} \left(\frac{\omega}{\omega_n} \right)^2 R_d \tag{3.3.3}$$

$$\ddot{u}_o = \frac{m_e e}{m} \omega^2 R_a = \frac{m_e e \omega_n^2}{m} \left(\frac{\omega}{\omega_n} \right)^2 R_a \tag{3.3.4}$$

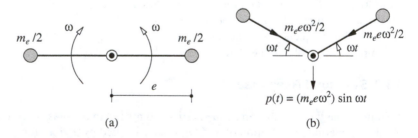

Figure 3.3.2 Vibration generator: (a) initial position; (b) position and forces at time t.

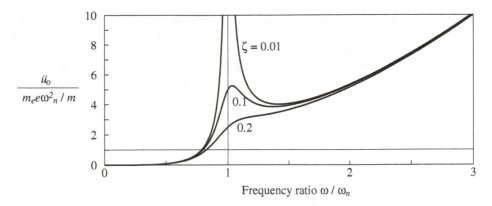

Figure 3.3.3

The acceleration amplitude of Eq. (3.3.4) is plotted as a function of the frequency ratio ω/ω_n in Fig. 3.3.3. For forcing frequencies ω greater than the natural frequency ω_n of the system, the acceleration increases rapidly with increasing ω because the amplitude of the exciting force, Eq. (3.3.1), is proportional to ω^2.

3.4 NATURAL FREQUENCY AND DAMPING FROM HARMONIC TESTS

The theory of forced harmonic vibration, presented in the preceding sections of this chapter, provides a basis to determine the natural frequency and damping of a structure from its measured response to a vibration generator. The measured damping provides data for an important structural property that cannot be computed from the design of the structure. The measured value of the natural frequency is the "actual" property of a structure against which values computed from the stiffness and mass properties of structural idealizations can be compared. Such research investigations have led to better procedures for developing structural idealizations that are representative of actual structures.

3.4.1 Resonance Testing

The concept of resonance testing is based on the result of Eq. (3.2.15), rewritten as

$$\zeta = \frac{1}{2}\frac{(u_{st})_o}{(u_o)_{\omega=\omega_n}} \tag{3.4.1}$$

The damping ratio ζ is calculated from experimentally determined values of $(u_{st})_o$ and of u_o at forcing frequency equal to the natural frequency of the system.[†] Usually, the acceleration amplitude is measured and $u_o = \ddot{u}_o/\omega^2$. This seems straightforward except that the true value ω_n of the natural frequency is unknown. The natural frequency is detected

---

[†]Strictly speaking, this is not the resonant frequency; see Section 3.2.5.

experimentally by utilizing the earlier result that the phase angle is 90° if $\omega = \omega_n$. Thus the structure is excited at forcing frequency ω, the phase angle is measured, and the exciting frequency is adjusted progressively until the phase angle is 90°.

If the displacement due to static force p_o—the amplitude of the harmonic force—can be obtained, Eq. (3.4.1) provides the damping ratio. As mentioned earlier, it is difficult for a vibration generator to produce a force at low frequencies and impractical to obtain a significant static force. An alternative is to measure the static response by some other means, such as by pulling on the structure. In this case, Eq. (3.4.1) should be modified to recognize any differences in the force applied in the static test relative to the amplitude of the harmonic force.

3.4.2 Frequency-Response Curve

Because of the difficulty in obtaining the static structural response using a vibration generator, the natural frequency and damping ratio of a structure are usually determined by obtaining the frequency-response curve experimentally. The vibration generator is operated at a selected frequency, the structural response is observed until the transient part damps out, and the amplitude of the steady-state acceleration is measured. The frequency of the vibration generator is adjusted to a new value and the measurements are repeated. The forcing frequency is varied over a range that includes the natural frequency of the system. A frequency-response curve in the form of acceleration amplitude versus frequency may be plotted directly from the measured data. This curve is for a force with amplitude proportional to ω^2 and would resemble the frequency-response curve of Fig. 3.3.3. If each measured acceleration amplitude is divided by ω^2, we obtain the frequency–acceleration curve

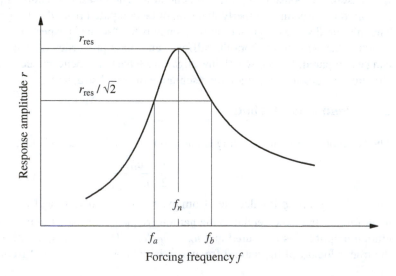

Figure 3.4.1   Evaluating damping from frequency-response curve.

for a constant-amplitude force. This curve from measured data would resemble a curve in Fig. 3.2.7c. If the measured accelerations are divided by ω^4, the resulting frequency–displacement curve for a constant-amplitude force would be an experimental version of the curve in Fig. 3.2.7a.

The natural frequency and damping ratio can be determined from any one of the experimentally obtained versions of the frequency-response curves of Figs. 3.3.3, 3.2.7c, and 3.2.7a. For the practical range of damping the natural frequency f_n is essentially equal to the forcing frequency at resonance. The damping ratio is calculated by Eq. (3.2.24) using the frequencies f_a and f_b, determined, as illustrated in Fig. 3.4.1, from the experimental curve shown schematically. Although this equation was derived from the frequency–displacement curve for a constant-amplitude harmonic force, it is approximately valid for the other response curves mentioned earlier as long as the structure is lightly damped.

Example 3.2

The plexiglass frame of Fig. 1.1.4 is mounted on a shaking table that can apply harmonic base motions of specified frequencies and amplitudes. At each excitation frequency ω, acceleration amplitudes $\ddot{u}_{go}$ and $\ddot{u}_o^t$ of the table and the top of the frame, respectively, are recorded. The transmissibility TR $= \ddot{u}_o^t / \ddot{u}_{go}$ is compiled and the data are plotted in Fig. E3.2. Determine the natural frequency and damping ratio of the plexiglass frame from these data.

Solution   The peak of the frequency-response curve occurs at 3.59 Hz. Assuming that the damping is small, the natural frequency $f_n = 3.59$ Hz.

The peak value of the transmissibility curve is 12.8. Now draw a horizontal line at $12.8/\sqrt{2} = 9.05$ as shown. This line intersects the frequency-response curve at $f_b = 3.74$ Hz and $f_a = 3.44$ Hz. Therefore, from Eq. (3.2.24),

$$\zeta = \frac{3.74 - 3.44}{2\,(3.59)} = 0.042 = 4.2\%$$

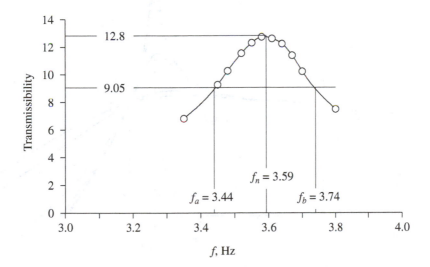

Figure E3.2

This damping value is slightly higher than the 3.96% determined from a free vibration test on the model (Example 2.5).

Note that we have used Eq. (3.2.24) to determine the damping ratio of the system from its transmissibility (TR) curve, whereas this equation had been derived from the frequency-displacement curve. This approximation is appropriate because at excitation frequencies in the range f_a to f_b, the numerical values of TR and R_d are close; this is left for the reader to verify after an equation for TR is presented in Section 3.6.

3.5 FORCE TRANSMISSION AND VIBRATION ISOLATION

Consider the mass–spring–damper system shown in the left inset in Fig. 3.5.1 subjected to a harmonic force. The force transmitted to the base is

$$f_T = f_S + f_D = ku(t) + c\dot{u}(t) \qquad (3.5.1)$$

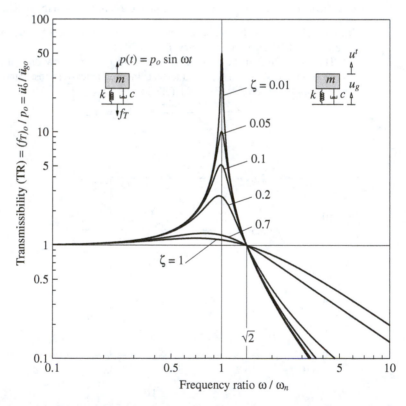

Figure 3.5.1 Transmissibility for harmonic excitation. Force transmissibility and ground motion transmissibility are identical.

Substituting Eq. (3.2.10) for $u(t)$ and Eq. (3.2.17) for $\dot{u}(t)$ and using Eq. (3.2.18) gives

$$f_T(t) = (u_{st})_o R_d [k\sin(\omega t - \phi) + c\omega\cos(\omega t - \phi)] \tag{3.5.2}$$

The maximum value of $f_T(t)$ over t is

$$(f_T)_o = (u_{st})_o R_d \sqrt{k^2 + c^2\omega^2}$$

which, after using $(u_{st})_o = p_o/k$ and $\zeta = c/2m\omega_n$, can be expressed as

$$\frac{(f_T)_o}{p_o} = R_d\sqrt{1 + \left(2\zeta\frac{\omega}{\omega_n}\right)^2}$$

Substituting Eq. (3.2.11) for R_d gives an equation for the ratio of the maximum transmitted force to the amplitude p_o of the applied force, known as the *transmissibility* (TR) of the system:

$$TR = \left\{\frac{1 + [2\zeta(\omega/\omega_n)]^2}{[1 - (\omega/\omega_n)^2]^2 + [2\zeta(\omega/\omega_n)]^2}\right\}^{1/2} \tag{3.5.3}$$

Note that if the spring is rigid, $\omega_n = \infty$ and TR $= 1$, implying that $(f_T)_0 = p_0$.

The transmissibility is plotted in Fig. 3.5.1 as a function of the frequency ratio ω/ω_n for several values of the damping ratio ζ. Logarithmic scales have been chosen to highlight the curves for large ω/ω_n, the region of interest. While damping decreases the amplitude of motion at all excitation frequencies (Fig. 3.2.6), damping decreases the transmitted force only if $\omega/\omega_n < \sqrt{2}$. For the transmitted force to be less than the applied force, i.e., TR < 1, the stiffness of the support system and hence the natural frequency should be small enough so that $\omega/\omega_n > \sqrt{2}$. No damping is desired in the support system because, in this frequency range, damping increases the transmitted force. This implies a trade-off between a soft spring to reduce the transmitted force and an acceptable static displacement.

If the applied force arises from a rotating machine, its frequency will vary as it starts to rotate and increases its speed to reach the operating frequency. In this case the choice of a flexible support system to minimize the transmitted force must be a compromise. It must have sufficient damping to limit the force transmitted while passing through resonance, but not enough to add significantly to the force transmitted at operating speeds. Luckily, natural rubber is a very satisfactory material and is often used for the isolation of vibration.

3.6 RESPONSE TO GROUND MOTION AND VIBRATION ISOLATION

In this section we determine the response of an SDF system (see the right inset in Fig. 3.5.1) to harmonic ground motion:

$$\ddot{u}_g(t) = \ddot{u}_{go}\sin\omega t \tag{3.6.1}$$

For this excitation the governing equation is Eq. (1.7.4), where the forcing function is $p_{eff}(t) = -m\ddot{u}_g(t) = -m\ddot{u}_{go}\sin\omega t$, the same as Eq. (3.2.1) for an applied harmonic force with p_o replaced by $-m\ddot{u}_{go}$. Making this substitution in Eqs. (3.1.9) and (3.2.10) gives

$$u(t) = \frac{-m\ddot{u}_{go}}{k} R_d \sin(\omega t - \phi) \tag{3.6.2}$$

The acceleration of the mass is

$$\ddot{u}^t(t) = \ddot{u}_g(t) + \ddot{u}(t) \tag{3.6.3}$$

Substituting Eq. (3.6.1) and the second derivative of Eq. (3.6.2) gives an equation for $\ddot{u}^t(t)$ from which the amplitude or maximum value $\ddot{u}_o^t$ can be determined (see Derivation 3.5):

$$\text{TR} = \frac{\ddot{u}_o^t}{\ddot{u}_{go}} = \left\{ \frac{1 + [2\zeta(\omega/\omega_n)]^2}{\left[1 - (\omega/\omega_n)^2\right]^2 + [2\zeta(\omega/\omega_n)]^2} \right\}^{1/2} \tag{3.6.4}$$

The ratio of acceleration $\ddot{u}_o^t$ transmitted to the mass and amplitude $\ddot{u}_{go}$ of ground acceleration is also known as the *transmissibility* (TR) of the system. From Eqs. (3.6.4) and (3.5.3) it is clear that the transmissibility for the ground excitation problem is the same as for the applied force problem.

Therefore, Fig. 3.5.1 also gives the ratio $\ddot{u}_o^t/\ddot{u}_{go}$ as a function of the frequency ratio ω/ω_n. If the excitation frequency ω is much smaller than the natural frequency ω_n of the system, $\ddot{u}_o^t \simeq \ddot{u}_{go}$ (i.e., the mass moves rigidly with the ground, both undergoing the same acceleration). If the excitation frequency ω is much higher than the natural frequency ω_n of the system, $\ddot{u}_o^t \simeq 0$ (i.e., the mass stays still while the ground beneath it moves). This is the basic concept underlying isolation of a mass from a moving base by using a very flexible support system. For example, buildings have been mounted on natural rubber bearings to isolate them from ground-borne vertical vibration—typically with frequencies that range from 25 to 50 Hz—due to rail traffic.

Before closing this section, we mention without derivation the results of a related problem. If the ground motion is defined as $u_g(t) = u_{go} \sin \omega t$, it can be shown that the amplitude u_o^t of the total displacement $u^t(t)$ of the mass is given by

$$\text{TR} = \frac{u_o^t}{u_{go}} = \left\{ \frac{1 + [2\zeta(\omega/\omega_n)]^2}{\left[1 - (\omega/\omega_n)^2\right]^2 + [2\zeta(\omega/\omega_n)]^2} \right\}^{1/2} \tag{3.6.5}$$

Comparing this with Eq. (3.6.4) indicates that the transmissibility for displacements and accelerations is identical.

Example 3.3

A sensitive instrument with weight 100 lb is to be installed at a location where the vertical acceleration is 0.1g at a frequency of 10 Hz. This instrument is mounted on a rubber pad of stiffness 80 lb/in. and damping such that the damping ratio for the system is 10%. **(a)** What acceleration is transmitted to the instrument? **(b)** If the instrument can tolerate only an acceleration of 0.005g, suggest a solution assuming that the same rubber pad is to be used. Provide numerical results.

Solution   **(a)** *Determine TR.*

$$\omega_n = \sqrt{\frac{80}{100/386}} = 17.58 \text{ rad/sec}$$

$$\frac{\omega}{\omega_n} = \frac{2\pi(10)}{17.58} = 3.575$$

Substituting these in Eq. (3.6.4) gives

$$TR = \frac{\ddot{u}_o^t}{\ddot{u}_{go}} = \sqrt{\frac{1 + [2(0.1)(3.575)]^2}{[1 - (3.575)^2]^2 + [2(0.1)(3.575)]^2}} = 0.104$$

Therefore, $\ddot{u}_o^t = (0.104)\ddot{u}_{go} = (0.104)0.1g = 0.01g$.

(b) *Determine the added mass to reduce acceleration.* The acceleration transmitted can be reduced by increasing ω/ω_n, which requires reducing ω_n by mounting the instrument on mass m_b. Suppose that we add a mass $m_b = 150$ lb/g; the total mass $= 250$ lb/g, and

$$\omega_n' = \sqrt{\frac{80}{250/386}} = 11.11 \text{ rad/sec} \qquad \frac{\omega}{\omega_n'} = 5.655$$

To determine the damping ratio for the system with added mass, we need the damping coefficient for the rubber pad:

$$c = \zeta(2m\omega_n) = 0.1(2)\left(\frac{100}{386}\right)17.58 = 0.911 \text{ lb-sec/in.}$$

Then

$$\zeta' = \frac{c}{2(m + m_b)\omega_n'} = \frac{0.911}{2(250/386)11.11} = 0.063$$

Substituting for ω/ω_n' and ζ' in Eq. (3.6.4) gives $\ddot{u}_o^t/\ddot{u}_{go} = 0.04$, $\ddot{u}_o^t = 0.004g$, which is satisfactory because it is less than 0.005g.

Instead of selecting an added mass by judgment, it is possible to set up a quadratic equation for the unknown mass, which will give $\ddot{u}_o^t = 0.005g$.

Example 3.4

An automobile is traveling along a multispan elevated roadway supported every 100 ft. Long-term creep has resulted in a 6-in. deflection at the middle of each span (Fig. E3.4a). The roadway profile can be approximated as sinusoidal with an amplitude of 3 in. and a period of 100 ft. The SDF system shown is a simple idealization of an automobile, appropriate for a "first approximation" study of the ride quality of the vehicle. When fully loaded, the weight of the automobile is 4 kips. The stiffness of the automobile suspension system is 800 lb/in., and its viscous damping coefficient is such that the damping ratio of the system is 40%. Determine

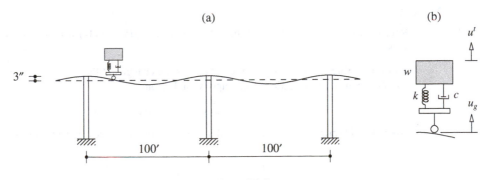

(a)   (b)

Figure E3.4

(a) the amplitude u_o^t of vertical motion $u^t(t)$ when the automobile is traveling at 40 mph, and
(b) the speed of the vehicle that would produce a resonant condition for u_o^t.

Solution Assuming that the tires are infinitely stiff and they remain in contact with the road, the problem can be idealized as shown in Fig. E3.4b. The vertical displacement of the tires is $u_g(t) = u_{go}\sin\omega t$, where $u_{go} = 3$ in. The forcing frequency $\omega = 2\pi/T$, where the forcing period $T = L/v$, the time taken by the automobile to cross the span; therefore, $\omega = 2\pi v/L$.

(a) *Determine* u_o^t.

$$v = 40\text{ mph} = 58.67\text{ ft/sec}\qquad \omega = \frac{2\pi(58.67)}{100} = 3.686\text{ rad/sec}$$

$$\omega_n = \sqrt{\frac{k}{m}} = \sqrt{\frac{800}{4000/386}} = 8.786\text{ rad/sec}\qquad \frac{\omega}{\omega_n} = 0.420$$

Substituting these data in Eq. (3.6.5) gives

$$\frac{u_o^t}{u_{go}} = \left\{\frac{1 + [2(0.4)(0.420)]^2}{[1 - (0.420)^2]^2 + [2(0.4)(0.420)]^2}\right\}^{1/2} = 1.186$$

$$u_o^t = 1.186u_{go} = 1.186(3) = 3.56\text{ in.}$$

(b) *Determine the speed at resonance.* If ζ were small, resonance would occur approximately at $\omega/\omega_n = 1$. However, automobile suspensions have heavy damping, to reduce vibration. In this case, $\zeta = 0.4$, and for such large damping the resonant frequency is significantly different from ω_n. By definition, resonance occurs for u_o^t when TR (or TR^2) is maximum over all ω. Substituting $\zeta = 0.4$ in Eq. (3.6.5) and introducing $\beta = \omega/\omega_n$ gives

$$TR^2 = \frac{1 + 0.64\beta^2}{(1 - 2\beta^2 + \beta^4) + 0.64\beta^2} = \frac{1 + 0.64\beta^2}{\beta^4 - 1.36\beta^2 + 1}$$

$$\frac{d(TR)^2}{d\beta} = 0 \Rightarrow \beta = 0.893 \Rightarrow \omega = 0.893\omega_n = 0.893(8.786) = 7.846\text{ rad/sec}$$

Resonance occurs at this forcing frequency, which implies a speed of

$$v = \frac{\omega L}{2\pi} = \frac{(7.846)100}{2\pi} = 124.9\text{ ft/sec} = 85\text{ mph}$$

Example 3.5

Repeat part (a) of Example 3.4 if the vehicle is empty (driver only) with a total weight of 3 kips.

Solution Since the damping coefficient c does not change but the mass m does, we need to recompute the damping ratio for an empty vehicle from

$$c = 2\zeta_f\sqrt{km_f} = 2\zeta_e\sqrt{km_e}$$

where the subscripts f and e denote full and empty conditions, respectively. Thus

$$\zeta_e = \zeta_f\left(\frac{m_f}{m_e}\right)^{1/2} = 0.4\left(\frac{4}{3}\right)^{1/2} = 0.462$$

For an empty vehicle

$$\omega_n = \sqrt{\frac{k}{m}} = \sqrt{\frac{800}{3000/386}} = 10.15 \text{ rad/sec}$$

$$\frac{\omega}{\omega_n} = \frac{3.686}{10.15} = 0.363$$

Substituting for ω/ω_n and ζ in Eq. (3.6.5) gives

$$\frac{u_o^t}{u_{go}} = \left\{ \frac{1 + [2(0.462)(0.363)]^2}{[1 - (0.363)^2]^2 + [2(0.462)(0.363)]^2} \right\}^{1/2} = 1.133$$

$$u_o^t = 1.133 u_{go} = 1.133(3) = 3.40 \text{ in.}$$

Derivation 3.5

Equation (3.6.2) is first rewritten as a linear combination of sine and cosine functions. This can be accomplished by substituting Eqs. (3.2.11) and (3.2.12) for the R_d and ϕ, respectively, or by replacing p_o in Eq. (3.2.4) with $-m\ddot{u}_{go}$ and substituting in Eq. (3.2.3). Either way the relative displacement is

$$u(t) = \frac{-m\ddot{u}_{go}}{k} \left\{ \frac{[1 - (\omega/\omega_n)^2] \sin \omega t - [2\zeta(\omega/\omega_n)] \cos \omega t}{[1 - (\omega/\omega_n)^2]^2 + [2\zeta(\omega/\omega_n)]^2} \right\} \qquad \text{(a)}$$

Differentiating this twice and substituting it in Eq. (3.6.3) together with Eq. (3.6.1) gives

$$\ddot{u}^t(t) = \ddot{u}_{go} (C_1 \sin \omega t + D_1 \cos \omega t) \qquad \text{(b)}$$

where

$$C_1 = \frac{1 - (\omega/\omega_n)^2 + 4\zeta^2(\omega/\omega_n)^2}{[1 - (\omega/\omega_n)^2]^2 + [2\zeta(\omega/\omega_n)]^2} \qquad D_1 = \frac{-2\zeta(\omega/\omega_n)^3}{[1 - (\omega/\omega_n)^2]^2 + [2\zeta(\omega/\omega_n)]^2} \qquad \text{(c)}$$

The acceleration amplitude is

$$\ddot{u}_o^t = \ddot{u}_{go} \sqrt{C_1^2 + D_1^2} \qquad \text{(d)}$$

This result, after substituting for C_1 and D_1 from Eq. (c) and some simplification, leads to Eq. (3.6.4).

3.7 VIBRATION-MEASURING INSTRUMENTS

Measurement of vibration is of great interest in many aspects of structural engineering. For example, measurement of ground shaking during an earthquake provides basic data for earthquake engineering, and records of the resulting motions of a structure provide insight into how structures respond during earthquakes. Although measuring instruments are highly developed and intricate, the basic element of these instruments is some form of a transducer. In its simplest form a transducer is a mass–spring–damper system mounted inside a rigid frame that is attached to the surface whose motion is to be measured. Figure 3.7.1 shows a schematic drawing of such an instrument to record the horizontal

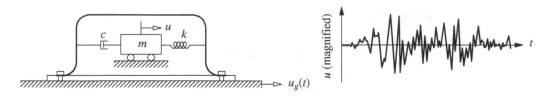

Figure 3.7.1   Schematic drawing of a vibration-measuring instrument and recorded motion.

motion of a support point; three separate transducers are required to measure the three components of motion. When subjected to motion of the support point, the transducer mass moves relative to the frame, and this relative displacement is recorded after suitable magnification. It is the objective of this brief presentation to discuss the principle underlying the design of vibration-measuring instruments so that the measured relative displacement provides the desired support motion—acceleration or displacement.

3.7.1 Measurement of Acceleration

The motion to be measured generally varies arbitrarily with time and may include many harmonic components covering a wide range of frequencies. It is instructive, however, to consider first the measurement of simple harmonic motion described by Eq. (3.6.1). The displacement of the instrument mass relative to the moving frame is given by Eq. (3.6.2), which can be rewritten as

$$u(t) = -\left(\frac{1}{\omega_n^2} R_d\right) \ddot{u}_g\left(t - \frac{\phi}{\omega}\right) \tag{3.7.1}$$

The recorded $u(t)$ is the base acceleration modified by a factor $-R_d/\omega_n^2$ and recorded with a time lag ϕ/ω. As shown in Fig. 3.2.6, R_d and ϕ vary with the forcing frequency ω, but ω_n^2 is an instrument constant independent of the support motion.

   The object of the instrument design is to make R_d and ϕ/ω as independent of excitation frequency as possible because then each harmonic component of acceleration will be recorded with the same modifying factor and the same time lag. Then, even if the motion to be recorded consists of many harmonic components, the recorded $u(t)$ will have the same shape as the support motion with a constant shift of time. This constant time shift simply moves the time scale a little, which is usually not important. According to Fig. 3.7.2 (which is a magnified plot of Fig. 3.2.6 with additional damping values), if $\zeta = 0.7$, then over the frequency range $0 \le \omega/\omega_n \le 0.50$, R_d is close to 1 (less than 2.5% error) and the variation of ϕ with ω is close to linear, implying that ϕ/ω is essentially constant. Thus an instrument with a natural frequency of 50 Hz and a damping ratio of 0.7 has a useful frequency range from 0 to 25 Hz with negligible error. These are the properties of modern, commercially available instruments designed to measure earthquake-induced ground acceleration. Because the measured amplitude of $u(t)$ is proportional to R_d/ω_n^2, a high-frequency instrument will result in a very small displacement that is substantially magnified in these instruments for proper measurement.

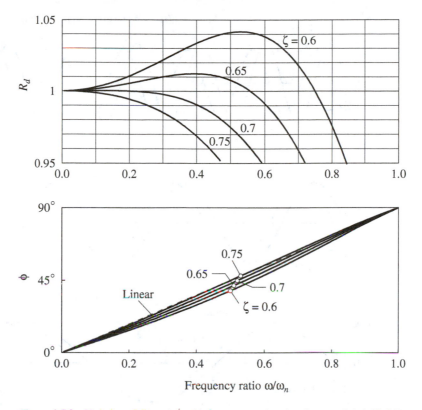

Figure 3.7.2   Variation of R_d and ϕ with frequency ratio ω/ω_n for $\zeta = 0.6, 0.65, 0.7,$ and 0.75.

Figure 3.7.3 shows a comparison of the actual ground acceleration $\ddot{u}_g(t) = 0.1\mathrm{g}\sin 2\pi f t$ and the measured relative displacement of $R_d\ddot{u}_g(t - \phi/\omega)$, except for the instrument constant $-1/\omega_n^2$ in Eq. (3.7.1). For excitation frequencies $f = 20$ and 10 Hz, the measured motion has accurate amplitude, but the error at $f = 40$ Hz is noticeable; and the time shift, although not identical for the three frequencies, is similar. If the ground acceleration is the sum of the three harmonic components, this figure shows that the recorded motion matches the ground acceleration in amplitude and shape reasonably well but not perfectly.

The accuracy of the recorded motion $u(t)$ can be improved by separating $u(t)$ into its harmonic components and correcting one component at a time, by calculating $\ddot{u}_g(t - \phi/\omega)$ from the measured $u(t)$ using Eq. (3.7.1) with R_d determined from Eq. (3.2.11) and known instrument properties ω_n and ζ. Such corrections are repeated for each harmonic component in $u(t)$, and the corrected components are then synthesized to obtain $\ddot{u}_g(t)$. These computations can be carried out by discrete Fourier transform procedures, which are introduced in Appendix A.

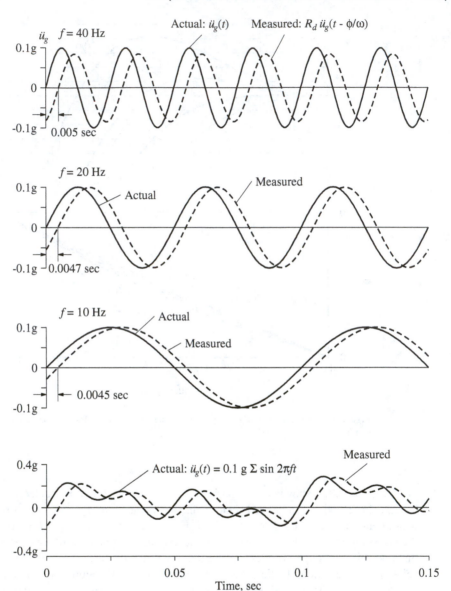

Figure 3.7.3 Comparison of actual ground acceleration and measured motion by an instrument with $f_n = 50$ Hz and $\zeta = 0.7$.

3.7.2 Measurement of Displacement

It is desired to design the transducer so that the relative displacement $u(t)$ measures the support displacement $u_g(t)$. This is achieved by making the transducer spring so flexible or the transducer mass so large, or both, that the mass stays still while the support beneath it moves. Such an instrument is unwieldy because of the heavy mass and soft spring, and because it must accommodate the anticipated support displacement, which may be as large as 36 in. during earthquakes.

To examine the basic concept further, consider harmonic support displacement

$$u_g(t) = u_{go} \sin \omega t \qquad (3.7.2)$$

With the forcing function $p_{eff}(t) = -m\ddot{u}_g(t) = m\omega^2 u_{go} \sin \omega t$, Eq. (1.7.4) governs the relative displacement of the mass; this governing equation is the same as Eq. (3.2.1) for applied harmonic force with p_o replaced by $m\omega^2 u_{go}$. Making this substitution in Eq. (3.2.10) and using Eqs. (3.1.9) and (3.2.20) gives

$$u(t) = R_a u_{go} \sin(\omega t - \phi) \qquad (3.7.3)$$

For excitation frequencies ω much higher than the natural frequency ω_n, R_a is close to unity (Fig. 3.2.7c) and ϕ is close to 180°, and Eq. (3.7.3) becomes

$$u(t) = -u_{go} \sin \omega t \qquad (3.7.4)$$

This recorded displacement is the same as the support displacement [Eq. (3.7.2)] except for the negative sign, which is usually inconsequential. Damping of the instrument is not a critical parameter because it has little effect on the recorded motion if ω/ω_n is very large.

3.8 ENERGY DISSIPATED IN VISCOUS DAMPING

Consider the steady-state motion of an SDF system due to $p(t) = p_o \sin \omega t$. The energy dissipated by viscous damping in one cycle of harmonic vibration is

$$E_D = \int f_D \, du = \int_0^{2\pi/\omega} (c\dot{u})\dot{u} \, dt = \int_0^{2\pi/\omega} c\dot{u}^2 \, dt$$

$$= c \int_0^{2\pi/\omega} [\omega u_o \cos(\omega t - \phi)]^2 \, dt = \pi c\omega u_o^2 = 2\pi \zeta \frac{\omega}{\omega_n} k u_o^2 \qquad (3.8.1)$$

The energy dissipated is proportional to the square of the amplitude of motion. It is not a constant value for any given amount of damping and amplitude since the energy dissipated increases linearly with excitation frequency.

In steady-state vibration, the energy input to the system due to the applied force is dissipated in viscous damping. The external force $p(t)$ inputs energy to the system, which for each cycle of vibration is

$$E_I = \int p(t) \, du = \int_0^{2\pi/\omega} p(t)\dot{u} \, dt$$

$$= \int_0^{2\pi/\omega} [p_o \sin \omega t][\omega u_o \cos(\omega t - \phi)] \, dt = \pi p_o u_o \sin \phi \qquad (3.8.2)$$

Utilizing Eq. (3.2.12) for phase angle, this equation can be rewritten as (see Derivation 3.6)

$$E_I = 2\pi \zeta \frac{\omega}{\omega_n} k u_o^2 \qquad (3.8.3)$$

Equations (3.8.1) and (3.8.3) indicate that $E_I = E_D$.

What about the potential energy and kinetic energy? Over each cycle of harmonic vibration the changes in potential energy (equal to the strain energy of the spring) and kinetic energy are zero. This can be confirmed as follows:

$$E_S = \int f_S \, du = \int_0^{2\pi/\omega} (ku)\dot{u} \, dt$$

$$= \int_0^{2\pi/\omega} k[u_o \sin(\omega t - \phi)][\omega u_o \cos(\omega t - \phi)] \, dt = 0$$

$$E_K = \int f_I \, du = \int_0^{2\pi/\omega} (m\ddot{u})\dot{u} \, dt$$

$$= \int_0^{2\pi/\omega} m[-\omega^2 u_o \sin(\omega t - \phi)][\omega u_o \cos(\omega t - \phi)] \, dt = 0$$

The preceding energy concepts help explain the growth of the displacement amplitude caused by harmonic force with $\omega = \omega_n$ until steady state is attained (Fig. 3.2.2). For $\omega = \omega_n$, $\phi = 90°$ and Eq. (3.8.2) gives

$$E_I = \pi p_o u_o \qquad (3.8.4)$$

The input energy varies linearly with the displacement amplitude (Fig. 3.8.1). In contrast, the dissipated energy varies quadratically with the displacement amplitude (Eq. 3.8.1). As shown in Fig. 3.8.1, before steady state is reached, the input energy per cycle exceeds the energy dissipated during the cycle by damping, leading to a larger amplitude of displacement in the next cycle. With growing displacement amplitude, the dissipated energy increases more rapidly than does the input energy. Eventually, the input and dissipated energies will match at the steady-state displacement amplitude u_o, which will be bounded no matter how small the damping. This energy balance provides an alternative means of

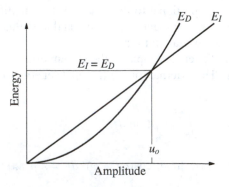

Figure 3.8.1   Input energy E_I and energy dissipated E_D in viscous damping.

finding u_o due to harmonic force with $\omega = \omega_n$; equating Eqs. (3.8.1) and (3.8.4) gives

$$\pi p_o u_o = \pi c \omega_n u_o^2 \qquad (3.8.5)$$

Solving for u_o leads to

$$u_o = \frac{p_o}{c \omega_n} \qquad (3.8.6)$$

This result agrees with Eq. (3.2.7), obtained by solving the equation of motion.

We will now present a graphical interpretation for the energy dissipated in viscous damping. For this purpose we first derive an equation relating the damping force f_D to the displacement u:

$$f_D = c\dot{u}(t) = c\omega u_o \cos(\omega t - \phi)$$

$$= c\omega \sqrt{u_o^2 - u_o^2 \sin^2(\omega t - \phi)}$$

$$= c\omega \sqrt{u_o^2 - [u(t)]^2}$$

This can be rewritten as

$$\left(\frac{u}{u_o}\right)^2 + \left(\frac{f_D}{c\omega u_o}\right)^2 = 1 \qquad (3.8.7)$$

which is the equation of the ellipse shown in Fig. 3.8.2a. Observe that the f_D–u curve is not a single-valued function but a loop known as a *hysteresis loop*. The area enclosed by the ellipse is $\pi(u_o)(c\omega u_o) = \pi c\omega u_o^2$, which is the same as Eq. (3.8.1). Thus the area within the hysteresis loop gives the dissipated energy.

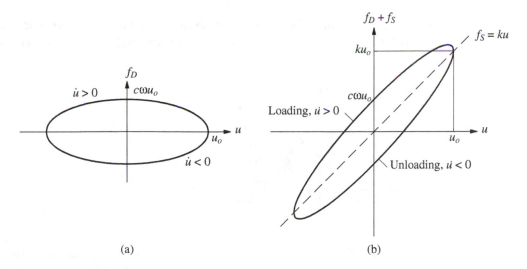

(a)   (b)

Figure 3.8.2 Hysteresis loops for (a) viscous damper; (b) spring and viscous damper in parallel.

It is of interest to examine the total (elastic plus damping) resisting force because this is the force that is measured in an experiment:

$$f_S + f_D = ku(t) + c\dot{u}(t)$$

$$= ku + c\omega\sqrt{u_o^2 - u^2} \tag{3.8.8}$$

A plot of $f_S + f_D$ against u is the ellipse of Fig. 3.8.2a rotated as shown in Fig. 3.8.2b because of the ku term in Eq. (3.8.8). The energy dissipated by damping is still the area enclosed by the ellipse because the area enclosed by the single-valued elastic force, $f_S = ku$, is zero.

The hysteresis loop associated with viscous damping is the result of *dynamic hysteresis* since it is related to the dynamic nature of the loading. The loop area is proportional to excitation frequency; this implies that the force–deformation curve becomes a single-valued curve (no hysteresis loop) if the cyclic load is applied slowly enough ($\omega = 0$). A distinguishing characteristic of dynamic hysteresis is that the hysteresis loops tend to be elliptical in shape rather than pointed, as in Fig. 1.3.1c, if they are associated with plastic deformations. In the latter case, the hysteresis loops develop even under static cyclic loads; this phenomenon is therefore known as *static hysteresis* because the force–deformation curve is insensitive to deformation rate.

In passing, we mention two measures of damping: *specific damping capacity* and the *specific damping factor*. The specific damping capacity, E_D/E_{So}, is that fractional part of the strain energy, $E_{So} = ku_o^2/2$, which is dissipated during each cycle of motion; both E_D and E_{So} are shown in Fig. 3.8.3. The specific damping factor, also known as the *loss factor,* is defined as

$$\xi = \frac{1}{2\pi}\frac{E_D}{E_{So}} \tag{3.8.9}$$

If the energy could be removed at a uniform rate during a cycle of simple harmonic motion (such a mechanism is not realistic), ξ could be interpreted as the energy loss per radian

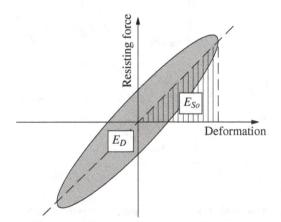

Figure 3.8.3 Definition of energy loss E_D in a cycle of harmonic vibration and maximum strain energy E_{So}.

divided by the strain energy, E_{So}. These two measures of damping are not often used in structural vibration since they are most useful for very light damping (e.g., they are useful in comparing the damping capacity of materials).

Derivation 3.6

Equation (3.8.2) gives the input energy per cycle where the phase angle, defined by Eq. (3.2.12), can be expressed as

$$\sin \phi = \left(2\zeta \frac{\omega}{\omega_n} \right) R_d = \left(2\zeta \frac{\omega}{\omega_n} \right) \frac{u_o}{p_o/k}$$

Substituting this in Eq. (3.8.2) gives Eq. (3.8.3).

3.9 EQUIVALENT VISCOUS DAMPING

As introduced in Section 1.4, damping in actual structures is usually represented by equivalent viscous damping. It is the simplest form of damping to use since the governing differential equation of motion is linear and hence amenable to analytical solution, as seen in earlier sections of this chapter and in Chapter 2. The advantage of using a linear equation of motion usually outweighs whatever compromises are necessary in the viscous damping approximation. In this section we determine the damping coefficient for viscous damping so that it is equivalent in some sense to the combined effect of all damping mechanisms present in the actual structure; these were mentioned in Section 1.4.

The simplest definition of equivalent viscous damping is based on the measured response of a system to harmonic force at exciting frequency ω equal to the natural frequency ω_n of the system. The damping ratio ζ_{eq} is calculated from Eq. (3.4.1) using measured values of u_o and $(u_{st})_o$. This is the equivalent viscous damping since it accounts for all the energy-dissipating mechanisms that existed in the experiments.

Another definition of equivalent viscous damping is that it is the amount of damping that provides the same bandwidth in the frequency-response curve as obtained experimentally for an actual system. The damping ratio ζ_{eq} is calculated from Eq. (3.2.24) using the excitation frequencies f_a, f_b, and f_n (Fig. 3.4.1) obtained from an experimentally determined frequency-response curve.

The most common method for defining equivalent viscous damping is to equate the energy dissipated in a vibration cycle of the actual structure and an equivalent viscous system. For an actual structure the force-displacement relation is obtained from an experiment under cyclic loading with displacement amplitude u_o; such a relation of arbitrary shape is shown schematically in Fig. 3.9.1. The energy dissipated in the actual structure is given by the area E_D enclosed by the hysteresis loop. Equating this to the energy dissipated in viscous damping given by Eq. (3.8.1) leads to

$$4\pi \zeta_{eq} \frac{\omega}{\omega_n} E_{So} = E_D \quad \text{or} \quad \zeta_{eq} = \frac{1}{4\pi} \frac{1}{\omega/\omega_n} \frac{E_D}{E_{So}} \qquad (3.9.1)$$

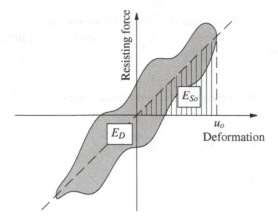

Figure 3.9.1 Energy dissipated E_D in a cycle of harmonic vibration determined from experiment.

where the strain energy, $E_{So} = ku_o^2/2$, is calculated from the stiffness k determined by experimentation.

The experiment leading to the force–deformation curve of Fig. 3.9.1 and hence E_D should be conducted at $\omega = \omega_n$, where the response of the system is most sensitive to damping. Thus Eq. (3.9.1) specializes to

$$\zeta_{eq} = \frac{1}{4\pi}\frac{E_D}{E_{So}} \tag{3.9.2}$$

The damping ratio ζ_{eq} determined from a test at $\omega = \omega_n$ would not be correct at any other exciting frequency, but it would be a satisfactory approximation (Section 3.10.2).

It is widely accepted that this procedure can be extended to model the damping in systems with many degrees of freedom. An equivalent viscous damping ratio is assigned to each natural vibration mode of the system (defined in Chapter 10) in such a way that the energy dissipated in viscous damping matches the actual energy dissipated in the system when the system vibrates in that mode at its natural frequency.

In this book the concept of equivalent viscous damping is restricted to systems vibrating at amplitudes within the linearly elastic limit of the overall structure. The energy dissipated in inelastic deformations of the structure has also been modeled as equivalent viscous damping in some research studies. This idealization is generally not satisfactory, however, for the large inelastic deformations of structures expected during strong earthquakes. We shall account for these inelastic deformations and the associated energy dissipation by nonlinear force–deformation relations, such as those shown in Fig. 1.3.4 (see Chapters 5 and 7).

Example 3.6

A body moving through a fluid experiences a resisting force that is proportional to the square of the speed, $f_D = \pm a\dot{u}^2$, where the positive sign applies to positive $\dot{u}$ and the negative sign to negative $\dot{u}$. Determine the equivalent viscous damping coefficient c_{eq} for such forces acting on an oscillatory system undergoing harmonic motion of amplitude u_o and frequency ω. Also find its displacement amplitude at $\omega = \omega_n$.

Solution   If time is measured from the position of largest negative displacement, the harmonic motion is

$$u(t) = -u_o \cos \omega t$$

The energy dissipated in one cycle of motion is

$$E_D = \int f_D \, du = \int_0^{2\pi/\omega} f_D \dot{u} \, dt = 2 \int_0^{\pi/\omega} f_D \dot{u} \, dt$$

$$= 2 \int_0^{\pi/\omega} (a\dot{u}^2) \dot{u} \, dt = 2a\omega^3 u_o^3 \int_0^{\pi/\omega} \sin^3 \omega t \, dt = \tfrac{8}{3} a\omega^2 u_o^3$$

Equating this to the energy dissipated in viscous damping [Eq. (3.8.1)] gives

$$\pi c_{\text{eq}} \omega u_o^2 = \frac{8}{3} a\omega^2 u_o^3 \quad \text{or} \quad c_{\text{eq}} = \frac{8}{3\pi} a\omega u_o \tag{a}$$

Substituting $\omega = \omega_n$ in Eq. (a) and the c_{eq} for c in Eq. (3.2.15) gives

$$u_o = \left(\frac{3\pi}{8a} \frac{p_o}{\omega_n^2} \right)^{1/2} \tag{b}$$

PART C: SYSTEMS WITH NONVISCOUS DAMPING

3.10 HARMONIC VIBRATION WITH RATE-INDEPENDENT DAMPING

3.10.1 Rate-Independent Damping

Experiments on structural metals indicate that the energy dissipated internally in cyclic straining of the material is essentially independent of the cyclic frequency. Similarly, forced vibration tests on structures indicate that the equivalent viscous damping ratio is roughly the same for all natural modes and frequencies. Thus we refer to this type of damping as *rate-independent linear damping*. Other terms used for this mechanism of internal damping are *structural damping, solid damping,* and *hysteretic damping.* We prefer not to use these terms because the first two are not especially meaningful, and the third is ambiguous because hysteresis is a characteristic of all materials or structural systems that dissipate energy.

Rate-independent damping is associated with static hysteresis due to plastic strain, localized plastic deformation, crystal plasticity, and plastic flow in a range of stresses within the apparent elastic limit. On the microscopic scale the inhomogeneity of stress distribution within crystals and stress concentration at crystal boundary intersections produce local stress high enough to cause local plastic strain even though the average (macroscopic) stress may be well below the elastic limit. This damping mechanism does not include the energy dissipation in macroscopic plastic deformations, which as mentioned earlier, is handled by a nonlinear relationship between force f_S and deformation u.

The simplest device that can be used to represent rate-independent linear damping during harmonic motion at frequency ω is to assume that the damping force is proportional to velocity and inversely proportional to frequency:

$$f_D = \frac{\eta k}{\omega}\dot{u} \tag{3.10.1}$$

where k is the stiffness of the structure and η is a damping coefficient. The energy dissipated by this type of damping in a cycle of vibration at frequency ω is independent of ω (Fig. 3.10.1). It is given by Eq. (3.8.1) with c replaced by $\eta k/\omega$:

$$E_D = \pi\eta k u_o^2 = 2\pi\eta E_{So} \tag{3.10.2}$$

In contrast, the energy dissipated in viscous damping [Eq. (3.8.1)] increases linearly with the forcing frequency as shown in Fig. 3.10.1.

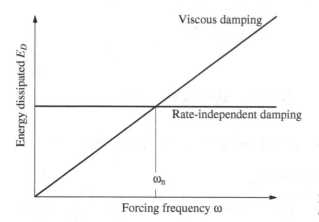

Figure 3.10.1   Energy dissipated in viscous damping and rate-independent damping.

Rate-independent damping is easily described if the excitation is harmonic and we are interested only in the steady-state response of this system. Difficulties arise in translating this damping mechanism back to the time domain. Thus it is most useful in the frequency-domain method of analysis (Appendix A).

3.10.2 Steady-State Response to Harmonic Force

The equation governing harmonic motion of an SDF system with rate-independent linear damping, denoted by a crossed box in Fig. 3.10.2, is Eq. (3.2.1) with the damping term replaced by Eq. (3.10.1):

$$m\ddot{u} + \frac{\eta k}{\omega}\dot{u} + ku = p(t) \tag{3.10.3}$$

The mathematical solution of this equation is quite complex for arbitrary $p(t)$. Here we consider only the steady-state motion due to a sinusoidal forcing function,

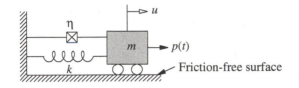

Friction-free surface

Figure 3.10.2   SDF system with rate-independent linear damping.

$p(t) = p_o \sin \omega t$, which is described by

$$u(t) = u_o \sin(\omega t - \phi) \qquad (3.10.4)$$

The amplitude u_o and phase angle ϕ are

$$u_o = (u_{st})_o \frac{1}{\sqrt{\left[1 - (\omega/\omega_n)^2\right]^2 + \eta^2}} \qquad (3.10.5)$$

$$\phi = \tan^{-1} \frac{\eta}{1 - (\omega/\omega_n)^2} \qquad (3.10.6)$$

These results are obtained by modifying the viscous damping ratio in Eqs. (3.2.11) and (3.2.12) to reflect the damping force associated with rate-independent damping, Eq. (3.10.1). In particular, ζ was replaced by

$$\zeta = \frac{c}{c_c} = \frac{\eta k/\omega}{2m\omega_n} = \frac{\eta}{2(\omega/\omega_n)} \qquad (3.10.7)$$

Shown in Fig. 3.10.3 by solid lines are plots of $u_o/(u_{st})_o$ and ϕ as a function of the frequency ratio ω/ω_n for damping coefficient $\eta = 0$, 0.2, and 0.4; the dashed lines are described in the next section. Comparing these results with those in Fig. 3.2.6 for viscous damping, two differences are apparent: First, resonance (maximum amplitude) occurs at $\omega = \omega_n$, not at $\omega < \omega_n$. Second, the phase angle for $\omega = 0$ is $\phi = \tan^{-1} \eta$ instead of zero for viscous damping; this implies that motion with rate-independent damping can never be in phase with the forcing function.

These differences between forced vibration with rate-independent damping and forced vibration with viscous damping are not significant, but they are the source of some difficulty in reconciling physical data. In most damped vibration, damping is not viscous, and to assume that it is without knowing its real physical characteristics is an assumption of some error. In the next section this error is shown to be small when the real damping is rate independent.

3.10.3 Solution Using Equivalent Viscous Damping

In this section an approximate solution for the steady-state harmonic response of a system with rate-independent damping is obtained by modeling this damping mechanism as equivalent viscous damping.

Matching dissipated energies at $\omega = \omega_n$ led to Eq. (3.9.2), where E_D is given by Eq. (3.10.2), leading to the equivalent viscous damping ratio:

$$\zeta_{eq} = \frac{\eta}{2} \qquad (3.10.8)$$

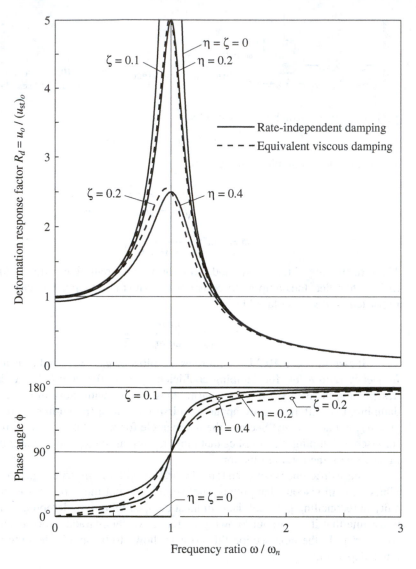

Figure 3.10.3 Response of system with rate-independent damping: exact solution and approximate solution using equivalent viscous damping.

Substituting this ζ_{eq} for ζ in Eqs. (3.2.10) to (3.2.12) gives the system response. The resulting amplitude u_o and phase angle ϕ are shown by the dashed lines in Fig. 3.10.3. This approximate solution matches the exact result at $\omega = \omega_n$ because that was the criterion used in selecting ζ_{eq} (Fig. 3.10.1). Over a wide range of excitation frequencies the approximate solution is seen to be accurate enough for many engineering applications. Thus Eq. (3.10.3)—which is difficult to solve for arbitrary force $p(t)$ that contains many harmonic components of different frequencies ω—can be replaced by the simpler Eq. (3.2.1) for a system with equivalent viscous damping defined by Eq. (3.10.8). This is the basic advantage of equivalent viscous damping.

3.11 HARMONIC VIBRATION WITH COULOMB FRICTION

3.11.1 Equation of Motion

Shown in Fig. 3.11.1 is a mass–spring system with Coulomb friction force $F = \mu N$ that opposes sliding of the mass. As defined in Section 2.4, the coefficients of static and kinetic friction are assumed to be equal to μ, and N is the normal force across the sliding surfaces. The equation of motion is obtained by including the exciting force in Eqs. (2.4.1) and (2.4.2) governing the free vibration of the system:

$$m\ddot{u} + ku \pm F = p(t) \qquad (3.11.1)$$

The sign of the friction force changes with the direction of motion; the positive sign applies if the motion is from left to right ($\dot{u} > 0$) and the negative sign is for motion from right to left ($\dot{u} < 0$). Each of the two differential equations is linear, but the overall problem is nonlinear because the governing equation changes every half-cycle of motion. Therefore, exact analytical solutions would not be possible except in special cases.

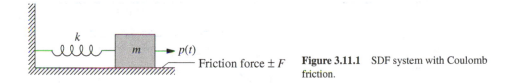

Friction force $\pm F$ **Figure 3.11.1** SDF system with Coulomb friction.

3.11.2 Steady-State Response to Harmonic Force

An exact analytical solution for the steady-state response of the system of Fig. 3.11.1 subjected to harmonic force was developed by J. P. Den Hartog in 1933. The analysis is not included here, but his results are shown by solid lines in Fig. 3.11.2; the dashed lines are described in the next section. The displacement amplitude u_o, normalized relative to $(u_{st})_o = p_o/k$, and the phase angle ϕ are plotted as a function of the frequency ratio ω/ω_n for three values of F/p_o. If there is no friction, $F = 0$ and $u_o/(u_{st})_o = (R_d)_{\zeta=0}$, the same as in Eq. (3.1.11) for an undamped system. The friction force reduces the displacement amplitude u_o, with the reduction depending on the frequency ratio ω/ω_n.

At $\omega = \omega_n$ the amplitude of motion is not limited by Coulomb friction if

$$\frac{F}{p_o} < \frac{\pi}{4} \qquad (3.11.2)$$

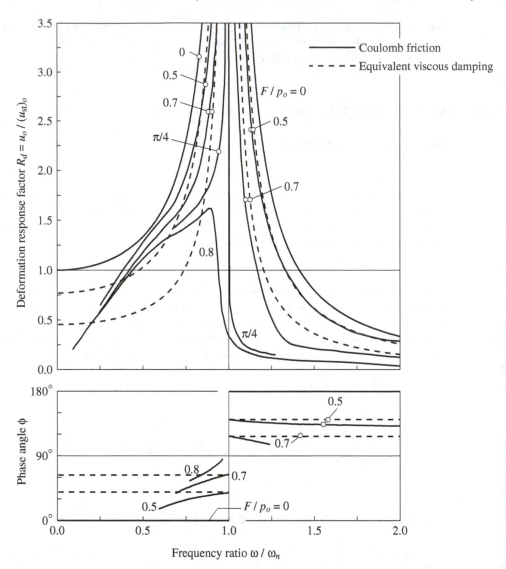

Figure 3.11.2 Deformation response factor and phase angle of a system with Coulomb friction excited by harmonic force. Exact solution from J. P. Den Hartog; approximate solution is based on equivalent viscous damping.

which is surprising since $F = (\pi/4)p_o$ represents a large friction force, but can be explained by comparing the energy E_F dissipated in friction against the input energy E_I. The energy dissipated by Coulomb friction in one cycle of vibration with displacement amplitude u_o is the area of the hysteresis loop enclosed by the friction force–displacement diagram (Fig. 3.11.3):

$$E_F = 4Fu_o \tag{3.11.3}$$

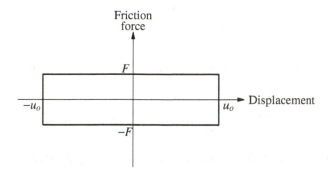

Figure 3.11.3   Hysteresis loop for Coulomb friction.

Observe that the dissipated energy in a vibration cycle is proportional to the amplitude of the cycle. The energy E_I input by the harmonic force applied at $\omega = \omega_n$ is also proportional to the displacement amplitude. If Eq. (3.11.2) is satisfied, it can be shown that

$$E_F < E_I$$

that is, the energy dissipated in friction per cycle is less than the input energy (Fig. 3.11.4). Therefore, the displacement amplitude would increase cycle after cycle and grow without bound. This behavior is quite different from that of systems with viscous damping or rate-independent damping. For these forms of damping, as shown in Section 3.8, the dissipated energy increases quadratically with displacement amplitude, and the displacement amplitude is bounded no matter how small the damping. In connection with the fact that infinite amplitudes occur at $\omega = \omega_n$ if Eq. (3.11.2) is satisfied, the phase angle shows a discontinuous jump at $\omega = \omega_n$ (Fig. 3.11.2).

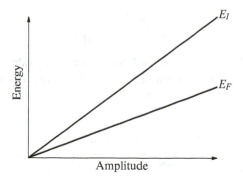

Figure 3.11.4   Input energy E_I and energy dissipated E_F by Coulomb friction.

3.11.3 Solution Using Equivalent Viscous Damping

In this section an approximate solution for the steady-state harmonic response of a system with Coulomb friction is obtained by modeling this damping mechanism by equivalent viscous damping. Substituting E_F, the energy dissipated by Coulomb friction given by Eq. (3.11.3), for E_D in Eq. (3.9.1) provides the equivalent viscous damping ratio:

$$\zeta_{eq} = \frac{2}{\pi} \frac{1}{\omega/\omega_n} \frac{u_F}{u_o} \tag{3.11.4}$$

where $u_F = F/k$. The approximate solution for the displacement amplitude u_o is obtained by substituting ζ_{eq} for ζ in Eq. (3.2.11):

$$\frac{u_o}{(u_{st})_o} = \frac{1}{\left\{\left[1 - (\omega/\omega_n)^2\right]^2 + \left[(4/\pi)(u_F/u_o)\right]^2\right\}^{1/2}}$$

This contains u_o on the right side also. Squaring and solving algebraically, the normalized displacement amplitude is

$$\frac{u_o}{(u_{st})_o} = \frac{\left\{1 - \left[(4/\pi)(F/p_o)\right]^2\right\}^{1/2}}{1 - (\omega/\omega_n)^2} \tag{3.11.5}$$

This approximate result is valid provided that $F/p_o < \pi/4$. The approximate solution cannot be used if $F/p_o > \pi/4$ because then the quantity under the radical is negative and the numerator is imaginary.

These approximate and exact solutions are compared in Fig. 3.11.2. If the friction force is small enough to permit continuous motion, this motion is practically sinusoidal and the approximate solution is close to the exact solution. If the friction force is large, discontinuous motion with stops and starts results, which is much distorted relative to a sinusoid, and the approximate solution is poor.

The approximate solution for the phase angle is obtained by substituting ζ_{eq} for ζ in Eq. (3.2.12):

$$\tan \phi = \frac{(4/\pi)(u_F/u_o)}{1 - (\omega/\omega_n)^2}$$

Substituting for u_o from Eq. (3.11.5) gives

$$\tan \phi = \pm \frac{(4/\pi)(F/p_o)}{\left\{1 - \left[(4/\pi)(F/p_o)\right]^2\right\}^{1/2}} \tag{3.11.6}$$

For a given value of F/p_o, the $\tan \phi$ is constant but with a positive value if $\omega/\omega_n < 1$ and a negative value if $\omega/\omega_n > 1$. This is shown in Fig. 3.11.2, where it is seen that the phase angle is discontinuous at $\omega = \omega_n$ for Coulomb friction.

Example 3.7

The structure of Example 2.7 with friction devices deflects 2 in. under a lateral force of $p = 500$ kips. What would be the approximate amplitude of motion if the lateral force is replaced by the harmonic force $p(t) = 500 \sin \omega t$, where the forcing period $T = 1$ sec?

Solution   The data (given and from Example 2.7) are

$$(u_{st})_o = \frac{p_o}{k} = 2 \text{ in.} \qquad u_F = 0.15 \text{ in.}$$

$$\frac{\omega}{\omega_n} = \frac{T_n}{T} = \frac{0.5}{1} = 0.5$$

Calculate u_o from Eq. (3.11.5).

$$\frac{F}{p_o} = \frac{F/k}{p_o/k} = \frac{u_F}{(u_{st})_o} = \frac{0.15}{2} = 0.075$$

Substituting for F/p_o in Eq. (3.11.5) gives

$$\frac{u_o}{(u_{st})_o} = \frac{\left\{1 - [(4/\pi)0.075]^2\right\}^{1/2}}{1 - (0.5)^2} = 1.327$$

$$u_o = 1.327(2) = 2.654 \text{ in.}$$

PART D: RESPONSE TO PERIODIC EXCITATION

A periodic function is one in which the portion defined over T_0 repeats itself indefinitely (Fig. 3.12.1). Many forces are periodic or nearly periodic. Under certain conditions, propeller forces on a ship, wave loading on an offshore platform, and wind forces induced by vortex shedding on tall, slender structures are nearly periodic. Earthquake ground motion usually has no resemblance to a periodic function. However, the base excitation arising from an automobile traveling on an elevated freeway that has settled because of long-term creep may be nearly periodic.

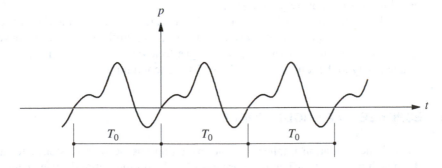

Figure 3.12.1   Periodic excitation.

We are interested in analyzing the response to periodic excitation for yet another reason. The analysis can be extended to arbitrary excitations utilizing discrete Fourier transform techniques. These are introduced in Appendix A.

3.12 FOURIER SERIES REPRESENTATION

A function $p(t)$ is said to be periodic with period T_0 if it satisfies the following relationship:

$$p(t + jT_0) = p(t) \qquad j = -\infty, \ldots, -3, -2, -1, 0, 1, 2, 3, \ldots, \infty$$

A periodic function can be separated into its harmonic components using the *Fourier series:*

$$p(t) = a_0 + \sum_{j=1}^{\infty} a_j \cos j\omega_0 t + \sum_{j=1}^{\infty} b_j \sin j\omega_0 t \qquad (3.12.1)$$

where the fundamental harmonic in the excitation has the frequency

$$\omega_0 = \frac{2\pi}{T_0} \qquad (3.12.2)$$

The coefficients in the Fourier series can be expressed in terms of $p(t)$ because the sine and cosine functions are orthogonal:

$$a_0 = \frac{1}{T_0} \int_0^{T_0} p(t) \, dt \qquad (3.12.3)$$

$$a_j = \frac{2}{T_0} \int_0^{T_0} p(t) \cos j\omega_0 t \, dt \qquad j = 1, 2, 3, \ldots \qquad (3.12.4)$$

$$b_j = \frac{2}{T_0} \int_0^{T_0} p(t) \sin j\omega_0 t \, dt \qquad j = 1, 2, 3, \ldots \qquad (3.12.5)$$

The coefficient a_0 is the average value of $p(t)$; coefficients a_j and b_j are the amplitudes of the jth harmonics of frequency $j\omega_0$.

Theoretically, an infinite number of terms are required for the Fourier series to converge to $p(t)$. In practice, however, a few terms are sufficient for good convergence. At a discontinuity, the Fourier series converges to a value that is the average of the values immediately to the left and to the right of the discontinuity.

3.13 RESPONSE TO PERIODIC FORCE

A periodic excitation implies that the excitation has been in existence for a long time, by which time the transient response associated with the initial displacement and velocity has decayed. Thus, we are interested in finding the steady-state response. Just as for harmonic excitation, the response of a linear system to a periodic force can be determined by combining the responses to individual excitation terms in the Fourier series.

The response of an undamped system to constant force $p(t) = a_0$ is given by Eq. (4.3.7), in which the oscillatory part decays because of damping, leaving the steady-state solution.[†]

$$u_0(t) = \frac{a_0}{k} \qquad (3.13.1)$$

The steady-state response of a viscously damped SDF system to harmonic cosine force $p(t) = a_j \cos(j\omega_0 t)$ is given by Eqs. (3.2.3) and (3.2.26) with ω replaced by $j\omega_0$:

$$u_j^c(t) = \frac{a_j}{k} \frac{2\zeta\beta_j \sin j\omega_0 t + (1 - \beta_j^2) \cos j\omega_0 t}{(1 - \beta_j^2)^2 + (2\zeta\beta_j)^2} \qquad (3.13.2)$$

where

$$\beta_j = \frac{j\omega_0}{\omega_n} \qquad (3.13.3)$$

Similarly, the steady-state response of the system to sinusoidal force $p(t) = b_j \sin j\omega_0 t$ is given by Eqs. (3.2.3) and (3.2.4) with ω replaced by $j\omega_0$:

$$u_j^s(t) = \frac{b_j}{k} \frac{(1 - \beta_j^2) \sin j\omega_0 t - 2\zeta\beta_j \cos j\omega_0 t}{(1 - \beta_j^2)^2 + (2\zeta\beta_j)^2} \qquad (3.13.4)$$

If $\zeta = 0$ and one of $\beta_j = 1$, the steady-state response is unbounded and not meaningful because the transient response never decays (see Section 3.1); in the following it is assumed that $\zeta \neq 0$ and $\beta_j \neq 1$.

The steady-state response of a system with damping to periodic excitation $p(t)$ is the combination of responses to individual terms in the Fourier series:

$$u(t) = u_0(t) + \sum_{j=1}^{\infty} u_j^c(t) + \sum_{j=1}^{\infty} u_j^s(t) \qquad (3.13.5)$$

Substituting Eqs. (3.13.1), (3.13.2), and (3.13.4) into (3.13.5) gives

$$u(t) = \frac{a_0}{k} + \sum_{j=1}^{\infty} \frac{1}{k} \frac{1}{(1 - \beta_j^2)^2 + (2\zeta\beta_j)^2} \left\{ \left[a_j(2\zeta\beta_j) + b_j(1 - \beta_j^2) \right] \sin j\omega_0 t \right.$$

$$\left. + \left[a_j(1 - \beta_j^2) - b_j(2\zeta\beta_j) \right] \cos j\omega_0 t \right\} \qquad (3.13.6)$$

The response $u(t)$ is a periodic function with period T_0.

The relative contributions of the various harmonic terms in Eq. (3.13.6) depend on two factors: (1) the amplitudes a_j and b_j of the harmonic components of the forcing function $p(t)$, and (2) the frequency ratio β_j. The response will be dominated by those harmonic components for which β_j is close to unity [i.e., the forcing frequency $j\omega_0$ is close to the natural frequency (see Fig. 3.2.6)].

[†]The notation u_0 used here includes the subscript zero consistent with a_0; this should not be confused with u_o with the subscript "oh" used earlier to denote the maximum value of $u(t)$.

Example 3.8

The periodic force shown in Fig. E3.8a is defined by

$$p(t) = \begin{cases} p_o & 0 \le t \le T_0/2 \\ -p_o & T_0/2 \le t \le T_0 \end{cases} \tag{a}$$

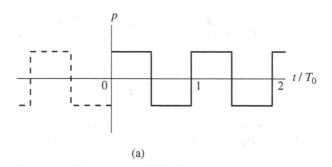

(a)

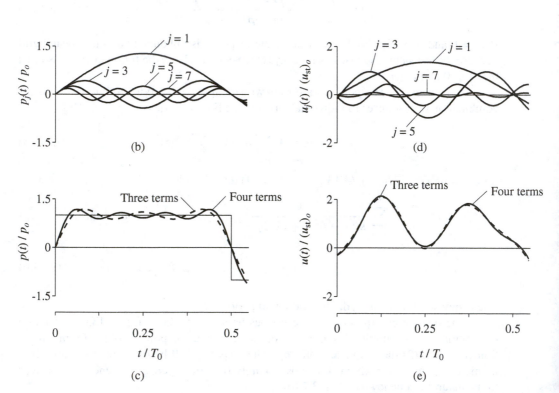

Figure E3.8

Substituting this in Eqs. (3.12.3) to (3.12.5) gives the Fourier series coefficients:

$$a_0 = \frac{1}{T_0} \int_0^{T_0} p(t)\, dt = 0 \tag{b}$$

$$a_j = \frac{2}{T_0} \int_0^{T_0} p(t) \cos j\omega_0 t\, dt$$

$$= \frac{2}{T_0} \left[p_o \int_0^{T_0/2} \cos j\omega_0 t\, dt + (-p_o) \int_{T_0/2}^{T_0} \cos j\omega_0 t\, dt \right] = 0 \tag{c}$$

$$b_j = \frac{2}{T_0} \int_0^{T_0} p(t) \sin j\omega_0 t\, dt$$

$$= \frac{2}{T_0} \left[p_o \int_0^{T_0/2} \sin j\omega_0 t\, dt + (-p_o) \int_{T_0/2}^{T_0} \sin j\omega_0 t\, dt \right]$$

$$= \begin{cases} 0 & j \text{ even} \\ 4p_o/j\pi & j \text{ odd} \end{cases} \tag{d}$$

Thus the Fourier series representation of $p(t)$ is

$$p(t) = \sum p_j(t) = \frac{4p_o}{\pi} \sum_{j=1,3,5}^{\infty} \frac{1}{j} \sin j\omega_0 t \tag{e}$$

The first four terms of this series are shown in Fig. E3.8b, where the frequencies and relative amplitudes—1, $\frac{1}{3}$, $\frac{1}{5}$, and $\frac{1}{7}$—of the four harmonics are apparent. The cumulative sum of the Fourier terms is shown in Fig. E3.8c, where four terms provide a reasonable representation of the forcing function. At $t = T_0/2$, where $p(t)$ is discontinuous, the Fourier series converges to zero, the average value of $p(T_0/2)$.

The response of an SDF system to the forcing function of Eq. (e) is obtained by substituting Eqs. (b), (c), and (d) in Eq. (3.13.6) to obtain

$$u(t) = (u_{st})_o \frac{4}{\pi} \sum_{j=1,3,5}^{\infty} \frac{1}{j} \frac{(1-\beta_j^2) \sin j\omega_0 t - 2\zeta\beta_j \cos j\omega_0 t}{(1-\beta_j^2)^2 + (2\zeta\beta_j)^2} \tag{f}$$

Shown in Fig. E3.8d are the responses of an SDF system with natural period $T_n = T_0/4$ and damping ratio $\zeta = 5\%$ to the first four loading terms in the Fourier series of Eq. (e). These are plots of individual terms in Eq. (f) with $\beta_j = j\omega_0/\omega_n = jT_n/T_0 = j/4$. The relative amplitudes of these terms are apparent. None of them is especially large because none of the β_j values is especially close to unity; note that $\beta_j = \frac{1}{4}, \frac{3}{4}, \frac{5}{4}, \frac{7}{4}$, and so on. The cumulative sum of the individual response terms of Eq. (f) is shown in Fig. E3.8e, where the contribution of the fourth term is seen to be small. The higher terms would be even smaller because the amplitudes of the harmonic components of $p(t)$ decrease with j and β_j would be even farther from unity.

FURTHER READING

Blake, R. E., "Basic Vibration Theory," Chapter 2 in *Shock and Vibration Handbook,* 3rd ed. (ed. C. M. Harris), McGraw-Hill, New York, 1988.

Hudson, D. E., *Reading and Interpreting Strong Motion Accelerograms,* Earthquake Engineering Research Institute, Berkeley, Calif., 1979.

Jacobsen, L. S., and Ayre, R. S., *Engineering Vibrations,* McGraw-Hill, New York, 1958, Section 5.8.

APPENDIX 3: FOUR-WAY LOGARITHMIC GRAPH PAPER

R_v is plotted as a function of ω/ω_n on log-log graph paper [i.e., $\log R_v$ is the ordinate and $\log(\omega/\omega_n)$ the abscissa]. Equation (3.2.21) gives

$$\log R_v = \log \frac{\omega}{\omega_n} + \log R_d \qquad (A3.1)$$

If R_d is a constant, Eq. (A3.1) represents a straight line with slope of $+1$. Grid lines showing constant R_d would therefore be straight lines of slope $+1$, and the R_d-axis would be perpendicular to them (Fig. A3.1). Equation (3.2.21) also gives

$$\log R_v = -\log \frac{\omega}{\omega_n} + \log R_a \qquad (A3.2)$$

If R_a is a constant, Eq. (A3.2) represents a straight line with slope of -1. Grid lines showing constant R_a would be straight lines of slope -1, and the R_a-axis would be perpendicular to them (Fig. A3.1).

　　With reference to Fig. A3.1, the scales are established as follows:

1. With the point $(R_v = 1, \omega/\omega_n = 1)$ as the origin, draw a vertical R_v-axis and a horizontal ω/ω_n-axis with *equal* logarithmic scales.

2. The mark A on the R_a-axis would be located at the point $(R_v = A^{1/2}, \omega/\omega_n = A^{1/2})$ in order to satisfy

$$R_a = \frac{\omega}{\omega_n} R_v \qquad (A3.3)$$

   R_v and ω/ω_n are taken to be equal because the R_a-axis has a slope of $+1$. This procedure is shown for $A = 9$, leading to the scale marks 3 on the R_v and ω/ω_n axes.

3. The mark D on the R_d-axis would be located at the point $(R_v = D^{1/2}, \omega/\omega_n = D^{-1/2})$ in order to satisfy

$$R_d = R_v \div \frac{\omega}{\omega_n} \qquad (A3.4)$$

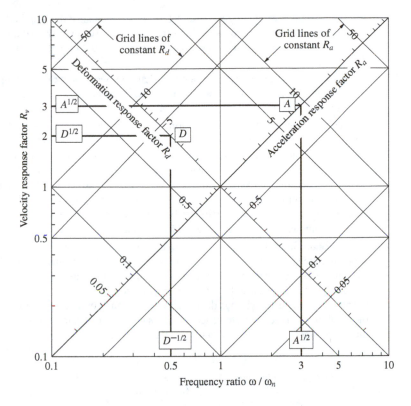

Figure A3.1   Construction of four-way logarithmic graph paper.

and the condition that the R_d-axis has a slope of -1. This procedure is shown for $D = 4$, leading to the scale mark 2 on the R_v-axis and to the scale mark $\frac{1}{2}$ on the ω/ω_n-axis.

The logarithmic scales along the R_d and R_a axes are equal but not the same as the R_v and ω/ω_n scales.

PROBLEMS

Part A

3.1   The mass m, stiffness k, and natural frequency ω_n of an undamped SDF system are unknown. These properties are to be determined by harmonic excitation tests. At an excitation frequency of 4 Hz, the response tends to increase without bound (i.e., a resonant condition). Next, a weight $\Delta w = 5$ lb is attached to the mass m and the resonance test is repeated. This time resonance occurs at $f = 3$ Hz. Determine the mass and the stiffness of the system.

3.2 An SDF system is excited by a sinusoidal force. At resonance the amplitude of displacement was measured to be 2 in. At an exciting frequency of one-tenth the natural frequency of the system, the displacement amplitude was measured to be 0.2 in. Estimate the damping ratio of the system.

3.3 In a forced vibration test under harmonic excitation it was noted that the amplitude of motion at resonance was exactly four times the amplitude at an excitation frequency 20% higher than the resonant frequency. Determine the damping ratio of the system.

3.4 A machine is supported on four steel springs for which damping can be neglected. The natural frequency of vertical vibration of the machine–spring system is 200 cycles per minute. The machine generates a vertical force $p(t) = p_0 \sin \omega t$. The amplitude of the resulting steady-state vertical displacement of the machine is $u_o = 0.2$ in. when the machine is running at 20 revolutions per minute (rpm), 1.042 in. at 180 rpm, and 0.0248 in. at 600 rpm. Calculate the amplitude of vertical motion of the machine if the steel springs are replaced by four rubber isolators that provide the same stiffness but introduce damping equivalent to $\zeta = 25\%$ for the system. Comment on the effectiveness of the isolators at various machine speeds.

3.5 An air-conditioning unit weighing 1200 lb is bolted at the middle of two parallel simply supported steel beams (Fig. P3.5). The clear span of the beams is 8 ft. The second moment of cross-sectional area of each beam is 10 in⁴. The motor in the unit runs at 300 rpm and produces an unbalanced vertical force of 60 lb at this speed. Neglect the weight of the beams and assume 1% viscous damping in the system; for steel $E = 30,000$ ksi. Determine the amplitudes of steady-state deflection and steady-state acceleration (in g's) of the beams at their midpoints which result from the unbalanced force.

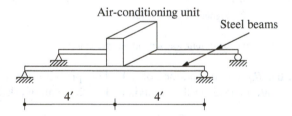

Figure P3.5

3.6 (a) Show that the steady-state response of an SDF system to a cosine force, $p(t) = p_0 \cos \omega t$, is given by

$$u(t) = \frac{p_o}{k} \frac{\left[1 - (\omega/\omega_n)^2\right] \cos \omega t + [2\zeta (\omega/\omega_n)] \sin \omega t}{\left[1 - (\omega/\omega_n)^2\right]^2 + [2\zeta (\omega/\omega_n)]^2}$$

(b) Show that the maximum deformation due to cosine force is the same as that due to sinusoidal force.

3.7 (a) Show that $\omega_r = \omega_n (1 - 2\zeta^2)^{1/2}$ is the resonant frequency for displacement amplitude of an SDF system.

(b) Determine the displacement amplitude at resonance.

3.8 (a) Show that $\omega_r = \omega_n (1 - 2\zeta^2)^{-1/2}$ is the resonant frequency for acceleration amplitude of an SDF system.

(b) Determine the acceleration amplitude at resonance.

3.9   (a) Show that $\omega_r = \omega_n$ is the resonant frequency for velocity amplitude of an SDF system.
(b) Determine the velocity amplitude at resonance.

Part B

3.10   A one-story reinforced concrete building has a roof mass of 500 kips/g, and its natural frequency is 4 Hz. This building is excited by a vibration generator with two weights, each 50 lb, rotating about a vertical axis at an eccentricity of 12 in. When the vibration generator runs at the natural frequency of the building, the amplitude of roof acceleration is measured to be 0.02g. Determine the damping of the structure.

3.11   The steady-state acceleration amplitude of a structure caused by an eccentric-mass vibration generator was measured for several excitation frequencies. These data are as follows:

Frequency (Hz)	Acceleration (10^{-3}g)	Frequency (Hz)	Acceleration (10^{-3}g)
1.337	0.68	1.500	7.10
1.378	0.90	1.513	5.40
1.400	1.15	1.520	4.70
1.417	1.50	1.530	3.80
1.438	2.20	1.540	3.40
1.453	3.05	1.550	3.10
1.462	4.00	1.567	2.60
1.477	7.00	1.605	1.95
1.487	8.60	1.628	1.70
1.493	8.15	1.658	1.50
1.497	7.60		

Determine the natural frequency and damping ratio of the structure.

3.12   Consider an industrial machine of mass m supported on spring-type isolators of total stiffness k. The machine operates at a frequency of f hertz with a force unbalance p_o.
(a) Determine an expression giving the fraction of force transmitted to the foundation as a function of the forcing frequency f and the static deflection $\delta_{st} = mg/k$. Consider only the steady-state response.
(b) Determine the static deflection δ_{st} for the force transmitted to be 10% of p_o if $f = 20$ Hz.

3.13   For the automobile in Example 3.4, determine the amplitude of the force developed in the spring of the suspension system when the automobile is traveling at 40 mph.

3.14   Determine the speed of the automobile in Example 3.4 that would produce a resonant condition for the spring force in the suspension system.

3.15   A vibration isolation block is to be installed in a laboratory so that the vibration from adjacent factory operations will not disturb certain experiments (Fig. P3.15). If the isolation block weighs 2000 lb and the surrounding floor and foundation vibrate at 1500 cycles per minute, determine the stiffness of the isolation system such that the motion of the isolation block is limited to 10% of the floor vibration; neglect damping.

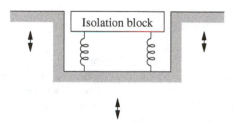

Figure P3.15

3.16 An SDF system is subjected to support displacement $u_g(t) = u_{go} \sin \omega t$. Show that the amplitude u_o^t of the total displacement of the mass is given by Eq. (3.6.5).

3.17 The natural frequency of an accelerometer is 50 Hz, and its damping ratio is 70%. Compute the recorded acceleration as a function of time if the input acceleration is $\ddot{u}_g(t) = 0.1g \sin 2\pi ft$ for $f = 10$, 20, and 40 Hz. A comparison of the input and recorded accelerations was presented in Fig. 3.7.3. The accelerometer is calibrated to read the input acceleration correctly at very low values of the excitation frequency. What would be the error in the measured amplitude at each of the given excitation frequencies?

3.18 An accelerometer has the natural frequency $f_n = 25$ Hz and damping ratio $\zeta = 60\%$. Write an equation for the response $u(t)$ of the instrument as a function of time if the input acceleration is $\ddot{u}_g(t) = \ddot{u}_{go} \sin 2\pi ft$. Sketch the ratio $\omega_n^2 u_o / \ddot{u}_{go}$ as a function of f/f_n. The accelerometer is calibrated to read the input acceleration correctly at very low values of the excitation frequency. Determine the range of frequencies for which the acceleration amplitude can be measured with an accuracy of $\pm 1\%$. Identify this frequency range on the above-mentioned plot.

3.19 The natural frequency of an accelerometer is $f_n = 50$ Hz, and its damping ratio is $\zeta = 70\%$. Solve Problem 3.18 for this accelerometer.

3.20 If a displacement-measuring instrument is used to determine amplitudes of vibration at frequencies very much higher than its own natural frequency, what would be the optimum instrument damping for maximum accuracy?

3.21 A displacement meter has a natural frequency $f_n = 0.5$ Hz and a damping ratio $\zeta = 0.6$. Determine the range of frequencies for which the displacement amplitude can be measured with an accuracy of $\pm 1\%$.

3.22 Repeat Problem 3.21 for $\zeta = 0.7$.

3.23 Show that the energy dissipated per cycle for viscous damping can be expressed by

$$E_D = \frac{\pi p_o^2}{k} \frac{2\zeta(\omega/\omega_n)}{\left[1 - (\omega/\omega_n)^2\right]^2 + [2\zeta(\omega/\omega_n)]^2}$$

3.24 Show that for viscous damping the loss factor ξ is independent of the amplitude and proportional to the frequency.

Part C

3.25 The properties of the SDF system of Fig. P2.20 are as follows: $w = 500$ kips, $F = 50$ kips, and $T_n = 0.25$ sec. Determine an approximate value for the displacement amplitude due to harmonic force with amplitude 100 kips and period 0.30 sec.

Part D

3.26 An SDF system with natural period T_n and damping ratio ζ is subjected to the periodic force shown in Fig. P3.26 with an amplitude p_o and period T_0.
(a) Expand the forcing function in its Fourier series.
(b) Determine the steady-state response of an undamped system. For what values of T_0 is the solution indeterminate?
(c) For $T_0/T_n = 2$, determine and plot the response to individual terms in the Fourier series. How many terms are necessary to obtain reasonable convergence of the series solution?

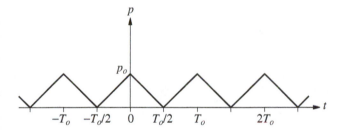

Figure P3.26

━━━◣ **4** ◢━━━

Response to Arbitrary, Step, and Pulse Excitations

PREVIEW

In many practical situations the dynamic excitation is neither harmonic nor periodic. Thus we are interested in studying the dynamic response of SDF systems to excitations varying arbitrarily with time. A general result for linear systems, Duhamel's integral, is derived in Part A of this chapter. This result is used in Part B to study the response of systems to step force, linearly increasing force, and step force with finite rise time. These results demonstrate how the dynamic response of the system is affected by the rise time.

   An important class of excitations that consist of essentially a single pulse is considered in Part C. The time variation of the response to three different force pulses is studied, and the concept of shock spectrum is introduced to present graphically the maximum response as a function of t_d/T_n, the ratio of pulse duration to the natural vibration period. It is then demonstrated that the response to short pulses is essentially independent of the pulse shape and that the response can be determined using only the pulse area. Most of the analyses and results presented are for systems without damping because the effect of damping on the response to a single pulse excitation is usually not important; this is demonstrated toward the end of the chapter.

PART A: RESPONSE TO ARBITRARILY TIME-VARYING FORCES

A general procedure is developed to analyze the response of an SDF system subjected to force $p(t)$ varying arbitrarily with time. This result will enable analytical evaluation of response to forces described by simple functions of time.

We seek the solution of the differential equation of motion

$$m\ddot{u} + c\dot{u} + ku = p(t)$$

subject to the initial conditions

$$u(0) = 0 \qquad \dot{u}(0) = 0$$

In developing the general solution, $p(t)$ is interpreted as a sequence of impulses of in-finitesimal duration, and the response of the system to $p(t)$ is the sum of the responses to individual impulses. These individual responses can conveniently be written in terms of the response of the system to a unit impulse.

4.1 RESPONSE TO UNIT IMPULSE

A very large force that acts for a very short time but with a time integral that is finite is called an *impulsive* force. Shown in Fig. 4.1.1 is the force $p(t) = 1/\varepsilon$, with time duration ε starting at the time instant $t = \tau$. As ε approaches zero the force becomes infinite; however, the *magnitude of the impulse*, defined by the time integral of $p(t)$, remains equal to unity. Such a force in the limiting case $\varepsilon \to 0$ is called the *unit impulse*. The *Dirac delta function* $\delta(t - \tau)$ mathematically defines a unit impulse centered at $t = \tau$.

According to Newton's second law of motion, if a force p acts on a body of mass m, the rate of change of momentum of the body is equal to the applied force, that is,

$$\frac{d}{dt}(m\dot{u}) = p \tag{4.1.1}$$

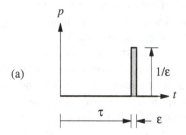

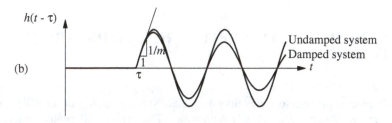

Figure 4.1.1   (a) Unit impulse; (b) response to unit impulse.

For constant mass, this equation becomes

$$p = m\ddot{u} \tag{4.1.2}$$

Integrating both sides with respect to t gives

$$\int_{t_1}^{t_2} p \, dt = m(\dot{u}_2 - \dot{u}_1) = m \, \Delta\dot{u} \tag{4.1.3}$$

The integral on the left side of this equation is the magnitude of the impulse. The product of mass and velocity is the *momentum*. Thus Eq. (4.1.3) states that the magnitude of the impulse is equal to the change in momentum.

This result is also applicable to an SDF mass–spring–damper system if the spring or damper has no effect. Such is the case because the impulsive force acts for an infinitesimally short duration. Thus a unit impulse at $t = \tau$ imparts to the mass, m, the velocity [from Eq. (4.1.3)]

$$\dot{u}(\tau) = \frac{1}{m} \tag{4.1.4}$$

but the displacement is zero prior to and up to the impulse:

$$u(\tau) = 0 \tag{4.1.5}$$

A unit impulse causes free vibration of the SDF system due to the initial velocity and displacement given by Eqs. (4.1.4) and (4.1.5). Substituting these in Eq. (2.2.4) gives the response of viscously damped systems:

$$h(t - \tau) \equiv u(t) = \frac{1}{m\omega_D} e^{-\zeta\omega_n(t-\tau)} \sin[\omega_D(t - \tau)] \qquad t \ge \tau \tag{4.1.6}$$

This *unit impulse-response function*, denoted by $h(t - \tau)$, is shown in Fig. 4.1.1b, together with the special case of $\zeta = 0$.

If the excitation is a unit impulse ground motion, based on Eq. (1.7.6), $p_{\text{eff}}(t) = -m\delta(t - \tau)$, then Eq. (4.1.4) becomes $\dot{u}(\tau) = -1$ and Eq. (4.1.6) changes to

$$h(t - \tau) = -\frac{1}{\omega_D} e^{-\zeta\omega_n(t-\tau)} \sin[\omega_D(t - \tau)] \qquad t \ge \tau \tag{4.1.7}$$

4.2 RESPONSE TO ARBITRARY FORCE

A force $p(t)$ varying arbitrarily with time can be represented as a sequence of infinitesimally short impulses (Fig. 4.2.1). The response of a linear dynamic system to one of these impulses, the one at time τ of magnitude $p(\tau) \, d\tau$, is this magnitude times the unit impulse-response function:

$$du(t) = [p(\tau) \, d\tau] h(t - \tau) \qquad t > \tau \tag{4.2.1}$$

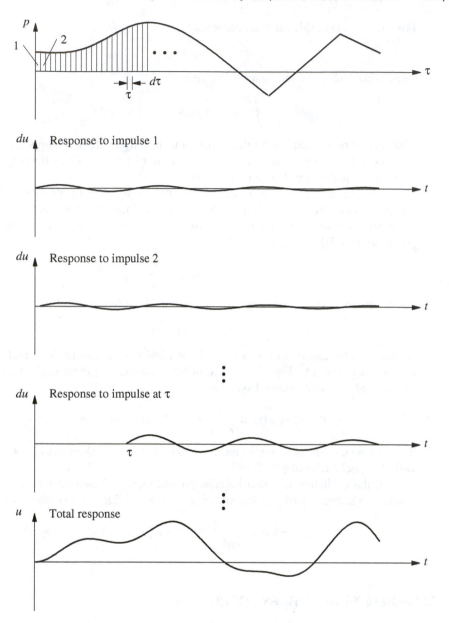

Figure 4.2.1   Schematic explanation of convolution integral.

The response of the system at time t is the sum of the responses to all impulses up to that time (Fig. 4.2.1). Thus

$$u(t) = \int_0^t p(\tau)h(t - \tau)\, d\tau \tag{4.2.2}$$

This is known as the *convolution integral*, a general result that applies to any linear dynamic system.

Specializing Eq. (4.2.2) for the SDF system by substituting Eq. (4.1.6) for the unit impulse response function gives *Duhamel's integral*:

$$u(t) = \frac{1}{m\omega_D} \int_0^t p(\tau) e^{-\zeta\omega_n(t-\tau)} \sin\left[\omega_D(t-\tau)\right] d\tau \tag{4.2.3}$$

For an undamped system this result simplifies to

$$u(t) = \frac{1}{m\omega_n} \int_0^t p(\tau) \sin\left[\omega_n(t-\tau)\right] d\tau \tag{4.2.4}$$

Implicit in this result are "at rest" initial conditions, $u(0) = 0$ and $\dot{u}(0) = 0$. If the initial displacement and velocity are $u(0)$ and $\dot{u}(0)$, the resulting free vibration response given by Eqs. (2.2.4) and (2.1.3) should be added to Eqs. (4.2.3) and (4.2.4), respectively. Recall that we had used Eq. (4.2.4) in Section 1.10.2, where four methods for solving the equation of motion were introduced.

Duhamel's integral provides a general result for evaluating the response of a linear SDF system to arbitrary force. This result is restricted to linear systems because it is based on the principle of superposition. Thus it does not apply to structures deforming beyond their linearly elastic limit. If $p(\tau)$ is a simple function, closed-form evaluation of the integral is possible and Duhamel's integral is an alternative to the classical method for solving differential equations (Section 1.10.1). If $p(\tau)$ is a complicated function that is described numerically, evaluation of the integral requires numerical methods. These will not be presented in this book, however, because they are not particularly efficient. More effective methods for numerical solution of the equation of motion are presented in Chapter 5.

PART B: RESPONSE TO STEP AND RAMP FORCES

4.3 STEP FORCE

A *step force* jumps suddenly from zero to p_o and stays constant at that value (Fig. 4.3.1b). It is desired to determine the response of an undamped SDF system (Fig. 4.3.1a) starting at rest to the step force:

$$p(t) = p_o \tag{4.3.1}$$

The equation of motion has been solved (Section 1.10.2) using Duhamel's integral to obtain

$$u(t) = (u_{\text{st}})_o(1 - \cos\omega_n t) = (u_{\text{st}})_o\left(1 - \cos\frac{2\pi t}{T_n}\right) \tag{4.3.2}$$

where $(u_{\text{st}})_o = p_o/k$, the static deformation due to force p_o.

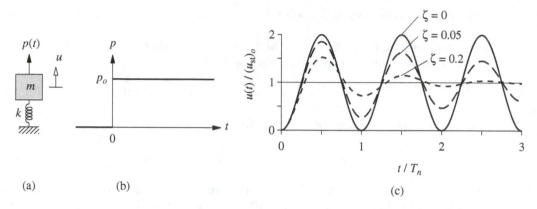

Figure 4.3.1   (a) SDF system; (b) step force; (c) dynamic response.

The normalized deformation or displacement, $u(t)/(u_{st})_o$, is plotted against normalized time, t/T_n, in Fig. 4.3.1c. It is seen that the system oscillates at its natural period about a new equilibrium position, which is displaced through $(u_{st})_o$ from the original equilibrium position of $u = 0$. The maximum displacement can be determined by differentiating Eq. (4.3.2) and setting $\dot{u}(t)$ to zero, which gives $\omega_n \sin \omega_n t = 0$. The values t_o of t that satisfy this condition are

$$\omega_n t_o = j\pi \quad \text{or} \quad t_o = \frac{j}{2} T_n \tag{4.3.3}$$

where j is an odd integer; even integers correspond to minimum values of $u(t)$. The maximum value u_o of $u(t)$ is given by Eq. (4.3.2) evaluated at $t = t_o$; these maxima are all the same:

$$u_o = 2(u_{st})_o \tag{4.3.4}$$

Thus a suddenly applied force produces twice the deformation it would have caused as a slowly applied force.

The response of a system with damping can be determined by substituting Eq. (4.3.1) in Eq. (4.2.3) and evaluating Duhamel's integral to obtain

$$u(t) = (u_{st})_o \left[1 - e^{-\zeta \omega_n t} \left(\cos \omega_D t + \frac{\zeta}{\sqrt{1 - \zeta^2}} \sin \omega_D t \right) \right] \tag{4.3.5}$$

For analysis of damped systems the classical method (Section 1.10.1) may be easier, however, than evaluating Duhamel's integral. The differential equation to be solved is

$$m\ddot{u} + c\dot{u} + ku = p_o \tag{4.3.6}$$

Its complementary solution is given by Eq. (f) of Derivation 2.2, the particular solution is $u_p = p_o/k$, and the complete solution is

$$u(t) = e^{-\zeta \omega_n t} (A \cos \omega_D t + B \sin \omega_D t) + \frac{p_o}{k} \tag{4.3.7}$$

where the constants A and B are to be determined from initial conditions. For a system starting from rest, $u(0) = \dot{u}(0) = 0$ and

$$A = -\frac{p_o}{k} \qquad B = -\frac{p_o}{k}\frac{\zeta}{\sqrt{1-\zeta^2}}$$

Substituting these constants in Eq. (4.3.7) gives the same result as Eq. (4.3.5). When specialized for undamped systems this result reduces to Eq. (4.3.2), already presented in Fig. 4.3.1c.

Equation (4.3.5) is plotted in Fig. 4.3.1c for two additional values of the damping ratio. With damping the overshoot beyond the static equilibrium position is smaller, and the oscillations about this position decay with time. The damping ratio determines the amount of overshoot and the rate at which the oscillations decay. Eventually, the system settles down to the static deformation, which is also the steady-state deformation.

4.4 RAMP OR LINEARLY INCREASING FORCE

In Fig. 4.4.1b, the applied force $p(t)$ increases linearly with time. Naturally, it cannot increase indefinitely, but our interest is confined to the time duration where $p(t)$ is still small enough that the resulting spring force is within the linearly elastic limit of the spring.

While the equation of motion can be solved by any one of several methods, we illustrate use of Duhamel's integral to obtain the solution. The applied force

$$p(t) = p_o\frac{t}{t_r} \tag{4.4.1}$$

is substituted in Eq. (4.2.4) to obtain

$$u(t) = \frac{1}{m\omega_n}\int_0^t \frac{p_o}{t_r}\tau \sin\omega_n(t-\tau)\,d\tau$$

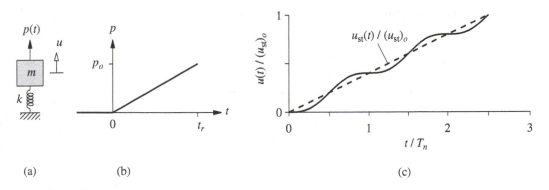

(a)   (b)   (c)

Figure 4.4.1   (a) SDF system; (b) ramp force; (c) dynamic and static responses.

This integral is evaluated and simplified to obtain

$$u(t) = (u_{st})_o \left(\frac{t}{t_r} - \frac{\sin \omega_n t}{\omega_n t_r} \right) = (u_{st})_o \left(\frac{t}{T_n} \frac{T_n}{t_r} - \frac{\sin 2\pi t/T_n}{2\pi t_r/T_n} \right) \tag{4.4.2}$$

where $(u_{st})_o = p_o/k$, the static deformation due to force p_o.

Equation (4.4.2) is plotted in Fig. 4.4.1c for $t_r/T_n = 2.5$, wherein the static deformation at each time instant,

$$u_{st}(t) = \frac{p(t)}{k} = (u_{st})_o \frac{t}{t_r} \tag{4.4.3}$$

is also shown; $u_{st}(t)$ varies with time in the same manner as $p(t)$ and the two differ by the scale factor $1/k$. It is seen that the system oscillates at its natural period T_n about the static solution.

4.5 STEP FORCE WITH FINITE RISE TIME

Since in reality a force can never be applied suddenly, it is of interest to consider a dynamic force that has a finite rise time, t_r, but remains constant thereafter, as shown in Fig. 4.5.1b:

$$p(t) = \begin{cases} p_o(t/t_r) & t \le t_r \\ p_o & t \ge t_r \end{cases} \tag{4.5.1}$$

The excitation has two phases: ramp or rise phase and constant phase.

For a system without damping starting from rest, the response during the ramp phase is given by Eq. (4.4.2), repeated here for convenience:

$$u(t) = (u_{st})_o \left(\frac{t}{t_r} - \frac{\sin \omega_n t}{\omega_n t_r} \right) \qquad t \le t_r \tag{4.5.2}$$

The response during the constant phase can be determined by evaluating Duhamel's integral after substituting Eq. (4.5.1) in Eq. (4.2.4). Alternatively, existing solutions for free vibration and step force could be utilized to express this response as

$$u(t) = u(t_r) \cos \omega_n (t - t_r) + \frac{\dot{u}(t_r)}{\omega_n} \sin \omega_n (t - t_r) + (u_{st})_o [1 - \cos \omega_n (t - t_r)] \tag{4.5.3}$$

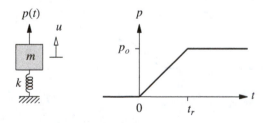

(a)   (b)

Figure 4.5.1 (a) SDF system; (b) step force with finite rise time.

The third term is the solution for a system at rest subjected to a step force starting at $t = t_r$; it is obtained from Eq. (4.3.2). The first two terms in Eq. (4.5.3) account for free vibration of the system resulting from its displacement $u(t_r)$ and velocity $\dot{u}(t_r)$ at the end of the ramp phase. Determined from Eq. (4.5.2), $u(t_r)$ and $\dot{u}(t_r)$ are substituted in Eq. (4.5.3) to obtain

$$u(t) = (u_{\text{st}})_o \left\{ 1 + \frac{1}{\omega_n t_r} \left[(1 - \cos \omega_n t_r) \sin \omega_n (t - t_r) \right. \right.$$

$$\left. \left. - \sin \omega_n t_r \cos \omega_n (t - t_r) \right] \right\} \qquad t \geq t_r \qquad (4.5.4a)$$

This equation can be simplified, using a trigonometric identity, to

$$u(t) = (u_{\text{st}})_o \left\{ 1 - \frac{1}{\omega_n t_r} \left[\sin \omega_n t - \sin \omega_n (t - t_r) \right] \right\} \qquad t \geq t_r \qquad (4.5.4b)$$

The normalized deformation, $u(t)/(u_{\text{st}})_o$, is a function of the normalized time, t/T_n, because $\omega_n t = 2\pi (t/T_n)$. This function depends only on the ratio t_r/T_n because $\omega_n t_r = 2\pi (t_r/T_n)$, not separately on t_r and T_n. Figure 4.5.2 shows $u(t)/(u_{\text{st}})_o$ plotted against t/T_n for several values of t_r/T_n, the ratio of the rise time to the natural period. Each plot is valid for all combinations of t_r and T_n with the same ratio t_r/T_n. Also plotted is $u_{\text{st}}(t) = p(t)/k$, the static deformation at each time instant. These results permit several observations:

1. During the force-rise phase the system oscillates at the natural period T_n about the static solution.
2. During the constant-force phase the system oscillates also at the natural period T_n about the static solution.
3. If the velocity $\dot{u}(t_r)$ is zero at the end of the ramp, the system does not vibrate during the constant-force phase.
4. For smaller values of t_r/T_n (i.e., short rise time), the response is similar to that due to a sudden step force; see Fig. 4.3.1c.
5. For larger values of t_r/T_n, the dynamic displacement oscillates close to the static solution, implying that the dynamic effects are small (i.e., a force increasing slowly— relative to T_n—from 0 to p_o affects the system like a static force).

The deformation attains its maximum value during the constant-force phase of the response. From Eq. (4.5.4a) the maximum value of $u(t)$ is

$$u_o = (u_{\text{st}})_o \left[1 + \frac{1}{\omega_n t_r} \sqrt{(1 - \cos \omega_n t_r)^2 + (\sin \omega_n t_r)^2} \right] \qquad (4.5.5)$$

Using trigonometric identities and $T_n = 2\pi/\omega_n$, Eq. (4.5.5) can be simplified to

$$R_d \equiv \frac{u_o}{(u_{\text{st}})_o} = 1 + \frac{|\sin(\pi t_r/T_n)|}{\pi t_r/T_n} \qquad (4.5.6)$$

The deformation response factor R_d depends only on t_r/T_n, the ratio of the rise time to the

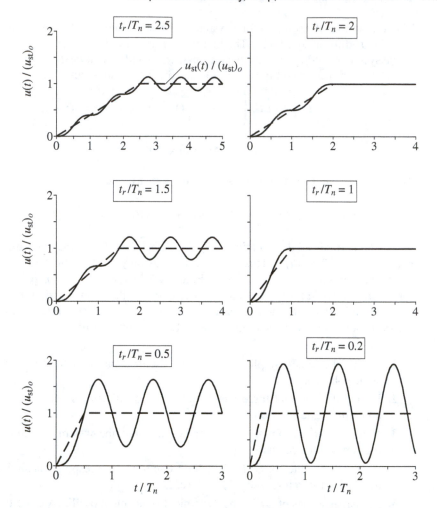

Figure 4.5.2 Dynamic response of undamped SDF system to step force with finite rise time; static solution is shown by dashed lines.

natural period. A graphical presentation of this relationship, as in Fig. 4.5.3, is called the *response spectrum* for the step force with finite rise time.

This response spectrum characterizes the problem completely. In this case it contains information on the normalized maximum response, $u_o/(u_{st})_o$, of all SDF systems (without damping) due to any step force p_o with any rise time t_r. The response spectrum permits several observations:

1. If $t_r < T_n/4$ (i.e., a relatively short rise time), $u_o \simeq 2(u_{st})_o$, implying that the structure "sees" this excitation like a suddenly applied force.

2. If $t_r > 3T_n$ (i.e., a relatively long rise time), $u_o \simeq (u_{st})_o$, implying that this excitation affects the structure like a static force.

3. If $t_r/T_n = 1, 2, 3, \ldots$, $u_o = (u_{st})_o$, because $\dot{u}(t_r) = 0$ at the end of the force-rise phase, and the system does not oscillate during the constant-force phase; see Fig. 4.5.2.

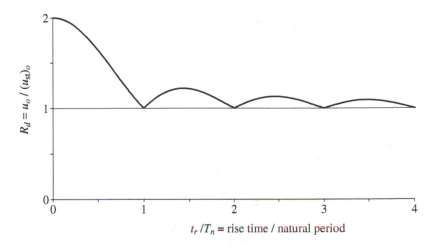

$$t_r/T_n = \text{rise time / natural period}$$

Figure 4.5.3   Response spectrum for step force with finite rise time.

PART C: RESPONSE TO PULSE EXCITATIONS

We next consider an important class of excitations that consist of essentially a single pulse, such as shown in Fig. 4.6.1. Air pressures generated on a structure due to aboveground blasts or explosions are essentially a single pulse and can usually be idealized by simple shapes such as those shown in the left part of Fig. 4.6.2. The dynamics of structures subjected to such excitations was the subject of much work during the cold war, especially in the 1950s and 1960s; interest in the topic has been revived because of recent terrorism acts.

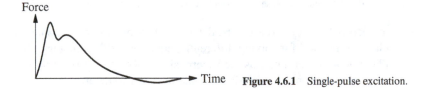

Figure 4.6.1   Single-pulse excitation.

4.6 SOLUTION METHODS

The response of the system to such pulse excitations does not reach a steady-state condition; the effects of the initial conditions must be considered. The response of the system

(a)

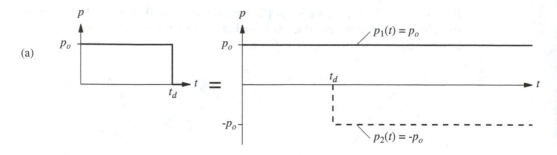

(b)

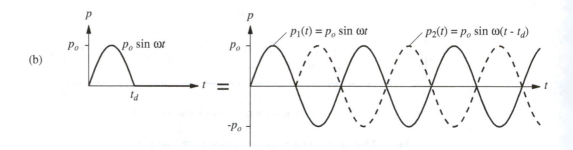

(c)

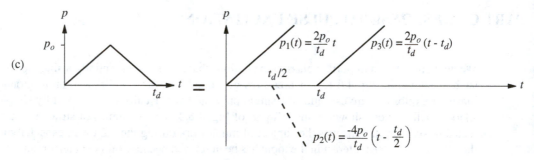

Figure 4.6.2 Expressing pulse force as superposition of simple functions: (a) rectangular pulse; (b) half-cycle sine pulse; (c) triangular pulse.

to such pulse excitations can be determined by one of several analytical methods: (1) the classical method for solving differential equations, (2) evaluating Duhamel's integral, and (3) expressing the pulse as the superposition of two or more simpler functions for which response solutions are already available or easier to determine.

The last of these approaches is illustrated in Fig. 4.6.2 for three pulse forces. For example, the rectangular pulse is the step function $p_1(t)$ plus the step function $p_2(t)$ of equal amplitude, but after a time interval t_d has passed. The desired response is the sum of the responses to each of these step functions, and these responses can be determined readily from the results of Section 4.3. A half-cycle sine pulse is the result of adding a sine function of amplitude p_o starting at $t = 0$ [$p_1(t)$ in Fig. 4.6.2b] and another sine function of the same frequency and amplitude starting at $t = t_d$ [$p_2(t)$ in Fig. 4.6.2b]. The desired response is the sum of the total (transient plus steady state) responses to the two sinusoidal forces, obtained using the results of Section 3.1. Similarly, the response to the symmetrical triangular pulse is the sum of the responses to the three ramp functions in Fig. 4.6.2c; the individual responses come from Section 4.4. Thus the third method involves adapting existing results and manipulating them to obtain the desired response.

We prefer to use the classical method in evaluating the response of SDF systems to pulse forces because it is closely tied to the dynamics of the system. Using the classical method the response to pulse forces will be determined in two phases. The first is the forced vibration phase, which covers the duration of the excitation. The second is the free vibration phase, which follows the end of the pulse force. Much of the presentation concerns systems without damping because, as will be shown in Section 4.11, damping has little influence on response to pulse excitations.

4.7 RECTANGULAR PULSE FORCE

We start with the simplest type of pulse, the rectangular pulse shown in Fig. 4.7.1. The equation to be solved is

$$m\ddot{u} + ku = p(t) = \begin{cases} p_o & t \le t_d \\ 0 & t \ge t_d \end{cases} \tag{4.7.1}$$

with at-rest initial conditions: $u(0) = \dot{u}(0) = 0$. The analysis is organized in two phases.

1. *Forced vibration phase.* During this phase, the system is subjected to a step force. The response of the system is given by Eq. (4.3.2), repeated for convenience:

$$\frac{u(t)}{(u_{\text{st}})_o} = 1 - \cos \omega_n t = 1 - \cos \frac{2\pi t}{T_n} \qquad t \le t_d \tag{4.7.2}$$

2. *Free vibration phase.* After the force ends at t_d, the system undergoes free vibration, defined by modifying Eq. (2.1.3) appropriately:

$$u(t) = u(t_d) \cos \omega_n (t - t_d) + \frac{\dot{u}(t_d)}{\omega_n} \sin \omega_n (t - t_d) \tag{4.7.3}$$

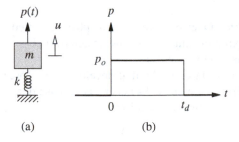

(a)   (b)

Figure 4.7.1   (a) SDF system; (b) rectangular pulse force.

This free vibration is initiated by the displacement and velocity of the mass at $t = t_d$, determined from Eq. (4.7.2):

$$u(t_d) = (u_{st})_o[1 - \cos \omega_n t_d] \qquad \dot{u}(t_d) = (u_{st})_o\omega_n \sin \omega_n t_d \qquad (4.7.4)$$

Substituting these in Eq. (4.7.3) gives

$$\frac{u(t)}{(u_{st})_o} = (1 - \cos \omega_n t_d) \cos \omega_n(t - t_d) + \sin \omega_n t_d \sin \omega_n(t - t_d) \qquad t \geq t_d$$

which can be simplified, using a trigonometric identity, to

$$\frac{u(t)}{(u_{st})_o} = \cos \omega_n(t - t_d) - \cos \omega_n t \qquad t \geq t_d$$

Expressing $\omega_n = 2\pi/T_n$ and using trigonometric identities enables us to rewrite these equations as

$$\frac{u(t)}{(u_{st})_o} = \left(2 \sin \frac{\pi t_d}{T_n}\right) \sin \left[2\pi \left(\frac{t}{T_n} - \frac{1}{2}\frac{t_d}{T_n}\right)\right] \qquad t \geq t_d \qquad (4.7.5)$$

Response history. The normalized deformation $u(t)/(u_{st})_o$ given by Eqs. (4.7.2) and (4.7.5) is a function of t/T_n. It depends only on t_d/T_n, the ratio of the pulse duration to the natural vibration period of the system, not separately on t_d or T_n, and has been plotted in Fig. 4.7.2 for several values of t_d/T_n. Also shown in dashed lines is the static solution $u_{st}(t) = p(t)/k$ at each time instant due to $p(t)$. The nature of the response is seen to vary greatly by changing just the duration t_d of the pulse. However, no matter how long the duration, the dynamic response is not close to the static solution, because the force is suddenly applied.

While the force is applied to the structure, the system oscillates about the shifted position, $(u_{st})_o = p_o/k$, at its own natural period T_n. After the pulse has ended, the system oscillates freely about the original equilibrium position at its natural period T_n, with no decay of motion because the system is undamped. If $t_d/T_n = 1, 2, 3, \ldots$, the system stays still in its original undeformed configuration during the free vibration phase, because the displacement and velocity of the mass are zero when the force ends.

Each response result of Fig. 4.7.2 is applicable to all combinations of systems and forces with fixed t_d/T_n. Implicit in this figure, however, is the presumption that the natural period T_n of the system is constant and the pulse duration t_d varies. By modifying the time scale, the results can be presented for a fixed value of t_d and varying values of T_n.

Maximum response. Over each of the two phases, forced vibration and free vibration, separately, the maximum value of response is determined next. The larger of the two maxima is the overall maximum response.

The number of local maxima or peaks that develop in the forced vibration phase depends on t_d/T_n (Fig. 4.7.2); more such peaks occur as the pulse duration lengthens. The first peak occurs at $t_o = T_n/2$ with the deformation

$$u_o = 2(u_{st})_o \qquad (4.7.6)$$

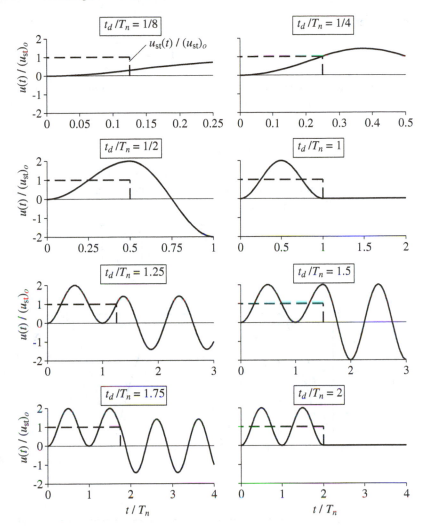

Figure 4.7.2 Dynamic response of undamped SDF system to rectangular pulse force; static solution is shown by dashed lines.

consistent with the results derived in Section 4.3. Thus t_d must be longer than $T_n/2$ for at least one peak to develop during the forced vibration phase. If more than one peak develops during this phase, they all have this same value and occur at $t_o = 3T_n/2, 5T_n/2$, and so on, again consistent with the results of Section 4.3.

As a corollary, if t_d is shorter than $T_n/2$, no peak will develop during the forced vibration phase (Fig. 4.7.2), and the response simply builds up from zero to $u(t_d)$. The displacement at the end of the pulse is given by Eq. (4.7.4a), rewritten to emphasize the parameter t_d/T_n:

$$u(t_d) = (u_{\mathrm{st}})_o \left(1 - \cos \frac{2\pi t_d}{T_n}\right) \qquad (4.7.7)$$

The maximum deformation during the forced vibration phase, Eqs. (4.7.6) and (4.7.7), can be expressed in terms of the *deformation response factor*:

$$R_d = \frac{u_o}{(u_{\mathrm{st}})_o} = \begin{cases} 1 - \cos(2\pi t_d/T_n) & t_d/T_n \le \frac{1}{2} \\ 2 & t_d/T_n \ge \frac{1}{2} \end{cases} \qquad (4.7.8)$$

This relationship is shown as "forced response" in Fig. 4.7.3a.

In the free vibration phase the system oscillates in simple harmonic motion, given by Eq. (4.7.3), with an amplitude

$$u_o = \sqrt{[u(t_d)]^2 + \left[\frac{\dot{u}(t_d)}{\omega_n}\right]^2} \qquad (4.7.9)$$

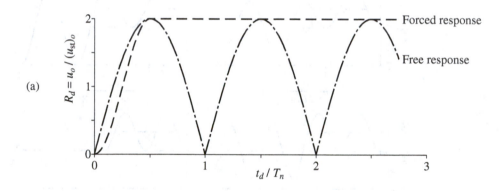

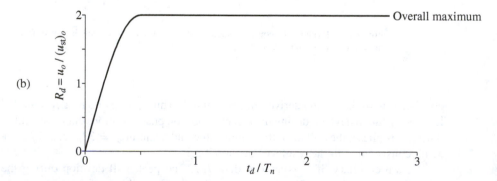

Figure 4.7.3 Response to rectangular pulse force: (a) maximum response during each of forced vibration and free vibration phases; (b) shock spectrum.

which after substituting Eq. (4.7.4) and some manipulation becomes

$$u_o = 2(u_{st})_o \left| \sin \frac{\pi t_d}{T_n} \right| \qquad (4.7.10)$$

The corresponding deformation response factor,

$$R_d \equiv \frac{u_o}{(u_{st})_o} = 2 \left| \sin \frac{\pi t_d}{T_n} \right| \qquad (4.7.11)$$

depends only on t_d/T_n and is shown as "free response" in Fig. 4.7.3a.

Having determined the maximum response during each of the forced and free vibration phases, we now determine the overall maximum. Figure 4.7.3a shows that if $t_d/T_n > \frac{1}{2}$, the overall maximum is the peak (or peaks because all are equal) in $u(t)$ that develops during the forced vibration phase because it will not be exceeded in free vibration; see Fig. 4.7.2 for $t_d/T_n = 1.25$. This observation can also be deduced from the mathematical results: the R_d of Eq. (4.7.11) for the free vibration phase can never exceed the $R_d = 2$, Eq. (4.7.8b), for the forced vibration phase.

If $t_d/T_n < \frac{1}{2}$, Fig. 4.7.3a shows that the overall maximum is the peak (or peaks because all are equal) in $u(t)$ that develops during the free vibration phase. In this case the response during the forced vibration phase has built up from zero at $t = 0$ to $u(t_d)$ at the end of the pulse, Eq. (4.7.7), and $\dot{u}(t_d)$ given by Eq. (4.7.4b) is positive; see Fig. 4.7.2 for $t_d/T_n = \frac{1}{8}$ or $\frac{1}{4}$. As a result, the first peak in free vibration is larger than $u(t_d)$.

Finally, if $t_d/T_n = \frac{1}{2}$, Fig. 4.7.3a shows that the overall maximum is given by either the forced-response maximum or the free-response maximum because the two are equal. The first peak occurs exactly at the end of the forced vibration phase (Fig. 4.7.2), the velocity $\dot{u}(t_d) = 0$, and the peaks in free vibration are the same as $u(t_d)$. This observation is consistent with Eqs. (4.7.8) and (4.7.11) because both of them give $R_d = 2$ for $t_d/T_n = \frac{1}{2}$.

In summary, the deformation response factor that defines the overall maximum response is

$$R_d = \frac{u_o}{(u_{st})_o} = \begin{cases} 2 \sin \pi t_d/T_n & t_d/T_n \leq \frac{1}{2} \\ 2 & t_d/T_n \geq \frac{1}{2} \end{cases} \qquad (4.7.12)$$

Clearly, R_d depends only on t_d/T_n, the ratio of the pulse duration to the natural period of the system, not separately on t_d or T_n. This relationship is shown in Fig. 4.7.3b.

Such a plot, which shows the maximum deformation of an SDF system as a function of the natural period T_n of the system (or a related parameter), is called a *response spectrum*. When the excitation is a single pulse, the terminology *shock spectrum* is also used for the response spectrum. Figure 4.7.3b then is the shock spectrum for a rectangular pulse force. The shock spectrum characterizes the problem completely.

The maximum deformation of an undamped SDF system having a natural period T_n to a rectangular pulse force of amplitude p_o and duration t_d can readily be determined if the shock spectrum for this excitation is available. Corresponding to the ratio t_d/T_n, the deformation response factor R_d is read from the spectrum, and the maximum deformation is computed from

$$u_o = (u_{st})_o R_d = \frac{p_o}{k} R_d \tag{4.7.13}$$

The maximum value of the equivalent static force (Section 1.8.2) is

$$f_{So} = k u_o = p_o R_d \tag{4.7.14}$$

that is, the applied force p_o multiplied by the deformation response factor. As mentioned in Section 1.8.2, static analysis of the structure subjected to f_{So} gives the internal forces and stresses.

Example 4.1

A one-story building, idealized as a 12-ft-high frame with two columns hinged at the base and a rigid beam, has a natural period of 0.5 sec. Each column is an American standard wide-flange steel section W8 × 18. Its properties for bending about its major axis are $I_x = 61.9$ in^4, $S = I_x/c = 15.2$ in^3; $E = 30,000$ ksi. Neglecting damping, determine the maximum response of this frame due to a rectangular pulse force of amplitude 4 kips and duration $t_d = 0.2$ sec. The response quantities of interest are displacement at the top of the frame and maximum bending stress in the columns.

Solution

    1. *Determine R_d.*

$$\frac{t_d}{T_n} = \frac{0.2}{0.5} = 0.4$$

$$R_d = \frac{u_o}{(u_{st})_o} = 2\sin\frac{\pi t_d}{T_n} = 2\sin(0.4\pi) = 1.902$$

    2. *Determine the lateral stiffness of the frame.*

$$k_{col} = \frac{3EI}{L^3} = \frac{3(30,000)61.9}{(12 \times 12)^3} = 1.865 \text{ kips/in.}$$

$$k = 2 \times 1.865 = 3.73 \text{ kips/in.}$$

    3. *Determine $(u_{st})_o$.*

$$(u_{st})_o = \frac{p_o}{k} = \frac{4}{3.73} = 1.07 \text{ in.}$$

    4. *Determine the maximum dynamic deformation.*

$$u_o = (u_{st})_o R_d = (1.07)(1.902) = 2.04 \text{ in.}$$

    5. *Determine the bending stress.* The resulting bending moments in each column are shown in Fig. E4.1c. At the top of the column the bending moment is largest and is given by

$$M = \frac{3EI}{L^2} u_o = \left[\frac{3(30,000)61.9}{(12 \times 12)^2}\right] 2.04 = 547.8 \text{ kip-in.}$$

Alternatively, we can find the bending moment from the equivalent static force:

$$f_{So} = p_o R_d = 4(1.902) = 7.61 \text{ kips}$$

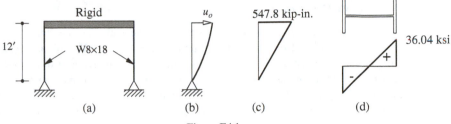

Figure E4.1

Because both columns are identical in cross section and length, the force f_{So} will be shared equally. The bending moment at the top of the column is

$$M = \frac{f_{So}}{2}h = \left(\frac{7.61}{2}\right)12 \times 12 = 547.8 \text{ kip-in.}$$

The bending stress is largest at the outside of the flanges at the top of the columns:

$$\sigma = \frac{M}{S} = \frac{547.8}{15.2} = 36.04 \text{ ksi}$$

The stress distribution is shown in Fig. E4.1d.

4.8 HALF-CYCLE SINE PULSE FORCE

The next pulse we consider is a half-cycle of sinusoidal force (Fig. 4.8.1b). The response analysis procedure for this pulse is the same as developed in Section 4.7 for a rectangular pulse, but the mathematical derivation becomes a little complicated. The solution of the governing equation

$$m\ddot{u} + ku = p(t) = \begin{cases} p_o \sin(\pi t/t_d) & t \le t_d \\ 0 & t \ge t_d \end{cases} \qquad (4.8.1)$$

with at-rest initial conditions is presented separately for (1) $\omega \equiv \pi/t_d \ne \omega_n$ or $t_d/T_n \ne \frac{1}{2}$ and (2) $\omega = \omega_n$ or $t_d/T_n = \frac{1}{2}$. For each case the analysis is organized in two phases: forced vibration and free vibration.

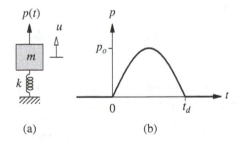

Figure 4.8.1 (a) SDF system; (b) half-cycle sine pulse force.

Case 1: $t_d/T_n \neq \frac{1}{2}$

Forced Vibration Phase. The force is the same as the harmonic force $p(t) = p_o \sin \omega t$ considered earlier with frequency $\omega = \pi/t_d$. The response of an undamped SDF system to such a force is given by Eq. (3.1.6b) in terms of ω and ω_n, the excitation and natural frequencies. The excitation frequency ω is not the most meaningful way of characterizing the pulse because, unlike a harmonic force, it is not a periodic function. A better characterization is the pulse duration t_d, which will be emphasized here. Using the relations $\omega = \pi/t_d$ and $\omega_n = 2\pi/T_n$, and defining $(u_{st})_o = p_o/k$, as before, Eq. (3.1.6b) becomes

$$\frac{u(t)}{(u_{st})_o} = \frac{1}{1-(T_n/2t_d)^2}\left[\sin\left(\pi\frac{t}{t_d}\right) - \frac{T_n}{2t_d}\sin\left(2\pi\frac{t}{T_n}\right)\right] \qquad t \leq t_d \qquad (4.8.2)$$

Free Vibration Phase. After the force pulse ends, the system vibrates freely with its motion described by Eq. (4.7.3). The displacement $u(t_d)$ and velocity $\dot{u}(t_d)$ at the end of the pulse are determined from Eq. (4.8.2). Substituting these in Eq. (4.7.3), using trigonometric identities and manipulating the mathematical quantities, we obtain

$$\frac{u(t)}{(u_{st})_o} = \frac{(T_n/t_d)\cos(\pi t_d/T_n)}{(T_n/2t_d)^2 - 1}\sin\left[2\pi\left(\frac{t}{T_n} - \frac{1}{2}\frac{t_d}{T_n}\right)\right] \qquad t \geq t_d \qquad (4.8.3)$$

Case 2: $t_d/T_n = \frac{1}{2}$

Forced Vibration Phase. The forced response is now given by Eq. (3.1.13b), repeated here for convenience:

$$\frac{u(t)}{(u_{st})_o} = \frac{1}{2}\left(\sin\frac{2\pi t}{T_n} - \frac{2\pi t}{T_n}\cos\frac{2\pi t}{T_n}\right) \qquad t \leq t_d \qquad (4.8.4)$$

Free Vibration Phase. After the force pulse ends at $t = t_d$, free vibration of the system is initiated by the displacement $u(t_d)$ and velocity $\dot{u}(t_d)$ at the end of the force pulse. Determined from Eq. (4.8.4), these are

$$\frac{u(t_d)}{(u_{st})_o} = \frac{\pi}{2} \qquad \dot{u}(t_d) = 0 \qquad (4.8.5)$$

The second equation implies that the displacement in the forced vibration phase reaches its maximum at the end of this phase. Substituting Eq. (4.8.5) in Eq. (4.7.3) gives the response of the system after the pulse has ended:

$$\frac{u(t)}{(u_{st})_o} = \frac{\pi}{2}\cos 2\pi\left(\frac{t}{T_n} - \frac{1}{2}\right) \qquad t \geq t_d \qquad (4.8.6)$$

Response history. The time variation of the normalized deformation, $u(t)/(u_{st})_o$, given by Eqs. (4.8.2) and (4.8.3) is plotted in Fig. 4.8.2 for several values of t_d/T_n. For the special case of $t_d/T_n = \frac{1}{2}$, Eqs. (4.8.4) and (4.8.6) describe the response of the system, and these are also plotted in Fig. 4.8.2. The nature of the response is seen to vary greatly by changing just the duration t_d of the pulse. Also plotted in Fig. 4.8.2 is $u_{st}(t) = p(t)/k$,

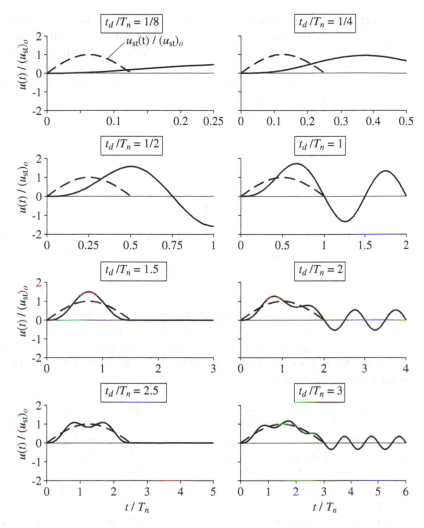

Figure 4.8.2   Dynamic response of undamped SDF system to half-cycle sine pulse force; static solution is shown by dashed lines.

the static solution. The difference between the two curves is an indication of the dynamic effects, which are seen to be small for $t_d = 3T_n$ because this implies that the force is varying slowly relative to the natural period T_n of the system.

The response during the force pulse contains both frequencies ω and ω_n and it is positive throughout. After the force pulse has ended, the system oscillates freely about its undeformed configuration with constant amplitude for lack of damping. If $t_d/T_n = 1.5, 2.5, \ldots$, the mass stays still after the force pulse ends because both the displacement and velocity of the mass are zero when the force pulse ends.

Maximum response. As in the preceding section, the maximum values of response over each of the two phases, forced vibration and free vibration, are determined separately. The larger of the two maxima is the overall maximum response.

During the forced vibration phase, the number of local maxima or peaks that develop depends on t_d/T_n (Fig. 4.8.2); more such peaks occur as the pulse duration lengthens. The time instants t_o when the peaks occur are determined by setting to zero the velocity associated with $u(t)$ of Eq. (4.8.2), leading to

$$\cos\frac{\pi t_o}{t_d} = \cos\frac{2\pi t_o}{T_n}$$

This equation is satisfied by

$$(t_o)_l = \frac{\mp 2l}{1 \mp 2(t_d/T_n)}t_d \qquad l = 1, 2, 3, \ldots \tag{4.8.7}$$

where the negative signs (numerator and denominator) are associated with local minima and the positive signs with local maxima. Thus the local maxima occur at time instants

$$(t_o)_l = \frac{2l}{1 + 2(t_d/T_n)}t_d \qquad l = 1, 2, 3, \ldots \tag{4.8.8}$$

While this gives an infinite number of $(t_o)_l$ values, only those that do not exceed t_d are relevant. For $t_d/T_n = 3$, Eq. (4.8.8) gives three relevant time instants: $t_o = \frac{2}{7}t_d$, $\frac{4}{7}t_d$, and $\frac{6}{7}t_d$; $l = 4$ gives $t_o = \frac{8}{7}t_d$, which is not valid because it exceeds t_d. Substituting in Eq. (4.8.2) the $(t_o)_l$ values of Eq. (4.8.8) gives the local maxima u_o, which can be expressed in terms of the deformation response factor:

$$R_d = \frac{u_o}{(u_{st})_o} = \frac{1}{1 - (T_n/2t_d)^2}\left(\sin\frac{2\pi l}{1 + 2t_d/T_n} - \frac{T_n}{2t_d}\sin\frac{2\pi l}{1 + T_n/2t_d}\right) \tag{4.8.9}$$

Figure 4.8.3a shows these peak values plotted as a function of t_d/T_n. For each t_d/T_n value the above-described computations were implemented and then repeated for many t_d/T_n values. If $0.5 \leq t_d/T_n \leq 1.5$, only one peak, $l = 1$, occurs during the force pulse. A second peak develops if $t_d/T_n > 1.5$, but it is smaller than the first peak if $1.5 < t_d/T_n < 2.5$. A third peak develops if $t_d/T_n > 2.5$. The second peak is larger than the first and third peaks if $2.5 < t_d/T_n < 4.5$. Usually, we will be concerned only with the largest peak because that controls the design of the system. The shock spectrum for the largest peak of the forced response is shown in Fig. 4.8.3b.

If $t_d/T_n < \frac{1}{2}$, no peak occurs during the forced vibration phase (Fig. 4.8.2). This becomes clear by examining the time of the first peak, Eq. (4.8.8), with $l = 1$:

$$t_o = \frac{2}{1 + 2t_d/T_n}t_d$$

If this t_o exceeds t_d, and it does for all $t_d < T_n/2$, no peak develops during the force pulse; the response builds up from zero to $u(t_d)$, obtained by evaluating Eq. (4.8.2) at $t = t_d$:

$$\frac{u(t_d)}{(u_{st})_o} = \frac{T_n/2t_d}{(T_n/2t_d)^2 - 1}\sin\left(2\pi\frac{t_d}{T_n}\right) \tag{4.8.10}$$

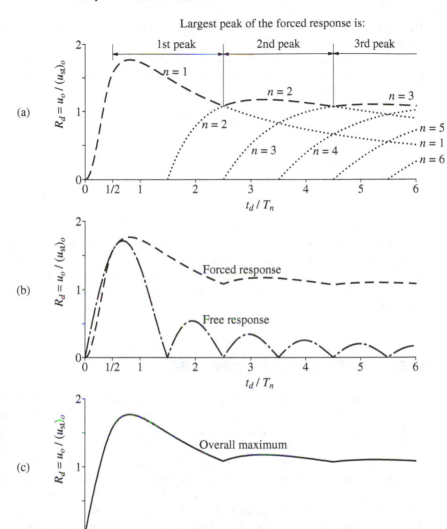

Largest peak of the forced response is:

Figure 4.8.3   Response to half-cycle sine pulse force: (a) response maxima during forced vibration phase; (b) maximum responses during each of forced vibration and free vibration phases; (c) shock spectrum.

This is the maximum response during the forced vibration phase and it defines the deformation response factor over the range $0 \le t_d/T_n < \frac{1}{2}$ in Fig. 4.8.3a and b.

In the free vibration phase the response of a system is given by the sinusoidal function of Eq. (4.8.3), and its amplitude is

$$R_d = \frac{u_o}{(u_{st})_o} = \frac{(T_n/t_d)\cos(\pi t_d/T_n)}{(T_n/2t_d)^2 - 1} \qquad (4.8.11)$$

This equation describes the maximum response during the free vibration phase and is plotted in Fig. 4.8.3b.

For the special case of $t_d/T_n = \frac{1}{2}$, the maximum response during each of the forced and free vibration phases can be determined from Eqs. (4.8.4) and (4.8.6), respectively; the two maxima are the same:

$$R_d = \frac{u_o}{(u_{st})_o} = \frac{\pi}{2} \qquad (4.8.12)$$

The overall maximum response is the larger of the two maxima determined separately for the forced and free vibration phases. Figure 4.8.3b shows that if $t_d > T_n/2$, the overall maximum is the largest peak that develops during the force pulse. On the other hand, if $t_d < T_n/2$, the overall maximum is given by the peak response during the free vibration phase. For the special case of $t_d = T_n/2$, as mentioned earlier, the two individual maxima are equal. The overall maximum response is plotted against t_d/T_n in Fig. 4.8.3c; for each t_d/T_n it is the larger of the two plots of Fig. 4.8.3b. This is the shock spectrum for the half-cycle sine pulse force. If it is available, the maximum deformation and equivalent static force can readily be determined using Eqs. (4.7.13) and (4.7.14).

4.9 SYMMETRICAL TRIANGULAR PULSE FORCE

Consider next an SDF system initially at rest and subjected to the symmetrical triangular pulse shown in Fig. 4.9.1. The response of an undamped SDF system to this pulse could be determined by any of the methods mentioned in Section 4.6. For example, the classical method could be implemented in three separate phases: $0 \le t \le t_d/2$, $t_d/2 \le t \le t_d$, and $t \ge t_d$. The classical method was preferred in Sections 4.8 and 4.9 because it is closely tied to the dynamics of the system, but is abandoned here for expedience. Perhaps the easiest way to solve the problem is to express the triangular pulse as the superposition of the three ramp functions shown in Fig. 4.6.2c. The response of the system to each of these ramp functions can readily be determined by appropriately adapting Eq. (4.4.2) to recognize the slope and starting time of each of the three ramp functions. These three individual responses are added to obtain the response to the symmetrical triangular pulse.

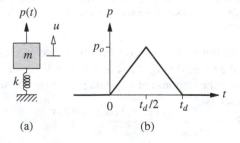

Figure 4.9.1   (a) SDF system; (b) triangular pulse force.

(a)   (b)

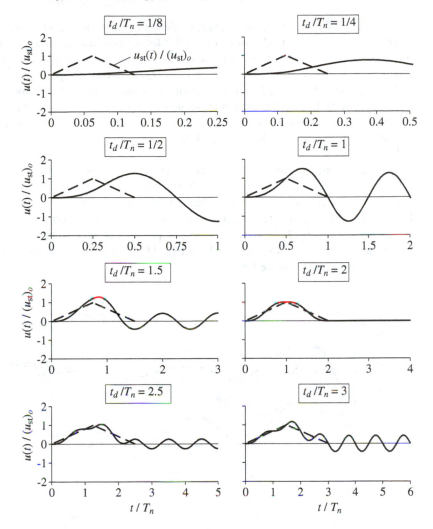

Figure 4.9.2   Dynamic response of undamped SDF system to triangular pulse force; static solution is shown by dashed lines.

The final result is

$$
\frac{u(t)}{(u_{st})_o} = \begin{cases}
2\left(\dfrac{t}{t_d} - \dfrac{T_n}{2\pi t_d}\sin 2\pi\dfrac{t}{T_n}\right) & 0 \le t \le \dfrac{t_d}{2} \quad (4.9.1a) \\[3ex]
2\left\{1 - \dfrac{t}{t_d} + \dfrac{T_n}{2\pi t_d}\left[2\sin\dfrac{2\pi}{T_n}\left(t - \dfrac{1}{2}t_d\right) - \sin 2\pi\dfrac{t}{T_n}\right]\right\} & \dfrac{t_d}{2} \le t \le t_d \quad (4.9.1b) \\[3ex]
2\left\{\dfrac{T_n}{2\pi t_d}\left[2\sin\dfrac{2\pi}{T_n}\left(t - \dfrac{1}{2}t_d\right) - \sin\dfrac{2\pi}{T_n}(t - t_d) - \sin 2\pi\dfrac{t}{T_n}\right]\right\} & t \ge t_d \quad (4.9.1c)
\end{cases}
$$

The variation of the normalized dynamic deformation $u(t)/(u_{st})_o$ and of the static solution $u_{st}(t)/(u_{st})_o$ with time is shown in Fig. 4.9.2 for several values of t_d/T_n. The dynamic effects are seen to decrease as the pulse duration t_d increases beyond $2T_n$. The first peak develops right at the end of the pulse if $t_d = T_n/2$, during the pulse if $t_d > T_n/2$, and after the pulse if $t_d < T_n/2$. The maximum response during free vibration (Fig. 4.9.3a) was obtained by finding the maximum value of Eq. (4.9.1c). The corresponding plot for maximum response during the forced vibration phase (Fig. 4.9.3a) was obtained by finding the largest of the local maxima of Eq. (4.9.1b), which is always larger than the maximum value of Eq. (4.9.1a).

The overall maximum response is the larger of the two maxima determined separately for the forced and free vibration phases. Figure 4.9.3a shows that if $t_d > T_n/2$, the overall maximum is the largest peak that develops during the force pulse. On the other hand, if $t_d < T_n/2$, the overall maximum is the peak response during the free vibration phase, and if $t_d = T_n/2$, the forced and free response maxima are equal. The overall maximum response is plotted against t_d/T_n in Fig. 4.9.3b. This is the shock spectrum for the symmetrical triangular pulse force.

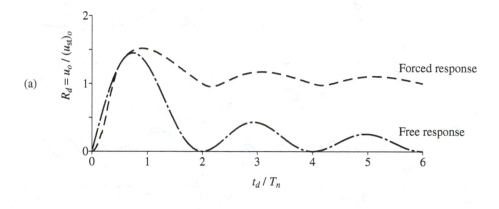

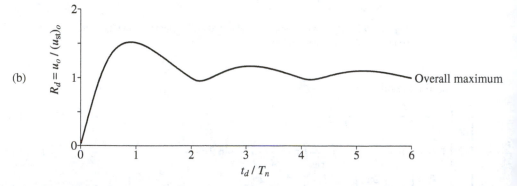

Figure 4.9.3 Response to triangular pulse force: (a) maximum response during each of forced vibration and free vibration phases; (b) shock spectrum.

4.10 EFFECTS OF PULSE SHAPE AND APPROXIMATE ANALYSIS FOR SHORT PULSES

The shock spectra for the three pulses of rectangular, half-cycle sine, and triangular shapes, each with the same value of maximum force p_o, are presented together in Fig. 4.10.1. As shown in the preceding sections, if the pulse duration t_d is longer than $T_n/2$, the overall maximum deformation occurs during the pulse. Then the pulse shape is of great significance. For the larger values of t_d/T_n, this overall maximum is influenced by the rapidity of the loading. The rectangular pulse in which the force increases suddenly from zero to p_o produces the largest deformation. The triangular pulse in which the increase in force is initially slowest among the three pulses produces the smallest deformation. The half-cycle sine pulse in which the force initially increases at an intermediate rate causes deformation that for many values of t_d/T_n is larger than the response to the triangular pulse.

If the pulse duration t_d is shorter than $T_n/2$, the overall maximum response of the system occurs during its free vibration phase and is controlled by the time integral of the pulse. This can be demonstrated by considering the limiting case as t_d/T_n approaches zero. As the pulse duration becomes extremely short compared to the natural period of the system, it becomes a pure impulse of magnitude

$$\mathcal{I} = \int_0^{t_d} p(t)\, dt \qquad (4.10.1)$$

The response of the system to this impulsive force is the unit impulse response of Eq. (4.1.6) times $\mathcal{I}$:

$$u(t) = \mathcal{I}\left(\frac{1}{m\omega_n}\sin \omega_n t\right) \qquad (4.10.2)$$

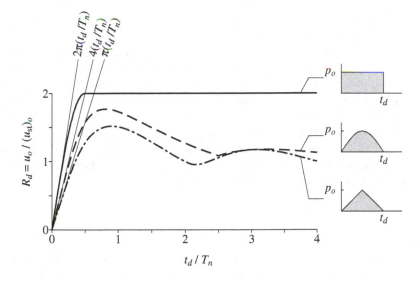

Figure 4.10.1   Shock spectra for three force pulses of equal amplitude.

The maximum deformation,

$$u_o = \frac{\mathcal{I}}{m\omega_n} = \frac{\mathcal{I}}{k}\frac{2\pi}{T_n} \qquad (4.10.3)$$

is proportional to the magnitude of the impulse.

Thus the maximum deformation due to the rectangular impulse of magnitude $\mathcal{I} = p_o t_d$ is

$$\frac{u_o}{(u_{st})_o} = 2\pi\frac{t_d}{T_n} \qquad (4.10.4)$$

that due to the half-cycle sine pulse with $\mathcal{I} = (2/\pi)p_o t_d$ is

$$\frac{u_o}{(u_{st})_o} = 4\frac{t_d}{T_n} \qquad (4.10.5)$$

and that due to the triangular pulse of magnitude $\mathcal{I} = p_o t_d/2$ is

$$\frac{u_o}{(u_{st})_o} = \pi\frac{t_d}{T_n} \qquad (4.10.6)$$

These pure impulse solutions, which vary linearly with t_d/T_n (Fig. 4.10.1), are exact if $t_d/T_n = 0$; for all other values of t_d, they provide an upper bound to the true maximum deformation since the effect of the pulse has been overestimated by assuming it to be concentrated at $t = 0$ instead of being spread out over 0 to t_d. Over the range $t_d/T_n < \frac{1}{4}$, the pure impulse solution is close to the exact response. The two solutions differ increasingly as t_d/T_n increases up to $\frac{1}{2}$. For larger values of t_d/T_n, the deformation attains its overall maximum during the pulse and the pure impulse solution is meaningless because it assumes that the maximum occurs in free vibration.

The preceding observations suggest that if the pulse duration is much shorter than the natural period, say $t_d < T_n/4$, the maximum deformation should be essentially controlled by the pulse area, independent of its shape. This expectation is confirmed by considering the rectangular pulse of amplitude $p_o/2$, the triangular pulse of amplitude p_o, and the half-cycle sine pulse of amplitude $(\pi/4)p_o$; these three pulses have the same area: $\frac{1}{2}p_o t_d$. For these three pulses, the shock spectra, determined by appropriately scaling the plots of Fig. 4.10.1, are presented in Fig. 4.10.2; observe that the quantity plotted now is $u_o \div p_o/k$, where p_o is the amplitude of the triangular pulse but not of the other two. Equation (4.10.3) with $\mathcal{I} = \frac{1}{2}p_o t_d$ gives the approximate result

$$\frac{u_o}{p_o/k} = \pi\frac{t_d}{T_n} \qquad (4.10.7)$$

which is also shown in Fig. 4.10.2. It is clear that for $t_d < T_n/4$, the shape of the pulse has little influence on the response and the response can be determined using only the pulse area. This conclusion is valid even if the pulse has a complicated shape.

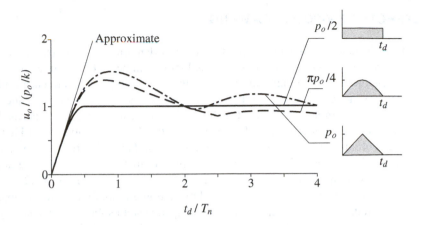

Figure 4.10.2   Shock spectra for three force pulses of equal area.

Example 4.2

The 80-ft-high full water tank of Example 2.7 is subjected to the force $p(t)$ shown in Fig. E4.2, caused by an aboveground explosion. Determine the maximum base shear and bending moment at the base of the tower supporting the tank.

Solution   For this water tank, from Example 2.7, weight $w = 100.03$ kips, $k = 8.2$ kips/in. $T_n = 1.12$ sec, and $\zeta = 1.23\%$. The ratio $t_d/T_n = 0.08/1.12 = 0.071$. Because $t_d/T_n < 0.25$, the forcing function may be treated as a pure impulse of magnitude

$$\mathcal{I} = \int_0^{0.08} p(t)\,dt = \frac{0.02}{2}[0 + 2(40) + 2(16) + 2(4) + 0] = 1.2 \text{ kip-sec}$$

where the integral is calculated by the trapezoidal rule. Neglecting the effect of damping, the maximum displacement is

$$u_o = \frac{\mathcal{I}}{k}\frac{2\pi}{T_n} = \frac{(1.2)2\pi}{(8.2)(1.12)} = 0.821 \text{ in.}$$

The equivalent static force f_{So} associated with this displacement is [from Eq. (1.8.1)]

$$f_{So} = ku_o = (8.2)0.821 = 6.73 \text{ kips}$$

The resulting shearing forces and bending moments over the height of the tower are shown in Fig. E4.2. The base shear and moment are $V_b = 6.73$ kips and $M_b = 538$ kip-ft.

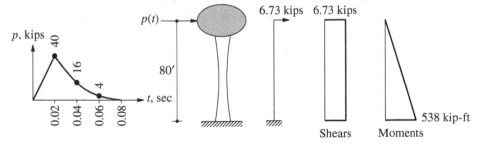

Figure E4.2

4.11 EFFECTS OF VISCOUS DAMPING

If the excitation is a single pulse, the effect of damping on the maximum response is usually not important unless the system is highly damped. This is in contrast to the results of Chapter 3, where damping was seen to have an important influence on the maximum steady-state response of systems to harmonic excitation at or near resonance.

For example, if the excitation frequency of harmonic excitation is equal to the natural frequency of the system, a tenfold increase in the damping ratio ζ, from 1% to 10%, results in a tenfold decrease in the deformation response factor R_d, from 50 to 5. Damping is so influential because of the cumulative energy dissipated in the many (the number depends on ζ) vibration cycles prior to attainment of steady state; see Figs. 3.2.2, 3.2.3, and 3.2.4.

In contrast, the energy dissipated by damping is small in systems subjected to pulse-type excitations. Consider a viscously damped system subjected to a half-cycle sine pulse with $t_d/T_n = \frac{1}{2}$ (which implies that $\omega = \omega_n$) and $\zeta = 0.1$. The variation of deformation with time (Fig. 4.11.1a) indicates that the maximum deformation (point b) is attained at

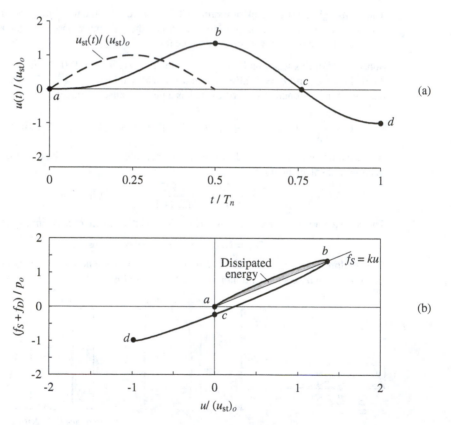

Figure 4.11.1   (a) Response of damped system ($\zeta = 0.1$) to a half-cycle sine pulse force with $t_d/T_n = \frac{1}{2}$; (b) force–deformation diagram showing energy dissipated in viscous damping.

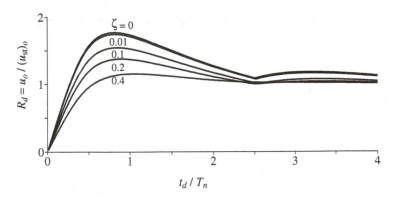

Figure 4.11.2   Shock spectra for a half-cycle sine pulse force for five damping values.

the end of the pulse before completion of a single vibration cycle. The total (elastic plus damping) force–deformation diagram of Fig. 4.11.1b indicates that before the maximum response is reached, the energy dissipated in viscous damping is only the small shaded area multiplied by p_o^2/k. Thus the influence of damping on maximum response is expected to be small.

This prediction is confirmed by the shock spectrum for the half-cycle sine pulse presented in Fig. 4.11.2. For $\zeta = 0$, this spectrum is the same as the spectrum of Fig. 4.8.3c for undamped systems. For $\zeta \neq 0$ and for each value of t_d/T_n, the dynamic response of the damped system was computed by a numerical time-stepping procedure (Chapter 5) and the maximum deformation was determined. In the case of the system acted upon by a half-cycle sine pulse of duration $t_d = T_n/2$, increase in the damping ratio from 1% to 10% reduces the maximum deformation by only 12%. Thus a conservative but not overly conservative estimate of the response of many practical structures with damping to pulse-type excitations may be obtained by neglecting damping and using the earlier results for undamped systems.

4.12 RESPONSE TO GROUND MOTION

The response spectrum characterizing the maximum response of SDF systems to ground motion $\ddot{u}_g(t)$ can be determined from the response spectrum for the applied force $p(t)$ with the same time variation as $\ddot{u}_g(t)$. This is possible because as shown in Eq. (1.7.6), the ground acceleration can be replaced by the effective force, $p_{\text{eff}}(t) = -m\ddot{u}_g(t)$.

The response spectrum for applied force $p(t)$ is a plot of $R_d = u_o/(u_{\text{st}})_o$, where $(u_{\text{st}})_o = p_o/k$, versus the appropriate system and excitation parameters: ω/ω_n for harmonic excitation and t_d/T_n for pulse-type excitation. Replacing p_o by $(p_{\text{eff}})_o$ gives

$$(u_{\text{st}})_o = \frac{(p_{\text{eff}})_o}{k} = \frac{m\ddot{u}_{go}}{k} = \frac{\ddot{u}_{go}}{\omega_n^2} \qquad (4.12.1)$$

where $\ddot{u}_{go}$ is the maximum value of $\ddot{u}_g(t)$ and the negative sign in $p_{\text{eff}}(t)$ has been dropped. Thus

$$R_d = \frac{u_o}{(u_{\text{st}})_o} = \frac{\omega_n^2 u_o}{\ddot{u}_{go}} \qquad (4.12.2)$$

Therefore, the response spectra presented in Chapters 3 and 4 showing the response $u_o/(u_{\text{st}})_o$ due to applied force also give the response $\omega_n^2 u_o/\ddot{u}_{go}$ to ground motion.

For undamped systems subjected to ground motion, Eqs. (1.7.4) and (1.7.3) indicate that the total acceleration of the mass is related to the deformation through $\ddot{u}^t(t) = -\omega_n^2 u(t)$. Thus the maximum values of the two responses are related by $\ddot{u}_o^t = \omega_n^2 u_o$. Substituting in Eq. (4.12.2) gives

$$R_d = \frac{\ddot{u}_o^t}{\ddot{u}_{go}} \qquad (4.12.3)$$

Thus the earlier response spectra showing the response $u_o/(u_{\text{st}})_o$ of undamped systems subjected to applied force also display the response $\ddot{u}_o^t/\ddot{u}_{go}$ to ground motion.

As an example, the response spectrum of Fig. 4.8.3c for a half-cycle sine pulse force also gives the maximum values of responses $\omega_n^2 u_o/\ddot{u}_{go}$ and $\ddot{u}_o^t/\ddot{u}_{go}$ due to ground acceleration described by a half-cycle sine pulse.

Example 4.3

Consider the SDF model of an automobile described in Example 3.4 running over the speed bump shown in Fig. E4.3 at velocity v. Determine the maximum force developed in the suspension spring and the maximum vertical acceleration of the mass if **(a)** $v = 5$ mph, and **(b)** $v = 10$ mph.

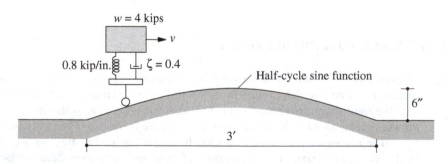

Figure E4.3

Solution

1. *Determine the system and excitation parameters.*

$$m = \frac{4000}{386} = 10.363 \text{ lb-sec}^2/\text{in.}$$

$k = 800$ lb/in.

$\omega_n = 8.786$ rad/sec $T_n = 0.715$ sec

$v = 5$ mph $= 7.333$ ft/sec $t_d = \dfrac{3}{7.333} = 0.4091$ sec $\dfrac{t_d}{T_n} = 0.572$

$v = 10$ mph $= 14.666$ ft/sec $t_d = \dfrac{3}{14.666} = 0.2046$ sec $\dfrac{t_d}{T_n} = 0.286$

The vertical ground displacement $u_g(t) = 6\sin(\pi t/t_d)$ for $0 \le t \le t_d$ and is zero for $t < 0$ and $t > t_d$. Two differentiations lead to an equation for ground acceleration: $\ddot{u}_g(t) = -(6\pi^2/t_d^2)\sin(\pi t/t_d)$ plus terms containing delta functions. The latter terms are due to the kink in the ground profile at the beginning and the end of the speed bump (Fig. E4.3). As a result, the ground acceleration is not a single pulse—in contrast to the ground displacement. Thus, the method presented in Section 4.12 is not applicable unless the latter terms are dropped, an approximation that may be appropriate at very low speeds. Such an approximate solution is presented to illustrate application of the method. With this approximation, $\ddot{u}_{go} = 6\pi^2/t_d^2$.

2. *Determine R_d for the t_d/T_n values above from Fig. 4.11.2.*

$$R_d = \begin{cases} 1.015 & v = 5 \text{ mph} \\ 0.639 & v = 10 \text{ mph} \end{cases}$$

Obviously, R_d cannot be read accurately to three or four significant digits; these values are from the numerical data used in plotting Fig. 4.11.2.

3. *Determine the maximum force, f_{So}.*

$$u_o = \frac{\ddot{u}_{go}}{\omega_n^2} R_d = 1.5 \left(\frac{T_n}{t_d}\right)^2 R_d$$

$$u_o = \begin{cases} 1.5\left(\dfrac{1}{0.572}\right)^2 1.015 = 4.65 \text{ in.} & v = 5 \text{ mph} \\ 1.5\left(\dfrac{1}{0.286}\right)^2 0.639 = 11.7 \text{ in.} & v = 10 \text{ mph} \end{cases}$$

$$f_{So} = ku_o = 0.8u_o = \begin{cases} 3.72 \text{ kips} & v = 5 \text{ mph} \\ 9.37 \text{ kips} & v = 10 \text{ mph} \end{cases}$$

Observe that the force in the suspension is much larger at the higher speed. The large deformation of the suspension suggests that it may deform beyond its linearly elastic limit.

4. *Determine the maximum acceleration, $\ddot{u}_o^t$.* Equation (4.12.3) provides a relation between $\ddot{u}_o^t$ and R_d that is exact for systems without damping but is approximate for damped

systems. These approximate results can readily be obtained for this problem:

$$\ddot{u}_o^t = \ddot{u}_{go} R_d = \frac{6\pi^2}{t_d^2} R_d$$

$$\ddot{u}_o^t = \begin{cases} \left[\dfrac{6\pi^2}{(0.4091)^2} \right] 1.015 = 359.1 \text{ in./sec}^2 & v = 5 \text{ mph} \\ \left[\dfrac{6\pi^2}{(0.2046)^2} \right] 0.639 = 903.7 \text{ in./sec}^2 & v = 10 \text{ mph} \end{cases}$$

Observe that the acceleration of the mass is much larger at the higher speed; in fact, it exceeds 1g, indicating that the SDF model would lift off from the road.

To evaluate the error in the approximate solution for $\ddot{u}_o^t$, a numerical solution of the equation of motion was carried out, leading to the "exact" value of $\ddot{u}_o^t = 422.7$ in./sec^2 for $v = 5$ mph.

FURTHER READING

Ayre, R. S., "Transient Response to Step and Pulse Functions," Chapter 8 in *Shock and Vibration Handbook,* 3rd ed. (ed. C. M. Harris), McGraw-Hill, New York, 1988.

Jacobsen, L. S., and Ayre, R. S., *Engineering Vibrations*, McGraw-Hill, New York, 1958, Chapters 3 and 4.

PROBLEMS

Part A

4.1 Show that the maximum deformation u_0 of an SDF system due to a unit impulse force, $p(t) = \delta(t)$, is

$$u_o = \frac{1}{m\omega_n} \exp\left(-\frac{\zeta}{\sqrt{1-\zeta^2}} \tan^{-1} \frac{\sqrt{1-\zeta^2}}{\zeta} \right)$$

Plot this result as a function of ζ. Comment on the influence of damping on the maximum response.

4.2 Consider the deformation response $g(t)$ of an SDF system to a unit step function $p(t) = 1$, $t \geq 0$, and $h(t)$ due to a unit impulse $p(t) = \delta(t)$. Show that $h(t) = \dot{g}(t)$.

4.3 An SDF undamped system is subjected to a force $p(t)$ consisting of a sequence of two impulses, each of magnitude I, as shown in Fig. P4.3.
(a) Plot the displacement response of the system for $t_d/T_n = \frac{1}{8}, \frac{1}{4}$, and 1. For each case show the response to individual impulses and the combined response.

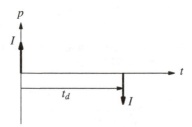

Figure P4.3

(b) Plot $u_o \div (I/m\omega_n)$ as a function of t_d/T_n. Indicate separately the maximum occurring at $t \leq t_d$ and $t \geq t_d$. Such a plot is called the response spectrum for this excitation.

4.4 Repeat Problem 4.3 for the case of the two impulses acting in the same direction.

4.5 **(a)** Show that the motion of an undamped system starting from rest due to a suddenly applied force p_o that decays exponentially with time (Fig. P4.5) is

$$\frac{u(t)}{(u_{st})_o} = \frac{1}{1+a^2/\omega_n^2} \left[\frac{a}{\omega_n} \sin \omega_n t - \cos \omega_n t + e^{-at} \right]$$

Note that a has the same units as ω_n.
(b) Plot this motion for selected values of $a/\omega_n = 0.01, 0.1,$ and 1.0.
(c) Show that the steady-state amplitude is

$$\frac{u_o}{(u_{st})_o} = \frac{1}{\sqrt{1+a^2/\omega_n^2}}$$

When is the steady-state motion reached?

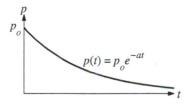

Figure P4.5

4.6 **(a)** Determine the motion of an undamped system starting from rest due to the force $p(t)$ shown in Fig. P4.6; $b > a$.
(b) Plot the motion for $b = 2a$ for three values of $a/\omega_n = 0.05, 0.1,$ and 0.5.

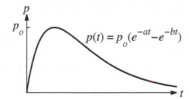

Figure P4.6

Part B

4.7 Using the classical method for solving differential equations, derive Eq. (4.4.2), which describes the response of an undamped SDF system to a linearly increasing force; the initial conditions are $u(0) = \dot{u}(0) = 0$.

4.8 An elevator is idealized as a weight of mass m supported by a spring of stiffness k. If the upper end of the spring begins to move with a steady velocity v, show that the distance u^t that the mass has risen in time t is governed by the equation

$$m\ddot{u}^t + ku^t = kvt$$

If the elevator starts from rest, show that the motion is

$$u^t(t) = vt - \frac{v}{\omega_n}\sin\omega_n t$$

Plot this result.

4.9 (a) Determine the maximum response of a damped SDF system to a step force.
(b) Plot the maximum response as a function of the damping ratio.

4.10 The deformation response of an undamped SDF system to a step force having finite rise time is given by Eqs. (4.5.2) and (4.5.4). Derive these results using Duhamel's integral.

4.11 Derive Eqs. (4.5.2) and (4.5.4) by considering the excitation as the sum of two ramp functions (Fig. P4.11). For $t \leq t_r$, $u(t)$ is the solution of the equation of motion for the first ramp function. For $t \geq t_r$, $u(t)$ is the sum of the responses to the two ramp functions.

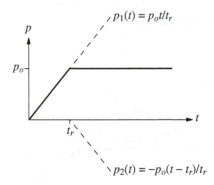

Figure P4.11

4.12 The elevated water tank of Fig. P4.12 weighs 100.03 kips when full with water. The tower has a lateral stiffness of 8.2 kips/in. Treating the water tower as an SDF system, estimate the maximum lateral displacement due to each of the two dynamic forces shown without any "exact" dynamic analysis. Instead, use your understanding of how the maximum response depends on the ratio of the rise time of the applied force to the natural vibration period of the system; neglect damping.

4.13 An SDF system with natural vibration period T_n is subjected to an alternating step force (Fig. P4.13). Note that $p(t)$ is periodic with period T_n.
(a) Determine the displacement as a function of time; the initial conditions are $u(0) = \dot{u}(0) = 0$.
(b) Plot the response.

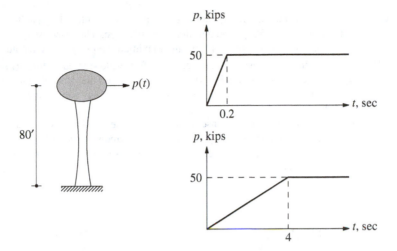

Figure P4.12

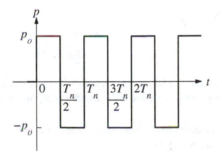

Figure P4.13

(c) Show that the displacement peaks are given by $u_n/(u_{st})_o = (-1)^{n-1}2n$, where n is the number of half cycles of $p(t)$.

Part C

4.14 Determine the response of an undamped system to a rectangular pulse force of amplitude p_o and duration t_d by considering the pulse as the superposition of two step excitations (Fig. 4.6.2).

4.15 Using Duhamel's integral, determine the response of an undamped system to a rectangular pulse force of amplitude p_o and duration t_d.

4.16 Determine the response of an undamped system to a half-cycle sine pulse force of amplitude p_o and duration t_d by considering the pulse as the superposition of two sinusoidal excitations (Fig. 4.6.2); $t_d/T_n \neq \frac{1}{2}$.

4.17 The one-story building of Example 4.1 is modified so that the columns are clamped at the base instead of hinged. For the same excitation determine the maximum displacement at the top of the frame and maximum bending stress in the columns. Comment on the effect of base fixity.

4.18 Determine the maximum response of the frame of Example 4.1 to a half-cycle sine pulse force of amplitude $p_o = 5$ kips and duration $t_d = 0.25$ sec. The response quantities of interest are: displacement at the top of the frame and maximum bending stress in columns.

4.19 An SDF undamped system is subjected to a full-cycle sine pulse force (Fig. P4.19).
(a) Derive equations describing $u(t)$ during the forced and free vibration phases.
(b) Plot the response $u(t)/(u_{st})_o$ versus t/T_n for various values of t_d/T_n; on the same plots show the static response $u_{st}(t)/(u_{st})_o$.
(c) Determine the peak response u_o, defined as the maximum of the absolute value of $u(t)$, during (i) the forced vibration phase, and (ii) the free vibration phase.
(d) Plot $R_d = u_o/(u_{st})_o$ for each of the two phases as a function of t_d/T_n.
(e) Plot the shock spectrum.

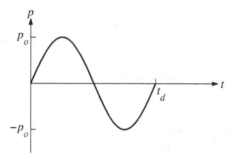

Figure P4.19

4.20 Derive equations (4.9.1) for the displacement response of an undamped SDF system to a symmetrical triangular pulse by considering the pulse as the superposition of three ramp functions (Fig. 4.6.2).

4.21 An undamped system is subjected to the triangular pulse in Fig. P4.21.
(a) Show that the displacement response is

$$
\frac{u(t)}{(u_{st})_o} =
\begin{cases}
\dfrac{t}{t_d} - \dfrac{1}{2\pi}\left(\dfrac{T_n}{t_d}\right)\sin\dfrac{2\pi t}{T_n} & 0 \le t \le t_d \\[3mm]
\cos\dfrac{2\pi}{T_n}(t - t_d) + \dfrac{1}{2\pi}\left(\dfrac{T_n}{t_d}\right)\sin\dfrac{2\pi}{T_n}(t - t_d) - \dfrac{1}{2\pi}\left(\dfrac{T_n}{t_d}\right)\sin\dfrac{2\pi t}{T_n} & t \ge t_d
\end{cases}
$$

Plot the response for two values of $t_d/T_n = \frac{1}{2}$ and 2.
(b) Derive equations for the deformation response factor R_d during (i) the forced vibration phase, and (ii) the free vibration phase.
(c) Plot R_d for the two phases against t_d/T_n. Also plot the shock spectrum.

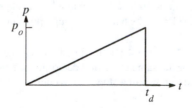

Figure P4.21

4.22 Derive equations for the deformation $u(t)$ of an undamped SDF system due to the force $p(t)$ shown in Fig. P4.22 for each of the following time ranges: $t \leq t_1, t_1 \leq t \leq 2t_1, 2t_1 \leq t \leq 3t_1$, and $t \geq 3t_1$.

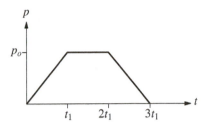

p

p_o

t_1 $2t_1$ $3t_1$ t

Figure P4.22

4.23 An SDF system is subjected to the force shown in Fig. P4.22. Determine the maximum response during free vibration of the system and the time instant the first peak occurs.

4.24 To determine the maximum response of an undamped SDF system to the force of Fig. P4.22 for a particular value of t_d/T_n, where $t_d = 3t_1$, you would need to identify the time range among the four time ranges mentioned in Problem 4.22 during which the overall maximum response would occur, and then find the value of that maximum. Such analyses would need to be repeated for many values of t_d/T_n to determine the complete shock spectrum. Obviously, this is time consuming but necessary if one wishes to determine the complete shock spectrum. However, the spectrum for small values of t_d/T_n can be determined by treating the force as an impulse. Determine the shock spectrum by this approach and plot it. What is the error in this approximate result for $t_d/T_n = \frac{1}{4}$?

4.25 **(a)** Determine the response of an undamped SDF system to the force shown in Fig. P4.25 for each of the following time intervals: (i) $0 \leq t \leq t_d/2$, (ii) $t_d/2 \leq t \leq t_d$, and (iii) $t \geq t_d$. Assume that $u(0) = \dot{u}(0) = 0$.
(b) Determine the maximum response u_o during free vibration of the system. Plot the deformation response factor $R_d = u_o/(u_{st})_o$ as a function of t_d/T_n over the range $0 \leq t_d/T_n \leq 4$.
(c) If $t_d \ll T_n$, can the maximum response be determined by treating the applied force as a pure impulse? State reasons for your answer.

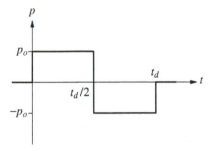

p

p_o

$t_d/2$ t_d t

$-p_o$

Figure P4.25

4.26 The 80-ft-high water tank of Examples 2.6 and 2.7 is subjected to the force $p(t)$ shown in Fig. E4.2a. The maximum response of the structure with the tank full (weight = 100.03 kips) was determined in Example 4.2.
(a) If the tank is empty (weight = 20.03 kips), calculate the maximum base shear and bending moment at the base of the tower supporting the tank.
(b) By comparing these results with those for the full tank (Example 4.2), comment on the effect of mass on the response to impulsive forces. Explain the reason.

5

Numerical Evaluation of Dynamic Response

PREVIEW

Analytical solution of the equation of motion for a single-degree-of-freedom system is usually not possible if the excitation—applied force $p(t)$ or ground acceleration $\ddot{u}_g(t)$—varies arbitrarily with time or if the system is nonlinear. Such problems can be tackled by numerical time-stepping methods for integration of differential equations. A vast body of literature, including major chapters of several books, exists about these methods for solving various types of differential equations that arise in the broad subject area of applied mechanics. The literature includes the mathematical development of these methods; their accuracy, convergence, and stability properties; and computer implementation.

Only a brief presentation of a very few methods that are especially useful in dynamic response analysis of SDF systems is included here, however. This presentation is intended to provide only the basic concepts underlying these methods and to provide a few computational algorithms. Although these would suffice for many practical problems and research applications, the reader should recognize that a wealth of knowledge exists on this subject.

5.1 TIME-STEPPING METHODS

For an inelastic system the equation of motion to be solved numerically is

$$ m\ddot{u} + c\dot{u} + f_S(u) = p(t) \qquad \text{or} \qquad -m\ddot{u}_g(t) \tag{5.1.1} $$

subject to the initial conditions

$$ u_0 = u(0) \qquad \dot{u}_0 = \dot{u}(0) $$

165

The system is assumed to have linear viscous damping, but other forms of damping, including nonlinear damping, could be considered, as will become obvious later. However, this is rarely done for lack of information on damping, especially at large amplitudes of motion. The applied force $p(t)$ is given by a set of discrete values $p_i = p(t_i)$, $i = 0$ to N (Fig. 5.1.1). The time interval

$$\Delta t_i = t_{i+1} - t_i \tag{5.1.2}$$

is usually taken to be constant, although this is not necessary. The response is determined at the discrete time instants t_i, denoted as time i; the displacement, velocity, and acceleration of the SDF system are u_i, $\dot{u}_i$, and $\ddot{u}_i$, respectively. These values, assumed to be known, satisfy Eq. (5.1.1) at time i:

$$m\ddot{u}_i + c\dot{u}_i + (f_S)_i = p_i \tag{5.1.3}$$

where $(f_S)_i$ is the resisting force at time i; $(f_S)_i = ku_i$ for a linearly elastic system but would depend on the prior history of displacement and on the velocity at time i if the system were nonlinear. The numerical procedures to be presented will enable us to determine the response quantities u_{i+1}, $\dot{u}_{i+1}$, and $\ddot{u}_{i+1}$ at time $i+1$ that satisfy Eq. (5.1.1) at

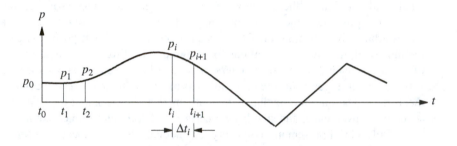

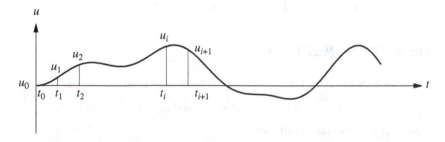

Figure 5.1.1   Notation for time-stepping methods.

time $i + 1$:

$$m\ddot{u}_{i+1} + c\dot{u}_{i+1} + (f_S)_{i+1} = p_{i+1} \tag{5.1.4}$$

When applied successively with $i = 0, 1, 2, 3, \ldots$, the time-stepping procedure gives the desired response at all time instants $i = 1, 2, 3, \ldots$. The known initial conditions, $u_0 = u(0)$ and $\dot{u}_0 = \dot{u}(0)$, provide the information necessary to start the procedure.

Stepping from time i to $i + 1$ is usually not an exact procedure. Many approximate procedures are possible that are implemented numerically. The three important requirements for a numerical procedure are (1) convergence—as the time step decreases, the numerical solution should approach the exact solution, (2) stability—the numerical solution should be stable in the presence of numerical round-off errors, and (3) accuracy—the numerical procedure should provide results that are close enough to the exact solution. These important issues are discussed briefly in this book; comprehensive treatments are available in books emphasizing numerical solution of differential equations.

Three types of time-stepping procedures are presented in this chapter: (1) methods based on interpolation of the excitation function, (2) methods based on finite difference expressions of velocity and acceleration, and (3) methods based on assumed variation of acceleration. Only one method is presented in each of the first two categories and two from the third group.

5.2 METHODS BASED ON INTERPOLATION OF EXCITATION

A highly efficient numerical procedure can be developed for linear systems by interpolating the excitation over each time interval and developing the exact solution using the methods of Chapter 4. If the time intervals are short, linear interpolation is satisfactory. Figure 5.2.1 shows that over the time interval $t_i \le t \le t_{i+1}$, the excitation function is given by

$$p(\tau) = p_i + \frac{\Delta p_i}{\Delta t_i}\tau \tag{5.2.1a}$$

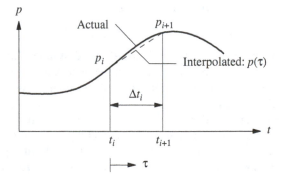

Figure 5.2.1   Notation for linearly interpolated excitation.

where

$$\Delta p_i = p_{i+1} - p_i \tag{5.2.1b}$$

and the time variable τ varies from 0 to Δt_i. For algebraic simplicity, we first consider systems without damping; later, the procedure will be extended to include damping. The equation to be solved is

$$m\ddot{u} + ku = p_i + \frac{\Delta p_i}{\Delta t_i}\tau \tag{5.2.2}$$

subject to initial conditions $u(0) = u_i$ and $\dot{u}(0) = \dot{u}_i$. The response $u(\tau)$ over the time interval $0 \le \tau \le \Delta t_i$ is the sum of three parts: (1) free vibration due to initial displacement u_i and velocity $\dot{u}_i$ at $\tau = 0$, (2) response to step force p_i with zero initial conditions, and (3) response to ramp force $(\Delta p_i/\Delta t_i)\tau$ with zero initial conditions. Adapting the available solutions for these three cases from Sections 2.1, 4.3, and 4.4, respectively, gives

$$u(\tau) = u_i \cos \omega_n \tau + \frac{\dot{u}_i}{\omega_n} \sin \omega_n \tau + \frac{p_i}{k}(1 - \cos \omega_n \tau) + \frac{\Delta p_i}{k}\left(\frac{\tau}{\Delta t_i} - \frac{\sin \omega_n \tau}{\omega_n \Delta t_i}\right) \tag{5.2.3a}$$

and differentiating $u(\tau)$ leads to

$$\frac{\dot{u}(\tau)}{\omega_n} = -u_i \sin \omega_n \tau + \frac{\dot{u}_i}{\omega_n} \cos \omega_n \tau + \frac{p_i}{k} \sin \omega_n \tau + \frac{\Delta p_i}{k}\frac{1}{\omega_n \Delta t_i}(1 - \cos \omega_n \tau) \tag{5.2.3b}$$

Evaluating these equations at $\tau = \Delta t_i$ gives the displacement u_{i+1} and velocity $\dot{u}_{i+1}$ at time $i + 1$:

$$u_{i+1} = u_i \cos(\omega_n \Delta t_i) + \frac{\dot{u}_i}{\omega_n} \sin(\omega_n \Delta t_i)$$

$$+ \frac{p_i}{k}[1 - \cos(\omega_n \Delta t_i)] + \frac{\Delta p_i}{k}\frac{1}{\omega_n \Delta t_i}[\omega_n \Delta t_i - \sin(\omega_n \Delta t_i)] \tag{5.2.4a}$$

$$\frac{\dot{u}_{i+1}}{\omega_n} = -u_i \sin(\omega_n \Delta t_i) + \frac{\dot{u}_i}{\omega_n} \cos(\omega_n \Delta t_i)$$

$$+ \frac{p_i}{k} \sin(\omega_n \Delta t_i) + \frac{\Delta p_i}{k}\frac{1}{\omega_n \Delta t_i}[1 - \cos(\omega_n \Delta t_i)] \tag{5.2.4b}$$

These equations can be rewritten after substituting Eq. (5.2.1b) as recurrence formulas:

$$u_{i+1} = Au_i + B\dot{u}_i + Cp_i + Dp_{i+1} \tag{5.2.5a}$$

$$\dot{u}_{i+1} = A'u_i + B'\dot{u}_i + C'p_i + D'p_{i+1} \tag{5.2.5b}$$

Repeating the derivation above for under-critically damped systems (i.e., $\zeta < 1$) shows that Eqs. (5.2.5) also apply to damped systems with the expressions for the coefficients $A, B, \ldots, D'$ given in Table 5.2.1. They depend on the system parameters ω_n, k, and ζ, and on the time interval $\Delta t \equiv \Delta t_i$.

Since the recurrence formulas are derived from exact solution of the equation of motion, the only restriction on the size of the time step Δt is that it permit a close approximation to the excitation function and that it provide response results at closely spaced time intervals so that the response peaks are not missed. This numerical procedure is especially

TABLE 5.2.1   COEFFICIENTS IN RECURRENCE FORMULAS ($\zeta < 1$)

$$A = e^{-\zeta \omega_n \Delta t} \left(\frac{\zeta}{\sqrt{1-\zeta^2}} \sin \omega_D \Delta t + \cos \omega_D \Delta t \right)$$

$$B = e^{-\zeta \omega_n \Delta t} \left(\frac{1}{\omega_D} \sin \omega_D \Delta t \right)$$

$$C = \frac{1}{k} \left\{ \frac{2\zeta}{\omega_n \Delta t} + e^{-\zeta \omega_n \Delta t} \left[\left(\frac{1-2\zeta^2}{\omega_D \Delta t} - \frac{\zeta}{\sqrt{1-\zeta^2}} \right) \sin \omega_D \Delta t - \left(1 + \frac{2\zeta}{\omega_n \Delta t} \right) \cos \omega_D \Delta t \right] \right\}$$

$$D = \frac{1}{k} \left[1 - \frac{2\zeta}{\omega_n \Delta t} + e^{-\zeta \omega_n \Delta t} \left(\frac{2\zeta^2 - 1}{\omega_D \Delta t} \sin \omega_D \Delta t + \frac{2\zeta}{\omega_n \Delta t} \cos \omega_D \Delta t \right) \right]$$

$$A' = -e^{-\zeta \omega_n \Delta t} \left(\frac{\omega_n}{\sqrt{1-\zeta^2}} \sin \omega_D \Delta t \right)$$

$$B' = e^{-\zeta \omega_n \Delta t} \left(\cos \omega_D \Delta t - \frac{\zeta}{\sqrt{1-\zeta^2}} \sin \omega_D \Delta t \right)$$

$$C' = \frac{1}{k} \left\{ -\frac{1}{\Delta t} + e^{-\zeta \omega_n \Delta t} \left[\left(\frac{\omega_n}{\sqrt{1-\zeta^2}} + \frac{\zeta}{\Delta t \sqrt{1-\zeta^2}} \right) \sin \omega_D \Delta t + \frac{1}{\Delta t} \cos \omega_D \Delta t \right] \right\}$$

$$D' = \frac{1}{k \Delta t} \left[1 - e^{-\zeta \omega_n \Delta t} \left(\frac{\zeta}{\sqrt{1-\zeta^2}} \sin \omega_D \Delta t + \cos \omega_D \Delta t \right) \right]$$

useful when the excitation is defined at closely spaced time intervals—as for earthquake ground acceleration—so that the linear interpolation is essentially perfect. If the time step Δt is constant, the coefficients A, B, ..., D' need to be computed only once.

The exact solution of the equation of motion required in this numerical procedure is feasible only for linear systems. It is conveniently developed for SDF systems, as shown above, but would be impractical for MDF systems unless their response is obtained by the superposition of modal responses (Chapters 12 and 13).

Example 5.1

An SDF system has the following properties: $m = 0.2533$ kip-sec^2/in., $k = 10$ kips/in., $T_n = 1$ sec ($\omega_n = 6.283$ rad/sec), and $\zeta = 0.05$. Determine the response $u(t)$ of this system to $p(t)$ defined by the half-cycle sine pulse force shown in Fig. E5.1 by (a) using piecewise linear interpolation of $p(t)$ with $\Delta t = 0.1$ sec, and (b) evaluating the theoretical solution.

Solution

1. *Initial calculations*

$$e^{-\zeta \omega_n \Delta t} = 0.9691 \qquad \omega_D = \omega_n \sqrt{1 - \zeta^2} = 6.275$$

$$\sin \omega_D \Delta t = 0.5871 \qquad \cos \omega_D \Delta t = 0.8095$$

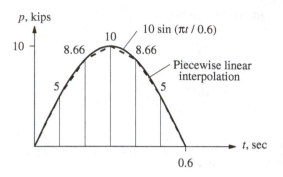

Figure E5.1

Substituting these in Table 5.2.1 gives

$$A = 0.8129 \qquad B = 0.09067 \qquad C = 0.01236 \qquad D = 0.006352$$

$$A' = -3.5795 \qquad B' = 0.7559 \qquad C' = 0.1709 \qquad D' = 0.1871$$

2. *Apply the recurrence equations (5.2.5).* The resulting computations are summarized in Tables E5.1a and E5.1b.

TABLE E5.1a   NUMERICAL SOLUTION USING LINEAR INTERPOLATION OF EXCITATION

t_i	p_i	Cp_i	Dp_{i+1}	$B\dot{u}_i$	$\dot{u}_i$	Au_i	u_i	Theoretical u_i
0.0	0.0000	0.0000	0.0318	0.0000	0.0000	0.0000	0.0000	0.0000
0.1	5.0000	0.0618	0.0550	0.0848	0.9354	0.0258	0.0318	0.0328
0.2	8.6602	0.1070	0.0635	0.2782	3.0679	0.1849	0.2274	0.2332
0.3	10.0000	0.1236	0.0550	0.4403	4.8558	0.5150	0.6336	0.6487
0.4	8.6603	0.1070	0.0318	0.4290	4.7318	0.9218	1.1339	1.1605
0.5	5.0000	0.0618	0.0000	0.1753	1.9336	1.2109	1.4896	1.5241
0.6	0.0000	0.0000	0.0000	−0.2735	−3.0159	1.1771	1.4480	1.4814
0.7	0.0000	0.0000	0.0000	−0.6767	−7.4631	0.7346	0.9037	0.9245
0.8	0.0000	0.0000	0.0000	−0.8048	−8.8765	0.0471	0.0579	0.0593
0.9	0.0000	0.0000	0.0000	−0.6272	−6.9177	−0.6160	−0.7577	−0.7751
1.0	0.0000				−2.5171		−1.2432	−1.2718

3. *Compute the theoretical response.* Equation (3.2.5)—valid for $t \leq 0.6$ sec, Eq. (2.2.4) modified appropriately—valid for $t \geq 0.6$ sec, and the derivatives of these two equations are evaluated for each t_i; the results are given in Tables E5.1a and E5.1b.

4. *Check the accuracy of the numerical results.* The numerical solution based on piecewise linear interpolation of the excitation agrees reasonably well with the theoretical solution. The discrepancy arises because the half-cycle sine curve has been replaced by the series of straight lines shown in Fig. E5.1. With a smaller Δt the piecewise linear approximation would be closer to the half-cycle sine curve, and the numerical solution would be more accurate.

TABLE E5.1b   NUMERICAL SOLUTION USING LINEAR INTERPOLATION OF EXCITATION

t_i	p_i	$C'p_i$	$D'p_{i+1}$	$A'u_i$	u_i	$B'\dot{u}_i$	$\dot{u}_i$	Theoretical u_i
0.0	0.0000	0.0000	0.9354	0.0000	0.0000	0.0000	0.0000	0.0000
0.1	5.0000	0.8544	1.6201	−0.1137	0.0318	0.7071	0.9354	0.9567
0.2	8.6602	1.4799	1.8707	−0.8140	0.2274	2.3192	3.0679	3.1383
0.3	10.0000	1.7088	1.6201	−2.2679	0.6336	3.6708	4.8558	4.9674
0.4	8.6603	1.4799	0.9354	−4.0588	1.1339	3.5771	4.7318	4.8408
0.5	5.0000	0.8544	0.0000	−5.3320	1.4896	1.4617	1.9336	1.9783
0.6	0.0000	0.0000	0.0000	−5.1832	1.4480	−2.2799	−3.0159	−3.0848
0.7	0.0000	0.0000	0.0000	−3.2347	0.9037	−5.6418	−7.4631	−7.6346
0.8	0.0000	0.0000	0.0000	−0.2074	0.0579	−6.7103	−8.8765	−9.0808
0.9	0.0000	0.0000	0.0000	2.7124	−0.7577	−5.2295	−6.9177	−7.0771
1.0	0.0000				−1.2432		−2.5171	−2.5754

5.3 CENTRAL DIFFERENCE METHOD

This method is based on a finite difference approximation of the time derivatives of displacement (i.e., velocity and acceleration). Taking constant time steps, $\Delta t_i = \Delta t$, the central difference expressions for velocity and acceleration at time i are

$$\dot{u}_i = \frac{u_{i+1} - u_{i-1}}{2\Delta t} \qquad \ddot{u}_i = \frac{u_{i+1} - 2u_i + u_{i-1}}{(\Delta t)^2} \tag{5.3.1}$$

Substituting these approximate expressions for velocity and acceleration into Eq. (5.1.3), specialized for linearly elastic systems, gives

$$m\frac{u_{i+1} - 2u_i + u_{i-1}}{(\Delta t)^2} + c\frac{u_{i+1} - u_{i-1}}{2\Delta t} + ku_i = p_i \tag{5.3.2}$$

In this equation u_i and u_{i-1} are assumed known (from implementation of the procedure for the preceding time steps). Transferring these known quantities to the right side leads to

$$\left[\frac{m}{(\Delta t)^2} + \frac{c}{2\Delta t}\right]u_{i+1} = p_i - \left[\frac{m}{(\Delta t)^2} - \frac{c}{2\Delta t}\right]u_{i-1} - \left[k - \frac{2m}{(\Delta t)^2}\right]u_i \tag{5.3.3}$$

or

$$\hat{k}u_{i+1} = \hat{p}_i \tag{5.3.4}$$

where

$$\hat{k} = \frac{m}{(\Delta t)^2} + \frac{c}{2\Delta t} \tag{5.3.5}$$

and

$$\hat{p}_i = p_i - \left[\frac{m}{(\Delta t)^2} - \frac{c}{2\Delta t}\right]u_{i-1} - \left[k - \frac{2m}{(\Delta t)^2}\right]u_i \tag{5.3.6}$$

TABLE 5.3.1   CENTRAL DIFFERENCE METHOD[†]

1.0 *Initial calculations*

  1.1 $\ddot{u}_0 = \dfrac{p_0 - c\dot{u}_0 - ku_0}{m}$.

  1.2 $u_{-1} = u_0 - \Delta t\,\dot{u}_0 + \dfrac{(\Delta t)^2}{2}\ddot{u}_0$.

  1.3 $\hat{k} = \dfrac{m}{(\Delta t)^2} + \dfrac{c}{2\Delta t}$.

  1.4 $a = \dfrac{m}{(\Delta t)^2} - \dfrac{c}{2\Delta t}$.

  1.5 $b = k - \dfrac{2m}{(\Delta t)^2}$.

2.0 *Calculations for time step i*

  2.1 $\hat{p}_i = p_i - au_{i-1} - bu_i$.

  2.2 $u_{i+1} = \dfrac{\hat{p}_i}{\hat{k}}$.

  2.3 If required: $\dot{u}_i = \dfrac{u_{i+1} - u_{i-1}}{2\Delta t}$; $\ddot{u}_i = \dfrac{u_{i+1} - 2u_i + u_{i-1}}{(\Delta t)^2}$.

3.0 *Repetition for the next time step*

  Replace i by $i + 1$ and repeat steps 2.1, 2.2, and 2.3 for the next time step.

[†] If the excitation is ground acceleration $\ddot{u}_g(t)$, according to Eq. (1.7.6), replace p_i by $-m\ddot{u}_{gi}$ in Table 5.3.1. The computed u_i, $\dot{u}_i$, and $\ddot{u}_i$ give response values relative to the ground. If needed, the total velocity and acceleration can be computed readily: $\dot{u}_i^t = \dot{u}_i + \dot{u}_{gi}$ and $\ddot{u}_i^t = \ddot{u}_i + \ddot{u}_{gi}$.

The unknown u_{i+1} is then given by

$$u_{i+1} = \frac{\hat{p}_i}{\hat{k}} \qquad\qquad (5.3.7)$$

The solution u_{i+1} at time $i + 1$ is determined from the equilibrium condition, Eq. (5.1.3), at time i without using the equilibrium condition, Eq. (5.1.4), at time $i + 1$. Such methods are called *explicit methods*.

Observe in Eq. (5.3.6) that known displacements u_i and u_{i-1} are used to compute u_{i+1}. Thus u_0 and u_{-1} are required to determine u_1; the specified initial displacement u_0 is known. To determine u_{-1}, we specialize Eq. (5.3.1) for $i = 0$ to obtain

$$\dot{u}_0 = \frac{u_1 - u_{-1}}{2\Delta t} \qquad \ddot{u}_0 = \frac{u_1 - 2u_0 + u_{-1}}{(\Delta t)^2} \qquad\qquad (5.3.8)$$

Solving for u_1 from the first equation and substituting in the second gives

$$u_{-1} = u_0 - \Delta t\,(\dot{u}_0) + \frac{(\Delta t)^2}{2}\ddot{u}_0 \qquad\qquad (5.3.9)$$

The initial displacement u_0 and initial velocity $\dot{u}_0$ are given, and the equation of motion at time 0 ($t_0 = 0$),

$$m\ddot{u}_0 + c\dot{u}_0 + ku_0 = p_0$$

provides the acceleration at time 0:

$$\ddot{u}_0 = \frac{p_0 - c\dot{u}_0 - ku_0}{m} \tag{5.3.10}$$

Table 5.3.1 summarizes the above-described procedure as it might be implemented on the computer.

The central difference method will "blow up," giving meaningless results, in the presence of numerical round-off if the time step chosen is not short enough. The specific requirement for stability is

$$\frac{\Delta t}{T_n} < \frac{1}{\pi} \tag{5.3.11}$$

This is never a constraint for SDF systems because a much smaller time step should be chosen to obtain results that are accurate. Typically, $\Delta t / T_n \leq 0.1$ to define the response adequately, and in most earthquake response analyses even a shorter time step, typically $\Delta t = 0.01$ to 0.02 sec, is chosen to define the ground acceleration $\ddot{u}_g(t)$ accurately.

Example 5.2

Solve Example 5.1 by the central difference method using $\Delta t = 0.1$ sec.

Solution

1.0 *Initial calculations*

$$m = 0.2533 \qquad k = 10 \qquad c = 0.1592$$

$$u_0 = 0 \qquad \dot{u}_0 = 0$$

1.1 $\ddot{u}_0 = \dfrac{p_0 - c\dot{u}_0 - ku_0}{m} = 0.$

1.2 $u_{-1} = u_0 - (\Delta t)\dot{u}_0 + \dfrac{(\Delta t)^2}{2}\ddot{u}_0 = 0.$

1.3 $\hat{k} = \dfrac{m}{(\Delta t)^2} + \dfrac{c}{2\Delta t} = 26.13.$

1.4 $a = \dfrac{m}{(\Delta t)^2} - \dfrac{c}{2\Delta t} = 24.53.$

1.5 $b = k - \dfrac{2m}{(\Delta t)^2} = -40.66.$

2.0 *Calculations for each time step*

2.1 $\hat{p}_i = p_i - au_{i-1} - bu_i = p_i - 24.53u_{i-1} + 40.66u_i.$

2.2 $u_{i+1} = \dfrac{\hat{p}_i}{\hat{k}} = \dfrac{\hat{p}_i}{26.13}.$

3.0 Computational steps 2.1 and 2.2 are repeated for $i = 0, 1, 2, 3, \ldots$ leading to Table E5.2, wherein the theoretical result (from Table E5.1a) is also included.

TABLE E5.2 NUMERICAL SOLUTION BY CENTRAL DIFFERENCE METHOD

t_i	p_i	u_{i-1}	u_i	$\hat{p}_i$ [Eq. (2.1)]	u_{i+1} [Eq. (2.2)]	Theoretical u_{i+1}
0.0	0.0000	0.0000	0.0000	0.0000	0.0000	0.0328
0.1	5.0000	0.0000	0.0000	5.0000	0.1914	0.2332
0.2	8.6602	0.0000	0.1914	16.4419	0.6293	0.6487
0.3	10.0000	0.1914	0.6293	30.8934	1.1825	1.1605
0.4	8.6603	0.6293	1.1825	41.3001	1.5808	1.5241
0.5	5.0000	1.1825	1.5808	40.2649	1.5412	1.4814
0.6	0.0000	1.5808	1.5412	23.8809	0.9141	0.9245
0.7	0.0000	1.5412	0.9141	−0.6456	−0.0247	0.0593
0.8	0.0000	0.9141	−0.0247	−23.4309	−0.8968	−0.7751
0.9	0.0000	−0.0247	−0.8968	−35.8598	−1.3726	−1.2718
1.0	0.0000	−0.8968	−1.3726	−33.8058	−1.2940	−1.2674

5.4 NEWMARK'S METHOD

5.4.1 Basic Procedure

In 1959, N. M. Newmark developed a family of time-stepping methods based on the following equations:

$$\dot{u}_{i+1} = \dot{u}_i + [(1 - \gamma)\,\Delta t]\,\ddot{u}_i + (\gamma\,\Delta t)\ddot{u}_{i+1} \tag{5.4.1a}$$

$$u_{i+1} = u_i + (\Delta t)\dot{u}_i + \left[(0.5 - \beta)(\Delta t)^2\right]\ddot{u}_i + \left[\beta(\Delta t)^2\right]\ddot{u}_{i+1} \tag{5.4.1b}$$

The parameters β and γ define the variation of acceleration over a time step and determine the stability and accuracy characteristics of the method. Typical selection for γ is $\frac{1}{2}$, and $\frac{1}{6} \leq \beta \leq \frac{1}{4}$ is satisfactory from all points of view, including that of accuracy. These two equations, combined with the equilibrium equation (5.1.4) at the end of the time step, provide the basis for computing u_{i+1}, $\dot{u}_{i+1}$, and $\ddot{u}_{i+1}$ at time $i + 1$ from the known u_i, $\dot{u}_i$, and $\ddot{u}_i$ at time i. Iteration is required to implement these computations because the unknown $\ddot{u}_{i+1}$ appears in the right side of Eq. (5.4.1).

For linear systems it is possible to modify Newmark's original formulation, however, to permit solution of Eqs. (5.4.1) and (5.1.4) without iteration. Before describing this modification, we demonstrate that two special cases of Newmark's method are the well-known constant average acceleration and linear acceleration methods.

5.4.2 Special Cases

For these two methods, Table 5.4.1 summarizes the development of the relationship between responses u_{i+1}, $\dot{u}_{i+1}$, and $\ddot{u}_{i+1}$ at time $i + 1$ to the corresponding quantities at time i. Equation (5.4.2) describes the assumptions that the variation of acceleration over a time step is constant, equal to the average acceleration, or varies linearly. Integration of $\ddot{u}(\tau)$ gives Eq. (5.4.3) for the variation $\dot{u}(\tau)$ of velocity over the time step in which

TABLE 5.4.1 AVERAGE ACCELERATION AND LINEAR ACCELERATION METHODS

Constant Average Acceleration	Linear Acceleration

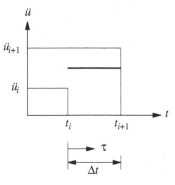

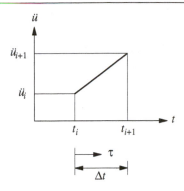

$$\ddot{u}(\tau) = \frac{1}{2}(\ddot{u}_{i+1} + \ddot{u}_i) \qquad\qquad \ddot{u}(\tau) = \ddot{u}_i + \frac{\tau}{\Delta t}(\ddot{u}_{i+1} - \ddot{u}_i) \qquad (5.4.2)$$

$$\dot{u}(\tau) = \dot{u}_i + \frac{\tau}{2}(\ddot{u}_{i+1} + \ddot{u}_i) \qquad \dot{u}(\tau) = \dot{u}_i + \ddot{u}_i\tau + \frac{\tau^2}{2\Delta t}(\ddot{u}_{i+1} - \ddot{u}_i) \qquad (5.4.3)$$

$$\dot{u}_{i+1} = \dot{u}_i + \frac{\Delta t}{2}(\ddot{u}_{i+1} + \ddot{u}_i) \qquad \dot{u}_{i+1} = \dot{u}_i + \frac{\Delta t}{2}(\ddot{u}_{i+1} + \ddot{u}_i) \qquad (5.4.4)$$

$$u(\tau) = u_i + \dot{u}_i\tau + \frac{\tau^2}{4}(\ddot{u}_{i+1} + \ddot{u}_i) \qquad u(\tau) = u_i + \dot{u}_i\tau + \ddot{u}_i\frac{\tau^2}{2} + \frac{\tau^3}{6\Delta t}(\ddot{u}_{i+1} - \ddot{u}_i) \qquad (5.4.5)$$

$$u_{i+1} = u_i + \dot{u}_i\,\Delta t + \frac{(\Delta t)^2}{4}(\ddot{u}_{i+1} + \ddot{u}_i) \qquad u_{i+1} = u_i + \dot{u}_i\,\Delta t + (\Delta t)^2\left(\frac{1}{6}\ddot{u}_{i+1} + \frac{1}{3}\ddot{u}_i\right) \qquad (5.4.6)$$

$\tau = \Delta t$ is substituted to obtain Eq. (5.4.4) for the velocity $\dot{u}_{i+1}$ at time $i+1$. Integration of $\dot{u}(\tau)$ gives Eq. (5.4.5) for the variation $u(\tau)$ of displacement over the time step in which $\tau = \Delta t$ is substituted to obtain Eq. (5.4.6) for the displacement u_{i+1} at time $i+1$. Comparing Eqs. (5.4.4) and (5.4.6) with Eq. (5.4.1) demonstrates that Newmark's equations with $\gamma = \frac{1}{2}$ and $\beta = \frac{1}{4}$ are the same as those derived assuming constant average acceleration, and those with $\gamma = \frac{1}{2}$ and $\beta = \frac{1}{6}$ correspond to the assumption of linear variation of acceleration.

5.4.3 Linear Systems

For linear systems, it is possible to modify Newmark's original formulation, to permit solution of Eqs. (5.4.1) and (5.1.4) without iteration. Specialized for linear systems, Eq. (5.1.4) becomes

$$m\ddot{u}_{i+1} + c\dot{u}_{i+1} + ku_{i+1} = p_{i+1} \qquad (5.4.7)$$

From Eq. (5.4.1b), $\ddot{u}_{i+1}$ can be expressed in terms of u_{i+1}:

$$\ddot{u}_{i+1} = \frac{1}{\beta(\Delta t)^2}(u_{i+1} - u_i) - \frac{1}{\beta\Delta t}\dot{u}_i - \left(\frac{1}{2\beta} - 1\right)\ddot{u}_i \qquad (5.4.8)$$

Substituting Eq. (5.4.8) into Eq. (5.4.1a) gives $\dot{u}_{i+1}$ in terms of u_{i+1}:

$$\dot{u}_{i+1} = \frac{\gamma}{\beta \Delta t}(u_{i+1} - u_i) + \left(1 - \frac{\gamma}{\beta}\right)\dot{u}_i + \Delta t\left(1 - \frac{\gamma}{2\beta}\right)\ddot{u}_i \qquad (5.4.9)$$

Next, Eqs. (5.4.8) and (5.4.9) are substituted into the governing equation (5.4.7) at time $i + 1$. This substitution gives

$$\hat{k}u_{i+1} = \hat{p}_{i+1} \qquad (5.4.10)$$

where

$$\hat{k} = k + \frac{\gamma}{\beta \Delta t}c + \frac{1}{\beta(\Delta t)^2}m \qquad (5.4.11)$$

and

$$\hat{p}_{i+1} = p_{i+1} + \left[\frac{1}{\beta(\Delta t)^2}m + \frac{\gamma}{\beta \Delta t}c\right]u_i + \left[\frac{1}{\beta \Delta t}m + \left(\frac{\gamma}{\beta} - 1\right)c\right]\dot{u}_i$$

$$+ \left[\left(\frac{1}{2\beta} - 1\right)m + \Delta t\left(\frac{\gamma}{2\beta} - 1\right)c\right]\ddot{u}_i \qquad (5.4.12)$$

With $\hat{k}$ and $\hat{p}_{i+1}$ known from the system properties m, k, and c, algorithm parameters γ and β, and the state of the system at time i defined by u_i, $\dot{u}_i$, and $\ddot{u}_i$, the displacement at time $i + 1$ is computed from

$$u_{i+1} = \frac{\hat{p}_{i+1}}{\hat{k}} \qquad (5.4.13)$$

Once u_{i+1} is known, the velocity $\dot{u}_{i+1}$ and acceleration $\ddot{u}_{i+1}$ can be computed from Eqs. (5.4.9) and (5.4.8), respectively.

The acceleration can also be obtained from the equation of motion at time $i + 1$:

$$\ddot{u}_{i+1} = \frac{p_{i+1} - c\dot{u}_{i+1} - ku_{i+1}}{m} \qquad (5.4.14)$$

rather than by Eq. (5.4.8). Equation (5.4.14) is needed to obtain $\ddot{u}_0$ to start the time-stepping computations [see Eq. (5.3.10)].

In Newmark's method, the solution at time $i + 1$ is determined from Eq. (5.4.7), the equilibrium condition at time $i + 1$. Such methods are called *implicit methods*. Although the resisting force is an implicit function of the unknown u_{i+1}, it was easy to calculate for linear systems.

Table 5.4.2 summarizes the time-stepping solution using Newmark's method as it might be implemented on the computer.

TABLE 5.4.2 NEWMARK'S METHOD: LINEAR SYSTEMS[†]

Special cases
  (1) Constant average acceleration method ($\gamma = \frac{1}{2}$, $\beta = \frac{1}{4}$)
  (2) Linear acceleration method ($\gamma = \frac{1}{2}$, $\beta = \frac{1}{6}$)

1.0 *Initial calculations*

  1.1 $\ddot{u}_0 = \dfrac{p_0 - c\dot{u}_0 - ku_0}{m}$.

  1.2 Select Δt.

  1.3 $a_1 = \dfrac{1}{\beta(\Delta t)^2}m + \dfrac{\gamma}{\beta \Delta t}c; \quad a_2 = \dfrac{1}{\beta \Delta t}m + \left(\dfrac{\gamma}{\beta} - 1\right)c; \quad$ and

$$a_3 = \left(\dfrac{1}{2\beta} - 1\right)m + \Delta t \left(\dfrac{\gamma}{2\beta} - 1\right)c.$$

  1.4 $\hat{k} = k + a_1$.

2.0 *Calculations for each time step, $i = 0, 1, 2, \ldots$*

  2.1 $\hat{p}_{i+1} = p_{i+1} + a_1 u_i + a_2 \dot{u}_i + a_3 \ddot{u}_i$.

  2.2 $u_{i+1} = \dfrac{\hat{p}_{i+1}}{\hat{k}}$.

  2.3 $\dot{u}_{i+1} = \dfrac{\gamma}{\beta \Delta t}(u_{i+1} - u_i) + \left(1 - \dfrac{\gamma}{\beta}\right)\dot{u}_i + \Delta t \left(1 - \dfrac{\gamma}{2\beta}\right)\ddot{u}_i$.

  2.4 $\ddot{u}_{i+1} = \dfrac{1}{\beta(\Delta t)^2}(u_{i+1} - u_i) - \dfrac{1}{\beta \Delta t}\dot{u}_i - \left(\dfrac{1}{2\beta} - 1\right)\ddot{u}_i$.

3.0 *Repetition for the next time step.* Replace i by $i + 1$ and implement steps 2.1 to 2.4 for the next time step.

[†]If the excitation is ground acceleration $\ddot{u}_g(t)$, according to Eq. (1.7.6), replace p_i by $-m\ddot{u}_{gi}$ in Table 5.4.2. The computed u_i, $\dot{u}_i$, and $\ddot{u}_i$ give response values relative to the ground. If needed, the total velocity and acceleration can be computed readily: $\dot{u}_i^t = \dot{u}_i + \dot{u}_{gi}$ and $\ddot{u}_i^t = \ddot{u}_i + \ddot{u}_{gi}$.

Newmark's method is stable if

$$\frac{\Delta t}{T_n} \leq \frac{1}{\pi\sqrt{2}}\frac{1}{\sqrt{\gamma - 2\beta}} \qquad (5.4.15)$$

For $\gamma = \frac{1}{2}$ and $\beta = \frac{1}{4}$ this condition becomes

$$\frac{\Delta t}{T_n} < \infty \qquad (5.4.16a)$$

This implies that the constant average acceleration method is stable for any Δt, no matter how large; however, it is accurate only if Δt is small enough, as discussed at the end of

Section 5.3. For $\gamma = \frac{1}{2}$ and $\beta = \frac{1}{6}$, Eq. (5.4.15) indicates that the linear acceleration method is stable if

$$\frac{\Delta t}{T_n} \leq 0.551 \tag{5.4.16b}$$

However, as in the case of the central difference method, this condition has little significance in the analysis of SDF systems because a much shorter time step than $0.551 T_n$ must be used to obtain an accurate representation of the excitation and response.

Example 5.3

Solve Example 5.1 by the constant average acceleration method using $\Delta t = 0.1$ sec.

Solution

1.0 *Initial calculations*

$$m = 0.2533 \qquad k = 10 \qquad c = 0.1592$$
$$u_0 = 0 \qquad \dot{u}_0 = 0 \qquad p_0 = 0$$

1.1 $\ddot{u}_0 = \dfrac{p_0 - c\dot{u}_0 - ku_0}{m} = 0.$

1.2 $\Delta t = 0.1.$

1.3 $a_1 = \dfrac{4}{(\Delta t)^2} m + \dfrac{2}{\Delta t} c = 104.5; \quad a_2 = \dfrac{4}{\Delta t} m + c = 10.29; \quad$ and

$\qquad a_3 = m = 0.2533.$

1.4 $\hat{k} = k + a_1 = 114.5.$

2.0 *Calculations for each time step,* $i = 0, 1, 2, \ldots$

2.1 $\hat{p}_{i+1} = p_{i+1} + a_1 u_i + a_2 \dot{u}_i + a_3 \ddot{u}_i = p_{i+1} + 104.5 u_i + 10.29 \dot{u}_i + 0.2533 \ddot{u}_i.$

2.2 $u_{i+1} = \dfrac{\hat{p}_{i+1}}{\hat{k}} = \dfrac{\hat{p}_{i+1}}{114.5}.$

2.3 $\dot{u}_{i+1} = \dfrac{2}{\Delta t}(u_{i+1} - u_i) - \dot{u}_i.$

2.4 $\ddot{u}_{i+1} = \dfrac{4}{(\Delta t)^2}(u_{i+1} - u_i) - \dfrac{4}{\Delta t}\dot{u}_i - \ddot{u}_i.$

3.0 *Repetition for the next time step.* Steps 2.1 to 2.4 are repeated for successive time steps and are summarized in Table E5.3, where the theoretical result (from Table E5.1a) is also included.

Example 5.4

Solve Example 5.1 by the linear acceleration method using $\Delta t = 0.1$ sec.

Solution

1.0 *Initial calculations*

$$m = 0.2533 \qquad k = 10 \qquad c = 0.1592$$
$$u_0 = 0 \qquad \dot{u}_0 = 0 \qquad p_0 = 0$$

TABLE E5.3 NUMERICAL SOLUTION BY CONSTANT AVERAGE ACCELERATION METHOD

t_i	p_i	$\hat{p}_i$ (Step 2.1)	$\ddot{u}_i$ (Step 2.4)	$\dot{u}_i$ (Step 2.3)	u_i (Step 2.2)	Theoretical u_i
0.0	0.0000		0.0000	0.0000	0.0000	0.0000
0.1	5.0000	5.0000	17.4666	0.8733	0.0437	0.0328
0.2	8.6603	26.6355	23.1801	2.9057	0.2326	0.2332
0.3	10.0000	70.0837	12.3719	4.6833	0.6121	0.6487
0.4	8.6603	123.9535	−11.5175	4.7260	1.0825	1.1605
0.5	5.0000	163.8469	−38.1611	2.2421	1.4309	1.5241
0.6	0.0000	162.9448	−54.6722	−2.3996	1.4230	1.4814
0.7	0.0000	110.1710	−33.6997	−6.8182	0.9622	0.9245
0.8	0.0000	21.8458	−2.1211	−8.6092	0.1908	0.0593
0.9	0.0000	−69.1988	28.4423	−7.2932	−0.6043	−0.7751
1.0	0.0000	−131.0066	47.3701	−3.5026	−1.1441	−1.2718

1.1 $\ddot{u}_0 = \dfrac{p_0 - c\dot{u}_0 - ku_0}{m} = 0.$

1.2 $\Delta t = 0.1.$

1.3 $a_1 = \dfrac{6}{(\Delta t)^2}m + \dfrac{3}{\Delta t}c = 156.8; \quad a_2 = \dfrac{6}{\Delta t}m + 2c = 15.52; \quad$ and

$a_3 = 2m + \dfrac{\Delta t}{2}c = 0.5146.$

1.4 $\hat{k} = k + a_1 = 166.8.$

2.0 *Calculations for each time step, $i = 0, 1, 2, \ldots$*

2.1 $\hat{p}_{i+1} = p_{i+1} + a_1 u_i + a_2 \dot{u}_i + a_3 \ddot{u}_i = p_{i+1} + 156.8u_i + 15.52\dot{u}_i + 0.5146\ddot{u}_i.$

2.2 $u_{i+1} = \dfrac{\hat{p}_{i+1}}{\hat{k}} = \dfrac{\hat{p}_{i+1}}{166.8}.$

2.3 $\dot{u}_{i+1} = \dfrac{3}{\Delta t}(u_{i+1} - u_i) - 2\dot{u}_i - \dfrac{\Delta t}{2}\ddot{u}_i.$

2.4 $\ddot{u}_{i+1} = \dfrac{6}{(\Delta t)^2}(u_{i+1} - u_i) - \dfrac{6}{\Delta t}\dot{u}_i - 2\ddot{u}_i.$

3.0 *Repetition for the next time step.* Steps 2.1 to 2.4 are repeated for successive time steps and are summarized in Table E5.4, where the theoretical result (from Table E5.1a) is also included.

Observe that the numerical results obtained by the linear acceleration method are closer to the theoretical solution (Table E5.4), hence more accurate, than those from the constant average acceleration method (Table E5.3).

TABLE E5.4   NUMERICAL SOLUTION BY LINEAR ACCELERATION METHOD

t_i	p_i	$\hat{p}_i$ (Step 2.1)	$\ddot{u}_i$ (Step 2.4)	$\dot{u}_i$ (Step 2.3)	u_i (Step 2.2)	Theoretical u_i
0.0	0.0000		0.0000	0.0000	0.0000	0.0000
0.1	5.0000	5.0000	17.9904	0.8995	0.0300	0.0328
0.2	8.6603	36.5748	23.6566	2.9819	0.2193	0.2332
0.3	10.0000	102.8221	12.1372	4.7716	0.6166	0.6487
0.4	8.6603	185.5991	−12.7305	4.7419	1.1130	1.1605
0.5	5.0000	246.4956	−39.9425	2.1082	1.4782	1.5241
0.6	0.0000	243.8733	−56.0447	−2.6911	1.4625	1.4814
0.7	0.0000	158.6538	−33.0689	−7.1468	0.9514	0.9245
0.8	0.0000	21.2311	0.4892	−8.7758	0.1273	0.0593
0.9	0.0000	−115.9590	31.9491	−7.1539	−0.6954	−0.7751
1.0	0.0000	−203.5678	50.1114	−3.0508	−1.2208	−1.2718

5.5 STABILITY AND COMPUTATIONAL ERROR

5.5.1 Stability

Numerical procedures that lead to bounded solutions if the time step is shorter than some stability limit are called *conditionally stable procedures*. Procedures that lead to bounded solutions regardless of the time-step length are called *unconditionally stable procedures*. The average acceleration method is unconditionally stable. The linear acceleration method is stable if $\Delta t/T_n < 0.551$, and the central difference method is stable if $\Delta t/T_n < 1/\pi$. Obviously, the latter two methods are conditionally stable.

The stability criteria are not restrictive (i.e., they do not dictate the choice of time step) in the analysis of SDF systems because $\Delta t/T_n$ must be considerably smaller than the stability limit (say, 0.1 or less) to ensure adequate accuracy in the numerical results. Stability of the numerical method is important, however, in the analysis of MDF systems, where it is often necessary to use unconditionally stable methods (Chapter 16).

5.5.2 Computational Error

Error is inherent in any numerical solution of the equation of motion. We do not discuss error analysis from a mathematical point of view. Rather, we examine two important characteristics of numerical solutions to develop a feel for the nature of the errors, and then mention a simple, useful way of managing error.

Consider the free vibration problem

$$m\ddot{u} + ku = 0 \qquad u(0) = 1 \quad \text{and} \quad \dot{u}(0) = 0$$

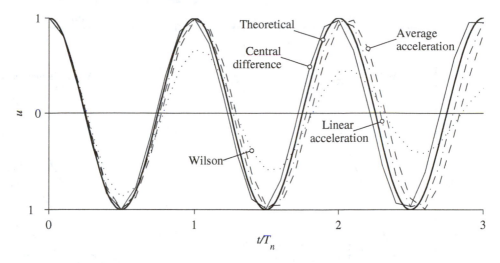

Figure 5.5.1   Free vibration solution by four numerical methods ($\Delta t/T_n = 0.1$) and the theoretical solution.

for which the theoretical solution is

$$u(t) = \cos \omega_n t \qquad (5.5.1)$$

This problem is solved by four numerical methods: central difference method, average acceleration method, linear acceleration method, and Wilson's method. The last of these methods is available elsewhere; see the references at the end of the chapter. The numerical results obtained using $\Delta t = 0.1 T_n$ are compared with the theoretical solution in Fig. 5.5.1. This comparison shows that some numerical methods may predict that the displacement amplitude decays with time, although the system is undamped, and that the natural period is elongated or shortened.

Figure 5.5.2 shows the amplitude decay AD and period elongation PE in the four numerical methods as a function of $\Delta t/T_n$; AD and PE are defined in part (b) of the figure. The mathematical analyses that led to these data are not presented, however. Three of the methods predict no decay of displacement amplitude. Wilson's method contains decay of amplitude, however, implying that this method introduces *numerical damping* in the system; the equivalent viscous damping ratio $\bar{\zeta}$ is shown in part (a) of the figure. Observe the rapid increase in the period error in the central difference method near $\Delta t/T_n = 1/\pi$, the stability limit for the method. The central difference method introduces the largest period error. In this sense it is the least accurate of the methods considered. For $\Delta t/T_n$ less than its stability limit, the linear acceleration method gives the least period elongation. This property, combined with no amplitude decay, makes this method the most suitable method (of the methods presented) for SDF systems. However,

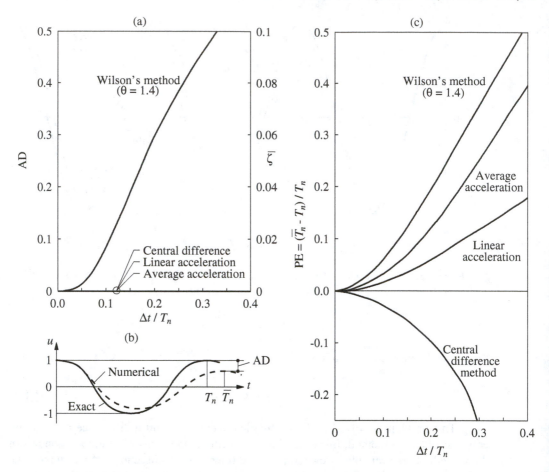

Figure 5.5.2 (a) Amplitude decay versus $\Delta t / T_n$; (b) definition of AD and PE; (c) period elongation versus $\Delta t / T_n$.

we shall arrive at a different conclusion for MDF systems because of stability requirements (Chapter 16).

The choice of time step also depends on the time variation of the dynamic excitation, in addition to the natural vibration period of the system. Figure 5.5.2 suggests that $\Delta t = 0.1 T_n$ would give reasonably accurate results. The time step should also be short enough to keep the distortion of the excitation function to a minimum. A very fine time step is necessary to describe numerically the highly irregular earthquake ground acceleration recorded during earthquakes; typically, $\Delta t = 0.02$ sec and the time step chosen for computing structural response should not be longer.

One useful, although unsophisticated technique for selecting the time step is to solve the problem with a time step that seems reasonable, then repeat the solution with a slightly smaller time step and compare the results, continuing the process until two successive solutions are close enough.

The preceding discussion of stability and accuracy applies strictly to linear systems. The reader should consult other references for how these issues affect nonlinear response analysis.

5.6 NONLINEAR SYSTEMS: CENTRAL DIFFERENCE METHOD

The dynamic response of a system beyond its linearly elastic range is generally not amenable to analytical solution even if the time variation of the excitation is described by a simple function. Numerical methods are therefore essential in the analysis of nonlinear systems. The central difference method can easily be adapted for solving the nonlinear equation of motion, Eq. (5.1.3), at time i. Substituting Eqs. (5.3.1), the central difference approximation for velocity and acceleration, gives Eq. (5.3.2) with ku_i replaced by $(f_S)_i$, which can be rewritten to obtain the following expression for response at time $i + 1$:

$$\hat{k}u_{i+1} = \hat{p}_i \tag{5.6.1}$$

where

$$\hat{k} = \frac{m}{(\Delta t)^2} + \frac{c}{2\Delta t} \tag{5.6.2}$$

and

$$\hat{p}_i = p_i - \left[\frac{m}{(\Delta t)^2} - \frac{c}{2\Delta t} \right] u_{i-1} + \frac{2m}{(\Delta t)^2} u_i - (f_S)_i \tag{5.6.3}$$

Comparing these equations with those for linear systems, it is seen that the only difference is in the definition for $\hat{p}_i$. With this modification Table 5.3.1 also applies to nonlinear systems.

The resisting force $(f_S)_i$ appears *explicitly*, as it depends only on the response at time i, not on the unknown response at time $i + 1$. Thus it is easily calculated, making the central difference method perhaps the simplest procedure for nonlinear systems.

5.7 NONLINEAR SYSTEMS: NEWMARK'S METHOD

In this section, Newmark's method described earlier for linear systems is extended to nonlinear systems. Recall that this method determines the solution at time $i + 1$ from the equilibrium condition at time $i + 1$, i.e., Eq. (5.1.4) for nonlinear systems. Because the resisting force $(f_S)_{i+1}$ is an implicit nonlinear function of the unknown u_{i+1}, iteration is required in this method. This requirement is typical of implicit methods. It is instructive first to develop the Newton–Raphson method of iteration for static analysis of a nonlinear SDF system.

5.7.1 Newton–Raphson Iteration

Dropping the inertia and damping terms in the equation of motion [Eq. (5.1.1)] gives the nonlinear equation to be solved in a static problem:

$$f_S(u) = p \tag{5.7.1}$$

The task is to determine the deformation u due to a given external force p, where the nonlinear force–deformation relation $f_S(u)$ is defined for the system to be analyzed.

Suppose that after j cycles of iteration, $u^{(j)}$ is an estimate of the unknown displacement and we are interested in developing an iterative procedure that provides an improved estimate $u^{(j+1)}$. For this purpose, expanding the resisting force $f_S^{(j+1)}$ in Taylor series about the known estimate $u^{(j)}$ gives

$$f_S^{(j+1)} = f_S^{(j)} + \left.\frac{\partial f_S}{\partial u}\right|_{u^{(j)}} \left(u^{(j+1)} - u^{(j)}\right) + \frac{1}{2} \left.\frac{\partial^2 f_S}{\partial u^2}\right|_{u^{(j)}} \left(u^{(j+1)} - u^{(j)}\right)^2 + \cdots \tag{5.7.2}$$

If $u^{(j)}$ is close to the solution, the change in u, $\Delta u^{(j)} = u^{(j+1)} - u^{(j)}$, will be small and the second- and higher-order terms can be neglected, leading to the linearized equation

$$f_S^{(j+1)} \simeq f_S^{(j)} + k_T^{(j)} \Delta u^{(j)} = p \tag{5.7.3}$$

or

$$k_T^{(j)} \Delta u^{(j)} = p - f_S^{(j)} = R^{(j)} \tag{5.7.4}$$

where $k_T^{(j)} = \left.\dfrac{\partial f_S}{\partial u}\right|_{u^{(j)}}$ is the tangent stiffness at $u^{(j)}$. Solving the linearized equation (5.7.4) gives $\Delta u^{(j)}$ and an improved estimate of the displacement:

$$u^{(j+1)} = u^{(j)} + \Delta u^{(j)} \tag{5.7.5}$$

The iterative procedure is described next with reference to Fig. 5.7.1. Associated with $u^{(j)}$ is the force $f_S^{(j)}$, which is not equal to the applied force p, and a residual force is defined: $R^{(j)} = p - f_S^{(j)}$. The additional displacement due to this residual force is determined from Eq. (5.7.4), leading to $u^{(j+1)}$. This new estimate of the solution is used to find a new value of the residual force $R^{(j+1)} = p - f_S^{(j+1)}$. The additional displacement $\Delta u^{(j+1)}$ due to this residual force is determined by solving

$$k_T^{(j+1)} \Delta u^{(j+1)} = R^{(j+1)} \tag{5.7.6}$$

This additional displacement is used to find a new value of the displacement:

$$u^{(j+2)} = u^{(j+1)} + \Delta u^{(j+1)} \tag{5.7.7}$$

and a new value of the residual force $R^{(j+2)}$, and the process is continued until convergence is achieved. This iterative process is known as the *Newton–Raphson method*.

Convergence rate. It can be proven that near the end of the iteration process the Newton–Raphson algorithm converges with quadratic rate to the exact solution u, i.e., $\left|u - u^{(j+1)}\right| \leq c \left|u - u^{(j)}\right|^2$, where c is a constant that depends on the second derivative

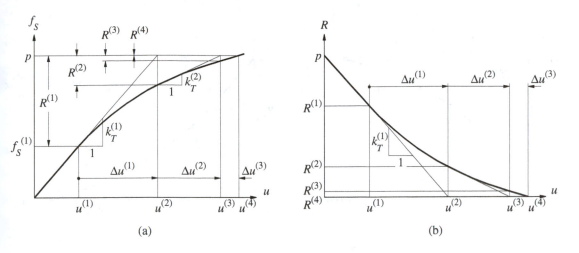

Figure 5.7.1   Newton–Raphson iteration: (a) applied and resisting forces; (b) residual force.

of the resisting force or the change in tangent stiffness. This result implies that near the solution the error in the $(j + 1)$th iterate (equal to the difference between u and $u^{(j+1)}$) is less than the square of the error in the previous iterate $u^{(j)}$.

Convergence criteria.   After each iteration the solution is checked and the iterative process is terminated when some measure of the error in the solution is less than a specified tolerance. Typically, one or more of the following convergence (or acceptance) criteria are enforced:

1. Residual force is less than a tolerance:

$$\left| R^{(j)} \right| \leq \varepsilon_R \tag{5.7.8a}$$

Conventional values for the tolerance ε_R range from 10^{-3} to 10^{-8}.

2. Change in displacement is less than a tolerance:

$$\left| \Delta u^{(j)} \right| \leq \varepsilon_u \tag{5.7.8b}$$

Conventional values for the tolerance ε_u range from 10^{-3} to 10^{-8}.

3. Incremental work done by the residual force acting through the change in displacement is less than a tolerance:

$$\tfrac{1}{2} \left| \Delta u^{(j)} R^{(j)} \right| \leq \varepsilon_w \tag{5.7.8c}$$

Tolerance ε_w must be at or near the computer (machine) tolerance because the left side is a product of small quantities.

Modified Newton–Raphson iteration.   To avoid computation of the tangent stiffness for each iteration, the initial stiffness at the beginning of a time step may be used

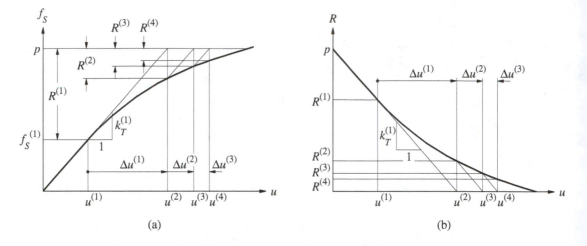

Figure 5.7.2 Modified Newton–Raphson iteration: (a) applied and resisting forces; (b) residual force.

as the constant stiffness for all iterations within the time step. This modified Newton–Raphson iteration is illustrated in Fig. 5.7.2, where it can be seen that convergence is now slower. At each iteration the residual force $R^{(j)}$ is now larger, as seen by comparing Figs. 5.7.1 and 5.7.2, and more iterations will be required to achieve convergence.

5.7.2 Newmark's Method

We have now developed Newton–Raphson iteration to solve a nonlinear equilibrium equation [e.g., Eq. (5.7.1)] that governs the static problem. In dynamic analysis the goal is to determine response quantities u_{i+1}, $\dot{u}_{i+1}$ and $\ddot{u}_{i+1}$ at time $i+1$ that satisfy Eq. (5.1.4), which can be written as

$$(\hat{f}_S)_{i+1} = p_{i+1} \tag{5.7.9}$$

where

$$(\hat{f}_S)_{i+1} = m\ddot{u}_{i+1} + c\dot{u}_{i+1} + (f_S)_{i+1} \tag{5.7.10}$$

By including the inertia and damping forces in defining the "resisting force" $\hat{f}_S$, the dynamic analysis equation (5.7.9) is of the same form as the static analysis equation (5.7.1).

Thus, we can adapt the Taylor series expansion of Eq. (5.7.2) to Eq. (5.7.9), interpret $(\hat{f}_S)_{i+1}$ as a function of u_{i+1}, and drop the second- and higher-order terms to obtain an equation analogous to Eq. (5.7.3):

$$(\hat{f}_S)_{i+1}^{(j+1)} \simeq (\hat{f}_S)_{i+1}^{(j)} + \frac{\partial \hat{f}_S}{\partial u_{i+1}} \Delta u^{(j)} = p_{i+1} \tag{5.7.11}$$

where

$$\Delta u^{(j)} = u_{i+1}^{(j+1)} - u_{i+1}^{(j)} \tag{5.7.12}$$

Differentiating Eq. (5.7.10) at the known displacement $u_{i+1}^{(j)}$ gives

$$\frac{\partial \hat{f}_S}{\partial u_{i+1}} = m\frac{\partial \ddot{u}}{\partial u_{i+1}} + c\frac{\partial \dot{u}}{\partial u_{i+1}} + \frac{\partial f_S}{\partial u_{i+1}}$$

where the derivatives in inertia and damping terms on the right side can be determined from Eqs. (5.4.8) and (5.4.9), respectively, which were derived from Newmark's equation (5.4.1):

$$\frac{\partial \ddot{u}}{\partial u_{i+1}} = \frac{1}{\beta(\Delta t)^2} \qquad \qquad \frac{\partial \dot{u}}{\partial u_{i+1}} = \frac{\gamma}{\beta \Delta t}$$

Putting together the preceding two equations and recalling the definition of tangent stiffness (Section 5.7.1) gives

$$(\hat{k}_T)_{i+1}^{(j)} \equiv \frac{\partial \hat{f}_S}{\partial u_{i+1}} = (\hat{k}_T)_{i+1}^{(j)} + \frac{\gamma}{\beta \Delta t}c + \frac{1}{\beta(\Delta t)^2}m \tag{5.7.13}$$

With the preceding definition of $(\hat{k}_T)_{i+1}^{(j)}$, Eq. (5.7.11) can be written as

$$(\hat{k}_T)_{i+1}^{(j)}\Delta u^{(j)} = p_{i+1} - (\hat{f}_S)_{i+1}^{(j)} \equiv \hat{R}_{i+1}^{(j)} \tag{5.7.14}$$

Substituting Eqs. (5.4.8) and (5.4.9) in Eq. (5.7.10) and then combining it with the right side of Eq. (5.7.14) leads to the following expression for the residual force:

$$\hat{R}_{i+1}^{(j)} = p_{i+1} - (f_S)_{i+1}^{(j)} - \left[\frac{1}{\beta(\Delta t)^2}m + \frac{\gamma}{\beta \Delta t}c\right]\left(u_{i+1}^{(j)} - u_i\right) + \left[\frac{1}{\beta \Delta t}m + \left(\frac{\gamma}{\beta} - 1\right)c\right]\dot{u}_i$$

$$+ \left[\left(\frac{1}{2\beta} - 1\right)m + \Delta t\left(\frac{\gamma}{2\beta} - 1\right)c\right]\ddot{u}_i \tag{5.7.15}$$

Note that the linearized equation (5.7.14) for the jth iteration in dynamic analysis is similar in form to the corresponding equation (5.7.4) in static analysis. However, there is an important difference in the two equations in that damping and inertia terms are now included in both the tangent stiffness $\hat{k}_T$ (Eq. 5.7.13) and the residual force $\hat{R}$ (Eq. 5.7.15). The first, fourth, and fifth terms on the right side of Eq. (5.7.15) do not change from one iteration to the next. The second and third terms need to be updated with every new estimate of displacement $u_{i+1}^{(j)}$ during iteration.

Equation (5.7.14) provides the basis for the Newton–Raphson iteration method, summarized in step 3.0 of Table 5.7.1. Once u_{i+1} is determined, the rest of the computation proceeds as for linear systems; in particular, $\ddot{u}_{i+1}$ and $\dot{u}_{i+1}$ are determined from Eqs. (5.4.8) and (5.4.9), respectively. Table 5.7.1 summarizes Newmark's algorithm as it might be implemented on the computer.

TABLE 5.7.1   NEWMARK'S METHOD: NONLINEAR SYSTEMS

Special cases

   (1) Average acceleration method ($\gamma = \frac{1}{2}, \beta = \frac{1}{4}$)

   (2) Linear acceleration method ($\gamma = \frac{1}{2}, \beta = \frac{1}{6}$)

1.0 *Initial calculations*

    1.1 State determination: $(f_S)_0$ and $(k_T)_0$.

    1.2 $\ddot{u}_0 = \dfrac{p_0 - c\dot{u}_0 - (f_S)_0}{m}$.

    1.3 Select Δt.

    1.4 $a_1 = \dfrac{1}{\beta(\Delta t)^2}m + \dfrac{\gamma}{\beta \Delta t}c; \quad a_2 = \dfrac{1}{\beta \Delta t}m + \left(\dfrac{\gamma}{\beta} - 1\right)c; \quad$ and

$$a_3 = \left(\dfrac{1}{2\beta} - 1\right)m + \Delta t\left(\dfrac{\gamma}{2\beta} - 1\right)c.$$

2.0 *Calculations for each time instant, $i = 0, 1, 2, \ldots$*

    2.1 Initialize $j = 1, u_{i+1}^{(j)} = u_i, (f_S)_{i+1}^{(j)} = (f_S)_i$, and $(k_T)_{i+1}^{(j)} = (k_T)_i$.

    2.2 $\hat{p}_{i+1} = p_{i+1} + a_1 u_i + a_2 \dot{u}_i + a_3 \ddot{u}_i$.

3.0 *For each iteration, $j = 1, 2, 3 \ldots$*

    3.1 $\hat{R}_{i+1}^{(j)} = \hat{p}_{i+1} - (f_S)_{i+1}^{(j)} - a_1 u_{i+1}^{(j)}$.

    3.2 Check convergence; If the acceptance criteria are not met, implement steps 3.3 to 3.7; otherwise, skip these steps and go to step 4.0.

    3.3 $(\hat{k}_T)_{i+1}^{(j)} = (k_T)_{i+1}^{(j)} + a_1$.

    3.4 $\Delta u^{(j)} = \hat{R}_{i+1}^{(j)} \div (\hat{k}_T)_{i+1}^{(j)}$.

    3.5 $u_{i+1}^{(j+1)} = u_{i+1}^{(j)} + \Delta u^{(j)}$.

    3.6 State determination: $(f_S)_{i+1}^{(j+1)}$ and $(k_T)_{i+1}^{(j+1)}$.

    Replace j by $j + 1$ and repeat steps 3.1 to 3.6; denote final value as u_{i+1}.

4.0 *Calculations for velocity and acceleration*

    4.1 $\dot{u}_{i+1} = \dfrac{\gamma}{\beta \Delta t}(u_{i+1} - u_i) + \left(1 - \dfrac{\gamma}{\beta}\right)\dot{u}_i + \Delta t\left(1 - \dfrac{\gamma}{2\beta}\right)\ddot{u}_i$.

    4.2 $\ddot{u}_{i+1} = \dfrac{1}{\beta(\Delta t)^2}(u_{i+1} - u_i) - \dfrac{1}{\beta \Delta t}\dot{u}_i - \left(\dfrac{1}{2\beta} - 1\right)\ddot{u}_i$.

5.0 *Repetition for next time step.* Replace i by $i + 1$ and implement steps 2.0 to 4.0 for the next time step.

Example 5.5

An SDF system has the same properties as in Example 5.1, except that the restoring force–deformation relation is elastoplastic with yield deformation $u_y = 0.75$ in. and yield force $f_y = 7.5$ kips (Fig. E5.5). Determine the response $u(t)$ of this system (starting from rest) to the half-cycle sine pulse force in Fig. E5.1 using the constant average acceleration method with $\Delta t = 0.1$ sec and Newton–Raphson iteration.

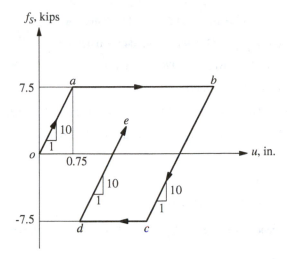

Figure E5.5

Solution

1.0 *Initial calculations*

$$m = 0.2533 \qquad k = 10 \qquad c = 0.1592$$

$$u_0 = 0 \qquad \dot{u}_0 = 0 \qquad p_0 = 0$$

1.1 State determination: $(f_S)_0 = 0$ and $(k_T)_0 = k = 10$.

1.2 $\ddot{u}_0 = \dfrac{p_0 - c\dot{u}_0 - (f_S)_0}{m} = 0.$

1.3 $\Delta t = 0.1.$

1.4 $a_1 = \dfrac{4}{(\Delta t)^2} m + \dfrac{2}{\Delta t} c = 104.5040; \qquad a_2 = \dfrac{4}{\Delta t} m + c = 10.2912; \quad$ and

$\qquad a_3 = m = 0.2533.$

As an example, the calculations of steps 2.0, 3.0 and 4.0 in Table 5.7.1, are implemented as follows for the time step that begins at 0.3 sec and ends at 0.4 sec.

2.0 *Calculations for* $i = 3$

2.1 Initialize $j = 1$

$\qquad u_{i+1}^{(1)} = u_i = 0.6121$, $(f_S)_{i+1}^{(1)} = (f_S)_i = 6.1206$, and $(k_T)_{i+1}^{(1)} = (k_T)_i = 10.$

2.2 $\hat{p}_{i+1} = p_{i+1} + 104.5u_i + 10.29\dot{u}_i + 0.2533\ddot{u}_i = 123.9535.$

3.0 *First iteration,* $j = 1$

3.1 $\hat{R}_{i+1}^{(1)} = \hat{p}_{i+1} - (f_S)_{i+1}^{(1)} - 104.5u_{i+1}^{(1)}$

$\qquad = 123.9535 - 6.1206 - 63.9630 = 53.8698.$

3.2 Check of convergence: Because $\left|\hat{R}_{i+1}^{(1)}\right| = 53.8698$ exceeds $\varepsilon_R = 10^{-3}$, chosen for this example, implement steps 3.3 to 3.7.

3.3 $(\hat{k}_T)_{i+1}^{(1)} = (k_T)_{i+1}^{(1)} + a_1 = 10 + 104.5040 = 114.5040.$

3.4 $\Delta u^{(1)} = \hat{R}_{i+1}^{(1)} \div (\hat{k}_T)_{i+1}^{(1)} = 53.8698 \div 114.5040 = 0.4705.$

3.5 $u_{i+1}^{(2)} = u_{i+1}^{(1)} + \Delta u^{(1)} = 0.6121 + 0.4705 = 1.0825.$

3.6 State determination: $(f_S)_{i+1}^{(2)}$ and $(k_T)_{i+1}^{(2)}$

$(f_S)_{i+1}^{(2)} = (f_S)_i + k(u_{i+1}^{(2)} - u_i) = 6.1206 + (10 \times 0.4705) = 10.8253.$

Because $(f_S)_{i+1}^{(2)} > f_y$, $(f_S)_{i+1}^{(2)} = f_y = 7.5$ and $(k_T)_{i+1}^{(2)} = 0.$

3.0 *Second iteration, $j = 2$*

3.1 $\hat{R}_{i+1}^{(2)} = \hat{p}_{i+1} - (f_S)_{i+1}^{(2)} - 104.5u_{i+1}^{(2)}$

$= 123.9535 - 7.5 - 113.1282 = 3.3253.$

3.2 Check of convergence: Because $\left| \hat{R}_{i+1}^{(2)} \right| = 3.3253$ exceeds ε_R, implement steps 3.3 to 3.7.

3.3 $(\hat{k}_T)_{i+1}^{(2)} = (k_T)_{i+1}^{(2)} + a_1 = 0 + 104.5040 = 104.5040.$

3.4 $\Delta u^{(2)} = \hat{R}_{i+1}^{(2)} \div (\hat{k}_T)_{i+1}^{(2)} = 3.3253 \div 104.5040 = 0.0318.$

3.5 $u_{i+1}^{(3)} = u_{i+1}^{(2)} + \Delta u^{(2)} = 1.0825 + 0.0318 = 1.1143.$

3.6 State determination: $(f_S)_{i+1}^{(3)}$ and $(k_T)_{i+1}^{(3)}$

$(f_S)_{i+1}^{(3)} = (f_S)_i + k(u_{i+1}^{(3)} - u_i) = 6.1206 + (10 \times 0.5023) = 11.1434.$

Because $(f_S)_{i+1}^{(3)} > f_y$, $(f_S)_{i+1}^{(3)} = f_y = 7.5$ and $(k_T)_{i+1}^{(3)} = 0.$

3.0 *Third iteration, $j = 3$*

3.1 $\hat{R}_{i+1}^{(3)} = \hat{p}_{i+1} - (f_S)_{i+1}^{(3)} - 104.5u_{i+1}^{(3)}$

$= 123.9535 - 7.5 - 116.4535 = 0.$

3.2 Check of convergence: Because $\left| \hat{R}_{i+1}^{(3)} \right| = 0$ is less than ε_R, skip steps 3.3 to 3.7;

set $u_4 = u_4^{(3)} = 1.1143.$

4.0 *Calculations for velocity and acceleration*

4.1 $\dot{u}_{i+1} = \dfrac{2}{\Delta t}(u_{i+1} - u_i) - \dot{u}_i = \dfrac{2}{0.1}(1.1143 - 0.6121) - 4.683 = 5.3624.$

4.2 $\ddot{u}_{i+1} = \dfrac{4}{(\Delta t)^2}(u_{i+1} - u_i) - \dfrac{4}{\Delta t}\dot{u}_i - \ddot{u}_i$

$= \dfrac{4}{(0.1)^2}(1.1143 - 0.6121) - \dfrac{4}{0.1}4.6833 - 12.3719 = 1.2103.$

These calculations for the time step 0.3 to 0.4 sec are summarized in Table E5.5.

5.0 *Repetition for next time step.* After replacing i by $i + 1$, steps 2.0 to 4.0 are repeated for successive time steps and are summarized in Table E5.5.

TABLE E5.5 NUMERICAL SOLUTION BY CONSTANT AVERAGE ACCELERATION METHOD WITH NEWTON–RAPHSON ITERATION

t_i	p_i	$\hat{R}_i$ or $\hat{R}_i^{(j)}$	$(k_T)_i$ or $(k_T)_i^{(j)}$	$(\hat{k}_T)_i$ or $(\hat{k}_T)_i^{(j)}$	$\Delta u^{(j)}$	u_i or $u_i^{(j+1)}$	$(f_S)_i^{(j+1)}$	$\dot{u}_i$	$\ddot{u}_i$
0.0	0.0000		10			0.0000		0.0000	0.0000
0.1	5.0000	5.0000	10	114.504	0.0437	0.0437	0.4367	0.8733	17.4666
0.2	8.6603	21.6355	10	114.504	0.1889	0.2326	2.3262	2.9057	23.1801
0.3	10.0000	43.4481	10	114.504	0.3794	0.6121	6.1206	4.6833	12.3719
0.4	8.6603	53.8698	10	114.504	0.4705	1.0825	7.5000		
		3.3253	0	104.504	0.0318	1.1143	7.5000	5.3624	1.2103
0.5	5.0000	55.9918	0	104.504	0.5071	1.6214	7.5000	4.7792	−12.8735
0.6	0.0000	38.4230	0	104.504	0.3677	1.9891	7.5000	2.5742	−31.2270
0.7	0.0000	11.0816	0	104.504	0.1060	2.0951	7.5000	−0.4534	−29.3242
0.8	0.0000	−19.5936	0	104.504	−0.1875	1.9076	5.6251		
		1.8749	10	114.504	0.0164	1.9240	5.7888	−2.9690	−20.9876
0.9	0.0000	−41.6593	10	114.504	−0.3638	1.5602	2.1506	−4.3075	−5.7830
1.0	0.0000	−47.9448	10	114.504	−0.4187	1.1415	−2.0366	−4.0668	10.5962

During the next three time steps (after 0.4 sec), the system is on the yielding branch *ab*. In other words, the stiffness $k_i = 0$ remains constant, and no iteration is necessary. Between 0.6 and 0.7 sec the velocity changes sign from positive to negative, implying that the deformation begins to decrease, the system begins to unload along the branch *bc*, and the stiffness $k_i = 10$. However, we have ignored this change during the time step, implying that the system stays on the branch *ab* and no iteration is necessary.

The computation for the time step starting at 0.6 sec can be made more accurate by finding, by a process of iteration, the time instant at which $\dot{u} = 0$. Then the calculations can be carried out with stiffness $k_i = 0$ over the first part of the time step and with $k_i = 10$ over the second part of the time step. Alternatively, a smaller time step can be used for improved accuracy.

Note that the solution over a time step is not exact because equilibrium is satisfied only at the beginning and end of the time step, not at all time instants within the time step. This implies that the energy balance equation (Chapter 7) is violated. The discrepancy in energy balance, usually calculated at the end of the excitation, is an indication of the error in the numerical solution.

Example 5.6

Repeat Example 5.5 using modified Newton–Raphson iteration within each time step of $\Delta t = 0.1$ sec.

Solution The procedure of Table 5.7.1 is modified to use the initial stiffness at the beginning of a time step as the constant stiffness for all iterations within the time step. The computations in steps 1.0 and 2.0 are identical to those presented in Example 5.5, but step 3.0 is now different. To illustrate these differences, step 3.0 in the modified Table 5.7.1 is implemented for the time step that begins at 0.3 sec and ends at 0.4 sec.

3.0 *First iteration, $j = 1$*

3.1 $\hat{R}^{(1)}_{i+1} = \hat{p}_{i+1} - (fs)^{(1)}_{i+1} - 104.5\, u^{(1)}_{i+1}$

$\qquad = 123.9535 - 6.1206 - 63.9630 = 53.8698.$

3.2 Check of convergence: Because $\left|\hat{R}^{(1)}_{i+1}\right| = 53.8698$ exceeds ε_R, implement steps 3.3 to 3.7.

3.3 $(\hat{k}_T)_{i+1} = (k_T)_{i+1} + a_1 = 10 + 104.5040 = 114.5040.$

3.4 $\Delta u^{(1)} = \hat{R}^{(1)}_{i+1} \div (\hat{k}_T)_{i+1} = 53.8698 \div 114.5040 = 0.4705.$

3.5 $u^{(2)}_{i+1} = u^{(1)}_{i+1} + \Delta u^{(1)} = 0.6121 + 0.4705 = 1.0825.$

3.6 State determination: $(fs)^{(2)}_{i+1}$

$\qquad (fs)^{(2)}_{i+1} = (fs)_i + k\left(u^{(2)}_{i+1} - u_i\right) = 6.1206 + (10 \times 0.4705) = 10.8253.$

$\qquad$ Because $(fs)^{(2)}_{i+1} > f_y$, $(fs)^{(2)}_{i+1} = f_y = 7.5.$

3.0 *Second iteration $j = 2$*

3.1 $\hat{R}^{(2)}_{i+1} = \hat{p}_{i+1} - (fs)^{(2)}_{i+1} - 104.5 u^{(2)}_{i+1}$

$\qquad = 123.9535 - 7.5 - 113.1282 = 3.3253.$

3.2 Check of convergence: Because $\left|\hat{R}^{(2)}_{i+1}\right| = 3.3253$ exceeds ε_R, implement steps 3.3 to 3.7.

3.3 $(\hat{k}_T)_{i+1} = 114.5040.$

3.4 $\Delta u^{(2)} = \hat{R}^{(2)}_{i+1} \div (\hat{k}_T)_{i+1} = 3.3253 \div 114.5040 = 0.0290.$

3.5 $u^{(3)}_{i+1} = u^{(2)}_{i+1} + \Delta u^{(2)} = 1.0825 + 0.0290 = 1.1116.$

3.6 State determination: $(fs)^{(3)}_{i+1}$

$\qquad (fs)^{(3)}_{i+1} = (fs)_i + k(u^{(3)}_{i+1} - u_i) = 6.1206 + (10 \times 0.5000) = 11.1157.$

$\qquad$ Because $(fs)^{(3)}_{i+1} > f_y$, $(fs)^{(3)}_{i+1} = f_y = 7.5.$

3.0 *Third iteration, $j = 3$*

3.1 $\hat{R}^{(3)}_{i+1} = \hat{p}_{i+1} - (fs)^{(3)}_{i+1} - 104.5 u^{(3)}_{i+1}$

$\qquad = 123.9535 - 7.5 - 116.1631 = 0.2904.$

3.2 Check of convergence: Because $\left|\hat{R}^{(3)}_{i+1}\right| = 0.2904$ exceeds ε_R, implement steps 3.3 to 3.7.

3.3 $(\hat{k}_T)_{i+1} = 114.5040.$

3.4 $\Delta u^{(3)} = \hat{R}^{(3)}_{i+1} \div (\hat{k}_T)_{i+1} = 0.2904 \div 114.5040 = 0.0025.$

3.5 $u_{i+1}^{(4)} = u_{i+1}^{(3)} + \Delta u^{(3)} = 1.1116 + 0.0025 = 1.1141.$

3.6 State determination: $(fs)_{i+1}^{(4)}$

$(fs)_{i+1}^{(4)} = (fs)_i + k(u_{i+1}^{(4)} - u_i) = 6.1206 + (10 \times 0.5020) = 11.1410.$

Because $(fs)_{i+1}^{(4)} > f_y$, $(fs)_{i+1}^{(4)} = f_y = 7.5.$

These calculations and those for additional iterations during the time step 0.3 to 0.4 sec are shown in Table E5.6.

TABLE E5.6 NUMERICAL SOLUTION BY CONSTANT AVERAGE ACCELERATION METHOD WITH MODIFIED NEWTON–RAPHSON ITERATION

t_i	p_i	$\hat{R}_i$ or $\hat{R}_i^{(j)}$	$(k_T)_i$ or $(k_T)_i^{(j)}$	$(\hat{k}_T)_i$ or $(\hat{k}_T)_i^{(j)}$	$\Delta u^{(j)}$	u_i or $u_i^{(j+1)}$	$(fs)_i^{(j+1)}$	$\dot{u}_i$	$\ddot{u}_i$
0.0	0.0000		10			0.0000		0.0000	0.0000
0.1	5.0000	5.0000	10	114.504	0.0437	0.0437	0.4367	0.8733	17.4666
0.2	8.6603	21.6355	10	114.504	0.1889	0.2326	2.3262	2.9057	23.1801
0.3	10.0000	43.4481	10	114.504	0.3794	0.6121	6.1206	4.6833	12.3719
0.4	8.6603	53.8698	10	114.504	0.4705	1.0825	7.5000		
		3.3253			0.02904	1.1116	7.5000		
		0.2904			2.536E-3	1.1141	7.5000		
		2.536E-2			2.215E-4	1.1143	7.5000		
		2.215E-3			1.934E-5	1.1143	7.5000	5.3623	1.2095
0.5	5.0000	55.9912	0	104.504	0.5071	1.6214	7.5000	4.7791	−12.8734
0.6	0.0000	38.4222	0	104.504	0.3677	1.9891	7.5000	2.5741	−31.2270
0.7	0.0000	11.0810	0	104.504	0.1060	2.0951	7.5000	−0.4534	−29.3242
0.8	0.0000	−19.5936	0	104.504	−0.1875	1.9076	5.6250		
		1.8750			1.794E-2	1.9256	5.8044		
		−0.1794			−1.717E-3	1.9238	5.7873		
		1.717E-2			1.643E-4	1.9240	5.7889		
		1.643E-3			−1.572E-5	1.9240	5.7888	−2.9690	−20.9879
0.9	0.0000	−41.6600	10	114.504	−0.3638	1.5602	2.1505	−4.3076	−5.7824
1.0	0.0000	−47.9451	10	114.504	−0.4187	1.1414	−2.0367	−4.0668	10.5969

The original Newton–Raphson iteration converges more rapidly than the modified Newton–Raphson iteration, as is apparent by comparing Tables E5.5 and E5.6 that summarize results from the two methods, respectively. Observe the following: (1) The results of the first iteration are identical in the two cases because both use the initial tangent stiffness. Consequently, the resisting force $(fs)_{i+1}^{(2)}$ and the residual force $\hat{R}_{i+1}^{(2)}$ are identical. (2) By using the current tangent stiffness $(k_T)_{i+1}^{(2)}$ and the associated value of $(\hat{k}_T)_{i+1}^{(2)}$ from Eq. (5.7.13) in the second iteration, the original Newton–Raphson method leads to a smaller residual force

$\hat{R}_{i+1}^{(3)} = 0$ (Example 5.5) compared to $\hat{R}_{i+1}^{(3)} = 0.2904$ from the modified Newton–Raphson method (Example 5.6). (3) Because at each iteration the residual force $\hat{R}_{i+1}^{(j)}$ is now smaller, convergence is achieved in fewer iterations; for this time step of this example, two iterations are required in the original Newton–Raphson method (Example 5.5) compared to five iterations in the modified Newton–Raphson method (Example 5.6).

FURTHER READING

Bathe, K.-J., *Finite Element Procedures,* Prentice Hall, Englewood Cliffs, N.J., 1996, Chapter 9.

Filippou, F. C., and Fenves, G. L., "Methods of Analysis for Earthquake-Resistant Structures," in: *Earthquake Engineering: From Engineering Seismology to Performance-Based Engineering* (eds. Y. Bozorgnia and V. V. Bertero), CRC Press, New York, 2004, Chapter 6.

Hughes, T. J. R., *The Finite Element Method,* Prentice Hall, Englewood Cliffs, N.J., 1987, Chapter 9.

Humar, J. L., *Dynamics of Structures,* 2nd ed., A. A. Balkema Publishers, Lisse, The Netherlands, 2002, Chapter 8.

Newmark, N. M., "A Method of Computation for Structural Dynamics," *Journal of the Engineering Mechanics Division, ASCE,* **85**, 1959, pp. 67–94.

PROBLEMS

5.1 In Section 5.2 we developed recurrence formulas for numerical solution of the equation of motion of a linear SDF system based on linear interpolation of the forcing function $p(t)$ over each time step. Develop a similar procedure using a piecewise-constant representation of the forcing function wherein the value of the force in the interval t_i to t_{i+1} is a constant equal to $\tilde{p}_i$ (Fig. P5.1). Show that the recurrence formulas for the response of an undamped system are

$$u_{i+1} = u_i \cos(\omega_n \, \Delta t_i) + \dot{u}_i \frac{\sin(\omega_n \, \Delta t_i)}{\omega_n} + \frac{\tilde{p}_i}{k}[1 - \cos(\omega_n \, \Delta t_i)]$$

$$\dot{u}_{i+1} = u_i[-\omega_n \sin(\omega_n \, \Delta t_i)] + \dot{u}_i \cos(\omega_n \, \Delta t_i) + \frac{\tilde{p}_i}{k}\omega_n \sin(\omega_n \, \Delta t_i)$$

Specialize the recurrence formulas for the following definition of the piecewise-constant force: $\tilde{p}_i = (p_i + p_{i+1})/2$. Write the recurrence formulas in the following form:

$$u_{i+1} = Au_i + B\dot{u}_i + Cp_i + Dp_{i+1}$$

$$\dot{u}_{i+1} = A'u_i + B'\dot{u}_i + C'p_i + D'p_{i+1}$$

with equations for the constants $A, B, C, \dots, D'$.

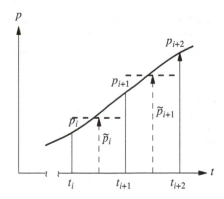

Figure P5.1

*5.2   Solve Example 5.1 using the piecewise-constant approximation of the forcing function; neglect damping in the SDF system.

*5.3   Solve the problem in Example 5.1 by the central difference method, implemented by a computer program in a language of your choice, using $\Delta t = 0.1$ sec. Note that this problem was solved as Example 5.2 and that the results were presented in Table E5.2.

*5.4   Repeat Problem 5.3 using $\Delta t = 0.05$ sec. How does the time step affect the accuracy of the solution?

*5.5   An SDF system has the same mass and stiffness as in Example 5.1, but the damping ratio is $\zeta = 20\%$. Determine the response of this system to the excitation of Example 5.1 by the central difference method using $\Delta t = 0.05$ sec. Plot the response as a function of time, compare with the solution of Problem 5.3, and comment on how damping affects the peak response.

*5.6   Solve the problem in Example 5.1 by the central difference method using $\Delta t = \frac{1}{3}$ sec. Carry out your solution to 2 sec, and comment on what happens to the solution and why.

*5.7   Solve the problem in Example 5.1 by the constant average acceleration method, implemented by a computer program in a language of your choice, using $\Delta t = 0.1$ sec. Note that this problem was solved as Example 5.3, and the results are presented in Table E5.3. Compare these results with those of Example 5.2, and comment on the relative accuracy of the constant average acceleration and central difference methods.

*5.8   Repeat Problem 5.7 using $\Delta t = 0.05$ sec. How does the time step affect the accuracy of the solution?

*5.9   Solve the problem in Example 5.1 by the constant average acceleration method using $\Delta t = \frac{1}{3}$ sec. Carry out the solution to 2 sec, and comment on the accuracy and stability of the solution.

*5.10   Solve the problem of Example 5.1 by the linear acceleration method, implemented by a computer program in a language of your choice, using $\Delta t = 0.1$ sec. Note that this problem was solved as Example 5.4 and that the results are presented in Table E5.4. Compare with the solution of Example 5.3, and comment on the relative accuracy of the constant average acceleration and linear acceleration methods.

*5.11   Repeat Problem 5.10 using $\Delta t = 0.05$ sec. How does the time step affect the accuracy of the solution?

*Denotes that a computer is necessary to solve this problem.

*5.12   Solve the problem of Example 5.5 by the central difference method, implemented by a computer program in a language of your choice, using $\Delta t = 0.05$ sec.

*5.13   Solve Example 5.5 by the constant average acceleration method with Newton–Raphson iteration, implemented by a computer program in a language of your choice. Note that this problem was solved as Example 5.5 and the results were presented in Table E5.5.

*5.14   Solve Example 5.6 by the constant average acceleration method with modified Newton–Raphson iteration, implemented by a computer program in a language of your choice. Note that this problem was solved as Example 5.6 and the results were presented in Table E5.6.

*5.15   Solve Example 5.5 by the linear acceleration method with Newton–Raphson iteration using $\Delta t = 0.1$ sec.

*5.16   Solve Example 5.5 by the linear acceleration method with modified Newton–Raphson iteration using $\Delta t = 0.1$ sec.

*Denotes that a computer is necessary to solve this problem.

6

Earthquake Response of Linear Systems

PREVIEW

One of the most important applications of the theory of structural dynamics is in analyzing the response of structures to ground shaking caused by an earthquake. In this chapter we study the earthquake response of linear SDF systems to earthquake motions. By definition, linear systems are elastic systems, and we shall also refer to them as linearly elastic systems to emphasize both properties. Because earthquakes can cause damage to many structures, we are also interested in the response of yielding or inelastic systems, the subject of Chapter 7.

The first part of this chapter is concerned with the earthquake response—deformation, internal element forces, stresses, and so on—of simple structures as a function of time and how this response depends on the system parameters. Then we introduce the response spectrum concept, which is central to earthquake engineering, together with procedures to determine the peak response of systems directly from the response spectrum. This is followed by a study of the characteristics of earthquake response spectra, which leads into the design spectrum for the design of new structures and safety evaluation of existing structures against future earthquakes. The important distinctions between design and response spectra are identified and the chapter closes with a discussion of two types of response spectra that are not used commonly.

6.1 EARTHQUAKE EXCITATION

For engineering purposes the time variation of ground acceleration is the most useful way of defining the shaking of the ground during an earthquake. The ground acceleration $\ddot{u}_g(t)$ appears on the right side of the differential equation (1.7.4) governing the response of structures to earthquake excitation. Thus, for given ground acceleration the problem to be solved is defined completely for an SDF system with known mass, stiffness, and damping properties.

The basic instrument to record three components of ground shaking during earthquakes is the strong-motion accelerograph (Fig. 6.1.1), which does not record continuously but is triggered into motion by the first waves of the earthquake to arrive. This is because even in earthquake-prone regions such as California and Japan, there may not be

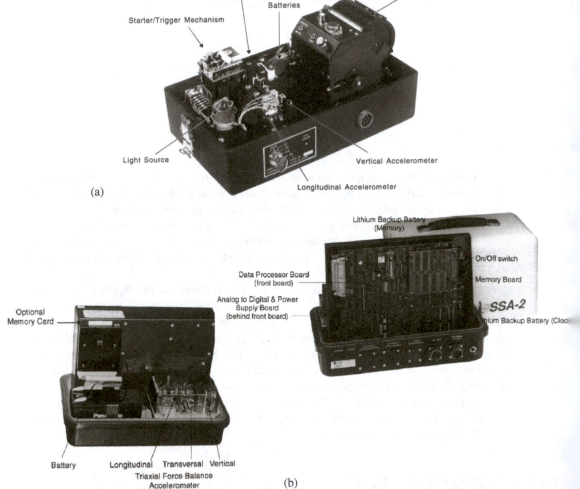

Figure 6.1.1 Strong motion accelerographs: (a) SMA-1, an analog-recording instrument with undamped natural frequency of 25 Hz and damping 60% of critical; (b) SSA-2, a digital recording instrument with undamped natural frequency of 50 Hz and damping 70% of critical. (Courtesy of Kinemetrics, Inc.)

any strong ground motion from earthquakes to record for months, or even years, at a time. Consequently, continual recordings of hundreds of such instruments would be a wasteful exercise. After triggering, the recording continues for some minutes or until the ground shaking falls again to imperceptible levels. Clearly, the instruments must be regularly maintained and serviced so that they produce a record when shaking occurs.

The basic element of an accelerograph is a transducer element, which in its simplest form is an SDF mass–spring–damper system (Section 3.7). Therefore, the transducer element is characterized by its natural frequency f_n and viscous damping ratio ζ; typically, $f_n = 25$ Hz and $\zeta = 60\%$ for modern analog accelerographs; and $f_n = 50$ Hz and $\zeta = 70\%$ in modern digital accelerographs.[†] These transducer parameters enable the digital instrument to record, without excessive distortion, acceleration–time functions containing frequencies from very low up to, say, 30 Hz; the analog instrument is accurate over a narrower frequency range, say, up to 15 Hz.

Ideally, many stations should be instrumented prior to an earthquake to record the ground motions. However, not knowing when and exactly where earthquakes will occur and having limited budgets for installation and maintenance of instruments, it is not always possible to obtain such recordings in the region of strongest shaking. For example, no strong-motion records were obtained from two earthquakes that caused much destruction: Killari, Maharashtra, India, September 30, 1993; and Guam, a U.S. territory, August 8, 1993; only one record resulted from the devastating earthquake in Haiti, January 12, 2010. In contrast, an earthquake in Japan or California, two well-instrumented regions, can be expected to produce a large number of records. For example, the magnitude 9.0 Tohoku earthquake on March 11, 2011, near the east coast of Honshu, Japan, produced several hundred records of strong shaking.

The first strong-motion accelerogram was recorded during the Long Beach earthquake of 1933, and as of April 2011, over 3000 records have now been obtained. As might be expected, most of these records are of small motion and only a fraction of them have acceleration of 20% g or more. The geographical distribution of these ground motion records is very uneven. A large majority of them are from California, Japan, and Taiwan; most of the intense records are from six earthquakes: the San Fernando earthquake of February 9, 1971, the Loma Prieta earthquake of October 17, 1989, and the Northridge earthquake of January 17, 1994, in California; the Kobe earthquake of January 16, 1995, and the Tohoku earthquake of March 11, 2011, in Japan; and the Chi-Chi earthquake of September 20, 1999 in Taiwan. The peak values of accelerations recorded at many different locations during the Loma Prieta earthquake are shown in Fig. 6.1.2. These acceleration values are largest near the epicenter of the earthquake and tend to decrease with distance from the fault causing the earthquake. However, the accelerations recorded at similar distances may vary significantly because of several factors, especially local soil conditions.

Figure 6.1.3 shows a collection of representative acceleration–time records of earthquake ground motions in the region of strong shaking. One horizontal component is given

---

[†]It should be noted that most if not all of the digital accelerographs use a force-balance type of transducer, for which two parameters will not completely define the instrument response, which is that of a higher-order (than a mass–spring–damper) system.

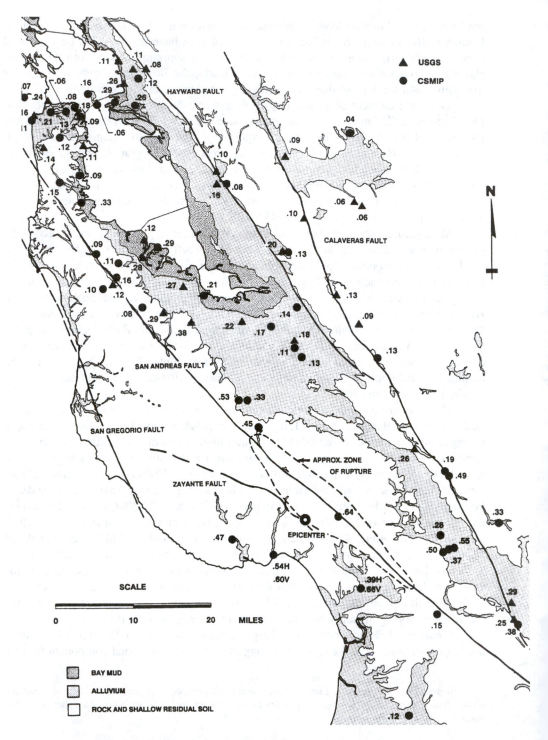

Figure 6.1.2 Peak horizontal ground accelerations recorded during the Loma Prieta earthquake of October 17, 1989. (Courtesy of R. B. Seed.)

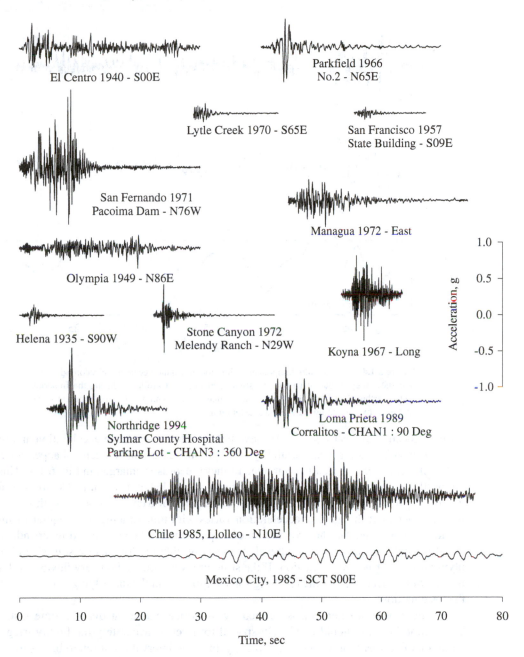

Figure 6.1.3 Ground motions recorded during several earthquakes. [Based in part on Hudson (1979).]

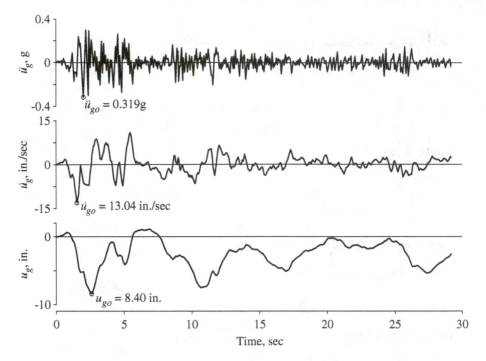

Figure 6.1.4 North–south component of horizontal ground acceleration recorded at the Imperial Valley Irrigation District substation, El Centro, California, during the Imperial Valley earthquake of May 18, 1940. The ground velocity and ground displacement were computed by integrating the ground acceleration.

for each location and earthquake. All have been plotted to the same acceleration and time scale. The wide and very real variability of amplitude, duration, and general appearance of different records can be clearly seen. One of these records is enlarged in Fig. 6.1.4. This is the north–south component of the ground motion recorded at a site in El Centro, California, during the Imperial Valley, California, earthquake of May 18, 1940.[†] At this scale it becomes apparent that ground acceleration varies with time in a highly irregular manner. No matter how irregular, the ground motion is presumed to be known and independent of the structural response. This is equivalent to saying that the foundation soil is rigid, implying no soil–structure interaction. If the structure were founded on very flexible soil, the motion of the structure and the resulting forces imposed on the underlying soil can modify the base motion.

    The ground acceleration is defined by numerical values at discrete time instants. These time instants should be closely spaced to describe accurately the highly irregular variation of acceleration with time. Typically, the time interval is chosen to be $\frac{1}{100}$ to $\frac{1}{50}$ of a second, requiring 1500 to 3000 ordinates to describe the ground motion of Fig. 6.1.4.

[†]This ground acceleration is used extensively in this book and, for brevity, will be called *El Centro ground motion*, although three components of motion have been recorded at the same site during several earthquakes after 1940.

The top curve in Fig. 6.1.4 shows the variation of El Centro ground acceleration with time. The peak ground acceleration $\ddot{u}_{go}$ is 0.319g. The second curve is the ground velocity, obtained by integrating the acceleration–time function. The peak ground velocity $\dot{u}_{go}$ is 13.04 in./sec. Integration of ground velocity provides the ground displacement, presented as the lowest trace. The peak ground displacement u_{go} is 8.40 in. It is difficult to determine accurately the ground velocity and displacement because analog accelerographs do not record the initial part—until the accelerograph is triggered—of the acceleration–time function, and thus the base (zero acceleration) line is unknown. Digital accelerographs overcome this problem by providing a short memory so that the onset of ground motion is measured.

In existence are several different versions of the El Centro ground motion. The variations among them arise from differences in (1) how the original analog trace of acceleration versus time was digitized into numerical data, and (2) the procedure chosen to introduce the missing baseline in the record. The version shown in Fig. 6.1.4 is used throughout this book and is tabulated in Appendix 6.

6.2 EQUATION OF MOTION

Equation (1.7.4) governs the motion of a linear SDF system (Fig. 6.2.1) subjected to ground acceleration $\ddot{u}_g(t)$. Dividing this equation by m gives

$$\ddot{u} + 2\zeta\omega_n\dot{u} + \omega_n^2 u = -\ddot{u}_g(t) \tag{6.2.1}$$

It is clear that for a given $\ddot{u}_g(t)$, the deformation response $u(t)$ of the system depends only on the natural frequency ω_n or natural period T_n of the system and its damping ratio, ζ; writing formally, $u \equiv u(t, T_n, \zeta)$. Thus any two systems having the same values of T_n and ζ will have the same deformation response $u(t)$ even though one system may be more massive than the other or one may be stiffer than the other.

Ground acceleration during earthquakes varies irregularly to such an extent (see Fig. 6.1.4) that analytical solution of the equation of motion must be ruled out. Therefore, numerical methods are necessary to determine the structural response, and any of the methods presented in Chapter 5 could be used. The response results presented in this chapter

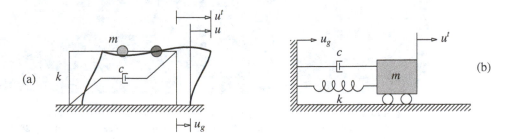

Figure 6.2.1   Single-degree-of-freedom systems.

were obtained by exact solution of the equation of motion for the ground motion varying linearly over every time step, $\Delta t = 0.02$ sec (Section 5.2).

6.3 RESPONSE QUANTITIES

Of greatest interest in structural engineering is the deformation of the system, or displacement $u(t)$ of the mass relative to the moving ground, to which the internal forces are linearly related. These are the bending moments and shears in the beams and columns of the one-story frame of Fig. 6.2.1a or the spring force in the system of Fig. 6.2.1b. Knowing the total displacement $u^t(t)$ of the mass would be useful in providing enough separation between adjacent buildings to prevent their pounding against each other during an earthquake. Pounding is the cause of damage to several buildings during almost every earthquake (see Fig. 6.3.1). Similarly, the total acceleration $\ddot{u}^t(t)$ of the mass would be needed if the structure is supporting sensitive equipment and the motion imparted to the equipment is to be determined.

The numerical solution of Eq. (6.2.1) can be implemented to provide results for relative quantities $u(t)$, $\dot{u}(t)$, and $\ddot{u}(t)$ as well as total quantities $u^t(t)$, $\dot{u}^t(t)$, and $\ddot{u}^t(t)$.

Figure 6.3.1 Two images of pounding damage to the Sanborns Building (shorter) and 33 Reforma Avenue Building (taller), Mexico City due to Mexico earthquake of July 28, 1957. (From the Steinbrugge Collection, National Information Service for Earthquake Engineering, University of California, Berkeley.)

6.4 RESPONSE HISTORY

For a given ground motion $\ddot{u}_g(t)$, the deformation response $u(t)$ of an SDF system depends only on the natural vibration period of the system and its damping ratio. Figure 6.4.1a shows the deformation response of three different systems due to El Centro ground acceleration. The damping ratio, $\zeta = 2\%$, is the same for the three systems, so that only the differences in their natural periods are responsible for the large differences in the deformation responses. It is seen that the time required for an SDF system to complete a cycle of vibration when subjected to this earthquake ground motion is very close to the natural period of the system. (This interesting result, valid for typical ground motions containing a wide range of frequencies, can be proven using random vibration theory, not included in this book.) The peak deformation [Eq. (1.11.1)] is also noted in each case. Observe that among these three systems, the longer the vibration period, the greater the peak deformation. As will be seen later, this trend is neither perfect nor valid over the entire range of periods.

      Figure 6.4.1b shows the deformation response of three systems to the same ground motion. The vibration period T_n is the same for the three systems, so that the differences

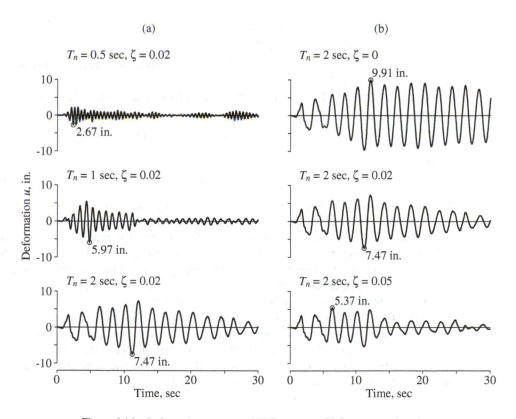

Figure 6.4.1 Deformation response of SDF systems to El Centro ground motion.

in their deformation responses are associated with their damping. We observe the expected trend that systems with more damping respond less than lightly damped systems. Because the natural period of the three systems is the same, their responses display a similarity in the time required to complete a vibration cycle and in the times the maxima and minima occur.

Once the deformation response history $u(t)$ has been evaluated by dynamic analysis of the structure, the internal forces can be determined by static analysis of the structure at each time instant. Two methods to implement such analysis were mentioned in Section 1.8. Between them, the preferred approach in earthquake engineering is based on the concept of the *equivalent static force* f_S (Fig. 6.4.2) because it can be related to earthquake forces specified in building codes; f_S was defined in Eq. (1.8.1), which is repeated here for convenience:

$$f_S(t) = ku(t) \tag{6.4.1}$$

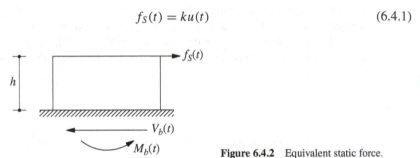

Figure 6.4.2   Equivalent static force.

where k is the lateral stiffness of the frame (Fig. 6.2.1a). Expressing k in terms of the mass m gives

$$f_S(t) = m\,\omega_n^2\,u(t) = mA(t) \tag{6.4.2}$$

where

$$A(t) = \omega_n^2\,u(t) \tag{6.4.3}$$

Observe that the equivalent static force is m times $A(t)$, the *pseudo-acceleration*, not m times the total acceleration $\ddot{u}^t(t)$. This distinction is discussed in Section 6.6.3.

The pseudo-acceleration response $A(t)$ of the system can readily be computed from the deformation response $u(t)$. For the three systems with $T_n = 0.5$, 1, and 2 sec, all having $\zeta = 0.02$, $u(t)$ is available in Fig. 6.4.1. Multiplying each $u(t)$ by the corresponding $\omega_n^2 = (2\pi/T_n)^2$ gives the pseudo-acceleration responses for these systems; they are presented in Fig. 6.4.3, where the peak value is noted for each system.

For the one-story frame the internal forces (e.g., the shears and moments in the columns and beam, or stress at any location) can be determined at a selected instant of time by static analysis of the structure subjected to the equivalent static lateral force $f_S(t)$ at the same time instant (Fig. 6.4.2). Thus a static analysis of the structure would be necessary at each time instant when the responses are desired. In particular, the base shear $V_b(t)$ and the base overturning moment $M_b(t)$ are

$$V_b(t) = f_S(t) \qquad M_b(t) = hf_S(t) \tag{6.4.4a}$$

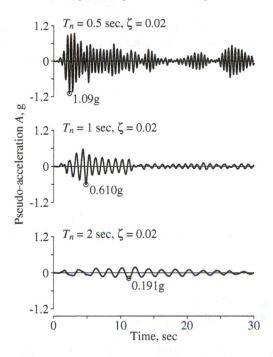

Figure 6.4.3   Pseudo-acceleration response of SDF systems to El Centro ground motion.

where h is the height of the mass above the base. We put Eq. (6.4.2) into these equations to obtain

$$V_b(t) = mA(t) \qquad M_b(t) = hV_b(t) \tag{6.4.4b}$$

If the SDF system is viewed as a mass–spring–damper system (Fig. 6.2.1b), the notion of equivalent static force is not necessary. One can readily visualize that the spring force is given by Eq. (6.4.1).

6.5 RESPONSE SPECTRUM CONCEPT

G. W. Housner was instrumental in the widespread acceptance of the concept of the earthquake response spectrum—initiated by M. A. Biot in 1932—as a practical means of characterizing ground motions and their effects on structures. Now a central concept in earthquake engineering, the response spectrum provides a convenient means to summarize the peak response of all possible linear SDF systems to a particular component of ground motion. It also provides a practical approach to applying the knowledge of structural dynamics to the design of structures and development of lateral force requirements in building codes.

A plot of the peak value of a response quantity as a function of the natural vibration period T_n of the system, or a related parameter such as circular frequency ω_n or cyclic frequency f_n, is called the *response spectrum* for that quantity. Each such plot is for SDF systems having a fixed damping ratio ζ, and several such plots for different values of ζ are included to cover the range of damping values encountered in actual structures. Whether the peak response is plotted against f_n or T_n is a matter of personal preference. We have chosen the latter because engineers prefer to use natural period rather than natural frequency because the period of vibration is a more familiar concept and one that is intuitively appealing.

A variety of response spectra can be defined depending on the response quantity that is plotted. Consider the following peak responses:

$$u_o(T_n, \zeta) \equiv \max_t |u(t, T_n, \zeta)|$$

$$\dot{u}_o(T_n, \zeta) \equiv \max_t |\dot{u}(t, T_n, \zeta)|$$

$$\ddot{u}_o^t(T_n, \zeta) \equiv \max_t |\ddot{u}^t(t, T_n, \zeta)|$$

The *deformation response spectrum* is a plot of u_o against T_n for fixed ζ. A similar plot for $\dot{u}_o$ is the *relative velocity response spectrum*, and for $\ddot{u}_o^t$ is the *acceleration response spectrum*.

6.6 DEFORMATION, PSEUDO-VELOCITY, AND PSEUDO-ACCELERATION RESPONSE SPECTRA

In this section the deformation response spectrum and two related spectra, the pseudo-velocity and pseudo-acceleration response spectra, are discussed. As shown in Section 6.4, only the deformation $u(t)$ is needed to compute internal forces. Obviously, then, the deformation spectrum provides all the information necessary to compute the peak values of deformation $D \equiv u_o$ and internal forces. The pseudo-velocity and pseudo-acceleration response spectra are included, however, because they are useful in studying characteristics of response spectra, constructing design spectra, and relating structural dynamics results to building codes.

6.6.1 Deformation Response Spectrum

Figure 6.6.1 shows the procedure to determine the deformation response spectrum. The spectrum is developed for El Centro ground motion, shown in part (a) of this figure. The time variation of the deformation induced by this ground motion in three SDF systems is presented in part (b). For each system the peak value of deformation $D \equiv u_o$ is determined from the deformation history. (Usually, the peak occurs during ground shaking; however, for lightly damped systems with very long periods the peak response may occur during the free vibration phase after the ground shaking has stopped.) The peak deformations are $D = 2.67$ in. for a system with natural period $T_n = 0.5$ sec and damping ratio $\zeta = 2\%$; $D = 5.97$ in. for a system with $T_n = 1$ sec and $\zeta = 2\%$; and $D = 7.47$ in. for a system with $T_n = 2$ sec and $\zeta = 2\%$. The D value so determined for each system provides one point on the deformation response spectrum; these three values of D are identified in Fig. 6.6.1c. Repeating such computations for a range of values of T_n while keeping ζ constant at 2% provides the deformation response spectrum shown in Fig. 6.6.1c. As we shall show

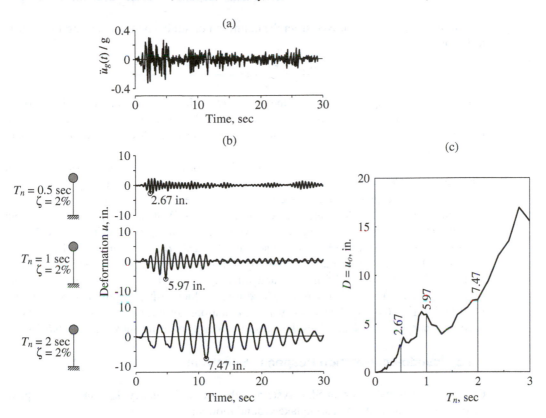

Figure 6.6.1 (a) Ground acceleration; (b) deformation response of three SDF systems with $\zeta = 2\%$ and $T_n = 0.5$, 1, and 2 sec; (c) deformation response spectrum for $\zeta = 2\%$.

later, the complete response spectrum includes such spectrum curves for several values of damping.

6.6.2 Pseudo-velocity Response Spectrum

Consider a quantity V for an SDF system with natural frequency ω_n related to its peak deformation $D \equiv u_o$ due to earthquake ground motion:

$$V = \omega_n D = \frac{2\pi}{T_n} D \qquad (6.6.1)$$

The quantity V has units of velocity. It is related to the peak value of strain energy E_{So} stored in the system during the earthquake by the equation

$$E_{So} = \frac{mV^2}{2} \qquad (6.6.2)$$

This relationship can be derived from the definition of strain energy and using Eq. (6.6.1) as follows:

$$E_{So} = \frac{ku_o^2}{2} = \frac{kD^2}{2} = \frac{k(V/\omega_n)^2}{2} = \frac{mV^2}{2}$$

The right side of Eq. (6.6.2) is the kinetic energy of the structural mass m with velocity V, called the peak *pseudo-velocity*. The prefix *pseudo* is used because V is not equal to the peak relative velocity $\dot{u}_o$, although it has the correct units. We return to this matter in Section 6.12.

The *pseudo-velocity response spectrum* is a plot of V as a function of the natural vibration period T_n, or natural vibration frequency f_n, of the system. For the ground motion of Fig. 6.6.1a, the peak pseudo-velocity V for a system with natural period T_n can be determined from Eq. (6.6.1) and the peak deformation D of the same system available from the response spectrum of Fig. 6.6.1c, which has been reproduced in Fig. 6.6.2a. As an example, for a system with $T_n = 0.5$ sec and $\zeta = 2\%$, $D = 2.67$ in.; from Eq. (6.6.1), $V = (2\pi/0.5)2.67 = 33.7$ in./sec. Similarly, for a system with $T_n = 1$ sec and the same ζ, $V = (2\pi/1)5.97 = 37.5$ in./sec; and for a system with $T_n = 2$ sec and the same ζ, $V = (2\pi/2)7.47 = 23.5$ in./sec. These three values of peak pseudo-velocity V are identified in Fig. 6.6.2b. Repeating such computations for a range of values of T_n while keeping ζ constant at 2% provides the pseudo-velocity spectrum shown in Fig. 6.6.2b.

6.6.3 Pseudo-acceleration Response Spectrum

Consider a quantity A for an SDF system with natural frequency ω_n related to its peak deformation $D \equiv u_o$ due to earthquake ground motion:

$$A = \omega_n^2 D = \left(\frac{2\pi}{T_n}\right)^2 D \tag{6.6.3}$$

The quantity A has units of acceleration and is related to the peak value of base shear V_{bo} [or the peak value of the equivalent static force f_{So}, Eq. (6.4.4a)]:

$$V_{bo} = f_{So} = mA \tag{6.6.4}$$

This relationship is simply Eq. (6.4.4b) specialized for the time of peak response with the peak value of $A(t)$ denoted by A. The peak base shear can be written in the form

$$V_{bo} = \frac{A}{g}w \tag{6.6.5}$$

where w is the weight of the structure and g the gravitational acceleration. When written in this form, A/g may be interpreted as the *base shear coefficient* or *lateral force coefficient*. It is used in building codes to represent the coefficient by which the structural weight is multiplied to obtain the base shear.

Observe that the base shear is equal to the inertia force associated with the mass m undergoing acceleration A. This quantity defined by Eq. (6.6.3) is generally different from the peak acceleration $\ddot{u}_o^t$ of the system. It is for this reason that we call A the *peak*

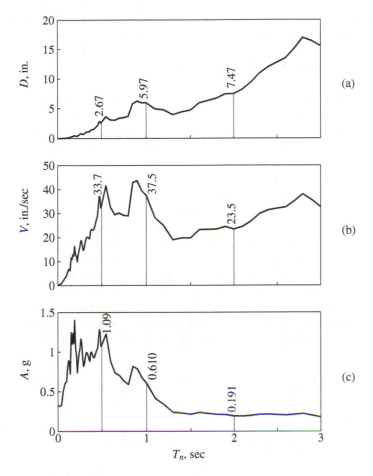

Figure 6.6.2 Response spectra ($\zeta = 0.02$) for El Centro ground motion: (a) deformation response spectrum; (b) pseudo-velocity response spectum; (c) pseudo-acceleration response spectrum.

pseudo-acceleration; the prefix *pseudo* is used to avoid possible confusion with the true peak acceleration $\ddot{u}_o^t$. We return to this matter in Section 6.12.

The *pseudo-acceleration response spectrum* is a plot of A as a function of the natural vibration period T_n, or natural vibration frequency f_n, of the system. For the ground motion of Fig. 6.6.1a, the peak pseudo-acceleration A for a system with natural period T_n and damping ratio ζ can be determined from Eq. (6.6.3), and the peak deformation D of the system from the spectrum of Fig. 6.6.2a. As an example, for a system with $T_n = 0.5$ sec and $\zeta = 2\%$, $D = 2.67$ in.; from Eq. (6.6.3), $A = (2\pi/0.5)^2 2.67 = 1.09$g, where g $= 386$ in./sec^2. Similarly, for a system with $T_n = 1$ sec and the same ζ, $A = (2\pi/1)^2 5.97 = 0.610$g; and for a system with $T_n = 2$ sec and the same ζ, $A = (2\pi/2)^2 7.47 = 0.191$g. Note that the same values for A are also available as the peak values of $A(t)$ presented in Fig. 6.4.3. These three values of peak pseudo-acceleration are identified in Fig. 6.6.2c.

Repeating such computations for a range of values of T_n while keeping ζ constant at 2% provides the pseudo-acceleration spectrum shown in Fig. 6.6.2c.

6.6.4 Combined *D–V–A* Spectrum

Each of the deformation, pseudo-velocity, and pseudo-acceleration response spectra for a given ground motion contains the same information, no more and no less. The three spectra are simply different ways of presenting the same information on structural response. Knowing one of the spectra, the other two can be obtained by algebraic operations using Eqs. (6.6.1) and (6.6.3).

   Why do we need three spectra when each of them contains the same information? One of the reasons is that each spectrum directly provides a physically meaningful quantity. The deformation spectrum provides the peak deformation of a system. The pseudo-velocity spectrum is related directly to the peak strain energy stored in the system during the earthquake; see Eq. (6.6.2). The pseudo-acceleration spectrum is related directly to the peak value of the equivalent static force and base shear; see Eq. (6.6.4). The second reason lies in the fact that the shape of the spectrum can be approximated more readily for design purposes with the aid of all three spectral quantities rather than any one of them alone; see Sections 6.8 and 6.9. For this purpose a combined plot showing all three of the spectral quantities is especially useful. This type of plot was developed for earthquake response spectra, apparently for the first time, by A. S. Veletsos and N. M. Newmark in 1960.

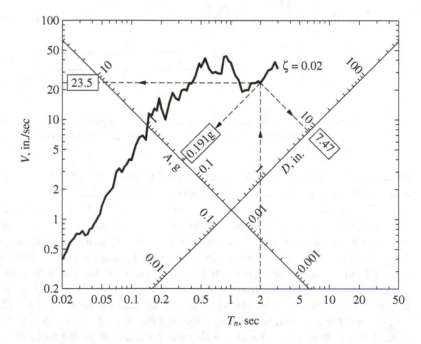

Figure 6.6.3   Combined *D–V–A* response spectrum for El Centro ground motion; $\zeta = 2\%$.

This integrated presentation is possible because the three spectral quantities are interrelated by Eqs. (6.6.1) and (6.6.3), rewritten as

$$\frac{A}{\omega_n} = V = \omega_n D \quad \text{or} \quad \frac{T_n}{2\pi} A = V = \frac{2\pi}{T_n} D \qquad (6.6.6)$$

Observe the similarity between these equations relating D, V, and A and Eq. (3.2.21) for the dynamic response factors R_d, R_v, and R_a for an SDF system subjected to harmonic excitation. Equation (3.2.21) permitted presentation of R_d, R_v, and R_a, all together, on four-way logarithmic paper (Fig. 3.2.8), constructed by the procedure described in Appendix 3 (Chapter 3). Similarly, the graph paper shown in Fig. A6.1 (Appendix 6) with four-way logarithmic scales can be constructed to display D, V, and A, all together. The vertical and horizontal scales for V and T_n are standard logarithmic scales. The two scales for D and A sloping at $+45°$ and $-45°$, respectively, to the T_n-axis are also logarithmic scales but not identical to the vertical scale; see Appendix 3.

Once this graph paper has been constructed, the three response spectra—deformation, pseudo-velocity, and pseudo-acceleration—of Fig. 6.6.2 can readily be combined into a single plot. The pairs of numerical data for V and T_n that were plotted in Fig. 6.6.2b on

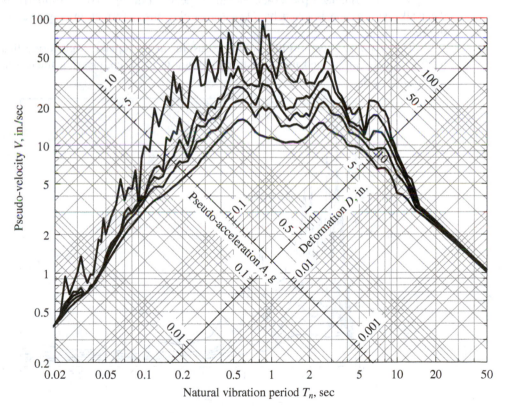

Figure 6.6.4   Combined D–V–A response spectrum for El Centro ground motion; $\zeta = 0$, 2, 5, 10, and 20%.

linear scales are replotted in Fig. 6.6.3 on logarithmic scales. For a given natural period T_n, the D and A values can be read from the diagonal scales. As an example, for $T_n = 2$ sec, Fig. 6.6.3 gives $D = 7.47$ in. and $A = 0.191$g. (Actually, these numbers cannot be read so accurately from the graph; in this case they were available from Fig. 6.6.2.) The four-way plot is a compact presentation of the three—deformation, pseudo-velocity, and pseudo-acceleration—response spectra, for a single plot of this form replaces the three plots of Fig. 6.6.2.

A response spectrum should cover a wide range of natural vibration periods and several damping values so that it provides the peak response of all possible structures. The period range in Fig. 6.6.3 should be extended because tall buildings and long-span bridges, among other structures, may have longer vibration periods (Fig. 2.1.2), and several damping values should be included to cover the practical range of $\zeta = 0$ to 20%. Figure 6.6.4 shows spectrum curves for $\zeta = 0, 2, 5, 10,$ and 20% over the period range 0.02 to 50 sec. This, then, is the response spectrum for the north–south component of ground motion recorded at one location during the Imperial Valley earthquake of May 18, 1940. Because the lateral force or base shear for an SDF system is related through Eq. (6.6.5) to A/g, we also plot this normalized pseudo-acceleration spectrum in Fig. 6.6.5. Similarly, because the peak deformation is given by D, we also plot this deformation response spectrum in Fig. 6.6.6.

The response spectrum has proven so useful in earthquake engineering that spectra for virtually all ground motions strong enough to be of engineering interest are now

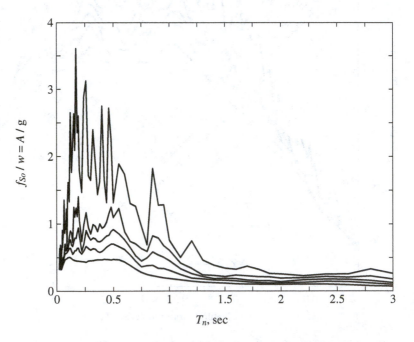

Figure 6.6.5 Normalized pseudo-acceleration, or base shear coefficient, response spectrum for El Centro ground motion; $\zeta = 0, 2, 5, 10,$ and 20%.

computed and published soon after they are recorded. Enough of them have been obtained to give us a reasonable idea of the kind of motion that is likely to occur in future earthquakes, and how response spectra are affected by distance to the causative fault, local soil conditions, and regional geology.

6.6.5 Construction of Response Spectrum

The response spectrum for a given ground motion component $\ddot{u}_g(t)$ can be developed by implementation of the following steps:

1. Numerically define the ground acceleration $\ddot{u}_g(t)$; typically, the ground motion ordinates are defined every 0.02 sec.
2. Select the natural vibration period T_n and damping ratio ζ of an SDF system.
3. Compute the deformation response $u(t)$ of this SDF system due to the ground motion $\ddot{u}_g(t)$ by any of the numerical methods described in Chapter 5. [In obtaining the

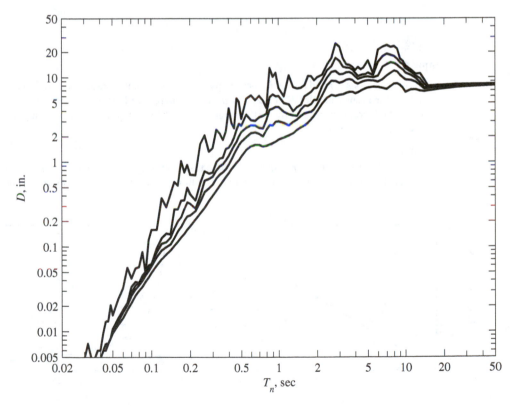

Figure 6.6.6 Deformation response spectrum for El Centro ground motion; $\zeta = 0, 2, 5, 10,$ and 20%.

responses shown in Fig. 6.6.1, the exact solution of Eq. (6.2.1) for ground motion assumed to be piecewise linear over every $\Delta t = 0.02$ sec was used; see Section 5.2.]

4. Determine u_o, the peak value of $u(t)$.
5. The spectral ordinates are $D = u_o$, $V = (2\pi/T_n)D$, and $A = (2\pi/T_n)^2 D$.
6. Repeat steps 2 to 5 for a range of T_n and ζ values covering all possible systems of engineering interest.
7. Present the results of steps 2 to 6 graphically to produce three separate spectra like those in Fig. 6.6.2 or a combined spectrum like the one in Fig. 6.6.4.

Considerable computational effort is required to generate an earthquake response spectrum. A complete dynamic analysis to determine the time variation (or history) of the deformation of an SDF system provides the data for one point on the spectrum corresponding to the T_n and ζ of the system. Each curve in the response spectrum of Fig. 6.6.4 was produced from such data for 112 values of T_n unevenly spaced over the range $T_n = 0.02$ to 50 sec.

Example 6.1

Derive equations for and plot deformation, pseudo-velocity, and pseudo-acceleration response spectra for ground acceleration $\ddot{u}_g(t) = \dot{u}_{go}\delta(t)$, where $\delta(t)$ is the Dirac delta function and $\dot{u}_{go}$ is the increment in velocity, or the magnitude of the acceleration impulse. Only consider systems without damping.

Solution

1. *Determine the response history.* The response of an SDF system to $p(t) = \delta(t - \tau)$ is available in Eq. (4.1.6). Adapting that solution to $p_{eff}(t) = -m\ddot{u}_g(t) = -m\dot{u}_{go}\delta(t)$

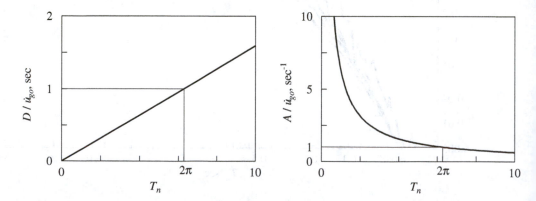

Figure E6.1

gives

$$u(t) = -\frac{\dot{u}_{go}}{\omega_n} \sin \omega_n t \qquad \text{(a)}$$

The peak value of $u(t)$ is

$$u_o = \frac{\dot{u}_{go}}{\omega_n} \qquad \text{(b)}$$

2. *Determine the spectral values.*

$$D \equiv u_o = \frac{\dot{u}_{go}}{\omega_n} = \frac{\dot{u}_{go}}{2\pi} T_n \qquad \text{(c)}$$

$$V = \omega_n D = \dot{u}_{go} \qquad A = \omega_n^2 D = \frac{2\pi \dot{u}_{go}}{T_n} \qquad \text{(d)}$$

Two of these response spectra are plotted in Fig. E6.1.

6.7 PEAK STRUCTURAL RESPONSE FROM THE RESPONSE SPECTRUM

If the response spectrum for a given ground motion component is available, the peak value of deformation or of an internal force in any linear SDF system can be determined readily. This is the case because the computationally intensive dynamic analyses summarized in Section 6.6.5 have already been completed in generating the response spectrum. Corresponding to the natural vibration period T_n and damping ratio ζ of the system, the values of D, V, or A are read from the spectrum, such as Fig. 6.6.6, 6.6.4, or 6.6.5. Now all response quantities of interest can be expressed in terms of D, V, or A and the mass or stiffness properties of the system. In particular, the peak deformation of the system is

$$u_o = D = \frac{T_n}{2\pi} V = \left(\frac{T_n}{2\pi}\right)^2 A \qquad (6.7.1)$$

and the peak value of the equivalent static force f_{So} is [from Eqs. (6.6.4) and (6.6.3)]

$$f_{So} = kD = mA \qquad (6.7.2)$$

Static analysis of the one-story frame subjected to lateral force f_{So} (Fig. 6.7.1) provides the internal forces (e.g., shears and moments in columns and beams). This involves application of well-known procedures of static structural analysis, as will be illustrated later

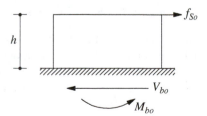

Figure 6.7.1   Peak value of equivalent static force.

by examples. We emphasize again that no further dynamic analysis is required beyond that necessary to determine $u(t)$. In particular, the peak values of shear and overturning moment at the base of the one-story structure are

$$V_{bo} = kD = mA \qquad M_{bo} = hV_{bo} \qquad\qquad (6.7.3)$$

We note that any one of these response spectra—deformation, pseudo-velocity, or pseudo-acceleration—is sufficient for computing the peak deformations and forces required in structural design. For such applications the velocity or acceleration spectra (defined in Section 6.5) are not required, but for completeness we discuss these spectra briefly at the end of this chapter.

Example 6.2

A 12-ft-long vertical cantilever, a 4-in.-nominal-diameter standard steel pipe, supports a 5200-lb weight attached at the tip as shown in Fig. E6.2. The properties of the pipe are: outside diameter, $d_o = 4.500$ in., inside diameter $d_i = 4.026$ in., thickness $t = 0.237$ in., and second moment of cross-sectional area, $I = 7.23$ in^4, elastic modulus $E = 29,000$ ksi, and weight $= 10.79$ lb/foot length. Determine the peak deformation and bending stress in the cantilever due to the El Centro ground motion. Assume that $\zeta = 2\%$.

Solution   The lateral stiffness of this SDF system is

$$k = \frac{3EI}{L^3} = \frac{3(29 \times 10^3)7.23}{(12 \times 12)^3} = 0.211 \text{ kip/in.}$$

The total weight of the pipe is $10.79 \times 12 = 129.5$ lb, which may be neglected relative to the lumped weight of 5200 lb. Thus

$$m = \frac{w}{g} = \frac{5.20}{386} = 0.01347 \text{ kip-sec}^2/\text{in.}$$

The natural vibration frequency and period of the system are

$$\omega_n = \sqrt{\frac{k}{m}} = \sqrt{\frac{0.211}{0.01347}} = 3.958 \text{ rad/sec} \qquad T_n = 1.59 \text{ sec}$$

From the response spectrum curve for $\zeta = 2\%$ (Fig. E6.2b), for $T_n = 1.59$ sec, $D = 5.0$ in. and $A = 0.20g$. The peak deformation is

$$u_o = D = 5.0 \text{ in.}$$

The peak value of the equivalent static force is

$$f_{So} = \frac{A}{g}w = 0.20 \times 5.2 = 1.04 \text{ kips}$$

The bending moment diagram is shown in Fig. E6.2d with the maximum moment at the base $= 12.48$ kip-ft. Points A and B shown in Fig. E6.2e are the locations of maximum bending stress:

$$\sigma_{\max} = \frac{Mc}{I} = \frac{(12.48 \times 12)(4.5/2)}{7.23} = 46.5 \text{ ksi}$$

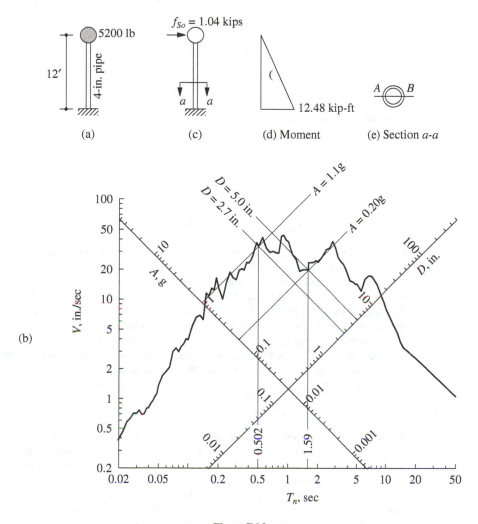

Figure E6.2

Thus, $\sigma = +46.5$ ksi at A and $\sigma = -46.5$ ksi at B, where $+$ denotes tension. The algebraic signs of these stresses are irrelevant because the direction of the peak force is not known, as the pseudo-acceleration spectrum is, by definition, positive.

Example 6.3

The stress computed in Example 6.2 exceeded the allowable stress and the designer decided to increase the size of the pipe to an 8-in.-nominal standard steel pipe. Its properties are $d_o = 8.625$ in., $d_i = 7.981$ in., $t = 0.322$ in., and $I = 72.5$ in^4. Comment on the advantages and disadvantages of using the larger pipe.

Solution

$$k = \frac{3(29 \times 10^3)72.5}{(12 \times 12)^3} = 2.112 \text{ kips/in.}$$

$$\omega_n = \sqrt{\frac{2.112}{0.01347}} = 12.52 \text{ rad/sec} \qquad T_n = 0.502 \text{ sec}$$

From the response spectrum (Fig. E6.2b): $D = 2.7$ in. and $A = 1.1g$. Therefore,

$$u_o = D = 2.7 \text{ in.}$$

$$f_{So} = 1.1 \times 5.2 = 5.72 \text{ kips}$$

$$M_{\text{base}} = 5.72 \times 12 = 68.64 \text{ kip-ft}$$

$$\sigma_{\text{max}} = \frac{(68.64 \times 12)(8.625/2)}{72.5} = 49.0 \text{ ksi}$$

Using the 8-in.-diameter pipe decreases the deformation from 5.0 in. to 2.7 in. However, contrary to the designer's objective, the bending stress increases slightly.

　　This example points out an important difference between the response of structures to earthquake excitation and to a fixed value of static force. In the latter case, the stress would decrease, obviously, by increasing the member size. In the case of earthquake excitation, the increase in pipe diameter shortens the natural vibration period from 1.59 sec to 0.50 sec, which for this response spectrum has the effect of increasing the equivalent static force f_{So}. Whether the bending stress decreases or increases by increasing the pipe diameter depends on the increase in section modulus, I/c, and the increase or decrease in f_{So}, depending on the response spectrum.

Example 6.4

A small one-story reinforced-concrete building is idealized for purposes of structural analysis as a massless frame supporting a total dead load of 10 kips at the beam level (Fig. E6.4a). The frame is 24 ft wide and 12 ft high. Each column and the beam has a 10-in.-square cross section. Assume that the Young's modulus of concrete is 3×10^3 ksi and the damping ratio for the building is estimated as 5%. Determine the peak response of this frame to the El Centro ground motion. In particular, determine the peak lateral deformation at the beam level and plot the diagram of bending moments at the instant of peak response.

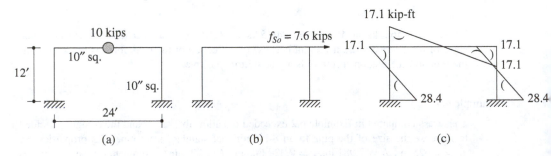

Figure E6.4　(a) Frame; (b) equivalent static force; (c) bending moment diagram.

Solution   The lateral stiffness of such a frame was calculated in Chapter 1: $k = 96EI/7h^3$, where EI is the flexural rigidity of the beam and columns and h is the height of the frame. For this particular frame,

$$k = \frac{96(3 \times 10^3)(10^4/12)}{7(12 \times 12)^3} = 11.48 \text{ kips/in.}$$

The natural vibration period is

$$T_n = \frac{2\pi}{\sqrt{k/m}} = 2\pi\sqrt{\frac{10/386}{11.48}} = 0.30 \text{ sec}$$

For $T_n = 0.3$ and $\zeta = 0.05$, we read from the response spectrum of Fig. 6.6.4: $D = 0.67$ in. and $A = 0.76g$. Peak deformation: $u_o = D = 0.67$ in. Equivalent static force: $f_{So} = (A/g)w = 0.76 \times 10 = 7.6$ kips. Static analysis of the frame for this lateral force, shown in Fig. E6.4b, gives the bending moments that are plotted in Fig. E6.4c.

Example 6.5

The frame of Example 6.4 is modified for use in a building to be located on sloping ground (Fig. E6.5). The beam is now made much stiffer than the columns and can be assumed to be rigid. The cross sections of the two columns are 10 in. square, as before, but their lengths are 12 ft and 24 ft, respectively. Determine the base shears in the two columns at the instant of peak response due to the El Centro ground motion. Assume the damping ratio to be 5%.

Solution
    1. *Compute the natural vibration period.*

$$k = \frac{12(3 \times 10^3)(10^4/12)}{(12 \times 12)^3} + \frac{12(3 \times 10^3)(10^4/12)}{(24 \times 12)^3}$$

$$= 10.05 + 1.26 = 11.31 \text{ kips/in.}$$

$$T_n = 2\pi\sqrt{\frac{10/386}{11.31}} = 0.30 \text{ sec}$$

    2. *Compute the shear force at the base of the short and long columns.*

$$u_o = D = 0.67 \text{ in.}, \qquad A = 0.76g$$

$$V_{\text{short}} = k_{\text{short}}u_o = (10.05)0.67 = 6.73 \text{ kips}$$

$$V_{\text{long}} = k_{\text{long}}u_o = (1.26)0.67 = 0.84 \text{ kip}$$

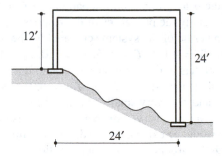

Figure E6.5

Observe that both columns go through equal deformation. Undergoing equal deformations, the stiffer column carries a greater force than the flexible column; the lateral force is distributed to the elements in proportion to their relative stiffnesses. Sometimes this basic principle has, inadvertently, not been recognized in building design, leading to unanticipated damage of the stiffer elements.

Example 6.6

For the three-span box-girder bridge of Example 1.3, determine the base shear in each of the six columns of the two bents due to El Centro ground motion applied in the longitudinal direction. Assume the damping ratio to be 5%.

Solution   The weight of the bridge deck was computed in Example 1.3: $w = 6919$ kips. The natural period of longitudinal vibration of the bridge was computed in Example 2.2: $T_n = 0.573$ sec. For $T_n = 0.573$ sec and $\zeta = 0.05$, we read from the response spectrum of Fig. 6.6.4: $D = 2.591$ in. and $A = 0.807g$.

All the columns have the same stiffness and they go through equal deformation $u_o = D = 2.591$ in. Thus, the base shear will be the same in all columns, which can be computed in one of two ways: The total equivalent static force on the bridge is [from Eq. (6.6.5)]

$$f_{so} = 0.807 \times 6919 = 5584 \text{ kips}$$

Base shear for one column, $V_b = 5584 \div 6 = 931$ kips. Alternatively, the base shear in each column is

$$V_b = k_{\text{col}}u_o = 4313 \times \frac{2.591}{12} = 931 \text{ kips}$$

6.8 RESPONSE SPECTRUM CHARACTERISTICS

We now study the important properties of earthquake response spectra. Figure 6.8.1 shows the response spectrum for El Centro ground motion together with $\ddot{u}_{go}$, $\dot{u}_{go}$, and u_{go}, the peak values of ground acceleration, ground velocity, and ground displacement, respectively, identified in Fig. 6.1.4. To show more directly the relationship between the response spectrum and the ground motion parameters, the data of Fig. 6.8.1 have been presented again in Fig. 6.8.2 using normalized scales: D/u_{go}, $V/\dot{u}_{go}$, and $A/\ddot{u}_{go}$. Figure 6.8.3 shows one of the spectrum curves of Fig. 6.8.2, the one for 5% damping, together with an idealized version shown in dashed lines; the latter will provide a basis for constructing smooth design spectra directly from the peak ground motion parameters (see Section 6.9). Based on Figs. 6.8.1 to 6.8.3, we first study the properties of the response spectrum over various ranges of the natural vibration period of the system separated by the period values at a, b, c, d, e, and f: $T_a = 0.035$ sec, $T_b = 0.125$, $T_c = 0.5$, $T_d = 3.0$, $T_e = 10$, and $T_f = 15$ sec. Subsequently, we identify the effects of damping on spectrum ordinates.

For systems with very short period, say $T_n < T_a = 0.035$ sec, the peak pseudo-acceleration A approaches $\ddot{u}_{go}$ and D is very small. This trend can be understood based on physical reasoning. For a fixed mass, a very short-period system is extremely stiff or essentially rigid. Such a system would be expected to undergo very little deformation and its mass would move rigidly with the ground; its peak acceleration should be approximately

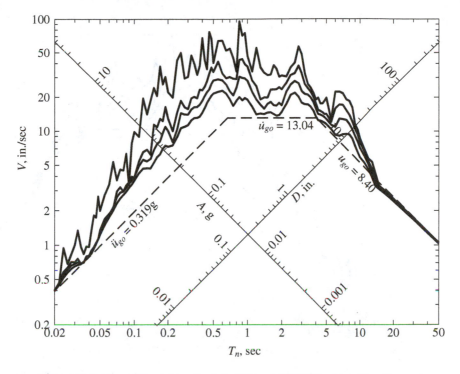

Figure 6.8.1 Response spectrum ($\zeta = 0$, 2, 5, and 10%) and peak values of ground acceleration, ground velocity, and ground displacement for El Centro ground motion.

equal to $\ddot{u}_{go}$ (Fig. 6.8.4d). This expectation is confirmed by Fig. 6.8.4, where the ground acceleration is presented in part (a), the total acceleration $\ddot{u}^t(t)$ of a system with $T_n = 0.02$ sec and $\zeta = 2\%$ in part (b), and the pseudo-acceleration $A(t)$ for the same system in part (c). Observe that $\ddot{u}^t(t)$ and $\ddot{u}_g(t)$ are almost identical functions and $\ddot{u}^t_o \simeq \ddot{u}_{go}$. Furthermore, for lightly damped systems $\ddot{u}^t(t) \simeq -A(t)$ and $\ddot{u}^t_o \simeq A$ (Section 6.12.2); therefore, $A \simeq \ddot{u}_{go}$.

For systems with a very long period, say $T_n > T_f = 15$ sec, D for all damping values approaches u_{go} and A is very small; thus the forces in the structure, which are related to mA, would be very small. This trend can again be explained by relying on physical reasoning. For a fixed mass, a very-long-period system is extremely flexible. The mass would be expected to remain essentially stationary while the ground below moves (Fig. 6.8.5c). Thus $\ddot{u}^t(t) \simeq 0$, implying that $A(t) \simeq 0$ (see Section 6.12.2); and $u(t) \simeq -u_g(t)$, implying that $D \simeq u_{go}$. This expectation is confirmed by Fig. 6.8.5, where the deformation response $u(t)$ of a system with $T_n = 30$ sec and $\zeta = 2\%$ to the El Centro ground motion is compared with the ground displacement $u_g(t)$. Observe that the peak values for u_o and u_{go} are close and the time variation of $u(t)$ is similar to that of $-u_g(t)$, but for rotation of the baseline. The discrepancy between the two arises, in part, from the loss of the initial portion of the recorded ground motion prior to triggering of the recording accelerograph.

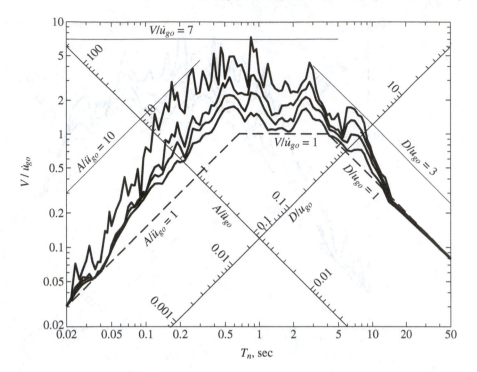

Figure 6.8.2   Response spectrum for El Centro ground motion plotted with normalized scales $A/\ddot{u}_{go}$, $V/\dot{u}_{go}$, and D/u_{go}; $\zeta = 0$, 2, 5, and 10%.

For short-period systems with T_n between $T_a = 0.035$ sec and $T_c = 0.50$ sec, A exceeds $\ddot{u}_{go}$, with the amplification depending on T_n and ζ. Over a portion of this period range, $T_b = 0.125$ sec to $T_c = 0.5$ sec, A may be idealized as constant at a value equal to $\ddot{u}_{go}$ amplified by a factor depending on ζ.

For long-period systems with T_n between $T_d = 3$ sec and $T_f = 15$ sec, D generally exceeds u_{go}, with the amplification depending on T_n and ζ. Over a portion of this period range, $T_d = 3.0$ sec to $T_e = 10$ sec, D may be idealized as constant at a value equal to u_{go} amplified by a factor depending on ζ.

For intermediate-period systems with T_n between $T_c = 0.5$ sec and $T_d = 3.0$ sec, V exceeds $\dot{u}_{go}$. Over this period range, V may be idealized as constant at a value equal to $\dot{u}_{go}$, amplified by a factor depending on ζ.

Based on these observations, it is logical to divide the spectrum into three period ranges (Fig. 6.8.3). The long-period region to the right of point d, $T_n > T_d$, is called the *displacement-sensitive region* because structural response is related most directly to ground displacement. The short-period region to the left of point c, $T_n < T_c$, is called the *acceleration-sensitive region* because structural response is most directly related to ground acceleration. The intermediate period region between points c and d, $T_c < T_n < T_d$, is called the *velocity-sensitive region* because structural response appears to be better related to ground velocity than to other ground motion parameters. For a particular ground motion,

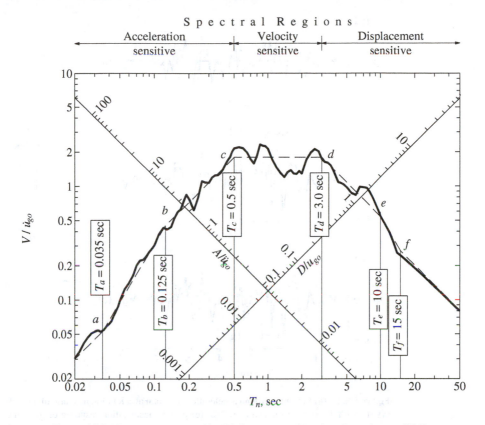

Figure 6.8.3 Response spectrum for El Centro ground motion shown by a solid line together with an idealized version shown by a dashed line; $\zeta = 5\%$.

the periods T_a, T_b, T_e, and T_f on the idealized spectrum are independent of damping, but T_c and T_d vary with damping.

The preceding observations and discussion have brought out the usefulness of the four-way logarithmic plot of the combined deformation, pseudo-velocity, and pseudo-acceleration response spectra. These observations would be difficult to glean from the three individual spectra.

Idealizing the spectrum by a series of straight lines a–b–c–d–e–f in the four-way logarithmic plot is obviously not a precise process. For a given ground motion, the period values associated with the points a, b, c, d, e, and f and the amplification factors for the segments b–c, c–d, and d–e are somewhat judgmental in the way we have approached them. However, formal curve-fitting techniques can be used to replace the actual spectrum by an idealized spectrum of a selected shape. In any case, the idealized spectrum in Fig. 6.8.3 is not a close approximation to the actual spectrum. This may not be visually apparent but becomes obvious when we note that the scales are logarithmic. As we shall see in the next section, the greatest benefit of the idealized spectrum is in constructing a design spectrum representative of many ground motions.

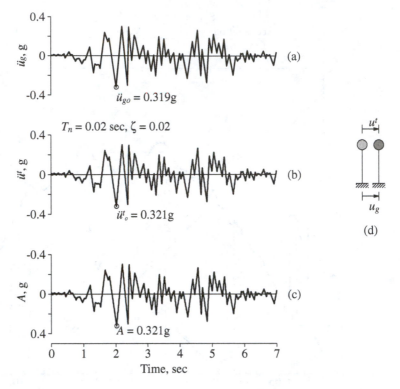

Figure 6.8.4 (a) El Centro ground acceleration; (b) total acceleration response of an SDF system with $T_n = 0.02$ sec and $\zeta = 2\%$; (c) pseudo-acceleration response of the same system; (d) rigid system.

The periods T_a, T_b, T_c, T_d, T_e, and T_f separating spectral regions and the amplification factors for the segments $b-c$, $c-d$, and $d-e$ depend on the time variation of ground motion, in particular, the relative values of peak ground acceleration, velocity, and displacement, as indicated by their ratios: $\dot{u}_{go}/\ddot{u}_{go}$ and $u_{go}/\dot{u}_{go}$. These ground motion characteristics depend on the earthquake magnitude, fault-to-site distance, source-to-site geology, and soil conditions at the site.

Ground motions recorded within the near-fault region of an earthquake at stations located toward the direction of the fault rupture are qualitatively quite different from the usual far-fault earthquake ground motions. The fault-normal component of a ground motion recorded in the near-fault region of the Northridge, California, earthquake of January 17, 1994 displays a long-period pulse in the acceleration history that appears as a coherent pulse in the velocity and displacements histories (Fig. 6.8.6a). Such a pronounced pulse does not exist in ground motions recorded at locations away from the near-fault region, such as the Taft record obtained from the Kern County, California, earthquake of July 21, 1952 (Fig. 6.8.6b).

The ratios $\dot{u}_{go}/\ddot{u}_{go}$ and $u_{go}/\dot{u}_{go}$ are very different between the fault normal components of near- and far-fault motions. As apparent from the peak values noted in Fig. 6.8.6,

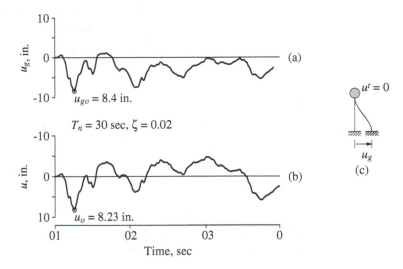

Figure 6.8.5   (a) El Centro ground displacement; (b) deformation response of SDF system with $T_n = 30$ sec and $\zeta = 2\%$; (c) very flexible system.

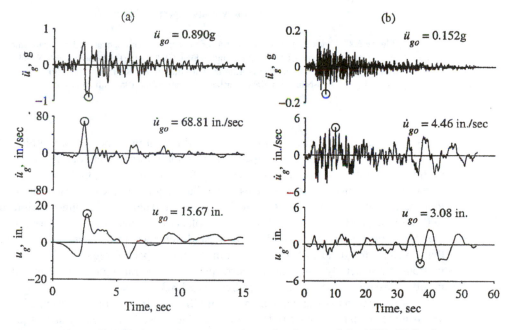

Figure 6.8.6   Fault-normal component of ground motions recorded at (a) Rinaldi Receiving Station, 1994 Northridge earthquake, and (b) Taft, 1952 Kern County earthquake.

the ratio $\dot{u}_{go}/\ddot{u}_{go}$ for near-fault motions is much larger than the ratio for far-fault motions, whereas the ratio $u_{go}/\dot{u}_{go}$ for near-fault motions is much smaller. As a result, the response

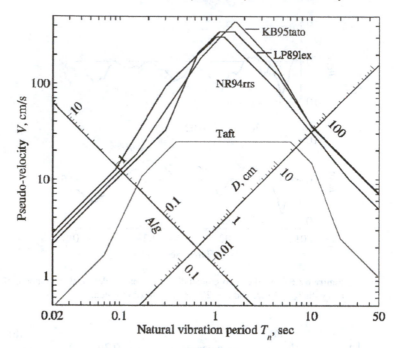

Figure 6.8.7 Idealized response spectra for fault-normal component of three near-fault ground motion records—LP89Lex: Lexington Dam, 1989 Loma Prieta earthquake; NR94rrs: Rinaldi Receiving Station, 1994 Northridge earthquake; and KB95tato: Takatori Station, 1994 Hygogo-Ken-Nanbu (or Kobe) earthquake—and of the 1952 Taft record; $\zeta = 5\%$.

spectra for near- and far-fault motions are very different in shape. Shown in Fig. 6.8.7 are the idealized versions of response spectra for the fault normal components of one far-fault motion—the Taft motion of Fig. 6.8.6a—and for three near-fault motions—including the one in Fig. 6.8.6b—from earthquakes of similar magnitudes. Comparing them indicates that the velocity-sensitive region is much narrower and shifted to a longer period for near-fault motions, and their acceleration- and displacement-sensitive regions are much wider than those for far-fault motions. Despite these differences, researchers have demonstrated that response trends identified earlier from the three spectral regions of far-fault ground motions are generally valid for the corresponding spectral regions of near-fault ground motions. We return to this assertion in Section 22.3.3.

We now turn to damping, which has significant influence on the earthquake response spectrum (Figs. 6.6.4 to 6.6.6). The zero damping curve is marked by abrupt jaggedness, which indicates that the response is very sensitive to small differences in the natural vibration period. The introduction of damping makes the response much less sensitive to the period.

Damping reduces the response of a structure, as expected, and the reduction achieved with a given amount of damping is different in the three spectral regions. In the limit as

$T_n \to 0$, damping does not affect the response because the structure moves rigidly with the ground. In the other limit as $T_n \to \infty$, damping again does not affect the response because the structural mass stays still while the ground underneath moves. Among the three period regions defined earlier, the effect of damping tends to be greatest in the velocity-sensitive region of the spectrum. In this spectral region the effect of damping depends on the ground motion characteristics. If the ground motion is nearly harmonic over many cycles (e.g., the record from Mexico City shown in Fig. 6.1.3), the effect of damping would be especially large for systems near "resonance" (Chapter 3). If the ground motion is short in duration with only a few major cycles (e.g., the record from Parkfield, California, shown in Fig. 6.1.3), the influence of damping would be small, as in the case of pulse excitations (Chapter 4).

Figure 6.8.8 shows the peak pseudo-acceleration $A(\zeta)$, normalized relative to $A(\zeta = 0)$, plotted as a function of ζ for several T_n values. These are some of the data from the response spectrum of Figs. 6.6.4 and 6.6.5 replotted in a different format. Observe that the effect of damping is stronger for smaller damping values. This means that if the damping ratio is increased from 0 to 2%, the reduction in response is greater than the response reduction, due to an increase in damping from 10% to 12%. The effect of damping in reducing the response depends on the period T_n of the system, but there is no clear trend from Fig. 6.8.8. This is yet another indication of the complexity of structural response to earthquakes.

The motion of a structure and the associated forces could be reduced by increasing the effective damping of the structure. The addition of dampers achieves this goal without significantly changing the natural vibration periods of the structure. Viscoelastic dampers have been introduced in many structures; for example, 10,000 dampers were

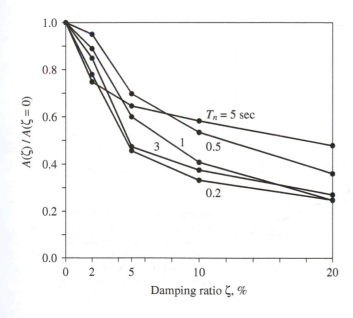

Figure 6.8.8 Variation of peak pseudo-acceleration with damping for systems with $T_n = 0.2, 0.5, 1, 3,$ and 5 sec; El Centro ground motion.

installed throughout the height of each tower of the World Trade Center in New York City to reduce wind-induced motion to within a comfortable range for the occupants. In recent years there is a growing interest in developing dampers suitable for structures in earthquake-prone regions. Because the inherent damping in most structures is relatively small, their earthquake response can be reduced significantly by the addition of dampers. These can be especially useful in improving the seismic safety of an existing structure. We will return to this topic in Chapter 7.

6.9 ELASTIC DESIGN SPECTRUM

In this section we introduce the concept of earthquake design spectrum for elastic systems and present a procedure to construct it from estimated peak values for ground acceleration, ground velocity, and ground displacement.

　　The design spectrum should satisfy certain requirements because it is intended for the design of new structures, or the seismic safety evaluation of existing structures, to resist future earthquakes. For this purpose the response spectrum for a ground motion recorded during a past earthquake is inappropriate. The jaggedness in the response spectrum, as seen in Fig. 6.6.4, is characteristic of that one excitation. The response spectrum for another ground motion recorded at the same site during a different earthquake is also jagged, but the peaks and valleys are not necessarily at the same periods. This is apparent from Fig. 6.9.1, where the response spectra for ground motions recorded at the same site during three past earthquakes are plotted. Similarly, it is not possible to predict the jagged response spectrum

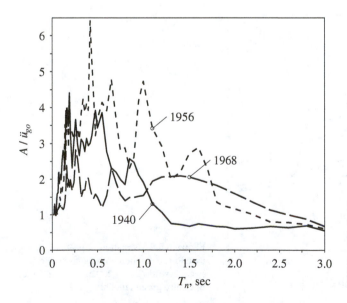

Figure 6.9.1 Response spectra for the north–south component of ground motions recorded at the Imperial Valley Irrigation District substation, El Centro, California, during earthquakes of May 18, 1940; February 9, 1956; and April 8, 1968. $\zeta = 2\%$.

in all its detail for a ground motion that may occur in the future. Thus the design spectrum should consist of a set of smooth curves or a series of straight lines with one curve for each level of damping.

The design spectrum should, in a general sense, be representative of ground motions recorded at the site during past earthquakes. If none have been recorded at the site, the design spectrum should be based on ground motions recorded at other sites under similar conditions. The factors that one tries to match in the selection include the magnitude of the earthquake, the distance of the site from the causative fault, the fault mechanism, the geology of the travel path of seismic waves from the source to the site, and the local soil conditions at the site. While this approach is feasible for some parts of the world, such as California and Japan, where numerous ground motion records are available, in many other regions it is hampered by the lack of a sufficient number of such records. In such situations compromises in the approach are necessary by considering ground motion records that were recorded for conditions different from those at the site. Detailed discussion of these issues is beyond the scope of this book. The presentation here is focused on the narrow question of how to develop the design spectrum that is representative of an available ensemble (or set) of recorded ground motions.

The design spectrum is based on statistical analysis of the response spectra for the ensemble of ground motions. Suppose that I is the number of ground motions in the ensemble, the ith ground motion is denoted by $\ddot{u}_g^i(t)$, and u_{go}^i, $\dot{u}_{go}^i$, and $\ddot{u}_{go}^i$ are its peak displacement, velocity, and acceleration, respectively. Each ground motion is normalized (scaled up or down) so that all ground motions have the same peak ground acceleration, say $\ddot{u}_{go}$; other bases for normalization can be chosen. The response spectrum for each normalized ground motion is computed by the procedures described in Section 6.6. At each period T_n there are as many spectral values as the number I of ground motion records in the ensemble: D^i, V^i, and A^i ($i = 1, 2, \ldots, I$), the deformation, pseudo-velocity, and pseudo-acceleration spectral ordinates. Such data were generated for an ensemble of 10 earthquake records, and selected aspects of the results are presented in Fig. 6.9.2. The quantities u_{go}, $\dot{u}_{go}$, and $\ddot{u}_{go}$ in the normalized scales of Fig. 6.9.2 are the average values of the peak ground displacement, velocity, and acceleration—averaged over the I ground motions. Statistical analysis of these data provide the probability distribution for the spectral ordinate, its mean value, and its standard deviation at each period T_n. The probability distributions are shown schematically at three selected T_n values, indicating that the coefficient of variation (= standard deviation ÷ mean value) varies with T_n. Connecting all the mean values gives the *mean response spectrum*. Similarly connecting all the mean-plus-one-standard-deviation values gives the *mean-plus-one-standard-deviation response spectrum*. Observe that these two response spectra are much smoother than the response spectrum for an individual ground motion (Fig. 6.6.4). As shown in Fig. 6.9.2, such a smooth spectrum curve lends itself to idealization by a series of straight lines much better than the spectrum for an individual ground motion (Fig. 6.8.3).

Researchers have developed procedures to construct such design spectra from ground motion parameters. One such procedure, which is illustrated in Fig. 6.9.3, will be summarized later. The recommended period values $T_a = \frac{1}{33}$ sec, $T_b = \frac{1}{8}$ sec, $T_e = 10$ sec, and $T_f = 33$ sec, and the amplification factors α_A, α_V, and α_D for the three spectral regions,

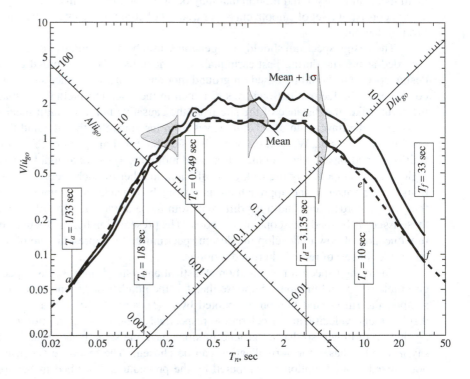

Figure 6.9.2 Mean and mean $+1\sigma$ spectra with probability distributions for V at $T_n = 0.25$, 1, and 4 sec; $\zeta = 5\%$. Dashed lines show an idealized design spectrum. (Based on numerical data from R. Riddell and N. M. Newmark, 1979.)

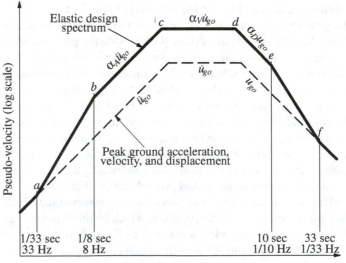

Figure 6.9.3 Construction of elastic design spectrum.

TABLE 6.9.1   AMPLIFICATION FACTORS: ELASTIC DESIGN SPECTRA

Damping, ζ (%)	Median (50th percentile)			One Sigma (84.1th percentile)		
	α_A	α_V	α_D	α_A	α_V	α_D
1	3.21	2.31	1.82	4.38	3.38	2.73
2	2.74	2.03	1.63	3.66	2.92	2.42
5	2.12	1.65	1.39	2.71	2.30	2.01
10	1.64	1.37	1.20	1.99	1.84	1.69
20	1.17	1.08	1.01	1.26	1.37	1.38

Source: N. M. Newmark and W. J. Hall, *Earthquake Spectra and Design*, Earthquake Engineering Research Institute, Berkeley, Calif., 1982, pp. 35 and 36.

TABLE 6.9.2   AMPLIFICATION FACTORS: ELASTIC DESIGN SPECTRA[a]

	Median (50th percentile)	One Sigma (84.1th percentile)
α_A	$3.21 - 0.68 \ln \zeta$	$4.38 - 1.04 \ln \zeta$
α_V	$2.31 - 0.41 \ln \zeta$	$3.38 - 0.67 \ln \zeta$
α_D	$1.82 - 0.27 \ln \zeta$	$2.73 - 0.45 \ln \zeta$

Source: N. M. Newmark and W. J. Hall, *Earthquake Spectra and Design*, Earthquake Engineering Research Institute, Berkeley, Calif., 1982, pp. 35 and 36.

[a] Damping ratio in percent.

were developed by the preceding analysis of a larger ensemble of ground motions recorded on firm ground (rock, soft rock, and competent sediments). The amplification factors for two different nonexceedance probabilities, 50% and 84.1%, are given in Table 6.9.1 for several values of damping and in Table 6.9.2 as a function of damping ratio. The 50% nonexceedance probability represents the median value of the spectral ordinates and the 84.1% approximates the mean-plus-one-standard-deviation value assuming lognormal probability distribution for the spectral ordinates.

Observe that the period values T_a, T_b, T_e, and T_f are fixed; the values in Fig. 6.9.3 are for firm ground. Period values T_c and T_d are determined by the intersections of the constant-A $(= \alpha_A \ddot{u}_{go})$, constant-$V$ $(= \alpha_V \dot{u}_{go})$, and constant-$D$ $(= \alpha_D u_{go})$ branches of the spectrum. Because α_A, α_V, and α_D are functions of ζ (Tables 6.9.1 and 6.9.2), T_c and T_d depend on the damping ratio.

Summary.   A procedure to construct a design spectrum is now summarized with reference to Fig. 6.9.3:

1. Plot the three dashed lines corresponding to the peak values of ground acceleration $\ddot{u}_{go}$, velocity $\dot{u}_{go}$, and displacement u_{go} for the design ground motion.
2. Obtain from Table 6.9.1 or 6.9.2 the values for α_A, α_V, and α_D for the ζ selected.

3. Multiply $\ddot{u}_{go}$ by the amplification factor α_A to obtain the straight line b–c representing a constant value of pseudo-acceleration A.

4. Multiply $\dot{u}_{go}$ by the amplification factor α_V to obtain the straight line c–d representing a constant value of pseudo-velocity V.

5. Multiply u_{go} by the amplification factor α_D to obtain the straight line d–e representing a constant value of deformation D.

6. Draw the line $A = \ddot{u}_{go}$ for periods shorter than T_a and the line $D = u_{go}$ for periods longer than T_f.

7. The transition lines a–b and e–f complete the spectrum.

We now illustrate use of this procedure by constructing the 84.1th percentile design spectrum for systems with 5% damping. For convenience, a peak ground acceleration $\ddot{u}_{go} = 1$g is selected; the resulting spectrum can be scaled by η to obtain the design spectrum corresponding to $\ddot{u}_{go} = \eta$g. Consider also that no specific estimates for peak ground velocity $\dot{u}_{go}$ and displacement u_{go} are provided; thus typical values $\dot{u}_{go}/\ddot{u}_{go} = 48$ in./sec/g and $\ddot{u}_{go} \times u_{go}/\dot{u}_{go}^2 = 6$, recommended for firm ground, are used. For $\ddot{u}_{go} = 1$g, these ratios give $\dot{u}_{go} = 48$ in./sec and $u_{go} = 36$ in.

The design spectrum shown in Fig. 6.9.4 is determined by the following steps:

1. The peak parameters for the ground motion: $\ddot{u}_{go} = 1$g, $\dot{u}_{go} = 48$ in./sec, and $u_{go} = 36$ in. are plotted.

2. From Table 6.9.1, the amplification factors for the 84.1th percentile spectrum and 5% damping are obtained: $\alpha_A = 2.71$, $\alpha_V = 2.30$, and $\alpha_D = 2.01$.

3–5. The ordinate for the constant-A branch is $A = 1$g $\times 2.71 = 2.71$g, for the constant-V branch: $V = 48 \times 2.30 = 110.4$, and for the constant-$D$ branch: $D = 36 \times 2.01 = 72.4$. The three branches are drawn as shown.

6. The line $A = 1$g is plotted for $T_n < \frac{1}{33}$ sec and $D = 36$ in. for $T_n > 33$ sec.

7. The transition line b–a is drawn to connect the point $A = 2.71$g at $T_n = \frac{1}{8}$ sec to $\ddot{u}_{go} = 1$g at $T_n = \frac{1}{33}$ sec. Similarly, the transition line e–f is drawn to connect the point $D = 72.4$ at $T_n = 10$ sec to $u_{go} = 36$ in. at $T_n = 33$ sec.

With the pseudo-velocity design spectrum known (Fig. 6.9.4), the pseudo-acceleration design spectrum and the deformation design spectrum are determined using Eq. (6.6.6) and plotted in Figs. 6.9.5 and 6.9.6, respectively. Observe that A approaches $\ddot{u}_{go} = 1$g at $T_n = 0$ and D tends to $u_{go} = 36$ in. at $T_n = 50$ sec. The design spectrum can be defined completely by numerical values for T_a, T_b, T_c, T_d, T_e, and T_f, and equations for $A(T_n)$, $V(T_n)$, or $D(T_n)$ for each branch of the spectrum. As mentioned before, some of these periods—T_a, T_b, T_e, and T_f—are fixed, but others—T_c and T_d—depend on damping. The intersections of $A = 2.71$g, $V = 110.4$ in./sec, and $D = 72.4$ in. are determined from Eq. (6.6.6): $T_c = 0.66$ sec and $T_d = 4.12$ sec for $\zeta = 5\%$. Equations describing various branches of the pseudo-acceleration design spectrum are given in Fig. 6.9.5.

Repeating the preceding construction of the design spectrum for additional values of the damping ratio leads to Figs. 6.9.7 to 6.9.10. This, then, is the design spectrum for

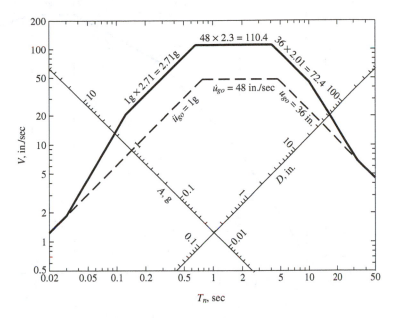

Figure 6.9.4 Construction of elastic design spectrum (84.1th percentile) for ground motions with $\ddot{u}_{go} = 1$g, $\dot{u}_{go} = 48$ in./sec, and $u_{go} = 36$ in.; $\zeta = 5\%$.

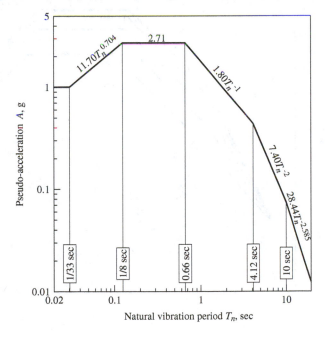

Figure 6.9.5 Elastic pseudo-acceleration design spectrum (84.1th percentile) for ground motions with $\ddot{u}_{go} = 1$g, $\dot{u}_{go} = 48$ in./sec, and $u_{go} = 36$ in.; $\zeta = 5\%$.

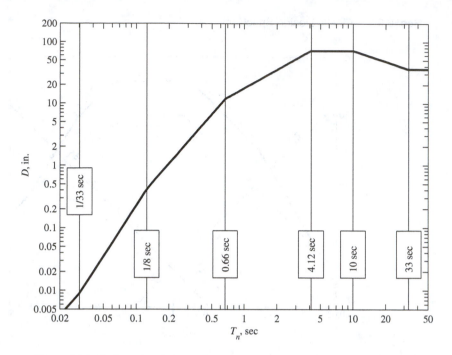

Figure 6.9.6 Deformation design spectrum (84.1th percentile) for ground motions with $\ddot{u}_{go} = 1$g, $\dot{u}_{go} = 48$ in./sec, and $u_{go} = 36$ in.; $\zeta = 5\%$.

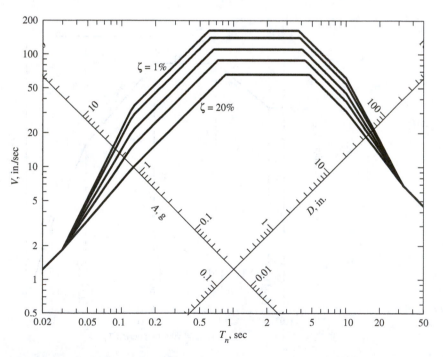

Figure 6.9.7 Pseudo-velocity design spectrum for ground motions with $\ddot{u}_{go} = 1$g, $\dot{u}_{go} = 48$ in./sec, and $u_{go} = 36$ in.; $\zeta = 1, 2, 5, 10,$ and 20%.

ground motions on firm ground with $\ddot{u}_{go} = 1g$, $\dot{u}_{go} = 48$ in./sec, and $u_{go} = 36$ in. in three different forms: pseudo-velocity, pseudo-acceleration, and deformation. Observe that the pseudo-acceleration design spectrum has been plotted in two formats: logarithmic scales (Fig. 6.9.8) and linear scales (Fig. 6.9.9).

The elastic design spectrum provides a basis for calculating the design force and deformation for SDF systems to be designed to remain elastic. For this purpose the design spectrum is used in the same way as the response spectrum was used to compute peak response; see Examples 6.2 to 6.6. The errors in reading spectral ordinates from a four-way logarithmic plot can be avoided, however, because simple functions of T_n define various branches of the spectrum in Figs. 6.9.4 to 6.9.6.

Parameters that enter into construction of the elastic design spectrum should be selected considering the factors that influence ground motion mentioned previously. Thus the selection of design ground motion parameter $\ddot{u}_{go}$, $\dot{u}_{go}$, and u_{go} should be based on

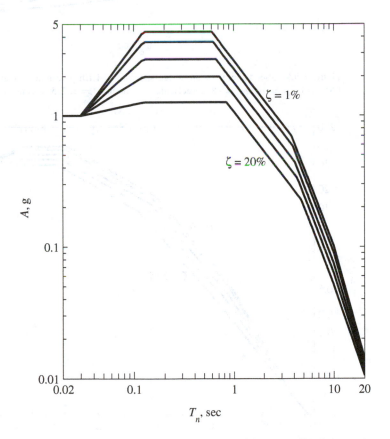

Figure 6.9.8 Pseudo-acceleration design spectrum (84.1th percentile) for ground motions with $\ddot{u}_{go} = 1g$, $\dot{u}_{go} = 48$ in./sec, and $u_{go} = 36$ in.; $\zeta = 1, 2, 5, 10,$ and 20%.

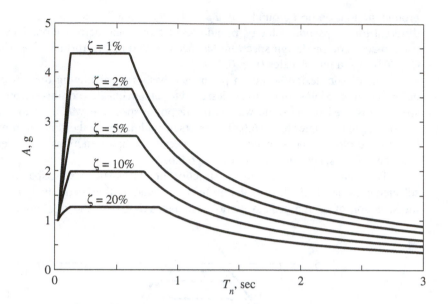

Figure 6.9.9 Pseudo-acceleration design spectrum (84.1th percentile) for ground motions with $\ddot{u}_{go} = 1$g, $\dot{u}_{go} = 48$ in./sec, and $u_{go} = 36$ in.; $\zeta = 1, 2, 5, 10,$ and 20%.

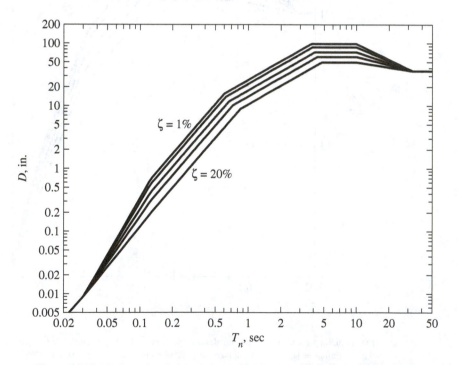

Figure 6.9.10 Deformation design spectrum (84.1th percentile) for ground motions with $\ddot{u}_{go} = 1$g, $\dot{u}_{go} = 48$ in./sec, and $u_{go} = 36$ in.; $\zeta = 1, 2, 5, 10,$ and 20%.

earthquake magnitude, distance to the earthquake fault, fault mechanism, wave-travel-path geology, and local soil conditions. Results of research on these factors and related issues are available; they are used to determine site-dependent design spectra for important projects. Similarly, numerical values for the amplification factors α_A, α_V, and α_D should be chosen consistent with the expected frequency content of the ground motion.

The selected values of $\ddot{u}_{go}$, $\dot{u}_{go}$, and u_{go} are consistent with $\dot{u}_{go}/\ddot{u}_{go} = 48$ in./sec/g and $\ddot{u}_{go} \times u_{go}/\dot{u}_{go}^2 = 6$. These ratios are considered representative of ground motions on firm ground. For such sites the resulting spectrum may be scaled to conform to the peak ground acceleration estimated for the site. Thus if this estimate is 0.4g, the spectrum of Figs. 6.9.7 to 6.9.9 multiplied by 0.4 gives the design spectrum for the site. Such a simple approach may be reasonable if a site-specific seismic hazard analysis is not planned.

6.10 COMPARISON OF DESIGN AND RESPONSE SPECTRA

It is instructive to compare the "standard" design spectrum developed in Section 6.9 for firm ground with an actual response spectrum for similar soil conditions. Figure 6.10.1 shows a standard design spectrum for $\ddot{u}_{go} = 0.319g$, the peak acceleration for the El Centro ground motion; the implied values for $\dot{u}_{go}$ and u_{go} are 15.3 in./sec and 11.5 in., respectively, based on the standard ratios mentioned in the preceding paragraph. Also shown in Fig. 6.10.1 is the response spectrum for the El Centro ground motion; recall that the actual peak values for this motion are $\dot{u}_{go} = 13.04$ in./sec and $u_{go} = 8.40$ in. The El Centro response spectrum agrees well with the design spectrum in the acceleration-sensitive region, largely because the peak accelerations for the two are matched. However, the two spectra are considerably different in the velocity-sensitive region because of the differences (15.3 in./sec versus 13.04 in./sec) in the peak ground velocity. Similarly, they are even more different in the displacement-sensitive region because of the larger differences (11.5 in. versus 8.4 in.) in the peak ground displacement.

The response spectrum for an individual ground motion differs from the design spectrum even if the peak values $\ddot{u}_{go}$, $\dot{u}_{go}$, and u_{go} for the two spectra are matched. In Fig. 6.10.2 the response spectrum for the El Centro ground motion is compared with the design spectrum for ground motion parameters $\ddot{u}_{go} = 0.319g$, $\dot{u}_{go} = 13.04$ in./sec, and $u_{go} = 8.40$ in.—the same as for the El Centro ground motion. Two design spectra are included: the 50th percentile spectrum and the 84.1th percentile spectrum. The agreement between the response and design spectra is now better because the ground motion parameters are matched. However, significant differences remain: over the acceleration-sensitive region the response spectrum is close to the 84.1th percentile design spectrum; over the velocity- and displacement-sensitive regions the response spectrum is between the two design spectra for some periods and below the median design spectrum for other periods.

Such differences are to be expected because the design spectrum is not intended to match the response spectrum for any particular ground motion but is constructed to represent the average characteristics of many ground motions. These differences are due to the inherent variability in ground motions as reflected in the probability distributions of the amplification factors and responses; see Fig. 6.9.2.

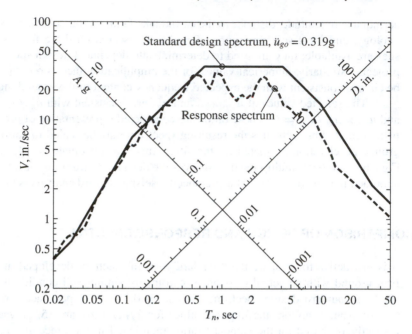

Figure 6.10.1 Comparison of standard design spectrum ($\ddot{u}_{go} = 0.319g$) with elastic response spectrum for El Centro ground motion; $\zeta = 5\%$.

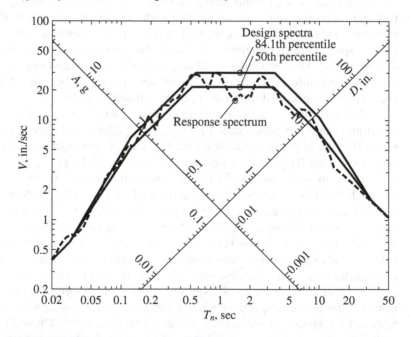

Figure 6.10.2 Comparison of design spectra ($\ddot{u}_{go} = 0.319g$, $\dot{u}_{go} = 13.04$ in./sec, $u_{go} = 8.40$ in.) with elastic response spectrum for El Centro ground motion; $\zeta = 5\%$.

6.11 DISTINCTION BETWEEN DESIGN AND RESPONSE SPECTRA

A design spectrum differs conceptually from a response spectrum in two important ways. First, the jagged response spectrum is a plot of the peak response of all possible SDF systems and hence is a description of a particular ground motion. The smooth design spectrum, however, is a specification of the level of seismic design force, or deformation, as a function of natural vibration period and damping ratio. This conceptual difference between the two spectra should be recognized, although in some situations, their shapes may be similar. Such is the case when the design spectrum is determined by statistical analysis of several comparable response spectra.

Second, for some sites a design spectrum is the envelope of two different elastic design spectra. Consider a site in southern California that could be affected by two different types of earthquakes: a Magnitude 6.5 earthquake originating on a nearby fault and a Magnitude 8.5 earthquake on the distant San Andreas fault. The design spectrum for each earthquake could be determined by the procedure developed in Section 6.9. The ordinates and shapes of the two design spectra would differ, as shown schematically in Fig. 6.11.1, because of the differences in earthquake magnitude and distance of the site from the earthquake fault. The design spectrum for this site is defined as the envelope of the design spectra for the two different types of earthquakes. Note that the short-period portion of the design spectrum is governed by the nearby earthquake, while the long-period portion of the design spectrum is controlled by the distant earthquake.

Before leaving the subject, we emphasize that this limited presentation on constructing elastic design spectra has been narrowly focused on methods that are directly related to structural dynamics that we have learned. In contrast, modern methods for constructing design spectra are based on probabilistic seismic hazard analysis, which considers the past rate of seismic activity on all faults that contribute to the seismic hazard at the site, leading to the uniform hazard spectrum.

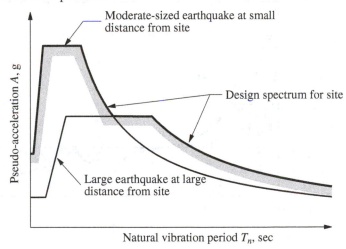

Figure 6.11.1   Design spectrum defined as the envelope of design spectra for earthquakes originating on two different faults.

6.12 VELOCITY AND ACCELERATION RESPONSE SPECTRA

We now return to the relative velocity response spectrum and the acceleration response spectrum that were introduced in Section 6.5. In one sense there is little motivation to study these "true" spectra because they are not needed to determine the peak deformations and forces in a system; for this purpose the pseudo-acceleration (or pseudo-velocity or deformation) response spectrum is sufficient. A brief discussion of these "true" spectra is included, however, because the distinction between them and "pseudo" spectra has not always been made in the early publications, and the two have sometimes been used interchangeably.

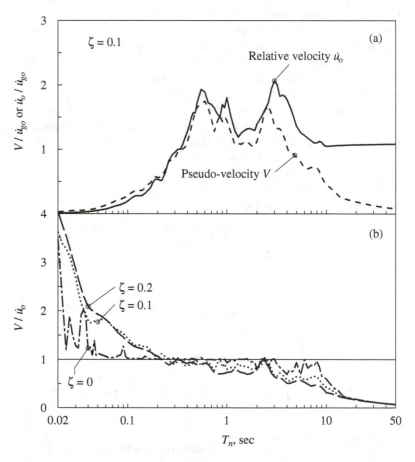

Figure 6.12.1   (a) Comparison between pseudo-velocity and relative-velocity response spectra; $\zeta = 10\%$; (b) ratio $V/\dot{u}_o$ for $\zeta = 0$, 10, and 20%.

To study the relationship between these spectra, we write them in mathematical form. The deformation response of a linear SDF system to an arbitrary ground motion with zero initial conditions is given by the convolution integral, Eq. (4.2.2), adapted for earthquake excitation:

$$u(t) = \int_0^t \ddot{u}_g(\tau) h(t - \tau) \, d\tau \tag{6.12.1}$$

where the unit impulse response function, $h(t - \tau)$, is given by Eq. (4.1.7). Thus,

$$u(t) = -\frac{1}{\omega_D} \int_0^t \ddot{u}_g(\tau) e^{-\zeta \omega_n (t-\tau)} \sin[\omega_D(t - \tau)] \, d\tau \tag{6.12.2}$$

Using theorems from calculus to differentiate under the integral sign leads to

$$\dot{u}(t) = -\zeta \omega_n u(t) - \int_0^t \ddot{u}_g(\tau) e^{-\zeta \omega_n (t-\tau)} \cos[\omega_D(t - \tau)] \, d\tau \tag{6.12.3}$$

An equation for the acceleration $\ddot{u}^t(t)$ of the mass can be obtained by differentiating Eq. (6.12.3) and adding the ground acceleration $\ddot{u}_g(t)$. However, the equation of motion for the system [Eq. (6.2.1)] provides a more convenient alternative:

$$\ddot{u}^t(t) = -\omega_n^2 u(t) - 2\zeta \omega_n \dot{u}(t) \tag{6.12.4}$$

As defined earlier, the relative-velocity spectrum and acceleration spectrum are plots of $\dot{u}_o$ and $\ddot{u}_o^t$, the peak values of $\dot{u}(t)$ and $\ddot{u}^t(t)$, respectively, as functions of T_n.

6.12.1 Pseudo-velocity and Relative-Velocity Spectra

In Fig. 6.12.1a the relative-velocity response spectrum is compared with the pseudo-velocity response spectrum, both for El Centro ground motion and systems with $\zeta = 10\%$. The latter spectrum is simply one of the curves of Fig. 6.6.4 presented in a different form. Each point on the relative-velocity response spectrum represents the peak velocity of an SDF system obtained from $\dot{u}(t)$ determined by the numerical methods of Chapter 5. The differences between the two spectra depend on the natural period of the system. For long-period systems, V is less than $\dot{u}_o$ and the differences between the two are large. This can be understood by recognizing that as T_n becomes very long, the mass of the system stays still while the ground underneath moves. Thus, as $T_n \to \infty$, $D \to u_{go}$ (see Section 6.8 and Fig. 6.8.5) and $\dot{u}_o \to \dot{u}_{go}$. Now $D \to u_{go}$ implies that $V \to 0$ because of Eq. (6.6.1). These trends are confirmed by the results presented in Fig. 6.12.1a. For short-period systems V exceeds $\dot{u}_o$, with the differences increasing as T_n becomes shorter. For medium-period systems, the differences between V and $\dot{u}_o$ are small over a wide range of T_n.

In Fig. 6.12.1b the ratio $V/\dot{u}_o$ is plotted for three damping values, $\zeta = 0$, 10, and 20%. The differences between the two spectra, as indicated by how much the ratio $V/\dot{u}_o$ differs from unity, are smallest for undamped systems and increase with damping. This can be explained from Eqs. (6.12.2) and (6.12.3) by observing that for $\zeta = 0$, $\dot{u}(t)$ and $\omega_n u(t)$ are the same except for the sine and cosine terms in the integrand. With damping, the first term in Eq. (6.12.3) contributes to $\dot{u}(t)$, suggesting that $\dot{u}(t)$ would differ from $\omega_n u(t)$ to a

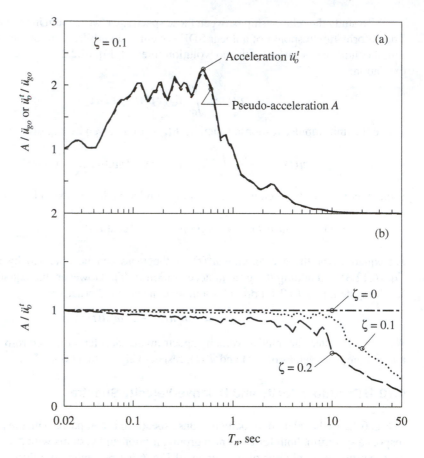

Figure 6.12.2 (a) Comparison between pseudo-acceleration and acceleration response spectra; $\zeta = 10\%$; (b) ratio $A/\ddot{u}_o^t$ for $\zeta = 0$, 10, and 20%.

greater degree. Over the medium-period range V can be taken as an approximation to $\dot{u}_o$ for the practical range of damping.

6.12.2 Pseudo-acceleration and Acceleration Spectra

The pseudo-acceleration and acceleration response spectra are identical for systems without damping. This is apparent from Eq. (6.12.4), which for undamped systems specializes to

$$\ddot{u}^t(t) = -\omega_n^2 u(t) \tag{6.12.5}$$

The peak values of the two sides are therefore equal, that is,

$$\ddot{u}_o^t = \omega_n^2 u_o = \omega_n^2 D = A \tag{6.12.6}$$

With damping, Eq. (6.12.5) is not valid at all times, but only at the time instants when $\dot{u}(t) = 0$, in particular when $u(t)$ attains its peak u_o. At this instant, $-\omega_n^2 u_o$ represents the true acceleration of the mass. The peak value $\ddot{u}_o^t$ of $\ddot{u}^t(t)$ does not occur at the same instant, however, unless $\zeta = 0$. The peak values $\ddot{u}_o^t$ and A occur at the same time and are equal only for $\zeta = 0$.

Equation (6.12.4) suggests that the differences between A and $\ddot{u}_o^t$ are expected to increase as the damping increases. This expectation is confirmed by the data presented in Fig. 6.12.2, where the pseudo-acceleration and the acceleration spectra for the El Centro ground motion are plotted for $\zeta = 10\%$, and the ratio $A/\ddot{u}_o^t$ is presented for three damping values. The difference between the two spectra is small for short-period systems and is of some significance only for long-period systems with large values of damping. Thus for a wide range of conditions the pseudo-acceleration may be treated as an approximation to the true acceleration.

As the natural vibration period T_n of a system approaches infinity, the mass of the system stays still while the ground underneath moves. Thus, as $T_n \rightarrow \infty$, $\ddot{u}_o^t \rightarrow 0$ and $D \rightarrow u_{go}$; the latter implies that $A \rightarrow 0$ because of Eq. (6.6.3). Both A and $(\ddot{u}_t)_o \rightarrow 0$ as $T_n \rightarrow \infty$, but at different rates, as evident from the ratio $A/\ddot{u}_o^t$ plotted as a function of T_n; $A \rightarrow 0$ at a much faster rate because of T_n^2 in the denominator of Eq. (6.6.3).

Another way of looking at the differences between the two spectra is by recalling that mA is equal to the peak value of the elastic-resisting force. In contrast, $m\ddot{u}_o^t$ is equal to the peak value of the sum of elastic and damping forces. As seen in Fig. 6.12.2b, the pseudo-acceleration is smaller than the true acceleration, because it is that part of the true acceleration which gives the elastic force.

Parenthetically, we note that the widespread adoption of the prefix *pseudo* is in one sense misleading. The literal meaning of *pseudo* (false) is not really appropriate since we are dealing with approximation rather than with concepts that are in any sense false or inappropriate. In fact, there is rarely the need to use the "pseudo"-spectra as approximations to the "true" spectra because the latter can be computed by the same numerical procedures as those used for the former. Furthermore, as emphasized earlier, the pseudo quantities provide the exact values of the desired deformation and forces.

FURTHER READING

Benioff, H., "The Physical Evaluation of Seismic Destructiveness," *Bulletin of the Seismological Society of America*, **24**, 1934, pp. 398–403.

Biot, M. A., "Theory of Elastic Systems Under Transient Loading with an Application to Earthquake Proof Buildings," *Proceedings, National Academy of Sciences*, **19**, 1933, pp. 262–268.

Biot, M. A., "A Mechanical Analyzer for the Prediction of Earthquake Stresses," *Bulletin of the Seismological Society of America*, **31**, 1941, pp. 151–171.

Bolt, B. A., *Earthquakes*, W.H. Freeman, New York, 1993, Chapters 1–7.

Clough, R. W., and Penzien, J., *Dynamics of Structures*, McGraw-Hill, New York, 1993, pp. 586–597.

Housner, G. W., "Calculating the Response of an Oscillator to Arbitrary Ground Motion," *Bulletin of the Seismological Society of America*, **31**, 1941, pp. 143–149.

Housner, G. W., and Jennings, P. C., *Earthquake Design Criteria*, Earthquake Engineering Research Institute, Berkeley, Calif., 1982, pp. 19–41 and 58–88.

Hudson, D. E., "Response Spectrum Techniques in Engineering Seismology," *Proceedings of the First World Conference in Earthquake Engineering*, Berkeley, Calif., 1956, pp. 4–1 to 4–12.

Hudson, D. E., *Reading and Interpreting Strong Motion Accelerograms*, Earthquake Engineering Research Institute, Berkeley, Calif., 1979, pp. 22–70 and 95–97.

Hudson, D. E., "A History of Earthquake Engineering," *Proceedings of the IDNDR International Symposium on Earthquake Disaster Reduction Technology—30th Anniversary of IISEE*, Tsukuba, Japan, 1992, pp. 3–13.

Mohraz, B., and Elghadamsi, F. E., "Earthquake Ground Motion and Response Spectra," Chapter 2 in *The Seismic Design Handbook* (ed. F. Naeim), Van Nostrand Reinhold, New York, 1989.

Newmark, N. M., and Hall, W. J., *Earthquake Spectra and Design*, Earthquake Engineering Research Institute, Berkeley, Calif., 1982, pp. 29–37.

Newmark, N. M., and Rosenblueth, E., *Fundamentals of Earthquake Engineering*, Prentice Hall, Englewood Cliffs, N.J., 1971, Chapter 7.

Riddell, R., and Newmark, N. M., "Statistical Analysis of the Response of Nonlinear Systems Subjected to Earthquakes," *Structural Research Series No. 468*, University of Illinois at Urbana-Champaign, Urbana, Ill., August 1979.

Rosenblueth, E., "Characteristics of Earthquakes," Chapter 1 in *Design of Earthquake Resistant Structures* (ed. E. Rosenblueth), Pentech Press, London, 1980.

Seed, H. B., and Idriss, I. M., *Ground Motions and Soil Liquefaction During Earthquakes*, Earthquake Engineering Research Institute, Berkeley, Calif., 1982, pp. 21–56.

Veletsos, A. S., "Maximum Deformation of Certain Nonlinear Systems," *Proceedings of the 4th World Conference on Earthquake Engineering*, Santiago, Chile, Vol. 1, 1969, pp. 155–170.

Veletsos, A. S., and Newmark, N. M., "Response Spectra for Single-Degree-of-Freedom Elastic and Inelastic Systems," *Report No. RTD-TDR-63-3096*, Vol. III, Air Force Weapons Laboratory, Albuquerque, N.M., June 1964.

Veletsos, A. S., Newmark, N. M., and Chelapati, C. V., "Deformation Spectra for Elastic and Elastoplastic Systems Subjected to Ground Shock and Earthquake Motion," *Proceedings of the 3rd World Conference on Earthquake Engineering*, New Zealand, Vol. II, 1965, pp. 663–682.

APPENDIX 6: EL CENTRO, 1940 GROUND MOTION

The north–south component of the ground motion recorded at a site in El Centro, California, during the Imperial Valley, California, earthquake of May 18, 1940 is shown in Fig. 6.1.4. This particular version of this record is used throughout this book, and is required in solving some of the end-of-chapter problems. Numerical values for the ground acceleration in units of g, the acceleration due to gravity, are presented in Table A6.1. This includes 1559 data points at equal time spacings of 0.02 sec, to be read row by row; the first value is at $t = 0.02$ sec; acceleration at $t = 0$ is zero. These data are also available electronically from the National Information Service for Earthquake Engineering (NISEE), University of California at Berkeley, on the World Wide Web at the following URL: <http://nisee.berkeley.edu/data/strong_motion/a.k.chopra/index.html>.

TABLE A6.1   GROUND ACCELERATION DATA

0.00630	0.00364	0.00099	0.00428	0.00758	0.01087	0.00682	0.00277
−0.00128	0.00368	0.00864	0.01360	0.00727	0.00094	0.00420	0.00221
0.00021	0.00444	0.00867	0.01290	0.01713	−0.00343	−0.02400	−0.00992
0.00416	0.00528	0.01653	0.02779	0.03904	0.02449	0.00995	0.00961
0.00926	0.00892	−0.00486	−0.01864	−0.03242	−0.03365	−0.05723	−0.04534
−0.03346	−0.03201	−0.03056	−0.02911	−0.02766	−0.04116	−0.05466	−0.06816
−0.08166	−0.06846	−0.05527	−0.04208	−0.04259	−0.04311	−0.02428	−0.00545
0.01338	0.03221	0.05104	0.06987	0.08870	0.04524	0.00179	−0.04167
−0.08513	−0.12858	−0.17204	−0.12908	−0.08613	−0.08902	−0.09192	−0.09482
−0.09324	−0.09166	−0.09478	−0.09789	−0.12902	−0.07652	−0.02401	0.02849
0.08099	0.13350	0.18600	0.23850	0.21993	0.20135	0.18277	0.16420
0.14562	0.16143	0.17725	0.13215	0.08705	0.04196	−0.00314	−0.04824
−0.09334	−0.13843	−0.18353	−0.22863	−0.27372	−0.31882	−0.25024	−0.18166
−0.11309	−0.04451	0.02407	0.09265	0.16123	0.22981	0.29839	0.23197
0.16554	0.09912	0.03270	−0.03372	−0.10014	−0.16656	−0.23299	−0.29941
−0.00421	0.29099	0.22380	0.15662	0.08943	0.02224	−0.04495	0.01834
0.08163	0.14491	0.20820	0.18973	0.17125	0.13759	0.10393	0.07027
0.03661	0.00295	−0.03071	−0.00561	0.01948	0.04458	0.06468	0.08478
0.10487	0.05895	0.01303	−0.03289	−0.07882	−0.03556	0.00771	0.05097
0.01013	−0.03071	−0.07156	−0.11240	−0.15324	−0.11314	−0.07304	−0.03294
0.00715	−0.06350	−0.13415	−0.20480	−0.12482	−0.04485	0.03513	0.11510
0.19508	0.12301	0.05094	−0.02113	−0.09320	−0.02663	0.03995	0.10653
0.17311	0.11283	0.05255	−0.00772	0.01064	0.02900	0.04737	0.06573
0.02021	−0.02530	−0.07081	−0.04107	−0.01133	0.00288	0.01709	0.03131
−0.02278	−0.07686	−0.13095	−0.18504	−0.14347	−0.10190	−0.06034	−0.01877
0.02280	−0.00996	−0.04272	−0.02147	−0.00021	0.02104	−0.01459	−0.05022
−0.08585	−0.12148	−0.15711	−0.19274	−0.22837	−0.18145	−0.13453	−0.08761
−0.04069	0.00623	0.05316	0.10008	0.14700	0.09754	0.04808	−0.00138
0.05141	0.10420	0.15699	0.20979	0.26258	0.16996	0.07734	−0.01527
−0.10789	−0.20051	−0.06786	0.06479	0.01671	−0.03137	−0.07945	−0.12753
−0.17561	−0.22369	−0.27177	−0.15851	−0.04525	0.06802	0.18128	0.14464
0.10800	0.07137	0.03473	0.09666	0.15860	0.22053	0.18296	0.14538
0.10780	0.07023	0.03265	0.06649	0.10033	0.13417	0.10337	0.07257
0.04177	0.01097	−0.01983	0.04438	0.10860	0.17281	0.10416	0.03551
−0.03315	−0.10180	−0.07262	−0.04344	−0.01426	0.01492	−0.02025	−0.05543
−0.09060	−0.12578	−0.16095	−0.19613	−0.14784	−0.09955	−0.05127	−0.00298
−0.01952	−0.03605	−0.05259	−0.04182	−0.03106	−0.02903	−0.02699	0.02515
0.01770	0.02213	0.02656	0.00419	−0.01819	−0.04057	−0.06294	−0.02417
0.01460	0.05337	0.02428	−0.00480	−0.03389	−0.00557	0.02274	0.00679
−0.00915	−0.02509	−0.04103	−0.05698	−0.01826	0.02046	0.00454	−0.01138
−0.00215	0.00708	0.00496	0.00285	0.00074	−0.00534	−0.01141	0.00361
0.01863	0.03365	0.04867	0.03040	0.01213	−0.00614	−0.02441	0.01375
0.01099	0.00823	0.00547	0.00812	0.01077	−0.00692	−0.02461	−0.04230
−0.05999	−0.07768	−0.09538	−0.06209	−0.02880	0.00448	0.03777	0.01773
−0.00231	−0.02235	0.01791	0.05816	0.03738	0.01660	−0.00418	−0.02496
−0.04574	−0.02071	0.00432	0.02935	0.01526	0.01806	0.02086	0.00793
−0.00501	−0.01795	−0.03089	−0.01841	−0.00593	0.00655	−0.02519	−0.05693
−0.04045	−0.02398	−0.00750	0.00897	0.00384	−0.00129	−0.00642	−0.01156
−0.02619	−0.04082	−0.05545	−0.04366	−0.03188	−0.06964	−0.05634	−0.04303

TABLE A6.1   GROUND ACCELERATION DATA (*Continued*)

−0.02972	−0.01642	−0.00311	0.01020	0.02350	0.03681	0.05011	0.02436
−0.00139	−0.02714	−0.00309	0.02096	0.04501	0.06906	0.05773	0.04640
0.03507	0.03357	0.03207	0.03057	0.03250	0.03444	0.03637	0.01348
−0.00942	−0.03231	−0.02997	−0.03095	−0.03192	−0.02588	−0.01984	−0.01379
−0.00775	−0.01449	−0.02123	0.01523	0.05170	0.08816	0.12463	0.16109
0.12987	0.09864	0.06741	0.03618	0.00495	0.00420	0.00345	0.00269
−0.05922	−0.12112	−0.18303	−0.12043	−0.05782	0.00479	0.06740	0.13001
0.08373	0.03745	0.06979	0.10213	−0.03517	−0.17247	−0.13763	−0.10278
−0.06794	−0.03310	−0.03647	−0.03984	−0.00517	0.02950	0.06417	0.09883
0.13350	0.05924	−0.01503	−0.08929	−0.16355	−0.06096	0.04164	0.01551
−0.01061	−0.03674	−0.06287	−0.08899	−0.05430	−0.01961	0.01508	0.04977
0.08446	0.05023	0.01600	−0.01823	−0.05246	−0.08669	−0.06769	−0.04870
−0.02970	−0.01071	0.00829	−0.00314	0.02966	0.06246	−0.00234	−0.06714
−0.04051	−0.01388	0.01274	0.00805	0.03024	0.05243	0.02351	−0.00541
−0.03432	−0.06324	−0.09215	−0.12107	−0.08450	−0.04794	−0.01137	0.02520
0.06177	0.04028	0.01880	0.04456	0.07032	0.09608	0.12184	0.06350
0.00517	−0.05317	−0.03124	−0.00930	0.01263	0.03457	0.03283	0.03109
0.02935	0.04511	0.06087	0.07663	0.09239	0.05742	0.02245	−0.01252
0.00680	0.02611	0.04543	0.01571	−0.01402	−0.04374	−0.07347	−0.03990
−0.00633	0.02724	0.06080	0.03669	0.01258	−0.01153	−0.03564	−0.00677
0.02210	0.05098	0.07985	0.06915	0.05845	0.04775	0.03706	0.02636
0.05822	0.09009	0.12196	0.10069	0.07943	0.05816	0.03689	0.01563
−0.00564	−0.02690	−0.04817	−0.06944	−0.09070	−0.11197	−0.11521	−0.11846
−0.12170	−0.12494	−0.16500	−0.20505	−0.15713	−0.10921	−0.06129	−0.01337
0.03455	0.08247	0.07576	0.06906	0.06236	0.08735	0.11235	0.13734
0.12175	0.10616	0.09057	0.07498	0.08011	0.08524	0.09037	0.06208
0.03378	0.00549	−0.02281	−0.05444	−0.04030	−0.02615	−0.01201	−0.02028
−0.02855	−0.06243	−0.03524	−0.00805	−0.04948	−0.03643	−0.02337	−0.03368
−0.01879	−0.00389	0.01100	0.02589	0.01446	0.00303	−0.00840	0.00463
0.01766	0.03069	0.04372	0.02165	−0.00042	−0.02249	−0.04456	−0.03638
−0.02819	−0.02001	−0.01182	−0.02445	−0.03707	−0.04969	−0.05882	−0.06795
−0.07707	−0.08620	−0.09533	−0.06276	−0.03018	0.00239	0.03496	0.04399
0.05301	0.03176	0.01051	−0.01073	−0.03198	−0.05323	0.00186	0.05696
0.01985	−0.01726	−0.05438	−0.01204	0.03031	0.07265	0.11499	0.07237
0.02975	−0.01288	0.01212	0.03711	0.03517	0.03323	0.01853	0.00383
0.00342	−0.02181	−0.04704	−0.07227	−0.09750	−0.12273	−0.08317	−0.04362
−0.00407	0.03549	0.07504	0.11460	0.07769	0.04078	0.00387	0.00284
0.00182	−0.05513	0.04732	0.05223	0.05715	0.06206	0.06698	0.07189
0.02705	−0.01779	−0.06263	−0.10747	−0.15232	−0.12591	−0.09950	−0.07309
−0.04668	−0.02027	0.00614	0.03255	0.00859	−0.01537	−0.03932	−0.06328
−0.03322	−0.00315	0.02691	0.01196	−0.00300	0.00335	0.00970	0.01605
0.02239	0.04215	0.06191	0.08167	0.03477	−0.01212	−0.01309	−0.01407
−0.05274	−0.02544	0.00186	0.02916	0.05646	0.08376	0.01754	−0.04869
−0.02074	0.00722	0.03517	−0.00528	−0.04572	−0.08617	−0.06960	−0.05303
−0.03646	−0.01989	−0.00332	0.01325	0.02982	0.01101	−0.00781	−0.02662
−0.00563	0.01536	0.03635	0.05734	0.03159	0.00584	−0.01992	−0.00201
0.01589	−0.01024	−0.03636	−0.06249	−0.04780	−0.03311	−0.04941	−0.06570
−0.08200	−0.04980	−0.01760	0.01460	0.04680	0.07900	0.04750	0.01600
−0.01550	−0.00102	0.01347	0.02795	0.04244	0.05692	0.03781	0.01870
−0.00041	−0.01952	−0.00427	0.01098	0.02623	0.04148	0.01821	−0.00506

TABLE A6.1   GROUND ACCELERATION DATA (*Continued*)

−0.00874	−0.03726	−0.06579	−0.02600	0.01380	0.05359	0.09338	0.05883
0.02429	−0.01026	−0.04480	−0.01083	−0.01869	−0.02655	−0.03441	−0.02503
−0.01564	−0.00626	−0.01009	−0.01392	0.01490	0.04372	0.03463	0.02098
0.00733	−0.00632	−0.01997	0.00767	0.03532	0.03409	0.03287	0.03164
0.02403	0.01642	0.00982	0.00322	−0.00339	0.02202	−0.01941	−0.06085
−0.10228	−0.07847	−0.05466	−0.03084	−0.00703	0.01678	0.01946	0.02214
0.02483	0.01809	−0.00202	−0.02213	−0.00278	0.01656	0.03590	0.05525
0.07459	0.06203	0.04948	0.03692	−0.00145	0.04599	0.04079	0.03558
0.03037	0.03626	0.04215	0.04803	0.05392	0.04947	0.04502	0.04056
0.03611	0.03166	0.00614	−0.01937	−0.04489	−0.07040	−0.09592	−0.07745
−0.05899	−0.04052	−0.02206	−0.00359	0.01487	0.01005	0.00523	0.00041
−0.00441	−0.00923	−0.01189	−0.01523	−0.01856	−0.02190	−0.00983	0.00224
0.01431	0.00335	−0.00760	−0.01856	−0.00737	0.00383	0.01502	0.02622
0.01016	−0.00590	−0.02196	−0.00121	0.01953	0.04027	0.02826	0.01625
0.00424	0.00196	−0.00031	−0.00258	−0.00486	−0.00713	−0.00941	−0.01168
−0.01396	−0.01750	−0.02104	−0.02458	−0.02813	−0.03167	−0.03521	−0.04205
−0.04889	−0.03559	−0.02229	−0.00899	0.00431	0.01762	0.00714	−0.00334
−0.01383	0.01314	0.04011	0.06708	0.04820	0.02932	0.01043	−0.00845
−0.02733	−0.04621	−0.03155	−0.01688	−0.00222	0.01244	0.02683	0.04121
0.05559	0.03253	0.00946	−0.01360	−0.01432	−0.01504	−0.01576	−0.04209
−0.02685	−0.01161	0.00363	0.01887	0.03411	0.03115	0.02819	0.02917
0.03015	0.03113	0.00388	−0.02337	−0.05062	−0.03820	−0.02579	−0.01337
−0.00095	0.01146	0.02388	0.03629	0.01047	−0.01535	−0.04117	−0.06699
−0.05207	−0.03715	−0.02222	−0.00730	0.00762	0.02254	0.03747	0.04001
0.04256	0.04507	0.04759	0.05010	0.04545	0.04080	0.02876	0.01671
0.00467	−0.00738	−0.00116	0.00506	0.01128	0.01750	−0.00211	−0.02173
−0.04135	−0.06096	−0.08058	−0.06995	−0.05931	−0.04868	−0.03805	−0.02557
−0.01310	−0.00063	0.01185	0.02432	0.03680	0.04927	0.02974	0.01021
−0.00932	−0.02884	−0.04837	−0.06790	−0.04862	−0.02934	−0.01006	0.00922
0.02851	0.04779	0.02456	0.00133	−0.02190	−0.04513	−0.06836	−0.04978
−0.03120	−0.01262	0.00596	0.02453	0.04311	0.06169	0.08027	0.09885
0.06452	0.03019	−0.00414	−0.03848	−0.07281	−0.05999	−0.04717	−0.03435
−0.03231	−0.03028	−0.02824	−0.00396	0.02032	0.00313	−0.01406	−0.03124
−0.04843	−0.06562	−0.05132	−0.03702	−0.02272	−0.00843	0.00587	0.02017
0.02698	0.03379	0.04061	0.04742	0.05423	0.03535	0.01647	0.01622
0.01598	0.01574	0.00747	−0.00080	−0.00907	0.00072	0.01051	0.02030
0.03009	0.03989	0.03478	0.02967	0.02457	0.03075	0.03694	0.04313
0.04931	0.05550	0.06168	−0.00526	−0.07220	−0.06336	−0.05451	−0.04566
−0.03681	−0.03678	−0.03675	−0.03672	−0.01765	0.00143	0.02051	0.03958
0.05866	0.03556	0.01245	−0.01066	−0.03376	−0.05687	−0.04502	−0.03317
−0.02131	−0.00946	0.00239	−0.00208	−0.00654	−0.01101	−0.01548	−0.01200
−0.00851	−0.00503	−0.00154	0.00195	0.00051	−0.00092	0.01135	0.02363
0.03590	0.04818	0.06045	0.07273	0.02847	−0.01579	−0.06004	−0.05069
−0.04134	−0.03199	−0.03135	−0.03071	−0.03007	−0.01863	−0.00719	0.00425
0.01570	0.02714	0.03858	0.02975	0.02092	0.02334	0.02576	0.02819
0.03061	0.03304	0.01371	−0.00561	−0.02494	−0.02208	−0.01923	−0.01638
−0.01353	−0.01261	−0.01170	−0.00169	0.00833	0.01834	0.02835	0.03836
0.04838	0.03749	0.02660	0.01571	0.00482	−0.00607	−0.01696	−0.00780
0.00136	0.01052	0.01968	0.02884	−0.00504	−0.03893	−0.02342	−0.00791
0.00759	0.02310	0.00707	−0.00895	−0.02498	−0.04100	−0.05703	−0.02920

TABLE A6.1 GROUND ACCELERATION DATA (*Continued*)

−0.00137	0.02645	0.05428	0.03587	0.01746	−0.00096	−0.01937	−0.03778
−0.02281	−0.00784	0.00713	0.02210	0.03707	0.05204	0.06701	0.08198
0.03085	−0.02027	−0.07140	−0.12253	−0.08644	−0.05035	−0.01426	0.02183
0.05792	0.09400	0.13009	0.03611	−0.05787	−0.04802	−0.03817	−0.02832
0.02970	0.03993	0.05017	0.06041	0.07065	0.08089	−0.00192	−0.08473
−0.01846	−0.00861	−0.03652	−0.06444	−0.06169	−0.05894	−0.05618	−0.06073
−0.06528	−0.04628	−0.02728	−0.00829	0.01071	0.02970	0.03138	0.03306
0.03474	0.03642	0.04574	0.05506	0.06439	0.07371	0.08303	0.03605
−0.01092	−0.05790	−0.04696	−0.03602	−0.02508	−0.01414	−0.03561	−0.05708
−0.07855	−0.06304	−0.04753	−0.03203	−0.01652	−0.00102	0.00922	0.01946
−0.07032	−0.05590	−0.04148	−0.05296	−0.06443	−0.07590	−0.08738	−0.09885
−0.06798	−0.03710	−0.00623	0.02465	0.05553	0.08640	0.11728	0.14815
0.08715	0.02615	−0.03485	−0.09584	−0.07100	−0.04616	−0.02132	0.00353
0.02837	0.05321	−0.00469	−0.06258	−0.12048	−0.09960	−0.07872	−0.05784
−0.03696	−0.01608	0.00480	0.02568	0.04656	0.06744	0.08832	0.10920
0.13008	0.10995	0.08982	0.06969	0.04955	0.04006	0.03056	0.02107
0.01158	0.00780	0.00402	0.00024	−0.00354	−0.00732	−0.01110	−0.00780
−0.00450	−0.00120	0.00210	0.00540	−0.00831	−0.02203	−0.03575	−0.04947
−0.06319	−0.05046	−0.03773	−0.02500	−0.01227	0.00046	0.00482	0.00919
0.01355	0.01791	0.02228	0.00883	−0.00462	−0.01807	−0.03152	−0.02276
−0.01401	−0.00526	0.00350	0.01225	0.02101	0.01437	0.00773	0.00110
0.00823	0.01537	0.02251	0.01713	0.01175	0.00637	0.01376	0.02114
0.02852	0.03591	0.04329	0.03458	0.02587	0.01715	0.00844	−0.00027
−0.00898	−0.00126	0.00645	0.01417	0.02039	0.02661	0.03283	0.03905
0.04527	0.03639	0.02750	0.01862	0.00974	0.00086	−0.01333	−0.02752
−0.04171	−0.02812	−0.01453	−0.00094	0.01264	0.02623	0.01690	0.00756
−0.00177	−0.01111	−0.02044	−0.02977	−0.03911	−0.02442	−0.00973	0.00496
0.01965	0.03434	0.02054	0.00674	−0.00706	−0.02086	−0.03466	−0.02663
−0.01860	−0.01057	−0.00254	−0.00063	0.00128	0.00319	0.00510	0.00999
0.01488	0.00791	0.00093	−0.00605	0.00342	0.01288	0.02235	0.03181
0.04128	0.02707	0.01287	−0.00134	−0.01554	−0.02975	−0.04395	−0.03612
−0.02828	−0.02044	−0.01260	−0.00476	0.00307	0.01091	0.00984	0.00876
0.00768	0.00661	0.01234	0.01807	0.02380	0.02953	0.03526	0.02784
0.02042	0.01300	−0.03415	−0.00628	−0.00621	−0.00615	−0.00609	−0.00602
−0.00596	−0.00590	−0.00583	−0.00577	−0.00571	−0.00564	−0.00558	−0.00552
−0.00545	−0.00539	−0.00532	−0.00526	−0.00520	−0.00513	−0.00507	−0.00501
−0.00494	−0.00488	−0.00482	−0.00475	−0.00469	−0.00463	−0.00456	−0.00450
−0.00444	−0.00437	−0.00431	−0.00425	−0.00418	−0.00412	−0.00406	−0.00399
−0.00393	−0.00387	−0.00380	−0.00374	−0.00368	−0.00361	−0.00355	−0.00349
−0.00342	−0.00336	−0.00330	−0.00323	−0.00317	−0.00311	−0.00304	−0.00298
−0.00292	−0.00285	−0.00279	−0.00273	−0.00266	−0.00260	−0.00254	−0.00247
−0.00241	−0.00235	−0.00228	−0.00222	−0.00216	−0.00209	−0.00203	−0.00197
−0.00190	−0.00184	−0.00178	−0.00171	−0.00165	−0.00158	−0.00152	−0.00146
−0.00139	−0.00133	−0.00127	−0.00120	−0.00114	−0.00108	−0.00101	−0.00095
−0.00089	−0.00082	−0.00076	−0.00070	−0.00063	−0.00057	−0.00051	−0.00044
−0.00038	−0.00032	−0.00025	−0.00019	−0.00013	−0.00006	0.00000	

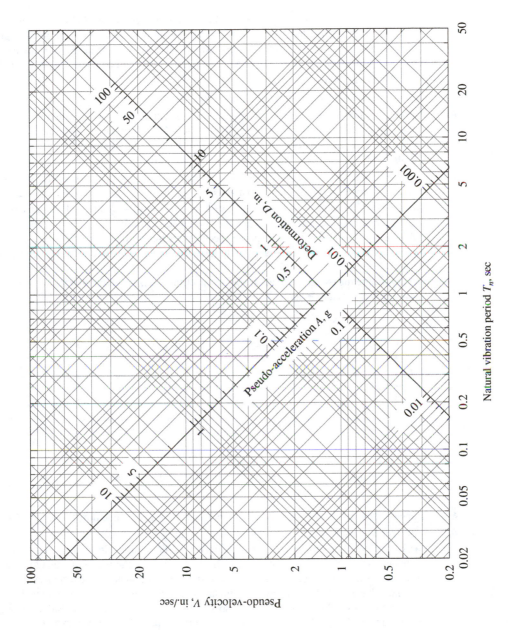

Natural vibration period T_n, sec

Pseudo-velocity V, in./sec

Figure A6.1   Graph paper with four-way logarithmic scales.

PROBLEMS

*6.1   Determine the deformation response $u(t)$ for $0 \le t \le 15$ sec for an SDF system with natural period $T_n = 2$ sec and damping ratio $\zeta = 0$ to El Centro 1940 ground motion. The ground acceleration values are available at every $\Delta t = 0.02$ sec in Appendix 6. Implement the numerical time-stepping algorithm of Section 5.2. Plot $u(t)$ and compare it with Fig. 6.4.1.

*6.2   Solve Problem 6.1 for $\zeta = 5\%$.

*6.3   Solve Problem 6.2 by the central difference method.

6.4   Derive equations for the deformation, pseudo-velocity, and pseudo-acceleration response spectra for ground acceleration $\ddot{u}_g(t) = \dot{u}_{go}\delta(t)$, where $\delta(t)$ is the Dirac delta function and $\dot{u}_{go}$ is the increment in velocity or the magnitude of the acceleration impulse. Plot the spectra for $\zeta = 0$ and 10%.

6.5   An SDF undamped system is subjected to ground motion $\ddot{u}_g(t)$ consisting of a sequence of two acceleration impulses, each with a velocity increment $\dot{u}_{go}$, as shown in Fig. P6.5.
(a) Plot the displacement response of the system $t_d/T_n = \frac{1}{8}, \frac{1}{4}, \frac{1}{2}$, and 1. For each case show the response to individual impulses and the combined response.
(b) Determine the deformation response spectrum for this excitation by plotting $u_o/(\dot{u}_{go}/\omega_n)$ as a function of t_d/T_n. Indicate separately the maximum occurring during $t \le t_d$ and during $t \ge t_d$.
(c) Determine the pseudo-velocity response spectrum for this excitation with $t_d = 0.5$ sec by plotting $V/\dot{u}_{go}$ as a function of $f_n = 1/T_n$.

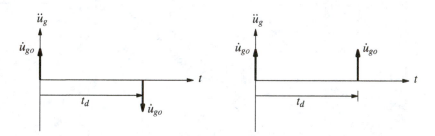

Figure P6.5   **Figure P6.6**

6.6   Repeat Problem 6.5 with the two velocity pulses acting in the same direction (Fig. P6.6).

6.7   Consider harmonic ground motion $\ddot{u}_g(t) = \ddot{u}_{go}\ \sin(2\pi t/T)$.
(a) Derive equations for A and for $\ddot{u}_o^t$ in terms of the natural vibration period T_n and the damping ratio ζ of the SDF system. A is the peak value for the pseudo-acceleration, and $\ddot{u}_o^t$ is the peak value of the true acceleration. Consider only the steady-state response.
(b) Show that A and $\ddot{u}_o^t$ are identical for undamped systems but different for damped systems.
(c) Graphically display the two response spectra by plotting the normalized values $A/\ddot{u}_{go}$ and $\ddot{u}_o^t/\ddot{u}_{go}$ against T_n/T, the ratio of the natural vibration period of the system and the period of the excitation.

*Denotes that a computer is necessary to solve this problem.

6.8– Certain types of near-fault ground motion can be represented by a full sinusoidal cycle of
6.9   ground acceleration (Fig. P6.8) or a full cosine cycle of ground acceleration (Fig. P6.9).
Assuming that the ground velocity and displacement are both zero at time zero, plot the ground
velocity and ground displacement as a function of time. Determine the pseudo-acceleration
response spectrum for undamped systems. Plot this spectrum against t_d/T_n. How will the
true-acceleration response spectrum differ?

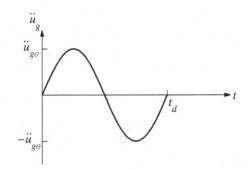

Figure P6.8

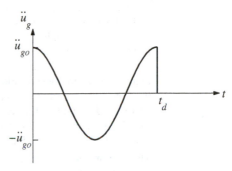

Figure P6.9

6.10 A 10-ft-long vertical cantilever made of a 6-in.-nominal-diameter standard steel pipe supports
a 3000-lb weight attached at the tip, as shown in Fig. P6.10. The properties of the pipe are:
outside diameter = 6.625 in., inside diameter = 6.065 in., thickness = 0.280 in., second moment
of cross-sectional area $I = 28.1$ in^4, Young's modulus $E = 29,000$ ksi, and weight = 18.97
lb/ft length. Determine the peak deformation and the bending stress in the cantilever due to
the El Centro ground motion; assume that $\zeta = 5\%$.

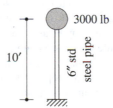

3000 lb

10′   6″ std
          steel pipe

Figure P6.10

6.11 (a) A full water tank is supported on an 80-ft-high cantilever tower. It is idealized as an SDF
system with weight $w = 100$ kips, lateral stiffness $k = 4$ kips/in., and damping ratio $\zeta = 5\%$.
The tower supporting the tank is to be designed for ground motion characterized by the design
spectrum of Fig. 6.9.5 scaled to 0.5g peak ground acceleration. Determine the design values
of lateral deformation and base shear.
(b) The deformation computed for the system in part (a) seemed excessive to the structural
designer, who decided to stiffen the tower by increasing its size. Determine the design values of deformation and base shear for the modified system if its lateral stiffness is 8 kips/in.;
assume that the damping ratio is still 5%. Comment on how stiffening the system has affected
the design requirements. What is the disadvantage of stiffening the system?

(c) If the stiffened tower were to support a tank weighing 200 kips, determine the design requirements; assume for purposes of this example that the damping ratio is still 5%. Comment on how the increased weight has affected the design requirements.

6.12 Solve Problem 6.11 modified as follows: $w = 16$ kips in part (a) and $w = 32$ kips in part (c).

6.13 Solve Problem 6.11 modified as follows: $w = 1600$ kips in part (a) and $w = 3200$ kips in part (c).

6.14 A one-story reinforced-concrete building is idealized for structural analysis as a massless frame supporting a dead load of 10 kips at the beam level. The frame is 24 ft wide and 12 ft high. Each column, clamped at the base, has a 10-in.-square cross section. The Young's modulus of concrete is 3×10^3 ksi, and the damping ratio of the building is estimated as 5%. If the building is to be designed for the design spectrum of Fig. 6.9.5 scaled to a peak ground acceleration of 0.5g, determine the design values of lateral deformation and bending moments in the columns for two conditions:
(a) The cross section of the beam is much larger than that of the columns, so the beam may be assumed as rigid in flexure.
(b) The beam cross section is much smaller than the columns, so the beam stiffness can be ignored. Comment on the influence of beam stiffness on the design quantities.

6.15 The columns of the frame of Problem 6.14 with condition (a) (i.e., rigid beam) are hinged at the base. For the same design earthquake, determine the design values of lateral deformation and bending moments on the columns. Comment on the influence of base fixity on the design deformation and bending moments.

6.16 Determine the peak response of the one-story industrial building of Example 1.2 to ground motion characterized by the design spectrum of Fig. 6.9.5 scaled to a peak ground motion acceleration of 0.25g.
(a) For north-south excitation determine the lateral displacement of the roof and the bending moments in the columns.
(b) For east-west excitation determine the lateral displacement of the roof and the axial force in each brace.

6.17 A small one-story reinforced-concrete building shown in Fig. P6.17 is idealized as a massless frame supporting a total dead load of 10 kips at the beam level. Each 10-in.-square column is hinged at the base; the beam may be assumed to be rigid in flexure; and $E = 3 \times 10^3$ ksi. Determine the peak response of this structure to ground motion characterized by the design spectrum of Fig. 6.9.5 scaled to 0.25g peak ground acceleration. The response quantities of interest are the displacement at the top of the frame and the bending moments in the two columns. Draw the bending moment diagram.

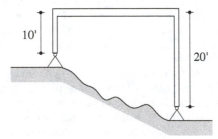

Figure P6.17

6.18   A one-story steel frame of 24-ft span and 12-ft height has the following properties: The second moments of cross-sectional area for beam and columns are $I_b = 160$ in^4 and $I_c = 320$ in^4, respectively; the elastic modulus for steel is 30×10^3 ksi. For purposes of dynamic analysis the frame is considered massless with a weight of 100 kips lumped at the beam level; the columns are clamped at the base; the damping ratio is estimated at 5%. Determine the peak values of lateral displacement at the beam level and bending moments throughout the frame due to the design spectrum of Fig. 6.9.5 scaled to a peak ground acceleration of 0.5g.

6.19   Solve Problem 6.18 assuming that the columns are hinged at the base. Comment on the influence of base fixity on the design deformation and bending moments.

6.20   The ash hopper in Fig. 6.20 consists of a bin mounted on a rigid platform supported by four columns 24 ft long. The weight of the platform, bin, and contents is 100 kips and may be taken as a point mass located 6 ft above the bottom of the platform. The columns are braced in the longitudinal direction, that is, normal to the plane of the paper, but are unbraced in the transverse direction. The column properties are: $A = 20$ in^2, $E = 29{,}500$ ksi, $I = 2000$ in^4, and $S = 170$ in^3. Taking the damping ratio to be 5%, find the peak lateral displacement and the peak stress in the columns due to gravity and the earthquake characterized by the design spectrum of Fig. 6.9.5 scaled to $\frac{1}{3}$g acting in the transverse direction. Take the columns to be clamped at the base and at the rigid platform. Neglect axial deformation of the column and gravity effects on the lateral stiffness.

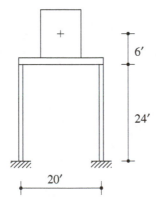

6'

24'

20'

Figure P6.20

6.21   The structure of Example 1.7 subjected to rotational acceleration $\ddot{u}_{g\theta} = \delta(t)$ of the foundation. Derive an equation for the rotation $u_\theta(t)$ of the roof slab in terms of I_O, k_x, k_y, b, and d. Neglect damping.

6.22   The peak response of the system described in Examples 1.7 and 2.4 due to rotational ground acceleration $\ddot{u}_{g\theta}$ (see Fig. E1.7) is to be determined; $\zeta = 5\%$. The design spectrum for translational ground acceleration $(b/2)\ddot{u}_{g\theta}$ is given by Fig. 6.9.5 scaled to a peak ground acceleration of 0.05g. Determine the displacement at each corner of the roof slab, the base torque, and the bending moments about the x and y axes at the base of each column.

6.23   For the design earthquake at a site, the peak values of ground acceleration, velocity, and displacement have been estimated: $\ddot{u}_{go} = 0.5g$, $\dot{u}_{go} = 24$ in./sec, and $u_{go} = 18$ in. For systems with 2% damping ratio, construct the 50th and 84.1th percentile design spectra.

(a) Plot both spectra, together, on four-way log paper.

(b) Plot the 84.1th percentile spectrum for pseudo-acceleration on log-log paper, and determine the equations for $A(T_n)$ for each branch of the spectrum and the period values at the intersections of the branches.

(c) Plot the spectrum of part (b) on a linear-linear graph (the T_n scale should cover the range 0 to 5 sec).

7

Earthquake Response of Inelastic Systems

PREVIEW

We have shown that the peak base shear induced in a linearly elastic system by ground motion is $V_b = (A/g)w$, where w is the weight of the system and A is the pseudo-acceleration spectrum ordinate corresponding to the natural vibration period and damping of the system (Chapter 6). Most buildings are designed, however, for base shear smaller than the elastic base shear associated with the strongest shaking that can occur at the site. This becomes clear from Fig. 7.1, wherein the base shear coefficient A/g

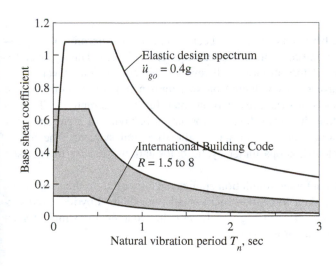

Figure 7.1 Comparison of base shear coefficients from elastic design spectrum and *International Building Code*.

from the design spectrum of Fig. 6.9.5, scaled by 0.4 to correspond to peak ground acceleration of 0.4g, is compared with the base shear coefficient specified in the 2009 *International Building Code*. This disparity implies that buildings designed for the code forces would be deformed beyond the limit of linearly elastic behavior when subjected to ground motions represented by the 0.4g design spectrum. Thus it should not be surprising that buildings suffer damage during intense ground shaking. However, if an earthquake causes damage that is too severe to be repaired economically (Figs. 7.2 and 7.3) or it causes a building to collapse (Fig. 7.4), the design was obviously flawed. The challenge to the engineer is to design the structure so that the damage is controlled to an acceptable degree.

The response of structures deforming into their inelastic range during intense ground shaking is therefore of central importance in earthquake engineering. This chapter is concerned with this important subject. After introducing the elastoplastic system and the parameters describing the system, the equation of motion is presented and the various parameters describing the system and excitation are identified. Then the earthquake response of elastic and inelastic systems is compared with the objective of understanding how yielding influences structural response. This is followed by a procedure to determine the response spectrum for yield force associated with specified values of the ductility factor, together with a discussion of how the spectrum can be used to determine the design force and deformation for inelastic systems. The chapter closes with a procedure to determine the design spectrum for inelastic systems from the elastic design spectrum, followed by a discussion of the important distinction between design and response spectra.

7.1 FORCE–DEFORMATION RELATIONS

7.1.1 Laboratory Tests

Since the 1960s hundreds of laboratory tests have been conducted to determine the force–deformation behavior of structural components for earthquake conditions. During an earthquake structures undergo oscillatory motion with reversal of deformation. Cyclic tests simulating this condition have been conducted on structural members, assemblages of members, reduced-scale models of structures, and on small full-scale structures. The experimental results indicate that the cyclic force–deformation behavior of a structure depends on the structural material (Fig. 7.1.1) and on the structural system. The force–deformation plots show hysteresis loops under cyclic deformations because of inelastic behavior.

Since the 1960s many computer simulation studies have focused on the earthquake response of SDF systems with their force–deformation behavior defined by idealized versions of experimental curves, such as in Fig. 7.1.1. For this chapter, the simplest such idealized force–deformation behavior is chosen.

(a)

(b)

Figure 7.2   The six-story Imperial County Services Building was overstrained by the Imperial Valley, California, earthquake of October 15, 1979. The building is located in El Centro, California, 9 km from the causative fault of the Magnitude 6.5 earthquake; the peak ground acceleration near the building was 0.23g. The first-story reinforced-concrete columns were overstrained top and bottom with partial hinging. The four columns at the right end were shattered at ground level, which dropped the end of the building about 6 in.; see detail. The building was demolished. (Courtesy of K. V. Steinbrugge Collection, Earthquake Engineering Research Center, University of California at Berkeley.)

(a)

(b)

(c)

(d)

Figure 7.3 The O'Higgin's Tower, built in 2009, is a 21-story reinforced-concrete building with an unsymmetric (in plan) shear wall and column-resisting system that is discontinuous and highly irregular over height. Located in Concepcion, 65 miles from the point of the initial rupture of the fault causing the Magnitude 8.8 Offshore Maule Region, Chile, earthquake of February 27, 2010, the building experienced very strong shaking. The damage was so extensive–including collapse of its 12th floor–that the building is slated to be demolished: (a) east face; (b) southeast face; (c) south face; and (d) southeast face: three upper floors and machine room. (Courtesy of Francisco Medina.)

(a)

(b)

Figure 7.4 Psychiatric Day Care Center: (a) before and (b) after the San Fernando, California, earthquake, Magnitude 6.4, February 9, 1971. The structural system for this two-story reinforced-concrete building was a moment-resisting frame. However, the masonry walls added in the second story increased significantly the stiffness and strength of this story. The first story of the building collapsed completely. (Photograph by V. V. Bertero in W. G. Godden collection, National Information Service for Earthquake Engineering, University of California, Berkeley.)

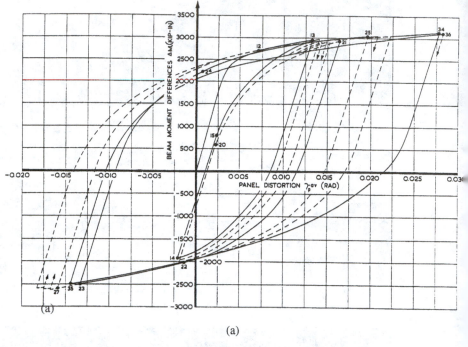

(a)

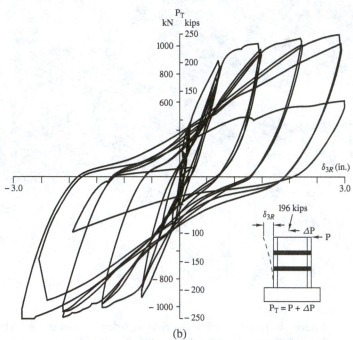

(b)

Figure 7.1.1 (*continues next page*)

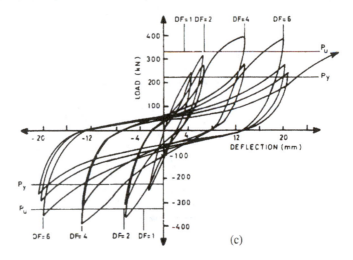

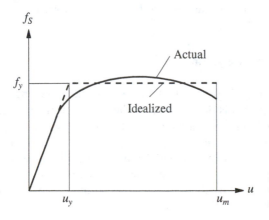

Figure 7.1.1   Force–deformation relations for structural components in different materials: (a) structural steel (from H. Krawinkler, V. V. Bertero, and E. P. Popov, "Inelastic Behavior of Steel Beam-to Column Subassemblages," *Report No. EERC 71-7,* University of California, Berkeley, 1971); (b) reinforced concrete [from E. P. Popov and V. V. Bertero, "On Seismic Behavior of Two R/C Structural Systems for Tall Buildings," in *Structural and Geotechnical Mechanics* (ed. W. J. Hall), Prentice Hall, Englewood Cliffs, N.J., 1977]; (c) masonry [from M. J. N. Priestley, "Masonry," in *Design of Earthquake Resistant Structures* (ed. E. Rosenblueth), Pentech Press, Plymouth, U.K., 1980].

7.1.2 Elastoplastic Idealization

Consider the force–deformation relation for a structure during its initial loading shown in Fig. 7.1.2. It is convenient to idealize this curve by an *elastic–perfectly plastic* (or *elastoplastic* for brevity) force–deformation relation because this approximation permits, as we will see later, the development of response spectra in a manner similar to linearly elastic systems. The elastoplastic approximation to the actual force–deformation curve is

Figure 7.1.2   Force–deformation curve during initial loading: actual and elastoplastic idealization.

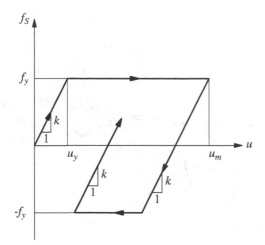

Figure 7.1.3 Elastoplastic force–deformation relation.

drawn, as shown in Fig. 7.1.2, so that the areas under the two curves are the same at the value selected for the maximum displacement u_m. On initial loading this idealized system is linearly elastic with stiffness k as long as the force is less than f_y. Yielding begins when the force reaches f_y, the *yield strength*. The deformation at which yielding begins is u_y, the *yield deformation*. Yielding takes place at constant force f_y (i.e., the stiffness is zero).

Figure 7.1.3 shows a typical cycle of loading, unloading, and reloading for an elastoplastic system. The yield strength is the same in the two directions of deformation. Unloading from a point of maximum deformation takes place along a path parallel to the initial elastic branch. Similarly, reloading from a point of minimum deformation takes place along a path parallel to the initial elastic branch. The cyclic force–deformation relation is path dependent; for deformation u at time t the resisting force f_S depends on the prior history of motion of the system and whether the deformation is currently increasing or decreasing. Thus, the resisting force is an implicit function of deformation: $f_s = f_s(u)$.

7.1.3 Corresponding Linear System

It is desired to evaluate the peak deformation of an elastoplastic system due to earthquake ground motion and to compare this deformation to the peak deformation caused by the same excitation in the *corresponding linear system*. This elastic system is defined to have the same stiffness as the stiffness of the elastoplastic system during its initial loading; see Fig. 7.1.4. Both systems have the same mass and damping. Therefore, the natural vibration period of the corresponding linear system is the same as the period of the elastoplastic system undergoing small ($u \le u_y$) oscillations. At larger amplitudes of motion the natural vibration period is not defined for inelastic systems.

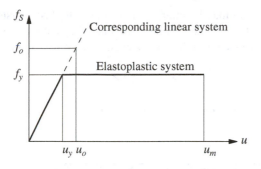

Figure 7.1.4   Elastoplastic system and its corresponding linear system.

7.2 NORMALIZED YIELD STRENGTH, YIELD STRENGTH REDUCTION FACTOR, AND DUCTILITY FACTOR

The *normalized yield strength* $\overline{f}_y$ of an elastoplastic system is defined as

$$\overline{f}_y = \frac{f_y}{f_o} = \frac{u_y}{u_o} \tag{7.2.1}$$

where f_o and u_o are the peak values of the earthquake-induced resisting force and deformation, respectively, in the corresponding linear system. (For brevity the notation f_o has been used instead of f_{So} employed in preceding chapters.) The second part of Eq. (7.2.1) is obvious because $f_y = ku_y$ and $f_o = ku_o$. We may interpret f_o as the minimum strength required for the structure to remain linearly elastic during the ground motion. Normalized yield strength less than unity implies that the yield strength of the system is less than the minimum strength required for the system to remain elastic during the ground motion. Such a system will yield and deform into the inelastic range. The normalized yield strength of a system that remains linearly elastic is equal to unity because such a system can be interpreted as an elastoplastic system with $f_y = f_o$. This system will deform exactly up to the yield deformation during the ground motion.

Alternatively, f_y can be related to f_o through a *yield strength reduction factor* R_y defined by

$$R_y = \frac{f_o}{f_y} = \frac{u_o}{u_y} \tag{7.2.2}$$

Obviously, R_y is the reciprocal of $\overline{f}_y$; R_y is equal to 1 for linearly elastic systems and R_y greater than 1 implies that the system is not strong enough to remain elastic during the ground motion. Such a system will yield and deform into the inelastic range.

The peak, or absolute (without regard to algebraic sign) maximum, deformation of the elastoplastic system due to the ground motion is denoted by u_m. It is meaningful to normalize u_m relative to the yield deformation of the system:

$$\mu = \frac{u_m}{u_y} \tag{7.2.3}$$

This dimensionless ratio is called the *ductility factor*. For systems deforming into the inelastic range, by definition, u_m exceeds u_y and the ductility factor is greater than unity.

The corresponding linear system may be interpreted as an elastoplastic system with $f_y = f_o$, implying that the ductility factor is unity. Later, we relate the peak deformations u_m and u_o of the elastoplastic and corresponding linear systems. Their ratio can be expressed as

$$\frac{u_m}{u_o} = \mu \overline{f}_y = \frac{\mu}{R_y} \tag{7.2.4}$$

This equation follows directly from Eqs. (7.2.1) to (7.2.3).

7.3 EQUATION OF MOTION AND CONTROLLING PARAMETERS

The governing equation for an inelastic system, Eq. (1.7.5), is repeated here for convenience:

$$m\ddot{u} + c\dot{u} + f_S(u) = -m\ddot{u}_g(t) \tag{7.3.1}$$

where the resisting force $f_S(u)$ for an elastoplastic system is shown in Fig. 7.1.3. Equation (7.3.1) will be solved numerically using the procedures of Chapter 5 to determine $u(t)$. The response results presented in this chapter were obtained by the constant average acceleration method using a time step $\Delta t = 0.02$ sec, which was further subdivided to detect the transition from elastic to plastic branches, and vice versa, in the force–deformation relation (Section 5.7).

For a given $\ddot{u}_g(t)$, $u(t)$ depends on three system parameters: ω_n, ζ, and u_y, in addition to the form of the force–deformation relation; here the elastoplastic form has been selected. To demonstrate this fact, Eq. (7.3.1) is divided by m to obtain

$$\ddot{u} + 2\zeta\omega_n\dot{u} + \omega_n^2 u_y \tilde{f}_S(u) = -\ddot{u}_g(t) \tag{7.3.2}$$

where

$$\omega_n = \sqrt{\frac{k}{m}} \qquad \zeta = \frac{c}{2m\,\omega_n} \qquad \tilde{f}_S(u) = \frac{f_S(u)}{f_y} \tag{7.3.3}$$

It is clear from Eq. (7.3.2) that $u(t)$ depends on ω_n, ζ, and u_y. The quantity ω_n is the natural frequency ($T_n = 2\pi/\omega_n$ is the natural period) of the inelastic system vibrating within its linearly elastic range (i.e., $u \leq u_y$). It is also the natural frequency of the corresponding linear system. We will also refer to ω_n and T_n as the small-oscillation frequency and small-oscillation period, respectively, of the inelastic system. Similarly, ζ is the damping ratio of the system based on the critical damping $2m\omega_n$ of the inelastic system vibrating within its linearly elastic range. It is also the damping ratio of the corresponding linear system. The function $\tilde{f}_S(u)$ describes the force–deformation relation in partially dimensionless form, as shown in Fig. 7.3.1a.

For a given $\ddot{u}_g(t)$, the ductility factor μ depends on three system parameters: ω_n, ζ, and $\overline{f}_y$; recall that $\overline{f}_y$ is the normalized yield strength of the elastoplastic system. This can be demonstrated as follows. First, Eq. (7.3.2) is rewritten in terms of $\mu(t) \equiv u(t)/u_y$. Substituting $u(t) = u_y\mu(t)$, $\dot{u}(t) = u_y\dot{\mu}(t)$, and $\ddot{u}(t) = u_y\ddot{\mu}(t)$ in Eq. (7.3.2) and dividing

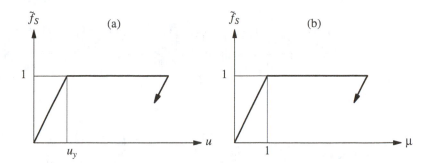

Figure 7.3.1   Force–deformation relations in normalized form.

by u_y gives

$$\ddot{\mu} + 2\zeta\omega_n\dot{\mu} + \omega_n^2\tilde{f}_S(\mu) = -\omega_n^2\frac{\ddot{u}_g(t)}{A_y} \qquad (7.3.4)$$

where $A_y = f_y/m$ may be interpreted as the acceleration of the mass necessary to produce the yield force f_y, and $\tilde{f}_S(\mu)$ is the force–deformation relation in dimensionless form (Fig. 7.3.1b). The acceleration ratio $\ddot{u}_g(t)/A_y$ is the ratio between the ground acceleration and a measure of the yield strength of the structure. Equation (7.3.4) indicates that doubling the ground accelerations $\ddot{u}_g(t)$ will produce the same response $\mu(t)$ as if the yield strength had been halved.

Second, we observe from Eq. (7.3.4) that for a given $\ddot{u}_g(t)$ and form for $\tilde{f}_S(\mu)$, say elastoplastic, $\mu(t)$ depends on ω_n, ζ, and A_y. In turn, A_y depends on ω_n, ζ, and $\overline{f}_y$; this can be shown by substituting Eq. (7.2.1) in the definition of $A_y = f_y/m$ to obtain $A_y = \omega_n^2 u_o \overline{f}_y$, and noting that the peak deformation u_o of the corresponding linear system depends on ω_n and ζ. We have now demonstrated that for a given $\ddot{u}_g(t)$, μ depends on ω_n, ζ, and $\overline{f}_y$.

7.4 EFFECTS OF YIELDING

To understand how the response of SDF systems is affected by inelastic action or yielding, in this section we compare the response of an elastoplastic system to that of its corresponding linear system. The excitation selected is the El Centro ground motion shown in Fig. 6.1.4.

7.4.1 Response History

Figure 7.4.1 shows the response of a linearly elastic system with weight w, natural vibration period $T_n = 0.5$ sec, and no damping. The time variation of deformation shows that the system oscillates about its undeformed equilibrium position and the peak deformation, $u_o = 3.34$ in.; this is also the deformation response spectrum ordinate for $T_n = 0.5$ sec and $\zeta = 0$ (Fig. 6.6.6). Also shown is the time variation of the elastic resisting force f_S; the peak value of this force f_o is given by $f_o/w = 1.37$. This is the minimum strength required

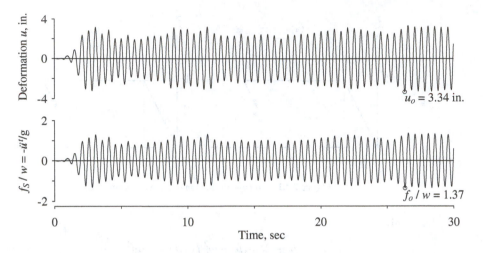

Figure 7.4.1   Response of linear system with $T_n = 0.5$ sec and $\zeta = 0$ to El Centro ground motion.

for the structure to remain elastic. In passing, note from Eq. (7.3.1) that for undamped systems, $f_S(t)/w = -\ddot{u}^t(t)/g$; recall that $\ddot{u}^t$ is the total acceleration of the mass. Thus the peak value of this acceleration is $\ddot{u}^t_o = 1.37g$; this is also the acceleration spectrum ordinate for $T_n = 0.5$ sec and $\zeta = 0$.

Figure 7.4.2 shows the response of an elastoplastic system having the same mass and initial stiffness as the linearly elastic system, with normalized strength $\overline{f}_y = 0.125$ (or yield strength reduction factor $R_y = 8$). The yield strength of this system is $f_y = 0.125 f_o$, where $f_o = 1.37w$ (Fig. 7.4.1); therefore, $f_y = 0.125(1.37w) = 0.171w$. To show more detail, only the first 10 sec of the response is shown in Fig. 7.4.2, which is organized in four parts: (a) shows the deformation $u(t)$; (b) shows the resisting force $f_S(t)$ and acceleration $\ddot{u}^t(t)$; (c) identifies the time intervals during which the system is yielding; and (d) shows the force–deformation relation for one cycle of motion. In the beginning, up to point b, the deformation is small, $f_S < f_y$, and the system is vibrating within its linearly elastic range. We now follow in detail a vibration cycle starting at point a when u and f_S are both zero. At this point the system is linearly elastic and remains so until point b. When the deformation reaches the yield deformation for the first time, identified as b, yielding begins. From b to c the system is yielding (Fig. c), the force is constant at f_y (Fig. b), and the system is on the plastic branch b–c of the force–deformation relation (Fig. d). At c, a local maximum of deformation, the velocity is zero, and the deformation begins to reverse (Fig. a); the system begins to unload elastically along c–d (Fig. d) and is not yielding during this time (Fig. c). Unloading continues until point d (Fig. d), when the resisting force reaches zero. Then the system begins to deform and load in the opposite direction and this continues until f_S reaches $-f_y$ at point e (Figs. b and d). Now yielding begins in the opposite direction and continues until point f (Fig. c); $f_S = -f_y$ during this time span (Fig. b) and the system is moving along the plastic branch e–f (Fig. d). At f a local minimum for deformation, the velocity is zero, and the

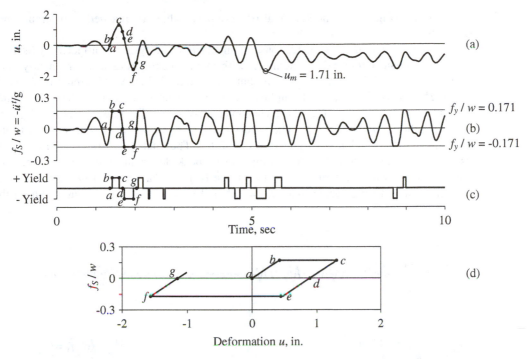

Figure 7.4.2   Response of elastoplastic system with $T_n = 0.5$ sec, $\zeta = 0$, and $\overline{f}_y = 0.125$ to El Centro ground motion: (a) deformation; (b) resisting force and acceleration; (c) time intervals of yielding; (d) force–deformation relation.

deformation begins to reverse (Fig. a); the system begins to reload elastically along f–g (Fig. d) and is not yielding during this time (Fig. c). Reloading brings the resisting force in the system to zero at g, and it continues along this elastic branch until the resisting force reaches $+f_y$.

The time variation of deformation of the yielding system differs from that of the elastic system. Unlike the elastic system (Fig. 7.4.1), the inelastic system after it has yielded does not oscillate about its initial equilibrium position. Yielding causes the system to drift from its initial equilibrium position, and the system oscillates around a new equilibrium position until this gets shifted by another episode of yielding. Therefore, after the ground has stopped shaking, the system will come to rest at a position generally different from its initial equilibrium position (i.e., permanent deformation remains). Thus a structure that has undergone significant yielding during an earthquake may not stand exactly vertical at the end of the motion. For example, the roof of the pergola shown in Fig. 1.1.1 was displaced by 9 in. relative to its original position at the end of the Caracas, Venezuela, earthquake of July 29, 1967; this permanent displacement resulted from yielding of the pipe columns. In contrast, a linear system returns to its initial equilibrium position following the decay of free vibration after the ground has stopped shaking. The peak deformation, 1.71 in., of the elastoplastic system is different from the peak deformation, 3.34 in., of the corresponding linear system (Figs. 7.4.1 and 7.4.2); also, these peak values are reached at different times in the two cases.

We next examine how the response of an elastoplastic system is affected by its yield strength. Consider four SDF systems all with identical properties in their linearly elastic range: $T_n = 0.5$ sec and $\zeta = 5\%$, but they differ in their yield strength: $\overline{f}_y = 1$, 0.5, 0.25, and 0.125. $\overline{f}_y = 1$ implies a linearly elastic system; it is the corresponding linear system for the other three elastoplastic systems. Decreasing values of $\overline{f}_y$ indicate smaller yield strength f_y. The deformation response of these four systems to the El Centro ground motion is presented in Fig. 7.4.3. The linearly elastic system $(\overline{f}_y = 1)$ oscillates around its equilibrium position and its peak deformation $u_o = 2.25$ in. The corresponding peak value of the resisting force is $f_o = ku_o = 0.919w$, the minimum strength required for a system with $T_n = 0.5$ and $\zeta = 5\%$ to remain elastic during the

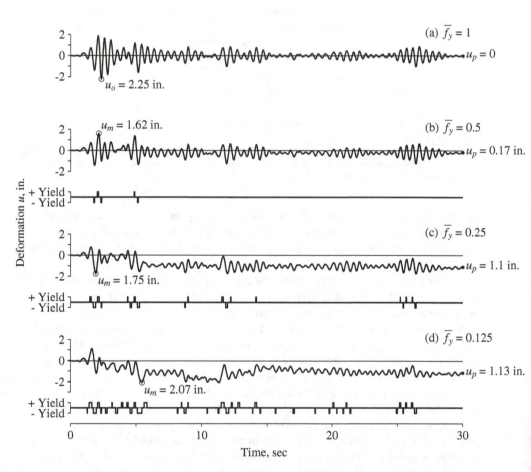

Figure 7.4.3 Deformation response and yielding of four systems due to El Centro ground motion; $T_n = 0.5$ sec, $\zeta = 5\%$; and $\overline{f}_y = 1, 0.5, 0.25,$ and 0.125.

selected ground motion. The other three systems with smaller yield strength—$f_y = 0.5 f_o$, $0.25 f_o$, and $0.125 f_o$, respectively—are therefore expected to deform into the inelastic range. This expectation is confirmed by Fig. 7.4.3, where the time intervals of yielding of these systems are identified. As might be expected intuitively, systems with lower yield strength yield more frequently and for longer intervals. With more yielding, the permanent deformation u_p of the structure after the ground stops shaking tends to increase, but this trend may not be perfect. For the values of T_n and ζ selected, the peak deformations u_m of the three elastoplastic systems are smaller than the peak deformation u_o of the corresponding linear system. This is not always the case, however, because the relative values of u_m and u_o depend on the natural vibration period T_n of the system and the characteristics of the ground motion, and to a lesser degree on the damping in the system.

The ductility factor for an elastoplastic system can be computed using Eq. (7.2.4). For example, the peak deformations of an elastoplastic system with $\overline{f}_y = 0.25$ and the corresponding linear system are $u_m = 1.75$ in. and $u_o = 2.25$ in., respectively. Substituting for u_m, u_o, and $\overline{f}_y$ in Eq. (7.2.4) gives the ductility factor: $\mu = (1/0.25)(1.75/2.25) = 3.11$. This is the *ductility demand* imposed on this elastoplastic system by the ground motion. It represents a requirement on the design of the system in the sense that its *ductility capacity* (i.e., the ability to deform beyond the elastic limit) should exceed the ductility demand.

7.4.2 Ductility Demand, Peak Deformations, and Normalized Yield Strength

In this section we examine how the ductility demand and the relationship between u_m and u_o depend on the natural vibration period T_n and on the normalized yield strength $\overline{f}_y$ or its reciprocal, the yield strength reduction factor R_y. Figure 7.4.4a is a plot of u_m as a function of T_n for four values of $\overline{f}_y = 1, 0.5, 0.25$, and 0.125; u_o is the same as u_m for $\overline{f}_y = 1$. (Note that u_o and u_m have been divided by the peak ground displacement $u_{go} = 8.4$ in.; see Fig. 6.1.4.) Figure 7.4.4b shows the ratio u_m/u_o. In Fig. 7.4.5 the ductility factor μ is plotted versus T_n for the same four values of $\overline{f}_y$; $\mu = 1$ if $\overline{f}_y = 1$. The response histories presented in Fig. 7.4.3 for systems with $T_n = 0.5$ sec and $\zeta = 5\%$ provide the value for $u_o = 2.25$ in., and $u_m = 1.62, 1.75$, and 2.07 in. for $\overline{f}_y = 0.5, 0.25$, and 0.125, respectively. Two of these four data points are identified in Fig. 7.4.4a. The ductility demands μ for the three elastoplastic systems are $1.44, 3.11$ (computed in Section 7.4.1), and 7.36, respectively. These three data points are identified in Fig. 7.4.5. Also identified in these plots are the period values T_a, T_b, T_c, T_d, T_e, and T_f that define the various spectral regions; these were introduced in Section 6.8.

We now study the trends for each spectral region based on the data in Figs. 7.4.4 and 7.4.5. For very-long-period systems ($T_n > T_f$) in the displacement-sensitive region of the spectrum, the deformation u_m of an elastoplastic system is independent of $\overline{f}_y$ and is essentially equal to the peak deformation u_o of the corresponding linear system; the ratio $u_m/u_o \simeq 1$. This observation can be explained as follows: For a fixed mass, such a system is very flexible and, as mentioned in Section 6.8, its mass stays still while the ground

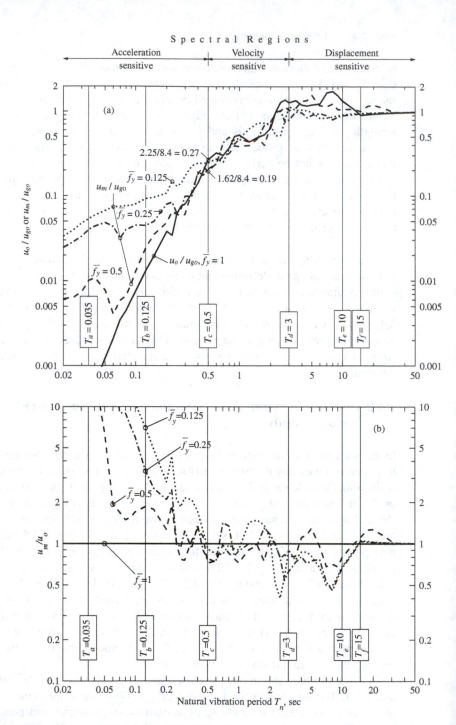

Figure 7.4.4 (a) Peak deformations u_m and u_o of elastoplastic systems and corresponding linear system due to El Centro ground motion; (b) ratio u_m/u_o. T_n is varied; $\zeta = 5\%$ and $\overline{f}_y = 1, 0.5, 0.25$, and 0.125.

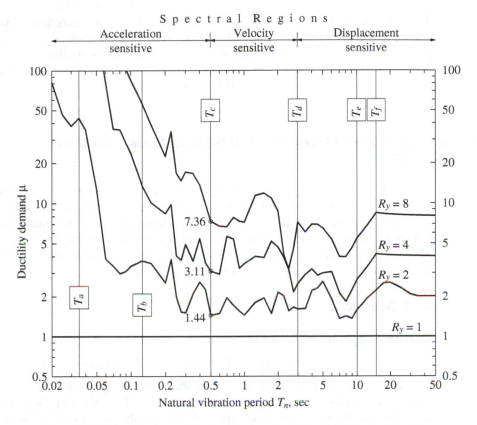

Figure 7.4.5   Ductility demand for elastoplastic system due to El Centro ground motion; $\zeta = 5\%$ and $\overline{f}_y = 1, 0.5, 0.25$, and 0.125, or $R_y = 1, 2, 4$, and 8.

beneath moves. It experiences a peak deformation equal to the peak ground displacement, independent of $\overline{f}_y$. Thus $u_m \simeq u_o \simeq u_{go}$ and Eq. (7.2.4) gives $\mu \simeq 1/\overline{f}_y$ or $\mu \simeq R_y$, a result confirmed by Fig. 7.4.5.

For systems with T_n in the velocity-sensitive region of the spectrum, u_m may be larger or smaller than u_o (i.e., u_m/u_o may or may not exceed 1); both are affected irregularly by variations in $\overline{f}_y$; the ductility demand μ may be larger or smaller than R_y; and the influence of $\overline{f}_y$, although small, is not negligible.

For systems in the acceleration-sensitive region of the spectrum, u_m is greater than u_o, and u_m/u_o increases with decreasing $\overline{f}_y$ (i.e., decreasing yield strength) and decreasing T_n. Therefore, according to Eq. (7.2.4), the ductility demand can be much larger than R_y, an observation confirmed by Fig. 7.4.5. This result implies that the ductility demand on very-short-period systems may be large even if their strength is only slightly below that required for the system to remain elastic.

In the preceding paragraphs we have examined the ductility demand and the relationship between the peak deformations u_m and u_o for elastoplastic and corresponding linear systems, and their dependence on T_n and $\overline{f}_y$ or R_y. Researchers have demonstrated that these relationships identified for the various regions of the response spectrum for one ground motion are valid for the corresponding spectral regions of other ground motions. The period values T_a, T_b, T_c, T_d, T_e, and T_f separating these regions vary from one ground motion to the next, however, as mentioned in Section 6.8.

7.5 RESPONSE SPECTRUM FOR YIELD DEFORMATION AND YIELD STRENGTH

For design purposes it is desired to determine the yield strength f_y (or yield deformation u_y) of the system necessary to limit the ductility demand imposed by the ground motion to a specified value. In their 1960 paper, A. S. Veletsos and N. M. Newmark developed a response spectrum for elastoplastic systems that readily provides the desired information. We present next a procedure to determine this spectrum, which is central to understanding the earthquake response and design of yielding structures.

7.5.1 Definitions

Response spectra are plotted for the quantities

$$D_y = u_y \qquad V_y = \omega_n u_y \qquad A_y = \omega_n^2 u_y \qquad (7.5.1)$$

Note that D_y is the yield deformation u_y of the elastoplastic system, *not* its peak deformation u_m. A plot of D_y against T_n for fixed values of the ductility factor μ is called the *yield–deformation response spectrum*. Following the definitions for linearly elastic systems (Section 6.6), similar plots of V_y and A_y are called the *pseudo-velocity response spectrum* and *pseudo-acceleration response spectrum*, respectively.

These definitions of D_y, V_y, and A_y for elastoplastic systems are consistent with the definitions of D, V, and A for linear systems. This becomes apparent by interpreting a linear system with peak deformation u_o as an elastoplastic system with yield deformation $u_y = u_o$. Then Eqs. (7.5.1) for the elastoplastic system are equivalent to Eqs. (6.6.1) and (6.6.3) for linear systems.

The quantities D_y, V_y, and A_y can be presented in a single four-way logarithmic plot in the same manner as for linear systems. This is possible because these quantities are related through

$$\frac{A_y}{\omega_n} = V_y = \omega_n D_y \qquad \text{or} \qquad \frac{T_n}{2\pi} A_y = V_y = \frac{2\pi}{T_n} D_y \qquad (7.5.2)$$

and these relations are analogous to Eq. (6.6.6) relating D, V, and A for linear systems.

The yield strength of an elastoplastic system is

$$f_y = \frac{A_y}{g} w \qquad (7.5.3)$$

where w is the weight of the system. This result can be derived using Eq. (7.5.1) as follows:

$$f_y = ku_y = m(\omega_n^2 u_y) = mA_y = \frac{A_y}{g} w$$

Observe that Eq. (7.5.3) is analogous to Eq. (6.7.2), repeated here for convenience:

$$f_o = \frac{A}{g} w \tag{7.5.4}$$

where A is the pseudo-acceleration response spectrum for linearly elastic systems.

7.5.2 Yield Strength for Specified Ductility

An interpolative procedure is necessary to obtain the yield strength of an elastoplastic system for a specified ductility factor since the response of a system with arbitrarily selected yield strength will seldom correspond to the desired ductility value. This becomes apparent considering the response results of Fig. 7.4.3 for four systems, all having the same $T_n = 0.5$ sec and $\zeta = 5\%$, but different yield strengths, as defined by the normalized yield strength $\bar{f}_y = 1, 0.5, 0.25$, and 0.125. The ductility factors for these four systems are $1, 1.44, 3.11$, and 7.36 (Section 7.4.2). Clearly, these results do not provide the $\bar{f}_y$ value corresponding to a specified ductility factor, say, 4.

These results provide the basis, however, to obtain the desired information. They lead to a plot showing $\bar{f}_y$ (or R_y) as functions of μ for a fixed T_n and ζ. The solid lines in Fig. 7.5.1 show such plots for several values of T_n and $\zeta = 5\%$. In the plot for $T_n = 0.5$ sec, three of the four pairs of $\bar{f}_y$ and μ values mentioned in the preceding paragraph are identified. To develop some insight into the trends, for each $\bar{f}_y$ two values of the ductility factor are shown: u_m^+/u_y, where u_m^+ is the maximum deformation in the positive direction, and u_m^-/u_y, where u_m^- is the absolute value of the largest deformation in the negative direction. The solid line represents μ, the larger of the two values of the ductility factor.

Contrary to intuition, the ductility factor μ does not always increase monotonically as the normalized strength $\bar{f}_y$ decreases. In particular, more than one yield strength is possible corresponding to a given μ. For example, the plot for $T_n = 2$ sec features two values of $\bar{f}_y$ corresponding to $\mu = 5$. This peculiar phenomenon occurs where the u_m^+/u_y and u_m^-/u_y curves cross (e.g., points a or b in Fig. 7.5.1). Such a point often corresponds to a local minimum of the ductility factor, which permits more than one value of $\bar{f}_y$ for a slightly larger value of μ. For each μ value, it is the largest $\bar{f}_y$, or the largest yield strength, that is relevant for design.

The yield strength f_y of an elastoplastic system for a specified ductility factor μ can be obtained using the corresponding $\bar{f}_y$ value and Eq. (7.2.1). To ensure accuracy in this $\bar{f}_y$ value it is obtained by an iterative procedure, not from a plot like Fig. 7.5.1. From the available data pairs $(\bar{f}_y, \mu)$ interpolation assuming a linear relation between $\log(\bar{f}_y)$ and $\log(\mu)$ leads to $\bar{f}_y$, corresponding to the μ specified. The response history of the system with this $\bar{f}_y$ is computed to determine the ductility factor. If this is close enough to the μ specified—say to within 1%—the $\bar{f}_y$ value is considered satisfactory; otherwise, it is modified until satisfactory agreement is reached.

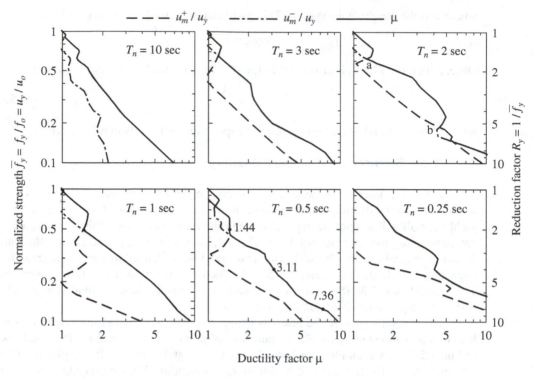

Figure 7.5.1 Relationship between normalized strength (or reduction factor) and ductility factor due to El Centro ground motion; $\zeta = 5\%$.

7.5.3 Construction of Constant-Ductility Response Spectrum

The procedure to construct the response spectrum for elastoplastic systems corresponding to specified levels of ductility factor is summarized as a sequence of steps:

1. Numerically define the ground motion $\ddot{u}_g(t)$.

2. Select and fix the damping ratio ζ for which the spectrum is to be plotted.

3. Select a value for T_n.

4. Determine the response $u(t)$ of the linear system with T_n and ζ equal to the values selected. From $u(t)$ determine the peak deformation u_o and the peak force $f_o = ku_o$. Such results for $T_n = 0.5$ sec and $\zeta = 5\%$ are shown in Fig. 7.4.3a.

5. Determine the response $u(t)$ of an elastoplastic system with the same T_n and ζ and yield force $f_y = \overline{f}_y f_o$, with a selected $\overline{f}_y < 1$. From $u(t)$ determine the peak deformation u_m and the associated ductility factor from Eq. (7.2.4). Repeat such an analysis for enough values of $\overline{f}_y$ to develop data points $(\overline{f}_y, \mu)$ covering the ductility range of interest. Such results are shown in Fig. 7.4.3 for $\overline{f}_y = 0.5, 0.25,$ and 0.125, which provide three data points for the $T_n = 0.5$ case in Fig. 7.5.1.

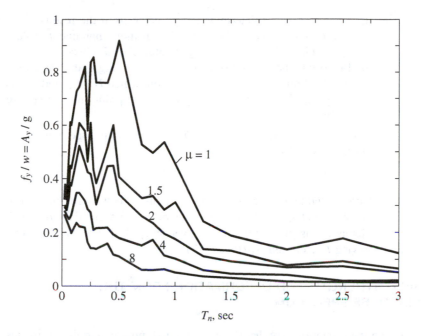

Figure 7.5.2   Constant-ductility response spectrum for elastoplastic systems and El Centro ground motion; $\mu = 1, 1.5, 2, 4,$ and 8; $\zeta = 5\%$.

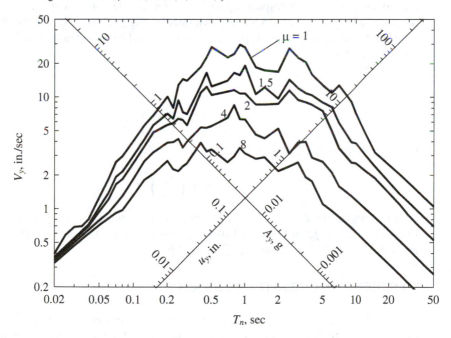

Figure 7.5.3   Constant-ductility response spectrum for elastoplastic systems and El Centro ground motion; $\mu = 1, 1.5, 2, 4,$ and 8; $\zeta = 5\%$.

6. **a.** For a selected μ determine the $\overline{f}_y$ value from the results of step 5 using the procedure described in Section 7.5.2. If more than one $\overline{f}_y$ value corresponds to a particular value of μ, the largest value of $\overline{f}_y$ is chosen.
   b. Determine the spectral ordinates corresponding to the value of $\overline{f}_y$ determined in step 6a. Equation (7.2.1) gives u_y, from which D_y, V_y, and A_y can be determined using Eq. (7.5.1). These data provide one point on the response spectrum plots of Figs. 7.5.2 and 7.5.3.

7. Repeat steps 3 to 6 for a range of T_n resulting in the spectrum valid for the μ value chosen in step 6a.

8. Repeat steps 3 to 7 for several values of μ.

Constructed by this procedure, the response spectrum for elastoplastic systems with $\zeta = 5\%$ subjected to the El Centro ground motion is presented for $\mu = 1, 1.5, 2, 4$, and 8 in two different forms: linear plot of A_y/g versus T_n (Fig. 7.5.2) and a four-way logarithmic plot showing D_y, V_y, and A_y (Fig. 7.5.3).

7.6 YIELD STRENGTH AND DEFORMATION FROM THE RESPONSE SPECTRUM

Given the excitation, say the El Centro ground motion, and the properties T_n and ζ of an SDF system, it is desired to determine the yield strength for the system consistent with a ductility factor μ. Corresponding to T_n, ζ, and μ, the value of A_y/g is read from the spectrum of Fig. 7.5.2 or 7.5.3 and substituted in Eq. (7.5.3) to obtain the desired yield strength f_y. An equation for the peak deformation can be derived in terms of A_y as follows. From Eq. (7.2.3):

$$u_m = \mu u_y \tag{7.6.1}$$

where

$$u_y = \frac{f_y}{k} = \left(\frac{T_n}{2\pi}\right)^2 A_y \tag{7.6.2}$$

Putting Eqs. (7.6.1) and (7.6.2) together gives

$$u_m = \mu \left(\frac{T_n}{2\pi}\right)^2 A_y \tag{7.6.3}$$

As an example, for $T_n = 0.5$ sec, $\zeta = 5\%$, and $\mu = 4$, Fig. 7.5.2 gives $A_y/g = 0.179$. From Eq. (7.5.3), $f_y = 0.179w$. From Eq. (7.6.2), $u_y = (0.5/2\pi)^2 0.179g = 0.438$ in., and Eq. (7.6.1) gives $u_m = 4(0.438) = 1.752$ in.

7.7 YIELD STRENGTH–DUCTILITY RELATION

The yield strength f_y required of an SDF system permitted to undergo inelastic deformation is less than the minimum strength necessary for the structure to remain elastic. Figure 7.5.2 shows that the required yield strength is reduced with increasing values of the

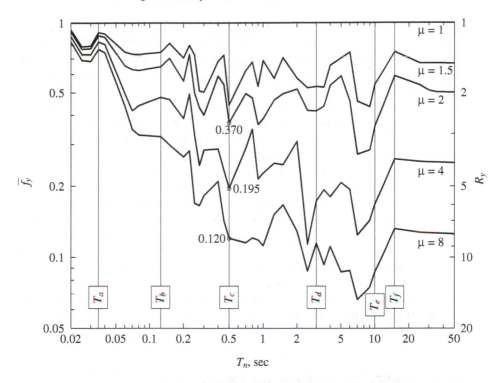

Figure 7.7.1 Normalized strength $\overline{f}_y$ of elastoplastic systems as a function of natural vibration period T_n for $\mu = 1, 1.5, 2, 4$, and 8; $\zeta = 5\%$; El Centro ground motion.

ductility factor. Even small amounts of inelastic deformation, corresponding to $\mu = 1.5$, produce a significant reduction in the required strength. Additional reductions are achieved with increasing values of μ but at a slower rate.

To study these reductions quantitatively, Fig. 7.7.1 shows the normalized yield strength $\overline{f}_y$ and yield strength reduction factor R_y of elastoplastic systems as a function of T_n for four values of μ. This is simply the data of Fig. 7.5.2 (or Fig. 7.5.3) plotted in a different form. From Fig. 7.5.2, for each value of T_n, the $\mu = 1$ curve gives f_o/w and the curve for another μ gives the corresponding f_y/w. The normalized strength $\overline{f}_y$ is then computed from Eq. (7.2.1). For example, consider systems with $T_n = 0.5$ sec; $f_o = 0.919w$ and $f_y = 0.179w$ for $\mu = 4$; the corresponding $\overline{f}_y = 0.195$. Such computations for $\mu = 1, 1.5, 2, 4$, and 8 give $\overline{f}_y = 1, 0.442, 0.370, 0.195$, and 0.120 (or 100, 44.2, 37.0, 19.5, and 12.0%), respectively; three of these data points are identified in Fig. 7.7.1. Repeating such computations for a range of T_n leads to Fig. 7.7.1, wherein the period values T_a, T_b, T_c, T_d, T_e, and T_f that define the various spectral regions are identified; these were introduced in Section 7.4.

The practical implication of these results is that a structure may be designed for earthquake resistance by making it strong, by making it ductile, or by designing it for economic combinations of both properties. Consider again an SDF system with

$T_n = 0.5$ sec and $\zeta = 5\%$ to be designed for the El Centro ground motion. If this system is designed for a strength $f_o = 0.919w$ or larger, it will remain within its linearly elastic range during this excitation; therefore, it need not be ductile. On the other hand, if it can develop a ductility factor of 8, it need be designed for only 12% of the strength f_o required for elastic behavior. Alternatively, it may be designed for strength equal to 37% of f_o and a ductility capacity of 2; or strength equal to 19.5% of f_o and a ductility capacity of 4. For some types of materials and structural members, ductility is difficult to achieve, and economy dictates designing for large lateral forces; for others, providing ductility is much easier than providing lateral strength, and the design practice reflects this. If the combination of strength and ductility provided is inadequate, the structure may be damaged to an extent that repair is not economical (see Fig. 7.2), or it may collapse (see Fig. 7.3).

The strength reduction permitted for a specified allowable ductility varies with T_n. As shown in Fig. 7.7.1, the normalized strength $\overline{f}_y$ tends to 1 (and the yield-strength reduction factor R_y tends to 1), implying no reduction, at the short-period end of the spectrum; and to $\overline{f}_y = 1/\mu$ (i.e., $R_y = \mu$) at the long-period end of the spectrum. In between, $\overline{f}_y$ determined for a single ground motion varies in an irregular manner. However, smooth curves can be developed for design purposes (Section 7.10).

The normalized strength for a specified ductility factor also depends on the damping ratio ζ, but this dependence is not strong. It is usually ignored, therefore, in design applications.

7.8 RELATIVE EFFECTS OF YIELDING AND DAMPING

Figure 7.8.1 shows the response spectra for linearly elastic systems for three values of viscous damping: $\zeta = 2$, 5, and 10%. For the same three damping values, response spectra for elastoplastic systems are presented for two different ductility factors: $\mu = 4$ and $\mu = 8$. From these results the relative effects of yielding and damping are identified in this section.

The effects of yielding and viscous damping are similar in one sense but different in another. They are similar in the sense that both mechanisms reduce the pseudo-acceleration A_y and hence the peak value of the lateral force for which the system should be designed. The relative effectiveness of yielding and damping is quite different, however, in the various spectral regions:

1. Damping has negligible influence on the response of systems with $T_n > T_f$ in the displacement-sensitive region of the spectrum, whereas for such systems the effects of yielding on the design force are very important, but on the peak deformation u_m they are negligible (Fig. 7.4.4).

2. Damping has negligible influence in the response of systems with $T_n < T_a$ in the acceleration-sensitive region of the spectrum, whereas for such systems the effects of yielding on the peak deformation and ductility demand are very important (Figs. 7.4.4 and 7.4.5), but on the design force they are small. In the limit as T_n tends to zero, the pseudo-acceleration A or A_y will approach the peak ground acceleration, implying that this response parameter is unaffected by damping or yielding.

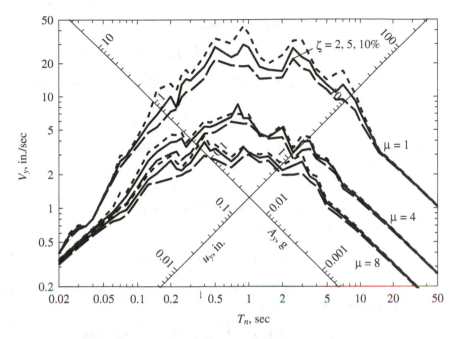

Figure 7.8.1 Response spectra for elastoplastic systems and El Centro ground motion; $\zeta = 2, 5,$ and 10% and $\mu = 1, 4,$ and 8.

3. Damping is most effective in reducing the response of systems with T_n in the velocity-sensitive region of the spectrum, where yielding is even more effective.

Thus, in general, the effects of yielding cannot be considered in terms of a fixed amount of equivalent viscous damping. If this were possible, the peak response of inelastic systems could be determined directly from the response spectrum for linearly elastic systems, which would have been convenient.

The effectiveness of damping in reducing the response is smaller for inelastic systems and decreases as inelastic deformations increase (Fig. 7.8.1). For example, averaged over the velocity-sensitive spectral region, the percentage reduction in response resulting from increasing the damping ratio from 2% to 10% for systems with $\mu = 4$ is about one-half of the reduction for linearly elastic systems. Thus the added viscoelastic dampers mentioned in Section 6.8 may be less beneficial in reducing the response of inelastic systems compared to elastic systems.

7.9 DISSIPATED ENERGY

The input energy imparted to an inelastic system by an earthquake is dissipated by both viscous damping and yielding. These energy quantities are defined and discussed in this section. The various energy terms can be defined by integrating the equation of motion of an inelastic system, Eq. (7.3.1), as follows:

$$\int_0^u m\ddot{u}(t)\,du + \int_0^u c\dot{u}(t)\,du + \int_0^u f_S(u)\,du = -\int_0^u m\ddot{u}_g(t)\,du \qquad (7.9.1)$$

The right side of this equation is the energy input to the structure since the earthquake excitation began:

$$E_I(t) = -\int_0^u m\ddot{u}_g(t)\,du \qquad (7.9.2)$$

This is clear by noting that as the structure moves through an increment of displacement du, the energy supplied to the structure by the effective force $p_{\text{eff}}(t) = -m\ddot{u}_g(t)$ is

$$dE_I = -m\ddot{u}_g(t)\,du$$

The first term on the left side of Eq. (7.9.1) is the kinetic energy of the mass associated with its motion relative to the ground:

$$E_K(t) = \int_0^u m\ddot{u}(t)\,du = \int_0^{\dot{u}} m\dot{u}(t)\,d\dot{u} = \frac{m\dot{u}^2}{2} \qquad (7.9.3)$$

The second term on the left side of Eq. (7.9.1) is the energy dissipated by viscous damping, defined earlier in Section 3.8:

$$E_D(t) = \int_0^u f_D(t)\,du = \int_0^u c\dot{u}(t)\,du \qquad (7.9.4)$$

The third term on the left side of Eq. (7.9.1) is the sum of the energy dissipated by yielding and the recoverable strain energy of the system:

$$E_S(t) = \frac{[f_S(t)]^2}{2k} \qquad (7.9.5)$$

where k is the initial stiffness of the inelastic system. Thus the energy dissipated by yielding is

$$E_Y(t) = \int_0^u f_S(u)\,du - E_S(t) \qquad (7.9.6)$$

Based on these energy quantities, Eq. (7.9.1) is a statement of energy balance for the system:

$$E_I(t) = E_K(t) + E_D(t) + E_S(t) + E_Y(t) \qquad (7.9.7)$$

Concurrent with the earthquake response analysis of a system these energy quantities can be computed conveniently by rewriting the integrals with respect to time. Thus

$$E_D(t) = \int_0^t c[\dot{u}(t)]^2\,dt$$

$$E_Y(t) = \left[\int_0^t \dot{u}\,f_S(u)\,dt\right] - E_S(t) \qquad (7.9.8)$$

The kinetic energy E_K and strain energy E_S at any time t can be computed conveniently from Eqs. (7.9.3) and (7.9.5), respectively.

   The foregoing energy analysis is for a structure whose mass is acted upon by a force $p_{\text{eff}}(t) = -m\ddot{u}_g(t)$, not for a structure whose base is excited by acceleration $\ddot{u}_g(t)$. Therefore, the input energy term in Eq. (7.9.1) represents the energy supplied by $p_{\text{eff}}(t)$, not by $\ddot{u}_g(t)$, and the kinetic energy term in Eq. (7.9.1) represents the energy of motion relative to the base rather than that due to the total motion. As it is the relative displacement and velocity that cause forces in a structure, an energy equation expressed in terms of the relative motion is more meaningful than one expressed in terms of absolute velocity and displacement. Furthermore, the energy dissipated in viscous damping or yielding depends only on the relative motion.

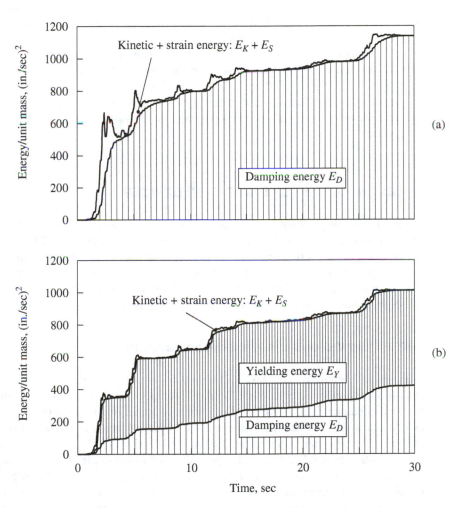

Figure 7.9.1 Time variation of energy dissipated by viscous damping and yielding, and of kinetic plus strain energy; (a) linear system, $T_n = 0.5$ sec, $\zeta = 5\%$; (b) elastoplastic system, $T_n = 0.5$ sec, $\zeta = 5\%$, $\overline{f}_y = 0.25$.

Shown in Fig. 7.9.1 is the variation of these energy quantities with time for two SDF systems subjected to the El Centro ground motion. The results presented are for a linearly elastic system with natural period $T_n = 0.5$ sec and damping ratio $\zeta = 0.05$, and for an elastoplastic system with the same properties in the elastic range and normalized strength $\bar{f}_y = 0.25$. Recall that the deformation response of these two systems was presented in Fig. 7.4.3.

The results of Fig. 7.9.1 show that eventually the structure dissipates by viscous damping and yielding all the energy supplied to it. This is indicated by the fact that the kinetic energy and recoverable strain energy diminish near the end of the ground shaking. Viscous damping dissipates less energy from the inelastic system, implying smaller velocities relative to the elastic system. Figure 7.9.1 also indicates that the energy input to a linear system and to an inelastic system, both with the same T_n and ζ, is not the same. Furthermore, the input energy varies with T_n for both systems.

The yielding energy shown in Fig. 7.9.1b indicates a demand imposed on the structure. If this much energy can be dissipated through yielding of the structure, it needs to be designed only for $\bar{f}_y = 0.25$ (i.e., one-fourth the force developed in the corresponding linear system). The repeated yielding that dissipates energy causes damage to the structure, however, and leaves it in a permanently deformed condition at the end of the earthquake.

7.10 SUPPLEMENTAL ENERGY DISSIPATION DEVICES

If part of this energy could be dissipated through supplemental devices that can easily be replaced, as necessary, after an earthquake, the structural damage could be reduced. Such devices may be cost-effective in the design of new structures and for seismic protection of existing structures. Available devices can be classified into three main categories: fluid viscous and viscoelastic dampers, metallic yielding dampers, and friction dampers. Only one of the several devices available in each category is described briefly here.

7.10.1 Fluid Viscous and Viscoelastic Dampers

In the most commonly used viscous damper for seismic protection of structures, a viscous fluid, typically silicone-based fluid, is forced to flow through small orifices within a closed container (Fig. 7.10.1a). Energy is dissipated due to friction between the fluid and orifice walls. The damper force–velocity relation, which is a function of the rate of loading, may be linear or nonlinear. Figure 7.10.1b shows an experimentally determined force–displacement relation for a damper subjected to sinusoidal force. An elliptical loop indicates a linear force–velocity relation, as demonstrated analytically in Section 3.10. Fluid viscous dampers are installed within the skeleton of a building frame, typically in line with diagonal bracing (Fig. 7.10.1c) or between the towers (or piers) and the deck of a bridge.

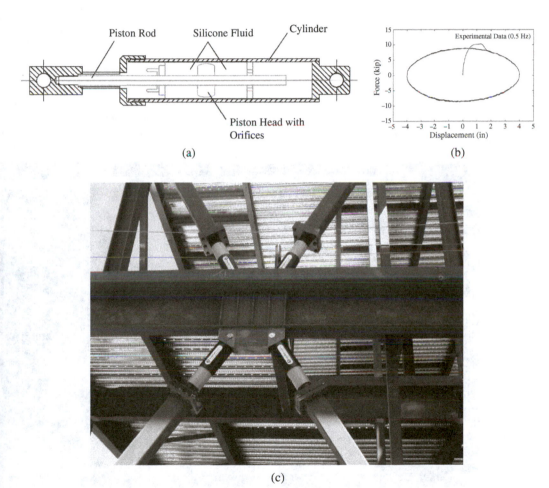

Figure 7.10.1   (a) Fluid viscous damper: schematic drawing; (b) force–displacement relation; and (c) diagonal bracing with fluid viscous damper. [Credits: (a) Cameron Black; (b) Cameron Black; and (c) Taylor Devices, Inc.]

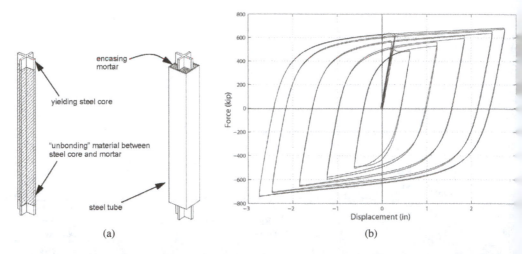

(a)   (b)

(c)

Figure 7.10.2   (a) Buckling restrained brace (BRB): schematic drawings; (b) force–displacement relation; and (c) diagonal bracing with BRB. [Credits: (a) Ian Aiken; (b) Cameron Black; and (c) Ian Aiken.]

7.10.2 Metallic Yielding Dampers

Metallic dampers dissipate energy through hysteretic behavior of metals when deformed into their inelastic range. A wide variety of devices have been developed and tested that dissipate energy in flexural, shear, or extensional deformation modes. Among them, the bucking restrained brace (BRB) has been widely used in buildings in Japan—where it was first developed—and in the United States. One such device consists of a steel core encased in a steel tube filled with mortar (Fig. 7.10.2a). The steel core carries the axial load, whereas the filler material provides lateral support to the core and prevents its buckling. Displaying stable hysteresis loops (Fig. 7.10.2b), BRBs are typically installed in Chevron-type bracing configurations (Fig. 7.10.2c).

7.10.3 Friction Dampers

Various types of energy-dissipating devices, utilizing friction as a means of energy dissipation, have been developed and tested by researchers. These devices increase the capacity of the structure to dissipate energy but do not change the natural vibration periods significantly—by about 10 to 20%. One of these devices is the slotted bolted connection (SBC). Figure 7.10.3 includes a schematic diagram of an SBC, the resulting almost rectangular hysteresis loop, the SBC connected to the top of chevron braces, and a test structure with 12 SBCs.

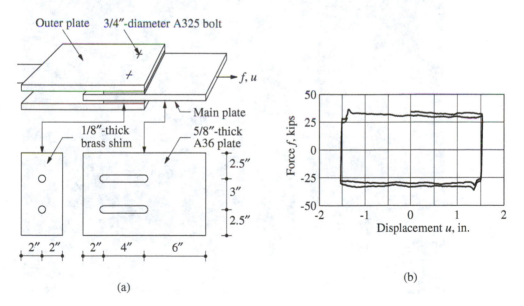

Figure 7.10.3a, b   (a) Schematic diagram of slotted bolted connection (SBC); (b) force–displacement diagram of an SBC. (Adapted from C. E. Grigorian and E. P. Popov, 1994.)

(c)

(d)

Figure 7.10.3c, d   (c) SBC at top of chevron brace in test structure; (d) test structure with 12 SBCs on the shaking table at the University of California at Berkeley. (Courtesy of K. V. Steinbrugge Collection, Earthquake Engineering Research Center, University of California at Berkeley.)

7.11 INELASTIC DESIGN SPECTRUM

In this section a procedure is presented for constructing the design spectrum for elasto-plastic systems for specified ductility factors. This could be achieved by constructing the constant-ductility response spectrum (Section 7.5.3) for many plausible ground motions for the site and, based on these data, the design spectrum associated with an exceedance probability could be established. A simpler approach is to develop a constant-ductility design spectrum from the elastic design spectrum (Section 6.9), multiplying it by the normalized strength $\overline{f}_y$ or dividing it by the yield strength reduction factor R_y.

7.11.1 R_y–μ–T_n Equations

Figure 7.11.1 shows the yield-strength reduction factor R_y for elastoplastic systems as a function of T_n for selected values of μ. These data for the El Centro ground motion, which are the reciprocal of the $\overline{f}_y$ values in Fig. 7.7.1, are shown in Fig. 7.11.1a and the median value over 20 ground motions in Fig. 7.11.1b. As noted in Section 7.7, the reduction in yield strength permitted for a specified ductility factor varies with T_n. At the short-period end of the spectrum, R_y tends to 1, implying no reduction. At the long-period end of the spectrum, R_y tends to μ. In between, R_y varies with T_n in an irregular manner for a single ground motion, but its median over the ground motion ensemble varies relatively

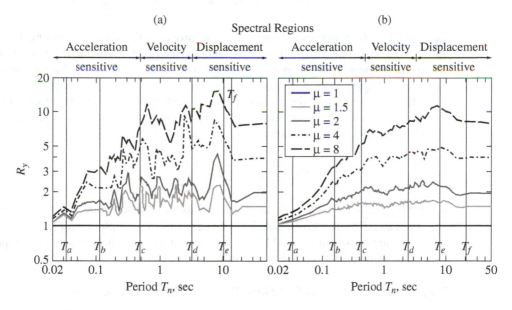

Figure 7.11.1   Yield-strength reduction factor R_y for elastoplastic systems as a function of T_n for $\mu = 1$, 1.5, 2, 4, and 8; $\zeta = 5\%$: (a) El Centro ground motion; (b) LMSR ensemble of ground motions (median values are presented).

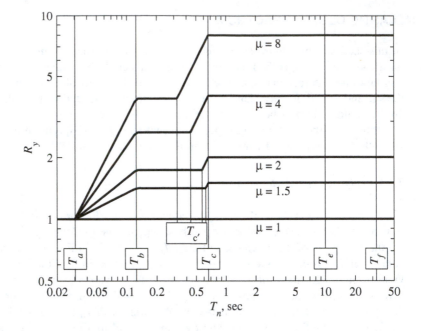

Figure 7.11.2   Design values of yield-strength reduction factor.

smoothly with T_n, generally increasing significantly with T_n over the acceleration-sensitive spectral region, but only slightly over the velocity-sensitive region and the T_d-to-T_f part of the displacement-sensitive region; in the period range longer than T_f, R_y decreases as T_n increases and approaches μ at very long periods.

Based on results similar to those presented in Fig. 7.11.1b, several researchers have proposed equations for the variation of R_y with T_n and μ. One of the early simpler proposals relates R_y to μ in different spectral regions as follows:

$$R_y = \begin{cases} 1 & T_n < T_a \\ \sqrt{2\mu - 1} & T_b < T_n < T_{c'} \\ \mu & T_n > T_c \end{cases} \qquad (7.11.1)$$

where the periods T_a, T_b, ..., T_f separating the spectral regions were defined in Figs. 6.9.2 and 6.9.3, and $T_{c'}$ will become clear later. Equation (7.11.1) is plotted for several values of μ in a log-log format in Fig. 7.11.2, where sloping straight lines are included to provide transitions among the three constant segments.

7.11.2 Construction of Constant-Ductility Design Spectrum

It is presumed that the elastic design spectrum a–b–c–d–e–f shown in Fig. 7.11.3 has been developed by the procedure described in Section 6.9. This elastic design spectrum is divided by R_y for a chosen value of ductility factor μ [Eq. (7.11.1) and Fig. 7.11.2] to construct the inelastic design spectrum a'–b'–c'–d'–e'–f' shown in Fig. 7.11.3. This implementation involves the following steps:

1. Divide the constant-A ordinate of segment b–c by $R_y = \sqrt{2\mu - 1}$ to locate the segment b'–c'.
2. Divide the constant-V ordinate of segment c–d by $R_y = \mu$ to locate the segment c'–d'.
3. Divide the constant-D ordinate of segment d–e by $R_y = \mu$ to locate the segment d'–e'.
4. Divide the ordinate at f by $R_y = \mu$ to locate f'. Join points f' and e'. Draw $D_y = u_{go}/\mu$ for $T_n > 33$ sec.
5. Take the ordinate a' of the inelastic spectrum at $T_n = \frac{1}{33}$ sec as equal to that of point a of the elastic spectrum. This is equivalent to $R_y = 1$. Join points a' and b'.
6. Draw $A_y = \ddot{u}_{go}$ for $T_n < \frac{1}{33}$ sec.

The period values associated with points a', b', e', and f' are fixed, as shown in Fig. 7.11.3, at the same values as the corresponding points of the elastic spectrum. For ground motions on firm ground, $T_a = \frac{1}{33}$ sec, $T_b = \frac{1}{8}$ sec, $T_e = 10$ sec, and $T_f = 33$ sec (Fig. 6.9.3). T_c and T_d depend on damping because they are determined by the amplification factors α_A, α_V, and α_D, which depend on damping (Table 6.9.1). The period values $T_{c'}$ and $T_{d'}$ depend on the values used to reduce the segments b–c, c–d, and d–e of the elastic design spectrum because the R_y vary with μ. With the selected R_y values of Eq. (7.11.1) for the three spectral regions, respectively, $T_{d'}$ is the same as T_d but $T_{c'}$ differs from T_c. $T_{d'}$ would differ from T_d if R_y was not the same for the c–d and d–e spectral regions. Observe that the c–d–e–f portion of the spectrum has been reduced by a constant factor μ.

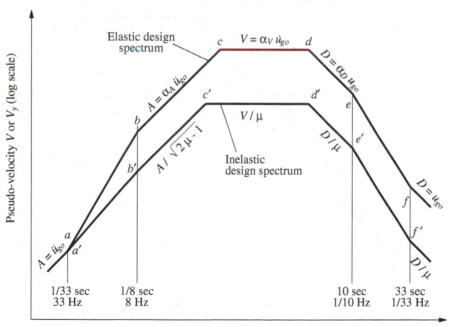

Figure 7.11.3   Construction of inelastic design spectrum.

Consider ground motions on firm ground with peak acceleration $\ddot{u}_{go} = 1g$, peak velocity $\dot{u}_{go} = 48$ in./sec, and peak displacement $u_{go} = 36$ in. The 84.1th percentile spectrum is desired for elastoplastic systems with damping ratio $\zeta = 5\%$ for a ductility factor of $\mu = 2$. The design spectrum for elastic systems with $\zeta = 5\%$ and the ground motion selected was presented in Fig. 6.9.4 and is reproduced in Fig. 7.11.4. The inelastic spectrum for $\mu = 2$ is determined by the following steps (with reference to Figs. 7.11.3 and 7.11.4):

1. The ordinate $A = 2.71g$ of the constant-A branch is divided by $R_y = \sqrt{2\mu - 1} = 1.732$ for $\mu = 2$ to obtain the ordinate $A_y = 1.56g$ for the segment $b'-c'$.
2. The ordinate $V = 110.4$ in./sec of the constant-V branch is divided by $R_y = \mu = 2$ to obtain the ordinate $V_y = 55.2$ in./sec for the segment $c'-d'$.
3. The ordinate $D = 72.4$ in. of the constant-D branch is divided by $R_y = \mu = 2$ to obtain the ordinate $D_y = 36.2$ for the segment $d'-e'$.
4. The ordinate $D = 36$ in. of the point f is divided by $R_y = \mu = 2$ to obtain the ordinate $D_y = 18$ in. for the point f'; points f' and e' are joined by a straight line. This D_y value also defines the spectrum for $T_n > 33$ sec.

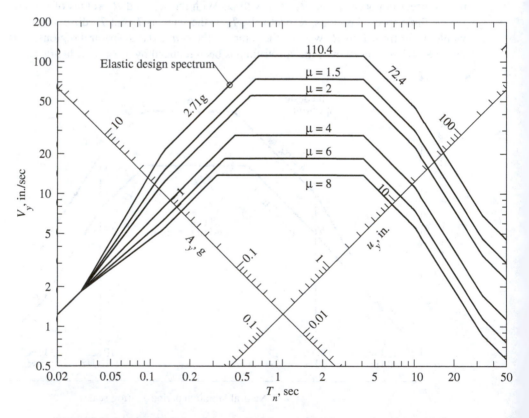

Figure 7.11.4 Inelastic design spectrum (84.1th percentile) for ground motions with $\ddot{u}_{go} = 1g$, $\dot{u}_{go} = 48$ in./sec, and $u_{go} = 36$ in.; $\mu = 1.5, 2, 4, 6,$ and 8; $\zeta = 5\%$.

5. Point a', the same as point a, is joined to point b'.

6. The line $A_y = \ddot{u}_{go} = 1\text{g}$ is drawn for $T_n < \frac{1}{33}$ sec.

The resulting inelastic design spectrum for $\mu = 2$ is shown in Fig. 7.11.4, together with design spectra for other values of $\mu = 1, 1.5, 4, 6$, and 8 constructed by the same procedure.

With the pseudo-velocity (V_y) design spectrum known (Fig. 7.11.4), the pseudo-acceleration (A_y) design spectrum is constructed by using Eq. (7.5.2), and the resulting spectrum has been plotted in two formats: logarithmic scales (Fig. 7.11.5) and linear scales (Fig. 7.11.6). Determined from the A_y data of Fig. 7.11.5 and Eq. (7.6.3), the deformation design spectrum is presented in Fig. 7.11.7. Shown herein is u_m versus T_n for $\mu = 1, 1.5, 2, 4, 6$, and 8. The $\mu = 1$ curve also gives the deformation u_o of the system if it were to remain elastic. Thus, the ratio u_m/u_o can be determined from Fig. 7.11.7; this is plotted against T_n in Fig. 7.11.8. Over a wide range of periods, $T_n > T_c$, the peak deformation of an inelastic system is independent of μ and equal to the peak deformation of the elastic (or corresponding linear) system. For shorter T_n, $T_n < T_c$, the peak deformation of an

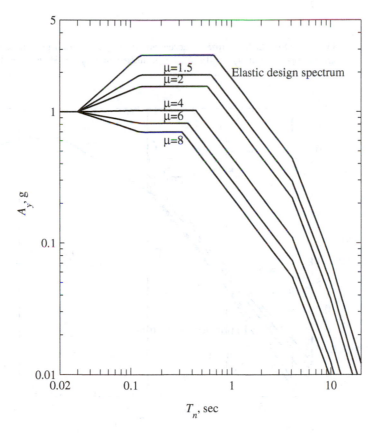

Figure 7.11.5   Inelastic (pseudo-acceleration) design spectrum (84.1th percentile) for ground motions with $\ddot{u}_{go} = 1\text{g}$, $\dot{u}_{go} = 48$ in./sec, and $u_{go} = 36$ in; $\mu = 1.5, 2, 4, 6,$ and 8; $\zeta = 5\%$.

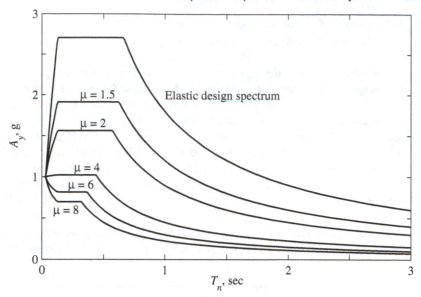

Figure 7.11.6 Inelastic (pseudo-acceleration) design spectrum (84.1th percentile) for ground motions with $\ddot{u}_{go} = 1g$, $\dot{u}_{go} = 48$ in./sec, and $u_{go} = 36$ in; $\mu = 1.5, 2, 4, 6,$ and 8; $\zeta = 5\%$.

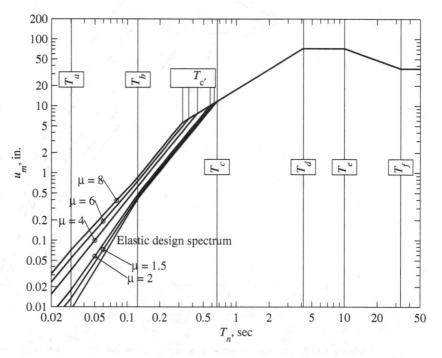

Figure 7.11.7 Inelastic (deformation) design spectrum (84.1th percentile) for ground motions with $\ddot{u}_{go} = 1g$, $\dot{u}_{go} = 48$ in./sec, and $u_{go} = 36$ in; $\mu = 1.5, 2, 4, 6,$ and 8; $\zeta = 5\%$.

inelastic system exceeds that of the elastic system; for fixed μ, the ratio u_m/u_o increases as T_n decreases; and for fixed T_n, the ratio u_m/u_o increases with μ.

Researchers have developed results for SDF systems with various inelastic force–deformation relations (see Section 7.1.1), similar to the data presented in this chapter for elastoplastic systems. In particular, they have demonstrated that the design spectrum for elastoplastic systems is generally conservative and may therefore be used for bilinear systems and stiffness degrading systems.

7.11.3 Equations Relating f_y to f_o and u_m to u_o

Using Eqs. (7.2.2), (7.2.4), and (7.11.1), simple relations can be developed between the peak deformations u_o and u_m and between the required yield strengths f_o and f_y for elastic and elastoplastic systems; these relations depend on the spectral regions (Fig. 7.11.2):

1. In the period region $T_n < T_a$, $R_y = 1$, which implies that

$$f_y = f_o \qquad u_m = \mu u_o \qquad\qquad (7.11.2)$$

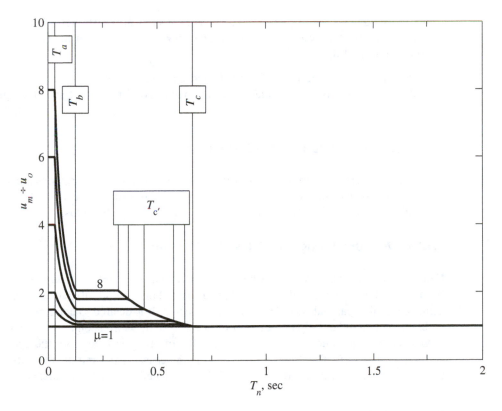

Figure 7.11.8 Ratio u_m/u_o of peak deformations u_m and u_o of elastoplastic system and corresponding linear system plotted against T_n; $\mu = 1, 1.5, 2, 4, 6,$ and 8.

Thus, the required strength for the elastoplastic system is the same for all μ, equal to the minimum strength required for the system to remain elastic (Fig. 7.11.5). The peak deformation of the elastoplastic system is μ times that of the elastic system (Fig. 7.11.8). If $f_y < f_o$ (i.e., less strength is provided), the ductility demand would be large (Fig. 7.4.5).

2. For the period region $T_b < T_n < T_{c'}$, $R_y = \sqrt{2\mu - 1}$, which implies that

$$f_y = \frac{f_o}{\sqrt{2\mu - 1}} \qquad u_m = \frac{\mu}{\sqrt{2\mu - 1}} u_o \qquad (7.11.3)^\dagger$$

Thus, the required strength for the elastoplastic system is the minimum strength required for the system to remain elastic divided by $\sqrt{2\mu - 1} = 1$, 2.65, and 3.87 for $\mu = 1$, 4, and 8, respectively. The peak deformation of the elastoplastic system is larger than that of the elastic system by the factor $\mu/\sqrt{2\mu - 1} = 1$, 1.51, and 2.06 for $\mu = 1$, 4, and 8, respectively.

3. For the period region $T_n > T_c$, $R_y = \mu$, which implies that

$$f_y = \frac{f_o}{\mu} \qquad u_m = u_o \qquad (7.11.4)$$

Thus, the required strength for the elastoplastic system is the strength demand for the elastic system divided by μ (Fig. 7.11.6). The peak deformations of the two systems are the same (Fig. 7.11.8).

Similar relations between f_y and f_o and between u_m and u_o can be developed to provide transitions between the relations developed above for the three spectral regions.

7.12 APPLICATIONS OF THE DESIGN SPECTRUM

The inelastic design spectrum, developed in the preceding section, provides a convenient basis to address questions that arise in the design of new structures and safety evaluation of existing structures. In this section we discuss three such applications.

7.12.1 Structural Design for Allowable Ductility

First, consider an SDF system to be designed for an *allowable ductility* μ, which has been decided based on the allowable deformation and on the ductility capacity that can be achieved for the materials and design details selected. It is desired to determine the design yield strength and the design deformation for the system. Corresponding to the allowable μ and the known values of T_n and ζ, the value of A_y/g is read from the spectrum of Fig. 7.11.5 or 7.11.6. The minimum yield strength necessary to limit the ductility demand to the allowable ductility is given by Eq. (7.5.3), repeated here for convenience:

[†]Although there is little rational basis for doing so, Eq. (7.11.3a) for $T_b < T_n < T_{c'}$ can be derived by equating the areas under the force–deformation relations for elastic and elastoplastic systems (Fig. 7.1.4).

$$f_y = \frac{A_y}{g}w \qquad (7.12.1)$$

The peak deformation is given by Eq. (7.6.3), which can be expressed in terms of A, the pseudo-acceleration design spectrum for elastic systems. For this purpose we use the relation

$$R_y = \frac{A}{A_y} \qquad (7.12.2)$$

which comes from substituting Eqs. (7.5.4) and (7.12.1) in Eq. (7.2.2). Putting Eqs. (7.6.3) and (7.12.2) together gives

$$u_m = \mu\frac{1}{R_y}\left(\frac{T_n}{2\pi}\right)^2 A \qquad (7.12.3)$$

The deformation of the inelastic system can be conveniently determined by Eq. (7.6.3) using the inelastic design spectrum, or by Eq. (7.12.3) using the elastic design spectrum. The R_y–μ–T_n relation needed in the latter case is available in Eq. (7.11.1) and Fig. 7.11.2.

Example 7.1

Consider a one-story frame with lumped weight w and natural vibration period in the linearly elastic range, $T_n = 0.25$ sec. Determine the lateral deformation and lateral force (in terms of w) for which the frame should be designed if (1) the system is required to remain elastic, (2) the allowable ductility factor is 4, and (3) the allowable ductility factor is 8. Assume that $\zeta = 5\%$ and elastoplastic force–deformation behavior. The design earthquake has a peak acceleration of 0.5g and its elastic design spectrum is given by Fig. 6.9.5 multiplied by 0.5.

Solution For a system with $T_n = 0.25$ sec, $A = (2.71g)\,0.5 = 1.355g$ from Fig. 6.9.5 and $R_y = \sqrt{2\mu - 1}$ from Eq. (7.11.1) or Fig. 7.11.2. Then, Eq. (7.12.2) gives $A_y = 1.355g/\sqrt{2\mu - 1}$ and Eq. (7.12.1) leads to

$$f_y = \frac{1.355w}{\sqrt{2\mu - 1}} \qquad (a)$$

Substituting known data in Eq. (7.12.3) gives

$$u_m = \frac{\mu}{\sqrt{2\mu - 1}}\left(\frac{0.25}{2\pi}\right)^2 1.355g = \frac{\mu}{\sqrt{2\mu - 1}}0.828 \qquad (b)$$

Observe that Eqs. (a) and (b) are equivalent to Eq. (7.11.3). Substituting $\mu = 1$, 4, and 8 in Eqs. (a) and (b) gives the following results.

μ	f_y/w	u_m (in.)
1	1.355	0.828
4	0.512	1.252
8	0.350	1.710

7.12.2 Evaluation of an Existing Structure

Consider the problem of estimating the deformation of an existing structure at which its performance should be evaluated. To illustrate application of the inelastic design spectrum to the solution of this problem, we consider the simplest possible structure, an SDF system. The mass m, initial stiffness k at small displacements, and the yield strength f_y of the structure are determined from its properties: dimensions, member sizes, and design details (reinforcement in reinforced-concrete structures, connections in steel structures, etc.) The small-oscillation period T_n is computed from the known k and m, and the damping ratio ζ is estimated by the approach presented in Chapter 11.

For a system with known T_n and ζ, A is read from the elastic design spectrum. From the known value of f_y, A_y is obtained by inverting Eq. (7.12.1):

$$A_y = \frac{f_y}{w/g} \tag{7.12.4}$$

With A and A_y known, R_y is calculated from Eq. (7.12.2). Corresponding to this R_y and T_n determined earlier, μ is calculated from Eq. (7.11.2) or Fig. 7.11.2. Substituting A, T_n, R_y, and μ in Eq. (7.12.3) provides an estimate of the peak deformation u_m.

Example 7.2

Consider a one-story frame with lumped weight w, $T_n = 0.25$ sec, and $f_y = 0.512w$. Assume that $\zeta = 5\%$ and elastoplastic force–deformation behavior. Determine the lateral deformation for the design earthquake defined in Example 7.1.

Solution   For a system with $T_n = 0.25$ sec, $A = (2.71g)0.5 = 1.355g$ from Fig. 6.9.5. Equation (7.12.4) gives

$$\frac{A_y}{g} = \frac{f_y}{w} = \frac{0.512w}{w} = 0.512$$

and Eq. (7.12.2) leads to

$$R_y = \frac{A}{A_y} = \frac{1.355g}{0.512g} = 2.646$$

Knowing R_y, μ can be computed from Eq. (7.11.1) for $T_n = 0.25$ sec:

$$\mu = \frac{1 + R_y^2}{2} = \frac{1 + (2.646)^2}{2} = 4$$

The desired u_m is calculated by substituting these values of μ and R_y in Eq. (7.12.3):

$$u_m = 4\frac{1}{2.646}\left(\frac{0.25}{2\pi}\right)^2 1.355g = 1.252 \text{ in.}$$

7.12.3 Displacement-Based Structural Design

The inelastic design spectrum is also useful for direct displacement-based design of structures. The goal is to determine the initial stiffness and yield strength of the structure necessary to limit the deformation to some acceptable value. Applied to an elastoplastic SDF system (Fig. 7.12.1), such a design procedure may be implemented as a sequence of the following steps:

1. Estimate the yield deformation u_y for the system.
2. Determine acceptable plastic rotation θ_p of the hinge at the base.
3. Determine the design displacement u_m from

$$u_m = u_y + h\theta_p \tag{7.12.5}$$

   and design ductility factor from Eq. (7.2.3).
4. Enter the deformation design spectrum (Fig. 7.11.7) with known u_m and μ to read T_n. Determine the initial elastic stiffness

$$k = \left(\frac{2\pi}{T_n}\right)^2 m \tag{7.12.6}$$

5. Determine the required yield strength

$$f_y = ku_y \tag{7.12.7}$$

6. Select member sizes and detailing (reinforcement in reinforced-concrete structures, connections in steel structures, etc.) to provide the strength determined from Eq. (7.12.7). For the resulting design of the structure, calculate its initial elastic stiffness k and yield deformation $u_y = f_y/k$.
7. Repeat steps 2 through 6 until a satisfactory solution is reached.

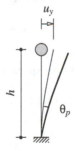

Figure 7.12.1   Idealized SDF system.

Example 7.3

   Consider a long reinforced-concrete viaduct that is part of a freeway. The total weight of the superstructure, 13 kips/ft, is supported on identical bents 30 ft high, uniformly spaced at

130 ft. Each bent consists of a single circular column 60 in. in diameter (Fig. E7.3a). Using the displacement-based design procedure, design the longitudinal reinforcement of the column for the design earthquake defined in Example 7.1.

Solution For transverse ground motion the viaduct can be idealized as an SDF system (Fig. E7.3b) with its lateral stiffness computed from

$$k = \frac{3EI}{h^3} \tag{a}$$

where E is the elastic modulus of concrete, I is the effective second moment of area of the reinforced-concrete cross section, and h is the column length. Based on the American Concrete Institute design provisions ACI 318-95, the effective EI for circular columns subjected to lateral force is given by

$$EI = E_c I_g \left(0.2 + 2\rho_t \gamma^2 \frac{E_s}{E_c} \right) \tag{b}$$

where I_g is the second moment of area of the gross cross section, E_c and E_s are the elastic moduli of concrete and reinforcing steel, respectively; ρ_t is the longitudinal reinforcement ratio, and γ is the ratio of the distances from the center of the column to the center of the outermost reinforcing bars and to the column edge.

We select the following system properties: concrete strength $= 4$ ksi, steel strength $= 60$ ksi, and $\gamma = 0.9$. The mass of the idealized SDF system is the tributary mass for one bent (i.e., the mass of 130-ft length of the superstructure):

$$m = \frac{w}{g} = \frac{1690}{386} = 4.378 \text{ kip-sec}^2/\text{in.} \tag{c}$$

The step-by-step procedure described earlier in this section is now implemented as follows:

1. An initial estimate of $u_y = 1.80$ in.
2. The plastic rotation acceptable at the base of the column is $\theta_p = 0.02$ rad.
3. The design displacement given by Eq. (7.12.5) is

$$u_m = u_y + h\theta_p = 1.80 + 360 \times 0.02 = 9.00 \text{ in.}$$

   and the design ductility factor is

$$\mu = \frac{u_m}{u_y} = \frac{9.00}{1.80} = 5$$

4. The deformation design spectrum for inelastic systems (Fig. 7.11.7) is shown in Fig. E7.3c for $\mu = 5$. Corresponding to $u_m = 9.00$ in., this spectrum gives $T_n = 1.024$ sec and k is computed by Eq. (7.12.6):

$$k = \left(\frac{2\pi}{1.024} \right)^2 4.378 = 164.9 \text{ kips/in.}$$

5. The yield strength is given by Eq. (7.12.7):

$$f_y = k u_y = 164.9 \times 1.80 = 296.9 \text{ kips}$$

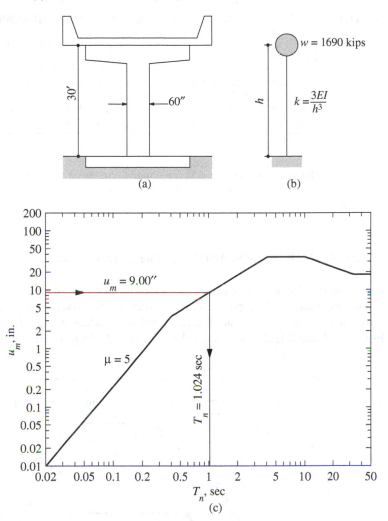

Figure E7.3

6. Using ACI 318-95, the circular column is then designed for axial force due to the dead load of 1690 kips due to the superstructure plus 127 kips due to self-weight of the column and the bending moment due to lateral force = f_y: $M = hf_y = 106{,}884$ kip-in. For the resulting column design, $\rho_t = 3.6\%$, flexural strength = 120,528 kip-in., and lateral strength = 334.8 kips. For $\rho_t = 3.6\%$, Eq. (b) gives $EI = 1.57 \times 10^9$ kip-in.^{2}; using this EI value, Eq. (a) gives $k = 101.1$ kips/in. The yield deformation is $u_y = f_y/k = 334.8/101.1 = 3.31$ in.

7. Since the yield deformation computed in step 6 differs significantly from the initial estimate of $u_y = 1.80$ in., iteration is necessary. The results of such iterations are summarized in Table E7.3.

TABLE E7.3 ITERATIVE PROCEDURE FOR DIRECT DISPLACEMENT-BASED DESIGN

Iteration	u_y (in.)	u_m (in.)	μ	T_n (sec)	k (kips/in.)	f_y (kips)	ρ_t (%)	Design f_y (kips)	Design k (kips/in.)	u_y (in.)
1	1.80	9.00	5.00	1.024	164.9	269.9	3.60	334.8	101.1	3.31
2	3.31	10.5	3.17	1.196	120.9	400.5	5.50	442.8	138.8	3.19
3	3.19	10.4	3.26	1.182	123.8	394.8	5.40	435.0	136.8	3.18
4	3.18	10.4	3.26	1.181	124.0	394.3	5.40	435.0	136.8	3.18

The procedure converged after four iterations, giving a column design with $\rho_t = 5.4\%$. This column has an initial stiffness $k = 136.8$ kips/in. and lateral yield strength $f_y = 435.0$ kips.

7.13 COMPARISON OF DESIGN AND RESPONSE SPECTRA

In this section the design spectrum presented in Section 7.11 is compared with the response spectrum for elastoplastic systems. Such a comparison for elastic systems was presented in Section 6.10, and the data presented in Figs. 6.10.1 and 6.10.2 are reproduced in Figs. 7.13.1 and 7.13.2. Also included now are (1) the inelastic design spectrum for $\mu = 8$

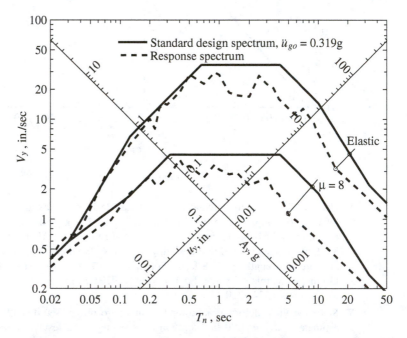

Figure 7.13.1 Comparison of standard design spectrum ($\ddot{u}_{go} = 0.319$g) with response spectrum for El Centro ground motion; $\mu = 1$ and 8; $\zeta = 5\%$.

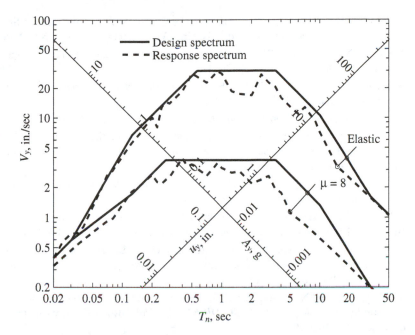

Figure 7.13.2 Comparison of design spectrum ($\ddot{u}_{go} = 0.319g$, $\dot{u}_{go} = 13.04$ in./sec, $u_{go} = 8.40$ in.) with response spectrum for El Centro ground motion; $\mu = 1$ and 8; $\zeta = 5\%$.

determined from the elastic design spectrum using the procedure described in Section 7.11, and (2) the actual spectrum for the El Centro ground motion for $\mu = 8$, reproduced from Fig. 7.5.3.

Observe that the differences between the design and response spectra for elastoplastic systems are greater than between the two spectra for elastic systems. For the latter case, the reasons underlying these differences were discussed in Section 6.10. Additional discrepancies arise in the two spectra for elastoplastic systems because the jagged variation of $\overline{f}_y$ with T_n (Fig. 7.7.1) was approximated by simple functions (Fig. 7.11.1). The errors introduced by this simplification are responsible for the additional discrepancies in the velocity- and displacement-sensitive regions of the spectrum.

FURTHER READING

Grigorian, C. E., and Popov, E. P., "Energy Dissipation with Slotted Bolted Connections," *Report No. UCB/EERC-94/02*, Earthquake Engineering Research Center, University of California at Berkeley, February 1994.

Mohraz, B., and Elghadamsi, F. E., "Earthquake Ground Motion and Response Spectra," Chapter 2 in *The Seismic Design Handbook* (ed. F. Naeim), Van Nostrand Reinhold, New York, 1989.

Newmark, N. M., and Hall, W. J., *Earthquake Spectra and Design*, Earthquake Engineering Research Institute, Berkeley, Calif., 1982, pp. 29–37.

Newmark, N. M., and Rosenblueth, E., *Fundamentals of Earthquake Engineering*, Prentice Hall, Englewood Cliffs, N.J., 1971, Chapter 11.

"Passive Energy Dissipation," *Earthquake Spectra*, **9**, 1993, pp. 319–636.

Riddell, R., and Newmark, N. M., "Statistical Analysis of the Response of Nonlinear Systems Subjected to Earthquakes," *Structural Research Series No. 468*, University of Illinois at Urbana-Champaign, Urbana, Ill., August 1979.

Soong, T. T., and Dargush, G. F., *Passive Energy Dissipation Systems in Structural Engineering*, Wiley, Chichester, U.K., 1997, Chapters 3 and 4.

Veletsos, A. S., "Maximum Deformation of Certain Nonlinear Systems," *Proceedings of the 4th World Conference on Earthquake Engineering*, Santiago, Chile, Vol. 1, 1969, pp. 155–170.

Veletsos, A. S., and Newmark, N. M., "Effect of Inelastic Behavior on the Response of Simple Systems to Earthquake Motions," *Proceedings of the 2nd World Conference on Earthquake Engineering*, Japan, Vol. 2, 1960, pp. 895–912.

Veletsos, A. S., and Newmark, N. M., "Response Spectra for Single-Degree-of-Freedom Elastic and Inelastic Systems," *Report No. RTD-TDR-63-3096*, Vol. III, Air Force Weapons Laboratory, Albuquerque, N.M., June 1964.

Veletsos, A. S., Newmark, N. M., and Chelapati, C. V., "Deformation Spectra for Elastic and Elastoplastic Systems Subjected to Ground Shock and Earthquake Motion," *Proceedings of the 3rd World Conference on Earthquake Engineering*, New Zealand, Vol. II, 1965, pp. 663–682.

PROBLEMS

7.1　　The lateral force–deformation relation of the system of Example 6.3 is idealized as elastic–perfectly plastic. In the linear elastic range of vibration this SDF system has the following properties: lateral stiffness, $k = 2.112$ kips/in., and $\zeta = 2\%$. The yield strength $f_y = 5.55$ kips and the lumped weight $w = 5200$ lb.

(a) Determine the natural period and damping ratio of this system vibrating at amplitudes smaller than u_y.

(b) Can these properties be defined for motions at larger amplitudes? Explain your answer.

(c) Determine the natural period and damping ratio of the corresponding linear system.

(d) Determine $\overline{f}_y$ and R_y for this system subjected to El Centro ground motion scaled up by a factor of 3.

*7.2　　Determine by the central difference method the deformation response $u(t)$ for $0 < t < 10$ sec of an elastoplastic undamped SDF system with $T_n = 0.5$ sec and $\overline{f}_y = 0.125$ to El Centro ground motion. Reproduce Fig. 7.4.2, showing the force–deformation relation in part (d) for the entire duration.

---

*Denotes that a computer is necessary to solve this problem.

*7.3   For a system with $T_n = 0.5$ sec and $\zeta = 5\%$ and El Centro ground motion, verify the following assertion: "doubling the ground acceleration $\ddot{u}_g(t)$ will produce the same response $\mu(t)$ as if the yield strength had been halved." Use the deformation–time responses available in Fig. 7.4.3a–c and generate similar results for an additional system and excitation as necessary.

*7.4   For a system with $T_n = 0.5$ sec and $\zeta = 5\%$ and El Centro ground motion, show that for $\overline{f}_y = 0.25$ the ductility factor μ is unaffected by scaling the ground motion.

7.5   From the response results presented in Fig. 7.4.3, compute the ductility demands for $\overline{f}_y = 0.5$, 0.25, and 0.125.

7.6   For the design earthquake at a site, the peak values of ground acceleration, velocity, and displacement have been estimated: $\ddot{u}_{go} = 0.5g$, $\dot{u}_{go} = 24$ in./sec, and $u_{go} = 18$ in. For systems with a 2% damping ratio and allowable ductility of 3, construct the 84.1th percentile design spectrum. Plot the elastic and inelastic spectra together on (a) four-way log paper, (b) log-log paper showing pseudo-acceleration versus natural vibration period, T_n, and (c) linear-linear paper showing pseudo-acceleration versus T_n from 0 to 5 sec. Determine equations $A(T_n)$ for each branch of the inelastic spectrum and the period values at intersections of branches.

7.7   Consider a vertical cantilever tower that supports a lumped weight w at the top; assume that the tower mass is negligible, $\zeta = 5\%$, and that the force–deformation relation is elastoplastic. The design earthquake has a peak acceleration of 0.5g, and its elastic design spectrum is given by Fig. 6.9.5 multiplied by 0.5. For three different values of the natural vibration period in the linearly elastic range, $T_n = 0.02, 0.2$, and 2 sec, determine the lateral deformation and lateral force (in terms of w) for which the tower should be designed if (i) the system is required to remain elastic, and (ii) the allowable ductility factor is 2, 4, or 8. Comment on how the design deformation and design force are affected by structural yielding.

7.8   Consider a vertical cantilever tower with lumped weight w, $T_n = 2$ sec, and $f_y = 0.112w$. Assume that $\zeta = 5\%$ and elastoplastic force–deformation behavior. Determine the lateral deformation for the elastic design spectrum of Fig. 6.9.5 scaled to a peak ground acceleration of 0.5g.

7.9   Solve Example 7.3 for an identical structure except for one change: The bents are 13 ft high.

*Denotes that a computer is necessary to solve this problem.

8

Generalized Single-Degree-of-Freedom Systems

PREVIEW

So far in this book we have considered single-degree-of-freedom systems involving a single point mass (Figs. 1.2.1 and 1.6.1) or the translation of a rigid distributed mass (Fig. 1.1.3a), which is exactly equivalent to a mass lumped at a single point. Once the stiffness of the system was determined, the equation of motion was readily formulated, and solution procedures were presented in Chapters 2 to 7.

In this chapter we develop the analysis of more complex systems treated as SDF systems, which we call generalized SDF systems. The analysis provides exact results for an assemblage of rigid bodies supported such that it can deflect in only one shape, but only approximate results for systems with distributed mass and flexibility. In the latter case, the approximate natural frequency is shown to depend on the assumed deflected shape. The same frequency estimate is also determined by the classical Rayleigh's method, based on the principle of conservation of energy; this method also provides insight into the error in the estimated natural frequency.

8.1 GENERALIZED SDF SYSTEMS

Consider, for example, the system of Fig. 8.1.1a, consisting of a rigid, massless bar supported by a hinge at the left end with two lumped masses, a spring and a damper, attached to it, subjected to a time-varying external force $p(t)$. Because the bar is rigid, its deflections can be related to a single *generalized displacement* $z(t)$ through a *shape function* $\psi(x)$ as

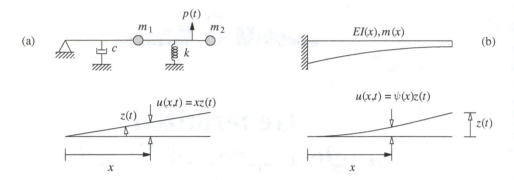

Figure 8.1.1   Generalized SDF systems.

shown, and can be expressed as

$$u(x, t) = \psi(x)z(t) \tag{8.1.1}$$

We have some latitude in choosing the displacement coordinate and, quite arbitrarily, we have chosen the rotation z of the bar. For this system $\psi(x) = x$ is known exactly from the configuration of the system and how it is constrained by a hinged support. This is an SDF system, but it is difficult to replace the two masses by an equivalent mass lumped at a single point.

Next consider, for example, the system of Fig. 8.1.1b consisting of a cantilever beam with distributed mass. This system can deflect in an infinite variety of shapes, and for exact analysis it must be treated as an infinite-degree-of-freedom system. Such exact analysis, developed in Chapter 17, shows that the system, unlike an SDF system, possesses an infinite number of natural vibration frequencies, each paired with a natural mode of vibration. It is possible to obtain approximate results that are accurate to a useful degree for the lowest (also known as *fundamental*) natural frequency, however, by restricting the deflections of the beam to a single shape function $\psi(x)$ that approximates the fundamental vibration mode. The deflections of the beam are then given by Eq. (8.1.1), where the generalized coordinate $z(t)$ is the deflection of the cantilever beam at a selected location—say the free end, as shown in Fig. 8.1.1b.

The two systems of Fig. 8.1.1 are called *generalized SDF systems* because in each case the displacements at all locations are defined in terms of the generalized coordinate $z(t)$ through the shape function $\psi(x)$. We will show that the equation of motion for a generalized SDF system is of the form

$$\tilde{m}\ddot{z} + \tilde{c}\dot{z} + \tilde{k}z = \tilde{p}(t) \tag{8.1.2}$$

where $\tilde{m}, \tilde{c}, \tilde{k}$, and $\tilde{p}(t)$ are defined as the *generalized mass, generalized damping, generalized stiffness*, and *generalized force* of the system; these generalized properties are associated with the generalized displacement $z(t)$ selected. Equation (8.1.2) is of the same form as the standard equation formulated in Chapter 1 for an SDF system with a single lumped mass. Thus the analysis procedures and response results presented in Chapters 2 to 7 can readily be adapted to determine the response $z(t)$ of generalized SDF systems. With $z(t)$

known, the displacements at all locations of the system are determined from Eq. (8.1.1). This analysis procedure leads to the exact results for the system of Fig. 8.1.1a because the shape function $\psi(x)$ could be determined exactly but provides only approximate results for the system of Fig. 8.1.1b because they are based on an assumed shape function.

The key step in the analysis outlined above that is new is the evaluation of the generalized properties $\tilde{m}$, $\tilde{c}$, $\tilde{k}$, and $\tilde{p}(t)$ for a given system. Procedures are developed to determine these properties for (1) assemblages of rigid bodies that permit exact evaluation of the deflected shape (Section 8.2), and (2) multi-degree-of-freedom systems with distributed mass or several lumped masses which require that a shape function be assumed that satisfies the displacement boundary conditions (Sections 8.3 and 8.4).

8.2 RIGID-BODY ASSEMBLAGES

In this section the equation of motion is formulated for an assemblage of rigid bodies with distributed mass supported by discrete springs and dampers subjected to time-varying forces. In formulating the equation of motion for such generalized SDF systems, application of Newton's second law of motion can be cumbersome, and it is simpler to use D'Alembert's principle and include inertia forces in the free-body diagram. The distributed inertia forces for a rigid body with distributed mass can be expressed in terms of the inertia force resultants at the center of gravity using the total mass and the moment of inertia of the body. These properties for rigid plates of three configurations are presented in Appendix 8.

Example 8.1

The system shown in Fig. E8.1a consists of a rigid bar supported by a fulcrum at O, with an attached spring and damper subjected to force $p(t)$. The mass m_1 of the part OB of the bar is distributed uniformly along its length. The portions OA and BC of the bar are massless, but a uniform circular plate of mass m_2 is attached at the midpoint of BC. Selecting the counterclockwise rotation about the fulcrum as the generalized displacement and considering small displacements, formulate the equation of motion for this generalized SDF system, determine the natural vibration frequency and damping ratio, and evaluate the dynamic response of the system without damping subjected to a suddenly applied force p_o. How would the equation of motion change with an axial force on the horizontal bar; what is the buckling load?

Solution

1. *Determine the shape function.* The L-shaped bar rotates about the fulcrum at O. Assuming small deflections, the deflected shape is shown in Fig. E8.1b.

2. *Draw the free-body diagram and write the equilibrium equation.* Figure E8.1c shows the forces in the spring and damper associated with the displacements of Fig. E8.1b, together with the inertia forces. Setting the sum of the moments of all forces about O to zero gives

$$I_1\ddot{\theta} + \left(m_1\frac{L}{2}\ddot{\theta}\right)\frac{L}{2} + I_2\ddot{\theta} + (m_2 L\ddot{\theta})L + \left(m_2\frac{L}{4}\ddot{\theta}\right)\frac{L}{4} + \left(c\frac{L}{2}\dot{\theta}\right)\frac{L}{2} + \left(k\frac{3L}{4}\theta\right)\frac{3L}{4} = p(t)\frac{L}{2}$$

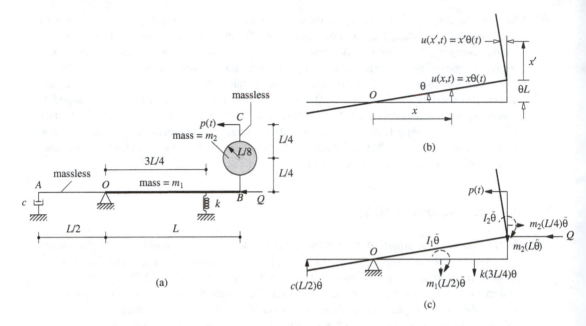

Figure E8.1

Substituting $I_1 = m_1 L^2/12$ and $I_2 = m_2(L/8)^2/2 = m_2 L^2/128$ (see Appendix 8) gives

$$\left(\frac{m_1 L^2}{3} + \frac{137}{128}m_2 L^2\right)\ddot{\theta} + \frac{cL^2}{4}\dot{\theta} + \frac{9kL^2}{16}\theta = p(t)\frac{L}{2} \tag{a}$$

The equation of motion is

$$\tilde{m}\ddot{\theta} + \tilde{c}\dot{\theta} + \tilde{k}\theta = \tilde{p}(t) \tag{b}$$

where

$$\tilde{m} = \left(\frac{m_1}{3} + \frac{137}{128}m_2\right)L^2 \qquad \tilde{c} = \frac{cL^2}{4} \qquad \tilde{k} = \frac{9kL^2}{16} \qquad \tilde{p}(t) = p(t)\frac{L}{2} \tag{c}$$

3. *Determine the natural frequency and damping ratio.*

$$\omega_n = \sqrt{\frac{\tilde{k}}{\tilde{m}}} \qquad \zeta = \frac{\tilde{c}}{2\sqrt{\tilde{k}\tilde{m}}} \tag{d}$$

4. *Solve the equation of motion.*

$$\tilde{p}(t) = \frac{p(t)L}{2} = \frac{p_o L}{2} \equiv \tilde{p}_o$$

By adapting Eq. (4.3.2), the solution of Eq. (b) with $c = 0$ is

$$\theta(t) = \frac{\tilde{p}_o}{\tilde{k}}(1 - \cos \omega_n t) = \frac{8p_o}{9kL}(1 - \cos \omega_n t) \tag{e}$$

5. *Determine the displacements.*

$$u(x, t) = x\theta(t) \qquad u(x', t) = x'\theta(t) \tag{f}$$

where $\theta(t)$ is given by Eq. (e).

6. *Include the axial force.* In the displaced position of the bar, the axial force Q introduces a counterclockwise moment $= QL\theta$. Thus Eq. (b) becomes

$$\tilde{m}\ddot{\theta} + \tilde{c}\dot{\theta} + (\tilde{k} - QL)\theta = \tilde{p}(t) \tag{g}$$

A compressive axial force decreases the stiffness of the system and hence its natural vibration frequency. These become zero if the axial force is

$$Q_{cr} = \frac{\tilde{k}}{L} = \frac{9kL}{16}$$

This is the critical or buckling axial load for the system.

8.3 SYSTEMS WITH DISTRIBUTED MASS AND ELASTICITY

As an illustration of approximating a system having an infinite number of degrees of freedom by a generalized SDF system, consider the cantilever tower shown in Fig. 8.3.1. This tower has mass $m(x)$ per unit length and flexural rigidity $EI(x)$, and the excitation is earthquake ground motion $u_g(t)$. In this section, first the equation of motion for this system without damping is formulated; damping is usually expressed by a damping ratio estimated based on experimental data from similar structures (Chapter 11). Then the equation of motion is solved to determine displacements and a procedure is developed to determine the internal forces in the tower. Finally, this procedure is applied to evaluation of the peak response of the system to earthquake ground motion.

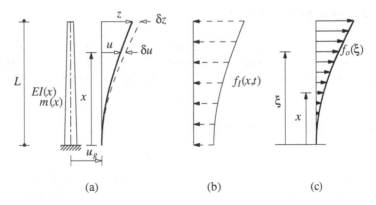

(a)   (b)   (c)

Figure 8.3.1   (a) Tower deflections and virtual displacements; (b) inertia forces; (c) equivalent static forces.

8.3.1 Assumed Shape Function

We assume that the displacement relative to the ground can be expressed by Eq. (8.1.1). The total displacement of the tower is

$$u^t(x, t) = u(x, t) + u_g(t) \tag{8.3.1}$$

The shape function $\psi(x)$ in Eq. (8.1.1) must satisfy the displacement boundary conditions (Fig. 8.3.1a). For this tower, these conditions at the base of the tower are $\psi(0) = 0$ and $\psi'(0) = 0$. Within these constraints a variety of shape functions could be chosen. One possibility is to determine the shape function as the deflections of the tower due to some static forces. For example, the deflections of a uniform tower with flexural rigidity EI due to a unit lateral force at the top are $u(x) = (3Lx^2 - x^3)/6EI$. If we select the generalized coordinate as the deflection of some convenient reference point, say the top of the tower, then $z = u(L) = L^3/3EI$, and

$$u(x) = \psi(x)z \qquad \psi(x) = \frac{3}{2}\frac{x^2}{L^2} - \frac{1}{2}\frac{x^3}{L^3} \tag{8.3.2a}$$

This $\psi(x)$ automatically satisfies the displacement boundary conditions at $x = 0$ because it was determined from static analysis of the system. The $\psi(x)$ of Eq. (8.3.2a) may also be used as the shape function for a nonuniform tower, although it was determined for a uniform tower. It is not necessary to select the shape function based on deflections due to static forces, and it could be assumed directly; possibilities are

$$\psi(x) = \frac{x^2}{L^2} \quad \text{and} \quad \psi(x) = 1 - \cos\frac{\pi x}{2L} \tag{8.3.2b}$$

The three shape functions above have $\psi(L) = 1$, although this is not necessary. The accuracy of the generalized SDF system formulation depends on the assumed shape function $\psi(x)$ in which the structure is constrained to vibrate. This issue will be discussed later together with how to select the shape function.

8.3.2 Equation of Motion

We now proceed to formulate the equation of motion for the tower. At each time instant the system is in equilibrium under the action of the internal resisting bending moments and the fictitious inertia forces (Fig. 8.3.1b), which by D'Alembert's principle are

$$f_I(x, t) = -m(x)\ddot{u}^t(x, t)$$

Substituting Eq. (8.3.1) for u^t gives

$$f_I(x, t) = -m(x)[\ddot{u}(x, t) + \ddot{u}_g(t)] \tag{8.3.3}$$

The equation of dynamic equilibrium of this generalized SDF system can be formulated conveniently only by work or energy principles. We prefer to use the principle of virtual displacements. This principle states that if the system in equilibrium is subjected to virtual displacements $\delta u(x)$, the external virtual work δW_E is equal to the internal virtual work δW_I:

$$\delta W_I = \delta W_E \qquad (8.3.4)$$

The external virtual work is due to the forces $f_I(x, t)$ acting through the virtual displacements $\delta u(x)$:

$$\delta W_E = \int_0^L f_I(x, t)\delta u(x)\, dx$$

which after substituting Eq. (8.3.3) becomes

$$\delta W_E = -\int_0^L m(x)\ddot{u}(x, t)\delta u(x)\, dx - \ddot{u}_g(t)\int_0^L m(x)\delta u(x)\, dx \qquad (8.3.5)$$

The internal virtual work is due to the bending moments $\mathcal{M}(x, t)$ acting through the curvature $\delta\kappa(x)$ associated with the virtual displacements:

$$\delta W_I = \int_0^L \mathcal{M}(x, t)\delta\kappa(x)\, dx$$

Substituting

$$\mathcal{M}(x, t) = EI(x)u''(x, t) \qquad \delta\kappa(x) = \delta[u''(x)]$$

where $u'' \equiv \partial^2 u/\partial x^2$ gives

$$\delta W_I = \int_0^L EI(x)u''(x, t)\delta[u''(x)]\, dx \qquad (8.3.6)$$

The internal and external virtual work is expressed next in terms of the generalized coordinate z and shape function $\psi(x)$. For this purpose, from Eq. (8.1.1) we obtain

$$u''(x, t) = \psi''(x)z(t) \qquad \ddot{u}(x, t) = \psi(x)\ddot{z}(t) \qquad (8.3.7)$$

The virtual displacement is selected consistent with the assumed shape function (Fig. 8.3.1a), giving Eq. (8.3.8a), and the virtual curvature is defined by Eq. (8.3.8b):

$$\delta u(x) = \psi(x)\, \delta z \qquad \delta[u''(x)] = \psi''(x)\, \delta z \qquad (8.3.8)$$

Substituting Eqs. (8.3.7b) and (8.3.8a) in Eq. (8.3.5) gives

$$\delta W_E = -\delta z\left[\ddot{z}\int_0^L m(x)[\psi(x)]^2\, dx + \ddot{u}_g(t)\int_0^L m(x)\psi(x)\, dx\right] \qquad (8.3.9)$$

Substituting Eqs. (8.3.7a) and (8.3.8b) in Eq. (8.3.6) gives

$$\delta W_I = \delta z\left[z\int_0^L EI(x)[\psi''(x)]^2\, dx\right] \qquad (8.3.10)$$

Having obtained the final expressions for δW_E and δW_I, Eq. (8.3.4) gives

$$\delta z\left[\tilde{m}\ddot{z} + \tilde{k}z + \tilde{L}\ddot{u}_g(t)\right] = 0 \qquad (8.3.11)$$

where

$$\tilde{m} = \int_0^L m(x)[\psi(x)]^2 \, dx$$

$$\tilde{k} = \int_0^L EI(x)[\psi''(x)]^2 \, dx \qquad (8.3.12)$$

$$\tilde{L} = \int_0^L m(x)\psi(x) \, dx$$

Because Eq. (8.3.11) is valid for every virtual displacement δz, we conclude that

$$\tilde{m}\ddot{z} + \tilde{k}z = -\tilde{L}\ddot{u}_g(t) \qquad (8.3.13a)$$

This is the equation of motion for the tower assumed to deflect according to the shape function $\psi(x)$. For this generalized SDF system, the generalized mass $\tilde{m}$, generalized stiffness $\tilde{k}$, and generalized excitation $-\tilde{L}\ddot{u}_g(t)$ are defined by Eq. (8.3.12). Dividing Eq. (8.3.13a) by $\tilde{m}$ gives

$$\ddot{z} + 2\zeta\omega_n\dot{z} + \omega_n^2 z = -\tilde{\Gamma}\ddot{u}_g(t) \qquad (8.3.13b)$$

where $\omega_n^2 = \tilde{k}/\tilde{m}$ and a damping term using an estimated damping ratio ζ has been included. This equation is the same as Eq. (6.2.1) for an SDF system, except for the factor

$$\tilde{\Gamma} = \frac{\tilde{L}}{\tilde{m}} \qquad (8.3.14)$$

8.3.3 Natural Vibration Frequency

Once the generalized properties $\tilde{m}$ and $\tilde{k}$ are determined, the natural vibration frequency of the system is given by

$$\omega_n^2 = \frac{\tilde{k}}{\tilde{m}} = \frac{\int_0^L EI(x)[\psi''(x)]^2 \, dx}{\int_0^L m(x)[\psi(x)]^2 \, dx} \qquad (8.3.15)$$

8.3.4 Response Analysis

The generalized coordinate response $z(t)$ of the system to specified ground acceleration can be determined by solving Eq. (8.3.13b) using the methods presented in Chapters 5 and 6. Equation (8.1.1) then gives the displacements $u(x, t)$ of the tower relative to the base.

The next step is to compute the internal forces—bending moments and shears—in the tower associated with the displacements $u(x, t)$. The second of the two methods described in Section 1.8 is used if we are working with deflected shape $\psi(x)$ that is assumed and not exact, as for generalized SDF systems. In this method internal forces are computed by static analysis of the structure subjected to *equivalent static forces*. Denoted by $f_S(x)$, these forces are defined as external forces that would cause displacements $u(x)$. Elementary beam theory gives

$$f_S(x) = \left[EI(x)u''(x) \right]'' \qquad (8.3.16)$$

Because u varies with time, so will f_S; thus

$$f_S(x, t) = \left[EI(x)u''(x, t)\right]'' \tag{8.3.17}$$

which after substituting Eq. (8.1.1) becomes

$$f_S(x, t) = \left[EI(x)\psi''(x)\right]''z(t) \tag{8.3.18}$$

These external forces, which depend on derivatives of the assumed shape function, will give internal forces that are usually less accurate than the displacements, because the derivatives of the assumed shape function are poorer approximations than the shape function itself.

The best estimate, best within the constraints of the assumed shape function, for equivalent static forces is

$$f_S(x, t) = \omega_n^2 m(x)\psi(x)z(t) \tag{8.3.19}$$

This can be shown to be identical to Eq. (8.3.18) if the assumed shape function is exact, as we shall see in Chapter 16. With an approximate shape function, the two sets of forces given by Eqs. (8.3.18) and (8.3.19) are not the same locally at all points along the length of the structure, but the two are globally equivalent (see Derivation 8.1). Furthermore, Eq. (8.3.19) does not involve the derivatives of the assumed $\psi(x)$, and is therefore a better approximation, relative to Eq. (8.3.18). Internal forces can be determined at each time instant by static analysis of the tower subjected to forces $f_S(x, t)$ determined from Eq. (8.3.19).

8.3.5 Peak Earthquake Response

Comparing Eq. (8.3.13b) to Eq. (6.2.1) for an SDF system and using the procedure of Section 6.7 gives the peak value of $z(t)$:

$$z_o = \tilde{\Gamma}D = \frac{\tilde{\Gamma}}{\omega_n^2}A \tag{8.3.20}$$

where D and A are the deformation and pseudo-acceleration ordinates, respectively, of the design spectrum at period $T_n = 2\pi/\omega_n$ for damping ratio ζ. In Eqs. (8.1.1) and (8.3.19), $z(t)$ is replaced by z_o of Eq. (8.3.20) to obtain the peak values of displacements and equivalent static forces:

$$u_o(x) = \tilde{\Gamma}D\psi(x) \qquad f_o(x) = \tilde{\Gamma}m(x)\psi(x)A \tag{8.3.21}$$

where the conventional subscript s has been dropped from f_{So} for brevity.

The internal forces—bending moments and shears—in the cantilever tower are obtained by static analysis of the structure subjected to the forces $f_o(x)$; see Fig. 8.3.1c. Thus the shear and bending moment at height x above the base are

$$V_o(x) = \int_x^L f_o(\xi)\,d\xi = \tilde{\Gamma}A\int_x^L m(\xi)\psi(\xi)\,d\xi \tag{8.3.22a}$$

$$M_o(x) = \int_x^L (\xi - x)f_o(\xi)\,d\xi = \tilde{\Gamma}A\int_x^L (\xi - x)m(\xi)\psi(\xi)\,d\xi \tag{8.3.22b}$$

In particular the shear and bending moment at the base of the tower are

$$\mathcal{V}_{bo} = \mathcal{V}_o(0) = \tilde{L}\tilde{\Gamma}A \qquad \mathcal{M}_{bo} = \mathcal{M}_o(0) = \tilde{L}^\theta \tilde{\Gamma}A \qquad (8.3.23)$$

where $\tilde{L}$ was defined in Eq. (8.3.12) and

$$\tilde{L}^\theta = \int_0^L xm(x)\psi(x)\,dx \qquad (8.3.24)$$

This completes the approximate evaluation of the earthquake response of a system with distributed mass and flexibility based on an assumed shape function $\psi(x)$.

8.3.6 Applied Force Excitation

If the excitation were external forces $p(x, t)$ instead of ground motion $\ddot{u}_g(t)$, the equation of motion could be derived following the methods of Section 8.3.2, leading to

$$\tilde{m}\ddot{z} + \tilde{k}z = \tilde{p}(t) \qquad (8.3.25)$$

where the generalized force

$$\tilde{p}(t) = \int_0^L p(x, t)\psi(x)\,dx \qquad (8.3.26)$$

Observe that the only difference in the two equations (8.3.25) and (8.3.13a) is in the excitation term.

Derivation 8.1

The equivalent static forces from elementary beam theory, Eq. (8.3.17), are written as

$$f_S(x, t) = \mathcal{M}''(x, t) \qquad \text{(a)}$$

where the internal bending moments

$$\mathcal{M}(x, t) = EI(x)u''(x, t) \qquad \text{(b)}$$

We seek lateral forces $\tilde{f}_S(x, t)$ that do not involve the derivatives of $\mathcal{M}(x, t)$ and at each time instant are in equilibrium with the internal bending moments; equilibrium is satisfied globally for the system (but not at every location x). Using the principle of virtual displacements, the external work done by the unknown forces $\tilde{f}_S(x, t)$ in acting through the virtual displacement $\delta u(x)$ equals the internal work done by the bending moments acting through the curvatures $\delta\kappa(x)$ associated with the virtual displacements:

$$\int_0^L \tilde{f}_S(x, t)\delta u(x)\,dx = \int_0^L \mathcal{M}(x, t)\delta\kappa(x)\,dx \qquad \text{(c)}$$

This equation is rewritten by substituting Eq. (8.3.8a) for $\delta u(x)$ in the left side and by using Eq. (8.3.10) for the integral on the right side; thus

$$\delta z \int_0^L \tilde{f}_S(x, t)\psi(x)\,dx = \delta z \left[z(t) \int_0^L EI(x)\left[\psi''(x)\right]^2 dx \right] \qquad \text{(d)}$$

Utilizing Eq. (8.3.15) and dropping δz, Eq. (d) can be rewritten as

$$\int_0^L \left[\tilde{f}_S(x,t) - \omega_n^2 m(x)\psi(x)z(t)\right]\psi(x)\,dx = 0 \tag{e}$$

Setting the quantity in brackets to zero gives

$$f_S(x,t) = \omega_n^2 m(x)\psi(x)z(t) \tag{f}$$

where the tilde above f_S has now been dropped. This completes the derivation of Eq. (8.3.19).

Example 8.2

A uniform cantilever tower of length L has mass per unit length $= m$ and flexural rigidity EI (Fig. E8.2). Assuming that the shape function $\psi(x) = 1 - \cos(\pi x/2L)$, formulate the equation of motion for the system excited by ground motion, and determine its natural frequency.

Solution

    1. *Determine the generalized properties.*

$$\tilde{m} = m\int_0^L \left(1 - \cos\frac{\pi x}{2L}\right)^2 dx = 0.227\ mL \tag{a}$$

$$\tilde{k} = EI\int_0^L \left(\frac{\pi^2}{4L^2}\right)^2 \cos^2\frac{\pi x}{2L}\,dx = 3.04\frac{EI}{L^3} \tag{b}$$

$$\tilde{L} = m\int_0^L \left(1 - \cos\frac{\pi x}{2L}\right)dx = 0.363\ mL \tag{c}$$

The computed $\tilde{k}$ is close to the stiffness of the tower under a concentrated lateral force at the top.

    2. *Determine the natural vibration frequency.*

$$\omega_n = \sqrt{\frac{\tilde{k}}{\tilde{m}}} = \frac{3.66}{L^2}\sqrt{\frac{EI}{m}} \tag{d}$$

This approximate result is close to the exact natural frequency, $\omega_{\text{exact}} = (3.516/L^2)\sqrt{EI/m}$, determined in Chapter 16. The error is only 4%.

    3. *Formulate the equation of motion.* Substituting $\tilde{L}$ and $\tilde{m}$ in Eq. (8.3.14) gives $\tilde{\Gamma} = 1.6$ and Eq. (8.3.13b) becomes

$$\ddot{z} + \omega_n^2 z = -1.6\ddot{u}_g(t) \tag{e}$$

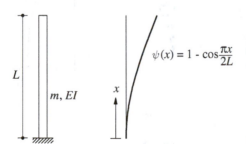

$$\psi(x) = 1 - \cos\frac{\pi x}{2L}$$

L

m, EI

x

Figure E8.2

Example 8.3

A reinforced-concrete chimney, 600 ft high, has a uniform hollow circular cross section with outside diameter 50 ft and wall thickness 2 ft 6 in. (Fig. E8.3a). For purposes of preliminary earthquake analysis, the chimney is assumed clamped at the base, the mass and flexural rigidity are computed from the gross area of the concrete (neglecting the reinforcing steel), and the damping is estimated as 5%. The unit weight of concrete is 150 lb/ft³ and its elastic modulus $E_c = 3600$ ksi.

Assuming the shape function as in Example 8.2, estimate the peak displacements, shear forces, and bending moments for the chimney due to ground motion characterized by the design spectrum of Fig. 6.9.5 scaled to a peak acceleration 0.25g.

Solution

1. *Determine the chimney properties.*

Length: $\qquad L = 600$ ft

Cross-sectional area: $\qquad A = \pi(25^2 - 22.5^2) = 373.1$ ft²

Mass/foot length: $\qquad m = \dfrac{150 \times 373.1}{32.2} = 1.738$ kip-sec²/ft²

Second moment of area: $\qquad I = \dfrac{\pi}{4}(25^4 - 22.5^4) = 105{,}507$ ft⁴

Flexural rigidity: $\qquad EI = 5.469 \times 10^{10}$ kip-ft²

2. *Determine the natural period.* From Example 8.2,

$$\omega_n = \frac{3.66}{L^2}\sqrt{\frac{EI}{m}} = 1.80 \text{ rad/sec}$$

$$T_n = \frac{2\pi}{\omega_n} = 3.49 \text{ sec}$$

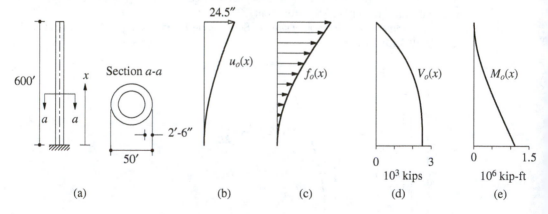

(a) (b) (c) (d) (e)

Figure E8.3

3. *Determine the peak value of* $z(t)$. For $T_n = 3.49$ sec and $\zeta = 0.05$, the design spectrum gives $A/g = 0.25(1.80/3.49) = 0.129$. The corresponding deformation is $D = A/\omega_n^2 = 15.3$ in. Equation (8.3.20) gives the peak value of $z(t)$:

$$z_o = 1.6D = 1.6 \times 15.3 = 24.5 \text{ in.}$$

4. *Determine the peak displacements* $u_o(x)$ *of the tower (Fig. E8.3b).*

$$u_o(x) = \psi(x)z_o = 24.5 \left(1 - \cos\frac{\pi x}{2L}\right) \text{ in.}$$

5. *Determine the equivalent static forces.*

$$f_o(x) = \tilde{\Gamma}m(x)\psi(x)A = (1.6)(1.738)\left(1 - \cos\frac{\pi x}{2L}\right)0.129g$$

$$= 11.58\left(1 - \cos\frac{\pi x}{2L}\right) \text{ kips/ft}$$

These forces are shown in Fig. E8.3c.

6. *Compute the shears and bending moments.* Static analysis of the chimney subjected to external forces $f_o(x)$ gives the shear forces and bending moments. The results using Eq. (8.3.22) are presented in Fig. E8.3d and e. If we were interested only in the forces at the base of the chimney, they could be computed directly from Eq. (8.3.23). In particular, the base shear is

$$V_{bo} = \tilde{L}\tilde{\Gamma}A = (0.363mL)(1.6)0.129g$$

$$= 0.0749mLg = 2518 \text{ kips}$$

This is 7.49% of the total weight of the chimney.

Example 8.4

A simply supported bridge with a single span of L feet has a deck of uniform cross section with mass m per foot length and flexural rigidity EI. A single wheel load p_o travels across the bridge at a uniform velocity of v, as shown in Fig. E8.4. Neglecting damping and assuming the shape function as $\psi(x) = \sin(\pi x/L)$, determine an equation for the deflection at midspan as a function of time. The properties of a prestressed-concrete box-girder elevated-freeway connector are $L = 200$ ft, $m = 11$ kips/g per foot, $I = 700$ ft^4, and $E = 576,000$ kips/ft^2. If $v = 55$ mph, determine the impact factor defined as the ratio of maximum deflection at midspan and the static deflection.

Solution We assume that the mass of the wheel load is small compared to the bridge mass, and it can be neglected.

1. *Determine the generalized mass, generalized stiffness, and natural frequency.*

$$\psi(x) = \sin\frac{\pi x}{L} \qquad \psi''(x) = -\frac{\pi^2}{L^2}\sin\frac{\pi x}{L}$$

$$\tilde{m} = \int_0^L m\sin^2\frac{\pi x}{L}\,dx = \frac{mL}{2} \tag{a}$$

$$\tilde{k} = \int_0^L EI\left(\frac{\pi^2}{L^2}\right)^2\sin^2\frac{\pi x}{L}\,dx = \frac{\pi^4 EI}{2L^3} \tag{b}$$

$$\omega_n = \sqrt{\frac{\tilde{k}}{\tilde{m}}} = \frac{\pi^2}{L^2}\sqrt{\frac{EI}{m}} \tag{c}$$

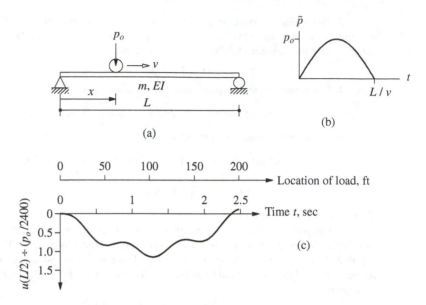

Figure E8.4

This happens to be the exact value of the lowest natural frequency of the bridge because the $\psi(x)$ selected is the exact shape of the fundamental natural vibration mode of a simply supported beam (see Section 16.3.1).

2. *Determine the generalized force.* A load p_o traveling with a velocity v takes time $t_d = L/v$ to cross the bridge. At any time t its position is as shown in Fig. E8.4a. Thus the moving load can be described mathematically as

$$p(x, t) = \begin{cases} p_o \delta(x - vt) & 0 \le t \le t_d \\ 0 & t \ge t_d \end{cases} \tag{d}$$

where $\delta(x - vt)$ is the Dirac delta function centered at $x = vt$; it is a mathematical description of the traveling concentrated load. From Eq. (8.3.26) the generalized force is

$$\tilde{p}(t) = \int_0^L p(x, t)\psi(x)\,dx = \begin{cases} \int_0^L p_o \delta(x - vt)\,\sin(\pi x/L)\,dx & 0 \le t \le t_d \\ 0 & t \ge t_d \end{cases}$$

$$= \begin{cases} p_o \sin(\pi vt/L) & 0 \le t \le t_d \\ 0 & t \ge t_d \end{cases}$$

$$= \begin{cases} p_o \sin(\pi t/t_d) & 0 \le t \le t_d \\ 0 & t \ge t_d \end{cases} \tag{e}$$

This generalized force is the half-cycle sine pulse shown in Fig. E8.4b.

3. *Solve the equation of motion.*

$$\tilde{m}\ddot{z} + \tilde{k}z = \tilde{p}(t) \tag{f}$$

Equations (4.8.2) and (4.8.3) describe the response of an SDF system to a half-cycle sine pulse. We will adapt this solution to the bridge problem by changing the notation from $u(t)$ to

$z(t)$ and noting that

$$t_d = \frac{L}{v} \qquad T_n = \frac{2\pi}{\omega_n} \qquad (z_{st})_o = \frac{p_o}{\tilde{k}} = \frac{2p_o}{mL\omega_n^2}$$

The results are

$$z(t) = \begin{cases} \dfrac{2p_o}{mL} \dfrac{1}{\omega_n^2 - (\pi v/L)^2} \left(\sin \dfrac{\pi vt}{L} - \dfrac{\pi v}{\omega_n L} \sin \omega_n t \right) & t \leq L/v \qquad \text{(g)} \\[3ex] -\dfrac{2p_o}{mL} \dfrac{(2\pi v/\omega_n L) \cos(\omega_n L/2v)}{\omega_n^2 - (\pi v/L)^2} \sin[\omega_n (t - L/2v)] & t \geq L/v \qquad \text{(h)} \end{cases}$$

The response is given by Eq. (g) while the moving load is on the bridge span and by Eq. (h) after the load has crossed the span. These equations are valid provided that $\omega_n \neq \pi v/L$ or $T_n \neq 2L/v$.

   4. *Determine the deflection at midspan.*

$$u(x, t) = z(t)\psi(x) = z(t) \sin \frac{\pi x}{L} \qquad \text{(i)}$$

At midspan, $x = L/2$ and

$$u\left(\frac{L}{2}, t\right) = z(t) \qquad \text{(j)}$$

Thus the deflection at midspan is also given by Eqs. (g) and (h).

   5. *Numerical results.* For the given prestressed-concrete structure and vehicle speed:

$$m = \frac{11}{32.2} = 0.3416 \text{ kip-sec}^2/\text{ft}^2$$

$$EI = 576,000 \times 700 = 4.032 \times 10^8 \text{ kip-ft}^2$$

$$\omega_n = \frac{\pi^2}{(200)^2} \sqrt{\frac{4.032 \times 10^8}{0.3416}} = 8.477 \text{ rad/sec}$$

$$T_n = 0.74 \text{ sec}$$

$$v = 55 \text{ mph} = 80.67 \text{ ft/sec}$$

$$\frac{\pi v}{L} = 1.267$$

$$\frac{L}{v} = \frac{200}{80.67} = 2.479 \text{ sec}$$

Because the duration of the half-cycle sine pulse $t_d = L/v$ is greater than $T_n/2$, the maximum response occurs while the moving load is on the bridge span. This phase of the response is given by Eqs. (g) and (j):

$$u(L/2, t) = \frac{2p_o}{(0.3416)200} \frac{1}{(8.477)^2 - (1.267)^2} \left(\sin 1.267t - \frac{1.267}{8.477} \sin 8.477t \right)$$

$$= \frac{p_o}{2400} (\sin 1.267t - 0.1495 \sin 8.477t) \qquad \text{(k)}$$

Equation (k) is valid for $0 \leq t \leq 2.479$ sec. The values of midspan deflections calculated from Eq. (k) at many values of t are shown in Fig. E8.4c; the maximum deflection is $u_o = (p_o/2400)(1.147)$. The static deflection is $u(L/2) = p_o L^3/48EI = p_o/2419$. The ratio of these deflections gives the impact factor: 1.156 (i.e., the static load should be increased by 15.6% to account for the dynamic effect).

Example 8.5

Determine the natural frequency of transverse vibration of the three-span, box-girder bridge of Example 1.3. Therein several of the properties of this structure were given. In addition, the second moment of area for transverse bending of the bridge deck is given: $I_y = 65,550$ ft^4. Neglect torsional stiffness of the bents.

Solution

1. *Select the shape function.* We select a function appropriate for a beam simply supported at both ends (Fig. E8.5):

$$\psi(x) = \sin \frac{\pi x}{L} \tag{a}$$

This shape function is shown in Fig. E8.5.

2. *Determine the generalized mass.*

$$\tilde{m} = \int_0^L m \sin^2 \frac{\pi x}{L} dx = \frac{mL}{2} \tag{b}$$

3. *Determine the generalized stiffness.*

$$\tilde{k} = \int_0^L EI_y[\psi''(x)]^2 dx + \sum k_{\text{bent}} \psi^2(x_{\text{bent}})$$

$$= \int_0^L EI_y \left(\frac{\pi^2}{L^2} \sin \frac{\pi x}{L} \right)^2 dx + k_{\text{bent}} \sin^2 \left(\frac{\pi L/3}{L} \right) + k_{\text{bent}} \sin^2 \left(\frac{\pi 2L/3}{L} \right)$$

$$= \frac{\pi^4 EI_y}{2L^3} + \frac{3}{4}k_{\text{bent}} + \frac{3}{4}k_{\text{bent}}$$

$$= \underbrace{\frac{\pi^4 EI_y}{2L^3}}_{\tilde{k}_{\text{deck}}} + \underbrace{\frac{3}{2}k_{\text{bent}}}_{\tilde{k}_{\text{bents}}} \tag{c}$$

4. *Determine numerical values for $\tilde{m}$ and $\tilde{k}$.* From Example 1.3, the weight of the bridge deck per unit length is 18.45 kips/ft.

$$\tilde{m} = \frac{1}{32.2} \frac{18.45 \times 375}{2} = 107.5 \text{ kip-sec}^2/\text{ft}$$

$$\tilde{k}_{\text{deck}} = \frac{\pi^4 (3000 \times 144)65,550}{2(375)^3} = 26,154 \text{ kips/ft}$$

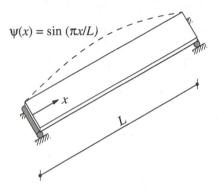

$\psi(x) = \sin(\pi x/L)$

Figure E8.5

From Example 1.3, the stiffness of each bent in the longitudinal direction is $k_{\text{bent}} = 12{,}940$ kips/ft. Because each column in the bent has a circular section, its second moment of area is the same for bending in the longitudinal or transverse directions. Thus, the transverse stiffness of each bent is also $k_{\text{bent}} = 12{,}940$ kips/ft.

$$\tilde{k}_{\text{bent}} = \frac{3}{2}k_{\text{bent}} = \frac{3}{2}(12{,}940) = 19{,}410 \text{ kips/ft}$$

$$\tilde{k} = \tilde{k}_{\text{deck}} + \tilde{k}_{\text{bents}} = 26{,}154 + 19{,}410 = 45{,}564 \text{ kips/ft}$$

5. *Determine the natural vibration period.*

$$\omega_n = \sqrt{\frac{\tilde{k}}{\tilde{m}}} = \sqrt{\frac{45{,}564}{107.5}} = 20.59 \text{ rad/sec}$$

$$T_n = \frac{2\pi}{\omega_n} = 0.305 \text{ sec}$$

8.4 LUMPED-MASS SYSTEM: SHEAR BUILDING

As an illustration of approximating a system having several degrees of freedom by a generalized SDF system, consider the frame shown in Fig. 8.4.1a and earthquake excitation. The mass of this N-story frame is lumped at the floor levels with m_j denoting the mass at the jth floor. This system has N degrees of freedom: u_1, u_2, ..., u_N. In this section, first the equation of motion for this system without damping is formulated; damping is

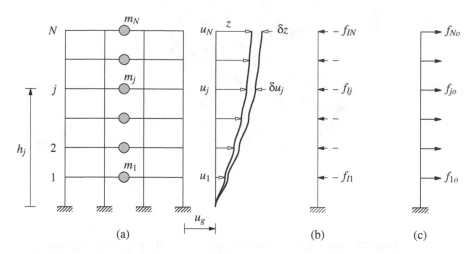

Figure 8.4.1 (a) Building displacements and virtual displacements; (b) inertia forces; (c) equivalent static forces.

usually defined by a damping ratio estimated from experimental data for similar structures (Chapter 11). Then the equation of motion is solved to determine the peak response—displacements and internal forces—of the structure to earthquake ground motion.

8.4.1 Assumed Shape Vector

We assume that the floor displacements relative to the ground can be expressed as

$$u_j(t) = \psi_j z(t) \qquad j = 1, 2, \ldots, N \tag{8.4.1a}$$

which in vector form is

$$\mathbf{u}(t) = \psi z(t) \tag{8.4.1b}$$

where ψ is an assumed shape vector that defines the deflected shape. The total displacement of the jth floor is

$$u_j^t(t) = u_j(t) + u_g(t) \tag{8.4.2}$$

8.4.2 Equation of Motion

Before we can formulate the equation of motion for this system, we must define how the internal forces are related to the displacements. This relationship is especially easy to develop if the beams are rigid axially as well as in flexure and this "shear building" assumption is adequate for our present objectives; however, realistic idealizations of multistory buildings will be developed in Chapter 9. For this shear building idealization, the shear V_j in the jth story (which is the sum of the shear in all columns) is related to the story drift $\Delta_j = u_j - u_{j-1}$ through the story stiffness k_j:

$$V_j = k_j \, \Delta_j = k_j(u_j - u_{j-1}) \tag{8.4.3}$$

The story stiffness is the sum of the lateral stiffnesses of all columns in the story:

$$k_j = \sum_{\text{columns}} \frac{12EI}{h^3} \tag{8.4.4}$$

where EI is the flexural rigidity of a column and h its height.

We now proceed to formulate the equation of motion for the shear building. At each time instant the system is in equilibrium under the action of the internal story shears $V_j(t)$ and the fictitious inertia forces (Fig. 8.4.1b), which by D'Alembert's principle are $f_{Ij} = -m_j \ddot{u}_j^t$. Substituting Eq. (8.4.2) for u_j^t gives

$$f_{Ij} = -m_j[\ddot{u}_j(t) + \ddot{u}_g(t)] \tag{8.4.5}$$

As before, the principle of virtual displacements provides a convenient approach for formulating the equation of motion. The procedure is similar to that developed in Section 8.3 for a beam. The external virtual work is due to the forces f_{Ij} acting through the virtual displacements δu_j:

$$\delta W_E = \sum_{j=1}^{N} f_{Ij}(t)\, \delta u_j$$

which after substituting Eq. (8.4.5) becomes

$$\delta W_E = -\sum_{j=1}^{N} m_j \ddot{u}_j(t)\, \delta u_j - \ddot{u}_g(t) \sum_{j=1}^{N} m_j\, \delta u_j \tag{8.4.6}$$

The internal virtual work is due to the story shears $V_j(t)$ acting through the story drifts associated with the virtual displacements:

$$\delta W_I = \sum_{j=1}^{N} V_j(t)(\delta u_j - \delta u_{j-1}) \tag{8.4.7}$$

where $V_j(t)$ is related to displacements by Eq. (8.4.3).

The internal and external virtual work can be expressed in terms of the generalized coordinate z and shape vector ψ by noting that the virtual displacements consistent with the assumed shape vector (Fig. 8.4.1a) are

$$\delta u_j = \psi_j\, \delta z \quad \text{or} \quad \delta \mathbf{u} = \psi\, \delta z \tag{8.4.8}$$

Proceeding as in Section 8.3 leads to the following equations of external and internal virtual work:

$$\delta W_E = -\delta z \left[\ddot{z} \sum_{j=1}^{N} m_j \psi_j^2 + \ddot{u}_g(t) \sum_{j=1}^{N} m_j \psi_j \right] \tag{8.4.9}$$

$$\delta W_I = \delta z \left[z \sum_{j=1}^{N} k_j (\psi_j - \psi_{j-1})^2 \right] \tag{8.4.10}$$

Having obtained these expressions for δW_E and δW_I, Eq. (8.3.4), after dropping δz (see Section 8.3.2), gives the equation of motion:

$$\tilde{m}\ddot{z} + \tilde{k}z = -\tilde{L}\ddot{u}_g(t) \qquad (8.4.11)$$

where

$$\tilde{m} = \sum_{j=1}^{N} m_j \psi_j^2 \qquad \tilde{k} = \sum_{j=1}^{N} k_j(\psi_j - \psi_{j-1})^2 \qquad \tilde{L} = \sum_{j=1}^{N} m_j \psi_j \qquad (8.4.12a)$$

We digress briefly to mention that the preceding derivation would have been easier if we used a matrix formulation, but we avoided this approach because the stiffness matrix **k** and the mass matrix **m** of a structure are not introduced until Chapter 9. However, most readers are expected to be familiar with the stiffness matrix of a structure, and as we shall see in Chapter 9, the mass matrix of the system of Fig. 8.4.1 is a diagonal matrix with $m_{jj} = m_j$. Using these matrices and the shape vector $\psi = \langle \psi_1 \quad \psi_2 \quad \cdots \quad \psi_N \rangle^T$, Eq. (8.4.12a) for the generalized properties becomes

$$\tilde{m} = \psi^T \mathbf{m} \psi \qquad \tilde{k} = \psi^T \mathbf{k} \psi \qquad \tilde{L} = \psi^T \mathbf{m} \mathbf{1} \qquad (8.4.12b)$$

where **1** is a vector with all elements unity. Equation (8.4.12b) for the generalized stiffness is a general result because, unlike Eq. (8.4.12a), it is not restricted to shear buildings, as long as **k** is determined for a realistic idealization of the structure.

Equation (8.4.11) governs the motion of the multistory shear frame assumed to deflect in the shape defined by the vector ψ. For this generalized SDF system, the generalized mass $\tilde{m}$, generalized stiffness $\tilde{k}$, and generalized excitation $-\tilde{L}\ddot{u}_g(t)$ are defined by Eq. (8.4.12). Dividing Eq. (8.4.11) by $\tilde{m}$ and including a damping term using an estimated modal damping ratio ζ gives Eqs. (8.3.13b) and (8.3.14), demonstrating that the same equation of motion applies to both—lumped mass or distributed mass—generalized SDF systems; the generalized properties $\tilde{m}$, $\tilde{k}$, and $\tilde{L}$ depend on the system, of course.

8.4.3 Response Analysis

The generalized SDF system can now be analyzed by the methods developed in preceding chapters for SDF systems. In particular, the natural vibration frequency of the system is given by

$$\omega_n^2 = \frac{\tilde{k}}{\tilde{m}} = \frac{\sum_{j=1}^{N} k_j(\psi_j - \psi_{j-1})^2}{\sum_{j=1}^{N} m_j \psi_j^2} \qquad (8.4.13a)$$

Rewriting this equation in matrix notation gives

$$\omega_n^2 = \frac{\psi^T \mathbf{k} \psi}{\psi^T \mathbf{m} \psi} \qquad (8.4.13b)$$

The generalized coordinate response $z(t)$ of the system to specified ground acceleration can be determined by solving Eq. (8.3.13b) using the methods of Chapters 5 and 6. Equation (8.4.1) then gives floor displacements relative to the base.

Suppose that it is desired to determine the peak response of the frame to earthquake excitation characterized by a design spectrum. The peak value of $z(t)$ is still given by Eq. (8.3.20), and the floor displacements relative to the ground are given by Eq. (8.4.1) with $z(t)$ replaced by z_o:

$$u_{jo} = \psi_j z_o = \tilde{\Gamma} D \psi_j \qquad j = 1, 2, \ldots, N \qquad (8.4.14)$$

The equivalent static forces associated with these floor displacements are given by Eq. (8.3.21b) modified for a lumped-mass system:

$$f_{jo} = \tilde{\Gamma} m_j \psi_j A \qquad j = 1, 2, \ldots, N \qquad (8.4.15)$$

Static analysis of the structure subjected to these floor forces (Fig. 8.4.1c) gives the shear force V_{io} in the ith story and overturning moment at the ith floor:

$$V_{io} = \sum_{j=i}^{N} f_{jo} \qquad M_{io} = \sum_{j=i}^{N} (h_j - h_i) f_{jo} \qquad (8.4.16)$$

where h_i is the height of the ith floor above the base. In particular, the shear and overturning moment at the base are

$$V_{bo} = \sum_{j=1}^{N} f_{jo} \qquad M_{bo} = \sum_{j=1}^{N} h_j f_{jo}$$

Substituting Eq. (8.4.15) gives

$$V_{bo} = \tilde{L} \tilde{\Gamma} A \qquad M_{bo} = \tilde{L}^\theta \tilde{\Gamma} A \qquad (8.4.17)$$

where $\tilde{L}$ was defined by Eq. (8.4.12) and

$$\tilde{L}^\theta = \sum_{j=1}^{N} h_j m_j \psi_j \qquad (8.4.18)$$

Observe that these equations for forces at the base of a lumped-mass system are the same as derived earlier for distributed-mass systems [Eqs. (8.3.23) and (8.3.24)]; the parameters $\tilde{L}$, $\tilde{L}^\theta$, and $\tilde{\Gamma}$ depend on the system, of course.

Example 8.6

The uniform five-story frame with rigid beams shown in Fig. E8.6a is subjected to ground acceleration $\ddot{u}_g(t)$. All the floor masses are m, and all stories have the same height h and stiffness k. Assuming the displacements to increase linearly with height above the base (Fig. E8.6b), formulate the equation of motion for the system and determine its natural frequency.

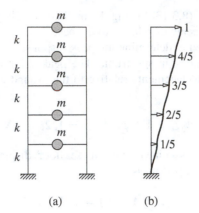

(a)   (b)   **Figure E8.6**

Solution

     1. *Determine the generalized properties.*

$$\tilde{m} = \sum_{j=1}^{5} m_j \psi_j^2 = m \frac{1^2 + 2^2 + 3^2 + 4^2 + 5^2}{5^2} = \frac{11}{5}m$$

$$\tilde{k} = \sum_{j=1}^{5} k_j (\psi_j - \psi_{j-1})^2 = k \frac{1^2 + 1^2 + 1^2 + 1^2 + 1^2}{5^2} = \frac{k}{5}$$

$$\tilde{L} = \sum_{j=1}^{5} m_j \psi_j = m \frac{1 + 2 + 3 + 4 + 5}{5} = 3m$$

     2. *Formulate the equation of motion.* Substituting for $\tilde{m}$ and $\tilde{L}$ in Eq. (8.3.14) gives $\tilde{\Gamma} = \frac{15}{11}$ and Eq. (8.3.13b) becomes

$$\ddot{z} + \omega_n^2 z = -\tfrac{15}{11} \ddot{u}_g(t)$$

where z is the lateral displacement at the location where $\psi_j = 1$, in this case the top of the frame.

     3. *Determine the natural vibration frequency.*

$$\omega_n = \sqrt{\frac{k/5}{11m/5}} = 0.302 \sqrt{\frac{k}{m}}$$

This is about 6% higher than $\omega_n = 0.285\sqrt{k/m}$, the exact frequency of the system determined in Chapter 12.

Example 8.7

Determine the peak displacements, story shears, and floor overturning moments for the frame of Example 8.6 with $m = 100$ kips/g, $k = 31.54$ kips/in., and $h = 12$ ft (Fig. E8.7a) due to the ground motion characterized by the design spectrum of Fig. 6.9.5 scaled to a peak ground acceleration of 0.25g.

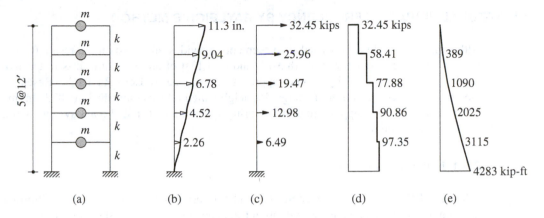

5@12′

m		
m	k	
m	k	
m	k	
m	k	
	k	

11.3 in.
9.04
6.78
4.52
2.26

32.45 kips
25.96
19.47
12.98
6.49

32.45 kips
58.41
77.88
90.86
97.35

389
1090
2025
3115
4283 kip-ft

(a)   (b)   (c)   (d)   (e)

Figure E8.7

Solution

1. *Compute the natural period.*

$$\omega_n = 0.302\sqrt{\frac{31.54}{100/386}} = 3.332$$

$$T_n = \frac{2\pi}{3.332} = 1.89 \text{ sec}$$

2. *Determine the peak value of $z(t)$.* For $T_n = 1.89$ sec and $\zeta = 0.05$, the design spectrum gives $A/g = 0.25(1.80/1.89) = 0.238$ and $D = A/\omega_n^2 = 8.28$ in. The peak value of $z(t)$ is

$$z_o = \tfrac{15}{11}D = \tfrac{15}{11}(8.28) = 11.3 \text{ in.}$$

3. *Determine the peak values u_{jo} of floor displacements.*

$$u_{jo} = \psi_j z_o \qquad \psi_j = \frac{j}{5}$$

Therefore, $u_{1o} = 2.26$, $u_{2o} = 4.52$, $u_{3o} = 6.78$, $u_{4o} = 9.04$, and $u_{5o} = 11.3$, all in inches (Fig. E8.7b).

4. *Determine the equivalent static forces.*

$$f_{jo} = \tilde{\Gamma} m_j \psi_j A = \tfrac{15}{11} m \psi_j (0.238g) = 32.45\psi_j \text{ kips}$$

These forces are shown in Fig. E8.7c.

5. *Compute the story shears and overturning moments.* Static analysis of the structure subjected to external floor forces f_{jo}, Eq. (8.4.16), gives the story shears (Fig. E8.7d) and overturning moments (Fig. E8.7e). If we were interested only in the forces at the base, they could be computed directly from Eq. (8.4.17). In particular, the base shear is

$$V_{bo} = \tilde{L}\tilde{\Gamma} A = (3m)\tfrac{15}{11}(0.238g)$$

$$= 0.195(5mg) = 97.35 \text{ kips}$$

This is 19.5% of the total weight of the building.

8.5 NATURAL VIBRATION FREQUENCY BY RAYLEIGH'S METHOD

Although the principle of virtual displacements provides an approximate result for the natural vibration frequency [Eqs. (8.3.15) and (8.4.13)] of any structure, it is instructive to obtain the same result by another approach, developed by Lord Rayleigh. Based on the principle of conservation of energy, Rayleigh's method was published in 1873. In this section this method is applied to a mass–spring system, a distributed-mass system, and a lumped-mass system.

8.5.1 Mass–Spring System

When an SDF system with lumped mass m and stiffness k is disturbed from its equilibrium position, it oscillates at its natural vibration frequency ω_n, and it was shown in Section 2.1 by solving the equation of motion that $\omega_n = \sqrt{k/m}$. Now we will obtain the same result using the principle of conservation of energy.

The simple harmonic motion of a freely vibrating mass–spring system, Eq. (2.1.3), can be described conveniently by defining a new time variable t' with its origin as shown in Fig. 8.5.1a:

$$u(t') = u_o \sin \omega_n t' \tag{8.5.1}$$

where the frequency ω_n is to be determined and the amplitude u_o of the motion is given by Eq. (2.1.9). The velocity of the mass, shown in Fig. 8.5.1b, is

$$\dot{u}(t') = \omega_n u_o \cos \omega_n t' \tag{8.5.2}$$

The potential energy of the system is the strain energy in the spring, which is proportional to the square of the spring deformation u (Eq. 2.3.2). Therefore, the strain energy is

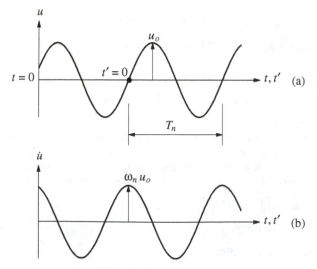

Figure 8.5.1   Simple harmonic motion of a freely vibrating system: (a) displacement; (b) velocity.

maximum at $t' = T_n/4$ (also at $t' = 3T_n/4, 5T_n/4, \ldots$) when $u(t) = u_o$ and is given by

$$E_{So} = \tfrac{1}{2}ku_o^2 \tag{8.5.3}$$

This is also the total energy of the system because at this t' the velocity is zero (Fig. 8.5.1b), implying that the kinetic energy is zero.

The kinetic energy of the system is proportional to the square of the velocity of the mass $\dot{u}$ [Eq. (2.3.2)]. Therefore, the kinetic energy is maximum at $t' = 0$ (also at $t' = T_n/2$, $3T_n/2, \ldots$) when the velocity $\dot{u}(t) = \omega_n u_o$ and is given by

$$E_{Ko} = \tfrac{1}{2}m\omega_n^2 u_o^2 \tag{8.5.4}$$

This is also the total energy of the system because at this t', the deformation is zero (Fig. 8.5.1a), implying that the strain energy is zero.

The principle of conservation of energy states that the total energy in a freely vibrating system without damping is constant (i.e., it does not vary with time), as shown by Eq. (2.3.5). Thus the two alternative expressions, E_{Ko} and E_{So}, for the total energy must be equal, leading to the important result:

$$\text{maximum kinetic energy, } E_{Ko} = \text{maximum potential energy, } E_{So} \tag{8.5.5}$$

Substituting Eqs. (8.5.3) and (8.5.4) gives

$$\omega_n = \sqrt{\frac{k}{m}} \tag{8.5.6}$$

This is the same result for the natural vibration frequency as Eq. (2.1.4) obtained by solving the equation of motion.

Rayleigh's method does not provide any significant advantage in obtaining the natural vibration frequency of a mass–spring system, but the underlying concept of energy conservation is useful for complex systems, as shown in the next two sections.

8.5.2 Systems with Distributed Mass and Elasticity

As an illustration of such a system, consider the cantilever tower of Fig. 8.3.1 vibrating freely in simple harmonic motion:

$$u(x, t') = z_o \sin \omega_n t' \psi(x) \tag{8.5.7}$$

where $\psi(x)$ is an assumed shape function that defines the form of deflections, z_o is the amplitude of the generalized coordinate $z(t)$, and the natural vibration frequency ω_n is to be determined. The velocity of the tower is

$$\dot{u}(x, t') = \omega_n z_o \cos \omega_n t' \psi(x) \tag{8.5.8}$$

The maximum potential energy of the system over a vibration cycle is equal to its strain energy associated with the maximum displacement $u_o(x)$:

$$E_{So} = \int_0^L \tfrac{1}{2}EI(x)[u_o''(x)]^2 \, dx \tag{8.5.9}$$

The maximum kinetic energy of the system over a vibration cycle is associated with the maximum velocity $\dot{u}_o(x)$:

$$E_{Ko} = \int_0^L \tfrac{1}{2} m(x)[\dot{u}_o(x)]^2 \, dx \qquad (8.5.10)$$

From Eqs. (8.5.7) and (8.5.8), $u_o(x) = z_o \psi(x)$ and $\dot{u}_o(x) = \omega_n z_o \psi(x)$. Substituting these in Eqs. (8.5.9) and (8.5.10) and equating E_{Ko} to E_{So} gives

$$\omega_n^2 = \frac{\int_0^L EI(x)[\psi''(x)]^2 \, dx}{\int_0^L m(x)[\psi(x)]^2 \, dx} \qquad (8.5.11)$$

This is known as *Rayleigh's quotient* for a system with distributed mass and elasticity; recall that the same result, Eq. (8.3.15), was obtained using the principle of virtual displacements. Rayleigh's quotient is valid for any natural vibration frequency of a multi-degree-of-freedom system, although its greatest utility is in determining the lowest or fundamental frequency.

8.5.3 Systems with Lumped Masses

As an illustration of such a system, consider the shear building of Fig. 8.4.1 vibrating freely in simple harmonic motion,

$$\mathbf{u}(t') = z_o \sin \omega_n t' \psi \qquad (8.5.12)$$

where the vector ψ is an assumed shape vector that defines the form of deflections, z_o is the amplitude of the generalized coordinate $z(t)$, and the natural vibration frequency ω_n is to be determined. The velocities of the lumped masses of the system are given by the vector

$$\dot{\mathbf{u}}(t') = \omega_n z_o \cos \omega_n t' \psi \qquad (8.5.13)$$

The maximum potential energy of the system over a vibration cycle is equal to its strain energy associated with the maximum displacements, $\mathbf{u}_o = \langle u_{1o} \quad u_{2o} \quad \cdots \quad u_{No} \rangle^T$:

$$E_{So} = \sum_{j=1}^N \tfrac{1}{2} k_j \left(u_{jo} - u_{j-1,o} \right)^2 \qquad (8.5.14)$$

The maximum kinetic energy of the system over a vibration cycle is associated with the maximum velocities, $\dot{\mathbf{u}}_o = \langle \dot{u}_{1o} \quad \dot{u}_{2o} \quad \cdots \quad \dot{u}_{No} \rangle^T$:

$$E_{Ko} = \sum_{j=1}^N \tfrac{1}{2} m_j \dot{u}_{jo}^2 \qquad (8.5.15)$$

From Eqs. (8.5.12) and (8.5.13), $u_{jo} = z_o \psi_j$ and $\dot{u}_{jo} = \omega_n z_o \psi_j$. Substituting these in Eqs. (8.5.14) and (8.5.15) and equating E_{Ko} to E_{So} gives

$$\omega_n^2 = \frac{\sum_{j=1}^N k_j (\psi_j - \psi_{j-1})^2}{\sum_{j=1}^N m_j \psi_j^2} \qquad (8.5.16a)$$

Rewriting this in matrix notation gives

$$\omega_n^2 = \frac{\psi^T k \psi}{\psi^T m \psi} \qquad (8.5.16b)$$

This is *Rayleigh's quotient* for the shear building with N lumped masses; recall that the same result, Eq. (8.4.13), was obtained using the principle of virtual displacements.

8.5.4 Properties of Rayleigh's Quotient

What makes Rayleigh's method especially useful for estimating the lowest or fundamental natural vibration frequency of a system are the properties of Rayleigh's quotient, presented formally in Section 10.12 but only conceptually here: First, the approximate frequency obtained from an assumed shape function is never smaller than the exact value. Second, Rayleigh's quotient provides excellent estimates of the fundamental frequency, even with a mediocre shape function.

Let us examine these properties in the context of a specific system, the cantilever tower considered in Example 8.2. Its fundamental frequency can be expressed as $\omega_n = \alpha_n \sqrt{EI/mL^4}$ [see Eq. (d) of Example 8.2]. Three different estimates of α_n using three different shape functions are summarized in Table 8.5.1. The second frequency estimate comes from Example 8.2. The same procedure leads to the results using the other two shape functions. The percentage error shown is relative to the exact value of $\alpha_n = 3.516$ (Chapter 16).

TABLE 8.5.1   NATURAL FREQUENCY ESTIMATES FOR A UNIFORM CANTILEVER

$\psi(x)$	α_n	% Error
$3x^2/2L^2 - x^3/2L^3$	3.57	1.5
$1 - \cos(\pi x/2L)$	3.66	4
x^2/L^2	4.47	27

Consistent with the properties of Rayleigh's quotient, the three estimates of the natural frequency are higher than its exact value. Even if we did not know the exact value, as would be the case for complex systems, we could say that the smallest value, $\alpha_n = 3.57$, is the best among the three estimates for the natural frequency. This concept can be used to determine the *exact* frequency of a two-DOF system by minimizing the Rayleigh's quotient over a shape function parameter.

Why such a large error in the third case of Table 8.5.1? The shape function $\psi(x) = x^2/L^2$ satisfies the displacement boundary conditions at the base of the tower but violates a force boundary condition at the free end. It implies a constant bending moment over the height of the tower, but a bending moment at the free end of a cantilever is unrealistic unless there is a mass at the free end with a moment of inertia. Thus, a shape function that satisfies only the geometric boundary conditions does not always ensure an accurate result for the natural frequency.

An estimate of the natural vibration frequency of a system obtained using Rayleigh's quotient can be improved by iterative methods. Such methods are developed in Chapter 10.

8.6 SELECTION OF SHAPE FUNCTION

The accuracy of the natural vibration frequency estimated using Rayleigh's quotient depends entirely on the shape function that is assumed to approximate the exact mode shape. In principle any shape may be selected that satisfies the displacement and force boundary conditions. In this section we address the question of how a reasonable shape function can be selected to ensure good results.

For this purpose it is useful to identify the properties of the exact mode shape. In free vibration the displacements are given by Eq. (8.5.7) and the associated inertia forces are

$$f_I(x, t) = -m(x)\ddot{u}(x, t') = \omega_n^2 z_o m(x)\psi(x) \sin \omega_n t'$$

If $\psi(x)$ were the exact mode shape, static application of these inertia forces at each time instant will produce deflections given by Eq. (8.5.7), a result that will become evident in Chapter 16. This concept is not helpful in evaluating the exact mode shape $\psi(x)$ because the inertia forces involve this unknown shape. However, it suggests that an approximate shape function $\psi(x)$ may be determined as the deflected shape due to static forces $p(x) = m(x)\tilde{\psi}(x)$, where $\tilde{\psi}(x)$ is any reasonable approximation of the exact mode shape.

In general, this procedure to select the shape function involves more computational effort than is necessary because, as mentioned earlier, Rayleigh's method gives excellent accuracy even if the shape function is mediocre. However, the preceding discussion does support the concept of determining the shape function from deflections due to a selected set of static forces. One common selection for these forces is the weight of the structure applied in an appropriate direction. For the cantilever tower it is the lateral direction (Fig. 8.6.1a). This selection is equivalent to taking $\tilde{\psi}(x) = 1$ in $p(x) = m(x)\tilde{\psi}(x)$. Another selection includes several concentrated forces as shown in Fig. 8.6.1b.

The displacement and force boundary conditions are satisfied automatically if the shape function is determined from the static deflections due to a selected set of forces. This choice of shape function has the additional advantage that the strain energy can be calculated as the work done by the static forces in producing the deflections, an approach that is usually simpler than Eq. (8.5.9). Thus, the maximum strain energy of the system associated with the forces $p(x)$ in Fig. 8.6.1a is

$$E_{So} = \frac{1}{2} \int_0^L p(x)u(x)\, dx$$

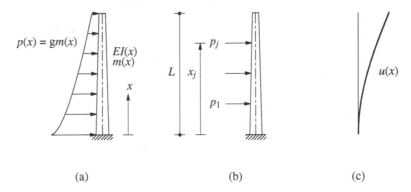

(a)   (b)   (c)

Figure 8.6.1   Shape function from deflections due to static forces.

Equating this E_{So} to E_{Ko} of Eq. (8.5.10) with $\dot{u}_o(x) = \omega_n u_o(x)$ and dropping the subscript "o" gives

$$\omega_n^2 = \frac{\int_0^L p(x)u(x)\,dx}{\int_0^L m(x)[u(x)]^2\,dx} \tag{8.6.1}$$

This equation with $p(x) = p_o$ (i.e., uniformly distributed forces) appears in the AASHTO (American Association of State Highway and Transportation Officials) code to estimate the fundamental natural frequency of a bridge.

For $p(x) = gm(x)$ in Fig. 8.6.1a, Eq. (8.6.1) becomes

$$\omega_n^2 = g\frac{\int_0^L m(x)u(x)\,dx}{\int_0^L m(x)[u(x)]^2\,dx} \tag{8.6.2}$$

Similarly, the maximum strain energy of the system associated with deflections $u(x)$ due to the forces of Fig. 8.6.1b is

$$E_{So} = \tfrac{1}{2}\sum p_j u(x_j)$$

Equating this E_{So} to E_{Ko} of Eq. (8.5.10) gives

$$\omega_n^2 = \frac{\sum p_j u(x_j)}{\int_0^L m(x)[u(x)]^2\,dx} \tag{8.6.3}$$

Although attractive in principle, selecting the shape function as the static deflections due to a set of forces can be cumbersome for the nonuniform (variable EI) tower shown in Fig. 8.6.1a. A convenient approach is to determine the static deflections of a uniform (constant EI) tower of the same length, and use the resulting shape function for the nonuniform tower. The reader is reminded, however, against doing complicated analysis to determine deflected shapes in the interest of obtaining an extremely accurate natural frequency. The principal attraction of Rayleigh's method lies in its ability to provide a useful estimate of the natural frequency from any reasonable assumption on the shape function that satisfies the displacement and force boundary conditions.

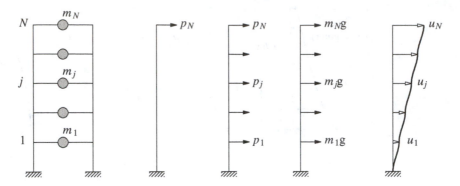

Figure 8.6.2   Shape function from deflections due to static forces.

The concept of using the shape function as the static deflections due to a selected set of forces is also useful for lumped-mass systems. Three sets of forces that may be used for a multistory building frame are shown in Fig. 8.6.2. The maximum strain energy of the system associated with deflections u_j in the three cases is

$$E_{So} = \tfrac{1}{2} p_N u_N \qquad E_{So} = \tfrac{1}{2} \sum_{j=1}^{N} p_j u_j \qquad E_{So} = \tfrac{1}{2} g \sum_{j=1}^{N} m_j u_j$$

Equating these E_{So} to E_{Ko} of Eq. (8.5.15) with $\dot{u}_{jo} = \omega_n u_{jo}$, dropping the subscript "o," and simplifying gives

$$\omega_n^2 = \frac{p_N u_N}{\sum m_j u_j^2} \qquad \omega_n^2 = \frac{\sum p_j u_j}{\sum m_j u_j^2} \qquad \omega_n^2 = \frac{g \sum m_j u_j}{\sum m_j u_j^2} \qquad (8.6.4)$$

respectively. Equations (8.6.4b) and (8.6.4c) appear in building codes to estimate the fundamental natural frequency of a building (Chapter 21). Unlike Eq. (8.5.16a), these results are not restricted to a shear building as long as the deflections are calculated using the actual stiffness properties of the frame.

It is important to recognize that the success of Rayleigh's method for estimating the lowest or fundamental natural frequency of a structure depends on the ability to visualize the corresponding natural mode of vibration that the shape function is intended to approximate. The fundamental mode of a multistory building or of a single-span beam is easy to visualize because the deflections in this mode are all of the same sign. However, the mode shape of more complex systems may not be easy to visualize, and even a shape function calculated from the static deflections due to the self-weight of the structure may not be appropriate. Consider, for example, a two-span continuous beam. Its symmetric deflected shape under its own weight, shown in Fig. 8.6.3a, is not appropriate for computing the lowest natural frequency because this frequency is associated with the antisymmetric mode shown in Fig. 8.6.3b. If this mode shape can be visualized, we can approximate it by the static deflections due to the self-weight of the beam applied downward in one span and upward in the other span (Fig. 8.6.3b).

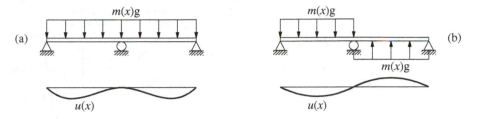

Figure 8.6.3   Shape functions resulting from self-weight applied in appropriate directions.

Example 8.8

Estimate the natural frequency of a uniform cantilever beam assuming the shape function obtained from static deflections due to a load p at the free end.

Solution

1. *Determine the deflections.* With the origin at the clamped end,

$$u(x) = \frac{p}{6EI}(3Lx^2 - x^3) \tag{a}$$

2. *Determine the natural frequency from Eq. (8.6.3).*

$$\sum_j p_j u(x_j) = pu(L) = p^2 \frac{L^3}{3EI} \tag{b}$$

$$\int_0^L m(x)[u(x)]^2\, dx = m\frac{p^2}{(6EI)^2}\int_0^L (3Lx^2 - x^3)^2\, dx = \frac{11p^2}{420}\frac{mL^7}{(EI)^2} \tag{c}$$

Substituting Eqs. (b) and (c) in Eq. (8.6.3) gives

$$\omega_n = \frac{3.57}{L^2}\sqrt{\frac{EI}{m}}$$

This is the first frequency estimate in Table 8.5.1.

Example 8.9

Estimate the fundamental natural frequency of the five-story frame in Fig. E8.9 assuming the shape function obtained from static deflections due to lateral forces equal to floor weights $w = mg$.

Solution

1. *Determine the deflections due to applied forces.* The static deflections are determined as shown in Fig. E8.9 by calculating the story shears and the resulting story drifts; and adding these drifts from the bottom to the top to obtain

$$\mathbf{u}^T = \frac{w}{k}\langle 5 \quad 9 \quad 12 \quad 14 \quad 15 \rangle^T$$

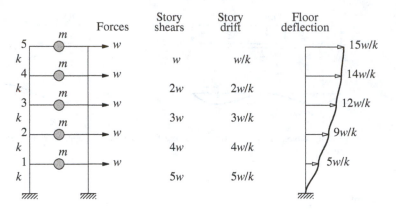

Figure E8.9

2. *Determine the natural frequency from Eq. (8.6.4c).*

$$\omega_n^2 = g\frac{w(w/k)(5 + 9 + 12 + 14 + 15)}{w(w/k)^2(25 + 81 + 144 + 196 + 225)} = \frac{55}{671}\frac{k}{m}$$

$$\omega_n = 0.286\sqrt{\frac{k}{m}}$$

This estimate is very close to the exact value, $\omega_{\text{exact}} = 0.285\sqrt{k/m}$, and better than from a linear shape function (Example 8.6).

FURTHER READING

Rayleigh, J. W. S., *Theory of Sound*, Dover, New York, 1945; published originally in 1894.

APPENDIX 8: INERTIA FORCES FOR RIGID BODIES

The inertia forces for a rigid bar and rectangular and circular rigid plates associated with accelerations $\ddot{u}_x$, $\ddot{u}_y$, and $\ddot{\theta}$ of the center of mass O (or center of gravity) are shown in Fig. A8.1. Each rigid body is of uniform thickness, and its total mass m is uniformly distributed; the moment of inertia I_O about the axis normal to the bar or plate and passing through O is as noted in the figure.

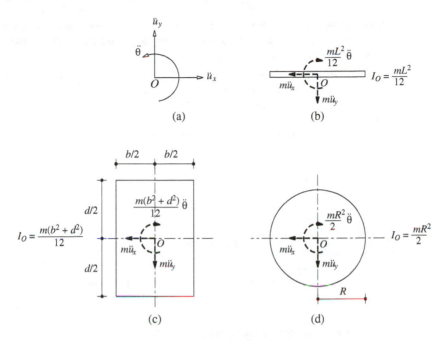

Figure A8.1   Inertia forces for rigid plates.

PROBLEMS

8.1 Repeat parts (a), (b), and (c) of Example 8.1 with one change: Use the horizontal displacement at C as the generalized coordinate. Show that the natural frequency, damping ratio, and displacement response are independent of the choice of generalized displacement.

8.2 For the rigid-body system shown in Fig. P8.2:
(a) Formulate the equation of motion governing the rotation at O.
(b) Determine the natural frequency and damping ratio.
(c) Determine the displacement response $u(x, t)$ to $p(t) = \delta(t)$, the Dirac delta function.

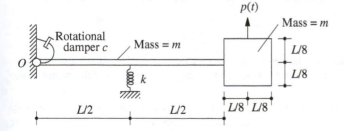

Figure P8.2

8.3   Solve Problem 8.2 with one change: Use the vertical displacement at the center of gravity of the square plate as the generalized displacement. Show that the results are independent of the choice of generalized displacement.

8.4   The rigid bar in Fig. P8.4 with a hinge at the center is bonded to a viscoelastic foundation, which can be modeled by stiffness k and damping coefficient c per unit of length. Using the rotation of the bar as the generalized coordinate:
(a) Formulate the equation of motion.
(b) Determine the natural vibration frequency and damping ratio.

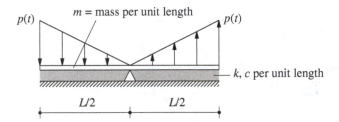

Figure P8.4

8.5   For the rigid-body system shown in Fig. P8.5:
(a) Choose a generalized coordinate.
(b) Formulate the equation of motion.
(c) Determine the natural vibration frequency and damping ratio.

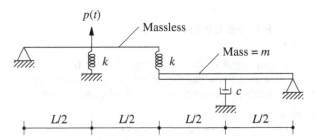

Figure P8.5

8.6   Solve Example 8.3 assuming the deflected shape function due to lateral force at the top:

$$\psi(x) = \frac{3}{2}\frac{x^2}{L^2} - \frac{1}{2}\frac{x^3}{L^3}$$

The shear forces and bending moments need to be calculated only at the base and midheight. (Note that these forces were determined in Example 8.3 throughout the height of the chimney.)

8.7   A reinforced-concrete chimney 600 ft high has a hollow circular cross section with outside diameter 50 ft at the base and 25 ft at the top; the wall thickness is 2 ft 6 in., uniform over the height (Fig. P8.7). Using the approximation that the wall thickness is small compared to the radius, the mass and flexural stiffness properties are computed from the gross area of concrete (neglecting reinforcing steel). The chimney is assumed to be clamped at the base, and its damping ratio is estimated to be 5%. The unit weight of concrete is 150 lb/ft^3, and its elastic modulus $E_c = 3600$ ksi. Assuming that the shape function is

$$\psi(x) = 1 - \cos \frac{\pi x}{2L}$$

where L is the length of the chimney and x is measured from the base, calculate the following quantities: **(a)** the shear forces and bending moments at the base and at the midheight, and **(b)** the top deflection due to ground motion defined by the design spectrum of Fig. 6.9.5, which is scaled to a peak acceleration of 0.25g.

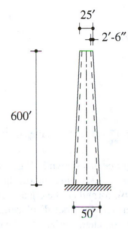

Figure P8.7

8.8   Solve Problem 8.7 asssuming that the shape function is

$$\psi(x) = \frac{3}{2}\frac{x^2}{L^2} - \frac{1}{2}\frac{x^3}{L^3}$$

8.9   Solve Problem 8.7 for a different excitation: a blast force varying linearly over height from zero at the base to $p(t)$ at the top, where $p(t)$ is given in Fig. P8.9.

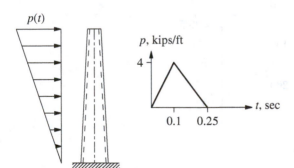

Figure P8.9

8.10– Three-story shear frames (rigid beams and flexible columns) in structural steel ($E = 29{,}000$ ksi)
8.11 are shown in Figs. P8.10 and P8.11; $w = 100$ kips; $I = 1400$ in^4; and the modal damping
ratios ζ_n are 5% for all modes. Assuming that the shape function is given by deflections due
to lateral forces that are equal to the floor weights, determine the floor displacements, story
shears, and overturning moments at the floors and base due to ground motion characterized
by the design spectrum of Fig. 6.9.5 scaled to a peak ground acceleration of 0.25g.

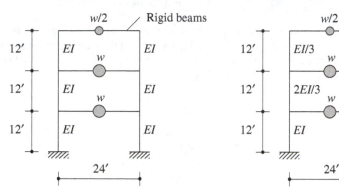

Figure P8.10   Figure P8.11

8.12– Solve Problems 8.10 and 8.11 using the shape function given by deflections due to a lateral
8.13 force at the roof level.

8.14 A five-story frame with rigid beams shown in Fig. P8.14a is subjected to ground acceleration
$\ddot{u}_g(t)$; k_j are story stiffnesses. Assuming the displacements to increase linearly with height
above the base (Fig. P8.14b), formulate the equation of motion for the system and determine
its natural frequency. Determine the floor displacements, story shears, and floor overturning
moments due to ground motion characterized by the design spectrum of Fig. 6.9.5 scaled to a
peak ground acceleration of 0.25g.

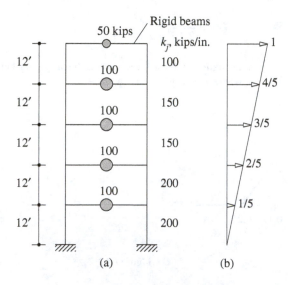

(a)   (b)   **Figure P8.14**

8.15   Solve Problem 8.14 using the shape function obtained from static deflections due to lateral forces equal to floor weights.

8.16   Solve Problem 8.14 using the shape function given by deflections due to a lateral force at the roof level.

8.17   Determine the natural vibration frequency of the inverted L-shaped frame shown in Fig. P8.17 using the shape function given by the deflections due to a vertical force at the free end. Neglect deformations due to shear and axial force. EI is constant.

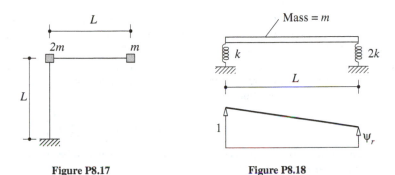

Figure P8.17   **Figure P8.18**

8.18   **(a)** By Rayleigh's method determine the natural vibration frequency of the rigid bar on two springs (Fig. P8.18) using the shape function shown. Note that the result involves the unknown ψ_r. Plot the value of ω_n^2 as a function of ψ_r.
(b) Using the properties of Rayleigh's quotient, determine the exact values of the two vibration frequencies and the corresponding vibration shapes.

8.19   The umbrella structure shown in Fig. P8.19 consists of a uniform column of flexural rigidity EI supporting a uniform slab of radius R and mass m. By Rayleigh's method determine the natural vibration frequency of the structure. Neglect the mass of the column and the effect of axial force on column stiffness. Assume that the slab is rigid in flexure and that the column is axially rigid.

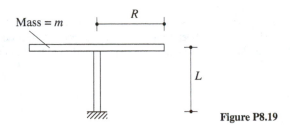

Figure P8.19

8.20   By Rayleigh's method determine the natural vibration frequency of the uniform beam shown in Fig. P8.20. Assume that the shape function is given by the deflections due to a force applied at the free end.

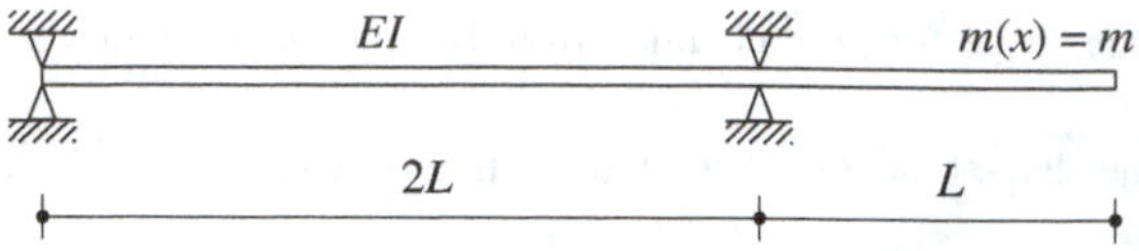

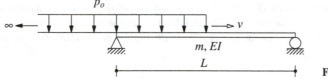

2L   L

EI   m(x) = m

Figure P8.20

8.21 By Rayleigh's method determine the natural vibration frequency of transverse vibration of the three-span, box-girder bridge of Example 8.5. Assume that the shape function is given by the deflections due to uniform force $p(x) = 1$ applied in the transverse direction. Neglect torsional stiffness of the bents.

8.22 Repeat Problem 8.21 using a simpler approach in which the deflections are assumed to be $u(x) = u_o \sin(\pi x/L)$, where u_o is the midspan deflection due to uniform force $p(x) = 1$ applied in the transverse direction.

8.23 Repeat Problem 8.21 using a simpler approach in which the deflections are assumed to be $u(x) = u_o \psi(x)$, where u_o is the midspan deflection due to uniform force $p(x) = 1$ applied in the transverse direction and

$$\psi(x) = \frac{16}{5}\left[\frac{x}{L} - 2\left(\frac{x}{L}\right)^3 + \left(\frac{x}{L}\right)^4\right]$$

Note that $\psi(x)$ is the deflected shape of a simply supported beam without bents subjected to transverse force $p(x) = 1$.

8.24 Repeat Problem 8.21 with one change: Consider torsional stiffness of the bents.

8.25 A simply supported bridge with a single span of L feet has a deck of uniform cross section with mass m per foot length and flexural rigidity EI. An infinitely long, uniformly distributed force p_o per foot length (that represents a very long train) travels across the bridge at a uniform velocity v (Fig. P8.25). Determine an equation for the deflection at midspan as a function of time. Neglect damping and assume the shape function to be $\psi(x) = \sin(\pi x/L)$.

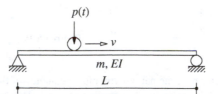

p_o

v

m, EI

L

Figure P8.25

8.26 A pulsating force $p(t) = p_o \cos \omega t$ travels across the bridge of Fig. P8.25 at a uniform velocity of v, as shown in Fig. P8.26. Determine an equation for the deflection at midspan as a function of time. Neglect damping and assume the shape function to be $\psi(x) = \sin(\pi x/L)$.

$p(t)$

v

m, EI

L

Figure P8.26

PART II

Multi-Degree-of-Freedom Systems

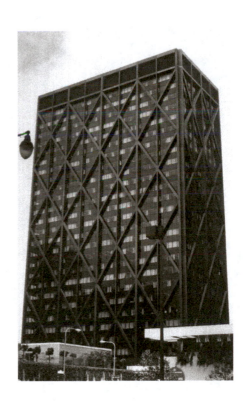

9

Equations of Motion, Problem Statement, and Solution Methods

PREVIEW

In this opening chapter of Part II the structural dynamics problem is formulated for structures discretized as systems with a finite number of degrees of freedom. The equations of motion are developed first for a simple multi-degree-of-freedom (MDF) system; a two-story shear frame is selected to permit easy visualization of elastic, damping, and inertia forces. Subsequently, a general formulation is presented for MDF systems subjected to external forces or earthquake-induced ground motion. This general formulation is then illustrated by several examples and applied to develop the equations of motion for multistory buildings, first for symmetric-plan buildings and then for unsymmetric-plan buildings. Subsequently, the formulation for earthquake response analysis is extended to systems subjected to spatially varying ground motion and to inelastic systems. The chapter ends with an overview of methods for solving the differential equations governing the motion of the structure and of how our study of dynamic analysis of MDF systems is organized.

9.1 SIMPLE SYSTEM: TWO-STORY SHEAR BUILDING

We first formulate the equations of motion for the simplest possible MDF system, a highly idealized two-story frame subjected to external forces $p_1(t)$ and $p_2(t)$ (Fig. 9.1.1a). In this idealization the beams and floor systems are rigid (infinitely stiff) in flexure, and several factors are neglected: axial deformation of the beams and columns, and the effect of axial force on the stiffness of the columns. This shear-frame or shear-building idealization, although unrealistic, is convenient for illustrating how the equations of motion for an MDF

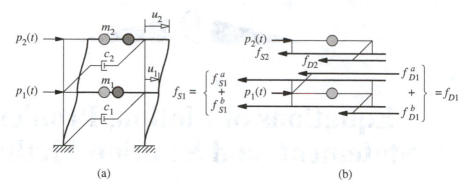

Figure 9.1.1   (a) Two-story shear frame; (b) forces acting on the two masses.

system are developed. Later, we extend the formulation to more realistic idealizations of buildings that consider beam flexure and joint rotations, and to structures other than buildings.

The mass is distributed throughout the building, but we will idealize it as concentrated at the floor levels. This assumption is generally appropriate for multistory buildings because most of the building mass is indeed at the floor levels.

Just as in the case of SDF systems (Chapter 1), we assume that a linear viscous damping mechanism represents the energy dissipation in a structure. If energy dissipation is associated with the deformational motions of each story, the viscous dampers may be visualized as shown.

The number of independent displacements required to define the displaced positions of all the masses relative to their original equilibrium position is called the number of degrees of freedom. The two-story frame of Fig. 9.1.1a, with lumped mass at each floor level, has two DOFs: the lateral displacements u_1 and u_2 of the two floors in the direction of the x-axis.

9.1.1 Using Newton's Second Law of Motion

The forces acting on each floor mass m_j are shown in Fig. 9.1.1b. These include the external force $p_j(t)$, the elastic (or inelastic) resisting force f_{Sj}, and the damping force f_{Dj}. The external force is taken to be positive along the positive direction of the x-axis. The elastic and damping forces are shown acting in the opposite direction because they are internal forces that resist the motions.

Newton's second law of motion then gives for each mass:

$$p_j - f_{Sj} - f_{Dj} = m_j \ddot{u}_j \quad \text{or} \quad m_j \ddot{u}_j + f_{Dj} + f_{Sj} = p_j(t) \qquad (9.1.1)$$

Equation (9.1.1) contains two equations for $j = 1$ and 2, and these can be written in matrix form:

$$\begin{bmatrix} m_1 & 0 \\ 0 & m_2 \end{bmatrix} \begin{Bmatrix} \ddot{u}_1 \\ \ddot{u}_2 \end{Bmatrix} + \begin{Bmatrix} f_{D1} \\ f_{D2} \end{Bmatrix} + \begin{Bmatrix} f_{S1} \\ f_{S2} \end{Bmatrix} = \begin{Bmatrix} p_1(t) \\ p_2(t) \end{Bmatrix} \qquad (9.1.2)$$

Equation (9.1.2) can be written compactly as

$$\mathbf{m\ddot{u}} + \mathbf{f}_D + \mathbf{f}_S = \mathbf{p}(t) \tag{9.1.3}$$

by introducing the following notation:

$$\mathbf{u} = \begin{Bmatrix} u_1 \\ u_2 \end{Bmatrix} \qquad \mathbf{m} = \begin{bmatrix} m_1 & 0 \\ 0 & m_2 \end{bmatrix} \qquad \mathbf{f}_D = \begin{Bmatrix} f_{D1} \\ f_{D2} \end{Bmatrix} \qquad \mathbf{f}_S = \begin{Bmatrix} f_{S1} \\ f_{S2} \end{Bmatrix} \qquad \mathbf{p} = \begin{Bmatrix} p_1 \\ p_2 \end{Bmatrix}$$

where $\mathbf{m}$ is the *mass matrix* for the two-story shear frame.

Assuming linear behavior, the elastic resisting forces $\mathbf{f}_S$ are next related to the floor displacements $\mathbf{u}$. For this purpose we introduce the lateral stiffness k_j of the jth story; it relates the story shear V_j to the story deformation or drift, $\Delta_j = u_j - u_{j-1}$, by

$$V_j = k_j \, \Delta_j \tag{9.1.4}$$

The story stiffness is the sum of the lateral stiffnesses of all columns in the story. For a story of height h and a column with modulus E and second moment of area I_c, the lateral stiffness of a column with clamped ends, implied by the shear-building idealization, is $12EI_c/h^3$. Thus the story stiffness is

$$k_j = \sum_{\text{columns}} \frac{12EI_c}{h^3} \tag{9.1.5}$$

With the story stiffnesses defined, we can relate the elastic resisting forces f_{S1} and f_{S2} to the floor displacements, u_1 and u_2. The force f_{S1} at the first floor is made up of two contributions: f_{S1}^a from the story above, and f_{S1}^b from the story below. Thus

$$f_{S1} = f_{S1}^b + f_{S1}^a$$

which, after substituting Eq. (9.1.4) and noting that $\Delta_1 = u_1$ and $\Delta_2 = u_2 - u_1$, becomes

$$f_{S1} = k_1 u_1 + k_2(u_1 - u_2) \tag{9.1.6a}$$

The force f_{S2} at the second floor is

$$f_{S2} = k_2(u_2 - u_1) \tag{9.1.6b}$$

Observe that f_{S1}^a and f_{S2} are equal in magnitude and opposite in direction because both represent the shear in the second story. In matrix form Eqs. (9.1.6a) and (9.1.6b) are

$$\begin{Bmatrix} f_{S1} \\ f_{S2} \end{Bmatrix} = \begin{bmatrix} k_1 + k_2 & -k_2 \\ -k_2 & k_2 \end{bmatrix} \begin{Bmatrix} u_1 \\ u_2 \end{Bmatrix} \qquad \text{or} \quad \mathbf{f}_S = \mathbf{ku} \tag{9.1.7}$$

Thus the elastic resisting force vector $\mathbf{f}_S$ and the displacement vector $\mathbf{u}$ are related through the *stiffness matrix* $\mathbf{k}$ for the two-story shear building.

The damping forces f_{D1} and f_{D2} are next related to the floor velocities $\dot{u}_1$ and $\dot{u}_2$. The jth story damping coefficient c_j relates the story shear V_j due to damping effects to the velocity $\dot{\Delta}_j$ associated with the story deformation by

$$V_j = c_j \, \dot{\Delta}_j \tag{9.1.8}$$

In a manner similar to Eq. (9.1.6), we can derive

$$f_{D1} = c_1 \dot{u}_1 + c_2(\dot{u}_1 - \dot{u}_2) \qquad f_{D2} = c_2(\dot{u}_2 - \dot{u}_1) \qquad (9.1.9)$$

In matrix form Eq. (9.1.9) is

$$\begin{Bmatrix} f_{D1} \\ f_{D2} \end{Bmatrix} = \begin{bmatrix} c_1 + c_2 & -c_2 \\ -c_2 & c_2 \end{bmatrix} \begin{Bmatrix} \dot{u}_1 \\ \dot{u}_2 \end{Bmatrix} \quad \text{or} \quad \mathbf{f}_D = \mathbf{c}\dot{\mathbf{u}} \qquad (9.1.10)$$

The damping resisting force vector $\mathbf{f}_D$ and the velocity vector $\dot{\mathbf{u}}$ are related through the *damping matrix* $\mathbf{c}$ for the two-story shear building.

We now substitute Eqs. (9.1.7) and (9.1.10) into Eq. (9.1.3) to obtain

$$\mathbf{m}\ddot{\mathbf{u}} + \mathbf{c}\dot{\mathbf{u}} + \mathbf{k}\mathbf{u} = \mathbf{p}(t) \qquad (9.1.11)$$

This matrix equation represents two ordinary differential equations governing the displacements $u_1(t)$ and $u_2(t)$ of the two-story frame subjected to external dynamic forces $p_1(t)$ and $p_2(t)$. Each equation contains both unknowns u_1 and u_2. The two equations are therefore coupled and in their present form must be solved simultaneously.

9.1.2 Dynamic Equilibrium

According to D'Alembert's principle (Chapter 1), with inertia forces included, a dynamic system is in equilibrium at each time instant. For the two masses in the system of Fig. 9.1.1a, Fig. 9.1.2 shows their free-body diagrams, including the inertia forces. Each inertia force is equal to the product of the mass times its acceleration and acts opposite to the direction of acceleration. From the free-body diagrams the condition of dynamic equilibrium also gives Eq. (9.1.3), which leads to Eq. (9.1.11), as shown in the preceding section.

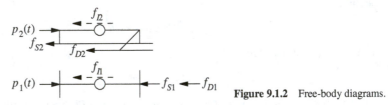

Figure 9.1.2   Free-body diagrams.

9.1.3 Mass–Spring–Damper System

We have introduced the linear two-DOF system by idealizing a two-story frame—an approach that should appeal to structural engineering students. However, the classic two-DOF system, shown in Fig. 9.1.3a, consists of two masses connected by linear springs and linear viscous dampers subjected to external forces $p_1(t)$ and $p_2(t)$. At any instant of time the forces acting on the two masses are as shown in their free-body diagrams (Fig. 9.1.3b). The resulting conditions of dynamic equilibrium also lead to Eq. (9.1.11), with $\mathbf{u}$, $\mathbf{m}$, $\mathbf{c}$, $\mathbf{k}$, and $\mathbf{p}(t)$ as defined earlier.

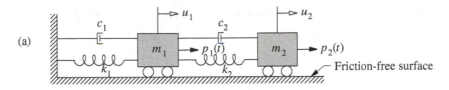

(a)

(b)

Figure 9.1.3   (a) Two-degree-of-freedom system; (b) free-body diagrams.

Example 9.1a

Formulate the equations of motion for the two-story shear frame shown in Fig. E9.1a.

Solution   Equation (9.1.11) is specialized for this system to obtain its equation of motion. To do so, we note that

$$m_1 = 2m \qquad m_2 = m$$

$$k_1 = 2\frac{12(2EI_c)}{h^3} = \frac{48EI_c}{h^3} \qquad k_2 = 2\frac{12(EI_c)}{h^3} = \frac{24EI_c}{h^3}$$

Substituting these data in Eqs. (9.1.2) and (9.1.7) gives the mass and stiffness matrices:

$$\mathbf{m} = m\begin{bmatrix} 2 & 0 \\ 0 & 1 \end{bmatrix} \qquad \mathbf{k} = \frac{24EI_c}{h^3}\begin{bmatrix} 3 & -1 \\ -1 & 1 \end{bmatrix}$$

Substituting these **m** and **k** in Eq. (9.1.11) gives the governing equations for this system without damping:

$$m\begin{bmatrix} 2 & 0 \\ 0 & 1 \end{bmatrix}\begin{Bmatrix} \ddot{u}_1 \\ \ddot{u}_2 \end{Bmatrix} + 24\frac{EI_c}{h^3}\begin{bmatrix} 3 & -1 \\ -1 & 1 \end{bmatrix}\begin{Bmatrix} u_1 \\ u_2 \end{Bmatrix} = \begin{Bmatrix} p_1(t) \\ p_2(t) \end{Bmatrix}$$

Observe that the stiffness matrix is nondiagonal, implying that the two equations are coupled, and in their present form must be solved simultaneously.

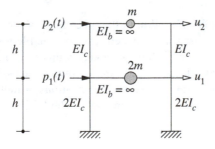

Figure E9.1a

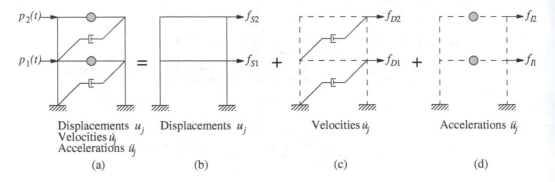

Displacements u_j Displacements u_j Velocities $\dot{u}_j$ Accelerations $\ddot{u}_j$
Velocities $\dot{u}_j$
Accelerations $\ddot{u}_j$

(a)   (b)   (c)   (d)

Figure 9.1.4   (a) System; (b) stiffness component; (c) damping component; (d) mass component.

9.1.4 Stiffness, Damping, and Mass Components

In this section the governing equations for the two-story shear frame are formulated based on an alternative viewpoint. Under the action of external forces $p_1(t)$ and $p_2(t)$ the state of the system at any time instant is described by displacements $u_j(t)$, velocities $\dot{u}_j(t)$, and accelerations $\ddot{u}_j(t)$; see Fig. 9.1.4a. Now visualize this system as the combination of three pure components: (1) stiffness component: the frame without damping or mass (Fig. 9.1.4b); (2) damping component: the frame with its damping property but no stiffness or mass (Fig. 9.1.4c); and (3) mass component: the floor masses without the stiffness or damping of the frame (Fig. 9.1.4d). The external forces f_{Sj} on the stiffness component are related to the displacements by Eq. (9.1.7). Similarly, the external forces f_{Dj} on the damping component are related to the velocities by Eq. (9.1.10). Finally, the external forces f_{Ij} on the mass component are related to the accelerations by $\mathbf{f}_I = \mathbf{m}\ddot{\mathbf{u}}$. The external forces $\mathbf{p}(t)$ on the system may therefore by visualized as distributed among the three components of the structure. Thus $\mathbf{f}_S + \mathbf{f}_D + \mathbf{f}_I$ must equal the applied forces $\mathbf{p}(t)$, leading to Eq. (9.1.3). This alternative viewpoint may seem unnecessary for the two-story shear frame, but it can be useful in visualizing the formulation of the equations of motion for complex MDF systems (Section 9.2).

9.2 GENERAL APPROACH FOR LINEAR SYSTEMS

The formulation of the equations of motion in the preceding sections, while easy to visualize for a shear building and other simple systems, is not suitable for complex structures. For this purpose a more general approach is presented in this section. We define the three types of forces—inertia forces, elastic forces, and damping forces—and use the line of reasoning presented in Section 9.1.4 to develop the equations of motion. Before defining the forces, we need to discretize the structure and define the DOFs.

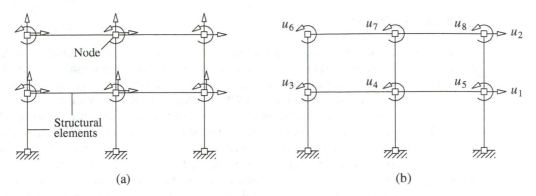

Figure 9.2.1   Degrees of freedom: (a) axial deformation included, 18 DOFs; (b) axial deformation neglected, 8 DOFs.

9.2.1 Discretization

A frame structure can be idealized as an assemblage of elements—beams, columns, walls—interconnected at nodal points or nodes (Fig. 9.2.1a). The displacements of the nodes are the degrees of freedom. In general, a node in a planar two-dimensional frame has three DOFs—two translations and one rotation. A node in a three-dimensional frame has six DOFs—three translations (the x, y, and z components) and three rotations (about the x, y, and z axes).

For example, a two-story, two-bay planar frame has six nodes and 18 DOFs (Fig. 9.2.1a). Axial deformations of beams can be neglected in analyzing most buildings, and axial deformations of columns need not be considered for low-rise buildings. With these assumptions the two-story, two-bay frame has eight DOFs (Fig. 9.2.1b). This is the structural idealization we use to illustrate a general approach for formulating equations of motion. The external dynamic forces are applied at the nodes (Fig. 9.2.2). The external moments $p_3(t)$ to $p_8(t)$ are zero in most practical cases.

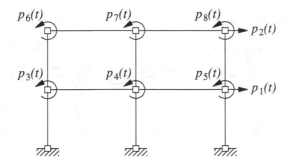

Figure 9.2.2   External dynamic forces, **p**(t).

9.2.2 Elastic Forces

We will relate the external forces f_{Sj} on the stiffness component of the structure to the resulting displacements u_j (Fig. 9.2.3a). For linear systems this relationship can be obtained by the method of superposition and the concept of stiffness influence coefficients.

We apply a unit displacement along DOF j, holding all other displacements to zero as shown; to maintain these displacements, forces must be applied generally along all DOFs. The *stiffness influence coefficient* k_{ij} is the force required along DOF i due to unit displacement at DOF j. In particular, the forces k_{i1} ($i = 1, 2, \ldots, 8$) shown in Fig. 9.2.3b are required to maintain the deflected shape associated with $u_1 = 1$ and all other $u_j = 0$. Similarly, the forces k_{i4} ($i = 1, 2, \ldots, 8$) shown in Fig. 9.2.3c are required to maintain the deflected shape associated with $u_4 = 1$ and all other $u_j = 0$. All forces in Fig. 9.2.3 are shown with their positive signs, but some of them may be negative to be consistent with the deformations imposed.

The force f_{Si} at DOF i associated with displacements u_j, $j = 1$ to N (Fig. 9.2.3a), is obtained by superposition:

$$f_{Si} = k_{i1}u_1 + k_{i2}u_2 + \cdots + k_{ij}u_j + \cdots + k_{iN}u_N \tag{9.2.1}$$

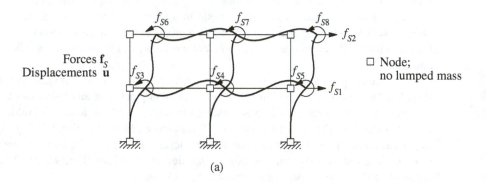

(a)

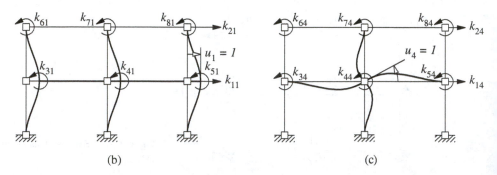

(b)   (c)

Figure 9.2.3   (a) Stiffness component of frame; (b) stiffness influence coefficients for $u_1 = 1$; (c) stiffness influence coefficients for $u_4 = 1$.

One such equation exists for each $i = 1$ to N. The set of N equations can be written in matrix form:

$$\begin{bmatrix} f_{S1} \\ f_{S2} \\ \vdots \\ f_{SN} \end{bmatrix} = \begin{bmatrix} k_{11} & k_{12} & \cdots & k_{1j} & \cdots & k_{1N} \\ k_{21} & k_{22} & \cdots & k_{2j} & \cdots & k_{2N} \\ \vdots & \vdots & & \vdots & & \vdots \\ k_{N1} & k_{N2} & \cdots & k_{Nj} & \cdots & k_{NN} \end{bmatrix} \begin{Bmatrix} u_1 \\ u_2 \\ \vdots \\ u_N \end{Bmatrix} \quad (9.2.2)$$

or

$$\mathbf{f}_S = \mathbf{k}\mathbf{u} \quad (9.2.3)$$

where $\mathbf{k}$ is the *stiffness matrix* of the structure; it is a symmetric matrix (i.e., $k_{ij} = k_{ji}$).

The stiffness matrix $\mathbf{k}$ for a discretized system can be determined by any one of several methods. The jth column of $\mathbf{k}$ can be obtained by calculating the forces k_{ij} ($i = 1, 2, \ldots, N$) required to produce $u_j = 1$ (with all other $u_i = 0$). The direct equilibrium method is feasible to implement such calculations for simple structures with a few DOFs; it is not practical, however, for complex structures or for computer implementation. The most commonly used method is the direct stiffness method, wherein the stiffness matrices of individual elements are assembled to obtain the structural stiffness matrix. This and other methods should be familiar to the reader. Therefore, these methods will not be developed in this book; we will use the simplest method appropriate for the problem to be solved.

9.2.3 Damping Forces

As mentioned in Section 1.4, the mechanisms by which the energy of a vibrating structure is dissipated can usually be idealized by equivalent viscous damping. With this assumption we relate the external forces f_{Dj} acting on the damping component of the structure to the velocities $\dot{u}_j$ (Fig. 9.2.4). We impart a unit velocity along DOF j, while the velocities in all other DOFs are kept zero. These velocities will generate internal damping forces that resist the velocities, and external forces would be necessary to equilibrate these forces. The *damping influence coefficient* c_{ij} is the external force in DOF i due to unit velocity in DOF j. The force f_{Di} at DOF i associated with velocities $\dot{u}_j$, $j = 1$ to N (Fig. 9.2.4), is obtained by superposition:

$$f_{Di} = c_{i1}\dot{u}_1 + c_{i2}\dot{u}_2 + \cdots + c_{ij}\dot{u}_j + \cdots + c_{iN}\dot{u}_N \quad (9.2.4)$$

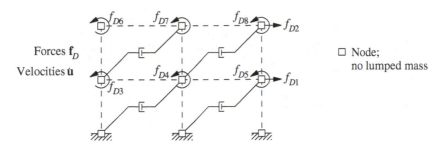

Figure 9.2.4   Damping component of frame.

Collecting all such equations for $i = 1$ to N and writing them in matrix form gives

$$
\begin{bmatrix} f_{D1} \\ f_{D2} \\ \vdots \\ f_{DN} \end{bmatrix} = \begin{bmatrix} c_{11} & c_{12} & \cdots & c_{1j} & \cdots & c_{1N} \\ c_{21} & c_{22} & \cdots & c_{2j} & \cdots & c_{2N} \\ \vdots & \vdots & & \vdots & & \vdots \\ c_{N1} & c_{N2} & \cdots & c_{Nj} & \cdots & c_{NN} \end{bmatrix} \begin{Bmatrix} \dot{u}_1 \\ \dot{u}_2 \\ \vdots \\ \dot{u}_N \end{Bmatrix}
$$
(9.2.5)

or

$$
\mathbf{f}_D = \mathbf{c}\dot{\mathbf{u}}
$$
(9.2.6)

where $\mathbf{c}$ is the *damping matrix* for the structure.

It is impractical to compute the coefficients c_{ij} of the damping matrix directly from the dimensions of the structure and the sizes of the structural elements. Therefore, damping for MDF systems is generally specified by numerical values for the damping ratios, as for SDF systems, based on experimental data for similar structures (Chapter 11). Methods are available to construct the damping matrix from known damping ratios (Chapter 11).

9.2.4 Inertia Forces

We will relate the external forces f_{Ij} acting on the mass component of the structure to the accelerations $\ddot{u}_j$ (Fig. 9.2.5a). We apply a unit acceleration along DOF j, while the

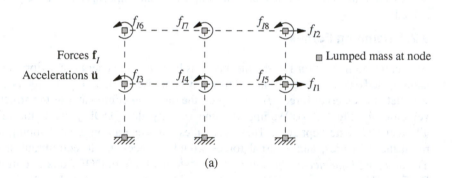

(a)

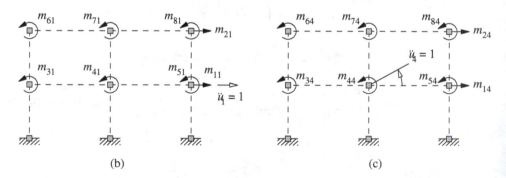

(b)   (c)

Figure 9.2.5   (a) Mass component of frame; (b) mass influence coefficients for $\ddot{u}_1 = 1$; (c) mass influence coefficients for $\ddot{u}_4 = 1$.

accelerations in all other DOFs are kept zero. According to D'Alembert's principle, the fictitious inertia forces oppose these accelerations; therefore, external forces will be necessary to equilibrate these inertia forces. The *mass influence coefficient* m_{ij} is the external force in DOF i due to unit acceleration along DOF j. In particular, the forces m_{i1} ($i = 1, 2, \ldots, 8$) shown in Fig. 9.2.5b are required in the various DOFs to equilibrate the inertia forces associated with $\ddot{u}_1 = 1$ and all other $\ddot{u}_j = 0$. Similarly, the forces m_{i4} ($i = 1, 2, \ldots, 8$) shown in Fig. 9.2.5c are associated with acceleration $\ddot{u}_4 = 1$ and all other $\ddot{u}_j = 0$. The force f_{Ii} at DOF i associated with accelerations $\ddot{u}_j$, $j = 1$ to N (Fig. 9.2.5a), is obtained by superposition:

$$f_{Ii} = m_{i1}\ddot{u}_1 + m_{i2}\ddot{u}_2 + \cdots + m_{ij}\ddot{u}_j + \cdots + m_{iN}\ddot{u}_N \qquad (9.2.7)$$

One such equation exists for each $i = 1$ to N. The set of N equations can be written in matrix form:

$$\begin{bmatrix} f_{I1} \\ f_{I2} \\ \vdots \\ f_{IN} \end{bmatrix} = \begin{bmatrix} m_{11} & m_{12} & \cdots & m_{1j} & \cdots & m_{1N} \\ m_{21} & m_{22} & \cdots & m_{2j} & \cdots & m_{2N} \\ \vdots & \vdots & & \vdots & & \vdots \\ m_{N1} & m_{N2} & \cdots & m_{Nj} & \cdots & m_{NN} \end{bmatrix} \begin{Bmatrix} \ddot{u}_1 \\ \ddot{u}_2 \\ \vdots \\ \ddot{u}_N \end{Bmatrix} \qquad (9.2.8)$$

or

$$\mathbf{f}_I = \mathbf{m}\ddot{\mathbf{u}} \qquad (9.2.9)$$

where $\mathbf{m}$ is the *mass matrix*. Just like the stiffness matrix, the mass matrix is symmetric (i.e., $m_{ij} = m_{ji}$).

The mass is distributed throughout an actual structure, but it can be idealized as lumped or concentrated at the nodes of the discretized structure; usually, such a lumped-mass idealization is satisfactory. The lumped mass at a node is determined from the portion of the weight that can reasonably be assigned to the node. Each structural element is replaced by point masses at its two nodes, with the distribution of the two masses being determined by static analysis of the element under its own weight. The lumped mass at a node of the structure is the sum of the mass contributions of all the structural elements connected to the node. This procedure is illustrated schematically in Fig. 9.2.6 for a two-story, two-bay frame where the beam mass includes the floor–slab mass it supports. The lumped masses m_a, m_b, and so on, at the various nodes are identified.

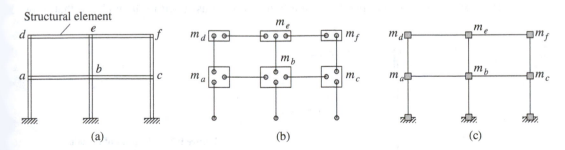

Figure 9.2.6   Lumping of mass at structural nodes.

Once the lumped masses at the nodes have been calculated, the mass matrix for the structure can readily be formulated. Consider again the two-story, two-bay frame of Fig. 9.2.1b. The external forces associated with acceleration $\ddot{u}_1 = 1$ (Fig. 9.2.5b) are $m_{11} = m_1$, where $m_1 = m_a + m_b + m_c$ (Fig. 9.2.6c), and $m_{i1} = 0$ for $i = 2, 3, \ldots, 8$. Similarly, the external forces m_{i4} associated with $\ddot{u}_4 = 1$ (Fig. 9.2.5c) are zero for all i, except possibly for $i = 4$. The coefficient m_{44} is equal to the rotational inertia of the mass lumped at the middle node at the first floor. This rotational inertia has negligible influence on the dynamics of practical structures; thus, we set $m_{44} = 0$.

In general, then, for a lumped-mass idealization, the mass matrix is diagonal:

$$m_{ij} = 0 \quad i \neq j \quad m_{jj} = m_j \quad \text{or} \quad 0 \qquad (9.2.10)$$

where m_j is the lumped mass associated with the jth translational DOF, and $m_{jj} = 0$ for a rotational DOF. The mass lumped at a node is associated with all the translational degrees of freedom of that node: (1) the horizontal (x) and vertical (z) DOFs for a two-dimensional frame, and (2) all three (x, y, and z) translational DOFs for a three-dimensional frame.

The mass representation can be simplified for multistory buildings because of the constraining effects of the floor slabs or floor diaphragms. Each floor diaphragm is usually assumed to be rigid in its own plane but is flexible in bending in the vertical direction, which is a reasonable representation of the true behavior of several types of floor systems (e.g., cast-in-place concrete). Introducing this assumption implies that both (x and y) horizontal DOFs of all the nodes at a floor level are related to the three rigid-body DOFs of the floor diaphragm in its own plane. For the jth floor diaphragm these three DOFs, defined at the center of mass, are translations u_{jx} and u_{jy} in the x and y directions and rotation $u_{j\theta}$ about the vertical axis (Fig. 9.2.7). Therefore, the mass needs to be defined only in these DOFs and need not be identified separately for each node. The diaphragm mass gives the mass associated with DOFs u_{jx} and u_{jy}, and the moment of inertia of the diaphragm about the vertical axis through O gives the mass associated with DOF $u_{j\theta}$. The diaphragm mass should include the contributions of the dead load and live load on the diaphragm and of the structural elements—columns, walls, etc.—and of the nonstructural elements—partition walls, architectural finishes, etc.—between floors.

The mass idealization for a multistory building becomes complicated if the floor diaphragm cannot be assumed as rigid in its own plane (e.g., floor system with wood framing and plywood sheathing). The diaphragm mass should then be assigned to individual nodes. The distributed dead and live loads at a floor level are assigned to the nodes at that floor

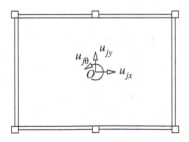

Figure 9.2.7 Degrees of freedom for in-plane-rigid floor diaphragm with distributed mass.

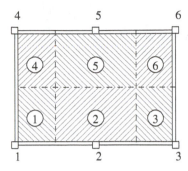

Figure 9.2.8 Tributary areas for distributing diaphragm mass to nodes.

in accordance with their respective tributary areas (Fig. 9.2.8). Similarly, the distributed weights of the structural and nonstructural elements between floors should be distributed to the nodes at the top and bottom of the story according to statics. The diaphragm flexibility must also be recognized in formulating the stiffness properties of the structure; the finite element method (Chapter 17) is effective in idealizing flexible diaphragms for this purpose.

9.2.5 Equations of Motion: External Forces

We now write the equations of motion for an MDF system subjected to external dynamic forces $p_j(t)$, $j = 1$ to N. The dynamic response of the structure to this excitation is defined by the displacements $u_j(t)$, velocities $\dot{u}_j(t)$, and accelerations $\ddot{u}_j(t)$, $j = 1$ to N. As mentioned in Section 9.1.4, the external forces $\mathbf{p}(t)$ may be visualized as distributed among the three components of the structure: $\mathbf{f}_S(t)$ to the stiffness components (Fig. 9.2.3a), $\mathbf{f}_D(t)$ to the damping component (Fig. 9.2.4), and $\mathbf{f}_I(t)$ to the mass component (Fig. 9.2.5a). Thus

$$\mathbf{f}_I + \mathbf{f}_D + \mathbf{f}_S = \mathbf{p}(t) \qquad (9.2.11)$$

Substituting Eqs. (9.2.3), (9.2.6), and (9.2.9) into Eq. (9.2.11) gives

$$\mathbf{m}\ddot{\mathbf{u}} + \mathbf{c}\dot{\mathbf{u}} + \mathbf{k}\mathbf{u} = \mathbf{p}(t) \qquad (9.2.12)$$

This is a system of N ordinary differential equations governing the displacements $\mathbf{u}(t)$ due to applied forces $\mathbf{p}(t)$. Equation (9.2.12) is the MDF equivalent of Eq. (1.5.2) for an SDF system; each scalar term in the SDF equation has become a vector or a matrix of order N, the number of DOFs in the MDF system.

Coupling of equations.   The off-diagonal terms in the coefficient matrices $\mathbf{m}$, $\mathbf{c}$, and $\mathbf{k}$ are known as *coupling terms*. In general, the equations have mass, damping, and stiffness coupling; however, the coupling in a system depends on the choice of degrees of freedom used to describe the motion. This is illustrated in Examples 9.2 and 9.3 for the same physical system with two different choices for the DOFs.

Example 9.1b

Formulate the equations of motion for the two-story shear frame in Fig. E9.1a using influence coefficients.

Solution

The two DOFs of this system are $\mathbf{u} = \langle u_1 \quad u_2 \rangle^T$.

$u_1 = 1, u_2 = 0$

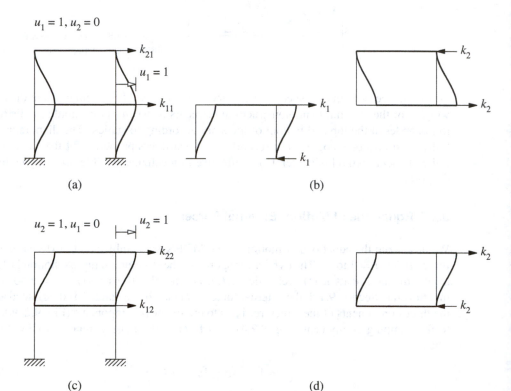

(a)   (b)

$u_2 = 1, u_1 = 0$

(c)   (d)

Figure E9.1b

    1. *Determine the stiffness matrix.* To obtain the first column of the stiffness matrix, we impose $u_1 = 1$ and $u_2 = 0$. The stiffness influence coefficients are k_{ij} (Fig. E9.1b). The forces necessary at the top and bottom of each story to maintain the deflected shape are expressed in terms of story stiffnesses k_1 and k_2 [part (b) of the figure], defined in Section 9.1.1 and determined in Example 9.1a:

$$k_1 = \frac{48EI_c}{h^3} \qquad k_2 = \frac{24EI_c}{h^3} \tag{a}$$

The two sets of forces in parts (a) and (b) of the figure are one and the same. Thus,

$$k_{11} = k_1 + k_2 = \frac{72EI_c}{h^3} \qquad k_{21} = -k_2 = -\frac{24EI_c}{h^3} \tag{b}$$

The second column of the stiffness matrix is obtained in a similar manner by imposing $u_2 = 1$ with $u_1 = 0$. The stiffness influence coefficients are k_{i2} [part (c) of the figure] and the forces necessary to maintain the deflected shape are shown in part (d) of the figure. The two sets of forces in parts (c) and (d) of the figure are one and the same. Thus,

$$k_{12} = -k_2 = -\frac{24EI_c}{h^3} \qquad k_{22} = k_2 = \frac{24EI_c}{h^3} \tag{c}$$

With the stiffness influence cofficients determined, the stiffness matrix is

$$\mathbf{k} = \frac{24EI_c}{h^3} \begin{bmatrix} 3 & -1 \\ -1 & 1 \end{bmatrix} \tag{d}$$

2. *Determine the mass matrix.* With the DOFs defined at the locations of the lumped masses, the diagonal mass matrix is given by Eq. (9.2.10):

$$\mathbf{m} = m \begin{bmatrix} 2 & 0 \\ 0 & 1 \end{bmatrix} \tag{e}$$

3. *Determine the equations of motion.* The governing equations are

$$\mathbf{m}\ddot{\mathbf{u}} + \mathbf{k}\mathbf{u} = \mathbf{p}(t) \tag{f}$$

where $\mathbf{m}$ and $\mathbf{k}$ are given by Eqs. (e) and (d), and $\mathbf{p}(t) = \langle\, p_1(t) \quad p_2(t) \,\rangle^T$.

Example 9.2

A uniform rigid bar of total mass m is supported on two springs k_1 and k_2 at the two ends and subjected to dynamic forces shown in Fig. E9.2a. The bar is constrained so that it can move only vertically in the plane of the paper; with this constraint the system has two DOFs.

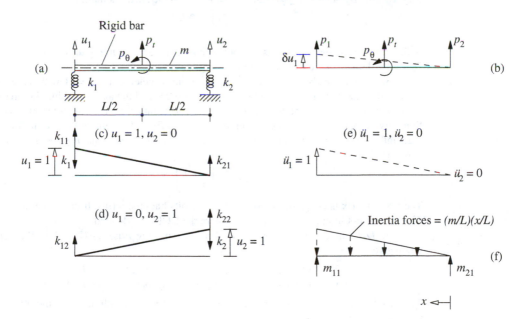

Figure E9.2

Formulate the equations of motion with respect to displacements u_1 and u_2 of the two ends as the two DOFs.

Solution

   1. *Determine the applied forces.* The external forces do not act along the DOFs and should therefore be converted to equivalent forces p_1 and p_2 along the DOFs (Fig. E9.2b) using equilibrium equations. This can also be achieved by the principle of virtual displacements. Thus if we introduce a virtual displacement δu_1 along DOF 1, the work done by the applied forces is

$$\delta W = p_t \frac{\delta u_1}{2} - p_\theta \frac{\delta u_1}{L} \tag{a}$$

Similarly, the work done by the equivalent forces is

$$\delta W = p_1 \delta u_1 + p_2(0) \tag{b}$$

Because the work done by the two sets of forces should be the same, we equate Eqs. (a) and (b) and obtain

$$p_1 = \frac{p_t}{2} - \frac{p_\theta}{L} \tag{c}$$

In a similar manner, by introducing a virtual displacement δu_2, we obtain

$$p_2 = \frac{p_t}{2} + \frac{p_\theta}{L} \tag{d}$$

   2. *Determine the stiffness matrix.* Apply a unit displacement $u_1 = 1$ with $u_2 = 0$ and identify the resulting elastic forces and the stiffness influence coefficients k_{11} and k_{21} (Fig. E9.2c). By statics, $k_{11} = k_1$ and $k_{21} = 0$. Now apply a unit displacement $u_2 = 1$ with $u_1 = 0$ and identify the resulting elastic forces and the stiffness influence coefficients (Fig. E9.2d). By statics, $k_{12} = 0$ and $k_{22} = k_2$. Thus the stiffness matrix is

$$\mathbf{k} = \begin{bmatrix} k_1 & 0 \\ 0 & k_2 \end{bmatrix} \tag{e}$$

In this case the stiffness matrix is diagonal (i.e., there are no coupling terms) because the two DOFs are defined at the locations of the springs.

   3. *Determine the mass matrix.* Impart a unit acceleration $\ddot{u}_1 = 1$ with $\ddot{u}_2 = 0$, determine the distribution of accelerations of (Fig. E9.2e) and the associated inertia forces, and identify mass influence coefficients (Fig. E9.2f). By statics, $m_{11} = m/3$ and $m_{21} = m/6$. Similarly, imparting a unit acceleration $\ddot{u}_2 = 1$ with $\ddot{u}_1 = 0$, defining the inertia forces and mass influence coefficients, and applying statics gives $m_{12} = m/6$ and $m_{22} = m/3$. Thus the mass matrix is

$$\mathbf{m} = \frac{m}{6} \begin{bmatrix} 2 & 1 \\ 1 & 2 \end{bmatrix} \tag{f}$$

The mass matrix is coupled, as indicated by the off-diagonal terms, because the mass is distributed and not lumped at the locations where the DOFs are defined.

   4. *Determine the equations of motion.* Substituting Eqs. (c)–(f) in Eq. (9.2.12) with $\mathbf{c} = \mathbf{0}$ gives

$$\frac{m}{6} \begin{bmatrix} 2 & 1 \\ 1 & 2 \end{bmatrix} \begin{bmatrix} \ddot{u}_1 \\ \ddot{u}_2 \end{bmatrix} + \begin{bmatrix} k_1 & 0 \\ 0 & k_2 \end{bmatrix} \begin{bmatrix} u_1 \\ u_2 \end{bmatrix} = \begin{bmatrix} (p_t/2) - (p_\theta/L) \\ (p_t/2) + (p_\theta/L) \end{bmatrix} \tag{g}$$

The two differential equations are coupled because of mass coupling due to the off-diagonal terms in the mass matrix.

Example 9.3

Formulate the equations of motion of the system of Fig. E9.2a with the two DOFs defined at the center of mass O of the rigid bar: translation u_t and rotation u_θ (Fig. E9.3a).

Solution

1. *Determine the stiffness matrix.* Apply a unit displacement $u_t = 1$ with $u_\theta = 0$ and identify the resulting elastic forces and k_{tt} and $k_{\theta t}$ (Fig. E9.3b). By statics, $k_{tt} = k_1 + k_2$ and $k_{\theta t} = (k_2 - k_1)L/2$. Now, apply a unit rotation $u_\theta = 1$ with $u_t = 0$ and identify the resulting elastic forces and $k_{t\theta}$ and $k_{\theta\theta}$ (Fig. E9.3c). By statics, $k_{t\theta} = (k_2 - k_1)L/2$ and $k_{\theta\theta} = (k_1 + k_2)L^2/4$. Thus the stiffness matrix is

$$\bar{\mathbf{k}} = \begin{bmatrix} k_1 + k_2 & (k_2 - k_1)L/2 \\ (k_2 - k_1)L/2 & (k_1 + k_2)L^2/4 \end{bmatrix} \tag{a}$$

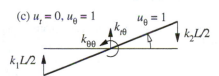

(a)

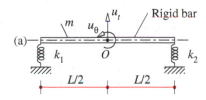

(b) $u_t = 1$, $u_\theta = 0$

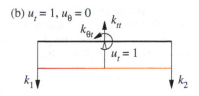

(c) $u_t = 0$, $u_\theta = 1$

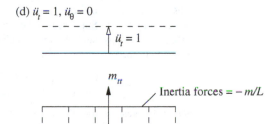

(d) $\ddot{u}_t = 1$, $\ddot{u}_\theta = 0$

Inertia forces $= -m/L$

(e)

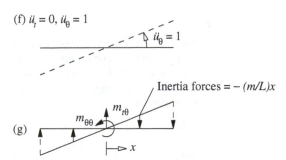

(f) $\ddot{u}_t = 0$, $\ddot{u}_\theta = 1$

Inertia forces $= -(m/L)x$

(g)

Figure E9.3

Observe that now the stiffness matrix has coupling terms because the DOFs chosen are not the displacements at the locations of the springs.

2. *Determine the mass matrix.* Impart a unit acceleration $\ddot{u}_t = 1$ with $\ddot{u}_\theta = 0$, determine the acceleration distribution (Fig. E9.3d) and the associated inertia forces, and identify m_{tt} and $m_{\theta t}$ (Fig. E9.3e). By statics, $m_{tt} = m$ and $m_{\theta t} = 0$. Now impart a unit rotational acceleration $\ddot{u}_\theta = 1$ with $\ddot{u}_t = 0$, determine the resulting accelerations (Fig. E9.3f) and the associated inertia forces, and identify $m_{t\theta}$ and $m_{\theta\theta}$ (Fig. E9.3g). By statics, $m_{t\theta} = 0$ and $m_{\theta\theta} = mL^2/12$. Note that $m_{\theta\theta} = I_O$, the moment of inertia of the bar about an axis that passes through O and is perpendicular to the plane of rotation. Thus the mass matrix is

$$\bar{\mathbf{m}} = \begin{bmatrix} m & 0 \\ 0 & mL^2/12 \end{bmatrix} \tag{b}$$

Now the mass matrix is diagonal (i.e., it has no coupling terms) because the DOFs of this rigid bar are defined at the mass center.

3. *Determine the equations of motion.* Substituting $\mathbf{u} = \langle u_t \quad u_\theta \rangle^T$, $\mathbf{p} = \langle p_t \quad p_\theta \rangle^T$, and Eqs. (a) and (b) in Eq. (9.2.12) gives

$$\begin{bmatrix} m & 0 \\ 0 & mL^2/12 \end{bmatrix} \begin{Bmatrix} \ddot{u}_t \\ \ddot{u}_\theta \end{Bmatrix} + \begin{bmatrix} k_1 + k_2 & (k_2 - k_1)L/2 \\ (k_2 - k_1)L/2 & (k_1 + k_2)L^2/4 \end{bmatrix} \begin{Bmatrix} u_t \\ u_\theta \end{Bmatrix} = \begin{Bmatrix} p_t \\ p_\theta \end{Bmatrix} \tag{c}$$

The two differential equations are now coupled through the stiffness matrix.

We should note that if the equations of motion for a system are available in one set of DOFs, they can be transformed to a different choice of DOF. This concept is illustrated for the system of Fig. E9.2a. Suppose that the mass and stiffness matrices and the applied force vector for the system are available for the first choice of DOF, $\mathbf{u} = \langle u_1 \quad u_2 \rangle^T$. These displacements are related to the second set of DOF, $\bar{\mathbf{u}} = \langle u_t \quad u_\theta \rangle^T$, by

$$\begin{Bmatrix} u_1 \\ u_2 \end{Bmatrix} = \begin{bmatrix} 1 & -L/2 \\ 1 & L/2 \end{bmatrix} \begin{Bmatrix} u_t \\ u_\theta \end{Bmatrix} \quad \text{or} \quad \mathbf{u} = \mathbf{a}\bar{\mathbf{u}} \tag{d}$$

where $\mathbf{a}$ denotes the coordinate transformation matrix. The stiffness and mass matrices and the applied force vector for the $\bar{\mathbf{u}}$ DOFs are given by

$$\bar{\mathbf{k}} = \mathbf{a}^T \mathbf{k} \mathbf{a} \qquad \bar{\mathbf{m}} = \mathbf{a}^T \mathbf{m} \mathbf{a} \qquad \bar{\mathbf{p}} = \mathbf{a}^T \mathbf{p} \tag{e}$$

Substituting for $\mathbf{a}$ from Eq. (d) and for $\mathbf{k}$, $\mathbf{m}$, and $\mathbf{p}$ from Example 9.2 into Eq. (e) leads to $\bar{\mathbf{k}}$ and $\bar{\mathbf{m}}$, which are identical to Eqs. (a) and (b) and to the $\bar{\mathbf{p}}$ in Eq. (c).

Example 9.4

A massless cantilever beam of length L supports two lumped masses $mL/2$ and $mL/4$ at the midpoint and free end as shown in Fig. E9.4a. The flexural rigidity of the uniform beam is EI. With the four DOFs chosen as shown in Fig. E9.4b and the applied forces $p_1(t)$ and $p_2(t)$, formulate the equations of motion of the system. Axial and shear deformations in the beam are neglected.

Solution

The beam consists of two beam elements and three nodes. The left node is constrained and each of the other two nodes has two DOFs (Fig. E9.4b). Thus, the displacement vector $\mathbf{u} = \langle u_1 \quad u_2 \quad u_3 \quad u_4 \rangle^T$.

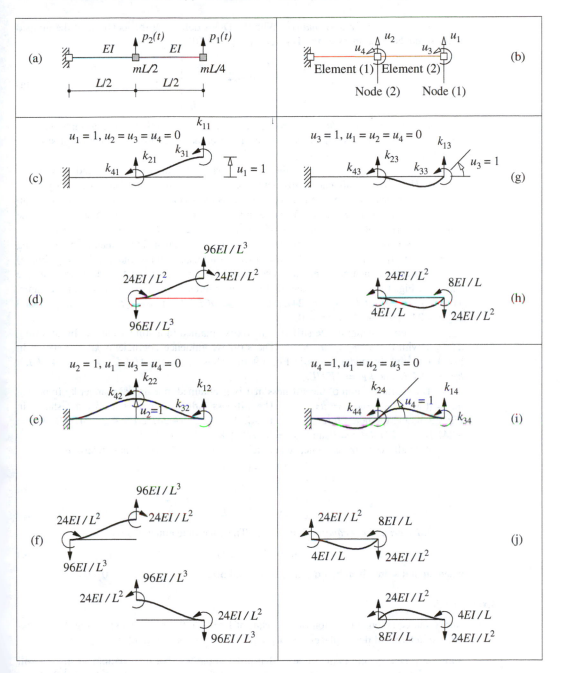

Figure E9.4

1. *Determine the mass matrix.* With the DOFs defined at the locations of the lumped masses, the diagonal mass matrix is given by Eq. (9.2.10):

$$
\mathbf{m} = \begin{bmatrix} mL/4 & & & \\ & mL/2 & & \\ & & 0 & \\ & & & 0 \end{bmatrix} \tag{a}
$$

2. *Determine the stiffness matrix.* Several methods are available to determine the stiffness matrix. Here we use the direct equilibrium method based on the definition of stiffness influence coefficients (Appendix 1).

To obtain the first column of the stiffness matrix, we impose $u_1 = 1$ and $u_2 = u_3 = u_4 = 0$. The stiffness influence coefficients are k_{i1} (Fig. E9.4c). The forces necessary at the nodes of each beam element to maintain the deflected shape are determined from the beam stiffness coefficients (Fig. E9.4d). The two sets of forces in figures (c) and (d) are one and the same. Thus $k_{11} = 96EI/L^3$, $k_{21} = -96EI/L^3$, $k_{31} = -24EI/L^2$, and $k_{41} = -24EI/L^2$.

The second column of the stiffness matrix is obtained in a similar manner by imposing $u_2 = 1$ with $u_1 = u_3 = u_4 = 0$. The stiffness influence coefficients are k_{i2} (Fig. E9.4e) and the forces on each beam element necessary to maintain the imposed displacements are shown in Fig. E9.4f. The two sets of forces in figures (e) and (f) are one and the same. Thus $k_{12} = -96EI/L^3$, $k_{32} = 24EI/L^2$, $k_{22} = 96EI/L^3 + 96EI/L^3 = 192EI/L^3$, and $k_{42} = -24EI/L^2 + 24EI/L^2 = 0$.

The third column of the stiffness matrix is obtained in a similar manner by imposing $u_3 = 1$ with $u_1 = u_2 = u_4 = 0$. The stiffness influence coefficients k_{i3} are shown in Fig. E9.4g and the nodal forces in Fig. E9.4h. Thus $k_{13} = -24EI/L^2$, $k_{23} = 24EI/L^2$, $k_{33} = 8EI/L$, and $k_{43} = 4EI/L$.

The fourth column of the stiffness matrix is obtained in a similar manner by imposing $u_4 = 1$ with $u_1 = u_2 = u_3 = 0$. The stiffness influence coefficients k_{i4} are shown in Fig. E9.4i, and the nodal forces in Fig. E9.4j. Thus $k_{14} = -24EI/L^2$, $k_{34} = 4EI/L$, $k_{24} = -24EI/L^2 + 24EI/L^2 = 0$, and $k_{44} = 8EI/L + 8EI/L = 16EI/L$.

With all the stiffness influence coefficients determined, the stiffness matrix is

$$
\mathbf{k} = \frac{8EI}{L^3} \begin{bmatrix} 12 & -12 & -3L & -3L \\ -12 & 24 & 3L & 0 \\ -3L & 3L & L^2 & L^2/2 \\ -3L & 0 & L^2/2 & 2L^2 \end{bmatrix} \tag{b}
$$

3. *Determine the equations of motion.* The governing equations are

$$
\mathbf{m\ddot{u} + ku = p}(t) \tag{c}
$$

where $\mathbf{m}$ and $\mathbf{k}$ are given by Eqs. (a) and (b), and $\mathbf{p}(t) = \langle\, p_1(t) \quad p_2(t) \quad 0 \quad 0\,\rangle^T$.

Example 9.5

Derive the equations of motion of the beam of Example 9.4 (also shown in Fig. E9.5a) expressed in terms of the displacements u_1 and u_2 of the masses (Fig. E9.5b).

Solution   This system is the same as that in Example 9.4, but its equations of motion will be formulated considering only the translational DOFs u_1 and u_2 (i.e., the rotational DOFs u_3 and u_4 will be excluded).

1. *Determine the stiffness matrix.* In a statically determinate structure such as the one in Fig. E9.5a, it is usually easier to calculate first the flexibility matrix and invert it to obtain

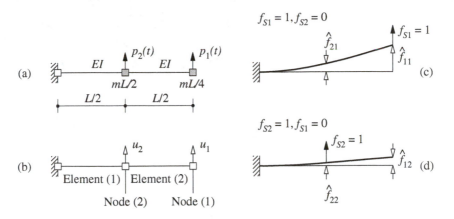

Figure E9.5

the stiffness matrix. The flexibility influence coefficient $\hat{f}_{ij}$ is the displacement in DOF i due to unit force applied in DOF j (Fig. E9.4c and d). The deflections are computed by standard procedures of structural analysis to obtain the flexibility matrix:

$$\hat{\mathbf{f}} = \frac{L^3}{48EI} \begin{bmatrix} 16 & 5 \\ 5 & 2 \end{bmatrix}$$

The off-diagonal elements $\hat{f}_{12}$ and $\hat{f}_{21}$ are equal, as expected, because of Maxwell's theorem of reciprocal deflections. By inverting $\hat{\mathbf{f}}$, the stiffness matrix is obtained:

$$\mathbf{k} = \frac{48EI}{7L^3} \begin{bmatrix} 2 & -5 \\ -5 & 16 \end{bmatrix} \tag{a}$$

2. *Determine the mass matrix.* This is a diagonal matrix because the lumped masses are located where the DOFs are defined:

$$\mathbf{m} = \begin{bmatrix} mL/4 & \\ & mL/2 \end{bmatrix} \tag{b}$$

3. *Determine the equations of motion.* Substituting $\mathbf{m}$, $\mathbf{k}$, and $\mathbf{p}(t) = \langle\, p_1(t) \quad p_2(t)\,\rangle^T$ in Eq. (9.2.12) with $\mathbf{c} = \mathbf{0}$ gives

$$\begin{bmatrix} mL/4 & \\ & mL/2 \end{bmatrix} \begin{Bmatrix} \ddot{u}_1 \\ \ddot{u}_2 \end{Bmatrix} + \frac{48EI}{7L^3} \begin{bmatrix} 2 & -5 \\ -5 & 16 \end{bmatrix} \begin{Bmatrix} u_1 \\ u_2 \end{Bmatrix} = \begin{Bmatrix} p_1(t) \\ p_2(t) \end{Bmatrix} \tag{c}$$

Example 9.6

Formulate the free vibration equations for the two-element frame of Fig. E9.6a. For both elements the flexural stiffness is EI, and axial deformations are to be neglected. The frame is massless with lumped masses at the two nodes as shown.

Solution   The two degrees of freedom of the frame are shown. The mass matrix is

$$\mathbf{m} = \begin{bmatrix} 3m & \\ & m \end{bmatrix} \tag{a}$$

Note that the mass corresponding to $\ddot{u}_1 = 1$ is $2m + m = 3m$ because both masses will undergo the same acceleration since the beam connecting the two masses is axially inextensible.

The stiffness matrix is formulated by first evaluating the flexibility matrix and then inverting it. The flexibility influence coefficients are identified in Fig. E9.6b and c, and the

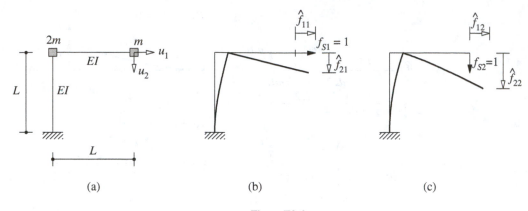

Figure E9.6

deflections are computed by standard procedures of structural analysis to obtain the flexibility matrix:

$$\hat{\mathbf{f}} = \frac{L^3}{6EI}\begin{bmatrix} 2 & 3 \\ 3 & 8 \end{bmatrix}$$

This matrix is inverted to determine the stiffness matrix:

$$\mathbf{k} = \frac{6EI}{7L^3}\begin{bmatrix} 8 & -3 \\ -3 & 2 \end{bmatrix}$$

Thus the equations in free vibration of the system (without damping) are

$$\begin{bmatrix} 3m & \\ & m \end{bmatrix}\begin{Bmatrix} \ddot{u}_1 \\ \ddot{u}_2 \end{Bmatrix} + \frac{6EI}{7L^3}\begin{bmatrix} 8 & -3 \\ -3 & 2 \end{bmatrix}\begin{Bmatrix} u_1 \\ u_2 \end{Bmatrix} = \begin{Bmatrix} 0 \\ 0 \end{Bmatrix}$$

Example 9.7

Formulate the equations of motion for the two-story frame in Fig. E9.7a. The flexural rigidity of the beams and columns and the lumped masses at the floor levels are as noted. The dynamic excitation consists of lateral forces $p_1(t)$ and $p_2(t)$ at the two floor levels. The story height is h and the bay width $2h$. Neglect axial deformations in the beams and the columns.

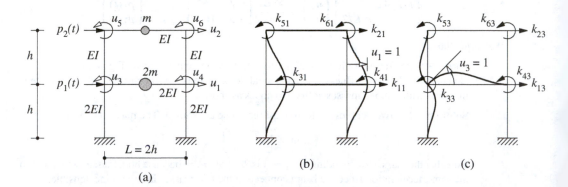

Figure E9.7

Solution   The system has six degrees of freedom shown in Fig. E9.7a: lateral displacements u_1 and u_2 of the floors and joint rotations u_3, u_4, u_5, and u_6. The displacement vector is

$$\mathbf{u} = \langle\, u_1 \quad u_2 \quad u_3 \quad u_4 \quad u_5 \quad u_6 \,\rangle^T \tag{a}$$

The mass matrix is given by Eq. (9.2.10):

$$\mathbf{m} = m \begin{bmatrix} 2 & & & & & \\ & 1 & & & & \\ & & 0 & & & \\ & & & 0 & & \\ & & & & 0 & \\ & & & & & 0 \end{bmatrix} \tag{b}$$

The stiffness influence coefficients are evaluated following the procedure of Example 9.4. A unit displacement is imposed, one at a time, in each DOF while constraining the other five DOFs, and the stiffness influence coefficients (e.g., shown in Fig. E9.7b and c for $u_1 = 1$ and $u_3 = 1$, respectively) are calculated by statics from the nodal forces for individual structural elements associated with the displacements imposed. These nodal forces are determined from the beam stiffness coefficients (Appendix 1). The result is

$$\mathbf{k} = \frac{EI}{h^3} \begin{bmatrix} 72 & -24 & 6h & 6h & -6h & -6h \\ -24 & 24 & 6h & 6h & 6h & 6h \\ 6h & 6h & 16h^2 & 2h^2 & 2h^2 & 0 \\ 6h & 6h & 2h^2 & 16h^2 & 0 & 2h^2 \\ -6h & 6h & 2h^2 & 0 & 6h^2 & h^2 \\ -6h & 6h & 0 & 2h^2 & h^2 & 6h^2 \end{bmatrix} \tag{c}$$

The dynamic forces applied are lateral forces $p_1(t)$ and $p_2(t)$ at the two floors without any moments at the nodes. Thus the applied force vector is

$$\mathbf{p}(t) = \langle\, p_1(t) \quad p_2(t) \quad 0 \quad 0 \quad 0 \quad 0 \,\rangle^T \tag{d}$$

The equations of motion are

$$\mathbf{m}\ddot{\mathbf{u}} + \mathbf{k}\mathbf{u} = \mathbf{p}(t) \tag{e}$$

where $\mathbf{u}$, $\mathbf{m}$, $\mathbf{k}$, and $\mathbf{p}(t)$ are given by Eqs. (a), (b), (c), and (d), respectively.

9.3 STATIC CONDENSATION

The static condensation method is used to eliminate from dynamic analysis those DOFs of a structure to which zero mass is assigned; however, all the DOFs are included in the static analysis. Consider the two-bay, two-story frame shown in Fig. 9.3.1. With axial deformations in structural elements neglected, the system has eight DOFs for formulating its stiffness matrix (Fig. 9.3.1a). As discussed in Section 9.2.4, typically the mass of the structure is idealized as concentrated in point lumps at the nodes (Fig. 9.3.1b), and the mass matrix contains zero diagonal elements in the rotational DOFs (see also Example 9.7). These are the DOFs that can be eliminated from the dynamic analysis of the structure provided that the dynamic excitation does not include any external forces in the rotational DOFs, as in the case of earthquake excitation (Section 9.4). Even if included in formulating the stiffness matrix, the vertical DOFs of the building can also be eliminated from dynamic analysis—because the inertial effects associated with the vertical DOFs of

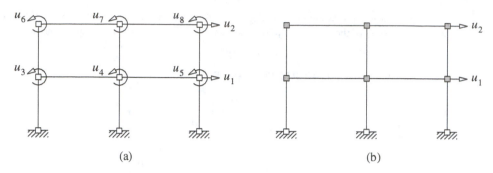

Figure 9.3.1 (a) Degrees of freedom (DOFs) for elastic forces—axial deformations neglected; (b) DOFs for inertia forces.

building frames are usually small—provided that the dynamic excitation does not include vertical forces at the nodes, as in the case of horizontal ground motion (Section 9.4).

The equations of motion for a system excluding damping [Eq. (9.2.12)] are written in partitioned form:

$$\begin{bmatrix} \mathbf{m}_{tt} & \mathbf{0} \\ \mathbf{0} & \mathbf{0} \end{bmatrix} \begin{Bmatrix} \ddot{\mathbf{u}}_t \\ \ddot{\mathbf{u}}_0 \end{Bmatrix} + \begin{bmatrix} \mathbf{k}_{tt} & \mathbf{k}_{t0} \\ \mathbf{k}_{0t} & \mathbf{k}_{00} \end{bmatrix} \begin{Bmatrix} \mathbf{u}_t \\ \mathbf{u}_0 \end{Bmatrix} = \begin{Bmatrix} \mathbf{p}_t(t) \\ \mathbf{0} \end{Bmatrix} \tag{9.3.1}$$

where $\mathbf{u}_0$ denotes the DOFs with zero mass and $\mathbf{u}_t$ the DOFs with mass, also known as dynamic DOFs; $\mathbf{k}_{t0} = \mathbf{k}_{0t}^T$. The two partitioned equations are

$$\mathbf{m}_{tt}\ddot{\mathbf{u}}_t + \mathbf{k}_{tt}\mathbf{u}_t + \mathbf{k}_{t0}\mathbf{u}_0 = \mathbf{p}_t(t) \qquad \mathbf{k}_{0t}\mathbf{u}_t + \mathbf{k}_{00}\mathbf{u}_0 = \mathbf{0} \tag{9.3.2}$$

Because no inertia terms or external forces are associated with $\mathbf{u}_0$, Eq. (9.3.2b) permits a static relationship between $\mathbf{u}_0$ and $\mathbf{u}_t$:

$$\mathbf{u}_0 = -\mathbf{k}_{00}^{-1}\mathbf{k}_{0t}\mathbf{u}_t \tag{9.3.3}$$

Substituting Eq. (9.3.3) in Eq. (9.3.2a) gives

$$\mathbf{m}_{tt}\ddot{\mathbf{u}}_t + \hat{\mathbf{k}}_{tt}\mathbf{u}_t = \mathbf{p}_t(t) \tag{9.3.4}$$

where $\hat{\mathbf{k}}_{tt}$ is the *condensed stiffness matrix* given by

$$\hat{\mathbf{k}}_{tt} = \mathbf{k}_{tt} - \mathbf{k}_{0t}^T\mathbf{k}_{00}^{-1}\mathbf{k}_{0t} \tag{9.3.5}$$

Solution of Eq. (9.3.4) provides the displacements $\mathbf{u}_t(t)$ in the dynamic DOFs, and at each instant of time the displacements $\mathbf{u}_0(t)$ in the condensed DOFs are determined from Eq. (9.3.3).

Henceforth, for notational convenience, Eq. (9.2.12) will also denote the equations of motion governing the dynamic DOFs [Eq. (9.3.4)], and it will be understood that only the dynamic DOFs have been retained. Before closing this section, note that an alternative method to determine $\hat{\mathbf{k}}_{tt}$ is by inverting the flexibility matrix $\hat{\mathbf{f}}_{tt}$. Each column of $\hat{\mathbf{f}}_{tt}$ is given by the displacements $\mathbf{u}_t$ due to a unit force applied successively in each DOF in $\mathbf{u}_t$, which can be determined by the force method. (This approach was employed in Examples 9.5 and 9.6.)

Example 9.8

Examples 9.4 and 9.5 were concerned with formulating the equations of motion for a cantilever beam with two lumped masses. The degrees of freedom chosen in Example 9.5 were the translational displacements u_1 and u_2 at the lumped masses; in Example 9.4 the four DOFs were u_1, u_2, and node rotations u_3 and u_4. Starting with the equations governing these four DOFs, derive the equations of motion in the two translational DOFs.

Solution The vector of four DOFs is partitioned in two parts: $\mathbf{u}_t = \langle\, u_1 \quad u_2\,\rangle^T$ and $\mathbf{u}_0 = \langle\, u_3 \quad u_4\,\rangle^T$. The equations of motion governing $\mathbf{u}_t$ are given by Eq. (9.3.4), where

$$\mathbf{m}_{tt} = \begin{bmatrix} mL/4 & \\ & mL/2 \end{bmatrix} \qquad \mathbf{p}_t(t) = \langle\, p_1(t) \quad p_2(t)\,\rangle^T \tag{a}$$

To determine $\hat{\mathbf{k}}_{tt}$, the 4×4 stiffness matrix determined in Example 9.4 is partitioned:

$$\mathbf{k} = \begin{bmatrix} \mathbf{k}_{tt} & \mathbf{k}_{t0} \\ \mathbf{k}_{0t} & \mathbf{k}_{00} \end{bmatrix} = \frac{8EI}{L^3} \left[\begin{array}{cc:cc} 12 & -12 & -3L & -3L \\ -12 & 24 & 3L & 0 \\ \hdashline -3L & 3L & L^2 & L^2/2 \\ -3L & 0 & L^2/2 & 2L^2 \end{array} \right] \tag{b}$$

Substituting these submatrices in Eq. (9.3.5) gives the condensed stiffness matrix:

$$\hat{\mathbf{k}}_{tt} = \frac{48EI}{7L^3} \begin{bmatrix} 2 & -5 \\ -5 & 16 \end{bmatrix} \tag{c}$$

This stiffness matrix of Eq. (c) is the same as that obtained in Example 9.5 by inverting the flexibility matrix corresponding to the two translational DOFs.

Substituting the stiffness submatrices in Eq. (9.3.3) gives the relation between the condensed DOF $\mathbf{u}_0$ and the dynamic DOF $\mathbf{u}_t$:

$$\mathbf{u}_0 = \mathbf{T}\mathbf{u}_t \qquad \mathbf{T} = \frac{1}{L} \begin{bmatrix} 2.57 & -3.43 \\ 0.857 & 0.857 \end{bmatrix} \tag{d}$$

The equations of motion are given by Eq. (9.3.4), where $\mathbf{m}_{tt}$ and $\mathbf{p}_t(t)$ are defined in Eq. (a) and $\hat{\mathbf{k}}_{tt}$ in Eq. (c). These are the same as Eq. (c) of Example 9.5.

Example 9.9

Formulate the equations of motion for the two-story frame of Example 9.7 governing the lateral floor displacements u_1 and u_2.

Solution The equations of motion for this system were formulated in Example 9.7 considering six DOFs which are partitioned into $\mathbf{u}_t = \langle\, u_1 \quad u_2\,\rangle^T$ and $\mathbf{u}_0 = \langle\, u_3 \quad u_4 \quad u_5 \quad u_6\,\rangle^T$.

The equations governing $\mathbf{u}_t$ are given by Eq. (9.3.4), where

$$\mathbf{m}_{tt} = m \begin{bmatrix} 2 & \\ & 1 \end{bmatrix} \qquad \mathbf{p}_t(t) = \langle\, p_1(t) \quad p_2(t)\,\rangle^T \tag{a}$$

To determine $\mathbf{k}_{tt}$, the 6×6 stiffness matrix determined in Example 9.7 is partitioned:

$$\mathbf{k} = \begin{bmatrix} \mathbf{k}_{tt} & \mathbf{k}_{t0} \\ \mathbf{k}_{0t} & \mathbf{k}_{00} \end{bmatrix} = \frac{EI}{h^3} \left[\begin{array}{cc:cccc} 72 & -24 & 6h & 6h & -6h & -6h \\ -24 & 24 & 6h & 6h & 6h & 6h \\ \hdashline 6h & 6h & 16h^2 & 2h^2 & 2h^2 & 0 \\ 6h & 6h & 2h^2 & 16h^2 & 0 & 2h^2 \\ -6h & 6h & 2h^2 & 0 & 6h^2 & h^2 \\ -6h & 6h & 0 & 2h^2 & h^2 & 6h^2 \end{array} \right] \tag{b}$$

Substituting these submatrices in Eq. (9.3.5) gives the condensed stiffness matrix:

$$\hat{\mathbf{k}}_{tt} = \frac{EI}{h^3} \begin{bmatrix} 54.88 & -17.51 \\ -17.51 & 11.61 \end{bmatrix} \tag{c}$$

This is called the *lateral stiffness matrix* because the DOFs are the lateral displacements of the floors. It enters into the earthquake analysis of buildings (Section 9.4).

Substituting the stiffness submatrices in Eq. (9.3.3) gives the relation between the condensed DOF $\mathbf{u}_0$ and the translational DOF $\mathbf{u}_t$:

$$\mathbf{u}_0 = \mathbf{T}\mathbf{u}_t \qquad \mathbf{T} = \frac{1}{h}\begin{bmatrix} -0.4426 & -0.2459 \\ -0.4426 & -0.2459 \\ 0.9836 & -0.7869 \\ 0.9836 & -0.7869 \end{bmatrix} \tag{d}$$

The equations of motion are given by Eq. (9.3.4), where $\mathbf{m}_{tt}$ and $\mathbf{p}_t$ are defined in Eq. (a) and $\hat{\mathbf{k}}_{tt}$ in Eq. (c):

$$m\begin{bmatrix} 2 & \\ & 1 \end{bmatrix}\begin{Bmatrix} \ddot{u}_1 \\ \ddot{u}_2 \end{Bmatrix} + \frac{EI}{h^3}\begin{bmatrix} 54.88 & -17.51 \\ -17.51 & 11.61 \end{bmatrix}\begin{Bmatrix} u_1 \\ u_2 \end{Bmatrix} = \begin{Bmatrix} p_1(t) \\ p_2(t) \end{Bmatrix} \tag{e}$$

9.4 PLANAR OR SYMMETRIC-PLAN SYSTEMS: GROUND MOTION

One of the important applications of structural dynamics is in predicting how structures respond to earthquake-induced motion of the base of the structure. In this and following sections equations of motion for MDF systems subjected to earthquake excitation are formulated. Planar systems subjected to translational and rotational ground motions are considered in Sections 9.4.1 and 9.4.3, symmetric-plan buildings subjected to translational and torsional excitations in Sections 9.4.2 and 9.6, and unsymmetric-plan buildings subjected to translational ground motion in Section 9.5. Systems excited by different prescribed motions at their multiple supports are the subject of Section 9.7.

9.4.1 Planar Systems: Translational Ground Motion

We start with the simplest case, where all the dynamic degrees of freedom are displacements in the same direction as the ground motion. Two such structures—a tower and a building frame—are shown in Fig. 9.4.1. The displacement of the ground is denoted by u_g, the total (or absolute) displacement of the mass m_j by u_j^t, and the relative displacement between this mass and the ground by u_j. At each instant of time these displacements are related by

$$u_j^t(t) = u_g(t) + u_j(t) \tag{9.4.1a}$$

Such equations for all the N masses can be combined in vector form:

$$\mathbf{u}^t(t) = u_g(t)\mathbf{1} + \mathbf{u}(t) \tag{9.4.1b}$$

where $\mathbf{1}$ is a vector of order N with each element equal to unity.

The equation of dynamic equilibrium, Eq. (9.2.11), developed earlier is still valid, except that $\mathbf{p}(t) = \mathbf{0}$ because no external dynamic forces are applied. Thus

$$\mathbf{f}_I + \mathbf{f}_D + \mathbf{f}_S = \mathbf{0} \tag{9.4.2}$$

Only the relative motions $\mathbf{u}$ between the masses and the base due to structural deformations produce elastic and damping forces (i.e., the rigid-body component of the displacement of the structure produces no internal forces). Thus for a linear system, Eqs. (9.2.3) and (9.2.6)

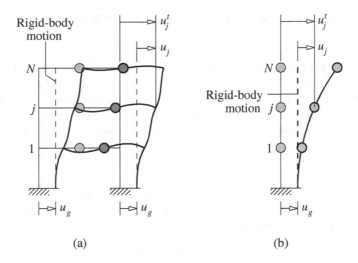

Figure 9.4.1   (a) Building frame; (b) tower.

are still valid. However, the inertia forces $\mathbf{f}_I$ are related to the total accelerations $\ddot{\mathbf{u}}^t$ of the masses, and Eq. (9.2.9) becomes

$$\mathbf{f}_I = \mathbf{m}\ddot{\mathbf{u}}^t \qquad (9.4.3)$$

Substituting Eqs. (9.2.3), (9.2.6), and (9.4.3) in Eq. (9.4.2) and using Eq. (9.4.1b) gives

$$\mathbf{m}\ddot{\mathbf{u}} + \mathbf{c}\dot{\mathbf{u}} + \mathbf{k}\mathbf{u} = -\mathbf{m}\mathbf{1}\ddot{u}_g(t) \qquad (9.4.4)$$

Equation (9.4.4) contains N differential equations governing the relative displacements $u_j(t)$ of a linearly elastic MDF system subjected to ground acceleration $\ddot{u}_g(t)$. The stiffness matrix in Eq. (9.4.4) refers to the horizontal displacements u_j and is obtained by the static condensation method (Section 9.3) to eliminate the rotational and vertical DOF of the nodes; hence this $\mathbf{k}$ is known as the *lateral stiffness matrix*.

Comparison of Eq. (9.4.4) with Eq. (9.2.12) shows that the equations of motion for the structure subjected to two separate excitations—ground acceleration $= \ddot{u}_g(t)$ and external forces $= -m_j\ddot{u}_g(t)$—are one and the same. As shown in Fig. 9.4.2, the ground motion

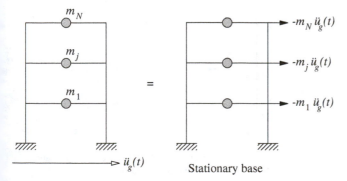

Stationary base

Figure 9.4.2   Effective earthquake forces.

can therefore be replaced by the *effective earthquake forces:*

$$\mathbf{p}_{\text{eff}}(t) = -\mathbf{m} \mathbf{1} \ddot{u}_g(t) \tag{9.4.5}$$

A generalization of the preceding derivation is useful if all the DOFs of the system are not in the direction of the ground motion (later in this section), or if the earthquake excitation is not identical at all the structural supports (Section 9.7). In this general approach the total displacement of each mass is expressed as its displacement u_j^s due to static application of the ground motion plus the dynamic displacement u_j relative to the quasi-static displacement:

$$u_j^t(t) = u_j^s(t) + u_j(t) \quad \text{or} \quad \mathbf{u}^t(t) = \mathbf{u}^s(t) + \mathbf{u}(t) \tag{9.4.6}$$

The quasi-static displacements can be expressed as $\mathbf{u}^s(t) = \boldsymbol{\iota} u_g(t)$, where the *influence vector* $\boldsymbol{\iota}$ represents the displacements of the masses resulting from static application of a unit ground displacement; thus Eq. (9.4.6b) becomes

$$\mathbf{u}^t(t) = \boldsymbol{\iota} u_g(t) + \mathbf{u}(t) \tag{9.4.7}$$

The equations of motion are obtained as before, except that Eq. (9.4.7) is used instead of Eq. (9.4.1b):

$$\mathbf{m}\ddot{\mathbf{u}} + \mathbf{c}\dot{\mathbf{u}} + \mathbf{k}\mathbf{u} = -\mathbf{m}\boldsymbol{\iota}\ddot{u}_g(t) \tag{9.4.8}$$

Now the effective earthquake forces are

$$\mathbf{p}_{\text{eff}}(t) = -\mathbf{m}\boldsymbol{\iota}\ddot{u}_g(t) \tag{9.4.9}$$

This generalization is of no special benefit in deriving the governing equations for the systems of Fig. 9.4.1. Static application of $u_g = 1$ to these systems gives $u_j = 1$ for all j (i.e., $\boldsymbol{\iota} = \mathbf{1}$), as shown in Fig. 9.4.3, where the masses are blank to emphasize that the displacements are static. Thus, Eqs. (9.4.8) and (9.4.9) become identical to Eqs. (9.4.4) and (9.4.5), respectively.

We next consider systems with not all the dynamic DOFs in the direction of the ground motion. An example is shown in Fig. 9.4.4a, where an inverted L-shaped frame with lumped masses is subjected to horizontal ground motion. Assuming the elements to be

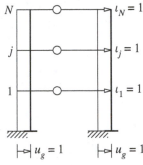

(a)

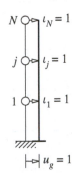

(b)

Figure 9.4.3 Influence vector $\boldsymbol{\iota}$: static displacements due to $u_g = 1$.

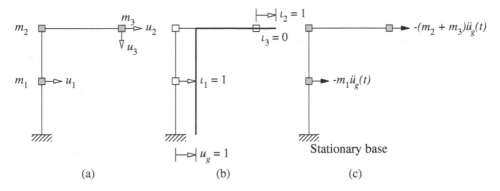

Figure 9.4.4   (a) L-shaped frame; (b) influence vector ι: static displacements due to $u_g = 1$; (c) effective earthquake forces.

axially rigid, the three DOFs are as shown; $\mathbf{u} = \langle u_1 \quad u_2 \quad u_3 \rangle^T$. Static application of $u_g = 1$ results in the displacements shown in Fig. 9.4.4b. Thus $\iota = \langle 1 \quad 1 \quad 0 \rangle^T$ in Eq. (9.4.8), and Eq. (9.4.9) becomes

$$\mathbf{p}_{\text{eff}}(t) = -\mathbf{m}\iota \ddot{u}_g(t) = -\ddot{u}_g(t) \begin{bmatrix} m_1 & & \\ & m_2 + m_3 & \\ & & m_3 \end{bmatrix} \begin{Bmatrix} 1 \\ 1 \\ 0 \end{Bmatrix} = -\ddot{u}_g(t) \begin{bmatrix} m_1 \\ m_2 + m_3 \\ 0 \end{bmatrix}$$

(9.4.10)

Note that the mass corresponding to $\ddot{u}_2 = 1$ is $m_2 + m_3$ because both masses will undergo the same acceleration since the connecting beam is axially rigid. The effective forces of Eq. (9.4.10) are shown in Fig. 9.4.4c. Observe that the effective force is zero in the vertical DOFs because the ground motion is horizontal.

9.4.2 Symmetric-Plan Buildings: Translational Ground Motion

Consider the N-story building shown in Fig. 9.4.5 having rigid floor diaphragms and several frames in each of the x and y directions; the planwise distribution of mass and stiffness is symmetric about the x and y axes. We show in Section 9.5 that such symmetric-plan buildings can be analyzed independently in the two lateral directions. The motion of the building due to ground motion along one of the two axes, say the x-axis, is also governed by Eq. (9.4.4) with appropriate interpretation of $\mathbf{m}$ and $\mathbf{k}$. The mass matrix is a diagonal matrix with diagonal elements $m_{jj} = m_j$, where m_j is the total mass lumped at the jth floor diaphragm (Section 9.2.4). The stiffness matrix $\mathbf{k}$ is the lateral stiffness matrix of the building for motion in the x-direction.

The lateral stiffness matrix of a building can be determined from the lateral stiffness matrices of the individual frames of the building. First, the lateral stiffness matrix $\mathbf{k}_{xi}$ of the ith frame oriented in the x-direction is determined by the static condensation procedure to condense out the joint rotations and vertical displacements at the joints (Section 9.3). This

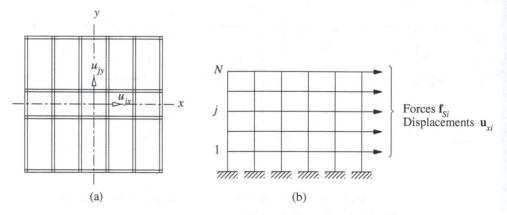

Figure 9.4.5   (a) jth floor plan with DOFs noted; (b) frame i, x-direction, with lateral forces and displacements shown.

lateral stiffness matrix provides the relation between the lateral forces $\mathbf{f}_{Si}$ on the ith frame and the lateral displacements $\mathbf{u}_{xi}$ of the frame (Fig. 9.4.5b):

$$\mathbf{f}_{Si} = \mathbf{k}_{xi}\mathbf{u}_{xi} \tag{9.4.11}$$

Because the floor diaphragms are assumed to be rigid, all frames undergo the same lateral displacements:

$$\mathbf{u}_{xi} = \mathbf{u}_x \tag{9.4.12}$$

where $\mathbf{u}_x^T = \langle u_{1x} \quad u_{2x} \quad \cdots \quad u_{jx} \quad \cdots \quad u_{Nx} \rangle$ are the lateral displacements of the floors defined at their centers of mass. Substituting Eq. (9.4.12) in Eq. (9.4.11) and adding the latter equations for all the frames gives

$$\mathbf{f}_S = \mathbf{k}_x\mathbf{u}_x \tag{9.4.13}$$

where $\mathbf{f}_S = \sum_i \mathbf{f}_{Si}$ is the vector of lateral forces at the floor centers of mass of the building and

$$\mathbf{k}_x = \sum_i \mathbf{k}_{xi} \tag{9.4.14}$$

is the x-lateral stiffness of the building. It is a matrix of order N for an N-story building. Equation (9.4.4) with $\mathbf{k} = \mathbf{k}_x$ governs the x-lateral motion of a multistory building due to ground motion in the x-direction.

9.4.3 Planar Systems: Rotational Ground Motion

Although the rotational components of ground motion are not measured during earthquakes, they can be estimated from the measured translational components, and it is of interest to apply the preceding concepts to this excitation. For this purpose, consider the frame of Fig. 9.4.6a subjected to base rotation $\theta_g(t)$. The total displacements $\mathbf{u}^t$ of the masses are made up of two parts: $\mathbf{u}$ associated with the structural deformations and a

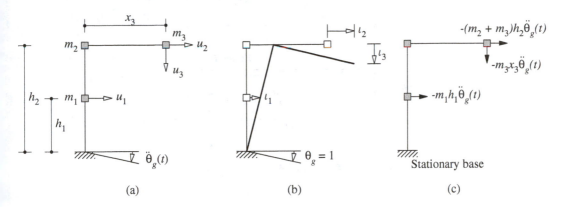

Figure 9.4.6 (a) Frame; (b) influence vector ι: static displacements due to $\theta_g = 1$; (c) effective earthquake forces.

rigid-body component $\mathbf{u}^s(t) = \iota\theta_g(t)$ due to static application of ground rotation θ_g:

$$\mathbf{u}^t(t) = \mathbf{u}(t) + \iota\theta_g(t) \tag{9.4.15}$$

Static application of $\theta_g = 1$ results in the displacements shown in Fig. 9.4.6b; thus $\iota = \langle\, h_1 \quad h_2 \quad x_3\,\rangle^T$. Equations (9.4.2) and (9.4.3) are still valid, but the total accelerations $\ddot{\mathbf{u}}^t(t)$ must now be determined from Eq. (9.4.15). Putting all these equations together leads to

$$\mathbf{m\ddot{u}} + \mathbf{c\dot{u}} + \mathbf{ku} = -\mathbf{m}\iota\ddot{\theta}_g(t) \tag{9.4.16}$$

The effective forces associated with ground rotation are shown in Fig. 9.4.6c:

$$\mathbf{p}_{\text{eff}}(t) = -\mathbf{m}\iota\ddot{\theta}_g(t) = -\ddot{\theta}_g(t)\begin{bmatrix} m_1 h_1 \\ (m_2 + m_3)h_2 \\ m_3 x_3 \end{bmatrix} \tag{9.4.17}$$

9.5 ONE-STORY UNSYMMETRIC-PLAN BUILDINGS

We now extend the development of the preceding sections to formulate the equations of motion for buildings with unsymmetrical plan. Such buildings, when subjected to, say, the y-component of ground motion, would simultaneously undergo lateral motion in two horizontal (x and y) directions and torsion about the vertical (z) axis. In this section the equations governing such coupled lateral-torsional motion are formulated—first for one-story systems, followed by multistory buildings.

9.5.1 Two-Way Unsymmetric System

System considered. Consider the idealized one-story building shown in Fig. 9.5.1, consisting of a roof diaphragm, assumed rigid in its own plane, supported on

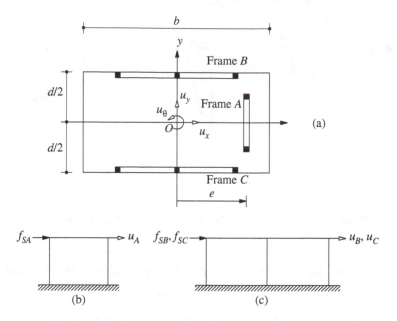

Figure 9.5.1   One-story system considered: (a) plan; (b) frame A; (c) frames B and C.

three frames: A, B, and C. Frame A is oriented in the y-direction, located at a distance e from the y-axis passing through the center of mass (CM) of the diaphragm. Frames B and C are oriented in the x-direction, located at the same distance $d/2$ on the two sides of the x-axis; for simplicity the frames are located at the edge of the diaphragm and we neglect the frame thickness. The motion of the roof mass will be described by three DOFs defined at the CM of the roof: displacements u_x in the x-direction and u_y in the y-direction, and torsional rotation u_θ about the vertical axis.

Force–displacement relation.   Let $\mathbf{f}_S$ represent the vector of statically applied external forces on the stiffness component of the structure and $\mathbf{u}$ the vector of resulting displacements both defined in terms of the three DOFs. The forces and displacements are related through

$$\begin{Bmatrix} f_{Sx} \\ f_{Sy} \\ f_{S\theta} \end{Bmatrix} = \begin{bmatrix} k_{xx} & k_{xy} & k_{x\theta} \\ k_{yx} & k_{yy} & k_{y\theta} \\ k_{\theta x} & k_{\theta y} & k_{\theta\theta} \end{bmatrix} \begin{bmatrix} u_x \\ u_y \\ u_\theta \end{bmatrix} \quad \text{or} \quad \mathbf{f}_S = \mathbf{k}\mathbf{u} \qquad (9.5.1)$$

The 3×3 stiffness matrix $\mathbf{k}$ of the structure can be determined by the direct equilibrium method (based on the definition of stiffness influence coefficients) or by the direct stiffness method.

For this purpose the lateral stiffness of each frame is defined. The lateral stiffness k_y of frame A relates the lateral force f_{SA} and displacement u_A (Fig. 9.5.1b):

$$f_{SA} = k_y u_A \qquad (9.5.2)$$

The lateral stiffnesses of frames B and C are k_{xB} and k_{xC}, respectively, and they relate the lateral forces and displacements shown in Fig. 9.5.1c:

$$f_{SB} = k_{xB}u_B \qquad f_{SC} = k_{xC}u_C \tag{9.5.3}$$

The lateral stiffness for each frame is determined by the static condensation procedure described in Section 9.3.

The stiffness matrix of the complete system is determined first by the direct equilibrium method. A unit displacement is imposed successively in each DOF, and the stiffness influence coefficients are determined by statics. The details are presented in Fig. 9.5.2 and should be self-explanatory. The resulting stiffness matrix for the structure is

$$\mathbf{k} = \begin{bmatrix} k_{xB} + k_{xC} & 0 & (d/2)(k_{xC} - k_{xB}) \\ 0 & k_y & ek_y \\ (d/2)(k_{xC} - k_{xB}) & ek_y & e^2 k_y + (d^2/4)(k_{xB} + k_{xC}) \end{bmatrix} \tag{9.5.4}$$

Observe that $k_{xy} = 0$ in Eq. (9.5.4) for the system of Fig. 9.5.1; in general, $k_{xy} \neq 0$.

Alternatively, the stiffness matrix of the structure may be formulated by the direct stiffness method implemented as follows: Determined first is the transformation matrix that relates the lateral displacement u_i of frame i to u_x, u_y, and u_θ, the global DOF of the system. This 1×3 matrix is denoted by $\mathbf{a}_{xi}$ if the frame is oriented in the x-direction, or by $\mathbf{a}_{yi}$ if in the y-direction. The lateral displacement of frame A, $u_A = u_y + eu_\theta$, or $u_A = \mathbf{a}_{yA}\mathbf{u}$, where $\mathbf{a}_{yA} = \langle 0 \quad 1 \quad e \rangle$. Similarly, the lateral displacement of frame B, $u_B = u_x - (d/2)u_\theta$, or $u_B = \mathbf{a}_{xB}\mathbf{u}$, where $\mathbf{a}_{xB} = \langle 1 \quad 0 \quad -d/2 \rangle$. Finally, the lateral displacement of frame C, $u_C = u_x + (d/2)u_\theta$, or $u_C = \mathbf{a}_{xC}\mathbf{u}$, where $\mathbf{a}_{xC} = \langle 1 \quad 0 \quad d/2 \rangle$.

Second, the stiffness matrix for frame i with respect to global DOF $\mathbf{u}$ is determined from the lateral stiffness k_{xi} or k_{yi} of frame i in local coordinates u_i from

$$\mathbf{k}_i = \mathbf{a}_{xi}^T k_{xi} \mathbf{a}_{xi} \quad \text{or} \quad \mathbf{k}_i = \mathbf{a}_{yi}^T k_{yi} \mathbf{a}_{yi} \tag{9.5.5}$$

The first equation applies to frames oriented in the x-direction and the second to frames in the y-direction. Substituting the appropriate $\mathbf{a}_{xi}$ or $\mathbf{a}_{yi}$ and k_{xi} or k_{yi} gives the stiffness matrices $\mathbf{k}_A$, $\mathbf{k}_B$, and $\mathbf{k}_C$ of the three frames:

$$\mathbf{k}_A = \begin{Bmatrix} 0 \\ 1 \\ e \end{Bmatrix} k_y \langle 0 \quad 1 \quad e \rangle = k_y \begin{bmatrix} 0 & 0 & 0 \\ 0 & 1 & e \\ 0 & e & e^2 \end{bmatrix} \tag{9.5.6}$$

$$\mathbf{k}_B = \begin{Bmatrix} 1 \\ 0 \\ -d/2 \end{Bmatrix} k_{xB} \langle 1 \quad 0 \quad -d/2 \rangle = k_{xB} \begin{bmatrix} 1 & 0 & -d/2 \\ 0 & 0 & 0 \\ -d/2 & 0 & d^2/4 \end{bmatrix} \tag{9.5.7}$$

$$\mathbf{k}_C = \begin{Bmatrix} 1 \\ 0 \\ d/2 \end{Bmatrix} k_{xC} \langle 1 \quad 0 \quad d/2 \rangle = k_{xC} \begin{bmatrix} 1 & 0 & d/2 \\ 0 & 0 & 0 \\ d/2 & 0 & d^2/4 \end{bmatrix} \tag{9.5.8}$$

Finally, the stiffness matrix of the system is

$$\mathbf{k} = \mathbf{k}_A + \mathbf{k}_B + \mathbf{k}_C \tag{9.5.9}$$

(a) $u_x = 1$, $u_y = u_\theta = 0$

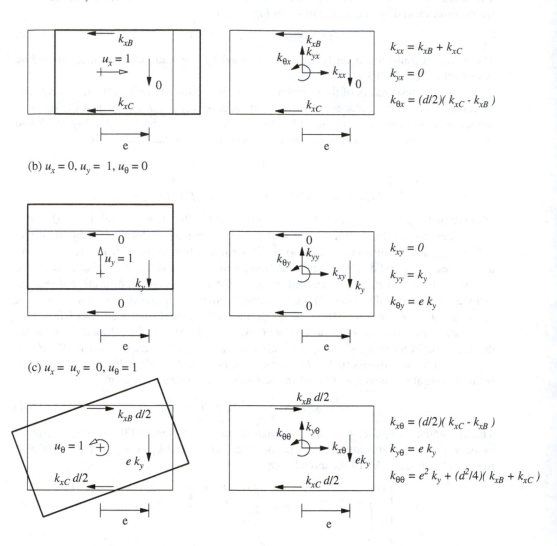

$k_{xx} = k_{xB} + k_{xC}$

$k_{yx} = 0$

$k_{\theta x} = (d/2)(k_{xC} - k_{xB})$

(b) $u_x = 0$, $u_y = 1$, $u_\theta = 0$

$k_{xy} = 0$

$k_{yy} = k_y$

$k_{\theta y} = e\,k_y$

(c) $u_x = u_y = 0$, $u_\theta = 1$

$k_{x\theta} = (d/2)(k_{xC} - k_{xB})$

$k_{y\theta} = e\,k_y$

$k_{\theta\theta} = e^2\,k_y + (d^2/4)(k_{xB} + k_{xC})$

Figure 9.5.2   Evaluation of stiffness matrix of a one-story, two-way unsymmetric system.

Substituting Eqs. (9.5.6), (9.5.7), and (9.5.8) gives

$$\mathbf{k} = \begin{bmatrix} k_{xB} + k_{xC} & 0 & (d/2)(k_{xC} - k_{xB}) \\ 0 & k_y & e k_y \\ (d/2)(k_{xC} - k_{xB}) & e k_y & e^2 k_y + (d^2/4)(k_{xB} + k_{xC}) \end{bmatrix} \qquad (9.5.10)$$

As expected, this stiffness matrix is the same as that determined earlier from the definition of stiffness influence coefficients.

Inertia forces.   Since the global DOFs selected are located at the center of mass O, the inertia forces on the mass component of the structure are

$$f_{Ix} = m\ddot{u}_x^t \qquad f_{Iy} = m\ddot{u}_y^t \qquad f_{I\theta} = I_O\ddot{u}_\theta^t \qquad (9.5.11)$$

where m is the diaphragm mass distributed uniformly over the plan, $I_O = m(b^2+d^2)/12$ is the moment of inertia of the diaphragm about the vertical axis passing through O, and $\ddot{u}_x^t$, $\ddot{u}_y^t$, and $\ddot{u}_\theta^t$ are the x, y, and θ components of the total acceleration of the center of mass. In matrix form the inertia forces and accelerations are related through the mass matrix:

$$\begin{Bmatrix} f_{Ix} \\ f_{Iy} \\ f_{I\theta} \end{Bmatrix} = \begin{bmatrix} m & & \\ & m & \\ & & I_O \end{bmatrix} \begin{Bmatrix} \ddot{u}_x^t \\ \ddot{u}_y^t \\ \ddot{u}_\theta^t \end{Bmatrix} \quad \text{or} \quad \mathbf{f}_I = \mathbf{m}\ddot{\mathbf{u}}^t \qquad (9.5.12)$$

Equations of motion.   Substituting Eqs. (9.5.12b) and (9.5.1b) into Eq. (9.4.2) and excluding damping forces gives

$$\mathbf{m}\ddot{\mathbf{u}}^t + \mathbf{k}\mathbf{u} = \mathbf{0} \qquad (9.5.13)$$

Consider the earthquake excitation defined by $\ddot{u}_{gx}(t)$ and $\ddot{u}_{gy}(t)$, the x and y components of the ground acceleration, and $\ddot{u}_{g\theta}(t)$, the rotational acceleration of the base of the building about the vertical axis. The total accelerations are

$$\begin{Bmatrix} \ddot{u}_x^t \\ \ddot{u}_y^t \\ \ddot{u}_\theta^t \end{Bmatrix} = \begin{Bmatrix} \ddot{u}_x \\ \ddot{u}_y \\ \ddot{u}_\theta \end{Bmatrix} + \begin{Bmatrix} \ddot{u}_{gx} \\ \ddot{u}_{gy} \\ \ddot{u}_{g\theta} \end{Bmatrix} \quad \text{or} \quad \ddot{\mathbf{u}}^t = \ddot{\mathbf{u}} + \ddot{\mathbf{u}}_g \qquad (9.5.14)$$

Substituting Eq. (9.5.14) into (9.5.13) and using $\mathbf{m}$ and $\mathbf{k}$ defined in Eqs. (9.5.12) and (9.5.10) gives

$$\begin{bmatrix} m & & \\ & m & \\ & & I_O \end{bmatrix} \begin{Bmatrix} \ddot{u}_x \\ \ddot{u}_y \\ \ddot{u}_\theta \end{Bmatrix} + \begin{bmatrix} k_{xx} & 0 & k_{x\theta} \\ 0 & k_{yy} & k_{y\theta} \\ k_{\theta x} & k_{\theta y} & k_{\theta\theta} \end{bmatrix} \begin{Bmatrix} u_x \\ u_y \\ u_\theta \end{Bmatrix} = -\begin{Bmatrix} m\ddot{u}_{gx}(t) \\ m\ddot{u}_{gy}(t) \\ I_O\ddot{u}_{g\theta}(t) \end{Bmatrix} \qquad (9.5.15)$$

where

$$k_{xx} = k_{xB} + k_{xC} \qquad k_{yy} = k_y \qquad k_{\theta\theta} = e^2 k_y + \frac{d^2}{4}(k_{xB} + k_{xC})$$

$$k_{x\theta} = k_{\theta x} = \frac{d}{2}(k_{xC} - k_{xB}) \qquad k_{y\theta} = k_{\theta y} = e k_y \qquad (9.5.16)$$

The three differential equations in Eq. (9.5.15) governing the three DOFs—u_x, u_y, and u_θ—are coupled through the stiffness matrix because the stiffness properties are not symmetric about the x or y axes. Thus the response of the system to the x (and y)-component of ground motion is not restricted to lateral displacement in the x (and y)-direction, but will include lateral motion in the transverse direction, y (and x), and torsion of the roof diaphragm about the vertical axis.

9.5.2 One-Way Unsymmetric System

We next consider a special case of the system of Fig. 9.5.1 for which the lateral stiffness of frames B and C is identical (i.e., $k_{xB} = k_{xC} = k_x$). This system is symmetric about the x-axis but not about the y-axis. For this one-way unsymmetric system, Eq. (9.5.15) specializes to

$$\begin{bmatrix} m & & \\ & m & \\ & & I_O \end{bmatrix} \begin{Bmatrix} \ddot{u}_x \\ \ddot{u}_y \\ \ddot{u}_\theta \end{Bmatrix} + \begin{bmatrix} 2k_x & 0 & 0 \\ 0 & k_y & ek_y \\ 0 & ek_y & e^2 k_y + (d^2/2)k_x \end{bmatrix} \begin{Bmatrix} u_x \\ u_y \\ u_\theta \end{Bmatrix} = - \begin{Bmatrix} m\ddot{u}_{gx}(t) \\ m\ddot{u}_{gy}(t) \\ 0 \end{Bmatrix}$$

$$(9.5.17)$$

where the rotational excitation has been dropped. The first of three equations,

$$m\ddot{u}_x + 2k_x u_x = -m\ddot{u}_{gx}(t) \tag{9.5.18}$$

is a familiar SDF equation of motion that governs the response u_x of the one-story system to ground motion in the x-direction; u_y and u_θ do not enter into this equation. This implies that motion in the x-direction occurs independent of the motion in the y-direction or of torsional motion. Such is the case because the system is symmetric about the x-axis.

The second and third equations can be rewritten as

$$\begin{bmatrix} m & \\ & I_O \end{bmatrix} \begin{Bmatrix} \ddot{u}_y \\ \ddot{u}_\theta \end{Bmatrix} + \begin{bmatrix} k_y & ek_y \\ ek_y & k_{\theta\theta} \end{bmatrix} \begin{Bmatrix} u_y \\ u_\theta \end{Bmatrix} = - \begin{bmatrix} m & \\ & I_O \end{bmatrix} \begin{Bmatrix} 1 \\ 0 \end{Bmatrix} \ddot{u}_{gy}(t) \tag{9.5.19}$$

These equations governing u_y and u_θ are coupled through the stiffness matrix because the stiffness properties are not symmetric about the y-axis. Thus the system response to the y-component of ground motion is not restricted to the lateral displacement in the y-direction but includes torsion about a vertical axis.

The separation of the governing equations into Eqs. (9.5.18) and (9.5.19) indicates that the earthquake response of a system with plan symmetric about the x-axis but unsymmetric about the y-axis can be determined by two independent analyses: (1) the response of the structure to ground motion in the x-direction can be determined by solving the SDF system equation (9.5.18) by the procedures of Chapter 6; and (2) the coupled lateral-torsional response of the structure to ground motion in the y-direction can be determined by solving the two-DOF-system equation (9.5.19) by the procedures of Chapter 13. In passing we observe that Eq. (9.5.19) can be interpreted as Eq. (9.4.8) without damping with the influence vector $\iota = \langle 1 \quad 0 \rangle^T$.

9.5.3 Symmetric System

We next consider a further special case of the system of Fig. 9.5.1 for which frames B and C are identical (i.e., $k_{xB} = k_{xC} = k_x$) and frame A is located at the center of mass (i.e., $e = 0$). For such systems Eq. (9.5.15) specializes to

$$\begin{bmatrix} m & & \\ & m & \\ & & I_O \end{bmatrix} \begin{Bmatrix} \ddot{u}_x \\ \ddot{u}_y \\ \ddot{u}_\theta \end{Bmatrix} + \begin{bmatrix} 2k_x & 0 & 0 \\ 0 & k_y & 0 \\ 0 & 0 & (d^2/2)k_x \end{bmatrix} \begin{Bmatrix} u_x \\ u_y \\ u_\theta \end{Bmatrix} = - \begin{Bmatrix} m\ddot{u}_{gx}(t) \\ m\ddot{u}_{gy}(t) \\ I_O\ddot{u}_{g\theta}(t) \end{Bmatrix} \tag{9.5.20}$$

The three equations are now uncoupled, and each is of the same form as the equation for an SDF system. This uncoupling of equations implies: (1) translational ground motion in the x (or y) direction would cause lateral motion of the system only in the x (or y) direction; (2) rotational ground motion would cause only torsional motion of the system; and (3) response to individual components of ground motion can be determined by solving only the corresponding equation in Eq. (9.5.20).

9.6 MULTISTORY UNSYMMETRIC-PLAN BUILDINGS

In this section, the equations of motion for a multistory building with plan unsymmetric about both x and y axes subjected to earthquake excitation are formulated. Figure 9.6.1 shows a schematic idealization of such a system, which consists of some frames oriented in the y-direction and others in the x-direction. The framing plan and hence stiffness properties are unsymmetric about both x and y axes; however, the mass distribution of each floor diaphragm is symmetric about both x and y axes, and the centers of mass O of all floor diaphragms lie on the same vertical axis. Each floor diaphragm, assumed to be rigid in its own plane, has three DOFs defined at the center of mass (Fig. 9.6.1a). The DOFs

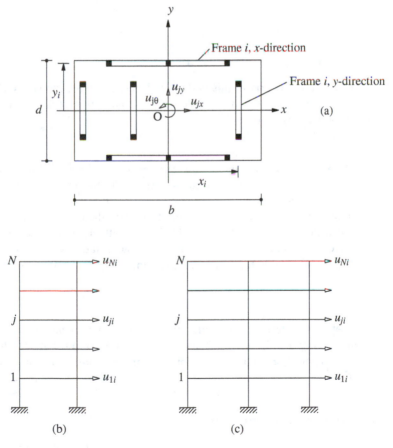

Figure 9.6.1 Multistory system: (a) plan; (b) frame i, y-direction; (c) frame i, x-direction.

for the jth floor are: translation u_{jx} along the x-axis, translation u_{jy} along the y-axis, and torsional rotation $u_{j\theta}$ about the vertical axis; u_{jx} and u_{jy} are defined relative to the ground.

The earthquake excitation is defined by $\ddot{u}_{gx}(t)$ and $\ddot{u}_{gy}(t)$, the x and y components of ground acceleration, and $\ddot{u}_{go}(t)$, the rotational ground acceleration about the vertical axis. Although rotational acceleration of the base of a building is not recorded by strong-motion accelerographs, in some cases it can be calculated from the translational accelerations recorded at two locations at the base (Section 13.4).

As suggested by the earlier formulation of the equations of motion for a one-story system, the multistory building would undergo coupled lateral-torsional motion described by $3N$ DOFs: u_{jx}, u_{jy}, and $u_{j\theta}$, $j = 1, 2, \ldots, N$. The displacement vector $\mathbf{u}$ of size $3N \times 1$ for the system is defined by

$$
\mathbf{u} = \left\{ \begin{array}{c} \mathbf{u}_x \\ \mathbf{u}_y \\ \mathbf{u}_\theta \end{array} \right\}
$$

where
$$
\mathbf{u}_x = \langle u_{1x} \quad u_{2x} \quad \cdots \quad u_{Nx} \rangle^T \qquad \mathbf{u}_y = \langle u_{1y} \quad u_{2y} \quad \cdots \quad u_{Ny} \rangle^T
$$

$$
\mathbf{u}_\theta = \langle u_{1\theta} \quad u_{2\theta} \quad \cdots \quad u_{N\theta} \rangle^T
$$

The stiffness matrix of this system with respect to the global DOFs $\mathbf{u}$ is formulated by the direct stiffness method by implementing four major steps [similar to Eqs. (9.5.5) to (9.5.10) for a one-story frame].

Step 1. Determine the lateral stiffness matrix for each frame. For the ith frame it is determined by the following steps: (a) Define the DOF for the ith frame: lateral displacements at floor levels, $\mathbf{u}_i = \langle u_{1i} \quad u_{2i} \quad \cdots \quad u_{Ni} \rangle^T$ (Fig. 9.5.3b and c), and vertical displacement and rotation of each node. (b) Obtain the complete stiffness matrix for the ith frame with reference to the frame DOF. (c) Statically condense all the rotational and vertical DOFs to obtain the $N \times N$ lateral stiffness matrix of the ith frame, denoted by $\mathbf{k}_{xi}$ if the frame is oriented in the x-direction, or by $\mathbf{k}_{yi}$ if the frame is parallel to the y-axis.

Step 2. Determine the displacement transformation matrix relating the lateral DOF $\mathbf{u}_i$ defined in step 1(a) for the ith frame to the global DOF $\mathbf{u}$ for the building. This $N \times 3N$ matrix is denoted by $\mathbf{a}_{xi}$ if the frame is oriented in the x-direction or $\mathbf{a}_{yi}$ if in the y-direction. Thus

$$
\mathbf{u}_i = \mathbf{a}_{xi}\mathbf{u} \quad \text{or} \quad \mathbf{u}_i = \mathbf{a}_{yi}\mathbf{u} \tag{9.6.1}
$$

These transformation matrices are

$$
\mathbf{a}_{xi} = [\mathbf{I} \quad \mathbf{O} \quad -y_i\mathbf{I}] \quad \text{or} \quad \mathbf{a}_{yi} = [\mathbf{O} \quad \mathbf{I} \quad x_i\mathbf{I}] \tag{9.6.2}
$$

where x_i and y_i define the location of the ith frame (Fig. 9.6.1a) oriented in the y and x directions, respectively, $\mathbf{I}$ is an identity matrix of order N, and $\mathbf{O}$ is a square matrix of order N with all elements equal to zero.

Step 3. Transform the lateral stiffness matrix for the ith frame to the building DOF $\mathbf{u}$ to obtain

$$\mathbf{k}_i = \mathbf{a}_{xi}^T \mathbf{k}_{xi} \mathbf{a}_{xi} \quad \text{or} \quad \mathbf{k}_i = \mathbf{a}_{yi}^T \mathbf{k}_{yi} \mathbf{a}_{yi} \tag{9.6.3}$$

The $3N \times 3N$ matrix $\mathbf{k}_i$ is the contribution of the ith frame to the building stiffness matrix.

Step 4. Add the stiffness matrices for all frames to obtain the stiffness matrix for the building:

$$\mathbf{k} = \sum_i \mathbf{k}_i \tag{9.6.4}$$

Substituting Eq. (9.6.2) into Eq. (9.6.3) and the latter into Eq. (9.6.4) leads to

$$\mathbf{k} = \begin{bmatrix} \mathbf{k}_{xx} & \mathbf{k}_{xy} & \mathbf{k}_{x\theta} \\ \mathbf{k}_{yx} & \mathbf{k}_{yy} & \mathbf{k}_{y\theta} \\ \mathbf{k}_{\theta x} & \mathbf{k}_{\theta y} & \mathbf{k}_{\theta\theta} \end{bmatrix} \tag{9.6.5}$$

where

$$\mathbf{k}_{xx} = \sum_i \mathbf{k}_{xi} \qquad \mathbf{k}_{yy} = \sum_i \mathbf{k}_{yi} \qquad \mathbf{k}_{\theta\theta} = \sum_i (x_i^2 \mathbf{k}_{yi} + y_i^2 \mathbf{k}_{xi})$$

$$\mathbf{k}_{xy} = \mathbf{0} \qquad \mathbf{k}_{x\theta} = \mathbf{k}_{\theta x}^T = \sum_i -y_i \mathbf{k}_{xi} \qquad \mathbf{k}_{y\theta} = \mathbf{k}_{\theta y}^T = \sum_i x_i \mathbf{k}_{yi} \tag{9.6.6}$$

The equations of undamped motion of the building are

$$\begin{bmatrix} \mathbf{m} & & \\ & \mathbf{m} & \\ & & \mathbf{I}_O \end{bmatrix} \begin{Bmatrix} \ddot{\mathbf{u}}_x \\ \ddot{\mathbf{u}}_y \\ \ddot{\mathbf{u}}_\theta \end{Bmatrix} + \begin{bmatrix} \mathbf{k}_{xx} & \mathbf{k}_{xy} & \mathbf{k}_{x\theta} \\ \mathbf{k}_{yx} & \mathbf{k}_{yy} & \mathbf{k}_{y\theta} \\ \mathbf{k}_{\theta x} & \mathbf{k}_{\theta y} & \mathbf{k}_{\theta\theta} \end{bmatrix} \begin{Bmatrix} \mathbf{u}_x \\ \mathbf{u}_y \\ \mathbf{u}_\theta \end{Bmatrix}$$

$$= - \begin{bmatrix} \mathbf{m} & & \\ & \mathbf{m} & \\ & & \mathbf{I}_O \end{bmatrix} \left(\begin{Bmatrix} \mathbf{1} \\ \mathbf{0} \\ \mathbf{0} \end{Bmatrix} \ddot{u}_{gx}(t) + \begin{Bmatrix} \mathbf{0} \\ \mathbf{1} \\ \mathbf{0} \end{Bmatrix} \ddot{u}_{gy}(t) + \begin{Bmatrix} \mathbf{0} \\ \mathbf{0} \\ \mathbf{1} \end{Bmatrix} \ddot{u}_{g\theta}(t) \right) \tag{9.6.7}$$

where $\mathbf{m}$ is a diagonal matrix of order N, with $m_{jj} = m_j$, the mass lumped at the jth floor diaphragm; $\mathbf{I}_O$ is a diagonal matrix of order N with $I_{jj} = I_{Oj}$, the moment of inertia of the jth floor diaphragm about the vertical axis through the center of mass; and $\mathbf{1}$ and $\mathbf{0}$ are vectors of dimension N with all elements equal to 1 and zero, respectively.

Considering one component of ground motion at a time, Eq. (9.6.7) indicates that ground motion in the x-direction can be replaced by effective earthquake forces $-m_j \ddot{u}_{gx}(t)$, ground motion in the y-direction by effective earthquake forces $-m_j \ddot{u}_{gy}(t)$, and motion in the θ-direction by effective earthquake forces $-I_{Oj} \ddot{u}_{g\theta}(t)$; note that these effective forces are along the direction of the ground motion component considered and are zero in the other two directions. Equation (9.6.7) can be interpreted as Eq. (9.4.8) with the influence vector ι associated with the x, y, and θ components of ground motion given by the three vectors on the right side of Eq. (9.6.7), respectively.

Because the three sets of DOFs—$\mathbf{u}_x$, $\mathbf{u}_y$, and $\mathbf{u}_\theta$— in Eq. (9.6.7) are coupled through the stiffness matrix, the system when subjected to any one component of ground motion will respond simultaneously in x-lateral, y-lateral, and torsional motion; such motion is referred to as coupled lateral-torsional motion.

9.6.1 One-Way Unsymmetric-Plan Buildings

We next consider a special case of the system of Fig. 9.6.1 that has stiffness properties symmetric about the x-axis. For such systems, the stiffness submatrices $\mathbf{k}_{xy} = \mathbf{k}_{x\theta} = \mathbf{0}$ and Eq. (9.6.7) may be written as

$$\mathbf{m}\ddot{\mathbf{u}}_x + \mathbf{k}_{xx}\mathbf{u}_x = -\mathbf{m1}\,\ddot{u}_{gx}(t) \tag{9.6.8a}$$

$$\begin{bmatrix} \mathbf{m} & \\ & \mathbf{I}_O \end{bmatrix} \begin{Bmatrix} \ddot{\mathbf{u}}_y \\ \ddot{\mathbf{u}}_\theta \end{Bmatrix} + \begin{bmatrix} \mathbf{k}_{yy} & \mathbf{k}_{y\theta} \\ \mathbf{k}_{\theta y} & \mathbf{k}_{\theta\theta} \end{bmatrix} \begin{Bmatrix} \mathbf{u}_y \\ \mathbf{u}_\theta \end{Bmatrix} = - \begin{bmatrix} \mathbf{m} & \\ & \mathbf{I}_O \end{bmatrix} \begin{Bmatrix} \mathbf{1} \\ \mathbf{0} \end{Bmatrix} \ddot{u}_{gy}(t) \tag{9.6.8b}$$

where the rotational excitation has been dropped temporarily.

Equation (9.6.8) permits the following observations: Ground motion in the x-direction, an axis of symmetry, would cause the building to undergo only lateral motion in the x-direction, and this response can be determined by solving the N-DOF system governed by Eq. (9.6.8a), which is similar to the equations of motion for planar systems [Eq. (9.4.8)]. Ground motion in the y-direction would cause coupled lateral (y)-torsional motion of the building and this response can be determined by solving the $2N$-DOF system governed by Eq. (9.6.8b).

9.6.2 Symmetric-Plan Buildings

We next consider a further special case of the system of Fig. 9.6.1 that has stiffness properties symmetric about both x and y axes. For such systems, the stiffness submatrices $\mathbf{k}_{xy} = \mathbf{k}_{x\theta} = \mathbf{k}_{y\theta} = \mathbf{0}$ and Eq. (9.6.7) may be written as

$$\mathbf{m}\ddot{\mathbf{u}}_x + \mathbf{k}_{xx}\mathbf{u}_x = -\mathbf{m1}\,\ddot{u}_{gx}(t) \tag{9.6.9a}$$

$$\mathbf{m}\ddot{\mathbf{u}}_y + \mathbf{k}_{yy}\mathbf{u}_y = -\mathbf{m1}\,\ddot{u}_{gy}(t) \tag{9.6.9b}$$

$$\mathbf{I}_O\ddot{\mathbf{u}}_\theta + \mathbf{k}_{\theta\theta}\mathbf{u}_\theta = -\mathbf{I}_O\mathbf{1}\,\ddot{u}_{g\theta}(t) \tag{9.6.9c}$$

It is clear from Eq. (9.6.9) that a symmetric-plan building subjected to x, y, or θ components of ground motion—one component at a time—will undergo only x-lateral, y-lateral, or torsional motion, respectively. As a corollary, a symmetric-plan system would experience no torsional motion unless the base motion includes rotation about a vertical axis; see Section 13.4 for further discussion of this topic. Response of a symmetric-plan building to an individual component of ground motion can be determined by solving an N-DOF system governed by Eq. (9.6.9a), Eq. (9.6.9b), or Eq. (9.6.9c), as appropriate.

9.7 MULTIPLE SUPPORT EXCITATION

So far, we have assumed that all supports where the structure is connected to the ground undergo identical motion that is prescribed. In this section we generalize the previous formulation of the equations of motion to allow for different—possibly even multicomponent—prescribed motions at the various supports. Such multiple-support excitation (or spatially varying excitation) may arise in several situations. First, consider the earthquake analysis of extended structures such as the Golden Gate Bridge, shown in Fig. 2.1.2. The ground motion generated by an earthquake on the nearby San Andreas fault is expected to vary significantly over the 6450-ft length of the structure. Therefore, different motions should be prescribed at the four supports: the base of the two towers and two ends of the bridge. Second, consider the dynamic analysis of piping in nuclear power plants. Although the piping may not be especially long, its ends are connected to different locations of the main structure and would therefore experience different motions during an earthquake.

For the analysis of such systems the formulation of Section 9.4 is extended to include the degrees of freedom at the supports (Fig. 9.7.1). The displacement vector now contains two parts: (1) $\mathbf{u}^t$ includes the N DOFs of the superstructure, where the superscript t denotes that these are total displacements; and (2) $\mathbf{u}_g$ contains the N_g components of support displacements. The equation of dynamic equilibrium for all the DOFs is written in partitioned form:

$$\begin{bmatrix} \mathbf{m} & \mathbf{m}_g \\ \mathbf{m}_g^T & \mathbf{m}_{gg} \end{bmatrix} \begin{Bmatrix} \ddot{\mathbf{u}}^t \\ \ddot{\mathbf{u}}_g \end{Bmatrix} + \begin{bmatrix} \mathbf{c} & \mathbf{c}_g \\ \mathbf{c}_g^T & \mathbf{c}_{gg} \end{bmatrix} \begin{Bmatrix} \dot{\mathbf{u}}^t \\ \dot{\mathbf{u}}_g \end{Bmatrix} + \begin{bmatrix} \mathbf{k} & \mathbf{k}_g \\ \mathbf{k}_g^T & \mathbf{k}_{gg} \end{bmatrix} \begin{Bmatrix} \mathbf{u}^t \\ \mathbf{u}_g \end{Bmatrix} = \begin{Bmatrix} \mathbf{0} \\ \mathbf{p}_g(t) \end{Bmatrix} \qquad (9.7.1)$$

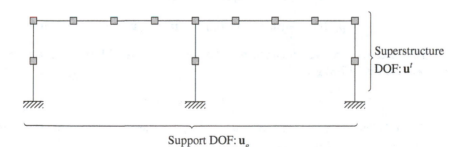

Figure 9.7.1   Definition of superstructure and support DOFs.

Observe that no external forces are applied along the superstructure DOFs. In Eq. (9.7.1) the mass, damping, and stiffness matrices can be determined from the properties of the structure using the procedures presented earlier in this chapter, while the support motions $\mathbf{u}_g(t)$, $\dot{\mathbf{u}}_g(t)$, and $\ddot{\mathbf{u}}_g(t)$ must be specified. It is desired to determine the displacements $\mathbf{u}^t$ in the superstructure DOF and the support forces $\mathbf{p}_g$.

To write the governing equations in a form familiar from the earlier formulation for a single excitation, we separate the displacements into two parts, similar to Eq. (9.4.6):

$$\left\{ \begin{array}{c} \mathbf{u}^t(t) \\ \mathbf{u}_g(t) \end{array} \right\} = \left\{ \begin{array}{c} \mathbf{u}^s(t) \\ \mathbf{u}_g(t) \end{array} \right\} + \left\{ \begin{array}{c} \mathbf{u}(t) \\ \mathbf{0} \end{array} \right\} \tag{9.7.2}$$

In this equation $\mathbf{u}^s$ is the vector of structural displacements due to static application of the prescribed support displacements $\mathbf{u}_g$ at each time instant. The two are related through

$$\begin{bmatrix} \mathbf{k} & \mathbf{k}_g \\ \mathbf{k}_g^T & \mathbf{k}_{gg} \end{bmatrix} \left\{ \begin{array}{c} \mathbf{u}^s(t) \\ \mathbf{u}_g(t) \end{array} \right\} = \left\{ \begin{array}{c} \mathbf{0} \\ \mathbf{p}_g^s(t) \end{array} \right\} \tag{9.7.3}$$

where $\mathbf{p}_g^s$ are the support forces necessary to statically impose displacements $\mathbf{u}_g$ that vary with time; obviously, $\mathbf{u}^s$ varies with time and is therefore known as the vector of quasi-static displacements. Observe that $\mathbf{p}_g^s = \mathbf{0}$ if the structure is statically determinate or if the support system undergoes rigid-body motion; for the latter condition an obvious example is identical horizontal motion of all supports. The remainder $\mathbf{u}$ of the structural displacements are known as dynamic displacements because a dynamic analysis is necessary to evaluate them.

With the total structural displacements split into quasi-static and dynamic displacements, Eq. (9.7.2), we return to the first of the two partitioned equations (9.7.1):

$$\mathbf{m}\ddot{\mathbf{u}}^t + \mathbf{m}_g\ddot{\mathbf{u}}_g + \mathbf{c}\dot{\mathbf{u}}^t + \mathbf{c}_g\dot{\mathbf{u}}_g + \mathbf{k}\mathbf{u}^t + \mathbf{k}_g\mathbf{u}_g = \mathbf{0} \tag{9.7.4}$$

Substituting Eq. (9.7.2) and transferring all terms involving $\mathbf{u}_g$ and $\mathbf{u}^s$ to the right side leads to

$$\mathbf{m}\ddot{\mathbf{u}} + \mathbf{c}\dot{\mathbf{u}} + \mathbf{k}\mathbf{u} = \mathbf{p}_{\text{eff}}(t) \tag{9.7.5}$$

where the vector of effective earthquake forces is

$$\mathbf{p}_{\text{eff}}(t) = -(\mathbf{m}\ddot{\mathbf{u}}^s + \mathbf{m}_g\ddot{\mathbf{u}}_g) - (\mathbf{c}\dot{\mathbf{u}}^s + \mathbf{c}_g\dot{\mathbf{u}}_g) - (\mathbf{k}\mathbf{u}^s + \mathbf{k}_g\mathbf{u}_g) \tag{9.7.6}$$

This effective force vector can be rewritten in a more useful form. The last term drops out because Eq. (9.7.3) gives

$$\mathbf{k}\mathbf{u}^s + \mathbf{k}_g\mathbf{u}_g = \mathbf{0} \tag{9.7.7}$$

This relation also enables us to express the quasi-static displacements $\mathbf{u}^s$ in terms of the specified support displacements $\mathbf{u}_g$:

$$\mathbf{u}^s = \iota\mathbf{u}_g \qquad \iota = -\mathbf{k}^{-1}\mathbf{k}_g \tag{9.7.8}$$

We call $\boldsymbol{\iota}$ the *influence matrix* because it describes the influence of support displacements on the structural displacements. Substituting Eqs. (9.7.8) and (9.7.7) in Eq. (9.7.6) gives

$$\mathbf{p}_{\text{eff}}(t) = -(\mathbf{m}\boldsymbol{\iota} + \mathbf{m}_g)\ddot{\mathbf{u}}_g(t) - (\mathbf{c}\boldsymbol{\iota} + \mathbf{c}_g)\dot{\mathbf{u}}_g(t) \tag{9.7.9}$$

If the ground (or support) accelerations $\ddot{\mathbf{u}}_g(t)$ and velocities $\dot{\mathbf{u}}_g(t)$ are prescribed, $\mathbf{p}_{\text{eff}}(t)$ is known from Eq. (9.7.9), and this completes the formulation of the governing equation [Eq. (9.7.5)].

Simplification of $\mathbf{p}_{\text{eff}}(t)$.   For many practical applications, further simplification of the effective force vector is possible on two counts. First, the damping term in Eq. (9.7.9) is zero if the damping matrices are proportional to the stiffness matrices (i.e., $\mathbf{c} = a_1\mathbf{k}$ and $\mathbf{c}_g = a_1\mathbf{k}_g$) because of Eq. (9.7.7); this stiffness-proportional damping will be shown in Chapter 11 to be unrealistic, however. While the damping term in Eq. (9.7.9) is not zero for arbitrary forms of damping, it is usually small relative to the inertia term and may therefore be dropped. Second, for structures with mass idealized as lumped at the DOFs, the mass matrix is diagonal, implying that $\mathbf{m}_g$ is a null matrix and $\mathbf{m}$ is diagonal. With these simplifications Eq. (9.7.9) reduces to

$$\mathbf{p}_{\text{eff}}(t) = -\mathbf{m}\boldsymbol{\iota}\ddot{\mathbf{u}}_g(t) \tag{9.7.10}$$

Observe that this equation for the effective earthquake forces associated with multiple-support excitation is a generalization of Eq. (9.4.9) valid for structures with single support (and for structures with identical motion at multiple supports). The $N \times N_g$ influence matrix $\boldsymbol{\iota}$ was previously an $N \times 1$ vector, and the $N_g \times 1$ vector $\ddot{\mathbf{u}}_g(t)$ of support motions was a scalar $\ddot{u}_g(t)$.

Interpretation of $\mathbf{p}_{\text{eff}}(t)$.   We will find it useful to use a different form of Eq. (9.7.8a):

$$\mathbf{u}^s(t) = \sum_{l=1}^{N_g} \boldsymbol{\iota}_l u_{gl}(t) \tag{9.7.11}$$

where $\boldsymbol{\iota}_l$, the lth column of the influence matrix $\boldsymbol{\iota}$, is the influence vector associated with the support displacement u_{gl}. It is the vector of static displacements in the structural DOFs due to $u_{gl} = 1$. By using Eqs. (9.7.8) and (9.7.11), the effective force vector, Eq. (9.7.10), can be expressed as

$$\mathbf{p}_{\text{eff}}(t) = -\sum_{l=1}^{N_g} \mathbf{m}\boldsymbol{\iota}_l \ddot{u}_{gl}(t) \tag{9.7.12}$$

The lth term in Eq. (9.7.12) that denotes the effective earthquake forces due to acceleration in the lth support DOF is of the same form as Eq. (9.4.9) for structures with single support (and for structures with identical motion at multiple supports). The two cases differ in an important sense, however: In the latter case, the influence vector can be determined by kinematics, but N algebraic equations [Eq. (9.7.7)] are solved to determine each influence vector $\boldsymbol{\iota}_l$ for multiple-support excitations.

Example 9.10

A uniform two-span continuous bridge with flexural stiffness EI is idealized as a lumped-mass system (Fig. E9.10a). Formulate the equations of motion for the bridge subjected to vertical motions u_{g1}, u_{g2}, and u_{g3} of the three supports. Consider only the translational degrees of freedom. Neglect damping.

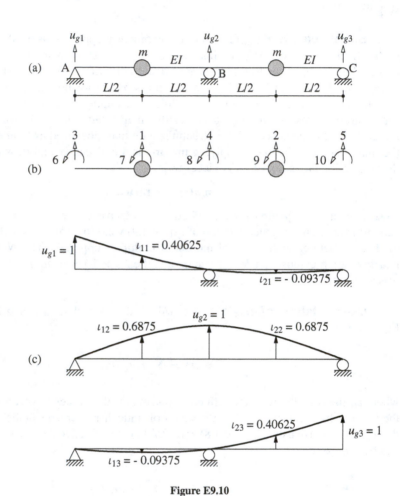

Figure E9.10

Solution

1. *Formulate the stiffness matrix.* With reference to the 10 DOFs identified in Fig. E9.10b, the stiffness matrix of the system is formulated by the procedure used in Example 9.7 for a two-story frame. Static condensation of the rotational DOFs using the procedure in Section 9.3 leads to the 5×5 stiffness matrix with reference to the five translational

DOFs:

$$\hat{\mathbf{k}} = \frac{EI}{L^3} \begin{bmatrix} 78.86 & 30.86 & -29.14 & -75.43 & -5.14 \\ 30.86 & 78.86 & -5.14 & -75.43 & -29.14 \\ -29.14 & -5.14 & 12.86 & 20.57 & 0.86 \\ -75.43 & -75.43 & 20.57 & 109.71 & 20.57 \\ -5.14 & -29.14 & 0.86 & 20.57 & 12.86 \end{bmatrix} \tag{a}$$

2. *Partition the stiffness matrix.* The vectors of the structural DOF and support DOF are

$$\mathbf{u} = \langle u_1 \quad u_2 \rangle^T \quad \text{and} \quad \mathbf{u}_g = \langle u_3 \quad u_4 \quad u_5 \rangle^T \tag{b}$$

The $\hat{\mathbf{k}}$ determined previously is partitioned:

$$\hat{\mathbf{k}} = \begin{bmatrix} \mathbf{k} & \mathbf{k}_g \\ \mathbf{k}_g^T & \mathbf{k}_{gg} \end{bmatrix} \tag{c}$$

where

$$\mathbf{k} = \frac{EI}{L^3} \begin{bmatrix} 78.86 & 30.86 \\ 30.86 & 78.86 \end{bmatrix} \tag{d1}$$

$$\mathbf{k}_g = \frac{EI}{L^3} \begin{bmatrix} -29.14 & -75.43 & -5.14 \\ -5.14 & -75.43 & -29.14 \end{bmatrix} \tag{d2}$$

$$\mathbf{k}_{gg} = \frac{EI}{L^3} \begin{bmatrix} 12.86 & 20.57 & 0.86 \\ 20.57 & 109.71 & 20.57 \\ 0.86 & 20.57 & 12.86 \end{bmatrix} \tag{d3}$$

3. *Formulate the mass matrix.* Relative to the DOFs u_1 and u_2, the mass matrix is

$$\mathbf{m} = m \begin{bmatrix} 1 & \\ & 1 \end{bmatrix} \tag{e}$$

4. *Determine the influence matrix.*

$$\iota = -\mathbf{k}^{-1}\mathbf{k}_g = \begin{bmatrix} 0.40625 & 0.68750 & -0.09375 \\ -0.09375 & 0.68750 & 0.40625 \end{bmatrix} \tag{f}$$

The influence vectors associated with each of the supports are

$$\iota_1 = \langle 0.40625 \quad -0.09375 \rangle^T \tag{g1}$$

$$\iota_2 = \langle 0.68750 \quad 0.68750 \rangle^T \tag{g2}$$

$$\iota_3 = \langle -0.09375 \quad 0.40625 \rangle^T \tag{g3}$$

The structural displacements described by each of the influence vectors are shown in Fig. E9.10c.

5. *Determine the equations of motion.*

$$\mathbf{m}\ddot{\mathbf{u}} + \mathbf{k}\mathbf{u} = \mathbf{p}_{\text{eff}}(t) \tag{h}$$

where $\mathbf{m}$ and $\mathbf{k}$ are defined by Eqs. (e) and (d1), respectively. The effective force vector is

$$\mathbf{p}_{\text{eff}}(t) = -\sum_{l=1}^{3} \mathbf{m}\iota_l \ddot{u}_{gl}(t) \tag{i}$$

where ι_l are given by Eq. (g) and $\ddot{u}_{gl}(t)$ are the support accelerations.

9.8 INELASTIC SYSTEMS

The force–deformation relation for a structural steel component undergoing cyclic defor-mations is shown in Fig. 1.3.1c. The initial loading curve is nonlinear at the larger ampli-tudes of deformation, and the unloading and reloading curves differ from the initial loading branch. Thus the relation between the resisting force vector $\mathbf{f}_S$ and displacement vector $\mathbf{u}$ is path dependent, i.e., it depends on whether the displacements are increasing or decreasing. Then $\mathbf{f}_S$ can be expressed as an implicit function of $\mathbf{u}$:

$$\mathbf{f}_S = \mathbf{f}_S(\mathbf{u}) \tag{9.8.1}$$

This general equation replaces Eq. (9.2.3) and Eq. (9.4.8) becomes

$$\mathbf{m}\ddot{\mathbf{u}} + \mathbf{c}\dot{\mathbf{u}} + \mathbf{f}_S(\mathbf{u}) = -\mathbf{m}\boldsymbol{\iota}\ddot{u}_g(t) \tag{9.8.2}$$

These are the equations of motion for inelastic MDF systems subjected to ground acceler-ation $\ddot{u}_g(t)$, the same at all support points.

Following the approach outlined in Section 1.4 for SDF systems, the damping matrix of an MDF system that models the energy dissipation arising from rate-dependent effects within the linearly elastic range of deformations (see Chapter 11) is also assumed to repre-sent this damping mechanism in the inelastic range of deformations. The additional energy dissipated due to inelastic behavior at larger deformations is accounted for by the hysteretic force–deformation relation used in the numerical time-stepping procedures for solving the equations of motion (Chapter 16).

These numerical procedures are based on linearizing the equations of motion over a time step t_i to $t_i + \Delta t$ and using Newton–Raphson iteration (Section 5.7). The struc-tural stiffness matrix at t_i is formulated by direct assembly of the element stiffness ma-trices. For each structural element—column, beam, or wall, etc.—the element stiffness matrix is determined for the state—displacements and velocities—of the system at t_i and the prescribed yielding mechanism of the material. The element stiffness matrices are then assembled. These procedures are not presented in this structural dynamics text because the reader is expected to be familiar with static analysis of inelastic systems. However, we discuss this issue briefly in Chapter 19 in the context of nonlinear analysis of simple idealizations of multistory buildings.

9.9 PROBLEM STATEMENT

Given the mass matrix $\mathbf{m}$, the stiffness matrix $\mathbf{k}$ of a linearly elastic system or the force–deformation relations $\mathbf{f}_S(\mathbf{u})$ for an inelastic system, the damping matrix $\mathbf{c}$, and the dynamic excitation—which may be external forces $\mathbf{p}(t)$ or ground acceleration $\ddot{u}_g(t)$—a fundamental problem in structural dynamics is to determine the response of the MDF structure.

The term *response* denotes any response quantity, such as displacement, velocity, and acceleration of each mass, and also an internal force or internal stress in the structural elements. When the excitation is a set of external forces, the displacements $\mathbf{u}(t)$, veloc-ities $\dot{\mathbf{u}}(t)$, and accelerations $\ddot{\mathbf{u}}(t)$ are of interest. For earthquake excitations the response quantities relative to the ground—$\mathbf{u}$, $\dot{\mathbf{u}}$, and $\ddot{\mathbf{u}}$—as well as the total responses—$\mathbf{u}^t$, $\dot{\mathbf{u}}^t$, and $\ddot{\mathbf{u}}^t$—may be needed. The relative displacements $\mathbf{u}(t)$ associated with deformations of the structure are the most important since the internal forces in the structure are directly related to $\mathbf{u}(t)$.

9.10 ELEMENT FORCES

Once the relative displacements $\mathbf{u}(t)$ have been determined by dynamic analysis, the element forces and stresses needed for structural design can be determined by static analysis of the structure at each time instant (i.e., no additional dynamic analysis is necessary). The static analysis of an MDF system can be visualized in one of two ways:

1. At each time instant the nodal displacements are known from $\mathbf{u}(t)$; if $\mathbf{u}(t)$ includes only the dynamic DOF, the displacements in the condensed DOF are given by Eq. (9.3.3). From the known displacements and rotations of the nodes of each structural element (beam and column), the element forces (bending moments and shears) can be determined through the element stiffness properties (Appendix 1), and stresses can be determined from the element forces.

2. The second approach is to introduce *equivalent static forces;* at any instant of time t these forces $\mathbf{f}_S$ are the external forces that will produce the displacements $\mathbf{u}$ at the same t in the stiffness component of the structure. Thus

$$\mathbf{f}_S(t) = \mathbf{k}\mathbf{u}(t) \qquad (9.10.1)$$

Element forces or stresses can be determined at each time instant by static analysis of the structure subjected to the forces $\mathbf{f}_S$. The repeated static analyses at many time instants can be implemented efficiently as described in Chapter 13.

For inelastic systems the element forces can be determined by appropriate modifications of these procedures to recognize that such systems are analyzed by time-stepping procedures with iteration within a time step (Chapter 16).

To keep the preceding problem statement simple, we have excluded systems subjected to spatially varying, multiple-support excitations (Section 9.7). Such dynamic response analyses involve additional considerations that are discussed in Section 13.5.

9.11 METHODS FOR SOLVING THE EQUATIONS OF MOTION: OVERVIEW

The dynamic response of linear systems with classical damping that is a reasonable model for many structures can be determined by classical modal analysis. Classical natural frequencies and modes of vibration exist for such systems (Chapter 10), and their equations of motion, when transformed to modal coordinates, become uncoupled (Chapters 12 and 13). Thus the response in each natural vibration mode can be computed independent of the others, and the modal responses can be combined to determine the total response. Each mode responds with its own particular pattern of deformation, the mode shape; with its own frequency, the natural frequency; and with its own damping. Each modal response can be computed as a function of time by analysis of an SDF system with the vibration properties—natural frequency and damping—of the particular mode. These SDF equations can be solved in closed form for excitations that can be described analytically (Chapters 3 and 4), or they can be solved by time-stepping methods for complicated excitations that are defined numerically (Chapter 5).

Classical modal analysis is not applicable to a structure consisting of subsystems with very different levels of damping. For such systems the classical damping model may not be appropriate, classical vibration modes do not exist, and the equations of motion cannot be uncoupled by transforming to modal coordinates of the system without damping. Such systems can be analyzed by (1) transforming the equations of motion to the eigenvectors of the complex eigenvalue problem that includes the damping matrix (Chapter 14); or (2) direct solution of the coupled system of differential equations (Chapter 16). The latter approach requires numerical methods because closed-form analytical solutions are not possible even if the dynamic excitation is a simple, analytically described function of time and also, of course, if the dynamic excitation is described numerically.

Classical modal analysis is also not applicable to inelastic systems irrespective of the damping model, classical or nonclassical. The standard approach is to solve directly the coupled equations in the original nodal displacements by numerical methods (Chapter 16). The overview of analysis procedures presented in this section is summarized in Fig. 9.11.1.

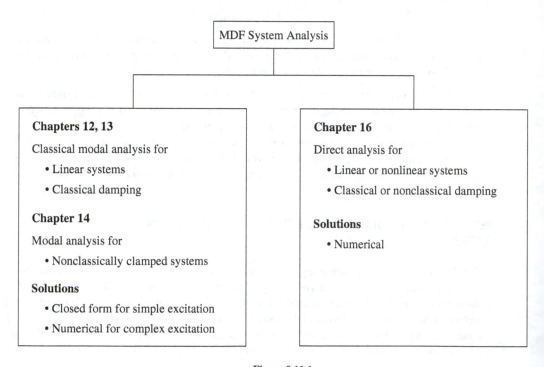

Figure 9.11.1

FURTHER READING

Clough, R. W., and Penzien, J., *Dynamics of Structures,* McGraw-Hill, New York, 1993, Chapters 9 and 10.

Craig, R. R., Jr. and Kurdila, A.J., *Fundamentals of Structural Dynamics,* 2nd ed., Wiley, New York, 2006, Chapter 8.

Humar, J. L., *Dynamics of Structures,* 2nd ed., A. A. Balkema Publishers, Lisse, The Netherlands, 2002, Chapter 3.

PROBLEMS

9.1   A uniform rigid bar of total mass m is supported on two springs k_1 and k_2 at the two ends and subjected to dynamic forces as shown in Fig. P9.1. The bar is constrained so that it can move only vertically in the plane of the paper. (*Note:* This is the system of Example 9.2.) Formulate the equations of motion with respect to the two DOFs defined at the left end of the bar.

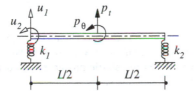

Figure P9.1

*9.2   A uniform simply supported beam of length L, flexural rigidity EI, and mass m per unit length has been idealized as the lumped-mass system shown in Fig. P9.2. The applied forces are also shown.
(**a**) Identify the DOFs to represent the elastic properties and determine the stiffness matrix. Neglect the axial deformations of the beam.
(**b**) Identify the DOFs to represent the inertial properties and determine the mass matrix.
(**c**) Formulate the equations governing the translational motion of the beam.

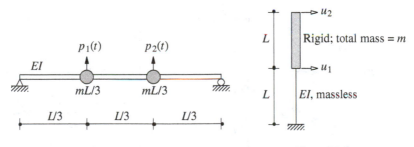

Figure P9.2   **Figure P9.4**

9.3   Derive the equations of motion of the beam of Fig. P9.2 governing the translational displacements u_1 and u_2 by starting directly with these two DOFs only.

9.4   A rigid bar is supported by a weightless column as shown in Fig. P9.4. Evaluate the mass, flexibility, and stiffness matrices of the system defined for the two DOFs shown. Do not use a lumped-mass approximation.

*Denotes that a computer is necessary to solve this problem.

9.5   Using the definition of stiffness and mass influence coefficients, formulate the equations of motion for the two-story shear frame with lumped masses shown in Fig. P9.5. The beams are rigid and the flexural rigidity of the columns is EI. Neglect axial deformations in all elements.

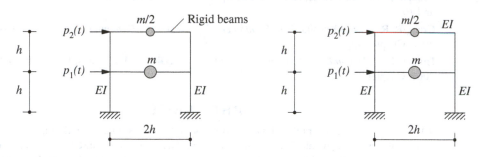

Figure P9.5   Figure P9.6

*9.6   Figure P9.6 shows a two-story frame with lumped masses subjected to lateral forces, together with some of its properties; in addition, the flexural rigidity is EI for all columns and beams.
(a) Identify the DOFs to represent the elastic properties and determine the stiffness matrix. Neglect axial deformations in all elements.
(b) Identify the DOFs to represent the inertial properties and determine the mass matrix. Assume the members to be massless and neglect their rotational inertia.
(c) Formulate the equations governing the motion of the frame in the DOFs in part (b).

9.7–   Using the definition of stiffness and mass influence coefficients, formulate the equations of
9.8   motion for the three-story shear frames with lumped masses shown in Figs. P9.7 and P9.8. The beams are rigid in flexure, and the flexural rigidity of the columns is as shown. Neglect axial deformations in all elements.

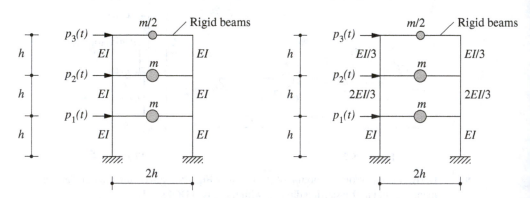

Figure P9.7   Figure P9.8

*Denotes that a computer is necessary to solve this problem.

*9.9– Figures P9.9–P9.12 show three-story frames with lumped masses subjected to lateral forces,
9.12 together with the flexural rigidity of columns and beams.

(a) Identify the DOFs to represent the elastic properties and determine the stiffness matrix.
Neglect the axial deformation of the members.

(b) Identify the DOFs to represent the inertial properties and determine the mass matrix. Assume the members to be massless and neglect their rotational inertia.

(c) Formulate the equations governing the motion of the frame in the DOFs in part (b).

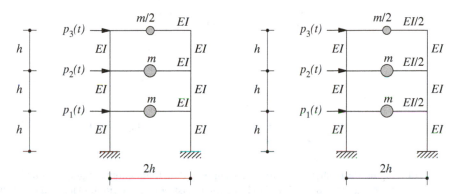

Figure P9.9   Figure P9.10

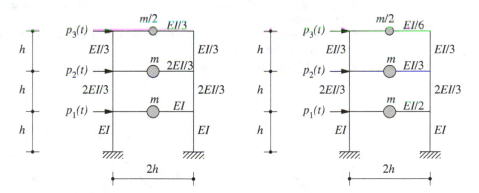

Figure P9.11   Figure P9.12

*9.13 An umbrella structure has been idealized as an assemblage of three flexural elements with
lumped masses at the nodes as shown in Fig. P9.13.

*Denotes that a computer is necessary to solve this problem.

(a) Identify the DOFs to represent the elastic properties and determine the stiffness matrix. Neglect axial deformations in all members.
(b) Identify the DOFs to represent the inertial properties and determine the mass matrix.
(c) Formulate the equations of motion governing the DOFs in part (b) when the excitation is (i) horizontal ground motion, (ii) vertical ground motion, (iii) ground motion in direction b–d, (iv) ground motion in direction b–c, and (v) rocking ground motion in the plane of the structure.

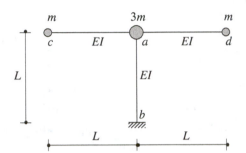

Figure P9.13

9.14 Figure P9.14 shows a uniform slab supported on four columns rigidly attached to the slab and clamped at the base. The slab has a total mass m and is rigid in plane and out of plane. Each column is of circular cross section, and its second moment of cross-sectional area about any diametrical axis is as noted. With the DOFs selected as u_x, u_y, and u_θ at the center of the slab, and using influence coefficients:
(a) Formulate the mass and stiffness matrices in terms of m and the lateral stiffness $k = 12EI/h^3$ of the smaller column; h is the height.
(b) Formulate the equations of motion for ground motion in (i) the x-direction, (ii) the y-direction, and (iii) the direction d–b.

9.15 Repeat Problem 9.14 using the second set of DOFs shown in Fig. P9.15.

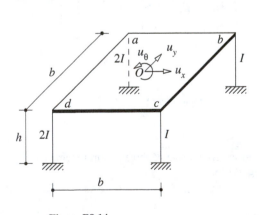

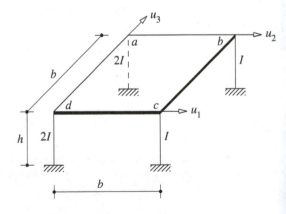

Figure P9.14   **Figure P9.15**

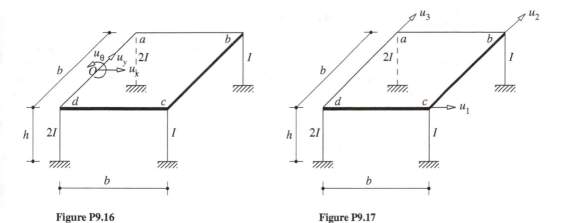

Figure P9.16   Figure P9.17

9.16 Repeat Problem 9.14 using the DOFs shown in Fig. P.9.16.

9.17 Repeat Problem 9.14 using the DOFs shown in Fig. P9.17.

9.18 Figure P9.18 shows a three-dimensional pipe *abcd* clamped at *a* with mass *m* at *d*. All members are made of the same material and have identical cross sections. Formulate the equations of motion governing the DOFs u_x, u_y, and u_z when the excitation is ground motion in (i) x-direction, (ii) y-direction, (iii) z-direction, and (iv) direction *a–d*. First express the flexibility matrix in terms of E, I, G, J, and L; then specialize it for $GJ = \frac{4}{5}EI$. Consider flexural and torsional deformations but neglect axial deformations.

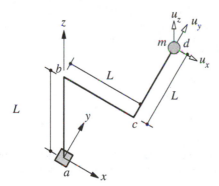

Figure P9.18

9.19 Formulate the equations of motion for the system shown in Fig. P9.19 subjected to support displacements $u_{g1}(t)$ and $u_{g2}(t)$. These equations governing the dynamic components of

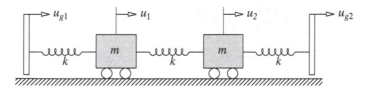

Figure P9.19

displacements u_1 and u_2 (total displacements minus quasistatic displacements) should be expressed in terms of m, k, $\ddot{u}_{g1}(t)$, and $\ddot{u}_{g2}(t)$.

9.20   Figure P9.20 shows a simply supported massless beam with a lumped mass at the center subjected to motions $u_{g1}(t)$ and $u_{g2}(t)$ at the two supports. Formulate the equation of motion governing the dynamic component of displacement u (= total displacement − quasi-static displacement) of the lumped mass. Express this equation in terms of m, EI, L, $\ddot{u}_{g1}(t)$, and $\ddot{u}_{g2}(t)$.

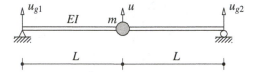

Figure P9.20

*9.21   Figure P9.21 shows a pipe in an industrial plant. The pipe is clamped at supports a and b and has a 90° bend at c. It supports two heavy valves of mass m as shown. Neglecting axial deformations and pipe mass, formulate the equations of motion for this system subjected to support displacements $u_{g1}(t)$ and $u_{g2}(t)$. These equations governing the dynamic component (= total displacement − quasi-static component) of the displacements u_1 and u_2 should be expressed in terms of m, EI, and L. How do these governing equations differ from the case of identical motion at both supports?

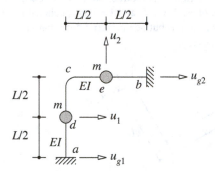

Figure P9.21

*9.22   Figure P9.22 shows a single-span bridge. Neglecting axial deformations, formulate the equations of motion for this system subjected to support displacements $u_{g1}(t)$ and $u_{g2}(t)$. These

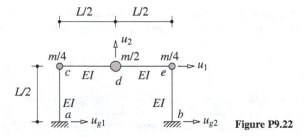

Figure P9.22

*Denotes that a computer is necessary to solve this problem.

equations governing the dynamic component (= total displacement − quasi-static component) of the displacements u_1 and u_2 should be expressed in terms of m, EI, and L. How do these governing equations differ from the case of identical motion at both supports?

*9.23   Figure P9.23 shows a uniform slab supported on four identical columns rigidly attached to the slab and clamped at the base. The slab has a total mass m and is rigid in plane and out of plane. Each column is of circular cross section, and its stiffness about any diametrical axis is as noted. With the DOFs selected as u_x, u_y, and u_θ, formulate the equations of motion for the system subjected to ground displacements $u_{ga}(t)$, $u_{gb}(t)$, $u_{gc}(t)$, and $u_{gd}(t)$ in the x-direction at the supports of columns a, b, c, and d, respectively. These equations governing the dynamic component (= total displacement − quasi-static component) of the displacements u_x, u_y, and u_θ should be expressed in terms of m, b, and $k = 12EI/h^3$ of the columns. How do these governing equations differ from the case of identical ground motion $u_g(t)$ at all column supports?

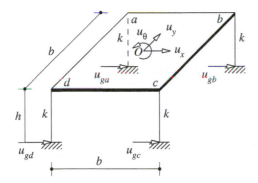

Figure P9.23

*9.24   Formulate the equations of motion for the system of Problem 9.14 subjected to ground displacements $u_{ga}(t)$, $u_{gb}(t)$, $u_{gc}(t)$, and $u_{gd}(t)$ in the x-direction at the supports of columns a, b, c, and d, respectively. These equations governing the dynamic component (= total displacement − quasi-static component) of the displacements u_x, u_y, and u_θ should be expressed in terms of m, b, h, and the lateral stiffness $k = 12EI/h^3$ of the smaller column. How do these governing equations differ from the case of identical ground motion $u_g(t)$ at all column supports?

*9.25   An intake–outlet tower fixed at the base is partially submerged in water and is accessible from the edge of the reservoir by a foot bridge that is axially rigid and pin-connected to the tower (Fig. P9.25). (In practice, sliding is usually permitted at the connection. The pin connection has been used here only for this hypothetical problem.) The 200-ft-high uniform tower has a hollow reinforced-concrete cross section with outside diameter = 25 ft and wall thickness = 1 ft 3 in. An approximate value of the flexural stiffness EI may be computed from the gross properties of the concrete section without the reinforcement; the elastic modulus of concrete $E = 3.6 \times 10^3$ ksi. For purposes of preliminary analysis the mass of the tower is lumped as shown at two equally spaced locations, where m is the mass per unit length and L the total length of the tower; the unit weight of concrete is 150 lb/ft^3. (The added mass of the surrounding water may be neglected here, but it should be considered in practical analysis.)

*Denotes that a computer is necessary to solve this problem.

It is desired to analyze the response of this structure to support motions $u_{g1}(t)$ and $u_{g2}(t)$. Formulate the equations of motion governing the dynamic components of displacements u_1 and u_2 (dynamic component = total displacement − quasi-static component).

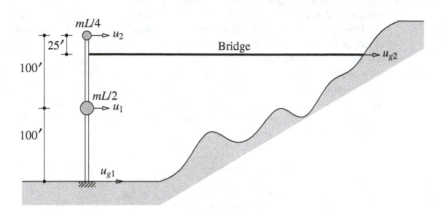

Figure P9.25

10

Free Vibration

PREVIEW

By *free vibration* we mean the motion of a structure without any dynamic excitation—external forces or support motion. Free vibration is initiated by disturbing the structure from its equilibrium position by some initial displacements and/or by imparting some initial velocities.

This chapter on free vibration of MDF systems is divided into three parts. In Part A we develop the notion of natural frequencies and natural modes of vibration of a structure; these concepts play a central role in the dynamic and earthquake analysis of linear systems (Chapters 12 and 13).

In Part B we describe the use of these vibration properties to determine the free vibration response of systems. Undamped systems are analyzed first. We then define systems with classical damping and systems with nonclassical damping. The analysis procedure is extended to systems with classical damping, recognizing that such systems possess the same natural modes as those of the undamped system.

Part C is concerned with numerical solution of the eigenvalue problem to determine the natural frequencies and modes of vibration. Vector iteration methods are effective in structural engineering applications, and we restrict this presentation to such methods. Only the basic ideas of vector iteration are included, without getting into subspace iteration or the Lanczos method. Although this limited treatment would suffice for many practical problems and research applications, the reader should recognize that a wealth of knowledge exists on the subject.

PART A: NATURAL VIBRATION FREQUENCIES AND MODES

10.1 SYSTEMS WITHOUT DAMPING

Free vibration of linear MDF systems is governed by Eq. (9.2.12) with $\mathbf{p}(t) = \mathbf{0}$, which for systems without damping is

$$\mathbf{m\ddot{u} + ku = 0} \qquad (10.1.1)$$

Equation (10.1.1) represents N homogeneous differential equations that are coupled through the mass matrix, the stiffness matrix, or both matrices; N is the number of DOFs. It is desired to find the solution $\mathbf{u}(t)$ of Eq. (10.1.1) that satisfies the initial conditions

$$\mathbf{u = u}(0) \qquad \mathbf{\dot{u} = \dot{u}}(0) \qquad (10.1.2)$$

at $t = 0$. A general procedure to obtain the desired solution for any MDF system is developed in Section 10.8. In this section the solution is presented in graphical form that enables us to understand free vibration of an MDF system in qualitative terms.

Figure 10.1.1 shows the free vibration of a two-story shear frame. The story stiffnesses and lumped masses at the floors are noted, and the free vibration is initiated by the deflections shown by curve a in Fig. 10.1.1b. The resulting motion u_j of the two masses is plotted in Fig. 10.1.1d as a function of time; T_1 will be defined later.

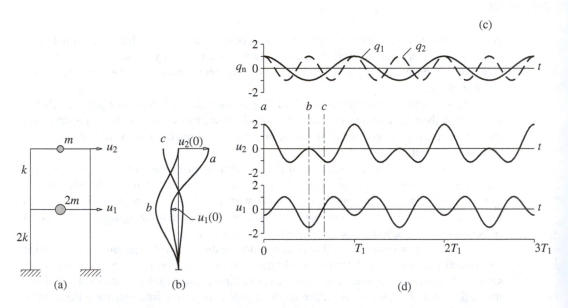

Figure 10.1.1 Free vibration of an undamped system due to arbitrary initial displacement: (a) two-story frame; (b) deflected shapes at time instants a, b, and c; (c) modal coordinates $q_n(t)$; (d) displacement history.

The deflected shapes of the structure at selected time instants a, b, and c are also shown; the $q_n(t)$ plotted in Fig.10.1.1c are discussed in Example 10.11. The displacement–time plot for the jth floor starts with the initial conditions $u_j(0)$ and $\dot{u}_j(0)$; the $u_j(0)$ are identified in Fig. 10.1.1b and $\dot{u}_j(0) = 0$ for both floors. Contrary to what we observed in Fig. 2.1.1 for SDF systems, the motion of each mass (or floor) is not a simple harmonic motion and the frequency of the motion cannot be defined. Furthermore, the deflected shape (i.e., the ratio u_1/u_2) varies with time, as is evident from the differing deflected shapes b and c, which are in turn different from the initial deflected shape a.

   An undamped structure would undergo simple harmonic motion without change of deflected shape, however, if free vibration is initiated by appropriate distributions of displacements in the various DOFs. As shown in Figs. 10.1.2 and 10.1.3, two characteristic deflected shapes exist for this two-DOF system such that if it is displaced in one of these shapes and released, it will vibrate in simple harmonic motion, maintaining the initial deflected shape. Both floors vibrate in the same phase, i.e., they pass through their equilibrium, maximum, or minimum positions at the same instant of time. Each characteristic deflected shape is called a *natural mode of vibration* of an MDF system.

   Observe that the displacements of both floors are in the same direction in the first mode but in opposite directions in the second mode. The point of zero displacement, called a *node*,[†] does not move at all (Fig. 10.1.3); as the mode number n increases, the number of nodes increases accordingly (see Fig. 12.8.2).

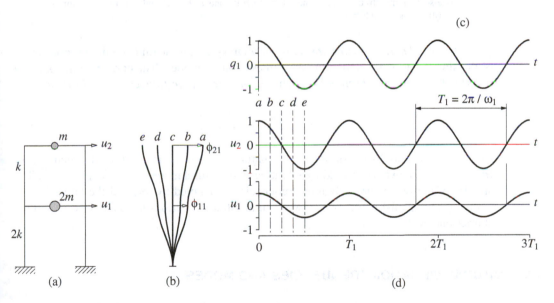

Figure 10.1.2 Free vibration of an undamped system in its first natural mode of vibration: (a) two-story frame; (b) deflected shapes at time instants a, b, c, d, and e; (c) modal coordinate $q_1(t)$; (d) displacement history.

[†]Recall that we have already used the term *node* for nodal points in the structural idealization; the two different uses of *node* should be clear from the context.

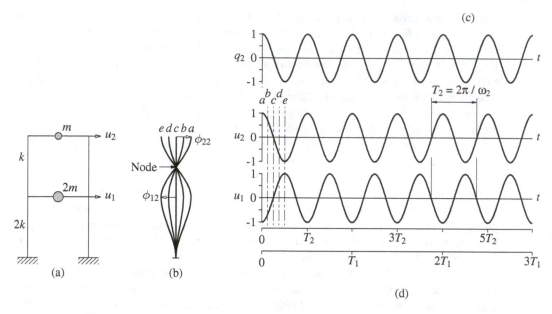

Figure 10.1.3 Free vibration of an undamped system in its second natural mode of vibration: (a) two-story frame; (b) deflected shapes at the time instants a, b, c, d, and e; (c) modal coordinate $q_2(t)$; (d) displacement history.

A *natural period of vibration* T_n of an MDF system is the time required for one cycle of the simple harmonic motion in one of these natural modes. The corresponding *natural circular frequency of vibration* is ω_n and the *natural cyclic frequency of vibration* is f_n, where

$$T_n = \frac{2\pi}{\omega_n} \qquad f_n = \frac{1}{T_n} \tag{10.1.3}$$

Figures 10.1.2 and 10.1.3 show the two natural periods T_n and natural frequencies ω_n ($n = 1, 2$) of the two-story building vibrating in its natural modes $\phi_n = \langle \phi_{1n} \quad \phi_{2n} \rangle^T$. The smaller of the two natural vibration frequencies is denoted by ω_1, and the larger by ω_2. Correspondingly, the longer of the two natural vibration periods is denoted by T_1 and the shorter one by T_2.

10.2 NATURAL VIBRATION FREQUENCIES AND MODES

In this section we introduce the eigenvalue problem whose solution gives the natural frequencies and modes of a system. The free vibration of an undamped system in one of its natural vibration modes, graphically displayed in Figs. 10.1.2 and 10.1.3 for a two-DOF system, can be described mathematically by

$$\mathbf{u}(t) = q_n(t)\phi_n \tag{10.2.1}$$

where the deflected shape ϕ_n does not vary with time. The time variation of the displacements is described by the simple harmonic function

$$q_n(t) = A_n \cos \omega_n t + B_n \sin \omega_n t \qquad (10.2.2)$$

where A_n and B_n are constants that can be determined from the initial conditions that initiate the motion. Combining Eqs. (10.2.1) and (10.2.2) gives

$$\mathbf{u}(t) = \phi_n \left(A_n \cos \omega_n t + B_n \sin \omega_n t \right) \qquad (10.2.3)$$

where ω_n and ϕ_n are unknown.

Substituting this form of $\mathbf{u}(t)$ in Eq. (10.1.1) gives

$$\left[-\omega_n^2 \mathbf{m} \phi_n + \mathbf{k} \phi_n \right] q_n(t) = \mathbf{0}$$

This equation can be satisfied in one of two ways. Either $q_n(t) = 0$, which implies that $\mathbf{u}(t) = \mathbf{0}$ and there is no motion of the system (this is the so-called trivial solution), or the natural frequencies ω_n and modes ϕ_n must satisfy the following algebraic equation:

$$\mathbf{k} \phi_n = \omega_n^2 \mathbf{m} \phi_n \qquad (10.2.4)$$

which provides a useful condition. This algebraic equation is called the *matrix eigenvalue problem.* When necessary it is called the real eigenvalue problem to distinguish it from the complex eigenvalue problem mentioned in Chapter 14 for systems with damping. The stiffness and mass matrices $\mathbf{k}$ and $\mathbf{m}$ are known; the problem is to determine the scalar ω_n^2 and vector ϕ_n.

To indicate the formal solution to Eq. (10.2.4), it is rewritten as

$$\left[\mathbf{k} - \omega_n^2 \mathbf{m} \right] \phi_n = \mathbf{0} \qquad (10.2.5)$$

which can be interpreted as a set of N homogeneous algebraic equations for the N elements ϕ_{jn} ($j = 1, 2, \ldots, N$). This set always has the trivial solution $\phi_n = \mathbf{0}$, which is not useful because it implies no motion. It has nontrivial solutions if

$$\det \left[\mathbf{k} - \omega_n^2 \mathbf{m} \right] = 0 \qquad (10.2.6)$$

When the determinant is expanded, a polynomial of order N in ω_n^2 is obtained. Equation (10.2.6) is known as the *characteristic equation* or *frequency equation.* This equation has N real and positive roots for ω_n^2 because $\mathbf{m}$ and $\mathbf{k}$, the structural mass and stiffness matrices, are symmetric and positive definite. The positive definite property of $\mathbf{k}$ is assured for all structures supported in a way that prevents rigid-body motion. Such is the case for civil engineering structures of interest to us, but not for unrestrained structures such as aircraft in flight—these are beyond the scope of this book. The positive definite property of $\mathbf{m}$ is also assured because the lumped masses are nonzero in all DOFs retained in the analysis after the DOFs with zero lumped mass have been eliminated by static condensation (Section 9.3).

The N roots, ω_n^2, of Eq. (10.2.6) determine the N natural frequencies ω_n ($n = 1, 2, \ldots, N$) of vibration, conventionally arranged in sequence from smallest to largest ($\omega_1 < \omega_2 < \cdots < \omega_N$). These roots of the characteristic equation are also known as *eigenvalues, characteristic values,* or *normal values.* When a natural frequency ω_n is known, Eq. (10.2.5) can be solved for the corresponding vector ϕ_n to within a multiplicative constant. The eigenvalue problem does not fix the absolute amplitude of the vectors ϕ_n, only the shape of the vector given by the relative values of the N displacements ϕ_{jn} ($j = 1, 2, \ldots, N$). Corresponding to the N natural vibration frequencies ω_n of an N-DOF system, there are N independent vectors ϕ_n, which are known as *natural modes of vibration*, or *natural mode shapes of vibration*. These vectors are also known as *eigenvectors, characteristic vectors,* or *normal modes*. The term *natural* is used to qualify each of these vibration properties to emphasize the fact that these are natural properties of the structure in free vibration, and they depend only on its mass and stiffness properties. The subscript n denotes the mode number, and the first mode ($n = 1$) is also known as the fundamental mode.

As mentioned earlier, during free vibration in each natural mode, an undamped system oscillates at its natural frequency with all DOFs of the system vibrating in the same phase, passing through their equilibrium, maximum, or minimum positions at the same instant of time. Because this type of natural mode was the subject of Lagrange's classical (1811) treatise on mechanics, we will refer to such modes as *classical natural modes*. In damped systems, this property is generally violated and classical natural modes may not exist, as we shall see later.

10.3 MODAL AND SPECTRAL MATRICES

The N eigenvalues and N natural modes can be assembled compactly into matrices. Let the natural mode ϕ_n corresponding to the natural frequency ω_n have elements ϕ_{jn}, where j indicates the DOFs. The N eigenvectors can then be displayed in a single square matrix, each column of which is a natural mode:

$$\mathbf{\Phi} = \left[\phi_{jn}\right] = \begin{bmatrix} \phi_{11} & \phi_{12} & \cdots & \phi_{1N} \\ \phi_{21} & \phi_{22} & \cdots & \phi_{2N} \\ \vdots & \vdots & \ddots & \vdots \\ \phi_{N1} & \phi_{N2} & \cdots & \phi_{NN} \end{bmatrix}$$

The matrix $\mathbf{\Phi}$ is called the *modal matrix* for the eigenvalue problem, Eq. (10.2.4). The N eigenvalues ω_n^2 can be assembled into a diagonal matrix $\mathbf{\Omega}^2$, which is known as the *spectral matrix* of the eigenvalue problem, Eq. (10.2.4):

$$\mathbf{\Omega}^2 = \begin{bmatrix} \omega_1^2 & & & \\ & \omega_2^2 & & \\ & & \ddots & \\ & & & \omega_N^2 \end{bmatrix}$$

Each eigenvalue and eigenvector satisfies Eq. (10.2.4), which can be rewritten as the relation

$$\mathbf{k}\phi_n = \mathbf{m}\phi_n\omega_n^2 \tag{10.3.1}$$

By using the modal and spectral matrices, it is possible to assemble all of such relations ($n = 1, 2, \ldots, N$) into a single matrix equation:

$$\mathbf{k}\mathbf{\Phi} = \mathbf{m}\mathbf{\Phi}\mathbf{\Omega}^2 \tag{10.3.2}$$

Equation (10.3.2) provides a compact presentation of the equations relating all eigenvalues and eigenvectors.

10.4 ORTHOGONALITY OF MODES

The natural modes corresponding to different natural frequencies can be shown to satisfy the following orthogonality conditions. When $\omega_n \neq \omega_r$,

$$\phi_n^T \mathbf{k} \phi_r = 0 \qquad \phi_n^T \mathbf{m} \phi_r = 0 \tag{10.4.1}$$

These important properties can be proven as follows: The nth natural frequency and mode satisfy Eq. (10.2.4); premultiplying it by ϕ_r^T, the transpose of ϕ_r, gives

$$\phi_r^T \mathbf{k} \phi_n = \omega_n^2 \phi_r^T \mathbf{m} \phi_n \tag{10.4.2}$$

Similarly, the rth natural frequency and mode satisfy Eq. (10.2.4); thus $\mathbf{k}\phi_r = \omega_r^2 \mathbf{m}\phi_r$. Premultiplying by ϕ_n^T gives

$$\phi_n^T \mathbf{k} \phi_r = \omega_r^2 \phi_n^T \mathbf{m} \phi_r \tag{10.4.3}$$

The transpose of the matrix on the left side of Eq. (10.4.2) will equal the transpose of the matrix on the right side of the equation; thus

$$\phi_n^T \mathbf{k} \phi_r = \omega_n^2 \phi_n^T \mathbf{m} \phi_r \tag{10.4.4}$$

wherein we have utilized the symmetry property of the mass and stiffness matrices. Subtracting Eq. (10.4.3) from (10.4.4) gives

$$(\omega_n^2 - \omega_r^2)\phi_n^T \mathbf{m} \phi_r = 0$$

Thus Eq. (10.4.1b) is true when $\omega_n^2 \neq \omega_r^2$, which for systems with positive natural frequencies implies that $\omega_n \neq \omega_r$. Substituting Eq. (10.4.1b) in (10.4.3) indicates that Eq. (10.4.1a) is true when $\omega_n \neq \omega_r$. This completes a proof for the orthogonality relations of Eq. (10.4.1).

We have established the orthogonality relations between modes with distinct frequencies (i.e., $\omega_n \neq \omega_r$). If the frequency equation (10.2.4) has a j-fold multiple root (i.e., the system has one frequency repeated j times), it is always possible to find j modes associated with this frequency that satisfy Eq. (10.4.1). If these j modes are included with the modes corresponding to the other frequencies, a set of N modes is obtained which satisfies Eq. (10.4.1) for $n \neq r$.

The orthogonality of natural modes implies that the following square matrices are diagonal:

$$\mathbf{K} \equiv \Phi^T \mathbf{k} \Phi \qquad \mathbf{M} \equiv \Phi^T \mathbf{m} \Phi \tag{10.4.5}$$

where the diagonal elements are

$$K_n = \phi_n^T \mathbf{k} \phi_n \qquad M_n = \phi_n^T \mathbf{m} \phi_n \tag{10.4.6}$$

Since $\mathbf{m}$ and $\mathbf{k}$ are positive definite, the diagonal elements of $\mathbf{K}$ and $\mathbf{M}$ are positive. They are related by

$$K_n = \omega_n^2 M_n \tag{10.4.7}$$

This can be demonstrated from the definitions of K_n and M_n as follows: Substituting Eq. (10.2.4) in (10.4.6a) gives

$$K_n = \phi_n^T(\omega_n^2 \mathbf{m}\phi_n) = \omega_n^2(\phi_n^T \mathbf{m}\phi_n) = \omega_n^2 M_n$$

10.5 INTERPRETATION OF MODAL ORTHOGONALITY

In this section we develop physically motivated interpretations of the orthogonality properties of natural modes. One implication of modal orthogonality is that the work done by the nth-mode inertia forces in going through the rth-mode displacements is zero. To demonstrate this result, consider a structure vibrating in the nth mode with displacements

$$\mathbf{u}_n(t) = q_n(t)\phi_n \qquad (10.5.1)$$

The corresponding accelerations are $\ddot{\mathbf{u}}_n(t) = \ddot{q}_n(t)\phi_n$ and the associated inertia forces are

$$(\mathbf{f}_I)_n = -\mathbf{m}\ddot{\mathbf{u}}_n(t) = -\mathbf{m}\phi_n\ddot{q}_n(t) \qquad (10.5.2)$$

Next, consider displacements of the structure in its rth natural mode:

$$\mathbf{u}_r(t) = q_r(t)\phi_r \qquad (10.5.3)$$

The work done by the inertia forces of Eq. (10.5.2) in going through the displacements of Eq. (10.5.3) is

$$(\mathbf{f}_I)_n^T \mathbf{u}_r = -\left(\phi_n^T \mathbf{m}\phi_r\right)\ddot{q}_n(t)q_r(t) \qquad (10.5.4)$$

which is zero because of the modal orthogonality relation of Eq. (10.4.1b). This completes the proof.

Another implication of the modal orthogonality properties is that the work done by the equivalent static forces associated with displacements in the nth mode in going through the rth-mode displacements is zero. These forces are

$$(\mathbf{f}_S)_n = \mathbf{k}\mathbf{u}_n(t) = \mathbf{k}\phi_n q_n(t)$$

and the work they do in going through the displacements of Eq. (10.5.3) is

$$(\mathbf{f}_S)_n^T \mathbf{u}_r = \left(\phi_n^T \mathbf{k}\phi_r\right)q_n(t)q_r(t)$$

which is zero because of the modal orthogonality relation of Eq. (10.4.1a). This completes the proof.

10.6 NORMALIZATION OF MODES

As mentioned earlier, the eigenvalue problem, Eq. (10.2.4), determines the natural modes to only within a multiplicative factor. If the vector ϕ_n is a natural mode, any vector proportional to ϕ_n is essentially the same natural mode because it also satisfies Eq. (10.2.4). Scale factors are sometimes applied to natural modes to standardize their elements associated with various DOFs. This process is called *normalization*. Sometimes it is convenient to normalize each mode so that its largest element is unity. Other times it may be advantageous to normalize each mode so that the element corresponding to a particular DOF, say the top floor of a multistory building, is unity. In theoretical discussions and computer programs it is common to normalize modes so that the M_n have unit values. In this case

$$M_n = \phi_n^T \mathbf{m} \phi_n = 1 \qquad \mathbf{\Phi}^T \mathbf{m} \mathbf{\Phi} = \mathbf{I} \tag{10.5.5}$$

where $\mathbf{I}$ is the identity matrix, a diagonal matrix with unit values along the main diagonal. Equation (10.5.5) states that the natural modes are not only orthogonal but are normalized with respect to $\mathbf{m}$. They are then called a mass *orthonormal set*. When the modes are normalized in this manner, Eqs. (10.4.6a) and (10.4.5a) become

$$K_n = \phi_n^T \mathbf{k} \phi_n = \omega_n^2 M_n = \omega_n^2 \qquad \mathbf{K} = \mathbf{\Phi}^T \mathbf{k} \mathbf{\Phi} = \mathbf{\Omega}^2 \tag{10.5.6}$$

Example 10.1

(a) Determine the natural vibration frequencies and modes of the system of Fig. E10.1a using the first set of DOFs shown.

(b) Repeat part (a) using the second set of DOFs in Fig. E10.1b.

(c) Show that the natural frequencies and modes determined using the two sets of DOFs are the same.

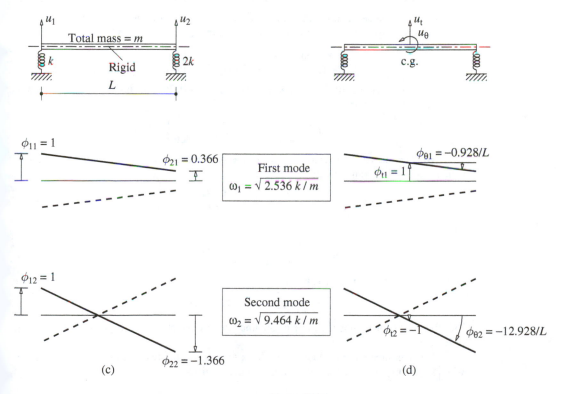

Figure E10.1

Solution (a) The mass and stiffness matrices for the system with the first set of DOFs were determined in Example 9.2:

$$\mathbf{m} = \frac{m}{6}\begin{bmatrix} 2 & 1 \\ 1 & 2 \end{bmatrix} \qquad \mathbf{k} = k\begin{bmatrix} 1 & 0 \\ 0 & 2 \end{bmatrix}$$

Then

$$\mathbf{k} - \omega_n^2 \mathbf{m} = \begin{bmatrix} k - m\omega_n^2/3 & -m\omega_n^2/6 \\ -m\omega_n^2/6 & 2k - m\omega_n^2/3 \end{bmatrix} \tag{a}$$

is substituted in Eq. (10.2.6) to obtain the frequency equation:

$$m^2\omega_n^4 - 12km\omega_n^2 + 24k^2 = 0$$

This is a quadratic equation in ω_n^2 that has the solutions

$$\omega_1^2 = (6 - 2\sqrt{3})\frac{k}{m} = 2.536\frac{k}{m} \qquad \omega_2^2 = (6 + 2\sqrt{3})\frac{k}{m} = 9.464\frac{k}{m} \tag{b}$$

Taking the square root of Eq. (b) gives the natural frequencies ω_1 and ω_2.

The natural modes are determined by substituting $\omega_n^2 = \omega_1^2$ in Eq. (a), and then Eq. (10.2.5) gives

$$k \begin{bmatrix} 0.155 & -0.423 \\ -0.423 & 1.155 \end{bmatrix} \begin{Bmatrix} \phi_{11} \\ \phi_{21} \end{Bmatrix} = \begin{Bmatrix} 0 \\ 0 \end{Bmatrix} \tag{c}$$

Now select any value for one unknown, say $\phi_{11} = 1$. Then the first or second of the two equations gives $\phi_{21} = 0.366$. Substituting $\omega_n^2 = \omega_2^2$ in Eq. (10.2.5) gives

$$k \begin{bmatrix} -2.155 & -1.577 \\ -1.577 & -1.155 \end{bmatrix} \begin{Bmatrix} \phi_{12} \\ \phi_{22} \end{Bmatrix} = \begin{Bmatrix} 0 \\ 0 \end{Bmatrix} \tag{d}$$

Selecting $\phi_{12} = 1$, either of these equations gives $\phi_{22} = -1.366$. In summary, the two modes plotted in Fig. E10.1c are

$$\phi_1 = \begin{Bmatrix} 1 \\ 0.366 \end{Bmatrix} \qquad \phi_2 = \begin{Bmatrix} 1 \\ -1.366 \end{Bmatrix} \tag{e}$$

(b) The mass and stiffness matrices of the system described by the second set of DOF were developed in Example 9.3:

$$\mathbf{m} = \begin{bmatrix} m & 0 \\ 0 & mL^2/12 \end{bmatrix} \qquad \mathbf{k} = \begin{bmatrix} 3k & kL/2 \\ kL/2 & 3kL^2/4 \end{bmatrix} \tag{f}$$

Then

$$\mathbf{k} - \omega_n^2 \mathbf{m} = \begin{bmatrix} 3k - m\omega_n^2 & kL/2 \\ kL/2 & (9k - m\omega_n^2)L^2/12 \end{bmatrix} \tag{g}$$

is substituted in Eq. (10.2.6) to obtain

$$m^2\omega_n^4 - 12km\omega_n^2 + 24k^2 = 0$$

This frequency equation is the same as obtained in part (a); obviously, it gives the ω_1 and ω_2 of Eq. (b).

To determine the nth mode we go back to either of the two equations of Eq. (10.2.5) with $\left[\mathbf{k} - \omega^2\mathbf{m}\right]$ given by Eq. (g). The first equation gives

$$\left(3k - m\omega_n^2\right)\phi_{tn} + \frac{kL}{2}\phi_{\theta n} = 0 \quad \text{or} \quad \phi_{\theta n} = -\frac{3k - m\omega_n^2}{kL/2}\phi_{tn} \tag{h}$$

Substituting for $\omega_1^2 = 2.536k/m$ and $\omega_2^2 = 9.464k/m$ in Eq. (h) gives

$$\frac{L}{2}\phi_{\theta 1} = -0.464\phi_{t1} \qquad \frac{L}{2}\phi_{\theta 2} = 6.464\phi_{t2}$$

If $\phi_{t1} = 1$, then $\phi_{\theta1} = -0.928/L$, and if $\phi_{t2} = -1$, then $\phi_{\theta2} = -12.928/L$. In summary, the two modes plotted in Fig. E10.1d are

$$\phi_1 = \left\{ \begin{matrix} 1 \\ -0.928/L \end{matrix} \right\} \qquad \phi_2 = \left\{ \begin{matrix} -1 \\ -12.928/L \end{matrix} \right\} \tag{i}$$

(c) The same natural frequencies were obtained using the two sets of DOFs. The mode shapes are given by Eqs. (e) and (i) for the two sets of DOFs. These two sets of results are plotted in Fig. E10.1c and d and can be shown to be equivalent on a graphical basis. Alternatively, the equivalence can be demonstrated by using the coordinate transformation from one set of DOFs to the other. The displacements $\mathbf{u} = \langle u_1 \quad u_2 \rangle^T$ are related to the second set of DOFs, $\bar{\mathbf{u}} = \langle u_t \quad u_\theta \rangle^T$ by

$$\left\{ \begin{matrix} u_1 \\ u_2 \end{matrix} \right\} = \begin{bmatrix} 1 & -L/2 \\ 1 & L/2 \end{bmatrix} \left\{ \begin{matrix} u_t \\ u_\theta \end{matrix} \right\} \qquad \text{or} \quad \mathbf{u} = \mathbf{a}\bar{\mathbf{u}} \tag{j}$$

The displacements $\bar{\mathbf{u}}$ in the first two modes are given by Eq. (i). Substituting the first mode in Eq. (j) leads to $\mathbf{u} = \langle 1.464 \quad 0.536 \rangle^T$. Normalizing the vector yields $\mathbf{u} = \langle 1 \quad 0.366 \rangle^T$, which is identical to ϕ_1 of Eq. (e). Similarly, substituting the second mode from Eq. (i) in Eq. (j) gives $\mathbf{u} = \langle 1 \quad -1.366 \rangle$, which is identical to ϕ_2 of Eq. (e).

Example 10.2

Determine the natural frequencies and modes of vibration of the system shown in Fig. E10.2a and defined in Example 9.5. Show that the modes satisfy the orthogonality properties.

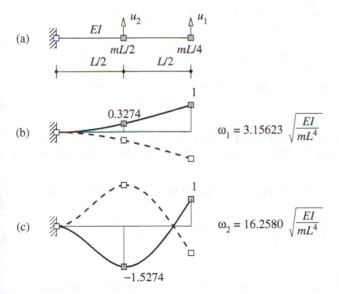

Figure E10.2

Solution   The stiffness and mass matrices were determined in Example 9.5 with reference to the translational DOFs u_1 and u_2:

$$\mathbf{m} = \begin{bmatrix} mL/4 & \\ & mL/2 \end{bmatrix} \qquad \mathbf{k} = \frac{48EI}{7L^3} \begin{bmatrix} 2 & -5 \\ -5 & 16 \end{bmatrix}$$

Then

$$\mathbf{k} - \omega^2\mathbf{m} = \frac{48EI}{7L^3} \begin{bmatrix} 2 - \lambda & -5 \\ -5 & 16 - 2\lambda \end{bmatrix} \qquad \text{(a)}$$

where

$$\lambda = \frac{7mL^4}{192EI}\omega^2 \qquad \text{(b)}$$

Substituting Eq. (a) in (10.2.6) gives the frequency equation

$$2\lambda^2 - 20\lambda + 7 = 0$$

which has two solutions: $\lambda_1 = 0.36319$ and $\lambda_2 = 9.6368$. The natural frequencies corresponding to the two values of λ are obtained from Eq. (b)[†]:

$$\omega_1 = 3.15623\sqrt{\frac{EI}{mL^4}} \qquad \omega_2 = 16.2580\sqrt{\frac{EI}{mL^4}} \qquad \text{(c)}$$

The natural modes are determined from Eq. (10.2.5) following the procedure shown in Example 10.1 to obtain

$$\phi_1 = \begin{Bmatrix} 1 \\ 0.3274 \end{Bmatrix} \qquad \phi_2 = \begin{Bmatrix} 1 \\ -1.5274 \end{Bmatrix} \qquad \text{(d)}$$

These natural modes are plotted in Fig. E10.2b and c.

With the modes known we compute the left side of Eq. (10.4.1):

$$\phi_1^T \mathbf{m}\phi_2 = \frac{mL}{4} \langle 1 \quad 0.3274 \rangle \begin{bmatrix} 1 & \\ & 2 \end{bmatrix} \begin{Bmatrix} 1 \\ -1.5274 \end{Bmatrix} = 0$$

$$\phi_1^T \mathbf{k}\phi_2 = \frac{48EI}{7L^3} \langle 1 \quad 0.3274 \rangle \begin{bmatrix} 2 & -5 \\ -5 & 16 \end{bmatrix} \begin{Bmatrix} 1 \\ -1.5274 \end{Bmatrix} = 0$$

This verifies that the natural modes computed for the system are orthogonal.

Example 10.3

Determine the natural frequencies and modes of vibration of the system shown in Fig. E10.3a and defined in Example 9.6. Normalize the modes to have unit vertical deflection at the free end.

Solution   The stiffness and mass matrices were determined in Example 9.6 with reference to DOFs u_1 and u_2:

$$\mathbf{m} = \begin{bmatrix} 3m & \\ & m \end{bmatrix} \qquad \mathbf{k} = \frac{6EI}{7L^3} \begin{bmatrix} 8 & -3 \\ -3 & 2 \end{bmatrix}$$

[†]Six significant digits are included so as to compare with the continuum model of a beam in Chapter 16.

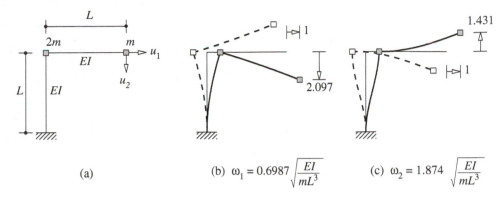

$$\text{(b)}\;\; \omega_1 = 0.6987\sqrt{\frac{EI}{mL^3}}\qquad\qquad \text{(c)}\;\; \omega_2 = 1.874\sqrt{\frac{EI}{mL^3}}$$

Figure E10.3

The frequency equation is Eq. (10.2.6), which, after substituting for **m** and **k**, evaluating the determinant, and defining

$$\lambda = \frac{7mL^3}{6EI}\omega^2 \tag{a}$$

can be written as

$$3\lambda^2 - 14\lambda + 7 = 0$$

The two roots are $\lambda_1 = 0.5695$ and $\lambda_2 = 4.0972$. The natural frequencies corresponding to the two values of λ are obtained from Eq. (a):

$$\omega_1 = 0.6987\sqrt{\frac{EI}{mL^3}}\qquad \omega_2 = 1.874\sqrt{\frac{EI}{mL^3}} \tag{b}$$

The natural modes are determined from Eq. (10.2.5) following the procedure used in Example 10.1 to obtain

$$\phi_1 = \begin{Bmatrix} 1 \\ 2.097 \end{Bmatrix}\qquad \phi_2 = \begin{Bmatrix} 1 \\ -1.431 \end{Bmatrix} \tag{c}$$

These modes are plotted in Fig. E10.3b and c.

In computing the natural modes the mode shape value for the first DOF had been arbitrarily set as unity. The resulting mode is normalized to unit value in DOF u_2 by dividing ϕ_1 in Eq. (c) by 2.097. Similarly, the second mode is normalized by dividing ϕ_2 in Eq. (c) by -1.431. Thus the normalized modes are

$$\phi_1 = \begin{Bmatrix} 0.4769 \\ 1 \end{Bmatrix}\qquad \phi_2 = \begin{Bmatrix} -0.6988 \\ 1 \end{Bmatrix} \tag{d}$$

Example 10.4

Determine the natural frequencies and modes of the system shown in Fig. E10.4a and defined in Example E9.1, a two-story frame idealized as a shear building. Normalize the modes so that $M_n = 1$.

Solution   The mass and stiffness matrices of the system, determined in Example 9.1, are

$$\mathbf{m} = \begin{bmatrix} 2m & \\ & m \end{bmatrix}\qquad \mathbf{k} = \begin{bmatrix} 3k & -k \\ -k & k \end{bmatrix} \tag{a}$$

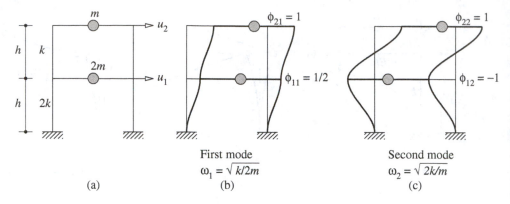

First mode
$\omega_1 = \sqrt{k/2m}$

Second mode
$\omega_2 = \sqrt{2k/m}$

(a)   (b)   (c)

Figure E10.4

where $k = 24EI_c/h^3$. The frequency equation is Eq. (10.2.6), which, after substituting for **m** and **k** and evaluating the determinant, can be written as

$$(2m^2)\omega^4 + (-5km)\omega^2 + 2k^2 = 0 \qquad\qquad (b)$$

The two roots are $\omega_1^2 = k/2m$ and $\omega_2^2 = 2k/m$, and the two natural frequencies are

$$\omega_1 = \sqrt{\frac{k}{2m}} \qquad \omega_2 = \sqrt{\frac{2k}{m}} \qquad\qquad (c)$$

Substituting for k gives

$$\omega_1 = 3.464\sqrt{\frac{EI_c}{mh^3}} \qquad \omega_2 = 6.928\sqrt{\frac{EI_c}{mh^3}} \qquad\qquad (d)$$

The natural modes are determined from Eq. (10.2.5) following the procedure used in Example 10.1 to obtain

$$\phi_1 = \left\{\begin{matrix} \frac{1}{2} \\ 1 \end{matrix}\right\} \qquad \phi_2 = \left\{\begin{matrix} -1 \\ 1 \end{matrix}\right\} \qquad\qquad (e)$$

These natural modes are shown in Fig. E10.4b and c.

To normalize the first mode, M_1 is calculated using Eq. (10.4.6), with ϕ_1 given by Eq. (e):

$$M_1 = \phi_1^T \mathbf{m}\phi_1 = m \left\langle \tfrac{1}{2} \ \ 1 \right\rangle \begin{bmatrix} 2 & \\ & 1 \end{bmatrix} \left\{\begin{matrix} \frac{1}{2} \\ 1 \end{matrix}\right\} = \frac{3}{2}m$$

To make $M_1 = 1$, divide ϕ_1 of Eq. (e) by $\sqrt{3m/2}$ to obtain the normalized mode,

$$\phi_1 = \frac{1}{\sqrt{6m}} \left\{\begin{matrix} 1 \\ 2 \end{matrix}\right\}$$

For this ϕ_1 it can be verified that $M_1 = 1$. The second mode can be normalized similarly.

Example 10.5

Determine the natural frequencies and modes of the system shown in Fig. E10.5a and defined earlier in Example 9.9. The story height $h = 10$ ft.

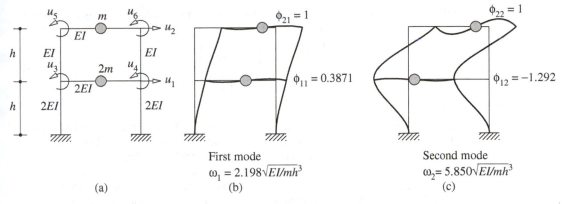

First mode
$\omega_1 = 2.198\sqrt{EI/mh^3}$

Second mode
$\omega_2 = 5.850\sqrt{EI/mh^3}$

(a)   (b)   (c)

Figure E10.5

Solution   With reference to the lateral displacements u_1 and u_2 of the two floors as the two DOFs, the mass matrix and the condensed stiffness matrix were determined in Example 9.9:

$$\mathbf{m}_{tt} = m\begin{bmatrix} 2 & \\ & 1 \end{bmatrix} \qquad \hat{\mathbf{k}}_{tt} = \frac{EI}{h^3}\begin{bmatrix} 54.88 & -17.51 \\ -17.51 & 11.61 \end{bmatrix} \qquad \text{(a)}$$

The frequency equation is

$$\det(\hat{\mathbf{k}}_{tt} - \omega^2 \mathbf{m}_{tt}) = 0 \qquad \text{(b)}$$

Substituting for $\mathbf{m}_{tt}$ and $\hat{\mathbf{k}}_{tt}$, evaluating the determinant, and obtaining the two roots just as in Example 10.4 leads to

$$\omega_1 = 2.198\sqrt{\frac{EI}{mh^3}} \qquad \omega_2 = 5.850\sqrt{\frac{EI}{mh^3}} \qquad \text{(c)}$$

It is of interest to compare these frequencies for a frame with flexible beams with those for the frame with flexurally rigid beams determined in Example 10.4. It is clear that beam flexibility has the effect of lowering the frequencies, consistent with intuition.

The natural modes are determined by solving

$$(\hat{\mathbf{k}}_{tt} - \omega_n^2 \mathbf{m}_{tt})\boldsymbol{\phi}_n = \mathbf{0} \qquad \text{(d)}$$

with ω_1 and ω_2 substituted successively from Eq. (c) to obtain

$$\boldsymbol{\phi}_1 = \begin{Bmatrix} 0.3871 \\ 1 \end{Bmatrix} \qquad \boldsymbol{\phi}_2 = \begin{Bmatrix} -1.292 \\ 1 \end{Bmatrix} \qquad \text{(e)}$$

These vectors define the lateral displacements of each floor. They are shown in Fig. E10.5b and c together with the joint rotations. The joint rotations associated with the first mode are determined by substituting $\mathbf{u}_t = \boldsymbol{\phi}_1$ from Eq. (e) in Eq. (d) of Example 9.9:

$$\begin{Bmatrix} u_3 \\ u_4 \\ u_5 \\ u_6 \end{Bmatrix} = \frac{1}{h}\begin{bmatrix} -0.4426 & -0.2459 \\ -0.4426 & -0.2459 \\ 0.9836 & -0.7869 \\ 0.9836 & -0.7869 \end{bmatrix}\begin{Bmatrix} 0.3871 \\ 1.0000 \end{Bmatrix} = \frac{1}{h}\begin{Bmatrix} -0.4172 \\ -0.4172 \\ -0.4061 \\ -0.4061 \end{Bmatrix} \qquad \text{(f)}$$

Similarly, the joint rotations associated with the second mode are obtained by substituting $\mathbf{u}_t = \boldsymbol{\phi}_2$ from Eq. (e) in Eq. (d) of Example 9.9:

$$\begin{Bmatrix} u_3 \\ u_4 \\ u_5 \\ u_6 \end{Bmatrix} = \frac{1}{h} \begin{Bmatrix} 0.3258 \\ 0.3258 \\ -2.0573 \\ -2.0573 \end{Bmatrix} \tag{g}$$

Example 10.6

Figure 9.5.1 shows the plan view of a one-story building. The structure consists of a roof, idealized as a rigid diaphragm, supported on three frames, A, B, and C, as shown. The roof weight is uniformly distributed and has a magnitude of 100 lb/ft.^{2} The lateral stiffnesses of the frames are $k_y = 75$ kips/ft for frame A, and $k_x = 40$ kips/ft for frames B and C. The plan dimensions are $b = 30$ ft and $d = 20$ ft, the eccentricity is $e = 1.5$ ft, and the height of the building is 12 ft. Determine the natural periods and modes of vibration of the structure.

Solution

Weight of roof slab: $w = 30 \times 20 \times 100$ lb $= 60$ kips

Mass: $m = w/g = 1.863$ kips-sec^2/ft

Moment of inertia: $I_O = \dfrac{m(b^2 + d^2)}{12} = 201.863$ kips-ft-sec^2

Lateral motion of the roof diaphragm in the x-direction is governed by Eq. (9.5.18):

$$m\ddot{u}_x + 2k_x u_x = 0 \tag{a}$$

Thus the natural frequency of x-lateral vibration is

$$\omega_x = \sqrt{\frac{2k_x}{m}} = \sqrt{\frac{2(40)}{1.863}} = 6.553 \text{ rad/sec}$$

The corresponding natural mode is shown in Fig. E10.6c.

The coupled lateral (u_y)-torsional (u_θ) motion of the roof diaphragm is governed by Eq. (9.5.19). Substituting for m and I_O gives

$$\mathbf{m} = \begin{bmatrix} 1.863 & \\ & 201.863 \end{bmatrix}$$

From Eqs. (9.5.16) and (9.5.19) the stiffness matrix has four elements:

$$k_{yy} = k_y = 75 \text{ kips/ft}$$

$$k_{y\theta} = k_{\theta y} = ek_y = 1.5 \times 75 = 112.5 \text{ kips}$$

$$k_{\theta\theta} = e^2 k_y + \frac{d^2}{2}k_x = 8168.75 \text{ kips-ft}$$

Hence,

$$\mathbf{k} = \begin{bmatrix} 75.00 & 112.50 \\ 112.50 & 8168.75 \end{bmatrix}$$

With $\mathbf{k}$ and $\mathbf{m}$ known, the eigenvalue problem for this two-DOF system is solved by standard procedures to obtain:

Natural frequencies (rad/sec): $\omega_1 = 5.878$; $\omega_2 = 6.794$

Natural modes: $\boldsymbol{\phi}_1 = \begin{Bmatrix} -0.5228 \\ 0.0493 \end{Bmatrix}$; $\boldsymbol{\phi}_2 = \begin{Bmatrix} -0.5131 \\ -0.0502 \end{Bmatrix}$

These mode shapes are plotted in Fig. E10.6a and b. The motion of the structure in each mode consists of translation of the rigid diaphragm coupled with torsion about the vertical axis through the center of mass.

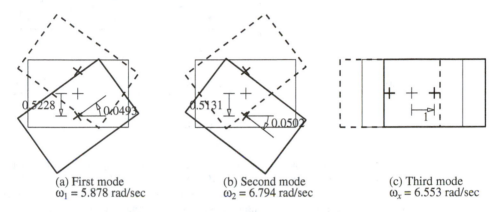

(a) First mode
$\omega_1 = 5.878$ rad/sec

(b) Second mode
$\omega_2 = 6.794$ rad/sec

(c) Third mode
$\omega_x = 6.553$ rad/sec

Figure E10.6

Example 10.7

Consider a special case of the system of Example 10.6 in which frame A is located at the center of mass (i.e., $e = 0$). Determine the natural frequencies and modes of this system.

Solution   Equation (9.5.20) specialized for free vibration of this system gives three equations of motion:

$$m\ddot{u}_x + 2k_x u_x = 0 \qquad m\ddot{u}_y + k_y u_y = 0 \qquad I_O \ddot{u}_\theta + \frac{d^2}{2} k_x u_\theta = 0 \qquad \text{(a)}$$

The first equation of motion indicates that translational motion in the x-direction would occur at the natural frequency

$$\omega_x = \sqrt{\frac{2k_x}{m}} = \sqrt{\frac{2(40)}{1.863}} = 6.553 \text{ rad/sec}$$

This motion is independent of lateral motion u_y or torsional motion u_θ (Fig. E10.7c). The second equation of motion indicates that translational motion in the y-direction would occur at the natural frequency

$$\omega_y = \sqrt{\frac{k_y}{m}} = \sqrt{\frac{75}{1.863}} = 6.344 \text{ rad/sec}$$

This motion is independent of the lateral motion u_x or torsional motion u_θ (Fig. E10.7b). The third equation of motion indicates that torsional motion would occur at the natural frequency

$$\omega_\theta = \sqrt{\frac{d^2 k_x}{2I_O}} = \sqrt{\frac{(20)^2 40}{2(201.863)}} = 6.295 \text{ rad/sec}$$

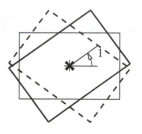
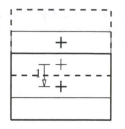
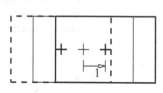

(a) First mode
$\omega_\theta = 6.295$ rad/sec

(b) Second mode
$\omega_y = 6.344$ rad/sec

(c) Third mode
$\omega_x = 6.553$ rad/sec

Figure E10.7

The roof diaphragm would rotate about the vertical axis through its center of mass without any translation of this point in the x or y directions (Fig. E10.7a).

Observe that the natural frequencies ω_1 and ω_2 of the unsymmetric-plan system (Example 10.6) are different from and more separated than the natural frequencies ω_y and ω_θ of the symmetric-plan system (Example 10.7).

10.7 MODAL EXPANSION OF DISPLACEMENTS

Any set of N independent vectors can be used as a basis for representing any other vector of order N. In the following sections the natural modes are used as such a basis. Thus, a modal expansion of any displacement vector $\mathbf{u}$ has the form

$$\mathbf{u} = \sum_{r=1}^{N} \phi_r q_r = \mathbf{\Phi q} \tag{10.7.1}$$

where q_r are scalar multipliers called *modal coordinates* or *normal coordinates* and $\mathbf{q} = \langle q_1 \quad q_2 \quad \cdots \quad q_n \rangle^T$. When the ϕ_r are known, for a given $\mathbf{u}$ it is possible to evaluate the q_r by multiplying both sides of Eq. (10.7.1) by $\phi_n^T \mathbf{m}$:

$$\phi_n^T \mathbf{mu} = \sum_{r=1}^{N} (\phi_n^T \mathbf{m}\phi_r) q_r$$

Because of the orthogonality relation of Eq. (10.4.1b), all terms in the summation above vanish except the $r = n$ term; thus

$$\phi_n^T \mathbf{mu} = (\phi_n^T \mathbf{m}\phi_n) q_n$$

The matrix products on both sides of this equation are scalars. Therefore,

$$q_n = \frac{\phi_n^T \mathbf{mu}}{\phi_n^T \mathbf{m}\phi_n} = \frac{\phi_n^T \mathbf{mu}}{M_n} \tag{10.7.2}$$

The modal expansion of the displacement vector **u**, Eq. (10.7.1), is employed in Section 10.8 to obtain solutions for the free vibration response of undamped systems. It also plays a central role in the analysis of forced vibration response and earthquake response of MDF systems (Chapters 12 and 13).

Example 10.8

For the two-story shear frame of Example 10.4, determine the modal expansion of the displacement vector $\mathbf{u} = \langle 1 \quad 1 \rangle^T$.

Solution   The displacement **u** is substituted in Eq. (10.7.2) together with $\phi_1 = \langle \frac{1}{2} \quad 1 \rangle^T$ and $\phi_2 = \langle -1 \quad 1 \rangle^T$, from Example 10.4, to obtain

$$q_1 = \frac{\langle \frac{1}{2} \ 1 \rangle \begin{bmatrix} 2m & \\ & m \end{bmatrix} \begin{Bmatrix} 1 \\ 1 \end{Bmatrix}}{\langle \frac{1}{2} \ 1 \rangle \begin{bmatrix} 2m & \\ & m \end{bmatrix} \begin{Bmatrix} \frac{1}{2} \\ 1 \end{Bmatrix}} = \frac{2m}{3m/2} = \frac{4}{3}$$

$$q_2 = \frac{\langle -1 \ 1 \rangle \begin{bmatrix} 2m & \\ & m \end{bmatrix} \begin{Bmatrix} 1 \\ 1 \end{Bmatrix}}{\langle -1 \ 1 \rangle \begin{bmatrix} 2m & \\ & m \end{bmatrix} \begin{Bmatrix} -1 \\ 1 \end{Bmatrix}} = \frac{-m}{3m} = -\frac{1}{3}$$

Substituting q_n in Eq. (10.7.1) gives the desired modal expansion, which is shown in Fig. E10.8.

$$\mathbf{u} \qquad = \qquad (4/3)\phi_1 \qquad + \qquad (-1/3)\phi_2 \qquad \textbf{Figure E10.8}$$

PART B: FREE VIBRATION RESPONSE

10.8 SOLUTION OF FREE VIBRATION EQUATIONS: UNDAMPED SYSTEMS

We now return to the problem posed by Eqs. (10.1.1) and (10.1.2) and find its solution. For the example structure of Fig. 10.1.1a, such a solution was shown in Fig. 10.1.1d. The differential equation (10.1.1) to be solved had led to the matrix eigenvalue problem of Eq. (10.2.4). Assuming that the eigenvalue problem has been solved for the natural frequencies and modes, the general solution of Eq. (10.1.1) is given by a superposition of the response in individual modes given by Eq. (10.2.3). Thus

$$\mathbf{u}(t) = \sum_{n=1}^{N} \boldsymbol{\phi}_n (A_n \cos \omega_n t + B_n \sin \omega_n t) \tag{10.8.1}$$

where A_n and B_n are $2N$ constants of integration. To determine these constants, we will also need the equation for the velocity vector, which is

$$\dot{\mathbf{u}}(t) = \sum_{n=1}^{N} \boldsymbol{\phi}_n \omega_n (-A_n \sin \omega_n t + B_n \cos \omega_n t) \tag{10.8.2}$$

Setting $t = 0$ in Eqs. (10.8.1) and (10.8.2) gives

$$\mathbf{u}(0) = \sum_{n=1}^{N} \boldsymbol{\phi}_n A_n \qquad \dot{\mathbf{u}}(0) = \sum_{n=1}^{N} \boldsymbol{\phi}_n \omega_n B_n \tag{10.8.3}$$

With the initial displacements $\mathbf{u}(0)$ and initial velocities $\dot{\mathbf{u}}(0)$ known, each of these two equation sets represents N algebraic equations in the unknowns A_n and B_n, respectively. Simultaneous solution of these equations is not necessary because they can be interpreted as a modal expansion of the vectors $\mathbf{u}(0)$ and $\dot{\mathbf{u}}(0)$. Following Eq. (10.7.1), we can write

$$\mathbf{u}(0) = \sum_{n=1}^{N} \boldsymbol{\phi}_n q_n(0) \qquad \dot{\mathbf{u}}(0) = \sum_{n=1}^{N} \boldsymbol{\phi}_n \dot{q}_n(0) \tag{10.8.4}$$

where, analogous to Eq. (10.7.2), $q_n(0)$ and $\dot{q}_n(0)$ are given by

$$q_n(0) = \frac{\boldsymbol{\phi}_n^T \mathbf{m} \mathbf{u}(0)}{M_n} \qquad \dot{q}_n(0) = \frac{\boldsymbol{\phi}_n^T \mathbf{m} \dot{\mathbf{u}}(0)}{M_n} \tag{10.8.5}$$

Equations (10.8.3) and (10.8.4) are equivalent, implying that $A_n = q_n(0)$ and $B_n = \dot{q}_n(0)/\omega_n$. Substituting these in Eq. (10.8.1) gives

$$\mathbf{u}(t) = \sum_{n=1}^{N} \boldsymbol{\phi}_n \left[q_n(0) \cos \omega_n t + \frac{\dot{q}_n(0)}{\omega_n} \sin \omega_n t \right] \tag{10.8.6}$$

or, alternatively,

$$\mathbf{u}(t) = \sum_{n=1}^{N} \boldsymbol{\phi}_n q_n(t) \tag{10.8.7}$$

where

$$q_n(t) = q_n(0) \cos \omega_n t + \frac{\dot{q}_n(0)}{\omega_n} \sin \omega_n t \tag{10.8.8}$$

is the time variation of modal coordinates, which is analogous to the free vibration response of SDF systems [Eq. (2.1.3)]. Equation (10.8.6) provides the displacement $\mathbf{u}$ as a function of time due to initial displacement $\mathbf{u}(0)$ and velocity $\dot{\mathbf{u}}(0)$; the $\mathbf{u}(t)$ is independent of how the modes are normalized, although $q_n(t)$ are not. Assuming that the natural frequencies ω_n and modes $\boldsymbol{\phi}_n$ are available, the right side of Eq. (10.8.6) is known with $q_n(0)$ and $\dot{q}_n(0)$ defined by Eq. (10.8.5).

Example 10.9

Determine the free vibration response of the two-story shear frame of Example 10.4 due to initial displacement $\mathbf{u}(0) = \langle \frac{1}{2} \quad 1 \rangle^T$.

Solution   The initial displacement and velocity vectors are

$$\mathbf{u}(0) = \left\{ \begin{matrix} \frac{1}{2} \\ 1 \end{matrix} \right\} \qquad \dot{\mathbf{u}}(0) = \left\{ \begin{matrix} 0 \\ 0 \end{matrix} \right\}$$

For the given $\mathbf{u}(0)$, $q_n(0)$ are calculated following the procedure of Example 10.8 and using ϕ_n from Eq. (e) of Example 10.4; the results are $q_1(0) = 1$ and $q_2(0) = 0$. Because the initial velocity $\dot{\mathbf{u}}(0)$ is zero, $\dot{q}_1(0) = \dot{q}_2(0) = 0$. Inserting $q_n(0)$ and $\dot{q}_n(0)$ in Eq. (10.8.8) gives the solution for modal coordinates

$$q_1(t) = 1 \cos \omega_1 t \qquad q_2(t) = 0$$

Substituting $q_n(t)$ and ϕ_n in Eq. (10.8.7) leads to

$$\left\{ \begin{matrix} u_1(t) \\ u_2(t) \end{matrix} \right\} = \left\{ \begin{matrix} \frac{1}{2} \\ 1 \end{matrix} \right\} \cos \omega_1 t$$

where $\omega_1 = \sqrt{k/2m}$ from Example 10.4. These solutions for $q_1(t)$, $u_1(t)$, and $u_2(t)$ had been plotted in Fig. 10.1.2c and d. Note that $q_2(t) = 0$ implies that the second mode has no contribution to the response, which is all due to the first mode. Such is the case because the initial displacement is proportional to the first mode and hence orthogonal to the second mode.

Example 10.10

Determine the free vibration response of the two-story shear frame of Example 10.4 due to initial displacement $\mathbf{u}(0) = \langle -1 \quad 1 \rangle^T$.

Solution   The calculations proceed as in Example 10.9, leading to $q_1(0) = 0$, $q_2(0) = 1$, and $\dot{q}_1(0) = \dot{q}_2(0) = 0$. Inserting these in Eq. (10.8.8) gives the solutions for modal coordinates:

$$q_1(t) = 0 \qquad q_2(t) = 1 \cos \omega_2 t$$

Substituting $q_n(t)$ and ϕ_n in Eq. (10.8.7) leads to

$$\left\{ \begin{matrix} u_1(t) \\ u_2(t) \end{matrix} \right\} = \left\{ \begin{matrix} -1 \\ 1 \end{matrix} \right\} \cos \omega_2 t$$

where $\omega_2 = \sqrt{2k/m}$ from Example 10.4. These solutions for $q_2(t)$, $u_1(t)$, and $u_2(t)$ had been plotted in Fig. 10.1.3c and d. Note that $q_1(t) = 0$ implies that the first mode has no contribution to the response and the response is due entirely to the second mode. Such is the case because the initial displacement is proportional to the second mode and hence orthogonal to the first mode.

Example 10.11

Determine the free vibration response of the two-story shear frame of Example 10.4 due to initial displacements $\mathbf{u}(0) = \langle -\frac{1}{2} \quad 2 \rangle^T$.

Solution   Following Example 10.8, $q_n(0)$ and $\dot{q}_n(0)$ are evaluated: $q_1(0) = 1$, $q_2(0) = 1$, and $\dot{q}_1(0) = \dot{q}_2(0) = 0$. Substituting these in Eq. (10.8.8) gives the solution for modal coordinates

$$q_1(t) = 1 \cos \omega_1 t \qquad q_2(t) = 1 \cos \omega_2 t$$

Substituting $q_n(t)$ and ϕ_n in Eq. (10.8.7) leads to

$$\begin{Bmatrix} u_1(t) \\ u_2(t) \end{Bmatrix} = \begin{Bmatrix} \frac{1}{2} \\ 1 \end{Bmatrix} \cos \omega_1 t + \begin{Bmatrix} -1 \\ 1 \end{Bmatrix} \cos \omega_2 t$$

These solutions for $q_n(t)$ and $u_j(t)$ had been plotted in Fig. 10.1.1c and d. Observe that both natural modes contribute to the response due to these initial displacements.

10.9 SYSTEMS WITH DAMPING

When damping is included, the free vibration response of the system is governed by Eq. (9.2.12) with $\mathbf{p}(t) = \mathbf{0}$:

$$\mathbf{m}\ddot{\mathbf{u}} + \mathbf{c}\dot{\mathbf{u}} + \mathbf{k}\mathbf{u} = \mathbf{0} \tag{10.9.1}$$

It is desired to find the solution $\mathbf{u}(t)$ of Eq. (10.9.1) that satisfies the initial conditions

$$\mathbf{u} = \mathbf{u}(0) \qquad \dot{\mathbf{u}} = \dot{\mathbf{u}}(0) \tag{10.9.2}$$

at $t = 0$. Procedures to obtain the desired solution differ depending on the form of damping: classical or nonclassical; these terms are defined next.

If the damping matrix of a linear system satisfies the identity

$$\mathbf{c}\mathbf{m}^{-1}\mathbf{k} = \mathbf{k}\mathbf{m}^{-1}\mathbf{c} \tag{10.9.3}$$

all the natural modes of vibration are real-valued and identical to those of the associated undamped system; they were determined by solving the real eigenvalue problem, Eq. (10.2.4). Such systems are said to possess *classical damping* because they have classical natural modes, defined in Section 10.2.

To state an important property of classically damped systems, we express the displacement $\mathbf{u}$ in terms of the natural modes of the associated undamped system; thus we substitute Eq. (10.7.1) in Eq. (10.9.1):

$$\mathbf{m}\boldsymbol{\Phi}\ddot{\mathbf{q}} + \mathbf{c}\boldsymbol{\Phi}\dot{\mathbf{q}} + \mathbf{k}\boldsymbol{\Phi}\mathbf{q} = \mathbf{0}$$

Premultiplying by $\boldsymbol{\Phi}^T$ gives

$$\mathbf{M}\ddot{\mathbf{q}} + \mathbf{C}\dot{\mathbf{q}} + \mathbf{K}\mathbf{q} = \mathbf{0} \tag{10.9.4}$$

where the diagonal matrices $\mathbf{M}$ and $\mathbf{K}$ were defined in Eq. (10.4.5) and

$$\mathbf{C} = \boldsymbol{\Phi}^T \mathbf{c} \boldsymbol{\Phi} \tag{10.9.5}$$

For classically damped systems, the square matrix $\mathbf{C}$ is diagonal. Then, Eq. (10.9.4) represents N uncoupled differential equations in modal coordinates q_n, and classical modal analysis is applicable to such systems. Such a procedure to solve Eq. (10.9.1) is presented in Section 10.10.

A linear system is said to possess *nonclassical damping* if its damping matrix does not satisfy Eq. (10.9.3). For such systems, the natural modes of vibration are not real-valued, and the square matrix $\mathbf{C}$ of Eq. (10.9.5) is no longer diagonal, thus they are not amenable to classical modal analysis. Analytical solutions of Eq. (10.9.1) for nonclassically damped systems are presented in Chapter 14, and numerical solution methods in Chapter 16.

10.10 SOLUTION OF FREE VIBRATION EQUATIONS: CLASSICALLY DAMPED SYSTEMS

For classically damped systems, each of the N uncoupled differential equations in Eq. (10.9.4) is of the form

$$M_n \ddot{q}_n + C_n \dot{q}_n + K_n q_n = 0 \tag{10.10.1}$$

where M_n and K_n were defined in Eq. (10.4.6) and

$$C_n = \phi_n^T \mathbf{c} \phi_n \tag{10.10.2}$$

Equation (10.10.1) is of the same form as Eq. (2.2.1a) for an SDF system with damping. Thus the damping ratio can be defined for each mode in a manner analogous to Eq. (2.2.2) for an SDF system:

$$\zeta_n = \frac{C_n}{2 M_n \omega_n} \tag{10.10.3}$$

Dividing Eq. (10.10.1) (t) by M_n gives

$$\ddot{q}_n + 2 \zeta_n \omega_n \dot{q}_n + \omega_n^2 q_n = 0 \tag{10.10.4}$$

This equation is of the same form as Eq. (2.2.1b) governing the free vibration of an SDF system with damping for which the solution is Eq. (2.2.4). Adapting this result, the solution for Eq. (10.10.4) is

$$q_n(t) = e^{-\zeta_n \omega_n t} \left[q_n(0) \cos \omega_{nD} t + \frac{\dot{q}_n(0) + \zeta_n \omega_n q_n(0)}{\omega_{nD}} \sin \omega_{nD} t \right] \tag{10.10.5}$$

where the nth natural frequency of the system with damping

$$\omega_{nD} = \omega_n \sqrt{1 - \zeta_n^2} \tag{10.10.6}$$

and ω_n is the nth natural frequency of the associated undamped system. The displacement response of the system is then obtained by substituting Eq. (10.10.5) for $q_n(t)$ in Eq. (10.8.7):

$$\mathbf{u}(t) = \sum_{n=1}^{N} \phi_n e^{-\zeta_n \omega_n t} \left[q_n(0) \cos \omega_{nD} t + \frac{\dot{q}_n(0) + \zeta_n \omega_n q_n(0)}{\omega_{nD}} \sin \omega_{nD} t \right] \tag{10.10.7}$$

This is the solution of the free vibration problem for classically damped MDF systems. It provides the displacement $\mathbf{u}$ as a function of time due to initial displacement $\mathbf{u}(0)$ and velocity $\dot{\mathbf{u}}(0)$. Assuming that the natural frequencies ω_n and modes ϕ_n of the system without damping are available together with the modal damping ratios ζ_n, the right side of Eq. (10.10.7) is known with $q_n(0)$ and $\dot{q}_n(0)$ defined by Eq. (10.8.5).

Damping influences the natural frequencies and periods of vibration of an MDF system according to Eq. (10.10.6), which is of the same form as Eq. (2.2.5) for an SDF system. Therefore, the effect of damping on the natural frequencies and periods of an MDF system is small for damping ratios ζ_n below 20% (Fig. 2.2.3), a range that includes most practical structures.

In a classically damped MDF system undergoing free vibration in its nth natural mode, the displacement amplitude at any DOF decreases with each vibration cycle. The rate of decay depends on the damping ratio ζ_n in that mode, in a manner similar to SDF systems. This similarity is apparent by comparing Eq. (10.10.5) with Eq. (2.2.4). Thus the ratio of two response peaks separated by j cycles of vibration is related to the damping ratio by Eq. (2.2.12) with appropriate change in notation.

Consequently, the damping ratio in a natural mode of an MDF system can be determined, in principle, from a free vibration test following the procedure presented in Section 2.2.4 for SDF systems. In such a test the structure would be deformed by pulling on it with a cable that is then suddenly released, thus causing the structure to undergo free vibration about its static equilibrium position. A difficulty in such tests is to apply the pull and release in such a way that the structure will vibrate in only one of its natural modes. For this reason this test procedure is not an effective means to determine damping except possibly for the fundamental mode. After the response contributions of the higher modes have damped out, the free vibration is essentially in the fundamental mode, and the damping ratio for this mode can be computed from the decay rate of vibration amplitudes.

Example 10.12

Determine the free vibration response of the two-story shear frame of Fig. E10.12.1a with $c = \sqrt{km/200}$ due to two sets of initial displacement (1) $\mathbf{u}(0) = \langle\, \tfrac{1}{2} \quad 1 \,\rangle^T$ and (2) $\mathbf{u}(0) = \langle -1 \quad 1 \rangle^T$.

Solution

Part 1 The $q_n(0)$ corresponding to this $\mathbf{u}(0)$ were determined in Example 10.9: $q_1(0) = 1$ and $q_2(0) = 0$; $\dot{q}_n(0) = 0$. The differential equations governing $q_n(t)$ are given by

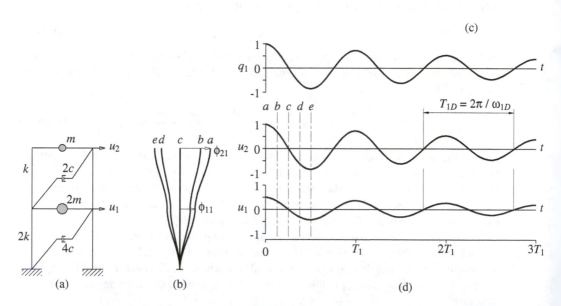

Figure E10.12.1 Free vibration of a classically damped system in the first natural mode of vibration: (a) two-story frame; (b) deflected shapes at time instants a, b, c, d, and e; (c) modal coordinate $q_1(t)$; (d) displacement history.

Eq. (10.10.4). Because $q_2(0)$ and $\dot{q}_2(0)$ are both zero, $q_2(t) = 0$ for all times. The response is given by the $n = 1$ term in Eq. (10.10.7). Substituting the aforementioned values for $q_1(0)$, $\dot{q}_1(0)$, and $\phi_1 = \langle \frac{1}{2} \quad 1 \rangle^T$ gives

$$\left\{ \begin{matrix} u_1(t) \\ u_2(t) \end{matrix} \right\} = \left\{ \begin{matrix} \frac{1}{2} \\ 1 \end{matrix} \right\} e^{-\zeta_1 \omega_1 t} \left(\cos \omega_{1D} t + \frac{\zeta_1}{\sqrt{1 - \zeta_1^2}} \sin \omega_{1D} t \right)$$

where $\omega_1 = \sqrt{k/2m}$ from Example 10.4 and $\zeta_1 = 0.05$ from Eq. (10.10.3).

Part 2 The $q_n(0)$ corresponding to this $\mathbf{u}(0)$ were determined in Example 10.10: $q_1(0) = 0$ and $q_2(0) = 1$; $\dot{q}_n(0) = 0$. The differential equations governing $q_n(t)$ are given by Eq. (10.10.4). Because $q_1(0)$ and $\dot{q}_1(0)$ are both zero, $q_1(t) = 0$ at all times. The response is given by the $n = 2$ term in Eq. (10.10.7). Substituting for $q_2(0)$, $\dot{q}_2(0)$, and $\phi_2 = \langle -1 \quad 1 \rangle^T$ gives

$$\left\{ \begin{matrix} u_1(t) \\ u_2(t) \end{matrix} \right\} = \left\{ \begin{matrix} -1 \\ 1 \end{matrix} \right\} e^{-\zeta_2 \omega_2 t} \left(\cos \omega_{2D} t + \frac{\zeta_2}{\sqrt{1 - \zeta_2^2}} \sin \omega_{2D} t \right)$$

where $\omega_2 = \sqrt{2k/m}$ from Example 10.4 and $\zeta_2 = 0.10$ from Eq. (10.10.3).

Observations The results for free vibration due to initial displacements $\mathbf{u}(0) = \phi_1$ are presented in Fig. E10.12.1, and for $\mathbf{u}(0) = \phi_2$ in Fig. E10.12.2, respectively. The solutions for $q_n(t)$ are presented in part (c) of these figures; the floor displacements $u_j(t)$ in part (d); and the deflected shapes at selected time instants are plotted in part (b) of these figures.

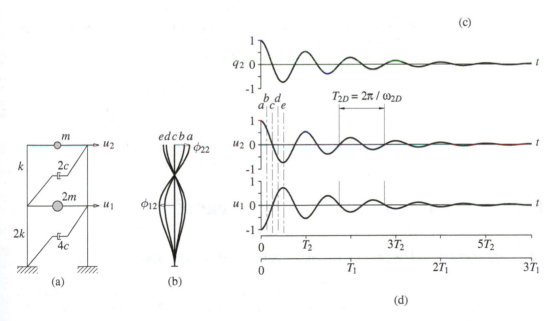

Figure E10.12.2   Free vibration of a classically damped system in the second natural mode of vibration: (a) two-story frame; (b) deflected shapes at time instants a, b, c, d, and e; (c) modal coordinate $q_2(t)$; (d) displacement history.

These results permit the following observations: First, if the initial displacement is proportional to the nth mode, the response is due entirely to that mode; the other mode has no contribution. Second, the initial deflected shape is maintained during free vibration, just as in the case of undamped systems (Figs. 10.1.2 and 10.1.3). Third, the system oscillates at the frequency ω_{nD} with all floors (or DOFs) vibrating in the same phase, passing through their equilibrium, maximum, or minimum positions at the same instant of time. Thus, the system possesses classical natural modes of vibration, defined first in Section 10.2, as expected of classically damped systems. Although based on results for a system with two DOFs, these observations are valid for classically damped systems with any number of DOFs.

Example 10.13

Determine the free vibration response of the two-story shear frame of Example 10.12 due to initial displacements $\mathbf{u}(0) = \left\langle -\frac{1}{2} \ 2 \right\rangle^T$.

Solution The $q_n(0)$ corresponding to this $\mathbf{u}(0)$ were determined in Example 10.11: $q_n(0) = 1$ and $q_n(0) = 1$; $\dot{q}_n(0) = 0$. Substituting them in Eq. (10.10.5) gives the solution for modal coordinates:

$$q_1(t) = e^{-\zeta_1 \omega_1 t} \left[\cos \omega_{1D} t + \frac{\zeta_1}{\sqrt{1 - \zeta_1^2}} \sin \omega_{1D} t \right] \tag{a}$$

$$q_2(t) = e^{-\zeta_2 \omega_2 t} \left[\cos \omega_{2D} t + \frac{\zeta_2}{\sqrt{1 - \zeta_2^2}} \sin \omega_{2D} t \right] \tag{b}$$

where, as determined earlier, $\omega_1 = \sqrt{k/2m}$ and $\omega_2 = \sqrt{2k/m}$; ω_{nD} are given by Eq. (10.10.6), and $\zeta_1 = 0.05$ and $\zeta_2 = 0.10$ from Eq. (10.10.3).

Substituting $q_n(t)$ and ϕ_n in Eq. (10.8.7) leads to

$$\begin{Bmatrix} u_1(t) \\ u_2(t) \end{Bmatrix} = \begin{Bmatrix} 1/2 \\ 1 \end{Bmatrix} e^{-\zeta_1 \omega_1 t} \left[\cos \omega_{1D} t + \frac{\zeta_1}{\sqrt{1 - \zeta_1^2}} \sin \omega_{1D} t \right]$$
$$+ \begin{Bmatrix} -1 \\ 1 \end{Bmatrix} e^{-\zeta_2 \omega_2 t} \left[\cos \omega_{2D} t + \frac{\zeta_2}{\sqrt{1 - \zeta_2^2}} \sin \omega_{2D} t \right] \tag{c}$$

PART C: COMPUTATION OF VIBRATION PROPERTIES

10.11 SOLUTION METHODS FOR THE EIGENVALUE PROBLEM

Finding the vibration properties—natural frequencies and modes—of a structure requires solution of the matrix eigenvalue problem of Eq. (10.2.4), which is repeated for convenience:

$$\mathbf{k}\phi = \lambda \mathbf{m}\phi \tag{10.11.1}$$

As mentioned earlier, the eigenvalues $\lambda_n \equiv \omega_n^2$ are the roots of the characteristic equation (10.2.6):

$$p(\lambda) = \det[\mathbf{k} - \lambda\mathbf{m}] = 0 \qquad (10.11.2)$$

where $p(\lambda)$ is a polynomial of order N, the number of DOFs of the system. This is not a practical method, especially for large systems (i.e., a large number of DOFs), because evaluation of the N coefficients of the polynomial requires much computational effort and the roots of $p(\lambda)$ are sensitive to numerical round-off errors in the coefficients.

Finding reliable and efficient methods to solve the eigenvalue problem has been the subject of much research, especially since development of the digital computer. Most of the methods available can be classified into three broad categories depending on which basic property is used as the basis of the solution algorithm: (1) Vector iteration methods work directly with the property of Eq. (10.11.1). (2) Transformation methods use the or-thogonality property of modes, Eqs. (10.4.1). (3) Polynomial iteration techniques work on the fact that $p(\lambda_n) = 0$. A number of solution algorithms have been developed within each of the foregoing three categories. Combination of two or more methods that belong to the same or to different categories have been developed to deal with large systems. Two ex-amples of such combined procedures are the determinant search method and the subspace iteration method.

All solution methods for eigenvalue problems must be iterative in nature because, basically, solving the eigenvalue problem is equivalent to finding the roots of the poly-nomial $p(\lambda)$. No explicit formulas are available for these roots when N is larger than 4, thus requiring an iterative solution. To find an eigenpair (λ_n, ϕ_n), only one of them is calculated by iteration; the other can be obtained without further iteration. For example, if λ_n is obtained by iteration, then ϕ_n can be evaluated by solving the algebraic equations $(\mathbf{k} - \lambda_n\mathbf{m})\phi_n = \mathbf{0}$. On the other hand, if ϕ_n is determined by iteration, λ_n can be ob-tained by evaluating Rayleigh's quotient (Section 10.12). Is it most economical to solve first for λ_n and then calculate ϕ_n (or vice versa), or to solve for both simultaneously? The answer to this question and hence the choice among the three procedure categories mentioned above depends on the properties of the mass and stiffness matrices—size N, bandwidth of $\mathbf{k}$, and whether $\mathbf{m}$ is diagonal or banded—and on the number of eigenpairs required.

In structural engineering we are usually analyzing systems with narrowly banded $\mathbf{k}$ and diagonal or narrowly banded $\mathbf{m}$ subjected to excitations that excite primarily the lower few (relative to N) natural modes of vibration. Inverse vector iteration methods are usually effective (i.e., reliable in obtaining accurate solutions and computationally efficient) for such situations, and this presentation is restricted to such methods. Only the basic ideas of vector iteration are included, without getting into subspace iteration or the Lanczos method. Similarly, transformation methods and polynomial iteration techniques are ex-cluded. In short, this is a limited treatment of solution methods for the eigenvalue problem arising in structural dynamics. This is sufficient for our purposes, but more comprehensive treatments are available in other books.

10.12 RAYLEIGH'S QUOTIENT

In this section Rayleigh's quotient is presented because it is needed in vector iteration methods; its properties are also presented. If Eq. (10.11.1) is premultiplied by ϕ^T, the following scalar equation is obtained:

$$\phi^T \mathbf{k} \phi = \lambda \phi^T \mathbf{m} \phi$$

The positive definiteness of $\mathbf{m}$ guarantees that $\phi^T \mathbf{m} \phi$ is nonzero, so that it is permissible to solve for the scalar λ:

$$\lambda = \frac{\phi^T \mathbf{k} \phi}{\phi^T \mathbf{m} \phi} \tag{10.12.1}$$

which obviously depends on the vector ϕ. This quotient is called *Rayleigh's quotient*. It may also be derived by equating the maximum value of kinetic energy to the maximum value of potential energy under the assumption that the vibrating system is executing simple harmonic motion at frequency ω with the deflected shape given by ϕ (Section 8.5.3).

Rayleigh's quotient has the following properties, presented without proof:

1. When ϕ is an eigenvector ϕ_n of Eq. (10.11.1), Rayleigh's quotient is equal to the corresponding eigenvalue λ_n.

2. If ϕ is an approximation to ϕ_n with an error that is a first-order infinitesimal, Rayleigh's quotient is an approximation to λ_n with an error which is a second-order infinitesimal λ, i.e., Rayleigh's quotient is *stationary* in the neighborhoods of the true eigenvectors. The stationary value is actually a minimum in the neighborhood of the first eigenvector and a maximum in the vicinity of the Nth eigenvector.

3. Rayleigh's quotient is bounded between $\lambda_1 \equiv \omega_1^2$ and $\lambda_N \equiv \omega_N^2$, the smallest and largest eigenvalues, i.e., it provides an upper bound for ω_1^2 and lower bound for ω_N^2.

A common engineering application of Rayleigh's quotient involves simply evaluating Eq. (10.12.1) for a trial vector ϕ that is selected on the basis of physical insight (Chapter 8). If the elements of an approximate eigenvector whose largest element is unity are correct to s decimal places, Rayleigh's quotient can be expected to be correct to about $2s$ decimal places. Several numerical procedures for solving eigenvalue problems make use of the stationary property of Rayleigh's quotient.

10.13 INVERSE VECTOR ITERATION METHOD

10.13.1 Basic Concept and Procedure

We restrict this presentation to systems with a stiffness matrix $\mathbf{k}$ that is positive definite, whereas the mass matrix $\mathbf{m}$ may be a banded mass matrix, or it may be a diagonal matrix with or without zero diagonal elements. The fact that vector iteration methods can handle zero diagonal elements in the mass matrix implies that these methods can be applied without requiring static condensation of the stiffness matrix (Section 9.3).

Our goal is to satisfy Eq. (10.11.1) by operating on it directly. We assume a trial vector for ϕ, say $\mathbf{x}_1$, and evaluate the right-hand side of Eq. (10.11.1). This we can do except for the eigenvalue λ, which is unknown. Thus we drop λ, which is equivalent to saying that we set $\lambda = 1$. Because eigenvectors can be determined only within a scale factor, the choice of λ will not affect the final result. With $\lambda = 1$ the right-hand side of Eq. (10.11.1) can be computed:

$$\mathbf{R}_1 = \mathbf{m}\mathbf{x}_1 \qquad (10.13.1)$$

Since $\mathbf{x}_1$ was an arbitrary choice, in general $\mathbf{k}\mathbf{x}_1 \neq \mathbf{R}_1$. (If by coincidence we find that $\mathbf{k}\mathbf{x}_1 = \mathbf{R}_1$, the $\mathbf{x}_1$ chosen is an eigenvector.) We now set up an equilibrium equation

$$\mathbf{k}\mathbf{x}_2 = \mathbf{R}_1 \qquad (10.13.2)$$

where $\mathbf{x}_2$ is the displacement vector corresponding to forces $\mathbf{R}_1$ and $\mathbf{x}_2 \neq \mathbf{x}_1$. Since we are using iteration to solve for an eigenvector, intuition may suggest that the solution of $\mathbf{x}_2$ of Eq. (10.13.2), obtained after one cycle of iteration, may be a better approximation to ϕ than was $\mathbf{x}_1$. This is indeed the case, as we shall demonstrate later, and by repeating the iteration cycle, we obtain an increasingly better approximation to the eigenvector. A corresponding value for the eigenvalue can be computed using Rayleigh's quotient, and the iteration can be terminated when two successive estimates of the eigenvalue are close enough. As the number of iterations increases, $\mathbf{x}_{i+1}$ approaches ϕ_1 and the eigenvalue approaches λ_1.

Thus the procedure starts with the assumption of a starting iteration vector $\mathbf{x}_1$ and consists of the following steps to be repeated for $j = 1, 2, 3, \ldots$ until convergence:

1. Determine $\bar{\mathbf{x}}_{j+1}$ by solving the algebraic equations:

$$\mathbf{k}\bar{\mathbf{x}}_{j+1} = \mathbf{m}\mathbf{x}_j \qquad (10.13.3)$$

2. Obtain an estimate of the eigenvalue by evaluating Rayleigh's quotient:

$$\lambda^{(j+1)} = \frac{\bar{\mathbf{x}}_{j+1}^T \mathbf{k}\bar{\mathbf{x}}_{j+1}}{\bar{\mathbf{x}}_{j+1}^T \mathbf{m}\bar{\mathbf{x}}_{j+1}} = \frac{\bar{\mathbf{x}}_{j+1}^T \mathbf{m}\mathbf{x}_j}{\bar{\mathbf{x}}_{j+1}^T \mathbf{m}\bar{\mathbf{x}}_{j+1}} \qquad (10.13.4)$$

3. Check convergence by comparing two successive values of λ:

$$\frac{|\lambda^{(j+1)} - \lambda^{(j)}|}{\lambda^{(j+1)}} \leq \text{tolerance} \qquad (10.13.5)$$

4. If the convergence criterion is not satisfied, normalize $\bar{\mathbf{x}}_{j+1}$:

$$\mathbf{x}_{j+1} = \frac{\bar{\mathbf{x}}_{j+1}}{(\bar{\mathbf{x}}_{j+1}^T \mathbf{m}\bar{\mathbf{x}}_{j+1})^{1/2}} \qquad (10.13.6)$$

   and go back to the first step and carry out another iteration using the next j.

5. Let l be the last iteration [i.e., the iteration that satisfies Eq. (10.13.5)]. Then

$$\lambda_1 \doteq \lambda^{(l+1)} \qquad \phi_1 \doteq \frac{\bar{\mathbf{x}}_{l+1}}{(\bar{\mathbf{x}}_{l+1}^T \mathbf{m}\bar{\mathbf{x}}_{l+1})^{1/2}} \qquad (10.13.7)$$

The basic step in the iteration is the solution of Eq. (10.13.3)—a set of N algebraic equations—which gives a better approximation to ϕ_1. The calculation in Eq. (10.13.4) gives an approximation to the eigenvalue λ_1 according to Rayleigh's quotient. It is this approximation to λ_1 that we use to determine convergence in the iteration. Equation (10.13.6) simply assures that the new vector satisfies the mass orthonormality relation

$$\mathbf{x}_{j+1}^T \mathbf{m} \mathbf{x}_{j+1} = 1 \qquad (10.13.8)$$

Although normalizing of the new vector does not affect convergence, it is numerically useful. If such normalizing is not included, the elements of the iteration vectors grow (or decrease) in each step, and this may cause numerical problems. Normalizing keeps the element values similar from one iteration to the next. The tolerance is selected depending on the accuracy desired. It should be 10^{-2s} or smaller when λ_1 is required to $2s$-digit accuracy. Then the eigenvector will be accurate to about s or more digits.

The inverse vector iteration algorithm can be organized a little differently for convenience in computer implementation, but such computational issues are not included in this presentation.

Example 10.14

The floor masses and story stiffnesses of the three-story frame, idealized as a shear frame, are shown in Fig. E10.14, where $m = 100$ kips/g and $k = 168$ kips/in. Determine the fundamental frequency ω_1 and mode shape ϕ_1 by inverse vector iteration.

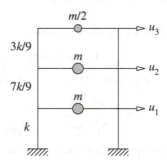

Figure E10.14

Solution   The mass and stiffness matrices for the system are

$$\mathbf{m} = m \begin{bmatrix} 1 & & \\ & 1 & \\ & & \frac{1}{2} \end{bmatrix} \qquad \mathbf{k} = \frac{k}{9} \begin{bmatrix} 16 & -7 & 0 \\ -7 & 10 & -3 \\ 0 & -3 & 3 \end{bmatrix}$$

where $m = 0.259$ kip-sec^2/in. and $k = 168$ kips/in.

The inverse iteration algorithm of Eqs. (10.13.3) to (10.13.7) is implemented starting with an initial vector $\mathbf{x}_1 = \langle 1 \quad 1 \quad 1 \rangle^T$ leading to Table E10.14. The final result is $\omega_1 \doteq \sqrt{144.14} = 12.006$ and $\phi_1 \doteq \langle 0.6377 \quad 1.2752 \quad 1.9122 \rangle^T$.

10.13.2 Convergence of Iteration

In the preceding section we have merely presented the inverse iteration scheme and stated that it converges to the first eigenvector associated with the smallest eigenvalue. We now

TABLE E10.14 INVERSE VECTOR ITERATION FOR THE FIRST EIGENPAIR

Iteration	$\mathbf{x}_j$	$\bar{\mathbf{x}}_{j+1}$	$\lambda^{(j+1)}$	$\mathbf{x}_{j+1}$
1	$\begin{bmatrix} 1 \\ 1 \\ 1 \end{bmatrix}$	$\begin{bmatrix} 0.0039 \\ 0.0068 \\ 0.0091 \end{bmatrix}$	147.73	$\begin{bmatrix} 0.7454 \\ 1.3203 \\ 1.7676 \end{bmatrix}$
2	$\begin{bmatrix} 0.7454 \\ 1.3203 \\ 1.7676 \end{bmatrix}$	$\begin{bmatrix} 0.0045 \\ 0.0089 \\ 0.0130 \end{bmatrix}$	144.29	$\begin{bmatrix} 0.6574 \\ 1.2890 \\ 1.8800 \end{bmatrix}$
3	$\begin{bmatrix} 0.6574 \\ 1.2890 \\ 1.8800 \end{bmatrix}$	$\begin{bmatrix} 0.0044 \\ 0.0089 \\ 0.0132 \end{bmatrix}$	144.15	$\begin{bmatrix} 0.6415 \\ 1.2785 \\ 1.9052 \end{bmatrix}$
4	$\begin{bmatrix} 0.6415 \\ 1.2785 \\ 1.9052 \end{bmatrix}$	$\begin{bmatrix} 0.0044 \\ 0.0089 \\ 0.0133 \end{bmatrix}$	144.14	$\begin{bmatrix} 0.6384 \\ 1.2758 \\ 1.9109 \end{bmatrix}$
5	$\begin{bmatrix} 0.6384 \\ 1.2758 \\ 1.9109 \end{bmatrix}$	$\begin{bmatrix} 0.0044 \\ 0.0088 \\ 0.0133 \end{bmatrix}$	144.14	$\begin{bmatrix} 0.6377 \\ 1.2752 \\ 1.9122 \end{bmatrix}$

demonstrate this convergence because the proof is instructive, especially in suggesting how to modify the procedure to achieve convergence to a higher eigenvector.

The modal expansion of vector $\mathbf{x}$ is [from Eqs. (10.7.1) and (10.7.2)]

$$\mathbf{x} = \sum_{n=1}^{N} \phi_n q_n = \sum_{n=1}^{N} \phi_n \frac{\phi_n^T \mathbf{mx}}{\phi_n^T \mathbf{m}\phi_n} \tag{10.13.9}$$

The nth term in this summation represents the nth modal component in $\mathbf{x}$.

The first iteration cycle involves solving the equilibrium equations (10.13.3) with $j = 1$: $\mathbf{k}\bar{\mathbf{x}}_2 = \mathbf{mx}_1$, where $\mathbf{x}_1$ is a trial vector. This solution can be expressed as $\bar{\mathbf{x}}_2 = \mathbf{k}^{-1}\mathbf{mx}_1$. Substituting the modal expansion of Eq. (10.13.9) for $\mathbf{x}_1$ gives

$$\bar{\mathbf{x}}_2 = \sum_{n=1}^{N} \mathbf{k}^{-1}\mathbf{m}\phi_n q_n \tag{10.13.10}$$

By rewriting Eq. (10.11.1) for the nth eigenpair as $\mathbf{k}^{-1}\mathbf{m}\phi_n = (1/\lambda_n)\phi_n$ and substituting it in Eq. (10.13.10), we get

$$\bar{\mathbf{x}}_2 = \sum_{n=1}^{N} \frac{1}{\lambda_n}\phi_n q_n = \frac{1}{\lambda_1} \sum_{n=1}^{N} \frac{\lambda_1}{\lambda_n}\phi_n q_n \tag{10.13.11}$$

The second iteration cycle involves solving Eq. (10.13.3) with $j = 2$: $\bar{\mathbf{x}}_3 = \mathbf{k}^{-1}\mathbf{m}\bar{\mathbf{x}}_2$, wherein we have used the unnormalized vector $\bar{\mathbf{x}}_2$ instead of the normalized vector $\mathbf{x}_2$. This is acceptable for the present purpose because convergence is unaffected by normalization and eigenvectors are arbitrary within a multiplicative factor. Following the derivation of

Eqs. (10.13.10) and (10.13.11), it can be shown that

$$\bar{\mathbf{x}}_3 = \frac{1}{\lambda_1^2} \sum_{n=1}^{N} \left(\frac{\lambda_1}{\lambda_n}\right)^2 \phi_n q_n \tag{10.13.12}$$

Similarly, the vector $\bar{\mathbf{x}}_{j+1}$ after j iteration cycles can be expressed as

$$\bar{\mathbf{x}}_{j+1} = \frac{1}{\lambda_1^j} \sum_{n=1}^{N} \left(\frac{\lambda_1}{\lambda_n}\right)^j \phi_n q_n \tag{10.13.13}$$

Since $\lambda_1 < \lambda_n$ for $n > 1$, $(\lambda_1/\lambda_n)^j \to 0$ as $j \to \infty$, and only the $n = 1$ term in Eq. (10.13.13) remains significant, indicating that

$$\bar{\mathbf{x}}_{j+1} \to \frac{1}{\lambda_1^j} \phi_1 q_1 \quad \text{as} \quad j \to \infty \tag{10.13.14}$$

Thus $\bar{\mathbf{x}}_{j+1}$ converges to a vector proportional to ϕ_1. Furthermore, the normalized vector $\mathbf{x}_{j+1}$ of Eq. (10.13.6) converges to ϕ_1, which is mass orthonormal.

The rate of convergence depends on λ_1/λ_2, the ratio that appears in the second term in the summation of Eq. (10.13.13). The smaller this ratio is, the faster is the convergence; this implies that convergence is very slow when λ_2 is nearly equal to λ_1. For such situations the convergence rate can be improved by the procedures of Section 10.14.

If only the first natural mode ϕ_1 and the associated natural frequency ω_1 are required, there is no need to proceed further. This is an advantage of the iteration method. It is unnecessary to solve the complete eigenvalue problem to obtain one or two of the modes.

10.13.3 Evaluation of Higher Modes

To continue the solution after ϕ_1 and λ_1 have been determined, the starting vector is modified to make the iteration procedure converge to the second eigenvector. The necessary modification is suggested by the proof presented in Section 10.13.2 to show that the iteration process converges to the first eigenvector. Observe that after each iteration cycle the other modal components are reduced relative to the first modal component because its eigenvalue λ_1 is smaller than all other eigenvalues λ_n. The iteration process converges to the first mode for the same reason because $(\lambda_1/\lambda_n)^j \to 0$ as $j \to \infty$. In general, the iteration procedure will converge to the mode with the lowest eigenvalue contained in a trial vector $\mathbf{x}$.

To make the iteration procedure converge to the second mode, a trial vector $\mathbf{x}$ should therefore be chosen so that it does not contain any first-mode component [i.e., q_1 should be zero in Eq. (10.13.9)] and $\mathbf{x}$ is said to be orthogonal to ϕ_1. It is not possible to start *a priori* with such an $\mathbf{x}$, however. We therefore start with an arbitrary $\mathbf{x}$ and make it orthogonal to ϕ_1 by the Gram–Schmidt orthogonalization process. This process can also be used to orthogonalize a trial vector with respect to the first n eigenvectors that have already been determined so that iteration on the purified trial vectors will converge to the $(n+1)$th mode, the mode with the next eigenvalue in ascending sequence.

In principle, the Gram–Schmidt orthogonalization process, combined with the inverse iteration procedure, provides a tool for computing the second and higher eigenvalues and eigenvectors. This tool is not effective, however, as a general computer method for two reasons. First, if $\mathbf{x}_1$ were made orthogonal to ϕ_1 [i.e., $q_1 = 0$ in Eq. (10.13.9)], theoretically the iteration will not converge to ϕ_1 but to ϕ_2 (or some other eigenvector—the one with the next-higher eigenvalue which is contained in the modal expansion of $\mathbf{x}_1$). However, in practice this never occurs since the inevitable round-off errors in finite-precision arithmetic continuously reintroduce small components of ϕ_1 which the iteration process magnifies. Second, convergence of the iteration process becomes progressively slower for the higher modes. It is for these reasons that this method is not developed in this book.

10.14 VECTOR ITERATION WITH SHIFTS: PREFERRED PROCEDURE

The inverse vector iteration procedure of Section 10.13, combined with the concept of "shifting" the eigenvalue spectrum (or scale), provides an effective means to improve the convergence rate of the iteration process and to make it converge to an eigenpair other than (λ_1, ϕ_1). Thus this is the preferred method, as it provides a practical tool for computing as many pairs of natural vibration frequencies and modes of a structure as desired.

10.14.1 Basic Concept and Procedure

The solutions of Eq. (10.11.1) are the eigenvalues λ_n and eigenvectors ϕ_n; the number of such pairs equals N, the order of $\mathbf{m}$ and $\mathbf{k}$. Figure 10.14.1a shows the eigenvalue spectrum (i.e., a plot of $\lambda_1, \lambda_2, \ldots$ along the eigenvalue axis). Introducing a shift μ in the origin of the eigenvalue axis (Fig. 10.14.1b) and defining $\check{\lambda}$ as the shifted eigenvalue measured from the shifted origin gives $\lambda = \check{\lambda} + \mu$. Substituting this in Eq. (10.11.1) leads to

$$\check{\mathbf{k}}\phi = \check{\lambda}\mathbf{m}\phi \qquad (10.14.1)$$

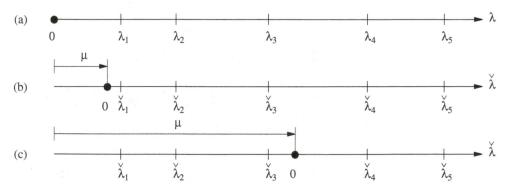

Figure 10.14.1   (a) Eigenvalue spectrum; (b) eigenvalue measured from a shifted origin; (c) location of shift point for convergence to λ_3.

where

$$\check{\mathbf{k}} = \mathbf{k} - \mu\mathbf{m} \qquad \check{\lambda} = \lambda - \mu \qquad (10.14.2)$$

The eigenvectors of the two eigenvalue problems—original Eq. (10.11.1) and shifted Eq. (10.14.1)—are the same. This is obvious because if a ϕ satisfies one equation, it will also satisfy the other. However, the eigenvalues $\check{\lambda}$ of the shifted problem differ from the eigenvalues λ of the original problem by the shift μ [Eq. (10.14.2)]. The spectrum of the shifted eigenvalues $\check{\lambda}$ is also shown in Fig. 10.14.1b with the origin at μ. If the inverse vector iteration method of Section 10.13.1 were applied to the eigenvalue problem of Eq. (10.14.1), it obviously will converge to the eigenvector having the smallest magnitude of the shifted eigenvalue $|\check{\lambda}_n|$ (i.e., the eigenvector with original eigenvalue λ_n closest to the shift value μ).

If μ were chosen as in Fig. 10.14.1b, the iteration will converge to the first eigenvector. The rate of convergence depends on the ratio $\check{\lambda}_1/\check{\lambda}_2 = (\lambda_1 - \mu)/(\lambda_2 - \mu)$. The convergence rate has improved because this ratio is smaller than the ratio λ_1/λ_2 for the original eigenvalue problem. If μ were chosen between λ_n and λ_{n+1}, and μ is closer to λ_n than λ_{n+1}, the iteration will converge to λ_n. On the other hand, if μ is closer to λ_{n+1} than λ_n, the iteration will converge to λ_{n+1}. Thus the "shifting" concept enables computation of any pair (λ_n, ϕ_n). In particular, if μ were chosen as in Fig. 10.14.1c, the iteration will converge to the third eigenvector.

Example 10.15

Determine the natural frequencies and modes of vibration of the system of Example 10.14 by inverse vector iteration with shifting.

Solution   Equation (10.14.1) with a selected shift μ is solved by inverse vector iteration. Selecting the shift $\mu_1 = 100$, $\check{\mathbf{k}}$ is calculated from Eq. (10.14.2) and the inverse vector iteration

TABLE E10.15a VECTOR ITERATION WITH SHIFT: FIRST EIGENPAIR

Iteration	$\mathbf{x}_j$	μ	$\bar{\mathbf{x}}_{j+1}$	$\lambda^{(j+1)}$	$\mathbf{x}_{j+1}$
1	$\begin{bmatrix}1\\1\\1\end{bmatrix}$	100	$\begin{bmatrix}0.0114\\0.0218\\0.0313\end{bmatrix}$	144.60	$\begin{bmatrix}0.6759\\1.2933\\1.8610\end{bmatrix}$
2	$\begin{bmatrix}0.6759\\1.2933\\1.8610\end{bmatrix}$	100	$\begin{bmatrix}0.0145\\0.0289\\0.0432\end{bmatrix}$	144.15	$\begin{bmatrix}0.6401\\1.2769\\1.9083\end{bmatrix}$
3	$\begin{bmatrix}0.6401\\1.2769\\1.9083\end{bmatrix}$	100	$\begin{bmatrix}0.0144\\0.0289\\0.0433\end{bmatrix}$	144.14	$\begin{bmatrix}0.6377\\1.2752\\1.9122\end{bmatrix}$
4	$\begin{bmatrix}0.6377\\1.2752\\1.9122\end{bmatrix}$	100	$\begin{bmatrix}0.0144\\0.0289\\0.0433\end{bmatrix}$	144.14	$\begin{bmatrix}0.6375\\1.2750\\1.9125\end{bmatrix}$

TABLE E10.15b   VECTOR ITERATION WITH SHIFT: SECOND EIGENPAIR

Iteration	$\mathbf{x}_j$	μ	$\bar{\mathbf{x}}_{j+1}$	$\lambda^{(j+1)}$	$\mathbf{x}_{j+1}$
1	$\begin{bmatrix} 1 \\ 1 \\ 1 \end{bmatrix}$	600	$\begin{bmatrix} 0.0044 \\ 0.0028 \\ -0.0133 \end{bmatrix}$	605.11	$\begin{bmatrix} 0.8030 \\ 0.5189 \\ -2.4277 \end{bmatrix}$
2	$\begin{bmatrix} 0.8030 \\ 0.5189 \\ -2.4277 \end{bmatrix}$	600	$\begin{bmatrix} 0.0197 \\ 0.0201 \\ -0.0373 \end{bmatrix}$	648.10	$\begin{bmatrix} 1.0062 \\ 1.0221 \\ -1.8994 \end{bmatrix}$
3	$\begin{bmatrix} 1.0062 \\ 1.0221 \\ -1.8994 \end{bmatrix}$	600	$\begin{bmatrix} 0.0201 \\ 0.0201 \\ -0.0405 \end{bmatrix}$	648.64	$\begin{bmatrix} 0.9804 \\ 0.9778 \\ -1.9717 \end{bmatrix}$
4	$\begin{bmatrix} 0.9804 \\ 0.9778 \\ -1.9717 \end{bmatrix}$	600	$\begin{bmatrix} 0.0202 \\ 0.0202 \\ -0.0404 \end{bmatrix}$	648.65	$\begin{bmatrix} 0.9827 \\ 0.9829 \\ -1.9642 \end{bmatrix}$

algorithm of Eqs. (10.13.3) to (10.13.7) is implemented starting with an initial vector of $\mathbf{x}_1 = \langle 1 \quad 1 \quad 1 \rangle^T$ leading to Table E10.15a. The final result is $\omega_1 = \sqrt{144.14} = 12.006$ and $\phi_1 = \langle 0.6375 \quad 1.2750 \quad 1.9125 \rangle^T$. This is obtained in one iteration cycle less than in the iteration without shift in Example 10.14.

Starting with the shift $\mu_1 = 600$ and the same $\mathbf{x}_1$, the inverse iteration algorithm leads to Table E10.15b. The final result is $\omega_2 = \sqrt{648.65} = 25.468$ and $\phi_2 = \langle 0.9827 \quad 0.9829 - 1.9642 \rangle^T$. Convergence is attained in four iteration cycles.

Starting with the shift $\mu_1 = 1500$ and the same $\mathbf{x}_1$, the inverse iteration algorithm leads to Table E10.15c. The final result is $\omega_3 = \sqrt{1513.5} = 38.904$ and $\phi_3 = \langle 1.5778 - 1.1270 \quad 0.4508 \rangle^T$. Convergence is attained in three cycles.

TABLE E10.15c   VECTOR ITERATION WITH SHIFT: THIRD EIGENPAIR

Iteration	$\mathbf{x}_j$	μ	$\bar{\mathbf{x}}_{j+1}$	$\lambda^{(j+1)}$	$\mathbf{x}_{j+1}$
1	$\begin{bmatrix} 1 \\ 1 \\ 1 \end{bmatrix}$	1500	$\begin{bmatrix} 0.0198 \\ -0.0156 \\ 0.0054 \end{bmatrix}$	1510.6	$\begin{bmatrix} 1.5264 \\ -1.2022 \\ 0.4148 \end{bmatrix}$
2	$\begin{bmatrix} 1.5264 \\ -1.2022 \\ 0.4148 \end{bmatrix}$	1500	$\begin{bmatrix} 0.1167 \\ -0.0832 \\ 0.0333 \end{bmatrix}$	1513.5	$\begin{bmatrix} 1.5784 \\ -1.1261 \\ 0.4509 \end{bmatrix}$
3	$\begin{bmatrix} 1.5784 \\ -1.1261 \\ 0.4509 \end{bmatrix}$	1500	$\begin{bmatrix} 0.1168 \\ -0.0834 \\ 0.0334 \end{bmatrix}$	1513.5	$\begin{bmatrix} 1.5778 \\ -1.1270 \\ 0.4508 \end{bmatrix}$

10.14.2 Rayleigh's Quotient Iteration

The inverse iteration method with shifts converges rapidly if a shift is chosen near enough to the eigenvalue of interest. However, selection of an appropriate shift is difficult without knowledge of the eigenvalue. Many techniques have been developed to overcome this difficulty; one of these is presented in this section.

The Rayleigh quotient calculated by Eq. (10.13.4) to estimate the eigenvalue provides an appropriate shift value, but it is not necessary to calculate and introduce a new shift at each iteration cycle. If this is done, however, the resulting procedure is called *Rayleigh's quotient iteration.*

This procedure starts with the assumption of a starting iteration vector $\mathbf{x}_1$ and starting shift $\lambda^{(1)}$ and consists of the following steps to be repeated for $j = 1, 2, 3, \ldots$ until convergence:

1. Determine $\bar{\mathbf{x}}_{j+1}$ by solving the algebraic equations:

$$[\mathbf{k} - \lambda^{(j)}\mathbf{m}]\bar{\mathbf{x}}_{j+1} = \mathbf{m}\mathbf{x}_j \qquad (10.14.3)$$

2. Obtain an estimate of the eigenvalue and the shift for the next iteration from

$$\lambda^{(j+1)} = \frac{\bar{\mathbf{x}}_{j+1}^T \mathbf{m}\mathbf{x}_j}{\bar{\mathbf{x}}_{j+1}^T \mathbf{m}\bar{\mathbf{x}}_{j+1}} + \lambda^{(j)} \qquad (10.14.4)$$

3. Normalize $\bar{\mathbf{x}}_{j+1}$:

$$\mathbf{x}_{j+1} = \frac{\bar{\mathbf{x}}_{j+1}}{(\bar{\mathbf{x}}_{j+1}^T \mathbf{m}\bar{\mathbf{x}}_{j+1})^{1/2}} \qquad (10.14.5)$$

This iteration converges to a particular eigenpair $(\lambda_n, \boldsymbol{\phi}_n)$ depending on the starting vector $\mathbf{x}_1$ and the initial shift $\lambda^{(1)}$. If $\mathbf{x}_1$ includes a strong contribution of the eigenvector $\boldsymbol{\phi}_n$ and $\lambda^{(1)}$ is close enough to λ_n, the iteration converges to the eigenpair $(\lambda_n, \boldsymbol{\phi}_n)$. The rate of convergence is faster than the standard vector iteration with shift described in Section 10.14.1, but at the expense of additional computation because a new $[\mathbf{k} - \lambda^{(j)}\mathbf{m}]$ has to be factorized in each iteration.

Example 10.16

Determine all three natural frequencies and modes of the system of Example 10.14 by inverse vector iteration with the shift in each iteration cycle equal to Rayleigh's quotient from the previous cycle.

Solution   The iteration procedure of Eqs. (10.14.3) to (10.14.5) is implemented with starting shifts of $\mu_1 = 100$, $\mu_2 = 600$, and $\mu_3 = 1500$, leading to Tables E10.16a, E10.16b, and E10.16c, respectively, where the final results are $\omega_1 = \sqrt{144.14} = 12.006$ and $\phi_1 = \langle 0.6375 \quad 1.2750 \quad 1.9125 \rangle^T$, $\omega_2 = \sqrt{648.65} = 25.468$ and $\phi_2 = \langle 0.9825 \quad 0.9825 \quad -1.9649 \rangle^T$, and $\omega_3 = \sqrt{1513.5} = 38.904$ and $\phi_3 = \langle 1.5778 \quad -1.1270 \quad 0.4508 \rangle^T$.

Observe that convergence is faster when a new shift is used in each iteration cycle. Only two cycles are required instead of four (Example 10.15) for the first mode, and three instead of four for the second mode.

TABLE E10.16a RAYLEIGH'S QUOTIENT ITERATION: FIRST EIGENPAIR

Iteration	$\mathbf{x}_j$	μ	$\bar{\mathbf{x}}_{j+1}$	$\lambda^{(j+1)}$	$\mathbf{x}_{j+1}$
1	$\begin{bmatrix} 1 \\ 1 \\ 1 \end{bmatrix}$	100	$\begin{bmatrix} 0.0114 \\ 0.0218 \\ 0.0313 \end{bmatrix}$	144.60	$\begin{bmatrix} 0.6759 \\ 1.2933 \\ 1.8610 \end{bmatrix}$
2	$\begin{bmatrix} 0.6759 \\ 1.2933 \\ 1.8610 \end{bmatrix}$	144.60	$\begin{bmatrix} -1.3947 \\ -2.7895 \\ -4.1845 \end{bmatrix}$	144.14	$\begin{bmatrix} -0.6375 \\ -1.2750 \\ -1.9126 \end{bmatrix}$
3	$\begin{bmatrix} -0.6375 \\ -1.2750 \\ -1.9126 \end{bmatrix}$	144.14	$\begin{bmatrix} 1.9738 \times 10^6 \\ 3.9476 \times 10^6 \\ 5.9214 \times 10^6 \end{bmatrix}$	144.14	$\begin{bmatrix} 0.6375 \\ 1.2750 \\ 1.9125 \end{bmatrix}$

TABLE E10.16b RAYLEIGH'S QUOTIENT ITERATION: SECOND EIGENPAIR

Iteration	$\mathbf{x}_j$	μ	$\bar{\mathbf{x}}_{j+1}$	λ^{j+1}	$\mathbf{x}_{j+1}$
1	$\begin{bmatrix} 1 \\ 1 \\ 1 \end{bmatrix}$	600	$\begin{bmatrix} 0.0044 \\ 0.0028 \\ -0.0133 \end{bmatrix}$	605.11	$\begin{bmatrix} 0.8030 \\ 0.5189 \\ -2.4277 \end{bmatrix}$
2	$\begin{bmatrix} 0.8030 \\ 0.5189 \\ -2.4277 \end{bmatrix}$	605.11	$\begin{bmatrix} 0.0220 \\ 0.0223 \\ -0.0418 \end{bmatrix}$	648.21	$\begin{bmatrix} 1.0036 \\ 1.0176 \\ -1.9070 \end{bmatrix}$
3	$\begin{bmatrix} 1.0036 \\ 1.0176 \\ -1.9070 \end{bmatrix}$	648.21	$\begin{bmatrix} 2.2624 \\ 2.2623 \\ -4.5249 \end{bmatrix}$	648.65	$\begin{bmatrix} 0.9825 \\ 0.9824 \\ -1.9650 \end{bmatrix}$
4	$\begin{bmatrix} 0.9825 \\ 0.9824 \\ -1.9650 \end{bmatrix}$	648.65	$\begin{bmatrix} 3.0372 \times 10^6 \\ 3.0372 \times 10^6 \\ -6.0745 \times 10^6 \end{bmatrix}$	648.65	$\begin{bmatrix} 0.9825 \\ 0.9825 \\ -1.9649 \end{bmatrix}$

TABLE E10.16c RAYLEIGH'S QUOTIENT ITERATION: THIRD EIGENPAIR

Iteration	$\mathbf{x}_j$	μ	$\bar{\mathbf{x}}_{j+1}$	λ^{j+1}	$\mathbf{x}_{j+1}$
1	$\begin{bmatrix} 1 \\ 1 \\ 1 \end{bmatrix}$	1500	$\begin{bmatrix} 0.0198 \\ -0.0156 \\ 0.0054 \end{bmatrix}$	1510.6	$\begin{bmatrix} 1.5264 \\ -1.2022 \\ 0.4148 \end{bmatrix}$
2	$\begin{bmatrix} 1.5264 \\ -1.2022 \\ 0.4148 \end{bmatrix}$	1510.6	$\begin{bmatrix} 0.5431 \\ -0.3879 \\ 0.1552 \end{bmatrix}$	1513.5	$\begin{bmatrix} 1.5779 \\ -1.1268 \\ 0.4508 \end{bmatrix}$
3	$\begin{bmatrix} 1.5779 \\ -1.1268 \\ 0.4508 \end{bmatrix}$	1513.5	$\begin{bmatrix} 9.7061 \times 10^4 \\ -6.9329 \times 10^4 \\ 2.7732 \times 10^4 \end{bmatrix}$	1513.5	$\begin{bmatrix} 1.5778 \\ -1.1270 \\ 0.4508 \end{bmatrix}$

Application to structural dynamics. In modal analysis of the dynamic response of structures, we are interested in the lower J natural frequencies and modes (Chapters 12 and 13); typically, J is much smaller than N, the number of degrees of freedom. Although Rayleigh's quotient iteration may appear to be an effective tool for the necessary computation, it may not always work. For example, with the starting vector $\mathbf{x}_1$ and starting shift $\lambda^{(1)} = 0$, Eq. (10.14.4) may provide a value for Rayleigh's quotient (which, according to Section 10.12, is always higher than the first eigenvalue), which is also the next shift, closer to the second eigenvalue than the first, resulting in the iteration converging to the second mode. Thus it is necessary to supplement Rayleigh's quotient iteration by another technique to assure convergence to the lowest eigenpair (λ_1, ϕ_1). One possibility is to use first the inverse iteration without shift, Eqs. (10.13.3) to (10.13.7), for a few cycles to obtain an iteration vector that is a good approximation (but has not converged) to ϕ_1, and then start with Rayleigh's quotient iteration.

Computer implementation of inverse vector iteration with shift should be reliable and efficient. By reliability we mean that it should give the desired eigenpair. Efficiency implies that with the fewest iterations and least computation, the method should provide results to the desired degree of accuracy. Both of these requirements are essential; otherwise, the computer program may skip a desired eigenpair, or the computations may be unnecessarily time consuming. The issues related to reliability and efficiency of computer methods for solving the eigenvalue problem are discussed further in other books.

10.15 TRANSFORMATION OF $\mathbf{k}\phi = \omega^2\mathbf{m}\phi$ TO THE STANDARD FORM

The standard eigenvalue problem $\mathbf{Ay} = \lambda\mathbf{y}$ arises in many situations in mathematics and in applications to problems in the physical sciences and engineering. It has therefore attracted much attention and many solution algorithms have been developed and are available in computer software libraries. These computer procedures could be used to solve the structural dynamics eigenvalue problem, $\mathbf{k}\phi = \omega^2\mathbf{m}\phi$, provided that it can be transformed to the standard form. Such a transformation is presented in this section.

We assume that $\mathbf{m}$ is positive definite; that is, it is either a diagonal matrix with nonzero masses or a banded matrix as in a consistent mass formulation (Chapter 17). If $\mathbf{m}$ is a diagonal matrix with zero mass in some degrees of freedom, these are first eliminated by static condensation (Section 9.3). Positive definiteness of $\mathbf{m}$ implies that $\mathbf{m}^{-1}$ can be calculated. Premultiplying the structural dynamics eigenvalue problem

$$\mathbf{k}\phi = \omega^2\mathbf{m}\phi \tag{10.15.1}$$

by $\mathbf{m}^{-1}$ gives the standard eigenvalue problem:

$$\mathbf{A}\phi = \lambda\phi \tag{10.15.2}$$

where

$$\mathbf{A} = \mathbf{m}^{-1}\mathbf{k} \qquad \lambda = \omega^2 \tag{10.15.3}$$

In general, $\mathbf{A}$ is not symmetric, although $\mathbf{m}$ and $\mathbf{k}$ are both symmetric matrices.

Because the computational effort could be greatly reduced if $\mathbf{A}$ were symmetric, we seek methods that yield a symmetric $\mathbf{A}$. Consider that $\mathbf{m} = \mathrm{diag}(m_j)$, a diagonal matrix with elements $m_{jj} = m_j$, and define $\mathbf{m}^{1/2} = \mathrm{diag}(m_j^{1/2})$ and $\mathbf{m}^{-1/2} = \mathrm{diag}(m_j^{-1/2})$. Then $\mathbf{m}$ and the identity matrix $\mathbf{I}$ can be expressed as

$$\mathbf{m} = \mathbf{m}^{1/2}\mathbf{m}^{1/2} \qquad \mathbf{I} = \mathbf{m}^{-1/2}\mathbf{m}^{1/2} \tag{10.15.4}$$

Using Eq. (10.15.4), Eq. (10.15.1) can be rewritten as

$$\mathbf{k}\mathbf{m}^{-1/2}\mathbf{m}^{1/2}\phi = \omega^2\mathbf{m}^{1/2}\mathbf{m}^{1/2}\phi$$

Premultiplying both sides by $\mathbf{m}^{-1/2}$ leads to

$$\mathbf{m}^{-1/2}\mathbf{k}\mathbf{m}^{-1/2}\mathbf{m}^{1/2}\phi = \omega^2\mathbf{m}^{-1/2}\mathbf{m}^{1/2}\mathbf{m}^{1/2}\phi$$

Utilizing Eq. (10.15.4b) to simplify the right-hand side of the equation above gives

$$\mathbf{A}\mathbf{y} = \lambda\mathbf{y} \tag{10.15.5}$$

where

$$\mathbf{A} = \mathbf{m}^{-1/2}\mathbf{k}\mathbf{m}^{-1/2} \qquad \mathbf{y} = \mathbf{m}^{1/2}\phi \qquad \lambda = \omega^2 \tag{10.15.6}$$

Equation (10.15.5) is the standard eigenvalue problem and $\mathbf{A}$ is now symmetric.

Thus if a computer program to solve $\mathbf{A}\mathbf{y} = \lambda\mathbf{y}$ were available, it could be utilized to determine the natural frequencies ω_n and modes ϕ_n of a system for which $\mathbf{m}$ and $\mathbf{k}$ were known as follows:

1. Compute $\mathbf{A}$ from Eq. (10.15.6a).
2. Determine the eigenvalues λ_n and eigenvectors $\mathbf{y}_n$ of $\mathbf{A}$ by solving Eq. (10.15.5).
3. Determine the natural frequencies and modes by

$$\omega_n = \sqrt{\lambda_n} \qquad \phi_n = \mathbf{m}^{-1/2}\mathbf{y}_n \tag{10.15.7}$$

The transformation of Eq. (10.15.6) can be generalized to situations where the mass matrix is not diagonal but is banded like the stiffness matrix; such banding is typical of finite element formulations (Chapter 17). Then, $\mathbf{A}$ is a full matrix, although $\mathbf{k}$ and $\mathbf{m}$ are banded. This is a major computational disadvantage for large systems. For such situations the transformation of $\mathbf{k}\phi = \omega^2\mathbf{m}\phi$ to $\mathbf{A}\mathbf{y} = \lambda\mathbf{y}$ may not be an effective approach, and the inverse iteration method, which works directly with $\mathbf{k}\phi = \omega^2\mathbf{m}\phi$, may be more efficient.

FURTHER READING

Bathe, K. J., *Finite Element Procedures,* Prentice Hall, Englewood Cliffs, N.J., 1996, Chapters 10 and 11.

Crandall, S. H., and McCalley, R. B., Jr., "Matrix Methods of Analysis," Chapter 28 in *Shock and Vibration Handbook* (ed. C. M. Harris), McGraw-Hill, New York, 1988.

Humar, J. L., *Dynamics of Structures,* 2nd ed., A. A. Balkema Publishers, Lisse, The Netherlands, 2002, Chapter 11.

Parlett, B. N., *The Symmetric Eigenvalue Problem,* Prentice Hall, Englewood Cliffs, N.J., 1980.

PROBLEMS

Parts A and B

10.1 Determine the natural vibration frequencies and modes of the system of Fig. P9.1 with $k_1 = k$ and $k_2 = 2k$ in terms of the DOFs in the figure. Show that these results are equivalent to those presented in Fig. E10.1.

10.2 For the system defined in Problem 9.2:
(**a**) Determine the natural vibration frequencies and modes; express the frequencies in terms of m, EI, and L. Sketch the modes and identify the associated natural frequencies.
(**b**) Verify that the modes satisfy the orthogonality properties.
(**c**) Normalize each mode so that the modal mass M_n has unit value. Sketch these normalized modes. Compare these modes with those obtained in part (a) and comment on the differences.

10.3 Determine the free vibration response of the system of Problem 9.2 (and Problem 10.2) due to each of the three sets of initial displacements: (**a**) $u_1(0) = 1$, $u_2(0) = 0$; (**b**) $u_1(0) = 1$, $u_2(0) = 1$; (**c**) $u_1(0) = 1$, $u_2(0) = -1$. Comment on the relative contribution of the modes to the response in the three cases. Neglect damping in the system.

10.4 Repeat Problem 10.3(a) considering damping in the system. For each mode the damping ratio is $\zeta_n = 5\%$.

10.5 For the system defined in Problem 9.4:
(**a**) Determine the natural vibration frequencies and modes. Express the frequencies in terms of m, EI, and L, and sketch the modes.
(**b**) Determine the displacement response due to an initial velocity $\dot{u}_2(0)$ imparted to the top of the system.

10.6 For the two-story shear building shown in Problem 9.5:
(**a**) Determine the natural vibration frequencies and modes; express the frequencies in terms of m, EI, and h.
(**b**) Verify that the modes satisfy the orthogonality properties.
(**c**) Normalize each mode so that the roof displacement is unity. Sketch the modes and identify the associated natural frequencies.
(**d**) Normalize each mode so that the modal mass M_n has unit value. Compare these modes with those obtained in part (c) and comment on the differences.

10.7 The structure of Problem 9.5 is modified so that the columns are hinged at the base. Determine the natural vibration frequencies and modes of the modified system, and compare them with the vibration properties of the original structure determined in Problem 10.6. Comment on the effect of the column support condition on the vibration properties.

10.8 Determine the free vibration response of the structure of Problem 10.6 (and Problem 9.5) if it is displaced as shown in Fig. P10.8a and b and released. Comment on the relative contributions of the two vibration modes to the response that was produced by the two initial displacements. Neglect damping.

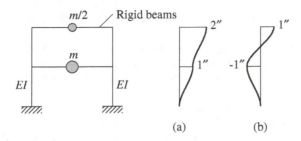

(a)   (b)

Figure P10.8

10.9   Repeat Problem 10.8 for the initial displacement of Fig. P10.8a, assuming that the damping ratio for each mode is 5%.

*__10.10__   Determine the natural vibration frequencies and modes of the system defined in Problem 9.6. Express the frequencies in terms of m, EI, and h and the joint rotations in terms of h. Normalize each mode to unit displacement at the roof and sketch it, identifying all DOFs.

10.11–   For the three-story shear buildings shown in Figs. P9.7 and P9.8:
10.12   **(a)** Determine the natural vibration frequencies and modes; express the frequencies in terms of m, EI, and h. Sketch the modes and identify the associated natural frequencies.
(b) Verify that the modes satisfy the orthogonality properties.
(c) Normalize each mode so that the modal mass M_n has unit value. Sketch these normalized modes. Compare these modes with those obtained in part (a) and comment on the differences.

10.13–   The structures of Figs. P9.7 and P9.8 are modified so that the columns are hinged at the base.
10.14   Determine the natural vibration frequencies and modes of the modified system, and compare them with the vibration properties of the original structures determined in Problems 10.11 and 10.12. Comment on the effect of the column support condition on the vibration properties.

10.15–   Determine the free vibration response of the structures of Problems 10.11 and 10.12 (and
10.16   Problems 9.7 and 9.8) if they are displaced as shown in Fig. P10.15–P10.16a, b, and c and released. Plot floor displacements versus t/T_1 and comment on the relative contributions of the three vibration modes to the response that was produced by each of the three initial displacements. Neglect damping.

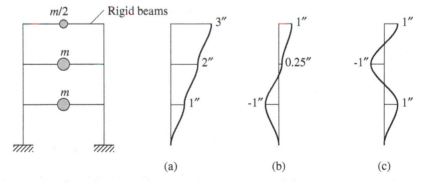

(a)   (b)   (c)

Figure P10.15–P10.16

*Denotes that a computer is necessary to solve this problem.

10.17– Repeat Problems 10.15 and 10.16 for the initial displacement of Fig. P10.15a, assuming that
10.18 the damping ratio for each mode is 5%.

*10.19– Determine the natural vibration frequencies and modes of the systems defined in Prob-
10.22 lems 9.9 to 9.12. Express the frequencies in terms of m, EI, and h and the joint rotations in
   terms of h. Normalize each mode to unit displacement at the roof and sketch it, including all
   DOFs.

10.23 **(a)** For the system in Problem 9.13, determine the natural vibration frequencies and modes.
   Express the frequencies in terms of m, EI, and L, and sketch the modes.
   (b) The structure is pulled through a lateral displacement $u_1(0) = 1$ and released. Determine
   the free vibration response.

10.24 For the system defined in Problem 9.14, $m = 90$ kips/g, $k = 1.5$ kips/in., and $b = 25$ ft.
   (a) Determine the natural vibration frequencies and modes.
   (b) Normalize each mode so that the modal mass M_n has unit value. Sketch these modes.

10.25 Repeat Problem 10.24 using a different set of DOFs—those defined in Problem 9.15. Show
   that the natural vibration frequencies and modes determined using the two sets of DOFs are
   the same.

10.26 Repeat Problem 10.24 using a different set of DOFs—those defined in Problem 9.16. Show
   that the natural vibration frequencies and modes determined using the two sets of DOFs are
   the same.

10.27 Repeat Problem 10.24 using a different set of DOFs—those defined in Problem 9.17. Show
   that the natural vibration frequencies and modes determined using the two sets of DOFs are
   the same.

10.28 For the structure defined in Problem 9.18, determine the natural frequencies and modes.
   Normalize each mode such that $\phi_n^T \phi_n = 1$.

Part C

*10.29 The floor weights and story stiffnesses of the three-story shear frame are shown in Fig. P10.29,
   where $w = 100$ kips and $k = 326.32$ kips/in. Determine the fundamental natural vibration
   frequency ω_1 and mode ϕ_1 by inverse vector iteration.

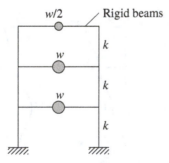

Figure P10.29

*10.30 For the system defined in Problem 10.29, there is concern for possible resonant vibrations
   due to rotating machinery mounted at the second-floor level. The operating speed of the

*Denotes that a computer is necessary to solve this problem.

motor is 430 rpm. Obtain the natural vibration frequency of the structure that is closest to the machine frequency.

*10.31 Determine the three natural vibration frequencies and modes of the system defined in Problem 10.29 by inverse vector iteration with shifting.

*10.32 Determine the three natural vibration frequencies and modes of the system defined in Problem 10.29 by inverse vector iteration with the shift in each iteration cycle equal to Rayleigh's quotient from the previous cycle.

*Denotes that a computer is necessary to solve this problem.

11

Damping in Structures

PREVIEW

Several issues that arise in defining the damping properties of structures are discussed in this chapter. It is impractical to determine the coefficients of the damping matrix directly from the structural dimensions, structural member sizes, and the damping properties of the structural materials used. Therefore, damping is generally specified by numerical values for the modal damping ratios, and these are sufficient for analysis of linear systems with classical damping. The experimental data that provide a basis for estimating these damping ratios are discussed in Part A of this chapter, which ends with recommended values for modal damping ratios. The damping matrix is needed, however, for analysis of linear systems with nonclassical damping and for analysis of nonlinear structures. Two procedures for constructing the damping matrix for a structure from the modal damping ratios are presented in Part B; classically damped systems as well as nonclassically damped systems are considered.

PART A: EXPERIMENTAL DATA AND RECOMMENDED MODAL DAMPING RATIOS

11.1 VIBRATION PROPERTIES OF MILLIKAN LIBRARY BUILDING

Chosen as an example to discuss damping, the Robert A. Millikan Library building is a nine-story reinforced-concrete building constructed in 1966–1967 on the campus of the California Institute of Technology in Pasadena, California. Figure 11.1.1 is a photograph of

Figure 11.1.1 Millikan Library, California Institute of Technology, Pasadena, California. (Courtesy of K. V. Steinbrugge Collection, Earthquake Engineering Research Center, University of California at Berkeley.)

the library building. It is 69 by 75 ft in plan and extends 144 ft above grade and 158 ft above the basement level. This includes an enclosed roof that houses air-conditioning equipment. Lateral forces in the north–south direction are resisted mainly by the 12-in. reinforced-concrete shear walls located at the east and west ends of the building. In the east–west direction the 12-in. reinforced-concrete walls of the central core, which houses the elevator and the emergency stairway, provide most of the lateral resistance. Precast concrete window wall panels are bolted in place on the north and south walls. These were intended to be architectural but provide stiffness in the east–west direction for low levels of vibration.

The vibration properties—natural periods, natural modes, and modal damping ratios—of the Millikan Library have been determined by forced harmonic vibration tests using the vibration generator shown in Fig. 3.3.1. Such a test leads to a frequency-response curve that shows a resonant peak corresponding to each natural frequency of the structure; e.g., the frequency-response curve near the fundamental natural frequency of vibration in the east–west direction is shown in Fig. 11.1.2. From such data the natural frequency and damping ratio for the fundamental vibration mode were determined by the methods of Section 3.4.2, and the results are presented in Table 11.1.1. The natural period for this

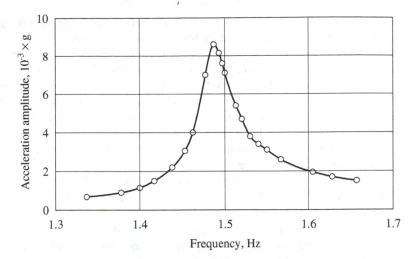

Figure 11.1.2 Frequency response curve for Millikan Library near its fundamental natural frequency of vibration in the east–west direction; acceleration measured is at the eighth floor. [Adapted from Jennings and Kuroiwa (1968).]

TABLE 11.1.1 NATURAL VIBRATION PERIODS AND MODAL DAMPING RATIOS OF MILLIKAN LIBRARY

Excitation	Roof Acceleration (g)	Fundamental Mode		Second Mode	
		Period (sec)	Damping (%)	Period (sec)	Damping (%)
North–South Direction					
Vibration generator	5×10^{-3} to 20×10^{-3}	0.51–0.53	1.2–1.8	[a]	[a]
Lytle Creek earthquake	0.05	0.52	2.9	0.12	1.0
San Fernando earthquake	0.312	0.62	6.4	0.13	4.7
East–West Direction					
Vibration generator	3×10^{-3} to 17×10^{-3}	0.66–0.68	0.7–1.5	[b]	[b]
Lytle Creek earthquake	0.035	0.71	2.2	0.18	3.6
San Fernando earthquake	0.348	0.98	7.0	0.20	5.9

[a] Not measured.

[b] Data not reliable.

mode of vibration in the east–west direction was 0.66 sec (observe that $f_n = 1.49$ Hz in Fig. 11.1.2). This value increased roughly 3% over the resonant amplitude range of testing: acceleration of 3×10^{-3}g to 17×10^{-3}g at the roof. The mode shape corresponding to this mode was found from measurements taken at various floors of the structure but is not presented here. In the vibration test the damping ratio in the fundamental east–west mode varied between 0.7 and 1.5%, increasing with the amplitude of response. In the north–south direction, the natural period of the fundamental mode was 0.51 sec, increasing roughly 4% over the resonant amplitude range of testing: acceleration of 5×10^{-3}g to 20×10^{-3}g at the roof. The damping ratio in this mode varied between 1.2 and 1.8%, again increasing with the amplitude of response.

The Millikan Library is located approximately 19 miles from the center of the Magnitude 6.4 San Fernando, California, earthquake of February 9, 1971. The strong motion accelerographs installed in the basement and the roof of the building recorded three components (two horizontal and one vertical) of accelerations. The recorded accelerations in the north–south direction given in Fig. 11.1.3 show that the peak acceleration of 0.202g at the basement amplified to 0.312g at the roof. Figure 11.1.4 shows that in the east–west direction the peak acceleration at the basement and roof were 0.185g and 0.348g, respectively. The accelerations at the roof represent the total motion of the building, which is composed of the motions of the building relative to the ground plus the motion of the ground. The total displacement at the roof of the building and the displacement of the basement were obtained by twice-integrating the recorded accelerations. The north–south and east–west components of the relative displacement of the roof, determined by subtracting the ground (basement) displacement from the total displacement at the roof, are presented in Fig. 11.1.5.

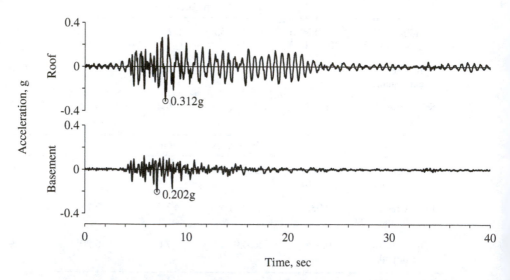

Figure 11.1.3 Accelerations in the north–south direction recorded at Millikan Library during the 1971 San Fernando, California, earthquake.

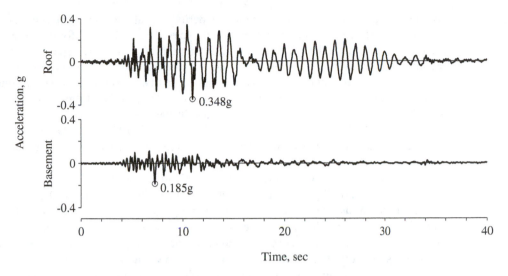

Figure 11.1.4 Accelerations in the east–west direction recorded at Millikan Library during the 1971 San Fernando, California, earthquake.

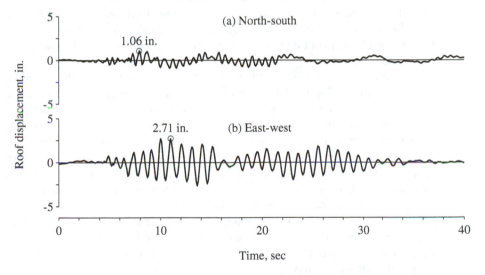

Figure 11.1.5 Relative displacement of the roof in (a) north–south direction; (b) east–west direction. [Adapted from Foutch, Housner, and Jennings (1975).]

It can be seen that the horizontal accelerations of the roof of the building are larger and their time variation is different from the ground (basement) accelerations. These differences arise because the building is flexible, not rigid. It is seen in the displacement plots that the displacement amplitude of the roof relative to the basement was 1.06 in. in the north–south direction and 2.71 in. in the east–west direction. The building vibrated in the north–south direction with a fundamental-mode period of approximately six-tenths of a second, while in the east–west direction this period was about 1 sec. These period values were estimated as the duration of a vibration cycle in Fig. 11.1.5. More accurate values for the first few natural periods and modal damping ratios can be determined from the recorded accelerations at the basement and the roof by system identification procedures (not presented in this book). The results for the first two modes in the north–south and east–west directions are presented in Table 11.1.1 for the Millikan Library building.

Acceleration records were also obtained in the basement and on the roof of this building from the Lytle Creek earthquake of September 12, 1970. The Magnitude 5.4 Lytle Creek earthquake, centered 40 miles from Millikan Library, produced a peak ground acceleration of approximately 0.02g and a roof acceleration of 0.05g in the building, fairly low levels for measured earthquake motion. System identification analysis of these records led to values for natural periods and damping ratios shown in Table 11.1.1. For the low-level vibrations due to the Lytle Creek earthquake, the fundamental periods of 0.52 and 0.71 sec in the north–south and east–west directions, respectively, were similar to—only slightly longer than—those determined in vibration generator tests. Similarly, the damping ratios were slightly increased relative to the vibration generator tests.

For the larger motions of the building during the San Fernando earthquake, the natural periods and damping ratios were increased significantly relative to the values from vibration generator tests. The fundamental period in the north–south direction increased from 0.51 to 0.62 sec and the damping ratio increased substantially, to 6.4%. In the east–west direction the building vibrated with a fundamental period of 0.98 sec, which is 50% longer than the period of 0.66 sec during vibration generator tests; the damping increased substantially, to 7.0%.

The lengthening of natural periods at the larger amplitudes of motion experienced by the building during the San Fernando earthquake implies a reduction in the stiffness of the structure. The stiffness in the east–west direction is reduced substantially, although except for the collapse of bookshelves and minor plaster cracking, the building suffered no observable damage. The apparent damage of the structure due to the earthquake is also the cause of the substantial increase in damping. Following the earthquake there is apparent recovery of the structural stiffness, as suggested by measured natural periods (not presented here) that are shorter than during the earthquake. Whether this recovery is complete or only partial appears to depend on how strongly the structure was excited by the earthquake. These are all indications of the complexity of the behavior of actual structures during earthquakes. We return to this issue in Chapter 13 (Section 13.6) after we have presented analytical procedures to compute the response of linearly elastic structures to specified ground motion.

11.2 ESTIMATING MODAL DAMPING RATIOS

It is usually not feasible to determine the damping properties or natural vibration periods of a structure to be analyzed in the way they were determined for the Millikan Library. If the seismic safety of an existing structure is to be evaluated, ideally we would like to determine experimentally the important properties of the structure, including its damping, but this is rarely done, for lack of budget and time. For a new building being designed, obviously its damping or other properties cannot be measured.

The modal damping ratios for a structure should therefore be estimated using measured data from similar structures. Although researchers have accumulated a substantial body of valuable data, it should be used with discretion because some of it is *not* directly applicable to earthquake analysis and design. It is clear from the Millikan Library data that the damping ratios determined from the low-amplitude forced vibration tests should not be used directly for the analysis of response to earthquakes that cause much larger motions of the structure, say, up to yielding of the structural materials. Modal damping ratios for such analysis should be based on data from recorded earthquake motions.

The data that are most useful but hard to come by are from structures shaken strongly but not deformed into the inelastic range. The damping ratios determined from structural motions that are small are not representative of the larger damping expected at higher amplitudes of structural motions. On the other hand, recorded motions of structures that have experienced significant yielding during an earthquake would provide damping ratios that also include the energy dissipation due to yielding of structural materials. These damping ratios would not be useful in dynamic analysis because the energy dissipation in yielding is accounted for separately through nonlinear force–deformation relationships (see Section 5.7).

Useful data on damping are slow to accumulate because relatively few structures are installed with permanent accelerographs ready to record motions when an earthquake occurs and strong earthquakes are infrequent. The bulk of records of earthquake-induced structural motions in the United States are from multistory buildings in California: more than 50 buildings in the greater Los Angeles area during the 1971 San Fernando earthquake; over 40 buildings in the Monterey Bay and San Francisco Bay areas during the 1989 Loma Prieta earthquake; and over 100 buildings in the greater Los Angeles area during the 1994 Northridge earthquake. Furthermore, the recorded motions of only some of these buildings have been analyzed to determine their natural periods and modal damping ratios.

Ideally, we would like to have data on damping determined from recorded earthquake motions of many structures of various types—buildings, bridges, dams, etc.—using different materials—steel, reinforced concrete, prestressed concrete, masonry, wood, etc. Such data would provide the basis for estimating the damping ratios for an existing structure to be evaluated for its seismic safety or for a new structure to be designed. Until we accumulate a sufficiently large database, selection of damping ratios is based on whatever data are available and expert opinion. Recommended damping values are given in Table 11.2.1 for two levels of motion: working stress levels or stress levels no more than one-half the yield point, and stresses at or just below the yield point. For each stress level, a range of damping values is given; the higher values of damping are to be used for ordinary structures, and the lower values are for special structures to be designed more conservatively. In addition to Table 11.2.1, recommended damping values are 3% for unreinforced masonry structures and 7% for reinforced masonry construction. Most building codes do not recognize the variation in damping with structural materials; and typically a 5% damping ratio is implicit in the code-specified earthquake forces and design spectrum.

TABLE 11.2.1 RECOMMENDED DAMPING VALUES

Stress Level	Type and Condition of Structure	Damping Ratio (%)
Working stress, no more than about $\frac{1}{2}$ yield point	Welded steel, prestressed concrete, well-reinforced concrete (only slight cracking)	2–3
	Reinforced concrete with considerable cracking	3–5
	Bolted and/or riveted steel, wood structures with nailed or bolted joints	5–7
At or just below yield point	Welded steel, prestressed concrete (without complete loss in prestress)	5–7
	Prestressed concrete with no prestress left	7–10
	Reinforced concrete	7–10
	Bolted and/or riveted steel, wood structures with bolted joints	10–15
	Wood structures with nailed joints	15–20

Source: N. M. Newmark, and W. J. Hall, *Earthquake Spectra and Design*, Earthquake Engineering Research Institute, Berkeley, Calif., 1982.

The recommended damping ratios can be used directly for the linearly elastic analysis of structures with classical damping. For such systems the equations of motion when transformed to natural vibration modes of the undamped system become uncoupled, and the estimated modal damping ratios are used directly in each modal equation. This concept was introduced in Section 10.10 and will be developed further in Chapters 12 and 13.

PART B: CONSTRUCTION OF DAMPING MATRIX

11.3 DAMPING MATRIX

When is the damping matrix needed? The damping matrix must be defined completely if classical modal analysis is not applicable. Such is the case for structures with nonclassical damping (see Section 11.5 for examples), even if our interest is confined to their linearly elastic response. Classical modal analysis is also not applicable to the analysis of nonlinear systems even if the damping is of classical form. One of the most important nonlinear problems of interest to us is calculating the response of structures beyond their linearly elastic range during earthquakes.

The damping matrix for practical structures should not be calculated from the structural dimensions, structural member sizes, and the damping of the structural materials used. One might think that it should be possible to determine the damping matrix for the structure from the damping properties of individual structural elements, just as the structural stiffness matrix is determined. However, it is impractical to determine the damping matrix in this manner because unlike the elastic modulus, which enters into the computation of stiffness, the damping properties of materials are not well established. Even if these properties were known, the resulting damping matrix would not account for a significant part of the energy dissipated in friction at steel connections, opening and closing of microcracks in concrete, stressing of nonstructural elements—partition walls, mechanical equipment, fireproofing, etc.—friction between the structure itself and nonstructural elements, and similar mechanisms, some of which are even difficult to identify.

Thus the damping matrix for a structure should be determined from its modal damping ratios, which account for all energy-dissipating mechanisms. As discussed in Section 11.2, the modal damping ratios should be estimated from available data on similar structures shaken strongly during past earthquakes but not deformed into the inelastic range; lacking such data, the values of Table 11.2.1 are recommended.

11.4 CLASSICAL DAMPING MATRIX

Classical damping is an appropriate idealization if similar damping mechanisms are distributed throughout the structure (e.g., a multistory building with a similar structural system and structural materials over its height). In this section we develop two procedures for constructing a classical damping matrix for a structure from modal damping ratios that have been estimated as described in Section 11.2. These two procedures are presented in the following three subsections.

11.4.1 Rayleigh Damping

Consider first mass-proportional damping and stiffness-proportional damping:

$$\mathbf{c} = a_0\mathbf{m} \quad \text{and} \quad \mathbf{c} = a_1\mathbf{k} \tag{11.4.1}$$

where the constants a_0 and a_1 have units of $\sec^{-1}$ and sec, respectively. For both of these damping matrices the matrix $\mathbf{C}$ of Eq. (10.9.4) is diagonal by virtue of the modal orthogonality properties of Eq. (10.4.1); therefore, these are classical damping matrices. Physically, they represent the damping models shown in Fig. 11.4.1 for a multistory building. Stiffness-proportional damping appeals to intuition because it can be interpreted to model the energy dissipation arising from story deformations. In contrast, mass-proportional damping is difficult to justify physically because the air damping it can be interpreted to model is negligibly small for most structures. Later we shall see that, by themselves, neither of the two damping models are appropriate for practical application.

We now relate the modal damping ratios for a system with mass-proportional damping to the coefficient a_0. The generalized damping for the nth mode, Eq. (10.9.10), is

$$C_n = a_0 M_n \tag{11.4.2}$$

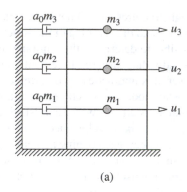 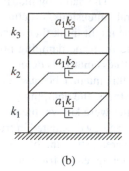

(a)   (b)

Figure 11.4.1   (a) Mass-proportional damping; (b) stiffness-proportional damping.

and the modal damping ratio, Eq. (10.9.11), is

$$\zeta_n = \frac{a_0}{2}\frac{1}{\omega_n} \tag{11.4.3}$$

The damping ratio is inversely proportional to the natural frequency (Fig. 11.4.2a). The coefficient a_0 can be selected to obtain a specified value of damping ratio in any one mode, say ζ_i for the ith mode. Equation (11.4.3) then gives

$$a_0 = 2\zeta_i\omega_i \tag{11.4.4}$$

With a_0 determined, the damping matrix $\mathbf{c}$ is known from Eq. (11.4.1a), and the damping ratio in any other mode, say the nth mode, is given by Eq. (11.4.3).

Similarly, the modal damping ratios for a system with stiffness-proportional damping can be related to the coefficient a_1. In this case

$$C_n = a_1\omega_n^2 M_n \quad \text{and} \quad \zeta_n = \frac{a_1}{2}\omega_n \tag{11.4.5}$$

wherein Eq. (10.2.4) is used. The damping ratio increases linearly with the natural frequency (Fig. 11.4.2a). The coefficient a_1 can be selected to obtain a specified value of the damping ratio in any one mode, say ζ_j for the jth mode. Equation (11.4.5b) then gives

$$a_1 = \frac{2\zeta_j}{\omega_j} \tag{11.4.6}$$

With a_1 determined, the damping matrix $\mathbf{c}$ is known from Eq. (11.4.1b), and the damping ratio in any other mode is given by Eq. (11.4.5b). Neither of the damping matrices defined by Eq. (11.4.1) is appropriate for practical analysis of MDF systems. The variations of modal damping ratios with natural frequencies they represent (Fig. 11.4.2a) are not consistent with experimental data that indicate roughly the same damping ratios for several vibration modes of a structure.

As a first step toward constructing a classical damping matrix somewhat consistent with experimental data, we consider *Rayleigh damping*:

$$\mathbf{c} = a_0 \mathbf{m} + a_1 \mathbf{k} \tag{11.4.7}$$

The damping ratio for the nth mode of such a system is

$$\zeta_n = \frac{a_0}{2}\frac{1}{\omega_n} + \frac{a_1}{2}\omega_n \tag{11.4.8}$$

The coefficients a_0 and a_1 can be determined from specified damping ratios ζ_i and ζ_j for the ith and jth modes, respectively. Expressing Eq. (11.4.8) for these two modes in matrix form leads to

$$\frac{1}{2}\begin{bmatrix} 1/\omega_i & \omega_i \\ 1/\omega_j & \omega_j \end{bmatrix}\begin{Bmatrix} a_0 \\ a_1 \end{Bmatrix} = \begin{Bmatrix} \zeta_i \\ \zeta_j \end{Bmatrix} \tag{11.4.9}$$

These two algebraic equations can be solved to determine the coefficients a_0 and a_1. If both modes are assumed to have the same damping ratio ζ, which is reasonable based on experimental data, then

$$a_0 = \zeta \frac{2\omega_i \omega_j}{\omega_i + \omega_j} \qquad a_1 = \zeta \frac{2}{\omega_i + \omega_j} \tag{11.4.10}$$

The damping matrix is then known from Eq. (11.4.7) and the damping ratio for any other mode, given by Eq. (11.4.8), varies with natural frequency, as shown in Fig. 11.4.2b.

In applying this procedure to a practical problem, the modes i and j with specified damping ratios should be chosen to ensure reasonable values for the damping ratios in all the modes contributing significantly to the response. Consider, for example, that five modes are to be included in the response analysis and roughly the same damping ratio ζ is desired for all modes. This ζ should be specified for the first mode and possibly for the fourth mode. Then Fig. 11.4.2b suggests that the damping ratio for the second and third

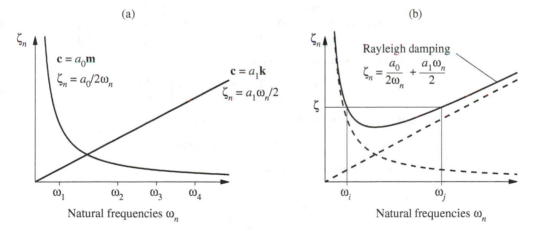

Figure 11.4.2   Variation of modal damping ratios with natural frequency: (a) mass-proportional damping and stiffness-proportional damping; (b) Rayleigh damping.

modes will be somewhat smaller than ζ and for the fifth mode it will be somewhat larger than ζ. The damping ratio for modes higher than the fifth will increase monotonically with frequency and the corresponding modal responses will be essentially eliminated because of their high damping.

Example 11.1

The properties of a three-story shear building are given in Fig. E11.1. These include the floor weights, story stiffnesses, natural frequencies, and modes. Derive a Rayleigh damping matrix such that the damping ratio is 5% for the first and second modes. Compute the damping ratio for the third mode.

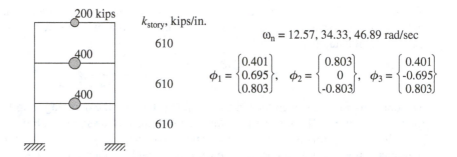

Figure E11.1

Solution

1. *Set up the mass and stiffness matrices.*

$$\mathbf{m} = \frac{1}{386}\begin{bmatrix} 400 & & \\ & 400 & \\ & & 200 \end{bmatrix} \qquad \mathbf{k} = 610\begin{bmatrix} 2 & -1 & 0 \\ -1 & 2 & -1 \\ 0 & -1 & 1 \end{bmatrix} \tag{a}$$

2. *Determine a_0 and a_1 from Eq. (11.4.9).*

$$\begin{bmatrix} 1/12.57 & 12.57 \\ 1/34.33 & 34.33 \end{bmatrix}\begin{Bmatrix} a_0 \\ a_1 \end{Bmatrix} = 2\begin{Bmatrix} 0.05 \\ 0.05 \end{Bmatrix} \tag{b}$$

These algebraic equations have the following solution:

$$a_0 = 0.9198 \qquad a_1 = 0.0021 \tag{c}$$

3. *Evaluate the damping matrix.*

$$\mathbf{c} = a_0\mathbf{m} + a_1\mathbf{k} = \begin{bmatrix} 3.55 & -1.30 & 0 \\ & 3.55 & -1.30 \\ (\text{sym}) & & 1.78 \end{bmatrix} \tag{d}$$

4. *Compute ζ_3 from Eq. (11.4.8).*

$$\zeta_3 = \frac{0.9198}{2(46.89)} + \frac{0.0021(46.89)}{2} = 0.0593 \tag{e}$$

11.4.2 Caughey Damping

If we wish to specify values for damping ratios in more than two modes, we need to consider the general form for a classical damping matrix (see Derivation 11.1), known as *Caughey damping*:

$$\mathbf{c} = \mathbf{m} \sum_{l=0}^{N-1} a_l [\mathbf{m}^{-1}\mathbf{k}]^l \tag{11.4.11}$$

where N is the number of degrees of freedom in the system and a_l are constants. The first three terms of the series are

$$a_0 \mathbf{m}(\mathbf{m}^{-1}\mathbf{k})^0 = a_0 \mathbf{m} \qquad a_1 \mathbf{m}(\mathbf{m}^{-1}\mathbf{k})^1 = a_1 \mathbf{k} \qquad a_2 \mathbf{m}(\mathbf{m}^{-1}\mathbf{k})^2 = a_2 \mathbf{k}\mathbf{m}^{-1}\mathbf{k} \tag{11.4.12}$$

Thus Eq. (11.4.11) with only the first two terms is the same as Rayleigh damping. Suppose that we wish to specify the damping ratios for J modes of an N-DOF system. Then J terms need to be included in the Caughey series; these could be any J of the N terms in Eq. (11.4.11). If the first J terms are included,

$$\mathbf{c} = \mathbf{m} \sum_{l=0}^{J-1} a_l [\mathbf{m}^{-1}\mathbf{k}]^l \tag{11.4.13}$$

and the modal damping ratio ζ_n is given by (see Derivation 11.2)

$$\zeta_n = \tfrac{1}{2} \sum_{l=0}^{J-1} a_l \omega_n^{2l-1} \tag{11.4.14}$$

The coefficients a_l can be determined from the damping ratios specified in any J modes, say the first J modes, by solving the J algebraic equations (11.4.14) for the unknowns $a_l, l = 0$ to $J - 1$. With a_l determined, the damping matrix $\mathbf{c}$ is known from Eq. (11.4.13), and the damping ratios for modes $n = J + 1, J + 2, \ldots, N$ are given by Eq. (11.4.14). It is recommended that these damping ratios be computed to ensure that their values are reasonable.

To illustrate that it is important to do so, we present results for an example structure for which the same damping ratio $\zeta = 5\%$ was specified for the first four modes, the first four terms were included in Eq. (11.4.11), and the values of a_l were determined as described above and substituted in Eq. (11.4.14) to determine damping ratio as a function of frequency. Plotted in Fig. 11.4.3, these results demonstrate that the damping ratio remains close to (slightly above or slightly below) the desired value ζ over the frequency range ω_1 to ω_4, being exactly equal to ζ at the first four natural frequencies, but increases monotonically with frequency higher than ω_4. As a result, the response contributions of the higher modes will be underestimated to a point that they are essentially excluded. On the other hand, when the damping ratio ζ was specified for only the first three modes, the same procedure led to a damping ratio that was close to the desired value over the frequency range ω_1 to ω_3, but decreased monotonically for modes higher than the third mode, eventually taking on negative values. These are obviously unrealistic because they imply free vibration that grows with time instead of decaying with time. In conclusion, Caughey damping

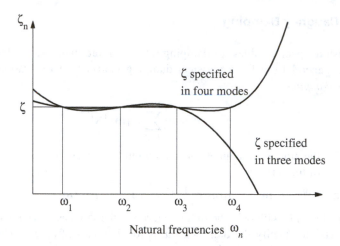

$$\zeta_n$$

ζ specified
in four modes

ζ specified
in three modes

ω_1 ω_2 ω_3 ω_4

Natural frequencies ω_n

Figure 11.4.3

should be defined such that modal damping ratios close to the desired value are achieved for all modes contributing significantly to the response, and none of the ζ_n values become negative.

While the general classical damping matrix given by Eq. (11.4.13) makes it possible to specify the damping ratios in any number of modes, there are two problems associated with its use. First, the algebraic equations (11.4.14) are numerically ill conditioned because the coefficients ω_n^{-1}, ω_n, ω_n^3, ω_n^5, ... can differ by orders of magnitude. Second, if more than two terms are included in the Caughey series, $\mathbf{c}$ is a full matrix, although $\mathbf{k}$ is a banded matrix, and for a lumped-mass system, $\mathbf{m}$ is a diagonal matrix. Since the computational effort for analyzing large systems increases significantly if the damping matrix is not banded, Rayleigh damping is often assumed in practical analyses.

Example 11.2

For the system of Fig. E11.1, evaluate the classical damping matrix if the damping ratio is 5% for all three modes.

Solution

1. *Caughey series for a 3-DOF system:*

$$\mathbf{c} = a_0\mathbf{m} + a_1\mathbf{k} + a_2\mathbf{km}^{-1}\mathbf{k} \tag{a}$$

2. *Determine a_0, a_1, and a_2 from Eq. (11.4.14):*

$$\zeta_n = \frac{a_0}{2}\frac{1}{\omega_n} + \frac{a_1}{2}\omega_n + \frac{a_2}{2}\omega_n^3 \qquad n = 1, 2, 3 \tag{b}$$

or

$$\begin{bmatrix} 1/12.57 & 12.57 & (12.57)^3 \\ 1/34.33 & 34.33 & (34.33)^3 \\ 1/46.89 & 46.89 & (46.89)^3 \end{bmatrix} \begin{Bmatrix} a_0 \\ a_1 \\ a_2 \end{Bmatrix} = 2 \begin{Bmatrix} 0.05 \\ 0.05 \\ 0.05 \end{Bmatrix} \tag{c}$$

These algebraic equations have the following solution:

$$a_0 = 0.8377 \qquad a_1 = 0.0027 \qquad a_2 = -4.416 \times 10^{-7} \tag{d}$$

3. *Evaluate* **c.** Substituting a_0, a_1, and a_2 from Eq. (d) in Eq. (a) gives

$$\mathbf{c} = \begin{bmatrix} 3.40 & -1.03 & -0.159 \\ & 3.08 & -1.03 \\ (\text{sym}) & & 1.62 \end{bmatrix} \tag{e}$$

Derivation 11.1

The natural frequencies ω_r and modes ϕ_r satisfy

$$\mathbf{k}\phi_r = \omega_r^2 \mathbf{m}\phi_r \tag{a}$$

Premultiplying both sides by $\phi_n^T \mathbf{km}^{-1}$ gives

$$\phi_n^T \left[\mathbf{km}^{-1}\mathbf{k} \right] \phi_r = \omega_r^2 \phi_n^T \mathbf{k}\phi_r = 0 \qquad n \neq r \tag{b}$$

wherein the second equality comes from the orthogonality equation (10.4.1a). Premultiplying both sides of Eq. (a) by $\phi_n^T (\mathbf{km}^{-1})^2$ gives

$$\phi_n^T \left[(\mathbf{km}^{-1})^2 \mathbf{k} \right] \phi_r = \omega_r^2 \phi_n^T \left[\mathbf{km}^{-1}\mathbf{km}^{-1}\mathbf{m} \right] \phi_r$$

$$= \omega_r^2 \phi_n^T \left[\mathbf{km}^{-1}\mathbf{k} \right] \phi_r = 0 \qquad n \neq r \tag{c}$$

wherein the second equality comes from Eq. (b). By repeated application of this procedure, a family of orthogonality relations can be obtained which can all be expressed in a compact form:

$$\phi_n^T \mathbf{c}_l \phi_r = 0 \qquad n \neq r \tag{d}$$

where

$$\mathbf{c}_l = \left[\mathbf{km}^{-1} \right]^l \mathbf{k} \qquad l = 0, 1, 2, 3, \ldots, \infty \tag{e}$$

The matrices $\mathbf{c}_l$ can be written in an alternative form by premultiplying Eq. (e) by the identity matrix, $\mathbf{I} = \mathbf{mm}^{-1}$:

$$\mathbf{c}_l = \mathbf{mm}^{-1}\mathbf{km}^{-1}\mathbf{km}^{-1} \cdots \mathbf{km}^{-1}\mathbf{k}$$

$$= \mathbf{m} \left[\mathbf{m}^{-1}\mathbf{k} \right]^l \qquad l = 0, 1, 2, 3, \ldots, \infty \tag{f}$$

Premultiplying Eq. (a) by $\phi_n^T \mathbf{mk}^{-1}$ and following the procedure above, it can be shown that Eq. (d) is satisfied by another infinite sequence of matrices:

$$\mathbf{c}_l = \mathbf{m} \left[\mathbf{m}^{-1}\mathbf{k} \right]^l \qquad l = -1, -2, -3, \ldots, -\infty \tag{g}$$

Combining Eqs. (f) and (g) gives

$$\mathbf{c} = \mathbf{m} \sum_{l=-\infty}^{\infty} a_l \left[\mathbf{m}^{-1}\mathbf{k} \right]^l \tag{h}$$

It can be shown that only N terms in this infinite series are independent, leading to Eq. (11.4.11) as the general form of classical damping matrices.

Derivation 11.2

For the nth mode the generalized damping is

$$C_n = \phi_n^T \mathbf{c}\phi_n = \sum_{l=0}^{N-1} \phi_n^T \mathbf{c}_l \phi_n \tag{a}$$

where c_l is given by Eq. (f) of Derivation 11.1; and the various terms in this series are

$$l = 0: \quad \phi_n^T c_o \phi_n = \phi_n^T (a_0 \mathbf{m}) \phi_n = a_o M_n$$

$$l = 1: \quad \phi_n^T c_1 \phi_n = \phi_n^T (a_1 \mathbf{k}) \phi_n = a_1 \omega_n^2 M_n$$

$$l = 2: \quad \phi_n^T c_2 \phi_n = \phi_n^T (a_2 \mathbf{km}^{-1}\mathbf{k}) \phi_n = a_2 \omega_n^2 \phi_n^T \mathbf{k} \phi_n = a_2 \omega_n^4 M_n$$

wherein Eq. (10.2.4) is used. Thus Eq. (a) becomes

$$C_n = \sum_{l=0}^{N-1} a_l \omega_n^{2l} M_n \tag{b}$$

The damping ratio for the nth mode, Eq. (10.9.11), is given by

$$\zeta_n = \frac{1}{2} \sum_{l=0}^{N-1} a_l \omega_n^{2l-1} \tag{c}$$

which is similar to Eq. (11.4.14).

11.4.3 Superposition of Modal Damping Matrices

An alternative procedure to determine a classical damping matrix from modal damping ratios can be derived starting with Eq. (10.9.4):

$$\mathbf{\Phi}^T \mathbf{c} \mathbf{\Phi} = \mathbf{C} \tag{11.4.15}$$

where $\mathbf{C}$ is a diagonal matrix with the nth diagonal element equal to the generalized modal damping:

$$C_n = \zeta_n (2 M_n \omega_n) \tag{11.4.16}$$

With ζ_n estimated as described in Section 11.2, $\mathbf{C}$ is known from Eq. (11.4.16) and Eq. (11.4.15) can be rewritten as

$$\mathbf{c} = \left(\mathbf{\Phi}^T\right)^{-1} \mathbf{C} \mathbf{\Phi}^{-1} \tag{11.4.17}$$

Using this equation to compute $\mathbf{c}$ may appear to be an inefficient procedure because it seems to require the inversion of two matrices of order N, the number of DOFs. However, the inverse of the modal matrix $\mathbf{\Phi}$ and of $\mathbf{\Phi}^T$ can be determined with little computation because of the orthogonality property of modes.

Starting with the orthogonality relationship of Eq. (10.4.5b),

$$\mathbf{\Phi}^T \mathbf{m} \mathbf{\Phi} = \mathbf{M} \tag{11.4.18}$$

it can be shown that

$$\mathbf{\Phi}^{-1} = \mathbf{M}^{-1} \mathbf{\Phi}^T \mathbf{m} \qquad \left(\mathbf{\Phi}^T\right)^{-1} = \mathbf{m} \mathbf{\Phi} \mathbf{M}^{-1} \tag{11.4.19}$$

Because $\mathbf{M}$ is a diagonal matrix of generalized modal masses M_n, $\mathbf{M}^{-1}$ is known immediately as a diagonal matrix with elements $= 1/M_n$. Thus $\mathbf{\Phi}^{-1}$ and $\left(\mathbf{\Phi}^T\right)^{-1}$ can be computed efficiently from Eq. (11.4.19).

Substituting Eq. (11.4.19) in Eq. (11.4.17) leads to

$$\mathbf{c} = (\mathbf{m}\boldsymbol{\Phi}\mathbf{M}^{-1})\mathbf{C}(\mathbf{M}^{-1}\boldsymbol{\Phi}^T\mathbf{m}) \qquad (11.4.20)$$

Since $\mathbf{M}$ and $\mathbf{C}$ are diagonal matrices, defined by Eqs. (11.4.18) and (11.4.15), respectively, Eq. (11.4.20) can be expressed as

$$\mathbf{c} = \mathbf{m}\left(\sum_{n=1}^{N} \frac{2\zeta_n\omega_n}{M_n}\boldsymbol{\phi}_n\boldsymbol{\phi}_n^T\right)\mathbf{m} \qquad (11.4.21)$$

The nth term in this summation is the contribution of the nth mode with its damping ratio ζ_n to the damping matrix $\mathbf{c}$; if this term is not included, the resulting $\mathbf{c}$ implies a zero damping ratio in the nth mode. It is reasonable to include in Eq. (11.4.21) only the first J modes that are expected to contribute significantly to the response. The lack of damping in modes $J + 1$ to N does not create numerical problems if an unconditionally stable time-stepping procedure is used to integrate the equations of motion; see Chapter 16.

Example 11.3

Determine a damping matrix for the system of Fig. E11.1 by superposing the damping matrices for the first two modes, each with $\zeta_n = 5\%$.

Solution

1. *Determine the individual terms in Eq. (11.4.21).*

$$c_1 = \frac{2(0.05)(12.57)}{1.0}\mathbf{m}\boldsymbol{\phi}_1\boldsymbol{\phi}_1^T\mathbf{m} \qquad\qquad c_2 = \frac{2(0.05)(34.33)}{1.0}\mathbf{m}\boldsymbol{\phi}_2\boldsymbol{\phi}_2^T\mathbf{m}$$

$$= \begin{bmatrix} 0.217 & 0.376 & 0.217 \\ & 0.651 & 0.376 \\ \text{(sym)} & & 0.217 \end{bmatrix} \qquad\qquad = \begin{bmatrix} 2.37 & 0 & -1.19 \\ & 0 & 0 \\ \text{(sym)} & & 0.593 \end{bmatrix}$$

2. *Determine* $\mathbf{c}$.

$$\mathbf{c} = c_1 + c_2 = \begin{bmatrix} 2.59 & 0.376 & -0.969 \\ & 0.651 & 0.376 \\ \text{(sym)} & & 0.810 \end{bmatrix}$$

Recall that this $\mathbf{c}$ implies a zero damping ratio for the third mode.

Example 11.4

Determine the damping matrix for the system of Fig. E11.1 by superposing the damping matrices for the three modes, each with $\zeta_n = 5\%$.

Solution

1. *Determine the individual terms in Eq. (11.4.21).* The first two terms, c_1 and c_2, are already computed in Example 11.3, and

$$c_3 = \frac{2(0.05)(46.89)}{1.0}\mathbf{m}\boldsymbol{\phi}_3\boldsymbol{\phi}_3^T\mathbf{m} = \begin{bmatrix} 0.809 & -1.40 & 0.810 \\ & 2.43 & -1.40 \\ \text{(sym)} & & 0.811 \end{bmatrix}$$

2. *Determine* $\mathbf{c}$.

$$\mathbf{c} = \sum_{n=1}^{3} c_n = \begin{bmatrix} 3.40 & -1.03 & -0.159 \\ & 3.08 & -1.03 \\ \text{(sym)} & & 1.62 \end{bmatrix}$$

Note that this **c** is the same as in Example 11.2 because $\zeta_n = 5\%$ for all three modes in both examples.

11.5 NONCLASSICAL DAMPING MATRIX

The assumption of classical damping is not appropriate if the system to be analyzed consists of two or more parts with significantly different levels of damping. One such example is a structure–soil system. While the underlying soil can be assumed as rigid in the analysis of many structures, soil–structure interaction should be considered in the analysis of structures with very short natural periods, such as the nuclear containment structure of Fig. 1.10.1. The modal damping ratio for the soil system would typically be much different than the structure, say 15 to 20% for the soil region compared to 3 to 5% for the structure. Therefore, the assumption of classical damping would not be appropriate for the combined structure–soil system, although it may be reasonable for the structure and soil regions separately. Another example is a concrete dam with water impounded behind the dam (Fig. 1.10.2). The damping of the water is negligible relative to damping for the dam, and classical damping is not an appropriate model for the dam–water system. While substructure methods (not developed in this book) are especially effective for the analysis of structure–soil and structure–fluid systems, these systems are also analyzed by standard methods, requiring the damping matrix for the complete system.

The damping matrix for the complete system is constructed by directly assembling the damping matrices for the two subsystems—structure and soil in the first case, dam and water in the second. As shown in Fig. 11.5.1, the stiffness and mass matrices of the combined structure–soil system are assembled from the corresponding matrices for the two subsystems. The portion of these matrices associated with the common DOFs at the interface (I) between the two subsystems includes contributions from both subsystems.

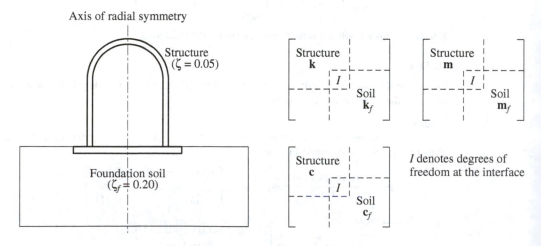

Figure 11.5.1   Assembly of subsystem matrices.

Thus all that remains to be described is the procedure to construct damping matrices for the individual subsystems, assumed to be classically damped.

In principle, these subsystem damping matrices could be constructed by any of the procedures developed in Section 11.4, but Rayleigh damping is perhaps most convenient for practical analyses. Thus the damping matrices for the structure and the foundation soil (denoted by subscript f) are

$$\mathbf{c} = a_0\mathbf{m} + a_1\mathbf{k} \qquad \mathbf{c}_f = a_{0f}\mathbf{m}_f + a_{1f}\mathbf{k}_f \tag{11.5.1}$$

The coefficients a_0 and a_1 are given by Eq. (11.4.10) using an appropriate damping ratio for the structure, say $\zeta = 0.05$, where ω_i and ω_j are selected as the frequencies of the ith and jth natural vibration modes of the combined system without damping. The coefficients a_{0f} and a_{1f} are determined similarly; they would be four times larger if the damping ratio for the foundation soil region is estimated as $\zeta_f = 0.20$.

The assumption of classical damping may not be appropriate either for structures with special energy-dissipating devices (Section 7.10) or on a base isolation system (Chapter 21), even if the structure itself has classical damping. The nonclassical damping matrix for the system is constructed by first evaluating the classical damping matrix $\mathbf{c}$ for the structure alone (without the special devices) from the damping ratios appropriate for the structure, using the procedures of Section 11.4. The damping contributions of the energy-dissipating devices are then assembled into $\mathbf{c}$ to obtain the damping matrix for the complete system.

FURTHER READING

Caughey, T. K., "Classical Normal Modes in Damped Linear Dynamic Systems," *Journal of Applied Mechanics, ASME*, **27**, 1960, pp 269–271.

Caughey, T. K., and O'Kelly, M. E. J., "Classical Normal Modes in Damped Linear Dynamic Systems," *Journal of Applied Mechanics, ASME*, **32**, 1965, pp. 583–588.

Foutch, D. A., Housner, G. W., and Jennings, P. C., "Dynamic Responses of Six Multistory Buildings during the San Fernando Earthquake," *Report No. EERL 75-02*, California Institute of Technology, Pasadena, Calif., October 1975.

Hart, G. C., and Vasudevan, R., "Earthquake Design of Buildings: Damping," *Journal of the Structural Division, ASCE*, **101**, 1975, pp. 11–30.

Hashimoto, P. S., Steele, L. K., Johnson, J. J., and Mensing, R. W., "Review of Structure Damping Values for Elastic Seismic Analysis of Nuclear Power Plants," *Report No. NUREG/CR-6011*, U.S. Nuclear Regulatory Commission, Washington, D.C., March 1993.

Jennings, P. C., and Kuroiwa, J. H., "Vibration and Soil–Structure Interaction Tests of a Nine-Story Reinforced Concrete Building," *Bulletin of the Seismological Society of America*, **58**, 1968, pp. 891–916.

McVerry, G. H., "Frequency Domain Identification of Structural Models from Earthquake Records," *Report No. EERL 79-02*, California Institute of Technology, Pasadena, Calif., October 1979.

Newmark, N. M., and Hall, W. J., *Earthquake Spectra and Design*, Earthquake Engineering Research Institute, Berkeley, Calif., 1982, pp. 53–54.

Rayleigh, Lord, *Theory of Sound*, Vol. 1, Dover Publications, New York, 1945; published originally in 1896.

Wilson, E. L., and Penzien, J., "Evaluation of Orthogonal Damping Matrices," *International Journal for Numerical Methods in Engineering*, **4**, 1972, pp. 5–10.

PROBLEMS

11.1   The properties of a three-story shear building are given in Fig. P11.1. These include the floor weights, story stiffnesses, natural vibration frequencies, and modes. Derive a Rayleigh damping matrix such that the damping ratio is 5% for the first and third modes. Compute the damping ratio for the second mode.

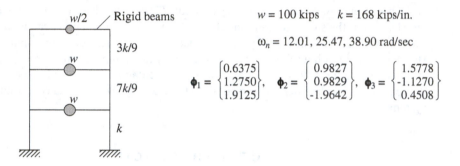

$w = 100$ kips $k = 168$ kips/in.

$\omega_n = 12.01, 25.47, 38.90$ rad/sec

$$\phi_1 = \begin{Bmatrix} 0.6375 \\ 1.2750 \\ 1.9125 \end{Bmatrix}, \quad \phi_2 = \begin{Bmatrix} 0.9827 \\ 0.9829 \\ -1.9642 \end{Bmatrix}, \quad \phi_3 = \begin{Bmatrix} 1.5778 \\ -1.1270 \\ 0.4508 \end{Bmatrix}$$

Figure P11.1

11.2   For the system of Fig. P11.1, use Caughey series to determine the classical damping matrix if the damping ratio is 5% for all three modes.

11.3   Determine a damping matrix for the system of Fig. P11.1 by superimposing the damping matrices for the first and third modes, each with $\zeta_n = 5\%$. Verify that the resulting damping matrix gives no damping in the second mode.

11.4   Determine the classical damping matrix for the system of Fig. P11.1 by superimposing the damping matrices for the three modes, each with $\zeta_n = 5\%$.

12

Dynamic Analysis and Response of Linear Systems

PREVIEW

Now that we have developed procedures to formulate the equations of motion for MDF systems subjected to dynamic forces (Chapters 9 and 11), we are ready to present the solution of these equations. In Part A of this chapter we show that the equations for a two-DOF system without damping subjected to harmonic forces can be solved analytically. Then we use these results to explain how a vibration absorber or tuned mass damper works to decrease or eliminate unwanted vibration. This simultaneous solution of the coupled equations of motion is not feasible in general, so in Part B we develop the classical modal analysis procedure. The equations of motion are transformed to modal coordinates, leading to an uncoupled set of modal equations; each modal equation is solved to determine the modal contributions to the response, and these modal responses are combined to obtain the total response. An understanding of the relative response contributions of the various modes is developed in Part C with the objective of deciding the number of modes to include in dynamic analysis. The chapter closes with Part D, which includes two analysis procedures useful in special situations: static correction method and mode acceleration method.

PART A: TWO-DEGREE-OF-FREEDOM SYSTEMS

12.1 ANALYSIS OF TWO-DOF SYSTEMS WITHOUT DAMPING

Consider the two-DOF systems shown in Fig. 12.1.1 excited by a harmonic force $p_1(t) = p_o \sin \omega t$ applied to the mass m_1. For both systems the equations of motion are

$$\begin{bmatrix} m_1 & 0 \\ 0 & m_2 \end{bmatrix} \begin{Bmatrix} \ddot{u}_1 \\ \ddot{u}_2 \end{Bmatrix} + \begin{bmatrix} k_1 + k_2 & -k_2 \\ -k_2 & k_2 \end{bmatrix} \begin{Bmatrix} u_1 \\ u_2 \end{Bmatrix} = \begin{Bmatrix} p_o \\ 0 \end{Bmatrix} \sin \omega t \qquad (12.1.1)$$

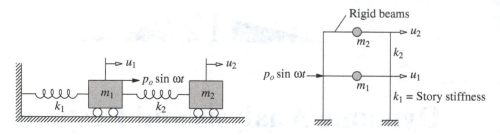

Figure 12.1.1   Two-degree-of-freedom systems.

Observe that the equations are coupled through the stiffness matrix. One equation cannot be solved independent of the other; that is, both equations must be solved simultaneously. Because the system is undamped, the steady-state solution can be assumed as

$$\begin{Bmatrix} u_1(t) \\ u_2(t) \end{Bmatrix} = \begin{Bmatrix} u_{1o} \\ u_{2o} \end{Bmatrix} \sin \omega t$$

Substituting this into Eq. (12.1.1), we obtain

$$\begin{bmatrix} k_1 + k_2 - m_1\omega^2 & -k_2 \\ -k_2 & k_2 - m_2\omega^2 \end{bmatrix} \begin{Bmatrix} u_{1o} \\ u_{2o} \end{Bmatrix} = \begin{Bmatrix} p_o \\ 0 \end{Bmatrix} \qquad (12.1.2)$$

or

$$[\mathbf{k} - \omega^2\mathbf{m}] \begin{Bmatrix} u_{1o} \\ u_{2o} \end{Bmatrix} = \begin{Bmatrix} p_o \\ 0 \end{Bmatrix}$$

Premultiplying by $[\mathbf{k} - \omega^2\mathbf{m}]^{-1}$ gives

$$\begin{Bmatrix} u_{1o} \\ u_{2o} \end{Bmatrix} = [\mathbf{k} - \omega^2\mathbf{m}]^{-1} \begin{Bmatrix} p_o \\ 0 \end{Bmatrix} = \frac{1}{\det[\mathbf{k} - \omega^2\mathbf{m}]} \, \text{adj} \, [\mathbf{k} - \omega^2\mathbf{m}] \begin{Bmatrix} p_o \\ 0 \end{Bmatrix} \qquad (12.1.3)$$

where $\det[\cdot]$ and $\text{adj}[\cdot]$ denote the determinant and adjoint of the matrix$[\cdot]$, respectively. The frequency equation [Eq. (10.2.6)]

$$\det[\mathbf{k} - \omega^2\mathbf{m}] = 0$$

can be solved for the natural frequencies ω_1 and ω_2 of the system. In terms of these frequencies, this determinant can be expressed as

$$\det[\mathbf{k} - \omega^2\mathbf{m}] = m_1 m_2(\omega^2 - \omega_1^2)(\omega^2 - \omega_2^2) \qquad (12.1.4)$$

Thus, Eq. (12.1.3) becomes

$$\begin{Bmatrix} u_{1o} \\ u_{2o} \end{Bmatrix} = \frac{1}{\det[\mathbf{k} - \omega^2\mathbf{m}]} \begin{bmatrix} k_2 - m_2\omega^2 & k_2 \\ k_2 & k_1 + k_2 - m_1\omega^2 \end{bmatrix} \begin{Bmatrix} p_o \\ 0 \end{Bmatrix} \qquad (12.1.5)$$

or

$$u_{1o} = \frac{p_o(k_2 - m_2\omega^2)}{m_1 m_2(\omega^2 - \omega_1^2)(\omega^2 - \omega_2^2)} \qquad u_{2o} = \frac{p_o k_2}{m_1 m_2(\omega^2 - \omega_1^2)(\omega^2 - \omega_2^2)} \qquad (12.1.6)$$

Example 12.1

Plot the frequency-response curve for the system shown in Fig. 12.1.1 with $m_1 = 2m$, $m_2 = m$, $k_1 = 2k$, and $k_2 = k$ subjected to harmonic force p_o applied on mass m_1.

Solution   Substituting the given mass and stiffness values in Eq. (12.1.6) gives

$$u_{1o} = \frac{p_o(k - m\omega^2)}{2m^2(\omega^2 - \omega_1^2)(\omega^2 - \omega_2^2)} \qquad u_{2o} = \frac{p_o k}{2m^2(\omega^2 - \omega_1^2)(\omega^2 - \omega_2^2)} \qquad (a)$$

where $\omega_1 = \sqrt{k/2m}$ and $\omega_2 = \sqrt{2k/m}$; these natural frequencies were obtained in Example 10.4. For given system parameters, Eq. (a) provides solutions for the response amplitudes u_{1o} and u_{2o}. It is instructive to rewrite them as

$$\frac{u_{1o}}{(u_{1\text{st}})_o} = \frac{1 - \frac{1}{2}(\omega/\omega_1)^2}{\left[1 - (\omega/\omega_1)^2\right]\left[1 - (\omega/\omega_2)^2\right]} \qquad \frac{u_{2o}}{(u_{2\text{st}})_o} = \frac{1}{\left[1 - (\omega/\omega_1)^2\right]\left[1 - (\omega/\omega_2)^2\right]} \qquad (b)$$

In these equations the response amplitudes have been divided $(u_{1\text{st}})_o = p_o/2k$ and $(u_{2\text{st}})_o = p_o/2k$, the maximum values of the *static displacements* (a concept introduced in Section 3.1), to obtain normalized or nondimensional responses that depend on frequency ratios ω/ω_1 and ω/ω_2, not separately on ω, ω_1, and ω_2.

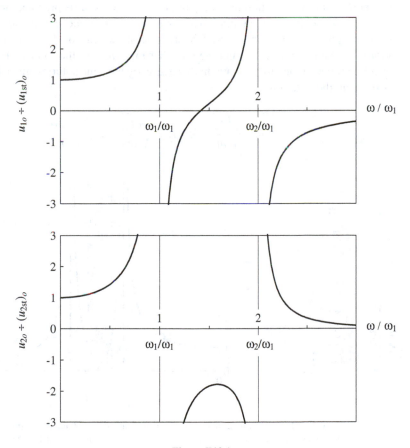

Figure E12.1

Figure E12.1 shows the normalized response amplitudes u_{1o} and u_{2o} plotted against the frequency ratio ω/ω_1. These frequency-response curves show two resonance conditions at $\omega = \omega_1$ and $\omega = \omega_2$; at these exciting frequencies the steady-state response is unbounded. At other exciting frequencies, the vibration is finite and could be calculated from Eq. (b). Note that there is an exciting frequency where the vibration of the first mass, where the exciting force is applied, is reduced to zero. This is the entire basis of the *dynamic vibration absorber* or *tuned mass damper* discussed next.

12.2 VIBRATION ABSORBER OR TUNED MASS DAMPER

The *vibration absorber* is a mechanical device used to decrease or eliminate unwanted vibration. The description *tuned mass damper* is often used in modern installation; this modern name has the advantage of showing its relationship to other types of dampers. In the brief presentation that follows, we restrict ourselves to the basic principle of a vibration absorber without getting into the many important aspects of its practical design.

In its simplest form, a vibration absorber consists of one spring and a mass. Such an absorber system is attached to a SDF system, as shown in Fig. 12.2.1a. The equations of motion for the main mass m_1 and the absorber mass m_2 are the same as Eq. (12.1.1). For harmonic force applied to the main mass we already have the solution given by Eq. (12.1.6). Introducing the notation

$$\omega_1^* = \sqrt{\frac{k_1}{m_1}} \qquad \omega_2^* = \sqrt{\frac{k_2}{m_2}} \qquad \mu = \frac{m_2}{m_1} \qquad (12.2.1)$$

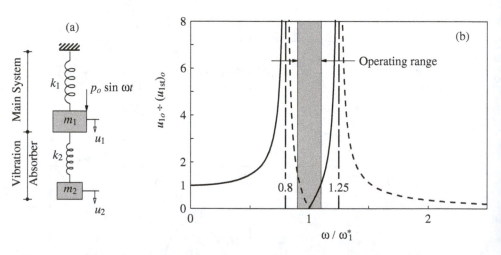

Figure 12.2.1   (a) Vibration absorber attached to an SDF system; (b) response amplitude versus exciting frequency (dashed curve indicates negative u_{1o} or phase opposite to excitation); $\mu = 0.2$ and $\omega_1^* = \omega_2^*$.

the available solution can be rewritten as

$$u_{1o} = \frac{p_o}{k_1} \frac{1 - (\omega/\omega_2^*)^2}{\left[1 + \mu \, (\omega_2^*/\omega_1^*)^2 - (\omega/\omega_1^*)^2 \right] \left[1 - (\omega/\omega_2^*)^2 \right] - \mu \, (\omega_2^*/\omega_1^*)^2} \qquad (12.2.2a)$$

$$u_{2o} = \frac{p_o}{k_1} \frac{1}{\left[1 + \mu \, (\omega_2^*/\omega_1^*)^2 - (\omega/\omega_1^*)^2 \right] \left[1 - (\omega/\omega_2^*)^2 \right] - \mu \, (\omega_2^*/\omega_1^*)^2} \qquad (12.2.2b)$$

At exciting frequency $\omega = \omega_2^*$, Eq. (12.2.2a) indicates that the motion of the main mass m_1 does not simply diminish, it ceases altogether. Figure 12.2.1b shows a plot of response amplitude $u_{1o} \div (u_{1st})_o$, where $(u_{1st})_o = p_o/k_1$, versus ω; for this example, the mass ratio $\mu = 0.2$ and $\omega_1^* = \omega_2^*$, the absorber being tuned to the natural frequency of the main system. Because the system has two DOFs, two resonant frequencies exist, and the response is unbounded at those frequencies. The operating frequency range where $u_{1o} \div (u_{1st})_o < 1$ is shown.

The usefulness of the vibration absorber becomes obvious if we compare the frequency-response function of Fig. 12.2.1b with the response of the main mass alone, without the absorber mass. At $\omega = \omega_1^*$ the response amplitude of the main mass alone is unbounded but is zero with the presence of the absorber mass. Thus, if the exciting frequency ω is close to the natural frequency ω_1^* of the main system, and operating restrictions make it impossible to vary either one, the vibration absorber can be used to reduce the response amplitude of the main system to near zero.

What should be the size of the absorber mass? To answer this question, we use Eq. (12.2.2b) to determine the motion of the absorber mass at $\omega = \omega_2^*$:

$$u_{2o} = -\frac{p_o}{k_2} \qquad (12.2.3)$$

The force acting on the absorber mass is

$$k_2 u_{2o} = \omega^2 m_2 u_{2o} = -p_o \qquad (12.2.4)$$

This implies that the absorber system exerts a force equal and opposite to the exciting force. Thus, the size of the absorber stiffness and mass, k_2 and m_2, depends on the allowable value of u_{2o}. There are other factors that affect the choice of the absorber mass. Obviously, a large absorber mass presents a practical problem. At the same time the smaller the mass ratio μ, the narrower will be the operating frequency range of the absorber.

The preceding presentation indicates that a vibration absorber has its greatest application to synchronous machinery, operating at nearly constant frequency, for it is tuned to one particular frequency and is effective only over a narrow band of frequencies. However, vibration absorbers are also used in situations where the excitation is not nearly harmonic.

The dumbbell-shaped devices that hang from highest-voltage transmission lines are vibration absorbers used to mitigate the fatiguing effects of wind-induced vibration. Vibration absorbers have also been used to reduce the wind-induced vibration of tall buildings when the motions have reached annoying levels for the occupants. An example of this is the 59-story Citicorp Center in midtown Manhattan; completed in 1977, this building has a 820-kip block of concrete installed on the 59th floor in a movable platform connected to the building by large hydraulic arms. When the building sways more than 1 foot a second, the computer directs the arms to move the block in the other direction. This action reduces such oscillation by 40%, considerably easing the discomfort of the building's occupants during high winds.

PART B: MODAL ANALYSIS

12.3 MODAL EQUATIONS FOR UNDAMPED SYSTEMS

The equations of motion for a linear MDF system without damping were derived in Chapter 9 and are repeated here:

$$\mathbf{m\ddot{u} + ku = p}(t) \tag{12.3.1}$$

The simultaneous solution of these coupled equations of motion that we have illustrated in Section 12.1 for a two-DOF system subjected to harmonic excitation is not efficient for systems with more DOFs, nor is it feasible for systems excited by other types of forces. Consequently, it is advantageous to transform these equations to modal coordinates, as we shall see next.

As mentioned in Section 10.7, the displacement vector $\mathbf{u}$ of an MDF system can be expanded in terms of modal contributions. Thus, the dynamic response of a system can be expressed as

$$\mathbf{u}(t) = \sum_{r=1}^{N} \phi_r q_r(t) = \mathbf{\Phi q}(t) \tag{12.3.2}$$

Using this equation, the coupled equations (12.3.1) in $u_j(t)$ can be transformed to a set of uncoupled equations with modal coordinates $q_n(t)$ as the unknowns. Substituting Eq. (12.3.2) in Eq. (12.3.1) gives

$$\sum_{r=1}^{N} \mathbf{m}\,\phi_r \ddot{q}_r(t) + \sum_{r=1}^{N} \mathbf{k}\,\phi_r q_r(t) = \mathbf{p}(t)$$

Premultiplying each term in this equation by ϕ_n^T gives

$$\sum_{r=1}^{N} \phi_n^T \mathbf{m}\phi_r \ddot{q}_r(t) + \sum_{r=1}^{N} \phi_n^T \mathbf{k}\phi_r q_r(t) = \phi_n^T \mathbf{p}(t)$$

Because of the orthogonality relations of Eq. (10.4.1), all terms in each of the summations vanish, except the $r = n$ term, reducing this equation to

$$(\phi_n^T \mathbf{m} \phi_n)\ddot{q}_n(t) + (\phi_n^T \mathbf{k} \phi_n)q_n(t) = \phi_n^T \mathbf{p}(t)$$

or

$$M_n \ddot{q}_n(t) + K_n q_n(t) = P_n(t) \qquad (12.3.3)$$

where

$$M_n = \phi_n^T \mathbf{m} \phi_n \qquad K_n = \phi_n^T \mathbf{k} \phi_n \qquad P_n(t) = \phi_n^T \mathbf{p}(t) \qquad (12.3.4)$$

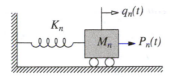

Figure 12.3.1   Generalized SDF system for the nth natural mode.

Equation (12.3.3) may be interpreted as the equation governing the response $q_n(t)$ of the SDF system shown in Fig. 12.3.1 with mass M_n, stiffness K_n, and exciting force $P_n(t)$. Therefore, M_n is called the *generalized mass* for the nth natural mode, K_n the *generalized stiffness* for the nth mode, and $P_n(t)$ the *generalized force* for the nth mode. These parameters depend only on the nth-mode ϕ_n. Thus, if we know only the nth mode, we can write the equation for q_n and solve it without even knowing the other modes. Dividing by M_n and using Eq. (10.4.7), Eq. (12.3.3) can be rewritten as

$$\ddot{q}_n + \omega_n^2 q_n = \frac{P_n(t)}{M_n} \qquad (12.3.5)$$

Equation (12.3.3) or (12.3.5) governs the nth modal coordinate $q_n(t)$, the only unknown in the equation, and there are N such equations, one for each mode. Thus, the set of N coupled differential equations (12.3.1) in nodal displacements $u_j(t)$—$j = 1, 2, \ldots, N$— has been transformed to the set of N uncoupled equations (12.3.3) in modal coordinates $q_n(t)$—$n = 1, 2, \ldots, N$. Written in matrix form the latter set of equations is

$$\mathbf{M}\ddot{\mathbf{q}} + \mathbf{K}\mathbf{q} = \mathbf{P}(t) \qquad (12.3.6)$$

where $\mathbf{M}$ is a diagonal matrix of the generalized modal masses M_n, $\mathbf{K}$ is a diagonal matrix of the generalized modal stiffnesses K_n, and $\mathbf{P}(t)$ is a column vector of the generalized modal forces $P_n(t)$. Recall that $\mathbf{M}$ and $\mathbf{K}$ were introduced in Section 10.4.

Example 12.2

Consider the systems and excitation of Example 12.1. By modal analysis determine the steady-state response of the system.

Solution   The natural vibration frequencies and modes of this system were determined in Example 10.4, from which the generalized masses and stiffnesses are calculated using

Eq. (12.3.4). These results are summarized next:

$$\omega_1 = \sqrt{\frac{k}{2m}} \qquad \omega_2 = \sqrt{\frac{2k}{m}}$$

$$\phi_1 = \left(\tfrac{1}{2} \quad 1\right)^T \qquad \phi_2 = \langle -1 \quad 1\rangle^T$$

$$M_1 = \frac{3m}{2} \qquad M_2 = 3m$$

$$K_1 = \frac{3k}{4} \qquad K_2 = 6k$$

1. *Compute the generalized forces.*

$$P_1(t) = \phi_1^T \mathbf{p}(t) = \underbrace{(p_o/2)}_{P_{1o}} \sin \omega t \qquad P_2(t) = \phi_2^T \mathbf{p}(t) = \underbrace{-p_o}_{P_{2o}} \sin \omega t \tag{a}$$

2. *Set up the modal equations.*

$$M_n \ddot{q}_n + K_n q_n = P_{no} \sin \omega t \tag{b}$$

3. *Solve the modal equations.* To solve Eq. (b) we draw upon the solution presented in Eq. (3.1.7) for an SDF system subjected to harmonic force. The governing equation is

$$m\ddot{u} + ku = p_o \sin \omega t \tag{c}$$

and its steady-state solution is

$$u(t) = \frac{p_o}{k} C \sin \omega t \qquad C = \frac{1}{1 - (\omega/\omega_n)^2} \tag{d}$$

where $\omega_n = \sqrt{k/m}$. Comparing Eqs. (c) and (b), the solution for Eq. (b) is

$$q_n(t) = \frac{P_{no}}{K_n} C_n \sin \omega t \tag{e}$$

where C_n is given by Eq. (d) with ω_n interpreted as the natural frequency of the nth mode. Substituting for P_{no} and K_n for $n = 1$ and 2 gives

$$q_1(t) = \frac{2p_o}{3k} C_1 \sin \omega t \qquad q_2(t) = -\frac{p_o}{6k} C_2 \sin \omega t \tag{f}$$

4. *Determine the modal responses.* The nth-mode contribution to displacements—from Eq. (12.3.2)—is $\mathbf{u}_n(t) = \phi_n q_n(t)$. Substituting Eq. (f) gives the displacement response due to the two modes:

$$\mathbf{u}_1(t) = \phi_1 \frac{2p_o}{3k} C_1 \sin \omega t \qquad \mathbf{u}_2(t) = \phi_2 \frac{-p_o}{6k} C_2 \sin \omega t \tag{g}$$

5. *Combine the modal responses.*

$$\mathbf{u}(t) = \mathbf{u}_1(t) + \mathbf{u}_2(t) \quad \text{or} \quad u_j(t) = u_{j1}(t) + u_{j2}(t) \qquad j = 1, 2 \tag{h}$$

Substituting Eq. (g) and for ϕ_1 and ϕ_2 gives

$$u_1(t) = \frac{p_o}{6k}(2\mathcal{C}_1 + \mathcal{C}_2)\sin \omega t \qquad u_2(t) = \frac{p_o}{6k}(4\mathcal{C}_1 - \mathcal{C}_2)\sin \omega t \qquad (i)$$

These results are equivalent to those obtained in Example 12.1 by solving the coupled equations (12.3.1) of motion.

12.4 MODAL EQUATIONS FOR DAMPED SYSTEMS

When damping is included, the equations of motion for an MDF system are

$$\mathbf{m\ddot{u} + c\dot{u} + ku = p}(t) \qquad (12.4.1)$$

Using the transformation of Eq. (12.3.2), where ϕ_r are the natural modes of the system without damping, these equations can be written in terms of the modal coordinates. Unlike the case of undamped systems (Section 12.3), these modal equations may be coupled through the damping terms. However, for certain forms of damping that are reasonable idealizations for many structures, the equations become uncoupled, just as for undamped systems. We shall demonstrate this next.

Substituting Eq. (12.3.2) in Eq. (12.4.1) gives

$$\sum_{r=1}^{N} \mathbf{m}\phi_r\ddot{q}_r(t) + \sum_{r=1}^{N} \mathbf{c}\phi_r\dot{q}_r(t) + \sum_{r=1}^{N} \mathbf{k}\phi_r q_r(t) = \mathbf{p}(t)$$

Premultiplying each term in this equation by ϕ_n^T gives

$$\sum_{r=1}^{N} \phi_n^T\mathbf{m}\phi_r\ddot{q}_r(t) + \sum_{r=1}^{N} \phi_n^T\mathbf{c}\phi_r\dot{q}_r(t) + \sum_{r=1}^{N} \phi_n^T\mathbf{k}\phi_r q_r(t) = \phi_n^T\mathbf{p}(t)$$

which can be rewritten as

$$M_n\ddot{q}_n(t) + \sum_{r=1}^{N} C_{nr}\dot{q}_r(t) + K_n q_n(t) = P_n(t) \qquad (12.4.2)$$

where M_n, K_n, and $P_n(t)$ were defined in Eq. (12.3.4) and

$$C_{nr} = \phi_n^T\mathbf{c}\phi_r \qquad (12.4.3)$$

Equation (12.4.2) exists for each $n = 1$ to N, and the set of N equations can be written in matrix form:

$$\mathbf{M\ddot{q} + C\dot{q} + Kq = P}(t) \qquad (12.4.4)$$

where $\mathbf{M}$, $\mathbf{K}$, and $\mathbf{P}(t)$ were introduced in Eq. (12.3.6) and $\mathbf{C}$ is a nondiagonal matrix of coefficients C_{nr}. These N equations in modal coordinates $q_n(t)$ are coupled through the damping terms because Eq. (12.4.2) contains more than one modal velocity.

The modal equations will be uncoupled if the system has classical damping. For such systems, as defined in Section 10.9, $C_{nr} = 0$ if $n \neq r$ and Eq. (12.4.2) reduces to

$$M_n \ddot{q}_n + C_n \dot{q}_n + K_n q_n = P_n(t) \tag{12.4.5}$$

where the generalized damping C_n is defined by Eq. (10.10.2). This equation governs the response of the SDF system shown in Fig. 12.4.1. Dividing Eq. (12.4.5) by M_n gives

$$\ddot{q}_n + 2\zeta_n \omega_n \dot{q}_n + \omega_n^2 q_n = \frac{P_n(t)}{M_n} \tag{12.4.6}$$

where ζ_n is the damping ratio for the nth mode. The damping ratio is usually not computed using Eq. (10.10.3) but is estimated based on experimental data for structures similar to the one being analyzed (Chapter 11). Equation (12.4.5) governs the nth modal coordinate $q_n(t)$, and the parameters M_n, K_n, C_n, and $P_n(t)$ depend only on the nth-mode ϕ_n, not on other modes. Thus, we have N uncoupled equations like Eq. (12.4.5), one for each natural mode. In summary, the set of N coupled differential equations (12.4.1) in nodal displacements $u_j(t)$ has been transformed to the set of N uncoupled equations (12.4.5) in modal coordinates $q_n(t)$.

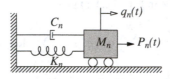

Figure 12.4.1   Generalized SDF system for the nth natural mode.

12.5 DISPLACEMENT RESPONSE

For given external dynamic forces defined by $\mathbf{p}(t)$, the dynamic response of an MDF system can be determined by solving Eq. (12.4.5) or (12.4.6) for the modal coordinate $q_n(t)$. Each modal equation is of the same form as the equation of motion for an SDF system. Thus, the solution methods and results available for SDF systems (Chapters 3 to 5) can be adapted to obtain solutions $q_n(t)$ for the modal equations. Once the modal coordinates $q_n(t)$ have been determined, Eq. (12.3.2) indicates that the contribution of the nth mode to the nodal displacements $\mathbf{u}(t)$ is

$$\mathbf{u}_n(t) = \phi_n q_n(t) \tag{12.5.1}$$

and combining these modal contributions gives the total displacements:

$$\mathbf{u}(t) = \sum_{n=1}^{N} \mathbf{u}_n(t) = \sum_{n=1}^{N} \phi_n q_n(t) \tag{12.5.2}$$

The resulting $\mathbf{u}(t)$ is independent of how the modes are normalized, although $q_n(t)$ are not.

   This procedure is known as *classical modal analysis* or the *classical mode superposition method* because individual (uncoupled) modal equations are solved to determine the modal coordinates $q_n(t)$ and the modal responses $\mathbf{u}_n(t)$, and the latter are combined to obtain the total response $\mathbf{u}(t)$. More precisely, this method is called the *classical mode displacement superposition method* because modal displacements are superposed. For brevity we usually refer to this procedure as *modal analysis*. This analysis method is restricted to linear systems with classical damping. The linearity of the system is implicit in using the principle of superposition, Eq. (12.3.2). Damping must be of the classical form in order to obtain modal equations that are uncoupled, a central feature of modal analysis.

12.6 ELEMENT FORCES

Two procedures described in Section 9.10 are available to determine the forces in various elements—beams, columns, walls, etc.—of the structure at time instant t from the displacements $\mathbf{u}(t)$ at the same time instant. In modal analysis it is instructive to determine the contributions of the individual modes to the element forces. In the first procedure, the nth-mode contribution $r_n(t)$ to an element force $r(t)$ is determined from modal displacements $\mathbf{u}_n(t)$ using element stiffness properties (Appendix 1). Then the element force considering contributions of all modes is

$$r(t) = \sum_{n=1}^{N} r_n(t) \qquad (12.6.1)$$

   In the second procedure, the equivalent static forces associated with the nth-mode response are defined using Eq. (9.10.1) with subscript s deleted: $\mathbf{f}_n(t) = \mathbf{k}\mathbf{u}_n(t)$. Substituting Eq. (12.5.1) and using Eq. (10.2.4) gives

$$\mathbf{f}_n(t) = \omega_n^2 \mathbf{m}\boldsymbol{\phi}_n q_n(t) \qquad (12.6.2)$$

Static analysis of the structure subjected to these external forces at each time instant gives the element force $r_n(t)$. Then the total force $r(t)$ is given by Eq. (12.6.1).

12.7 MODAL ANALYSIS: SUMMARY

The dynamic response of an MDF system to external forces $\mathbf{p}(t)$ can be computed by modal analysis, summarized next as a sequence of steps:

1. Define the structural properties.
   a. Determine the mass matrix $\mathbf{m}$ and stiffness matrix $\mathbf{k}$ (Chapter 9).
   b. Estimate the modal damping ratios ζ_n (Chapter 11).
2. Determine the natural frequencies ω_n and modes $\boldsymbol{\phi}_n$ (Chapter 10).

3. Compute the response in each mode by the following steps:
   a. Set up Eq. (12.4.5) or (12.4.6) and solve for $q_n(t)$.
   b. Compute the nodal displacements $\mathbf{u}_n(t)$ from Eq. (12.5.1).
   c. Compute the element forces associated with the nodal displacements $\mathbf{u}_n(t)$ by implementing one of the two methods described in Section 12.6 for the desired values of t and the element forces of interest.
4. Combine the contributions of all the modes to determine the total response. In particular, the nodal displacements $\mathbf{u}(t)$ are given by Eq. (12.5.2) and element forces by Eq. (12.6.1).

Example 12.3

Consider the systems and excitation of Example 12.1. Determine the spring forces $V_j(t)$ for the system of Fig. 12.1.1a, or story shears $V_j(t)$ in the system of Fig. 12.1.1b, without introducing equivalent static forces. Consider only the steady-state response.

Solution   Steps 1, 2, 3a, and 3b of the analysis summary of Section 12.7 have already been completed in Example 12.2.

Step 3c: The spring forces in the system of Fig. 12.1.1a or the story shears in the system of Fig. 12.1.1b are

$$V_{1n}(t) = k_1 u_{1n}(t) = k_1 \phi_{1n} q_n(t) \tag{a}$$

$$V_{2n}(t) = k_2 [u_{2n}(t) - u_{1n}(t)] = k_2(\phi_{2n} - \phi_{1n}) q_n(t) \tag{b}$$

Substituting Eq. (f) of Example 12.2 in Eqs. (a) and (b) with $n = 1$, $k_1 = 2k$, $k_2 = k$, $\phi_{11} = \frac{1}{2}$, and $\phi_{21} = 1$ gives the forces due to the first mode:

$$V_{11}(t) = \frac{2p_o}{3} C_1 \sin \omega t \qquad V_{21}(t) = \frac{p_o}{3} C_1 \sin \omega t \tag{c}$$

Substituting Eq. (f) of Example 12.2 in Eqs. (a) and (b) with $n = 2$, $\phi_{12} = -1$, and $\phi_{22} = 1$ gives the second-mode forces:

$$V_{12}(t) = \frac{p_o}{3} C_2 \sin \omega t \qquad V_{22}(t) = -\frac{p_o}{3} C_2 \sin \omega t \tag{d}$$

Step 4b: Substituting Eqs. (c) and (d) in $V_j(t) = V_{j1}(t) + V_{j2}(t)$ gives

$$V_1(t) = \frac{p_o}{3} (2C_1 + C_2) \sin \omega t \qquad V_2(t) = \frac{p_o}{3} (C_1 - C_2) \sin \omega t \tag{e}$$

Equation (e) gives the time variation of spring forces and story shears. For a given p_o and ω and the ω_n already determined, all quantities on the right side of these equations are known; thus $V_j(t)$ can be computed.

Example 12.4

Repeat Example 12.3 using equivalent static forces.

Solution   From Eq. (12.6.2), for a lumped-mass system the equivalent static force in the jth DOF due to the nth mode is

$$f_{jn}(t) = \omega_n^2 m_j \phi_{jn} q_n(t) \tag{a}$$

Step 3c: In Eq. (a) with $n = 1$, substitute $m_1 = 2m$, $m_2 = m$, $\phi_{11} = \frac{1}{2}$, $\phi_{21} = 1$, $\omega_1^2 = k/2m$, and $q_1(t)$ from Eq. (f) of Example 12.2 to obtain

$$f_{11}(t) = \frac{p_o}{3} C_1 \sin \omega t \qquad f_{21}(t) = \frac{p_o}{3} C_1 \sin \omega t \qquad \text{(b)}$$

In Eq. (a) with $n = 2$, substituting $m_1 = 2m$, $m_2 = m$, $\phi_{12} = -1$, $\phi_{22} = 1$, $\omega_2^2 = 2k/m$, and $q_2(t)$ from Eq. (f) of Example 12.2 gives

$$f_{12}(t) = \frac{2p_o}{3} C_2 \sin \omega t \qquad f_{22}(t) = -\frac{p_o}{3} C_2 \sin \omega t \qquad \text{(c)}$$

Static analysis of the systems of Fig. E12.4 subjected to forces $f_{jn}(t)$ gives the two spring forces and story shears due to the nth mode:

$$V_{1n}(t) = f_{1n}(t) + f_{2n}(t) \qquad V_{2n}(t) = f_{2n}(t) \qquad \text{(d)}$$

Substituting Eq. (b) in Eq. (d) with $n = 1$ gives the first mode forces that are identical to Eq. (c) of Example 12.3. Similarly, substituting Eq. (c) in Eq. (d) with $n = 2$ gives the second-mode results that are identical to Eq. (d) of Example 12.3.

Step 4: Proceed as in step 4b of Example 12.3.

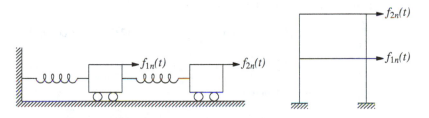

Figure E12.4

Example 12.5

Consider the system and excitation of Example 12.1 with modal damping ratios ζ_n. Determine the steady-state displacement amplitudes of the system.

Solution   Steps 1 and 2 of the analysis summary have been completed in Example 12.2.

Step 3: The modal equations without damping were developed in Example 12.2. Now including damping they become

$$M_n \ddot{q}_n + C_n \dot{q}_n + K_n q_n = P_{no} \sin \omega t \qquad \text{(a)}$$

where M_n, K_n, and P_{no} are available and C_n is known in terms of ζ_n.

To solve Eq. (a), we draw upon the solution presented in Eq. (3.2.3) for an SDF system with damping subjected to harmonic force. The governing equation is

$$m\ddot{u} + c\dot{u} + ku = p_o \sin \omega t \qquad \text{(b)}$$

and its steady-state solution is

$$u(t) = \frac{p_o}{k} (C \sin \omega t + D \cos \omega t) \qquad \text{(c)}$$

with

$$C = \frac{1 - (\omega/\omega_n)^2}{[1 - (\omega/\omega_n)^2]^2 + (2\zeta \,\omega/\omega_n)^2} \qquad D = \frac{-2\zeta \,\omega/\omega_n}{[1 - (\omega/\omega_n)^2]^2 + (2\zeta \,\omega/\omega_n)^2} \tag{d}$$

where $\omega_n = \sqrt{k/m}$ and $\zeta = c/2m\omega_n$.

Comparing Eqs. (b) and (a), the solution for the latter is

$$q_n(t) = \frac{P_{no}}{K_n} (C_n \sin \omega t + D_n \cos \omega t) \tag{e}$$

where C_n and D_n are given by Eq. (d) with ω_n interpreted as the natural frequency of the nth mode and $\zeta = \zeta_n$, the damping ratio for the nth mode. Substituting for P_{no} and K_n for $n = 1$ and 2 gives

$$q_1(t) = \frac{2p_o}{3k} (C_1 \sin \omega t + D_1 \cos \omega t) \tag{f}$$

$$q_2(t) = -\frac{p_o}{6k} (C_2 \sin \omega t + D_2 \cos \omega t) \tag{g}$$

Steps 3b and 4: Substituting ϕ_n in Eqs. (12.5.2) gives the nodal displacements:

$$u_1(t) = \tfrac{1}{2} q_1(t) - q_2(t) \qquad u_2(t) = q_1(t) + q_2(t)$$

Substituting Eqs. (f) and (g) for $q_n(t)$ gives

$$u_1(t) = \frac{p_o}{6k} [(2C_1 + C_2) \sin \omega t + (2D_1 + D_2) \cos \omega t] \tag{h}$$

$$u_2(t) = \frac{p_o}{6k} [(4C_1 - C_2) \sin \omega t + (4D_1 - D_2) \cos \omega t] \tag{i}$$

The displacement amplitudes are

$$u_{1o} = \frac{p_o}{6k} \sqrt{(2C_1 + C_2)^2 + (2D_1 + D_2)^2} \tag{j}$$

$$u_{2o} = \frac{p_o}{6k} \sqrt{(4C_1 - C_2)^2 + (4D_1 - D_2)^2} \tag{k}$$

These u_{jo} can be computed when the amplitude p_o and frequency ω of the exciting force are known together with system properties k, ω_n, and ζ_n.

It can be shown that Eqs. (h) and (i), specialized for $\zeta_n = 0$, are identical to the results for the system without damping obtained in Example 12.2.

Example 12.6

The dynamic response of the system of Fig. E12.6a to the excitation shown in Fig. E12.6b is desired. Determine **(a)** displacements $u_1(t)$ and $u_2(t)$; **(b)** bending moments and shears at sections a, b, c, and d as functions of time; **(c)** the shearing force and bending moment diagrams at $t = 0.18$ sec. The system and excitation parameters are $E = 29{,}000$ ksi, $I = 100$ in^4, $L = 120$ in., $mL = 0.1672$ kip-sec^2/in., and $p_o = 5$ kips. Neglect damping.

Solution The mass and stiffness matrices are available from Example 9.5. The natural frequencies and modes of this system were determined in Example 10.2. They are $\omega_1 = 3.156\sqrt{EI/mL^4}$ and $\omega_2 = 16.258\sqrt{EI/mL^4}$; $\phi_1 = \langle 1 \ 0.3274 \rangle^T$ and $\phi_2 = \langle 1 \ -1.5274 \rangle^T$. Substituting for E, I, m, and L gives $\omega_1 = 10.00$ and $\omega_2 = 51.51$ rad/sec.

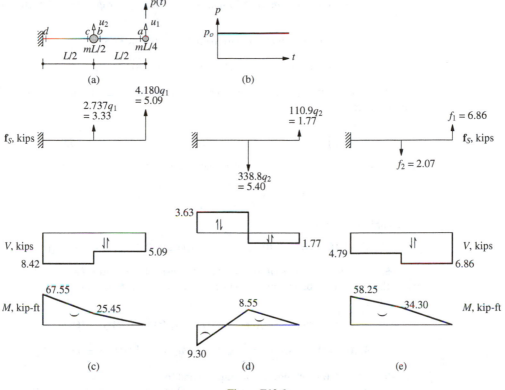

Figure E12.6

1. *Set up the modal equations.*

$$M_1 = \phi_1^T \mathbf{m} \phi_1 = 0.0507 \qquad M_2 = \phi_2^T \mathbf{m} \phi_2 = 0.2368 \text{ kip-sec}^2/\text{in.}$$

$$P_1(t) = \phi_1^T \left\{ \begin{array}{c} p_o \\ o \end{array} \right\} = 5 \qquad P_2(t) = \phi_2^T \left\{ \begin{array}{c} p_o \\ o \end{array} \right\} = 5 \text{ kips}$$

The modal equations (12.4.6) are

$$\ddot{q}_1 + 10^2 q_1 = \frac{5}{0.0507} = 98.62 \qquad \ddot{q}_2 + (51.51)^2 q_2 = \frac{5}{0.2368} = 21.12 \tag{a}$$

2. *Solve the modal equations.* Adapting the SDF system result, Eq. (4.3.2), to Eq. (a) gives

$$q_1(t) = \frac{98.62}{10^2}(1 - \cos 10t) = 0.986(1 - \cos 10t)$$

$$q_2(t) = \frac{21.12}{(51.51)^2}(1 - \cos 51.51t) = 0.008(1 - \cos 51.51t) \tag{b}$$

3. *Determine the displacement response.* Substituting for ϕ_1, ϕ_2, $q_1(t)$, and $q_2(t)$ in Eq. (12.5.2) gives

$$u_1(t) = 0.994 - 0.986\cos 10t - 0.008\cos 51.51t$$

$$u_2(t) = 0.311 - 0.323\cos 10t + 0.012\cos 51.51t$$

(c)

4. *Determine the equivalent static forces.* Substituting for ω_1^2, $\mathbf{m}$, and ϕ_1 in Eq. (12.6.2) gives the forces shown in Fig. E12.6c:

$$\mathbf{f}_1(t) = \begin{bmatrix} f_1(t) \\ f_2(t) \end{bmatrix}_1 = 10^2 \begin{bmatrix} 0.0418 & \\ & 0.0836 \end{bmatrix} \begin{Bmatrix} 1 \\ 0.3274 \end{Bmatrix} q_1(t) = \begin{Bmatrix} 4.180 \\ 2.737 \end{Bmatrix} q_1(t) \qquad (d)$$

Similarly substituting ω_2^2, $\mathbf{m}$, and ϕ_2 gives the forces shown in Fig. E12.6d:

$$\mathbf{f}_2(t) = \begin{bmatrix} f_1(t) \\ f_2(t) \end{bmatrix}_2 = \begin{bmatrix} 110.9 \\ -338.8 \end{bmatrix} q_2(t) \qquad (e)$$

The combined forces are

$$f_1(t) = 4.180\,q_1(t) + 110.9\,q_2(t) \qquad f_2(t) = 2.737\,q_1(t) - 338.8\,q_2(t) \qquad (f)$$

5. *Determine the internal forces.* Static analysis of the cantilever beam of Fig. E12.6e gives the shearing forces and bending moments at the various sections a, b, c, and d:

$$V_a(t) = V_b(t) = f_1(t) \qquad V_c(t) = V_d(t) = f_1(t) + f_2(t) \qquad (g)$$

$$M_a(t) = 0 \qquad M_b(t) = \frac{L}{2}f_1(t) \qquad M_d(t) = Lf_1(t) + \frac{L}{2}f_2(t) \qquad (h)$$

where $f_1(t)$ and $f_2(t)$ are known from Eqs. (f) and (b).

6. *Determine the internal forces at* $t = 0.18$ *sec.* At $t = 0.18$ sec, from Eq. (b), $q_1 = 1.217$ in. and $q_2 = 0.0159$ in. Substituting these in Eqs. (d) and (e) gives numerical values for the equivalent static forces shown in Fig. E12.6c and d, wherein the shearing forces and bending moments due to each mode are plotted. The combined values of these element forces are shown in Fig. E12.6e.

PART C: MODAL RESPONSE CONTRIBUTIONS

12.8 MODAL EXPANSION OF EXCITATION VECTOR $\mathbf{p}(t) = \mathbf{s}p(t)$

We now consider a common loading case in which the applied forces $p_j(t)$ have the same time variation $p(t)$, and their spatial distribution is defined by $\mathbf{s}$, independent of time. Thus

$$\mathbf{p}(t) = \mathbf{s}p(t) \qquad (12.8.1)$$

A central idea of this formulation, which we will find instructive, is to expand the vector $\mathbf{s}$ as

$$\mathbf{s} = \sum_{r=1}^{N} \mathbf{s}_r = \sum_{r=1}^{N} \Gamma_r\,\mathbf{m}\,\phi_r \qquad (12.8.2)$$

Premultiplying both sides of Eq. (12.8.2) by ϕ_n^T and utilizing the orthogonality property of modes gives

$$\Gamma_n = \frac{\phi_n^T \mathbf{s}}{M_n} \tag{12.8.3}$$

The contribution of the nth mode to **s** is

$$\mathbf{s}_n = \Gamma_n \mathbf{m} \phi_n \tag{12.8.4}$$

which is independent of how the modes are normalized. This should be clear from the structure of Eqs. (12.8.3) and (12.8.4).

Equation (12.8.2) may be viewed as an expansion of the distribution **s** of applied forces in terms of inertia force distributions $\mathbf{s}_n$ associated with natural modes. This interpretation becomes apparent by considering the structure vibrating in its nth mode with accelerations $\ddot{\mathbf{u}}_n(t) = \ddot{q}_n(t)\,\phi_n$. The associated inertia forces are

$$(\mathbf{f}_I)_n = -\mathbf{m}\ddot{\mathbf{u}}_n(t) = -\mathbf{m}\,\phi_n\,\ddot{q}_n(t)$$

and their spatial distribution, given by the vector $\mathbf{m}\phi_n$, is the same as that of $\mathbf{s}_n$.

The expansion of Eq. (12.8.2) has two useful properties: (1) the force vector $\mathbf{s}_n\,p(t)$ produces response only in the nth mode but no response in any other mode; and (2) the dynamic response in the nth mode is due entirely to the partial force vector $\mathbf{s}_n\,p(t)$ (see Derivation 12.1).

To study the modal expansion of the force vector $\mathbf{s}\,p(t)$ further, we consider the structure of Fig. 12.8.1: a five-story shear building (i.e., flexurally rigid floor beams and slabs) with lumped mass m at each floor, and same story stiffness k for all stories.

Floor Mass   Story Stiffness

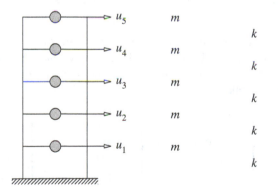

Figure 12.8.1   Uniform five-story shear building.

The mass and stiffness matrices of the structure are

$$\mathbf{m} = m \begin{bmatrix} 1 & & & & \\ & 1 & & & \\ & & 1 & & \\ & & & 1 & \\ & & & & 1 \end{bmatrix} \qquad \mathbf{k} = k \begin{bmatrix} 2 & -1 & & & \\ -1 & 2 & -1 & & \\ & -1 & 2 & -1 & \\ & & -1 & 2 & -1 \\ & & & -1 & 1 \end{bmatrix}$$

Determined by solving the eigenvalue problem, the natural frequencies are

$$\omega_n = \alpha_n \left(\frac{k}{m}\right)^{1/2}$$

where $\alpha_1 = 0.285$, $\alpha_2 = 0.831$, $\alpha_3 = 1.310$, $\alpha_4 = 1.682$, and $\alpha_5 = 1.919$. For a structure with $m = 100$ kips/g, the natural vibration modes, which have been normalized to obtain $M_n = 1$, are (Fig. 12.8.2)

$$\phi_1 = \begin{Bmatrix} 0.334 \\ 0.641 \\ 0.895 \\ 1.078 \\ 1.173 \end{Bmatrix} \quad \phi_2 = \begin{Bmatrix} -0.895 \\ -1.173 \\ -0.641 \\ 0.334 \\ 1.078 \end{Bmatrix} \quad \phi_3 = \begin{Bmatrix} 1.173 \\ 0.334 \\ -1.078 \\ -0.641 \\ 0.895 \end{Bmatrix} \quad \phi_4 = \begin{Bmatrix} -1.078 \\ 0.895 \\ 0.334 \\ -1.173 \\ 0.641 \end{Bmatrix} \quad \phi_5 = \begin{Bmatrix} 0.641 \\ -1.078 \\ 1.173 \\ -0.895 \\ 0.334 \end{Bmatrix}$$

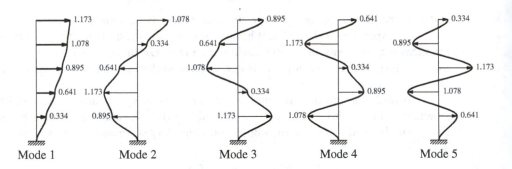

Figure 12.8.2 Natural modes of vibration of uniform five-story shear building.

Consider two different sets of applied forces: $\mathbf{p}(t) = \mathbf{s}_a p(t)$ and $\mathbf{p}(t) = \mathbf{s}_b p(t)$, where $\mathbf{s}_a^T = \langle 0 \; 0 \; 0 \; 0 \; 1 \rangle$ and $\mathbf{s}_b^T = \langle 0 \; 0 \; 0 \; -1 \; 2 \rangle$; note that the resultant force is unity in both cases (Fig. 12.8.3). Substituting for $\mathbf{m}$, ϕ_n, and $\mathbf{s} = \mathbf{s}_a$ in Eqs. (12.8.4) and (12.8.3) gives the modal contributions $\mathbf{s}_n$:

$$\mathbf{s}_1 = \begin{Bmatrix} 0.101 \\ 0.195 \\ 0.272 \\ 0.327 \\ 0.356 \end{Bmatrix} \quad \mathbf{s}_2 = \begin{Bmatrix} -0.250 \\ -0.327 \\ -0.179 \\ 0.093 \\ 0.301 \end{Bmatrix} \quad \mathbf{s}_3 = \begin{Bmatrix} 0.272 \\ 0.077 \\ -0.250 \\ -0.149 \\ 0.208 \end{Bmatrix} \quad \mathbf{s}_4 = \begin{Bmatrix} -0.179 \\ 0.149 \\ 0.055 \\ -0.195 \\ 0.106 \end{Bmatrix} \quad \mathbf{s}_5 = \begin{Bmatrix} 0.055 \\ -0.093 \\ 0.101 \\ -0.077 \\ 0.029 \end{Bmatrix}$$

Similarly, for $\mathbf{s} = \mathbf{s}_b$, the $\mathbf{s}_n$ vectors are

$$\mathbf{s}_1 = \begin{Bmatrix} 0.110 \\ 0.210 \\ 0.294 \\ 0.354 \\ 0.385 \end{Bmatrix} \quad \mathbf{s}_2 = \begin{Bmatrix} -0.423 \\ -0.553 \\ -0.302 \\ 0.157 \\ 0.508 \end{Bmatrix} \quad \mathbf{s}_3 = \begin{Bmatrix} 0.739 \\ 0.210 \\ -0.679 \\ -0.403 \\ 0.564 \end{Bmatrix} \quad \mathbf{s}_4 = \begin{Bmatrix} -0.685 \\ 0.569 \\ 0.212 \\ -0.746 \\ 0.407 \end{Bmatrix} \quad \mathbf{s}_5 = \begin{Bmatrix} 0.259 \\ -0.436 \\ 0.475 \\ -0.363 \\ 0.135 \end{Bmatrix}$$

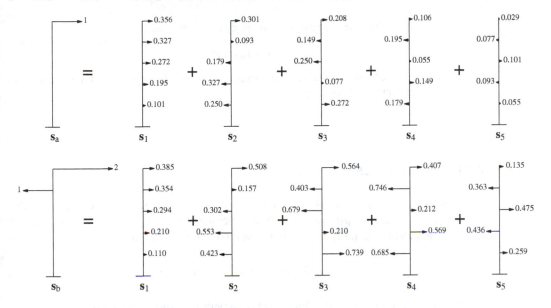

Figure 12.8.3   Modal expansion of excitation vectors s_a and s_b.

Both sets of vectors are displayed in Fig. 12.8.3. The contributions of the higher modes to s are larger for s_b than for s_a, suggesting that these modes may contribute more to the response if the force distribution is s_b than if it is s_a. We will return to this observation in Section 12.11.

Derivation 12.1

The first property can be demonstrated from the generalized force for the rth mode:

$$P_r(t) = \phi_r^T s_n p(t) = \Gamma_n (\phi_r^T m \phi_n) p(t) \tag{a}$$

Because of Eq. (10.4.1b), the orthogonality property of modes,

$$P_r(t) = 0 \qquad r \neq n \tag{b}$$

indicating that the excitation vector $s_n p(t)$ produces no generalized force and hence no response in the rth mode, $r \neq n$. Equation (a) for $r = n$ is

$$P_n(t) = \Gamma_n M_n p(t) \tag{c}$$

which is nonzero, implying that $s_n p(t)$ produces a response only in the nth mode.

The second property becomes obvious by examining the generalized force for the nth mode associated with the total force vector:

$$P_n(t) = \phi_n^T s p(t)$$

Substituting Eq. (12.8.2) for s gives

$$P_n(t) = \sum_{r=1}^{N} \Gamma_r \left(\phi_n^T m \phi_r \right) p(t)$$

which, after utilizing the orthogonality property of modes, reduces to

$$P_n(t) = \Gamma_n M_n p(t) \tag{d}$$

This generalized force for the complete force vector $\mathbf{s}p(t)$ is the same as Eq. (c) associated with the partial force vector $\mathbf{s}_n p(t)$.

12.9 MODAL ANALYSIS FOR $\mathbf{p}(t) = \mathbf{s}p(t)$

The dynamic analysis of an MDF system subjected to forces $\mathbf{p}(t)$ is specialized in this section for the excitation $\mathbf{p}(t) = \mathbf{s}p(t)$. The generalized force $P_n(t) = \Gamma_n M_n p(t)$ for the nth mode is substituted in Eq. (12.4.6) to obtain the modal equation:

$$\ddot{q}_n + 2\zeta_n \omega_n \dot{q}_n + \omega_n^2 q_n = \Gamma_n p(t) \tag{12.9.1}$$

The factor Γ_n that multiplies the force $p(t)$ is sometimes called a modal participation factor, but we avoid this terminology because Γ_n has two disadvantages; it is not independent of how the mode is normalized, or a measure of the contribution of the mode to a response quantity. Both these drawbacks are overcome by modal contribution factors that are defined in the next section.

We will write the solution $q_n(t)$ in terms of the response of an SDF system. Consider such a system with unit mass, and vibration properties—natural frequency ω_n and damping ratio ζ_n—of the nth mode of the MDF system excited by the force $p(t)$. The response of this nth-*mode SDF system* is governed by Eq. (1.5.2) with $m = 1$ and $\zeta = \zeta_n$, which is repeated here with u replaced by D_n to emphasize its connection with the nth mode:

$$\ddot{D}_n + 2\zeta_n \omega_n \dot{D}_n + \omega_n^2 D_n = p(t) \tag{12.9.2}$$

Comparing Eqs. (12.9.2) and (12.9.1) gives

$$q_n(t) = \Gamma_n D_n(t) \tag{12.9.3}$$

Thus $q_n(t)$ is readily available once Eq. (12.9.2) has been solved for $D_n(t)$, utilizing the available results for SDF systems subjected to, for example, harmonic, step, and impulsive forces (Chapters 3 and 4).

Then the contribution of the nth mode to nodal displacements $\mathbf{u}(t)$, Eq. (12.5.1), is

$$\mathbf{u}_n(t) = \Gamma_n \, \phi_n D_n(t) \tag{12.9.4}$$

Substituting Eq. (12.9.3) in (12.6.2) gives the equivalent static forces:

$$\mathbf{f}_n(t) = \mathbf{s}_n\left[\omega_n^2 D_n(t)\right] \tag{12.9.5}$$

The nth-mode contribution $r_n(t)$ to any response quantity $r(t)$ is determined by static analysis of the structure subjected to forces $\mathbf{f}_n(t)$. If r_n^{st} denotes the *modal static response*, the static value (indicated by superscript "st") of r due to external forces $\mathbf{s}_n$,[†] then

$$r_n(t) = r_n^{st} \left[\omega_n^2 D_n(t) \right] \tag{12.9.6}$$

Combining the response contributions of all the modes gives the total response:

$$r(t) = \sum_{n=1}^{N} r_n(t) = \sum_{n=1}^{N} r_n^{st} \left[\omega_n^2 D_n(t) \right] \tag{12.9.7}$$

The modal analysis procedure just presented, a special case of the one presented in Section 12.7, has the advantage of providing a basis for identifying and understanding the factors that influence the relative modal contributions to the response, as we shall see in Section 12.11.

Interpretation of modal analysis. In the first phase of the modal analysis procedure, the vibration properties—natural frequencies and modes—of the structure are computed, and the force distribution $\mathbf{s}$ is expanded into its modal components $\mathbf{s}_n$. The rest of the procedure is shown schematically in Fig. 12.9.1 to emphasize the underlying concepts. The contribution $r_n(t)$ of the nth mode to the dynamic response is obtained by multiplying the results of two analyses: (1) static analysis of the structure subjected to external forces $\mathbf{s}_n$, and (2) dynamic analysis of the nth-mode SDF system excited by the force $p(t)$. Thus, modal analysis requires static analysis of the structure for N sets of external forces, $\mathbf{s}_n$, $n = 1, 2, \ldots, N$, and dynamic analysis of N different SDF systems. Combining the modal responses gives the dynamic response of the structure.

12.10 MODAL CONTRIBUTION FACTORS

The contribution r_n of the nth mode to response quantity r, Eq. (12.9.6), can be expressed as

$$r_n(t) = r^{st} \bar{r}_n \left[\omega_n^2 D_n(t) \right] \tag{12.10.1}$$

where r^{st} is the static value of r due to external forces $\mathbf{s}$ and the nth *modal contribution factor*:

$$\bar{r}_n = \frac{r_n^{st}}{r^{st}} \tag{12.10.2}$$

[†]Although we loosely refer to $\mathbf{s}_n$ as forces, they are dimensionless because $p(t)$ has units of force. Thus, r_n^{st} does not have the same units as r, but Eq. (12.9.6) gives the correct units for r_n and Eq. (12.9.7) for r.

Mode	Static Analysis of Structure	Dynamic Analysis of SDF System	Modal Contribution to Dynamic Response
1	Forces s_1 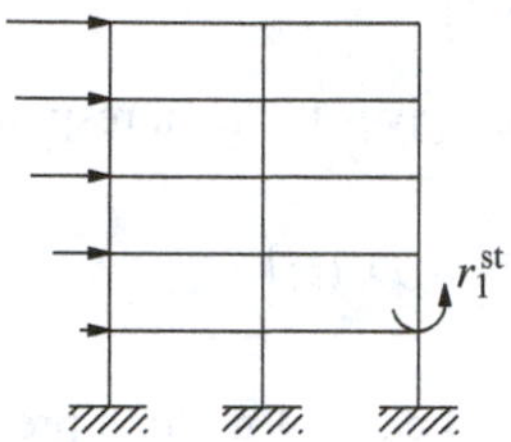 r_1^{st}	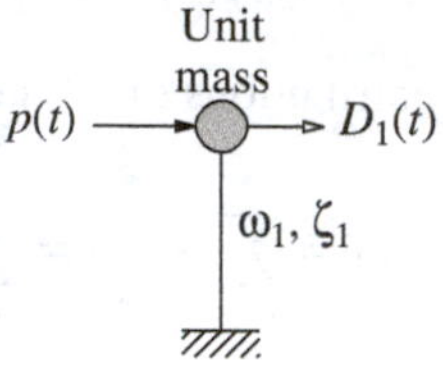 Unit mass $p(t) \longrightarrow \bullet \longrightarrow D_1(t)$ ω_1, ζ_1	$r_1(t) = r_1^{st} [\omega_1^2 D_1(t)]$
2	Forces s_2 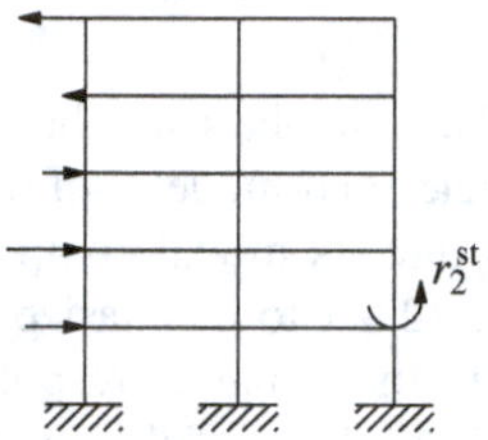 r_2^{st}	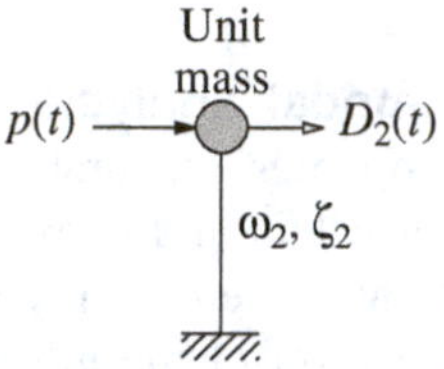 Unit mass $p(t) \longrightarrow \bullet \longrightarrow D_2(t)$ ω_2, ζ_2	$r_2(t) = r_2^{st} [\omega_2^2 D_2(t)]$
. . .	.	.	. . .
N	Forces s_N r_N^{st} 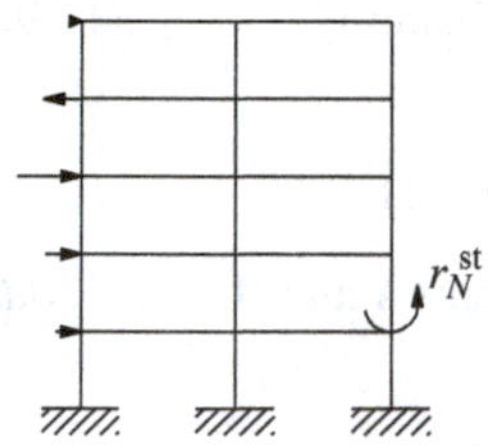	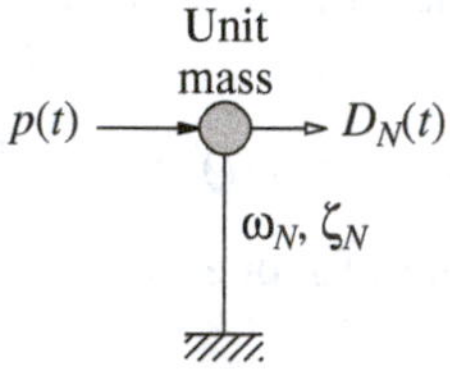 Unit mass $p(t) \longrightarrow \bullet \longrightarrow D_N(t)$ ω_N, ζ_N	$r_N(t) = r_N^{st} [\omega_N^2 D_N(t)]$
Total response			$r(t) = \sum_{n=1}^{N} r_n(t)$

Figure 12.9.1 Conceptual explanation of modal analysis.

These modal contribution factors $\bar{r}_n$ have three important properties:

1. They are dimensionless.
2. They are independent of how the modes are normalized.
3. The sum of the modal contribution factors over all modes is unity, that is,

$$\sum_{n=1}^{N} \bar{r}_n = 1 \qquad (12.10.3)$$

The first property is obvious from their definition [Eq. (12.10.2)]. The second property becomes obvious by noting that r_n^{st} is the static effect of $\mathbf{s}_n$, which does not depend on the normalization, and the modal properties do not enter into r^{st}. Equation (12.10.3) can be proven by recognizing that $\mathbf{s} = \sum \mathbf{s}_n$ [Eq. (12.8.2)], which implies that $r^{\text{st}} = \sum r_n^{\text{st}}$. Dividing by r^{st} gives the desired result.

12.11 MODAL RESPONSES AND REQUIRED NUMBER OF MODES

Consider the displacement $D_n(t)$ of the nth-mode SDF system and define its peak value $D_{no} \equiv \max_t |D_n(t)|$. The corresponding value of $r_n(t)$, Eq. (12.10.1), is

$$r_{no} = r^{\text{st}} \bar{r}_n \omega_n^2 D_{no} \qquad (12.11.1)$$

We shall rewrite this equation in terms of the dynamic response factor introduced in Chapters 3 and 4. For the nth-mode SDF system governed by Eq. (12.9.2), this factor is $R_{dn} = D_{no}/(D_{n,\text{st}})_o$, where $(D_{n,\text{st}})_o$ is the peak value of $D_{n,\text{st}}(t)$, the static response. Obtained by dropping the $\dot{D}_n$ and $\ddot{D}_n$ terms in Eq. (12.9.2), $D_{n,\text{st}}(t) = p(t)/\omega_n^2$ and its peak value is $(D_{n,\text{st}})_o = p_o/\omega_n^2$. Therefore, Eq. (12.11.1) becomes

$$r_{no} = p_o r^{\text{st}} \bar{r}_n R_{dn} \qquad (12.11.2)$$

The algebraic sign of r_{no} is the same as that of the modal static response $r_n^{\text{st}} = r^{\text{st}} \bar{r}_n$ because D_{no} is positive by definition [Eq. (12.11.1)]. Although it has an algebraic sign, r_{no} will be referred to as the peak value of the contribution of the nth mode to response r or, for brevity, the peak modal response because it corresponds to the peak value of $D_n(t)$.

Equation (12.11.2) indicates that the peak modal response is the product of four quantities: (1) the dimensionless dynamic response factor R_{dn} for the nth-mode SDF system excited by force $p(t)$; (2) the dimensionless modal contribution factor $\bar{r}_n$ for the response quantity r; (3) r^{st}, the static value of r due to the external forces $\mathbf{s}$; and (4) p_o, the peak value of $p(t)$. The quantities r^{st} and $\bar{r}_n$ depend on the spatial distribution $\mathbf{s}$ of the applied forces but are independent of the time variation $p(t)$ of the forces; on the other hand, R_{dn} depends on $p(t)$, but is independent of $\mathbf{s}$.

The modal contribution factor $\bar{r}_n$ and the dynamic response factor R_{dn} influence the relative response contributions of the various vibration modes, and hence the minimum number of modes that should be included in dynamic analysis. This becomes obvious by noting that among the four quantities that enter in Eq. (12.11.2), p_o and r^{st} are independent of the mode number, but $\bar{r}_n$ and R_{dn} vary with n.

How many modes should one include in modal analysis? The response contributions of all the modes must obviously be included in order to obtain the exact value of the response, but few modes can usually provide sufficiently accurate results. Typically, in analyzing an N-DOF system, the first J modes are included, where J may be much smaller than N, and the modal summation of Eq. (12.9.7) is truncated accordingly. Thus, we need to compute the natural frequencies, natural modes, and the response $D_n(t)$ only for the first J modes, leading to computational savings.

Before studying the relative response contributions of the various vibration modes and the number of modes to be included for specific examples, we make a couple of general observations. If only the first J modes are included, the error in the static response is

$$e_J = 1 - \sum_{n=1}^{J} \bar{r}_n \tag{12.11.3}$$

For a fixed J the error e_J depends on the spatial distribution $\mathbf{s}$ of the applied forces. For any $\mathbf{s}$ the error e_J will be zero when all the modes are included ($J = N$) because of Eq. (12.10.3), and the error will be unity when no modes are included ($J = 0$). Thus, modal analysis can be truncated when $|e_J|$, the absolute value of e_J, becomes sufficiently small for the response quantity r of interest. However, the relative values of R_{dn} for various modes also influence the minimum number of modes to be included. R_{dn} depends on the time variation $p(t)$ of the applied forces and on the vibration properties ω_n and ζ_n of the nth mode.

Next, we will study the relative contributions of various modes to the response and the number of modes that should be included in dynamic analysis with reference to the structure of Fig. 12.8.1 and the applied force distributions $\mathbf{s}_a$ and $\mathbf{s}_b$ in Fig. 12.8.3. For this purpose, we must investigate the two influencing factors: the modal contribution factor, which depends on the spatial distribution of forces, and the dynamic response factor, which is controlled by the time variation of forces.

12.11.1 Modal Contribution Factors

In this section we first describe a procedure to determine the modal contribution factors for the base shear V_b and roof (Nth floor) displacement u_N of a multistory building. Figure 12.11.1 shows the external forces $\mathbf{s}$ and $\mathbf{s}_n$. The latter is defined by Eq. (12.8.4); in particular, the lateral force at the jth floor level is the jth element of $\mathbf{s}_n$:

$$s_{jn} = \Gamma_n m_j \phi_{jn} \tag{12.11.4}$$

where m_j is the lumped mass and ϕ_{jn} the nth-mode shape value at the jth floor. As a result of the static forces $\mathbf{s}_n$ (Fig. 12.11.1b), the base shear is

$$V_{bn}^{st} = \sum_{j=1}^{N} s_{jn} = \Gamma_n \sum_{j=1}^{N} m_j \phi_{jn} \tag{12.11.5}$$

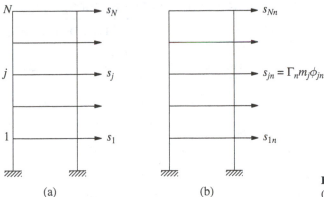

Figure 12.11.1   External forces: (a) **s**; (b) **s**$_n$.

(a)   (b)

and the floor displacements are $\mathbf{u}_n^{st} = \mathbf{k}^{-1}\mathbf{s}_n$. Inserting Eq. (12.8.4) and using Eq. (10.2.4) gives $\mathbf{u}_n^{st} = \left(\Gamma_n/\omega_n^2\right)\boldsymbol{\phi}_n$. In particular, the roof displacement is

$$u_{Nn}^{st} = \frac{\Gamma_n}{\omega_n^2}\phi_{Nn} \tag{12.11.6}$$

Equations (12.11.5) and (12.11.6) define the modal static responses r_n^{st} for base shear and roof displacement. With r_n^{st} known, the modal contribution factor is obtained from Eq. (12.10.2), where r^{st} is computed by static analysis of the building subjected to forces **s** (Fig. 12.11.1a).

Now, we study how the modal contribution factors depend on the spatial distribution of the applied forces. The procedure just described is used to determine the modal contribution factors for roof displacement and base shear of the five-story shear frame (Fig. 12.8.1) for the two different force distribution vectors $\mathbf{s}_a$ and $\mathbf{s}_b$ introduced in Section 12.8. These computations require the modal expansions of $\mathbf{s}_a$ and $\mathbf{s}_b$ and the natural modes of the system, which are available in Figs. 12.8.2 and 12.8.3. The modal contribution factors and their cumulative values considering the first J modes are presented in Table 12.11.1. Consistent with Eq. (12.10.3), the sum of modal contribution factors over all modes is unity, although the convergence may or may not be monotonic. For the structure and force distributions considered, Table 12.11.1 indicates that the convergence is monotonic for roof displacement but not for base shear.

The data of Table 12.11.1 permit two useful observations pertaining to relative values of the modal responses:

1. For a particular spatial distribution of forces, the modal contribution factors for higher modes are larger for base shear than for roof displacement, suggesting that the higher modes contribute more to base shear (and other element forces) than to roof displacement (and other floor displacements).

2. For a particular response quantity, the modal contribution factors for higher modes are larger for force distribution $\mathbf{s}_b$ than for $\mathbf{s}_a$, suggesting that the higher modes contribute more to a response in the $\mathbf{s}_b$ case. Recall that the modal expansion of $\mathbf{s}_a$ and $\mathbf{s}_b$ in Section 12.8 had suggested the same conclusion.

TABLE 12.11.1   MODAL AND CUMULATIVE CONTRIBUTION FACTORS

| | Force Distribution, s_a | | | | Force Distribution, s_b | | | |
| | Roof Displacement | | Base Shear | | Roof Displacement | | Base Shear | |
Mode n or Number of Modes, J	$\bar{u}_{5n}$	$\sum\limits_{n=1}^{J} \bar{u}_{5n}$	$\bar{V}_{bn}$	$\sum\limits_{n=1}^{J} \bar{V}_{bn}$	$\bar{u}_{5n}$	$\sum\limits_{n=1}^{J} \bar{u}_{5n}$	$\bar{V}_{bn}$	$\sum\limits_{n=1}^{J} \bar{V}_{bn}$
1	0.880	0.880	1.252	1.252	0.792	0.792	1.353	1.353
2	0.087	0.967	−0.362	0.890	0.123	0.915	−0.612	0.741
3	0.024	0.991	0.159	1.048	0.055	0.970	0.431	1.172
4	0.008	0.998	−0.063	0.985	0.024	0.994	−0.242	0.930
5	0.002	1.000	0.015	1.000	0.006	1.000	0.070	1.000

How many modes should be included in modal analysis? We first examine how the number of modes required to keep the error in static response below some selected value is influenced by the spatial distribution **s** of the applied forces. If the objective is to keep $|e_J| < 0.05$ (5%) for the base shear, the data of Table 12.11.1 indicate that three modes suffice for the force distribution s_a, whereas all five modes need to be included in the case of s_b. For the same accuracy in the roof displacement, two modes suffice for the force distribution s_a, but three modes are needed in the case of s_b. More modes need to be included for the force distribution s_b than for s_a because, as mentioned earlier, the modal contribution factors for higher modes are larger for s_b than for s_a.

We next examine how the number of modes required is influenced by the response quantity of interest. If the objective is to keep $|e_J| < 0.05$ (5%), three modes need to be included to determine the base shear for force distribution s_a, whereas two modes would suffice for roof displacement. To achieve the same accuracy for force distribution s_b, all five modes are needed for base shear, whereas three modes would suffice for roof displacement. More modes need to be included for base shear than for roof displacement because, as mentioned earlier, the modal contribution factors for higher modes are larger for base shear than for roof displacement.

It is not necessary to repeat the preceding analysis for all response quantities. Instead, some of the key response quantities, especially those that are likely to be sensitive to higher modes, should be identified for deciding the number of modes to be included in modal analysis.

12.11.2 Dynamic Response Factor

We now study how the modal response contributions depend on the time variation of the excitation. The dynamic response to $p(t)$ is characterized by the dynamic response factor R_{dn} in Eq. (12.11.2). This factor was derived in Chapter 3 for harmonic excitation (Fig. 3.2.6) and in Chapter 4 for various excitations, including the half-cycle sine pulse (Fig. 4.8.3c). In Fig. 12.11.2, R_d for harmonic force of period T is plotted against T_n/T

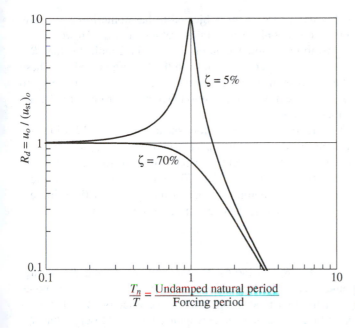

Figure 12.11.2   Dynamic response factor for harmonic force; $\zeta = 5\%$ and 70%.

for SDF systems with natural period T_n and two damping ratios: $\zeta = 5$ and 70%; R_d for a half-cycle sine pulse force of duration t_d is plotted against T_n/t_d in Fig. 12.11.3 for undamped SDF systems.

How R_{dn}, the value of R_d for the nth mode, for a given excitation $p(t)$ varies with n depends on where the natural periods T_n fall on the period scale. In the case of pulse excitation, Fig. 12.11.3 shows that R_{dn} varies over a narrow range for a wide range of T_n and could have similar values for several modes. Thus, several modes would generally have to be included in modal analysis with their relative response contributions, Eq. (12.11.2), determined primarily by the relative values of $\bar{r}_n$, the modal contribution factors. The same

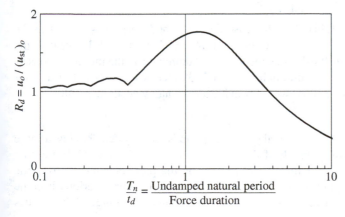

Figure 12.11.3   Dynamic response factor for half-cycle sine pulse force; $\zeta = 0$.

conclusion also applies to highly damped systems subjected to harmonic force because, as seen in Fig. 12.11.2, several modes could have similar values of R_{dn}. However, for lightly damped systems subjected to harmonic excitation, Fig. 12.11.2 indicates that R_{dn} is especially large for modes with natural period T_n close to the forcing period T. These modes would contribute most to the response and are perhaps the only modes that need to be included in modal analysis unless the modal contribution factors $\bar{r}_n$ for these modes are much smaller than for some other modes.

To explore these ideas further, consider the five-story shear frame (Fig. 12.8.1) without damping subjected to harmonic forces $\mathbf{p}(t) = \mathbf{s}_b p(t)$, where $p(t) = p_o \sin(2\pi t/T)$. Further consider four different values of the forcing period T relative to the fundamental natural period of the system: $T_1/T = 0.75, 2.75, 3.50,$ and 4.30. For each of these forcing periods the R_{dn} values for the five natural periods of the system (defined in Section 12.8) are identified in Fig. 12.11.4. These data permit the following observations for each of the four cases:

1. $T_1/T = 0.75$ (Fig. 12.11.4a): R_{dn} is largest for the first mode, and R_{d1} is significantly larger than other R_{dn}. The larger R_{d1} combined with the larger modal contribution factor $\bar{V}_{b1}$ for the first mode, compared to these factors $\bar{V}_{bn}$ for other modes (Table 12.11.1), will make the first-mode response largest. Observe that R_{dn} for higher modes are close to 1, indicating that the response in these modes is essentially static.

2. $T_1/T = 2.75$ (Fig. 12.11.4b): For this case $T_2/T = 0.943$, and therefore $R_{d2} = 8.89$ is much larger than the other R_{dn}. Therefore, the second-mode response will dominate all the modal responses, even the first-mode response, although $\bar{V}_{b1} = 1.353$ is more than twice $\bar{V}_{b2} = -0.612$. The second mode alone would provide a reasonably accurate result.

3. $T_1/T = 3.5$ (Fig. 12.11.4c): R_{d2} and R_{d3} are similar in magnitude, much larger than R_{d1}, and significantly larger than R_{d4} and R_{d5}. This suggests that the second- and third-mode contributions to the response would be the larger ones. The second-mode response would exceed the third-mode response because the magnitude of $\bar{V}_{b2} = -0.612$ is larger than $\bar{V}_{b3} = 0.431$.

4. $T_1/T = 4.3$ (Fig. 12.11.4d): For this case the third-mode period is closest to the exciting period ($T_3/T = 0.935$), and therefore $R_{d3} = 7.89$ is much larger than the other R_{dn}. Therefore, the third-mode response will dominate all the modal responses, even the first-mode response, although $\bar{V}_{b1} = 1.353$ exceeds $\bar{V}_{b3} = 0.431$ by a factor of over 3. The third mode alone would provide a reasonably accurate result.

It is not necessary to compute all the natural periods of a system having a large number of DOFs in ascertaining which of the R_{dn} values are significant. Only the first few natural periods need to be calculated and located on the plot showing the dynamic response factor. Then the approximate locations of the higher natural periods become readily apparent, thus providing sufficient information to estimate the range of R_{dn} values

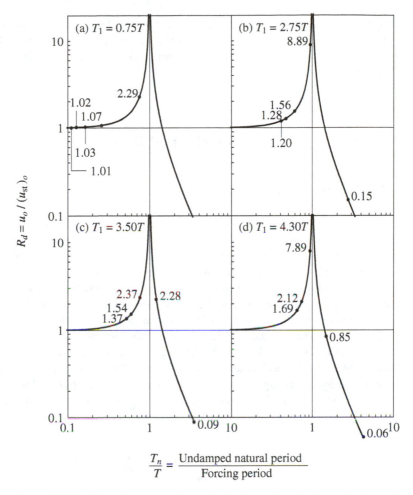

$$\frac{T_n}{T} = \frac{\text{Undamped natural period}}{\text{Forcing period}}$$

Figure 12.11.4 Dynamic response factors R_{dn} for five modes of undamped system and four forcing periods T: (a) $T_1 = 0.75T$; (b) $T_1 = 2.75T$; (c) $T_1 = 3.50T$; (d) $T_1 = 4.30T$.

and to make a preliminary decision on the modes that may contribute significant response. Precise values of R_{dn} can then be calculated for these modes to be included in modal analysis.

In judging the contribution of a natural mode to the dynamic response of a structure, it is necessary to consider the combined effects of the modal contribution factor $\bar{r}_n$ and the dynamic response factor R_{dn}. The two factors have been discussed separately in this and the preceding sections because $\bar{r}_n$ depends on the spatial distribution s of the applied forces, whereas R_{dn} depends on the time variation $p(t)$ of the excitation. However, they both enter into the modal response, Eq. (12.11.2). By retaining only the first few modes with significant values of $\bar{r}_n$ or R_{dn}, or both, the computational effort can be reduced. The resulting computational savings may not be significant in dynamic analysis of systems with a few dynamic DOFs, such as the five-story shear frame considered here. However, substantial reduction in computation can be achieved for practical, complex structures that may require hundreds or thousands of DOFs for their idealization.

PART D: SPECIAL ANALYSIS PROCEDURES

12.12 STATIC CORRECTION METHOD

In Section 12.11 we have shown that the dynamic response factor R_{dn} for some of the higher modes of a structure may be only slightly larger than unity. For the five-story shear frame subjected to harmonic excitation with $T_1/T = 0.75$, $R_{dn} = 1.07$, 1.03, 1.02, and 1.01 for the second, third, fourth, and fifth modes, respectively (Fig. 12.11.4a). Such is the case when the higher-mode period T_n is much shorter than the period T of the harmonic excitation or the duration t_d of an impulsive excitation (Fig. 12.11.3). The response in such a higher mode could be determined by the easier static analysis instead of dynamic analysis. This is the essence of the *static correction method*, developed next.

Suppose that we include all N modes in the analysis but divide them into two parts: (1) the first N_d modes with natural periods T_n such that the dynamic effects are significant, as indicated by R_{dn} being significantly different than 1; and (2) modes $N_d + 1$ to N with natural periods T_n such that R_{dn} is close to 1. Then the modal contributions to the response can be divided into two parts:

$$r(t) = \sum_{n=1}^{N_d} r_n(t) + \sum_{n=N_d+1}^{N} r_n(t) \tag{12.12.1}$$

The nth-mode response $r_n(t)$ is given by Eq. (12.10.1), where $D_n(t)$ is the dynamic response of the nth-mode SDF system governed by Eq. (12.9.2). A quasi-static solution of this equation gives $D_n(t)$ for modes $N_d + 1$ to N; dropping the velocity $\dot{D}_n(t)$ and acceleration $\ddot{D}_n(t)$ leads to

$$\omega_n^2 D_n(t) = p(t) \tag{12.12.2}$$

Substituting Eqs. (12.10.1) and (12.12.2) in Eq. (12.12.1) gives

$$r(t) = r^{st} \sum_{n=1}^{N_d} \bar{r}_n \left[\omega_n^2 D_n(t)\right] + r^{st} p(t) \sum_{n=N_d+1}^{N} \bar{r}_n \tag{12.12.3}$$

Thus, Eq. (12.9.2) needs to be solved by dynamic analysis procedures (e.g., analytical solution or numerical integration of the differential equation) only for the first N_d modes.

However, Eq. (12.12.3) suggests that the modal contribution factors $\bar{r}_n$ are still needed for the higher modes $N_d + 1$ to N and, as seen earlier, the natural frequencies and modes of the system are required to compute $\bar{r}_n$.

Therefore, we look for a way to circumvent this requirement by rewriting

$$\sum_{n=N_d+1}^{N} \bar{r}_n = \sum_{n=1}^{N} \bar{r}_n - \sum_{n=1}^{N_d} \bar{r}_n = 1 - \sum_{n=1}^{N_d} \bar{r}_n \tag{12.12.4}$$

wherein we have utilized Eq. (12.10.3). Substituting Eq. (12.12.4) in (12.12.3) gives

$$r(t) = r^{\text{st}} \left[\sum_{n=1}^{N_d} \bar{r}_n \left(\omega_n^2 D_n(t) \right) + \left(1 - \sum_{n=1}^{N_d} \bar{r}_n \right) p(t) \right] \tag{12.12.5}$$

or

$$r(t) = r^{\text{st}} \left[p(t) + \sum_{n=1}^{N_d} \bar{r}_n \left(\omega_n^2 D_n(t) - p(t) \right) \right] \tag{12.12.6}$$

When written in either of these forms, only the first N_d natural frequencies and modes are needed in computing the dynamic response. In Eq. (12.12.5) the second term in brackets is the quasi-static response solution for the higher modes, $n = N_d + 1$ to N, which may be considered as the static correction to the dynamic response solution given by the first term. This method is therefore known as the *static correction method*.

The static correction method is effective in analyses where many higher modes must be included to represent satisfactorily the spatial distribution **s** of the applied forces, but where the exciting force $p(t)$ is such that the dynamic response factor for only a few lower modes is significantly different than 1. In these situations the combined dynamic response of these few modes together with the static correction term will give results comparable to a classical modal analysis, including many more modes. If $p(t)$ is defined numerically, the static correction method requires substantially less computational effort by avoiding the numerical time-stepping solution for the higher, statically responding modes. The computational savings can be substantial because the time step to be used in numerical solution of the higher modal equations must be very short (Chapters 5 and 16).

Example 12.7

Compute the base shear response of the uniform five-story shear frame of Fig. 12.8.1 to $\mathbf{p}(t) = \mathbf{s}_b p(t)$, where $\mathbf{s}_b^T = \langle 0\ 0\ 0\ -1\ 2 \rangle$ and $p(t) = p_o \sin \omega t$ by two methods: (a) classical modal analysis, and (b) static correction method; $\omega/\omega_1 = T_1/T = 0.75$ and the system is undamped. Compute only the steady-state response.

Solution (a) *Classical modal analysis.* The nth-mode response is given by Eq. (12.10.1). The steady-state solution of Eq. (12.9.2) with $\zeta_n = 0$ and harmonic $p(t)$ is

$$D_n(t) = \frac{p_o}{\omega_n^2} R_{dn} \sin \omega t \qquad R_{dn} = \frac{1}{1 - (\omega/\omega_n)^2} \tag{a}$$

Note that this R_{dn} may be positive or negative, in contrast to the absolute value used in Chapter 3 and in Fig. 12.11.4. Retaining the algebraic sign is useful when dealing with the steady-state response of undamped systems to harmonic excitation. Substituting Eq. (a) in Eq. (12.10.1) gives the response in the nth mode:

$$r_n(t) = p_o r^{\text{st}} \bar{r}_n R_{dn} \sin \omega t \tag{b}$$

Combining the contributions of all modes gives the total response:

$$r(t) = p_o r^{\text{st}} \left(\sum_{n=1}^{N} \bar{r}_n R_{dn} \right) \sin \omega t \tag{c}$$

and its maximum value is

$$r_o = p_o r^{\text{st}} \sum_{n=1}^{N} \bar{r}_n R_{dn} \tag{d}$$

Equation (d) specialized for the base shear can be expressed as

$$\frac{V_{bo}}{p_o V_b^{\text{st}}} = \sum_{n=1}^{N} \bar{V}_{bn} R_{dn} \tag{e}$$

For the system of Fig. 12.8.1, $N = 5$ and the static base shear due to forces s_b (Fig. 12.8.3) is $V_b^{\text{st}} = 1$. Therefore, Eq. (e) becomes

$$\frac{V_{bo}}{p_o} = \sum_{n=1}^{5} \bar{V}_{bn} R_{dn} \tag{f}$$

Substituting for modal contribution factors $\bar{V}_{bn}$ from Table 12.11.1 and calculating R_{dn} from Eq. (a) leads to

$$\frac{V_{bo}}{p_o} = \bar{V}_{b1} R_{d1} + \bar{V}_{b2} R_{d2} + \bar{V}_{b3} R_{d3} + \bar{V}_{b4} R_{d4} + \bar{V}_{b5} R_{d5}$$

$$= (1.353)(2.29) + (-0.612)(1.07) + (0.431)(1.03) + (-0.242)(1.02)$$

$$+ (0.070)(1.01)$$

$$= 2.71$$

(b) *Static correction method.* Specializing Eq. (12.12.6) for base shear:

$$V_b(t) = V_b^{\text{st}} \left[p(t) + \sum_{n=1}^{N_d} \bar{V}_{bn} \left(\omega_n^2 D_n(t) - p(t) \right) \right] \tag{g}$$

Substituting for V_b^{st}, p_o, and $D_n(t)$ from Eq. (a) gives

$$V_b(t) = p_o \sin \omega t \left[1 + \sum_{n=1}^{N_d} \bar{V}_{bn} (R_{dn} - 1) \right] \tag{h}$$

The peak value V_{bo} of the base shear is given by

$$\frac{V_{bo}}{p_o} = 1 + \sum_{n=1}^{N_d} \bar{V}_{bn} (R_{dn} - 1) \tag{i}$$

Suppose that we consider the dynamic response only in the first mode. Then $N_d = 1$ and Eq. (i) gives

$$\frac{V_{bo}}{p_o} = 1 + \overline{V}_{b1}\,(R_{d1} - 1) = 1 + 1.353(2.29 - 1) = 2.75$$

This result is close to the exact result of 2.71 obtained from dynamic response solutions for all modes. This good agreement was anticipated in Section 12.11.2, where we noted that R_{dn} for $n > 1$ were only slightly larger than 1.

Note that both $p(t)$ and $D_n(t)$ vary as $\sin \omega t$ in this special case of steady-state response of undamped systems to harmonic excitation. In general, however, the time variation of $D_n(t)$ would differ from that of $p(t)$.

12.13 MODE ACCELERATION SUPERPOSITION METHOD

Another method that can provide the same general effect as the static correction method, called the *mode acceleration superposition method*, can be derived readily from the modal equations of motion. The total response is the sum of the modal response contributions, Eq. (12.10.1):

$$r(t) = r^{\mathrm{st}} \sum_{n=1}^{N} \overline{r}_n \left[\omega_n^2 D_n(t)\right] \qquad (12.13.1)$$

where $D_n(t)$ is governed by Eq. (12.9.2), which can be rewritten as

$$\omega_n^2 D_n(t) = p(t) - \ddot{D}_n(t) - 2\zeta_n\,\omega_n \dot{D}_n(t) \qquad (12.13.2)$$

Substituting Eq. (12.13.2) in Eq. (12.13.1) gives

$$r(t) = r^{\mathrm{st}} \sum_{n=1}^{N} \overline{r}_n \left[p(t) - \ddot{D}_n(t) - 2\zeta_n\omega_n \dot{D}_n(t) \right]$$

which, by using Eq. (12.10.3), can be expressed as

$$r(t) = r^{\mathrm{st}} p(t) - r^{\mathrm{st}} \sum_{n=1}^{N} \overline{r}_n [\ddot{D}_n(t) + 2\zeta_n\omega_n \dot{D}_n(t)] \qquad (12.13.3)$$

This can be interpreted as the quasi-static solution given by the first term on the right side, modified by the second term to obtain the dynamic response of the system. If the response in a higher mode, say the nth mode, is essentially static, the nth term in the summation will be negligible. Thus, if the response in all modes higher than the first N_d modes is essentially static, the summation can be truncated accordingly to obtain

$$r(t) = r^{\mathrm{st}} \left\{ p(t) - \sum_{n=1}^{N_d} \overline{r}_n \left[\ddot{D}_n(t) + 2\zeta_n\omega_n \dot{D}_n(t)\right] \right\} \qquad (12.13.4)$$

This method is often referred as the *mode acceleration superposition method* since

Eq. (12.13.4) involves the superposition of modal accelerations $\ddot{D}_n$ (and velocities $\dot{D}_n$) rather than modal displacements D_n.

The mode acceleration superposition method is equivalent to the static correction method, which becomes obvious by comparing Eqs. (12.13.4) and (12.12.6) in light of Eq. (12.9.2). Thus, the two methods should provide identical results except for minor differences arising from their numerical implementation. The choice between the two methods is often dictated by ease of implementation in a computer code. In this regard the static correction method is usually more convenient because it requires simple modification of classical modal analysis (or the classical mode displacement superposition method). In the classical procedure the first term in Eq. (12.12.5) is computed anyway for the specified number N_d of modes. For these modes the modal contribution factors $\bar{r}_n$ are available, so the second term can be computed with little additional effort.

12.14 MODE ACCELERATION SUPERPOSITION METHOD: ARBITRARY EXCITATION

Now that we understand the concepts underlying the static correction and mode acceleration superposition methods, two equivalent methods, the latter method is presented for arbitrary forces $\mathbf{p}(t)$, i.e., forces not restricted to $\mathbf{p}(t) = sp(t)$, together with a comprehensive example. Wave forces on offshore drilling platforms are an example of such excitation.

The displacement response of an N-DOF system is given by Eq. (12.5.2), repeated here for convenience:

$$\mathbf{u}(t) = \sum_{n=1}^{N} \mathbf{u}_n(t) = \sum_{n=1}^{N} \phi_n q_n(t) \tag{12.14.1}$$

where the modal coordinate $q_n(t)$ is governed by Eq. (12.4.6):

$$\ddot{q}_n + 2\zeta_n \omega_n \dot{q}_n + \omega_n^2 q_n = \frac{P_n(t)}{M_n} \tag{12.14.2}$$

where $P_n(t)$ and M_n were defined in Eq. (12.3.4). In the mode acceleration superposition method, the response due to the lower N_d modes is determined by dynamic analysis, i.e., solving Eq. (12.14.2), whereas the contributions of modes $N_d + 1$ to N are determined by static analysis. Dropping the velocity $\dot{q}_n(t)$ and acceleration $\ddot{q}_n(t)$ terms in Eq. (12.14.2) gives the quasi-static solution $q_n(t) = P_n(t)/K_n$, where the nth-mode generalized stiffness $K_n = \omega_n^2 M_n$; see Eq. (10.4.7). Substituting this $q_n(t)$ in the terms associated with $n = N_d + 1$ to N in Eq. (12.14.1) gives

$$\mathbf{u}(t) = \sum_{n=1}^{N_d} \phi_n q_n(t) + \sum_{n=N_d+1}^{N} \phi_n \frac{P_n(t)}{K_n} \tag{12.14.3}$$

where an individual term in the second summation represents the contribution of the nth mode to the static response. Rewriting the second summation as

$$\sum_{n=N_d+1}^{N} \phi_n \frac{P_n(t)}{K_n} = \sum_{n=1}^{N} \phi_n \frac{P_n(t)}{K_n} - \sum_{n=1}^{N_d} \phi_n \frac{P_n(t)}{K_n}$$

and noting that the first summation includes all modes and thus represents the (total) quasi-static response of the structure to $\mathbf{p}(t)$, the preceding equation becomes

$$\sum_{n=N_d+1}^{N} \phi_n \frac{P_n(t)}{K_n} = \mathbf{k}^{-1}\mathbf{p}(t) - \sum_{n=1}^{N_d} \phi_n \frac{P_n(t)}{K_n} \tag{12.14.4}$$

Substituting Eq. (12.14.4) in Eq. (12.14.3) gives

$$\mathbf{u}(t) = \mathbf{k}^{-1}\mathbf{p}(t) + \sum_{n=1}^{N_d} \phi_n \left[q_n(t) - \frac{P_n(t)}{K_n} \right] \tag{12.14.5}$$

The analysis procedure embodied in Eq. (12.14.5), which is a generalized version of Eq. (12.12.6), is known as the *static correction method*.

To express Eq. (12.14.5) in terms of modal velocity $\dot{q}_n$ and acceleration $\ddot{q}_n$, we divide Eq. (12.14.2) by ω_n^2 and rewrite it as

$$q_n(t) - \frac{P_n(t)}{K_n} = -\left[\frac{2\zeta_n}{\omega_n} \dot{q}_n(t) + \frac{1}{\omega_n^2} \ddot{q}_n(t) \right] \tag{12.14.6}$$

Substituting Eq. (12.14.6) in Eq. (12.14.5) gives

$$\mathbf{u}(t) = \mathbf{k}^{-1}\mathbf{p}(t) - \sum_{n=1}^{N_d} \phi_n \left[\frac{2\zeta_n}{\omega_n} \dot{q}_n(t) + \frac{1}{\omega_n^2} \ddot{q}_n(t) \right] \tag{12.14.7}$$

The analysis procedure embodied in Eq. (12.14.7), which is a generalized version of Eq. (12.13.4), is known as the *mode acceleration superposition method*. The first term in Eq. (12.14.7) represents the quasi-static solution, and the summation represents a correction applied to the quasi-static response to obtain the dynamic response.

It is obvious from the preceding derivation that the mode acceleration superposition method is equivalent to the static correction method. For either method it is necessary to decide N_d, the number of modes for which the response is to be determined by dynamic analysis. This decision is straightforward if the exciting forces are harmonic with frequency ω. Referring to Fig. 3.2.6, it is clear that dynamic analysis is necessary for all modes for which ω/ω_n is such that the deformation response factor R_d differs significantly from 1.0 or the phase angle ϕ differs significantly from zero. It is difficult to state similarly simple criteria for nonharmonic excitations that vary arbitrarily in time.

12.14.1 EXAMPLE

The system considered is a uniform cantilever tower of length L, mass per unit length m, flexural rigidity EI, and damping ratio of 5% in all natural vibration modes (Fig. 12.14.1a). For purposes of dynamic analysis, the system is discretized with 10 lumped masses, as shown in Fig. 12.14.1b, where $m_o = mL/10$ and the DOFs are identified. Harmonic forces applied at DOFs u_9 and u_{10} are $p_9(t) = -2p(t)$ and $p_{10}(t) = p(t)$, where

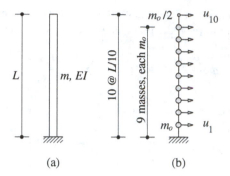

Figure 12.14.1 (a) Cantilever tower (b) Lumped mass system.

(a)   (b)

$p(t) = p_o \sin \omega t$ with $\omega/\omega_1 = 1.5$, where ω_1 is the fundamental natural frequency of the system[†]; no forces are applied at any of the other DOFs. Thus the forces are of the form $\mathbf{p}(t) = \mathbf{s}\, p(t)$, where their spatial distribution is defined by $\mathbf{s} = <0\ 0\ 0\ 0\ 0\ 0\ 0\ 0\ -2\ 1>$. The modal expansion of $\mathbf{s}$ (Section 12.8) leads to the modal contributions $\mathbf{s}_n$, displayed in Fig. 12.14.2, where we observe that the contributions of modes 4 to 9 are larger than those of modes 1 to 3, suggesting that these higher modes may contribute significantly to the dynamic response.

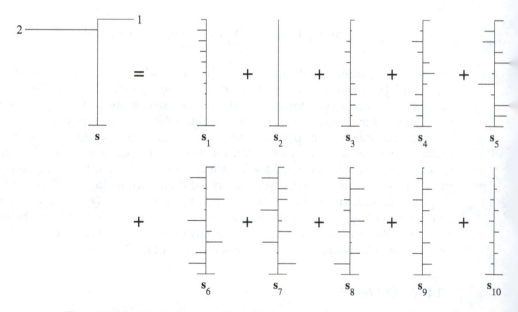

Figure 12.14.2 Modal expansion of force distribution $\mathbf{s}$; forces $\mathbf{s}_n$, shown without arrowheads, are drawn to the same scale as $\mathbf{s}$, but numerical values are not included.

[†]Although these forces belong to the restricted class $\mathbf{p}(t) = \mathbf{s}p(t)$, we will utilize Eqs. (12.14.7), (12.5.2), and (12.6.1) to implement dynamic analysis of the system for the restricted class, not the procedures of Sections 12.9 and 12.13.

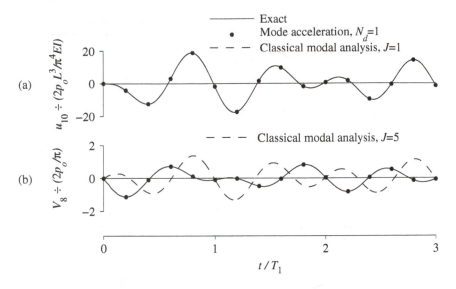

Figure 12.14.3 Response history determined by two approximate methods—classical modal analysis and mode acceleration superposition—compared with exact response; (a) $u_{10}(t)$ and (b) $V_8(t)$.

Approximate results for the dynamic response of the system obtained by two methods—the mode acceleration superposition method (Eq. 12.14.7) with $N_d = 1$ and classical modal analysis, including the contributions of the first $J = 1$, 3, or 5 modes in Eqs. (12.5.2) and (12.6.1)—are compared with the exact result, which can be obtained by including contributions of all modes in classical modal analysis or choosing $N_d = 10$ in Eq. (12.14.7); both approaches give identical results. The time variation of two response quantities—displacement $u_{10}(t)$ at the free end and shear force $V_8(t)$ at the section just above DOF u_8—are presented in Fig. 12.14.3. In obtaining these results, the combined steady-state and transient parts (see Section 3.2.1) of $q_n(t)$, starting from "at rest" initial conditions, were included. The height-wise distribution of the peak values of bending moments and shears considering only the steady-state part of the response are presented in Fig. 12.14.4. Spline functions were fitted to values of these internal forces determined at the locations of the 10 DOFs to facilitate visualization of results.

These results permit the following observations: The mode acceleration method with $N_d = 1$ provides essentially the exact results for displacements and forces. Such is the case because values of R_d, defined by Eq. (3.2.11), for ω/ω_n associated with modes 2, 3, and 4 are 1.061, 1.007, and 1.002, and the R_d values associated with higher modes are even closer to 1.0, implying that the response contributions of modes 2 to 10 may be determined by static analysis; i.e., the response due only to the first mode need be determined by dynamic analysis. In contrast, classical modal analysis including one mode provides excellent results for displacement, but even when as many as five modes are included, the moment and shears are inaccurate to an unacceptable degree; as many as nine modes had to be included to obtain accurate results for shears. The fact that higher modes contribute

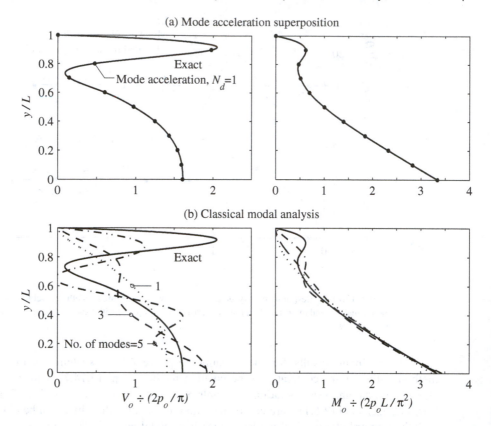

Figure 12.14.4 Heightwise distributions of peak shears and moments due to harmonic excitation with $\omega/\omega_1 = 1.5$: (a) mode acceleration superposition; (b) classical modal analysis.

more to shears than to moments, and more to moments than to displacements, is supported by analysis (not presented here) similar to Section 12.11 of the modal contribution factors, determined from Eq. (12.10.2).

The preceding analysis was repeated for exciting forces $\mathbf{p}(t) = \mathbf{s}\,p(t)$, where $\mathbf{s}$ was defined above and $p(t)$ is a step force that jumps suddenly from zero to p_o and stays constant (Fig. 4.3.1b). The heightwise distributions of the peak values of bending moments and shears are presented in Fig. 12.14.5. For this excitation, the mode acceleration method with $N_d = 1$ provides good results for moments and shears in the lower part of the tower, but it is unable to reproduce the rapid variation of forces in the upper part of the tower. As N_d is increased, the results improve, and $N_d = 5$ gives essentially the exact solution, implying that dynamic analysis is necessary to determine the response contributions of the first five modes. Such is the case because the value of R_d associated with all modes is $1.85^{\dagger}$, implying that the contributions of these modes cannot be determined by static analysis.

$^{\dagger}R_d = u_o \big/ (u_{\text{st}})_o$, where u_o is the peak value of $u\,(t)$ defined by Eq. (4.3.5).

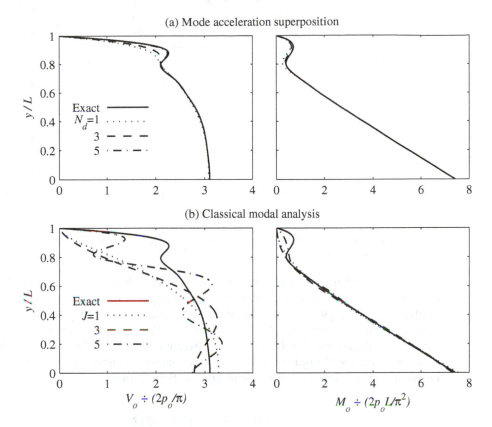

Figure 12.14.5 Heightwise distributions of peak shears and moments due to step force excitation: (a) mode acceleration superposition; (b) classical modal analysis.

The fact that peak values of forces determined by the mode acceleration method do not change significantly as N_d is increased beyond 5 suggests that the static response in modes 6 to 10, which we already included in the first term of Eq. (12.14.7), is adequate. However, this suggestion is not supported by the fact that the R_d value associated with modes 6 to 10 is 1.85, which indicates that dynamic analysis is necessary to determine the response contributions of these modes also. Such analysis is indeed necessary to obtain an accurate description of the complete response history; however, by coincidence, it had very little effect on the peak response.

Turning now to classical modal analysis, as the number of modes included increases, Figure 12.4.5 shows that the peak response approaches the exact result at a much slower rate compared to the convergence achieved by the mode acceleration method as N_d increases. Even with five modes included, the bending moment in the upper part of the tower and shear forces over its entire height are inaccurate to an unacceptable degree.

The mode acceleration method is more accurate because it does not lose any details of the height-wise distribution of applied forces; they are fully considered in the static solution represented by the first term in Eq. (12.14.7), which considers the exact **s** vector. In

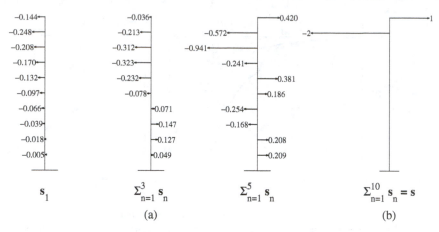

Figure 12.14.6 (a) Partial force distributions $\sum_{n=1}^{J} \mathbf{s}_n$, $J = 1, 3$, and 5; (b) complete force distribution $\mathbf{s}$.

contrast, classical modal analysis does not retain the complete force distribution. Including only the first J modes in classical modal analysis is equivalent to determining the response to $\mathbf{p}(t) = p(t)\sum_{n=1}^{J} \mathbf{s}_n$, i.e., considering only the force distribution $\sum_{n=1}^{J} \mathbf{s}_n$. This partial force distribution is shown for $J = 1, 3, 5$, and 10 in Fig. 12.14.6, where it is clear that the partial force distribution for $J = 5$ is very different than the complete force distribution $\mathbf{s}$. Thus, it is not surprising that classical modal analysis including the first five modes is unable to produce accurate results for responses (Fig. 12.14.5).

FURTHER READING

Bisplinghoff, R. L., Ashley, H., and Halfman, R. L., *Aeroelasticity*, Addison-Wesley, Reading, Mass., 1955.

Craig, R. R., Jr. and Kurdilla, A. J. *Fundamentals of Structural Dynamics*, 2nd ed., Wiley, New York, 2006, Section 11.4.

Crandall, S. H., and McCalley, R. B., Jr., "Matrix Methods of Analysis," Chapter 28 in *Shock and Vibration Handbook* (ed. C. M. Haris), McGraw-Hill, New York, 1988.

Den Hartog, J. P., *Mechanical Vibrations*, McGraw-Hill, New York, 1956, pp. 87–105.

Humar, J. L., *Dynamics of Structures,* 2nd ed., A. A. Balkema Publishers, Lisse, The Netherlands, 2002, Chapter 10.

PROBLEMS

Part A

12.1   Figure P12.1 shows a shear frame (i.e., rigid beams) and its floor masses and story stiffnesses. This structure is subjected to harmonic horizontal force $p(t) = p_o \sin \omega t$ at the top floor.

(a) Derive equations for the steady-state displacements of the structure by two methods: (i) direct solution of coupled equations, and (ii) modal analysis.
(b) Show that both methods give equivalent results.
(c) Plot on the same graph the two displacement amplitudes u_{1o} and u_{2o} as functions of the excitation frequency. Use appropriate normalizations of the displacement and frequency scales. Neglect damping.

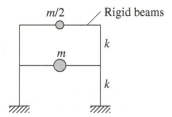

Figure P12.1

12.2   For the system and excitation of Problem 12.1, derive equations for story shears (considering steady-state response only) by two methods: (a) directly from displacements (without introducing equivalent static forces), and (b) using equivalent static forces. Show that the two methods give equivalent results.

Part B

12.3   Consider the system of Fig. P12.1 with modal damping ratios ζ_n subjected to the same excitation. Derive equations for the steady-state displacement amplitudes of the system.

12.4   The undamped system of Fig. P12.1 is subjected to an impulsive force at the first-floor mass: $p_1(t) = p_o\delta(t)$. Derive equations for the lateral floor displacements as functions of time.

12.5   The undamped system of Fig. P12.1 is subjected to a suddenly applied force at the first-floor mass: $p_1(t) = p_o, t \geq 0$. Derive equations for (a) the lateral floor displacements as functions of time, and (b) the story drift (or deformation) in the second story as a function of time.

12.6   The undamped system of Fig. P12.1 is subjected to a rectangular pulse force at the first floor. The pulse has an amplitude p_o and duration $t_d = T_1/2$, where T_1 is the fundamental vibration period of the system. Derive equations for the floor displacements as functions of time.

12.7   Figure P12.7 shows a shear frame (i.e., rigid beams) and its floor weights and story stiffnesses. This structure is subjected to harmonic force $p(t) = p_o \sin \omega t$ at the top floor.
(a) Determine the steady-state displacements as functions of ω by two methods: (i) direct solution of coupled equations, and (ii) modal analysis.
(b) Show that both methods give the same results.
(c) Plot on the same graph the three displacement amplitudes as a function of the excitation frequency over the frequency range 0 to $5\omega_1$. Use appropriate normalizations of the displacement and frequency scales.

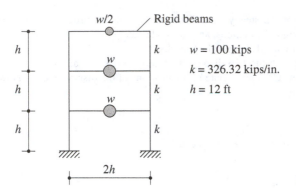

Figure P12.7

12.8   For the system and excitation of Problem 12.7 determine the story shears (considering steady-state response only) by two methods: (**a**) directly from displacements (without introducing equivalent static forces), and (**b**) using equivalent static forces. Show that the two methods give identical results.

12.9   An eccentric mass shaker is mounted on the roof of the system of Fig. P12.7. The shaker has two counterrotating weights, each 20 lb, at an eccentricity of 12 in. with respect to the vertical axis of rotation. Determine the steady-state amplitudes of displacement and acceleration at the roof as a function of excitation frequency. Plot the frequency response curves over the frequency range 0 to 15 Hz. Assume that the modal damping ratios ζ_n are 5%.

12.10   The undamped system of Fig. P12.7 is subjected to an impulsive force at the second-floor mass: $p_2(t) = p_o\delta(t)$, where $p_o = 20$ kips. Derive equations for the lateral floor displacements as functions of time.

12.11   The undamped system of Fig. P12.7 is subjected to a suddenly applied force at the first-floor mass: $p_1(t) = p_o, t \geq 0$, where $p_o = 200$ kips. Derive equations for (**a**) the lateral floor displacements as functions of time, and (**b**) the story drift (or deformation) in the second story as a function of time.

12.12   The undamped system of Fig. P12.7 is subjected to a rectangular pulse force at the third floor. The pulse has an amplitude $p_o = 200$ kips and duration $t_d = T_1/2$, where T_1 is the fundamental vibration period of the system. Derive equations for the floor displacements as functions of time.

12.13   Figure P12.13 shows a structural steel beam with $E = 30,000$ ksi, $I = 100$ in^4, $L = 150$ in., and $mL = 0.864$ kip-sec^2/in. Determine the displacement response of the system to an impulsive force $p_1(t) = p_o\delta(t)$ at the left mass, where $p_o = 10$ kips and $\delta(t)$ is the Dirac delta function. Plot as functions of time the displacements u_j due to each vibration mode separately and combined.

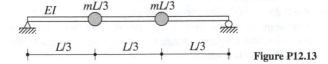

Figure P12.13

12.14 Determine the displacement response of the system of Problem 12.13 to a suddenly applied force of 100 kips applied at the left mass. Plot as functions of time the displacements u_j due to each vibration mode separately and combined.

12.15 Determine the displacement response of the system of Problem 12.13 to force $p(t)$, which is shown in Fig. P12.15 and applied at the left mass. Plot as functions of time the displacements $u_j(t)$ due to each vibration mode separately and combined.

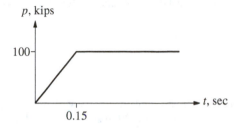

Figure P12.15

12.16 Determine the displacement response of the system of Problem 12.13 to force $p(t)$, which is shown in Fig. P12.16 and applied at the right mass. Plot as functions of time the displacements $u_j(t)$ due to each vibration mode separately and combined.

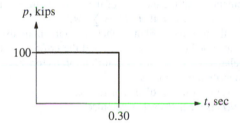

Figure P12.16

12.17 Repeat part (c) of Example 12.6 without using equivalent static forces. In other words, determine the shears and bending moments directly from the displacements and rotations.

12.18 For the system and excitation of Problem 12.14, determine the shears and bending moments at sections a, b, c, d, e, and f (Fig. P12.18) at $t = 0.1$ sec by using equivalent static forces. Draw the shear force and bending moment diagrams due to each mode separately and combined.

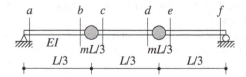

Figure P12.18

12.19 The system of Problem 12.13 is subjected to a harmonic force at the left mass: $p(t) = p_o \sin \omega t$, where $p_o = 100$ kips and $\omega = 25$ rad/sec. Neglecting damping, determine the forced (or steady-state) response of the system. In particular, determine (a) the displacements and accelerations of the two masses as functions of time, and (b) the amplitudes of displacements and accelerations.

12.20 For the system and excitation of Problem 12.19, determine the amplitude of the forced (or steady-state) bending moment at the location of each mass by using equivalent static forces.

12.21 Solve Problem 12.19 assuming modal damping ratios of 10% for the system.

Part C

*__12.22__ Figure P12.22 shows a massless simply supported beam with three lumped masses and the following properties: $L = 150$ in., $m = 0.192$ kip-sec^2/in., $E = 30,000$ ksi, and $I = 100$ in^4. We are interested in studying the dynamic response of the beam to two sets of applied forces: $\mathbf{p}(t) = \mathbf{s}p(t)$, $\mathbf{s}_a^T = \langle 1 \quad 0 \quad 0 \rangle$, and $\mathbf{s}_b^T = \langle 2 \quad 0 \quad -1 \rangle$.
(a) Determine the modal expansion of the vectors $\mathbf{s}_a$ and $\mathbf{s}_b$ that define the spatial distribution of forces. Show these modal expansions graphically and comment on the relative contributions of the various modes to $\mathbf{s}_a$ and $\mathbf{s}_b$, and how these contributions differ between $\mathbf{s}_a$ and $\mathbf{s}_b$.
(b) For the bending moment M_1 at the location of the u_1 DOF, determine the modal static responses M_{1n}^{st} for both $\mathbf{s}_a$ and $\mathbf{s}_b$. Show that $M_1^{st} = \sum M_{1n}^{st}$.
(c) Calculate and tabulate the modal contribution factors, their cumulative values for the various numbers of modes included ($J = 1, 2,$ or 3), and the error e_J in the static response. Comment on how the relative values of modal contribution factors and the error e_J are influenced by the spatial distribution of forces.
(d) Determine the peak values $(M_{1n})_o$ of modal responses $M_{1n}(t)$ due to $\mathbf{p}(t) = \mathbf{s}p(t)$, where $\mathbf{s} = \mathbf{s}_a$ or $\mathbf{s}_b$ and $p(t)$ is the half-cycle sine pulse:

$$p(t) = \begin{cases} p_o \sin(\pi t/t_d) & t \leq t_d \\ 0 & t \geq t_d \end{cases}$$

The duration of the pulse t_d is T_1, the fundamental period of the system. Figure 12.11.3 gives the shock spectrum for a half-cycle sine pulse with numerical ordinates $R_d = 1.73$, 1.14, and 1.06 for $T_1/t_d = 1$, $T_2/t_d = 0.252$, and $T_3/t_d = 0.119$, respectively. It should be convenient to organize your computations in a table with the following column headings: mode n, T_n/t_d, R_{dn}, $\bar{M}_{1n}$, and $(M_{1n})_o/p_o M_1^{st}$.
(e) Comment on how the peak modal responses determined in part (d) depend on the modal static responses M_{1n}^{st}, modal contribution factors $\bar{M}_{1n}$, dynamic response factors R_{dn}, and the force distributions $\mathbf{s}_a$ and $\mathbf{s}_b$.
(f) Can you determine the peak value of the total (considering all modes) response from the peak modal responses? Justify your answer.

---

*Denotes that a computer is necessary to solve this problem.

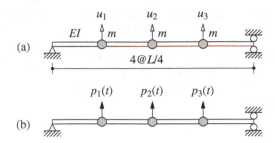

(a)

(b)

Figure P12.22

*12.23   The structure of Fig. P9.13 has the following properties: $L = 100$ in., $m = 0.192$ kip-sec²/in., $E = 30,000$ ksi, and $I = 150$ in⁴. We are interested in studying the dynamic response of the structure to three sets of applied forces: $\mathbf{p}(t) = \mathbf{s}p(t)$, $\mathbf{s}_a = \langle 1 \quad -1 \quad 1 \rangle^T$, $\mathbf{s}_b = \langle 1 \quad 1 \quad -1 \rangle^T$, $\mathbf{s}_c = \langle 1 \quad 2 \quad 2 \rangle^T$.

(a) Determine the modal expansion of the vectors $\mathbf{s}_a$, $\mathbf{s}_b$, and $\mathbf{s}_c$ that define the spatial distribution of forces. Show these modal expansions graphically and comment on the relative contributions of the various modes to $\mathbf{s}_a$, $\mathbf{s}_b$, and $\mathbf{s}_c$, and how these contributions differ among $\mathbf{s}_a$, $\mathbf{s}_b$, and $\mathbf{s}_c$.

(b) For the bending moment M_1 at the base b and the bending moment M_2 to the right-hand side of point a, determine the modal static responses M_{1n}^{st} and M_{2n}^{st} for $\mathbf{s}_a$, $\mathbf{s}_b$, and $\mathbf{s}_c$. Show that $M_1^{st} = \sum M_{1n}^{st}$ and $M_2^{st} = \sum M_{2n}^{st}$.

(c) Calculate and tabulate the modal contribution factors, their cumulative values for various numbers of modes included ($J = 1, 2,$ or 3), and the error e_J in the static response. Comment on how the relative values of modal contribution factors and error e_J are influenced by the spatial distribution of forces.

(d) Determine the peak values $(M_{1n})_o$ and $(M_{2n})_o$ of modal responses M_{1n} and M_{2n} due to $\mathbf{p}(t) = \mathbf{s}p(t)$, where $\mathbf{s} = \mathbf{s}_a$, $\mathbf{s}_b$, or $\mathbf{s}_c$, and $p(t)$ is a rectangular pulse:

$$p(t) = \begin{cases} p_0 & t \le t_d \\ 0 & t > t_d \end{cases}$$

The duration of the pulse $t_d = 0.2T_1$, where T_1 is the fundamental period of the system. Figure 4.7.3b gives the shock spectrum for a rectangular pulse with numerical ordinates $R_d = 0.691, 2.0,$ and 2.0 for $T_1/t_d = 5.0$, $T_2/t_d = 1.63$, and $T_3/t_d = 1.51$, respectively. It should be efficient to organize your computations in a table with the following column headings for M_1: mode n, T_n/t_d, R_{dn}, $\overline{M}_{1n}$, and $(M_{1n})_o/p_0 M_1^{st}$, and with similar column headings for M_2.

(e) Comment on how the peak modal responses determined in part (d) depend on the modal contribution factors $\overline{M}_{1n}$ and $\overline{M}_{2n}$ and on dynamic response factor R_{dn}.

(f) Can you determine the peak value of the total (considering all modes) response from the peak modal responses? Justify your answer.

Part D

12.24   The undamped system of Fig. P12.22 with its properties defined in Problem 12.22 is subjected to dynamic forces $\mathbf{p}(t) = \mathbf{s}_b p(t)$, where $\mathbf{s}_b^T = \langle 2 \quad 0 \quad -1 \rangle$ and $p(t)$ is the half-cycle

---

*Denotes that a computer is necessary to solve this problem.

sine pulse defined in part (d) of Problem 12.22 with the pulse duration t_d the same as T_1, the fundamental vibration period of the system. Determine the bending moment $M(t)$ at the location of the u_1 DOF as a function of time by two methods: **(a)** classical modal analysis, and **(b)** the static correction method with dynamic response determined only in the first mode. For part (a) plot the individual modal contributions to $M(t)$ and the total response. On a separate plot compare this total response with the results of part (b). Comment on the accuracy of the static correction method.

13

Earthquake Analysis of Linear Systems

PREVIEW

In this chapter procedures for earthquake analysis of structures, idealized as lumped-mass systems, are developed. The presentation is organized in two parts. Part A is concerned with the calculation of structural response as a function of time when the system is subjected to a given ground acceleration $\ddot{u}_g(t)$. This *response history analysis* (RHA) procedure is first presented for an arbitrary structural configuration and subsequently specialized for multistory buildings with a symmetric plan, and for multistory buildings with an unsymmetric plan. A brief discussion of the torsional response of symmetric-plan buildings is also included. Part A is devoted to a single component of ground motion, typically one of the horizontal components. Combining the structural responses determined from such independent analyses for each excitation component gives the response of linear systems to multiple-component excitation. Also developed is a procedure to analyze the response of a structure subjected to different prescribed motions at its various supports.

Part B is concerned with procedures to compute the peak response of a structure during an earthquake directly from the earthquake response (or design) spectrum without the need for response history analysis of the structure. Known as *response spectrum analysis* (RSA), this procedure is not an exact predictor of peak response, but it provides an estimate that is sufficiently accurate for structural design applications. The procedure is first presented for structures subjected to individual translational components of ground motion. Subsequently, rules to combine the three individual peaks to estimate the peak response to multicomponent excitation are presented. Also included in this chapter are response-spectrum-based equations to determine an envelope that bounds the joint response trajectory of all simultaneously acting forces that control the seismic design of a structural element.

PART A: RESPONSE HISTORY ANALYSIS

13.1 MODAL ANALYSIS

In this section we develop the modal analysis procedure to determine the response of a structure to earthquake-induced ground motion $\ddot{u}_g(t)$, identical at all support points of the structure.

13.1.1 Equations of Motion

The differential equations (9.4.8) governing the response of an MDF system to earthquake-induced ground motion are repeated:

$$\mathbf{m}\ddot{\mathbf{u}} + \mathbf{c}\dot{\mathbf{u}} + \mathbf{k}\mathbf{u} = \mathbf{p}_{\text{eff}}(t) \tag{13.1.1}$$

where

$$\mathbf{p}_{\text{eff}}(t) = -\mathbf{m}\iota\ddot{u}_g(t) \tag{13.1.2}$$

The mass and stiffness matrices, $\mathbf{m}$ and $\mathbf{k}$, and the influence vector ι are determined by the methods of Chapter 9. The damping matrix $\mathbf{c}$ would not be needed in modal analysis of earthquake response; instead, modal damping ratios suffice and their numerical values can be estimated as discussed in Chapter 11. The modal analysis procedure developed in Chapter 12 to solve Eq. (12.4.1) is applicable to the solution of Eq. (13.1.1).

13.1.2 Modal Expansion of Displacements and Forces

The displacement $\mathbf{u}$ of an N-DOF system can be expressed, as in Eq. (12.3.2), as the superposition of the modal contributions:

$$\mathbf{u}(t) = \sum_{n=1}^{N} \phi_n q_n(t) \tag{13.1.3}$$

The spatial distribution of the effective earthquake forces $\mathbf{p}_{\text{eff}}(t)$ is defined by $\mathbf{s} = \mathbf{m}\iota$. This force distribution can be expanded as a summation of modal inertia force distributions $\mathbf{s}_n$ (Section 12.8):

$$\mathbf{m}\iota = \sum_{n=1}^{N} \mathbf{s}_n = \sum_{n=1}^{N} \Gamma_n \mathbf{m}\phi_n \tag{13.1.4}$$

where

$$\Gamma_n = \frac{L_n}{M_n} \qquad L_n = \phi_n^T \mathbf{m}\iota \qquad M_n = \phi_n^T \mathbf{m}\phi_n \tag{13.1.5}$$

Equation (13.1.5) for the coefficient Γ_n can be derived by premultiplying both sides of Eq. (13.1.4) by ϕ_r^T and using the orthogonality property of modes, or by specializing Eq. (12.8.3) for $\mathbf{s} = \mathbf{m}\iota$. The contribution of the nth mode to $\mathbf{m}\iota$ is

$$\mathbf{s}_n = \Gamma_n \mathbf{m}\phi_n \tag{13.1.6}$$

which is independent of how the modes are normalized.

13.1.3 Modal Equations

Equation (12.4.6) is specialized for earthquake excitation by replacing $\mathbf{p}(t)$ in Eq. (12.3.4) by $\mathbf{p}_{\text{eff}}(t)$ to obtain

$$\ddot{q}_n + 2\zeta_n\omega_n\dot{q}_n + \omega_n^2 q_n = -\Gamma_n\ddot{u}_g(t) \qquad (13.1.7)$$

The solution $q_n(t)$ can readily be obtained by comparing Eq. (13.1.7) to the equation of motion for the nth-mode SDF system, an SDF system with vibration properties—natural frequency ω_n and damping ratio ζ_n—of the nth mode of the MDF system. Equation (6.2.1) with $\zeta = \zeta_n$, which governs the motion of this SDF system subjected to ground acceleration $\ddot{u}_g(t)$, is repeated here with u replaced by D_n to emphasize its connection to the nth mode:

$$\ddot{D}_n + 2\zeta_n\omega_n\dot{D}_n + \omega_n^2 D_n = -\ddot{u}_g(t) \qquad (13.1.8)$$

Comparing Eq. (13.1.8) to (13.1.7) gives the relation between q_n and D_n:

$$q_n(t) = \Gamma_n D_n(t) \qquad (13.1.9)$$

Thus $q_n(t)$ is readily available once Eq. (13.1.8) has been solved for $D_n(t)$, utilizing numerical time-stepping methods for SDF systems (Chapter 5).

The factor Γ_n [defined in Eq. (13.1.5a)] that multiples $\ddot{u}_g(t)$ in Eq. (13.1.7) is the same as the coefficient in the modal expansion (Section 10.7) of the influence vector:

$$\iota = \sum_{n=1}^{N} \Gamma_n\phi_n$$

It is usually referred to as a modal participation factor, implying that it is a measure of the degree to which the nth mode participates in the response. This terminology is misleading, however, because Γ_n is not independent of how the mode is normalized, nor a measure of the modal contribution to a response quantity. Both these drawbacks are overcome by modal contribution factors that were introduced in Section 12.10 and will be utilized later (Chapter 19) to investigate earthquake response of buildings.

13.1.4 Modal Responses

The contribution of the nth mode to the nodal displacements $\mathbf{u}(t)$ is

$$\mathbf{u}_n(t) = \phi_n q_n(t) = \Gamma_n\phi_n D_n(t) \qquad (13.1.10)$$

Two static analysis procedures described in Section 9.10 are available to determine the forces in various structural elements—beams, columns, walls, etc.—from the displacements $\mathbf{u}_n(t)$. The second of these procedures, using equivalent static forces, is preferred in earthquake analysis because it facilitates comparison of dynamic analysis procedures with the earthquake design forces specified in building codes (Chapter 21). The equivalent static forces associated with the nth-mode response are $\mathbf{f}_n(t) = \mathbf{k}\mathbf{u}_n(t)$, where $\mathbf{u}_n(t)$ is given by Eq. (13.1.10). Putting these equations together and using Eqs. (10.2.4) and (13.1.6) leads to

$$\mathbf{f}_n(t) = \mathbf{s}_n A_n(t) \tag{13.1.11}$$

where, similar to Eq. (6.4.3),

$$A_n(t) = \omega_n^2 D_n(t) \tag{13.1.12}$$

The equivalent static forces $\mathbf{f}_n(t)$ are the product of two quantities: (1) the nth-mode contribution $\mathbf{s}_n$ to the spatial distribution $\mathbf{m}\iota$ of $\mathbf{p}_{\text{eff}}(t)$, and (2) the pseudo-acceleration response of the nth-mode SDF system to $\ddot{u}_g(t)$.

The nth-mode contribution $r_n(t)$ to any response quantity $r(t)$ that can be expressed as a linear combination of the structural displacements $\mathbf{u}(t)$ is determined by static analysis of the structure subjected to external forces $\mathbf{f}_n(t)$. If r_n^{st} denotes the modal static response, the static value (indicated by superscript "st") of r due to external forces[†] $\mathbf{s}_n$, then

$$r_n(t) = r_n^{\text{st}} A_n(t) \tag{13.1.13}$$

Observe that r_n^{st} may be positive or negative and is independent of how the mode is normalized. Equation (13.1.13) also applies to the displacement response, although its derivation had been motivated by the desire to compute forces from the displacements. The static displacements due to forces $\mathbf{s}_n$ satisfy $\mathbf{k}\mathbf{u}_n^{\text{st}} = \mathbf{s}_n$. Substituting Eq. (13.1.6) for $\mathbf{s}_n$ and using Eq. (10.2.4) gives

$$\mathbf{u}_n^{\text{st}} = \mathbf{k}^{-1}(\Gamma_n \mathbf{m} \phi_n) = \frac{\Gamma_n}{\omega_n^2} \phi_n$$

Substituting this in Eq. (13.1.13) gives

$$\mathbf{u}_n(t) = \frac{\Gamma_n}{\omega_n^2} \phi_n A_n(t) \tag{13.1.14}$$

which is equivalent to Eq. (13.1.10) because of Eq. (13.1.12).

13.1.5 Total Response

Combining the response contributions of all the modes gives the total response of the structure to the ground motion. Thus the nodal displacements are

$$\mathbf{u}(t) = \sum_{n=1}^{N} \mathbf{u}_n(t) = \sum_{n=1}^{N} \Gamma_n \phi_n D_n(t) \tag{13.1.15}$$

wherein Eq. (13.1.10) has been substituted for $\mathbf{u}_n(t)$. Using Eq. (13.1.13) gives a general result valid for any response quantity:

$$r(t) = \sum_{n=1}^{N} r_n(t) = \sum_{n=1}^{N} r_n^{\text{st}} A_n(t) \tag{13.1.16}$$

[†] Although we loosely refer to $\mathbf{s}_n$ as forces, they have units of mass. Thus, r_n^{st} does not have the same units as r, but Eq. (13.1.13) gives the correct units for r_n.

The response contributions of some of the higher modes may, under appropriate circumstances, be determined by the simpler static analysis, instead of dynamic analysis. As shown in Fig. 6.8.4, for SDF systems with very short periods the pseudo-acceleration $A(t) \simeq -\ddot{u}_g(t)$. For the design spectrum of Fig. 6.9.3, $A = \ddot{u}_{go}$ for $T_n \leq \frac{1}{33}$ sec. If this period range includes the natural periods of modes $N_d + 1$ to N, then Eq. (13.1.16) can be expressed as

$$r(t) = \sum_{n=1}^{N_d} r_n^{st} A_n(t) - \ddot{u}_g(t) \left(r^{st} - \sum_{n=1}^{N_d} r_n^{st} \right) \tag{13.1.17}$$

where r^{st} is the static value of r due to external forces $\mathbf{s}$ (Section 12.10) and $r^{st} = \sum_{n=1}^{N} r_n^{st}$ because $\mathbf{s} = \sum \mathbf{s}_n$ [Eq. (13.1.4)]. This solution is in two parts: the first term is the dynamic response considering the first N_d modes and the second is the response of the higher modes determined by static analysis. Equation (13.1.17) is the static correction method that can also be derived following Section 12.12 or rewritten in the form of the mode acceleration method (Section 12.13). These methods are usually not useful in seismic analysis of structures because earthquake ground motions typically contain a wide band of frequencies that includes structural frequencies and higher-mode components in $\mathbf{s} = \mathbf{m}\boldsymbol{\iota}$ are small.

13.1.6 Interpretation of Modal Analysis

In the first phase of this dynamic analysis procedure, the vibration properties—natural frequencies and modes—of the structure are computed and the force distribution vector $\mathbf{m}\boldsymbol{\iota}$ is expanded into its modal components $\mathbf{s}_n$. The rest of the analysis procedure is shown schematically in Fig. 13.1.1 to emphasize the underlying concepts. The contribution of the nth mode to the dynamic response is obtained by multiplying the results of two analyses: (1) static analysis of the structure with applied forces $\mathbf{s}_n$, and (2) dynamic analysis of the nth-mode SDF system excited by $\ddot{u}_g(t)$. Thus modal analysis requires static analysis of the structure for N sets of forces: $\mathbf{s}_n, n = 1, 2, \ldots, N$; and dynamic analysis of N different SDF systems. Combining the modal responses gives the earthquake response of the structure.

Example 13.1

Determine the response of the inverted L-shaped frame of Fig. E9.6a to horizontal ground motion.

Solution   Assuming the two elements to be axially rigid, the DOFs are u_1 and u_2 (Fig. E9.6a). The equations of motion are given by Eqs. (13.1.1) and (13.1.2), where the influence vector $\boldsymbol{\iota} = \langle 1 \quad 0 \rangle^T$ (Fig. 9.4.4) and the mass and stiffness matrices (from Example 9.6) are

$$\mathbf{m} = \begin{bmatrix} 3m & \\ & m \end{bmatrix} \qquad \mathbf{k} = \frac{6EI}{7L^3} \begin{bmatrix} 8 & -3 \\ -3 & 2 \end{bmatrix} \tag{a}$$

The effective earthquake forces are

$$\mathbf{p}_{\text{eff}}(t) = -\mathbf{m}\boldsymbol{\iota}\ddot{u}_g(t) = -\begin{bmatrix} 3m & \\ & m \end{bmatrix} \begin{Bmatrix} 1 \\ 0 \end{Bmatrix} \ddot{u}_g(t) = -\begin{Bmatrix} 3m \\ 0 \end{Bmatrix} \ddot{u}_g(t) \tag{b}$$

The force in the vertical DOF is zero because the ground motion is horizontal.

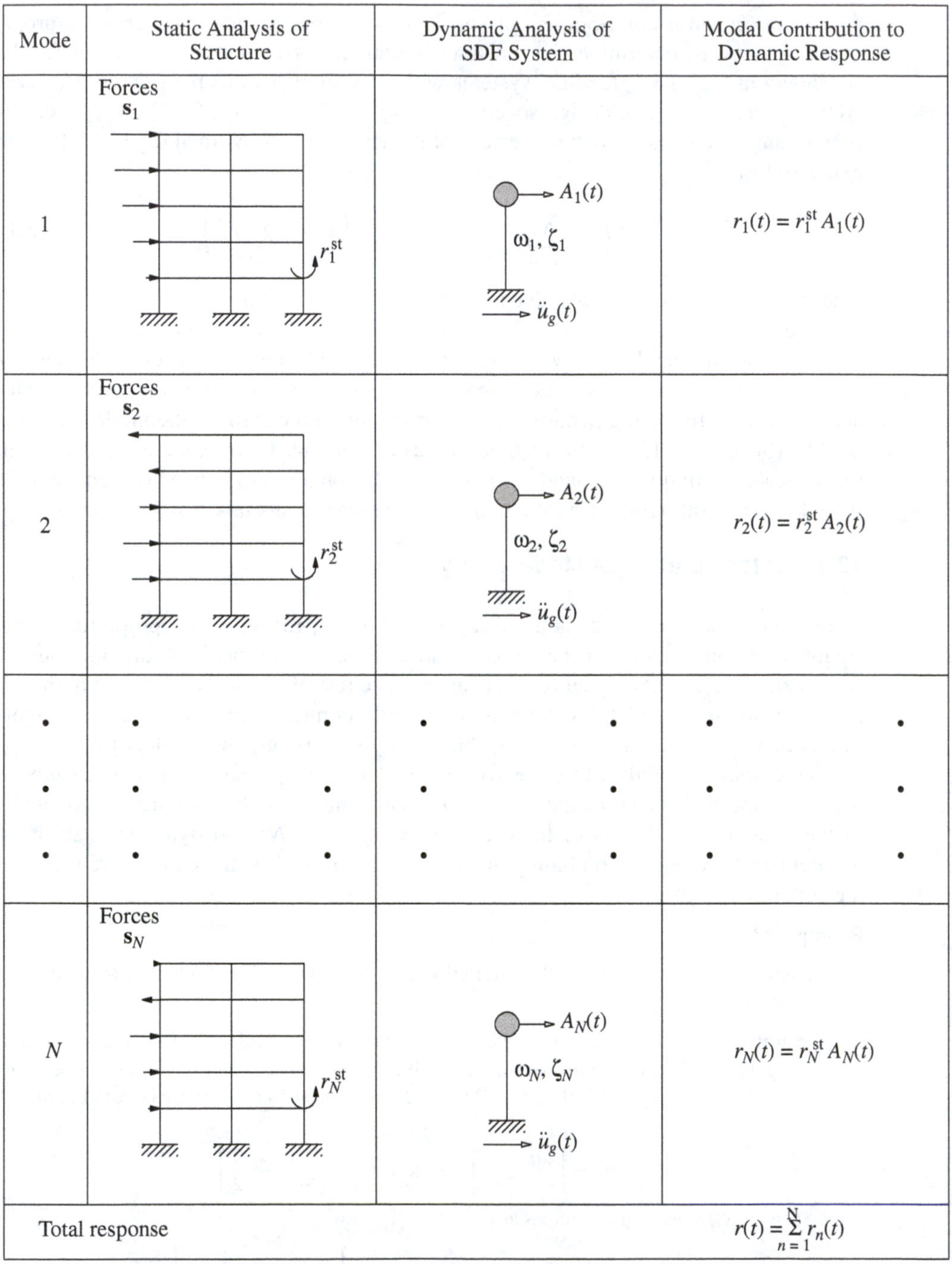

Mode	Static Analysis of Structure	Dynamic Analysis of SDF System	Modal Contribution to Dynamic Response
1	Forces $\mathbf{s}_1$ r_1^{st}	$A_1(t)$ ω_1, ζ_1 $\ddot{u}_g(t)$	$r_1(t) = r_1^{st} A_1(t)$
2	Forces $\mathbf{s}_2$ r_2^{st}	$A_2(t)$ ω_2, ζ_2 $\ddot{u}_g(t)$	$r_2(t) = r_2^{st} A_2(t)$
. . .	. . .	. . .	. . .
N	Forces $\mathbf{s}_N$ r_N^{st}	$A_N(t)$ ω_N, ζ_N $\ddot{u}_g(t)$	$r_N(t) = r_N^{st} A_N(t)$
Total response			$r(t) = \sum_{n=1}^{N} r_n(t)$

Figure 13.1.1 Conceptual explanation of modal analysis.

518

The natural frequencies and modes of the system are (from Example 10.3)

$$\omega_1 = 0.6987\sqrt{\frac{EI}{mL^3}} \qquad \omega_2 = 1.874\sqrt{\frac{EI}{mL^3}} \tag{c}$$

$$\phi_1 = \left\{\begin{matrix}1\\2.097\end{matrix}\right\} \qquad \phi_2 = \left\{\begin{matrix}1\\-1.431\end{matrix}\right\} \tag{d}$$

Substituting for $\mathbf{m}$ and ι in Eq. (13.1.5) gives the first-mode quantities:

$$L_1 = \phi_1^T \mathbf{m}\iota = \langle 1 \quad 2.097\rangle \begin{bmatrix}3m & \\ & m\end{bmatrix}\left\{\begin{matrix}1\\0\end{matrix}\right\} = 3m$$

$$M_1 = \phi_1^T \mathbf{m}\phi_1 = \langle 1 \quad 2.097\rangle \begin{bmatrix}3m & \\ & m\end{bmatrix}\left\{\begin{matrix}1\\2.097\end{matrix}\right\} = 7.397m$$

$$\Gamma_1 = \frac{L_1}{M_1} = \frac{3m}{7.397m} = 0.406$$

Similar calculations for the second mode give $L_2 = 3m$, $M_2 = 5.048m$, and $\Gamma_2 = 0.594$. Substituting Γ_n, $\mathbf{m}$, and ϕ_n in Eq. (13.1.6) gives

$$\mathbf{s}_1 = \Gamma_1 \mathbf{m}\phi_1 = 0.406\begin{bmatrix}3m & \\ & m\end{bmatrix}\left\{\begin{matrix}1\\2.097\end{matrix}\right\} = m\left\{\begin{matrix}1.218\\0.851\end{matrix}\right\} \tag{e}$$

$$\mathbf{s}_2 = \Gamma_2 \mathbf{m}\phi_2 = 0.594\begin{bmatrix}3m & \\ & m\end{bmatrix}\left\{\begin{matrix}1\\-1.431\end{matrix}\right\} = m\left\{\begin{matrix}1.782\\-0.851\end{matrix}\right\} \tag{f}$$

Then Eq. (13.1.4) specializes to

$$m\left\{\begin{matrix}3\\0\end{matrix}\right\} = m\left\{\begin{matrix}1.218\\0.851\end{matrix}\right\} + m\left\{\begin{matrix}1.782\\-0.851\end{matrix}\right\} \tag{g}$$

This modal expansion of the spatial distribution of effective forces is shown in Fig. E13.1. Observe that the forces along the vertical DOF in the two modes cancel each other because the effective earthquake force in this DOF is zero.

Substituting for Γ_n and ϕ_n in Eq. (13.1.10) gives the first-mode displacements

$$\mathbf{u}_1(t) = \left\{\begin{matrix}u_1(t)\\u_2(t)\end{matrix}\right\}_1 = \Gamma_1\phi_1 D_1(t) = 0.406\left\{\begin{matrix}1\\2.097\end{matrix}\right\} D_1(t) = \left\{\begin{matrix}0.406\\0.851\end{matrix}\right\} D_1(t) \tag{h}$$

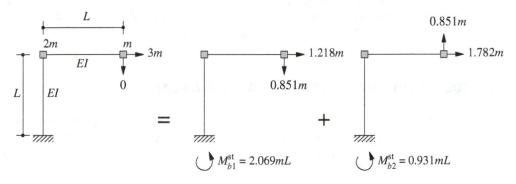

Figure E13.1

and the second-mode displacements

$$\mathbf{u}_2(t) = \left\{ \begin{matrix} u_1(t) \\ u_2(t) \end{matrix} \right\}_2 = \Gamma_2\boldsymbol{\phi}_2 D_2(t) = 0.594 \left\{ \begin{matrix} 1 \\ -1.431 \end{matrix} \right\} D_2(t) = \left\{ \begin{matrix} 0.594 \\ -0.851 \end{matrix} \right\} D_2(t) \tag{i}$$

Combining Eqs. (h) and (i) gives the total displacements:

$$u_1(t) = 0.406D_1(t) + 0.594D_2(t) \qquad u_2(t) = 0.851D_1(t) - 0.851D_2(t) \tag{j}$$

The earthquake-induced bending moment M_b at the base of the column due to the nth mode [from Eq. (13.1.13)] is

$$M_{bn}(t) = M_{bn}^{st} A_n(t) \tag{k}$$

Static analyses of the frame for the forces $\mathbf{s}_1$ and $\mathbf{s}_2$ give M_{b1}^{st} and M_{b2}^{st} as shown in Fig. E13.1. Substituting for M_{bn}^{st} and combining modal contributions gives the total bending moment:

$$M_b(t) = \sum_{n=1}^{2} M_{bn}(t) = 2.069mLA_1(t) + 0.931mLA_2(t) \tag{l}$$

The three response quantities considered have been, and other responses can be, expressed in terms of $D_n(t)$ and $A_n(t)$. These responses of the nth-mode SDF system to given ground acceleration $\ddot{u}_g(t)$ can be determined by numerical time-stepping methods (Chapter 5).

13.1.7 Analysis of Response to Base Rotation

The modal analysis procedure is applicable after slight modification when the excitation is base rotation. As shown in Section 9.4.3, the motion of a structure due to rotational acceleration $\ddot{\theta}_g(t)$ of the base (Fig. 9.4.6a) is governed by Eq. (13.1.1), with

$$\mathbf{p}_{\text{eff}}(t) = -\mathbf{m}\boldsymbol{\iota}\ddot{\theta}_g(t) \tag{13.1.18}$$

where $\boldsymbol{\iota}$ is the vector of static displacements in all the DOFs due to unit base rotation, $\theta_g = 1$. For the system of Fig. 9.4.6a, this influence vector is $\boldsymbol{\iota} = \langle h_1 \quad h_2 \quad x_3 \rangle^T$. With $\boldsymbol{\iota}$ determined, the structural response due to base rotation is calculated by the procedures of Sections 13.1.1 to 13.1.5 with $\ddot{u}_g(t)$ replaced by $\ddot{\theta}_g(t)$.

13.2 MULTISTORY BUILDINGS WITH SYMMETRIC PLAN

In this section the modal analysis of Section 13.1 is specialized for multistory buildings with rigid floor diaphragms and plans having two orthogonal axes of symmetry subjected to horizontal ground motion along one of those axes. The equations of motion for this system, Eq. (9.4.4), are repeated:

$$\mathbf{m}\ddot{\mathbf{u}} + \mathbf{c}\dot{\mathbf{u}} + \mathbf{k}\mathbf{u} = -\mathbf{m}\mathbf{1}\ddot{u}_g(t) \tag{13.2.1}$$

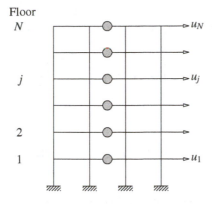

Floor
N

j

2

1

u_N

u_j

u_1

Figure 13.2.1   Dynamic degrees of freedom of a multistory frame: lateral displacements relative to the ground.

where **u** is the vector of lateral floor displacements relative to the ground (Fig. 13.2.1); **m** is a diagonal matrix with elements $m_{jj} = m_j$, the lumped mass at the jth floor level; **k** is the lateral stiffness matrix of the building defined in Section 9.4.2; and each element of **1** is unity. The modal analysis procedure developed in Section 13.1 is applicable to the multistory building problem because its governing equations are the same as Eq. (13.1.1) with the influence vector $\iota = 1$. For convenience, we present the analysis procedure with reference to a single frame (Fig. 13.2.1), although it applies to a building with several frames (see Section 9.4.2).

13.2.1 Modal Expansion of Effective Earthquake Forces

Substituting $\iota = 1$ in Eqs. (13.1.4) and (13.1.5) gives the modal expansion of the spatial distribution of effective earthquake forces:

$$\mathbf{m1} = \sum_{n=1}^{N} \mathbf{s}_n = \sum_{n=1}^{N} \Gamma_n \mathbf{m} \phi_n \qquad (13.2.2)$$

where

$$\Gamma_n = \frac{L_n^h}{M_n} \qquad L_n^h = \sum_{j=1}^{N} m_j \phi_{jn} \qquad M_n = \sum_{j=1}^{N} m_j \phi_{jn}^2 \qquad (13.2.3)$$

In Eq. (13.2.2) the contribution of the nth mode to **m1** is $\mathbf{s}_n$, a vector of lateral forces s_{jn} at the various floor levels:

$$\mathbf{s}_n = \Gamma_n \mathbf{m} \phi_n \qquad s_{jn} = \Gamma_n m_j \phi_{jn} \qquad (13.2.4)$$

Example 13.2

A two-story shear frame has the mass and story stiffnesses properties shown in Fig. E13.2a. Determine the modal expansion of the effective earthquake force distribution associated with horizontal ground acceleration $\ddot{u}_g(t)$.

Solution   The stiffness and mass matrices (from Example 9.1) are

$$\mathbf{k} = k \begin{bmatrix} 3 & -1 \\ -1 & 1 \end{bmatrix} \qquad \mathbf{m} = m \begin{bmatrix} 2 & 0 \\ 0 & 1 \end{bmatrix}$$

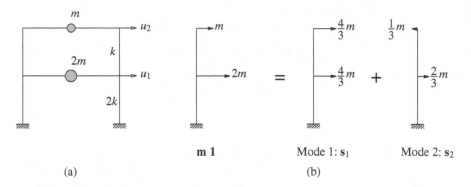

(a)   (b)

Figure E13.2   (a) Two-story shear frame; (b) modal expansion of **m1**.

where $k = 24EI_c/h^3$, and the natural frequencies and modes (from Example 10.4) are

$$\omega_1 = \sqrt{\frac{k}{2m}} \qquad \omega_2 = \sqrt{\frac{2k}{m}}$$

$$\phi_1 = \left\{ \begin{matrix} \frac{1}{2} \\ 1 \end{matrix} \right\} \qquad \phi_2 = \left\{ \begin{matrix} -1 \\ 1 \end{matrix} \right\}$$

The modal properties M_n, L_n^h, and Γ_n are computed from Eq. (13.2.3). For the first mode:
$M_1 = 2m \left(\frac{1}{2}\right)^2 + m(1)^2 = 3m/2$; $L_1^h = 2m \left(\frac{1}{2}\right) + m(1) = 2m$; $\Gamma_1 = L_1^h/M_1 = \frac{4}{3}$. Similarly,
for the second mode: $M_2 = 3m$, $L_2^h = -m$, and $\Gamma_2 = -\frac{1}{3}$. Substituting for Γ_n, **m**, and ϕ_n in
Eq. (13.2.4) gives

$$\mathbf{s}_1 = \frac{4}{3}m \begin{bmatrix} 2 & \\ & 1 \end{bmatrix} \left\{ \begin{matrix} \frac{1}{2} \\ 1 \end{matrix} \right\} = \frac{4}{3}m \left\{ \begin{matrix} 1 \\ 1 \end{matrix} \right\}$$

$$\mathbf{s}_2 = -\frac{1}{3}m \begin{bmatrix} 2 & \\ & 1 \end{bmatrix} \left\{ \begin{matrix} -1 \\ 1 \end{matrix} \right\} = -\frac{1}{3}m \left\{ \begin{matrix} -2 \\ 1 \end{matrix} \right\}$$

The modal expansion of **m1** is displayed in Fig. E13.2b.

13.2.2 Modal Responses

The differential equation governing the nth modal coordinate is Eq. (13.1.7) with Γ_n defined by Eq. (13.2.3). Using this Γ_n, Eq. (13.1.10) gives the contribution $\mathbf{u}_n(t)$ of the nth
mode to the lateral displacements $\mathbf{u}(t)$. In particular, the lateral displacement of the jth
floor of the building is

$$u_{jn}(t) = \Gamma_n \phi_{jn} D_n(t) \qquad (13.2.5)$$

The drift, or deformation, in story j is given by the difference of displacements of the
floors above and below:

$$\Delta_{jn}(t) = u_{jn}(t) - u_{j-1,n}(t) = \Gamma_n(\phi_{jn} - \phi_{j-1,n})D_n(t) \qquad (13.2.6)$$

The equivalent static forces $\mathbf{f}_n(t)$ for the nth mode [from Eq. (13.1.11)] are

$$\mathbf{f}_n(t) = \mathbf{s}_n A_n(t) \qquad f_{jn}(t) = s_{jn} A_n(t) \qquad (13.2.7)$$

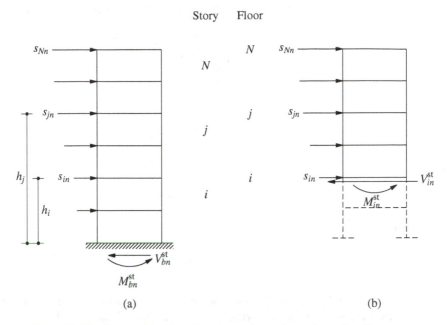

Figure 13.2.2 Computation of modal static responses of story forces from force vector $\mathbf{s}_n$: (a) base shear and base overturning moment; (b) ith story shear and ith floor overturning moment.

where f_{jn} is the lateral force at the jth floor level. Then the response $r_n(t)$ due to the nth mode is given by Eq. (13.1.13), repeated here for convenience:

$$r_n(t) = r_n^{\text{st}} A_n(t) \qquad (13.2.8)$$

The modal static response r_n^{st} is determined by static analysis of the building due to external forces $\mathbf{s}_n$ (Fig. 13.2.2). In applying these forces to the structure, the direction of forces is controlled by the algebraic sign of ϕ_{jn}. Hence these forces for the fundamental mode will all act in the same direction, as shown in Fig. 13.2.2a, but for the second and higher modes they will change direction as one moves up the structure.

The modal static responses are presented in Table 13.2.1 for six response quantities: the shear V_i in the ith story, the overturning moment M_i at the ith floor, the base shear V_b, the base overturning moment M_b, floor displacements u_j, and story drifts Δ_j. The first four equations come from static analysis of the problem in Fig. 13.2.2, which also provides modal static responses for internal forces—bending moments, shears, etc.—in structural elements: beams, columns, walls, etc. The results for u_j and Δ_j are obtained by comparing Eqs. (13.2.5) and (13.2.6) to Eq. (13.2.8). Parts of the equations for V_{bn}^{st} and M_{bn}^{st} are obtained by substituting Eq. (13.2.4) for s_{jn}, Eq. (13.2.3) for L_n^h, and defining M_n^* and h_n^*:

$$M_n^* = \Gamma_n L_n^h = \frac{(L_n^h)^2}{M_n} \qquad h_n^* = \frac{L_n^\theta}{L_n^h} \qquad (13.2.9a)$$

TABLE 13.2.1   MODAL STATIC RESPONSES

Response, r	Modal Static Response, r_n^{st}
V_i	$V_{in}^{st} = \sum_{j=i}^{N} s_{jn}$
M_i	$M_{in}^{st} = \sum_{j=i}^{N} (h_j - h_i)s_{jn}$
V_b	$V_{bn}^{st} = \sum_{j=1}^{N} s_{jn} = \Gamma_n L_n^h \equiv M_n^*$
M_b	$M_{bn}^{st} = \sum_{j=1}^{N} h_j s_{jn} = \Gamma_n L_n^\theta \equiv h_n^* M_n^*$
u_j	$u_{jn}^{st} = (\Gamma_n/\omega_n^2)\phi_{jn}$
Δ_j	$\Delta_{jn}^{st} = (\Gamma_n/\omega_n^2)(\phi_{jn} - \phi_{j-1,n})$

where

$$L_n^\theta = \sum_{j=1}^{N} h_j m_j \phi_{jn} \qquad (13.2.9b)$$

and h_j is the height of the jth floor above the base. Observe that M_n^* and h_n^* are independent of how the mode is normalized, unlike M_n, L_n^h, and Γ_n. Physically meaningful interpretations of M_n^* and h_n^* are presented in Section 13.2.5.

13.2.3 Total Response

Combining the response contributions of all the modes gives the earthquake response of the multistory building:

$$r(t) = \sum_{n=1}^{N} r_n(t) = \sum_{n=1}^{N} r_n^{st} A_n(t) \qquad (13.2.10)$$

wherein Eq. (13.2.8) has been substituted for $r_n(t)$, the nth-mode response.

The modal analysis procedure can also provide floor accelerations, although these are not necessary to compute earthquake-induced forces in the structure. The floor accelerations can be computed from

$$\ddot{u}_j^t(t) = \ddot{u}_g(t) + \sum_{n=1}^{N} \Gamma_n \phi_{jn} \ddot{D}_n(t) \qquad (13.2.11)$$

using the values of $\ddot{D}_n$ available at each time step from the numerical time-stepping procedure used to solve Eq. (13.1.8) for $D_n(t)$.

13.2.4 Summary

The response of an N-story building with plan symmetric about two orthogonal axes to earthquake ground motion along an axis of symmetry can be computed as a function of time by the procedure just developed, which is summarized next in step-by-step form:

1. Define the ground acceleration $\ddot{u}_g(t)$ numerically at every time step Δt.
2. Define the structural properties.
   a. Determine the mass matrix $\mathbf{m}$ and lateral stiffness matrix $\mathbf{k}$ (Section 9.4).
   b. Estimate the modal damping ratios ζ_n (Chapter 11).
3. Determine the natural frequencies ω_n (natural periods $T_n = 2\pi/\omega_n$) and natural modes ϕ_n of vibration (Chapter 10).
4. Determine the modal components $\mathbf{s}_n$ [Eq. (13.2.4)] of the effective earthquake force distribution.
5. Compute the response contribution of the nth mode by the following steps, which are repeated for all modes, $n = 1, 2, \dots, N$:
   a. Perform static analysis of the building subjected to lateral forces $\mathbf{s}_n$ to determine r_n^{st}, the modal static response for each desired response quantity r (Table 13.2.1).
   b. Determine the pseudo-acceleration response $A_n(t)$ of the nth-mode SDF system to $\ddot{u}_g(t)$, using numerical time-stepping methods (Chapter 5).
   c. Determine $r_n(t)$ from Eq. (13.2.8).
6. Combine the modal contributions $r_n(t)$ to determine the total response using Eq. (13.2.10).

As will be shown later, usually only the lower few modes contribute significantly to the response. Therefore, steps 3, 4, and 5 need to be implemented only for these modes, and the modal summation of Eq. (13.2.10) truncated accordingly.

Example 13.3

Derive equations for **(a)** the floor displacements and **(b)** the story shears for the shear frame of Example 13.2 subjected to ground motion $\ddot{u}_g(t)$.

Solution   Steps 1 to 4 of the procedure summary have already been implemented in Example 13.2.

(a) *Floor displacements.* Substituting Γ_n and ϕ_{jn} from Example 13.2 in Eq. (13.2.5) gives the floor displacements due to the each mode:

$$\left\{ \begin{matrix} u_1(t) \\ u_2(t) \end{matrix} \right\}_1 = \frac{4}{3} \left\{ \begin{matrix} \frac{1}{2} \\ 1 \end{matrix} \right\} D_1(t) \qquad \left\{ \begin{matrix} u_1(t) \\ u_2(t) \end{matrix} \right\}_2 = -\frac{1}{3} \left\{ \begin{matrix} -1 \\ 1 \end{matrix} \right\} D_2(t) \tag{a}$$

Combining the contributions of the two modes gives the floor displacements:

$$u_1(t) = u_{11}(t) + u_{12}(t) = \tfrac{2}{3} D_1(t) + \tfrac{1}{3} D_2(t) \tag{b}$$

$$u_2(t) = u_{21}(t) + u_{22}(t) = \tfrac{4}{3} D_1(t) - \tfrac{1}{3} D_2(t) \tag{c}$$

(b) *Story shears.* Static analysis of the frame for external floor forces $\mathbf{s}_n$ gives V_{in}^{st}, $i = 1$ and 2, shown in Fig. E13.3. Substituting these results in Eq. (13.2.8) gives

$$V_{11}(t) = \tfrac{8}{3} m A_1(t) \qquad V_{21}(t) = \tfrac{4}{3} m A_1(t) \tag{d}$$

$$V_{12}(t) = \tfrac{1}{3} m A_2(t) \qquad V_{22}(t) = -\tfrac{1}{3} m A_2(t) \tag{e}$$

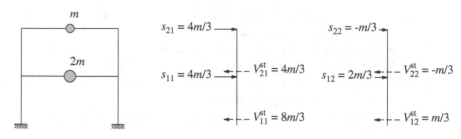

Figure E13.3

Combining the contributions of two modes gives the story shears

$$V_1(t) = V_{11}(t) + V_{12}(t) = \tfrac{8}{3}mA_1(t) + \tfrac{1}{3}mA_2(t) \tag{f}$$

$$V_2(t) = V_{21}(t) + V_{22}(t) = \tfrac{4}{3}mA_1(t) - \tfrac{1}{3}mA_2(t) \tag{g}$$

The floor displacements and story shears have been expressed in terms of $D_n(t)$ and $A_n(t)$. These responses of the nth-mode SDF system to prescribed $\ddot{u}_g(t)$ can be determined by numerical time-stepping methods (Chapter 5).

Example 13.4

Derive equations for **(a)** the floor displacements and **(b)** the element forces for the two-story frame of Fig. E13.4a due to horizontal ground motion $\ddot{u}_g(t)$.

Solution Equation (9.3.4) with $\mathbf{p}_t(t) = -\mathbf{m}_{tt}\mathbf{1}\ddot{u}_g(t)$ governs the displacement vector $\mathbf{u}_t = \langle u_1 \quad u_2 \rangle$; where $\mathbf{m}_{tt}$ and $\mathbf{k}_{tt}$, determined in Example 9.9, are

$$\mathbf{m}_{tt} = m\begin{bmatrix} 2 & \\ & 1 \end{bmatrix} \qquad \hat{\mathbf{k}}_{tt} = \frac{EI}{h^3}\begin{bmatrix} 54.88 & -17.51 \\ -17.51 & 11.61 \end{bmatrix} \tag{a}$$

where $h = 10$ ft. The natural frequencies and modes of the system, determined in Example 10.5, are

$$\omega_1 = 2.198\sqrt{\frac{EI}{mh^3}} \qquad \omega_2 = 5.850\sqrt{\frac{EI}{mh^3}} \tag{b}$$

$$\phi_1 = \begin{Bmatrix} 0.3871 \\ 1 \end{Bmatrix} \qquad \phi_2 = \begin{Bmatrix} -1.292 \\ 1 \end{Bmatrix} \tag{c}$$

Thus steps 1 to 3 of Section 13.2.4 have already been implemented.

(a) *Floor displacements and joint rotations.* The floor displacements are given by Eq. (13.2.5), where Γ_n are computed from Eq. (13.2.3): $M_1 = 2m(0.3871)^2 + m(1)^2 = 1.300m$, $L_1^h = 2m(0.3871) + m(1) = 1.774m$, and $\Gamma_1 = 1.774m/1.300m = 1.365$. Similarly, $M_2 = 4.337m$, $L_2^h = -1.583m$, and $\Gamma_2 = -0.365$. Substituting these in Eq. (13.2.5) with $n = 1$ gives the floor displacements due to the first mode:

$$\mathbf{u}_1(t) = \begin{Bmatrix} u_1(t) \\ u_2(t) \end{Bmatrix}_1 = 1.365\begin{Bmatrix} 0.3871 \\ 1 \end{Bmatrix}D_1(t) = \begin{Bmatrix} 0.5284 \\ 1.365 \end{Bmatrix}D_1(t) \tag{d}$$

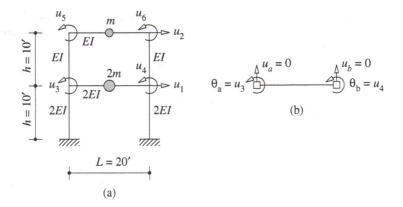

Figure E13.4

The joint rotations associated with these floor displacements are determined from Eq. (d) of Example 9.9 by substituting $\mathbf{u}_1$ from Eq. (d) for $\mathbf{u}_t$:

$$\mathbf{u}_{01}(t) = \begin{Bmatrix} u_3(t) \\ u_4(t) \\ u_5(t) \\ u_6(t) \end{Bmatrix}_1 = \frac{1}{h} \begin{bmatrix} -0.4426 & -0.2459 \\ -0.4426 & -0.2459 \\ 0.9836 & -0.7869 \\ 0.9836 & -0.7869 \end{bmatrix} \begin{Bmatrix} 0.5284 \\ 1.365 \end{Bmatrix} D_1(t)$$

$$= \frac{1}{h} \begin{Bmatrix} -0.5696 \\ -0.5696 \\ -0.5544 \\ -0.5544 \end{Bmatrix} D_1(t) \qquad (e)$$

Similarly, the floor displacements $\mathbf{u}_2(t)$ and joint rotations $\mathbf{u}_{02}(t)$ due to the second mode are determined:

$$\mathbf{u}_2(t) = \begin{Bmatrix} u_1(t) \\ u_2(t) \end{Bmatrix}_2 = \begin{Bmatrix} 0.4716 \\ -0.3651 \end{Bmatrix} D_2(t)$$

$$\mathbf{u}_{02}(t) = \begin{Bmatrix} u_3(t) \\ u_4(t) \\ u_5(t) \\ u_6(t) \end{Bmatrix}_2 = \frac{1}{h} \begin{Bmatrix} -0.1189 \\ -0.1189 \\ 0.7511 \\ 0.7511 \end{Bmatrix} D_2(t) \qquad (f)$$

Combining the contributions of the two modes gives the floor displacements and joint rotations:

$$\mathbf{u}(t) = \mathbf{u}_1(t) + \mathbf{u}_2(t) \qquad \mathbf{u}_0(t) = \mathbf{u}_{01}(t) + \mathbf{u}_{02}(t) \qquad (g)$$

(b) *Element forces.* Instead of implementing step 5 of the procedure (Section 13.2.4), we will illustrate the computation of element forces from the floor displacements and joint rotations by using the beam stiffness coefficients (Appendix 1). For example, the bending moment at the left end of the first-floor beam (Fig. E13.4b) is

$$M_a = \frac{4EI}{L}\theta_a + \frac{2EI}{L}\theta_b + \frac{6EI}{L^2}u_a - \frac{6EI}{L^2}u_b \qquad (h)$$

The vertical displacements u_a and u_b are zero because the columns are assumed to be axially rigid; joint rotations $\theta_a = u_3$ and $\theta_b = u_4$, where u_3 and u_4 are known from Eqs. (e), (f), and (g); thus

$$\theta_a(t) = -\frac{0.5696}{h}D_1(t) - \frac{0.1189}{h}D_2(t) \qquad \theta_b(t) = \theta_a(t) \tag{i}$$

Substituting for u_a, u_b, θ_a, and θ_b in Eq. (h), replacing EI by $2EI$, and using $D_n(t) = A_n(t)/\omega_n^2$ gives

$$M_a(t) = mh[-0.7077A_1(t) - 0.0209A_2(t)] \qquad M_b(t) = M_a(t) \tag{j}$$

Equations for forces in all beams and columns can be obtained similarly.

Comparing the two terms in Eq. (j) for $M_a(t)$ with Eq. (13.2.8) indicates that $M_{a1}^{st} = -0.7077mh$ and $M_{a2}^{st} = -0.0209mh$. These modal static responses could have been obtained by static analysis of the structure due to s_n determined from Eq. (13.2.4).

The various response quantities have been expressed in terms of $D_n(t)$ and $A_n(t)$; these responses of the nth-mode SDF system to given $\ddot{u}_g(t)$ can be determined by numerical time-stepping methods (Chapter 5).

13.2.5 Effective Modal Mass and Modal Height

In this section physically motivated interpretations of M_n^* and h_n^*, introduced in Eq. (13.2.9a), are presented. The base shear due to the nth mode (Fig. 13.2.3a) is obtained by specializing Eq. (13.2.8) for V_b:

$$V_{bn}(t) = V_{bn}^{st} A_n(t) \tag{13.2.12a}$$

which after substituting for V_{bn}^{st} from Table 13.2.1 becomes

$$V_{bn}(t) = M_n^* A_n(t) \tag{13.2.12b}$$

In contrast to Eq. (13.2.12b), the base shear in a one-story system (Fig. 6.2.1a) with lumped mass m is given by Eq. (6.7.3). Defining the natural frequency of this system as ω_n and its damping ratio as ζ_n—the same as the vibration properties of the nth mode of the multistory building—Eq. (6.7.3) becomes

$$V_b(t) = m A_n(t) \tag{13.2.13}$$

Comparing Eqs. (13.2.12b) and (13.2.13) indicates that if the mass of this SDF system were M_n^* (Fig. 13.2.3b), its base shear would be the same as V_{bn}, the nth-mode base shear in a multistory system with its mass distributed among the various floor levels. Thus M_n^* is called the *base shear effective modal mass* or, for brevity, *effective modal mass*.

Equation (13.2.13) implies that the total mass m of a single-mass system is effective in producing the base shear. This is so because the mass and hence the equivalent static force are concentrated at one location. In contrast, only the portion M_n^* of the mass of a multistory building is effective in producing the base shear due to the nth mode because the building mass is distributed among the various floor levels (Fig. 13.2.3a) and the equivalent static forces [Eq. (13.1.11)] vary over the height as $m_j\phi_{jn}$ [Eq. (13.2.4)]. This

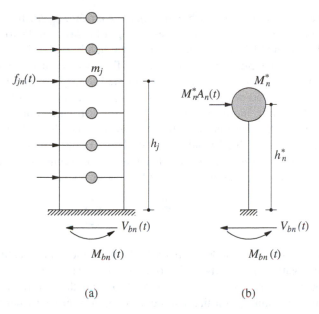

Figure 13.2.3   (a) Equivalent static forces and base shear in the nth mode; (b) one-story system with effective modal mass and effective modal height.

portion depends on the distribution of the mass of the building over its height and on the shape of the mode, as indicated by Eqs. (13.2.9a) and (13.2.3). As intuition might suggest, the sum of the effective modal masses M_n^* over all the modes is equal to the total mass of the building (see Derivation 13.1):

$$\sum_{n=1}^{N} M_n^* = \sum_{j=1}^{N} m_j \qquad (13.2.14)$$

Now we compare the base overturning moment equations for multistory and one-story systems. The base overturning moment in a multistory building due to its nth mode is obtained by specializing Eq. (13.2.8) for M_b:

$$M_{bn}(t) = M_{bn}^{st} A_n(t) \qquad (13.2.15a)$$

which after substituting for M_{bn}^{st} from Table 13.2.1 becomes

$$M_{bn}(t) = h_n^* V_{bn}(t) \qquad (13.2.15b)$$

In contrast, the base overturning moment in a one-story system with mass m lumped at height h above the base is given by Eq. (6.7.3), repeated here for convenience:

$$M_b(t) = h V_b(t) \qquad (13.2.16)$$

Comparing Eqs. (13.2.15b) and (13.2.16) indicates that if the mass of this SDF system were M_n^* and it were lumped at height h_n^* (Fig. 13.2.3b), its base overturning moment would be the same as M_{bn}, the nth-mode base overturning moment, in a multistory building with its mass distributed among the various floor levels. Thus h_n^* is called the *base-moment* *effective modal height* or, for brevity, *effective modal height*. It may also be interpreted as the height of the resultant of the forces $\mathbf{s}_n$ (Fig. 13.2.2a) or of the forces $f_{jn}(t)$ (Fig. 13.2.3a).

Equation (13.2.16) implies that the total height h of a single mass system is effective in producing the base overturning moment. This is so because the mass of the structure and hence the equivalent static force is concentrated at height h above the base. In contrast, the *effective modal height* h_n^* is less than the total height of the building because the building mass and hence the equivalent static forces are distributed among the various floor levels; h_n^* depends on the distribution of the mass over the height of the building and on the shape of the mode [Eqs. (13.2.9) and (13.2.3)]. The sum of the first moments about the base of the effective modal masses M_n^* located at effective heights h_n^* is equal to the first moment of the floor masses about the base (see Derivation 13.2):

$$\sum_{n=1}^{N} h_n^* M_n^* = \sum_{j=1}^{N} h_j m_j \qquad (13.2.17)$$

For some of the modes higher than the fundamental mode, the effective modal height may be negative. A negative value of h_n^* implies that the modal static base shear V_{bn}^{st} and the modal static base overturning moment M_{bn}^{st} for the nth mode have opposite algebraic signs; the M_{b1}^{st} and V_{b1}^{st} for the first mode are both positive, by definition.

Derivation 13.1

Premultiplying both sides of Eq. (13.2.2) by $\mathbf{1}^T$ gives

$$\mathbf{1}^T \mathbf{m} \mathbf{1} = \sum_{n=1}^{N} \Gamma_n (\mathbf{1}^T \mathbf{m} \boldsymbol{\phi}_n)$$

Noting that $\mathbf{m}$ is a diagonal matrix with $m_{jj} = m_j$, this can be rewritten as

$$\sum_{j=1}^{N} m_j = \sum_{n=1}^{N} \Gamma_n L_n^h$$

This provides a proof for Eq. (13.2.14) because the nth term on the right side is M_n^* (Table 13.2.1).

Derivation 13.2

A modal expansion of the force vector $\mathbf{m}\mathbf{h}$ where $\mathbf{h} = \langle h_1 \quad h_2 \quad \cdots \quad h_N \rangle^T$ is obtained by substituting $\mathbf{s} = \mathbf{m}\mathbf{h}$ in Eqs. (12.8.2) and (12.8.3):

$$\mathbf{m}\mathbf{h} = \sum_{n=1}^{N} \frac{L_n^\theta}{M_n} \mathbf{m} \boldsymbol{\phi}_n$$

Premultiplying both sides by $\mathbf{1}^T$ gives

$$\mathbf{1}^T \mathbf{m}\mathbf{h} = \sum_{n=1}^{N} \frac{L_n^\theta}{M_n} \mathbf{1}^T \mathbf{m} \boldsymbol{\phi}_n$$

Noting that **m** is a diagonal matrix with $m_{jj} = m_j$, this can be rewritten as

$$\sum_{j=1}^{N} m_j h_j = \sum_{n=1}^{N} \frac{L_n^\theta}{M_n} L_n^h = \sum_{n=1}^{N} h_n^* M_n^*$$

wherein Eq. (13.2.9) has been used. This provides a proof for Eq. (13.2.17).

Example 13.5

Determine the effective modal masses and effective modal heights for the two-story shear frame of Example 13.2. The height of each story is h.

Solution   In Example 13.2 the **m**, **k**, ω_n, and ϕ_n for this system were presented, and L_n^h and M_n for each of the two modes computed. These are listed next, together with the new computations for M_n^* and h_n^*. For the first mode: $L_1^h = 2m$, $M_1 = 3m/2$, $M_1^* = (L_1^h)^2/M_1 = \frac{8}{3}m$, $L_1^\theta = h(2m)\frac{1}{2} + 2h(m)1 = 3hm$, and $h_1^* = L_1^\theta/L_1^h = 3hm/2m = 1.5h$. Similarly, for the second mode: $L_2^h = -m$, $M_2 = 3m$, $M_2^* = (L_2^h)^2/M_2 = \frac{1}{3}m$, $L_2^\theta = h(2m)(-1) + 2h(m)1 = 0$, and $h_2^* = L_2^\theta/L_2^h = 0$.

Observe that $M_1^* + M_2^* = 3m$, the total mass of the frame, confirming that Eq. (13.2.14) is satisfied; also note that the effective height for the second mode is zero, implying that the base overturning moment $M_{b2}(t)$ due to that mode will be zero at all t. This is an illustration of a more general result developed in Example 13.6.

Example 13.6

Show that the base overturning moment in a multistory building due to the second and higher modes is zero if the first mode shape is linear (i.e., the floor displacements are proportional to floor heights above the base).

Solution   Equation (13.2.15) gives the nth-mode contribution to the base overturning moment. A linear first mode implies that $\phi_{j1} = h_j/h_N$, where h_j is the height of the jth floor above the base and h_N is the total height of the building. Substituting $h_j = h_N \phi_{j1}$ in (13.2.9b) gives

$$L_n^\theta = \sum_{j=1}^{N} h_j m_j \phi_{jn} = h_N \phi_1^T \mathbf{m} \phi_n$$

and this is zero for all $n \neq 1$ because of the orthogonality property of modes. Therefore, for all $n \neq 1$, $h_n^* = 0$ from Eq. (13.2.9a) and $M_{bn}(t) = 0$ from Eq. (13.2.15).

13.2.6 Example: Five-Story Shear Frame

In this section the earthquake analysis procedure summarized in Section 13.2.4 is implemented for the five-story shear frame of Fig. 12.8.1, subjected to the El Centro ground motion shown in Fig. 6.1.4. The results presented are accompanied by interpretive comments that should assist us in developing an understanding of the response behavior of multistory buildings.

System properties.   The lumped mass $m_j = m = 100$ kips/g at each floor, the lateral stiffness of each story is $k_j = k = 31.54$ kips/in., and the height of each story is 12 ft. The damping ratio for all natural modes is $\zeta_n = 5\%$. The mass matrix **m**, stiffness matrix **k**, natural frequencies, and natural modes of this system were presented in Section 12.8. For the given k and m, the natural periods are $T_n = 2.0, 0.6852, 0.4346, 0.3383,$ and 0.2966 sec. (These natural periods, which are much longer than for typical five-story buildings, were chosen to accentuate the contributions of the second through fifth modes to the structural response.) Thus steps 1, 2, and 3 of the analysis procedure (Section 13.2.4) have already been completed.

Modal expansion of m1.   To implement step 4 of the analysis procedure (Section 13.2.4), the modal properties M_n, L_n^h, and L_n^θ are computed from Eqs. (13.2.3) and (13.2.9b) using the known modes ϕ_n (Table 13.2.2). The Γ_n are computed from Eq. (13.2.3)

TABLE 13.2.2   MODAL PROPERTIES

Mode	M_n	L_n^h	L_n^θ/h
1	1.000	1.067	3.750
2	1.000	−0.336	0.404
3	1.000	0.177	0.135
4	1.000	−0.099	0.059
5	1.000	0.045	0.023

and substituted in Eq. (13.2.4), together with values for m_j and ϕ_{jn}, to obtain the $\mathbf{s}_n$ vectors shown in Fig. 13.2.4. Observe that the direction of forces $\mathbf{s}_n$ is controlled by the algebraic sign of ϕ_{jn} (Fig. 12.8.2). Hence, these forces for the fundamental mode act in the same direction, but for the second and higher modes they change direction as one moves up the structure. The contribution of the fundamental mode to the force distribution $\mathbf{s} = \mathbf{m1}$ of the effective earthquake forces is the largest, and the modal contributions to these forces decrease progressively for higher modes.

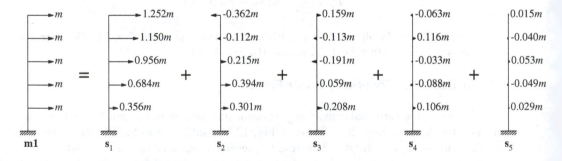

Figure 13.2.4   Modal expansion of **m1**.

Modal static responses.   Table 13.2.3 gives the results for four response quantities—base shear V_b, fifth-story shear V_5, base overturning moment M_{bn}, and roof

TABLE 13.2.3   MODAL STATIC RESPONSES

Mode	V_{bn}^{st}/m	V_{5n}^{st}/m	M_{bn}^{st}/mh	u_{5n}^{st}
1	4.398	1.252	15.45	0.127
2	0.436	−0.362	−0.525	−0.004
3	0.121	0.159	0.092	0.0008
4	0.037	−0.063	−0.022	−0.0002
5	0.008	0.015	0.004	0.00003

displacement u_5—obtained using the equations in Table 13.2.1 and the known s_{jn}, ϕ_{5n}, and ω_n^2 (step 5a of Section 13.2.4). Observe that the modal static responses are largest for the first mode and decrease progressively for higher modes. The effective modal masses $M_n^* = V_{bn}^{st}$ and effective modal heights $h_n^* = M_{bn}^{st}/V_{bn}^{st}$ are shown schematically in Fig. 13.2.5; note that h_n^* are plotted without their algebraic signs. Observe that $\sum M_n^* = 5m$, confirming that Eq. (13.2.14) is satisfied. Also note that $\sum h_n^* M_n^* = 15mh$; this is the same as $\sum h_j m_j = 15mh$, confirming that Eq. (13.2.17) is satisfied.

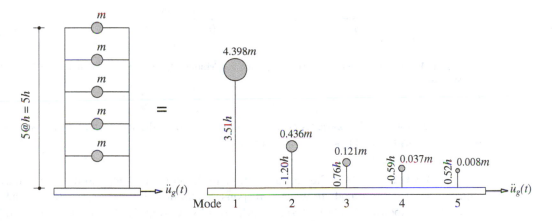

Figure 13.2.5   Effective modal masses and effective modal heights.

Earthquake excitation.   The ground acceleration $\ddot{u}_g(t)$ is defined by its numerical values at time instants equally spaced at every Δt. This time step $\Delta t = 0.01$ sec is chosen to be small enough to define $\ddot{u}_g(t)$ accurately and to determine accurately the response of SDF systems with natural periods T_n, the shortest of which is 0.2966 sec.

Response of SDF systems.   The deformation response $D_n(t)$ of the nth-mode SDF system with natural period T_n and damping ratio ζ_n to the ground motion is determined (step 5b of Section 13.2.4). The time-stepping linear acceleration method (Chapter 5) was implemented to obtain discrete values of D_n at every Δt. For convenience, however, we continue to denote these discrete values as $D_n(t)$. At each time instant the pseudo-acceleration is calculated from $A_n(t) = \omega_n^2 D_n(t)$. These computations are

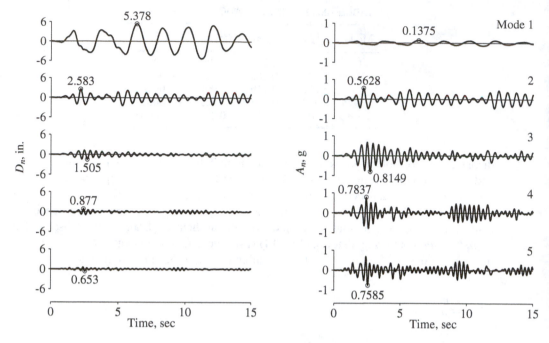

Figure 13.2.6 Displacement $D_n(t)$ and pseudo-acceleration $A_n(t)$ responses of modal SDF systems.

implemented for the SDF systems corresponding to each of the five modes of the structure, and the results are presented in Fig. 13.2.6.

Modal responses.   Step 5c of Section 13.2.4 is implemented to determine the contribution of the nth mode to selected response quantities: V_b, V_5, M_{bn}, and u_5. The modal static responses (Table 13.2.3) are multiplied by A_n (Fig. 13.2.6) at each time step to obtain the results presented in Figs. 13.2.7 and 13.2.8.

These results give us a first impression of the relative values of the response contributions of the various modes. The modal static responses (Table 13.2.3) had suggested that the response will be largest in the fundamental mode and will tend to decrease in the higher modes. Such is the case in this example for roof displacement, base shear, and base overturning moment but not for the fifth-story shear. How the relative modal responses depend on the response quantity and on the building properties is discussed in Chapter 19.

Total responses.   The total responses, determined by combining the modal contributions $r_n(t)$ (step 6 of Section 13.2.4) according to Eq. (13.2.10), are shown in Figs. 13.2.7 and 13.2.8. The results presented indicate that it is not necessary to include the contributions of all the modes in computing the response of a multistory building; the

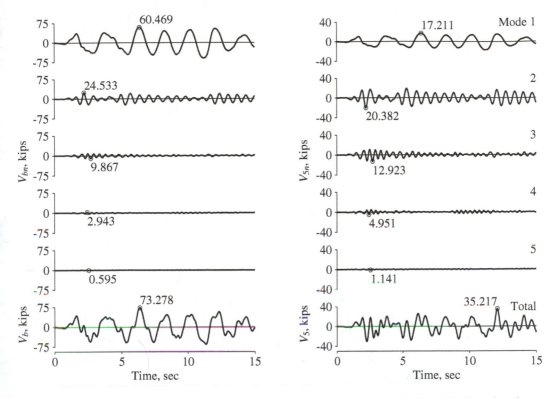

Figure 13.2.7 Base shear and fifth-story shear: modal contributions, $V_{bn}(t)$ and $V_{5n}(t)$, and total responses, $V_b(t)$ and $V_5(t)$.

lower few modes may suffice and the modal summations can be truncated accordingly. In this particular example, the contribution of the fourth and fifth modes could be neglected; the results would still be accurate enough for use in structural design. How many modes should be included depends on the earthquake ground motion and building properties. This issue is addressed in Chapter 19.

Before leaving this example, we make three additional observations that will be especially useful in Part B of this chapter. First, as seen in Chapter 6, the peak values of $D_n(t)$ and $A_n(t)$, noted in Fig. 13.2.6, can be determined directly from the response spectrum for the ground motion. This fact will enable us to determine the peak value of the nth-mode contribution to any response quantity directly from the response spectrum. Second, the contribution of the nth mode to every response quantity attains its peak value at the same time as $A_n(t)$ does. Third, the peak value of the total response occurs at a time instant different from when the individual modal peaks are attained. Furthermore, the peak values of the total responses for the four response quantities occur at different time instants because the relative values of the modal contributions vary with the response quantity.

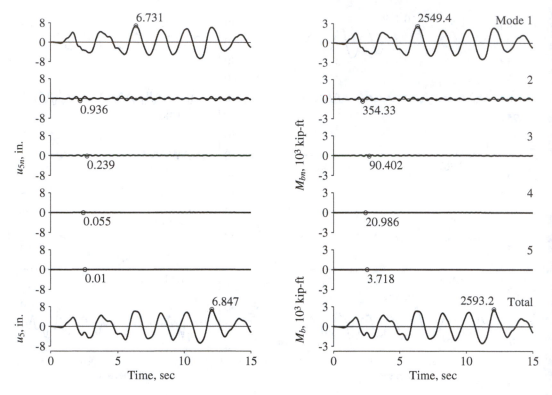

Figure 13.2.8 Roof displacement and base overturning moment: modal contributions, $u_{5n}(t)$ and $M_{bn}(t)$, and total responses, $u_5(t)$ and $M_b(t)$.

13.2.7 Example: Four-Story Frame with an Appendage

This section is concerned with the earthquake analysis and response of a four-story building with a light appendage—a penthouse, a small housing for mechanical equipment, an advertising billboard, or the like. This example is presented because it brings out certain special response features representative of a system with two natural frequencies that are close.

System properties.   The lumped masses at the first four floors are $m_j = m$, the appendage mass $m_5 = 0.01m$, and $m = 100$ kips/g. The lateral stiffness of each of the first four stories is $k_j = k$, the appendage stiffness $k_5 = 0.0012k$, and $k = 22.599$ kips/in. The height of each story and the appendage is 12 ft. The damping ratio for all natural modes is $\zeta_n = 5\%$. The response of this system to the El Centro ground motion is determined. The analysis procedure and its implementation are identical to Section 13.2.6; therefore, only a summary of the results is presented.

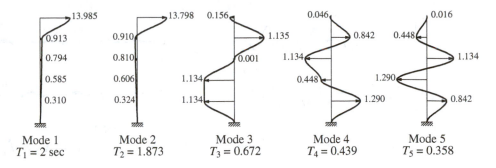

Figure 13.2.9   Natural periods and modes of vibration of building with appendage.

TABLE 13.2.4   MODAL STATIC RESPONSES

	Mode				
	1	2	3	4	5
V_{bn}^{st}/m	1.951	1.633	0.333	0.078	0.015
V_{5n}^{st}/m_5	9.938	−8.979	0.046	−0.007	0.0001

Summary of results.   The natural periods T_n and modes ϕ_n of this system are presented in Fig. 13.2.9. Observe that T_1 and T_2 are close and the corresponding modes show large deformations in the appendage. Table 13.2.4 gives the modal static responses for the base shear V_b and appendage shear V_5. Observe that V_{bn}^{st} for the first two modes are similar in magnitude and of the same algebraic sign; V_{5n}^{st} for the first two modes are also of similar magnitude but of opposite signs. The responses $D_n(t)$ and $A_n(t)$ of the SDF systems corresponding to the five modes of the system are shown in Fig. 13.2.10. Note that $D_n(t)$—also $A_n(t)$—for the first two modes are essentially in phase because the two natural periods are close; the peak values are similar because of similar periods and identical damping in the two modes.

The modal contributions to the base shear and to the appendage shear together with the total response are presented in Fig. 13.2.11. Observe that the response contributions of the first two modes are similar in magnitude because the modal static responses are about the same and the $A_n(t)$ are similar. In the case of base shear, the two modal static responses are of the same algebraic sign, implying that the two modal contributions are essentially in phase [because so are $A_1(t)$ and $A_2(t)$], and hence the combined base shear is much larger than the individual modal responses. In contrast, the modal static responses for the appendage shear are of opposite algebraic sign, indicating that the two modal contributions are essentially out of phase, and hence the combined appendage shear is much smaller than the individual modal responses. However, it is very large, being almost equal to its own weight. As a result, significant damage to appendages at the top of essentially undamaged structures has been observed during earthquakes. Two examples are shown in Figures 13.2.12 and 13.2.13.

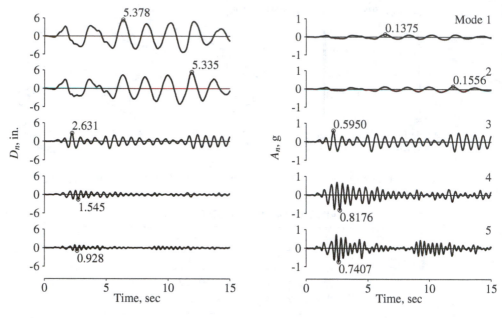

Figure 13.2.10 Displacement $D_n(t)$ and pseudo-acceleration $A_n(t)$ responses of modal SDF systems.

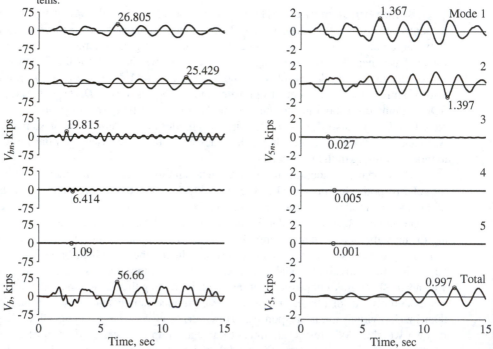

Figure 13.2.11 Base shear and appendage shear: modal contributions, $V_{bn}(t)$ and $V_{5n}(t)$, and total responses, $V_b(t)$ and $V_5(t)$.

Figure 13.2.12   The Hill Building in Anchorage, Alaska, after the Prince William Sound earthquake, March 27, 1964, Magnitude 9.2. Except for the damage to the penthouse wall (close-up), the building was essentially undamaged. (From the Steinbrugge Collection, National Information Service for Earthquake Engineering, University of California, Berkeley.)

Figure 13.2.13   The 1091 ft-high Tokyo Tower performed well during the Tohoku, Japan, earthquake, March 11, 2011, Magnitude 9.0, except that the top part bent permanently. Shown is the tower before and after, with a second focus on the bend near the top. (Courtesy of C. D. James and C. Bodnar-Anderson, National Information Service for Earthquake Engineering, University of California, Berkeley.)

13.3 MULTISTORY BUILDINGS WITH UNSYMMETRIC PLAN

In this section the modal analysis of Section 13.1 is specialized for multistory buildings with their plans symmetric about the x-axis but unsymmetric about the y-axis subjected to ground motion $\ddot{u}_{gy}(t)$ in the y-direction. Equation (9.6.8b) governs the motion of the $2N$ DOFs of the system. If the floor diaphragms have the same radius of gyration r (i.e., $I_{Oj} = r^2 m_j$), Eq. (9.6.8b) specializes to

$$\begin{bmatrix} \mathbf{m} & 0 \\ 0 & r^2\mathbf{m} \end{bmatrix} \begin{Bmatrix} \ddot{u}_y \\ \ddot{u}_\theta \end{Bmatrix} + \begin{bmatrix} \mathbf{k}_{yy} & \mathbf{k}_{y\theta} \\ \mathbf{k}_{\theta y} & \mathbf{k}_{\theta\theta} \end{bmatrix} \begin{Bmatrix} u_y \\ u_\theta \end{Bmatrix} = -\begin{bmatrix} \mathbf{m} & 0 \\ 0 & r^2\mathbf{m} \end{bmatrix} \begin{Bmatrix} 1 \\ 0 \end{Bmatrix} \ddot{u}_{gy}(t) \qquad (13.3.1)$$

The general analysis procedure developed in Section 13.1 is applicable to unsymmetric–plan buildings because Eq. (13.3.1) is of the same form as Eq. (13.1.1).

13.3.1 Modal Expansion of Effective Earthquake Forces

The effective earthquake forces $\mathbf{p}_{\text{eff}}(t)$ are defined by the right side of Eq. (13.3.1):

$$\mathbf{p}_{\text{eff}}(t) = -\begin{Bmatrix} \mathbf{m}1 \\ 0 \end{Bmatrix} \ddot{u}_{gy}(t) \equiv -\mathbf{s}\ddot{u}_{gy}(t) \qquad (13.3.2)$$

The spatial distribution $\mathbf{s}$ of these effective earthquake forces can be expanded as a summation of modal inertia force distributions $\mathbf{s}_n$ (Section 12.8):

$$\begin{Bmatrix} \mathbf{m}1 \\ 0 \end{Bmatrix} = \sum_{n=1}^{2N} \mathbf{s}_n = \sum_{n=1}^{2N} \Gamma_n \begin{Bmatrix} \mathbf{m}\phi_{yn} \\ r^2\mathbf{m}\phi_{\theta n} \end{Bmatrix} \qquad (13.3.3)$$

In this equation ϕ_{yn} includes the translations and $\phi_{\theta n}$ the rotations of the N floors about a vertical axis in the nth mode (i.e., $\phi_n^T = \langle \phi_{yn}^T \quad \phi_{\theta n}^T \rangle$):

$$\Gamma_n = \frac{L_n^h}{M_n} \qquad (13.3.4)$$

where

$$L_n^h = \langle \phi_{yn}^T \quad \phi_{\theta n}^T \rangle \begin{Bmatrix} \mathbf{m}1 \\ 0 \end{Bmatrix} = \phi_{yn}^T \mathbf{m}1 = \sum_{j=1}^{N} m_j \phi_{jyn} \qquad (13.3.5)$$

and

$$M_n = \langle \phi_{yn}^T \quad \phi_{\theta n}^T \rangle \begin{bmatrix} \mathbf{m} & \\ & r^2\mathbf{m} \end{bmatrix} \begin{Bmatrix} \phi_{yn} \\ \phi_{\theta n} \end{Bmatrix}$$

or

$$M_n = \phi_{yn}^T \mathbf{m}\phi_{yn} + r^2 \phi_{\theta n}^T \mathbf{m}\phi_{\theta n} = \sum_{j=1}^{N} m_j \phi_{jyn}^2 + r^2 \sum_{j=1}^{N} m_j \phi_{j\theta n}^2 \qquad (13.3.6)$$

where j denotes the floor number and m_j the floor mass. Equation (13.3.5) differs from Eq. (13.2.3b) for symmetric-plan systems because ϕ_{yn} is not necessarily the same as ϕ_n. Equations (13.3.4) to (13.3.6) for Γ_n can be derived by premultiplying both sides of Eq. (13.3.3) by ϕ_r^T and using the orthogonality property of modes. In Eq. (13.3.3) the nth-mode contribution to the spatial distribution of effective earthquake forces is

$$\mathbf{s}_n = \left\{ \begin{matrix} \mathbf{s}_{yn} \\ \mathbf{s}_{\theta n} \end{matrix} \right\} = \Gamma_n \left\{ \begin{matrix} \mathbf{m}\phi_{yn} \\ r^2\mathbf{m}\phi_{\theta n} \end{matrix} \right\} \tag{13.3.7}$$

The jth element of these subvectors gives the lateral force s_{jyn} and torque $s_{j\theta n}$ at the jth floor level:

$$s_{jyn} = \Gamma_n m_j \phi_{jyn} \qquad s_{j\theta n} = \Gamma_n r^2 m_j \phi_{j\theta n} \tag{13.3.8}$$

Premultiplying each submatrix equation in Eq. (13.3.3) by $\mathbf{1}^T$, two interesting results can be derived:

$$\sum_{n=1}^{2N} M_n^* = \sum_{j=1}^{N} m_j \qquad \sum_{n=1}^{2N} I_{On}^* = 0 \tag{13.3.9}$$

where

$$M_n^* = \frac{(L_n^h)^2}{M_n} \qquad I_{On}^* = r^2 \Gamma_n \mathbf{1}^T \mathbf{m}\phi_{\theta n} \tag{13.3.10}$$

Although this equation for M_n^* for unsymmetric-plan systems appears to be the same as Eq. (13.2.9a) for symmetric-plan systems, the two may not be identical because the mode shapes ϕ_{yn} and ϕ_n in the two cases are not necessarily the same. We shall see later that M_n^* is the base shear effective modal mass for the nth mode and also the modal static response for base shear. As for symmetric-plan buildings, Eq. (13.3.9a) implies that the sum of the effective modal masses over all modes is equal to the total mass of the building. As we shall see later, I_{On}^* is the modal static response for base torque; their sum over all modes is zero according to Eq. (13.3.9b).

Example 13.7

Determine the modal expansion for the distribution of the effective earthquake forces for the system of Example 10.6 subjected to ground motion in the y-direction. Also compute the modal static responses for base shear and base torque, and verify Eq. (13.3.9)

Solution   The DOFs are the lateral displacement u_y and rotation u_θ of the roof. With reference to these DOFs, the natural frequencies and modes were determined in Example 10.6:

$$\omega_1 = 5.878 \qquad \omega_2 = 6.794 \text{ rad/sec}$$

$$\phi_1 = \left\{ \begin{matrix} -0.5228 \\ 0.0493 \end{matrix} \right\} \qquad \phi_2 = \left\{ \begin{matrix} -0.5131 \\ -0.0502 \end{matrix} \right\}$$

Specializing Eqs. (13.3.6), (13.3.5), and (13.3.4) to a one-story system with roof mass m and radius of gyration r yields

$$M_n = m(\phi_{yn}^2 + r^2\phi_{\theta n}^2) \qquad L_n^h = m\phi_{yn} \qquad \Gamma_n = \frac{L_n^h}{M_n} \tag{a}$$

For this system, $m = 1.863$ kip-sec^2/ft and $r^2 = (b^2 + d^2)/12 = (30^2 + 20^2)/12 = 108.\overline{3}$ ft^2 (see Example 10.6). In Eq. (a) with $n = 1$, substituting known values of m, r, ϕ_{y1}, and $\phi_{\theta 1}$, we obtain $M_1 = 1.863\left[(-0.5228)^2 + 108.\overline{3}(0.0493)^2\right] = 1.0$, $L_1^h = 1.863(-0.5228) =$

-0.974, and $\Gamma_1 = -0.974$. Similarly, in Eq. (a) with $n = 2$, substituting for m, r, ϕ_{y2}, and $\phi_{\theta 2}$, we obtain $M_2 = 1.0$, $L_2^h = -0.956$, and $\Gamma_2 = -0.956$.

The modal expansion of the effective earthquake force distribution is obtained from Eq. (13.3.3) by specializing it to a one-story system:

$$\left\{ \begin{array}{c} m \\ 0 \end{array} \right\} = \sum_{n=1}^{2} \Gamma_n \left\{ \begin{array}{c} m\phi_{yn} \\ r^2 m\phi_{\theta n} \end{array} \right\} \tag{b}$$

Substituting numerical values for Γ_n, ϕ_{yn}, $\phi_{\theta n}$, and r gives

$$m \left\{ \begin{array}{c} 1 \\ 0 \end{array} \right\} = m \left\{ \begin{array}{c} 0.509 \\ -5.203 \end{array} \right\} + m \left\{ \begin{array}{c} 0.491 \\ 5.203 \end{array} \right\} \tag{c}$$

This modal expansion is shown on the structural plan in Fig. E13.7.

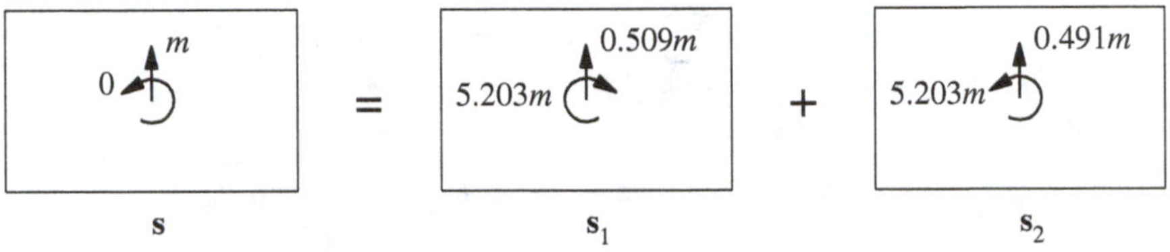

Figure E13.7   Modal expansion of effective force vector shown on plan view of the building.

Static analysis of the structure subjected to forces $\mathbf{s}_n$ (Fig. E13.7) gives the modal static responses for base shear and base torque: $V_{b1}^{st} = 0.509\,m$ and $V_{b2}^{st} = 0.491\,m$; $T_{b1}^{st} = -5.203\,m$ and $T_{b2}^{st} = 5.203\,m$.

Specializing Eq. (13.3.10) to a one-story system gives

$$M_n^* = \frac{(L_n^h)^2}{M_n} \qquad I_{On}^* = \Gamma_n r^2 m\phi_{\theta n} \tag{d}$$

Substituting numerical values for Γ_n, L_n^h, M_n, r, m, and $\phi_{\theta n}$ gives

$$M_1^* = 0.509\,m \qquad I_{O1}^* = -5.203\,m \qquad M_2^* = 0.491\,m \qquad I_{O2}^* = 5.203\,m$$

Observe that these data show that $M_1^* + M_2^* = m$ and $I_{O1}^* + I_{O2}^* = 0$, which provides a numerical confirmation of Eq. (13.3.9) for this one-story ($N = 1$) system. Note that, as expected, $V_{bn}^{st} = M_n^*$ and $T_{bn}^{st} = I_{On}^*$.

13.3.2 Modal Responses

Displacements.   The differential equation governing the nth modal coordinate is Eq. (13.1.7), with $\ddot{u}_g(t)$ replaced by $\ddot{u}_{gy}(t)$ and Γ_n defined by Eqs. (13.3.4) to (13.3.6). Using this Γ_n, Eq. (13.1.10) gives the contribution $\mathbf{u}_n(t)$ of the nth mode to the displacement $\mathbf{u}(t)$. The lateral displacement $\mathbf{u}_{yn}$ and torsional displacements $\mathbf{u}_{\theta n}$ are

$$\mathbf{u}_{yn}(t) = \Gamma_n \phi_{yn} D_n(t) \qquad \mathbf{u}_{\theta n}(t) = \Gamma_n \phi_{\theta n} D_n(t) \tag{13.3.11}$$

In particular, the lateral and torsional displacements of the jth floor are

$$u_{jyn}(t) = \Gamma_n \phi_{jyn} D_n(t) \qquad u_{j\theta n}(t) = \Gamma_n \phi_{j\theta n} D_n(t) \tag{13.3.12}$$

Building forces.   The equivalent static forces $\mathbf{f}_n(t)$ associated with displacements $\mathbf{u}_n(t)$ include lateral forces $\mathbf{f}_{yn}(t)$ and torques $\mathbf{f}_{\theta n}(t)$. These forces are given by a generalization of Eq. (13.1.11):

$$\begin{Bmatrix} \mathbf{f}_{yn}(t) \\ \mathbf{f}_{\theta n}(t) \end{Bmatrix} = \begin{Bmatrix} \mathbf{s}_{yn} \\ \mathbf{s}_{\theta n} \end{Bmatrix} A_n(t) \tag{13.3.13}$$

The lateral force and torque at the jth floor level are

$$f_{jyn}(t) = s_{jyn} A_n(t) \qquad f_{j\theta n}(t) = s_{j\theta n} A_n(t) \tag{13.3.14}$$

Then the response r due to the nth mode is given by Eq. (13.1.13), repeated here for convenience:

$$r_n(t) = r_n^{\text{st}} A_n(t) \tag{13.3.15}$$

The modal static response r_n^{st} is determined by static analysis of the building due to external forces $\mathbf{s}_{yn}$ and $\mathbf{s}_{\theta n}$. In applying these forces to the structure, the direction of forces is controlled by the algebraic sign of ϕ_{jyn} and $\phi_{j\theta n}$. In particular for the fundamental mode, the lateral forces all act in the same direction, and the torques all act in the same direction (Fig. 13.3.1). However, for the second and higher modes, the lateral forces or torques, or both, will change direction as one moves up the structure.

The modal static responses are presented in Table 13.3.1 for eight response quantities: the shear V_i and torque T_i in the ith story, the overturning moment M_i at the ith floor, the base shear V_b, the base torque T_b, and the base overturning moment M_b, floor translations u_{jy}, and floor rotations $u_{j\theta}$. The equations for forces are determined by static analysis of the building subjected to lateral forces $\mathbf{s}_{yn}$ and torques $\mathbf{s}_{\theta n}$ (Fig. 13.3.1); and the results for u_{jy} and $u_{j\theta}$ are obtained by writing Eqs. (13.3.12) in a form similar to Eq. (13.3.15).

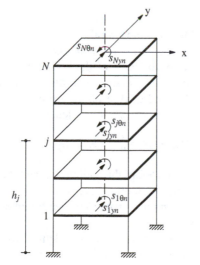

Figure 13.3.1   External forces s_{jyn} and $s_{j\theta n}$ for nth mode.

TABLE 13.3.1   MODAL STATIC RESPONSES

Response, r	Modal Static Response, r_n^{st}
V_i	$V_{in}^{st} = \sum_{j=i}^{N} s_{jyn}$
M_i	$M_{in}^{st} = \sum_{j=i}^{N} (h_j - h_i)s_{jyn}$
T_i	$T_{in}^{st} = \sum_{j=i}^{N} s_{j\theta n}$
V_b	$V_{bn}^{st} = \sum_{j=1}^{N} s_{jyn} = \Gamma_n L_n^h = M_n^*$
M_b	$M_{bn}^{st} = \sum_{j=1}^{N} h_j s_{jyn} = \Gamma_n L_n^\theta = h_n^* M_n^*$
T_b	$T_{bn}^{st} = \sum_{j=1}^{N} s_{j\theta n} = I_{On}^*$
u_{jy}	$u_{jyn}^{st} = (\Gamma_n/\omega_n^2)\phi_{jyn}$
$u_{j\theta}$	$u_{j\theta n}^{st} = (\Gamma_n/\omega_n^2)\phi_{j\theta n}$

Parts of the equations for V_{bn}^{st}, M_{bn}^{st}, and T_{bn}^{st} are obtained by substituting Eq. (13.3.8) for s_{jyn} and $s_{j\theta n}$, Eq. (13.3.5) for L_n^h, Eq. (13.3.10) for M_n^* and I_{On}^*, and Eq. (13.2.9a) for h_n^* with ϕ_{jn} replaced by ϕ_{jyn} in Eq. (13.2.9b).

Specializing Eq. (13.3.15) for V_b, M_b, and T_b and substituting for V_{bn}^{st}, M_{bn}^{st}, and T_{bn}^{st} from Table 13.3.1 gives

$$V_{bn}(t) = M_n^* A_n(t) \qquad T_{bn}(t) = I_{On}^* A_n(t) \qquad M_{bn}(t) = h_n^* M_n^* A_n(t) \qquad (13.3.16)$$

For reasons mentioned in Section 13.2.5, M_n^* is called the *effective modal mass* and h_n^* the *effective modal height*.

Frame forces.   In addition to the overall story forces for the building, it is desired to determine the element forces—bending moments, shears, etc.—in structural elements—beams, columns, walls, etc.—of each frame of the building. For this purpose, the lateral displacements $\mathbf{u}_{in}$ of the ith frame associated with displacements $\mathbf{u}_n$ in the floor DOFs of the building are determined from Eq. (9.6.1). Substituting Eq. (9.6.2) for $\mathbf{a}_{xi}$ and $\mathbf{a}_{yi}$, $\mathbf{u}_n^T = \langle \mathbf{u}_{yn}^T \quad \mathbf{u}_{\theta n}^T \rangle$, and Eq. (13.3.11) for $\mathbf{u}_{yn}$ and $\mathbf{u}_{\theta n}$ leads to

$$\mathbf{u}_{in}(t) = \Gamma_n(-y_i\phi_{\theta n})D_n(t) \qquad \mathbf{u}_{in}(t) = \Gamma_n(\phi_{yn} + x_i\phi_{\theta n})D_n(t) \qquad (13.3.17)$$

The first equation is for frames oriented in the x-direction and the second for frames in the y-direction. At each time instant, the internal forces in elements of frame i can be determined from these displacements and joint rotations (see Example 13.4) using the element stiffness properties (Appendix 1).

Alternatively, equivalent static forces $\mathbf{f}_{in}$ can be defined for the ith frame with lateral stiffness matrix $\mathbf{k}_{xi}$ if the frame is oriented in the x-direction or $\mathbf{k}_{yi}$ for a frame along the

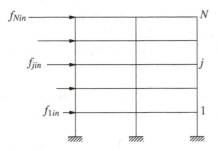

Figure 13.3.2   Equivalent static forces for the ith frame associated with response of the building in its nth natural mode.

y-direction. Thus

$$\mathbf{f}_{in}(t) = \mathbf{k}_{xi}\mathbf{u}_{in}(t) = (\Gamma_n/\omega_n^2)\mathbf{k}_{xi}(-y_i\boldsymbol{\phi}_{\theta n})A_n(t) \qquad (13.3.18a)$$

$$\mathbf{f}_{in}(t) = \mathbf{k}_{yi}\mathbf{u}_{in}(t) = (\Gamma_n/\omega_n^2)\mathbf{k}_{yi}(\boldsymbol{\phi}_{yn} + x_i\boldsymbol{\phi}_{\theta n})A_n(t) \qquad (13.3.18b)$$

where Eqs. (13.3.17) and (13.1.12) are used to obtain the second part of these equations. At each instant of time the element forces are determined by static analysis of the ith frame subjected to the lateral forces $\mathbf{f}_{in}(t)$ shown in Fig. 13.3.2.

13.3.3 Total Response

Combining the response contributions of all the modes gives the total response of the unsymmetric-plan building to earthquake excitation:

$$r(t) = \sum_{n=1}^{2N} r_n(t) = \sum_{n=1}^{2N} r_n^{st} A_n(t) \qquad (13.3.19)$$

wherein Eq. (13.3.15) has been substituted for $r_n(t)$, the nth-mode response.

13.3.4 Summary

The response history of an N-story building with plan unsymmetric about the y-axis to earthquake ground motion in the y-direction can be computed by the procedure just developed, which is summarized next in step-by-step form:

1. Define the ground acceleration $\ddot{u}_{gy}(t)$ numerically at every time step Δt.
2. Define the structural properties.
   a. Determine the mass and stiffness matrices from Eqs. (13.3.1) and (9.5.26).
   b. Estimate the modal damping ratios ζ_n (Chapter 11).
3. Determine the natural frequencies ω_n (natural periods $T_n = 2\pi/\omega_n$) and natural modes of vibration (Chapter 10).
4. Determine the modal components $\mathbf{s}_n^T = \langle \mathbf{s}_{yn}^T \quad \mathbf{s}_{\theta n}^T \rangle$—defined by Eqs. (13.3.7) and (13.3.8)—of the effective force distribution.

5. Compute the response contribution of the nth mode by the following steps, which are repeated for all modes, $n = 1, 2, \ldots, 2N$:

   a. Perform static analysis of the building subjected to lateral forces $\mathbf{s}_{yn}$ and torques $\mathbf{s}_{\theta n}$ to determine r_n^{st}, the modal static response for each desired response quantity r (Table 13.3.1).

   b. Determine the deformation response $D_n(t)$ and pseudo-acceleration response $A_n(t)$ of the nth-mode SDF system to $\ddot{u}_{gy}(t)$, using numerical time-stepping methods (Chapter 5).

   c. Determine $r_n(t)$ from Eq. (13.3.15). This equation may also be used to determine the element forces in the ith frame provided that the modal static responses are derived for these response quantities. Alternatively, these internal forces can be determined by static analysis of the frame subjected to the lateral forces of Eq. (13.3.18).

6. Combine the modal contributions $r_n(t)$ to determine the total response using Eq. (13.3.19).

Only the modes with significant response contributions need to be included in modal analysis. The system considered has coupled lateral-torsional motion in $2N$ modes or N pairs of modes. For many buildings both modes in a pair have similar natural frequencies and responses of similar magnitude (see Example 13.8). Usually, only the lower few pairs of modes contribute significantly to the response. Therefore, steps 3, 4, and 5 need to be implemented only for these modal pairs, and the modal summation in step 6 truncated accordingly.

Extension to arbitrary-plan buildings. The procedure just summarized is for earthquake analysis of multistory buildings with plan unsymmetric about one axis, say the y-axis, but symmetric about the other axis, the x-axis, subjected to ground motion in the y-direction. This procedure can be extended to multistory buildings with arbitrary plan that has no axis of symmetry. In this case the system with $3N$ dynamic degrees of freedom is governed by Eq. (9.6.7) and will respond in coupled x-lateral, y-lateral, and torsional motions when excited by ground motion in the x or y directions.

Example 13.8

Determine the response of the system of Examples 13.7 and 10.6 with modal damping ratios $\zeta_n = 5\%$ to the El Centro ground motion acting along the y-axis. The response quantities of interest are floor displacements, base shear, and base torque in the building, and base shear in frames A and B.

Solution Steps 1 to 4 of the analysis procedure (Section 13.3.4) have already been implemented in Examples 10.6 and 13.7.

Step 5a: The modal static responses for base shear and base torque are (from Example 13.7): $V_{b1}^{\text{st}} = 0.509m$ and $V_{b2}^{\text{st}} = 0.491m$; $T_{b1}^{\text{st}} = -5.203m$ and $T_{b2}^{\text{st}} = 5.203m$. The modal static responses for the lateral displacement u_y and rotation u_θ of the roof are obtained by specializing the equations in Table 13.3.1 for this one-story system:

$$u_{yn}^{\text{st}} = \frac{\Gamma_n}{\omega_n^2}\phi_{yn} \qquad u_{\theta n}^{\text{st}} = \frac{\Gamma_n}{\omega_n^2}\phi_{\theta n} \tag{a}$$

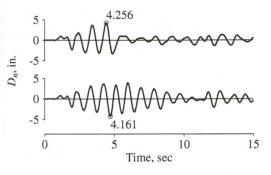
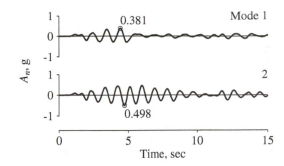

Figure E13.8a Displacement $D_n(t)$ and pseudo-acceleration $A_n(t)$ responses of modal SDF systems.

Substituting numerical values for Γ_n, ϕ_{yn}, and $\phi_{\theta n}$ (from Example 13.7) for the first mode in Eq. (a) gives $u_{y1}^{st} = 0.509/\omega_1^2$ and $u_{\theta 1}^{st} = -0.0480/\omega_1^2$. Similarly, for the second mode, $u_{y2}^{st} = 0.491/\omega_2^2$ and $u_{\theta 2}^{st} = 0.0480/\omega_2^2$. Observe that the modal static responses u_{yn}^{st} and V_{bn}^{st} for the two modes are similar in magnitude and of the same algebraic sign; $u_{\theta n}^{st}$ and T_{bn}^{st} for the two modes are also of similar (identical for a one-story system) magnitude but of opposite signs.

Step 5b: Response analysis of the first-mode SDF system ($T_1 = 2\pi/\omega_1 = 2\pi/5.878 = 1.069$ sec and $\zeta_1 = 5\%$) and the second-mode SDF system ($T_2 = 2\pi/\omega_2 = 2\pi/6.794 = 0.9248$ sec and $\zeta_2 = 5\%$) to the El Centro ground motion gives the $D_n(t)$ and $A_n(t)$ shown in Fig. E13.8a. Observe that $D_n(t)$—also $A_n(t)$—for the two modes are similar and roughly in phase because their natural periods are similar.

Step 5c: Substituting V_{bn}^{st} and T_{bn}^{st} from step 5a in Eq. (13.3.15) gives the contributions of the nth mode to the base shear and base torque:

$$V_{b1}(t) = 0.509m\,A_1(t) \qquad T_{b1}(t) = -5.203m\,A_1(t) \tag{b}$$

$$V_{b2}(t) = 0.491m\,A_2(t) \qquad T_{b2}(t) = 5.203m\,A_2(t) \tag{c}$$

Figure E13.8b shows $V_{bn}(t)$ and $T_{bn}(t)$ computed from Eqs. (b) and (c) using $m = 1.863$ kip–sec^2/ft and the $A_n(t)$ in Fig. E13.8a.

Substituting u_{yn}^{st} and $u_{\theta n}^{st}$ from step 5a in Eq. (13.3.15) give the contributions of the nth mode to roof displacements:

$$u_{y1}(t) = 0.509D_1(t) \qquad u_{\theta 1}(t) = -0.0480D_1(t) \tag{d}$$

$$u_{y2}(t) = 0.491D_2(t) \qquad u_{\theta 2}(t) = 0.0480D_2(t) \tag{e}$$

where $D_n(t)$ is in units of feet. Figure E13.8c shows $u_{yn}(t)$ and $(b/2)u_{\theta n}(t)$ computed from Eqs. (d) and (e) using the $D_n(t)$ in Fig. E13.8a.

The lateral force for frame A is given by Eq. (13.3.18b) specialized for a one-story frame:

$$f_{An}(t) = k_A \left[u_{yn}(t) + x_A u_{\theta n}(t) \right] \tag{f}$$

Substituting for $k_A = 75$ kips/ft, $x_A = 1.5$ ft, $u_{yn}(t)$ and $u_{\theta n}(t)$ from Eqs. (d) and (e) gives

$$f_{A1}(t) = 32.80D_1(t) \text{ kips} \qquad f_{A2}(t) = 42.19D_2(t) \text{ kips}$$

where $D_n(t)$ are in feet. The base shear in a one-story frame is equal to the lateral force; thus

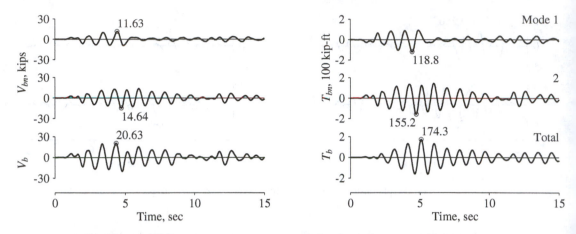

Figure E13.8b Base shear and base torque: modal contributions, $V_{bn}(t)$ and $T_{bn}(t)$, and total responses, $V_b(t)$ and $T_b(t)$.

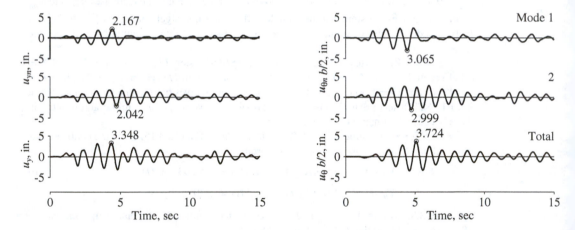

Figure E13.8c Lateral displacement and $b/2$ times roof rotation: modal contributions, $u_{yn}(t)$ and $(b/2)u_{\theta_n}(t)$, and total responses, $u_y(t)$ and $(b/2)u_\theta(t)$.

the base shear due to the two modes is

$$V_{bA1}(t) = 32.80D_1(t) \qquad V_{bA2}(t) = 42.19D_2(t) \tag{g}$$

These base shears are computed using the known $D_n(t)$ from Fig. E13.8a and are shown in Fig. E13.8d. Note that the base shears for frame A alone and the building are identical because the system has only frame A in the y-direction and this frame carries the entire force.

The lateral force for frame B is given by Eq. (13.3.18a) specialized for a one-story frame:

$$f_{Bn}(t) = k_B \left[-y_B u_{\theta n}(t)\right] \tag{h}$$

Substituting for $k_B = 40$ kips/ft, $y_B = 10$ ft, $u_{\theta n}(t)$ from Eqs. (d) and (e) gives

$$f_{B1}(t) = 19.2D_1(t) \qquad f_{B2}(t) = -19.2D_2(t)$$

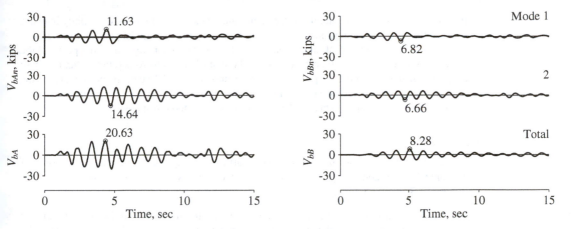

Figure E13.8d Base shears in frames A and B: modal contributions, V_{bAn} and V_{bBn}, and total responses, V_{bA} and V_{bB}.

where $D_n(t)$ are in feet. The base shear due to the two modes is

$$V_{bB1}(t) = 19.2 D_1(t) \qquad V_{bB2}(t) = -19.2 D_2(t)$$

These base shears in frame B are computed using the known $D_n(t)$ from Fig. E13.8a and are shown in Fig. E13.8d.

Figure E13.8b–d show that the two modes contribute similarly to the response of this one-story system. This is typical of unsymmetric-plan systems where pairs of modes in a structure with one axis of symmetry (or triplets of modes if the system has no axis of symmetry) may have similar response contributions.

Step 6: Combining the modal contributions gives the total response for this two-DOF system:

$$r(t) = r_1(t) + r_2(t)$$

The combined values of lateral displacement, rotation, base shear, and base torque for the building, and base shear for frames A and B are shown in Fig. E13.8b–d. Observe that the combined response attains its peak value at a time instant different from when the modal peaks are attained.

Interpretive Comments. Observe that the modal contributions to lateral displacement (and to base shear) are similar in magnitude because the modal static responses are about the same and the $D_n(t)$ and $A_n(t)$ are similar for the two modes (Fig. E13.8a). The modal contributions are roughly in phase because $D_n(t)$—also $A_n(t)$—for the two modes are roughly in phase and the two modal static responses are of the same algebraic sign. The peak of the total response is therefore much larger than the peaks of the modal responses. In contrast, the modal contributions to roof rotation (and to base torque), although similar in magnitude, are out of phase and the two modal static responses are of opposite algebraic sign. Therefore, the peak of the total response is only slightly larger than the peaks of the modal responses.

Consider another one-story unsymmetric-plan system similar to the one analyzed in Example 13.8 (Fig. 9.5.1) but with a smaller e, say $e = 0.5$ ft instead of 1.5 ft. The two natural periods will now be much closer than for the structure analyzed in Example 13.8, and the $D_n(t)$—also $A_n(t)$—for the two modes will be essentially in phase. For such a system, the modal contributions to lateral displacement (and to base shear) will be essentially in phase because the modal static responses are of the same sign; the two modal peaks will be almost directly additive, and the peak of the total response will be much larger than the peaks of the modal responses. On the other hand, the modal contributions to roof rotation (and to base torque) will have essentially opposite phase because the modal static responses are of the opposite sign; the two modal peaks will tend to cancel each other, and the peak of the total response will then be much smaller than the peaks of the modal responses.

Example 13.9

Identify the effects of plan asymmetry on the earthquake response of the one-story system of Example 13.8 by comparing its response with that of the symmetric-plan one-story system defined in Section 9.5.3.

Solution   The response of the symmetric-plan system to ground motion in the y-direction is governed by the second of the three differential equations (9.5.20). Dividing this equation by m and introducing damping gives the familiar equation for an SDF system:

$$\ddot{u}_y + 2\zeta_y \omega_y \dot{u}_y + \omega_y^2 u_y = -\ddot{u}_{gy}(t) \tag{a}$$

where $\omega_y = \sqrt{k_y/m}$. As mentioned in Section 9.5.3, the y-component of ground motion will only produce lateral response in the y-direction without any torsion about a vertical axis or displacements in the x-direction. The lateral displacement in the y-direction is

$$u_y(t) = D(t, \omega_y, \zeta_y) \tag{b}$$

and the associated base shear in frame A is

$$V_{bA} = m A(t, \omega_y, \zeta_y) \tag{c}$$

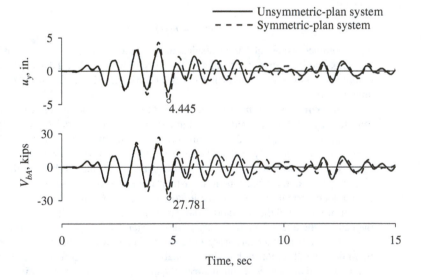

Figure E13.9   Responses of unsymmetric-plan system and symmetric-plan system.

where $D(t, \omega_y, \zeta)$ and $A(t, \omega_y, \zeta)$ denote the deformation and pseudo-acceleration responses, respectively, of an SDF system with natural frequency ω_y and damping ratio ζ to ground acceleration $\ddot{u}_{gy}(t)$. Frames B and C would experience no forces.

For the symmetric-plan system associated with Example 13.8, $\omega_y = 6.344$ (see Example 10.7) and the damping ratio is the same, $\zeta = 5\%$. The response of this SDF system is computed from Eqs. (a) to (c) and shown in Fig. E13.9, where it is also compared with the response of the unsymmetric-plan system (Example 13.8). It is clear that plan asymmetry has the effect of (1) modifying the lateral displacement and base shear in frame A, and (2) causing torsion in the system and forces in frames B and C that do not exist if the building plan is symmetric. In this particular case, the base shear in frame A is reduced because of plan asymmetry, but such is not always the case, depending on the natural period of the structure, ground motion characteristics, and the location of the frame in the building plan.

13.4 TORSIONAL RESPONSE OF SYMMETRIC-PLAN BUILDINGS

In this section the torsional response of multistory buildings with their plans nominally symmetric about two orthogonal axis is discussed briefly. Such structures may undergo "accidental" torsional motions for mainly two reasons: the building is usually not *perfectly* symmetric, and the spatial variations in ground motion may cause rotation (about the vertical axis) of the building's base, which will induce torsional motion of the building even if its plan is perfectly symmetric.

Consider first the analysis of torsional response of a building with a perfectly symmetric plan due to rotation of its base. For a given rotational excitation $\ddot{u}_{g\theta}(t)$, the governing equations (9.6.9c) can be solved by the modal analysis procedure, considering only the purely torsional vibration modes of the building. This procedure could be developed along the lines of Section 13.3. It is not presented, however, for two reasons: (1) it is straightforward; and (2) in structural engineering practice, buildings are not analyzed for rotational excitation. Therefore, in this brief section we present the results of such analysis and compare them with building torsion recorded during an earthquake.

Consider the building shown in Fig. 13.4.1, located in Pomona, California. This reinforced-concrete frame building has two stories, a partial basement, and a light penthouse structure. For all practical and code design purposes, the building has a nominally symmetric floor plan, as indicated by its framing plan in Fig. 13.4.2. The lateral force-resisting system in the building consists of peripheral columns interconnected by longitudinal and transverse beams, but the L-shaped exterior corner columns as well as the interior columns in the building are not designed especially for earthquake resistance. The floor decking system is formed by a 6-in.-thick concrete slab. The building also includes walls in the stairwell system—concrete walls in the basement and masonry walls in upper stories. Foundations of columns and interior walls are supported on piles.

The accelerograph channels located as shown in Fig. 13.4.3 recorded the motion of the building during the Upland (February 28, 1990) earthquake, including three channels of horizontal motion at each of three levels: roof, second floor, and basement. The peak accelerations of the basement were 0.12g and 0.13g in the x and y directions, respectively.

Figure 13.4.1 First Federal Savings building, a two-story reinforced-concrete building (with a partial basement) in Pomona, California. (Courtesy of California Strong Motion Instrumentation Program.)

These motions were amplified to 0.24g in the x-direction and 0.22g in the y-direction at the roof. The building experienced no structural damage during this earthquake.

Some of the recorded motions are shown in Fig. 13.4.4. These include the x-translational accelerations at two locations at the basement of the building and at two locations at the roof level. By superimposing the motions at two locations on the roof in Fig. 13.4.5 it is clear that this building experienced some torsion; otherwise, these two motions would have been identical. Assuming the base to be rigid, its rotational acceleration is computed as the difference between the two x-translational records at the basement of the building divided by the distance between the two locations. This rotational base acceleration is multiplied by $b/2$, where the building-plan dimension $b = 109.75$ ft, and plotted in Fig. 13.4.6. The peak value of $(b/2)\ddot{u}_{g\theta}(t)$ is 0.029g, compared with the peak acceleration of 0.12g in the x-direction.

The torsional response of the building to the rotational motion of the basement, Fig. 13.4.6, is determined by modal solution of Eq. (9.6.9c) with modal damping ratios of 5%. These damping ratios were estimated from the recorded motions at the roof and basement using some of the procedures mentioned in Chapter 11, Part A. The response history of the shear force in a selected column of the building is presented in Fig. 13.4.7. This is only a part of the element force due to the actual torsional motion of the building during the earthquake, as will be demonstrated next.

Approximate values of the element forces due to recorded torsion can be determined at each instant of time by static analysis of the building subjected to floor inertia torques $I_{Oj}\ddot{u}_{j\theta}^t(t)$ at all floors ($j = 1, 2, \ldots, N$), where I_{Oj} is the moment of inertia of the jth floor mass about the vertical axis through O, the center of mass (CM) of the floor, and

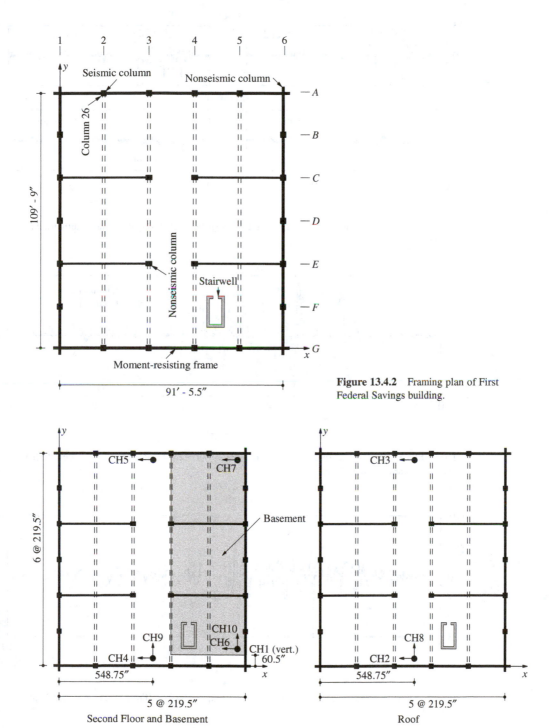

Figure 13.4.2 Framing plan of First Federal Savings building.

Figure 13.4.3 Accelerograph channels in First Federal Savings building.

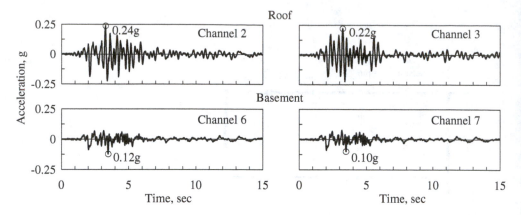

Figure 13.4.4   Motions recorded at First Federal Savings building during the Upland earthquake of February 28, 1990.

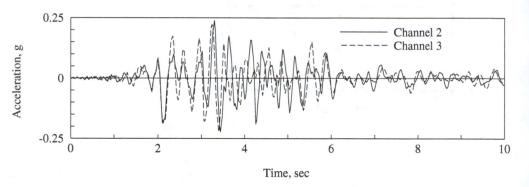

Figure 13.4.5   Motions recorded at two locations on the roof of First Federal Savings building during the Upland earthquake of February 28, 1990.

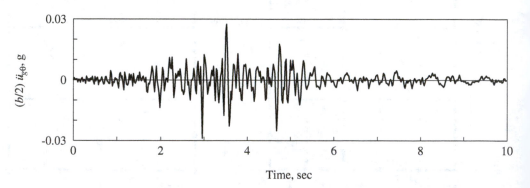

Figure 13.4.6   Rotational acceleration of basement multiplied by $b/2$. [From De la Llera and Chopra (1994).]

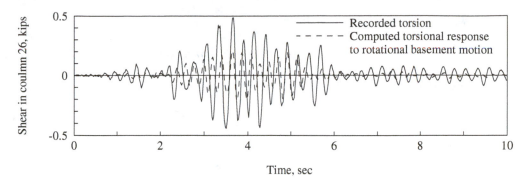

Figure 13.4.7   Comparison of shear force (x component) in column 26 due to (1) recorded torsion of the building, and (2) computed torsional response of the building to rotational basement motion (After De la Llera and Chopra, 1994).

$\ddot{u}^t_{j\theta}$ is the torsional acceleration of the jth floor diaphragm. By using these inertia forces as equivalent static forces, we have included the damping forces, and this is a source of approximation (see the last paragraph of Section 1.8.2). The results of these static analyses for the shear force in the same column are also presented in Fig. 13.4.7.

This figure shows that the peak force due to rotational basement motion is about 45% of the peak force due to the actual torsional motion of the building. The remaining 55% of the force arises, in part, because this building is not *perfectly* symmetric due to several factors, the most obvious of them being the stairwell system shown in Fig. 13.4.2, and because the basement, which is under one-half of the floor plan, is not symmetrically located.

Torsional motion of buildings with nominally symmetric plan, such as the building of Fig. 13.4.1, is usually called *accidental torsion*. Such motion contributes a small fraction of the total earthquake forces in the structure. For the building and earthquake considered, accidental torsion contributed about 4% of the total force (results not presented here), but larger contributions have been identified in the earthquake response of other buildings. The structural response associated with accidental torsion is not amenable to calculation in structural design for two reasons: (1) the rotational base motion is not defined, and (2) it is not practical to identify and analyze the effect of each source of asymmetry in a building with nominally symmetric plan. Therefore, building codes include a simple design provision to account for accidental torsion in symmetric and unsymmetric buildings; in the latter case it is considered in addition to torsion arising from plan asymmetry (Section 13.3). Research has demonstrated deficiencies in this code provision.

13.5 RESPONSE ANALYSIS FOR MULTIPLE SUPPORT EXCITATION

In this section the modal analysis procedure of Section 13.1 is extended to MDF systems excited by prescribed motions $\ddot{u}_{gl}(t)$ at the various supports ($l = 1, 2, \ldots, N_g$) of the structure. In Section 9.7 the governing equations were shown to be the same as Eq. (13.1.1), with the effective earthquake forces

$$\mathbf{p}_{\text{eff}}(t) = -\sum_{l=1}^{N_g} \mathbf{m}\iota_l \ddot{u}_{gl}(t) \tag{13.5.1}$$

instead of Eq. (13.1.2). The modal Eq. (13.1.7) now becomes

$$\ddot{q}_n + 2\zeta_n\omega_n\dot{q}_n + \omega_n^2 q_n = -\sum_{l=1}^{N_g} \Gamma_{nl}\ddot{u}_{gl}(t) \tag{13.5.2}$$

where

$$\Gamma_{nl} = \frac{L_{nl}}{M_n} \qquad L_{nl} = \phi_n^T \mathbf{m}\iota_l \qquad M_n = \phi_n^T \mathbf{m}\phi_n \tag{13.5.3}$$

The solution of Eq. (13.5.2) can be written as a generalization of Eq. (13.1.9):

$$q_n(t) = \sum_{l=1}^{N_g} \Gamma_{nl} D_{nl}(t) \tag{13.5.4}$$

where $D_{nl}(t)$ is the deformation response of the nth-mode SDF system to support acceleration $\ddot{u}_{gl}(t)$.

The displacement response of the structure, Eq. (9.7.2), contains two parts:

1. The dynamic displacements are obtained by combining Eqs. (13.1.3) and (13.5.4):

$$\mathbf{u}(t) = \sum_{l=1}^{N_g}\sum_{n=1}^{N} \Gamma_{nl}\phi_n D_{nl}(t) \tag{13.5.5}$$

2. The quasi-static displacements $\mathbf{u}^s$ are given by Eq. (9.7.11).

Combining the two parts gives the total displacements in the structural DOFs:

$$\mathbf{u}^t(t) = \sum_{l=1}^{N_g} \iota_l u_{gl}(t) + \sum_{l=1}^{N_g}\sum_{n=1}^{N} \Gamma_{nl}\phi_n D_{nl}(t) \tag{13.5.6}$$

The forces in structural elements can be obtained from the structural displacements $\mathbf{u}^t(t)$ and prescribed support displacements $\mathbf{u}_g(t)$ without additional dynamic analyses by using either of the two procedures mentioned in Section 9.10. In the first method, the element forces are calculated from the known nodal displacements using the element stiffness properties. This method is usually preferred in computer implementation of force calculations for multiple support excitation. It is instructive, however, to generalize the second method based on equivalent static forces. The rest of this section is devoted to this development.

The equivalent static forces in the structural DOF are given by the last term on the left side of Eq. (9.7.1):

$$\mathbf{f}_S = \mathbf{k}\mathbf{u}^t + \mathbf{k}_g\mathbf{u}_g \tag{13.5.7}$$

Substituting Eq. (9.7.2) for $\mathbf{u}^t$ and using Eq. (9.7.7) gives

$$\mathbf{f}_S(t) = \mathbf{k}\mathbf{u}(t) \tag{13.5.8}$$

These forces depend only on the dynamic displacements, Eq. (13.5.5). Therefore,

$$\mathbf{f}_S(t) = \sum_{l=1}^{N_g} \sum_{n=1}^{N} \Gamma_{nl} \mathbf{k} \phi_n D_{nl}(t) \tag{13.5.9}$$

which can be written in terms of the mass matrix by utilizing Eq. (10.2.4):

$$\mathbf{f}_S(t) = \sum_{l=1}^{N_g} \sum_{n=1}^{N} \Gamma_{nl} \mathbf{m} \phi_n A_{nl}(t) \tag{13.5.10}$$

where

$$A_{nl}(t) = \omega_n^2 D_{nl}(t) \tag{13.5.11}$$

is the pseudo-acceleration response of the nth-mode SDF system to support acceleration $\ddot{u}_{gl}(t)$.

The equivalent static forces along the support DOF are also given by the last term on the left side of Eq. (9.7.1):

$$\mathbf{f}_{Sg} = \mathbf{k}_g^T \mathbf{u}^t + \mathbf{k}_{gg} \mathbf{u}_g \tag{13.5.12}$$

Substituting Eq. (9.7.2) for $\mathbf{u}^t$ and using Eq. (9.7.3) for the quasi-static support forces $\mathbf{p}_g^s(t)$ gives

$$\mathbf{f}_{Sg}(t) = \mathbf{k}_g^T \mathbf{u}(t) + \mathbf{p}_g^s(t) \tag{13.5.13}$$

Observe that the support forces $\mathbf{f}_{Sg}$ depend on the displacements in the structural DOFs as well as on support displacements, and can no longer be obtained by statics from the force vector $\mathbf{f}_S$. This is different from Section 13.1, where for a structure excited at its only support, or excited by identical motion at all supports, the base shear could be determined from $\mathbf{f}_S$. By utilizing Eqs. (9.7.11) and (13.5.5), the support forces can be expressed as

$$\mathbf{f}_{Sg}(t) = \sum_{l=1}^{N_g} \left(\mathbf{k}_g^T \iota_l + \mathbf{k}_{gg}^l \right) u_{gl}(t) + \sum_{l=1}^{N_g} \sum_{n=1}^{N} \Gamma_{nl} \mathbf{k}_g^T \phi_n D_{nl}(t) \tag{13.5.14}$$

where $\mathbf{k}_{gg}^l$ is the lth column of $\mathbf{k}_{gg}$.

The element forces at each time instant are evaluated by static analysis of the structure subjected to the forces $\mathbf{f}_S(t)$ and $\mathbf{f}_{Sg}(t)$, given by Eqs. (13.5.10) and (13.5.14), respectively. Although this procedure was presented to show that the equivalent static force concept can be generalized to structures excited by multiple support excitation, as mentioned earlier, in computer analysis it is usually preferable to evaluate the element forces directly from the nodal displacements using the element stiffness properties.

Example 13.10

In the two-span continuous bridge of Example 9.10, support A undergoes vertical motion $u_g(t)$, support B describes the same motion as A, but it does so t' seconds later, and support

C undergoes the same motion $2t'$ after support A. Determine the following responses as a function of time: displacement of the two masses; bending moments at the midpoint of each span; and bending moment at the center support. Express all results in terms of $D_n(t)$ and $A_n(t)$, the displacement and pseudo-acceleration responses of the nth-mode SDF system to $\ddot{u}_g(t)$. Compare the preceding results with the response of the bridge if all supports undergo identical motion $u_g(t)$.

Solution

1. *Evaluate the natural frequencies and modes.* The eigenvalue problem to be solved is

$$\mathbf{k}\boldsymbol{\phi} = \omega^2 \mathbf{m}\boldsymbol{\phi}$$

where Eqs. (c) and (e) of Example 9.10 give

$$\mathbf{k} = \frac{EI}{L^3}\begin{bmatrix} 78.86 & 30.86 \\ 30.86 & 78.86 \end{bmatrix} \qquad \mathbf{m} = m\begin{bmatrix} 1 & \\ & 1 \end{bmatrix} \tag{a}$$

Solution of the eigenvalue problem gives

$$\omega_1 = 6.928\sqrt{\frac{EI}{mL^3}} \qquad \omega_2 = 10.47\sqrt{\frac{EI}{mL^3}} \tag{b}$$

$$\boldsymbol{\phi}_1 = \begin{Bmatrix} -1 \\ 1 \end{Bmatrix} \qquad \boldsymbol{\phi}_2 = \begin{Bmatrix} 1 \\ 1 \end{Bmatrix} \tag{c}$$

2. *Determine $\Gamma_{nl} = L_{nl}/M_n$.*

$$L_{nl} = \boldsymbol{\phi}_n^T \mathbf{m}\boldsymbol{\iota}_l \qquad l = 1, 2, 3, \quad n = 1, 2$$

Substituting for $\boldsymbol{\phi}_n$ and $\mathbf{m}$ from Eqs. (c) and (a), respectively, and for $\boldsymbol{\iota}_l$ from Eq. (g) of Example 9.10 gives

$$\mathbf{L} = [L_{nl}] = \begin{bmatrix} -0.5000m & 0 & 0.5000m \\ 0.3125m & 1.375m & 0.3125m \end{bmatrix} \begin{matrix} \leftarrow \text{mode 1} \\ \leftarrow \text{mode 2} \end{matrix} \tag{d}$$

$$\begin{matrix} \uparrow & \uparrow & \uparrow \\ \ddot{u}_{g1} & \ddot{u}_{g2} & \ddot{u}_{g3} \end{matrix}$$

$$M_n = \boldsymbol{\phi}_n^T \mathbf{m}\boldsymbol{\phi}_n \qquad n = 1, 2$$

Substituting for $\boldsymbol{\phi}_n$ and $\mathbf{m}$ gives $M_n = 2m, n = 1, 2$. Then $\Gamma_{nl} = L_{nl}/M_n$ gives

$$\boldsymbol{\Gamma} = [\Gamma_{nl}] = \begin{bmatrix} -0.25000 & 0 & 0.25000 \\ 0.15625 & 0.6875 & 0.15625 \end{bmatrix} \begin{matrix} \leftarrow \text{mode 1} \\ \leftarrow \text{mode 2} \end{matrix} \tag{e}$$

$$\begin{matrix} \uparrow & \uparrow & \uparrow \\ \ddot{u}_{g1} & \ddot{u}_{g2} & \ddot{u}_{g3} \end{matrix}$$

3. *Determine the response of the nth-mode SDF system to $\ddot{u}_{gl}(t)$.* Given $\ddot{u}_{g1}(t) = \ddot{u}_g(t)$, $\ddot{u}_{g2}(t) = \ddot{u}_g(t-t')$, $\ddot{u}_{g3}(t) = \ddot{u}_g(t-2t')$. Then

$$D_{n1}(t) = D_n(t) \qquad D_{n2}(t) = D_n(t-t') \qquad D_{n3}(t) = D_n(t-2t') \tag{f}$$

$$A_{n1}(t) = A_n(t) \qquad A_{n2}(t) = A_n(t-t') \qquad A_{n3}(t) = A_n(t-2t') \tag{g}$$

4. *Determine the displacement response.* In Eq. (13.5.6) with $N = 2$ and $N_g = 3$, substituting for Γ_{nl}, $\boldsymbol{\phi}_n$, and D_{nl} from Eqs. (e), (c), and (f), respectively, and for $\boldsymbol{\iota}_l$ from

Eq. (g) of Example 9.10 gives

$$\begin{Bmatrix} u_1^t(t) \\ u_2^t(t) \end{Bmatrix} = \begin{Bmatrix} 0.40625 \\ -0.09375 \end{Bmatrix} u_g(t) + \begin{Bmatrix} 0.6875 \\ 0.6875 \end{Bmatrix} u_g(t - t') + \begin{Bmatrix} -0.09375 \\ 0.40625 \end{Bmatrix} u_g(t - 2t')$$

$$- 0.25 \begin{Bmatrix} -1 \\ 1 \end{Bmatrix} D_1(t) + 0 \begin{Bmatrix} -1 \\ 1 \end{Bmatrix} D_1(t - t') + 0.25 \begin{Bmatrix} -1 \\ 1 \end{Bmatrix} D_1(t - 2t')$$

$$+ 0.15625 \begin{Bmatrix} 1 \\ 1 \end{Bmatrix} D_2(t) + 0.6875 \begin{Bmatrix} 1 \\ 1 \end{Bmatrix} D_2(t - t') + 0.15625 \begin{Bmatrix} 1 \\ 1 \end{Bmatrix} D_2(t - 2t') \qquad \text{(h)}$$

5. *Compute the equivalent static forces.* In Eq. (13.5.10) with $N = 2$ and $N_g = 3$, substituting for $\mathbf{m}$, $\boldsymbol{\phi}_n$, $\boldsymbol{\Gamma}_{nl}$, and $A_{nl}(t)$ from Eqs. (a), (c), (e), and (g), respectively, gives

$$\mathbf{f}_S(t) = -0.25 \begin{Bmatrix} -1 \\ 1 \end{Bmatrix} m A_1(t) + 0 \begin{Bmatrix} -1 \\ 1 \end{Bmatrix} m A_1(t - t') + 0.25 \begin{Bmatrix} -1 \\ 1 \end{Bmatrix} m A_1(t - 2t')$$

$$+ 0.15625 \begin{Bmatrix} 1 \\ 1 \end{Bmatrix} m A_2(t) + 0.6875 \begin{Bmatrix} 1 \\ 1 \end{Bmatrix} m A_2(t - t') + 0.15625 \begin{Bmatrix} 1 \\ 1 \end{Bmatrix} m A_2(t - 2t') \qquad \text{(i)}$$

6. *Compute the equivalent static support forces.* In Eq. (13.5.12) substituting for $\mathbf{k}_g$ and $\mathbf{k}_{gg}$ from Eq. (d) of Example 9.10 and $\mathbf{u}^t(t)$ from Eq. (h) gives

$$\mathbf{f}_{Sg}(t) = \begin{Bmatrix} -0.125 \\ 0 \\ 0.125 \end{Bmatrix} m A_1(t) + \begin{Bmatrix} 0 \\ 0 \\ 0 \end{Bmatrix} m A_1(t - t') + \begin{Bmatrix} 0.125 \\ 0 \\ -0.125 \end{Bmatrix} m A_1(t - 2t')$$

$$+ \begin{Bmatrix} -0.0488 \\ -0.2148 \\ -0.0488 \end{Bmatrix} m A_2(t) + \begin{Bmatrix} -0.2149 \\ -0.9454 \\ -0.2149 \end{Bmatrix} m A_2(t - t') + \begin{Bmatrix} -0.0488 \\ -0.2148 \\ -0.0488 \end{Bmatrix} m A_2(t - 2t')$$

$$+ \begin{Bmatrix} 1.5 \\ -3.0 \\ 1.5 \end{Bmatrix} \frac{EI}{L^3} u_g(t) + \begin{Bmatrix} -3 \\ 6 \\ -3 \end{Bmatrix} \frac{EI}{L^3} u_g(t - t') + \begin{Bmatrix} 1.5 \\ -3 \\ 1.5 \end{Bmatrix} u_g(t - 2t') \qquad \text{(j)}$$

where Eq. (b) was used to express EI/L^3 in terms of ω_n and Eq. (13.5.11) to express D_{nl} in terms of A_{nl}. The equivalent static forces given by Eqs. (i) and (j) are shown in Fig. E13.10. Observe that at each time instant, these forces defined by Eqs. (i) and (j) are in equilibrium.

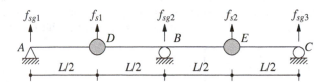

Figure E13.10

7. *Compute the bending moments.* The bending moments M_D, M_E, and M_B at the locations of the left mass, right mass, and support B, respectively, are obtained by static analysis of the system subjected to the forces shown in Fig. E13.10:

$$M_D = mL\left(-0.0625A_1(t) + 0A_1(t - t') + 0.0625A_1(t - 2t')\right)$$

$$+ mL\left(-0.0244A_2(t) - 0.1074A_2(t - t') - 0.0244A_2(t - 2t')\right)$$

$$+ \frac{EI}{L^2}\left(0.75u_g(t) - 1.50u_g(t - t') + 0.75u_g(t - 2t')\right) \tag{k}$$

$$M_E = mL\left(0.0625A_1(t) + 0A_1(t - t') - 0.0625A_1(t - 2t')\right)$$

$$+ mL\left(-0.0244A_2(t) - 0.1074A_2(t - t') - 0.0244A_2(t - 2t')\right)$$

$$+ \frac{EI}{L^2}\left(0.75u_g(t) - 1.50u_g(t - t') + 0.75u_g(t - 2t')\right) \tag{l}$$

$$M_B = mL\left(0.0293A_2(t) + 0.1289A_2(t - t') + 0.0293A_2(t - 2t')\right)$$

$$+ \frac{EI}{L^2}\left(1.5u_g(t) - 3.0u_g(t - t') + 1.5u_g(t - 2t')\right) \tag{m}$$

Observe that the first mode does not contribute to M_B because B is a point of inflection for this mode.

8. Identical support motions. If all the supports undergo identical motion $u_g(t)$, the motion of the structure is given by Eq. (13.1.15), where Γ_n is defined by Eq. (13.1.5) with $\iota = 1$. For this system

$$\Gamma_1 = 0 \qquad \Gamma_2 = 1$$

When we substitute these data, Eq. (13.1.15) gives

$$\mathbf{u}(t) = \Gamma_2\phi_2 D_2(t) = \begin{Bmatrix} 1 \\ 1 \end{Bmatrix} D_2(t) \tag{n}$$

Observe that the first mode, which is antisymmetric, is not excited by the symmetric excitation, and all the response is due to the second mode.

The equivalent static forces are given by Eq. (13.1.11):

$$\mathbf{f}_S(t) = \Gamma_2\mathbf{m}\phi_2 A_2(t) = m \begin{Bmatrix} 1 \\ 1 \end{Bmatrix} A_2(t) \tag{o}$$

The support forces can be obtained by static analysis of the bridge subjected to the external forces of Eq. (o). Alternatively, the support forces are given by Eq. (j), specialized by substituting $A_2(t) = A_2(t - t') = A_2(t - 2t')$ and $u_g(t) = u_g(t - t') = u_g(t - 2t')$. Either method gives

$$\mathbf{f}_{Sg}(t) = \begin{Bmatrix} -0.3125 \\ -1.3750 \\ -0.3125 \end{Bmatrix} mA_2(t) \tag{p}$$

Determined by static analysis of the structure due to the forces of Eq. (o), the bending moments are

$$M_D = -0.15625mLA_2(t) \qquad M_E = -0.15625mLA_2(t) \qquad M_B = 0.1875mLA_2(t) \tag{q}$$

9. *Comparison.* If the support motions are identical, the quasi-static forces $\mathbf{p}_g^s(t)$ in Eq. (13.5.13) are zero, there is no quasi-static component in the bending moments, and all support forces and internal forces can be computed directly (by statics) from the equivalent static forces in the structural DOFs. In contrast, if the support motions are different, the calculation of forces is more involved. In particular, the quasi-static forces associated with the different displacements of the supports must be included, and support forces cannot be obtained from only the equivalent static forces in the structural DOF.

13.6 STRUCTURAL IDEALIZATION AND EARTHQUAKE RESPONSE

With the development of earthquake analysis procedures presented in this chapter and the availability of modern computers, it is now possible to determine the linearly elastic response of an idealization (mathematical model) of any structure to prescribed ground motion. How well the computed response agrees with the actual response of a structure during an earthquake depends primarily on the quality of the structural idealization.

To illustrate this concept, we return to the natural periods and damping ratios for the Millikan Library building. Presented in Chapter 11 were these data from low-amplitude forced vibration tests, and from the Lytle Creek and San Fernando earthquakes, which caused roof accelerations of approximately 0.05g and 0.31g, respectively. These results demonstrated that with increasing levels of motion the natural periods lengthen and the damping ratios increase. The loss of stiffness indicated by this period change is believed to be primarily the result of cracking and other types of degradation of the nonstructural elements during the higher-level earthquake responses, especially from the San Fernando earthquake. A nonlinear structural idealization having stiffness and damping properties varying with deformation level would be necessary to reproduce this period change and to describe the behavior of a structure through the complete range of deformation amplitudes.

However, if the structure experiences no structural damage, good estimates of the response during the earthquake can usually be computed from an equivalent linear model with viscous damping. If the computed natural periods and modes and the estimated damping ratios represent the properties of the structure during the earthquake, the modal analysis procedure (Sections 13.1 to 13.3) will accurately predict "linear" response. This has been demonstrated by numerous analyses of recorded motions of structures during earthquakes; one such example is the response of the Millikan Library building during the San Fernando earthquake (Figs. 11.1.3 and 11.1.4). Using the natural periods and damping ratios of this building determined from these recorded motions and system identification procedures (Table 11.1.1), the displacement response of this building to the basement motion calculated by modal analysis was shown to agree almost perfectly with the displacements (relative to the ground) shown in Fig. 11.1.5, which were determined from the accelerations recorded at the roof and at the basement.

The usual situation, however, is different in the sense that the natural periods and modes are computed from an idealization of the structure. It is the quality of this idealization that determines the accuracy of response. Therefore, only those structural and nonstructural elements that contribute to the mass and stiffness of the structure at the amplitudes of motion expected during the earthquake should be included in the structural idealization; and their stiffness properties should be determined using realistic assumptions. Similarly, as discussed in Chapter 11, selection of damping values for analysis of a structure should be based on available data from recorded earthquake responses of similar structures.

PART B: RESPONSE SPECTRUM ANALYSIS

13.7 PEAK RESPONSE FROM EARTHQUAKE RESPONSE SPECTRUM

The response history analysis (RHA) procedure presented in Part A provides structural response $r(t)$ as a function of time, but structural design is usually based on the peak values of forces and deformations over the duration of the earthquake-induced response. Can the peak response be determined directly from the response spectrum for the ground motion without carrying out a response history analysis? For SDF systems the answer to this question is yes (Chapter 6). However, for MDF systems the answer is a qualified yes. The peak response of MDF systems can be calculated from the response spectrum, but the result is not exact—in the sense that it is not identical to the RHA result; the estimate obtained is accurate enough for structural design applications, however. In Part B we first present such response spectrum analysis (RSA) procedures for structures excited by a single component of ground motion; subsequently, procedures to estimate the peak response to three components of ground motion acting simultaneously are presented.

13.7.1 Peak Modal Responses

The peak value r_{no} of the nth-mode contribution $r_n(t)$ to response $r(t)$ can be obtained from the earthquake response spectrum or design spectrum. This becomes evident from Eq. (13.1.13) by recalling that the peak value of $A_n(t)$ is available from the pseudo-acceleration spectrum as its ordinate $A(T_n, \zeta_n)$, denoted as A_n, for brevity. Therefore,

$$r_{no} = r_n^{st} A_n \tag{13.7.1}$$

The algebraic sign of r_{no} is the same as that of r_n^{st} because A_n is positive by definition. Although it has an algebraic sign, r_{no}[†] will be referred to as the *peak modal response* because it corresponds to the peak value of $A_n(t)$. This algebraic sign must be retained because it can be important, as will be seen in Section 13.7.2. All response quantities $r_n(t)$ associated with a particular mode, say the nth mode, reach their peak values at the same

[†]The notation r_{no} should not be confused with the use of a subscript o in Chapter 6 to denote the maximum (over time) of the absolute value of the response quantity, which is positive by definition.

time instant as $A_n(t)$ reaches its peak (see Figs. 13.2.6 to 13.2.8, 13.2.10, 13.2.11, and E13.8a–d).

13.7.2 Modal Combination Rules

How do we combine the peak modal responses r_{no} $(n = 1, 2, \ldots, N)$ to determine the peak value $r_o \equiv \max_t |r(t)|$ of the total response? It will not be possible to determine the exact value of r_o from r_{no} because, in general, the modal responses $r_n(t)$ attain their peaks at different time instants and the combined response $r(t)$ attains its peak at yet a different instant. This phenomenon can be observed in Fig. 13.2.7b, where results for the shear in the top story of a five-story frame are presented. The individual modal responses $V_{5n}(t), n = 1, 2, \ldots, 5$, are shown together with the total response $V_5(t)$.

Approximations must be introduced in combining the peak modal responses r_{no} determined from the earthquake response spectrum because no information is available when these peak modal values occur. The assumption that all modal peaks occur at the same time and their algebraic sign is ignored provides an upper bound to the peak value of the total response:

$$r_o \leq \sum_{n=1}^{N} |r_{no}| \tag{13.7.2}$$

This upper-bound value is usually too conservative, as we shall see in example computations to be presented later. Therefore, this *absolute sum* (ABSSUM) modal combination rule is not popular in structural design applications.

The *square-root-of-sum-of-squares* (SRSS) rule for modal combination, developed in E. Rosenblueth's Ph.D. thesis (1951), is

$$r_o \simeq \left(\sum_{n=1}^{N} r_{no}^2 \right)^{1/2} \tag{13.7.3}$$

The peak response in each mode is squared, the squared modal peaks are summed, and the square root of the sum provides an estimate of the peak total response. As will be seen later, this modal combination rule provides good estimates of response for structures with well-separated natural frequencies. This limitation has not always been recognized in applying this rule to practical problems, and at times it has been misapplied to systems with closely spaced natural frequencies, such as piping systems in nuclear power plants and multistory buildings with unsymmetric plan.

The *complete quadratic combination* (CQC) rule for modal combination is applicable to a wider class of structures as it overcomes the limitations of the SRSS rule. According to the CQC rule,

$$r_o \simeq \left(\sum_{i=1}^{N} \sum_{n=1}^{N} \rho_{in} r_{io} r_{no} \right)^{1/2} \tag{13.7.4}$$

Each of the N^2 terms on the right side of this equation is the product of the peak responses in the ith and nth modes and the correlation coefficient ρ_{in} for these two modes; ρ_{in} varies between 0 and 1 and $\rho_{in} = 1$ for $i = n$. Thus Eq. (13.7.4) can be rewritten as

$$r_o \simeq \left(\sum_{n=1}^{N} r_{no}^2 + \underbrace{\sum_{i=1}^{N} \sum_{n=1}^{N} \rho_{in} r_{io} r_{no}}_{i \neq n} \right)^{1/2} \tag{13.7.5}$$

to show that the first summation on the right side is identical to the SRSS combination rule of Eq. (13.7.3); each term in this summation is obviously positive. The double summation includes all the cross ($i \neq n$) terms; each of these terms may be positive or negative. A cross term is negative when the modal static responses r_i^{st} and r_n^{st} assume opposite signs—for the algebraic sign of r_{no} is the same as that of r_n^{st} because A_n is positive by definition. Thus the estimate for r_o obtained by the CQC rule may be larger or smaller than the estimate provided by the SRSS rule. [It can be shown that the double summation inside the parentheses of Eq. (13.7.4) is always positive.]

Starting in the late 1960s and continuing through the 1970s and early 1980s, several formulations for the peak response to earthquake excitation were published. Some of these are identical or similar to Eq. (13.7.4) but differ in the mathematical expressions given for the correlation coefficient. Here we include two: one due to E. Rosenblueth and J. Elorduy for historical reasons because it was apparently the earliest (1969) result; and a second (1981) due to A. Der Kiureghian because it is now widely used.

The 1971 textbook *Fundamentals of Earthquake Engineering* by N. M. Newmark and E. Rosenblueth gives the Rosenblueth–Elorduy equations for the correlation coefficient:

$$\rho_{in} = \frac{1}{1 + \epsilon_{in}^2} \tag{13.7.6}$$

where

$$\epsilon_{in} = \frac{\omega_i \sqrt{1 - \zeta_i^2} - \omega_n \sqrt{1 - \zeta_n^2}}{\zeta_i' \omega_i + \zeta_n' \omega_n} \qquad \zeta_n' = \zeta_n + \frac{2}{\omega_n s} \tag{13.7.7}$$

and s is the duration of the strong phase of the earthquake excitation. Equations (13.7.6) and (13.7.7) show that $\rho_{in} = \rho_{ni}$; $0 \leq \rho_{in} \leq 1$; and $\rho_{in} = 1$ for $i = n$ or for two modes with equal frequencies and equal damping ratios. It is instructive to specialize Eq. (13.7.6) for systems with the same damping ratio in all modes subjected to earthquake excitation with duration s long enough to replace Eq. (13.7.7b) by $\zeta_n' = \zeta_n$. We substitute $\zeta_i = \zeta_n = \zeta$ in Eq. (13.7.7a), introduce $\beta_{in} = \omega_i / \omega_n$, and insert Eq. (13.7.7a) in Eq. (13.7.6) to obtain

$$\rho_{in} = \frac{\zeta^2 (1 + \beta_{in})^2}{(1 - \beta_{in})^2 + 4\zeta^2 \beta_{in}} \tag{13.7.8}$$

The equation for the correlation coefficient due to Der Kiureghian is

$$\rho_{in} = \frac{8\sqrt{\zeta_i \zeta_n}(\beta_{in}\zeta_i + \zeta_n)\beta_{in}^{3/2}}{(1 - \beta_{in}^2)^2 + 4\zeta_i \zeta_n \beta_{in}(1 + \beta_{in}^2) + 4(\zeta_i^2 + \zeta_n^2)\beta_{in}^2} \tag{13.7.9}$$

This equation also implies that $\rho_{in} = \rho_{ni}$, $\rho_{in} = 1$ for $i = n$ or for two modes with equal frequencies and equal damping ratios. For equal modal damping $\zeta_i = \zeta_n = \zeta$ this equation simplifies to

$$\rho_{in} = \frac{8\zeta^2(1+\beta_{in})\beta_{in}^{3/2}}{(1-\beta_{in}^2)^2 + 4\zeta^2\beta_{in}(1+\beta_{in})^2} \tag{13.7.10}$$

Figure 13.7.1 shows Eqs. (13.7.8) and (13.7.10) for the correlation coefficient ρ_{in} plotted as a function of $\beta_{in} = \omega_i/\omega_n$ for four damping values: $\zeta = 0.02, 0.05, 0.10,$ and 0.20. Observe that the two expressions give essentially identical values for ρ_{in}, especially in the neighborhood of $\beta_{in} = 1$, where ρ_{in} is the most significant.

This figure also provides an understanding of the correlation coefficient. Observe that this coefficient diminishes rapidly as the two natural frequencies ω_i and ω_n move farther apart. This is especially the case at small damping values that are typical of structures. In other words, it is only in a narrow range of β_{in} around $\beta_{in} = 1$ that ρ_{in} has significant values; and this range depends on damping. For example, $\rho_{in} > 0.1$ for systems with 5% damping over the frequency ratio range $1/1.35 \leq \beta_{in} \leq 1.35$. If the damping is 2%, this range is reduced to $1/1.13 \leq \beta_{in} \leq 1.13$. For structures with well-separated natural frequencies the coefficients ρ_{in} vanish; as a result all cross ($i \neq n$) terms in the CQC rule, Eq. (13.7.5), can be neglected and it reduces to the SRSS rule, Eq. (13.7.3). It is now clear that the SRSS rule applies to structures with well-separated natural frequencies of those modes that contribute significantly to the response.

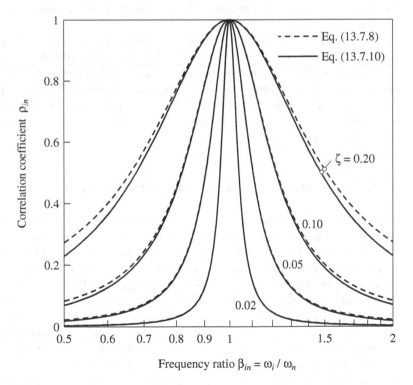

Figure 13.7.1   Variation of correlation coefficient ρ_{in} with modal frequency ratio, $\beta_{in} = \omega_i/\omega_n$, as given by two different equations for four damping values; abcissa scale is logarithmic.

The SRSS and CQC rules for combination of peak modal responses have been presented without the underlying derivations based on random vibration theory, a subject beyond the scope of this book. It is important, however, to recognize the implications of the assumptions behind the derivations. These assumptions indicate that the modal combination rules would be most accurate for earthquake excitations that contain a wide band of frequencies with long phases of strong shaking, which are several times longer than the fundamental periods of the structures, which are not too lightly damped ($\zeta_n > 0.005$). In particular, these modal combination rules will become less accurate for short-duration impulsive ground motions and are not recommended for ground motions that contain many cycles of essentially harmonic excitation.

Considering that the SRSS and CQC modal combination rules are based on random vibration theory, r_o should be interpreted as the mean of the peak values of response to an ensemble of earthquake excitations. Thus the modal combination rules are intended for use when the excitation is characterized by a smooth response (or design) spectrum, based on the response spectra for many earthquake excitations. The smooth spectrum may be the mean or median of the individual response spectra or it may be a more conservative spectrum, such as the mean-plus-one-standard-deviation spectrum (Section 6.9). The CQC or SRSS modal combination rule (as appropriate depending on the closeness of natural frequencies) when used in conjunction with, say, the mean spectrum provides an estimate of the peak response that is reasonably close to the mean of the peak values of response to individual excitations.

This estimate of the peak response generally, but not always, errs on the unconservative side. The magnitude of the error depends on the vibration properties—periods and modes—of the structure and spectrum shape. Over a range of buildings analyzed, researchers have observed errors up to 25%, especially in estimating local response quantities such as story drifts in the upper stories. The error may be larger or smaller if modal combination rules are used to estimate the peak response to a single ground motion characterized by a jagged response spectrum.

13.7.3 Interpretation of Response Spectrum Analysis

The response spectrum analysis (RSA) described in the preceding section is a procedure for dynamic analysis of a structure subjected to earthquake excitation, but it reduces to a series of static analyses. For each mode considered, static analysis of the structure subjected to forces s_n provides the modal static response r_n^{st}, which is multiplied by the spectral ordinate A_n to obtain the peak modal response r_{no} [Eq. (13.7.1)]. Thus the RSA procedure avoids the dynamic analysis of SDF systems necessary for response history analysis (Fig. 13.1.1). However, the RSA is still a dynamic analysis procedure, because it uses the vibration properties—natural frequencies, natural modes, and modal damping ratios—of the structure and the dynamic characteristics of the ground motion through its response (or design) spectrum. It is just that the user does not have to carry out any response history calculations; somebody has already done these in developing the earthquake response spectrum or the earthquake excitation has been characterized by a smooth design spectrum.

13.8 MULTISTORY BUILDINGS WITH SYMMETRIC PLAN

13.8.1 Response Spectrum Analysis Procedure

In this section the response spectrum analysis procedure of Section 13.7 is specialized for multistory buildings with their plans having two axes of symmetry subjected to horizontal ground motion along one of these axes. The peak value[†] of the nth-mode contribution $r_n(t)$ to a response quantity is given by Eq. (13.7.1). The modal static response r_n^{st} is calculated by static analysis of the building subjected to lateral forces $\mathbf{s}_n$ of Eq. (13.2.4). Equations for r_n^{st} for several response quantities are available in Table 13.2.1. Substituting these formulas for floor displacement u_j, story drift Δ_j, base shear V_b, and base overturning moment M_b in Eq. (13.7.1) gives

$$u_{jn} = \Gamma_n \phi_{jn} D_n \qquad \Delta_{jn} = \Gamma_n(\phi_{jn} - \phi_{j-1,n})D_n \qquad (13.8.1a)$$

$$V_{bn} = M_n^* A_n \qquad M_{bn} = h_n^* M_n^* A_n \qquad (13.8.1b)$$

where $D_n \equiv D(T_n, \zeta_n)$, the deformation spectrum ordinate corresponding to natural period T_n and damping ratio ζ_n; $D_n = A_n/\omega_n^2$.

Equations (13.8.1) for the peak modal responses are equivalent to static analysis of the building subjected to the equivalent static forces associated with the nth-mode peak response:

$$\mathbf{f}_n = \mathbf{s}_n A_n \qquad f_{jn} = \Gamma_n m_j \phi_{jn} A_n \qquad (13.8.2)$$

where $\mathbf{f}_n$ is the vector of forces f_{jn} at the various floor levels, $j = 1, 2, \ldots, N$ (Fig. 13.8.1); $\mathbf{s}_n$ is defined by Eq. (13.2.4). The force vector $\mathbf{f}_n$ is the peak value of $\mathbf{f}_n(t)$, obtained by

Floor

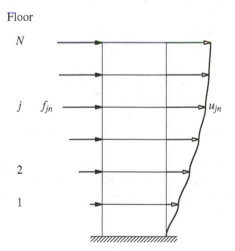

Figure 13.8.1   Peak values of lateral displacements and equivalent static lateral forces associated with the nth mode.

[†]From now on, the subscript o is dropped from r_o for brevity [i.e., r will denote the peak value of $r(t)$].

replacing $A_n(t)$ in Eq. (13.2.7) by the spectral ordinate A_n. Because only one static analysis is required for each mode, it is more direct to do so for the forces $\mathbf{f}_n$ instead of $\mathbf{s}_n$ and then multiplying the latter results by A_n. In contrast, the use of the modal static response r_n^{st} was emphasized in response history analysis because it highlighted the fact that the static analysis for forces $\mathbf{s}_n$ was needed only once even though the response was computed at many time instants.

Thus the peak value r_n of the nth-mode contribution to a response quantity r is determined by static analysis of the building due to lateral forces $\mathbf{f}_n$; the direction of forces f_{jn} is controlled by the algebraic sign of ϕ_{jn}. Hence these forces for the fundamental mode will act in the same direction (Fig. 13.8.1), but for the second and higher modes they will change direction as one moves up the building. Observe that this static analysis is not necessary to determine floor displacements or story drifts; Eq. (13.8.1a) provides the more convenient alternative. The peak value of the total response is estimated using the modal combination rules of Eq. (13.7.3) or (13.7.4), as appropriate, including all modes that contribute significantly to the response.

Summary. The procedure to compute the peak response of an N-story building with plan symmetric about two orthogonal axes to earthquake ground motion along an axis of symmetry, characterized by a response spectrum or design spectrum, is summarized in step-by-step form:

1. Define the structural properties.
   a. Determine the mass matrix $\mathbf{m}$ and lateral stiffness matrix $\mathbf{k}$ (Section 9.4).
   b. Estimate the modal damping ratios ζ_n (Chapter 11).
2. Determine the natural frequencies ω_n (natural periods $T_n = 2\pi/\omega_n$) and natural modes ϕ_n of vibration (Chapter 10).
3. Compute the peak response in the nth mode by the following steps to be repeated for all modes, $n = 1, 2, \ldots, N$:
   a. Corresponding to natural period T_n and damping ratio ζ_n, read D_n and A_n, the deformation and pseudo-acceleration, from the earthquake response spectrum or the design spectrum.
   b. Compute the floor displacements and story drifts from Eq. (13.8.1a).
   c. Compute the equivalent static lateral forces $\mathbf{f}_n$ from Eq. (13.8.2).
   d. Compute the story forces—shear and overturning moment—and element forces—bending moments and shears—by static analysis of the structure subjected to lateral forces $\mathbf{f}_n$.
4. Determine an estimate for the peak value r of any response quantity by combining the peak modal values r_n according to the SRSS rule, Eq. (13.7.3), if the natural frequencies are well separated. The CQC rule, Eq. (13.7.4), should be used if the natural frequencies are closely spaced.

Usually, only the lower modes contribute significantly to the response. Therefore, steps 2 and 3 need to be implemented for only these modes and the modal combinations of Eqs. (13.7.3) and (13.7.4) truncated accordingly.

Example 13.11

The peak response of the two-story frame of Example 13.4, shown in Fig. E13.11a, to ground motion characterized by the design spectrum of Fig. 6.9.5 scaled to 0.5g peak ground acceleration is to be determined. This reinforced-concrete frame has the following properties: $m = 200$ kips/g, $E = 3 \times 10^3$ ksi, $I = 1000$ in^4, $h = 10$ ft, $L = 20$ ft. Determine the lateral displacements of the frame and bending moments at both ends of each beam and column.

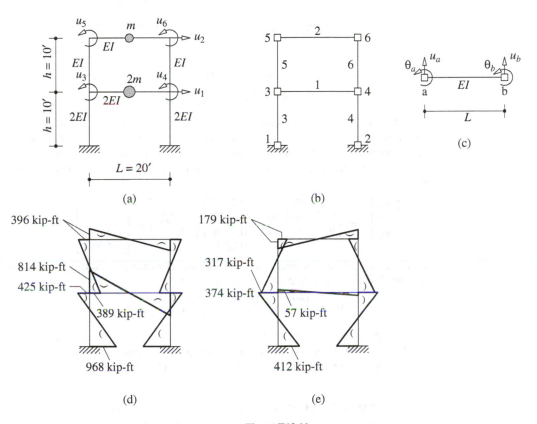

Figure E13.11

Solution   Steps 1 and 2 of the summary have already been implemented and the results are available in Examples 10.5 and 13.4. Substituting for E, I, and h in Eq. (b) of Example 13.4 gives ω_n and $T_n = 2\pi/\omega_n$:

$$\omega_1 = 4.023 \qquad \omega_2 = 10.71 \text{ rad/sec}$$

$$T_1 = 1.562 \qquad T_2 = 0.5868 \text{ sec}$$

Step 3a: Corresponding to these periods, the spectral ordinates are $D_1 = 13.72$ in. and $D_2 = 4.578$ in.

1. *Determine the floor displacements.*

Step 3b: Using Eq. (13.8.1a) with numerical values for Γ_n and ϕ_{jn} from Example 13.4 and D_n from step 3(a) gives the peak displacements $\mathbf{u}_n$ due to the two modes:

$$\mathbf{u}_1 = \left\{ \begin{array}{c} u_1 \\ u_2 \end{array} \right\}_1 = 1.365 \left\{ \begin{array}{c} 0.3871 \\ 1 \end{array} \right\} 13.72 = \left\{ \begin{array}{c} 7.252 \\ 18.73 \end{array} \right\} \text{in.}$$

$$\mathbf{u}_2 = \left\{ \begin{array}{c} u_1 \\ u_2 \end{array} \right\}_2 = -0.365 \left\{ \begin{array}{c} -1.292 \\ 1 \end{array} \right\} 4.578 = \left\{ \begin{array}{c} 2.159 \\ -1.672 \end{array} \right\} \text{in.}$$

Step 4: Using the SRSS rule for modal combination, estimates for the peak values of the floor displacements are

$$u_1 \simeq \sqrt{(7.252)^2 + (2.159)^2} = 7.566 \text{ in.}$$

$$u_2 \simeq \sqrt{(18.73)^2 + (-1.672)^2} = 18.81 \text{ in.}$$

2. *Determine the element forces.* Instead of implementing steps 3c and 3d as described in the summary, here we illustrate the computation of element forces from the floor displacements and joint rotations. The elements and nodes are numbered as shown in Fig. E13.11b.

First mode. Joint rotations are obtained from Eq. (d) of Example 9.9 with $\mathbf{u}_t$ replaced by $\mathbf{u}_1$:

$$\mathbf{u}_{01} = \left\{ \begin{array}{c} u_3 \\ u_4 \\ u_5 \\ u_6 \end{array} \right\}_1 = \frac{1}{120} \left[\begin{array}{cc} -0.4426 & -0.2459 \\ -0.4426 & -0.2459 \\ 0.9836 & -0.7869 \\ 0.9836 & -0.7869 \end{array} \right] \left[\begin{array}{c} 7.252 \\ 18.73 \end{array} \right] = \left[\begin{array}{c} -6.514 \\ -6.514 \\ -6.340 \\ -6.340 \end{array} \right] \times 10^{-2}$$

From $\mathbf{u}_1$ and $\mathbf{u}_{01}$ all element forces can be calculated. For example, the bending moment at the left end of the first floor beam (Fig. E13.11c) is

$$M_a = \frac{4EI}{L}\theta_a + \frac{2EI}{L}\theta_b + \frac{6EI}{L^2}u_a - \frac{6EI}{L^2}u_b$$

Substituting $E = 3 \times 10^3$ ksi, $I = 2000$ in^4, $L = 240$ in., $\theta_a = u_3$, $\theta_b = u_4$, $u_a = u_b = 0$ gives $M_a = -9770$ kip-in. $= -814$ kip-ft. Bending moments in all elements can be calculated similarly. The results are summarized in Table E13.11 and in Fig. E13.11d.

TABLE E13.11   PEAK BENDING MOMENTS (KIP-FT)

Element	Node	Mode 1	Mode 2	SRSS
Beam 1	3	−814	− 57	816
	4	−814	− 57	816
Beam 2	5	−396	179	435
	6	−396	179	435
Column 3	3	425	374	566
	1	968	412	1052
Column 5	5	396	−179	435
	3	389	−317	502

Second mode. Joint rotations $\mathbf{u}_{02}$ are obtained from Eq. (d) of Example 9.9 with $\mathbf{u}_t$ replaced by $\mathbf{u}_2$. Computations for the element forces parallel those shown for the first mode, but using $\mathbf{u}_2$ and $\mathbf{u}_{02}$, leading to the results in Table E13.11 and in Fig. E13.11c.

Step 4: The peak value of each element force is estimated by combining its peak modal values by the SRSS rule. The results are shown in Table E13.11. Note that the algebraic signs of the bending moments are lost in the total values; therefore, it is not meaningful to draw the bending moment diagram, and the total moments do not satisfy equilibrium at joints.

13.8.2 Example: Five-Story Shear Frame

In this section the RSA procedure is implemented for the five-story shear frame of Fig. 12.8.1. The complete history of this structure's response to the El Centro ground motion was determined in Section 13.2.6. We now estimate its peak response directly from the response spectrum for this excitation (i.e., without computing its response history).

Presented in Sections 12.8 and 13.2.6 were the mass and stiffness matrices and the natural vibration periods and modes of this structure. From these data, the modal properties M_n and L_n^h were computed (Table 13.2.2). The damping ratios are estimated as $\zeta_n = 5\%$.

Response spectrum ordinates. The response spectrum for the El Centro ground motion for 5% damping gives the values of D_n and A_n noted in Fig. 13.8.2 corresponding to the natural periods T_n. These are the precise values for the spectral ordinates, the peak values of $D_n(t)$ and $A_n(t)$ in Fig. 13.2.6, thus eliminating any errors in reading spectral ordinates. Such errors are inherent in practical implementation of the RSA procedure with a jagged response spectrum, but are eliminated if a smooth design spectrum, such as Fig. 6.9.5, is used.

Peak modal responses. The floor displacements are determined from Eq. (13.8.1a) using known values of ϕ_{jn} (Section 12.8), of L_n^h (Table 13.2.2) and $\Gamma_n = L_n^h$ (because $M_n = 1$), and of D_n (Fig. 13.8.2). For example, the floor displacements due to the first mode are computed as follows:

$$\mathbf{u}_1 = \Gamma_1 \phi_1 D_1 = 1.067 \begin{Bmatrix} 0.334 \\ 0.641 \\ 0.895 \\ 1.078 \\ 1.173 \end{Bmatrix} 5.378 = \begin{Bmatrix} 1.916 \\ 3.677 \\ 5.139 \\ 6.188 \\ 6.731 \end{Bmatrix} \text{in.}$$

These displacements are shown in Fig. 13.8.3a. The equivalent static forces for the nth mode are computed from Eq. (13.8.2a), multiplying known values of $\mathbf{s}_n$ (Fig.13.2.4) by A_n (Fig. 13.8.2). For example, the forces associated with the first mode are computed as follows:

$$\mathbf{f}_1 = \mathbf{s}_1 A_1 = \begin{Bmatrix} 0.356\,m \\ 0.684\,m \\ 0.956\,m \\ 1.150\,m \\ 1.252\,m \end{Bmatrix} 0.1375g = \begin{Bmatrix} 4.899 \\ 9.401 \\ 13.141 \\ 15.817 \\ 17.211 \end{Bmatrix} \text{kips}$$

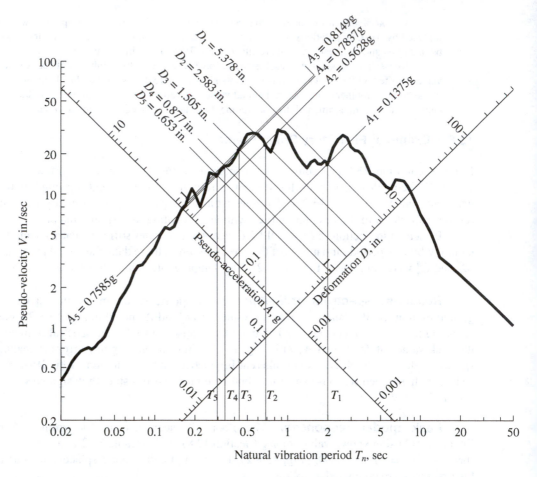

Figure 13.8.2 Earthquake response spectrum with natural vibration periods T_n of example structure shown together with spectral values D_n and A_n.

These forces are also shown in Fig. 13.8.3a. Repeating these computations for modes $n = 2, 3, 4,$ and 5 leads to the remaining results of Fig. 13.8.3. Observe that the equivalent static forces for the first mode all act in the same direction, but for the second and higher modes they change direction as one moves up the building; the direction of forces is controlled by the algebraic sign of ϕ_{jn} (Fig. 12.8.2).

For each mode the peak value of any story force or element force is computed by static analysis of the structure subjected to the equivalent static lateral forces $\mathbf{f}_n$. Table 13.8.1 summarizes these peak values for the base shear V_b, top-story shear V_5, and base overturning moment M_b. The earlier data for roof displacement u_5 are also included. These peak modal values are exact because the errors in reading spectral ordinates had been eliminated in this example. This is apparent by comparing the data in Table 13.8.1

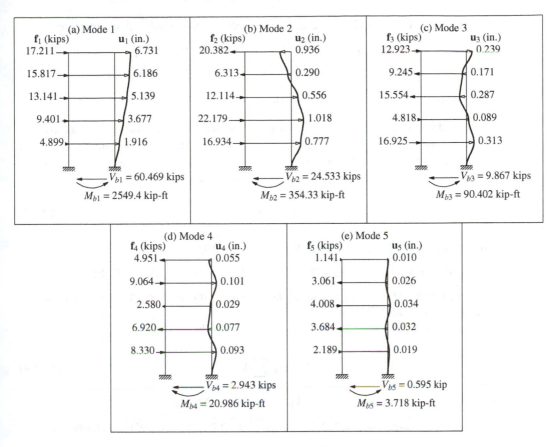

Figure 13.8.3   Peak values of displacements and equivalent static lateral forces due to the five natural vibration modes.

and the peak modal values from response history analysis in Figs. 13.2.7 and 13.2.8. The two sets of data agree except possibly for their algebraic signs because the peak values D_n and A_n are positive by definition.

Alternatively, Eq. (13.7.1) could have been used for computing the peak modal response. For example, the modal static responses V_{bn}^{st} and M_{bn}^{st} are available from

TABLE 13.8.1   PEAK MODAL RESPONSES

Mode	V_b (kips)	V_5 (kips)	M_b (kip-ft)	u_5 (in.)
1	60.469	17.211	2549.4	6.731
2	24.533	−20.382	−354.33	−0.936
3	9.867	12.923	90.402	0.239
4	2.943	−4.951	−20.986	−0.055
5	0.595	1.141	3.718	0.010

Table 13.2.3 and A_n from Fig. 13.8.2. For example, the first-mode calculations are

$$V_{b1} = V_{b1}^{st} A_1 = [4.398(100/g)]0.1375g = 60.469 \text{ kips}$$

$$M_{b1} = M_{b1}^{st} A_1 = [(15.45)(100/g)12]0.1375g = 2549.4 \text{ kip-ft}$$

As expected, these are the same as the data in Table 13.8.1.

Modal combination. The peak value r of the total response $r(t)$ is estimated by combining the peak modal responses according to the ABSSUM, SRSS, and CQC rules of Eqs. (13.7.2) to (13.7.4). Their use is illustrated for one response quantity, the base shear. The ABSSUM rule of Eq. (13.7.2) is specialized for the base shear:

$$V_b \leq \sum_{n=1}^{5} |V_{bn}| \tag{13.8.3}$$

Substituting for the known values of V_{bn} from Table 13.8.1 gives

$$V_b \leq 60.469 + 24.533 + 9.867 + 2.943 + 0.595 \quad \text{or} \quad V_b \leq 98.407 \text{ kips}$$

As expected, the ABSSUM estimate of 98.407 kips is much larger than the exact value of 73.278 kips (Fig. 13.2.7). The SRSS rule of Eq. (13.7.3) is specialized for the base shear:

$$V_b \simeq \left(\sum_{n=1}^{5} V_{bn}^2 \right)^{1/2} \tag{13.8.4}$$

Substituting for the known values of V_{bn} from Table 13.8.1 gives

$$V_b \simeq \sqrt{(60.469)^2 + (24.533)^2 + (9.867)^2 + (2.943)^2 + (0.595)^2} = 66.066 \text{ kips}$$

Observe that the contributions of modes higher than the second are small. The CQC rule of Eq. (13.7.4) is specialized for the base shear:

$$V_b \simeq \left(\sum_{i=1}^{5} \sum_{n=1}^{5} \rho_{in} V_{bi} V_{bn} \right)^{1/2} \tag{13.8.5}$$

Needed in this equation are the correlation coefficients ρ_{in}, which depend on the frequency ratios $\beta_{in} = \omega_i/\omega_n$, computed from the known natural frequencies (Section 13.2.6) and repeated in Table 13.8.2 for convenience.

TABLE 13.8.2 NATURAL FREQUENCY RATIOS β_{in}

Mode, i	$n = 1$	$n = 2$	$n = 3$	$n = 4$	$n = 5$	ω_i (rad/sec)
1	1.000	0.343	0.217	0.169	0.148	3.1416
2	2.919	1.000	0.634	0.494	0.433	9.1703
3	4.602	1.576	1.000	0.778	0.683	14.4561
4	5.911	2.025	1.285	1.000	0.877	18.5708
5	6.742	2.310	1.465	1.141	1.000	21.1810

TABLE 13.8.3   CORRELATION COEFFICIENTS ρ_{in}

Mode, i	$n = 1$	$n = 2$	$n = 3$	$n = 4$	$n = 5$
1	1.000	0.007	0.003	0.002	0.001
2	0.007	1.000	0.044	0.018	0.012
3	0.003	0.044	1.000	0.136	0.062
4	0.002	0.018	0.136	1.000	0.365
5	0.001	0.012	0.062	0.365	1.000

TABLE 13.8.4   INDIVIDUAL TERMS IN CQC RULE: BASE SHEAR V_b

Mode, i	$n = 1$	$n = 2$	$n = 3$	$n = 4$	$n = 5$
1	3656.476	10.172	1.615	0.306	0.049
2	10.172	601.844	10.687	1.284	0.178
3	1.615	10.687	97.354	3.943	0.365
4	0.306	1.284	3.943	8.658	0.639
5	0.049	0.178	0.365	0.639	0.354

For each β_{in} value in Table 13.8.2, ρ_{in} is determined from Eq. (13.7.10) for $\zeta = 0.05$ and presented in Table 13.8.3. Observe that the cross-correlation coefficients ρ_{in} ($i \neq n$) are small because the natural frequencies of the five-story shear frame are well separated.

The 25 terms in the double summation of Eq. (13.8.5), computed using the known values of ρ_{in} (Table 13.8.3) and V_{bn} (Table 13.8.1), are given in Table 13.8.4. Adding these 25 terms and taking the square root gives $V_b \simeq 66.507$ kips. It is clear that only the $i = n$ terms are significant and the cross-terms ($i \neq n$) are small because the cross-correlation coefficients are small. Note that the contributions of modes higher than the second mode could be neglected, thus reducing the computational effort.

Comparison of RSA and RHA results.   The RSA estimates of peak response obtained from the ABSSUM, SRSS, and CQC rules are summarized in Table 13.8.5 together with the RHA results from Figs. 13.2.7 to 13.2.8. In the preceding section, computational details for estimating the peak base shear by RSA were presented; similarly, results for V_5, M_b, and u_5 were obtained. These data permit several observations. First, the ABSSUM rule can be excessively conservative and should therefore not be used. Second, the SRSS and CQC rules give essentially the same estimates of peak response because the cross-correlation coefficients are small for this structure with well-separated natural frequencies. Third, the peak responses estimated by SRSS or CQC rules are smaller than the RHA values; this is not a general trend, however, and larger values can also be estimated when using a jagged response spectrum for a single excitation. Fourth, the error in SRSS (or CQC) estimates of peak response, expressed as a percentage of the RHA value, vary with the response quantity. It is about 15% for the top-story shear V_5, 10% for the base shear V_b, and less than 1% for the base overturning moment M_b and top-floor displacement

TABLE 13.8.5 RSA AND RHA VALUES OF PEAK RESPONSE

	V_b (kips)	V_5 (kips)	M_b (kip-ft)	u_5 (in.)
ABSSUM	98.407	56.608	3018.8	7.971
SRSS	66.066	30.074	2575.6	6.800
CQC	66.507	29.338	2572.7	6.793
RHA	73.278	35.217	2593.2	6.847

u_5. The error is largest for V_5 because the responses due to the higher modes are most significant (compared to other response quantities considered) relative to the first mode (Table 13.8.1). Similarly, the error is smallest for M_b because the higher-mode responses are a very small fraction of the first-mode response (Table 13.8.1).

Now consider a typical application of the RSA procedure in which the peak response is estimated for excitations characterized by a smooth design spectrum, say the mean (or median) spectrum derived from individual spectra for many ground motions (Section 6.9). Modal combination rules are more dependable when used in conjunction with such a smooth spectrum, because the variability of above-noted errors from excitation to excitation is "averaged" out.

Avoid a pitfall. Observe that the peak value r of each response quantity was determined by combining the peak values r_n of the modal contributions to the same response quantity. This is the correct way of estimating the peak value of a response quantity.

On the other hand, it is wrong to compute the combined peak value of one response quantity from the combined peak values of other response quantities. For example, it is desired to determine Δ_5, the drift in the fifth story of the building just analyzed. It is incorrect to determine its peak value from $\Delta_5 = u_5 - u_4$, where u_5 and u_4 have been determined by combining their modal peaks u_{5n} and u_{4n}, respectively. The correct procedure to determine Δ_5 is by combining the peak modal values, $\Delta_{5n} = u_{5n} - u_{4n}$.

Similarly, it is erroneous to compute the combined peak value of an internal force from the combined peak values of other forces. In particular, it is incorrect to determine the story shears or story overturning moments from the combined peak values of the equivalent static forces. The SRSS combination of the peak values of the equivalent static forces f_{jn}

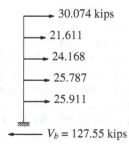

30.074 kips
21.611
24.168
25.787
25.911

$V_b = 127.55$ kips

Figure 13.8.4 Wrong procedure for computing internal forces.

for each mode of the five-story shear building (Fig. 13.8.3) is shown in Fig. 13.8.4. Static analysis of the structure with these external forces gives the base shear $V_b = 127.55$ kips, which is almost twice the correct SRSS value presented in Table 13.8.5. This erroneous value is much larger because the algebraic signs of f_{jn} (Fig. 13.8.3) are lost in the SRSS combination and the forces shown in Fig. 13.8.4 are all in the same direction.

13.8.3 Example: Four-Story Frame with an Appendage

This section is concerned with the four-story frame with a light appendage of Section 13.2.7, where its response history due to El Centro ground motion was presented. In this section the peak responses of the same structure are estimated by the RSA procedure directly from the response spectrum for the ground motion. The analysis procedure and the details of its implementation are identical to those described in Section 13.8.2. Therefore, only a summary of the results is presented.

Table 13.8.6 shows the natural periods T_n and the associated spectral ordinates for 5% damping together with the peak modal responses for two response quantities: base

TABLE 13.8.6 SPECTRAL VALUES AND PEAK MODAL RESPONSES

Mode	T_n (sec)	D_n (in.)	A_n/g	V_b (kips)	V_5 (kips)
1	2.000	5.378	0.1375	26.805	1.367
2	1.873	5.335	0.1556	25.429	-1.397
3	0.672	2.631	0.5950	19.816	0.027
4	0.439	1.545	0.8176	6.414	-0.005
5	0.358	0.928	0.7407	1.090	0.001

shear V_b and appendage shear V_5. The ratios β_{in} of natural frequencies are given in Table 13.8.7. The correlation coefficients computed by Eq. (13.7.10) for each β_{in} value are listed in Table 13.8.8.

TABLE 13.8.7 NATURAL FREQUENCY RATIOS β_{in}

Mode, i	$n=1$	$n=2$	$n=3$	$n=4$	$n=5$	ω_i (rad/sec)
1	1.000	0.936	0.336	0.219	0.179	3.142
2	1.068	1.000	0.359	0.234	0.191	3.355
3	2.974	2.785	1.000	0.653	0.532	9.344
4	4.556	4.266	1.532	1.000	0.815	14.314
5	5.589	5.233	1.879	1.227	1.000	17.558

TABLE 13.8.8   CORRELATION COEFFICIENTS ρ_{in}

Mode, i	$n = 1$	$n = 2$	$n = 3$	$n = 4$	$n = 5$
1	1.000	0.698	0.007	0.003	0.002
2	0.698	1.000	0.008	0.003	0.002
3	0.007	0.008	1.000	0.050	0.023
4	0.003	0.003	0.050	1.000	0.192
5	0.002	0.002	0.023	0.192	1.000

The 25 terms in the double summation of Eq. (13.8.5) for V_b are presented in Table 13.8.9; similar data for V_5 appear in Table 13.8.10. The cross ($i \neq n$) terms in Table 13.8.9 are all positive because the modal static responses V_{bn}^{st} for base shear are all positive. Some of the cross terms in Table 13.8.10 are negative because all the modal static responses V_{5n}^{st} for the appendage shear do not have the same algebraic sign; a cross term is negative when V_{5i}^{st} and V_{5n}^{st} assume opposite signs. Finally, estimates for the peak values of the two response quantities obtained by ABSSUM, SRSS, and CQC procedures are presented in Table 13.8.11.

These results bring out several response features of systems with two modes having close natural frequencies and contributing significantly to the response (e.g., the first two modes of the four-story building with an appendage). The cross-correlation coefficient for these two modes is 0.698 (Table 13.8.8), which is significant relative to its largest possible value of unity. As a result, the 1–2 cross terms for V_5 and V_b are comparable in magnitude to the individual modal (1–1 or 2–2) terms (Tables 13.8.9 and 13.8.10). Therefore,

TABLE 13.8.9   INDIVIDUAL TERMS IN CQC RULE: BASE SHEAR V_b

Mode, i	$n = 1$	$n = 2$	$n = 3$	$n = 4$	$n = 5$
1	718.516	475.836	3.491	0.474	0.056
2	475.836	646.646	3.850	0.510	0.059
3	3.491	3.850	392.654	6.385	0.488
4	0.474	0.510	6.385	41.144	1.341
5	0.056	0.059	0.488	1.341	1.188

TABLE 13.8.10   INDIVIDUAL TERMS IN CQC RULE: APPENDAGE SHEAR V_5

Mode, i	$n = 1$	$n = 2$	$n = 3$	$n = 4$	$n = 5$
1	1.868	−1.333	0.000	0.000	0.000
2	−1.333	1.951	0.000	0.000	0.000
3	0.000	0.000	0.001	0.000	0.000
4	0.000	0.000	0.000	0.000	0.000
5	0.000	0.000	0.000	0.000	0.000

TABLE 13.8.11   RSA AND RHA VALUES OF PEAK RESPONSE

	V_b (kips)	V_5 (kips)
ABSSUM	79.554	2.797
SRSS	42.428	1.954
CQC	52.774	1.074
RHA	56.660	0.997

the SRSS and CQC modal combination rules provide very different estimates of peak responses (Table 13.8.11). The CQC rule gives a base shear that is larger than its value from the SRSS rule because all the cross ($i \neq n$) terms are positive (Table 13.8.9). For the appendage shear, however, the significant cross-term associated with the first two modes is negative (Table 13.8.10). Therefore, the CQC rule gives an appendage shear that is smaller than that obtained from the SRSS rule. Table 13.8.11 shows that only the CQC modal combination rule provides estimates of peak response that are close to the RHA results of Fig. 13.2.11. The errors in the SRSS estimates are unacceptably large; and they are even larger in the ABSSUM results.

An examination of the RHA results reveals the reasons for these large errors in the SRSS combination rule. Observe that the SDF system responses $A_n(t)$ for the first two modes are highly correlated, as they are essentially in phase because the two natural periods are close (Fig. 13.2.10); the peak values of the two $A_n(t)$ are similar because their natural periods are close and their damping ratios are identical. As a result and because the modal static responses are similar for the first two modes (Table 13.2.4), the response contributions of the first two modes are similar in magnitude (Fig. 13.2.11). These modal contributions to the base shear are almost directly additive because they are essentially in phase (Fig. 13.2.11a). This feature of the response is not represented by the SRSS rule, whereas it is recognized in the CQC rule by the significant cross term (between modes 1 and 2) with positive value (Table 13.8.9). In contrast, the two modal contributions to the appendage shear tend to cancel each other because they have essentially opposite phase (Fig. 13.2.11b). This feature of the response is again not represented by the SRSS rule, whereas it is recognized in the CQC rule by the significant cross term (between modes 1 and 2) with negative value (Table 13.8.10). It is clear from this discussion that the SRSS modal combination rule should not be used for systems with closely spaced natural frequencies.

13.9 MULTISTORY BUILDINGS WITH UNSYMMETRIC PLAN

In this section the response spectrum analysis procedure of Section 13.7 is specialized for multistory buildings with their plans symmetric about the x-axis but unsymmetric about the y-axis subjected to ground motion in the y-direction. The peak value of the nth-mode contribution $r_n(t)$ to a response quantity is given by Eq. (13.7.1). The modal static response r_n^{st} is calculated by static analysis of the building subjected to lateral forces $\mathbf{s}_{yn}$ and torques $\mathbf{s}_{\theta n}$ of Eq. (13.3.7). Equations for the modal static response r_n^{st} for several response quantities are available in Table 13.3.1. Substituting these formulas for floor translation u_{jy}, floor rotation $u_{j\theta}$, base shear V_b, base overturning moment M_b, and base torque T_b in Eq. (13.7.1) gives

$$u_{jyn} = \Gamma_n \phi_{jyn} D_n \qquad u_{j\theta n} = \Gamma_n \phi_{j\theta n} D_n \tag{13.9.1a}$$

$$V_{bn} = M_n^* A_n \qquad M_{bn} = h_n^* M_n^* A_n \qquad T_{bn} = I_{On}^* A_n \tag{13.9.1b}$$

Equations (13.9.1) for the peak modal responses are equivalent to static analysis of the building subjected to the equivalent static forces associated with the nth-mode peak response; the lateral forces $\mathbf{f}_{yn}$ and torques $\mathbf{f}_{\theta n}$ are

$$\left\{ \begin{matrix} \mathbf{f}_{yn} \\ \mathbf{f}_{\theta n} \end{matrix} \right\} = \left\{ \begin{matrix} \mathbf{s}_{yn} \\ \mathbf{s}_{\theta n} \end{matrix} \right\} A_n \tag{13.9.2}$$

The lateral force and torque at the jth floor level (Fig. 13.9.1) are

$$f_{jyn} = \Gamma_n m_j \phi_{jyn} A_n \qquad f_{j\theta n} = \Gamma_n r^2 m_j \phi_{j\theta n} A_n \tag{13.9.3}$$

For reasons mentioned in Section 13.8, it is more direct to do the static analysis for the forces $\mathbf{f}_{yn}$ and $\mathbf{f}_{\theta n}$ instead of $\mathbf{s}_{yn}$ and $\mathbf{s}_{\theta n}$ and then multiplying the latter results by A_n.

For any response quantity, therefore, the peak value of the nth-mode response is determined by static analysis of the building subjected to lateral forces $\mathbf{f}_{yn}$ and torques $\mathbf{f}_{\theta n}$; the direction of forces f_{jyn} and $f_{j\theta n}$ is controlled by the algebraic signs of ϕ_{jyn} and $\phi_{j\theta n}$. Observe that this static analysis is not necessary to determine floor displacements or rotations; Eq. (13.9.1a) provides the more convenient alternative. Although such a three-dimensional static analysis of an unsymmetric-plan building provides the element forces in all frames

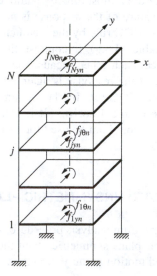

Figure 13.9.1   Peak values of equivalent static forces: lateral forces and torques.

of the building, it may be useful to recognize that the element forces in an individual (ith) frame can also be determined by planar analysis of the frame subjected to lateral forces:

$$\mathbf{f}_{in} = (\Gamma_n/\omega_n^2)\mathbf{k}_{xi}(-y_i\boldsymbol{\phi}_{\theta n})A_n \qquad \mathbf{f}_{in} = (\Gamma_n/\omega_n^2)\mathbf{k}_{yi}(\boldsymbol{\phi}_{yn} + x_i\boldsymbol{\phi}_{\theta n})A_n \qquad (13.9.4)$$

The first of these equations applies to frames in the x-direction and the second to frames in the y-direction. They are obtained from Eq. (13.3.18) with $A_n(t)$ replaced by the corresponding spectral value A_n.

Once these peak modal responses have been determined for all the modes that contribute significantly to the total response, they can be combined using the CQC rule, Eq. (13.7.4), with N replaced by $2N$—the number of DOFs for the unsymmetric-plan building—in both summations, to obtain an estimate of the peak total response. The SRSS rule for modal combination should not be used because many unsymmetric-plan buildings have pairs (or triplets) of closely spaced natural frequencies.

Summary. The procedure to compute the peak response of an N-story building with its plan symmetric about the x-axis but unsymmetric about the y-axis subjected to the y-component of ground motion, characterized by a response spectrum or design spectrum, is summarized in step-by-step form:

1. Define the structural properties.
   a. Determine the mass and stiffness matrices from Eqs. (13.3.1) and (9.5.26).
   b. Estimate the modal damping ratios ζ_n (Chapter 11).
2. Determine the natural frequencies ω_n (natural periods $T_n = 2\pi/\omega_n$) and natural modes $\boldsymbol{\phi}_n$ of vibration (Chapter 10).
3. Compute the peak response in the nth mode by the following steps, to be repeated for all modes, $n = 1, 2, \ldots, 2N$:
   a. Corresponding to the natural period T_n and damping ratio ζ_n, read D_n and A_n, the deformation and pseudo-acceleration, from the earthquake response spectrum or the design spectrum.
   b. Compute the lateral displacements and rotations of the floors from Eq. (13.9.1a).
   c. Compute the equivalent static forces: lateral forces $\mathbf{f}_{yn}$ and torques $\mathbf{f}_{\theta n}$ from Eq. (13.9.2) or (13.9.3).
   d. Compute the story forces—shear, torque, and overturning moment—and element forces—bending moments and shears—by three-dimensional static analysis of the structure subjected to external forces $\mathbf{f}_{yn}$ and $\mathbf{f}_{\theta n}$. Alternatively, the element forces in the ith frame can be calculated by planar static analysis of this frame subjected to the lateral forces of Eq. (13.9.4).
4. Determine an estimate for the peak value r of any response quantity by combining the peak modal values r_n. The CQC rule for modal combination should be used because unsymmetric-plan buildings usually have pairs of closely spaced frequencies.

Usually, only the lower pairs of modes contribute significantly to the response. Therefore, steps 2 and 3 need to be implemented for only these modes and the double summations in the CQC rule truncated accordingly.

Example 13.12

Determine the peak values of the response of the one-story unsymmetric-plan system of Examples 13.7 and 10.6 with modal damping ratios $\zeta_n = 5\%$ to the El Centro ground motion in the y-direction, directly from the response spectrum for this ground motion.

Solution   Steps 1 and 2 of the procedure summary just presented have already been implemented in Example 10.6.

Step 3a: Corresponding to the known T_n and $\zeta_n = 5\%$, Fig. 6.6.4 gives the ordinates D_n and A_n. For $T_1 = 1.069$ sec: $D_1 = 4.256$ in. $= 0.3547$ ft and $A_1/g = 0.381$. For $T_2 = 0.9248$ sec: $D_2 = 4.161$ in. $= 0.3468$ ft and $A_2/g = 0.497$. (Obviously, numbers cannot be read to four significant figures from the response spectrum; they were obtained from the numerical data used in plotting Fig. 6.6.4; see also Fig. E13.8a.)

Step 3b: The peak values of roof displacement and rotation are obtained by specializing Eq. (13.9.1a) for the one-story system:

$$u_{yn} = \Gamma_n \phi_{yn} D_n \qquad u_{\theta n} = \Gamma_n \phi_{\theta n} D_n \tag{a}$$

Substituting numerical values for Γ_n, ϕ_{yn}, and $\phi_{\theta n}$ (from Example 13.7) in Eq. (a) with $n = 1$ gives the first-mode peak responses:

$$u_{y1} = (-0.974)(-0.5228)(0.3547) = 0.1806 \text{ ft} = 2.168 \text{ in.}$$

$$\frac{b}{2}u_{\theta 1} = \frac{30}{2}(-0.974)(0.0493)(0.3547) = -0.2555 \text{ ft} = -3.065 \text{ in.} \tag{b}$$

where $(b/2)u_{\theta 1}$ represents the lateral displacement at the edge of the plan due to floor rotation. Similarly, the second-mode peak responses are

$$u_{y2} = (-0.956)(-0.5131)(0.3468) = 0.1701 \text{ ft} = 2.042 \text{ in.}$$

$$\frac{b}{2}u_{\theta 2} = \frac{30}{2}(-0.956)(-0.0502)(0.3468) = 0.2497 \text{ ft} = 2.999 \text{ in.} \tag{c}$$

Step 3c: The peak values of f_{yn} and $f_{\theta n}$, the lateral force and torque, are obtained by specializing Eq. (13.9.3) for this one-story frame:

$$f_{yn} = \Gamma_n m \phi_{yn} A_n \qquad f_{\theta n} = \Gamma_n r^2 m \phi_{\theta n} A_n \tag{d}$$

By statics, the base shear and base torque are

$$V_{bn} = f_{yn} \qquad T_{bn} = f_{\theta n} \tag{e}$$

Alternatively, Eq. (13.7.1) can be used for computing the peak modal response. For example, the modal static responses V_{bn}^{st} and M_{bn}^{st} are available from Example 13.8. Substituting $V_{b1}^{st} = 0.509m$, $T_{b1}^{st} = -5.203m$, $m = 60$ kips/g, and $A_1 = 0.381g$ in Eq. (13.7.1) gives

$$V_{b1} = [0.509(60/g)]0.381g = 11.63 \text{ kips}$$

$$T_{b1} = [-5.203(60/g)](0.381g) = -118.8 \text{ kip-ft} \tag{f}$$

Substituting $V_{b2}^{st} = 0.491m$, $T_{b2}^{st} = 5.203m$, and $A_2 = 0.497g$ in Eq. (13.7.1) gives

$$V_{b2} = [0.491(60/g)]0.497g = 14.64 \text{ kips}$$

$$T_{b2} = [5.203(60/g)](0.497g) = 155.2 \text{ kip-ft} \tag{g}$$

Step 3d: The peak lateral force for frame A is given by Eq. (13.9.4b) specialized for a one-story frame:

$$f_{An} = k_A \Gamma_n (\phi_{yn} + x_A \phi_{\theta n}) D_n \qquad \text{(h)}$$

Substituting $k_A = 75$ kips/ft, $x_a = 1.5$ ft, and numerical values for Γ_n, ϕ_{yn}, $\phi_{\theta n}$, and D_n gives

$$f_{A1} = 75(-0.974)[-0.5228 + 1.5(0.0493)]0.3547 = 11.63 \text{ kips}$$

$$f_{A2} = 75(-0.956)[-0.5131 + 1.5(-0.0502)]0.3468 = 14.64 \text{ kips}$$

The base shear in a one-story frame is equal to the lateral force; thus

$$V_{bA1} = 11.63 \text{ kips} \qquad V_{bA2} = 14.64 \text{ kips} \qquad \text{(i)}$$

The lateral force for frame B is given by Eq. (13.9.4a) specialized for a one-story frame:

$$f_{Bn} = k_B \Gamma_n (-y_B \phi_{\theta n}) D_n \qquad \text{(j)}$$

Substituting $k_B = 40$ kips/ft, $y_B = 10$ ft, and numerical values for Γ_n, $\phi_{\theta n}$, and D_n gives

$$f_{B1} = 40(-0.974)[-10(0.0493)]0.3547 = 6.814 \text{ kips}$$

$$f_{B2} = 40(-0.956)[-10(-0.0502)]0.3468 = -6.662 \text{ kips}$$

The corresponding base shears are

$$V_{bB1} = 6.814 \text{ kips} \qquad V_{bB2} = -6.662 \text{ kips} \qquad \text{(k)}$$

The results for peak modal responses are presented in Table E13.12a.

TABLE E13.12a   PEAK MODAL RESPONSES

Mode	u_y (in.)	$b/2u_\theta$ (in.)	V_b (kips)	T_b (kip-ft)	V_{bA} (kips)	V_{bB} (kips)
1	2.168	−3.065	11.63	−118.8	11.63	6.814
2	2.042	2.999	14.64	155.2	14.64	−6.662

Step 4: For this system with two modes, the ABSSUM, SRSS, and CQC rules, Eqs. (13.7.2)–(13.7.4), specialize to

$$r \leq |r_1| + |r_2| \qquad r \simeq (r_1^2 + r_2^2)^{1/2} \qquad r \simeq (r_1^2 + r_2^2 + 2\rho_{12} r_1 r_2)^{1/2} \qquad \text{(l)}$$

For this system, $\beta_{12} = \omega_1/\omega_2 = 5.878/6.794 = 0.865$. For this value of β_{12} and $\zeta = 0.05$, Eq. (13.7.10) gives $\rho_{12} = 0.322$. The results from Eq. (l) are summarized in Table E13.12b, wherein the peak values of total responses determined by RHA are also included. These were computed using the results of Example 13.8, where $D_n(t)$ and $A_n(t)$ were computed by dynamic analysis of the nth-mode SDF system.

As expected, the ABSSUM estimate is always larger than the RHA value. The SRSS estimate is better, but the CQC estimate is the best because it accounts for the cross-correlation term in the modal combination, which is significant in this example because the natural frequencies are close, a situation common for unsymmetric-plan systems.

	u_y (in.)	$(b/2)u_\theta$ (in.)	V_b (kips)	T_b (kip-ft)	V_{bA} (kips)	V_{bB} (kips)
ABSSUM	4.210	6.064	26.27	274.0	26.27	13.48
SRSS	2.978	4.289	18.70	195.5	18.70	9.530
CQC	3.423	3.532	21.43	162.3	21.43	7.848
RHA	3.349	3.724	20.63	174.3	20.63	8.275

Example 13.13

Figure E13.13a–c shows a two-story building consisting of rigid diaphragms supported by three frames, A, B, and C. The lumped weights at the first and second floor levels are 120 and 60 kips, respectively. The lateral stiffness matrices of these frames, each idealized as a shear frame, are

$$\mathbf{k}_{yA} = \mathbf{k}_y = \begin{bmatrix} 225 & -75 \\ -75 & 75 \end{bmatrix} \qquad \mathbf{k}_{xB} = \mathbf{k}_{xC} = \mathbf{k}_x = \begin{bmatrix} 120 & -40 \\ -40 & 40 \end{bmatrix}$$

The design spectrum for $\zeta_n = 5\%$ is given by Fig. 6.9.5 scaled to 0.5g peak ground acceleration. Determine the peak value of the base shear in frame A.

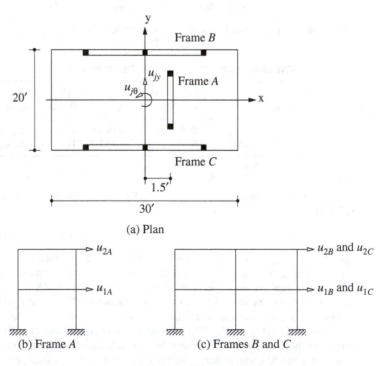

(a) Plan

(b) Frame A (c) Frames B and C

Figure E13.13a–c

Solution   This system has four DOFs: u_{yj} and $u_{\theta j}$ (Fig. E13.13a); $j = 1$ and 2. The stiffness matrix of Eqs. (9.5.25) and (9.5.26) is specialized for this system with three frames:

$$\mathbf{k} = \begin{bmatrix} \mathbf{k}_y & e\mathbf{k}_y \\ e\mathbf{k}_y & e^2\mathbf{k}_y + (d^2/2)\mathbf{k}_x \end{bmatrix}$$

Substituting for $\mathbf{k}_x$, $\mathbf{k}_y$, $e = 1.5$ ft, $d = 20$ ft, gives

$$\mathbf{k} = \begin{bmatrix} 225.0 & -75.00 & 337.5 & -112.5 \\ & 75.00 & -112.5 & 112.5 \\ & \text{(sym)} & 24{,}506 & -8169 \\ & & & 8169 \end{bmatrix}$$

The floor masses are $m_1 = 120/g = 3.727$ kip-sec^2/ft and $m_2 = 60/g = 1.863$ kip-sec^2/ft, and the floor moments of inertia are $I_{Oj} = m_j(b^2 + d^2)/12 = m_j(30^2 + 20^2)/12 = 1300m_j/12$. Substituting these data in the mass matrix of Eq. (9.5.27) gives

$$\mathbf{m} = \begin{bmatrix} 3.727 & & & \\ & 1.863 & & \\ & & 403.7 & \\ & & & 201.9 \end{bmatrix}$$

The eigenvalue problem is solved to determine the natural periods T_n and modes ϕ_n shown in Fig. E13.13d. Observe that each mode includes lateral and torsional motion. In the first mode the two floors displace in the same lateral direction and the two floors rotate in the same direction. In the second mode the two floors rotate in the same direction, which is

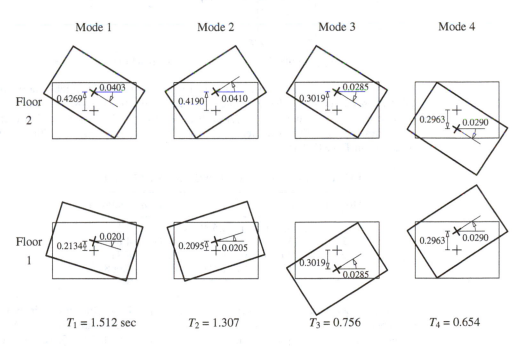

Figure E13.13d

opposite to the first mode. In the third and fourth modes the lateral displacements at the two floors are in opposite directions; the same is true for the rotations of the two floors.

The Γ_n are computed from Eqs. (13.3.4) to (13.3.6): $\Gamma_1 = 1.591$, $\Gamma_2 = 1.561$, $\Gamma_3 = -0.562$, and $\Gamma_4 = 0.552$.

For $T_n = 1.512, 1.307, 0.756$, and 0.654 sec, the design spectrum gives $A_1/g = 0.595$, $A_2/g = 0.688$, $A_3/g = 1.191$, and $A_4/g = 1.355$.

The peak values of the equivalent static lateral forces for frame A are [from Eq. (13.9.4b)]

$$\mathbf{f}_{An} = (\Gamma_n/\omega_n^2)\mathbf{k}_y(\boldsymbol{\phi}_{yn} + e\boldsymbol{\phi}_{\theta n})A_n$$

Substituting for Γ_1, $\omega_1(= 4.156)$, $\mathbf{k}_y$, A_1, $\boldsymbol{\phi}_{y1}$, and $\boldsymbol{\phi}_{\theta 1}$ gives the lateral forces associated with the first mode:

$$\left\{\begin{matrix} f_{A1} \\ f_{A2} \end{matrix}\right\}_1 = \frac{1.591}{(4.156)^2}(0.595 \times 32.2)\begin{bmatrix} 225 & -75 \\ -75 & -75 \end{bmatrix}\left(\left\{\begin{matrix} 0.2134 \\ 0.4269 \end{matrix}\right\} + 1.5\left\{\begin{matrix} -0.0201 \\ -0.0403 \end{matrix}\right\}\right) = \left\{\begin{matrix} 24.2 \\ 24.2 \end{matrix}\right\}$$

Static analysis of the frame subjected to these lateral forces (Fig. E13.13e) gives the internal forces. In particular, the base shear is $V_{bA1} = f_{11} + f_{21} = 48.4$ kips. Similar computations lead to the peak base shear due to the second, third, and fourth modes: $V_{bA2} = 53.9$, $V_{bA3} = 12.1$, and $V_{bA4} = 13.3$ kips.

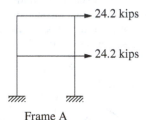

24.2 kips

24.2 kips

Frame A   **Figure E13.13e**

The peak value r of the total response $r(t)$ will be estimated by combining the peak modal responses according to the CQC rule, Eq. (13.7.4). For this purpose it is necessary to determine the frequency ratios $\beta_{in} = \omega_i/\omega_n$; these are given in Table E13.13a. For each of the β_{in} values the correlation coefficient ρ_{in} is computed from Eq. (13.7.10) with $\zeta = 0.05$ and presented in Table E13.3b.

TABLE E13.13a NATURAL FREQUENCY RATIOS β_{in}

Mode, i	$n = 1$	$n = 2$	$n = 3$	$n = 4$	ω_i (rad/sec)
1	1.000	0.865	0.500	0.433	4.157
2	1.156	1.000	0.578	0.500	4.804
3	2.000	1.730	1.000	0.865	8.313
4	2.312	2.000	1.156	1.000	9.608

Substituting the peak modal values V_{bAn} and the correlation coefficients ρ_{in} in the CQC rule, we obtain the 16 terms in the double summation of Eq. (13.7.4) (Table E13.13c). Adding the 16 terms and taking the square root gives $V_{bA} = 86.4$ kips. Table E13.13c shows that the terms with significant values are the $i = n$ terms, and the cross terms between modes 1 and 2

TABLE E13.13b   CORRELATION COEFFICIENTS ρ_{in}

Mode, i	$n = 1$	$n = 2$	$n = 3$	$n = 4$
1	1.000	0.322	0.018	0.012
2	0.322	1.000	0.030	0.018
3	0.018	0.030	1.000	0.322
4	0.012	0.018	0.322	1.000

TABLE E13.13c   INDIVIDUAL TERMS IN CQC RULE: BASE SHEAR V_{bA} IN FRAME A

Mode, i	$n = 1$	$n = 2$	$n = 3$	$n = 4$
1	2344.039	839.912	10.833	7.839
2	839.913	2905.669	19.748	13.250
3	10.833	19.748	146.502	51.797
4	7.839	13.250	51.797	176.807

and between modes 3 and 4. The cross terms between modes 1 and 3, 1 and 4, 2 and 3, or 2 and 4 are small because those frequencies are well separated. The square root of the sum of the four $i = n$ terms in Table E13.3c gives the SRSS estimate: $V_{bA} = 74.7$ kips. This is less accurate.

13.10 A RESPONSE-SPECTRUM-BASED ENVELOPE FOR SIMULTANEOUS RESPONSES

The seismic design of a structural element may be controlled by the simultaneous action of two or more responses. For example, a column in a three-dimensional frame must be designed to resist an axial force and bending moments about two axes that act concurrently and vary in time. We limit this section to two response quantities: $r_a(t)$ and $r_b(t)$; for a column $r_a(t)$ represents the bending moment $M(t)$ about a cross-sectional axis and $r_b(t)$ represents the axial force $P(t)$. The values of r_a and r_b at a time instant t are denoted by one point in the two-dimensional response space (Fig. 13.10.1a), and the response

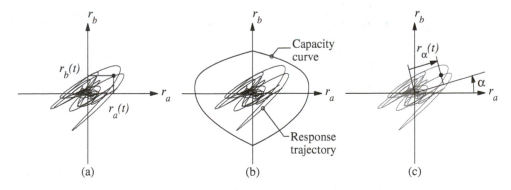

Figure 13.10.1

trajectory—i.e., combinations of r_a and r_b throughout the earthquake—shows how this pair of response quantities evolves in time. A structural element will be considered as designed adequately if the response trajectory stays within the safe region of the response space, defined by the relevant capacity curve for the element (Fig. 13.10.1b); e.g., the capacity curve for a column is commonly known as the *P-M* interaction diagram.

How can such a comparison between seismic demands and structural element capacity be achieved within the context of the response spectrum analysis (RSA) procedure presented in Section 13.7? The RSA procedure provides an estimate of the peak value of each response quantity, but these peaks generally do not occur at the same instant. Recognizing this fact, we present a set of response-spectrum-based equations to determine an envelope that bounds the response trajectory. This bounding envelope can then be compared to the capacity curve for the element to determine whether or not it is adequately designed.

The peak values of $r_a(t)$ and $r_b(t)$ estimated by the RSA procedure are identified as r_{ao} and r_{bo} in Fig. 13.10.2a. They can be interpreted as estimates of the peak values of the projections of the response trajectory on the r_a and r_b axes, respectively. Using these response-spectrum-based estimates of peak values, r_{ao} and r_{bo}, a rectangular "envelope" that "bounds" the response trajectory is shown in Fig. 13.10.2a[†]; visually this envelope seems overly conservative. It is indeed conservative because it requires that the structural element should be designed for all four combinations of the peak responses: $\pm r_{ao}$ and $\pm r_{bo}$, implying that the peaks occur at the same instant; both algebraic signs are considered because the sign is lost in the RSA procedure, which always gives a positive value. Our goal is to construct a tighter bounding envelope.

For this purpose, consider an arbitrary direction in the response space that is rotated α radians counterclockwise from the r_a axis (Fig. 13.10.1c). The projection of the response

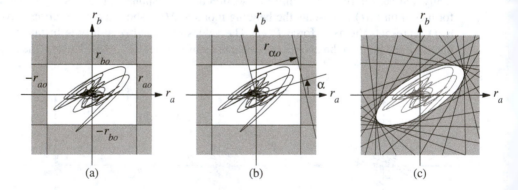

Figure 13.10.2

[†]The quotation marks are included to emphasize the fact that this envelope may not strictly bound the response trajectory because, as mentioned in Section 13.7, an RSA estimate of peak response may provide an underestimate relative to the exact value from RHA.

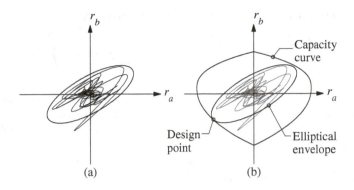

Figure 13.10.3

trajectory along this direction is given by[†]

$$r_\alpha(t) = r_a(t)\cos\alpha + r_b(t)\sin\alpha \qquad (13.10.1)$$

The peak value $r_{\alpha o}$ of $r_\alpha(t)$ can be estimated by the RSA procedure, wherein the required modal static responses $r_{\alpha n}^{\text{st}}$ can be determined from r_{an}^{st} and r_{bn}^{st} following Eq. (13.10.1):

$$r_{\alpha n}^{\text{st}} = r_{an}^{\text{st}}\cos\alpha + r_{bn}^{\text{st}}\sin\alpha \qquad (13.10.2)$$

As shown in Fig. 13.10.2b, the response trajectory is bounded in the direction α by $r_{\alpha o}$, just like it was bounded by r_{ao} and r_{bo} in directions $\alpha = 0°$ and $90°$, respectively.

We can repeat the preceding calculation to determine $r_{\alpha o}$ for several directions α and plot the resulting bounds (Fig. 13.10.2c), thus gradually tightening the envelope. The resulting envelope can be shown to be an ellipse (Fig. 13.10.3a) with its coordinates defined by

$$\begin{Bmatrix} r_a \\ r_b \end{Bmatrix} = \begin{Bmatrix} \left(r_{ao}^2\cos\alpha + r_{abo}\sin\alpha\right)/r_{\alpha o} \\ \left(r_{bo}^2\sin\alpha + r_{abo}\cos\alpha\right)/r_{\alpha o} \end{Bmatrix} \qquad 0 \le \alpha < 2\pi \text{ radians} \qquad (13.10.3)$$

where the cross term r_{abo} remains to be defined. In passing, we note that this cross term determines the orientation of the principal axes of the ellipse relative to the r_a and r_b axes of the response space. The coordinates given by Eq. (13.10.3) of a point on the elliptical envelope do not necessarily lie in direction α away from the origin of the response space; in fact, the direction is $\tan^{-1}(r_b/r_a)$. Recall that α defines the direction along which the response trajectory is projected. It can be shown that the elliptical envelope is inscribed within the conservative rectangular envelope.

Independent of the modal combination rule used to estimate peak responses, the preceding presentation is now specialized for the CQC and SRSS rules. Based on the CQC

[†]In general, r_α does not represent any physical response quantity; however, in some cases it does. For example, if r_a and r_b denote the bending moments in a column about two orthogonal axes a and b of its cross section, then r_α represents the bending moment about a cross-sectional axis rotated α radians counterclockwise from the a-axis.

modal combination rule, r_{ao}, r_{bo}, and r_{abo} are given by

$$r_{ao} \simeq \left(\sum_{i=1}^{N} \sum_{n=1}^{N} \rho_{in} r_{aio} r_{ano} \right)^{1/2} \qquad r_{bo} \simeq \left(\sum_{i=1}^{N} \sum_{n=1}^{N} \rho_{in} r_{bio} r_{bno} \right)^{1/2} \qquad r_{abo} \simeq \sum_{i=1}^{N} \sum_{n=1}^{N} \rho_{in} r_{aio} r_{bno}$$

$$(13.10.4)$$

where the first two are familiar equations (see Section 13.7) for peak values of responses $r_a(t)$ and $r_b(t)$, and the cross term defined by Eq. (13.10.4c) is similar in appearance to Eqs. (13.10.4a) and (13.10.4b), but it involves both responses; recall that $\rho_{in} = 1$ for $i = n$ (Section 13.7).

If the natural vibration frequencies of the structure are well separated, $\rho_{in} \simeq 0$ for $i \neq n$, and Eq. (13.10.4) simplifies to

$$r_{ao} \simeq \left(\sum_{n=1}^{N} r_{ano}^2 \right)^{1/2} \qquad r_{bo} \simeq \left(\sum_{n=1}^{N} r_{bno}^2 \right)^{1/2} \qquad r_{abo} \simeq \sum_{n=1}^{N} r_{ano} r_{bno} \qquad (13.10.5)$$

Equations (13.10.5a) and (13.10.5b) are the familiar equations (see Section 13.7) for peak values of responses $r_a(t)$ and $r_b(t)$ based on the SRSS modal combination rule, and Eq. (13.10.5c) is a special case of Eq. (13.10.4c).

Based on the RSA procedure, Eq. (13.10.3) combined with Eq. (13.10.4) or (13.10.5), as appropriate, defines the elliptical bounding envelope. Instead of the complete response trajectory (Fig. 13.10.1b), this envelope can be compared to the capacity curve of the structural element to determine whether or not it is adequately designed (Fig. 13.10.3b). In particular, an element is considered to be adequately designed if the elliptical envelope, which envelopes all combinations of the responses occurring during an earthquake, is within the safe region of the response space defined by the capacity surface of the element. It is often observed that the critical response combination, denoted by the design point, does not include the peak value of either of the two responses, as shown in Fig. 13.10.3b.

Researchers have developed a general theory leading to a bounding envelope for a set of more than two response quantities, e.g., axial force and bending moments about cross-sectional axes x and y acting on a structural element. In this case, the bounding envelope, which is an ellipse in the two-dimensional response space, generalizes to an ellipsoid in the three-dimensional response space. Researchers have also developed strategies for comparing three-dimensional response envelopes and capacity surfaces to judge whether or not a structural element is adequately designed.

Contribution of static forces.   In the preceding presentation of the elliptical envelope, only time-varying responses to earthquake excitation were considered. However, initial static (dead and live) loads acting on the structure cause time-invariant components r_{as} and r_{bs}. For linearly elastic systems, the initial static and time-varying components of the response can be added to yield the total response that varies in time around the starting point $<r_{as}, r_{bs}>$ in the $r_a - r_b$ response space (Fig. 13.10.4). The size and orientation of the elliptical envelope is unaffected by the static components, and the center of the envelope is translated from the origin of the response space to $<r_{as}, r_{bs}>$, as shown in Fig.13.10.4.

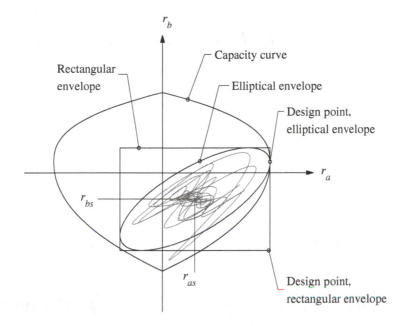

Figure 13.10.4

The coordinates of this translated envelope are

$$\begin{Bmatrix} r_a \\ r_b \end{Bmatrix} = \begin{Bmatrix} (r_{ao}^2 \cos\alpha + r_{abo}\sin\alpha)/r_{ao} \\ (r_{bo}^2 \sin\alpha + r_{abo}\cos\alpha)/r_{\alpha o} \end{Bmatrix} + \begin{Bmatrix} r_{as} \\ r_{bs} \end{Bmatrix} \qquad 0 \le \alpha < 2\pi \text{ radians} \quad (13.10.6)$$

It is clear from Fig. 13.10.4 that the critical design point, which is the point on the response envelope that governs the design of a structural element, based on the rectangular envelope lies outside the capacity curve, indicating that the structural element does not satisfy the design requirements. However, the elliptical envelope is encompassed by the capacity surface of the element, indicating that the element is adequately designed. Construction of an appropriate envelope is obviously important to arrive at the correct decision.

Example 13.14

Consider the inverted L-shaped frame of Fig. E13.1a subjected to the El Centro ground motion. Determine the response trajectory and bounding envelopes for two simultaneously acting forces at the base of the column: bending moment M_b and axial force P_b. Given $m = 50$ kips/g, $EI = 7.8 \times 10^6$ kip-in^2, $L = 12$ ft, and assume that $\zeta_n = 5\%$.

Solution Substituting the given values of m, EI, and L in Eq. (c) of Example 13.1 gives natural vibration frequencies $\omega_1 = 3.139$ and $\omega_2 = 8.420$ rad/s; the corresponding natural vibration periods are $T_1 = 2.0$ and $T_2 = 0.746$ sec.

Response History Analysis. Response analysis of the first-mode SDF system ($T_1 = 2$ sec and $\zeta_1 = 5\%$) and the second-mode SDF system ($T_2 = 0.746$ sec and $\zeta_2 = 5\%$) to the El Centro ground motion gives the $D_n(t)$ and $A_n(t)$ shown in Fig. E13.14a.

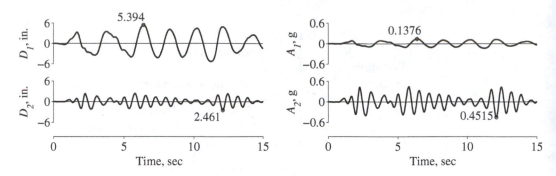

Figure E13.14a

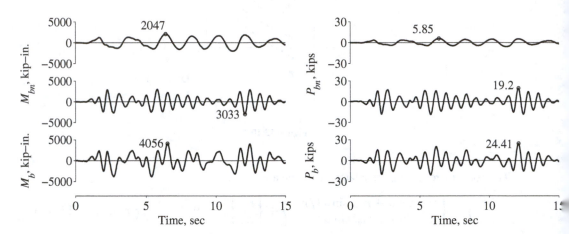

Figure E13.14b

The modal static responses M_{bn}^{st} and P_n^{st} are as follows: $M_{b1}^{st} = 2.069mL$ and $M_{b2}^{st} = 0.931mL$ (from Example 13.1); $P_{b1}^{st} = 0.851m$ and $P_{b2}^{st} = -0.851m$ (from Fig. E13.1). Substituting these in Eq. (13.1.16) gives the response history:

$$M_b(t) = 2.069mLA_1(t) + 0.931mLA_2(t)$$

$$P_b(t) = 0.851mA_1(t) - 0.851mA_2(t)$$

(a)

The base moment and axial force, computed from Eq. (a) using the known $A_n(t)$ from Fig. E13.14a, are shown in Fig. E13.14b. Presented are contributions of each mode separately and the combined, total responses. The peak values are $M_{bo} = 4056$ kip-in. and $P_{bo} = 24.41$ kips, as noted.

Plotting $M_b(t)$ and $P_b(t)$ together in the response space (Fig. E13.14c) provides the response trajectory. Observe that the peak values of the projections of the response trajectory on the M_b and P axes are equal to M_{bo} and P_{bo} determined above (Fig. E13.14c).

Response Spectrum Analysis. The response spectrum for the El Centro ground motion (Fig. 6.6.4) for 5% damping gives $A_1 = 0.1376g$ and $A_2 = 0.4515g$, corresponding to T_1 and T_2, respectively. These are the precise values for spectral ordinates, same as the peak values of $A_n(t)$ noted in Fig. E13.14a.

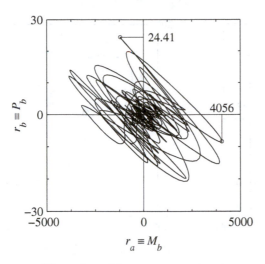

$$r_a \equiv M_b$$

Figure E13.14c

The peak modal responses are given by

$$M_{bno} = M_{bn}^{st} A_n \qquad P_{bno} = P_n^{st} A_n \qquad \text{(b)}$$

Substituting for A_n, the modal static responses, m, and L leads to the peak modal values for the bending moment at the base of the column:

$$M_{b1o} = 2.069mL(0.1376g) = 2047 \text{ kip-in.}$$
$$M_{b2o} = 0.931mL(0.4514g) = 3033 \text{ kip-in.}$$

and for the axial force in the column:

$$P_{b1o} = 0.851m(0.1376g) = 5.85 \text{ kips}$$
$$P_{b2o} = -0.851m(0.4514g) = -19.2 \text{ kips}$$

Because the natural frequencies of this system are well separated, we can combine the peak modal response by the SRSS rule:

$$M_{bo} = \sqrt{(2047)^2 + (3033)^2} = 3659 \text{ kip-in.}$$
$$P_{bo} = \sqrt{(5.85)^2 + (-19.2)^2} = 20.07 \text{ kips}$$

These estimates of the peak responses are shown in the $r_a - r_b$ response space in Fig. E13.14d. Because these estimates of the peak responses are smaller than the exact peak values from RHA (Fig. E13.14b), they do not strictly bound the response trajectory (Fig. E13.14d).

The modal static responses for r_α are computed from Eq. (13.10.2):

$$r_{\alpha n}^{st} = M_{bn}^{st} \cos \alpha + P_{bn}^{st} \sin \alpha$$

For $\alpha = 30°$, these modal static responses are

$$r_{\alpha 1}^{st} = 2.069mL \cos 30° + 0.851m \sin 30°$$
$$= (1.792L + 0.4255)m$$
$$r_{\alpha 2}^{st} = 0.931mL \cos 30° - 0.851m \sin 30°$$
$$= (0.8062L - 0.4255)m$$

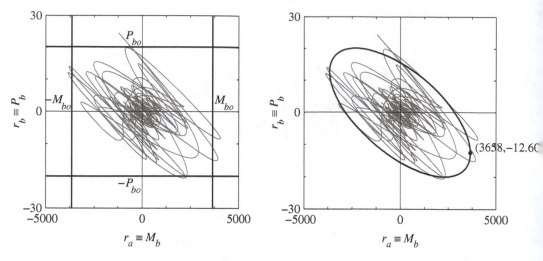

Figure E13.14d, e

Substituting these $r_{\alpha n}^{\text{st}}$ and the A_n values in $r_{\alpha n o} = r_{\alpha n}^{\text{st}} A_n$ gives the peak modal responses

$$r_{\alpha 1 o} = (1.792L + 0.4255)m(0.1376g) = 1776$$

$$r_{\alpha 2 o} = (0.8062L - 0.4255)m(0.4515g) = 2617$$

Combining these peak modal responses gives

$$r_{\alpha o} = \sqrt{(1776)^2 + (2617)^2} = 3163$$

Note that r_{α} does not represent a physical quantity, hence units are not shown.

Bounding Envelope   The bounding envelope is given by Eq. (13.10.3) where $r_{ao}, r_{bo},$ and $r_{\alpha o}$ have already been determined and r_{abo} is given by Eq. (13.10.5c):

$$r_{abo} = \sum_{n=1}^{N} r_{ano} r_{bno} = (2047)(5.85) + (3033)(-19.2)$$

$$= -46{,}260$$

With $r_{ao}, r_{bo}, r_{\alpha o},$ and r_{abo} known, calculating the coordinates of the bounding envelope is illustrated for $\alpha = 30°$:

$$r_a = [(3659)^2 \cos 30° + (-46260) \sin 30°]/3163 = 3658$$

$$r_b = [(20.07)^2 \sin 30° + (-46260) \cos 30°]/3163 = -12.60$$

These are the coordinates of one point on the bounding envelope (Fig. E13.14e). The elliptical envelope is bounded in the direction α by $r_{\alpha o} = 3163$. This bound will be tangential to the envelope at the point (3658, −12.60). However, this assertion cannot be verified in Fig. E13.14e because the scales along r_a and r_b axes are not the same, causing distortion of angles. Repeating such calculations for a large number of α values leads to the bounding ellipse (Fig. E13.14e).

13.11 PEAK RESPONSE TO MULTICOMPONENT GROUND MOTION

Ground motion at a location is defined by six components: translation along orthogonal axes, x, y, and z and rotations about these axes; the x and y axes are horizontal and the z-axis is vertical. We will restrict consideration to three translational components because these are the only components recorded during earthquakes. Let us temporarily assume that x, y, and z axes coincide with the axes defining the structure. The dynamic response of the structure within its linearly elastic range of behavior to one component of ground motion can be determined by the response history analysis (RHA) procedure (Chapter 13, Part A), and the combined response to three components acting simultaneously is obtained by superposition.

How can the combined response be determined within the context of the response spectrum analysis (RSA) procedure presented in Section 13.7? The RSA procedure provides an estimate of the peak values of the responses to individual components of excitation. Recognizing that these peaks generally do not occur simultaneously, we present procedures to combine the three individual peaks to estimate the peak response to multicomponent ground motion. For this purpose, it is necessary first to introduce the concept of principal axes of ground motion, which in turn requires new terminology and definitions.

13.11.1 Earthquake Excitation

Consider three orthogonal components—x, y, and z—of recorded ground acceleration $a_x(t)$,[†] $a_y(t)$, and $a_z(t)$ and define the 3×3 covariance matrix μ by its elements:

$$\mu_{ij} = \frac{1}{t_d} \int_0^{t_d} a_i(t) a_j(t) \, dt \qquad i, j = x, y, z \qquad (13.11.1)$$

where t_d is the duration of ground motion. The diagonal terms of this matrix represent the mean square intensities of the three components, and the off-diagonal terms represent cross correlations between pairs of components. In general, μ_{ij} for $i \neq j$ is nonzero and the components are said to be correlated.

Researchers have developed three concepts: First, ground motion can be transformed to a new orthogonal set of axes 1, 2, and 3 such that ground accelerations $a_1(t)$, $a_2(t)$, and $a_3(t)$ along these axes are uncorrelated; i.e., $\mu_{ij} = 0$, $i \neq j$, where $i, j = 1, 2, 3$; the corresponding covariance matrix is obviously a diagonal matrix. Second, these axes are defined as the *principal axes of ground motion*. Ordering the components starting from the largest mean square intensity to the lowest mean square intensity, i.e., $\mu_{11} > \mu_{22} > \mu_{33}$, $a_1(t)$ is called the major principal component, $a_2(t)$ the intermediate principal component, and $a_3(t)$ the minor principal component. Third, the major principal axis is horizontal and

---

[†]In this section ground acceleration is denoted by $a(t)$ instead of $\ddot{u}_g(t)$, to simplify notation.

directed roughly toward the epicenter of the earthquake; the intermediate principal axis is also horizontal but orthogonal to the first; and the minor principal axis is roughly vertical. This simple model is reasonable for ground motions recorded not too close to the causative fault.

The excitation is defined in terms of design spectra associated with the principal components of ground motion. The pseudo-acceleration spectra are denoted as $A(T_n)$ for the major principal component, $\gamma A(T_n)$ for the intermediate principal component, and $A_3(T_n)$ for the minor principal component. Note that the design spectra in the two horizontal directions have the same shape but differ by the factor γ, where $0 \leq \gamma \leq 1$, whereas the design spectrum $A_3(T_n)$ for vertical ground motion has a different shape, assumptions that are consistent with response spectra for recorded motions.

13.11.2 Structural Response-Incident Angle Relation

Figure 13.11.1 presents a schematic plan for a structure with two sets of axes: one (axes 1 and 2) defining the principal directions of ground motion, and the other (x and y axes) defining the structure; the third axis for the structure (z-axis) and the minor principal axis (axis 3) for the ground motion are both vertical. The relative orientation of the two coordinate systems is defined by the angle θ between the two sets of horizontal axes. Defined as the *seismic incident angle*, θ in the counterclockwise direction is taken to be positive. Components of ground motion along the structural axes are naturally correlated.

Consider a response quantity $r(t)$ that can be expressed as a linear combination of structural displacements. The CQC3 rule has been developed to estimate the peak value r_o of the combined response due to simultaneous application of the principal components of ground motion. This rule gives the peak response as a function of the seismic incident angle, presented without derivation:

$$r(\theta) \simeq \left\{ \left[r_x^2 + (\gamma r_y)^2 \right] \cos^2 \theta + \left[(\gamma r_x)^2 + r_y^2 \right] \sin^2 \theta + 2(1 - \gamma^2) r_{xy} \sin \theta \cos \theta + r_z^2 \right\}^{1/2}$$

$$(13.11.2)$$

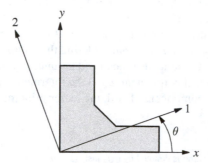

Figure 13.11.1   Relative orientation of two coordinate systems; structural axes x, y, and z; and principal axes 1, 2, and 3 of ground motion (vertical axes z and 3 not shown).

where the subscript "o" that denotes the peak value has been dropped, and r_x is the peak response to a single component of ground motion, defined by the spectrum $A(T_n)$ of the major principal component, applied along the structural axis x; r_y is the peak response to a single component of ground motion, defined by the spectrum $A(T_n)$, applied along the structural axis y; r_z equals the peak response to vertical ground motion defined by the spectrum $A_z(T_n)$. Note that r_y is not the actual response to the intermediate principal component of ground motion, which was defined by the spectrum $\gamma A(T_n)$; r_y equals the actual response divided by γ. If the intensities of the two horizontal components of ground motion are equal, i.e., $\gamma = 1$, it can readily be seen from Eq. (13.11.2) that the peak response is independent of the seismic incident angle θ. However, recorded ground motions do not support the hypothesis of $\gamma = 1$.

The peak response r_k to an individual component of ground motion is given by the CQC rule [see Eq. (13.7.4)]:

$$r_k \simeq \left(\sum_{i=1}^{N} \sum_{n=1}^{N} \rho_{in} r_{ik} r_{nk} \right)^{1/2} \qquad k = x, y, z \qquad (13.11.3)$$

where r_{nk} is the peak response due to the nth natural mode of vibration and ρ_{in} is the modal correlation coefficient for modes i and n; note that the first subscript in r_{nk} denotes the mode number and the second refers to the direction of application of ground motion. Rewriting Eq. (13.7.1),

$$r_{nk} = r_{nk}^{\text{st}} A_{nk} \qquad (13.11.4)$$

where r_{nk}^{st} is the nth modal static response and A_{nk} is the ordinate of the pseudo-acceleration spectrum at the nth-mode period and damping ratio, both associated with the k-direction of application of ground motion. The term r_{xy} in Eq. (13.11.2) is a cross term between modal responses contributing to r_x and r_y that arises from correlation between the ground motion components along the structural axes:

$$r_{xy} = \sum_{i=1}^{N} \sum_{n=1}^{N} \rho_{in} r_{ix} r_{ny} \qquad (13.11.5)$$

Note that r_{xy} involves the same terms that enter in Eq. (13.11.3) for peak responses r_x and r_y, but it involves modal responses to both components of ground motion.

If the natural vibration frequencies of modes contributing significantly to the response are well separated, the modal correlation coefficients $\rho_{in} \simeq 0$ for $i \neq n$, and r_k may be determined by the simpler SRSS modal combination rule [Eq. (13.7.3)]:

$$r_k \simeq \left(\sum_{n=1}^{N} r_{nk}^2 \right)^{1/2} \qquad (13.11.6)$$

Similarly, Eq. (13.11.5) simplifies to

$$r_{xy} \simeq \sum_{n=1}^{N} r_{nx} r_{ny} \tag{13.11.7}$$

where we have used the property that $\rho_{in} = 1$ for $i = n$.

13.11.3 Critical Response

The *critical response* r_{cr} is defined as the largest of the responses $r(\theta)$ for all possible seismic incident angles θ. It is of interest because in practical situations the location of the earthquake epicenter is not known; hence θ is unknown. Differentiating Eq. (13.11.2) with respect to θ and setting the derivative equal to zero gives the *critical incident angle*:

$$\theta_{cr} = \frac{1}{2} \tan^{-1}\left(\frac{2r_{xy}}{r_x^2 - r_y^2}\right) \tag{13.11.8}$$

This equation leads to two values of θ between 0 and π rad, corresponding to the maximum and minimum values of $r(\theta)$: r_{max} and r_{min}; the two values of θ are separated by $\pi/2$ rad. Observe that θ_{cr} is independent of the intensity ratio γ between the horizontal components of ground motion and is not influenced by the vertical component of ground motion.

Numerical values for r_{max} and r_{min} can be determined by substituting the two numerical values of θ_{cr} obtained from Eq. (13.11.8) into Eq. (13.11.2). It is possible, however, to derive explicit equations for r_{max} and r_{min}, and the one for maximum response is

$$r_{cr} \equiv r_{max} \simeq \left[(1+\gamma^2)\left(\frac{r_x^2 + r_y^2}{2}\right) + (1-\gamma^2)\sqrt{\left(\frac{r_x^2 - r_y^2}{2}\right)^2 + r_{xy}^2 + r_z^2} \right]^{1/2} \tag{13.11.9}$$

Thus the CQC3 estimate of the critical response has been expressed in terms of the peak responses r_x and r_y to the major principal component of ground motion, characterized by the design spectrum $A(T_n)$, applied along the x and y axes of the structure, respectively; the peak response r_z to the vertical component of ground motion, characterized by the design spectrum $A_z(T_n)$; and the cross term r_{xy} that arises from correlation between the ground motion components along the structural axes.

13.11.4 Other Multicomponent Combination Rules

If the principal axes of ground motion coincide with the structural axes, the response is given by Eq. (13.11.2) with $\theta = 0$ if the major principal axis is oriented in the x-direction and with $\theta = \pi/2$ rad if the major principal axis is oriented in the y-direction:

$$r(0) = \left[r_x^2 + (\gamma r_y)^2 + r_z^2\right]^{1/2} \qquad r(\pi/2) = \left[(\gamma r_x)^2 + r_y^2 + r_z^2\right]^{1/2} \tag{13.11.10}$$

According to these equations, the peak total response is given by the square-root-of-the-sum-of-squares of the peak responses to individual components of ground motion. Although the design spectrum for the y-component is less intense than that for the x-component by the factor $\gamma < 1$, r_y can be larger than r_x, depending on the relative values of the modal static responses associated with the two components; see Eq. (13.11.4). Therefore, the SRSS rule is defined as

$$r_{cr} \simeq \max[r\,(0)\,,\,r\,(\pi/2)] \tag{13.11.11}$$

If the two horizontal components of ground motion are equal in intensity, i.e., $\gamma = 1$, the CQC3 rule [Eq. (13.11.2)] reduces to the SRSS rule:

$$r_{cr} \simeq \left(r_x^2 + r_y^2 + r_z^2\right)^{1/2} \tag{13.11.12}$$

Note that Eq. (13.11.12) is the same as Eq. (13.11.11), specialized for $\gamma = 1$. Some building codes define the same design spectrum for both horizontal components of ground motion and specify the SRSS rule to determine the peak response to multicomponent excitation.

A percent rule to determine the peak response to multicomponent excitation also appears in some building codes. This rule is also based on the assumption that the principal axes of ground motion coincide with the structural axes and that both horizontal components of ground motion have the same intensity. According to this rule, r_{cr} is estimated as the sum of the response due to excitation in one direction and some fraction, α percent, of the responses due to excitations in the other two directions. Structural design is based on the combination that yields the largest estimate of the total response. Thus, three cases must be considered:

$$r_{cr} \simeq \max[r_x + \alpha r_y + \alpha r_z, \quad \alpha r_x + r_y + \alpha r_z, \quad \alpha r_x + \alpha r_y + r_z] \tag{13.11.13}$$

Some design codes specify $\alpha = 30\%$, whereas $\alpha = 40\%$ in other codes.

13.11.5 Example: Three-Dimensional System

The analysis procedures presented in the preceding sections are implemented for a simple three-dimensional system subjected to two horizontal components of ground motion, applied simultaneously. Peak values of selected response quantities are estimated by the various multicomponent combination rules. The results presented are accompanied by interpretive comments that should assist us in developing an understanding of these rules.

System and excitation.   The three-dimensional pipe of Fig. E13.11.2 is made of 3-in.-nominal diameter standard steel pipe. Its properties are $I = 3.017\,\text{in}^4$, $J = 6.034\,\text{in}^4$, $E = 30,000$ ksi, $G = 12,000$ ksi, $m = 1.0$ kips/g, and $L = 36$ in. This system is subjected to horizontal ground motion defined by its two principal components, characterized by $A(T_n)$ and $\gamma A(T_n)$, $0 \le \gamma \le 1$, respectively, where $A(T_n)$ is the design spectrum of Fig. 6.9.5 ($\zeta = 5\%$) scaled to 0.20g peak ground acceleration.

Response to individual components.   Response quantities selected are the bending moments about the x and y axes and torque at the clamped end a of the pipe. The peak responses due to a single component of ground motion characterized by the design spectrum $A\,(T_n)$ applied first along the x-axis and subsequently along the y-axis are

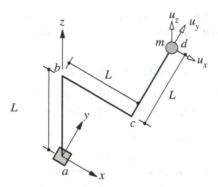

Figure E13.11.2

determined by the RSA procedure of Section 13.7. These calculations are not presented in detail (they are left to the reader as Problems 13.59 and 13.60), but the salient results needed to determine the response to multicomponent excitation are summarized next.

The modal correlation coefficients ρ_{in} are needed in Eq. (13.11.3) to estimate the peak response to individual components of ground motion. For this purpose it is necessary to compute the frequency ratios $\beta_{in} = \omega_i/\omega_n$ from the natural frequencies $\omega_n = 13.24$, 13.66, and 49.59 rad/sec. The corresponding values of ρ_{in}, computed from Eq. (13.7.10) with $\zeta = 0.05$, are presented in Table 13.11.1.

TABLE 13.11.1 MODAL CORRELATION COEFFICIENTS ρ_{in}

Mode, i	$n = 1$	$n = 2$	$n = 3$
1	1	0.909	0.006
2	0.909	1	0.007
3	0.006	0.007	1

The peak modal responses for M_x, M_y, and T_z together with the CQC estimate [Eq. (13.11.3)] of the peak value of the combined response are presented in Table 13.11.2.

TABLE 13.11.2 PEAK MODAL RESPONSES AND CQC ESTIMATE (KIP-IN.)

Excitation	x-direction			y-direction		
Mode	M_x	M_y	T_z	M_x	M_y	T_z
1	−1.509	−17.725	19.234	0.956	11.235	−12.191
2	−5.228	2.804	2.424	9.721	−5.214	−4.507
3	6.737	−4.591	−2.146	8.835	−6.021	−2.814
CQC	9.421	15.925	21.555	13.842	9.097	16.653

Cross term r_{xy}. Subsequent calculations are illustrated for M_x, i.e., $r \equiv M_x$. Table 13.11.3 gives $r_x = 9.421$ kips-in. and $r_y = 13.842$ kips-in. To determine the cross term defined by Eq. (13.11.5), the nine terms in the double summation, computed using the known values of ρ_{in}, r_{ix}, and r_{ny}, are presented in Table 13.11.3.

TABLE 13.11.3 INDIVIDUAL TERMS IN EQ. (13.11.5)

Mode, i	$n = 1$	$n = 2$	$n = 3$
1	−1.443	−13.333	−0.081
2	−4.543	−50.821	−0.305
3	0.039	0.432	59.521

Adding the nine terms gives $r_{xy} = -10.538$.

Response-incident angle relation. Now that r_x, r_y, and r_{xy} are known, they are substituted in Eq. (13.11.2) to obtain the response as a function of incident angle; e.g., for $\theta = \pi/4$ and $\gamma = 0.67$, $r\,(\pi/4) = 14.046$. Such calculations are repeated for many values of θ to obtain M_x as a function of θ for four values of γ. Similar calculations are implemented to obtain M_y and T_z as functions of θ.

The bending moments M_x and M_y, and torque T_z, estimated by the CQC3 rule, are presented in Fig. 13.11.3 for all possible orientations θ of the principal axes of ground motion, relative to the structural axes. These results for four values of γ permit the following observations: (1) the CQC3 estimate of peak response varies moderately with θ, increasingly so for the smaller values of γ; and (2) for the limiting case of $\gamma = 1$, the CQC3 estimate is independent of θ, a numerical confirmation of the earlier observation from Eq. (13.11.2).

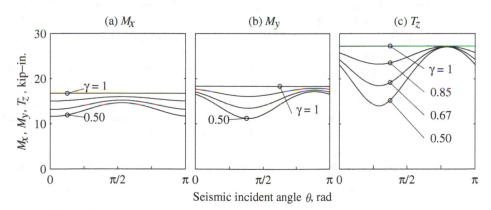

Figure 13.11.3 Peak responses M_x, M_y, and T_z for all orientations of principal axes of ground motion.

Critical response. Substituting $r_x = 9.421$, $r_y = 13.842$, $r_{xy} = -10.538$, and $\gamma = 0.67$ in Eq. (13.11.9) leads to $r_{cr} = 15.232$ kips-in. This is the critical value of M_x (Fig. 13.11.3); i.e., the largest value of M_x over all possible seismic incident angles, θ; the critical incident angle $\theta_{cr} = 95.8°$ is calculated from Eq. (13.11.8).

Comparison of CQC3 and SRSS rules. Estimates of the critical values of M_x, M_y, and T_z obtained from CQC3 and SRSS rules are presented in Table 13.11.4 for four values of γ. The CQC3 estimate is obtained, as illustrated earlier for M_x for $\gamma = 0.67$. The SRSS estimate is obtained using Eqs. (13.11.10) and (13.11.11), as is illustrated next for response M_x, i.e., $r \equiv M_x$. Substituting $r_x = 9.421$, $r_y = 13.842$, and $\gamma = 0.67$ in Eq. (13.11.10) gives $r(0) = 13.220$ and $r(\pi/2) = 15.213$; according to Eq. (13.11.11), $r_{cr} = \max(13.220, 15.213) = 15.213$ kip-in.

TABLE 13.11.4   SRSS AND CQC3 ESTIMATES OF CRITICAL RESPONSE

	M_x(kip-in.)		M_y(kip-in.)		T_z(kip-in.)	
Intensity Ratio γ	SRSS	CQC3	SRSS	CQC3	SRSS	CQC3
0.5	14.62	14.65	16.56	17.16	23.11	27.05
0.67	15.21	15.23	17.05	17.48	24.27	27.10
0.85	15.99	16.00	17.70	17.91	25.79	27.17
1	16.74	16.74	18.34	18.34	27.24	27.24

Observe that the SRSS estimate is smaller than the CQC3 estimate for all response quantities and for all values of $\gamma < 1$; for the limit case of $\gamma = 1$, the two estimates are identical.

Comments on SRSS and percent rules in building codes. We first illustrate estimation of the critical value of response M_x according to the SRSS and percent rules specified in building codes, which are based on the assumption that $\gamma = 1$. Substituting $r_x = 9.421$ and $r_y = 13.842$ (and dropping the r_z term) in Eq. (13.11.12) gives the SRSS estimate $r_{cr} = 16.744$ kip-in. According to the percent rules, the critical value of response is estimated from Eq. (13.11.13). Substituting $r_x = 9.421$ and $r_y = 13.842$, the 30% rule gives $r_{cr} = \max(16.668, 13.573) = 16.668$ kip-in., and the 40% rule gives $r_{cr} = \max(17.610, 14.958) = 17.610$ kip-in.

The critical values of M_x, M_y, and T_z estimated by the SRSS and percent rules in building codes are presented in Table 13.11.5, together with the CQC3 estimate (from Table 13.11.4); the latter varies with γ. Comparing these estimates permits the following

TABLE 13.11.5   BUILDING CODE AND CQC3 ESTIMATES OF CRITICAL RESPONSE

	M_x(kip-in.)				M_y(kip-in.)				T_z(kip-in.)			
γ	CQC3	SRSS	40% Rule	30% Rule	CQC3	SRSS	40% Rule	30% Rule	CQC3	SRSS	40% Rule	30% Rule
0.5	14.65				17.16				27.05			
0.67	15.23				17.48				27.10			
0.85	16.00				17.91				27.17			
1.0	16.74	16.74	17.61	16.67	18.34	18.34	19.56	18.65	27.24	27.24	28.22	26.55

observations: For all three response quantities the SRSS estimate (with $\gamma = 1$) of the critical value of response exceeds the CQC3 estimate for all values of γ, except $\gamma = 1$ when the two are identical. This overestimation increases as γ decreases. The preceding comments also apply to the estimate of critical response determined by the 40% rule, which is even more conservative than the SRSS rule. In contrast, the 30% rule may or may not be conservative, depending on the response quantity of interest and on γ.

13.11.6 A Response-Spectrum-Based Envelope for Simultaneous Responses

In Section 13.10 we presented a response-spectrum-based envelope for two or more responses of a structure excited by one component of ground motion. Researchers have developed such an elliptical envelope for multicomponent excitation when the principal directions of ground motion are known. For the case where the principal directions of ground motion are not known in advance, researchers have developed a *supreme envelope* that represents the union of the elliptical envelopes for all orientations of the principal axes. Comparing this supreme envelope for a structural element against its capacity surface to determine whether or not the element is adequately designed is even more challenging than was alluded to in Section 13.10, and strategies have been developed for this purpose.

FURTHER READING

Anastassiadis, K., Avramidis, I. E., and Panetsos, P., "Concurrent Design Forces in Structures Under Three-Component Orthotropic Seismic Excitation," *Earthquake Spectra*, **18**, 2002, pp. 1–17.

De la Llera, J. C., and Chopra, A. K., "Evaluation of Code Accidental Torsional Provisions from Building Records," *Journal of Structural Engineering*, ASCE, **120**, 1994, pp. 597–616.

Der Kiureghian, A., "A Response Spectrum Method for Random Vibration Analysis of MDF Systems," *Earthquake Engineering and Structural Dynamics*, **9**, 1981, pp. 419–435.

Gupta, A. K., *Response Spectrum Method in Seismic Analysis and Design of Structures*, Blackwell, Cambridge, Mass, 1990.

Gupta, A. K., and Singh, M. P., "Design of Column Sections Subjected to Three Components of Earthquake," *Nuclear Engineering and Design,* **41**, 1977, pp. 129–133.

Lopez, O. A., Chopra, A. K., and Hernandez, J. J., "Critical Response of Structures to Multicomponent Earthquake Excitation," *Earthquake Engineering and Structural Dynamics*, **29**, 2000, pp. 1759–1778.

Lopez, O. A., Chopra, A. K., and Hernandez, J. J., "Evaluation of Combination Rules for Maximum Response Calculation in Multicomponent Seismic Analysis," *Earthquake Engineering and Structural Dynamics*, **30**, 2001, pp. 1379–1398.

Menun, C., "Strategies for Identifying Critical Response Combinations," *Earthquake Spectra*, **20**, 2004, pp. 1139–1165.

Menun, C., and Der Kiureghian, A., "A Replacement for the 30%, 40%, and SRSS Rules for Multicomponent Seismic Analysis," *Earthquake Spectra*, **14**, 1998, pp. 153–156.

Menun, C., and Der Kiureghian, A., "Envelopes for Seismic Response Vectors I: Theory," *Journal of Structural Engineering*, ASCE, **126**, 2000, pp. 467–473.

Menun, C., and Der Kiureghian, A., "Envelopes for Seismic Response Vectors II: Application," *Journal of Structural Engineering, ASCE,* **126**, 2000, pp. 474–481.

Newmark, N. M., and Rosenblueth E., *Fundamentals of Earthquake Engineering,* Prentice Hall, Englewood Cliffs, N.J., 1971, pp. 308–312.

Penzien, J., and Watabe, M., "Characteristics of 3-Dimensional Earthquake Ground Motion, *Earthquake Engineering and Structural Dynamics,* **3**, 1975, pp. 365–374.

Rosenblueth, E., "A Basis for Aseismic Design," Ph.D. thesis, University of Illinois, Urbana, Ill., 1951.

Rosenblueth, E., and Contreras, H., "Approximate Design for Multicomponent Earthquakes," *Journal of Engineering Mechanics,* ASCE, **103**, 1977, pp. 895–911.

Rosenblueth, E., and Elorduy, J., "Responses of Linear Systems to Certain Transient Disturbances," *Proceedings of the 4th World Conference on Earthquake Engineering,* Santiago, Chile, Vol. I, 1969, pp. 185–196.

PROBLEMS

Part A: Sections 13.1–13.4

13.1   For the two-story shear frame of Fig. P13.1 (also of Problems 9.5 and 10.6) excited by horizontal ground motion $\ddot{u}_g(t)$, determine **(a)** the modal expansion of effective earthquake forces, **(b)** the floor displacement response in terms of $D_n(t)$, **(c)** the story shear response in terms of $A_n(t)$, and **(d)** the first-floor and base overturning moments in terms of $A_n(t)$.

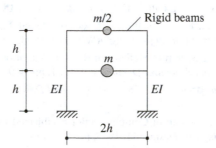

Figure P13.1

*13.2   The response of the two-story shear frame of Fig. P13.1 (also of Problems 9.5 and 10.6) to El Centro ground motion is to be computed as a function of time. The properties of the frame are $h = 12$ ft, $m = 100$ kips/g, $I = 727$ in^4, $E = 29{,}000$ ksi, and $\zeta_n = 5\%$. The ground acceleration data are available in Appendix 6 at every $\Delta t = 0.02$ sec.

(a) Determine the SDF system responses $D_n(t)$ and $A_n(t)$ using a numerical time-stepping method of your choice with an appropriate Δt; plot $D_n(t)$ and $A_n(t)$.

(b) For each natural mode calculate as a function of time the following response quantities: (i) the displacements at each floor, (ii) the story shears, and (iii) the floor and base overturning moments.

*Denotes that a computer is necessary to solve this problem.

(c) At each instant of time, combine the modal contributions to each of the response quantities to obtain the total response; determine the peak value of the total responses. For selected response quantities plot as a function of time the modal responses and total response.

13.3   Determine the effective modal masses and effective modal heights for the two-story shear frame of Fig. P13.1 (also of Problems 9.5 and 10.6); the height of each story is h. Display this information on the SDF systems for the modes. Verify that Eqs. (13.2.14) and (13.2.17) are satisfied.

*13.4   Figure P13.4 shows a two-story frame (the same as that in Problems 9.6 and 10.10) with flexural rigidity EI for beams and columns. Determine the dynamic response of this structure to horizontal ground motion $\ddot{u}_g(t)$. Express (a) the floor displacements and joint rotations in terms of $D_n(t)$, and (b) the bending moments in a first-story column and in the second-floor beam in terms of $A_n(t)$.

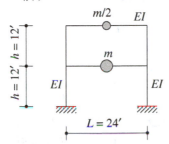

Figure P13.4

13.5–   For the three-story shear frames of Figs. P13.5–P13.6 (also of Problems 9.7–9.8 and 10.11–
13.6   10.12) excited by horizontal ground motion $\ddot{u}_g(t)$, determine (a) the modal expansion of effective earthquake forces, (b) the floor displacement response in terms of $D_n(t)$, (c) the story shear response in terms of $A_n(t)$, and (d) the base overturning moment in terms of $A_n(t)$.

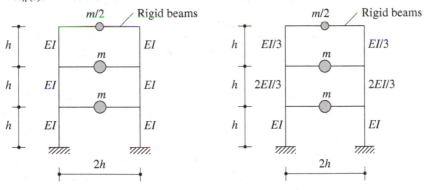

Figure P13.5   **Figure P13.6**

*13.7–   The response of the three-story shear frames of Figs. P13.5–P13.6 (also of Problems 9.7–9.8
13.8   and 10.11–10.12) to El Centro ground motion is to be computed as a function of time. The properties of the frame are $h = 12$ ft, $m = 100$ kips/g, $I = 1400$ in^4, $E = 29{,}000$ ksi, and $\zeta_n = 5\%$. The ground acceleration data are available in Appendix 6 at every $\Delta t = 0.02$ sec.

*Denotes that a computer is necessary to solve this problem.

(a) Determine the SDF system responses $D_n(t)$ and $A_n(t)$ using a numerical time-stepping method of your choice with an appropriate Δt; plot $D_n(t)$ and $A_n(t)$.

(b) For each natural mode, calculate as a function of time the following response quantities: (i) the roof displacement, (ii) the story shears, and (iii) the base overturning moment.

(c) At each instant of time combine the modal contributions to each of the response quantities to obtain the total response; determine the peak value of the total responses. For selected response quantities, plot as a function of time the modal responses and total response.

13.9–   Determine the effective modal masses and effective modal heights for the three-story shear
13.10   frames of Figs. P13.5–P13.6; the height of each story is h. Display this information on the SDF systems for the modes. Verify that Eqs. (13.2.14) and (13.2.17) are satisfied.

*13.11– Figures P13.11–P13.14 show three-story frames (the same as those in Problems 9.9–9.12 and
13.14   10.19–10.22) together with flexural rigidity for beams and columns. Determine the dynamic

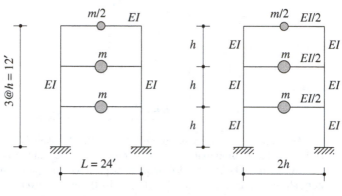

Figure P13.11   Figure P13.12

response of this three-story frame to horizontal ground motion $\ddot{u}_g(t)$. Express (a) the floor displacements and joint rotations in terms of $D_n(t)$, and (b) the bending moments in a first-story column and in the second-floor beam in terms of $A_n(t)$.

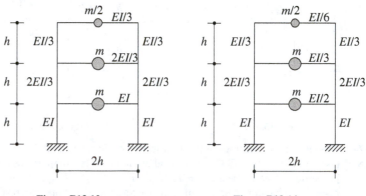

Figure P13.13   Figure P13.14

*Denotes that a computer is necessary to solve this problem.

13.15   For the inverted L-shaped frame of Fig. E9.6a excited by vertical ground motion $\ddot{u}_g(t)$, determine (a) the modal expansion of effective earthquake forces, (b) the displacement response in terms of $D_n(t)$, and (c) the bending moment at the base of the column in terms of $A_n(t)$.

13.16   Solve Problem 13.15 for the ground motion shown in Fig. P13.16.

*13.17   For the umbrella structure of Fig. P13.17 (also of Problems 9.13 and 10.23) excited by horizontal ground motion $\ddot{u}_g(t)$, determine (a) the modal expansion of effective earthquake forces, (b) the displacement response in terms of $D_n(t)$, and (c) the bending moments at the base of the column and at location a of the beam in terms of $A_n(t)$.

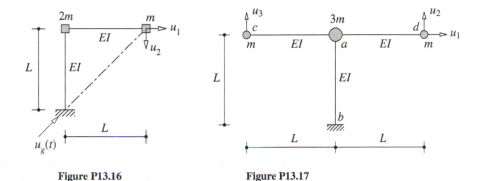

Figure P13.16   Figure P13.17

*13.18   Solve Problem 13.17 for vertical ground motion.

*13.19   Solve Problem 13.17 for ground motion in the direction b–d.

*13.20   Solve Problem 13.17 for ground motion in the direction b–c.

*13.21   Solve Problem 13.17 for rocking ground motion in the plane of the structure.

*13.22   A cantilever tower is shown in Fig. P13.22 with three lumped masses and its flexural stiffness properties; $m = 0.486$ kip-sec^2/in., $EI/L^3 = 56.26$ kips/in., and $EI'/L^3 = 0.0064$ kip/in. Note that the top mass and its supporting element are an appendage to the main tower. Damping is defined by modal damping ratios, with $\zeta_n = 5\%$ for all modes.

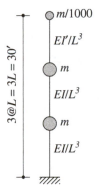

Figure P13.22

*Denotes that a computer is necessary to solve this problem.

(**a**) Determine the natural vibration periods and modes; sketch the modes.

(**b**) Expand the effective earthquake forces into their modal components and show this expansion graphically.

(**c**) Compute the modal static responses for three quantities: (i) the displacement of the appendage mass, (ii) the shear force at the base of the appendage, and (iii) the shear force at the base of the tower.

(**d**) What can you predict about the relative values of modal contributions to each of the response quantity from the results of parts (a) and (c)?

*13.23   The response of the tower with appendage of Fig. P13.22 to El Centro ground motion is to be computed as a function of time. The ground acceleration is available in Appendix 6 at every $\Delta t = 0.02$ sec. The damping of the structure is defined by the modal damping ratios $\zeta_n = 5\%$ for all modes.

(**a**) Determine the SDF system responses $D_n(t)$ and $A_n(t)$ using a numerical time-stepping method of your choice with an appropriate Δt.

(**b**) For each vibration mode calculate and plot as a function of time the following response quantities: (i) the displacement of the appendage mass, (ii) the shear force in the appendage, and (iii) the shear force at the base of the tower. Determine the peak value of each modal response.

(**c**) Calculate and plot as a function of time the total values of the three response quantities determined in part (b); determine the peak values of the total responses.

(**d**) Compute the seismic coefficients (defined as the shear force normalized by the weight) for the appendage and the tower. Why is the seismic coefficient for the appendage much larger than for the tower?

13.24   The one-story, unsymmetric-plan system of Fig. P13.24 (the same as that defined in Problem 9.14 for which the natural vibration frequencies and modes were to be determined in Problem 10.24) is excited by ground motion $\ddot{u}_{gy}(t)$ in the y-direction. Formulate the equations of motion for this 3DOF system and:

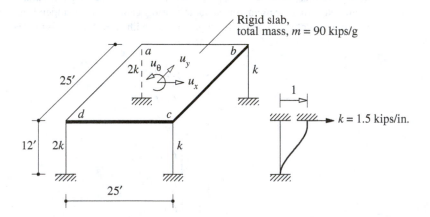

Figure P13.24

*Denotes that a computer is necessary to solve this problem.

(a) Expand the effective earthquake forces in terms of their modal components and show this expansion graphically.

(b) Verify that Eq. (13.3.9), generalized from a 2N-DOF system to a 3N-DOF system, is satisfied.

(c) Determine the displacement u_y and rotation u_θ of the slab in terms of $D_n(t)$.

(d) Determine the base shear and base torque in terms of $A_n(t)$.

13.25   The one-story, unsymmetric-plan system of Fig. P13.24 (the same as that defined in Problem 9.14 for which the natural vibration frequencies and modes were to be determined in Problem 10.24) is excited by ground motion $\ddot{u}_g(t)$ along the diagonal d–b. Formulate the equations of motion for this 3DF system and:

(a) Expand the effective earthquake forces in terms of their modal components and show this expansion graphically.

(b) Verify that Eq. (13.3.9), generalized from a 2N-DOF system to a 3N-DOF system, is satisfied.

(c) Determine the displacement u_y and rotation u_θ of the slab in terms of $D_n(t)$.

(d) Determine the x and y components of the base shear and base torque in terms of $A_n(t)$.

*13.26   The response history of the system of Problem 13.24 (the same as that in Problem 9.14 for which the natural vibration frequencies and modes were to be determined in Problem 10.24) to El Centro ground motion along the y-direction is to be determined. In addition to the system properties given in Fig. P13.24, $\zeta_n = 5\%$ for all natural vibration modes. The ground acceleration is available in Appendix 6 at every $\Delta t = 0.02$ sec.

(a) Determine the SDF system responses $D_n(t)$ and $A_n(t)$ using a numerical time-stepping method of your choice with an appropriate Δt; plot $D_n(t)$ and $A_n(t)$.

(b) For each vibration mode, calculate and plot as a function of time the following response quantities: u_y, $b/2u_\theta$, base shear V_b, and base torque T_b.

(c) Calculate and plot as a function of time the total responses; determine the peak values of the total responses.

13.27   For the system of Fig. P13.27 (also of Problems 9.18 and 10.28), which is subjected to ground motion in the x-direction: (a) expand the effective earthquake forces in terms of their modal components and show this expansion graphically; (b) determine the displacements u_x, u_y, and u_z of the mass in terms of $D_n(t)$; and (c) determine the bending moments about the x and y axes and the torque at the clamped end a in terms of $A_n(t)$.

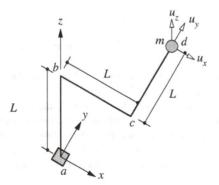

Figure P13.27

*Denotes that a computer is necessary to solve this problem.

13.28   Solve Problem 13.27 for ground motion in the y-direction.

13.29   Solve Problem 13.27 for ground motion in the z-direction.

13.30   Solve Problem 13.27 for ground motion in the direction a–d.

Part A: Section 13.5

13.31   The system of Fig. P13.31 (and of Problem 9.19) is subjected to support motions $u_{g1}(t)$ and $u_{g2}(t)$. Determine the motion of the two masses as a function of time for two excitations: (a) $u_{g1}(t) = -u_{g2}(t) = u_g(t)$, and (b) $u_{g2}(t) = u_{g1}(t) = u_g(t)$; express all results in terms of $D_n(t)$, the deformation response of the nth-mode SDF system to $\ddot{u}_g(t)$. Comment on how the response to the two excitations differs and why.

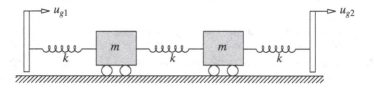

Figure P13.31

13.32   The undamped system of Fig. P13.32 (and of Problem 9.20), with $L = 50$ ft, $m = 0.2$ kip-sec²/in., and $EI = 5 \times 10^8$ kip-in², is subjected to support motions $u_{g1}(t)$ and $u_{g2}(t)$. Determine the steady-state motion of the lumped mass and the steady-state value of the bending moment at the midspan due to two harmonic excitations: (i) $u_{g1}(t) = u_{go} \sin \omega t$, $u_{g2}(t) = 0$; and (ii) $u_{g1}(t) = u_{g2}(t) = u_{go} \sin \omega t$. The excitation frequency ω is $0.8\omega_n$, where ω_n is the natural vibration frequency of the system. Express your results in terms of u_{go}. Comment on (a) the relative contributions of the quasi-static and dynamic components in each response quantity due to each excitation case, and (b) how the responses to the two excitations differ and why.

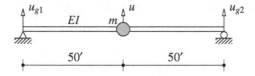

Figure P13.32

***13.33**   The equations governing the motion of the system in Fig. P9.21 due to support motions were formulated in Problem 9.21.

(a) Support a undergoes motion $u_g(t)$ in the x-direction and support b undergoes the same motion, but t' seconds later. Determine the following responses as a function of time: (i) the displacements u_1 and u_2 of the valves, and (ii) the bending moments at a, b, c, d, and e. Express the displacements in terms of $u_g(t)$ and $D_n(t)$, and forces in terms of $u_g(t)$ and $A_n(t)$, where $D_n(t)$ and $A_n(t)$ are the deformation and pseudo-acceleration responses of the nth-mode SDF system to $\ddot{u}_g(t)$.

*Denotes that a computer is necessary to solve this problem.

(b) Compare the preceding results with the response of the system if both supports undergo identical motion $u_g(t)$. Comment on how the responses in the two cases differ and why.

*13.34 The equations governing the motion of the system in Fig. P9.22 due to spatially varying ground motion in the x-direction were formulated in Problem 9.22.

(a) Support a undergoes motion $u_g(t)$ in the x-direction and support b undergoes the same motion, but t' seconds later. Determine the following responses as a function of time: (i) the displacements u_1 and u_2, and (ii) the bending moments at a, b, c, d, and e. Express the displacements in terms of $u_g(t)$ and $D_n(t)$, and forces in terms of $u_g(t)$ and $A_n(t)$, where $D_n(t)$ and $A_n(t)$ are the deformation and pseudo-acceleration response of the nth-mode SDF system to $\ddot{u}_g(t)$.

(b) Compare the preceding results with the response of the system if both supports undergo identical motion $u_g(t)$. Comment on how the responses in the two cases differ and why.

*13.35 For the system defined in Problem 9.23, the equations governing its motion due to spatially varying ground motion in the x-direction were formulated in Problem 9.23.

(a) The supports of columns a and b undergo motion $u_g(t)$ in the x-direction and the supports of columns c and d undergo the same motion, but t' seconds later. Determine the following responses as a function of time: (i) the displacements u_x, u_y, and u_θ of the roof slab, and (ii) the shear in each column. Express the displacements in terms of $u_g(t)$ and $D_n(t)$, and forces in terms of $u_g(t)$ and $A_n(t)$, where $D_n(t)$ and $A_n(t)$ are the deformation and pseudo-acceleration response of the nth-mode SDF system to $\ddot{u}_g(t)$.

(b) Compare the preceding results with the response of the structure if all column supports undergo identical motion $u_g(t)$. Comment on how the responses in the two cases differ and why.

*13.36 For the system of Fig. P13.24 the natural vibration frequencies and modes were determined in Problem 10.24, and the equations governing its motion due to spatially varying ground motion in the x-direction were formulated in Problem 9.24.

(a) The supports of columns a and b undergo motion $u_g(t)$ in the x-direction and the supports of columns c and d undergo the same motion, but t' seconds later. Determine the following responses as a function of time: (i) the displacements u_x, u_y, and u_θ of the roof slab, and (ii) the shear in each column. Express the displacements in terms of $u_g(t)$ and $D_n(t)$, and forces in terms of $u_g(t)$ and $A_n(t)$, where $D_n(t)$ and $A_n(t)$ are the deformation and pseudo-acceleration response of the nth-mode SDF system to $\ddot{u}_g(t)$.

(b) Compare the preceding results with the response of the structure if all column supports undergo identical motion $u_g(t)$. Comment on how the responses in the two cases differ and why.

*13.37 Implement numerically the solution to Problem 13.36 for El Centro ground motion with $t' = 0.1$ sec. In addition to the system properties given in Fig. P13.24, $\zeta_n = 5\%$ for all modes. The ground acceleration is available in Appendix 6 at every $\Delta t = 0.02$ sec. Plot (i) $D_n(t)$, and (ii) the modal contributions and the total response for each response quantity. Determine the peak values of the total response. Comment on the influence of spatial variations in the excitation.

*13.38 **(a)** In the intake tower of Problem 9.25, the base of the tower undergoes horizontal motion $u_g(t)$, and the right end of the bridge undergoes the same motion as the base, but it does so t' seconds later. Determine the following responses as a function of time: (i) the displacement at the top of the tower, (ii) the shear and bending moment at the tower base, and (iii) the

*Denotes that a computer is necessary to solve this problem.

axial force in the bridge. Express the displacements in terms of $D_n(t)$ and forces in terms of $A_n(t)$, where $D_n(t)$ and $A_n(t)$ are the deformation and pseudo-acceleration responses of the nth-mode SDF system to $\ddot{u}_g(t)$.

(b) Compare the preceding results with the response of the tower if both supports undergo identical motion $u_g(t)$. Comment on how the responses in the two cases differ and why.

Part B

*13.39   Figure P13.4 shows a two-story frame (the same as that in Problems 9.6 and 10.10) with $m = 100$ kips/g, $I = 727$ in^4 for beams and columns, and $E = 29,000$ ksi. Determine the response of this frame to ground motion characterized by the design spectrum of Fig. 6.9.5 (for 5% damping) scaled to $\frac{1}{3}$g peak ground acceleration. Compute (a) the floor displacements, and (b) the bending moments in a first-story column and in the second-floor beam.

13.40   The two-story shear frame of Fig. P13.1 (also of Problems 9.5 and 10.6) has the following properties: $h = 12$ ft, $m = 100$ kips/g, $I = 727$ in^4 for columns, $E = 29,000$ ksi, and $\zeta_n = 5\%$. The peak response of this structure to El Centro ground motion is to be estimated by response spectrum analysis (RSA) and compared with the results of Problem 13.2 from response history analysis (RHA). For the purposes of this comparison the RSA is to be implemented as follows.

(a) Determine the spectral ordinates D_n and A_n for the nth-mode SDF system as the peak values of $D_n(t)$ and $A_n(t)$, respectively, determined in part (a) of Problem 13.2. [We are doing so to avoid errors inherent in reading D_n and A_n from the response spectrum. However, in the standard application of RSA, $D_n(t)$ or $A_n(t)$ would not be available and D_n or A_n will be read from the response or design spectrum.]

(b) For each mode calculate the peak values of the following response quantities: (i) the floor displacements, (ii) the story shears, and (iii) the floor and base overturning moments.

(c) Combine the peak modal responses using an appropriate modal combination rule to obtain the peak value of the total response for each response quantity in part (b).

(d) Comment on the accuracy of the modal combination rule by comparing the RSA results from part (c) with the RHA results of Problem 13.2.

*13.41–   Figures P13.11–P13.14 show three-story frames (the same as those in Problems 9.9–9.12 and
13.44   10.19–10.22) with $m = 100$ kips/g, $I = 1400$ in^4, $E = 29,000$ ksi, and $h = 12'$. Determine the response of this frame to ground motion characterized by the design spectrum of Fig. 6.9.5 (for 5% damping) scaled to $\frac{1}{3}$g peak ground acceleration. Compute (a) the floor displacements, and (b) the bending moments in a first-story column and in the second-floor beam.

13.45–   The three-story shear frames of Figs. P13.5–P13.6 (also of Problems 9.7–9.8 and 10.11–
13.46   10.12) have the following properties: $h = 12$ ft, $m = 100$ kips/g, $I = 1400$ in^4, $E = 29,000$ ksi, and $\zeta_n = 5\%$. The peak response of this structure to El Centro ground motion is to be estimated by response spectrum analysis (RSA) and compared with the results of Problems 13.7–13.8 from response history analysis (RHA). For the purposes of this comparison the RSA is to be implemented as follows.

(a) Determine the spectral ordinates D_n and A_n for the nth-mode SDF system as the peak values of $D_n(t)$ and $A_n(t)$, respectively, determined in part (a) of Problems 13.7–13.8. [We are doing so to avoid errors inherent in reading D_n and A_n from the response spectrum.

*Denotes that a computer is necessary to solve this problem.

However, in the standard application of RSA, $D_n(t)$ or $A_n(t)$ would not be available, and D_n or A_n will be read from the response or design spectrum.]

(**b**) For each mode calculate the peak values of the following response quantities: (i) the floor displacements, (ii) the story shears, and (iii) the floor and base overturning moments.

(**c**) Combine the peak modal responses using an appropriate modal combination rule to obtain the peak value of the total response for each response quantity in part (b).

(**d**) Comment on the accuracy of the modal combination rule by comparing the RSA results from part (c) with the RHA results of Problems 13.7–13.8.

13.47   Determine the response (displacements and base moment) of the inverted L-shaped frame of Fig. E9.6a to horizontal ground motion characterized by the design spectrum of Fig. 6.9.5 scaled to 0.20g peak ground acceleration, given that $L = 10$ ft, $m = 1.5$ kips/g, $E = 29{,}000$ ksi, and $I = 28.1$ in^4; the given value of I is for a 6-in. standard steel pipe.

13.48   Solve Problem 13.47 for vertical ground motion.

13.49   Solve Problem 13.47 for the ground motion shown in Fig. P13.16.

13.50   The umbrella structure of Fig. P13.17 (also of Problems 9.13 and 10.23) is made of 6-in.-nominal diameter standard steel pipe. Its properties are: $I = 28.1$ in^4, $E = 29{,}000$ ksi, weight = 18.97 lb/ft, $m = 1.5$ kips/g, and $L = 10$ ft. Determine the peak response of this structure to horizontal ground motion characterized by the design spectrum of Fig. 6.9.5 (for 5% damping) scaled to 0.20g peak ground acceleration. Compute (**a**) displacements u_1, u_2, and u_3, and (**b**) the bending moments at the base of the column and at location a of the beam. Comment on the differences between the results from the SRSS and CQC modal combination rules.

13.51   Solve Problem 13.50 if the excitation is vertical ground motion characterized by the design spectrum of Fig. 6.9.5 (for 5% damping) scaled to 0.20g peak ground acceleration.

13.52   Solve Problem 13.50 if the excitation is ground motion in the direction b–d, characterized by the design spectrum of Fig. 6.9.5 (for 5% damping) scaled to 0.20g peak ground acceleration.

13.53   Solve Problem 13.50 if the excitation is ground motion in the direction b–c, characterized by the design spectrum of Fig. 6.9.5 (for 5% damping) scaled to 0.20g peak ground acceleration.

13.54   The peak earthquake response of the tower with the appendage of Fig. P13.22 is to be determined. The ground motion is characterized by the design spectrum of Fig. 6.9.5 (for 5% damping), scaled to $\frac{1}{3}$g peak ground acceleration.

(**a**) Using the SRSS and CQC modal combination rules, calculate the peak values of the following response quantities: (i) the displacement of the appendage mass, (ii) the shear force at the base of the appendage, and (iii) the shear force at the base of the tower.

(**b**) Comment on the differences between the results from the two modal combination rules and the reasons for these differences. Which of the two methods is accurate?

13.55   The peak response of the tower with the appendage of Fig. P13.22 to El Centro ground motion is to be estimated by response spectrum analysis (RSA) and compared with the results of Problem 13.23 from response history analysis (RHA). For the purposes of this comparison the RSA is to be implemented as follows.

(**a**) Determine the spectral ordinates D_n and A_n for the nth-mode SDF system as the peak values of $D_n(t)$ and $A_n(t)$, respectively, determined in part (a) of Problem 13.23. [We are doing so to avoid errors inherent in reading D_n and A_n from the response spectrum. However, in the standard application of RSA, $D_n(t)$ or $A_n(t)$ would not be available and D_n or A_n will be read from the response or design spectrum.]

(b) For each mode calculate the peak values of the following response quantities: (i) the displacement of the appendage mass, (ii) the shear force in the appendage, and (iii) the shear force at the base of the tower.

(c) Using the CQC method, combine the modal peak to determine the peak value of each of the response quantities of part (b). Which of the modal correlation terms must be retained and which could be dropped from CQC calculations, and why?

(d) Repeat part (c) using the SRSS method.

(e) Comment on the accuracy of the CQC and SRSS modal combination rules by comparing the RSA results from parts (c) and (d) with the RHA results by solving Problem 13.23.

13.56 The peak response of the one-story, unsymmetric-plan system of Fig. P13.24 with $\zeta_n = 5\%$ is to be estimated by response spectrum analysis (RSA) and compared with the results of Problem 13.26 from response history analysis (RHA). For purposes of this comparison the RSA is implemented as follows.

(a) Determine the spectral ordinates D_n and A_n for the nth-mode SDF system as the peak values of $D_n(t)$ and $A_n(t)$, respectively, determined in part (a) of Problem 13.26. [We are doing so to avoid errors inherent in reading D_n or A_n from the response spectrum. However, in the standard application of RSA, $D_n(t)$ or $A_n(t)$ would not be available, and D_n or A_n will be read from the response or design spectrum.]

(b) For each mode calculate the peak values of the following response quantities: u_y, $(b/2)u_\theta$, the base shear V_b, and the base torque T_b.

(c) Using the SRSS and CQC modal combination rules, compute the peak value for each response quantity.

(d) Comment on the accuracy of the SRSS and CQC methods by comparing the RSA results from part (c) with the RHA results of Problem 13.26.

13.57 Determine the peak response of the one-story, unsymmetric-plan system of Fig. P13.24 to ground motion along the y-direction. The excitation is characterized by the design spectrum of Fig. 6.9.5 (for 5% damping), scaled to 0.5g peak ground acceleration:

(a) Using the SRSS and CQC modal combination rules, calculate the peak values of the following response quantities: u_x, u_y, $b/2u_\theta$, the base shears in the x and y directions and the base torque, and the bending moments about the x and y axes at the base of each column.

(b) Comment on the differences between the results from the two modal combination rules and the reasons for these differences. Which of the two methods is accurate?

13.58 Determine the peak response of the one-story, unsymmetric-plan system of Fig. P13.24 to ground motion along the diagonal d–b. The excitation is characterized by the design spectrum of Fig. 6.9.5 (for 5% damping), scaled to 0.5g peak ground acceleration.

(a) Using the SRSS and CQC modal combination rules, calculate the peak values of the following response quantities: (i) u_x, (ii) u_y, (iii) $b/2u_\theta$, (iv) the base shears in the x and y directions and the base torque, and (v) the bending moments about the x and y axes at the base of each column.

(b) Comment on the differences between the results from the two modal combination rules and the reasons for these differences. Which of the two methods is accurate?

13.59 The three-dimensional pipe of Fig. P13.27 is made of 3-in.-nominal diameter standard steel pipe. Its properties are $I = 3.017$ in^4, $J = 6.034$ in^4, $E = 30{,}000$ ksi, $G = 12{,}000$ ksi, $m = 1.0$ kip/g, and $L = 36$ in. Determine the peak response of the system to ground motion in the x-direction characterized by the design spectrum of Fig. 6.9.5 ($\zeta = 5\%$) scaled to 0.20g peak ground acceleration. Using the SRSS and CQC modal combination rules, calculate the peak values of (a) the displacements u_x, u_y, and u_z of the mass, and

(b) the bending moments about the x and y axes and the torque at a. Comment on the differences between the results from the two modal combination rules.

13.60 Solve Problem 13.59 for ground motion in the y-direction.

13.61 Solve Problem 13.59 for ground motion in the z-direction.

13.62 Solve Problem 13.59 for ground motion in the direction a–d.

13.63 For the structure and ground motion defined in Problem 13.59, estimate the peak bending moment at a about an axis oriented at an angle $\alpha = 30°$ counterclockwise from the x-axis. Comment on the differences between the SRSS and CQC estimates.

13.64 Solve Problem 13.63 for ground motion in the y-direction.

13.65 Solve Problem 13.63 for ground motion in the z-direction.

13.66 Solve Problem 13.63 for ground motion in the direction a–d.

13.67 **(a)** For the structure defined in Problem 13.59 and ground motion in the x-direction, estimate the peak bending moment at a about an axis oriented at an arbitrary angle α counterclockwise from the x-axis.
(b) Compute the maximum value of the peak bending moment in the pipe at a and the corresponding value of α. Comment on the differences between the SRSS and CQC estimates.

13.68 Solve Problem 13.67 for ground motion in the y-direction.

13.69 Solve Problem 13.67 for ground motion in the z-direction.

13.70 Solve Problem 13.67 for ground motion in the direction a–d.

14

Analysis of Nonclassically Damped Linear Systems

PREVIEW

Now that we have developed the modal analysis procedure for structural systems with classical damping subjected to earthquake excitation, we are ready to deal with the more challenging analysis of nonclassically damped systems that arise in several practical situations mentioned in Section 11.5. In Part A of this chapter, we revisit classically damped systems and recast the analysis procedure developed in Chapters 10 and 13 for free vibration analysis and earthquake analysis of underdamped systems ($\zeta_n < 1$ in all modes) in a form that facilitates its extension to the more general case. Part B, which occupies most of this chapter, is devoted to response history analysis of nonclassically damped systems subjected to earthquake excitation. Presented first is the theory for free vibration analysis that is specialized to obtain the response to unit impulse excitation. The convolution integral approach is then used to develop the procedure for analysis of response to arbitrary ground motion.

PROBLEM STATEMENT

The response of an MDF system to ground acceleration $\ddot{u}_g(t)$ is governed by Eqs. (13.1.1) and (13.1.2), repeated here for convenience:

$$\mathbf{m}\ddot{\mathbf{u}} + \mathbf{c}\dot{\mathbf{u}} + \mathbf{k}\mathbf{u} = -\mathbf{m}\iota\ddot{u}_g(t) \tag{14.1}$$

Limiting our interest to "restrained systems"[†]—i.e., systems that do not permit any rigid body modes of natural vibration[‡]—the mass matrix $\mathbf{m}$, the stiffness matrix $\mathbf{k}$, and the damping matrix $\mathbf{c}$ of the system are real-valued and symmetric; furthermore, $\mathbf{m}$ and $\mathbf{k}$ are positive definite, whereas $\mathbf{c}$ is semipositive definite. Our objective is to develop a procedure for analyzing the response of the system without imposing any further restriction on the form of the damping matrix; in particular, for analyzing the response of nonclassically damped systems, defined in Section 10.9.

PART A: CLASSICALLY DAMPED SYSTEMS: REFORMULATION

14.1 NATURAL VIBRATION FREQUENCIES AND MODES

Free vibration of classically damped systems (defined in Section 10.9) is governed by Eq. (10.10.4), repeated here for convenience:

$$\ddot{q}_n + 2\zeta_n \omega_n \dot{q}_n + \omega_n^2 q_n = 0 \tag{14.1.1}$$

where ω_n is the undamped natural circular frequency and ζ_n is the damping ratio for the nth mode of vibration. The solution of this differential equation has the form

$$q_n(t) = e^{\lambda_n t} \tag{14.1.2}$$

Substituting Eq. (14.1.2) in Eq. (14.1.1) and proceeding as in Derivation 2.2 (Chapter 2) leads to the characteristic equation, which can be solved to determine its two eigenvalues, which are a complex-conjugate pair:

$$\lambda_n, \bar{\lambda}_n = -\zeta_n \omega_n \pm i\omega_{nD} \tag{14.1.3}$$

where $i = \sqrt{-1}$ is the unit imaginary quantity; ζ_n, ω_n, and ω_{nD} are real positive scalars;

$$\omega_{nD} = \omega_n \sqrt{1 - \zeta_n^2} \tag{14.1.4}$$

is the damped natural frequency (first introduced in Section 10.10), and the overbar denotes a complex conjugate, i.e., $\bar{\lambda}_n$ is the complex conjugate of λ_n.

Note that the natural frequency ω_n of the associated undamped system and the damping ratio ζ_n are related to the eigenvalue λ_n as follows:

$$\omega_n = |\lambda_n| \qquad \zeta_n = -\frac{\text{Re}(\lambda_n)}{|\lambda_n|} \tag{14.1.5}$$

[†] Such systems are sometimes referred to as passive systems.

[‡] In contrast, airplanes in flight have rigid-body modes.

where $|\cdot|$ denotes the modulus, and $\text{Re}(\cdot)$ represents the real part, of the complex-valued quantity enclosed in $(\cdot)$; similarly, $\text{Im}(\cdot)$ will represent the imaginary part of the complex-valued quantity. Associated with the two eigenvalues λ_n and $\bar{\lambda}_n$ is a real-valued eigenvector ϕ_n. Although a computed eigenvector may be complex valued, its real and imaginary subvectors are proportional; hence, it can be normalized as a real-valued vector.

14.2 FREE VIBRATION

Derived in Chapter 10, the solution for free vibration of classical damped systems due to initial displacements and velocities is given by Eq. (10.10.7). In this section we derive this result using an alternative approach that will permit its generalization to nonclassically damped systems.

Following Eq. (g) in Derivation 2.2 (Chapter 2), the general solution of Eq. (14.1.1) is

$$q_n(t) = B_n e^{\lambda_n t} + \bar{B}_n e^{\bar{\lambda}_n t}$$

and substituting it in Eq. (10.2.1) gives the contribution of the nth mode to the displacement response of the system:

$$\mathbf{u}_n(t) = B_n \phi_n e^{\lambda_n t} + \bar{B}_n \phi_n e^{\bar{\lambda}_n t} \tag{14.2.1}$$

in which B_n is a complex-valued constant and $\bar{B}_n$ is its complex conjugate; ϕ_n is the nth natural mode of vibration. Since the second term on the right side of Eq. (14.2.1) is the complex conjugate of the first, the two imaginary parts cancel each other, resulting in

$$\mathbf{u}_n(t) = 2\text{Re}(B_n \phi_n e^{\lambda_n t}) \tag{14.2.2}$$

Observe that the modal solution given by Eq. (14.2.1) or Eq. (14.2.2) is associated with the pair of eigenvalues λ_n and $\bar{\lambda}_n$, and their common eigenvector ϕ_n.

The response of the system to arbitrary initial excitation is given by the superposition of the modal solutions [Eqs. (14.2.1) or (14.2.2)]:

$$\mathbf{u}(t) = \sum_{n=1}^{N} B_n \phi_n e^{\lambda_n t} + \bar{B}_n \phi_n e^{\bar{\lambda}_n t} = 2\sum_{n=1}^{N} \text{Re}(B_n \phi_n e^{\lambda_n t}) \tag{14.2.3}$$

where the complex-valued constants B_n are determined from the given initial displacements $\mathbf{u}(0)$ and initial velocities $\dot{\mathbf{u}}(0)$ by invoking the orthogonality properties of modes (Appendix 14.1):

$$B_n = \frac{1}{2}\left[q_n(0) - i\frac{\dot{q}_n(0) + \zeta_n \omega_n q_n(0)}{\omega_{nD}}\right] \tag{14.2.4}$$

where

$$q_n(0) = \frac{\phi_n^T \mathbf{m}\mathbf{u}(0)}{M_n} \qquad \dot{q}_n(0) = \frac{\phi_n^T \mathbf{m}\dot{\mathbf{u}}(0)}{M_n} \qquad M_n = \phi_n^T \mathbf{m}\phi_n \tag{14.2.5}$$

which are identical to Eq. (10.8.5). Substituting Eq. (14.2.4) in Eq. (14.2.2), substituting Eq. (14.1.3) for λ_n, and finally, using Euler's relations, $e^{ix} = \cos x + i \sin x$ and $e^{-ix} = \cos x - i \sin x$, leads to the contribution of the nth vibration mode to the free vibration response:

$$\mathbf{u}_n(t) = \boldsymbol{\phi}_n e^{-\zeta_n \omega_n t} \left[q_n(0) \cos \omega_{nD} t + \frac{\dot{q}_n(0) + \zeta_n \omega_n q_n(0)}{\omega_{nD}} \sin \omega_{nD} t \right] \quad (14.2.6)$$

Superposition of the modal responses $\mathbf{u}_n(t)$, $n = 1, 2, \ldots, N$, leads to Eq. (10.10.7) for the total response.

In preparation for the analysis of nonclassically damped systems, we express Eq. (14.2.2) in an alternative form. For this purpose, we first evaluate the product

$$2 B_n \boldsymbol{\phi}_n = \boldsymbol{\beta}_n + i \boldsymbol{\gamma}_n \quad (14.2.7)$$

in which $\boldsymbol{\beta}_n$ and $\boldsymbol{\gamma}_n$ are real-valued vectors, then substitute Eq. (14.2.7) into Eq. (14.2.2) and Eq. (14.1.3) for λ_n, and finally, use Euler's relations to obtain

$$\mathbf{u}_n(t) = e^{-\zeta_n \omega_n t} [\boldsymbol{\beta}_n \cos \omega_{nD} t - \boldsymbol{\gamma}_n \sin \omega_{nD} t] \quad (14.2.8)$$

Since Eqs. (14.2.8) and Eq. (14.2.6) are equivalent,

$$\boldsymbol{\beta}_n = q_n(0) \boldsymbol{\phi}_n \qquad \boldsymbol{\gamma}_n = -\frac{\dot{q}_n(0) + \zeta_n \omega_n q_n(0)}{\omega_{nD}} \boldsymbol{\phi}_n \quad (14.2.9)$$

Superposing the modal responses defined by Eq. (14.2.8) gives an alternative expression for the total response:

$$\mathbf{u}(t) = \sum_{n=1}^{N} e^{-\zeta_n \omega_n t} [\boldsymbol{\beta}_n \cos \omega_{nD} t - \boldsymbol{\gamma}_n \sin \omega_{nD} t] \quad (14.2.10)$$

14.3 UNIT IMPULSE RESPONSE

A unit impulse ground acceleration, $\ddot{u}_g(t) = \delta(t - \tau)$, imparts to an SDF system the initial velocity $\dot{u}(\tau) = -1$ and initial displacement $u(\tau) = 0$. The resulting free vibration response was described by Eq. (4.1.7), which is repeated here for convenience after specializing it for $\tau = 0$:

$$h(t) = -\frac{1}{\omega_D} e^{-\zeta \omega_n t} \sin \omega_D t \quad (14.3.1)$$

Extending the preceding concepts to MDF systems, a unit impulse ground acceleration, $\ddot{u}_g(t) = \delta(t)$, imparts to the system the initial velocities $\dot{\mathbf{u}}(0) = -\boldsymbol{\iota}$, but no initial displacements, i.e., $\mathbf{u}(0) = \mathbf{0}$; the influence vector $\boldsymbol{\iota}$ was first defined in Section 9.4. Substituting these initial displacement and velocity vectors in Eq. (14.2.5) gives the initial conditions for the modal coordinates: $q_n(0) = 0$ and $\dot{q}_n(0) = -\Gamma_n$ [see Eq. (13.1.5) for the definition of Γ_n]; then Eq. (14.2.4) specializes to

$$B_n^g = \frac{i}{2} \frac{\Gamma_n}{\omega_{nD}} \quad (14.3.2)$$

which is substituted in the general solution for free vibration, Eq. (14.2.2), to obtain the unit impulse response; note that superscript "g" has been added in B_n^g to emphasize that these constants are associated with ground acceleration. To express such a result in a form similar to Eq. (14.2.10), we first express the product $2B_n^g\phi_n$ as in Eq. (14.2.7):

$$2B_n^g\phi_n = \beta_n^g + i\gamma_n^g \tag{14.3.3}$$

where

$$\beta_n^g = 0 \qquad \gamma_n^g = \frac{\Gamma_n}{\omega_{nD}}\phi_n \tag{14.3.4}$$

are real-valued vectors that are independent of how the modes are normalized.

Thus the response is given by Eq. (14.2.10) with an obvious change of notation:

$$\mathbf{h}(t) = \sum_{n=1}^{N} e^{-\zeta_n\omega_n t}\left[\beta_n^g \cos\omega_{nD}t - \gamma_n^g \sin\omega_{nD}t\right] \tag{14.3.5}$$

where $\mathbf{h}(t)$ denotes the vector of unit impulse response functions for displacements $\mathbf{u}(t)$ of the system.

Recognizing that β_n^g and γ_n^g are given by Eq. (14.3.4), the preceding result can be expressed as

$$\mathbf{h}(t) = \sum_{n=1}^{N} \Gamma_n h_n(t)\phi_n \tag{14.3.6}$$

in which

$$h_n(t) = -\frac{1}{\omega_{nD}}e^{-\zeta_n\omega_n t}\sin\omega_{nD}t \tag{14.3.7}$$

is the unit impulse response function for deformation of the nth-mode SDF system, an SDF system with vibration properties—natural frequency ω_n and ζ_n—of the nth mode of the MDF system. This becomes apparent by comparing Eqs. (14.3.7) and (14.3.1).

14.4 EARTHQUAKE RESPONSE

Extending the convolution integral concept for SDF systems (Sections 4.2 and 6.12), the response of an MDF system to arbitrary ground acceleration may be expressed as

$$\mathbf{u}(t) = \int_0^t \ddot{u}_g(\tau)\mathbf{h}(t-\tau)\,d\tau \tag{14.4.1}$$

in which $\mathbf{h}(t)$ is substituted from Eq. (14.3.6) to obtain

$$\mathbf{u}(t) = \sum_{n=1}^{N} \Gamma_n D_n(t)\phi_n \tag{14.4.2}$$

where

$$D_n(t) = \int_0^t \ddot{u}_g(\tau)h_n(t-\tau)\,d\tau \tag{14.4.3}$$

represents the deformation response of the nth-mode SDF system to ground acceleration $\ddot{u}_g(t)$, introduced in Section 13.1.3. This becomes apparent by substituting Eq. (14.3.7) for $h_n(t)$ and comparing the resulting equation with Eq. (6.12.1). Note that Eq. (14.4.2) is identical to Eq. (13.1.15).

PART B: NONCLASSICALLY DAMPED SYSTEMS

Now that we have reformulated the analysis of classically damped systems, we return to the original problem of analyzing the response of nonclassically damped systems, defined in Section 10.9.

14.5 NATURAL VIBRATION FREQUENCIES AND MODES

Free vibration of an MDF system is governed by Eq. (10.9.1), which is repeated for convenience:

$$\mathbf{m\ddot{u} + c\dot{u} + ku = 0} \qquad (14.5.1)$$

This equation admits a solution of the form

$$\mathbf{u}(t) = \psi e^{\lambda t} \qquad (14.5.2)$$

Substituting this form of $\mathbf{u}(t)$ in Eq. (14.5.1) leads to the quadratic eigenvalue problem (also known as the complex eigenvalue problem):

$$(\lambda^2 \mathbf{m} + \lambda \mathbf{c} + \mathbf{k})\psi = \mathbf{0} \qquad (14.5.3)$$

Although Eq. (14.5.3) can be solved directly for an eigenvalue λ and the associated eigenvector ψ, these may be determined more conveniently by first reducing the N second-order differential equations (14.1) to a system of $2N$ first-order differential equations (Appendix 14.2). The corresponding eigenvalue problem of order $2N$ [Eq. (A14.2.8)] can be solved by well-established procedures and computer algorithms. The $2N$ roots of λ are either real valued or they occur in complex-conjugate pairs (Appendix A14.2). Response analysis of systems with all complex-valued roots is developed in Sections 14.5 to 14.9; whereas systems with some real-valued roots are deferred until Section 14.10.

Provided the amount of damping is not very high—i.e., small enough to ensure oscillatory free vibration in all modes—the eigenvalues occur in complex-conjugate pairs with negative or zero real parts, just as in the case of classically damped systems [see Eq. (14.1.3)]. For an N-DOF system, there are N pairs of eigenvalues, and to each such pair corresponds a complex-conjugate pair of eigenvectors.

A complex-conjugate pair of eigenvalues, denoted by λ_n and $\bar{\lambda}_n$, may be expressed in the same form as Eq. (14.1.3) for classically damped systems:

$$\lambda_n, \bar{\lambda}_n = -\zeta_n \omega_n \pm i \omega_{nD} \qquad (14.5.4)$$

where

$$\omega_{nD} = \omega_n \sqrt{1 - \zeta_n^2} \tag{14.5.5}$$

Note that ω_n and ζ_n are related to the eigenvalues as follows:

$$\omega_n = |\lambda_n| \qquad \zeta_n = -\frac{\text{Re}(\lambda_n)}{|\lambda_n|} \tag{14.5.6}$$

The associated pair of complex-valued eigenvectors is separated into its real and imaginary parts:

$$\psi_n, \bar{\psi}_n = \phi_n \pm i\chi_n \tag{14.5.7}$$

in which ϕ_n and χ_n are real-valued vectors of N elements each. The values of ω_n are numbered in ascending order and the values of λ_n and ψ_n are numbered in the order corresponding to ω_n.

Considering that Eqs. (14.5.4) and (14.5.5) for nonclassically damped systems are identical to the corresponding Eqs. (14.1.4) and (14.1.5) for classically damped systems, ω_n will be referred to as the nth pseudo-undamped natural circular frequency of the system, ω_{nD} as the corresponding frequency with damping, and ζ_n as the modal damping ratio. The prefix "pseudo" has been included to denote that for nonclassically damped systems, ω_n is a function of the amount of system damping and, hence, differs from the corresponding frequency of the associated undamped system; where confusion may arise, the latter frequency will be denoted by ω_n^o. Because ψ_n for nonclassically damped systems is akin to ϕ_n for classically damped systems, ψ_n will be referred to as the nth natural mode of vibration.

Studies on the effect of light damping on the natural frequencies of MDF systems have demonstrated that (1) the natural frequency of the highest mode of a damped system is always less than or equal to the corresponding undamped frequency, no matter whether damping is classical or nonclassical; and (2) the damped natural frequency of the lowest mode may be higher than the corresponding undamped frequency, depending on the form of the damping matrix and on separation of ω_1 and ω_2.

For the special case of classically damped systems, as noted in Section 14.1, the eigenvalues also occur in complex-conjugate pairs; the modulus ω_n of each pair of eigenvalues is equal to the natural frequency of the associated undamped system ω_n^o; and the damped natural frequencies ω_{nD} are always lower than the corresponding undamped frequencies ω_n^o. However, the eigenvectors are real valued and equal to those of the associated undamped system, i.e., $\chi_n = \mathbf{0}$, $\psi_n = \bar{\psi}_n = \phi_n$.

14.6 ORTHOGONALITY OF MODES

A pair of eigenvectors corresponding to distinct eigenvalues satisfies the following orthogonality conditions (see Appendix 14.3 for proof):

$$(\lambda_n + \lambda_r)\,\psi_n^T \mathbf{m}\psi_r + \psi_n^T \mathbf{c}\,\psi_r = 0 \tag{14.6.1}$$

$$\psi_n^T \mathbf{k}\psi_r - \lambda_n \lambda_r \psi_n^T \mathbf{m}\,\psi_r = 0 \tag{14.6.2}$$

These orthogonality relations are also valid for a complex-conjugate pair of eigenvectors because their eigenvalues are distinct.

For classically damped systems, it can be shown that Eqs. (14.6.1) and (14.6.2) reduce to the familiar orthogonality relations of Eq. (10.4.1) (see Appendix 14.3).

Example 14.1

Determine the natural vibration frequencies and modes and modal damping ratios for the two-story shear frame of Fig. E10.12.1a with $c = \sqrt{km/200}$, a classically damped system. Use the theory for nonclassically damped systems, developed in Section 14.5, to solve the problem.

Solution The mass and stiffness matrices of the system, determined in Example 9.1, are

$$\mathbf{m} = \begin{bmatrix} 2m & \\ & m \end{bmatrix} \quad \mathbf{c} = \begin{bmatrix} 6c & -2c \\ -2c & 2c \end{bmatrix} \quad \mathbf{k} = \begin{bmatrix} 3k & -k \\ -k & k \end{bmatrix} \tag{a}$$

The damping matrix satisfies Eq. (10.9.3), implying that the system is classically damped. The eigenvalue problem to be solved is defined by Eq. (A14.2.8), repeated here for convenience:

$$\lambda \mathbf{a}\kappa + \mathbf{b}\kappa = \mathbf{0} \tag{b}$$

where the matrices $\mathbf{a}$ and $\mathbf{b}$, defined in Eq. (A14.2.5), for this system are

$$\mathbf{a} = \begin{bmatrix} \mathbf{0} & \mathbf{m} \\ \mathbf{m} & \mathbf{c} \end{bmatrix} = \begin{bmatrix} 0 & 0 & 2m & 0 \\ 0 & 0 & 0 & m \\ 2m & 0 & 6c & -2c \\ 0 & m & -2c & 2c \end{bmatrix} \tag{c}$$

$$\mathbf{b} = \begin{bmatrix} -\mathbf{m} & \mathbf{0} \\ \mathbf{0} & \mathbf{k} \end{bmatrix} = \begin{bmatrix} -2m & 0 & 0 & 0 \\ 0 & -m & 0 & 0 \\ 0 & 0 & 3k & -k \\ 0 & 0 & -k & k \end{bmatrix} \tag{d}$$

The eigenvalue problem can be solved numerically using an appropriate algorithm, e.g., the Matlab function *eig*($\mathbf{b}, -\mathbf{a}$), resulting in the eigenvalues

$$\lambda_1, \bar{\lambda}_1 = \sqrt{\frac{k}{m}}(-0.0354 \pm 0.7062i) \tag{e.1}$$

$$\lambda_2, \bar{\lambda}_2 = \sqrt{\frac{k}{m}}(-0.1414 \pm 1.4071i) \tag{e.2}$$

From these eigenvalues, ω_n and ζ_n can be determined using Eq. (14.5.6):

$$\omega_1 = |\lambda_1| = 0.7071\sqrt{\frac{k}{m}} \quad \omega_2 = |\lambda_2| = 1.4142\sqrt{\frac{k}{m}} \tag{f}$$

$$\zeta_1 = -\frac{\text{Re}(\lambda_1)}{|\lambda_1|} = 0.05 \quad \zeta_2 = -\frac{\text{Re}(\lambda_2)}{|\lambda_2|} = 0.10 \tag{g}$$

Observe that ω_n in Eq. (f) are the same as the natural frequencies of the associated undamped system, determined by solving the real eigenvalue problem in Problem 10.4, and ζ_n in Eq. (g) are identical to the values determined using Eq. (10.10.3).

From the eigenvalues of Eq. (e), the damped frequencies are determined from their definition in Eq. (14.5.5):

$$\omega_{1D} = \text{Im}(\lambda_1) = 0.7062\sqrt{\frac{k}{m}} \qquad \omega_{2D} = \text{Im}(\lambda_2) = 1.4071\sqrt{\frac{k}{m}} \tag{h}$$

Note that the damped frequencies ω_{nD} [Eq. (h)] are lower than the undamped frequencies ω_n [Eq. (f)]. Solution of the eigenvalue problem [Eq. (b)] also provides the eigenvectors:

$$\psi_1 = \left\{ \begin{matrix} \frac{1}{2} \\ 1 \end{matrix} \right\} \qquad \psi_2 = \left\{ \begin{matrix} -1 \\ 1 \end{matrix} \right\} \tag{i}$$

Note that the eigenvectors are real valued, as expected for a classically damped system, and identical to the natural modes of the associated undamped system determined in Example 10.4.

Example 14.2

Determine the natural frequencies and modes of vibration, and the modal damping ratios for the system shown in Fig. E14.2a, a two-story frame idealized as a shear building with a damper only in the first story with $c = \sqrt{km}$. Show that the eigenvectors satisfy the orthogonality properties.

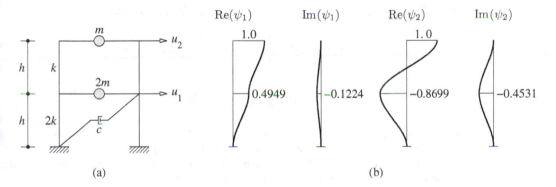

Figure E14.2   (a) Nonclassically damped system; (b) real and imaginary parts of eigenvectors ψ_1 and ψ_2.

Solution   The mass, damping, and stiffness matrices of the system are

$$\mathbf{m} = \begin{bmatrix} 2m & \\ & m \end{bmatrix} \qquad \mathbf{c} = \begin{bmatrix} c & \\ & 0 \end{bmatrix} \qquad \mathbf{k} = \begin{bmatrix} 3k & -k \\ -k & k \end{bmatrix} \tag{a}$$

The damping matrix does not satisfy Eq. (10.9.3), implying that the system is nonclassically damped. The eigenvalue problem to be solved is defined by Eq. (A14.2.8), repeated here for convenience:

$$\lambda \mathbf{a}\kappa + \mathbf{b}\kappa = 0 \tag{b}$$

where the matrices **a** and **b**, defined in Eq. (A14.2.5), for this system are

$$\mathbf{a} = \begin{bmatrix} \mathbf{0} & \mathbf{m} \\ \mathbf{m} & \mathbf{c} \end{bmatrix} = \begin{bmatrix} 0 & 0 & 2m & 0 \\ 0 & 0 & 0 & m \\ 2m & 0 & c & 0 \\ 0 & m & 0 & 0 \end{bmatrix} \tag{c}$$

$$\mathbf{b} = \begin{bmatrix} -\mathbf{m} & \mathbf{0} \\ \mathbf{0} & \mathbf{k} \end{bmatrix} = \begin{bmatrix} -2m & 0 & 0 & 0 \\ 0 & -m & 0 & 0 \\ 0 & 0 & 3k & -k \\ 0 & 0 & -k & k \end{bmatrix} \tag{d}$$

The eigenvalue problem can be solved numerically using an appropriate algorithm, e.g., the Matlab function eig (**b**, $-$**a**), resulting in the eigenvalues

$$\lambda_1, \bar{\lambda}_1 = \sqrt{\frac{k}{m}}(-0.0855 \pm 0.7159i) \tag{e.1}$$

$$\lambda_2, \bar{\lambda}_2 = \sqrt{\frac{k}{m}}(-0.1645 \pm 1.3773i) \tag{e.2}$$

From these eigenvalues, ω_n and ζ_n can be determined using Eq. (14.5.6):

$$\omega_1 = |\lambda_1| = 0.7209\sqrt{\frac{k}{m}} \qquad \omega_2 = |\lambda_2| = 1.3871\sqrt{\frac{k}{m}} \tag{f}$$

$$\zeta_1 = -\frac{\text{Re}(\lambda_1)}{|\lambda_1|} = 0.1186 \qquad \zeta_2 = -\frac{\text{Re}(\lambda_2)}{|\lambda_2|} = 0.1186 \tag{g}$$

Observe that the pseudo-undamped frequencies ω_n in Eq. (f) are slightly different than the frequencies of the associated undamped system ω_n^o given by Eq. (f) in Example 14.1.

From the eigenvalues of Eq. (e), the corresponding frequencies ω_{nD} of the damped system are determined from their definition in Eq. (14.5.5):

$$\omega_{1D} = \text{Im}(\lambda_1) = 0.7159\sqrt{\frac{k}{m}} \qquad \omega_{2D} = \text{Im}(\lambda_2) = 1.3773\sqrt{\frac{k}{m}} \tag{h}$$

Note that the damped natural frequency ω_{1D} of the lowest (or first) mode is higher than the undamped frequency ω_1^o; however, the damped frequency ω_{2D} of the highest (or second) mode is lower than the undamped frequency ω_2^o.

Solution of the eigenvalue problem [Eq. (b)] also provides the 4×1 eigenvectors, but only the third and fourth components [see Eq. (A14.2.7)] are relevant and shown below:

$$\psi_1 = \begin{Bmatrix} 0.4949 - 0.1224i \\ 1 \end{Bmatrix} \qquad \psi_2 = \begin{Bmatrix} -0.8699 - 0.4531i \\ 1 \end{Bmatrix} \tag{i}$$

Note that the eigenvectors are now complex valued, as expected for a nonclassically damped system. Their real and imaginary parts are plotted in Fig. E14.2b.

To verify that the eigenvectors ψ_n are orthogonal, we compute the individual terms in the left side of Eqs. (14.6.1) and (14.6.2):

$$\psi_1^T \mathbf{m}\psi_2 = \left\{ \begin{matrix} 0.4949 - 0.1224i \\ 1 \end{matrix} \right\}^T \left[\begin{matrix} 2m & \\ & m \end{matrix} \right] \left\{ \begin{matrix} -0.8699 - 0.4531i \\ 1 \end{matrix} \right\} = m\,(0.0281 - 0.2355i)$$

$$\psi_1^T \mathbf{k}\psi_2 = \left\{ \begin{matrix} 0.4949 - 0.1224i \\ 1 \end{matrix} \right\}^T \left[\begin{matrix} 3k & -k \\ -k & k \end{matrix} \right] \left\{ \begin{matrix} -0.8699 - 0.4531i \\ 1 \end{matrix} \right\} = k\,(-0.0828 + 0.2223i)$$

$$\psi_1^T \mathbf{c}\psi_2 = \left\{ \begin{matrix} 0.4949 - 0.1224i \\ 1 \end{matrix} \right\}^T \left[\begin{matrix} c & \\ & 0 \end{matrix} \right] \left\{ \begin{matrix} -0.8699 - 0.4531i \\ 1 \end{matrix} \right\} = \sqrt{km}\,(-0.4859 - 0.1178i)$$

Substituting individual terms in the left side of Eqs. (14.6.1) and (14.6.2) gives

$$(\lambda_1 + \lambda_2)\,\psi_1^T \mathbf{m}\psi_2 + \psi_1^T \mathbf{c}\psi_2 = \sqrt{km}\,(-0.2500 + 2.0931i)\,(0.0281 - 0.2355i)$$
$$+ \sqrt{km}\,(-0.4859 - 0.1178i) = 0$$

$$\psi_1^T \mathbf{k}\psi_2 - \lambda_1\lambda_2\psi_1^T \mathbf{m}\psi_2 = k\,(-0.0828 + 0.2223i)$$
$$+ k(0.9719 + 0.2355i)\,(0.0281 - 0.2355i) = 0$$

This verifies that the eigenvectors computed for the system are orthogonal.

14.7 FREE VIBRATION

The modal solution associated with the complex-conjugate pair of eigenvalues λ_n and $\bar{\lambda}_n$ and their eigenvectors ψ_n and $\bar{\psi}_n$ (derived in Appendix A14.4) is given by

$$\mathbf{u}_n(t) = B_n\psi_n e^{\lambda_n t} + \bar{B}_n\bar{\psi}_n e^{\bar{\lambda}_n t} \tag{14.7.1}$$

which may be viewed as a generalization of Eq. (14.2.1) for nonclassically damped systems. Since the second term on the right-hand side of Eq. (14.7.1) is the complex conjugate of the first, the two imaginary parts cancel each other, resulting in

$$\mathbf{u}_n(t) = 2\text{Re}\left(B_n\psi_n e^{\lambda_n t}\right) \tag{14.7.2}$$

The response of the system to arbitrary initial excitation is given by the superposition of the modal solutions [Eq. (14.7.2)]:

$$\mathbf{u}(t) = 2\sum_{n=1}^{N} \text{Re}\left(B_n\psi_n e^{\lambda_n t}\right) \tag{14.7.3}$$

By invoking the orthogonality properties of modes [Eqs. (14.6.1) and (14.6.2)], the complex-valued constants are determined (see Appendix 14.4):

$$B_n = \frac{\lambda_n\psi_n^T \mathbf{m}\mathbf{u}(0) + \psi_n^T \mathbf{c}\mathbf{u}(0) + \psi_n^T \mathbf{m}\dot{\mathbf{u}}(0)}{2\lambda_n\psi_n^T \mathbf{m}\psi_n + \psi_n^T \mathbf{c}\psi_n} \tag{14.7.4}$$

Following the derivation of the alternative form of the free vibration solution for classically damped systems (Section 14.2), we first evaluate the product:

$$2B_n\psi_n = \beta_n + i\gamma_n \tag{14.7.5}$$

in which $\boldsymbol{\beta}_n$ and $\boldsymbol{\gamma}_n$ are real-valued vectors, then substitute Eq. (14.7.5) into Eq. (14.7.2), and finally relate the exponential functions to trigonometric functions, leading to

$$\mathbf{u}_n(t) = e^{-\zeta_n \omega_n t}[\boldsymbol{\beta}_n \cos \omega_{nD}t - \boldsymbol{\gamma}_n \sin \omega_{nD}t] \tag{14.7.6}$$

which is identical to Eq. (14.2.8) for classically damped systems, but the vectors $\boldsymbol{\beta}_n$ and $\boldsymbol{\gamma}_n$ will no longer be defined by Eq. (14.2.9). Superposition of the modal solutions, as expressed in Eq. (14.7.6), provides an alternative form for the free vibration response:

$$\mathbf{u}(t) = \sum_{n=1}^{N} e^{-\zeta_n \omega_n t}[\boldsymbol{\beta}_n \cos \omega_{nD}t - \boldsymbol{\gamma}_n \sin \omega_{nD}t] \tag{14.7.7}$$

For classically damped systems, Eq. (14.7.4) reduces to Eq. (14.2.4), $\boldsymbol{\beta}_n$ and $\boldsymbol{\gamma}_n$ are given by Eqs. (14.2.9) (see Appendix 14.4), and Eq. (14.7.7) becomes equivalent to Eq. (14.2.10).

Example 14.3

Determine the free vibration response of the two-story shear frame of Fig. E10.12.1a with $c = \sqrt{km/200}$, a classically damped system, due to initial displacements $\mathbf{u}(0) = \langle -\frac{1}{2} \quad 2 \rangle^T$. Use the theory for nonclassically damped systems, developed in Section 14.7, to solve the problem.

Solution The initial displacement and velocity vectors are

$$\mathbf{u}(0) = \left\{ \begin{matrix} -\frac{1}{2} \\ 2 \end{matrix} \right\} \qquad \dot{\mathbf{u}}(0) = \left\{ \begin{matrix} 0 \\ 0 \end{matrix} \right\} \tag{a}$$

Substituting them in Eq. (14.7.4) together with $\mathbf{m}$, $\mathbf{c}$, λ_n, and ψ_n determined in Example 14.1 gives

$$B_1 = \frac{\lambda_1 \psi_1^T \mathbf{mu}(0) + \psi_1^T \mathbf{cu}(0) + \psi_1^T \mathbf{m\dot{u}}(0)}{2\lambda_1 \psi_1^T \mathbf{m}\psi_1 + \psi_1^T \mathbf{c}\psi_1} = 0.5000 - 0.0250i \tag{b.1}$$

$$B_2 = \frac{\lambda_2 \psi_2^T \mathbf{mu}(0) + \psi_2^T \mathbf{cu}(0) + \psi_2^T \mathbf{m\dot{u}}(0)}{2\lambda_2 \psi_2^T \mathbf{m}\psi_2 + \psi_2^T \mathbf{c}\psi_2} = 0.5000 - 0.0503i \tag{b.2}$$

Using the B_n in Eq. (b) and ψ_n from Example 14.1, $\boldsymbol{\beta}_n$ and $\boldsymbol{\gamma}_n$ are determined from Eq. (14.7.5) as follows:

$$\boldsymbol{\beta}_1 = \mathrm{Re}(2B_1\psi_1) = \left\{ \begin{matrix} 0.5000 \\ 1.0000 \end{matrix} \right\} \qquad \boldsymbol{\beta}_2 = \mathrm{Re}(2B_2\psi_2) = \left\{ \begin{matrix} -1.0000 \\ 1.0000 \end{matrix} \right\} \tag{c.1}$$

$$\boldsymbol{\gamma}_1 = \mathrm{Im}(2B_1\psi_1) = \left\{ \begin{matrix} -0.0250 \\ -0.0501 \end{matrix} \right\} \qquad \boldsymbol{\gamma}_2 = \mathrm{Im}(2B_2\psi_2) = \left\{ \begin{matrix} 0.1005 \\ -0.1005 \end{matrix} \right\} \tag{c.2}$$

Substituting the $\boldsymbol{\beta}_n$ and $\boldsymbol{\gamma}_n$ from Eq. (c) into Eq. (14.7.7) gives the free-vibration response:

$$\begin{aligned} \mathbf{u}(t) = e^{-0.05\omega_1 t}&\left[\left\{ \begin{matrix} 0.5000 \\ 1.0000 \end{matrix} \right\} \cos \omega_{1D}t + \left\{ \begin{matrix} 0.0250 \\ 0.0501 \end{matrix} \right\} \sin \omega_{1D}t \right] \\ + e^{-0.1\omega_2 t}&\left[\left\{ \begin{matrix} -1.0000 \\ 1.0000 \end{matrix} \right\} \cos \omega_{2D}t + \left\{ \begin{matrix} -0.1005 \\ 0.1005 \end{matrix} \right\} \sin \omega_{2D}t \right] \end{aligned} \tag{d}$$

Observe that this solution is identical to the result obtained by the classical method in Example 10.13.

Example 14.4

Determine the free vibration response of the two-story shear frame of Fig. E14.2a with $c = \sqrt{km}$ due to initial displacements $\mathbf{u}(0) = \langle -\frac{1}{2}\ \ 2 \rangle^T$.

Solution   The initial displacement and velocity vectors are

$$\mathbf{u}(0) = \left\{ \begin{array}{c} -\frac{1}{2} \\ 2 \end{array} \right\} \qquad \dot{\mathbf{u}}(0) = \left\{ \begin{array}{c} 0 \\ 0 \end{array} \right\} \tag{a}$$

Substituting them in Eq. (14.7.4) together with $\mathbf{m}$, $\mathbf{c}$, λ_n, and ψ_n determined in Example 14.2 gives

$$B_1 = \frac{\lambda_1 \psi_1^T \mathbf{m} \mathbf{u}(0) + \psi_1^T \mathbf{c} \mathbf{u}(0) + \psi_1^T \mathbf{m} \dot{\mathbf{u}}(0)}{2\lambda_1 \psi_1^T \mathbf{m} \psi_1 + \psi_1^T \mathbf{c} \psi_1} = 0.5101 + 0.3137i \tag{b.1}$$

$$B_2 = \frac{\lambda_2 \psi_2^T \mathbf{m} \mathbf{u}(0) + \psi_2^T \mathbf{c} \mathbf{u}(0) + \psi_2^T \mathbf{m} \dot{\mathbf{u}}(0)}{2\lambda_2 \psi_2^T \mathbf{m} \psi_2 + \psi_2^T \mathbf{c} \psi_2} = 0.4899 - 0.2532i \tag{b.2}$$

Using the B_n in Eq. (b) and ψ_n from Example 14.2, β_n and γ_n are determined from Eq. (14.7.5) as follows:

$$\beta_1 = \text{Re}\,(2B_1\psi_1) = \left\{ \begin{array}{c} 0.5817 \\ 1.0203 \end{array} \right\} \qquad \beta_2 = \text{Re}(2B_2\psi_2) = \left\{ \begin{array}{c} -1.0817 \\ 0.9797 \end{array} \right\} \tag{c.1}$$

$$\gamma_1 = \text{Im}(2B_1\psi_1) = \left\{ \begin{array}{c} 0.1856 \\ 0.6274 \end{array} \right\} \qquad \gamma_2 = \text{Im}\,(2B_2\psi_2) = \left\{ \begin{array}{c} -0.0034 \\ -0.5065 \end{array} \right\} \tag{c.2}$$

Substituting the β_n and γ_n from Eq. (c) into Eq. (14.7.7) gives the free-vibration response:

$$\mathbf{u}(t) = e^{-0.1186\,\omega_1 t}\left[\left\{ \begin{array}{c} 0.5817 \\ 1.0203 \end{array} \right\} \cos\omega_{1D}t - \left\{ \begin{array}{c} 0.1856 \\ 0.6274 \end{array} \right\} \sin\omega_{1D}t\right]$$

$$+ e^{-0.1186\,\omega_2 t}\left[\left\{ \begin{array}{c} -1.0817 \\ 0.9797 \end{array} \right\} \cos\omega_{2D}t + \left\{ \begin{array}{c} 0.0034 \\ 0.5065 \end{array} \right\} \sin\omega_{2D}t\right] \tag{d}$$

Example 14.5

Determine the free vibration responses of the two-story shear frame of Fig. E14.2a with $c = \sqrt{km}$ due to two sets of initial displacements: (1) $\mathbf{u}(0) = \phi_1$ and (2) $\mathbf{u}(0) = \phi_2$, where ϕ_n is the nth natural vibration mode of the associated undamped system; $\phi_1 = \langle \frac{1}{2}\ \ 1 \rangle^T$ and $\phi_2 = \langle -1\ \ 1 \rangle^T$.

Solution   **Part 1**   Substituting the first $\mathbf{u}(0)$ in Eq. (14.7.4) together with $\mathbf{m}$, $\mathbf{c}$, λ_n, and ψ_n determined in Example 14.2 gives

$$B_1 = \frac{\lambda_1 \psi_1^T \mathbf{m} \mathbf{u}(0) + \psi_1^T \mathbf{c} \mathbf{u}(0) + \psi_1^T \mathbf{m} \dot{\mathbf{u}}(0)}{2\lambda_1 \psi_1^T \mathbf{m} \psi_1 + \psi_1^T \mathbf{c} \psi_1} = 0.5103 - 0.0200i \tag{a.1}$$

$$B_2 = \frac{\lambda_2 \psi_2^T \mathbf{m} \mathbf{u}(0) + \psi_2^T \mathbf{c} \mathbf{u}(0) + \psi_2^T \mathbf{m} \dot{\mathbf{u}}(0)}{2\lambda_2 \psi_2^T \mathbf{m} \psi_2 + \psi_2^T \mathbf{c} \psi_2} = -0.0103 - 0.0200i \tag{a.2}$$

Using the B_n in Eq. (a) and ψ_n from Example 14.2, β_n and γ_n are determined from Eq. (14.7.5) as follows:

$$\beta_1 = \text{Re}(2B_1\psi_1) = \left\{ \begin{array}{c} 0.5002 \\ 1.0207 \end{array} \right\} \qquad \beta_2 = \text{Re}(2B_2\psi_2) = \left\{ \begin{array}{c} -0.0002 \\ -0.0207 \end{array} \right\} \tag{b.1}$$

$$\gamma_1 = \text{Im}(2B_1\psi_1) = \left\{ \begin{array}{c} -0.1448 \\ -0.0401 \end{array} \right\} \qquad \gamma_2 = \text{Im}(2B_2\psi_2) = \left\{ \begin{array}{c} 0.0442 \\ -0.0401 \end{array} \right\} \tag{b.2}$$

Substituting the β_n and γ_n from Eq. (b) into Eq. (14.7.7) gives the free-vibration response:

$$\mathbf{u}(t) = e^{-0.1186\,\omega_1 t} \left[\begin{Bmatrix} 0.5002 \\ 1.0207 \end{Bmatrix} \cos \omega_{1D} t + \begin{Bmatrix} 0.1448 \\ 0.0401 \end{Bmatrix} \sin \omega_{1D} t \right]$$
$$+ e^{-0.1186\,\omega_2 t} \left[\begin{Bmatrix} -0.0002 \\ -0.0207 \end{Bmatrix} \cos \omega_{2D} t - \begin{Bmatrix} 0.0442 \\ -0.0401 \end{Bmatrix} \sin \omega_{2D} t \right] \tag{c}$$

Part 2   Proceeding as in Part 1, the free vibration response due to the second $\mathbf{u}(0)$ can be determined:

$$\mathbf{u}(t) = e^{-0.1186\,\omega_1 t} \left[\begin{Bmatrix} 0.0815 \\ -0.0004 \end{Bmatrix} \cos \omega_{1D} t - \begin{Bmatrix} 0.3304 \\ 0.6675 \end{Bmatrix} \sin \omega_{1D} t \right]$$
$$+ e^{-0.1186\,\omega_2 t} \left[\begin{Bmatrix} -1.0815 \\ 1.0004 \end{Bmatrix} \cos \omega_{2D} t + \begin{Bmatrix} 0.0476 \\ 0.4664 \end{Bmatrix} \sin \omega_{2D} t \right] \tag{d}$$

Observations   The displacements $\mathbf{u}(t)$ are expressed as a linear combination of ϕ_n, the natural vibration modes of the associated undamped system:

$$\mathbf{u}(t) = \sum_{n=1}^{2} \phi_n q_n(t) \tag{e}$$

where the modal coordinates are given by

$$q_n(t) = \frac{\phi_n^T \mathbf{m} \mathbf{u}(t)}{\phi_n^T \mathbf{m} \phi_n} \tag{f}$$

which is a generalization of Eq. (10.7.2). Substituting the known ϕ_n and Eq. (c) for $\mathbf{u}(t)$ in Eq. (f) leads to

$$q_1(t) = e^{-0.1186\,\omega_1 t}(1.0139 \cos \omega_{1D} t + 0.1232 \sin \omega_{1D} t)$$
$$+ e^{-0.1186\,\omega_2 t}(0.0068 \cos \omega_{2D} t - 0.0832 \sin \omega_{2D} t) \tag{g}$$

$$q_2(t) = e^{-0.1186\,\omega_1 t}(-0.0139 \cos \omega_{1D} t - 0.0028 \sin \omega_{1D} t)$$
$$+ e^{-0.1186\,\omega_2 t}(-0.0068 \cos \omega_{2D} t + 0.0428 \sin \omega_{2D} t) \tag{h}$$

These are the $q_n(t)$ associated with Eq. (c), the free vibration response due to the first set of initial displacements.

Similarly, substituting the known ϕ_n and Eq. (d) for $\mathbf{u}(t)$, Eq. (f) leads to

$$q_1(t) = e^{-0.1186\,\omega_1 t}(0.0541 \cos \omega_{1D} t - 0.6652 \sin \omega_{1D} t)$$
$$+ e^{-0.1186\,\omega_2 t}(-0.0545 \cos \omega_{2D} t - 0.0023 \sin \omega_{2D} t) \tag{j}$$

$$q_2(t) = e^{-0.1186\,\omega_1 t}(-0.0541 \cos \omega_{1D} t + 0.3427 \sin \omega_{1D} t)$$
$$+ e^{-0.1186\,\omega_2 t}(1.0545 \cos \omega_{2D} t + 0.1237 \sin \omega_{2D} t) \tag{k}$$

These are the $q_n(t)$ associated with Eq. (d), the free vibration response to the second set of initial displacements.

These results, presented in Figs. E14.5.1 and E14.5.2, are for a system disturbed from its equilibrium position by imposing initial displacements that are proportional to a natural vibration mode ϕ_n of the associated undamped system. The solutions for $q_n(t)$ are presented in part (c) of these figures; the floor displacements in part (d); and the deflected shapes at selected time instants—a, b, c, d, and e—in part (b). These results permit three observations that contrast with what we observed earlier for undamped systems (Figs. 10.1.2 and 10.1.3) and for classically damped systems (Figs. E10.12.1 and E10.12.2): First, $q_2(t) \neq 0$ in Fig. E14.5.1c and $q_1(t) \neq 0$ in Fig. E14.5.2c. Second, the initial deflected shape is not maintained in free vibration; see Figs. E14.5.1b and E14.5.2b. Third, all floors (or DOFs) of the system do not vibrate in the same phase; they do not pass through zero, maximum, or minimum positions at the same instant of time; see Fig. E14.5.2d.

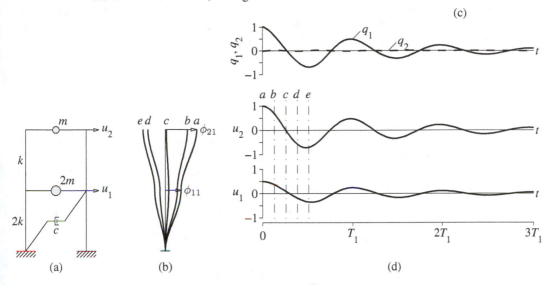

Figure E14.5.1 Free vibration of a nonclassically damped system due to initial displacement in the first natural mode of the undamped system: (a) two-story frame; (b) deflected shapes at time instants a, b, c, d, and e; (c) modal coordinates $q_n(t)$; (d) displacement history.

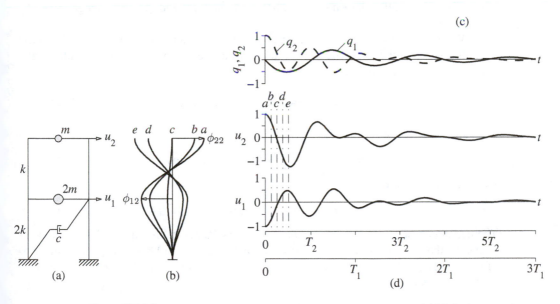

Figure E14.5.2 Free vibration of a nonclassically damped system due to initial displacement in the second natural mode of the undamped system: (a) two-story frame; (b) deflected shapes at time instants a, b, c, d, and e; (c) modal coordinates $q_n(t)$; (d) displacement history.

14.8 UNIT IMPULSE RESPONSE

As first mentioned in Section 14.3, a unit impulse ground acceleration $\ddot{u}_g(t) = \delta(t)$ imparts to an MDF system the initial velocities $\dot{\mathbf{u}}(0) = -\boldsymbol{\iota}$, but no initial displacements, i.e., $\mathbf{u}(0) = \mathbf{0}$. Substituting these initial conditions into Eq. (14.7.4) provides the complex-valued constant

$$B_n^g = \frac{-\boldsymbol{\psi}_n^T \mathbf{m} \boldsymbol{\iota}}{2\lambda_n \boldsymbol{\psi}_n^T \mathbf{m} \boldsymbol{\psi}_n + \boldsymbol{\psi}_n^T \mathbf{c} \boldsymbol{\psi}_n} \tag{14.8.1}$$

As before, the superscript "g" has been added in B_n^g to emphasize that these constants are associated with ground acceleration. Substituting Eq. (14.8.1) in the general solution for free vibration, Eq. (14.7.3) leads to the unit impulse response.

To express this result in a form similar to Eq. (14.7.6), we first express the product $2B_n^g \boldsymbol{\psi}_n$ as in Eq. (14.7.5):

$$2B_n^g \boldsymbol{\psi}_n = \boldsymbol{\beta}_n^g + i \boldsymbol{\gamma}_n^g \tag{14.8.2}$$

Thus, the response is given by Eq. (14.7.7) with an obvious change of notation:

$$\mathbf{h}(t) \equiv \mathbf{u}(t) = \sum_{n=1}^{N} e^{-\zeta_n \omega_n t} \left[\boldsymbol{\beta}_n^g \cos \omega_{nD} t - \boldsymbol{\gamma}_n^g \sin \omega_{nD} t \right] \tag{14.8.3}$$

where $\mathbf{h}(t)$ denotes the vector of unit impulse response functions for displacements of the system. Note that Eqs. (14.8.2) and (14.8.3) have the same form as Eqs. (14.3.3) and (14.3.5) for classically damped systems, but the vectors $\boldsymbol{\beta}_n^g$ and $\boldsymbol{\gamma}_n^g$ are no longer given by Eq. (14.3.4).

We prefer to express Eq. (14.8.3) in terms of the unit impulse response function $h_n(t)$ for deformation of the nth-mode SDF system, just as we did in Eq. (14.3.6) for classically damped systems. Recall that

$$h_n(t) = -\frac{1}{\omega_{nD}} e^{-\zeta_n \omega_n t} \sin \omega_{nD} t \tag{14.8.4}$$

and its first derivative is given by

$$\dot{h}_n(t) = -e^{-\zeta_n \omega_n t} \cos \omega_{nD} t - \zeta_n \omega_n h_n(t) \tag{14.8.5}$$

which can be rewritten as

$$-e^{-\zeta_n \omega_n t} \cos \omega_{nD} t = \dot{h}_n(t) + \zeta_n \omega_n h_n(t) \tag{14.8.6}$$

The trigonometric functions multiplying the vectors $\boldsymbol{\beta}_n^g$ and $\boldsymbol{\gamma}_n^g$ in Eq. (14.8.3) are now replaced by the corresponding expressions obtained from Eqs. (14.8.4) and (14.8.6) in terms of $h_n(t)$ and $\dot{h}_n(t)$ to obtain the vector of unit impulse response functions for displacements $\mathbf{u}(t)$ of the system:

$$\mathbf{h}(t) = -\sum_{n=1}^{N} \left[\boldsymbol{\alpha}_n^g \omega_n h_n(t) + \boldsymbol{\beta}_n^g \dot{h}_n(t) \right] \tag{14.8.7}$$

in which

$$\alpha_n^g = \zeta_n \beta_n^g - \sqrt{1 - \zeta_n^2}\, \gamma_n^g \tag{14.8.8}$$

where β_n^g and γ_n^g are known from Eqs. (14.8.2) and (14.8.1).

For classically damped systems, Eq. (14.8.7) specializes to Eq. (14.3.6). This can be verified by substituting Eq. (14.3.4) in Eq. (14.8.8) to obtain $\alpha_n^g = -(\Gamma_n/\omega_n)\phi_n$ and substituting it together with $\beta_n^g = \mathbf{0}$ in Eq. (14.8.7).

Properties of α_n^g and β_n^g. The vectors α_n^g and β_n^g that appear in Eq. (14.8.3) satisfy the following relations:

$$\sum_{n=1}^N \beta_n^g = 0 \qquad \sum_{n=1}^N \omega_n \left[\alpha_n^g - 2\zeta_n \beta_n^g \right] = -\boldsymbol{\iota} \tag{14.8.9}$$

These equations can be derived by recalling that Eq. (14.8.3) must satisfy the initial conditions associated with unit impulse ground acceleration: $\mathbf{u}(0) = \mathbf{0}$ and $\dot{\mathbf{u}}(0) = -\boldsymbol{\iota}$. Specializing Eq. (14.8.3) for $t = 0$ and imposing the first initial condition leads to Eq. (14.9.9a). Similarly, differentiating Eq. (14.8.3) to obtain an equation for $\dot{\mathbf{u}}(t)$, specializing it for $t = 0$, and imposing the second initial condition leads to Eq. (14.8.9b); see Appendix A14.5.

For classically damped systems, $\beta_n^g = \mathbf{0}$ for all modes [Eq. (14.3.4a)], implying that Eq. (14.8.9a) is satisfied. Furthermore, for such systems $\alpha_n^g = -(\Gamma_n/\omega_n)\,\phi_n$, and Eq. (14.8.9b) reduces to

$$\sum_{n=1}^N \Gamma_n \phi_n = \boldsymbol{\iota} \tag{14.8.10}$$

This is the modal expansion of the influence vector $\boldsymbol{\iota}$, which first appeared in Section 13.1.3.

Example 14.6

Determine the response of the two-story shear frame of Fig. E10.12.1a with $c = \sqrt{km/200}$, a classically damped system, due to unit impulse ground acceleration, $\ddot{u}_g(t) = \delta(t)$. Use the theory for nonclassically damped systems developed in Section 14.8 to solve the problem.

Solution B_n^g are determined by substituting $\mathbf{m}$, $\mathbf{c}$, and ψ_n from Example 14.1 in Eq. (14.8.1):

$$B_1^g = \frac{-\psi_1^T \mathbf{m}\boldsymbol{\iota}}{2\lambda_1 \psi_1^T \mathbf{m}\psi_1 + \psi_1^T \mathbf{c}\psi_1} = 0.9440\sqrt{\frac{m}{k}}\, i \tag{a.1}$$

$$B_2^g = \frac{-\psi_2^T \mathbf{m}\boldsymbol{\iota}}{2\lambda_2 \psi_2^T \mathbf{m}\psi_2 + \psi_2^T \mathbf{c}\psi_2} = -0.1184\sqrt{\frac{m}{k}}\, i \tag{a.2}$$

Using the B_n^g in Eq. (a) and ψ_n, β_n^g and γ_n^g are determined from Eq. (14.8.2) as follows:

$$\beta_1^g = \mathrm{Re}(2B_1^g \psi_1) = \begin{Bmatrix} 0 \\ 0 \end{Bmatrix} \qquad \beta_2^g = \mathrm{Re}(2B_2^g \psi_2) = \begin{Bmatrix} 0 \\ 0 \end{Bmatrix} \tag{b.1}$$

$$\gamma_1^g = \mathrm{Im}(2B_1^g \psi_1) = \begin{Bmatrix} 0.9440 \\ 1.8880 \end{Bmatrix}\sqrt{\frac{m}{k}} \qquad \gamma_2^g = \mathrm{Im}(2B_2^g \psi_2) = \begin{Bmatrix} 0.2369 \\ -0.2369 \end{Bmatrix}\sqrt{\frac{m}{k}} \tag{b.2}$$

Substituting the β_n^g and γ_n^g from Eq. (b) into Eq. (14.8.3) gives the desired response:

$$\mathbf{u}(t) = \sqrt{\frac{m}{k}} \left[e^{-\zeta_1 \omega_1 t} \left\{ \begin{matrix} -0.9440 \\ -1.8880 \end{matrix} \right\} \sin \omega_{1D} t + e^{-\zeta_2 \omega_2 t} \left\{ \begin{matrix} -0.2369 \\ 0.2369 \end{matrix} \right\} \sin \omega_{2D} t \right] \qquad \text{(c)}$$

These solutions for $u_j(t)$ are plotted in Fig. E14.6.

The response can also be expressed in terms of the unit impulse response functions $h_n(t)$. For this purpose Eq. (14.8.8) is used to obtain

$$\alpha_1^g = \left\{ \begin{matrix} -0.9428 \\ -1.8856 \end{matrix} \right\} \sqrt{\frac{m}{k}} \qquad \alpha_2^g = \left\{ \begin{matrix} -0.2357 \\ 0.2357 \end{matrix} \right\} \sqrt{\frac{m}{k}} \qquad \text{(d)}$$

which are substituted in Eq. (14.8.7) to obtain

$$\mathbf{u}(t) = \left\{ \begin{matrix} 0.6667 \\ 1.3333 \end{matrix} \right\} h_1(t) + \left\{ \begin{matrix} 0.3333 \\ -0.3333 \end{matrix} \right\} h_2(t) \qquad \text{(e)}$$

where $h_n(t)$ is given by Eq. (14.8.4).

The vectors α_n^g and β_n^g satisfy Eq. (14.8.9):

$$\beta_1^g + \beta_2^g = \left\{ \begin{matrix} 0 \\ 0 \end{matrix} \right\} \qquad \text{(f)}$$

$$\sum_{n=1}^{N} \omega_n \left[\alpha_n^g - 2\zeta_n \beta_n^g \right] = \omega_1 \sqrt{\frac{m}{k}} \left\{ \begin{matrix} -0.9428 \\ -1.8856 \end{matrix} \right\} + \omega_2 \sqrt{\frac{m}{k}} \left\{ \begin{matrix} -0.2357 \\ 0.2357 \end{matrix} \right\} = -\left\{ \begin{matrix} 1 \\ 1 \end{matrix} \right\} \qquad \text{(g)}$$

which can be verified by substituting for ω_1 and ω_2 from Eq. (f) of Example 14.2.

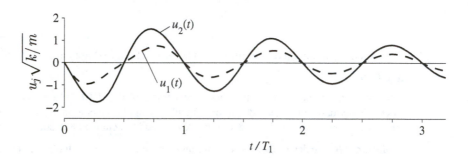

Figure E14.6   Unit impulse response functions for a classically damped system.

Example 14.7

Determine the response of the two-story shear frame of Fig. E14.2a with $c = \sqrt{km}$ due to unit impulse ground acceleration, $\ddot{u}_g(t) = \delta(t)$.

Solution B_n^g are determined by substituting $\mathbf{m}$, $\mathbf{c}$, and ψ_n from Example 14.2 in Eq. (14.8.1):

$$B_1^g = \frac{-\psi_1^T \mathbf{m} \iota}{2\lambda_1 \psi_1^T \mathbf{m} \psi_1 + \psi_1^T \mathbf{c} \psi_1} = \sqrt{\frac{m}{k}} (-0.0383 + 0.9835i) \qquad \text{(a.1)}$$

$$B_2^g = \frac{-\psi_2^T \mathbf{m} \iota}{2\lambda_2 \psi_2^T \mathbf{m} \psi_2 + \psi_2^T \mathbf{c} \psi_2} = \sqrt{\frac{m}{k}} (0.0383 - 0.1504i) \qquad \text{(a.2)}$$

Using the B_n^g in Eq. (a) and ψ_n, β_n^g and γ_n^g are determined from Eq. (14.8.2) as follows:

$$\beta_1^g = \text{Re}\left(2B_1^g \psi_1\right) = \left\{ \begin{array}{c} 0.2029 \\ -0.0766 \end{array} \right\} \sqrt{\frac{m}{k}} \qquad \beta_2^g = \text{Re}\left(2B_2^g \psi_2\right) = \left\{ \begin{array}{c} -0.2029 \\ 0.0766 \end{array} \right\} \sqrt{\frac{m}{k}} \qquad \text{(b.1)}$$

$$\gamma_1^g = \text{Im}\left(2B_1^g \psi_1\right) = \left\{ \begin{array}{c} 0.9828 \\ 1.9671 \end{array} \right\} \sqrt{\frac{m}{k}} \qquad \gamma_2^g = \text{Im}\left(2B_2^g \psi_2\right) = \left\{ \begin{array}{c} 0.2269 \\ -0.3007 \end{array} \right\} \sqrt{\frac{m}{k}} \qquad \text{(b.2)}$$

Substituting the β_n^g and γ_n^g from Eq. (b) into Eq. (14.8.3) gives the total response of the system:

$$\mathbf{u}(t) = e^{-\zeta_1 \omega_1 t}\left[\left\{ \begin{array}{c} 0.2029 \\ -0.0766 \end{array} \right\} \cos \omega_{1D}t + \left\{ \begin{array}{c} -0.9828 \\ -1.9671 \end{array} \right\} \sin \omega_{1D}t\right]\sqrt{\frac{m}{k}}$$

$$+ e^{-\zeta_2 \omega_2 t}\left[\left\{ \begin{array}{c} -0.2029 \\ 0.0766 \end{array} \right\} \cos \omega_{2D}t + \left\{ \begin{array}{c} -0.2269 \\ 0.3007 \end{array} \right\} \sin \omega_{2D}t\right]\sqrt{\frac{m}{k}} \qquad \text{(c)}$$

These solutions for $u_j(t)$ are plotted in Fig. E14.7.

The response can also be expressed in terms of the unit impulse response functions $h_n(t)$. For this purpose Eq. (14.8.8) is used to obtain

$$\alpha_1^g = \left\{ \begin{array}{c} -0.9518 \\ -1.9623 \end{array} \right\} \sqrt{\frac{m}{k}} \qquad \alpha_2^g = \left\{ \begin{array}{c} -0.2493 \\ 0.3077 \end{array} \right\} \sqrt{\frac{m}{k}} \qquad \text{(d)}$$

which are substituted together with β_n^g from Eq. (b.1) in Eq. (14.8.8) to obtain

$$\mathbf{u}(t) = \left\{ \begin{array}{c} 0.6862 \\ 1.4147 \end{array} \right\} h_1(t) + \left\{ \begin{array}{c} -0.2029 \\ 0.0766 \end{array} \right\} \sqrt{\frac{m}{k}} \, \dot{h}_1(t) + \left\{ \begin{array}{c} 0.3459 \\ -0.4268 \end{array} \right\} h_2(t) + \left\{ \begin{array}{c} 0.2029 \\ -0.0766 \end{array} \right\} \sqrt{\frac{m}{k}} \, \dot{h}_2(t) \quad \text{(e)}$$

where $h_n(t)$ and $\dot{h}_n(t)$ are given by Eq. (14.8.4) and (14.8.5), respectively.

The vectors α_n^g and β_n^g satisfy Eq. (14.8.9):

$$\beta_1^g + \beta_2^g = \sqrt{\frac{m}{k}}\left[\left\{ \begin{array}{c} 0.2029 \\ -0.0766 \end{array} \right\} + \left\{ \begin{array}{c} -0.2029 \\ 0.0766 \end{array} \right\}\right] = \left\{ \begin{array}{c} 0 \\ 0 \end{array} \right\} \qquad \text{(f)}$$

$$\sum_{n=1}^{N} \omega_n\left[\alpha_n^g - 2\zeta_n \beta_n^g\right] = \omega_1 \sqrt{\frac{m}{k}}\left[\left\{ \begin{array}{c} -0.9518 \\ -1.9623 \end{array} \right\} - 2\zeta_1 \left\{ \begin{array}{c} 0.2029 \\ -0.0766 \end{array} \right\}\right]$$

$$+ \omega_2 \sqrt{\frac{m}{k}}\left[\left\{ \begin{array}{c} -0.2493 \\ 0.3077 \end{array} \right\} - 2\zeta_2 \left\{ \begin{array}{c} -0.2029 \\ 0.0766 \end{array} \right\}\right] = \left\{ \begin{array}{c} -1 \\ -1 \end{array} \right\} \qquad \text{(g)}$$

which can be verified by substituting for ω_1 and ω_2 from Eq. (f) of Example 14.2.

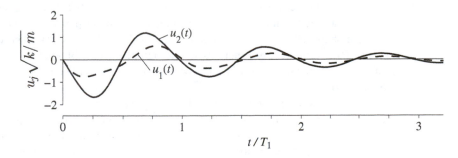

Figure E14.7   Unit impulse response functions for a nonclassically damped system.

14.9 EARTHQUAKE RESPONSE

The displacements (relative to the ground) of the system at time t due to arbitrary ground acceleration $\ddot{u}_g(t)$ are given by the convolution integral introduced in Eq. (14.4.1):

$$\mathbf{u}(t) = \int_0^t \ddot{u}_g(\tau)\mathbf{h}(t-\tau)\,d\tau \tag{14.9.1}$$

in which $\mathbf{h}(t)$ is substituted from Eq. (14.8.7) to obtain

$$\mathbf{u}(t) = -\sum_{n=1}^{N}\left[\alpha_n^g \omega_n D_n(t) + \beta_n^g \dot{D}_n(t)\right] \tag{14.9.2}$$

where

$$D_n(t) = \int_0^t \ddot{u}_g(\tau)\, h_n(t-\tau)\,d\tau \tag{14.9.3}$$

and

$$\dot{D}_n(t) = \int_0^t \ddot{u}_g(\tau)\,\dot{h}_n(t-\tau)\,d\tau \tag{14.9.4}$$

The unit impulse response functions $h_n(t)$ and $\dot{h}_n(t)$ for the nth-mode SDF system were presented in Eqs. (14.8.4) and (14.8.5), respectively.

In Eqs. (14.9.2) and (14.9.3), $D_n(t)$ represents the deformation response of the nth-mode SDF system, an SDF system with vibration properties—natural frequency ω_n and damping ratio ζ_n—of the nth mode of the MDF system, to ground acceleration $\ddot{u}_g(t)$; $D_n(t)$ was introduced in Section 13.1.3 and also appeared in the response of classically damped systems (Section 14.4). The quantity $\dot{D}_n(t)$ represents the corresponding relative velocity response of the nth-mode SDF system, introduced in Section 6.12. These SDF system responses, $D_n(t)$ and $\dot{D}_n(t)$, are usually computed by one of the numerical methods presented in Chapter 5, not by numerical evaluation of the convolution integrals in Eqs. (14.9.3) and (14.9.4) because the latter approach is numerically inefficient.

The response of a nonclassically damped MDF system has been expressed as a linear combination of N pairs of terms; the nth pair represents the nth-modal solution, associated with the nth pair of eigenvalues λ_n and $\bar{\lambda}_n$ and their associated eigenvectors ψ_n and $\bar{\psi}_n$. The first part in the nth such pair represents a motion in the deflected shape α_n^g with its temporal variation defined by $D_n(t)$, whereas the second part represents a motion in the deflected shape β_n^g with its temporal variation defined by $\dot{D}_n(t)$. The deflected shapes α_n^g and β_n^g are obviously functions of the natural modes of vibration of the system, as is apparent from Eqs. (14.8.2) and (14.8.8); these are independent of $\ddot{u}_g(t)$ and satisfy Eq. (14.8.9).

How does the response of nonclassically damped systems differ from that of classically damped systems? Both $D_n(t)$ and $\dot{D}_n(t)$ appear in Eq. (14.9.2), which defines the earthquake response of nonclassically damped systems; however, only $D_n(t)$ enters into Eq. (14.4.2), the corresponding result for classically damped systems. For classically damped systems, the deflected shape ϕ_n associated with the contribution of the nth mode of vibration remains invariant over time [Eq. (14.4.2)], but it varies with time in the case of nonclassically damped systems [Eq. (14.9.2)].

Once the displacements $\mathbf{u}(t)$ have been computed, internal forces in the structure can be determined as described earlier for classically damped systems. Forces in viscous dampers (that may be part of a supplemental damping system) can be determined from the damper properties and $\dot{\mathbf{u}}(t)$. Differentiating Eq. (14.9.2) gives a formal equation for these velocities that involve $\dot{D}_n(t)$ and $\ddot{D}_n(t)$. Both of these quantities are computed in the process of numerically evaluating $D_n(t)$ using one of the numerical methods presented in Chapter 5.

Example 14.8

Derive equations for the floor displacements of the shear frame of Fig. E10.12.1a with $c = \sqrt{km/200}$, a classically damped system, subjected to ground acceleration $\ddot{u}_g(t)$. Use the theory for nonclassically damped systems developed in Section 14.9 to solve the problem.

Solution   Substituting the α_n^g and β_n^g determined in Example 14.6 into Eq. (14.9.2) provides the equations for the floor displacements:

$$\mathbf{u}(t) = \left\{ \begin{array}{c} 2/3 \\ 4/3 \end{array} \right\} D_1(t) + \left\{ \begin{array}{c} 1/3 \\ -1/3 \end{array} \right\} D_2(t)$$

Note that this result is identical to that obtained in Example 13.3 by classical methods.

Example 14.9

Derive equations for the floor displacements of the shear frame of Example 14.7 subjected to ground acceleration $\ddot{u}_g(t)$.

Solution   Substituting the α_n^g and β_n^g determined in Example 14.7 into Eq. (14.9.2) provides the equation for the floor displacements:

$$\mathbf{u}(t) = \left\{ \begin{array}{c} 0.6862 \\ 1.4147 \end{array} \right\} D_1(t) + \left\{ \begin{array}{c} -0.2029 \\ 0.0766 \end{array} \right\} \sqrt{\frac{m}{k}} \dot{D}_1(t) + \left\{ \begin{array}{c} 0.3459 \\ -0.4268 \end{array} \right\} D_2(t) + \left\{ \begin{array}{c} 0.2029 \\ -0.0766 \end{array} \right\} \sqrt{\frac{m}{k}} \dot{D}_2(t)$$

$$\text{(a)}$$

where $D_n(t)$ and $\dot{D}_n(t)$ represent the deformation and relative velocity response of the nth-mode SDF system; see Eqs. (14.9.3) and (14.9.4).

Example 14.10

Determine the displacement response of the two-story shear frame of Fig. E14.2a with $c = \sqrt{km}$, $m = 100$ kips/g, and $k = 2\pi^2 m$, due to the El Centro ground motion.

Solution   The equation describing the displacement response of this system, derived in Example 14.9, contains $D_n(t)$ and $\dot{D}_n(t)$, the deformation and relative velocity responses of the nth-mode SDF system [Eqs. (14.9.3) and (14.9.4)]. The natural frequencies ω_n and damping ratios ζ_n of the two modal SDF systems are given in terms of m and k by by Eqs. (f) and (g) of Example 14.2.

Substituting $m = 0.2591$ kip-sec^2/in. and $k = 5.1138$ kips/in. gives

$$\omega_1 = 3.203 \text{ rad/sec} \qquad \omega_2 = 6.163 \text{ rad/sec} \qquad \text{(a)}$$

$$T_1 = 1.962 \text{ sec} \qquad T_2 = 1.020 \text{ sec} \qquad \text{(b)}$$

$$\zeta_1 = 0.1186 \qquad \zeta_2 = 0.1186 \qquad \text{(c)}$$

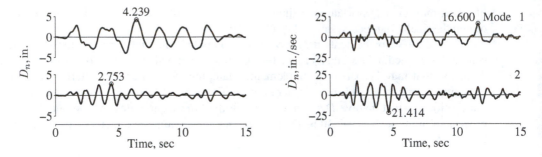

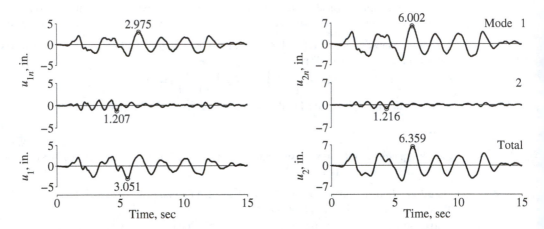

Figure E14.10a   Deformation $D_n(t)$ and relative velocity $\dot{D}_n(t)$ responses of modal SDF systems.

Figure E14.10b   Floor displacements: modal contributions, $u_{1n}(t)$ and $u_{2n}(t)$, and total responses $u_1(t)$ and $u_2(t)$.

Numerical values of $D_n(t)$ and $\dot{D}_n(t)$ for the nth-mode SDF system with natural period T_n and damping ratio ζ_n to the El Centro ground are determined by the linear acceleration method (Chapter 5) to obtain discrete values of D_n and $\dot{D}_n$ at every Δt. These computations were implemented for the two modal SDF systems with properties given by Eqs. (b) and (c), and the results are plotted in Fig. E14.10a.

Substituting these numerical values of $D_n(t)$ and $\dot{D}_n(t)$ into Eq. (a) of Example 14.9 provides numerical values of the floor displacements; the individual modal contributions are combined to obtain the total response shown in Fig. E14.10b.

14.10 SYSTEMS WITH REAL-VALUED EIGENVALUES

The preceding development led to the complete solution for the earthquake analysis of non-classically damped systems for which all eigenvalues λ_n and the associated eigenvectors are complex valued; the modal damping ratios ζ_n for such modes are less than unity. In general, a damped system—classical or nonclassical—may have an even number of real-valued, negative eigenvalues, each associated with a real-valued eigenvector; the modal damping ratios ζ_n for such modes are greater than unity. Extension of the preceding analysis to determine the earthquake response of such overdamped modes is presented next; although these modal responses are expected to be small, this extension is included for completeness.

14.10.1 Free Vibration[†]

Consider a pair of real-valued eigenvalues λ_n and λ_r, such that $|\lambda_r| > |\lambda_n|$, and the associated real eigenvectors ψ_n and ψ_r, determined by solving the eigenvalue problem of Eq. (A14.2.8). We express this pair of eigenvalues in a form motivated by the eigenvalues of an overdamped SDF system (Appendix 14.6) as

$$\lambda_n = -\zeta_n\omega_n + \omega_{nD} \qquad \lambda_r = -\zeta_n\omega_n - \omega_{nD} \qquad (14.10.1)$$

in which[‡]

$$\omega_{nD} = \omega_n\sqrt{\zeta_n^2 - 1} \qquad (14.10.2)$$

and ω_n and ζ_n are real-valued, positive quantities that can be determined from the known eigenvalues λ_n and λ_r as follows.

Multiplying Eqs. (14.10.1a) and (14.10.1b) and making use of Eq. (14.10.2) leads to

$$\omega_n = \sqrt{\lambda_n\lambda_r} \qquad (14.10.3)$$

Adding Eqs. (14.10.1a) and (14.10.1b) leads to

$$\zeta_n = -\frac{\lambda_n + \lambda_r}{2\omega_n} = -\frac{\lambda_n + \lambda_r}{2\sqrt{\lambda_n\lambda_r}} \qquad (14.10.4)$$

Finally, subtracting Eq. (14.10.1b) from (14.10.1a) gives

$$\omega_{nD} = \frac{\lambda_n - \lambda_r}{2} \qquad (14.10.5)$$

The motion represented by a linear combination of two real-valued eigenvectors ψ_n and ψ_r is given by

$$\mathbf{u}_n(t) = B_n\psi_n e^{\lambda_n t} + B_r\psi_r e^{\lambda_r t} \qquad (14.10.6)$$

[†]This terminology is used for consistency with Section 14.7, although the system returns to its equilibrium position without oscillating (Section 2.2.1).

[‡]Note that the notation ω_{nD} here represents a different quantity compared to Eq. (14.5.6).

in which B_n and B_r in this case are real-valued constants [in contrast to Eq. (14.7.1), where these constants were complex valued] that can be determined from Eq. (14.7.4). Substituting Eq. (14.10.1) and expressing exponential functions in terms of hyperbolic functions, Eq. (14.10.6) may be rewritten as (see Appendix 14.7)

$$\mathbf{u}_n(t) = e^{-\zeta_n \omega_n t}[\beta_n \cosh \omega_{nD} t - \gamma_n \sinh \omega_{nD} t] \tag{14.10.7}$$

where

$$\beta_n = B_r \psi_r + B_n \psi_n \tag{14.10.8}$$

$$\gamma_n = B_r \psi_r - B_n \psi_n \tag{14.10.9}$$

Note the similarity between modal solutions associated with a pair of real-valued eigenvalues [Eq. (14.10.7)] and with a complex-conjugate pair of complex-valued eigenvalues [Eq. (14.7.6)], with the only difference being that hyperbolic functions appear in the former case, in contrast to trigonometric functions in the latter case.

Example 14.11

Determine the free vibration characteristics of the system shown in Fig. E14.11a, a two-story frame idealized as a shear building with a damper in the second story with $c = 2\sqrt{km}$. Show that the eigenvectors satisfy the orthogonality properties.

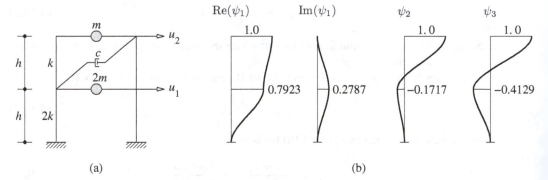

Figure E14.11 (a) Nonclassically damped system; (b) real and imaginary parts of eigenvector ψ_1; and real-valued eigenvectors ψ_2 and ψ_3.

Solution The mass, damping, and stiffness matrices of the system are

$$\mathbf{m} = \begin{bmatrix} 2m & \\ & m \end{bmatrix} \qquad \mathbf{c} = \begin{bmatrix} c & -c \\ -c & c \end{bmatrix} \qquad \mathbf{k} = \begin{bmatrix} 3k & -k \\ -k & k \end{bmatrix} \tag{a}$$

The eigenvalue problem to be solved is defined by Eq. (A14.2.8), repeated here for convenience:

$$\lambda \mathbf{a}\kappa + \mathbf{b}\kappa = 0 \tag{b}$$

where the matrices $\mathbf{a}$ and $\mathbf{b}$, defined in Eq. (A14.2.5), for this system are

$$\mathbf{a} = \begin{bmatrix} \mathbf{0} & \mathbf{m} \\ \mathbf{m} & \mathbf{c} \end{bmatrix} = \begin{bmatrix} 0 & 0 & 2m & 0 \\ 0 & 0 & 0 & m \\ 2m & 0 & c & -c \\ 0 & m & -c & c \end{bmatrix} \tag{c}$$

$$\mathbf{b} = \begin{bmatrix} -\mathbf{m} & \mathbf{0} \\ \mathbf{0} & \mathbf{k} \end{bmatrix} = \begin{bmatrix} -2m & 0 & 0 & 0 \\ 0 & -m & 0 & 0 \\ 0 & 0 & 3k & -k \\ 0 & 0 & -k & k \end{bmatrix} \qquad (d)$$

The eigenvalue problem can be solved numerically using an appropriate algorithm, e.g., the Matlab function $eig(\mathbf{b}, -\mathbf{a})$, resulting in the eigenvalues

$$\lambda_1, \bar{\lambda}_1 = \sqrt{\frac{k}{m}}(-0.0501 \pm 0.7955i) \qquad (e.1)$$

$$\lambda_2 = -0.7232\sqrt{\frac{k}{m}} \qquad \lambda_3 = -2.1766\sqrt{\frac{k}{m}} \qquad (e.2)$$

Note that two of the eigenvalues of the system are a complex-conjugate pair, whereas two are real and negative valued, with $|\lambda_3| > |\lambda_2|$

From the eigenvalues λ_1 and $\bar{\lambda}_1$, ω_1 and ζ_1 can be determined using Eq. (14.5.6):

$$\omega_1 = |\lambda_1| = 0.7971\sqrt{\frac{k}{m}} \qquad \zeta_1 = -\frac{\mathrm{Re}(\lambda_1)}{|\lambda_1|} = 0.0629 \qquad (f)$$

Substituting eigenvalues λ_2 and λ_3 in Eqs. (14.10.3) and (14.10.4) gives

$$\omega_2 = \sqrt{\lambda_2\lambda_3} = 1.2546\sqrt{\frac{k}{m}} \qquad \zeta_2 = -\frac{\lambda_2 + \lambda_3}{2\omega_2} = 1.1556 \qquad (g)$$

From the eigenvalue λ_1 of Eq. (e.1), the corresponding frequency ω_{1D} of the damped system is determined from its definition in Eq. (14.5.5):

$$\omega_{1D} = \mathrm{Im}(\lambda_1) = 0.7955\sqrt{\frac{k}{m}} \qquad (h)$$

However, ω_{2D} should be determined from Eq. (14.10.5):

$$\omega_{2D} = \frac{\lambda_2 - \lambda_3}{2} = 0.7267\sqrt{\frac{k}{m}} \qquad (i)$$

Solution of the eigenvalue problem [Eq. (b)] also provides the 4×1 eigenvectors, but only the third and fourth components [see Eq. (A14.2.7)] are relevant and shown below:

$$\psi_1 = \left\{ \begin{array}{c} 0.7923 + 0.2787i \\ 1 \end{array} \right\} \qquad \psi_2 = \left\{ \begin{array}{c} -0.1717 \\ 1 \end{array} \right\} \qquad \psi_3 = \left\{ \begin{array}{c} -0.4129 \\ 1 \end{array} \right\} \qquad (j)$$

Note that ψ_1 is complex valued, but ψ_2 and ψ_3 are real valued (Fig. E14.11b).

To verify that the eigenvectors ψ_1 and ψ_2 are orthogonal, we compute the individual terms in the left side of Eqs. (14.6.1) and (14.6.2):

$$\psi_1^T \mathbf{m} \psi_2 = \left\{ \begin{array}{c} 0.7923 + 0.2787i \\ 1 \end{array} \right\}^T \begin{bmatrix} 2m & \\ & m \end{bmatrix} \left\{ \begin{array}{c} -0.1717 \\ 1 \end{array} \right\} = m\ (0.7279 - 0.0957i)$$

$$\psi_1^T \mathbf{k} \psi_2 = \left\{ \begin{array}{c} 0.7923 + 0.2787i \\ 1 \end{array} \right\}^T \begin{bmatrix} 3k & -k \\ -k & k \end{bmatrix} \left\{ \begin{array}{c} -0.1717 \\ 1 \end{array} \right\} = k\ (-0.0287 - 0.4222i)$$

$$\psi_1^T \mathbf{c} \psi_2 = \left\{ \begin{array}{c} 0.7923 + 0.2787i \\ 1 \end{array} \right\}^T \begin{bmatrix} c & -c \\ -c & c \end{bmatrix} \left\{ \begin{array}{c} -0.1717 \\ 1 \end{array} \right\} = \sqrt{km}\ (0.4868 - 0.6531i)$$

Substituting these individual terms in the left side of Eqs. (14.6.1) and (14.6.2) gives

$$(\lambda_1 + \lambda_2)\,\psi_1^T \mathbf{m}\psi_2 + \psi_1^T \mathbf{c}\psi_2 = \sqrt{km}\,(-0.7733 + 0.7955i)\,(0.7279 - 0.0957i)$$
$$+\sqrt{km}\,(0.4868 - 0.6531i) = 0$$

$$\psi_1^T \mathbf{k}\psi_2 - \lambda_1\lambda_2\psi_1^T \mathbf{m}\psi_2 = k\,(-0.0287 - 0.4222i)$$
$$- k(0.0362 - 0.5753i)\,(0.7279 - 0.0957i) = 0$$

This verifies that the eigenvectors ψ_1 and ψ_2 computed for the system are orthogonal; similar calculations show that other pairs of eigenvectors are also orthogonal.

In passing, we rewrite Eqs. (h) and (i) to facilitate an interesting observation:

$$\omega_{1D} = 0.9743\sqrt{\frac{2k}{3m}} \qquad \omega_{2D} = 0.8900\sqrt{\frac{2k}{3m}} \tag{k}$$

If the second story were rigid, the system of Fig. E14.11a would reduce to an SDF system with natural vibration frequency = $\sqrt{2k/3m}$. Note that the frequencies of the damped system [Eq. (k)] are close to this reference frequency because the damper coefficient is so large that the damper provides much resistance to deformation in the second story.

Example 14.12

Determine the free vibration response of the two-story shear frame of Fig. E14.11a with $c = 2\sqrt{km}$ due to initial displacements $\mathbf{u}(0) = \langle -\frac{1}{2}\ \ 2\rangle^T$.

Solution The initial displacement and velocity vectors are

$$\mathbf{u}(0) = \begin{Bmatrix} -\frac{1}{2} \\ 2 \end{Bmatrix} \qquad \dot{\mathbf{u}}(0) = \begin{Bmatrix} 0 \\ 0 \end{Bmatrix} \tag{a}$$

Substituting them in Eq. (14.7.3) together with $\mathbf{m}$, $\mathbf{c}$, λ_n, and ψ_n determined in Example 14.11 gives

$$B_1 = \frac{\lambda_1\psi_1^T \mathbf{m}\mathbf{u}(0) + \psi_1^T \mathbf{c}\mathbf{u}(0) + \psi_1^T \mathbf{m}\dot{\mathbf{u}}(0)}{2\lambda_1\psi_1^T \mathbf{m}\psi_1 + \psi_1^T \mathbf{c}\psi_1} = -0.2747 - 0.2440i \tag{b.1}$$

$$B_2 = \frac{\lambda_2\psi_2^T \mathbf{m}\mathbf{u}(0) + \psi_2^T \mathbf{c}\mathbf{u}(0) + \psi_2^T \mathbf{m}\dot{\mathbf{u}}(0)}{2\lambda_2\psi_2^T \mathbf{m}\psi_2 + \psi_2^T \mathbf{c}\psi_2} = 3.5317 \tag{b.2}$$

$$B_3 = \frac{\lambda_3\psi_3^T \mathbf{m}\mathbf{u}(0) + \psi_3^T \mathbf{c}\mathbf{u}(0) + \psi_3^T \mathbf{m}\dot{\mathbf{u}}(0)}{2\lambda_3\psi_3^T \mathbf{m}\psi_3 + \psi_3^T \mathbf{c}\psi_3} = -0.9824 \tag{b.3}$$

Using the B_n in Eq. (b) and ψ_n from Example 14.11, β_1 and γ_1 are determined from Eq. (14.7.5), and β_2 and γ_2 from Eqs. (14.10.8) and (14.10.9) as follows:

$$\beta_1 = \text{Re}(2B_1\psi_1) = \begin{Bmatrix} -0.2992 \\ -0.5493 \end{Bmatrix} \qquad \gamma_1 = \text{Im}(2B_1\psi_1) = \begin{Bmatrix} -0.5397 \\ -0.4880 \end{Bmatrix} \tag{c.1}$$

$$\beta_2 = B_2\psi_2 + B_3\psi_3 = \begin{Bmatrix} -0.2008 \\ 2.5493 \end{Bmatrix} \qquad \gamma_2 = B_3\psi_3 - B_2\psi_2 = \begin{Bmatrix} 1.0120 \\ -4.5142 \end{Bmatrix} \tag{c.2}$$

The free-vibration response is given by

$$\mathbf{u}(t) = \mathbf{u}_1(t) + \mathbf{u}_2(t) \tag{d}$$

where $\mathbf{u}_1(t)$ is determined by substituting β_1 and γ_1 from Eq. (c.1) into the $n = 1$ term on the right side of Eq. (14.7.5):

$$\mathbf{u}_1(t) = e^{-0.0629\,\omega_1 t} \left[\left\{ \begin{matrix} -0.2992 \\ -0.5493 \end{matrix} \right\} \cos \omega_{1D} t - \left\{ \begin{matrix} -0.5397 \\ -0.4880 \end{matrix} \right\} \sin \omega_{1D} t \right] \tag{e}$$

and $\mathbf{u}_2(t)$ is determined by substituting β_2 and γ_2 from Eq. (c.2) into Eq. (14.10.7):

$$\mathbf{u}_2(t) = e^{-1.1556\,\omega_2 t} \left[\left\{ \begin{matrix} -0.2008 \\ 2.5493 \end{matrix} \right\} \cosh \omega_{2D} t - \left\{ \begin{matrix} 1.0120 \\ -4.5142 \end{matrix} \right\} \sinh \omega_{2D} t \right] \tag{f}$$

Substituting Eqs. (e) and (f) into Eq. (d) gives the total response $\mathbf{u}(t)$.

14.10.2 Unit Impulse Response

Recall that unit impulse ground acceleration $\ddot{u}_g(t) = \delta(t)$ imparts to the system the initial velocities $\dot{\mathbf{u}}(0) = -\boldsymbol{\iota}$ but no initial displacements, i.e., $\mathbf{u}(0) = \mathbf{0}$ (Section 14.3). For these initial conditions, the constants B_n^g are given by Eq. (14.8.1) and Eq. (14.10.7) can be expressed in a form analogous to the nth-mode term in Eq. (14.8.3):

$$\mathbf{u}_n(t) = e^{-\zeta_n \omega_n t} \left[\boldsymbol{\beta}_n^g \cosh \omega_{nD} t - \boldsymbol{\gamma}_n^g \sinh \omega_{nD} t \right] \tag{14.10.10}$$

where

$$\boldsymbol{\beta}_n^g = B_r^g \boldsymbol{\psi}_r + B_n^g \boldsymbol{\psi}_n \tag{14.10.11}$$

$$\boldsymbol{\gamma}_n^g = B_r^g \boldsymbol{\psi}_r - B_n^g \boldsymbol{\psi}_n \tag{14.10.12}$$

As in Section 14.8, we prefer to express Eq. (14.10.10) in terms of the unit impulse response function $h_n(t)$ for deformation of an overdamped SDF system with undamped natural frequency ω_n defined by Eq. (14.10.3) and damping ratio ζ_n by Eq. (14.10.4), respectively (see Appendix 14.6):

$$h_n(t) = -\frac{1}{\omega_{nD}} e^{-\zeta_n \omega_n t} \sinh \omega_{nD} t \tag{14.10.13}$$

and its first derivative

$$\dot{h}_n(t) = -e^{-\zeta_n \omega_n t} \cosh \omega_{nD} - \zeta_n \omega_n h_n(t) \tag{14.10.14}$$

which can be rewritten as

$$-e^{-\zeta_n \omega_n t} \cosh \omega_{nD} = \dot{h}_n(t) + \zeta_n \omega_n h_n(t) \tag{14.10.15}$$

The hyperbolic functions multiplying the vectors $\boldsymbol{\beta}_n^g$ and $\boldsymbol{\gamma}_n^g$ in Eq. (14.10.10) are now replaced by the corresponding expressions obtained from Eqs. (14.10.13) and (14.10.15), in terms of $h_n(t)$ and $\dot{h}_n(t)$ to obtain the vector of unit impulse response for displacements $\mathbf{u}_n(t)$:

$$\mathbf{h}_n(t) = -\left[\boldsymbol{\alpha}_n^g \omega_n h_n(t) + \boldsymbol{\beta}_n^g \dot{h}_n(t) \right] \tag{14.10.16}$$

where

$$\alpha_n^g = \zeta_n \beta_n^g - \sqrt{\zeta_n^2 - 1}\, \gamma_n^g \tag{14.10.17}$$

The vectors β_n^g and γ_n^g are given by Eqs. (14.10.11) and (14.10.12), wherein B_n^g are defined by Eq. (14.8.1). Observe that the form of Eq. (14.10.16) is identical to Eq. (14.8.7), but Eq. (14.10.17) is a modified version of Eq. (14.8.8).

Example 14.13

Determine the response of the two-story shear frame of Fig. E14.11a with $c = 2\sqrt{km}$ due to unit impulse ground acceleration, $\ddot{u}_g(t) = \delta(t)$.

Solution B_n^g are determined by substituting $\mathbf{m}$, $\mathbf{c}$, and ψ_n from Example 14.11 in Eq. (14.8.1):

$$B_1^g = \frac{-\psi_1^T \mathbf{m}\iota}{2\lambda_1 \psi_1^T \mathbf{m}\psi_1 + \psi_1^T \mathbf{c}\psi_1} = \sqrt{\frac{m}{k}}\,(0.2232 + 0.7311i) \tag{a.1}$$

$$B_2^g = \frac{-\psi_2^T \mathbf{m}\iota}{2\lambda_2 \psi_2^T \mathbf{m}\psi_2 + \psi_2^T \mathbf{c}\psi_2} = -0.5408\sqrt{\frac{m}{k}} \tag{a.2}$$

$$B_3^g = \frac{-\psi_3^T \mathbf{m}\iota}{2\lambda_3 \psi_3^T \mathbf{m}\psi_3 + \psi_3^T \mathbf{c}\psi_3} = 0.0945\sqrt{\frac{m}{k}} \tag{a.3}$$

Using the B_n^g in Eq. (a) and ψ_n in Eq. (i) of Example 14.10, β_1^g and γ_1^g are determined from Eq. (14.8.2), and β_2^g and γ_2^g from Eqs. (14.10.11) and (14.10.12) as follows:

$$\beta_1^g = \mathrm{Re}\big(2B_1^g\psi_1\big) = \left\{\begin{array}{c} -0.0539 \\ 0.4463 \end{array}\right\}\sqrt{\frac{m}{k}} \qquad \gamma_1^g = \mathrm{Im}\big(2B_1^g\psi_1\big) = \left\{\begin{array}{c} 1.2828 \\ 1.4621 \end{array}\right\}\sqrt{\frac{m}{k}} \tag{b.1}$$

$$\beta_2^g = B_2^g\psi_2 + B_3^g\psi_3 = \left\{\begin{array}{c} 0.0539 \\ -0.4463 \end{array}\right\}\sqrt{\frac{m}{k}} \qquad \gamma_2^g = B_3^g\psi_3 - B_2^g\psi_2 = \left\{\begin{array}{c} 0.1319 \\ -0.6353 \end{array}\right\}\sqrt{\frac{m}{k}} \tag{b.2}$$

The response of the system is given by

$$\mathbf{u}(t) = \mathbf{u}_1(t) + \mathbf{u}_2(t) \tag{c}$$

where $\mathbf{u}_1(t)$ is determined by substituting β_1^g and γ_1^g from Eq. (b.1) into the $n = 1$ term on the right side of Eq. (14.8.3):

$$\mathbf{u}_1(t) = e^{-\zeta_1\omega_1 t}\left[\left\{\begin{array}{c} -0.0539 \\ 0.4463 \end{array}\right\}\cos\omega_{1D}t - \left\{\begin{array}{c} 1.2828 \\ 1.4621 \end{array}\right\}\sin\omega_{1D}t\right]\sqrt{\frac{m}{k}} \tag{d}$$

and $\mathbf{u}_2(t)$ is determined by substituting β_2^g and γ_2^g from Eq. (b.2) into Eq. (14.10.10):

$$\mathbf{u}_2(t) = e^{-\zeta_2\omega_2 t}\left[\left\{\begin{array}{c} 0.0539 \\ -0.4463 \end{array}\right\}\cosh\omega_{2D}t - \left\{\begin{array}{c} 0.1319 \\ -0.6353 \end{array}\right\}\sinh\omega_{2D}t\right]\sqrt{\frac{m}{k}} \tag{e}$$

Substituting Eqs. (d) and (e) into Eq. (c) gives the total response $\mathbf{u}(t)$. The individual terms [Eqs. (d) and (e)] and the total response are plotted in Fig. E14.13; note that $\mathbf{u}_2(t)$ is plotted at a scale different from that for $\mathbf{u}_1(t)$.

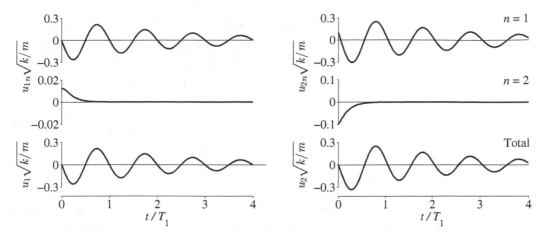

Figure E14.13 Floor displacements of a nonclassically damped system due to unit impulse ground acceleration: modal contributions $u_{1n}(t)$ and $u_{2n}(t)$, and total responses $u_1(t)$ and $u_2(t)$.

The response can also be expressed in terms of the unit impulse response functions $h_n(t)$. For this purpose, Eqs. (14.8.8) and (14.10.17) give

$$\alpha_1^g = \zeta_1 \beta_1^g - \sqrt{1 - \zeta_1^2}\, \gamma_1^g = \left\{ \begin{matrix} -1.2836 \\ -1.4312 \end{matrix} \right\} \sqrt{\frac{m}{k}} \tag{f.1}$$

$$\alpha_2^g = \zeta_2 \beta_2^g - \sqrt{\zeta_2^2 - 1}\, \gamma_2^g = \left\{ \begin{matrix} 0.1386 \\ -0.8838 \end{matrix} \right\} \sqrt{\frac{m}{k}} \tag{f.2}$$

The response of the system can be determined from

$$\mathbf{u}(t) = \mathbf{u}_1(t) + \mathbf{u}_2(t) \tag{g}$$

The first term $\mathbf{u}_1(t)$ is determined by substituting α_1^g from Eq. (f.1) and β_1^g from Eq. (b.1) into the $n = 1$ term on the right side of Eq. (14.8.7):

$$\mathbf{u}_1(t) = \left\{ \begin{matrix} 1.0231 \\ 1.1407 \end{matrix} \right\} h_1(t) + \left\{ \begin{matrix} 0.0539 \\ -0.4463 \end{matrix} \right\} \sqrt{\frac{m}{k}}\, \dot{h}_1(t) \tag{h}$$

where $h_1(t)$ and $\dot{h}_1(t)$ are given by Eqs. (14.8.4) and (14.8.5), respectively. The second term $\mathbf{u}_2(t)$ in Eq. (g) is determined by substituting α_2^g from Eq. (f.2) and β_2^g from Eq. (b.2) into Eq. (14.10.16):

$$\mathbf{u}_2(t) = \left\{ \begin{matrix} -0.1739 \\ 1.1088 \end{matrix} \right\} h_2(t) + \left\{ \begin{matrix} -0.0539 \\ 0.4463 \end{matrix} \right\} \sqrt{\frac{m}{k}}\, \dot{h}_2(t) \tag{i}$$

where $h_2(t)$ and $\dot{h}_2(t)$ are given by Eqs. (14.10.13) and (14.10.14), respectively.

The vectors β_n^g and γ_n^g satisfy Eq. (14.8.9):

$$\beta_1^g + \beta_2^g = \sqrt{\frac{m}{k}}\left[\left\{\begin{matrix}-0.0539\\0.4463\end{matrix}\right\} + \left\{\begin{matrix}0.0539\\-0.4463\end{matrix}\right\}\right] = \left\{\begin{matrix}0\\0\end{matrix}\right\} \tag{j}$$

$$\sum_{n=1}^{N}\omega_n\left[\alpha_n^g - 2\zeta_n\beta_n^g\right] = \omega_1\sqrt{\frac{m}{k}}\left[\left\{\begin{matrix}-1.2836\\-1.4312\end{matrix}\right\} - 2\zeta_1\left\{\begin{matrix}-0.0539\\0.4463\end{matrix}\right\}\right]$$

$$+ \omega_2\sqrt{\frac{m}{k}}\left[\left\{\begin{matrix}0.1386\\-0.8838\end{matrix}\right\} - 2\zeta_2\left\{\begin{matrix}0.0539\\-0.4463\end{matrix}\right\}\right] = \left\{\begin{matrix}-1\\-1\end{matrix}\right\} \tag{k}$$

Equation (k) can be verified by substituting for ω_1 and ω_2 from Example 14.11.

14.10.3 Earthquake Response

For a system subjected to arbitrary ground acceleration $\ddot{u}_g(t)$ the response associated with a linear combination of two real-valued eigenvectors ψ_n and ψ_r, and the associated real-valued eigenvalues λ_n and λ_r is still described by the nth term in the summation of Eq. (14.9.2). This becomes apparent if $\mathbf{h}(t)$ in Eq. (14.9.1) is replaced by $\mathbf{h}_n(t)$ defined in Eq. (14.10.16). Given by Eqs. (14.9.3) and (14.9.4), $D_n(t)$ and $\dot{D}_n(t)$ represent the deformation and relative velocity response of an overdamped SDF system with undamped natural frequency ω_n defined by Eq. (14.10.3) and damping ratio by Eq. (14.10.4), and the vectors α_n^g and β_n^g are given by Eqs. (14.10.17) and (14.10.11), respectively. Note that the SDF system responses $D_n(t)$ and $\dot{D}_n(t)$ are usually computed by one of the numerical methods presented in Chapter 5, not by evaluating the convolution integrals of Eqs. (14.9.3) and (14.9.4).

Example 14.14

Derive equations for the floor displacements of the shear frame of Fig. E14.11a with $c = 2\sqrt{km}$ subjected to ground acceleration $\ddot{u}_g(t)$.

Solution   Substituting the α_n^g and β_n^g determined in Example 14.13 into Eq. (14.9.2) provides equations for the floor displacements:

$$\mathbf{u}(t) = \left\{\begin{matrix}1.0231\\1.1407\end{matrix}\right\}D_1(t) + \left\{\begin{matrix}0.0539\\-0.4463\end{matrix}\right\}\sqrt{\frac{m}{k}}\,\dot{D}_1(t) + \left\{\begin{matrix}-0.1739\\1.1088\end{matrix}\right\}D_2(t) + \left\{\begin{matrix}-0.0539\\0.4463\end{matrix}\right\}\sqrt{\frac{m}{k}}\,\dot{D}_2(t)$$

where $D_n(t)$ and $\dot{D}_n(t)$ represent the deformation and relative velocity response of the nth-mode SDF system; see Eqs. (14.9.3) and (14.9.4). Recall that in these equations, $h_1(t)$ and $\dot{h}_1(t)$ are given by Eq. (14.8.4) and (14.8.5), respectively, whereas $h_2(t)$ and $\dot{h}_2(t)$ are given by Eq. (14.10.13) and (14.10.14), respectively.

14.11 RESPONSE SPECTRUM ANALYSIS

In Part B of Chapter 13, we presented the response spectrum analysis (RSA) procedure for classically damped systems to estimate their peak response directly from the earthquake response (or design) spectrum. Researchers have developed RSA procedures also for non-classically damped systems. The rationale for not including them here is as follows: These procedures require two response spectra: (1) pseudo-acceleration (or pseudo-velocity or deformation) response spectrum, which was needed for RSA procedures presented in Chapter 13; and (2) relative velocity response spectrum. Because the latter is not readily available or specified as part of structural design criteria, it has often been approximated by the pseudo-velocity response spectrum. As mentioned in Section 6.12, this approximation is valid only over a limited range of vibration periods and damping ratios.

14.12 SUMMARY

The preceding analysis of the earthquake response of an N-DOF nonclassically damped system with known mass, stiffness, and damping matrices $\mathbf{m}$, $\mathbf{k}$, and $\mathbf{c}$ to ground acceleration $\ddot{u}_g(t)$ is summarized as a sequence of steps:

1. Compute the eigenvalues λ_n and the associated eigenvectors ψ_n by solving the eigenvalue problem [Eq. (A14.2.8)].
2. Determine the damped and pseudo-undamped natural frequencies ω_{nD} and ω_n, and the modal damping ratios ζ_n, from Eqs. (14.5.5) and (14.5.6) for a complex-conjugate pair of eigenvalues, λ_n and $\bar{\lambda}_n$, or from Eqs. (14.10.3) through (14.10.5) for a pair of real-valued eigenvalues, λ_n and λ_r.
3. Determine the complex-valued constants B_n^g [Eq. (14.8.1)].
4. Determine β_n^g, γ_n^g, and α_n^g as follows: For a complex-conjugate pair of eigenvalues λ_n and $\bar{\lambda}_n$, compute the complex-valued product $2B_n^g \psi_n$, then determine β_n^g and γ_n^g from Eq. (14.8.2) and α_n^g from Eq. (14.8.8). For a pair of real eigenvalues λ_n and λ_r, compute β_n^g, γ_n^g, and α_n^g from Eqs. (14.10.11), (14.10.12), and (14.10.17), respectively.
5. Compute the deformation response $D_n(t)$ and relative velocity response $\dot{D}_n(t)$ of the nth-mode SDF system with ω_n and ζ_n determined in step 2 to prescribed ground acceleration $\ddot{u}_g(t)$ by one of the numerical time-stepping methods (Chapter 5).
6. Compute the displacements $\mathbf{u}(t)$ from Eq. (14.9.2) using the α_n^g and β_n^g appropriate for each term in the summation; see step 4.

The analysis of nonclassically damped systems differs from that of classically damped systems (Section 13.2.4) in two principal ways:

1. The eigenvalue problem to be solved [Eq. (A14.2.8)] is now of order $2N$.
2. In addition to the deformation $D_n(t)$ of the nth-mode SDF system, its relative velocity $\dot{D}_n(t)$ is now required. However, in a step-by-step numerical evaluation of the response of an SDF system (Chapter 5), $\dot{D}_n(t)$ is normally computed in the process of obtaining $D_n(t)$.

FURTHER READING

Caughey, T. K., and O'Kelly, M. E. J., "Effect of Damping on the Natural Frequencies of Linear Dynamic Systems," *Journal of the Acoustical Society of America*, **33**, 1961, pp. 1458–1461.

Caughey, T. K., and O'Kelly, M. E. J., "Classical Normal Modes in Damped Linear Dynamic Systems," *ASME Journal of Applied Mechanics*, **32**, 1965, pp. 583–588.

Veletsos, A. S., and Ventura, C. E., "Modal Analysis of Non-classically Damped Linear Dynamic Systems," *Earthquake Engineering and Structural Dynamics*, **14**, 1986, pp. 217–243.

APPENDIX 14: DERIVATIONS

A14.1 Complex-Valued Constants: Free Vibration of Classically Damped Systems

The displacements given by Eq. (14.2.3) are repeated herein for convenience:

$$\mathbf{u}(t) = \sum_{r=1}^{N}\left(B_r\boldsymbol{\phi}_r e^{\lambda_r t} + \bar{B}_r\boldsymbol{\phi}_r e^{\bar{\lambda}_r t}\right) = 2\sum_{r=1}^{N}\text{Re}\left(B_r\boldsymbol{\phi}_r e^{\lambda_r t}\right) \tag{A14.1.1}$$

These displacements can be differentiated to obtain the velocity response:

$$\dot{\mathbf{u}}(t) = \sum_{r=1}^{N}\left(\lambda_r B_r\boldsymbol{\phi}_r e^{\lambda_r t} + \bar{\lambda}_r \bar{B}_r\boldsymbol{\phi}_r e^{\bar{\lambda}_r t}\right) \tag{A14.1.2}$$

Setting $t = 0$ in Eqs. (A14.1.1) and (A14.1.2) gives

$$\mathbf{u}(0) = \sum_{r=1}^{N}\left(B_r + \bar{B}_r\right)\boldsymbol{\phi}_r \qquad \dot{\mathbf{u}}(0) = \sum_{r=1}^{N}\left(\lambda_r B_r + \bar{\lambda}_r \bar{B}_r\right)\boldsymbol{\phi}_r \tag{A14.1.3}$$

With the initial displacements $\mathbf{u}(0)$ and initial velocities $\dot{\mathbf{u}}(0)$ known, each of these two equations represents N algebraic equations in the $2N$ unknowns $\text{Re}(B_r)$ and $\text{Im}(B_r)$. Premultiplying both sides of Eqs. (A14.1.3a) and (A14.1.3b) by $\boldsymbol{\phi}_n^T\mathbf{m}$ and utilizing the orthogonality property of modes [Eq. (10.4.1)] gives

$$\boldsymbol{\phi}_n^T\mathbf{m}\mathbf{u}(0) = \left(B_n + \bar{B}_n\right)\boldsymbol{\phi}_n^T\mathbf{m}\boldsymbol{\phi}_n \qquad \boldsymbol{\phi}_n^T\mathbf{m}\dot{\mathbf{u}}(0) = \left(\lambda_n B_n + \bar{\lambda}_n \bar{B}_n\right)\boldsymbol{\phi}_n^T\mathbf{m}\boldsymbol{\phi}_n \tag{A14.1.4}$$

which may be written as

$$B_n + \bar{B}_n = q_n(0) \qquad \lambda_n B_n + \bar{\lambda}_n \bar{B}_n = \dot{q}_n(0) \tag{A14.1.5}$$

where Eq. (10.8.5), which defines the initial displacement $q_n(0)$ and initial velocity $\dot{q}_n(0)$ of the modal coordinate $q_n(t)$, has been utilized. The two equations (A14.1.5) are solved to determine the real and imaginary parts of B_n:

$$\text{Re}(B_n) = \frac{q_n(0)}{2} \qquad \text{Im}(B_n) = -\frac{\dot{q}_n(0) + \zeta_n\omega_n q_n(0)}{2\omega_{nD}} \tag{A14.1.6}$$

leading to Eq. (14.2.4) for the complex-valued constant B_n.

A14.2 First-Order Equations of Motion and Eigenvalue Problem

Equation (14.1) is written in augmented form:

$$m\dot{u} - m\dot{u} = 0 \tag{A14.2.1}$$

$$m\ddot{u} + c\dot{u} + ku = -m\iota\ddot{u}_g(t) \tag{A14.2.2}$$

The preceding two sets of equations can be combined to obtain a state-space formulation of the equations of motion:

$$a\dot{\hat{u}} + b\hat{u} = e(t) \tag{A14.2.3}$$

in which $\hat{u}$ and $e(t)$ are vectors of $2N$ elements defined as

$$\hat{u} = \begin{Bmatrix} \dot{u} \\ u \end{Bmatrix} \qquad e(t) = \begin{Bmatrix} 0 \\ -m\iota\ddot{u}_g(t) \end{Bmatrix} \tag{A14.2.4}$$

and a and b are square matrices of order $2N$ given by

$$a = \begin{bmatrix} 0 & m \\ m & c \end{bmatrix} \qquad b = \begin{bmatrix} -m & 0 \\ 0 & k \end{bmatrix} \tag{A14.2.5}$$

Thus, the system of N second-order differential equations [Eq. (14.1)] has been reduced to a system of 2N first-order differential equations (A14.2.3).

The solution of the homogeneous form of Eq. (A14.2.3) is of the form

$$\hat{u}(t) = \kappa e^{\lambda t} \tag{A14.2.6}$$

where λ is an eigenvalue and κ is the associated eigenvector of $2N$ elements. The lower N elements of κ represent the desired modal displacements ψ and the upper N elements represent the corresponding modal velocities $\lambda\psi$; i.e.,

$$\kappa = \begin{Bmatrix} \lambda\psi \\ \psi \end{Bmatrix} \tag{A14.2.7}$$

Substituting Eq. (A14.2.6) into the homogeneous form of Eq. (A14.2.3), i.e., the equation governing free vibration of nonclassically damped systems, leads to the eigenvalue problem:

$$\lambda a\kappa + b\kappa = 0 \tag{A14.2.8}$$

The 2N roots of λ are either real valued and negative or they occur in complex-conjugate pairs with negative (or zero) real parts; the latter fact can be demonstrated as follows: If λ_n and κ_n are an eigenvalue–eigenvector pair, they satisfy

$$(\lambda_n a + b)\kappa_n = 0 \tag{A14.2.9}$$

Taking the conjugate of both sides and noting that a and b are real valued, i.e., $\bar{a} = a$ and $\bar{b} = b$, gives

$$\left(\bar{\lambda}_n a + b\right)\bar{\kappa}_n = 0 \tag{A14.2.10}$$

which implies that $\bar{\lambda}_n$ and $\bar{\kappa}_n$ also satisfy Eq. (A14.2.8), and hence represent an eigenvalue–eigenvector pair.

[†]Other versions of state-space analysis, which are mathematically equivalent to the ones presented here, are popular in mathematics, physics, electrical engineering, and control.

A14.3 Orthogonality of Modes

Since $\mathbf{a}$ and $\mathbf{b}$ in Eq. (A14.2.3) are real-valued symmetric matrices, the eigenvectors κ_n and κ_r corresponding to any pair of distinct eigenvalues λ_n and λ_r can be shown to satisfy the orthogonality relations

$$\kappa_n^T \mathbf{a} \kappa_r = 0 \qquad \kappa_n^T \mathbf{b} \kappa_r = 0 \tag{A14.3.1}$$

These relations are also valid for a complex-conjugate pair κ_n and $\bar{\kappa}_n$ since the associated eigenvalues λ_n and $\bar{\lambda}_n$ are different. Substituting Eqs. (A14.2.5) and (A14.2.7) into Eq. (A14.3.1) leads to the orthogonality relations expressed by Eqs. (14.6.1) and (14.6.2).

Special Case.   For a classically damped system $\psi_n = \bar{\psi}_n = \phi_n$, $\omega_n = \omega_n^o$,

$$\lambda_n = -\zeta_n \omega_n^o + i\omega_n^o \sqrt{1 - \zeta^2} \tag{A14.3.2}$$

and the eigenvalue problem is

$$\mathbf{k}\phi_n = \left(\omega_n^o\right)^2 \mathbf{m}\phi_n \tag{A14.3.3}$$

Specializing Eq. (14.5.3) for the nth eigenvalue and eigenvector pair, and replacing ψ_n by ϕ_n gives

$$\lambda_n^2 \mathbf{m}\phi_n + \lambda_n \mathbf{c}\phi_n + \mathbf{k}\phi_n = \mathbf{0} \tag{A14.3.4}$$

Substituting Eq. (A14.3.3) for the last term on the left side of Eq. (A14.3.4), combining it with the first term, and making use of Eq. (A14.3.2) leads to the relation

$$\mathbf{c}\phi_r = 2\zeta_r \omega_r^o \mathbf{m}\phi_r \tag{A14.3.5}$$

where the subscript n has been replaced by r. Premultiplying both sides of Eq. (A14.3.5) by ϕ_n^T and recognizing that $\psi_n = \phi_n$, we obtain

$$\psi_n^T \mathbf{c}\psi_r = \phi_n^T \mathbf{c}\phi_r = 2\zeta_r \omega_r^o \phi_n^T \mathbf{m}\phi_r \tag{A14.3.6}$$

On making use of Eq. (A14.3.6), the first orthogonality condition given by Eq. (14.6.1) reduces to

$$\phi_n^T \mathbf{m}\phi_r = 0 \tag{A14.3.7}$$

Furthermore, substituting Eq. (A14.3.7) into Eq. (14.6.2) leads to the second orthogonality condition:

$$\phi_n^T \mathbf{k}\phi_r = 0 \tag{A14.3.8}$$

Equations (A14.3.7) and (A14.3.8) are the familiar orthogonality relationships for classically damped systems, introduced in Section 10.4.

A14.4 Complex-Valued Constants: Free Vibration of Nonclassically Damped Systems

The vector $\hat{\mathbf{u}}$ can be expanded in terms of the $2N$ eigenvectors

$$\hat{\mathbf{u}}(t) = \sum_{n=1}^{2N} \kappa_n \hat{q}_n(t) = \kappa \hat{q}(t) \tag{A14.4.1}^\dagger$$

---

† $\hat{}$ appears in $\hat{q}_n$ and $\hat{q}$ to distinguish them from the modal coordinates for classically damped systems, introduced in Section 10.7.

where $\kappa = [\kappa_1 \quad \kappa_2 \quad \cdots \quad \kappa_{2N}]$, $\hat{q}_n$ are scalar multipliers, and $\hat{\mathbf{q}} = <\hat{q}_1 \quad \hat{q}_2 \quad \cdots \quad \hat{q}_{2N}>^T$. The $\hat{q}_n$ are to be determined for prescribed initial conditions. Substituting Eq. (A14.4.1) in the homogeneous form of Eq. (A14.2.3) gives

$$\mathbf{a}\kappa\hat{\mathbf{q}} + \mathbf{b}\kappa\hat{\mathbf{q}} = \mathbf{0}$$

Premultiplying each term by κ^T gives

$$(\kappa^T \mathbf{a}\kappa)\hat{\mathbf{q}} + (\kappa^T \mathbf{b}\kappa)\hat{\mathbf{q}} = \mathbf{0}$$

Because of the orthogonality relations of Eq. (14.3.1), the two coefficient matrices are diagonal matrices, resulting in a set of $2N$ uncoupled equations governing $\hat{q}_n(t)$:

$$A_{nn}\hat{\dot{q}}_n + B_{nn}\hat{q}_n = 0 \tag{A14.4.2}$$

where

$$A_{nn} = \kappa_n^T \mathbf{a}\kappa_n \qquad B_{nn} = \kappa_n^T \mathbf{b}\kappa_n \tag{A14.4.3}$$

Premultiplying Eq. (A14.2.9) by κ_n^T gives $B_{nn} = -\lambda_n A_{nn}$, which is substituted in Eq. (A14.4.2) to obtain

$$\hat{\dot{q}}_n - \lambda_n \hat{q}_n = 0 \tag{A14.4.4}$$

which is the equation governing $\hat{q}_n(t)$ associated with the eigenvector κ_n (and eigenvalue λ_n). A companion equation exists for the complex-conjugate eigenvector $\bar{\kappa}_n$. It can be demonstrated that

$$\bar{A}_{nn} = \bar{\kappa}_n^T \mathbf{a}\bar{\kappa}_n \qquad \bar{B}_{nn} = \bar{\kappa}_n^T \mathbf{b}\bar{\kappa}_n \tag{A14.4.5}$$

and $\bar{B}_{nn} = -\bar{\lambda}_n \bar{A}_{nn}$. From these relationships it can be shown that the companion equation is

$$\hat{\dot{q}}_n - \bar{\lambda}_n \hat{q}_n = 0 \tag{A14.4.6}$$

The general solutions of Eqs. (A14.4.4) and (A14.4.6) are

$$\hat{q}_n(t) = B_n e^{\lambda_n t} \qquad \hat{q}_n(t) = \bar{B}_n e^{\bar{\lambda}_n t} \tag{A14.4.7}$$

respectively, where the constants B_n and $\bar{B}_n$ are to be determined from the prescribed initial displacements and velocities that initiate free vibration. Thus the response associated with the eigenvector pair κ_n and $\bar{\kappa}_n$ is given by the two corresponding terms in the summation of Eq. (A14.4.1):

$$\hat{\mathbf{u}}_n(t) = B_n \kappa_n e^{\lambda_n t} + \bar{B}_n \bar{\kappa}_n e^{\bar{\lambda}_n t} \tag{A14.4.8}$$

where the lower N equations are the same as Eq. (14.7.1), which becomes apparent after substituting Eqs. (A14.2.4a) and (A14.2.7). Combining such response contributions for the N pairs of eigenvectors gives the complete solution of the homogeneous form of Eq. (A14.2.3):

$$\hat{\mathbf{u}}(t) = \sum_{r=1}^{N} B_r \kappa_r e^{\lambda_r t} + \sum_{r=1}^{N} \bar{B}_r \bar{\kappa}_r e^{\bar{\lambda}_r t} \tag{A14.4.9}$$

The complex-valued constants B_n are to be determined from the initial conditions

$$\hat{\mathbf{u}}(0) = \left\{ \begin{array}{c} \dot{\mathbf{u}}(0) \\ \mathbf{u}(0) \end{array} \right\} \tag{A14.4.10}$$

where $\mathbf{u}(0)$ and $\dot{\mathbf{u}}(0)$ are the vectors of initial displacements and velocities, respectively. Specializing Eq. (A14.4.9) for $t = 0$ gives

$$\hat{\mathbf{u}}(0) = \sum_{r=1}^{N} B_r \boldsymbol{\kappa}_r + \sum_{r=1}^{N} \bar{B}_r \bar{\boldsymbol{\kappa}}_r \qquad (A14.4.11)$$

Premultiplying both sides of Eq. (A14.4.11) by $\boldsymbol{\kappa}_n^T \mathbf{a}$ gives

$$\boldsymbol{\kappa}_n^T \mathbf{a} \hat{\mathbf{u}}(0) = \sum_{r=1}^{N} B_r \boldsymbol{\kappa}_n^T \mathbf{a} \boldsymbol{\kappa}_r + \sum_{r=1}^{N} \bar{B}_r \boldsymbol{\kappa}_n^T \mathbf{a} \bar{\boldsymbol{\kappa}}_r \qquad (A14.4.12)$$

Because of the orthogonality condition of Eq. (A14.3.1a), all terms in both summations vanish except the $r = n$ term in the first summation; thus

$$\boldsymbol{\kappa}_n^T \mathbf{a} \hat{\mathbf{u}}(0) = \left(\boldsymbol{\kappa}_n^T \mathbf{a} \boldsymbol{\kappa}_n \right) B_n$$

The matrix products on both sides of this equation are scalars. Therefore,

$$B_n = \frac{\boldsymbol{\kappa}_n^T \mathbf{a} \hat{\mathbf{u}}(0)}{\boldsymbol{\kappa}_n^T \mathbf{a} \boldsymbol{\kappa}_n} \qquad (A14.4.13)$$

which, on substituting Eqs. (A14.2.5), (A14.2.7), and (A14.4.10), reduces to Eq. (14.7.4).

Special case: classically damped systems. For classically damped systems, $\psi_n = \phi_n$, $\omega_n = \omega_n^o$, λ_n is given by Eq. (A14.3.2), Eq. (A14.3.5) is premultiplied by ϕ_r^T, and the subscript r is replaced by n to obtain

$$\phi_n^T \mathbf{c} \, \phi_n = 2\zeta_n \omega_n^o \phi_n^T \mathbf{m} \phi_n \qquad (A14.4.14)$$

and the following generalized version of Eq. (A14.3.6) is valid:

$$\phi_n^T \mathbf{c} \mathbf{u}(0) = 2\zeta_n \omega_n^o \phi_n^T \mathbf{m} \, \mathbf{u}(0) \qquad (A14.4.15)$$

On making use of these results and of Eqs. (A14.3.2) and (A14.3.6), Eq. (14.7.4) reduces to Eq. (14.2.4).

A14.5 Derivation of Eq. (14.8.9)

Differentiating Eq. (14.8.3) gives the velocity vector $\dot{\mathbf{u}}(t)$ that is specialized for $t = 0$ to obtain

$$\dot{\mathbf{u}}(0) = \sum_{n=1}^{N} -\omega_n \left[\zeta_n \beta_n^g + \sqrt{1 - \zeta_n^2} \gamma_n^g \right] \qquad (A14.5.1)$$

Expressing the second term on the right side in terms of α_n^g and β_n^g by using Eq. (14.8.8) yields

$$\dot{\mathbf{u}}(0) = \sum_{n=1}^{N} \omega_n \left[\alpha_n - 2\zeta_n \beta_n^g \right] \qquad (A14.5.2)$$

Imposing the initial condition $\dot{\mathbf{u}}(0) = -\boldsymbol{\iota}$ in this equation leads to Eq. (14.8.9b).

A14.6 Overdamped SDF System

The characteristic equation in Derivation 2.2 (Chapter 2) is rewritten for damping ratio $\zeta > 1$ as

$$\lambda_{1,2} = -\zeta\omega_n \pm \omega_D \tag{A14.6.1}$$

where

$$\omega_D = \omega_n\sqrt{\zeta^2 - 1} \tag{A14.6.2}$$

The solution of Eq. (2.2.1b) is

$$u(t) = a_1 e^{\lambda_1 t} + a_2 e^{\lambda_2 t} \tag{A14.6.3}$$

which after substituting Eq. (A14.6.1) becomes

$$u(t) = e^{-\zeta\omega_n t}\left(a_1 e^{\omega_D t} + a_2 e^{-\omega_D t}\right) \tag{A14.6.4}$$

The constants of integration a_1 and a_2 are determined by imposing the requirement that $u(t)$ must satisfy the given initial displacement and initial velocity at $t = 0$.

A unit impulse ground acceleration $\ddot{u}_g(t) = \delta(t)$ applied at $t = 0$ imparts to the system the initial velocity $\dot{u}(0) = -1$ and initial displacement $u(0) = 0$. For these initial conditions, the constants are

$$a_1 = -\frac{1}{2\omega_D} \qquad a_2 = \frac{1}{2\omega_D} \tag{A14.6.5}$$

Substituting Eq. (A14.6.5) in Eq. (A14.6.4) leads to the resulting free vibration response:

$$u(t) = -\frac{1}{2\omega_D} e^{-\zeta\omega_n t}\left(e^{\omega_D t} - e^{-\omega_D t}\right) \tag{A14.6.6}$$

The exponential functions within the parentheses are related to hyperbolic functions as follows:

$$e^{\omega_D t} = \cosh\omega_D t + \sinh\omega_D t \qquad e^{-\omega_D t} = \cosh\omega_D t - \sinh\omega_D t \tag{A14.6.7}$$

Substituting these relations into Eq. (A14.6.6) leads to

$$h(t) = -\frac{1}{\omega_D} e^{-\zeta\omega_n t} \sinh\omega_D t \tag{A14.6.8}$$

where $h(t)$ denotes the unit impulse response of an overdamped SDF system.

Interpreting ω_n as given by Eq. (14.10.3) and changing the notation from ω_D to ω_{nD} [Eq. (14.10.2)], and from ζ to ζ_n [Eq. (14.10.4)], Eq. (A14.6.8) can be rewritten as Eq. (14.10.13).

A14.7 Derivation of Eq. (14.10.7)

Substituting Eq. (14.10.1) into Eq. (14.10.6) leads to

$$\mathbf{u}_n(t) = e^{-\zeta_n\omega_n t}\left(B_n\psi_n e^{\omega_{nD} t} + B_r\psi_r e^{-\omega_{nD} t}\right) \tag{A14.7.1}$$

The exponential functions within the parentheses are related to hyperbolic functions as follows:

$$e^{\omega_{nD} t} = \cosh\omega_{nD} t + \sinh\omega_{nD} t \qquad e^{-\omega_{nD} t} = \cosh\omega_{nD} t - \sinh\omega_{nD} t \tag{A14.7.2}$$

Substituting these relations into Eq. (A14.7.1) and collecting the terms containing $\cosh \omega_{nD} t$ and those where $\sinh \omega_{nD} t$ appears gives

$$\mathbf{u}_n(t) = e^{-\zeta_n \omega_n t}[(B_r \psi_r + B_n \psi_n) \cosh \omega_{nD} t - (B_r \psi_r - B_n \psi_n) \sinh \omega_{nD} t] \qquad \text{(A14.7.3)}$$

With β_n and γ_n defined by Eqs. (14.10.8) and (14.10.9), Eq. (A14.7.3) is equivalent to Eq. (14.10.7).

PROBLEMS

14.1 Determine the natural frequencies, natural modes, and modal damping ratios for the two-story shear frame of Figure P9.5 with damping. The Rayleigh damping matrix provides a damping ratio of 5% in both modes. Use the theory for nonclassically damped systems, developed in Section 14.5, to solve this problem for a classically damped system. Verify that the eigenvectors are orthogonal. Verify that the results match the solution of Problem 10.6 by classical modal analysis.

14.2 Determine the natural frequencies, natural modes, and modal damping ratios for the two-story frame of Figure P9.5, with a damper only in the first story with damping coefficient $c_1 = 0.4\sqrt{km}$, where $k = 24EI/h^3$ is the story stiffness; express frequencies in terms of m and k. Show that the natural modes satisfy the orthogonality properties.

14.3 Determine the free vibration response of the two-story shear frame of Problem 14.1, a classically damped system, due to initial displacements of Figure P10.8a. Use the theory for nonclassically damped systems developed in Section 14.7, to solve the problem. Verify that the results match the solution of Problem 10.9 by classical modal analysis.

14.4 Determine the free vibration response of the two-story shear frame of Problem 14.2 due to initial displacements of Fig. P10.8a.

14.5 Determine the response of the two-story shear frame of the system of Problem 14.1, a classically damped system, due to unit impulse ground acceleration, $\ddot{u}_g(t) = \delta(t)$. Use the theory for nonclassically damped systems developed in Section 14.8, to solve the problem. Compare the result with the solution given by Eqs. (14.3.6) and (14.3.7) for classically damped systems.

14.6 Determine the free vibration response of the two-story shear frame of Problem 14.2 due to unit impulse ground acceleration, $\ddot{u}_g(t) = \delta(t)$. Verify that Eq. (14.8.9) is satisfied.

14.7 For the two-story shear frame of Problem 14.1, a classically damped system, excited by horizontal ground motion $\ddot{u}_g(t)$, determine the floor displacement response in terms of $D_n(t)$. Use the theory for nonclassically damped systems developed in Section 14.9 to solve the problem. Compare the result with that determined in Problem 13.1.

14.8 For the two-story shear frame of Problem 14.2 excited by horizontal ground motion $\ddot{u}_g(t)$, determine the floor displacement response in terms of $D_n(t)$ and $\dot{D}_n(t)$.

14.9 Determine the natural frequencies, natural modes, and modal damping ratios for the two-story frame of Fig. P9.5, with dampers $c_1 = 0.6\sqrt{km}$ in the first story and $c_2 = 1.2\sqrt{km}$ in the second story, where $k = 24EI/h^3$ is the story stiffness; express frequencies in terms of m and k. Show that the natural modes satisfy the orthogonality properties.

14.10 Determine the free vibration response of the two-story shear frame of Problem 14.9 due to initial displacements of Fig. P10.8a.

14.11   Determine the free vibration response of the two-story shear frame of Problem 14.9 due to unit impulse ground acceleration, $\ddot{u}_g(t) = \delta(t)$. Verify that Eq. (14.8.9) is satisfied.

14.12   For the two-story shear frame of Problem 14.9 excited by horizontal ground motion $\ddot{u}_g(t)$, determine the floor displacement response in terms of $D_n(t)$ and $\dot{D}_n(t)$

14.13   Consider a one-story building with mass m, lateral stiffness k, and damping coefficient c (Fig. 21.2.1a). On a fixed base, this SDF system has the natural frequency ω_f, natural period $T_f = 0.4$ sec, and damping ratio $\zeta_f = 2\%$; the subscript f is chosen instead of n to emphasize that these are properties of the structure on a fixed base. As shown in Fig. 21.2.1b. this one-story building is mounted on a base slab of mass $m_b = 2m/3$, which in turn is supported on a base isolation system with lateral stiffness k_b and linear viscous damping c_b. The isolation system is characterized by two parameters:

$$T_b = 2\pi \div \sqrt{\frac{k_b}{m + m_b}} \qquad \zeta_b = \frac{c_b}{2(m + m_b)\,\omega_b}$$

which are given: $T_b = 2.0$ sec, $\zeta_b = 10\%$.

Determine the response of this nonclassically damped system to the El Centro ground motion by three methods:

1. Solving the coupled equations of motion.

2. Using the theory developed in Sections 14.5 to 14.9. Verify that these results agree with those from Method 1.

3. Modal analysis of the system approximated as classically damped by neglecting the off-diagonal terms in $\mathbf{C}$, the damping matrix in modal coordinates [see Eq. (10.9.5)].

15

Reduction of Degrees of Freedom

PREVIEW

Although our objective in this book is the analysis of structures for dynamic excitation, we recognize that in practice a dynamic analysis is usually preceded by static analysis for dead and live loads. The structural idealization for the static analysis is dictated by the complexity of the structure, and several hundred to a few thousand DOFs may be necessary for accurate evaluation of the internal element forces and stresses in a complex structure.

The same refined idealization may be used for dynamic analysis of the structure, but this may be unnecessarily refined and drastically fewer DOFs could suffice. Such is the case because the dynamic response of many structures can be represented well by the first few natural vibration modes, and these modes can be determined accurately from a structural idealization with drastically fewer DOFs than required for static analysis. Thus we are interested in reducing the number of DOFs as much as reasonably possible before proceeding with computation of natural frequencies and modes, which is perhaps the most demanding phase of dynamic analysis.

Presented in this chapter are two approaches to reducing the number of DOFs: mass lumping in selected DOFs and the Rayleigh–Ritz method. Before presenting these procedures we mention how kinematic constraints based on structural properties can be used to reduce the number of DOFs in the structural idealization for static analysis; this idealization is the starting point for dynamic analysis.

15.1 KINEMATIC CONSTRAINTS

The configuration and properties of a structure may suggest kinematic constraints that express the displacements of many DOFs in terms of a smaller set of displacements. For example, the floor diaphragms (or slabs) of a multistory building, although flexible in the vertical direction, are usually very stiff in their own plane and can be assumed as rigid without introducing significant error. With this assumption the horizontal displacements of all the joints at one floor level are related to the three rigid-body DOFs of the diaphragm in its own plane: the two horizontal components of displacement and rotation about a vertical axis.

As a result of this kinematic constraint, the number of DOFs that would be considered in a static analysis can be reduced almost by half. Consider, for example, the 20-story building shown in Fig. 15.1.1, consisting of eight frames in the y-direction and four in the x-direction. With 640 joints and six DOFs (three translations and three rotations) per joint, the system has 3840 DOFs. Assuming the floor diaphragms to be rigid in their own planes, the system has only 1980 DOFs. These include the vertical displacement and two rotations (in xz and yz planes) of each joint and three rigid-body DOFs per floor.

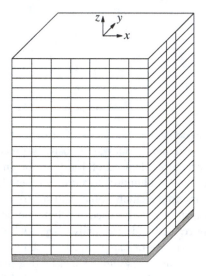

Figure 15.1.1   Twenty-story building.

Another kinematic constraint sometimes assumed in building analysis is that the columns are axially rigid. This assumption should be used with discretion because it may be reasonable only in special circumstances: for example, buildings that are *not* slender. If justifiable, the assumption leads to further reduction in the number of DOFs; for static analysis of the multistory building of Fig. 15.1.1 this number reduces to 1340.

Once the structural idealization has been established for static analysis after considering kinematic constraints appropriate to the structure, the number of DOFs can be reduced for dynamic analysis by the procedures presented next.

15.2 MASS LUMPING IN SELECTED DOFs

As mentioned in Section 9.2.4, the mass is distributed throughout an actual structure, but it can be idealized as lumped at the nodes of the discretized structure with the mass in rotational DOFs usually set to zero. Procedures were then described to determine the lumped masses. The DOFs in the structural idealization established for static analysis are subdivided into two parts: $\mathbf{u}_t$, which have mass, and the remaining $\mathbf{u}_0$, which have zero mass and no external dynamic force, but are necessary for accurate representation of the stiffness properties of the structure. The DOFs $\mathbf{u}_0$ are related to $\mathbf{u}_t$ by Eq. (9.3.3) and, as indicated by Eq. (9.3.4), the equations of motion can be formulated in terms of only $\mathbf{u}_t$, the dynamic DOFs. This is the static condensation method developed in Section 9.3 by which the number of DOFs is reduced to only the dynamic DOFs.

The static condensation method is especially effective in earthquake analysis of multistory buildings subjected to horizontal ground motion because of three special features of this class of structures and excitations. First, the floor diaphragms (or floor slabs) are usually assumed to be rigid in their own plane. Second, the effective earthquake forces [Eq. (9.4.9)] associated with rotations and vertical displacements of the joints are zero. Third, the inertial effects associated with these same DOFs are usually not significant in the lower vibration modes that contribute dominantly to structural response. Assigning zero mass to these DOFs leaves only the three rigid-body DOFs of each floor diaphragm for dynamic analysis. For the 20-story building of Fig. 15.1.1, this method reduces the number of degrees of freedom from 1980 to 60.

The reduction in actual computational effort may be much less significant, however, than the reduction in the number of DOFs. This is because the efficiency of computation permitted by the narrow banding of the stiffness matrix $\mathbf{k}$ in Eq. (9.2.12) is in part lost in using the fully populated condensed stiffness matrix $\hat{\mathbf{k}}_{tt}$ in Eq. (9.3.4).

The relationship between $\mathbf{u}_0$ and $\mathbf{u}_t$, Eq. (9.3.3), although exact only if the $\mathbf{u}_0$ DOFs have zero mass, can also be used if this condition is not satisfied. In such cases Eq. (9.3.3) provides a basis to select displacement shapes for use in the Rayleigh–Ritz method described in the next section.

15.3 RAYLEIGH–RITZ METHOD

A most general technique for reducing the number of DOFs and finding approximations to the lower natural frequencies and modes is the Rayleigh–Ritz method. It is an extension of Rayleigh's method suggested by W. Ritz in 1909. Originally developed for systems with distributed mass and elasticity (see Chapter 17), the method is presented next for discretized systems.

15.3.1 Reduced Equations of Motion

The equations of motion for a system with N DOFs subjected to forces $\mathbf{p}(t) = \mathbf{s}p(t)$ are

$$\mathbf{m}\ddot{\mathbf{u}} + \mathbf{c}\dot{\mathbf{u}} + \mathbf{k}\mathbf{u} = \mathbf{s}p(t) \qquad (15.3.1)$$

In Rayleigh's method we expressed the structural displacements as $\mathbf{u}(t) = z(t)\psi$, where ψ was an assumed shape vector; this reduced the system to one with a single degree of freedom and led to an approximate value for the fundamental natural frequency. In the Rayleigh–Ritz method, the displacements are expressed as a linear combination of several shape vectors ψ_j:

$$\mathbf{u}(t) = \sum_{j=1}^{J} z_j(t)\psi_j = \mathbf{\Psi}\mathbf{z}(t) \tag{15.3.2}$$

where $z_j(t)$ are called the generalized coordinates, and the Ritz vectors ψ_j—$j = 1, 2, \ldots,$ J—must be linearly independent vectors satisfying the geometric boundary conditions. They are selected appropriate for the system to be analyzed, as discussed in Section 15.4. The vectors ψ_j make up the columns of the $N \times J$ matrix $\mathbf{\Psi}$ in Eq. (15.3.2) and $\mathbf{z}$ is the vector of the J generalized coordinates.

Substituting the Ritz transformation of Eq. (15.3.2) in Eq. (15.3.1) gives

$$\mathbf{m}\mathbf{\Psi}\ddot{\mathbf{z}} + \mathbf{c}\mathbf{\Psi}\dot{\mathbf{z}} + \mathbf{k}\mathbf{\Psi}\mathbf{z} = \mathbf{s}p(t)$$

Each term is premultiplied by $\mathbf{\Psi}^T$ to obtain

$$\tilde{\mathbf{m}}\ddot{\mathbf{z}} + \tilde{\mathbf{c}}\dot{\mathbf{z}} + \tilde{\mathbf{k}}\mathbf{z} = \tilde{\mathbf{L}}p(t) \tag{15.3.3}$$

where

$$\tilde{\mathbf{m}} = \mathbf{\Psi}^T\mathbf{m}\mathbf{\Psi} \qquad \tilde{\mathbf{c}} = \mathbf{\Psi}^T\mathbf{c}\mathbf{\Psi} \qquad \tilde{\mathbf{k}} = \mathbf{\Psi}^T\mathbf{k}\mathbf{\Psi} \qquad \tilde{\mathbf{L}} = \mathbf{\Psi}^T\mathbf{s} \tag{15.3.4}$$

Equation (15.3.3) is a system of J differential equations in the J generalized coordinates $\mathbf{z}(t)$.

We now make two observations: (1) Equation (15.3.3) in generalized coordinates is similar to Eq. (12.4.4) in modal coordinates. (2) Equation (15.3.4) defining $\tilde{\mathbf{m}}$, $\tilde{\mathbf{c}}$, and $\tilde{\mathbf{k}}$ is of the same form as Eqs. (12.3.4) and (12.4.3) for $\mathbf{M}$, $\mathbf{C}$, and $\mathbf{K}$. Obtained by transforming Eq. (15.3.1), both sets of equations differ, however, in an important sense that the Ritz vectors are used for transformation in one case, whereas the natural vibration modes in the other. Because the Ritz vectors are generally different from the natural modes, $\tilde{\mathbf{m}}$ and $\tilde{\mathbf{k}}$ are not diagonal matrices, whereas $\mathbf{M}$ and $\mathbf{K}$ are diagonal; see Eq. (12.3.6).

In summary, the Ritz transformation of Eq. (15.3.2) has made it possible to reduce the original set of N equations (15.3.1) in the nodal displacements $\mathbf{u}$ to a smaller set of J equations (15.3.3) in the generalized coordinates $\mathbf{z}$.

15.3.2 "Best" Approximation

The reduced system of equations we have just derived represents a powerful procedure because they are rooted in Rayleigh's principle (Section 10.12). The approximations to the natural modes of the system determined by solving the eigenvalue problem associated with Eq. (15.3.3) represent the "best" solution among all possible solutions that are linear combinations of the selected Ritz vectors. In this section we use Rayleigh's principle to demonstrate that this solution is "best" in the sense that the associated natural frequencies of the system are closest to the true frequencies among all approximate values possible with the selected Ritz vectors.

For this purpose, we first determine Rayleigh's quotient, Eq. (10.12.1), for a vector $\tilde{\phi}$ defined as a linear combination of the Ritz vectors, consistent with Eq. (15.3.2):

$$\tilde{\phi} = \Psi\chi \tag{15.3.5}$$

Substituting Eq. (15.3.5) in Eq. (10.12.1) gives

$$\lambda(\chi) = \frac{\chi^T \tilde{\mathbf{k}}\chi}{\chi^T \tilde{\mathbf{m}}z} \equiv \frac{\tilde{k}(\chi)}{\tilde{m}(\chi)} \tag{15.3.6}$$

where $\tilde{k}(\chi)$ and $\tilde{m}(\chi)$ are scalar quantities; and $\tilde{\mathbf{k}}$ and $\tilde{\mathbf{m}}$ are the $J \times J$ matrices defined by Eq. (15.3.4) with their typical elements given by

$$\tilde{k}_{ij} = \psi_i^T \mathbf{k}\psi_j \qquad \tilde{m}_{ij} = \psi_i^T \mathbf{m}\psi_j \tag{15.3.7}$$

and λ has been replaced by $\lambda(\chi)$ to emphasize its dependence on χ. Eq. (15.3.6) can be rewritten as

$$\lambda(\chi) = \frac{\sum_{i=1}^{J} \sum_{j=1}^{J} \chi_i \chi_j \tilde{k}_{ij}}{\sum_{i=1}^{J} \sum_{j=1}^{J} \chi_i \chi_j \tilde{m}_{ij}} \tag{15.3.8}$$

Rayleigh's quotient cannot be determined from Eq. (15.3.8) because the generalized coordinates χ_n are unknown. However, from Section 10.12 it is known that

$$\omega_1^2 \le \lambda(\chi) \le \omega_N^2 \tag{15.3.9}$$

where ω_1 and ω_N are the smallest and largest natural vibration frequencies.

To proceed further we invoke *Rayleigh's stationary condition*, the property that Rayleigh's quotient is stationary in the neighborhood of the true modes (or true values of χ); see Section 10.12 for detail. Because χ_i are the only variables, the necessary condition for $\lambda(\chi)$ to be stationary is

$$\frac{\partial \lambda}{\partial \chi_i} = 0 \qquad i = 1, 2, \ldots, J \tag{15.3.10}$$

For the λ given by Eq. (15.3.8),

$$\frac{\partial \lambda}{\partial \chi_i} = \frac{2\tilde{m} \sum_{j=1}^{J} \chi_j \tilde{k}_{ij} - 2\tilde{k} \sum_{j=1}^{J} \chi_j \tilde{m}_{ij}}{\tilde{m}^2}$$

This condition can be rewritten by substituting $\lambda = \tilde{k}/\tilde{m}$ from Eq. (15.3.8):

$$\sum_{j=1}^{J} (\tilde{k}_{ij} - \lambda \tilde{m}_{ij})\chi_j = 0 \qquad i = 1, 2, \ldots, J$$

Writing these J equations in matrix form gives the reduced eigenvalue problem

$$\tilde{\mathbf{k}}\chi = \lambda \tilde{\mathbf{m}}\chi \tag{15.3.11}$$

where $\tilde{\mathbf{k}}$ and $\tilde{\mathbf{m}}$ are the $J \times J$ matrices defined by Eqs. (15.3.4) and (15.3.7), and χ is the vector of generalized coordinates that remain to be determined. Observe that Eq. (15.3.11), derived using Rayleigh's stationary condition, is also the eigenvalue problem associated with Eq. (15.3.3). This proves the assertion at the beginning of this section.

15.3.3 Approximate Frequencies and Modes

The solution to Eq. (15.3.11) by methods of Chapter 10 yields J eigenvalues λ_n—$n = 1$, $2, \ldots, J$—and the corresponding eigenvectors

$$\chi_n = \langle \chi_{1n}, \chi_{2n}, \ldots, \chi_{Jn} \rangle^T \tag{15.3.12}$$

The eigenvalues provide

$$\tilde{\omega}_n = \sqrt{\rho_n} \qquad n = 1, 2, \ldots, J \tag{15.3.13}$$

which are approximations to the natural frequencies ω_n. The eigenvectors χ_n substituted in Eq. (15.3.5) provide the vectors

$$\tilde{\phi}_n = \boldsymbol{\Psi}\chi_n \qquad n = 1, 2, \ldots, J \tag{15.3.14}$$

which are approximations to the natural modes ϕ_n. The accuracy of these approximate results is generally better for the lower modes than for higher modes. Therefore, more Ritz vectors should be included than the number of modes desired.

In light of the properties of Rayleigh's quotient (Section 10.12), the approximate frequencies are never lower than the fundamental frequency and never higher than the highest frequency, that is,

$$\tilde{\omega}_i \geq \omega_i \qquad \tilde{\omega}_J \leq \omega_N \tag{15.3.15}$$

Furthermore, an approximate frequency approaches the exact value as the number J of Ritz vectors is increased.

15.3.4 Orthogonality of Approximate Modes

In this section we demonstrate that the vectors $\tilde{\phi}_n$ satisfy the orthogonality conditions

$$\tilde{\phi}_n^T \mathbf{k} \tilde{\phi}_r = 0 \qquad \tilde{\phi}_n^T \mathbf{m} \tilde{\phi}_r = 0 \qquad n \neq r \tag{15.3.16}$$

This result is by no means obvious because the vectors $\tilde{\phi}_n$ are only approximations of the natural modes ϕ_n, which are known to satisfy Eq. (10.4.1).

The eigenvectors χ_n of Eq. (15.3.11) obviously satisfy the orthogonality conditions:

$$\chi_n^T \tilde{\mathbf{k}} \chi_r = 0 \qquad \chi_n^T \tilde{\mathbf{m}} \chi_r = 0 \qquad n \neq r \tag{15.3.17}$$

Using this property and Eq. (15.3.14), the first orthogonality condition in Eq. (15.3.16) can be proven as follows:

$$\tilde{\phi}_n^T \mathbf{k} \tilde{\phi}_r = \chi_n^T \boldsymbol{\Psi}^T \mathbf{k} \boldsymbol{\Psi} \chi_r = \chi_n^T \tilde{\mathbf{k}} \chi_r = 0 \qquad n \neq r$$

The second orthogonality condition in Eq. (15.3.16) can be demonstrated similarly.

If the eigenvectors χ_n were made to be mass orthonormal (Section 10.6), then

$$\chi_n^T \tilde{\mathbf{m}} \chi_n = 1 \qquad \chi_n^T \tilde{\mathbf{k}} \chi_n = \tilde{\omega}_n^2 \tag{15.3.18}$$

This implies, as can easily be demonstrated, that the approximate modes $\tilde{\phi}_n$ are also mass orthonormal:

$$\tilde{\phi}_n^T \mathbf{m} \tilde{\phi}_n = 1 \qquad \tilde{\phi}_n^T \mathbf{k} \tilde{\phi}_n = \tilde{\omega}_n^2 \tag{15.3.19}$$

Because the approximate modes $\tilde{\phi}_n$ satisfy the orthogonality conditions of Eq. (15.3.16), they can be used in classical modal solution of Eq. (15.3.1). Therefore, in the rest of this chapter we do not distinguish between the approximate values $(\tilde{\omega}_n, \tilde{\phi}_n)$ and the exact values (ω_n, ϕ_n).

15.4 SELECTION OF RITZ VECTORS

The success of the Rayleigh–Ritz method depends on how well linear combinations of Ritz vectors can approximate the natural modes of vibration. Therefore, it is important that the Ritz vectors be selected judiciously. In this section we present two very different approaches; the first is based on physical insight into shapes of natural modes, and the second is a formal computational procedure.

15.4.1 Physical Insight into Natural Mode Shapes

If we can visualize the shapes of the first few natural vibration modes of a structure, the Ritz vectors can be selected as approximations to these modes. In particular, the nth Ritz vector ψ_n is selected to approximate the nth natural mode ϕ_n of the structure. For example, based on what we have learned by solving several examples in Chapters 10 and 12, we can visualize the first two natural modes in planar vibration of a multistory frame. Thus the two Ritz vectors shown in Fig. 15.4.1 could be used in the Rayleigh–Ritz method to determine approximations to the first two natural frequencies and modes of this structure.

This approach may not be possible for complex systems because it may be difficult to visualize their mode shapes if we have never determined the natural modes of similar structures. Such visualization can be especially difficult if the natural mode includes two- or three-dimensional motions. A general procedure to select Ritz vectors that does not depend on physical visualization of the natural modes is therefore developed in the next section. This systematic procedure is suitable for implementation on a computer.

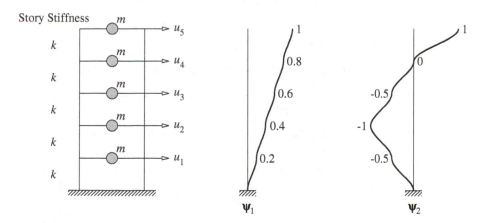

Figure 15.4.1   Ritz vectors for a five-story frame.

Example 15.1

By the Rayleigh–Ritz method, determine the first two natural frequencies and modes of a uniform five-story shear frame with story stiffnesses k and lumped floor masses m. Use the two Ritz vectors shown in Fig. 15.4.1.

Solution

1. *Formulate the stiffness and mass matrices.*

$$\mathbf{k} = k \begin{bmatrix} 2 & -1 & 0 & 0 & 0 \\ -1 & 2 & -1 & 0 & 0 \\ 0 & -1 & 2 & -1 & 0 \\ 0 & 0 & -1 & 2 & -1 \\ 0 & 0 & 0 & -1 & 1 \end{bmatrix} \qquad \mathbf{m} = m \begin{bmatrix} 1 & & & & \\ & 1 & & & \\ & & 1 & & \\ & & & 1 & \\ & & & & 1 \end{bmatrix}$$

2. *Compute $\tilde{\mathbf{k}}$ and $\tilde{\mathbf{m}}$.*

$$\mathbf{\Psi} = [\boldsymbol{\psi}_1 \quad \boldsymbol{\psi}_2] = \begin{bmatrix} 0.2 & -0.5 \\ 0.4 & -1.0 \\ 0.6 & -0.5 \\ 0.8 & 0 \\ 1.0 & 1.0 \end{bmatrix}$$

$$\tilde{\mathbf{k}} = \mathbf{\Psi}^T \mathbf{k} \mathbf{\Psi} = k \begin{bmatrix} 0.2 & 0.2 \\ 0.2 & 2.0 \end{bmatrix} \qquad \tilde{\mathbf{m}} = \mathbf{\Psi}^T \mathbf{m} \mathbf{\Psi} = m \begin{bmatrix} 2.2 & 0.2 \\ 0.2 & 2.5 \end{bmatrix}$$

3. *Solve the reduced eigenvalue problem, Eq. (15.3.11).* Substituting $\tilde{\mathbf{m}}$ and $\tilde{\mathbf{k}}$, Eq. (15.3.11) gives

$$\begin{bmatrix} 0.2 & 0.2 \\ 0.2 & 2.0 \end{bmatrix} \begin{bmatrix} \chi_1 \\ \chi_2 \end{bmatrix} = \left(\rho \frac{m}{k} \right) \begin{bmatrix} 2.2 & 0.2 \\ 0.2 & 2.5 \end{bmatrix} \begin{bmatrix} \chi_1 \\ \chi_2 \end{bmatrix} \tag{a}$$

The eigenvalue problem of Eq. (a) is solved to obtain

$$\lambda_1 = 0.08238(k/m) \qquad \lambda_2 = 0.8004(k/m)$$

$$\chi_1 = \left\{ \begin{matrix} 1.329 \\ -0.1360 \end{matrix} \right\} \qquad \chi_2 = \left\{ \begin{matrix} 0.03170 \\ 1.240 \end{matrix} \right\}$$

4. *Determine the approximate frequencies and modes.*

$$\tilde{\omega}_n = \sqrt{\lambda_n} \qquad \tilde{\phi}_n = \mathbf{\Psi} \chi_n$$

The results are presented in Table E15.1.

TABLE E15.1 COMPARISON OF APPROXIMATE AND EXACT RESULTS

Approximate	Exact
$\tilde{\omega}_1 = 0.2870\sqrt{k/m}$	$\omega_1 = 0.2846\sqrt{k/m}$
$\tilde{\omega}_2 = 0.8947\sqrt{k/m}$	$\omega_2 = 0.8308\sqrt{k/m}$
$\tilde{\Phi} = \begin{bmatrix} 0.3338 & -0.6135 \\ 0.6676 & -1.227 \\ 0.8654 & -0.6008 \\ 1.063 & 0.02536 \\ 1.193 & 1.271 \end{bmatrix}$	$\Phi = \begin{bmatrix} 0.3338 & -0.8954 \\ 0.6405 & -1.173 \\ 0.8954 & -0.6411 \\ 1.078 & 0.3338 \\ 1.173 & 1.078 \end{bmatrix}$

5. *Compare with the exact results.* The approximate values of the natural frequencies and modes are compared in Table E15.1 with their exact values obtained in Section 12.8. The errors in the approximate frequencies and modes are less than 1% in the first frequency, 8% in the second frequency, and 4% in the first mode. However, the second mode is so much in error that it may be useless.

15.4.2 Force-Dependent Ritz Vectors

It is desired to determine Ritz vectors appropriate for analysis of a structure subjected to external dynamic forces:

$$\mathbf{p}(t) = \mathbf{s}p(t) \tag{15.4.1}$$

The spatial distribution of forces defined by the vector $\mathbf{s}$ does not vary with time, and the time dependence of all forces is given by the same scalar function $p(t)$. Using the vector $\mathbf{s}$, a procedure is presented next to generate a sequence of mass-orthonormal Ritz vectors.

The first Ritz vector ψ_1 is defined as the static displacements due to applied forces $\mathbf{s}$. It is determined by solving

$$\mathbf{k}\mathbf{y}_1 = \mathbf{s} \tag{15.4.2}$$

The vector $\mathbf{y}_1$ is normalized to be mass orthonormal; thus

$$\psi_1 = \frac{\mathbf{y}_1}{\left(\mathbf{y}_1^T \mathbf{m}\mathbf{y}_1\right)^{1/2}} \tag{15.4.3}$$

The second Ritz vector ψ_2 is determined from the vector $\mathbf{y}_2$ of static displacements due to applied forces given by the inertia force distribution associated with the first Ritz vector ψ_1. The vector $\mathbf{y}_2$ is obtained by solving

$$\mathbf{k}\mathbf{y}_2 = \mathbf{m}\psi_1 \tag{15.4.4}$$

The vector $\mathbf{y}_2$ will in general contain a component of the previous vector, ψ_1. It can therefore be expressed as

$$\mathbf{y}_2 = \hat{\psi}_2 + a_{12}\psi_1 \tag{15.4.5}$$

where $\hat{\psi}_2$ is a "pure" vector, which is orthogonal to the previous vector, and $a_{12}\psi_1$ is the component of the previous vector present in $\mathbf{y}_2$. The coefficient a_{12} is determined by premultiplying both sides of Eq. (15.4.5) by $\psi_1^T \mathbf{m}$ to obtain

$$\psi_1^T \mathbf{m}\mathbf{y}_2 = \psi_1^T \mathbf{m}\hat{\psi}_2 + a_{12}(\psi_1^T \mathbf{m}\psi_1)$$

Note that $\psi_1^T \mathbf{m}\hat{\psi}_2 = 0$ by definition of $\hat{\psi}_2$, and $\psi_1^T \mathbf{m}\psi_1 = 1$ from Eq. (15.4.3). Thus

$$a_{12} = \psi_1^T \mathbf{m}\mathbf{y}_2 \tag{15.4.6}$$

The pure vector $\hat{\psi}_2$ is given by

$$\hat{\psi}_2 = \mathbf{y}_2 - a_{12}\psi_1 \tag{15.4.7}$$

where a_{12} is known from Eq. (15.4.6). Finally, the vector $\hat{\psi}_2$ is normalized so that it is mass orthonormal to obtain the second Ritz vector:

$$\psi_2 = \frac{\hat{\psi}_2}{\left(\hat{\psi}_2^T \mathbf{m} \hat{\psi}_2\right)^{1/2}} \tag{15.4.8}$$

Generalizing this procedure, the nth Ritz vector ψ_n is determined from the vector $\mathbf{y}_n$ of the static displacements due to applied forces given by the inertia force distribution associated with the $(n-1)$th Ritz vector ψ_{n-1}. The vector $\mathbf{y}_n$ is determined by solving

$$\mathbf{k}\mathbf{y}_n = \mathbf{m}\psi_{n-1} \tag{15.4.9}$$

The vector $\mathbf{y}_n$ will in general contain components of previous Ritz vectors ψ_j and can therefore be expressed as

$$\mathbf{y}_n = \hat{\psi}_n + \sum_{j=1}^{n-1} a_{jn}\psi_j \tag{15.4.10}$$

where $\hat{\psi}_n$ is a "pure" vector, which is orthogonal to the previous vectors, and $a_{jn}\psi_j$ are the components of the previous vectors present in $\mathbf{y}_n$. The coefficient a_{in} is determined by premultiplying both sides of Eq. (15.4.10) by $\psi_i^T \mathbf{m}$:

$$\psi_i^T \mathbf{m}\mathbf{y}_n = \psi_i^T \mathbf{m}\hat{\psi}_n + \sum_{j=1}^{n-1} a_{jn}\left(\psi_i^T \mathbf{m}\psi_j\right)$$

Observe that $\psi_i^T \mathbf{m}\hat{\psi}_n = 0$ by definition of $\hat{\psi}_n$, $\psi_i^T \mathbf{m}\psi_j = 0$ for $i \neq j$ and $\psi_i^T \mathbf{m}\psi_i = 1$ because all previous vectors are mass orthonormal. Thus

$$a_{in} = \psi_i^T \mathbf{m}\mathbf{y}_n \qquad i = 1, 2, \ldots, n-1 \tag{15.4.11}$$

The pure vector $\hat{\psi}_n$ is given by

$$\hat{\psi}_n = \mathbf{y}_n - \sum_{i=1}^{n-1} a_{in}\psi_i \tag{15.4.12}$$

where a_{in} are known from Eq. (15.4.11). Finally, the vector $\hat{\psi}_n$ is normalized so that it is mass orthonormal to obtain the nth Ritz vector:

$$\psi_n = \frac{\hat{\psi}_n}{\left(\hat{\psi}_n^T \mathbf{m} \hat{\psi}_n\right)^{1/2}} \tag{15.4.13}$$

The sequence of vectors $\psi_1, \psi_2, \ldots, \psi_J$ are mutually mass-orthonormal and hence they satisfy the linear independence requirement of the Rayleigh–Ritz method.

While the Gram–Schmidt orthogonalization procedure of Eqs. (15.4.11) and (15.4.12) should theoretically mass-orthogonalize the new vector with respect to all previous vectors, the actual computer implementation may be fraught with loss-of-orthogonality problems due to numerical round-off errors. To overcome these difficulties the Gram–Schmidt procedure is modified as follows. After computation of *each* a_{in} from Eq. (15.4.11), an improved vector $\hat{\psi}_n$ is calculated from Eq. (15.4.12), which is used instead of $\mathbf{y}_n$ in Eq. (15.4.11) to calculate the next a_{in}. Including this modification, the procedure to generate the force-dependent Ritz vectors is summarized in Table 15.4.1 as it might be implemented on the computer.

TABLE 15.4.1   GENERATION OF FORCE-DEPENDENT RITZ VECTORS

1. Determine the first vector, ψ_1.

   a. Determine $\mathbf{y}_1$ by solving: $\mathbf{ky}_1 = \mathbf{s}$.
   b. Normalize $\mathbf{y}_1$: $\psi_1 = \mathbf{y}_1 \div (\mathbf{y}_1^T \mathbf{m} \mathbf{y}_1)^{1/2}$.

2. Determine additional vectors, ψ_n, $n = 2, 3, \ldots, J$.

   a. Determine $\mathbf{y}_n$ by solving: $\mathbf{ky}_n = \mathbf{m}\psi_{n-1}$.
   b. Orthogonalize $\mathbf{y}_n$ with respect to previous $\psi_1, \psi_2, \ldots, \psi_{n-1}$ by repeating the following steps for $i = 1, 2, \ldots, n - 1$:

   - $a_{in} = \psi_i^T \mathbf{m} \mathbf{y}_n$.
   - $\hat{\psi}_n = \mathbf{y}_n - a_{in}\psi_i$.
   - $\mathbf{y}_n = \hat{\psi}_n$.

   c. Normalize $\hat{\psi}_n$: $\psi_n = \hat{\psi}_n \div (\hat{\psi}_n^T \mathbf{m} \hat{\psi}_n)^{1/2}$.

The procedure to generate these vectors is reminiscent of the vector sequence $\mathbf{x}_1$, $\mathbf{k}^{-1}\mathbf{m}\mathbf{x}_1$, $(\mathbf{k}^{-1}\mathbf{m})^2\mathbf{x}_1, \ldots$ generated in the inverse iteration procedure (Section 10.13). When obtained without making the vectors orthogonal, this vector sequence converges to the lowest natural mode. With Gram–Schmidt orthogonalization, as in Table 15.4.1, this sequence provides the force-dependent Ritz vectors.

Example 15.2

The vibration properties of the uniform five-story shear frame of Example 15.1 with $m = 100$ kips/g $= 0.2591$ kip-sec²/in. and $k = 31.56$ kips/in. are to be determined by the Rayleigh–Ritz method using Ritz vectors determined from a force distribution $\mathbf{s} = \langle m \quad m \quad m \quad m \quad m \rangle^T$. Using two force-dependent vectors, determine the first two natural frequencies and modes of vibration.

Solution

1. The stiffness and mass matrices, $\mathbf{k}$ and $\mathbf{m}$, are given in Example 15.1 with $k = 31.56$ kips/in. and $m = 0.2591$ kip-sec²/in.

2. *Determine the first Ritz vector, ψ_1.*

- Solve $\mathbf{ky}_1 = m\mathbf{1}$ to obtain $\mathbf{y}_1 = \langle 0.0410 \quad 0.0739 \quad 0.0985 \quad 0.1149 \quad 0.1231 \rangle^T$.
- Divide $\mathbf{y}_1$ by $(\mathbf{y}_1^T \mathbf{m} \mathbf{y}_1)^{1/2} = 0.1082$ to obtain the normalized vector:

$$\psi_1 = \langle 0.3792 \quad 0.6826 \quad 0.9102 \quad 1.062 \quad 1.138 \rangle^T$$

3. *Determine the second Ritz vector* ψ_2.

- Solve $\mathbf{k}\mathbf{y}_2 = \mathbf{m}\psi_1$ to obtain $\mathbf{y}_2 = \langle 0.0342 \quad 0.0654 \quad 0.0909 \quad 0.1090 \quad 0.1183 \rangle^T$.
- Orthogonalize $\mathbf{y}_2$ with respect to ψ_1:

$$a_{12} = \psi_1^T \mathbf{m}\mathbf{y}_2 = 0.1012$$

$$\hat{\psi}_2 = \mathbf{y}_2 - 0.1012\psi_1$$

$$= 10^{-2}\langle -0.4134 \quad -0.3705 \quad -0.1204 \quad 0.1500 \quad 0.3164 \rangle^T$$

- Divide $\hat{\psi}_2$ by $(\hat{\psi}_2^T \mathbf{m}\hat{\psi}_2)^{1/2} = 0.3396 \times 10^{-2}$ to get the normalized vector:

$$\psi_2 = \langle -1.217 \quad -1.091 \quad -0.3546 \quad 0.4418 \quad 0.9316 \rangle^T$$

4. *Compute* $\tilde{\mathbf{k}}$ *and* $\tilde{\mathbf{m}}$.

$$\mathbf{\Psi} = [\psi_1 \ \psi_2]$$

$$\tilde{\mathbf{k}} = \mathbf{\Psi}^T \mathbf{k} \mathbf{\Psi} = \begin{bmatrix} 9.986 & -3.086 \\ -3.086 & 91.95 \end{bmatrix} \qquad \tilde{\mathbf{m}} = \mathbf{\Psi}^T \mathbf{m} \mathbf{\Psi} = \begin{bmatrix} 1.0 & \\ & 1.0 \end{bmatrix}$$

5. *Solve the reduced eigenvalue problem, Eq. (15.3.11).*

$$\tilde{\omega}_1 = 3.142 \qquad \tilde{\omega}_2 = 9.595$$

$$\chi_1 = \begin{Bmatrix} 0.9993 \\ 0.0376 \end{Bmatrix} \qquad \chi_2 = \begin{Bmatrix} -0.0376 \\ 0.9993 \end{Bmatrix}$$

6. *Determine the natural modes.* Substituting $\mathbf{\Psi}_n$ and χ_n in Eq. (15.3.14) gives

$$\tilde{\phi}_1 = \langle 0.3332 \quad 0.6412 \quad 0.8962 \quad 1.078 \quad 1.172 \rangle^T$$

$$\tilde{\phi}_2 = \langle -1.230 \quad -1.116 \quad -0.3886 \quad 0.4016 \quad 0.8882 \rangle^T$$

7. *Compare with the exact results.* Table E15.1 gives the exact modes and frequencies; the latter, after substituting for k and m, are

$$\omega_1 = 3.142 \qquad \omega_2 = 9.170 \text{ rad/sec}$$

The approximate frequencies $\tilde{\omega}_n$ and modes from this example, using force-dependent Ritz vectors, are better than those determined in Example 15.1 from assumed vectors.

15.5 DYNAMIC ANALYSIS USING RITZ VECTORS

Now that we have developed procedures to generate Ritz vectors, we return to the solution of Eq. (15.3.3), the reduced system of equations. With J Ritz vectors included, these J equations are coupled because in general the matrices $\tilde{\mathbf{m}}$, $\tilde{\mathbf{c}}$, and $\tilde{\mathbf{k}}$ in Eq. (15.3.3) are not diagonal. However, if the force-dependent Ritz vectors of Section 15.4.2 are used, $\tilde{\mathbf{m}} = \mathbf{I}$, the identity matrix. The set of J coupled equations can be solved for the unknowns $z_j(t)$— $j = 1, 2, \ldots, J$—by numerical time-stepping methods (Chapter 16). Then at each time instant, the nodal displacement vector $\mathbf{u}$ is determined from Eq. (15.3.2) and the element forces by the methods of Section 9.10. This method is quite general in the sense that it applies to classically damped systems as well as nonclassically damped systems. For systems with Rayleigh damping, an alternative procedure is presented at the end of this section.

The number of force-dependent Ritz vectors included in the dynamic analysis should be sufficient to represent accurately the vector $\mathbf{s}$ that defines the spatial distribution of forces. Because these Ritz vectors are mass-orthonormal, following Eqs. (12.8.2) to (12.8.4), the vector $\mathbf{s}$ can be expanded as follows:

$$\mathbf{s} = \sum_{n=1}^{N} \tilde{\Gamma}_n \mathbf{m} \psi_n \qquad \text{where} \qquad \tilde{\Gamma}_n = \psi_n^T \mathbf{s} \qquad (15.5.1)$$

The J Ritz vectors included in dynamic analysis provide an approximation to $\mathbf{s}$, and an error vector can be defined as

$$\mathbf{e}_J = \mathbf{s} - \sum_{n=1}^{J} \tilde{\Gamma}_n \mathbf{m} \psi_n \qquad (15.5.2)$$

Considering that a logical norm for the vector $\mathbf{s}$ is its length $(\mathbf{s}^T \mathbf{s})^{1/2}$, an error norm e_J is defined as

$$e_J = \frac{\mathbf{s}^T \mathbf{e}_J}{\mathbf{s}^T \mathbf{s}} \qquad (15.5.3)$$

This error e_J will be zero when all N Ritz vectors are included ($J = N$) because of Eq. (15.5.1), and e_J will equal unity when no Ritz vectors are included ($J = 0$). Thus, enough Ritz vectors should be included so that e_J is sufficiently small.

To illustrate these concepts, the error is computed for the five-story uniform shear frame of Example 15.1. The results presented in Fig. 15.5.1 are for three different force distributions: $\mathbf{s}_a = \langle 0 \ \ 0 \ \ 0 \ \ 0 \ \ 1 \rangle^T$, $\mathbf{s}_b = \langle 0 \ \ 0 \ \ 0 \ \ -2 \ \ 1 \rangle^T$, and $\mathbf{s}_c = \langle 1 \ \ 1 \ \ 1 \ \ 1 \ \ 1 \rangle^T$. For a given force distribution, the error decreases as more Ritz vectors are included, and is zero when all five Ritz vectors are included. For a fixed number of Ritz vectors, the error is smallest for the force distribution $\mathbf{s}_c$, largest for $\mathbf{s}_b$, and has an intermediate value for $\mathbf{s}_a$.

Figure 15.5.1 also provides a comparison of the error e_J if J Ritz vectors are included in the analysis versus the error e_J if J natural vibration modes of the system are considered. The latter was calculated from formulas similar to Eqs. (15.5.2) and (15.5.3), with the Ritz vectors ψ_n replaced by the natural modes ϕ_n. The error is smaller when Ritz vectors are used because they are derived from the force distribution. While this result would indicate that Ritz vectors are preferable to natural modes, the latter lead to uncoupled modal equations, which have advantages: Solving uncoupled modal equations, each of the same form as the governing equation for an SDF system, is easier than dealing with the coupled equations in Ritz coordinates. Also, the modal equations permit estimation of the peak value of the earthquake response of a structure by response spectrum analysis (Chapter 13, Part B).

Do we need a static correction term (see Section 12.12) to supplement the response obtained by dynamic analysis using a truncated set of Ritz vectors? This is not necessary

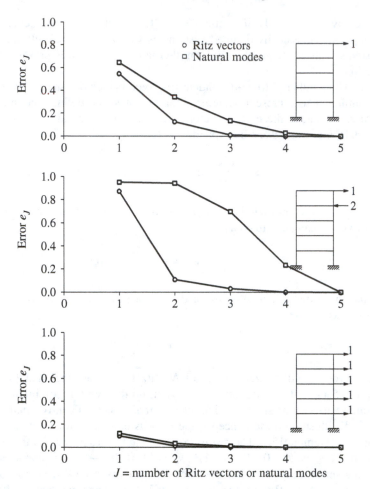

Figure 15.5.1   Variation of error e_J with the number J of Ritz vectors and of natural modes for three distributions of lateral forces.

because the static correction effect is contained in the first Ritz vector because it is obtained from the static displacements due to the applied forces.

For dynamic analysis of systems with Rayleigh damping, classical modal analysis (Section 12.9) of Eq. (15.3.1) may be preferable over solution of the coupled equations (15.3.3) in Ritz coordinates, especially if the natural frequencies and modes of the system are desired. The Rayleigh–Ritz concept is still useful, however, because these vibration properties are obtained by solving Eq. (15.3.11), a smaller eigenvalue problem of order J, instead of the original eigenvalue problem of size N [Eq. (10.2.4)]. Because the resulting approximate modes $\tilde{\phi}_n$ are orthogonal with respect to the mass and stiffness matrices $\mathbf{m}$ and $\mathbf{k}$ (Section 15.3.4), they can be used just like the exact modes in classical modal analysis of the system.

FURTHER READING

Clough, R. W., and Penzien, J., *Dynamics of Structures*, McGraw-Hill, New York, 1993, pp. 314–323.

Humar, J. L., *Dynamics of Structures*, 2nd ed., A. A. Balkema Publishers, Lisse, The Netherlands, 2002, pp. 702–747.

Leger, P., Wilson, E. L., and Clough, R. W., "The Use of Load Dependent Vectors for Dynamic and Earthquake Analysis," *Report No. UCB/EERC 86–04*, Earthquake Engineering Research Center, University of California, Berkeley, Calif., 1986.

PROBLEMS

15.1 By the Rayleigh–Ritz method, determine the first two natural vibration frequencies and modes of the system in Fig. 15.4.1 using the following two Ritz vectors:

$$\psi_1 = \langle\, 0.3 \quad 0.6 \quad 0.8 \quad 0.9 \quad 1 \,\rangle^T$$

$$\psi_2 = \langle\, -1 \quad -1 \quad -0.5 \quad 0.5 \quad 1 \,\rangle^T$$

Compare these results with those obtained in Example 15.1 and the exact values presented in Section 12.8 and comment on how the selected Ritz vectors influence the accuracy of the results.

*15.2 Resolve Example 15.2 using Ritz vectors determined from the force distribution $s = \langle\, 0 \quad 0 \quad 0 \quad 0 \quad 1 \,\rangle^T$. Comment on the accuracy of the results and how the force distribution used in generating Ritz vectors influences the accuracy.

15.3 By the Rayleigh–Ritz method, determine the first two natural vibration frequencies and modes of the five-story shear frame in Fig. P15.3 using the following two Ritz vectors:

$$\psi_1 = \langle\, 0.2 \quad 0.4 \quad 0.6 \quad 0.8 \quad 1 \,\rangle^T$$

$$\psi_2 = \langle\, -0.5 \quad -1 \quad -0.5 \quad 0 \quad 1 \,\rangle^T$$

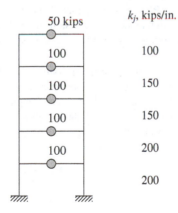

	k_j, kips/in.
50 kips	
100	100
100	150
100	150
100	200
100	200

Figure P15.3

*Denotes that a computer is necessary to solve this problem.

*15.4 For the five-story shear frame in Fig. P15.3, determine the first two natural vibration frequencies and modes using two force-dependent Ritz vectors determined from the force distribution $\mathbf{s} = \langle 1 \quad 1 \quad 1 \quad 1 \quad 0.5 \rangle^T$. Comment on the relative accuracy of the results of Problems 15.3 and 15.4.

*15.5 Solve Problem 15.4 using the force distribution $\mathbf{s} = \langle 0 \quad 0 \quad 0 \quad 0 \quad 1 \rangle^T$. Comment on the relative accuracy of the solutions of Problems 15.4 and 15.5.

*15.6 Solve Problem 15.4 using the force distribution $\mathbf{s} = \langle 0 \quad 0 \quad 0 \quad -2 \quad 1 \rangle^T$. Comment on the relative accuracy of the solutions of Problems 15.4, 15.5, and 15.6.

*15.7 Compute the error e_J where J is the number of Ritz vectors included in dynamic analysis of the five-story shear frame in Fig. P15.3 subjected to dynamic forces with three different force distributions:

$$\mathbf{s}_a = \langle 0 \quad 0 \quad 0 \quad 0 \quad 1 \rangle^T$$

$$\mathbf{s}_b = \langle 0 \quad 0 \quad 0 \quad -2 \quad 1 \rangle^T$$

$$\mathbf{s}_c = \langle 1 \quad 1 \quad 1 \quad 1 \quad 0.5 \rangle^T$$

Plot e_J against J and comment on how, for a given $\mathbf{s}$, this error depends on J, and how, for a given J, this error depends on $\mathbf{s}$.

*15.8 (a) Compute the error e_J where J is the number of natural vibration modes of the system included in dynamic analysis of the five-story shear frame in Fig. P15.3 subjected to dynamic forces with three different force distributions:

$$\mathbf{s}_a = \langle 0 \quad 0 \quad 0 \quad 0 \quad 1 \rangle^T$$

$$\mathbf{s}_b = \langle 0 \quad 0 \quad 0 \quad -2 \quad 1 \rangle^T$$

$$\mathbf{s}_c = \langle 1 \quad 1 \quad 1 \quad 1 \quad 0.5 \rangle^T$$

(b) Plot e_J against J and comment on how, for a given $\mathbf{s}$, this error depends on J, and how, for a given J, this error depends on $\mathbf{s}$.
(c) Compare the error e_J if J Ritz vectors are included in the analysis (Problem 15.7) versus the error e_J if J natural vibration modes of the system are included. Discuss the pros and cons of using the two sets of vectors in response history analysis and response spectrum analysis of earthquake response.

*15.9 (a) Determine the steady-state response of the five-story shear frame in Fig. P15.3 to ground motion $\ddot{u}_g(t) = 0.2g \sin 15t$ using two force-dependent Ritz vectors determined from the force distribution $\mathbf{s} = \langle 1 \quad 1 \quad 1 \quad 1 \quad 0.5 \rangle^T$. Neglect damping.
(b) Compare these results with those from modal analysis, including (i) the first two natural vibration modes, and (ii) all five modes.

Numerical Evaluation of Dynamic Response

PREVIEW

So far, we have been concerned primarily with modal analysis of MDF systems with classical damping responding within their linearly elastic range; see Fig. 9.11.1. The uncoupled modal equations could be solved in closed form if the excitation were a simple function (Chapter 12), but the numerical methods of Chapter 5 were necessary for complex excitations such as earthquake ground motion (Chapter 13). Uncoupling of modal equations is not possible if the system has nonclassical damping or it responds into the nonlinear range. For such systems, coupled equations of motion in nodal, modal, or Ritz coordinates—Eqs. (9.8.2), (12.4.4), or (15.3.3), respectively—need to be solved by numerical methods. A vast body of literature, including major chapters of several books, exists about these methods. This chapter includes only a few methods, however, that build upon the procedures presented in Chapter 5 for SDF systems. It provides the basic concepts underlying these methods and the computational algorithms needed to implement the methods.

16.1 TIME-STEPPING METHODS

The objective is to solve numerically the system of differential equations governing the response of MDF systems:

$$\mathbf{m}\ddot{\mathbf{u}} + \mathbf{c}\dot{\mathbf{u}} + \mathbf{f}_S(\mathbf{u}) = \mathbf{p}(t) \quad \text{or} \quad -\mathbf{m}\boldsymbol{\iota}\ddot{u}_g(t) \qquad (16.1.1)$$

with the initial conditions

$$\mathbf{u} = \mathbf{u}(0) \quad \text{and} \quad \dot{\mathbf{u}} = \dot{\mathbf{u}}(0) \tag{16.1.2}$$

at $t = 0$. The solution will provide the displacement vector $\mathbf{u}(t)$ as a function of time.

As in Chapter 5, the time scale is divided into a series of time steps, usually of constant duration Δt. The excitation is given at discrete time instants $t_i = i \, \Delta t$; at t_i, denoted as time i, the excitation vector is $\mathbf{p}_i \equiv \mathbf{p}(t_i)$. The response will be determined at the same time instants and is denoted by $\mathbf{u}_i \equiv \mathbf{u}(t_i)$, $\dot{\mathbf{u}}_i \equiv \dot{\mathbf{u}}(t_i)$, and $\ddot{\mathbf{u}}_i \equiv \ddot{\mathbf{u}}(t_i)$.

Starting with the known response of the system at time i that satisfies Eq. (16.1.1) at time i,

$$\mathbf{m}\ddot{\mathbf{u}}_i + \mathbf{c}\dot{\mathbf{u}}_i + (\mathbf{f}_S)_i = \mathbf{p}_i \tag{16.1.3}$$

time-stepping methods enable us to step ahead to determine the response $\mathbf{u}_{i+1}$, $\dot{\mathbf{u}}_{i+1}$, and $\ddot{\mathbf{u}}_{i+1}$ of the system at time $i + 1$ that satisfies Eq. (16.1.1) at time $i + 1$:

$$\mathbf{m}\ddot{\mathbf{u}}_{i+1} + \mathbf{c}\dot{\mathbf{u}}_{i+1} + (\mathbf{f}_S)_{i+1} = \mathbf{p}_{i+1} \tag{16.1.4}$$

When applied successively with $i = 0, 1, 2, 3, \ldots$, the time-stepping procedure gives the desired response at all time instants $i = 1, 2, 3, \ldots$. The known initial conditions at time $i = 0$, Eq. (16.1.2), provide the information necessary to start the procedure.

The numerical procedure requires three matrix equations to determine the three unknown vectors $\mathbf{u}_{i+1}$, $\dot{\mathbf{u}}_{i+1}$, and $\ddot{\mathbf{u}}_{i+1}$. Two of these equations are derived from either finite difference equations for the velocity and acceleration vectors or from an assumption on how the acceleration varies during a time step. The third is Eq. (16.1.1) at a selected time instant. If it is the current time i, the method of integration is said to be an *explicit method*. If the time $i + 1$ at the end of the time step is used, the method is known as an *implicit method;* see Chapter 5.

As mentioned in Chapter 5, for a numerical procedure to be useful, it should (1) converge to the exact solution as Δt decreases, (2) be stable in the presence of numerical round-off errors, and (3) be accurate (i.e., the computational errors should be within an acceptable limit). The stability criteria were shown not to be restrictive in the response analysis of SDF systems because Δt must be considerably smaller than the stability limit to ensure adequate accuracy in the numerical results. Stability of the numerical method is a critical consideration, however, in the analysis of MDF systems, as we shall see in this chapter. In particular, conditionally stable procedures can be used effectively for analysis of linear response of large MDF systems, but unconditionally stable procedures are generally necessary for nonlinear response analysis of such systems. In the following sections we present some of the numerical methods for each type of response analysis.

16.2 LINEAR SYSTEMS WITH NONCLASSICAL DAMPING

The N differential equations (16.1.1) to be solved for the nodal displacements $\mathbf{u}$, when specialized for linear systems, are

$$\mathbf{m\ddot{u}} + \mathbf{c\dot{u}} + \mathbf{ku} = \mathbf{p}(t) \quad \text{or} \quad \mathbf{m}\iota\ddot{u}_g(t) \tag{16.2.1}$$

In this section, we present an alternative to the generalized modal analysis procedure (Chapter 14) for solving Eq. (16.2.1). If the system has a few DOFs, it may be appropriate to solve these equations in their present form. For large systems, however, it is usually advantageous to transform Eq. (16.2.1) to a smaller set of equations by expressing the displacements in terms of the first few natural vibration modes ϕ_n of the undamped system (Chapter 12) or an appropriate set of Ritz vectors (Chapter 15). In this section we use the modal transformation; extension of the concepts to use Ritz vector transformation is straightforward.

Thus, the nodal displacements of the system are approximated by a linear combination of the first J natural modes:

$$\mathbf{u}(t) \simeq \sum_{n=1}^{J} \phi_n \mathbf{q}_n(t) = \mathbf{\Phi q}(t) \tag{16.2.2}$$

where J can be selected using the concepts and procedure developed in Section 12.11. Using this transformation, as shown in Section 12.4, Eq. (16.2.1) becomes

$$\mathbf{M\ddot{q}} + \mathbf{C\dot{q}} + \mathbf{Kq} = \mathbf{P}(t) \tag{16.2.3}$$

where

$$\mathbf{M} = \mathbf{\Phi}^T \mathbf{m} \mathbf{\Phi} \qquad \mathbf{C} = \mathbf{\Phi}^T \mathbf{c} \mathbf{\Phi} \qquad \mathbf{K} = \mathbf{\Phi}^T \mathbf{k} \mathbf{\Phi} \qquad \mathbf{P}(t) = \mathbf{\Phi}^T \mathbf{p}(t) \tag{16.2.4}$$

with $\mathbf{M}$ and $\mathbf{K}$ being diagonal matrices. Equation (16.2.3) is a system of J equations in the unknowns $q_n(t)$, and if J is much smaller than N, it may be advantageous to solve them numerically instead of Eq. (16.2.1). The resulting computational savings can more than compensate for the additional computational effort necessary to determine the first J modes.

The J equations (16.2.3) may be coupled or uncoupled depending on the form of the damping matrix. They are uncoupled for systems with classical damping, and each modal equation can be solved numerically by the methods of Chapter 5. For systems with nonclassical damping, $\mathbf{C}$ is not a diagonal matrix and the equations are coupled. In this section numerical methods are presented for solving such coupled equations for linear systems. Although these methods are presented with reference to Eq. (16.2.3), they can be extended to the reduced set of equations (15.3.3) using Ritz vectors.

Conditionally stable numerical methods can be used to solve Eq. (16.2.3); that is, we need not insist on an unconditionally stable procedure (see Section 5.5.1). The time step Δt should be chosen so that $\Delta t / T_n$ is small enough to ensure an accurate solution for each of the modes included, $n = 1, 2, \ldots, J$; T_n is the natural period of the nth mode of the undamped system. The choice for Δt would be dictated by the period of the Jth mode because it has the shortest period; thus $\Delta t / T_J$ should be small, say less than 0.1. This choice implies that $\Delta t / T_n$ for all the lower modes is even smaller, ensuring an accurate solution for all the modes included. The Δt chosen to satisfy the accuracy requirement, say

$\Delta t < 0.1T_J$, would obviously satisfy the stability requirement. For example, $\Delta t = 0.1T_J$ is much smaller than the stability limits of T_J/π and $0.551T_J$ for the central difference method and the linear acceleration method, respectively (Sections 5.3 and 5.4).

Direct solution of Eq. (16.2.1)—without transforming to modal coordinates—may be preferable for systems with few DOFs or for systems and excitations where most of the modes contribute significantly to the response, because in these situations there is little to be gained by modal transformation. The numerical methods presented next are readily adaptable to such direct solution as long as the time step Δt is chosen to satisfy the stability requirement relative to the shortest natural period T_N of the undamped system.

Two conditionally stable procedures are presented next for linear response analysis of MDF systems. These are the central difference method and Newmark's method.

16.2.1 Central Difference Method

Developed in Section 5.3 for SDF systems, the central difference method can readily be extended to MDF systems. The scalar equations (5.3.1) that relate the response quantities at time $i + 1$ to those at times i and $i - 1$, and the scalar equation (5.1.3) of equilibrium at time i, all now become matrix equations. The other new feature arises from the need to transform the initial conditions on nodal displacements, Eq. (16.1.2), to modal coordinates, and to transform back the solution of Eq. (16.2.3) in modal coordinates to nodal displacements. Putting all these ideas together in Table 5.3.1 leads to Table 16.2.1, where the central difference method is presented as it might be implemented on the computer.

Two observations regarding the central difference method may be useful. First, the algebraic equations to be solved in step 1.3 to determine $\ddot{\mathbf{q}}_0$ are uncoupled because $\mathbf{M}$ is a diagonal matrix when modal coordinates or force-dependent Ritz vectors are used. Second, step 2.3 is based on equilibrium at time i, and the stiffness matrix $\mathbf{K}$ does not enter into the system of algebraic equations solved to determine $\mathbf{q}_{i+1}$ at time $i + 1$, implying that the central difference method is an explicit method.

The central difference method can also be used for direct solution of the original equations in nodal displacements, Eq. (16.2.1), without transforming them to modal coordinates, by modifying Table 16.2.1 as follows: Delete steps 1.1, 1.2, 2.1, and 2.5. Replace (1) $\mathbf{q}$, $\dot{\mathbf{q}}$, and $\ddot{\mathbf{q}}$ by $\mathbf{u}$, $\dot{\mathbf{u}}$, and $\ddot{\mathbf{u}}$; (2) $\mathbf{M}$, $\mathbf{C}$, and $\mathbf{K}$ by $\mathbf{m}$, $\mathbf{c}$, and $\mathbf{k}$; (3) $\mathbf{P}$ by $\mathbf{p}$; and (4) $\hat{\mathbf{K}}$ and $\hat{\mathbf{P}}$ by $\hat{\mathbf{k}}$ and $\hat{\mathbf{p}}$.

16.2.2 Newmark's Method

Developed in Section 5.4 for SDF systems, Newmark's method can readily be extended to MDF systems. The scalar equations (5.4.9) that relate the response—displacement, velocity, and acceleration—increments over the time step i to $i + 1$ to each other and the response values at time i, and the scalar equation (5.4.12) of incremental equilibrium, all now become matrix equations. Implementing this change in Table 5.4.2 together with transformation of initial conditions to modal coordinates and of modal solutions to nodal displacements, as in Section 16.2.1, leads to Table 16.2.2, where the time-stepping solution using Newmark's method is summarized as it might be implemented on the computer.

TABLE 16.2.1   CENTRAL DIFFERENCE METHOD: LINEAR SYSTEMS

1.0   *Initial calculations*

1.1 $(q_n)_0 = \dfrac{\phi_n^T \mathbf{m} \mathbf{u}_0}{\phi_n^T \mathbf{m} \phi_n}$; $(\dot{q}_n)_0 = \dfrac{\phi_n^T \mathbf{m} \dot{\mathbf{u}}_0}{\phi_n^T \mathbf{m} \phi_n}$

$\mathbf{q}_0^T = \langle (q_1)_0, \dots, (q_J)_0 \rangle$ $\dot{\mathbf{q}}_0^T = \langle (\dot{q}_1)_0, \dots, (\dot{q}_J)_0 \rangle$.

1.2 $\mathbf{P}_0 = \mathbf{\Phi}^T \mathbf{p}_0$.

1.3   Solve: $\mathbf{M}\ddot{\mathbf{q}}_0 = \mathbf{P}_0 - \mathbf{C}\dot{\mathbf{q}}_0 - \mathbf{K}\mathbf{q}_0 \Rightarrow \ddot{\mathbf{q}}_0$.

1.4   Select Δt.

1.5 $\mathbf{q}_{-1} = \mathbf{q}_0 - \Delta t\, \dot{\mathbf{q}}_0 + \dfrac{(\Delta t)^2}{2}\ddot{\mathbf{q}}_0$.

1.6 $\hat{\mathbf{K}} = \dfrac{1}{(\Delta t)^2}\mathbf{M} + \dfrac{1}{2\Delta t}\mathbf{C}$.

1.7 $\mathbf{a} = \dfrac{1}{(\Delta t)^2}\mathbf{M} - \dfrac{1}{2\Delta t}\mathbf{C}$; $\mathbf{b} = \mathbf{K} - \dfrac{2}{(\Delta t)^2}\mathbf{M}$.

2.0   *Calculations for each time step i*

2.1 $\mathbf{P}_i = \mathbf{\Phi}^T \mathbf{p}_i$.

2.2 $\hat{\mathbf{P}}_i = \mathbf{P}_i - \mathbf{a}\mathbf{q}_{i-1} - \mathbf{b}\mathbf{q}_i$.

2.3   Solve: $\hat{\mathbf{K}}\mathbf{q}_{i+1} = \hat{\mathbf{P}}_i \Rightarrow \mathbf{q}_{i+1}$.

2.4   If required:

$$\dot{\mathbf{q}}_i = \frac{1}{2\Delta t}(\mathbf{q}_{i+1} - \mathbf{q}_{i-1})   \qquad \ddot{\mathbf{q}}_i = \frac{1}{(\Delta t)^2}(\mathbf{q}_{i+1} - 2\mathbf{q}_i + \mathbf{q}_{i-1})$$

2.5 $\mathbf{u}_{i+1} = \mathbf{\Phi}\mathbf{q}_{i+1}$.

3.0   *Repetition for the next time step.* Replace i by $i + 1$ and repeat steps 2.1 to 2.5 for the next time step.

The two special cases of Newmark's method that are commonly used are (1) $\gamma = \frac{1}{2}$ and $\beta = \frac{1}{4}$, which gives the constant average acceleration method, and (2) $\gamma = \frac{1}{2}$ and $\beta = \frac{1}{6}$, corresponding to the linear acceleration method. The constant average acceleration method is unconditionally stable, whereas the linear acceleration method is conditionally stable for $\Delta t \leq 0.551 T_J$. For a given time step that is much smaller than this stability limit, the linear acceleration method is more accurate than the average acceleration method. Therefore, it is especially useful for linear systems because the Δt chosen to obtain accurate response in the highest mode included would satisfy the stability requirements. Observe that step 2.3 is based on equilibrium at time $i + 1$, and the stiffness matrix $\mathbf{K}$ enters into the system of algebraic equations solved to determine q_{i+1} at time $i + 1$, implying that Newmark's method is an implicit method.

Newmark's method can also be used for direct solution of the original equations in nodal displacements, Eq. (16.2.1), without transforming them to modal coordinates, by modifying Table 16.2.2 appropriately, as indicated at the end of Section 16.2.1.

TABLE 16.2.2 NEWMARK'S METHOD: LINEAR SYSTEMS

Special cases

    (1) Constant average acceleration method ($\gamma = \frac{1}{2}$, $\beta = \frac{1}{4}$)

    (2) Linear acceleration method ($\gamma = \frac{1}{2}$, $\beta = \frac{1}{6}$)

1.0 *Initial calculations*

    1.1 $(q_n)_0 = \dfrac{\phi_n^T \mathbf{m}\, \mathbf{u}_0}{\phi_n^T \mathbf{m}\, \phi_n}$; $(\dot{q}_n)_0 = \dfrac{\phi_n^T \mathbf{m}\, \dot{\mathbf{u}}_0}{\phi_n^T \mathbf{m}\, \phi_n}$

        $\mathbf{q}_0^T = \langle (q_1)_0, \ldots, (q_J)_0 \rangle$ $\dot{\mathbf{q}}_0^T = \langle (\dot{q}_1)_0, \ldots, (\dot{q}_J)_0 \rangle$.

    1.2 $\mathbf{P}_0 = \mathbf{\Phi}^T \mathbf{p}_0$.

    1.3 Solve $\mathbf{M}\ddot{\mathbf{q}}_0 = \mathbf{P}_0 - \mathbf{C}\dot{\mathbf{q}}_0 - \mathbf{K}\mathbf{q}_0 \Rightarrow \ddot{\mathbf{q}}_0$.

    1.4 Select Δt.

    1.5 $\mathbf{a}_1 = \dfrac{1}{\beta(\Delta t)^2}\mathbf{M} + \dfrac{\gamma}{\beta \Delta t}\mathbf{C}$; $\mathbf{a}_2 = \dfrac{1}{\beta \Delta t}\mathbf{M} + \left(\dfrac{\gamma}{\beta} - 1\right)\mathbf{C}$; and

        $\mathbf{a}_3 = \left(\dfrac{1}{2\beta} - 1\right)\mathbf{M} + \Delta t\left(\dfrac{\gamma}{2\beta} - 1\right)\mathbf{C}$.

    1.6 $\hat{\mathbf{K}} = \mathbf{K} + \mathbf{a}_1$.

2.0 *Calculations for each time step, $i = 0, 1, 2, \ldots$*

    2.1 $\hat{\mathbf{P}}_{i+1} = \mathbf{\Phi}^T \mathbf{p}_{i+1} + \mathbf{a}_1 \mathbf{q}_i + \mathbf{a}_2 \dot{\mathbf{q}}_i + \mathbf{a}_3 \ddot{\mathbf{q}}_i$.

    2.2 Solve $\hat{\mathbf{K}}\mathbf{q}_{i+1} = \hat{\mathbf{P}}_{i+1} \Rightarrow \mathbf{q}_{i+1}$.

    2.3 $\dot{\mathbf{q}}_{i+1} = \dfrac{\gamma}{\beta \Delta t}(\mathbf{q}_{i+1} - \mathbf{q}_i) + \left(1 - \dfrac{\gamma}{\beta}\right)\dot{\mathbf{q}}_i + \Delta t\left(1 - \dfrac{\gamma}{2\beta}\right)\ddot{\mathbf{q}}_i$.

    2.4 $\ddot{\mathbf{q}}_{i+1} = \dfrac{1}{\beta(\Delta t)^2}(\mathbf{q}_{i+1} - \mathbf{q}_i) - \dfrac{1}{\beta \Delta t}\dot{\mathbf{q}}_i - \left(\dfrac{1}{2\beta} - 1\right)\ddot{\mathbf{q}}_i$.

    2.5 $\mathbf{u}_{i+1} = \mathbf{\Phi}\mathbf{q}_{i+1}$.

3.0 *Repetition for the next time step.* Replace i by $i+1$ and implement steps 2.1 to 2.5 for the next time step.

Example 16.1

The five-story shear building of Fig. 12.8.1 (repeated for convenience as Fig. E.16.1a) is subjected to a full sinusoidal cycle of ground acceleration $\ddot{u}_g(t) = \ddot{u}_{go} \sin 2\pi t$ (Fig. E.16.1b) with $t_d = 1$ sec; $\ddot{u}_{go} = 0.5g$; $m = 100$ kips/g; $k = 100$ kips/in.; and modal damping ratios $\zeta_n = 5\%$ for all modes. Solve the equations of motion after transforming them to the first two modes by the linear acceleration method with $\Delta t = 0.1$ sec.

Solution First, we set up modal equations. The mass and stiffness matrices are available in Section 12.8 and the effective earthquake forces in Eq. (13.1.2):

$$\mathbf{m} = m\begin{bmatrix} 1 & & & & \\ & 1 & & & \\ & & 1 & & \\ & & & 1 & \\ & & & & 1 \end{bmatrix} \quad \mathbf{k} = k\begin{bmatrix} 2 & -1 & & & \\ -1 & 2 & -1 & & \\ & -1 & 2 & -1 & \\ & & -1 & 2 & -1 \\ & & & -1 & 1 \end{bmatrix} \quad \mathbf{p}(t) = -m\begin{bmatrix} 1 \\ 1 \\ 1 \\ 1 \\ 1 \end{bmatrix}\ddot{u}_g(t)$$

                                                                                                                                                               (a)

where the subscript "eff" in Eq. (13.1.2) has been dropped.

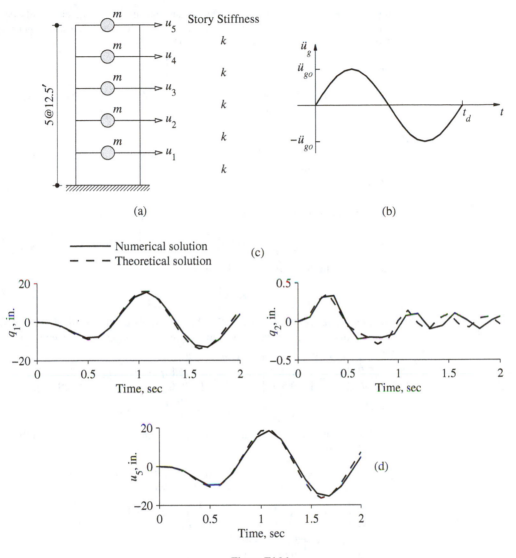

Figure E16.1

Solving the eigenvalue problem gives the first two natural frequencies and modes:

$$\mathbf{\Omega} = \begin{bmatrix} 5.592 & \\ & 16.32 \end{bmatrix} \qquad \mathbf{\Phi} = \begin{bmatrix} 0.334 & -0.895 \\ 0.641 & -1.173 \\ 0.895 & -0.641 \\ 1.078 & 0.334 \\ 1.173 & 1.078 \end{bmatrix} \tag{b}$$

Substituting $\mathbf{m}$, $\mathbf{k}$, and $\mathbf{\Phi}$ in Eq. (16.2.4) gives

$$\mathbf{M} = \begin{bmatrix} 1 & \\ & 1 \end{bmatrix} \qquad \mathbf{K} = \begin{bmatrix} 31.27 & \\ & 266.4 \end{bmatrix} \qquad \mathbf{P}(t) = \begin{bmatrix} -1.067 \\ 0.336 \end{bmatrix} \ddot{u}_g(t) \tag{c}$$

The two equations (16.2.3) in modal coordinates are uncoupled for the classically damped system, and $\mathbf{C}$ is a diagonal matrix with the nth diagonal element equal to the generalized modal damping $C_n = \zeta_n(2M_n\omega_n)$; see Eq. (10.9.11):

$$\mathbf{C} = \begin{bmatrix} 0.559 & \\ & 1.632 \end{bmatrix} \tag{d}$$

For generality, this uncoupling property is not used in solving this example, however. The procedure of Table 16.2.2 is implemented as follows:

1.0 *Initial calculations*

    1.1 Since the system starts from rest, $\mathbf{u}_0 = \dot{\mathbf{u}}_0 = \mathbf{0}$; therefore, $\mathbf{q}_0 = \dot{\mathbf{q}}_0 = \mathbf{0}$.

    1.2 $\mathbf{p}_0 = \mathbf{0}$; therefore, $\mathbf{P}_0 = \mathbf{0}$.

    1.3 $\ddot{\mathbf{q}}_0 = \mathbf{0}$.

    1.4 $\Delta t = 0.1$ sec.

    1.5 Substituting $\mathbf{M}$, $\mathbf{C}$, Δt, $\gamma = \frac{1}{2}$, and $\beta = \frac{1}{6}$ in step 1.5 gives

$$\mathbf{a}_1 = \begin{bmatrix} 616.8 & \\ & 649.0 \end{bmatrix} \quad \mathbf{a}_2 = \begin{bmatrix} 61.12 & \\ & 63.27 \end{bmatrix} \quad \mathbf{a}_3 = \begin{bmatrix} 2.028 & \\ & 2.082 \end{bmatrix}$$

    1.6 Substituting $\mathbf{K}$ and $\mathbf{a}_3$ in step 1.6 gives

$$\hat{\mathbf{K}} = \begin{bmatrix} 648.0 & \\ & 915.4 \end{bmatrix}$$

2.0 *Calculations for each time step, i.* For the parameters of this example, computational steps 2.1 through 2.5 are specialized and implemented for each time step i as follows:

    2.1 $\hat{\mathbf{P}}_{i+1} = \mathbf{\Phi}^T \mathbf{p}_{i+1} + \mathbf{a}_1 \mathbf{q}_i + \mathbf{a}_2\dot{\mathbf{q}}_i + \mathbf{a}_3 \ddot{\mathbf{q}}_i$.

$$\begin{bmatrix} \hat{P}_1 \\ \hat{P}_2 \end{bmatrix}_{i+1} = \begin{bmatrix} -1.067 \\ 0.336 \end{bmatrix} (\ddot{u}_g)_{i+1} + \begin{bmatrix} 616.8\,q_1 + 61.12\,\dot{q}_1 + 2.028\,\ddot{q}_1 \\ 649.0\,q_2 + 63.26\,\dot{q}_2 + 2.028\,\ddot{q}_2 \end{bmatrix}_i$$

    2.2 Solve $\begin{bmatrix} 648.0 & \\ & 915.4 \end{bmatrix} \begin{bmatrix} q_1 \\ q_2 \end{bmatrix}_{i+1} = \begin{bmatrix} \hat{P}_1 \\ \hat{P}_2 \end{bmatrix}_{i+1} \Rightarrow \mathbf{q}_{i+1}$.

The modal displacements $\mathbf{q}_i$ for the first 20 time steps are shown in Table E16.1 and Fig. E16.1c.

    2.3 $\begin{bmatrix} \dot{q}_1 \\ \dot{q}_2 \end{bmatrix}_{i+1} = 30 \left(\begin{bmatrix} q_1 \\ q_2 \end{bmatrix}_{i+1} - \begin{bmatrix} q_1 \\ q_2 \end{bmatrix}_i \right) - 2\begin{bmatrix} \dot{q}_1 \\ \dot{q}_2 \end{bmatrix}_i - 0.05\begin{bmatrix} \ddot{q}_1 \\ \ddot{q}_2 \end{bmatrix}_i$.

    2.4 $\begin{bmatrix} \ddot{q}_1 \\ \ddot{q}_2 \end{bmatrix}_{i+1} = 600 \left(\begin{bmatrix} q_1 \\ q_2 \end{bmatrix}_{i+1} - \begin{bmatrix} q_1 \\ q_2 \end{bmatrix}_i \right) - 60\begin{bmatrix} \dot{q}_1 \\ \dot{q}_2 \end{bmatrix}_i - 2\begin{bmatrix} \ddot{q}_1 \\ \ddot{q}_2 \end{bmatrix}_i$.

    2.5 $\begin{bmatrix} u_1 \\ u_2 \\ u_3 \\ u_4 \\ u_5 \end{bmatrix}_{i+1} = \begin{bmatrix} 0.334 & -0.895 \\ 0.641 & -1.173 \\ 0.895 & -0.641 \\ 1.078 & 0.334 \\ 1.173 & 1.078 \end{bmatrix} \begin{bmatrix} q_1 \\ q_2 \end{bmatrix}_{i+1}$.

These displacements are also presented in Table E16.1, and u_5 is plotted in Fig. E16.1d as a function of time.

Comparison with theoretical solution. The modal equations defined by Eqs. E16.1c–d can also be solved analytically, extending the procedure of Section 4.8 to damped systems. Considering the first two modes of the system, such theoretical results were derived. They were computed at every 0.1 sec and are presented as the dashed lines in Fig. E16.1c and d.

TABLE E16.1 NUMERICAL SOLUTION OF MODAL EQUATIONS BY THE LINEAR ACCELERATION METHOD

t_i	q_1	q_2	u_1	u_2	u_3	u_4	u_5
0.1	-0.1868	-0.1868	-0.0997	-0.1685	-0.1940	-0.1875	-0.1742
0.2	-1.3596	-1.3596	-0.6688	-1.1524	-1.3711	-1.3851	-1.3357
0.3	-3.7765	-3.7765	-1.5977	-2.8605	-3.6226	-3.9442	-4.0229
0.4	-6.6733	-6.6733	-2.4239	-4.5317	-6.1156	-7.1185	-7.5893
0.5	-8.5377	-8.5377	-2.7869	-5.3864	-7.5996	-9.2244	-10.0877
0.6	-7.8337	-7.8337	-2.4301	-4.7759	-6.8820	-8.5111	-9.4087
0.7	-3.8483	-3.8483	-1.1041	-2.2286	-3.3166	-4.2146	-4.7301
0.8	2.7434	2.7434	1.1162	2.0198	2.5998	2.8818	2.9758
0.9	9.8980	9.8980	3.5110	6.6113	9.0106	10.5897	11.3579
1.0	14.8661	14.8661	5.0228	9.6017	13.3542	15.9984	17.3602
1.1	15.4597	15.4597	5.0214	9.7205	13.7429	16.7125	18.2966
1.2	11.5465	11.5465	3.7794	7.2981	10.2851	12.4714	13.6304
1.3	4.4929	4.4929	1.6127	3.0259	4.1037	4.7998	5.1327
1.4	-3.4964	-3.4964	-1.0838	-2.1305	-3.0710	-3.7990	-4.2003
1.5	-10.0597	-10.0597	-3.4465	-6.5597	-9.0707	-10.8081	-11.6901
1.6	-13.3706	-13.3706	-4.5502	-8.6786	-12.0342	-14.3768	-15.5745
1.7	-12.6389	-12.6389	-4.1522	-8.0085	-11.2691	-13.6456	-14.9016
1.8	-8.2858	-8.2858	-2.6779	-5.1924	-7.3562	-8.9622	-9.8223
1.9	-1.7591	-1.7591	-0.6336	-1.1876	-1.6083	-1.8784	-2.0069
2.0	4.9390	4.9390	1.5634	3.0520	4.3614	5.3545	5.8944

The dashed line in Fig. E16.1d also represents the theoretical solution including all five modes, indicating that the response contributions of the third, fourth, and fifth modes are negligible. The numerical results for q_1 are accurate because the chosen time step $\Delta t = 0.1$ sec and natural period $T_1 = 2\pi/5.592 = 1.12$ sec, implying a small $\Delta t/T_1 = 0.089$. However, the same Δt implies that $\Delta t/T_2 = 0.16$, which is not small enough to provide good accuracy for q_2. The numerical solution for u_5 is quite accurate, however, because the contribution of the second mode is small.

16.3 NONLINEAR SYSTEMS

To numerically evaluate the dynamic response of systems responding beyond their linearly elastic range, the N equations for an N-DOF system are usually solved in their original form, Eq. (16.1.1), because classical modal analysis is not applicable to nonlinear systems (Fig. 9.11.1). However, even the displacements of a nonlinear system can always be expressed as a combination of the natural modes of the undamped system vibrating within the range of its linear behavior:

$$\mathbf{u}(t) = \sum_{n=1}^{N} \phi_n q_n(t) \qquad (16.3.1)$$

Numerical solutions

——————— Direct solution by linear acceleration method
— — — Two–mode solution (Example 16.1) using $\Delta t = 0.12$ sec

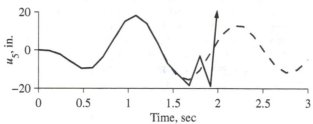

Time, sec

Figure 16.3.1

Direct solution of Eq. (16.1.1) is equivalent to including all the N modes in the analysis, although only the first J terms in Eq. (16.3.1) may be sufficient to represent accurately the structural response. It would seem that the choice of Δt should be based on the accuracy requirements for the Jth mode, say $\Delta t = T_J/10$, where T_J is the period of the Jth mode of undamped linear vibration. Direct solution of Eq. (16.1.1) with this choice of Δt will give $\mathbf{u}(t)$ such that the higher-mode ($J + 1$ to N) terms in Eq. (16.3.1) would be inaccurate, but this should not be of concern because we had concluded that these higher-mode contributions to the response were negligible. Although this choice of Δt would seem to provide accurate results, it may not be sufficiently small to ensure stability of the numerical procedure. Accuracy is required only for the first J modes, but stability must be ensured for all modes because even if the response in the higher modes is insignificant, it will "blow up" if the stability requirements are not satisfied relative to these modes. This problem is illustrated in Fig. 16.3.1, where the response of the shear building of Example 16.1 to one sinusoidal cycle of ground motion is presented as obtained by two numerical methods. The dashed curve shows the results determined by solving the first two modal equations by the linear acceleration method, as in Example 16.1 but now using $\Delta t = 0.12$ sec. When the original equations (16.1.1) are solved by the same method using the same time step, this direct solution (shown by the solid curve) "blows up" around $t = 2$ sec.

Requiring stability for all modes imposes very severe restrictions on Δt, as illustrated by the following example. Consider a system in which the highest mode with significant response contribution has a period $T_J = 0.10$ sec, whereas the period of the highest mode of the system is $T_N = 0.001$ sec. To ensure stability of the numerical procedure, Δt should be less than T_N/π (i.e., $\Delta t < 0.00032$ sec) for the central difference method, and $0.551 T_N$ (i.e. $\Delta t < 0.00055$ sec) for the linear acceleration method.

Presented in this section are one conditionally stable method: the central difference method, which is an explicit method (Section 16.3.1); and one unconditionally stable method: the average acceleration method, which is an implicit method. For implicit methods, there is little difference between nonlinear static and nonlinear dynamic analysis. The approach adopted here is first to present Newton–Raphson iteration for nonlinear static analysis (Section 16.3.2), then use Newmark's equations (5.4.8 and 5.4.9)—adapted to MDF systems—to extend it to nonlinear dynamic analysis (Section 16.3.3).

The constant average acceleration method has the drawback that it provides no numerical damping (Fig. 5.5.2). This is a disadvantage because it is desirable to filter out the response contributions of modes higher than the J significant modes because these higher modes and their frequencies, which have been calculated from an idealization of the structure, are usually not accurate relative to the actual properties of the structure. One approach for achieving this goal is to define the damping matrix consistent with increasing damping ratio for modes higher than the Jth mode (see Section 11.4). Researchers have also been interested in formulating numerical time-stepping algorithms which, in some sense, have optimal numerical damping.

16.3.1 Central Difference Method

The central difference method for nonlinear SDF systems (Section 5.6) can readily be adapted for MDF systems. Each scalar equation in the procedure for SDF systems (Eqs. 5.6.1 to 5.6.3) now becomes a matrix equation for MDF systems. Table 16.3.1 summarizes the procedure as it might be implemented on the computer.

The resisting forces $(\mathbf{f}_S)_i$ required in step 2.1 can be evaluated explicitly, as they depend only on the known state of the system at time i. Thus, $(\mathbf{f}_S)_i$ is easily calculated, making the central difference method perhaps the simplest procedure for analysis of nonlinear MDF systems.

TABLE 16.3.1   CENTRAL DIFFERENCE METHOD: NONLINEAR SYSTEMS

1.0   *Initial calculations*

    1.1   State determination for initial $\mathbf{u} = \mathbf{u}_0 : (\mathbf{f}_S)_0$.

    1.2   Solve $\mathbf{m}\ddot{\mathbf{u}}_0 = \mathbf{p}_0 - \mathbf{c}\dot{\mathbf{u}}_0 - (\mathbf{f}_S)_0 \Rightarrow \ddot{\mathbf{u}}_0$.

    1.3   Select Δt.

    1.4 $\mathbf{u}_{-1} = \mathbf{u}_0 - \Delta t\,\dot{\mathbf{u}}_0 + \dfrac{(\Delta t)^2}{2}\ddot{\mathbf{u}}_0$.

    1.5 $\hat{\mathbf{k}} = \dfrac{1}{(\Delta t)^2}\mathbf{m} + \dfrac{1}{2\Delta t}\mathbf{c}$.

    1.6 $\mathbf{a} = \dfrac{1}{(\Delta t)^2}\mathbf{m} - \dfrac{1}{2\Delta t}\mathbf{c}$ and $\mathbf{b} = -\dfrac{2}{(\Delta t)^2}\mathbf{m}$.

2.0   *Calculations for each time step, $i = 0, 1, 2, \ldots$*

    2.1 $\hat{\mathbf{p}}_i = \mathbf{p}_i - \mathbf{a}\,\mathbf{u}_{i-1} - \mathbf{b}\,\mathbf{u}_i - (\mathbf{f}_S)_i$.

    2.2   Solve $\hat{\mathbf{k}}\mathbf{u}_{i+1} = \hat{\mathbf{p}}_i \Rightarrow \mathbf{u}_{i+1}$.

    2.3   State determination: $(\mathbf{f}_S)_{i+1}$.

    2.4   If required:

$$\dot{\mathbf{u}}_i = \frac{1}{2\Delta t}(\mathbf{u}_{i+1} - \mathbf{u}_{i-1}) \quad \text{and} \quad \ddot{\mathbf{u}}_i = \frac{1}{(\Delta t)^2}(\mathbf{u}_{i+1} - 2\mathbf{u}_i + \mathbf{u}_{i-1})$$

3.0   *Repetition for next time step.* Replace i by $i + 1$ and repeat steps 2.1–2.4 for the next time step.

Explicit methods, typically the central difference method, are used in software such as LS-DYNA, ABAQUS-Explicit, and OpenSees. Because the time step required is very small, these methods are impractical for analysis of large systems using conventional computers (one to four computer processors). However, explicit methods have the advantage that they can be conveniently programmed for parallel computing using a large number of computer processors. Most applications to large systems diagonalize the damping matrix **c** so that $\hat{\mathbf{k}}$ is diagonal and the equations in step 2.2 can be solved efficiently. Researchers have developed various approximate models for damping to achieve this goal.

16.3.2 Nonlinear Static Analysis

Nonlinear static analysis is used to investigate the force–deformation behavior of a structure for a specified distribution of forces, typically lateral forces. With certain assumption of force distribution, nonlinear static analysis is called *pushover analysis* (see Chapters 19 and 22).

Dropping the inertia and damping terms in the equations of motion [Eq. (16.1.1)] gives the system of nonlinear equations to be solved in a static problem:

$$\mathbf{f}_S(\mathbf{u}) = \mathbf{p}(t) \tag{16.3.2}$$

Before examining multiple-force steps necessary in a pushover analysis, consider the solution of the equilibrium equations for a single set of forces:

$$\mathbf{f}_S(\mathbf{u}) = \mathbf{p} \tag{16.3.3}$$

The task is to determine the displacements **u** due to a set of given external forces **p**, where the nonlinear force–deformation relation $\mathbf{f}_S(\mathbf{u})$ is known for the system to be analyzed.

Suppose that after j cycles of iteration, $\mathbf{u}^{(j)}$ is an estimate of the unknown displacements and we are interested in developing an iterative procedure that provides an improved estimate $\mathbf{u}^{(j+1)}$. Expanding the resisting forces $\mathbf{f}_S^{(j+1)}(\mathbf{u})$ in Taylor series about the known estimate $\mathbf{u}^{(j)}$, and dropping the second- and higher-order terms, leads to the linearized equation (see Section 5.7.1)

$$\mathbf{k}_T^{(j)} \Delta \mathbf{u}^{(j)} = \mathbf{p} - \mathbf{f}_S^{(j)} = \mathbf{R}^{(j)} \tag{16.3.4}$$

where

$$\mathbf{k}_T^{(j)} = \left. \frac{\partial \mathbf{f}_S(\mathbf{u})}{\partial \mathbf{u}} \right|_{\mathbf{u}^{(j)}} \tag{16.3.5}$$

is the tangent stiffness matrix at $\mathbf{u}^{(j)}$; $(k_T)_{i,j}$ = the change in force at DOF i due to unit change in displacement at DOF j at the current state of the system. Solving the linearized system of equations (16.3.4) gives $\Delta \mathbf{u}^{(j)}$ and an improved estimate of displacements:

$$\mathbf{u}^{(j+1)} = \mathbf{u}^{(j)} + \Delta \mathbf{u}^{(j)} \tag{16.3.6}$$

This is the essence of the Newton–Raphson method for iterative solution of the nonlinear equations (16.3.3). As shown in Section 5.7.1, this iterative method converges with quadratic rate to the exact solution.

The preceding description of the Newton–Raphson procedure for a single force step can be generalized for multiple force steps. For this purpose, the forces are represented by a reference spatial distribution $\mathbf{p}_{\text{ref}}$ and a variable scalar λ_i; thus

$$\mathbf{p}_i = \lambda_i \mathbf{p}_{\text{ref}} \qquad (16.3.7)$$

The nonlinear equilibrium equations for each force level are solved by Newton–Raphson iteration starting with the initial estimate of the solution as the displacements at the previous force level. Table 16.3.2 summarizes such a procedure for nonlinear static analysis as it might be implemented on the computer.

TABLE 16.3.2 NONLINEAR STATIC ANALYSIS

1.0 *State determination for* $\mathbf{u} = \mathbf{u}_0^{\dagger}$: $(\mathbf{f}_S)_0$ and $(\mathbf{k}_T)_0$.

2.0 *Calculations for each force step,* $i = 0, 1, 2, \ldots$
    2.1 Initialize $j = 1$, $\mathbf{u}_{i+1}^{(j)} = \mathbf{u}_i$, $(\mathbf{f}_S)_{i+1}^{(j)} = (\mathbf{f}_S)_i$, and $(\mathbf{k}_T)_{i+1}^{(j)} = (\mathbf{k}_T)_i$.
    2.2 $\mathbf{p}_{i+1} = \lambda_{i+1}\mathbf{p}_{\text{ref}}$.

3.0 *For each iteration,* $j = 1, 2, 3, \ldots$
    3.1 $\mathbf{R}_{i+1}^{(j)} = \mathbf{p}_{i+1} - (\mathbf{f}_S)_{i+1}^{(j)}$.
    3.2 Check convergence; if the acceptance criteria are not met, implement steps 3.3 to 3.6; otherwise, skip these steps and go to step 4.0.
    3.3 Solve $(\mathbf{k}_T)_{i+1}^{(j)} \Delta\mathbf{u}^{(j)} = \mathbf{R}_{i+1}^{(j)} \Rightarrow \Delta\mathbf{u}^{(j)}$.
    3.4 $\mathbf{u}_{i+1}^{(j+1)} = \mathbf{u}_{i+1}^{(j)} + \Delta\mathbf{u}^{(j)}$.
    3.5 State determination: $(\mathbf{f}_S)_{i+1}^{(j+1)}$ and $(\mathbf{k}_T)_{i+1}^{(j+1)}$.
    3.6 Replace j by $j + 1$ and repeat steps 3.1 to 3.5; denote final value as $\mathbf{u}_{i+1}$.

4.0 *Repetition for next force step.* Replace i by $i + 1$ and implement steps 2.0 and 3.0 for the next force step.

---

† $\mathbf{u}_0$ may be nonzero if initial gravity load effects are included in the analysis.

In step 3.2, the solution is checked and the iterative process is terminated when some measure of the error in the solution falls below a specified tolerance. Typically, one or more of the following convergence (or acceptance) criteria are enforced:

1. Residual force is less than a tolerance:

$$\left\| \mathbf{R}^{(j)} \right\| \leq \varepsilon_R \qquad (16.3.8a)$$

where $\| \cdot \|$ denotes the Euclidean norm of the vector. Conventional values for the tolerance ε_R range from 10^{-3} to 10^{-8}.

2. Change in displacement is less than a tolerance:

$$\left\| \Delta\mathbf{u}^{(j)} \right\| \leq \varepsilon_u \qquad (16.3.8b)$$

Conventional values for the tolerance ε_u range from 10^{-3} to 10^{-8}.

3. Incremental work done by the residual force acting through the change in displacement is less than a tolerance:

$$\tfrac{1}{2} \left\| [\Delta\mathbf{u}^{(j)}]^{\mathrm{T}} \mathbf{R}^{(j)} \right\| \leq \varepsilon_w \qquad (16.3.8c)$$

Tolerance ε_w must be at or near the computer (machine) tolerance because the left side is a product of small quantities.

Although the examples presented subsequently use the preceding criteria, for large MDF systems it is better to use relative force or displacement measures:

$$\frac{\left\| \mathbf{R}^{(j)} \right\|}{\left\| \mathbf{P}_{\text{ref}} \right\|} \leq \varepsilon_R' \qquad \frac{\left\| \Delta \mathbf{u}^{(j)} \right\|}{\left\| \mathbf{u}^{(j)} \right\|} \leq \varepsilon_u' \qquad (16.3.9a)$$

where the recommended value for tolerances ε_R' and ε_u' is 10^{-3} to 10^{-6}. For frames, the displacement vector contains translations and rotations (and the force vectors contains forces and moments) whose magnitudes may be vastly different. For such situations, we recommend use of relative incremental work to check convergence. The convergence criterion then is

$$\frac{\left\| \left[\Delta \mathbf{u}^{(j)} \right]^T \mathbf{R}^{(j)} \right\|}{\left\| \left[\Delta \mathbf{u}^{(1)} \right]^T \mathbf{R}^{(1)} \right\|} \leq \varepsilon_w' \qquad (16.3.9b)$$

where the recommended value for ε_w' is on the order of 10^{-16}.

Example 16.2

The five-story shear building of Example 16.1 is subjected to monotonically increasing lateral forces with the invariant distribution presented in Fig. E16.2a. The story shear–drift $(V_j - \delta_j)$ relationship is identical for all stories; it is bilinear with initial stiffness $k = 100$ kips/in., post yield stiffness ratio $\alpha = 0.05$, and yield shear $V_{jy} = 125$ kips (Fig. E16.2b). Conduct nonlinear static analysis of the building for the $\mathbf{p}_{\text{ref}}$ shown and force factors:

$$\boldsymbol{\lambda}^T = \langle 0 \quad 1.0 \quad 1.1 \quad 1.2 \quad 1.3 \quad 1.4 \quad 1.5 \quad 1.6 \rangle$$

Note that the base shear associated with $\mathbf{p}_{\text{ref}}$ is $V_b = 125$ kips, the yield value of base shear.

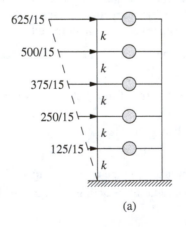

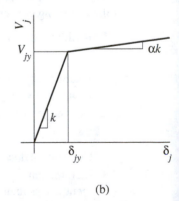

(a)   (b)

Figure E16.2 a, b

Solution   The procedure of Table 16.3.2 is implemented as follows:

1.0 *State determination for* $\mathbf{u}_0 = \mathbf{0}$

Gravity load effects would not produce any lateral displacements; thus $\mathbf{u}_0 = \mathbf{0}$, $(\mathbf{f}_S)_0 = \mathbf{0}$ and

$$(\mathbf{k}_T)_0 = k \begin{bmatrix} 2 & -1 & & & \\ -1 & 2 & -1 & & \\ & -1 & 2 & -1 & \\ & & -1 & 2 & -1 \\ & & & -1 & 1 \end{bmatrix}$$

To demonstrate calculations of steps 2.0 and 3.0 in Table 16.3.1, they are implemented for the force step from $i = 1$ to $i = 2$.

2.0 *Calculations for* $i = 1$

2.1 Initialize $j = 1$

$\mathbf{u}_{i+1}^{(1)} = \mathbf{u}_i = \langle 1.250 \quad 2.417 \quad 3.417 \quad 4.167 \quad 4.583 \rangle^T$; see Table E16.2.
State determination at $i = 1(\lambda = 1.0)$ leads to
$(\mathbf{f}_S)_{i+1}^{(1)} = (\mathbf{f}_S)_i = \langle 8.333 \quad 16.67 \quad 25.00 \quad 33.33 \quad 41.67 \rangle^T$

$$(\mathbf{k}_T)_{i+1}^{(1)} = (\mathbf{k}_T)_i = k \begin{bmatrix} 1.05 & -1 & & & \\ -1 & 2 & -1 & & \\ & -1 & 2 & -1 & \\ & & -1 & 2 & -1 \\ & & & -1 & 1 \end{bmatrix}$$

2.2 $\mathbf{p}_{i+1} = \lambda_{i+1}\mathbf{p}_{\text{ref}}$, where $\mathbf{p}_{\text{ref}}$ is shown in Fig. E16.2a.
$\mathbf{p}_{i+1} = \langle 9.167 \quad 18.33 \quad 27.50 \quad 36.67 \quad 45.83 \rangle^T$.

3.0 *First iteration,* $j = 1$

3.1 $\mathbf{R}_{i+1}^{(1)} = \mathbf{p}_{i+1} - (\mathbf{f}_S)_{i+1}^{(1)} = \langle 0.833 \quad 1.667 \quad 2.500 \quad 3.333 \quad 4.167 \rangle^T$.

3.2 Check convergence: Because $\left\| \mathbf{R}_{i+1}^{(1)} \right\| = 6.180$ exceeds $\varepsilon_R = 10^{-3}$, chosen for this example, implement steps 3.3 to 3.6.

3.3 Solve $(\mathbf{k}_T)_{i+1}^{(1)} \Delta\mathbf{u}^{(1)} = \mathbf{R}_{i+1}^{(1)} \Rightarrow \Delta\mathbf{u}^{(1)} = \langle 2.500 \quad 2.617 \quad 2.717 \quad 2.792 \quad 2.833 \rangle^T$.

3.4 $\mathbf{u}_{i+1}^{(2)} = \mathbf{u}_{i+1}^{(1)} + \Delta\mathbf{u}^{(1)} = \langle 3.750 \quad 5.033 \quad 6.133 \quad 6.958 \quad 7.417 \rangle^T$.

3.5 State determination.

First determine story drifts:

$$\boldsymbol{\delta}_{i+1}^{(2)} = \langle 3.750 \quad 1.283 \quad 1.100 \quad 0.825 \quad 0.458 \rangle^T$$

Knowing story drift, the story stiffness and story shear force can be obtained from the story shear–story drift relationship (Fig. E16.2b); e.g., the first story drift, $\delta_1 = 3.750$ in. is larger than the yield drift, $\delta_{1y} = 1.25$ in.; therefore, the story stiffness is $0.05k$ and the shear force is $125 + 0.05\,k\,(3.75 - 1.25) = 137.5$ kips. Knowing the stiffness of each story, the stiffness matrix can be obtained using the procedure of Example 9.1b.

$$(\mathbf{k}_T)_{i+1}^{(2)} = k \begin{bmatrix} 0.10 & -0.05 & & & \\ -0.05 & 1.05 & -1 & & \\ & -1 & 2 & -1 & \\ & & -1 & 2 & -1 \\ & & & -1 & 1 \end{bmatrix}$$

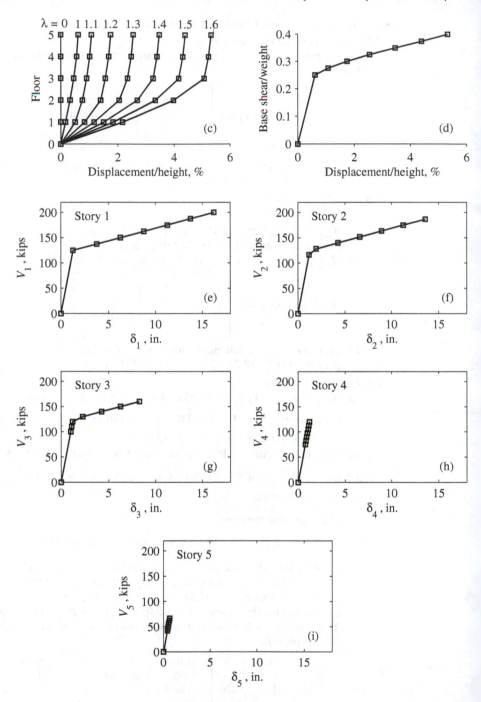

Figure E16.2 (c)–(i)

The resisting force f_{S1} at the first floor is the difference between story shears in the first and second stories; f_{Sj} at other stories are determined similarly to obtain

$$(\mathbf{f}_S)_{i+1}^{(2)} = \begin{bmatrix} 137.50 - 125.17 \\ 125.17 - 110.00 \\ 110.00 - 82.50 \\ 82.50 - 45.83 \\ 45.83 \end{bmatrix} = \begin{bmatrix} 12.33 \\ 15.17 \\ 27.50 \\ 36.67 \\ 45.83 \end{bmatrix}$$

3.0 *Second iteration, $j = 2$*

3.1 $\mathbf{R}_{i+1}^{(2)} = \mathbf{p}_{i+1} - (\mathbf{f}_S)_{i+1}^{(2)} = \langle -3.167 \quad 3.167 \quad 0 \quad 0 \quad 0 \rangle^T.$

3.2 Check convergence: Because $\left\| \mathbf{R}_{i+1}^{(2)} \right\| = 4.478$ exceeds ε_R, implement steps 3.3 to 3.6.

3.3 Solve $(\mathbf{k}_T)_{i+1}^{(2)} \Delta \mathbf{u}^{(2)} = \mathbf{R}_{i+1}^{(2)} \Rightarrow \Delta \mathbf{u}^{(2)} = \langle 0 \quad 0.633 \quad 0.633 \quad 0.633 \quad 0.633 \rangle^T.$

3.4 $\mathbf{u}_{i+1}^{(3)} = \mathbf{u}_{i+1}^{(2)} + \Delta \mathbf{u}^{(2)} = \langle 3.750 \quad 5.667 \quad 6.767 \quad 7.592 \quad 8.050 \rangle^T.$

3.5 State determination:

$$\boldsymbol{\delta}_{i+1}^{(3)} = \langle 3.750 \quad 1.917 \quad 1.100 \quad 0.825 \quad 0.458 \rangle^T$$

$$(\mathbf{f}_S)_{i+1}^{(3)} = \begin{bmatrix} 9.167 \\ 18.33 \\ 27.50 \\ 36.67 \\ 45.83 \end{bmatrix} \qquad (\mathbf{k}_T)_{i+1}^{(3)} = k \begin{bmatrix} 0.10 & -0.05 & & & \\ -0.05 & 1.05 & -1 & & \\ & -1 & 2 & -1 & \\ & & -1 & 2 & -1 \\ & & & -1 & 1 \end{bmatrix}$$

3.0 *Third iteration, $j = 3$*

3.1 $\mathbf{R}_{i+1}^{(3)} = \mathbf{p}_{i+1} - (\mathbf{f}_S)_{i+1}^{(3)} \simeq 0.$

3.2 Check convergence: Because $\left\| \mathbf{R}_{i+1}^{(3)} \right\| = 0$ is less than ε_R, iteration is terminated and the final value is denoted as $\mathbf{u}_{i+1}$; see row 3 of Table E16.2.

4.0 *Repetition for the next time step*: Steps 2.0 and 3.0 are implemented for $i = 0, 1, 2, 3, \ldots$ to obtain the floor displacements u_1, u_2, u_3, u_4, and u_5 presented in Table E16.2 and plotted in Fig. E16.2c. The base shear is plotted as a function of the roof displacement u_5 in Fig. E16.2d. Story shear–story drift relationships are presented in Figs. E16.2e–i for the five stories of the building.

TABLE E16.2   RESULTS OF NONLINEAR STATIC ANALYSIS

i	λ_i	u_1	u_2	u_3	u_4	u_5
0	0	0.0000	0.0000	0.0000	0.0000	0.0000
1	1.0	1.2500	2.4167	3.4167	4.1667	4.5833
2	1.1	3.7500	5.6667	6.7667	7.5917	8.0500
3	1.2	6.2500	10.5000	11.7000	12.6000	13.1000
4	1.3	8.7500	15.3333	17.5833	18.5583	19.1000
5	1.4	11.2500	20.1667	24.4167	25.4667	26.0500
6	1.5	13.7500	25.0000	31.2500	32.3750	33.0000
7	1.6	16.2500	29.8333	38.0833	39.2833	39.9500

Example 16.3

The five-story shear building of Example 16.2 is subjected to a full sinusoidal cycle of ground acceleration $\ddot{u}_g(t) = \ddot{u}_{go} \sin 2\pi t$ (Fig. E16.1b) with $t_d = 1$ sec and $\ddot{u}_{go} = 0.5g$. Solve the static equilibrium equations at every time step $\Delta t = 0.05$ sec.

Solution   The static equilibrium equations are

$$\mathbf{f}_S(\mathbf{u}) = -\mathbf{m}\iota\ddot{u}_g(t) = -\mathbf{m}\iota\ddot{u}_{go} \sin 2\pi t \qquad (a)$$

The reference force distribution $\mathbf{p}_{\text{ref}} = -\mathbf{m}\iota\ddot{u}_{go}$ and the load factor is $\lambda_i = \sin 2\pi t_i$, where $t_i = i\Delta t$. Thus, $\mathbf{p}_{\text{ref}} = -50\langle 1 \quad 1 \quad 1 \quad 1 \quad 1 \rangle^T$ and $\lambda = \langle 0 \quad 0.309 \quad 0.588 \quad 0.809 \quad 0.951 \quad 1.0 \rangle^T$. Computational step 1.0 is identical to Example 16.2. Computational steps 2.0 and 3.0 are implemented for $i = 0, 1, 2, 3, \ldots$ to obtain the displacements u_1, u_2, u_3, u_4, and u_5 presented in Table E16.3 and plotted in Figure E16.3a at $i = 0, 2, 4, \ldots$. The base shear is plotted as a function of the roof displacement u_5 in Fig. E16.3b. Story shear–story drift relationships are presented in Fig. E16.3c–g for the five stories of the building.

The computations in step 3.5 must recognize that the force–deformation relation of inelastic systems is path dependent; see Section 16.3.3.

TABLE E16.3   RESULTS OF NONLINEAR STATIC ANALYSIS

i	t_i	λ_i	u_1	u_2	u_3	u_4	u_5
0	0.00	0.0000	0.0000	0.0000	0.0000	0.0000	0.0000
1	0.05	0.3090	−0.7725	−1.3906	−1.8541	−2.1631	−2.3176
2	0.10	0.5878	−5.6393	−6.8148	−7.6965	−8.2843	−8.5782
3	0.15	0.8090	−16.7008	−25.3115	−26.5251	−27.3341	−27.7386
4	0.20	0.9511	−23.8028	−38.0951	−42.8768	−43.8278	−44.3034
5	0.25	1.0000	−26.2500	−42.5000	−48.7500	−49.7500	−50.2500
6	0.30	0.9511	−26.1276	−42.2798	−48.4563	−49.4074	−49.8829
7	0.35	0.8090	−25.7725	−41.6406	−47.6041	−48.4131	−48.8176
8	0.40	0.5878	−25.2195	−40.6450	−46.2767	−46.8645	−47.1584
9	0.45	0.3090	−24.5225	−39.3906	−44.6041	−44.9131	−45.0676
10	0.50	0.0000	−23.7500	−38.0000	−42.7500	−42.7500	−42.7500
11	0.55	−0.3090	−8.2992	−19.6885	−23.9749	−23.6659	−23.5114
12	0.60	−0.5878	5.6393	5.4007	1.5324	2.1201	2.4140
13	0.65	−0.8090	16.7008	25.3115	25.8320	26.6411	27.0456
14	0.70	−0.9511	23.8028	38.0951	42.8768	43.8278	44.3034
15	0.75	−1.0000	26.2500	42.5000	48.7500	49.7500	50.2500
16	0.80	−0.9511	26.1276	42.2798	48.4563	49.4074	49.8829
17	0.85	−0.8090	25.7725	41.6406	47.6041	48.4131	48.8176
18	0.90	−0.5878	25.2195	40.6450	46.2767	46.8645	47.1584
19	0.95	−0.3090	24.5225	39.3906	44.6041	44.9131	45.0676
20	1.00	0.0000	23.7500	38.0000	42.7500	42.7500	42.7500

16.3.3 Constant Average Acceleration Method

The constant average acceleration method with Newton–Raphson iteration has already been presented for analysis of nonlinear SDF systems; it is the procedure summarized in

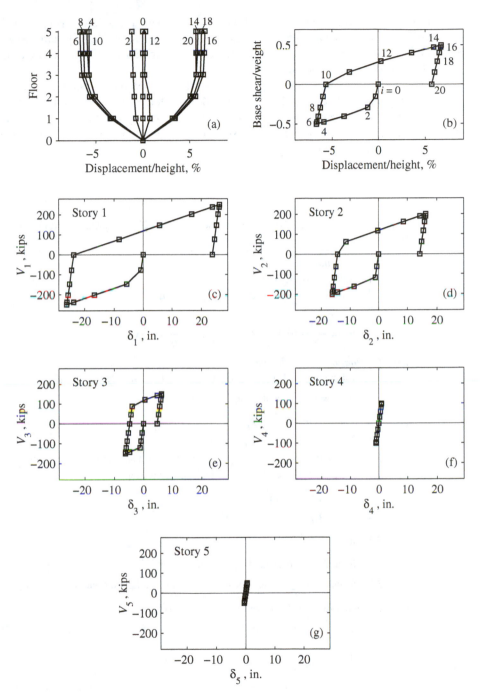

Figure E16.3

Table 5.7.1, specialized for $\gamma = \frac{1}{2}$ and $\beta = \frac{1}{4}$. This procedure carries over directly to MDF systems with each scalar equation in the procedure for SDF systems now becoming a matrix equation for MDF systems. Table 16.3.3 summarizes the procedure as it might be implemented on the computer.

To avoid computation of a new tangent stiffness matrix for each iteration—which can be computationally demanding for large MDF systems—the initial stiffness matrix at the beginning of a time step may be used for all iterations within the time step. This modified Newton–Raphson iteration results in slower convergence, i.e., it requires more iterations to achieve convergence (see Figs. 5.7.1 and 5.7.2).

TABLE 16.3.3 CONSTANT AVERAGE ACCELERATION METHOD: NONLINEAR SYSTEMS

1.0 *Initial calculations*

    1.1   State determination: $(\mathbf{f}_S)_0$ and $(\mathbf{k}_T)_0$.

    1.2   Solve $\mathbf{m}\ddot{\mathbf{u}}_0 = \mathbf{p}_0 - \mathbf{c}\dot{\mathbf{u}}_0 - (\mathbf{k}_T)_0\,\mathbf{u}_0 \Rightarrow \ddot{\mathbf{u}}_0$.

    1.3   Select Δt.

    1.4 $\mathbf{a}_1 = \dfrac{4}{(\Delta t)^2}\mathbf{m} + \dfrac{2}{\Delta t}\mathbf{c}$ and $\mathbf{a}_2 = \dfrac{4}{\Delta t}\mathbf{m} + \mathbf{c}$.

2.0 *Calculations for each time instant, $i = 0, 1, 2, \ldots$*

    2.1   Initialize $j = 1$, $\mathbf{u}_{i+1}^{(j)} = \mathbf{u}_i$, $(\mathbf{f}_S)_{i+1}^{(j)} = (\mathbf{f}_S)_i$, and $(\mathbf{k}_T)_{i+1}^{(j)} = (\mathbf{k}_T)_i$.

    2.2 $\hat{\mathbf{p}}_{i+1} = \mathbf{p}_{i+1} + \mathbf{a}_1\mathbf{u}_i + \mathbf{a}_2\dot{\mathbf{u}}_i + \mathbf{m}\ddot{\mathbf{u}}_i$.

3.0 *For each iteration, $j = 1, 2, 3, \ldots$*

    3.1 $\hat{\mathbf{R}}_{i+1}^{(j)} = \hat{\mathbf{p}}_{i+1} - (\mathbf{f}_S)_{i+1}^{(j)} - \mathbf{a}_1\mathbf{u}_{i+1}^{(j)}$.

    3.2   Check convergence; if the acceptance criteria are not met, implement steps 3.3 to 3.7; otherwise, skip these steps and go to step 4.0.

    3.3 $(\hat{\mathbf{k}}_T)_{i+1}^{(j)} = (\mathbf{k}_T)_{i+1}^{(j)} + \mathbf{a}_1$.

    3.4   Solve $(\hat{\mathbf{k}}_T)_{i+1}^{(j)}\Delta\mathbf{u}^{(j)} = \hat{\mathbf{R}}_{i+1}^{(j)} \Rightarrow \Delta\mathbf{u}^{(j)}$.

    3.5 $\mathbf{u}_{i+1}^{(j+1)} = \mathbf{u}_{i+1}^{(j)} + \Delta\mathbf{u}^{(j)}$.

    3.6   State determination: $(\mathbf{f}_S)_{i+1}^{(j+1)}$ and $(\mathbf{k}_T)_{i+1}^{(j+1)}$.

    3.7   Replace j by $j+1$ and repeat steps 3.1 to 3.6; denote final value as $\mathbf{u}_{i+1}$.

4.0 *Calculations for velocity and acceleration vectors*

    4.1 $\dot{\mathbf{u}}_{i+1} = \dfrac{2}{\Delta t}(\mathbf{u}_{i+1} - \mathbf{u}_i) - \dot{\mathbf{u}}_i$.

    4.2 $\ddot{\mathbf{u}}_{i+1} = \dfrac{4}{(\Delta t)^2}(\mathbf{u}_{i+1} - \mathbf{u}_i) - \dfrac{4}{\Delta t}\dot{\mathbf{u}}_i - \ddot{\mathbf{u}}_i$.

5.0 *Repetition for next time step.* Replace i by $i+1$ and implement steps 2.0 to 4.0 for the next time step.

Path dependence.   One of the important operations in the computations is state determination (step 3.6). It must recognize that the force–deformation relation of inelastic systems is path dependent, i.e., it depends on whether the deformation is increasing or decreasing during the time step. This feature must be considered in calculating the tangent stiffness matrix and the resisting forces associated with the displacements $\mathbf{u}_i$ at each time i. Such procedures are available in textbooks on static structural analysis and are not included here [see, e.g., Filippou and Fenves (2004)].

Example 16.4

The five-story shear building of Example 16.2 is subjected to a full sinusoidal cycle of ground acceleration $\ddot{u}_g(t) = \ddot{u}_{go} \sin 2\pi t$ (Fig. E16.1b); $\ddot{u}_{go} = 0.5g$. Solve the equations of motion by the constant average acceleration method with Newton–Raphson iteration using a time step of $\Delta t = 0.1$ sec; assume zero initial conditions and modal damping ratios $\zeta_n = 5\%$ for all modes.

Solution   The 5×5 mass and initial stiffness matrices were defined in Example 16.1; the damping matrix corresponding to given modal damping ratios $\zeta_n = 5\%$ is determined by superposing modal damping matrices (Section 11.4.3).

$$
\mathbf{c} = \begin{bmatrix}
69.01 & -19.81 & -3.395 & -1.370 & -0.873 \\
 & 65.70 & -21.18 & -4.268 & -2.243 \\
 & & 64.83 & -22.05 & -5.638 \\
(\text{sym}) & & & 63.46 & -25.45 \\
 & & & & 43.65
\end{bmatrix} \times 10^{-2}
$$

1.0 *Initial calculations*

   1.1 State determination for $\mathbf{u}_0 = \mathbf{0}$:

$$(\mathbf{f}_S)_0 = \mathbf{0}$$

$$
(\mathbf{k}_T)_0 = k \begin{bmatrix}
2 & -1 & & & \\
-1 & 2 & -1 & & \\
 & -1 & 2 & -1 & \\
 & & -1 & 2 & -1 \\
 & & & -1 & 1
\end{bmatrix}
$$

   1.2 Solve $\mathbf{m}\ddot{\mathbf{u}}_0 = \mathbf{p}_0 - \mathbf{c}\dot{\mathbf{u}}_0 - (\mathbf{k}_T)_0 \mathbf{u}_0 \Rightarrow \ddot{\mathbf{u}}_0 = \mathbf{0}$.

   1.3 $\Delta t = 0.1$.

   1.4 Matrices $\mathbf{a}_1$ and $\mathbf{a}_2$:

$$
\mathbf{a}_1 = \frac{4}{(\Delta t)^2}\mathbf{m} + \frac{2}{\Delta t}\mathbf{c} = 400\mathbf{m} + 20\mathbf{c} = \begin{bmatrix}
117.4 & -3.962 & -0.679 & -0.274 & -0.175 \\
 & 116.8 & -4.236 & -0.854 & -0.449 \\
 & & 116.6 & -4.411 & -1.128 \\
(\text{sym}) & & & 116.3 & -5.090 \\
 & & & & 112.4
\end{bmatrix}
$$

$$
\mathbf{a}_2 = \frac{4}{\Delta t}\mathbf{m} + \mathbf{c} = 40\mathbf{m} + \mathbf{c} = \begin{bmatrix}
110.5 & -1.981 & -0.340 & -0.137 & -0.087 \\
 & 110.2 & -2.118 & -0.427 & -0.224 \\
 & & 110.1 & -2.205 & -0.564 \\
(\text{sym}) & & & 110.0 & -2.545 \\
 & & & & 108.0
\end{bmatrix} \times 10^{-1}
$$

2.0 *Calculations for each time step, i*

Computational steps 2.0 to 5.0 are implemented for $i = 0, 1, 2, 3, \ldots$ to obtain the displacements $u_1, u_2, u_3, u_4,$ and u_5 presented in Table E16.4. The roof displacement u_5 is plotted as a function of time in Fig. E16.4a, and the variation of floor displacements over height at selected time instants $i = 0, 2, 4, \ldots$ are plotted in Fig. E16.4b.

TABLE E16.4 NUMERICAL SOLUTION BY CONSTANT AVERAGE
ACCELERATION METHOD

i	t_i	u_1	u_2	u_3	u_4	u_5
0	0.0	0.0000	0.0000	0.0000	0.0000	0.0000
1	0.1	−0.1708	−0.2359	−0.2613	−0.2712	−0.2746
2	0.2	−0.7701	−1.1762	−1.3750	−1.4663	−1.5015
3	0.3	−1.8807	−2.8973	−3.5514	−3.9154	−4.0737
4	0.4	−3.5344	−5.1266	−6.2603	−7.0473	−7.4383
5	0.5	−5.3831	−7.5152	−8.7127	−9.6905	−10.2388
6	0.6	−6.4439	−9.0716	−10.0489	−10.7549	−11.1525
7	0.7	−5.9863	−8.4266	−9.1988	−9.5634	−9.6810
8	0.8	−4.3618	−5.5450	−5.7476	−5.9381	−6.0052
9	0.9	−2.0815	−1.8558	−0.9983	−0.6832	−0.6541
10	1.0	0.7305	1.6946	3.3099	4.3848	4.9299
11	1.1	3.4782	5.0342	6.7247	7.8941	8.6981
12	1.2	5.5469	7.8168	9.2078	9.7899	10.0600
13	1.3	6.7492	9.1381	10.2318	10.4083	10.2899
14	1.4	6.6858	8.6380	9.6715	10.0370	10.2689
15	1.5	5.5610	7.2111	8.2729	8.9549	9.4278
16	1.6	4.8642	6.0410	6.8150	7.1914	7.3539
17	1.7	4.9813	5.5452	5.5219	5.1092	4.8836
18	1.8	4.7428	5.1217	4.5283	3.6914	3.2563
19	1.9	4.0458	4.3222	4.0478	3.4673	3.0939
20	2.0	3.7573	4.0322	4.1602	4.0642	4.0387

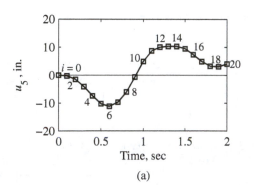

(a)

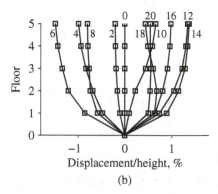

(b)

Figure E16.4

FURTHER READING

Bathe, K.-J., *Finite Element Procedures,* Prentice Hall, Englewood Cliffs, N.J., 1996, Chapter 9.

Filippou, F. C., and Fenves, G. L., "Methods of Analysis for Earthquake-Resistant Structures," in: *Earthquake Engineering: from Engineering Seismology to Performance-Based Engineering* (eds. Y. Bozorgnia and V. V. Bertero), CRC Press, New York, 2004, Chapter 6.

Hughes, T. J. R., *The Finite Element Method,* Prentice Hall, Englewood Cliffs, N.J., 1987, Chapter 9.

Humar, J. L., *Dynamics of Structures,* 2nd ed., A. A. Balkema Publishers, Lisse, The Netherlands, 2002, pp. 748–765.

Newmark, N. M., "A Method of Computation for Structural Dynamics," *Journal of the Engineering Mechanics Division, ASCE,* **85**, 1959, pp. 67–94.

PROBLEMS

*16.1 Solve the problem in Example 16.1 by the central difference method, implemented by a computer program in a language of your choice using $\Delta t = 0.1$ sec.

*16.2 Repeat Problem 16.1 using $\Delta t = 0.05$ sec. How does the time step affect the accuracy of the solution?

*16.3 Solve the problem in Example 16.1 by the constant average acceleration method, implemented by a computer program in a language of your choice using $\Delta t = 0.1$ sec. Based on these results and those from Problem 16.1, comment on the relative accuracy of the average acceleration and central difference methods.

*16.4 Repeat Problem 16.3 using $\Delta t = 0.05$ sec. How does the time step affect the accuracy of the solution?

*16.5 Solve the problem in Example 16.1 by the linear acceleration method, implemented by a computer program in a language of your choice using $\Delta t = 0.1$ sec. Based on these results and those from Problem 16.3, comment on the relative accuracy of the linear acceleration and constant average acceleration methods. Note that this problem was solved as Example 16.1 and the results were presented in Table E16.1.

*16.6 Repeat Problem 16.5 using $\Delta t = 0.05$ sec. How does the time step affect the accuracy of the solution?

*16.7 Solve the problem of Example 16.4 by the central difference method, implemented by a computer program in a language of your choice using a time step of 0.05 sec.

*16.8 Solve the problem in Example 16.2, implemented by a computer program in a language of your choice.

*16.9 Solve Problem 16.8 with a uniform distribution of lateral forces.

*16.10 Solve the problem in Example 16.3, implemented by a computer program in a language of your choice.

*Denotes that a computer is necessary to solve this problem.

*16.11   Solve the problem in Example 16.4, implemented by a computer program in a language of your choice.

*16.12   Solve the problem in Example 16.4 using modified Newton–Raphson iteration. Compare the number of iterations required for convergence using Newton–Raphson iteration (Problem 16.11) and modified Newton–Raphson iteration (Problem 16.12).

*Denotes that a computer is necessary to solve this problem.

Systems with Distributed Mass and Elasticity

PREVIEW

So far in this book we have focused on discretized systems, typically with lumped masses; such a system is an assemblage of rigid elements having mass (e.g., the floor diaphragms of a multistory building) and massless elements that are flexible (e.g., the beams and columns of a building). A major part of this book is devoted to lumped-mass discretized systems, for two reasons. First, such systems can effectively idealize many classes of structures, especially multistory buildings. Second, effective methods that are ideal for computer implementation are available to solve the system of ordinary differential equations governing the motion of such systems. However, a lumped-mass idealization, although applicable, is not a natural approach for certain types of structures, such as a bridge (Fig. 2.1.2e), a chimney (Fig. 2.1.2f), an arch dam (Fig. 1.10.2), or a nuclear containment structure (Fig. 1.10.1).

In this chapter we formulate the structural dynamics problem for one-dimensional systems with distributed mass, such as a beam or a tower, and solutions are presented for simple systems (e.g., a uniform beam and a uniform tower). The solutions presented for these simple cases provide insight into the dynamics of distributed-mass systems that have an infinite number of DOFs and how they differ from lumped-mass systems with a finite number of DOFs. The chapter ends with a discussion of why this infinite-DOF approach is not feasible for practical systems, pointing to the need for discretized methods for distributed-mass systems. The results presented for the simple systems provide the exact solution against which results from discretized methods can be compared (Chapter 18).

17.1 EQUATION OF UNDAMPED MOTION: APPLIED FORCES

In this section we develop the equation governing the transverse vibration of a straight beam without damping subjected to external force. Figure 17.1.1a shows such a beam with flexural rigidity $EI(x)$ and mass $m(x)$ per unit length, both of which may vary with position x. The external forces $p(x, t)$, which may vary with position and time, cause motion of the beam described by the transverse displacement $u(x, t)$ (Fig. 17.1.1b). The equation of motion to be developed will be valid for support conditions other than the simple supports shown and for beams with intermediate supports.

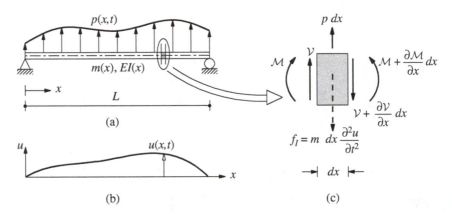

Figure 17.1.1   System with distributed mass and elasticity: (a) beam and applied force; (b) displacement; (c) forces on element.

The system has an infinite number of DOFs because its mass is distributed. Therefore, we consider a differential element of the beam, isolated by two adjoining sections. The forces on the element are shown in Fig. 17.1.1c, where an inertia force has been included following D'Alembert's principle (Section 1.5.2); $V(x, t)$ is the transverse shear force and $\mathcal{M}(x, t)$ is the bending moment. Equilibrium of forces in the y-direction gives

$$\frac{\partial V}{\partial x} = p - m \frac{\partial^2 u}{\partial t^2} \qquad (17.1.1)$$

Without the inertia force this equation is the familiar relation between the shear force in a beam and external transverse force. The inertia force modifies the external force in recognition of the dynamics of the problem. If the inertial moment associated with angular acceleration of the element is neglected, rotational equilibrium of the element gives the standard relation

$$V = \frac{\partial \mathcal{M}}{\partial x} \qquad (17.1.2)$$

We now use Eqs. (17.1.1) and (17.1.2) to write the equation governing the transverse displacement $u(x, t)$. With shear deformation neglected, the moment–curvature relation is

$$M = EI\frac{\partial^2 u}{\partial x^2} \qquad (17.1.3)$$

Substituting Eqs. (17.1.3) and (17.1.2) into Eq. (17.1.1) gives

$$m(x)\frac{\partial^2 u}{\partial t^2} + \frac{\partial^2}{\partial x^2}\left[EI(x)\frac{\partial^2 u}{\partial x^2}\right] = p(x, t) \qquad (17.1.4)$$

This is the partial differential equation governing the motion $u(x, t)$ of the beam subjected to external dynamic forces $p(x, t)$. To obtain a unique solution to this equation, we must specify two boundary conditions at each end of the beam and the initial displacement $u(x, 0)$ and initial velocity $\dot{u}(x, 0)$.

17.2 EQUATION OF UNDAMPED MOTION: SUPPORT EXCITATION

Consider two simple cases: a cantilever beam subjected to horizontal base motion (Fig. 17.2.1a) or a beam with multiple supports subjected to identical motion in the vertical direction (Fig. 17.2.1b). The total displacement of the beam is

$$u^t(x, t) = u_g(t) + u(x, t) \qquad (17.2.1)$$

where the beam displacement $u(x, t)$, measured relative to the support motion $u_g(t)$, results from the deformations of the beam.

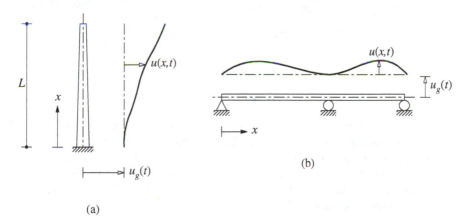

Figure 17.2.1   (a) Cantilever beam subjected to base excitation; (b) continuous beam subjected to identical motion at all supports.

For these simple cases of a beam excited by support motion the derivation of the equation of motion is only slightly different than that for applied forces. Recognizing that the inertia forces are now related to the total accelerations and that external forces $p(x, t)$ do not exist, Eq. (17.1.1) becomes

$$\frac{\partial \mathcal{V}}{\partial x} = -m\frac{\partial^2 u^t}{\partial t^2} = -m\frac{\partial^2 u}{\partial t^2} - m\frac{d^2 u_g}{dt^2} \tag{17.2.2}$$

wherein Eq. (17.2.1) has been used to obtain the second half of the equation. Substituting Eqs. (17.1.3) and (17.1.2) into Eq. (17.2.2) gives

$$m(x)\frac{\partial^2 u}{\partial t^2} + \frac{\partial^2}{\partial x^2}\left[EI(x)\frac{\partial^2 u}{\partial x^2}\right] = -m(x)\ddot{u}_g(t) \tag{17.2.3}$$

By comparing Eqs. (17.2.3) and (17.1.4), it is clear that the deformation response $u(x, t)$ of the beam to support acceleration $\ddot{u}_g(t)$ will be identical to the response of the system with stationary supports due to external forces $= -m(x)\ddot{u}_g(t)$. The support excitation can therefore be replaced by effective forces (Fig. 17.2.2):

$$p_{\text{eff}}(x, t) = -m(x)\ddot{u}_g(t) \tag{17.2.4}$$

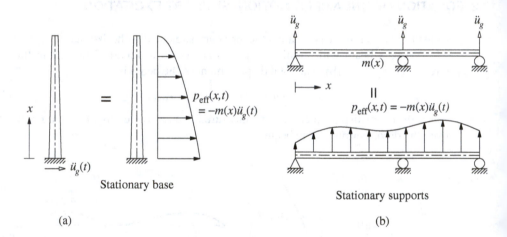

(a)   (b)

Figure 17.2.2 Effective forces $p_{\text{eff}}(x, t)$.

This formulation can be generalized to include the possibility of different motions of the various supports of a structure. Such multiple support excitation may exist in several practical situations (Section 9.7), but is not included in this chapter because it is usually not possible to analyze such practical problems as infinite-DOF systems. They are usually discretized by the finite element method (Chapter 18) and analyzed by extensions of the procedures of Section 13.5.

17.3 NATURAL VIBRATION FREQUENCIES AND MODES

For the case of free vibration, Eqs. (17.1.4) and (17.2.3) become

$$m(x)\frac{\partial^2 u}{\partial t^2} + \frac{\partial^2}{\partial x^2}\left[EI(x)\frac{\partial^2 u}{\partial x^2}\right] = 0 \tag{17.3.1}$$

We attempt a solution of the form

$$u(x, t) = \phi(x)q(t) \tag{17.3.2}$$

Then

$$\frac{\partial^2 u}{\partial t^2} = \phi(x)\ddot{q}(t) \qquad \frac{\partial^2 u}{\partial x^2} = \phi''(x)q(t) \tag{17.3.3}$$

where overdots denote a time derivative and primes denote an x derivative; thus $\dot{q}(t) \equiv dq/dt$, $\ddot{q}(t) = d^2q/dt^2$, and $\phi''(x) = d^2\phi/dx^2$. Substituting Eq. (17.3.3) in Eq. (17.3.1) leads to

$$m(x)\phi(x)\ddot{q}(t) + q(t)\left[EI(x)\phi''(x)\right]'' = 0$$

which, when divided by $m(x)\phi(x)q(t)$, becomes

$$\frac{-\ddot{q}(t)}{q(t)} = \frac{[EI(x)\phi''(x)]''}{m(x)\phi(x)} \tag{17.3.4}$$

The expression on the left is a function of t only and the one on the right depends only on x. For Eq. (17.3.4) to be valid for all values of x and t, the two expressions must therefore be constant, say ω^2. Thus the partial differential equation (17.3.1) becomes two ordinary differential equations, one governing the time function $q(t)$ and the other governing the spatial function $\phi(x)$:

$$\ddot{q} + \omega^2 q = 0 \tag{17.3.5}$$

$$\left[EI(x)\phi''(x)\right]'' - \omega^2 m(x)\phi(x) = 0 \tag{17.3.6}$$

Equation (17.3.5) has the same form as the equation governing free vibration of an SDF system with natural frequency ω. For any given stiffness and mass functions, $EI(x)$ and $m(x)$, respectively, there is an infinite set of frequencies ω and associated modes $\phi(x)$ that satisfy the eigenvalue problem defined by Eq. (17.3.6) and the support (boundary) conditions for the beam.

For the special case of a uniform beam, $EI(x) = EI$ and $m(x) = m$, and Eq. (17.3.6) becomes

$$EI\phi^{IV}(x) - \omega^2 m\phi(x) = 0 \quad \text{or} \quad \phi^{IV}(x) - \beta^4\phi(x) = 0 \tag{17.3.7}$$

where

$$\beta^4 = \frac{\omega^2 m}{EI} \tag{17.3.8}$$

The general solution of Eq. (17.3.7) is (see Derivation 17.1)

$$\phi(x) = C_1 \sin \beta x + C_2 \cos \beta x + C_3 \sinh \beta x + C_4 \cosh \beta x \tag{17.3.9}$$

This solution contains four unknown constants, C_1, C_2, C_3, and C_4, and the eigenvalue parameter β. Application of the four boundary conditions for a single-span beam, two at each end of the beam, will provide a solution for β and hence for the natural frequency ω [from Eq. (17.3.8)] and for three constants in terms of the fourth, resulting in the natural mode of

Eq. (17.3.9). This procedure is illustrated next by two examples: a simply supported beam and a cantilever beam. Results are also available for other boundary conditions but are not included in this book.

17.3.1 Uniform Simply Supported Beam

The natural frequencies and modes of vibration of a uniform beam simply supported at both ends are determined next. At $x = 0$ and $x = L$, the displacement and bending moment are zero. Thus, using Eqs. (17.3.2), (17.1.3), and (17.3.9) at $x = 0$ gives

$$u(0, t) = 0 \Rightarrow \phi(0) = 0 \Rightarrow C_2 + C_4 = 0 \tag{17.3.10a}$$

$$M(0, t) = 0 \Rightarrow EI\phi''(0) = 0 \Rightarrow \beta^2(-C_2 + C_4) = 0 \tag{17.3.10b}$$

These two equations give $C_2 = C_4 = 0$ and the general solution reduces to

$$\phi(x) = C_1 \sin \beta x + C_3 \sinh \beta x \tag{17.3.11}$$

Then at $x = L$,

$$u(L, t) = 0 \Rightarrow \phi(L) = 0 \Rightarrow C_1 \sin \beta L + C_3 \sinh \beta L = 0 \tag{17.3.12a}$$

$$M(L, t) = 0 \Rightarrow EI\phi''(L) = 0 \Rightarrow \beta^2(-C_1 \sin \beta L + C_3 \sinh \beta L) = 0 \tag{17.3.12b}$$

Adding these two equations after dropping β^2 from the second equation gives

$$C_3 \sinh \beta L = 0$$

Since $\sinh \beta L$ cannot be zero (otherwise, ω will be zero, a trivial solution implying no vibration at all), so C_3 must be zero. This leads to the frequency equation:

$$C_1 \sin \beta L = 0 \tag{17.3.13}$$

This equation can be satisfied by selecting $C_1 = 0$, which gives $\phi(x) = 0$, a trivial solution. Therefore, $\sin \beta L$ must be zero, from which

$$\beta L = n\pi \qquad n = 1, 2, 3, \ldots \tag{17.3.14}$$

Equation (17.3.8) then gives the natural vibration frequencies:

$$\omega_n = \frac{n^2\pi^2}{L^2} \sqrt{\frac{EI}{m}} \qquad n = 1, 2, 3, \ldots \tag{17.3.15}$$

The natural vibration mode corresponding to ω_n is obtained by substituting Eq. (17.3.14) in Eq. (17.3.11) with $C_3 = 0$ as determined earlier:

$$\phi_n(x) = C_1 \sin \frac{n\pi x}{L} \tag{17.3.16}$$

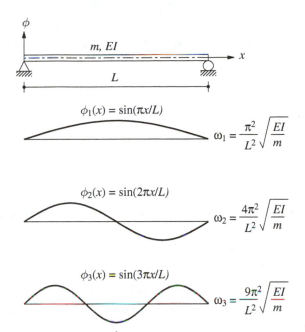

$\phi_1(x) = \sin(\pi x/L)$

$$\omega_1 = \frac{\pi^2}{L^2}\sqrt{\frac{EI}{m}}$$

$\phi_2(x) = \sin(2\pi x/L)$

$$\omega_2 = \frac{4\pi^2}{L^2}\sqrt{\frac{EI}{m}}$$

$\phi_3(x) = \sin(3\pi x/L)$

$$\omega_3 = \frac{9\pi^2}{L^2}\sqrt{\frac{EI}{m}}$$

Figure 17.3.1   Natural vibration modes and frequencies of uniform simply supported beams.

The value of C_1 is arbitrary; $C_1 = 1$ will make the maximum value of $\phi_n(x)$ equal to unity. These natural modes are shown in Fig. 17.3.1.

For a simply supported uniform beam, we have determined an infinite series of modes each with its vibration frequency. Equations (17.3.15) and (17.3.16) and Fig. 17.3.1 tell us that the first mode is a half sine wave and that its frequency $\omega_1 = \pi^2(EI/mL^4)^{1/2}$. The second mode is a complete sine wave with frequency $\omega_2 = 4\omega_1$; the third is one and a half sine waves with frequency $\omega_3 = 9\omega_1$; and so on.

17.3.2 Uniform Cantilever Beam

In this section the natural vibration frequencies and modes of a uniform cantilever beam are determined. At the clamped end, $x = 0$, the displacement and slope are zero. Thus Eq. (17.3.9) gives

$$u(0, t) = 0 \Rightarrow \phi(0) = 0 \Rightarrow C_2 + C_4 = 0 \Rightarrow C_4 = -C_2 \qquad (17.3.17a)$$

$$u'(0, t) = 0 \Rightarrow \phi'(0) = 0 \Rightarrow \beta(C_1 + C_3) = 0 \Rightarrow C_3 = -C_1 \qquad (17.3.17b)$$

At the free end, $x = L$, of the cantilever the bending moment and shear are both zero. Thus, from Eqs. (17.3.9) and after using Eq. (17.3.17), we obtain

$$\mathcal{M}(L,t) = 0 \Rightarrow EI\phi''(L) = 0$$

$$\Rightarrow C_1(\sin \beta L + \sinh \beta L) + C_2(\cos \beta L + \cosh \beta L) = 0 \qquad (17.3.18a)$$

$$\mathcal{V}(L,t) = 0 \Rightarrow EI\phi'''(L) = 0$$

$$\Rightarrow C_1(\cos \beta L + \cosh \beta L) + C_2(-\sin \beta L + \sinh \beta L) = 0 \quad (17.3.18b)$$

Rewriting Eqs. (17.3.18a) and (17.3.18b) in matrix form yields

$$\begin{bmatrix} \sin \beta L + \sinh \beta L & \cos \beta L + \cosh \beta L \\ \cos \beta L + \cosh \beta L & -\sin \beta L + \sinh \beta L \end{bmatrix} \begin{bmatrix} C_1 \\ C_2 \end{bmatrix} = \begin{bmatrix} 0 \\ 0 \end{bmatrix} \qquad (17.3.19)$$

Equation (17.3.19) can be satisfied by selecting both C_1 and C_2 equal to zero, but this would give a trivial solution of no vibration at all. For either or both of C_1 and C_2 to be nonzero, the coefficient matrix in Eq. (17.3.19) must be singular (i.e., its determinant must be zero). This leads to the frequency equation:

$$1 + \cos \beta L \cosh \beta L = 0 \qquad (17.3.20)$$

No simple solution is available for βL, so Eq. (17.3.20) is solved numerically to obtain

$$\beta_n L = 1.8751, \ 4.6941, \ 7.8548, \ \text{and} \ 10.996 \qquad (17.3.21)$$

for $n = 1, 2, 3,$ and 4. For $n > 4$, $\beta_n L \simeq (2n - 1)\pi/2$. Equation (17.3.8) then gives the first four natural frequencies:

$$\omega_1 = \frac{3.516}{L^2} \sqrt{\frac{EI}{m}} \qquad \omega_2 = \frac{22.03}{L^2} \sqrt{\frac{EI}{m}} \qquad \omega_3 = \frac{61.70}{L^2} \sqrt{\frac{EI}{m}} \qquad \omega_4 = \frac{120.9}{L^2} \sqrt{\frac{EI}{m}}$$

$$(17.3.22)$$

Corresponding to each value of $\beta_n L$, the natural vibration mode is

$$\phi_n(x) = C_1 \left[\cosh \beta_n x - \cos \beta_n x - \frac{\cosh \beta_n L + \cos \beta_n L}{\sinh \beta_n L + \sin \beta_n L} (\sinh \beta_n x - \sin \beta_n x) \right]$$

$$(17.3.23)$$

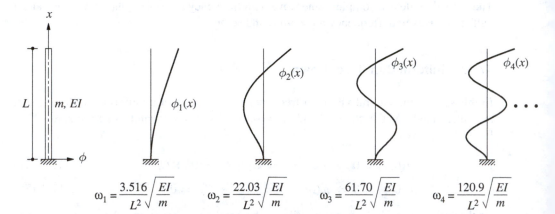

Figure 17.3.2 Natural vibration modes and frequencies of uniform cantilever beams.

where C_1 is an arbitrary constant. To arrive at Eq. (17.3.23), C_2 is expressed in terms of C_1 from Eq. (17.3.18a) and substituted in the general solution, Eq. (17.3.9), and Eq. (17.3.17) is used. The first four natural vibration modes are shown in Fig. 17.3.2.

Derivation 17.1

The solution of the fourth-order ordinary differential equation (17.3.7) is of the form

$$\phi(x) = Ae^{ax} \tag{a}$$

where A is an arbitrary constant. Substituting for $\phi(x)$ and its fourth derivative in Eq. (a) yields the characteristic equation

$$a^4 - \beta^4 = 0 \quad \text{or} \quad (a^2 - \beta^2)(a^2 + \beta^2) = 0 \tag{b}$$

which gives $a = \pm\beta$ and $a = \pm i\beta$. Thus the general solution of Eq. (17.3.7) is

$$\phi(x) = A_1 e^{i\beta x} + A_2 e^{-i\beta x} + A_3 e^{\beta x} + A_4 e^{-\beta x} \tag{c}$$

Equation (c) can be rewritten as Eq. (17.3.9) because

$$e^{\pm\beta x} = \cosh \beta x \pm \sinh \beta x \qquad e^{\pm i\beta x} = \cos \beta x \pm i \sin \beta x \tag{d}$$

17.3.3 Shear Deformation and Rotational Inertia

In the preceding derivation of the equation of motion for the transverse vibration of a beam, the inertial moment associated with rotation of the beam sections was ignored in Eq. (17.1.2), and only the deflection associated with bending stress in the beam was included in Eq. (17.1.3), thus ignoring the deflection due to shear stress in the beam. The analysis of beam vibration, including both the effects of rotational inertia and shear deformation, is called the *Timoshenko beam theory*.

The following equation governs such free vibration of a uniform beam with $m(x) = m$ and $EI(x) = EI$:

$$m\frac{\partial^2 u}{\partial t^2} + EI\frac{\partial^4 u}{\partial x^4} - mr^2 \left(1 + \frac{E}{\kappa G}\right) \frac{\partial^4 u}{\partial x^2 \partial t^2} + \frac{m^2 r^2}{\kappa G A}\frac{\partial^4 u}{\partial t^4} = 0 \tag{17.3.24}$$

where G is the modulus of rigidity, $r = \sqrt{I/A}$ is the radius of gyration of the beam cross section, A is the area of cross section, and κ is a constant that depends on the cross-sectional shape and accounts for the nonuniform distribution of shear stress across the section. The constant κ is derived for various cross-sectional shapes in textbooks on solid mechanics (e.g., κ is $\frac{5}{6}$ for rectangular cross section and $\frac{9}{10}$ for circular cross section).

Consider a beam with both ends simply supported. Assuming a solution of the form $u(x, t) = C \sin(n\pi x/L) \sin \omega'_n t$, which satisfies the necessary end conditions, the frequency equation is obtained. Denoting a natural frequency of the beam by ω'_n if shear and rotational inertia effects are included, by ω_n if these effects are neglected [Eq. (17.3.15)], and defining $\Omega_n = \omega'_n/\omega_n$, this frequency equation can be written as

$$(1 - \Omega_n^2) - \Omega_n^2 \left(\frac{n\pi r}{L}\right)^2 \left(1 + \frac{E}{\kappa G}\right) + \Omega_n^4 \left(\frac{n\pi r}{L}\right)^4 \frac{E}{\kappa G} = 0 \qquad (17.3.25)$$

If it is assumed that $nr/L \ll 1$, the $(n\pi r/L)^4$ term may be dropped and Eq. (17.3.25) reduces to

$$\omega'_n = \omega_n \frac{1}{\sqrt{1 + (n\pi r/L)^2(1 + E/\kappa G)}} \qquad (17.3.26)$$

which implies that $\omega'_n < \omega_n$. The correction due to rotational inertia is represented by the term $(n\pi r/L)^2$ in the denominator, whereas the shear deformation correction appears as $(n\pi r/L)^2(E/\kappa G)$. Thus the correction term for shear deformation is $E/\kappa G$ times larger than the rotational inertia correction term. For steel beams of rectangular cross section, $E/\kappa G$ is approximately 3.2. Values of $\Omega_n = \omega'_n/\omega_n$ are plotted in Fig. 17.3.3 using the solution of Eq. (17.3.25), a quadratic equation in Ω_n^2, for three values of $E/\kappa G$; these results are valid for all natural frequencies because n is included in the abscissa scale. Similar results are presented in Fig. 17.3.4 for the first five natural frequencies of a beam with $E/\kappa G = 3.2$. Also included is the approximate value of the frequency

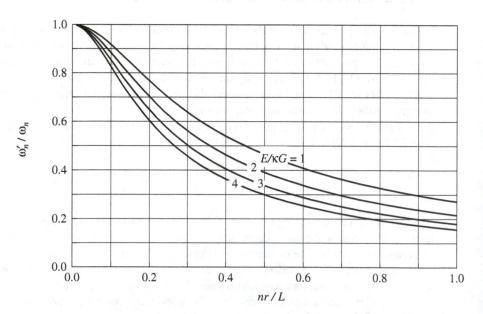

Figure 17.3.3   Influence of shear deformation and rotational inertia on natural frequencies of simply supported beams.

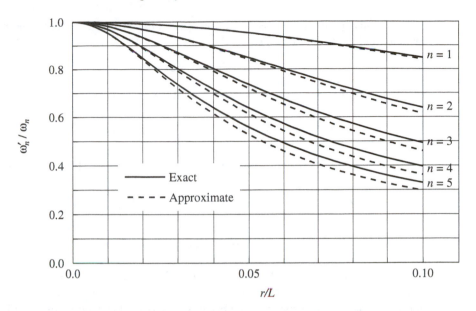

Figure 17.3.4   Influence of shear deformation and rotational inertia on natural frequencies of simply supported beams.

given by Eq. (17.3.26). When $nr/L < 0.2$ the error in the approximate equation is less than 5%.

Shear deformation and rotational inertia have the effect of lowering the natural frequencies, as shown in Figs. 17.3.3 and 17.3.4. For a fixed value of the slenderness ratio L/r of the beam, the frequency reduction due to shear deformation and rotational inertia increases with the $E/\kappa G$ value and with mode number. The latter observation implies that while the corrections due to shear deformation and rotational inertia may be unimportant for the fundamental natural frequency, they could be significant for the higher frequencies. For a fixed value of $E/\kappa G$ and mode number, the frequency reduction increases with r/L, implying its significance for less slender or stubby beams. From the results presented one can estimate whether these corrections need to be included in a particular problem. For earthquake response analysis of many practical structures, these corrections are not significant, but it is important to realize that these corrections do exist. If significant, they can be included in the finite element formulation for practical structures which are not amenable to solution as infinite-DOF systems.

17.4 MODAL ORTHOGONALITY

In this section we derive the orthogonality properties of natural vibration modes of systems with distributed mass and elasticity. For convenience, the derivation is restricted to single-span beams with hinged, clamped, or free ends and without any lumped mass at the ends, although the final result applies in general.

The starting point for this derivation is Eq. (17.3.6), which governs the natural frequencies and modes; for mode r,

$$\left[EI(x)\phi_r''(x)\right]'' = \omega_r^2 m(x)\phi_r(x) \tag{17.4.1}$$

Multiplying both sides by $\phi_n(x)$ and integrating from 0 to L gives

$$\int_0^L \phi_n(x)\left[EI(x)\phi_r''(x)\right]'' dx = \omega_r^2 \int_0^L m(x)\phi_n(x)\phi_r(x)\, dx \tag{17.4.2}$$

The left side of this equation is integrated by parts; applying this procedure twice leads to

$$\int_0^L \phi_n(x)\left[EI(x)\phi_r''(x)\right]'' dx = \left\{\phi_n(x)[EI(x)\phi_r''(x)]'\right\}_0^L - \left\{\phi_n'(x)[EI(x)\phi_r''(x)]\right\}_0^L$$

$$+ \int_0^L EI(x)\phi_n''(x)\phi_r''(x)\, dx \tag{17.4.3}$$

It is easy to see that the quantities enclosed in $\{\cdots\}$ are zero at $x = 0$ and L if the ends of the beam are free, simply supported, or clamped. This is true at a clamped end because $\phi = 0$ and $\phi' = 0$, at a simply supported end because $\phi = 0$ and the bending moment is zero (i.e., $EI\phi'' = 0$), and at a free end because the bending moment is zero (i.e., $EI\phi'' = 0$) and the shear force is zero [i.e., $(EI\phi'')' = 0$]. With the quantities in $\{\cdots\}$ set to zero, Eq. (17.4.3) substituted in Eq. (17.4.2) gives

$$\int_0^L EI(x)\phi_n''(x)\phi_r''(x)\, dx = \omega_r^2 \int_0^L m(x)\phi_n(x)\phi_r(x)\, dx \tag{17.4.4}$$

Similarly, starting with Eq. (17.3.6) written for mode n, multiplying both sides by $\phi_r(x)$, integrating from 0 to L, and using integration by parts twice leads to

$$\int_0^L EI(x)\phi_n''(x)\phi_r''(x)\, dx = \omega_n^2 \int_0^L m(x)\phi_n(x)\phi_r(x)\, dx \tag{17.4.5}$$

Subtracting Eq. (17.4.4) from Eq. (17.4.5) gives

$$(\omega_n^2 - \omega_r^2)\int_0^L m(x)\phi_n(x)\phi_r(x)\, dx = 0$$

Therefore, if $\omega_n \neq \omega_r$,

$$\int_0^L m(x)\phi_n(x)\phi_r(x)\, dx = 0 \tag{17.4.6a}$$

and this substituted in Eq. (17.4.2) leads to

$$\int_0^L \phi_n(x)\left[EI(x)\phi_r''(x)\right]'' dx = 0 \tag{17.4.6b}$$

Equations (17.4.6a) and (17.4.6b) are the orthogonality relations for the natural vibration modes. If a system has repeated frequencies, modes $\phi_n(x)$ still exist such that any two modes, $n \neq r$, satisfy the orthogonality relations even if $\omega_n = \omega_r$.

17.5 MODAL ANALYSIS OF FORCED DYNAMIC RESPONSE

We now return to the partial differential equation (17.1.4), which is to be solved for a given applied force $p(x, t)$. Assuming that the associated eigenvalue problem of Eq. (17.3.6) has been solved for the natural frequencies and modes, the displacement due to each mode is given by Eq. (17.3.2) and the total displacement by

$$u(x, t) = \sum_{r=1}^{\infty} \phi_r(x) q_r(t) \tag{17.5.1}$$

Thus the response $u(x, t)$ has been expressed as the superposition of the contributions of the individual modes; the rth term in the series of Eq. (17.5.1) is the contribution of the rth mode to the response.

We will see next that Eq. (17.1.4) can be transformed to an infinite set of ordinary differential equations, each of which has one modal coordinate $q_n(t)$ as the unknown. Substituting Eq. (17.5.1) in Eq. (17.1.4) gives

$$\sum_{r=1}^{\infty} m(x)\phi_r(x)\ddot{q}_r(t) + \sum_{r=1}^{\infty} \left[EI(x)\phi_r''(x) \right]'' q_r(t) = p(x, t)$$

Now we multiply each term by $\phi_n(x)$, integrate it over the length of the beam, and interchange the order of integration and summation to get

$$\sum_{r=1}^{\infty} \ddot{q}_r(t) \int_0^L m(x)\phi_n(x)\phi_r(x)\, dx + \sum_{r=1}^{\infty} q_r(t) \int_0^L \phi_n(x) \left[EI(x)\phi_r''(x) \right]'' dx$$

$$= \int_0^L p(x, t)\phi_n(x)\, dx$$

By virtue of the orthogonality properties of modes given by Eq. (17.4.6), all terms in each of the summations on the left side vanish except the one term for which $r = n$, leaving

$$\ddot{q}_n(t) \int_0^L m(x) \left[\phi_n(x) \right]^2 dx + q_n(t) \int_0^L \phi_n(x) \left[EI(x)\phi_n''(x) \right]'' dx = \int_0^L p(x, t)\phi_n(x)\, dx$$

This equation can be rewritten as

$$M_n \ddot{q}_n(t) + K_n q_n(t) = P_n(t) \tag{17.5.2}$$

where

$$M_n = \int_0^L m(x) \left[\phi_n(x) \right]^2 dx \qquad K_n = \int_0^L \phi_n(x) \left[EI(x)\phi_n''(x) \right]'' dx$$

$$P_n(t) = \int_0^L p(x, t)\phi_n(x)\, dx \tag{17.5.3}$$

If each end of the beam is free, hinged, or clamped, Eq. (17.4.3) and subsequent discussion gives an alternative equation for K_n:

$$K_n = \int_0^L EI(x) \left[\phi_n''(x)\right]^2 dx \tag{17.5.4}$$

The *generalized mass* M_n and *generalized stiffness* K_n for the nth mode are related:

$$K_n = \omega_n^2 M_n \tag{17.5.5}$$

This relation can be derived by writing Eq. (17.4.1) for the nth mode, multiplying both sides by $\phi_n(x)$, integrating over 0 to L, and utilizing the definitions of M_n and K_n. The term $P_n(t)$ in Eq. (17.5.2) is called the *generalized force* for the nth mode. Equation (17.5.2) governs the nth modal coordinate $q_n(t)$, and the generalized properties M_n, K_n, and $P_n(t)$ depend only on the nth mode $\phi_n(x)$.

Thus we have an infinite number of equations like Eq. (17.5.2), one for each mode. The partial differential equation (17.1.4) in the unknown function $u(x, t)$ has been transformed to an infinite set of ordinary differential equations (17.5.2) in unknowns $q_n(t)$. Recall that the same equations (12.3.3), N in number, were obtained for N-DOF systems.

For applied dynamic forces defined by $p(x, t)$, the motion $u(x, t)$ of the system can be determined by solving the modal equations for $q_n(t)$. The equation for each mode is independent of the equations for all other modes and can therefore be solved separately. Furthermore, each modal equation is of the same form as the equation of motion for an SDF system. Thus the results obtained in Chapters 3 and 4 for the response of SDF systems to various dynamic forces—harmonic force, impulsive force, etc.—can be adapted to obtain solutions $q_n(t)$ for the modal equations.

Once the $q_n(t)$ have been determined, the contribution of the nth mode to the displacement $u(x, t)$ is given by

$$u_n(x, t) = \phi_n(x)q_n(t) \tag{17.5.6}$$

The total displacement is the combination of the contributions of all the modes:

$$u(x, t) = \sum_{n=1}^{\infty} u_n(x, t) = \sum_{n=1}^{\infty} \phi_n(x)q_n(t) \tag{17.5.7}$$

The bending moment and shear force at any section along the length of the beam are related to the displacements $u(x)$ as follows:

$$\mathcal{M}(x) = EI(x)u''(x) \qquad \mathcal{V}(x) = \left[EI(x)u''(x)\right]' \tag{17.5.8}$$

These static relationships apply at each instant of time with $u(x)$ replaced by $u_n(x, t)$, which is given by Eq. (17.5.6). Thus the contribution of the nth mode to the internal forces is given by

$$\mathcal{M}_n(x, t) = EI(x)\phi_n''(x)q_n(t) \qquad \mathcal{V}_n(x, t) = \left[EI(x)\phi_n''(x)\right]' q_n(t) \tag{17.5.9}$$

Combining the contributions of all modes gives the total internal forces:

$$\mathcal{M}(x,t) = \sum_{n=1}^{\infty} \mathcal{M}_n(x,t) = \sum_{n=1}^{\infty} EI(x)\phi_n''(x)q_n(t) \tag{17.5.10a}$$

$$\mathcal{V}(x,t) = \sum_{n=1}^{\infty} \mathcal{V}_n(x,t) = \sum_{n=1}^{\infty} \left[EI(x)\phi_n''(x) \right]' q_n(t) \tag{17.5.10b}$$

Example 17.1

Derive mathematical expressions for the dynamic response—displacement and bending moments—of a uniform simply supported beam to a step-function force p_o at distance ξ from the left end (Fig. E17.1). Specialize the results for the force applied at midspan.

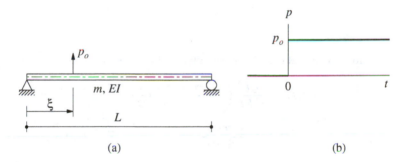

Figure E17.1

Solution

1. *Determine the natural vibration frequencies and modes.*

$$\omega_n = \frac{n^2\pi^2}{L^2}\sqrt{\frac{EI}{m}} \qquad \phi_n(x) = \sin\frac{n\pi x}{L} \tag{a}$$

2. *Set up the modal equations.* Substituting $\phi_n(x)$ in Eq. (17.5.3a) gives M_n, which is substituted in Eq. (17.5.5) together with ω_n^2 to get K_n:

$$M_n = \frac{mL}{2} \qquad K_n = \frac{n^4\pi^4 EI}{2L^3} \tag{b}$$

Substituting $p(x,t) = p_o\delta(x - \xi)$, where $\delta(x - \xi)$ is the Dirac delta function centered at ξ, in Eq. (17.5.3c) gives

$$P_n(t) = p_o\phi_n(\xi) \tag{c}$$

Then the nth modal equation is

$$M_n\ddot{q}_n(t) + K_n q_n(t) = p_o\phi_n(\xi) \tag{d}$$

3. *Solve the modal equations.* Equation (4.3.2) describes the response of an SDF system to a step force. We will adapt this solution to Eq. (d) by changing the notation $u(t)$ to $q_n(t)$ and noting that $(u_{st})_o = p_o \phi_n(\xi)/K_n$. Thus

$$q_n(t) = \frac{p_o \phi_n(\xi)}{K_n}(1 - \cos \omega_n t) = \frac{2 p_o L^3}{\pi^4 E I} \frac{\phi_n(\xi)}{n^4}(1 - \cos \omega_n t) \qquad \text{(e)}$$

Substituting Eq. (e) in Eq. (17.5.7) and noting that $\phi_n(x)$ is known from Eq. (a), we obtain the displacement response $u(x, t)$.

4. *Specialize for $\xi = L/2$.* Substituting $\xi = L/2$ in Eq. (e) and the latter in Eq. (17.5.7) gives

$$u(x, t) = \frac{2 p_o L^3}{\pi^4 E I} \sum_{n=1}^{\infty} \frac{\phi_n(L/2)}{n^4}(1 - \cos \omega_n t) \sin \frac{n \pi x}{L} \qquad \text{(f)}$$

where

$$\phi_n\left(\frac{L}{2}\right) = \begin{cases} 0 & n = 2, 4, 6, \ldots \\ 1 & n = 1, 5, 9, \ldots \\ -1 & n = 3, 7, 11, \ldots \end{cases} \qquad \text{(g)}$$

from Eq. (a) and Fig. 17.3.1. Substituting Eq. (g) in Eq. (f) gives

$$u(x, t) = \frac{2 p_o L^3}{\pi^4 E I} \left(\frac{1 - \cos \omega_1 t}{1} \sin \frac{\pi x}{L} - \frac{1 - \cos \omega_3 t}{81} \sin \frac{3 \pi x}{L} \right.$$
$$\left. + \frac{1 - \cos \omega_5 t}{625} \sin \frac{5 \pi x}{L} - \frac{1 - \cos \omega_7 t}{2401} \sin \frac{7 \pi x}{L} + \cdots \right) \qquad \text{(h)}$$

The displacement at midspan is

$$u\left(\frac{L}{2}, t\right) = \frac{2 p_o L^3}{\pi^4 E I} \left(\frac{1 - \cos \omega_1 t}{1} + \frac{1 - \cos \omega_3 t}{81} + \frac{1 - \cos \omega_5 t}{625} + \frac{1 - \cos \omega_7 t}{2401} + \cdots \right) \qquad \text{(i)}$$

The coefficients 1, 81, 625, 2401, and so on, in the denominator suggest that the first-mode contribution is dominant and that the series converges rapidly.

The bending moments are obtained by substituting Eq. (h) in Eq. (17.5.9a):

$$M(x, t) = -\frac{2 p_o L}{\pi^2} \left(\frac{1 - \cos \omega_1 t}{1} \sin \frac{\pi x}{L} - \frac{1 - \cos \omega_3 t}{9} \sin \frac{3 \pi x}{L} \right.$$
$$\left. + \frac{1 - \cos \omega_5 t}{25} \sin \frac{5 \pi x}{L} - \frac{1 - \cos \omega_7 t}{49} \sin \frac{7 \pi x}{L} + \cdots \right) \qquad \text{(j)}$$

The bending moment at midspan is

$$M\left(\frac{L}{2}, t\right) = -\frac{2 p_o L}{\pi^2} \left(\frac{1 - \cos \omega_1 t}{1} + \frac{1 - \cos \omega_3 t}{9} + \frac{1 - \cos \omega_5 t}{25} + \frac{1 - \cos \omega_7 t}{49} + \cdots \right) \qquad \text{(k)}$$

This series with n^2 in the denominator converges slowly compared to Eq. (i) with n^4 in the denominator. This difference implies that higher modes contribute more significantly to forces than to displacements, a result consistent with the conclusions of Chapters 12 and 13 for discretized systems.

Example 17.2

A simply supported bridge with a single span of length L has a deck of uniform cross section with mass m per feet length and flexural rigidity EI. A single wheel load p_o travels across the bridge at uniform velocity of v, as shown in Fig. E17.2a. Neglecting damping, determine an equation for the deflection at midspan as a function of time.

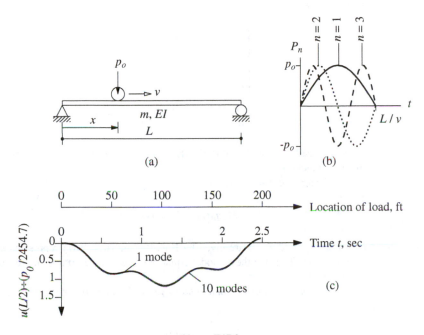

Figure E17.2

The properties of a prestressed concrete box girder elevated freeway connector are $L = 200$ ft, $m = 11$ kips/g per foot, $I = 700$ ft⁴, and $E = 576,000$ kips/ft². If $v = 55$ mph, determine an equation for the deflection at midspan as a function of time. Also determine the maximum value of deflection over time.

Solution   We assume that the mass of the wheel load is small compared to the bridge mass, and it can be neglected.

  1. *Determine the natural vibration frequencies and modes.*

$$\omega_n = \frac{n^2\pi^2}{L^2}\sqrt{\frac{EI}{m}} \qquad \phi_n(x) = \sin\frac{n\pi x}{L} \tag{a}$$

  2. *Determine the generalized mass and stiffness.* Substituting $\phi_n(x)$ in Eq. (17.5.3a) gives M_n, which is substituted in Eq. (17.5.5) together with ω_n^2 to get K_n:

$$M_n = \frac{mL}{2} \qquad K_n = \frac{n^4\pi^4 EI}{2L^3} \tag{b}$$

3. *Determine the generalized force.* A load p_o traveling with a velocity v takes time $t_d = L/v$ to cross the bridge. At any time t the position is as shown in Fig. E17.2a. Thus the moving load can be described mathematically as

$$p(x, t) = \begin{cases} p_o\delta(x - vt) & 0 \le t \le t_d \\ 0 & t \ge t_d \end{cases} \tag{c}$$

where $\delta(x - vt)$ is the Dirac delta function centered at $x = vt$. Substituting Eq. (c) in Eq. (17.5.3c) gives

$$\begin{aligned} P_n(t) &= \begin{cases} \int_0^L p_o\delta(x - vt)\phi_n(x)dx & 0 \le t \le t_d \\ 0 & t \ge t_d \end{cases} \\ &= \begin{cases} p_o\phi_n(vt) & 0 \le t \le t_d \\ 0 & t \ge t_d \end{cases} \\ &= \begin{cases} p_o \sin(n\pi t/t_d) & 0 \le t \le t_d \\ 0 & t \ge t_d \end{cases} \end{aligned} \tag{d}$$

This generalized force is shown in Fig. E17.2b; for the nth mode, it consists of n half-cycles of the sine function.

4. *Set up modal equations.*

$$M_n\ddot{q}_n(t) + K_n q_n(t) = P_n(t) \tag{e}$$

where M_n, K_n, and $P_n(t)$ are given by Eqs. (b) and (d). $P_n(t)$ represents n half-cycles of a sine function. To solve these modal equations, we first determine the response of an SDF system to such an excitation.

5. *Response of SDF system to n half-cycles of* $p(t) = p_o \sin \omega t$. The equation of motion is

$$m\ddot{u} + ku = \begin{cases} p_o \sin(n\pi t/t_d) & t \le t_d \\ 0 & t \ge t_d \end{cases} \tag{f}$$

During $t \le t_d$, the force is the same as the harmonic force $p(t) = p_o \sin \omega t$ considered earlier with frequency:

$$\omega = \frac{n\pi}{t_d} = \frac{n\pi v}{L} \tag{g}$$

The response is given by Eq. (3.1.6b), which is repeated here for convenience, with $(u_{st})_o \equiv p_o/k$:

$$\frac{u(t)}{(u_{st})_o} = \frac{1}{1 - (\omega/\omega_n)^2}\left(\sin \omega t - \frac{\omega}{\omega_n}\sin \omega_n t\right) \quad t \le t_d \tag{h}$$

After the force ends (i.e., $t \ge t_d$) the system vibrates freely with its motion described by Eq. (4.7.3). The displacement $u(t_d)$ and velocity $\dot{u}(t_d)$ at the end of the excitation are determined from Eq. (h):

$$\frac{u(t_d)}{(u_{st})_o} = \frac{-\omega/\omega_n}{1-(\omega/\omega_n)^2} \sin \omega_n t_d \tag{i1}$$

$$\frac{\dot{u}(t_d)}{(u_{st})_o} = \frac{\omega}{1-(\omega/\omega_n)^2} \left[(-1)^n - \cos \omega_n t_d\right] \tag{i2}$$

Substituting Eqs. (i) in Eq. (4.7.3) gives

$$\frac{u(t)}{(u_{st})_o} = \frac{\omega/\omega_n}{1-(\omega/\omega_n)^2} \left[(-1)^n \sin \omega_n (t-t_d) - \sin \omega_n t\right] \qquad t \geq t_d \tag{j}$$

6. *Solve modal equations.* The solution of Eq. (f) is given by Eqs. (h) and (j). We will adapt this solution to the modal equations (e) by changing the notation $u(t)$ to $q_n(t)$ and noting that

$$t_d = \frac{L}{v} \qquad (u_{st})_o \equiv \frac{p_o}{k} = \frac{P_{no}}{K_n} = \frac{2p_o}{mL\omega_n^2}$$

where ω_n and ω are given by Eqs. (a) and (g), respectively. The results are

$$q_n(t) = \frac{2p_o}{mL} \frac{1}{\omega_n^2 - (n\pi v/L)^2} \left(\sin \frac{n\pi vt}{L} - \frac{n\pi v}{\omega_n L} \sin \omega_n t\right) \qquad t \leq L/v \tag{k}$$

$$q_n(t) = \frac{2p_o}{mL} \frac{1}{\omega_n^2 - (n\pi v/L)^2} \frac{n\pi v}{\omega_n L} \left[(-1)^n \sin \omega_n (t-L/v) - \sin \omega_n t\right] \qquad t \geq L/v \tag{l}$$

This solution is valid provided that $\omega_n \neq n\pi v/L$ or $T_n \neq 2L/nv$. Note that when specialized for $n = 1$, Eqs. (k) and (l) reduce to Eqs. (g) and (h) of Example 8.4.

7. *Determine the total response.* The displacement response of the beam is given by Eq. (17.5.7):

$$u(x,t) = \sum_{n=1}^{\infty} \phi_n(x) q_n(t) \tag{m}$$

where $\phi_n(x)$ is given by Eq. (a) and $q_n(t)$ by Eqs. (k) and (l).

8. *Determine deflection at midspan.* Substituting $x = L/2$ in Eq. (m) gives

$$u(L/2, t) = \sum_{n=1}^{\infty} \phi_n(L/2) q_n(t) \tag{n}$$

where

$$\phi_n\left(\frac{L}{2}\right) = \begin{cases} 0 & n = 2, 4, 6, \ldots \\ 1 & n = 1, 5, 9, \ldots \\ -1 & n = 3, 7, 11, \ldots \end{cases} \tag{o}$$

Equation (o) indicates that the even-numbered modes, the antisymmetric modes, do not contribute to the midspan deflection.

9. *Numerical results.* For the given prestressed concrete structure and vehicle speed:

$$m = \frac{11}{32.2} = 0.3416 \text{ kip-sec}^2/\text{ft}^2$$

$$EI = 576{,}000 \times 700 = 4.032 \times 10^8 \text{ kip-ft}^2$$

$$\omega_1 = \frac{\pi^2}{200^2} \sqrt{\frac{4.032 \times 10^8}{0.3416}} = 8.477 \text{ rad/sec}$$

$$T_1 = 0.74 \text{ sec}$$

$$\omega_n = n^2 \omega_1$$

$$v = 55 \text{ mph} = 80.67 \text{ ft/sec}$$

$$\frac{\pi v}{L} = 1.267$$

$$t_d = \frac{L}{v} = \frac{200}{80.67} = 2.479 \text{ sec}$$

Because the duration of the excitation $t_d = L/v$ is greater than $T_n/2$ for all n, the maximum response occurs while the moving load is on the bridge span. This phase of the response is given by Eqs. (k) and (n):

$$u(L/2, t) = \frac{2p_o}{(0.3416)200} \sum_{n=1}^{\infty} \frac{\phi_n(L/2)}{(8.477n^2)^2 - (1.267n)^2} \left(\sin 1.267nt - \frac{1.267n}{8.477n^2} \sin 8.477n^2 t \right)$$

$$= \frac{p_o}{2454.7} \sum_{n=1}^{\infty} \frac{\phi_n(L/2)}{n^4(1 - 0.02234/n^2)} \left(\sin 1.267nt - \frac{0.1495}{n} \sin 8.477n^2 t \right) \qquad \text{(p)}$$

Equation (p) is valid for $0 \le t \le 2.479$ sec. The values of midspan deflection calculated from Eq. (p) at many values of t are shown in Fig. E17.2c; the maximum deflection $u_o = 1.1869 p_o/2454.7$ is attained when the moving load is near midspan. This deflection is only 17% larger than the static deflection ($= p_o L^3/48EI$) due to a stationary load p_o at midspan, implying that the dynamic effects of a moving load are small.

Also shown is the result considering only the contribution of the first vibration mode [i.e., the $n = 1$ term in Eq. (p)]. (Recall that this was the result obtained in Example 8.4.) It is clear that the response contributions of higher modes are negligible.

17.6 EARTHQUAKE RESPONSE HISTORY ANALYSIS

As shown earlier, when the excitation is acceleration $\ddot{u}_g(t)$ of the supports, the equation of motion for a beam is the same as for applied force $p(x, t)$ except that this force is replaced by $p_{\text{eff}}(x, t)$ given by Eq. (17.2.4). Thus the modal analysis procedure of Section 17.5 can readily be extended to the earthquake problem.

From Eq. (17.2.4) the effective earthquake forces are

$$p_{\text{eff}}(x, t) = -m(x)\ddot{u}_g(t) \qquad (17.6.1)$$

The spatial distribution of these forces is defined by $m(x)$. This force distribution can be expanded as a summation of inertia force distributions $s_n(x)$ associated with natural vibration modes (see Section 12.8):

$$m(x) = \sum_{r=1}^{\infty} s_r(x) = \sum_{r=1}^{\infty} \Gamma_r m(x) \phi_r(x) \qquad (17.6.2)$$

Premultiplying both sides by $\phi_n(x)$, integrating over the length of the beam, and utilizing modal orthogonality with respect to the mass distribution, Eq. (17.4.6a), leads to

$$\Gamma_n = \frac{L_n^h}{M_n} \qquad \text{where} \qquad L_n^h = \int_0^L m(x) \phi_n(x)\, dx \qquad (17.6.3)$$

The contribution of the nth mode to $m(x)$ is

$$s_n(x) = \Gamma_n m(x) \phi_n(x) \qquad (17.6.4)$$

Observe that these modal expansion equations for a distributed-mass system are similar to the corresponding equations (13.2.2) and (13.2.4) for lumped-mass systems.

For uniform cantilever towers with mass m per unit length the modal expansion of Eq. (17.6.2) is as shown in Fig. 17.6.1. The functions $s_n(x)$ were evaluated from Eq. (17.6.4) using Eqs. (17.6.3) and (17.5.3a) and the $\phi_n(x)$ given by Eqs. (17.3.23) and (17.3.21).

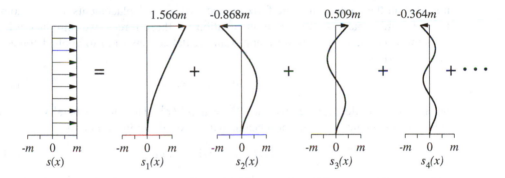

Figure 17.6.1   Modal expansion of effective earthquake forces for uniform towers.

Returning now to the modal analysis procedure of Section 17.5, $p(x, t)$ in Eq. (17.5.3c) is replaced by $p_{\text{eff}}(x, t)$ given by Eq. (17.6.1), to obtain

$$P_n(t) = -L_n^h \ddot{u}_g(t) \qquad (17.6.5)$$

Substituting Eq. (17.6.5) in Eq. (17.5.2), dividing by M_n, and using Eqs. (17.5.5) and (17.6.2a) gives the modal equations of an undamped tower subjected to earthquake excitation:

$$\ddot{q}_n + \omega_n^2 q_n = -\Gamma_n \ddot{u}_g(t) \qquad (17.6.6)$$

For classically damped systems, Eq. (17.6.6) becomes

$$\ddot{q}_n + 2\zeta_n \omega_n \dot{q}_n + \omega_n^2 q_n = -\Gamma_n \ddot{u}_g(t) \tag{17.6.7}$$

where ζ_n is the damping ratio for the nth mode. This is the same as Eq. (13.1.7) derived earlier for N-DOF systems, except that L_n^h and M_n that enter into Γ_n are now given by Eqs. (17.6.3) and (17.5.3), respectively. As shown in Section 13.1.3, the solution of Eq. (17.6.7) is

$$q_n(t) = \Gamma_n D_n(t) \tag{17.6.8}$$

where $D_n(t)$ is the deformation response of the nth-*mode SDF system*. This is an SDF system with vibration properties—natural frequency ω_n and damping ratio ζ_n—of the nth mode of the distributed-mass system. Thus $q_n(t)$ is readily available once the SDF system response has been determined by the methods of Chapter 6. The contribution of the nth mode to the earthquake response of the tower can be expressed in terms of $D_n(t)$. Substituting Eq. (17.6.8) in Eqs. (17.5.6) and (17.5.9) gives the displacements, bending moments, and shear forces due to the nth mode:

$$u_n(x, t) = \Gamma_n \phi_n(x) D_n(t) \tag{17.6.9}$$

$$\mathcal{M}_n(x, t) = \Gamma_n EI(x)\phi_n''(x) D_n(t) \qquad V_n(x, t) = \Gamma_n \left[EI(x)\phi_n''(x) \right]' D_n(t) \tag{17.6.10}$$

Alternatively, as in Chapter 13 for an N-DOF system, the internal forces can be determined from the equivalent static forces associated with displacements $u_n(x, t)$ computed from dynamic analysis. To derive an equation for these forces, we introduce a familiar equation from elementary beam theory relating deflections $u(x)$ to applied forces $f(x)$. For a uniform beam

$$EIu^{\text{IV}}(x) = f(x) \tag{17.6.11}$$

where EI is the flexural rigidity and $u^{\text{IV}} = d^4u/dx^4$. The more general version of this equation applicable to nonuniform beams with flexural rigidity $EI(x)$ is

$$\left[EI(x)u''(x) \right]'' = f(x) \tag{17.6.12}$$

Replacing $u(x)$ by the time-varying displacements $u_n(x, t)$ from Eq. (17.6.9) gives

$$f_n(x, t) = \Gamma_n \left[EI(x)\phi_n''(x) \right]'' D_n(t) \tag{17.6.13}$$

which, by using Eq. (17.4.1) rewritten for the nth mode, becomes

$$f_n(x, t) = s_n(x) A_n(t) \tag{17.6.14}$$

where $s_n(x)$ is given by Eq. (17.6.4), and $A_n(t)$, the pseudo-acceleration response of the nth-mode SDF system, is given by Eq. (13.1.12), which is repeated:

$$A_n(t) = \omega_n^2 D_n(t) \tag{17.6.15}$$

Observe the similarity between Eqs. (17.6.14) and (13.2.7) for a lumped-mass system. At any time instant the contribution $r_n(t)$ of the nth mode to any response quantity $r(t)$—deflection, shear force, or bending moment at any location of the beam—is determined by static analysis of the beam subjected to external forces $f_n(x, t)$, and can be expressed as

$$r_n(t) = r_n^{\text{st}} A_n(t) \qquad (17.6.16)$$

The modal static response r_n^{st} is determined by static analysis of the tower due to external forces $s_n(\xi)$ (Fig. 17.6.2). As shown in Fig. 17.6.1, these forces due to the fundamental mode all act in the same direction, but for the second and higher modes they will change direction as one moves up the tower.

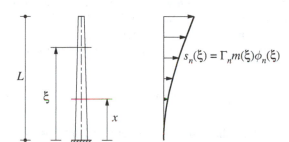

$$s_n(\xi) = \Gamma_n m(\xi) \phi_n(\xi)$$

Figure 17.6.2   Static problem to be solved to determine modal static responses.

The modal static responses are presented in Table 17.6.1 for five response quantities: the shear $V(x)$ at location x, the bending moment $M(x)$ at location x, the base shear $V_b = V(0)$, the base moment $M_b = M(0)$, and deflection $u(x)$. The first four equations come from static analysis of the system in Fig. 17.6.2. The result for $u(x)$ is obtained by comparing Eqs. (17.6.9) and (17.6.16) and using Eq. (17.6.15). Parts of the equations for V_{bn}^{st} and M_{bn}^{st} are obtained by substituting Eq. (17.6.4) for $s_n(\xi)$, using Eq. (17.6.3) for L_n^h, and defining

$$M_n^* = \Gamma_n L_n^h \qquad h_n^* = \frac{L_n^\theta}{L_n^h} \qquad L_n^\theta = \int_0^L x m(x) \phi_n(x)\, dx \qquad (17.6.17)$$

TABLE 17.6.1   MODAL STATIC RESPONSES

Response, r	Modal Static Response, r_n^{st}
$V(x)$	$V_n^{\text{st}}(x) = \int_x^L s_n(\xi)\, d\xi$
$M(x)$	$M_n^{\text{st}}(x) = \int_x^L (\xi - x) s_n(\xi)\, d\xi$
V_b	$V_{bn}^{\text{st}} = \int_0^L s_n(\xi)\, d\xi = \Gamma_n L_n^h = M_n^*$
M_b	$M_{bn}^{\text{st}} = \int_0^L \xi s_n(\xi)\, d\xi = \Gamma_n L_n^\theta = h_n^* M_n^*$
$u(x)$	$u_n^{\text{st}}(x) = (\Gamma_n / \omega_n^2) \phi_n(x)$

Observe the similarity between the equations in Table 17.6.1 and those for a lumped-mass system in Table 13.2.1. The approach symbolized by Eq. (17.6.16) to determine shear and moment is preferable over Eq. (17.6.10) because it avoids computation of the second and third derivatives of the mode shapes; obviously, both methods will give identical results.

The base shear $V_{bn}(t)$ and base moment $\mathcal{M}_{bn}(t)$ due to the nth mode are obtained by specializing Eq. (17.6.16) for V_b and $\mathcal{M}_b$ and substituting for V_{bn}^{st} and $\mathcal{M}_{bn}^{st}$ from Table 17.6.1:

$$V_{bn}(t) = M_n^* A_n(t) \qquad \mathcal{M}_{bn}(t) = h_n^* V_{bn}(t) \qquad (17.6.18)$$

Because Eq. (17.6.18) is identical to Eqs. (13.2.12b) and (13.2.15b) for lumped-mass systems, following Section 13.2.5, M_n^* and h_n^* may be interpreted as the *effective modal mass* and *effective modal height* for the nth mode. Observe that Eq. (17.6.17a and b) is identical to Eq. (13.2.9a) for lumped-mass systems; the definitions of M_n, L_n^h, and L_n^θ differ, however, between distributed-mass and lumped-mass systems.

The sum of the effective modal masses over all modes is equal to the total mass of the tower:

$$\sum_{n=1}^{\infty} M_n^* = \int_0^L m(x)\, dx \qquad (17.6.19)$$

and the sum of the first moments about the base of the effective modal masses M_n^* located at heights h_n^* is equal to the first moment of the distributed mass about the base:

$$\sum_{n=1}^{\infty} h_n^* M_n^* = \int_0^L x m(x)\, dx \qquad (17.6.20)$$

These relations can be proven in the same manner as the analogous equations (13.2.14) and (13.2.17) for a lumped-mass system. In particular, Eq. (17.6.19) can be proven by integrating Eq. (17.6.2) over the height of the tower and using Eq. (17.6.3b). Similarly, Eq. (17.6.20) can be derived from the modal expansion of forces $xm(x)$.

The effective modal masses M_n^* and effective modal heights h_n^* for a uniform cantilever tower are shown in Fig. 17.6.3; note that h_n^* are plotted without their algebraic signs. M_n^* and h_n^* were determined from Eq. (17.6.17) using the known modes (Fig. 17.3.2). Observe that the sum of M_n^* for the first four modes gives 90% of the total mass of the tower.

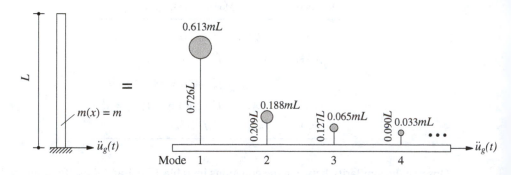

Figure 17.6.3   Effective modal masses and effective modal heights.

Combining the response contributions of all the modes the earthquake response of the system:

$$r(t) = \sum_{n=1}^{\infty} r_n(t) = \sum_{n=1}^{\infty} r_n^{st} A_n(t) \qquad (17.6.21)$$

where Eq. (17.6.16) has been used for $r_n(t)$. This nth-mode contribution to the response can be determined from the modal static response (Table 17.6.1) and $A_n(t)$, the pseudo-acceleration response of the nth-mode SDF system, just as for N-DOF systems (Fig. 13.1.1).

17.7 EARTHQUAKE RESPONSE SPECTRUM ANALYSIS

The peak response of a distributed-mass system, such as a cantilever tower, can be estimated from the earthquake response (or design) spectrum by procedures analogous to those developed in Chapter 13, Part B for lumped-mass systems.

The exact peak value of the nth-mode response $r_n(t)$ is

$$r_{no} = r_n^{st} A_n \qquad (17.7.1)$$

where $A_n \equiv A(T_n, \zeta_n)$ is the ordinate of the pseudo-acceleration spectrum corresponding to natural period T_n and damping ratio ζ_n. Alternatively, r_{no} may be viewed as the result of static analysis of the tower subjected to external forces

$$f_{no}(x) = s_n(x) A_n \qquad (17.7.2)$$

which are the peak values of the equivalent static forces $f_n(x, t)$ defined in Eq. (17.6.14).

The peak value r_o of the total response $r(t)$ can be estimated by combining the modal peaks r_{no} according to one of the modal combination rules presented in Section 13.7.2. Because the natural frequencies of transverse vibration of a beam are well separated, the SRSS combination rule is satisfactory. Thus

$$r_o \simeq \left(\sum_{n=1}^{\infty} r_{no}^2 \right)^{1/2} \qquad (17.7.3)$$

Example 17.3

A reinforced-concrete chimney, 600 ft high, has a uniform hollow circular cross section with outside diameter 50 ft and wall thickness 2 ft 6 in. (Fig. E17.3a). For purposes of earthquake analysis, the chimney is assumed clamped at the base and its mass and flexural stiffness are computed from the gross area of the concrete (neglecting the reinforcing steel). The elastic modulus for concrete $E_c = 3600$ ksi, and its unit weight is 150 lb/ft^3. Modal damping ratios are estimated as 5%. Determine the displacements, shear forces, and bending moments due to an earthquake characterized by the design spectrum of Fig. 6.9.5 scaled to a peak ground acceleration of 0.25g. Neglect shear deformations and rotational inertia.

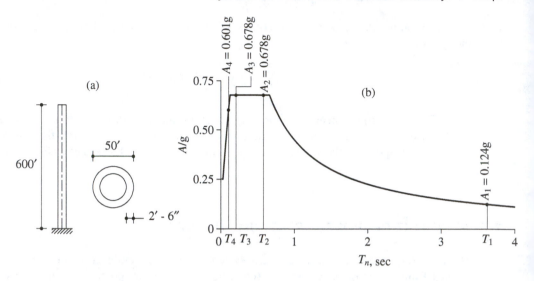

Figure E17.3a, b

Solution

1. *Determine the chimney properties.*

$$m = \pi \frac{[25^2 - (22.5)^2]0.15}{32.2} = 1.738 \text{ kip-sec}^2/\text{ft}^2$$

$$EI = (3600 \times 144)\frac{\pi}{4}\left[25^4 - (22.5)^4\right] = 5.469 \times 10^{10} \text{ kip-ft}^2$$

2. *Determine the natural vibration periods and modes.* Equation (17.3.22) gives the natural frequencies of vibration, and the corresponding periods, in seconds, are $T_1 = 3.626$, $T_2 = 0.5787$, $T_3 = 0.2067$, $T_4 = 0.1055$, and so on. The natural modes, given by Eq. (17.3.23) with $\beta_n L$ defined by Eq. (17.3.21), were evaluated numerically for many values of x and are shown in Fig. 17.3.2, normalized to unit value at the top.

3. *Compute the modal properties.* With the mode shapes known, the properties M_n, L_n^h, L_n^θ, M_n^*, and h_n^* were obtained by numerically evaluating their respective integrals, and are presented in Table E17.3.

4. *Read the design spectrum ordinates.* The design spectrum of Fig. 6.9.5 scaled to a peak acceleration of 0.25g is shown in Fig. E17.3b, wherein the pseudo-acceleration ordinates corresponding to the first four periods are noted: $A_n/g = 0.124, 0.678, 0.678$, and 0.601.

5. *Compute the displacements.* The peak displacements $u_{no}(x)$ due to the nth mode are given by Eq. (17.7.1), where the modal static response r_n^{st} becomes $u_n^{\text{st}}(x)$ given in Table 17.6.1. Substituting known values of Γ_n, $\phi_n(x)$, ω_n^2, and A_n leads to $u_{no}(x)$, $n = 1, 2, 3$, and 4, shown in Fig. E17.3c. At each location x these peak modal displacements are combined according to the SRSS rule, Eq. (17.7.3), to obtain an estimate of the total displacements $u_o(x)$, which are also shown. Observe that the total displacements are due primarily to the first mode.

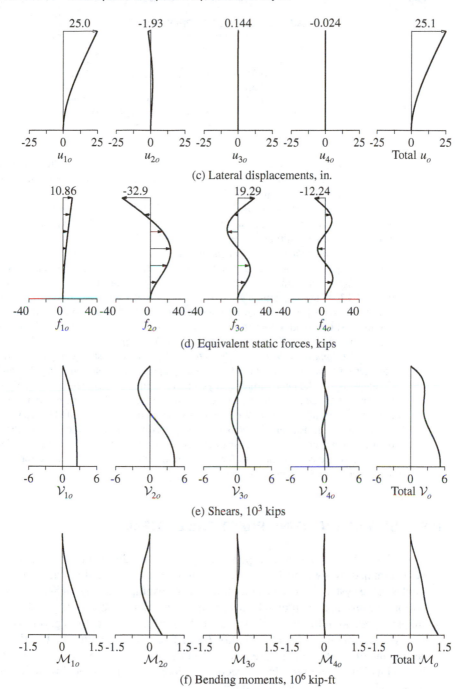

(c) Lateral displacements, in.

(d) Equivalent static forces, kips

(e) Shears, 10^3 kips

(f) Bending moments, 10^6 kip-ft

Figure E17.3c–f

TABLE E17.3 MODAL PROPERTIES

Mode	M_n/mL	L_n^h/mL	Γ_n	L_n^θ/mL^2
1	0.2500	0.3915	1.5660	0.2844
2	0.2500	−0.2170	−0.8679	−0.0454
3	0.2500	0.1272	0.5089	0.0162
4	0.2498	−0.0909	−0.3637	−0.0082

6. *Determine the modal expansion of $m(x)$, Eq. (17.6.2).* With $\phi_n(x)$ known from Fig. 17.3.2 and Γ_n from Table E17.3, the functions $s_n(x)$ are determined from Eq. (17.6.4). Actually, these were presented in Fig. 17.6.1.

7. *Compute the equivalent static forces for the nth mode.* These forces $f_{no}(x)$ are determined from Eq. (17.7.2) using the $s_n(x)$ of Fig. 17.6.1 and the A_n values of Fig. E17.3b. The results for the first four modes are shown in Fig. E17.3d.

8. *Compute the shears and bending moments.* For each mode the peak values of shears and bending moments at location x are computed by static analysis of the chimney subjected to forces $f_{no}(x)$. The resulting shears and bending moments due to the first four modes are shown in Fig. E17.3e and f. At each section x, these modal responses are combined by the SRSS rule [Eq. (17.7.3)] to obtain an estimate of the total forces, which are also shown. Observe that the first two modes contribute significantly to the total response, with the second-mode contribution more significant for the shears than for moments.

9. *Compare with Rayleigh's method.* It is of interest to compare the results above considering response in four modes with the approximate solution using Rayleigh's method (Example 8.3). The approximate analysis predicts the displacements reasonably well but not the bending moments or shears. There are two reasons for the larger errors in forces: (a) The approximate results differ from the exact response due to the first mode because the assumed shape function in Rayleigh's method is an approximation to this mode; this discrepancy introduces larger errors in forces than in displacements. (b) The second and higher modes, whose response contributions to forces are more significant than they are for displacements, are neglected in Rayleigh's method.

17.8 DIFFICULTY IN ANALYZING PRACTICAL SYSTEMS

It is evident that the dynamic response of systems with distributed mass and elasticity can be determined by the modal analysis procedure once the natural vibration frequencies and modes of the system have been determined. Both examples solved in Section 17.3 involved uniform beams, and we found the natural frequencies and modes analytically, although the frequency equation for the cantilever had to be solved numerically. This classical approach is rarely feasible if the flexural rigidity EI or mass m vary along the length of the beam, several intermediate supports are involved, or the system is an assemblage of several members with distributed mass. In this section we identify some of the difficulties in obtaining analytical solutions for the above-mentioned systems.

Consider a single-span beam with mass $m(x)$ and flexural stiffness $EI(x)$. To determine the natural frequencies and modes, we need to solve Eq. (17.3.6), which can be rewritten as

$$EI(x)\phi^{IV}(x) + 2EI'(x)\phi'''(x) + EI''(x)\phi''(x) - \omega^2 m(x)\phi(x) = 0 \qquad (17.8.1)$$

Because the coefficients $EI(x)$, $EI'(x)$, $EI''(x)$, and $m(x)$ of this fourth-order differential equation vary with x, an analytical solution is rarely feasible for ω^2 and $\phi(x)$. Therefore, it is not practical to use the classical approach for practical problems in which $EI(x)$ and $m(x)$ may be complicated functions.

In finding the natural frequencies and modes of a beam on multiple supports, the uniform segment between each pair of supports is considered as a separate beam with its origin at the left end of the segment. Equation (17.3.9) applies to each segment, there is one such equation for each segment, and the necessary boundary conditions are:

1. At each end of the beam the usual boundary conditions are applicable, depending on the type of support.
2. At each intermediate support the deflection is zero, and since the beam is continuous, the slope and the moment just to the left and to the right of the support are the same.

This process quickly becomes unmanageable because of the four constants in Eq. (17.3.9), which must be evaluated in each segment. An analytical solution is rarely feasible for ω^2 and $\phi(x)$, especially if the span lengths vary and $m(x)$ and $EI(x)$ vary within each segment, as would often be the case for a multispan bridge.

Consider the two-member frame shown in Fig. 17.8.1. Each member is axially rigid and has uniform properties—flexural rigidity and mass—as indicated; however, they may differ from one member to the other. Each member is considered as a separate beam with its origin at one end. Equation (17.3.9) applies to each uniform member, there is one such equation for each member, and the necessary end and joint conditions are:

1. At the supports of the frame the usual boundary conditions are applicable, depending on the type of support, resulting in four equations for the frame of Fig. 17.8.1:

$$\phi_{(1)}(L_1) = 0 \qquad \phi'_{(1)}(L_1) = 0 \qquad \phi_{(2)}(0) = 0 \qquad \phi'_{(2)}(0) = 0$$

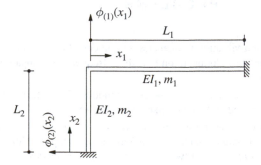

Figure 17.8.1

2. At the joint the end displacements of the joining members should be compatible; this condition for axially rigid members gives

$$\phi_{(1)}(0) = 0 \qquad \phi_{(2)}(L_2) = 0$$

3. At the joint the end slopes of the joining members should be compatible; thus

$$\phi'_{(1)}(0) = \phi'_{(2)}(L_2)$$

4. At the joint the bending moments should be in equilibrium; thus

$$EI_1\phi''_{(1)}(0) + EI_2\phi''_{(2)}(L_2) = 0$$

A simple two-member frame requires setting up these eight conditions and the evaluation of eight constants. The process becomes unmanageable for a frame with many members.

    It should now be evident that the classical procedure to determine the natural frequencies and modes of a distributed-mass system with infinite number of DOF, is not feasible for practical structures. Such problems can be analyzed by discretizing them as systems with a finite number of DOFs, as discussed in the next chapter.

FURTHER READING

Clough, R. W., and Penzien, J., *Dynamics of Structures*, McGraw-Hill, New York, 1993, Chapters 17–19.

Humar, J. L., *Dynamics of Structures*, 2nd ed., A. A. Balkema Publishers, Lisse, The Netherlands, 2002, Chapters 14–17.

Stokey, W. F., "Vibration of Systems Having Distributed Mass and Elasticity," Chapter 8 in *Shock and Vibration Handbook* (ed. C. M. Harris), McGraw-Hill, New York, 1988.

Timoshenko, S., Young, D. H., and Weaver, W., Jr., *Vibration Problems in Engineering*, Wiley, New York, 1974.

PROBLEMS

17.1   Find the first three natural vibration frequencies and modes of a uniform beam clamped at both ends. Sketch the modes. Comment on how these frequencies compare with those of a simply supported beam.

17.2   Find the first three natural vibration frequencies and modes of a uniform beam clamped at one end and simply supported at the other. Sketch the modes.

17.3   Find the first five natural vibration frequencies and modes of a uniform beam free at both ends. Sketch the modes. Comment on how these frequencies compare with those for a beam clamped at both ends. (*Hint*: The first two modes are rigid-body modes.)

17.4   A weight W is suspended from the midspan of a simply supported beam as shown in Fig. P17.4. If the wire by which the weight is suspended suddenly snaps, describe the subsequent vibration of the beam. Specialize the general result to obtain the deflection at midspan. Neglect damping.

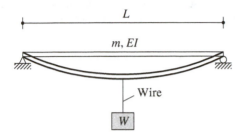

Wire

W

Figure P17.4

17.5   Derive mathematical expressions for the displacement response of a simply supported uniform beam to the force distribution shown in Fig. P17.5; the time variation of the force is a step function. Express the displacements $u(x, t)$ in terms of the natural vibration modes of the beam. Identify the modes that do not contribute to the response. Specialize the general result to obtain the deflection at midspan. Neglect damping.

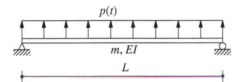

Figure P17.5

17.6   Derive mathematical expressions for the displacement response of a simply supported uniform beam to the force distribution shown in Fig. P17.6; the time variation of the force is a step function. Express the displacements $u(x, t)$ in terms of the natural vibration modes of the beam. Identify the modes that do not contribute to the response. Specialize the general result to obtain the deflection at quarter span. Neglect damping.

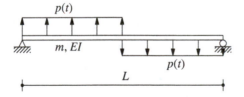

Figure P17.6

17.7   Solve Problem 8.25 considering all natural vibration modes of the bridge.

17.8   Solve Problem 8.26 considering all natural vibration modes of the bridge.

17.9   Prove that Eq. (17.6.19) is valid for a vertical cantilever beam.

17.10   Prove that Eq. (17.6.20) is valid for a vertical cantilever beam.

17.11 A free-standing intake–outlet tower 200 ft high has a uniform hollow circular cross section with outside diameter 25 ft and wall thickness 1 ft 3 in. Assume that the tower is clamped at the base and that its mass and flexural stiffness are computed from the gross area of the concrete (neglecting reinforcing steel). The elastic modulus for concrete is 3600 ksi, and its unit weight is 150 lb/ft^3. Modal damping ratios are estimated as 5%. Determine the top displacement, base shear, and base overturning moment due to an earthquake characterized by the design spectrum of Fig. 6.9.5 scaled to the peak ground acceleration of $\frac{1}{3}$g. Neglect shear deformations and rotational inertia.

<h1>18</h1>

Introduction to the
Finite Element Method

PREVIEW

The classical analysis of distributed-mass systems with infinite number of DOFs is not feasible for practical structures, for reasons mentioned in Chapter 17. In this chapter, two methods are presented for discretizing one-dimensional distributed-mass systems: the Rayleigh–Ritz method and the finite element method. As a result, the governing partial differential equation is replaced by a system of ordinary differential equations, as many as the DOFs in the discretized system, which can be solved by the methods presented in Chapters 10 to 16. The consistent mass matrix concept is introduced and the accuracy and convergence of the approximate natural frequencies of a cantilever beam, determined by the finite element method using consistent or lumped-mass matrices, is demonstrated. The chapter ends with a short discussion on application of the finite element method to the dynamic analysis of structural continua.

PART A: RAYLEIGH–RITZ METHOD

18.1 FORMULATION USING CONSERVATION OF ENERGY

Developed in Chapter 15 for lumped-mass systems, the Rayleigh–Ritz method is also applicable to systems with distributed mass and elasticity. It was for the latter class of systems that the method was originally developed by W. Ritz in 1909. The method, applicable to any one-, two-, or three-dimensional system with distributed mass and elasticity, reduces the system with an infinite number of DOFs to one with a finite number of

DOFs. In this section we present the Rayleigh–Ritz method for the transverse vibration of a straight beam.

Consider such a beam with flexural rigidity $EI(x)$ and mass $m(x)$ per unit length, both of which may vary arbitrarily with position x. The deflections $u(x)$ of the system are expressed as a linear combination of several trial functions $\psi_j(x)$:

$$u(x) = \sum_{j=1}^{N} z_j \psi_j(x) = \mathbf{\Psi}(x)\mathbf{z} \qquad (18.1.1)$$

where z_j are the generalized coordinates, which vary with time in a dynamic problem, $\mathbf{z}$ is the $N \times 1$ vector of generalized coordinates, and the $1 \times N$ matrix of trial functions is

$$\mathbf{\Psi}(x) = [\psi_1(x) \quad \psi_2(x) \quad \cdots \quad \psi_N(x)]$$

Each trial function $\psi_j(x)$—also known as Ritz function or shape function—must be admissible: that is, continuous and have a continuous first derivative, and satisfy the displacement boundary conditions on the system. All the trial functions must be linearly independent and are selected appropriate for the system to be analyzed.

The starting point for formulation of the Rayleigh–Ritz method is the Rayleigh's quotient [Eq. (8.5.11)] for a function $\tilde{\phi}(\chi)$ defined consistent with Eq. (18.1.1):

$$\tilde{\phi}(x) = \mathbf{\Psi}(x)\chi \qquad (18.1.2)$$

Replacing $\psi(x)$ in Eq. (8.5.11) by $\tilde{\phi}(x)$ and substituting Eq. (18.1.2) leads to

$$\rho(\chi) = \frac{\sum_{i=1}^{N}\sum_{j=1}^{N} \chi_i \chi_j \tilde{k}_{ij}}{\sum_{i=1}^{N}\sum_{j=1}^{N} \chi_i \chi_j \tilde{m}_{ij}} \qquad (18.1.3)$$

where

$$\tilde{k}_{ij} = \int_0^L EI(x)\psi_i''(x)\psi_j''(x)\,dx \qquad \tilde{m}_{ij} = \int_0^L m(x)\psi_i(x)\psi_j(x)\,dx \qquad (18.1.4)$$

Rayleigh's quotient cannot be determined from Eq. (18.1.3) because the N generalized coordinates z_i are unknown. Our objective is to find the values that provide the "best" approximate solution for the natural vibration frequencies and modes of the system.

For this purpose we invoke the property that Rayleigh's quotient is stationary in the neighborhood of the true modes (or true values of χ), which implies that $\partial \rho / \partial \chi_i = 0$— $i = 1, 2, \ldots, N$. We do not need to go through the details of the derivation because the $\rho(\chi)$ for distributed-mass systems, Eq. (18.1.3), is of the same form as that for discretized systems, Eq. (15.3.8). Thus the stationary condition on Eq. (18.1.3) leads to the eigenvalue problem of Eq. (15.3.11), which is repeated here for convenience:

$$\tilde{\mathbf{k}}\chi = \rho \tilde{\mathbf{m}}\chi \qquad (18.1.5)$$

where $\tilde{\mathbf{k}}$ and $\tilde{\mathbf{m}}$ are square matrices of order N, the number of Ritz functions used to represent the deflections $u(x)$ in Eq. (18.1.1), with their elements given by Eq. (18.1.4). The original eigenvalue problem for distributed-mass systems, Eq. (17.3.6), has been reduced to a matrix eigenvalue problem of order N by using Rayleigh's stationary condition.

The solution of Eq. (18.1.5), obtained by the methods of Chapter 10, yields N eigenvalues $\rho_1, \rho_2, \ldots, \rho_N$ and the corresponding eigenvectors

$$X_n = \langle X_{1n} \quad X_{2n} \quad \cdots \quad X_{Nn} \rangle^T \qquad n = 1, 2, \ldots, N \qquad (18.1.6)$$

The eigenvalues provide

$$\tilde{\omega}_n = \sqrt{\rho_n} \qquad (18.1.7)$$

When arranged in increasing order of magnitude, $\tilde{\omega}_n$ are upper bound approximations to the true natural frequencies ω_n of the system, that is,

$$\tilde{\omega}_n \geq \omega_n \qquad n = 1, 2, \ldots, N \qquad (18.1.8)$$

Furthermore, an approximate frequency approaches the exact value from above as the number N of Ritz functions is increased. The eigenvectors X_n substituted in Eq. (18.1.2) provide the functions:

$$\tilde{\phi}_n(x) = \sum_{j=1}^{N} X_{jn} \psi_j(x) = \Psi(x) X_n \qquad n = 1, 2, \ldots, N \qquad (18.1.9)$$

which are approximations to the true natural modes $\phi_n(x)$ of the system. The quality of these approximate results is generally better for the lower modes than for higher modes. Therefore, more Ritz functions should be included than the number of modes desired for dynamic response analysis of the system. Although the natural modes $\tilde{\phi}_n(x)$ are approximate, they satisfy the orthogonality properties of Eq. (17.4.6)—as demonstrated for discretized systems (Section 15.3.4)—and can therefore be used in modal analysis of the system as described in Sections 17.5 to 17.7.

Example 18.1

Find approximations for the first two natural frequencies and modes of lateral vibration of a uniform cantilever beam of Fig. E18.1a by the Rayleigh–Ritz method using the shape functions shown in Fig. E18.1b:

$$\psi_1(x) = 1 - \cos \frac{\pi x}{2L} \qquad \psi_2(x) = 1 - \cos \frac{3\pi x}{2L} \qquad (a)$$

(a)   (b)   (c)

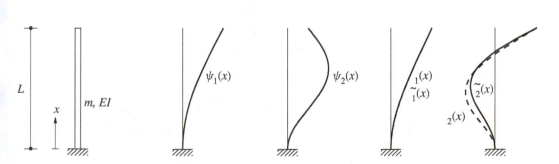

Figure E18.1

These functions are admissible because they are continuous, have a continuous first derivative, and satisfy the two displacement boundary conditions at the clamped end. However, they do not satisfy one of the force boundary conditions at the free end. Note that the two shape functions satisfy the requirement of linear independence.

Solution

1. Set up $\tilde{k}$ and $\tilde{m}$. For the selected Ritz functions of Eq. (a), the stiffness coefficients are computed from Eq. (18.1.4a):

$$\tilde{k}_{11} = \frac{1}{16} \frac{\pi^4 EI}{L^4} \int_0^L \left(\cos \frac{\pi x}{2L} \right)^2 dx = \frac{1}{32} \frac{\pi^4 EI}{L^3} \tag{b1}$$

$$\tilde{k}_{12} = \frac{9}{16} \frac{\pi^4 EI}{L^4} \int_0^L \cos \frac{\pi x}{2L} \cos \frac{3\pi x}{2L} dx = 0 \tag{b2}$$

$$\tilde{k}_{21} = \tilde{k}_{12} = 0 \tag{b3}$$

$$\tilde{k}_{22} = \frac{81}{16} \frac{\pi^4 EI}{L^4} \int_0^L \left(\cos \frac{3\pi x}{2L} \right)^2 dx = \frac{81}{32} \frac{\pi^4 EI}{L^3} \tag{b4}$$

Similarly, the mass coefficients are determined from Eq. (18.1.4b):

$$\tilde{m}_{11} = m \int_0^L \left(1 - \cos \frac{\pi x}{2L} \right)^2 dx = 0.2268 mL \tag{c1}$$

$$\tilde{m}_{12} = m \int_0^L \left(1 - \cos \frac{\pi x}{2L} \right) \left(1 - \cos \frac{3\pi x}{2L} \right) dx = 0.5756 mL \tag{c2}$$

$$\tilde{m}_{21} = \tilde{m}_{12} = 0.5756 mL \tag{c3}$$

$$\tilde{m}_{22} = m \int_0^L \left(1 - \cos \frac{3\pi x}{2L} \right)^2 dx = 1.9244 mL \tag{c4}$$

Substituting Eqs. (b) and (c) in Eq. (18.1.5) gives

$$\begin{bmatrix} 1 & 0 \\ 0 & 81 \end{bmatrix} \left\{ \begin{matrix} X_1 \\ X_2 \end{matrix} \right\} = \bar{\rho} \begin{bmatrix} 0.2268 & 0.5756 \\ 0.5756 & 1.9244 \end{bmatrix} \left\{ \begin{matrix} X_1 \\ X_2 \end{matrix} \right\} \tag{d}$$

where

$$\bar{\rho} = \frac{32 mL^4}{\pi^4 EI} \rho \tag{e}$$

2. Solve the reduced eigenvalue problem.

$$\bar{\rho}_1 = 4.0775 \qquad \bar{\rho}_2 = 188.87$$

$$X_1 = \begin{bmatrix} X_{11} \\ X_{21} \end{bmatrix} = \begin{bmatrix} 1 \\ 0.0321 \end{bmatrix} \qquad X_2 = \begin{bmatrix} X_{12} \\ X_{22} \end{bmatrix} = \begin{bmatrix} 1 \\ -0.3848 \end{bmatrix}$$

3. Determine the approximate frequencies from Eqs. (e) and (18.1.7).

$$\tilde{\omega}_1 = \frac{3.523}{L^2} \sqrt{\frac{EI}{m}} \qquad \tilde{\omega}_2 = \frac{23.978}{L^2} \sqrt{\frac{EI}{m}} \tag{f}$$

4. *Determine the approximate modes from Eq. (18.1.9).*

$$\tilde{\phi}_1(x) = \left(1 - \cos\frac{\pi x}{2L}\right) + 0.0321\left(1 - \cos\frac{3\pi x}{2L}\right) = 1.0321 - \cos\frac{\pi x}{2L} - 0.0321\cos\frac{3\pi x}{2L} \quad \text{(g1)}$$

$$\tilde{\phi}_2(x) = \left(1 - \cos\frac{\pi x}{2L}\right) - 0.3848\left(1 - \cos\frac{3\pi x}{2L}\right) = 0.6152 - \cos\frac{\pi x}{2L} + 0.3848\cos\frac{3\pi x}{2L} \quad \text{(g2)}$$

These approximate modes are plotted in Fig. E18.1c.

5. *Compare with the exact solution.* The exact values for natural frequencies and modes of a cantilever beam were determined in Section 17.3.2. The exact frequencies are

$$\omega_1 = \frac{3.516}{L^2}\sqrt{\frac{EI}{m}} \qquad \omega_2 = \frac{22.03}{L^2}\sqrt{\frac{EI}{m}} \quad \text{(h)}$$

Both approximate frequencies are higher than the corresponding exact values, and the error is larger in the second frequency. The exact modes are also plotted in Fig. E18.1c. It is clear that the approximate solution is excellent for the first mode but not as good for the second mode.

The approximate value for the fundamental frequency in Eq. (h) is lower and hence better than the result obtained in Example 8.2 by Rayleigh's method with $\psi_1(x)$ as the only trial function.

18.2 FORMULATION USING VIRTUAL WORK

In this section the equation governing the transverse vibration of a straight beam due to external forces will be formulated using the principle of virtual displacements. At each time instant the system is in equilibrium under the action of the external forces $p(x, t)$, internal resisting bending moments $M(x, t)$, and the fictitious inertia forces, which by D'Alembert's principle are

$$f_I(x, t) = -m(x)\ddot{u}(x, t) \quad \text{(18.2.1)}$$

If the system in equilibrium is subjected to virtual displacements $\delta u(x)$, the external virtual work δW_E is equal to the internal virtual work δW_I:

$$\delta W_I = \delta W_E \quad \text{(18.2.2)}$$

Based on the development of Section 8.3.2, these work quantities are

$$\delta W_E = -\int_0^L m(x)\ddot{u}(x, t)\delta u(x)\, dx + \int_0^L p(x, t)\delta u(x)\, dx \quad \text{(18.2.3a)}$$

$$\delta W_I = \int_0^L EI(x)u''(x, t)\delta[u''(x)]\, dx \quad \text{(18.2.3b)}$$

The displacements $u(x, t)$ are given by Eq. (18.1.1) and $\delta u(x)$ is any admissible virtual displacement:

$$\delta u(x) = \psi_i(x)\delta z_i \qquad i = 1, 2, \ldots, N \quad \text{(18.2.4)}$$

Substituting Eqs. (18.1.1) and (18.2.4) in Eq. (18.2.3) leads to

$$\delta W_E = -\delta z_i \sum_{j=1}^{N} \ddot{z}_j \tilde{m}_{ij} + \delta z_i \tilde{p}_i(t) \qquad (18.2.5a)$$

$$\delta W_I = \delta z_i \sum_{j=1}^{N} z_j \tilde{k}_{ij} \qquad (18.2.5b)$$

where

$$\tilde{m}_{ij} = \int_0^L m(x)\psi_i(x)\psi_j(x)\,dx$$

$$\tilde{k}_{ij} = \int_0^L EI(x)\psi_i''(x)\psi_j''(x)\,dx \qquad (18.2.6)$$

$$\tilde{p}_i(t) = \int_0^L p(x,t)\psi_i(x)\,dx$$

Substituting Eq. (18.2.5) in Eq. (18.2.2) gives

$$\delta z_i \left(\sum_{j=1}^{N} \ddot{z}_j \tilde{m}_{ij} + \sum_{j=1}^{N} z_j \tilde{k}_{ij} \right) = \delta z_i \tilde{p}_i(t) \qquad (18.2.7)$$

and δz_i can be dropped from both sides because this equation is valid for any δz_i.

Corresponding to the N independent virtual displacements of Eq. (18.2.4), there are N equations like Eq. (18.2.7). Together they can be expressed in matrix notation:

$$\tilde{\mathbf{m}}\ddot{\mathbf{z}} + \tilde{\mathbf{k}}\mathbf{z} = \tilde{\mathbf{p}}(t) \qquad (18.2.8)$$

where $\mathbf{z}$ is the vector of N generalized coordinates, $\tilde{\mathbf{m}}$ the generalized mass matrix with its elements defined by Eq. (18.2.6a), $\tilde{\mathbf{k}}$ the generalized stiffness matrix whose elements are given by Eq. (18.2.6b), and $\tilde{\mathbf{p}}(t)$ the generalized applied force vector with its elements defined by Eq. (18.2.6c). It is obvious from Eq. (18.2.6) that $\tilde{\mathbf{m}}$ and $\tilde{\mathbf{k}}$ are symmetric matrices. A damping matrix can also be included in the virtual work formulation if the damping mechanisms can be defined. The system of coupled differential equations (18.2.8) can be solved for the unknowns $z_j(t)$ using the numerical procedures presented in Chapter 16. Then at each time instant the displacement $u(x)$ is determined from Eq. (18.1.1). This is an alternative to the classical modal analysis of the system mentioned at the end of Section 18.1.

The mass and stiffness matrices obtained using the principle of virtual displacements are identical to those derived in Section 18.1 from the principle of energy conservation, the concept underlying the original development of the Rayleigh–Ritz method. We will draw upon the virtual work approach when we introduce the finite element method in Part B of this chapter.

18.3 DISADVANTAGES OF RAYLEIGH–RITZ METHOD

The Rayleigh–Ritz method leads to natural frequencies $\tilde{\omega}_n$ and natural modes $\tilde{\phi}_n(x)$ that approximate the true values ω_n and $\phi_n(x)$, respectively, best among the admissible class of functions described by Eq. (18.1.1). However, the method is not practical for general application and automated computer implementation to analyze complex structures for several reasons: (1) It is difficult to select the Ritz trial functions because they should be suitable for the particular system and its boundary conditions. (2) It is not clear how to select additional functions to improve the accuracy of an approximate solution obtained using fewer functions. (3) It may be difficult to evaluate the integrals of Eq. (18.2.6) over the entire structure, especially with higher-order trial functions. (4) It is computationally demanding to work with the matrices $\tilde{\mathbf{m}}$ and $\tilde{\mathbf{k}}$ because they are full matrices. (5) It is difficult to interpret the generalized coordinates, as they do not necessarily represent displacements at physical locations on the structure. These difficulties are overcome by the finite element method introduced next.

PART B: FINITE ELEMENT METHOD

The finite element method is one of the most important developments in applied mechanics. Although the method is applicable to a wide range of problems, only an introduction is included here with reference to systems that can be idealized as an assemblage of one-dimensional structural finite elements. The presentation, although self-contained, is based on the presumption that the reader is familiar with the finite element method for analysis of static structural problems.

18.4 FINITE ELEMENT APPROXIMATION

In the finite element method the trial functions are selected in a special way to overcome the aforementioned difficulties of the Rayleigh–Ritz method. To illustrate this concept, consider the cantilever beam shown in Fig. 18.4.1, which is subdivided into a number of segments, called *finite elements*. Their size is arbitrary; they may be all of the same size or all different. The elements are interconnected only at *nodes* or *nodal points*. In this simple case the nodal points are the ends of the element, and each node has two DOFs, transverse displacement and rotation. In the finite element method nodal displacements are selected as the generalized coordinates, and the equations of motion are formulated in terms of these physically meaningful displacements.

The deflection of the beam is expressed in terms of the nodal displacements through trial functions $\hat{\psi}_i(x)$ shown in Fig. 18.4.1. Corresponding to each DOF, a trial function is selected with the following properties: unit value at the DOF; zero value at all other DOFs; continuous function with continuous first derivative. No trial functions are shown for the node at the clamped end because the displacement and slope are both zero. These trial functions satisfy the requirements of admissibility because they are linearly independent, continuous with continuous first derivative, and consistent with the geometric boundary conditions. The deflection of the beam is expressed as

$$u(x) = \sum_i u_i \hat{\psi}_i(x) \qquad (18.4.1)$$

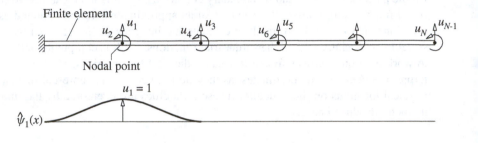

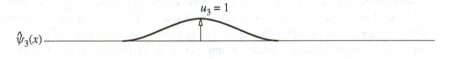

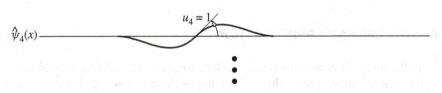

Figure 18.4.1

where u_i is the nodal displacement in the ith DOF and $\hat{\psi}_i(x)$ is the associated trial function. Because these functions define the displacements between nodal points (in contrast to global displacements of the structure in the Rayleigh–Ritz method), they are called *interpolation functions*.

The finite element method offers several important advantages over the Rayleigh–Ritz method. Stated in the same sequence as the disadvantages of the Rayleigh–Ritz method mentioned at the end of Section 18.3, the advantages of the finite element method are: (1) Simple interpolation functions can be chosen for each finite element. (2) Accuracy of the solution can be improved by increasing the number of finite elements in the structural idealization. (3) Computation of the integrals of Eq. (18.2.6) is much easier because the interpolation functions are simple, and the same functions may be chosen for each finite element. (4) The structural stiffness and mass matrices developed by the finite element method are narrowly banded, a property that reduces the computational effort necessary to solve the equations of motion. (5) The generalized displacements are physically meaningful, as they give the nodal displacements directly.

18.5 ANALYSIS PROCEDURE

The formulation of the equations of motion for a structure by the finite element method may be summarized as a sequence of the following steps:

1. Idealize the structure as an assemblage of finite elements interconnected only at nodes (Fig. 18.5.1a); define the DOF **u** at these nodes (Fig. 18.5.1b).
2. For each finite element form the element stiffness matrix $\mathbf{k}_e$, the element mass matrix $\mathbf{m}_e$, and the element (applied) force vector $\mathbf{p}_e(t)$ with reference to the DOF for the element (Fig. 18.5.1c). For each element the force–displacement relation and the

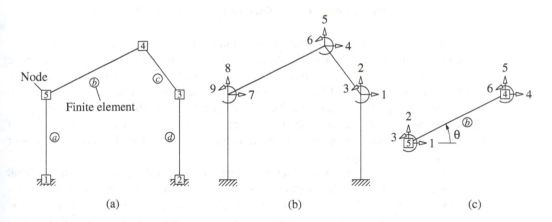

(a)   (b)   (c)

Figure 18.5.1   (a) Finite elements and nodes; (b) assemblage DOF **u**; (c) element DOF $\mathbf{u}_e$.

inertia force–acceleration relation are

$$(\mathbf{f}_S)_e = \mathbf{k}_e\mathbf{u}_e \qquad (\mathbf{f}_I)_e = \mathbf{m}_e\ddot{\mathbf{u}}_e \tag{18.5.1}$$

In the finite element formulation these relations are obtained by assuming the displacement field over the element, expressed in terms of nodal displacements.

3. Form the transformation matrix $\mathbf{a}_e$ that relates the displacements $\mathbf{u}_e$ and forces $\mathbf{p}_e$ for the element to the displacements $\mathbf{u}$ and forces $\mathbf{p}$ for the finite element assemblage:

$$\mathbf{u}_e = \mathbf{a}_e\mathbf{u} \qquad \mathbf{p}(t) = \mathbf{a}_e^T\mathbf{p}_e(t) \tag{18.5.2}$$

where $\mathbf{a}_e$ is a Boolean matrix consisting of zeros and ones. It simply locates the elements of $\mathbf{k}_e$, $\mathbf{m}_e$, and $\mathbf{p}_e$ at the proper locations in the mass matrix, stiffness matrix, and (applied) force vector for the finite element assemblage. Therefore, it is not necessary to carry out the transformations: $\hat{\mathbf{k}}_e = \mathbf{a}_e^T\mathbf{k}_e\mathbf{a}_e$, $\hat{\mathbf{m}}_e = \mathbf{a}_e^T\mathbf{m}_e\mathbf{a}_e$, or $\hat{\mathbf{p}}_e(t) = \mathbf{a}_e^T\mathbf{p}_e(t)$ to transform the element stiffness and mass matrices and applied force vector to the nodal displacements for the assemblage.

4. Assemble the element matrices to determine the stiffness and mass matrices and the applied force vector for the assemblage of finite elements:

$$\mathbf{k} = \mathcal{A}_{e=1}^{N_e}\mathbf{k}_e \qquad \mathbf{m} = \mathcal{A}_{e=1}^{N_e}\mathbf{m}_e \qquad \mathbf{p}(t) = \mathcal{A}_{e=1}^{N_e}\mathbf{p}_e(t) \tag{18.5.3}$$

The operator $\mathcal{A}$ denotes the *direct assembly procedure* for assembling according to the matrix $\mathbf{a}_e$, the element stiffness matrix, element mass matrix, and the element force vector—for each element $e = 1$ to N_e, where N_e is the number of elements— into the assemblage stiffness matrix, assemblage mass matrix, and assemblage force vector, respectively.

5. Formulate the equations of motion for the finite element assemblage:

$$\mathbf{m}\ddot{\mathbf{u}} + \mathbf{c}\dot{\mathbf{u}} + \mathbf{k}\mathbf{u} = \mathbf{p}(t) \tag{18.5.4}$$

where the damping matrix $\mathbf{c}$ is established by the methods of Chapter 11.

The governing equations (18.5.4) for a finite element system are of the same form as formulated in Chapter 9 for frame structures. It should be clear from the outline above that the only difference between the displacement method for analysis of frame structures and the finite element method is in the formulation of the element mass and stiffness matrices. Therefore, Eq. (18.5.4) can be solved for $\mathbf{u}(t)$ by the methods developed in preceding chapters. The classical modal analysis procedure of Chapters 12 and 13 is applicable if the system has classical damping and the direct methods of Chapter 16 enable analysis of nonclassically damped systems.

In the subsequent sections, which are restricted to assemblages of one-dimensional finite elements, we define the interpolation functions and develop the element stiffness matrix $\mathbf{k}_e$, the element mass matrix $\mathbf{m}_e$, and the element (applied) force vector $\mathbf{p}_e(t)$. Assembly of these element matrices to construct the corresponding matrices for the finite element assemblage is illustrated by an example.

18.6 ELEMENT DEGREES OF FREEDOM AND INTERPOLATION FUNCTIONS

Consider a straight-beam element of length L, mass per unit length $m(x)$, and flexural rigidity $EI(x)$. The two nodes by which the finite element can be assembled into a structure are located at its ends. If only planar displacements are considered, each node has two DOFs: the transverse displacement and rotation (Fig. 18.6.1a). The displacement of the beam element is related to its four DOFs:

$$u(x, t) = \sum_{i=1}^{4} u_i(t)\psi_i(x) \tag{18.6.1}$$

where the function $\psi_i(x)$ defines the displacement of the element due to unit displacement u_i while constraining other DOFs to zero. Thus $\psi_i(x)$ satisfies the following boundary conditions:

$$i = 1: \quad \psi_1(0) = 1, \ \psi_1'(0) = \psi_1(L) = \psi_1'(L) = 0 \tag{18.6.2a}$$

$$i = 2: \quad \psi_2'(0) = 1, \ \psi_2(0) = \psi_2(L) = \psi_2'(L) = 0 \tag{18.6.2b}$$

$$i = 3: \quad \psi_3(L) = 1, \ \psi_3(0) = \psi_3'(0) = \psi_3'(L) = 0 \tag{18.6.2c}$$

$$i = 4: \quad \psi_4'(L) = 1, \ \psi_4(0) = \psi_4'(0) = \psi_4(L) = 0 \tag{18.6.2d}$$

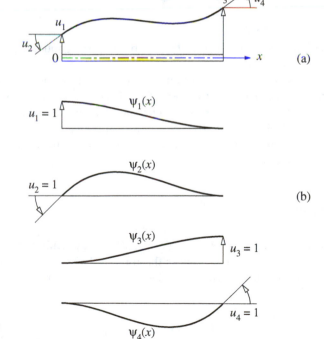

Figure 18.6.1   (a) Degrees of freedom for a beam element; (b) interpolation functions.

Introduction to the Finite Element Method

These interpolation functions could be any arbitrary shapes satisfying the boundary conditions. One possibility is the exact deflected shapes of the beam element due to the imposed boundary conditions, but these are difficult to determine if the flexural rigidity varies over the length of the element. However, they can conveniently be obtained for a uniform beam, as illustrated next.

Neglecting shear deformations, the equilibrium equation for a beam loaded only at its ends is

$$EI\frac{d^4u}{dx^4} = 0 \tag{18.6.3}$$

The general solution of Eq. (18.6.3) for a uniform beam is a cubic polynomial

$$u(x) = a_1 + a_2\left(\frac{x}{L}\right) + a_3\left(\frac{x}{L}\right)^2 + a_4\left(\frac{x}{L}\right)^3 \tag{18.6.4}$$

The constants a_i can be determined for each of the four sets of boundary conditions, Eq. (18.6.2), to obtain

$$\psi_1(x) = 1 - 3\left(\frac{x}{L}\right)^2 + 2\left(\frac{x}{L}\right)^3 \tag{18.6.5a}$$

$$\psi_2(x) = L\left(\frac{x}{L}\right) - 2L\left(\frac{x}{L}\right)^2 + L\left(\frac{x}{L}\right)^3 \tag{18.6.5b}$$

$$\psi_3(x) = 3\left(\frac{x}{L}\right)^2 - 2\left(\frac{x}{L}\right)^3 \tag{18.6.5c}$$

$$\psi_4(x) = -L\left(\frac{x}{L}\right)^2 + L\left(\frac{x}{L}\right)^3 \tag{18.6.5d}$$

These interpolation functions, illustrated in Fig. 18.6.1b, can be used in formulating the element matrices for nonuniform elements.

The foregoing approach is possible for a beam finite element—because the governing differential equation (18.6.3) could be solved for a uniform beam—but not for two- or three-dimensional finite elements. Therefore, the finite element method is based on assumed relationships between the displacements at interior points of the element and the displacements at the nodes. Proceeding in this manner makes the problem tractable but introduces approximations in the solution.

18.7 ELEMENT STIFFNESS MATRIX

Consider a beam element of length L with flexural rigidity $EI(x)$. By definition, the stiffness influence coefficient k_{ij} of the beam element is the force in DOF i due to unit displacement in DOF j. Using the principle of virtual displacement in a manner similar to Section 18.2, we can derive a general equation for k_{ij}:

$$k_{ij} = \int_0^L EI(x)\psi_i''(x)\psi_j''(x)\,dx \tag{18.7.1}$$

The symmetric form of this equation shows that the element stiffness matrix is symmetric; $k_{ij} = k_{ji}$. Observe that this result for an element has the same form as Eq. (18.2.6) for the structure. Equation (18.7.1) is a general result in the sense that it is applicable to elements with arbitrary variation of flexural rigidity $EI(x)$, although the interpolation functions of Eq. (18.6.5) are exact only for uniform elements. The associated errors can be reduced to any desired degree by reducing the element size and increasing the number of finite elements in the structural idealization.

For a uniform finite element with $EI(x) = EI$, the integral of Eq. (18.7.1) can be evaluated analytically for $i, j = 1, 2, 3$ and 4, resulting in the element stiffness matrix:

$$\bar{\mathbf{k}}_e = \frac{EI}{L^3} \begin{bmatrix} 12 & 6L & -12 & 6L \\ 6L & 4L^2 & -6L & 2L^2 \\ -12 & -6L & 12 & -6L \\ 6L & 2L^2 & -6L & 4L^2 \end{bmatrix} \qquad (18.7.2)$$

These stiffness coefficients are the exact values for a uniform beam, neglecting shear deformation, because the interpolation functions of Eq. (18.6.5) are the true deflection shapes for this case. Observe that the stiffness matrix of Eq. (18.7.2) is equivalent to the force–displacement relations for a uniform beam that are familiar from classical structural analysis (see Chapter 1, Appendix 1). For nonuniform elements, such as a haunched beam, approximate values for the stiffness coefficients can be determined by numerically evaluating Eq. (18.7.1).

The 4×4 element stiffness matrix $\bar{\mathbf{k}}_e$ of Eq. (18.7.2) in local element coordinates (Fig. 18.6.1a) is transformed to the 6×6 $\mathbf{k}_e$ of Eq. (18.5.1) in global element coordinates (Fig. 18.5.1c). Before carrying out this transformation, $\bar{\mathbf{k}}_e$ is expanded to a 6×6 matrix that includes the stiffness coefficients associated with the axial DOF at each node. The transformation matrix that depends on the orientation θ of the member (Fig. 18.5.1c) should be familiar to the reader.

18.8 ELEMENT MASS MATRIX

As defined in Section 9.2.4, the mass influence coefficient m_{ij} for a structure is the force in the ith DOF due to unit acceleration in the jth DOF. Applying this definition to a beam element with distributed mass $m(x)$ and using the principle of virtual displacement along the lines of Section 18.2, a general equation for m_{ij} can be derived:

$$m_{ij} = \int_0^L m(x)\psi_i(x)\psi_j(x)\,dx \qquad (18.8.1)$$

The symmetric form of this equation shows that the mass matrix is symmetric; $m_{ij} = m_{ji}$. Observe that the result for an element has the same form as Eq. (18.2.6) for the structure.

If we use the same interpolation functions in Eq. (18.8.1) as were used to derive the element stiffness matrix, the result obtained is known as the *consistent mass matrix*. The integrals of Eq. (18.8.1) are evaluated numerically or analytically depending on the function $m(x)$. For an element with uniform mass [i.e., $m(x) = m$], the integrals can be evaluated analytically to obtain the element (consistent) mass matrix:

$$\overline{\mathbf{m}}_e = \frac{mL}{420} \begin{bmatrix} 156 & 22L & 54 & -13L \\ 22L & 4L^2 & 13L & -3L^2 \\ 54 & 13L & 156 & -22L \\ -13L & -3L^2 & -22L & 4L^2 \end{bmatrix} \qquad (18.8.2)$$

Observe that the consistent-mass matrix is not diagonal, whereas the lumped-mass approximation leads to a diagonal matrix, as we shall see next.

The mass matrix of a finite element can be simplified by assuming that the distributed mass of the element can be lumped as point masses along the translational DOF u_1 and u_3 at the ends (Fig. 18.6.1), with the two masses being determined by static analysis of the beam under its own weight. For example, if the mass of a uniform element is m per unit length, a point mass of $mL/2$ will be assigned to each end, leading to

$$\overline{\mathbf{m}}_e = mL \begin{bmatrix} \frac{1}{2} & 0 & 0 & 0 \\ 0 & 0 & 0 & 0 \\ 0 & 0 & \frac{1}{2} & 0 \\ 0 & 0 & 0 & 0 \end{bmatrix} \qquad (18.8.3)$$

Observe that for a lumped-mass idealization of the finite element, the mass matrix is diagonal. The off-diagonal terms m_{ij} of this matrix are zero because an acceleration of any point mass produces an inertia force only in the same DOF. The diagonal terms m_{ii} associated with the rotational degrees of freedom are zero because of the idealization that the mass is lumped in points that have no rotational inertia.

The 4×4 element mass matrix $\overline{\mathbf{m}}_e$ given by Eq. (18.8.2) or (18.8.3) in local element coordinates (Fig. 18.6.1a) is transformed to the 6×6 $\mathbf{m}_e$ of Eq. (18.5.1) in global element coordinates (Fig. 18.5.1c). The procedure and transformation matrix are the same as described earlier for the stiffness matrix.

The dynamic analysis of a consistent-mass system requires considerably more computational effort than does a lumped-mass idealization, for two reasons: (1) The lumped-mass matrix is diagonal, whereas the consistent-mass matrix has off-diagonal terms; and (2) the rotational DOF can be eliminated by static condensation (see Section 9.3) from the equations of motion for a lumped-mass system, whereas all DOFs must be retained in a consistent-mass system.

However, the consistent-mass formulation has two advantages. First, it leads to greater accuracy in the results and rapid convergence to the exact results with an increasing number of finite elements, as we shall see later by an example, but in practice the improvement is often only slight because the inertia forces associated with node rotations are generally not significant in many structural earthquake engineering problems. Second, with a consistent-mass approach, the potential energy and kinetic energy quantities are evaluated in a consistent manner, and therefore we know how the computed values of the natural frequencies relate to the exact values (see Part A of this chapter).

The second advantage seldom outweighs the additional computational effort required to achieve a slight increase in accuracy, and therefore the lumped-mass idealization is widely used.

18.9 ELEMENT (APPLIED) FORCE VECTOR

If the external forces $p_i(t)$, $i = 1, 2, 3$ and 4, are applied along the four DOFs at the two nodes of the finite element, the element force vector can be written directly. On the other hand, if the external forces include distributed force $p(x, t)$ and concentrated forces $p'_j(t)$ at locations x_j, the nodal force in the ith DOF is

$$p_i(t) = \int_0^L p(x, t)\psi_i(x)\,dx + \sum_j p'_j(t)\psi_i(x_j) \tag{18.9.1}$$

This equation can be obtained by the principle of virtual displacements, following the derivation of the similar equation (18.2.6c) for the complete structure. If we use the same interpolation functions in Eq. (18.9.1) as were used to derive the element stiffness matrix, the results obtained are called *consistent nodal forces*.

A simpler, less accurate approach is to use linear interpolation functions:

$$\psi_1(x) = 1 - \frac{x}{L} \qquad \psi_3(x) = \frac{x}{L} \tag{18.9.2}$$

Then Eq. (18.9.1) gives the nodal forces $p_1(t)$ and $p_3(t)$ in the translational DOF; the forces $p_2(t)$ and $p_4(t)$ in the rotational DOF are zero unless external moments are applied directly to the nodes.

The 4×1 element force vector $\bar{\mathbf{p}}_e$ given by Eq. (18.9.1) in local element coordinates is transformed to the 6×1 $\mathbf{p}_e$ of Eq. (18.5.2) in global element coordinates. The transformation matrix is the transpose of the one described earlier for the stiffness matrix.

Example 18.2

Determine the natural frequencies and modes of vibration of a uniform cantilever beam, idealized as an assemblage of two finite elements (Fig. E18.2a) using the consistent mass matrix. The flexural rigidity is EI and mass per unit length is m.

Solution

1. *Identify the assemblage and element DOFs.* The six DOFs for the finite element assemblage are shown in Fig. E18.2b, and the two finite elements and their local DOFs in Fig. E18.2c.

2. *Form the element stiffness matrices.* Replacing L in Eq. (18.7.2) by $L/2$, the length of each finite element in Fig. E18.2a, gives the stiffness matrices $\bar{\mathbf{k}}_1$ and $\bar{\mathbf{k}}_2$ for the two finite elements in their local DOFs. Since both local element DOFs and assemblage DOFs are defined along the same set of Cartesian coordinates, no transformation of coordinates is required.

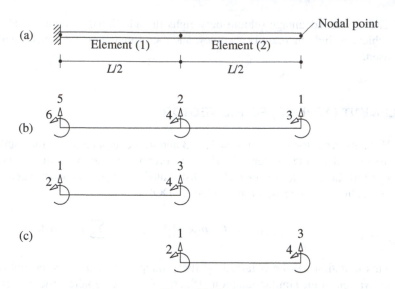

Figure E18.2

Therefore, $\mathbf{k}_1 = \bar{\mathbf{k}}_1$ and $\mathbf{k}_2 = \bar{\mathbf{k}}_2$; thus

$$
\mathbf{k}_1 = \frac{8EI}{L^3}
\begin{array}{c}
\begin{array}{cccc} (5) & (6) & (2) & (4) \end{array} \\
\left[
\begin{array}{cccc}
12 & 3L & -12 & 3L \\
3L & L^2 & -3L & L^2/2 \\
-12 & -3L & 12 & -3L \\
3L & L^2/2 & -3L & L^2
\end{array}
\right]
\begin{array}{c} (5) \\ (6) \\ (2) \\ (4) \end{array}
\end{array}
$$

$$
\mathbf{k}_2 = \frac{8EI}{L^3}
\begin{array}{c}
\begin{array}{cccc} (2) & (4) & (1) & (3) \end{array} \\
\left[
\begin{array}{cccc}
12 & 3L & -12 & 3L \\
3L & L^2 & -3L & L^2/2 \\
-12 & -3L & 12 & -3L \\
3L & L^2/2 & -3L & L^2
\end{array}
\right]
\begin{array}{c} (2) \\ (4) \\ (1) \\ (3) \end{array}
\end{array}
$$

The numbers in parentheses alongside rows and columns of $\mathbf{k}_e$ refer to the assemblage DOFs that correspond to the element DOFs. This information enables $\mathbf{k}_e$ to be assembled.

3. *Transform the element stiffness matrices to assemblage DOFs.* This step is not required for solving this problem but is included to assist in better understanding of the procedure. The element stiffness matrix with reference to the nodal displacements of the finite element assemblage is given by

$$ \hat{\mathbf{k}}_e = \mathbf{a}_e^T \mathbf{k}_e \mathbf{a}_e \tag{a} $$

The nodal displacements of elements (1) and (2) are related to the assemblage displacements by

$$ (u_1)_1 = u_5 \quad (u_2)_1 = u_6 \quad (u_3)_1 = u_2 \quad (u_4)_1 = u_4 $$

$$ (u_1)_2 = u_2 \quad (u_2)_2 = u_4, \quad (u_3)_2 = u_1 \quad (u_4)_2 = u_3 $$

These relationships for the two elements can be expressed as

$$\mathbf{u}_1 = \mathbf{a}_1\mathbf{u} \qquad \mathbf{u}_2 = \mathbf{a}_2\mathbf{u} \qquad\qquad (b)$$

where the transformation matrices are

$$\mathbf{a}_1 = \begin{bmatrix} 0 & 0 & 0 & 0 & 1 & 0 \\ 0 & 0 & 0 & 0 & 0 & 1 \\ 0 & 1 & 0 & 0 & 0 & 0 \\ 0 & 0 & 0 & 1 & 0 & 0 \end{bmatrix} \qquad \mathbf{a}_2 = \begin{bmatrix} 0 & 1 & 0 & 0 & 0 & 0 \\ 0 & 0 & 0 & 1 & 0 & 0 \\ 1 & 0 & 0 & 0 & 0 & 0 \\ 0 & 0 & 1 & 0 & 0 & 0 \end{bmatrix}$$

Observe that $a_{ij} = 1$ indicates that the element DOF i corresponds to (and becomes renumbered as) the assemblage DOF j. The same information is available through the numbers in parentheses alongside rows and columns of $\mathbf{k}_1$ and $\mathbf{k}_2$. Thus the matrices $\mathbf{a}_1$ and $\mathbf{a}_2$ simply locate the elements of $\mathbf{k}_1$ and $\mathbf{k}_2$ in the stiffness matrix for the assemblage. This implies that it is not necessary to carry out the transformation of Eq. (a), because $\mathbf{a}_1$ and $\mathbf{a}_2$ consist of only ones and zeros.

4. *Assemble the element stiffness matrices.* The stiffness matrix $\mathbf{k}$ for the finite element system is determined by assembling $\mathbf{k}_1$ and $\mathbf{k}_2$ by locating the elements of $\mathbf{k}_1$ and $\mathbf{k}_2$ in $\mathbf{k}$ according to $\mathbf{a}_1$ and $\mathbf{a}_2$, respectively.

$$\mathbf{k} = \mathcal{A}_{i=1}^2 \mathbf{k}_i$$

$$= \frac{8EI}{L^3} \left[\begin{array}{cccc|cc} 12 & -12 & -3L & -3L & 0 & 0 \\ -12 & 24 & 3L & 0 & -12 & -3L \\ -3L & 3L & L^2 & L^2/2 & 0 & 0 \\ -3L & 0 & L^2/2 & 2L^2 & 3L & L^2/2 \\ \hline 0 & -12 & 0 & 3L & 12 & 3L \\ 0 & -3L & 0 & L^2/2 & 3L & L^2 \end{array} \right] \Big\} \text{ support DOFs}$$

$$\underbrace{\qquad\qquad\qquad}_{\text{support DOFs}}$$

5. *Form the element mass matrices.* Replacing L in Eq. (18.8.2) by $L/2$ gives the element mass matrices $\overline{\mathbf{m}}_1$ and $\overline{\mathbf{m}}_2$ in their local DOF; and as for the stiffness matrices, $\mathbf{m}_1 = \overline{\mathbf{m}}_1$ and $\mathbf{m}_2 = \overline{\mathbf{m}}_2$. Thus

$$\mathbf{m}_1 = \frac{mL}{840} \begin{array}{c} \\ \end{array} \begin{matrix} (5) & (6) & (2) & (4) \\ \begin{bmatrix} 156 & 11L & 54 & -6.5L \\ 11L & L^2 & 6.5L & -0.75L^2 \\ 54 & 6.5L & 156 & -11L \\ -6.5L & -0.75L^2 & -11L & L^2 \end{bmatrix} & \begin{matrix} (5) \\ (6) \\ (2) \\ (4) \end{matrix} \end{matrix}$$

$$\mathbf{m}_2 = \frac{mL}{840} \begin{matrix} (2) & (4) & (1) & (3) \\ \begin{bmatrix} 156 & 11L & 54 & -6.5L \\ 11L & L^2 & 6.5L & -0.75L^2 \\ 54 & 6.5L & 156 & -11L \\ -6.5L & -0.75L^2 & -11L & L^2 \end{bmatrix} & \begin{matrix} (2) \\ (4) \\ (1) \\ (3) \end{matrix} \end{matrix}$$

6. *Assemble the element mass matrices.* The mass matrix for the finite element system is determined by assembling $\mathbf{m}_1$ and $\mathbf{m}_2$ in a manner analogous to the stiffness matrix assembly:

$$\mathbf{m} = \mathcal{A}_{i=1}^{2}\,\mathbf{m}_i$$

$$= \frac{mL}{840}\begin{bmatrix} 156 & 54 & -11L & 6.5L & 0 & 0 \\ 54 & 312 & -6.5L & 0 & 54 & 6.5L \\ -11L & -6.5L & L^2 & -0.75L^2 & 0 & 0 \\ 6.5L & 0 & -0.75L^2 & 2L^2 & -6.5L & -0.75L^2 \\ \hline 0 & 54 & 0 & -6.5L & 156 & 11L \\ 0 & 6.5L & 0 & -0.75L^2 & 11L & L^2 \end{bmatrix} \Big\}\ \text{support DOFs}$$

$$\underbrace{\qquad\qquad\qquad}_{\text{support DOFs}}$$

7. *Formulate the equations of motion.* Before writing the equation of motion, the support conditions must be imposed. For the cantilever beam of Fig. E18.2a, $u_5 = u_6 = 0$. Thus the fifth and sixth rows and columns are deleted from the matrices $\mathbf{m}$ and $\mathbf{k}$ to obtain

$$\mathbf{m\ddot{u}} + \mathbf{ku} = \mathbf{0} \tag{c}$$

or

$$\frac{mL}{840}\begin{bmatrix} 156 & 54 & -11L & 6.5L \\ & 312 & -6.5L & 0 \\ (\text{sym}) & & L^2 & -0.75L^2 \\ & & & 2L^2 \end{bmatrix}\begin{bmatrix} \ddot{u}_1 \\ \ddot{u}_2 \\ \ddot{u}_3 \\ \ddot{u}_4 \end{bmatrix}$$

$$+ \frac{8EI}{L^3}\begin{bmatrix} 12 & -12 & -3L & -3L \\ & 24 & 3L & 0 \\ (\text{sym}) & & L^2 & 0.5L^2 \\ & & & 2L^2 \end{bmatrix}\begin{bmatrix} u_1 \\ u_2 \\ u_3 \\ u_4 \end{bmatrix} = \begin{bmatrix} 0 \\ 0 \\ 0 \\ 0 \end{bmatrix} \tag{d}$$

Note that the stiffness matrix in Eq. (d) is the same as the one obtained by classical methods in Example 9.4.

8. *Solve the eigenvalue problem.* The natural frequencies are determined by solving $\mathbf{k}\phi = \omega^2\mathbf{m}\phi$:

$$\omega_1 = 3.51772\sqrt{\frac{EI}{mL^4}} \qquad \omega_2 = 22.2215\sqrt{\frac{EI}{mL^4}}$$

$$\omega_3 = 75.1571\sqrt{\frac{EI}{mL^4}} \qquad \omega_4 = 218.138\sqrt{\frac{EI}{mL^4}} \tag{e}$$

Example 18.3

Repeat Example 18.2 using the lumped-mass approximation.

Solution The only change is in formulation of the mass matrix. Using the lumped-mass matrix of Eq. (18.8.3) for each element and proceeding as in steps 5, 6, and 7 of Example 18.2 leads to

$$\mathbf{m\ddot{u}} + \mathbf{ku} = \mathbf{0} \tag{a}$$

where $\mathbf{u}$ and $\mathbf{k}$ are the same as in Eq. (d) of Example 18.2 but $\mathbf{m}$ is different:

$$\mathbf{m} = \begin{bmatrix} mL/4 & 0 & 0 & 0 \\ & mL/2 & 0 & 0 \\ \text{(sym)} & & 0 & 0 \\ & & & 0 \end{bmatrix} \tag{b}$$

Because the mass associated with the rotational DOFs u_3 and u_4 is zero, they can be eliminated from the stiffness matrix by static condensation. The resulting 2×2 stiffness matrix in terms of the translational DOF was presented in Example 9.5. The eigenvalue problem was solved in Example 10.2 to obtain the natural frequencies:

$$\omega_1 = 3.15623\sqrt{\frac{EI}{mL^4}} \qquad \omega_2 = 16.2580\sqrt{\frac{EI}{mL^4}} \tag{c}$$

18.10 COMPARISON OF FINITE ELEMENT AND EXACT SOLUTIONS

In this section the approximate values for the natural frequencies of a uniform cantilever beam determined by the finite element method are compared with the exact solutions presented in Chapter 17. The approximate results are obtained by discretizing the beam into N finite elements of equal length and analyzing it by the finite element method. Such results for the coefficient α_n in $\omega_n = \alpha_n\sqrt{EI/mL^4}$ obtained for $N_e = 1, 2, 3, 4,$ and 5 finite elements and using the consistent-mass matrix are presented in Table 18.10.1. Example 18.2 provides the results for $N_e = 2$ and those for other N_e were obtained similarly.

Observe from these results that the accuracy of the natural frequencies deteriorates for the higher modes of a particular N_e-element system, but the accuracy is improved by increasing N_e and hence the number of DOFs. The accuracy is quite good for a number of

TABLE 18.10.1 NATURAL FREQUENCIES OF A UNIFORM CANTILEVER BEAM: CONSISTENT-MASS FINITE ELEMENT AND EXACT SOLUTIONS

Mode	\multicolumn{5}{c}{Number of Finite Elements, N_e}	Exact
	1	2	3	4	5	
1	3.53273	3.51772	3.51637	3.51613	3.51606	3.51602
2	34.8069	22.2215	22.1069	22.0602	22.0455	22.0345
3		75.1571	62.4659	62.1749	61.9188	61.6972
4		218.138	140.671	122.657	122.320	120.902
5			264.743	228.137	203.020	199.860
6			527.796	366.390	337.273	298.556
7				580.849	493.264	416.991
8				953.051	715.341	555.165
9					1016.20	713.079
10					1494.88	890.732

Source: R. R. Craig, Jr. and A.J. Kurdila, *Fundamentals of Structural Dynamics*, 2nd ed., Wiley, New York, 2006, Section 14.8.

TABLE 18.10.2 NATURAL FREQUENCIES OF A UNIFORM CANTILEVER BEAM: LUMPED-MASS FINITE ELEMENT AND EXACT SOLUTIONS

| Mode | Number of Finite Elements, N_e | | | | | Exact |
	1	2	3	4	5	
1	2.44949	3.15623	3.34568	3.41804	3.45266	3.51602
2		16.2580	18.8859	20.0904	20.7335	22.0345
3			47.0284	53.2017	55.9529	61.6972
4				92.7302	104.436	120.902
5					153.017	199.860

Source: R. R. Craig, Jr. and A.J. Kurdila, *Fundamentals of Structural Dynamics*, 2nd ed., Wiley, New York, 2006, Section 14.8.

modes equal to the number of elements, but the frequencies of the higher modes are poor. As expected from Rayleigh–Ritz or consistent finite element formulations, the frequencies converge from above to the exact solution.

The natural frequencies obtained by a lumped-mass idealization of the finite element system and using the condensed stiffness matrix with $N_e = 1, 2, 3, 4$, and 5 finite elements are presented in Table 18.10.2. Example 18.3 provides the results for $N_e = 2$, and those for other N_e were obtained similarly.

For the particular mass-lumping procedure employed (i.e., half the element mass is distributed to each node), the lumped-mass approximation gives frequencies that converge slowly from below. Observe that for a given number N_e of elements the consistent-mass formulation provides a better result then does the lumped-mass approximation. However, this improved accuracy comes at the expense of increased computational effort because the size of the eigenvalue problem solved is doubled if consistent mass is used.

18.11 DYNAMIC ANALYSIS OF STRUCTURAL CONTINUA

The finite element method is one of the most important developments in structural analysis. Any structural continuum with an infinite number of degrees of freedom can be idealized as an assemblage of finite elements with a finite number of DOFs. Thus the partial differential equations governing the motion of the structural continuum are reduced to a system of ordinary differential equations, as many as the DOFs in the finite element idealization. Because the equations of motion for a finite element idealization of a structure [Eq. (18.5.4)] are of the same form as formulated in Chapter 9 for frame structures, they can be solved by the methods developed in this book.

The formulation of the governing equations outlined in Section 18.5 is also applicable to assemblages of two- or three-dimensional finite elements. Introductory comments regarding such applications are included in this paragraph with reference to a body in the state of plane stress (Fig. 18.11.1a) idealized as an assemblage of two-dimensional (quadrilateral in this case) elements (Fig. 18.11.1b), each with eight DOFs (Fig. 18.11.1c). The number of elements chosen for the assemblage depends on the accuracy desired. With properly formulated finite elements, the results converge to the exact solution with decreasing element size; accordingly, the larger the number of elements, the more accurate

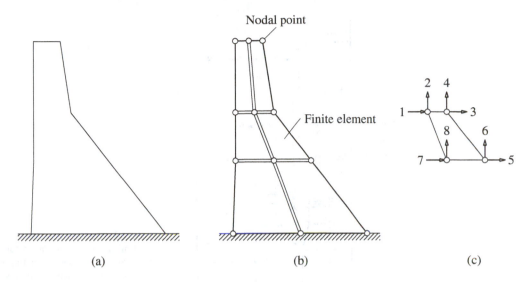

Figure 18.11.1   (a) Body in plane stress; (b) finite element idealization; (c) DOF $\mathbf{u}_e$.

will be the solution. Unlike assemblages of one-dimensional finite elements, compatibility at the nodes does not always ensure compatibility across the element boundaries. For example, as external forces are applied to the finite element assemblage of Fig. 18.11.1b, the boundaries of adjacent elements between the nodes would tend to open up or overlap. To avoid such discontinuities, interpolation functions over the element are assumed in such a fashion that the common boundaries will deform together; such elements are called *compatible elements*. With these additional considerations, the finite element method of Section 18.5 is applicable to structural continua. Solution of Eq. (18.5.4) will give the time variation of nodal displacements. At each time instant, the state of stress within each finite element is determined from the nodal displacements using interpolation functions, strain–displacement relations, and constitutive properties of the material. The finite element method differs from the displacement method for frame structures, as mentioned earlier, primarily in the formulation of the element mass and stiffness matrices.

An explosive growth of research on the finite element method took place beginning in the early 1960s, leading to the development of finite elements appropriate for idealizing different types of structural continua and their application to practical problems. These developments and applications are documented in thousands of published papers and dozens of textbooks and are too numerous to describe here. We mention just two applications.

Figure 18.11.2 shows the finite element idealization for the reactor building of a nuclear power plant. The idealization includes shell and solid elements, both axially symmetric, and includes a portion of the foundation soil. In the dynamic analysis conducted to satisfy U.S. Nuclear Regulatory Commission requirements for licensing the power plant, the prescribed earthquake motion was applied at the lower boundary of the system shown.

Figure 18.11.3 shows Koyna Dam, a 338-ft-high and 2800-ft-long concrete dam, in India. The Koyna earthquake of December 11, 1967, with peak acceleration around 0.5g

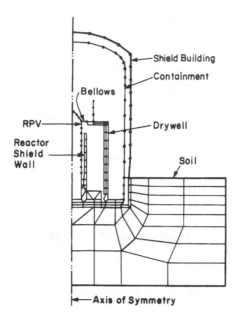

Figure 18.11.2 Axisymmetric finite element idealization of a nuclear reactor building. (From A. K. Chopra, "Earthquake Analysis of Complex Structures," in *Applied Mechanics in Earthquake Engineering,* ASME, New York, 1974.)

Figure 18.11.3 Koyna Dam, a 338-ft-high and 2800-ft-long concrete gravity dam near Poona, India. This major dam was damaged by the Koyna earthquake of December 11, 1967, a Magnitude 6.5 event centered 3 miles from the dam. [From A. K. Chopra, "Earthquake Response Analysis of Concrete Dams," in *Advanced Dam Engineering for Design, Construction and Rehabilitation* (ed. R. B. Jansen), Van Nostrand Reinhold, New York, 1988.]

Figure 18.11.4   Addition of buttresses to Koyna Dam after the damage caused by the Koyna earthquake of December 11, 1967. [From A. K. Chopra, "Earthquake Response Analysis of Concrete Dams," in *Advanced Dam Engineering for Design, Construction and Rehabilitation* (ed. R. B. Jansen), Van Nostrand Reinhold, New York, 1988.]

in the transverse direction, caused significant cracking in the dam. Although the dam survived the earthquake without any sudden release of water, the cracking appeared serious enough that it was decided to strengthen the dam by providing concrete buttresses on the downstream face of the nonoverflow monoliths; Fig. 18.11.4 shows the buttresses under construction. The finite element method was used to explain why the dam was damaged. Figure 18.11.5a shows the finite element idealization for the tallest monolith of Koyna Dam, including 136 quadrilateral plane-stress elements interconnected at 162 nodes. From the equations of motion formulated by the finite element method, the natural periods and modes of vibration of this structure were determined; the first four are shown in Fig. 18.11.5b. Assuming linear behavior, the governing equations were solved to predict the dynamic response of the dam to the ground motion recorded at the dam site (Fig. 6.1.3). The results of this computer analysis included the time variation of the displacement of each node and of the stress in each finite element. The peak value of tensile stress in each finite element over the duration of the earthquake was noted and the contours of equal stress are shown in Fig. 18.11.5c. These results indicate larger tensile stresses in upper parts of the dam, especially around the elevation where the slope of the downstream

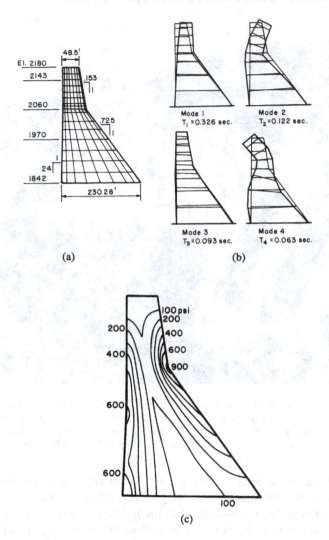

Figure 18.11.5 (a) Finite element idealization, (b) first four natural vibration modes and periods, and (c) stresses in the dam due to the Koyna earthquake, computed by linear analysis. [From A. K. Chopra, "Earthquake Response Analysis of Concrete Dams," in *Advanced Dam Engineering for Design, Construction and Rehabilitation* (ed. R. B. Jansen), Van Nostrand Reinhold, New York, 1988.]

face changes. These stresses, which exceed 600 psi on the upstream face and 900 psi on the downstream face, are approximately two to three times the tensile strength (350 psi) of the concrete used in the upper parts of the dam. Hence, based on the finite element analysis results and concrete strength data, it was possible to identify the locations in the dam where significant cracking of concrete can be expected. These are consistent with the locations where the dam was damaged by the Koyna earthquake.

FURTHER READING

Bathe, K. J., *Finite Element Procedures in Engineering Analysis*, Prentice Hall, Englewood Cliffs, N.J., 1982.

Chopra, A. K., "Earthquake Analysis of Complex Structures," in *Applied Mechanics in Earthquake Engineering*, AMD, Vol. 8 (ed. by W. D. Iwan), ASME, New York, 1974, pp. 163–204.

Chopra, A. K., "Earthquake Response Analysis of Concrete Dams," Chapter 15 in *Advanced Dam Engineering for Design, Construction and Rehabilitation* (ed. R. B. Jansen), Van Nostrand Reinhold, New York, 1988, pp. 416–465.

Clough, R. W., "The Finite Element Method in Plane Stress Analysis," *Proceedings of the 2nd ASCE Conference on Electronic Computation*, Pittsburgh, Pa., September 1960.

Cook, D., *Concepts and Applications of Finite Element Analysis*, Wiley, New York, 1981.

Hughes, T. J. R., *The Finite Element Method: Linear Static and Dynamic Finite Element Analysis*, Prentice Hall, Englewood Cliffs, N.J., 1987.

Zienkiewicz, O. C., and Taylor, R. L., *The Finite Element Method*, Vols. 1 and 2, 4th ed., McGraw-Hill, London, 1989.

PROBLEMS

Part A

18.1 A chimney of height L has been idealized as a cantilever beam with mass per unit length varying linearly from m at the base to $m/2$ at the top, and with second moment of cross-sectional area varying linearly from I at the base to $I/2$ at the top. Estimate the first two natural frequencies and modes of lateral vibration of the chimney using the shape functions $\psi_1(x) = 1 - \cos(\pi x/2L)$ and $\psi_2(x) = 1 - \cos(3\pi x/2L)$, where x is measured from the base.

18.2 A simply supported beam of length L has constant flexural rigidity EI and a mass distribution

$$m(x) = \begin{cases} 2m_o(x/L) & 0 \le x \le L/2 \\ 2m_o(1 - x/L) & L/2 \le x \le L \end{cases}$$

Determine the first two natural frequencies and modes of symmetric vibration of this nonuniform beam by the Rayleigh–Ritz method using two modes of the uniform beam as the shape functions.

Part B

*18.3 Determine the natural vibration frequencies and modes of a simply supported uniform beam, idealized as an assemblage of two finite elements (Fig. P18.3), using **(a)** the consistent-mass matrix, and **(b)** the lumped-mass matrix. Compare these results with the exact solutions obtained in Example 17.1.

---

*Denotes that a computer is necessary to solve this problem.

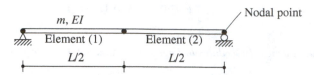

Figure P18.3

18.4 Determine the natural vibration frequencies and modes of a uniform beam clamped at both ends, idealized as an assemblage of two finite elements (Fig. P18.4), using (a) the consistent-mass matrix, and (b) the lumped-mass matrix. Compare these results with the exact solutions obtained by solving Problem 17.1.]

Figure P18.4

18.5 Determine the natural vibration frequencies and modes of a uniform beam clamped at one end and simply supported at the other, idealized as an assemblage of two finite elements (Fig. P18.5), using (a) the consistent-mass matrix, and (b) the lumped-mass matrix. Compare these results with the exact solutions obtained by solving Problem 17.2.

Figure P18.5

*18.6 Figure P18.6 shows a one-story, one-bay frame with mass per unit length and second moment of cross-sectional area given for each member. The frame, idealized as an assemblage of three finite elements, has the three DOFs shown if axial deformations are neglected in all elements.
(a) Using influence coefficients, formulate the stiffness matrix and the consistent-mass matrix. Express these matrices in terms of m, EI, and h.
(b) Determine the natural vibration frequencies and modes of the frame; express rotations in terms of h. Sketch the modes showing translations and rotations of the nodes.

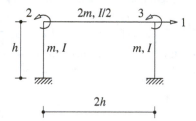

Figure P18.6

18.7 Repeat Problem 18.6 using the lumped-mass approximation. Comment on the effects of mass lumping on the vibration properties.

18.8 Repeat part (a) of Problem 18.6 starting with the stiffness and mass matrices for each element and using the direct assembly procedure of Section 18.5.

*Denotes that a computer is necessary to solve this problem.

PART III

Earthquake Response, Design, and Evaluation of Multistory Buildings

755

19

Earthquake Response of Linearly Elastic Buildings

PREVIEW

Developed in Chapter 13 were two procedures—response history analysis (RHA) and response spectrum analysis (RSA)—for calculating the earthquake response of any structure described as a linearly elastic system with a finite number of degrees of freedom. Determined by RSA, the earthquake response of multistory buildings to excitations characterized by a design spectrum is presented in this chapter for a wide range of the two key parameters: fundamental natural vibration period and beam-to-column stiffness ratio. Based on these results, we develop an understanding of how these parameters affect the earthquake response of buildings and how they affect the relative response contributions of the different natural vibration modes. These results also enable us to identify the conditions under which the first mode or the first two modes are sufficient to provide a useful approximation to the total response. The understanding we develop of the significance of the higher modes in building response will be useful in Chapter 22, where we evaluate the equivalent static force procedure in seismic building codes in light of the results of dynamic analyses.

19.1 SYSTEMS ANALYZED, DESIGN SPECTRUM, AND RESPONSE QUANTITIES

19.1.1 Systems Analyzed

The systems analyzed are single-bay, five-story frames with constant story height $= h$ and bay width $= 2h$ (Fig. 19.1.1). All the beams have the same flexural rigidity, EI_b, and the

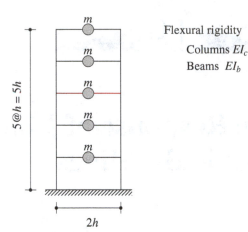

Flexural rigidity

Columns EI_c

Beams EI_b

Figure 19.1.1   Properties of uniform five-story frames.

column rigidity, EI_c, does not vary with height. The building is idealized as a lumped-mass system with the same mass m at all the floor levels. The damping ratio for all five natural vibration modes is assumed to be 5%.

Only two additional parameters are needed to define the system completely: the fundamental natural vibration period T_1 and the *beam-to-column stiffness ratio ρ*. The latter parameter is based on the properties of the beams and columns in the story closest to the midheight of the frame:

$$\rho = \frac{\sum_{\text{beams}} EI_b/L_b}{\sum_{\text{columns}} EI_c/L_c} \tag{19.1.1}$$

where L_b and L_c are the lengths of the beams and columns and the summations include all the beams and columns in the midheight story. For the uniform, one-bay frame defined in the preceding paragraph, Eq. (19.1.1) reduces to

$$\rho = \frac{I_b}{4I_c} \tag{19.1.2}$$

which was introduced in Section 1.3 for a one-story frame. This parameter is a measure of the relative beam-to-column stiffness and indicates how much the system may be expected to behave as a frame. For $\rho = 0$ the beams impose no restraint on joint rotations, and the frame behaves as a flexural beam (Fig. 19.1.2a). For $\rho = \infty$ the beams restrain the joint rotations completely, and the structure behaves as a shear beam with double-curvature bending of the columns in each story (Fig. 19.1.2c). An intermediate value of ρ represents a frame in which beams and columns undergo bending deformation with joint rotation (Fig. 19.1.2b). As an example for the frame of Fig. 19.1.1, $\rho = \frac{1}{8}$ represents $I_b = I_c/2$, which implies a frame with columns stiffer than the beams, typical of earthquake-resistant construction.

The parameter ρ controls several properties of the frame: the fundamental natural period, the relative closeness or separation of the natural periods, and the shapes of the natural modes. These vibration properties of the frame of Fig. 19.1.1 are calculated by

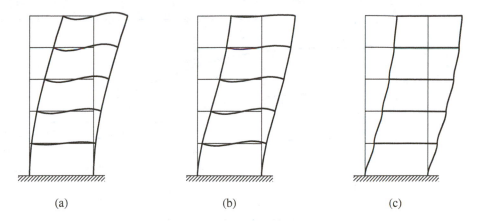

(a)   (b)   (c)

Figure 19.1.2   Deflected shapes: (a) $\rho = 0$; (b) $\rho = \frac{1}{8}$; (c) $\rho = \infty$.

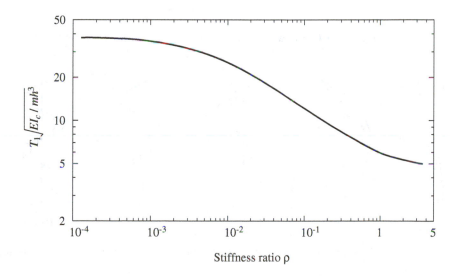

Stiffness ratio ρ

Figure 19.1.3   Fundamental natural vibration period of uniform five-story frames.

the procedures of Chapters 9 and 10. The variation of the fundamental period with ρ is shown in Fig. 19.1.3, which indicates that for a fixed column stiffness EI_c and floor mass m, the fundamental period is reduced by a factor of over 8 as ρ increases from 0 to ∞. The ratios of the natural periods are independent of T_1 but depend strongly on ρ, especially the higher-mode periods, as shown in Fig. 19.1.4. As a result, the natural periods of a frame with small ρ are more separated from each other than if ρ is large. The shapes of the natural modes depend significantly on ρ, as shown in Fig. 19.1.5. It is clear from these results that the stiffness ratio ρ must have great importance in determining the dynamic (and static) behavior of the frame. We demonstrate this in the sections that follow.

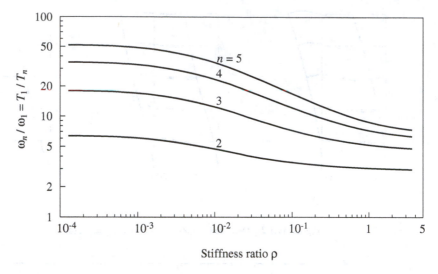

Figure 19.1.4 Natural vibration period ratios for uniform five-story frames. (After Roehl, 1971.)

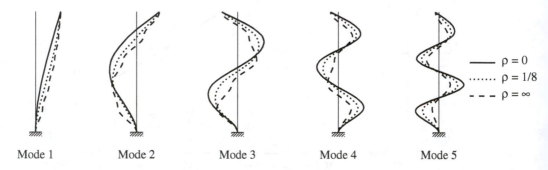

Figure 19.1.5 Natural vibration modes of uniform five-story frame for three values of ρ.

The fundamental period T_1 will be varied over a range that is much wider than is reasonable for five-story frames. However, this is appropriate for the objectives of this chapter, where we are studying the influence of T_1 on the response of buildings. The response behavior is controlled by T_1 primarily and affected only secondarily by the number of stories. Hence the observations we glean from the results presented are not restricted to five-story buildings.

19.1.2 Design Spectrum

The earthquake excitation is characterized by the design spectrum of Fig. 19.1.6 (identical to Fig. 6.9.5), multiplied by 0.5, so that it applies to ground motions with a peak ground acceleration $\ddot{u}_{go} = 0.5g$, velocity $\dot{u}_{go} = 24$ in./sec, and displacement $u_{go} = 18$ in. In the

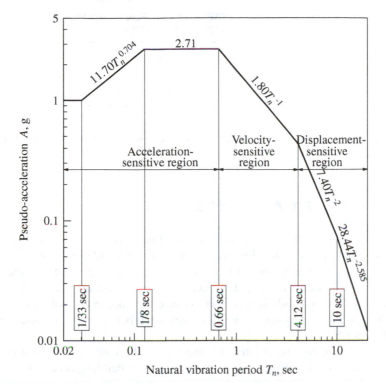

Figure 19.1.6 Design spectrum for ground motions with $\ddot{u}_{go} = 1g$, $\dot{u}_{go} = 48$ in./sec, and $u_{go} = 36$ in.; $\zeta = 5\%$.

design spectrum shown (Fig. 19.1.6) for 5% damping, the acceleration-sensitive, velocity-sensitive, and displacement-sensitive regions (defined in Chapter 6) are identified.

19.1.3 Response Quantities

The peak values of the responses of a frame described in Section 19.1.1 with specified T_1 and ρ to ground motion characterized by the design spectrum of Fig. 19.1.6 are determined by the RSA procedure. Such analyses were repeated for three values of ρ—0, $\frac{1}{8}$, and ∞—and many values of T_1. Among the many response quantities, we will examine four of them: top-floor displacement u_5 relative to the ground, base shear V_b, base overturning moment M_b, and top-story shear V_5. The first three will be normalized as follows: (1) u_5/u_{go}, where u_{go} is the peak ground displacement; (2) V_b/W_1^*, where $W_1^* = M_1^* g$ and M_1^* is the effective modal mass for the first mode; and (3) $M_b/W_1^* h_1^*$, where h_1^* is the effective modal height for the first mode. The values of W_1^* and h_1^* are computed using Eq. (13.2.9) and the shape of the first mode (Fig. 19.1.5). Presented in Table 19.1.1 are (1) W_1^*/W, where W is the total weight of the frame, and (2) $h_1^*/5h$, where $5h$ is the total

TABLE 19.1.1   FUNDAMENTAL MODE PROPERTIES

	$\rho = 0$	$\rho = \frac{1}{8}$	$\rho = \infty$
W_1^*/W	0.679	0.796	0.880
$h_1^*/5h$	0.794	0.742	0.703

height of the frame. It is clear that W_1^* and h_1^* depend on the beam-to-column stiffness ratio ρ.

19.2 INFLUENCE OF T_1 AND ρ ON RESPONSE

Shown in Fig. 19.2.1 are three normalized responses of the frame plotted against its fundamental period T_1 for three values of ρ. Over a wide range of T_1 values, the top-floor displacement varies very little with ρ (i.e., it is not sensitive to variations in the beam-to-column stiffness ratio). For very-long-period systems the top-floor displacement approaches the ground displacement because the floor masses of such a system remain stationary while the ground beneath moves; such behavior of a one-story frame is shown in Fig. 6.8.5.

The shear and overturning moment at the base of the frame are of special interest because their design values are specified in building codes; they are also the forces

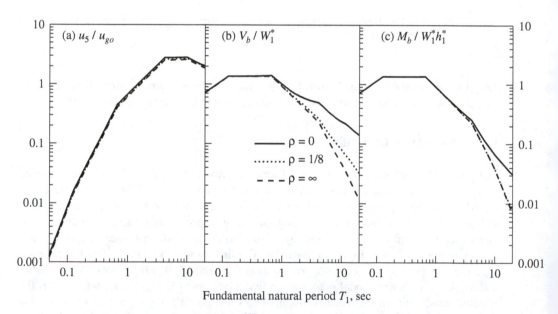

Figure 19.2.1   Normalized values of top-floor displacement u_5, base shear V_b, and base overturning moment M_b in uniform five-story frames for three values of ρ.

needed in the design of the foundation system. When these normalized forces are plotted against T_1, as in Fig. 19.2.1, the curves have the general appearance of the pseudo-acceleration spectrum of Fig. 19.1.6. Thus the individual curves tend to the peak ground acceleration, 0.5g, for short T_1, and to zero for long T_1, as does the pseudo-acceleration spectrum.

The normalized base shear and base overturning moment vary significantly with ρ for buildings with T_1 in the velocity- or displacement-sensitive regions of the spectrum, with the variation in M_b not as great as in V_b. However, these normalized responses do not vary appreciably with ρ in the acceleration-sensitive region of the spectrum. Observe that W_1^* and h_1^* themselves vary with ρ, as shown in Table 19.1.1, and hence they influence the values of the actual (in contrast to normalized) responses V_b and M_b and how they depend on ρ.

The variation in the normalized responses with ρ is closely related to the significance of the response contributions of the second and higher modes, which generally increase with decreasing ρ (Section 19.5) and—for the design spectrum selected—generally increase with increasing T_1 (Section 19.4). To study individual modal responses, we use the modal contribution factors introduced in Chapter 12, Part C.

19.3 MODAL CONTRIBUTION FACTORS

The peak value of the nth-mode contribution to a response quantity r is given by Eq. (13.7.1), repeated here for convenience:

$$r_n = r_n^{\text{st}} A_n \tag{19.3.1}$$

where A_n is the ordinate of the pseudo-acceleration response (or design) spectrum corresponding to natural period T_n and damping ratio ζ_n of the nth mode; and r_n^{st} is the modal static response. As defined in Section 13.2.2, r_n^{st} is the static value of response quantity r due to external forces $\mathbf{s}_n$, given by Eq. (13.2.4). These modal static responses are presented in Table 13.2.1 and repeated here for base shear V_b, top-story shear V_5, base overturning moment M_b, and top-floor displacement u_5:

$$V_{bn}^{\text{st}} = M_n^* \qquad V_{5n}^{\text{st}} = \Gamma_n m \phi_{5n} \qquad M_{bn}^{\text{st}} = h_n^* M_n^* \qquad u_{5n}^{\text{st}} = \frac{\Gamma_n}{\omega_n^2} \phi_{5n} \tag{19.3.2}$$

Alternatively, Eq. (19.3.1) can be expressed as

$$r_n = r^{\text{st}} \bar{r}_n A_n \tag{19.3.3}$$

where

$$r^{\text{st}} = \sum_{n=1}^{N} r_n^{\text{st}} \quad \text{and} \quad \bar{r}_n = \frac{r_n^{\text{st}}}{r^{\text{st}}} \tag{19.3.4}$$

As demonstrated in Section 12.10, r^{st} is also the static value of r due to external forces $\mathbf{s} = \mathbf{m1}$. The modal contribution factors $\bar{r}_n$ for a response quantity r are dimensionless, are independent of how the modes are normalized, and add up to unity when summed over all modes:

$$\sum_{n=1}^{N} \bar{r}_n = 1 \tag{19.3.5}$$

We first study how the modal contribution factors depend on the beam-to-column stiffness ratio ρ and on the response quantity. For the four response quantities mentioned in Section 19.1.3, the modal contribution factors $\bar{r}_n$ are calculated from Eqs. (19.3.2) and (19.3.4) using the known system properties and computed natural frequencies and modes. The results presented in Tables 19.3.1a and 19.3.1b for $\rho = 0$, $\frac{1}{8}$, and ∞ are independent of T_1. Consistent with Eq. (19.3.5) for each response quantity and each ρ, the sum of modal contribution factors over all modes is unity, although the convergence may or may not be monotonic. For the class of structures considered, the convergence is monotonic for base shear, but not for the other response quantities. The data of Tables 19.3.1a and 19.3.1b permit three useful observations that have a bearing on relative values of the modal responses.

1. For a fixed value of ρ and each of the response quantities, the modal contribution factor $\bar{r}_1$ for the first mode is larger than the factors $\bar{r}_n$ for the higher modes, suggesting that the fundamental mode should have the largest contribution to each of these responses.

2. For a fixed value of ρ, the absolute values of $\bar{r}_n$ for the second and higher modes are larger for V_5 than for V_b, and the values for V_b in turn are larger than those for M_b and u_5. This observation suggests that the second- and higher-mode response contributions are expected to be more significant for base shear V_b than for the base overturning moment M_b or top-floor displacement u_5. Among the story shears the higher-mode responses should be more significant for the fifth-story shear than for the base shear.

3. As ρ decreases, the absolute values of the higher-mode contribution factors $\bar{r}_n$ for V_5, V_b, and M_b, increase (but for minor exceptions), especially in the second mode. This observation suggests that the higher-mode contributions to any of these forces should become a larger fraction of the total response as ρ decreases and should be largest for a flexural beam with $\rho = 0$.

TABLE 19.3.1a MODAL CONTRIBUTION FACTORS FOR V_b AND V_5

	Base Shear V_b			Top-Story Shear V_5		
Mode	$\rho = 0$	$\rho = \frac{1}{8}$	$\rho = \infty$	$\rho = 0$	$\rho = \frac{1}{8}$	$\rho = \infty$
1	0.679	0.796	0.879	1.38	1.30	1.25
2	0.206	0.117	0.087	−0.528	−0.441	−0.362
3	0.070	0.051	0.024	0.204	0.211	0.159
4	0.033	0.026	0.007	−0.080	−0.089	−0.063
5	0.012	0.009	0.002	0.020	0.023	0.015

TABLE 19.3.1b   MODAL CONTRIBUTION FACTORS FOR M_b AND u_5

	Base Overturning Moment M_b			Top-Floor Displacement u_5		
Mode	$\rho = 0$	$\rho = \frac{1}{8}$	$\rho = \infty$	$\rho = 0$	$\rho = \frac{1}{8}$	$\rho = \infty$
1	0.898	0.985	1.030	1.009	1.027	1.030
2	0.078	−0.003	−0.035	−0.009	−0.030	−0.035
3	0.016	0.014	0.006	0.0005	0.003	0.006
4	0.006	0.003	−0.001	−0.00005	−0.0005	−0.001
5	0.002	0.001	0.0003	0.000005	0.00007	0.0003

19.4 INFLUENCE OF T_1 ON HIGHER-MODE RESPONSE

In this section we use the preceding concepts and data to predict how the modal response contributions depend on the fundamental natural period T_1 of the structure. For this purpose we examine the three factors that enter into Eq. (19.3.3) for the peak modal response: (1) The static value r^{st} of r is a common factor in all modal responses and therefore does not influence the relative values of the modal responses. (2) As mentioned in Section 19.3, for a fixed ρ the modal contribution factors $\bar{r}_n$ are independent of T_1. (3) The pseudo-acceleration spectrum ordinate A_n is the only factor in Eq. (19.3.3) that depends on the fundamental period T_1 and period ratios T_n/T_1, which, for a fixed ρ, do not depend on T_1 (see Section 19.1). Thus the variation in higher-mode response with increasing T_1 must be related to the shape of the design spectrum. This is illustrated for the selected design spectrum in parts (a) and (b) of Fig. 19.4.1, wherein the natural periods T_n of two shear frames ($\rho = \infty$) with fundamental natural periods $T_1 = 0.5$ and 3.0 sec, respectively, are identified. For the building with $T_1 = 3$ sec, the A_n values for the higher modes are larger than A_1 for the fundamental mode, whereas for the building with $T_1 = 0.5$ sec, the A_n ($n \geq 2$) values are either equal to or smaller than A_1. Thus the higher-mode response, expressed as a percentage of the total response, should be larger for the building with $T_1 = 3$ sec than for the $T_1 = 0.5$ sec building. In general, for the spectrum selected, as T_1 increases within the velocity- and displacement-sensitive regions of the spectrum, the higher-mode response will become an increasing percentage of the total response.

This prediction is confirmed by the results of dynamic analysis. The peak values of the responses of a frame with specified T_1 and ρ are determined by considering (1) all five modes, and (2) only the first mode. Such analyses were repeated for three values of ρ—0, $\frac{1}{8}$, and ∞—and many values of T_1. The results for normalized base shear are plotted in Fig. 19.4.2. The one-mode curves are independent of ρ and identical to the design spectrum of Fig. 19.1.6 because Eqs. (19.3.1) and (19.3.2a) give

$$V_{b1} = \frac{A_1}{g} W_1^* \quad \text{or} \quad \frac{V_{b1}}{W_1^*} = \frac{A_1}{g} \tag{19.4.1}$$

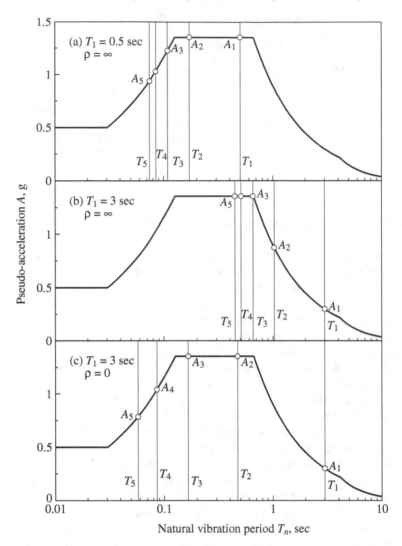

Figure 19.4.1   Natural periods and spectral ordinates for three cases: (a) $T_1 = 0.5$ sec, $\rho = \infty$; (b) $T_1 = 3$ sec, $\rho = \infty$; and (c) $T_1 = 3$ sec, $\rho = 0$.

The difference between the two results for the peak value is the higher-mode response (i.e., the combined response due to all modes higher than the first mode). The higher-mode response, expressed as a percentage of the total response, is presented in Fig. 19.4.3 for the four response quantities. The data for base shear are obtained from the results of Fig. 19.4.2, and those for other response quantities are obtained similarly. The higher-mode response of buildings is negligible for T_1 in the acceleration-sensitive region of the spectrum and increases with increasing T_1 in the velocity- and displacement-sensitive

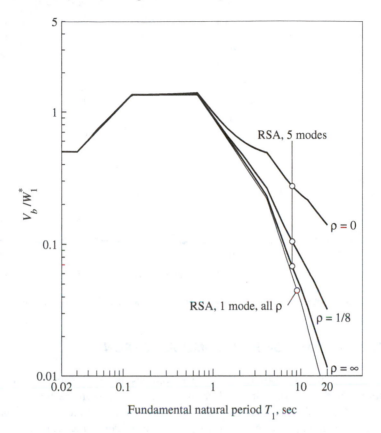

Figure 19.4.2   Normalized base shear in uniform five-story frames for three values of ρ. Results were obtained by response spectrum analysis (RSA), including one or five modes.

regions. Such results are useful in evaluating the lateral force provisions in building codes (Chapter 22).

Based on modal contribution factors, we had predicted earlier how the significance of the higher-mode response would depend on the response quantity (Section 19.3). This prediction is confirmed by the results of dynamic analysis; Fig. 19.4.3 demonstrates:

1. The higher-mode response is more significant for forces (e.g., V_5, V_b, and M_b) than for displacements (e.g., u_5). However, the higher-mode contributions to u_5 and M_b are identical for shear frames (Fig. 19.4.3c) because the modal contribution factors are identical (Table 19.3.1b).

2. The higher-mode response is more significant for base shear than for base overturning moment.

3. The higher-mode response is more significant for top-story shear than for base shear.

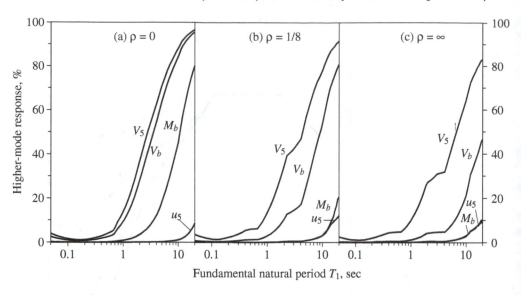

Figure 19.4.3   Higher-mode response in V_b, V_5, M_b, and u_5 for uniform five-story frames for three values of ρ.

19.5 INFLUENCE OF ρ ON HIGHER-MODE RESPONSE

In this section we predict how the modal response contributions depend on the beam-to-column stiffness ratio ρ. For this purpose we examine three factors that enter into Eq. (19.3.3) for the peak modal response: (1) The static value r^{st} is a common factor in all modal responses and therefore does not influence the relative values of the modal responses. (2) As ρ decreases, the higher-mode contribution factors $\bar{r}_n$ for the base shear and top-story shear increase especially in the second mode (Table 19.3.1a). (3) The pseudo-acceleration ordinates depend on T_1 and on T_1/T_n; the latter becomes larger as ρ decreases (Fig. 19.1.4) and the T_n values are spread out over a wider period range of the design spectrum. This is illustrated in parts (b) and (c) of Fig. 19.4.1. Both frames have the same fundamental period, $T_1 = 3$ sec, but they differ in ρ—one is a shear beam ($\rho = \infty$), and the other a flexural beam ($\rho = 0$). As a result, the ratio A_2 for the second mode—generally the most significant of the higher modes—to A_1 for the first mode is larger for buildings with $\rho = 0$ than for the $\rho = \infty$ case. Thus, putting the second and third observations together, both the modal contribution factor $\bar{r}_n$ and the spectral ordinate A_n for the second mode are larger for the $\rho = 0$ frame; therefore, the higher-mode response is more significant in this case than for the frame with $\rho = \infty$. In general, for the design spectrum selected and for T_1 within the velocity- and displacement-sensitive regions of the spectrum, the ratio A_n/A_1 increases (or more precisely, does not decrease) with decreasing ρ, and this trend should lead to increased higher-mode response.

    This prediction is confirmed by the results of dynamic analyses in Fig. 19.4.3, which demonstrate that for each response quantity the higher-mode response is least significant for systems behaving like shear beams ($\rho = \infty$), becomes increasingly significant as ρ decreases, and is largest for systems deforming like flexural beams ($\rho = 0$).

19.6 HEIGHTWISE VARIATION OF HIGHER-MODE RESPONSE

In this section we examine how the response contributions of modes higher than the first mode to the story shears and floor overturning moments vary over building height. Compared in Figs. 19.6.1 and 19.6.2 are these forces due to the first mode only and the total forces considering all five modes. The higher mode response, given by the difference between the two sets of forces, is expressed as a percentage of the total force and presented in Figs. 19.6.3 and 19.6.4. These results indicate, as before, that (1) for a fixed ρ, the higher-mode response is more significant for longer-period buildings, and (2) for a fixed T_1, the higher-mode response is more significant for frames with smaller ρ.

The results presented in Figs. 19.6.3 and 19.6.4 provide information on how the higher modes affect the forces in different stories of a building. For a particular building with fixed values of T_1 and ρ, the percentage contribution of the higher-mode response tends to increase as one moves up the building, although the trend is not perfect in all cases. This contribution is small in the overturning moment at the base but can be significant in the upper stories, especially for frames with longer-period T_1 and smaller ρ. While

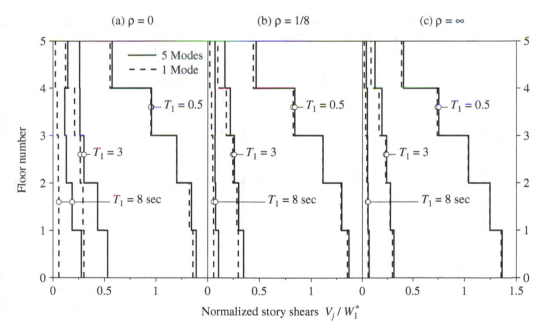

Figure 19.6.1 Normalized story shears in uniform five-story frames for three values of ρ and three values of T_1. Results were obtained by RSA, including one or five modes.

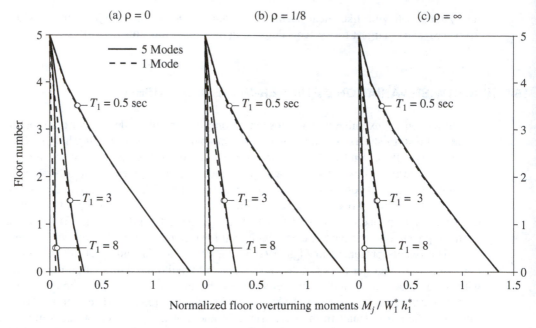

Figure 19.6.2 Normalized floor overturning moments in uniform five-story frames for three values of ρ and three values of T_1. Results were obtained by RSA, including one or five modes.

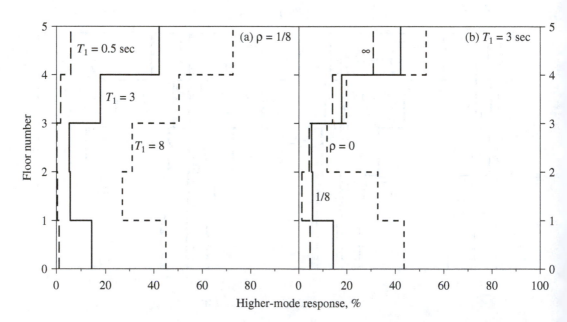

Figure 19.6.3 Higher-mode response in story shears for uniform five-story frames.

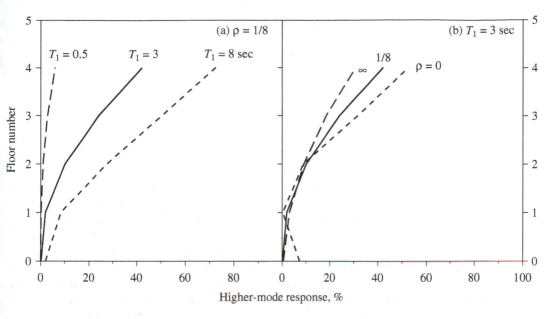

Figure 19.6.4 Higher-mode response in floor overturning moments for uniform five-story frames.

the above-noted trend is essentially consistent for story overturning moments, it is not always consistent for story shears because the higher-mode response tends to increase for the forces in the stories near the bottom of the building in addition to the stories near the top of the building. This lack of perfect consistency in the above-noted trend is an indication of the complex dependence of the earthquake response of buildings on the system parameters and on the earthquake excitation.

19.7 HOW MANY MODES TO INCLUDE

The response contributions of all the natural modes of vibration must be included if the "exact" value of the structural response to earthquake excitation is desired, but the first few modes can usually provide sufficiently accurate results. The number of modes to be included depends on two factors, modal contribution factor $\bar{r}_n$ and spectral ordinate A_n, that enter into the modal response equation (19.3.3).

Recall the important result that the sum of the modal contribution factors over all the modes is unity, Eq. (19.3.5). If only the first J modes are included, the error in the static response is

$$e_J = 1 - \sum_{n=1}^{J} \bar{r}_n \tag{19.7.1}$$

We first examine the error e_2 in static response if two modes are included. Table 19.7.1 shows this error computed from Eq. (19.7.1) and numerical values for the modal

TABLE 19.7.1 $e_2 = 1 - \sum_{n=1}^{2} \bar{r}_n$

Response	$\rho = 0$	$\rho = \frac{1}{8}$	$\rho = \infty$
V_5	0.144	0.144	0.110
V_b	0.115	0.086	0.033
M_b	0.024	0.018	0.005
u_5	0.0004	0.003	0.005

contribution factors in Table 19.3.1. The error e_2 is below 0.15 or 15% for the four response quantities. For a fixed ρ the error varies with the response quantity. It is smaller in the base overturning moment M_b relative to the base shear V_b, and in V_b compared to the top-story shear V_5. The error is much smaller for the top-floor displacement u_5, and it is less than 3% if the first mode alone is considered. For a particular response quantity the error e_2 varies with ρ, being smallest for $\rho = \infty$ (i.e., shear beams) and largest for $\rho = 0$ (i.e., flexural beams). The top-floor displacement displays trends opposite to the forces in the sense that e_2 increases with increasing ρ, but e_2 is so small that higher modes are of little consequence in displacements (Fig. 19.4.3). These data suggest that the first one or two modes may provide a good approximation to the total response, with the accuracy depending on the response quantity and on ρ.

We next examine how the spectral ordinates A_n influence the number of modes that should be included in the analysis. For a fixed ρ and T_1 in the velocity- or displacement-sensitive regions of the spectrum, the ratio A_n/A_1 is larger for frames with longer fundamental period T_1 (Fig. 19.4.1a and b). Thus for the same desired accuracy, more modes should be included in the analysis of buildings with longer T_1 than the number of modes necessary for shorter-period buildings. For a fixed T_1 in the velocity-sensitive or displacement-sensitive regions of the spectrum, the ratio A_n/A_1 is larger for frames with smaller ρ (Fig. 19.4.1b and c). Thus, for the same desired accuracy, more modes should be included in the analysis of buildings with smaller ρ compared to the number of modes necessary for buildings with larger ρ; in particular, more modes should be included in the analysis of flexural frames ($\rho = 0$) than for shear frames ($\rho = \infty$).

These expectations regarding how T_1 and ρ influence the number of modes that should be included in earthquake response analysis are confirmed by the results of Fig. 19.7.1, where, for each ρ value, five response curves for base shear are identified by indicating the number of modes included in the analysis. It is clear that the first two modes provide a reasonably accurate value for the base shear in frames with T_1 in the velocity-sensitive region of the spectrum, and one mode is sufficient in the acceleration-sensitive region. This conclusion is also valid for shears in all the stories and overturning moments at all floors. The first mode alone provides accurate results for u_5 over the entire range of T_1, and for all ρ values, as indicated by Fig. 19.4.3.

In light of the preceding observations, it is instructive to examine the 90% rule for participating mass specified in some building codes. Because the effective modal mass is equal to the modal static response V_{bn}^{st} for base shear (Section 13.2.5), the prior rule implies that enough—say J—modes should be included so that e_J for base shear is less than 10%.

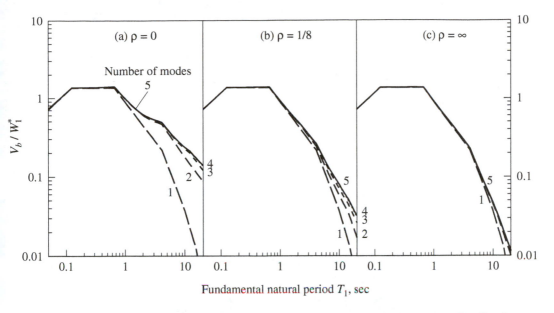

V_b / W_1^*

Fundamental natural period T_1, sec

Figure 19.7.1   Normalized base shear in uniform five-story frames for three values of ρ. Results were determined by RSA considering one, two, three, four, or five modes.

However, as noted earlier, e_J varies with the response quantity, and therefore this error may exceed 10% for other response quantities such as shears in upper stories and bending moments and shears in some structural elements. Furthermore, even if the error e_J in the static response is less than 10%, the error in the dynamic response may exceed 10% for buildings with longer T_1 and smaller ρ.

FURTHER READING

Blume, J. A., "Dynamic Characteristics of Multistory Buildings," *Journal of the Structural Division, ASCE,* **94**, (ST2), February 1968, pp. 337–402.

Cruz, E. F., and Chopra, A. K., "Elastic Earthquake Response of Building Frames," *Journal of Structural Engineering, ASCE,* **112**, 1986, pp. 443–459.

Cruz, E. F., and Chopra, A. K., "Simplified Procedures for Earthquake Analysis of Buildings," *Journal of Structural Engineering, ASCE,* **112**, 1986, pp. 461-480.

Roehl, J. L., "Dynamic Response of Ground-Excited Building Frames," Ph.D. thesis, Rice University, Houston, Tex., October 1971.

20

Earthquake Analysis and Response of Inelastic Buildings

PREVIEW

As mentioned in Chapter 7, most buildings are expected to deform beyond the limit of linearly elastic behavior when subjected to strong ground shaking. Thus the earthquake response of buildings deforming into their inelastic range is of central importance in earthquake engineering. This chapter covers selected aspects of this vast subject, organized into two parts.

Rigorous nonlinear response history analysis (RHA) is discussed in Part A. Mentioned are the governing equations of motion and differences in methodologies for solving these equations for inelastic MDF systems compared to elastic systems. Next, we demonstrate that the inelastic response of buildings is strongly influenced by assumptions in idealizing or modeling the structure, by second-order $P-\Delta$ effects of gravity loads acting on the laterally deformed state of the structure, and by the detailed variation of ground motion with time. These factors are much more influential on the response of structures deforming into their inelastic range than for those remaining elastic. Then we demonstrate that the story ductility demands and their heightwise variation depend on the relative yield strengths of beams versus columns and on the relative yield strengths of various stories. Identified next are the differences between the ductility demands imposed by earthquake excitation on multistory buildings and on an SDF system, both designed for the same base shear, and culminating in a quantitative discussion of the increase in strength necessary to limit the story ductility demands in a multistory building below the SDF-system ductility factor.

Recognizing that at the present time, nonlinear RHA is an onerous task for several reasons and an unreasonable requirement for every building—no matter how simple—and of every structural engineering office—no matter how small—approximate analysis procedures are developed in Part B of the chapter. Utilizing the modal expansion of the effective earthquake forces, two procedures for approximate analysis of inelastic buildings are developed: uncoupled modal response history analysis (UMRHA) and modal pushover analysis (MPA). Not intended for practical application, the UMRHA procedure is developed only to provide a rationale for the MPA procedure. In the latter procedure, the seismic demands due to individual terms in the modal expansion of the effective earthquake forces are determined by nonlinear static (or pushover) analyses using the modal inertia force distributions, and the peak "modal" responses are combined by modal combination rules to estimate the total response. The principal approximations in the MPA procedure are identified and the accuracy of the procedure is evaluated by comparing the estimated demands against results of nonlinear RHA for several buildings. Finally, MPA is simplified for practical application to estimate seismic demands for evaluation of existing buildings.

PART A: NONLINEAR RESPONSE HISTORY ANALYSIS

20.1 EQUATIONS OF MOTION: FORMULATION AND SOLUTION

The resisting force term in the equations of motion for an elastic multistory building [Eq. (13.2.1)] is modified to recognize the inelastic behavior of the building. The force–deformation relation for each structural member undergoing cyclic deformations is now nonlinear and hysteretic. The initial loading curve is nonlinear at larger amplitudes of deformation, and the unloading and reloading curves differ from the initial loading branch. Experiments on structural components have provided force–deformation relations appropriate for various types of structural elements (beams, columns, walls, braces, etc.) using a variety of structural materials (steel, reinforced concrete, masonry, wood, etc.) (see Fig. 7.1.1).

For inelastic systems the nonlinear relationship between lateral forces $\mathbf{f}_S$ at the N floor levels and resulting lateral floor displacements $\mathbf{u}$ is path dependent, i.e., it depends on whether deformation is increasing or decreasing during the time step:

$$\mathbf{f}_S = \mathbf{f}_S(\mathbf{u}) \tag{20.1.1}$$

With this generalization for inelastic systems, Eq. (13.2.1) becomes

$$\mathbf{m}\ddot{\mathbf{u}} + \mathbf{c}\dot{\mathbf{u}} + \mathbf{f}_S(\mathbf{u}) = -\mathbf{m}\iota\ddot{u}_g(t) \tag{20.1.2}$$

where $\mathbf{m}, \mathbf{c}$, and ι are as defined in Section 13.2. This matrix equation represents N non-linear differential equations for the N floor displacements $u_j(t)$, $j = 1, 2, \ldots, N$. Given the structural mass matrix $\mathbf{m}$, damping matrix $\mathbf{c}$, inelastic force–deformation relation $\mathbf{f}_S(\mathbf{u})$, and ground acceleration $\ddot{u}_g(t)$, nonlinear response history analysis requires numerical solution of Eq. (20.1.2) to obtain the displacement response of the structure, and internal forces can be determined from the displacements.

Formulation of the nonlinear differential equations, in particular the $\mathbf{f}_S(\mathbf{u})$ term, is computationally demanding. The structural stiffness matrix must be reformulated at each time instant from the element tangent stiffness matrices corresponding to the present deformation and its path dependence—whether it is on the initial loading, unloading, or reloading branches of the element force–deformation relation—and for a large structure this process must be repeated for thousands of structural elements. In formulating these equations, nonlinear geometry should be considered because structures subjected to intense ground motions may undergo large displacements. In earthquake engineering, the nonlinear equilibrium and compatibility relations are generally approximated by an approach referred to as $P{-}\Delta$ *analysis*. Detailed formulation of the governing equations is beyond the scope of this book, and the reader is referred to other sources (e.g., Filippou and Fenves, 2004).

Numerical solution of Eq. (20.1.2) is computationally demanding for large (number of DOFS) inelastic systems because these coupled differential equations must be solved simultaneously; for inelastic systems they cannot be uncoupled by transforming to modal coordinates, as will be demonstrated later. Such numerical solutions must be repeated at every time step Δt, which must be very short, short enough to ensure that the numerical procedure converges, remains stable, and gives accurate results. In Chapter 16 we developed selected numerical methods that are commonly used in earthquake engineering to solve these nonlinear differential equations.

20.2 COMPUTING SEISMIC DEMANDS: FACTORS TO BE CONSIDERED

Several factors should be recognized to obtain meaningful results for the inelastic response of a structure. Three of these factors—$P{-}\Delta$ effects, structural modeling (or idealization) assumptions, and ground motion characteristics—are discussed in this section based on results presented for a perimeter frame of the SAC–Los Angeles 20-story steel moment-resisting frame building.[†] The computed values of response quantities—floor displacements, story drifts, and plastic hinge rotations—represent the demands imposed on the structure by the design earthquake.

20.2.1 $P{-}\Delta$ Effects

The second-order effect of the downward gravity loads acting on the laterally deformed state of a structure, known as $P{-}\Delta$ *effects*, can profoundly influence the earthquake

[†]SAC was a joint venture of three nonprofit organizations: Structural Engineers Association of California (SEAOC), Applied Technology Council (ATC), and California Universities for Research in Earthquake Engineering (CUREE). Prompted by unexpected damage to the steel frames of many buildings during the 1994 Northridge earthquake, SAC was organized to conduct an extensive program of applied research. SAC commissioned three consulting firms to design 3-, 9-, and 20-story special moment-resisting frame buildings according to local code requirements in three cities: Los Angeles (UBC, 1994), Seattle (UBC, 1994), and Boston (BOCA, 1993). Square in plan, these buildings have identical properties in both lateral directions. Descriptions of their dimensions in plan and elevation, member sizes, and other properties are available in several publications (e.g., Gupta and Krawinkler, 1999). Perimeter frames of seven of these nine buildings are used as examples in this chapter. Ground motions selected for these examples are the 20 ground motion records assembled in the SAC project to represent 2% probability of exceedance in 50 years (a return period of 2475 years).

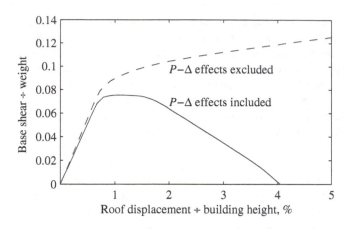

Figure 20.2.1   Pushover curves for the SAC–Los Angeles 20-story building with and without $P\text{–}\Delta$ effects. (From Gupta and Krawinkler, 2000b.)

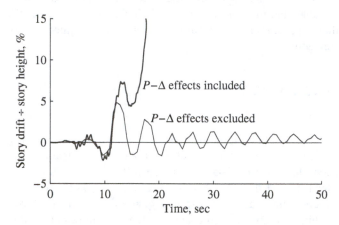

Figure 20.2.2   Importance of $P\text{–}\Delta$ effects on the second-story drift of the SAC–Los Angeles 20-story building due to LA30 ground motion. (Adapted from Gupta and Krawinkler, 2000b.)

response of buildings in their inelastic range. With or without these effects, Fig. 20.2.1 shows plots of the base shear V_b (normalized by the total weight W) against roof displacement (normalized by building height)—commonly known as a *pushover curve*—determined by nonlinear static analysis of the frame subjected to lateral forces with specified heightwise distribution that are gradually increased to push the building to large displacements. $P\text{–}\Delta$ effects reduce the initial elastic stiffness of a structure slightly and will therefore have little influence on the earthquake response of a structure if it remained elastic during the design ground motion. However, $P\text{–}\Delta$ effects have a profound influence on the postyield response, which now displays a short constant-strength plateau at a reduced yield strength, followed by a rapid decrease in lateral force resistance represented by the negative stiffness, culminating in zero lateral resistance at a roof displacement equal to 4% of the building height; in contrast, the postyield stiffness remains positive if $P\text{–}\Delta$ effects are ignored.

These profound differences in the postyield static behavior of a building suggest that $P\text{–}\Delta$ effects should also be important in the building's response to earthquake excitation. This expectation is confirmed in Fig. 20.2.2, where the response history of interstory drift or relative deformation (normalized by story height) in the second story of the building due to one of the SAC ground motions—the LA30 Tabas record—is presented for two cases: $P-\Delta$ effects included or excluded. When these effects are included, after the first episode of major yielding the story drift grows in one direction without any reversal toward the opposite lateral direction, resulting in dynamic instability. In contrast, analysis excluding these effects predicts oscillatory response that remains bounded. Clearly, it is essential to include $P-\Delta$ effects in predicting the earthquake response of buildings deforming significantly into their inelastic range.

20.2.2 Modeling Assumptions

The earthquake response of a building can be influenced significantly by the assumptions in modeling (or idealizing) the structure for computer analysis. To demonstrate this possibility, three different planar idealizations of the frame selected are considered: (1) model M1, a basic centerline model in which the panel zone size, strength, and flexibility are not represented; (2) model M2, a model that explicitly incorporates the strength and flexibility properties of panel zones; and (3) model M2A, an enhanced version of model M2, which includes the interior gravity columns, shear connections, and floor slabs.

The earthquake response of a building may be affected profoundly by the differences in these analytical models, as demonstrated by the response history of second-story drift due to the same ground motion (Fig. 20.2.3). The M1 model predicts that after the first large inelastic excursion, the story drift will not reverse direction and will continue to grow rapidly so that the building would become dynamically unstable within the first 20 sec of the excitation. This early instability does not occur in the M2 model, but the subsequent (after 20 sec) ground motion, although weaker, causes the drift to grow to near dynamic instability of the frame. When other sources of stiffness and strength are considered (model M2A), the response is radically different. After the first large inelastic excursion, the story drift now recovers partially and oscillates about a shifted position, showing no signs of dynamic instability of the structure. It is evident that dynamic response is extremely sensitive

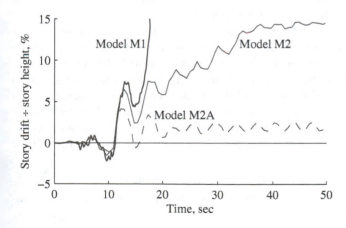

Figure 20.2.3   Influence of modeling assumptions on second-story drift in the SAC–Los Angeles 20-story building due to LA30 ground motion. (Adapted from Gupta and Krawinkler, 2000b.)

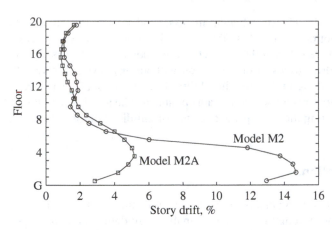

Figure 20.2.4 Influence of modeling assumptions on the story drift demands for the SAC–Los Angeles 20-story building due to LA30 ground motion; results are shown for models M2 and M2A, but model M1 predicted collapse of the building. (Adapted from Gupta and Krawinkler, 2000a.)

to modeling assumptions once P–Δ effects become important and a story is deformed into the range of negative postyield stiffness.

Consequently, the story drift demands for a building may be affected profoundly by the modeling assumptions. This is evident in Fig. 20.2.4, where the peak values of story drifts are presented for the same frame due to the same ground motion. No results are shown for model M1 because it predicted collapse of the building. Model M2 predicts story drifts approaching 15%, which are so large that the performance of the building would not be acceptable. However, the most realistic model (M2A) predicts much smaller story drifts, with the largest drift among all stories near 5%.

20.2.3 Response Variability with Ground Motion

The story drift demands are also sensitive to the time variation of ground acceleration, hence they vary from one ground motion to the next. This is evident from the peak values of story drifts in the M2 model of the same 20-story building due to 20 SAC ground motion records (Fig. 20.2.5). The story drift demands imposed by the 20 individual records vary widely, implying that the response to any one excitation should not be the basis for designing new buildings or evaluating existing buildings.

The seismic demands due to a large enough number of ground motions must be determined and their record-to-record variability considered in selecting demand values to be considered in the design and evaluation of structures. The median[†] and 84th percentile[‡] values of the story drift demands for the M2 model of the structure are presented in Fig. 20.2.6 for three ensembles of 20 ground motions for Los Angeles representing different probabilities of exceedance: 2% in 50 years (a return period of 2475 years), 10% in 50 years (a return period of 475 years), and 50% in 50 years (a return period of 72 years). As the intensity of the ground motion ensemble increases, both the median and 84th percentile

---

[†]*Median* refers to the exponent of the mean of the natural log values of the data set.

[‡]The *84th percentile* is the median multiplied by the exponent of the dispersion, where dispersion is determined as the standard deviation of the natural log values of the data set.

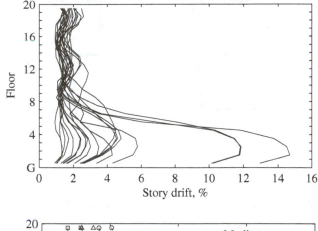

Figure 20.2.5 Story drift demands for the SAC–Los Angeles 20-story building due to 20 SAC ground motions. (Data provided by Akshay Gupta.)

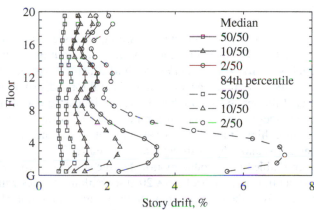

Figure 20.2.6 Median and 84th percentile values of the story drift demands for the SAC–Los Angeles 20-story building for three ensembles of ground motions. (From Gupta and Krawinkler, 2000a.)

values of story drift demands become larger, as expected; more important, dispersion of the demand increases. The excitation-to-excitation variability in response is larger for the most intense ensemble of ground motions because they cause story drifts large enough for P–Δ effects to be important.

20.3 STORY DRIFT DEMANDS

20.3.1 Influence of Plastic Hinge Mechanism

The heightwise variation in ductility demands on multistory buildings depends, in part, on the relative yield strengths of the beams and columns and the relative yield strengths of various stories. To demonstrate this important concept, we present the ductility demands for the three types of structures shown in Fig. 20.3.1. Designated as the *beam-hinge model*, the first structural type is a frame with strong columns and weak beams in which a complete mechanism forms with plastic hinges in all beams as lateral forces with a code-specified distribution are increased (Fig. 20.3.1a). Designated as the *column-hinge model*,

Figure 20.3.1   (a) Beam-hinge, (b) column-hinge, and (c) weak-story models. (From Krawinkler and Nassar, 1991.)

the second structural type is a system with weak columns and strong beams in which all columns develop plastic hinges as lateral forces with a code-specified distribution are increased (Fig. 20.3.1b). Designated as the *weak-story model*, the third structural type develops a story mechanism only in the first story under code-specified lateral force distribution (Fig. 20.3.1c); the yield strengths of the second and higher stories are increased considerably relative to the column-hinge model to ensure that they remain elastic. Thus the first story is no weaker than in the column-hinge model; it is weak only relative to the second and higher stories, implying a large discontinuity of strength across the first floor.

The mean values (over an ensemble of 15 ground motions) of the story ductility factor will be presented for the aforementioned models of three 20-story building frames with base shear yield strength determined from the corresponding SDF system. In the linearly elastic range, the natural period and damping ratio of the corresponding SDF system are the same as the fundamental mode properties T_1 and ζ_1 of the multistory frame. The weight of the corresponding SDF system is the same as the total weight W of the multistory frame. The base-shear yield strength for this SDF system corresponding to a selected ductility factor μ is given by Eq. (7.12.1) with appropriate change in notation:

$$V_{by} = \frac{A_y}{g} W \qquad (20.3.1)$$

where A_y is the pseudo-acceleration corresponding to the μ selected (in this example $\mu = 8$) and known T_1 and ζ_1. The pseudo-acceleration is determined from the mean inelastic response spectrum of an ensemble of 15 ground motions.

The story ductility demands differ considerably among the three structural types (Fig. 20.3.2). These ductility factors may be unrealistically large, but this example is chosen to illustrate trends. Observe that the story ductility demand for the structures considered is largest in the bottom story, which is generally—but not always—true. Among the three structural types, the ductility demands in the lower stories are largest for the weak-story model, smallest for the beam-hinge model, and in between for the column-hinge model. It is also apparent that the structural type has a great influence on the

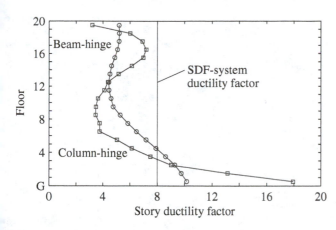

Figure 20.3.2   Mean story ductility demands for beam-hinge, column-hinge, and weak-story models of a 20-story frame due to an ensemble of 15 ground motions, compared with the SDF-system ductility factor $\mu = 8$. For the weak-story model, the ductility demand is about 50 for the first story but all other stories remained elastic. (Data from Nassar and Krawinkler, 1991.)

heightwise variation of story ductility demands. Although this variation is significant in all three cases, it is smallest for the beam-hinge model and greatest for the weak-story model; in the latter case, the upper stories remain essentially elastic. Thus all the energy that was dissipated through yielding of the upper stories in the beam–hinge and column–hinge models must be dissipated by the weak first story, resulting in the very large ductility demand of about 50.

In actual buildings, if the first story is relatively weak, it is usually also relatively flexible because stiffness and strength are often interrelated. The behavior of such a soft-first-story building is similar to that of a weak-first-story building: The upper stories remain essentially elastic, with yielding confined to the first story, resulting in large ductility demands in this story.

A well-known example of a building with a soft first story is the Olive View Hospital building. This was a six-story reinforced-concrete building with its first story partially underground. The lateral force-resisting system included large walls in the upper four stories that did not extend down to the lower two stories (Fig. 20.3.3). These discontinuous shear walls created a large discontinuity in strength and stiffness at the second-floor level. During the February 9, 1971 San Fernando earthquake, this structure behaved as suggested by the preceding results from dynamic analysis of a hypothetical building. The upper four stories of this building escaped with minor damage, with the damage decreasing toward the top. Most of the damage was concentrated in the partially underground story and the first aboveground story, with permanent drift in the latter story exceeding 30 in. (Fig. 20.3.4). This large drift imposed very severe deformation and ductility demands on the first-story columns. As a result, the tied columns failed in a brittle manner (Fig. 20.3.5); however, the ductile behavior of the spirally reinforced columns prevented the collapse of the building (Fig. 20.3.6). This building, completed only a few months prior to the earthquake, was damaged so severely that it had to be demolished. There are many such examples of severe damage to buildings with a soft first story even during recent earthquakes.

Although soft-first-story buildings are obviously not appropriate for earthquake-prone regions, their response during past earthquakes suggests the possibility of

Figure 20.3.3 Olive View Hospital building. The shear walls in the upper four stories did not extend to the lower two stories. (From K. V. Steinbrugge Collection, courtesy of the Earthquake Engineering Research Center, University of California at Berkeley.)

reducing the damage to a building by a base isolation system that acts like a soft first story. This topic is discussed in Chapter 21.

20.3.2 Influence of Inelastic Behavior

The distribution of story drifts over the height of a multistory frame also depends on how far the frame deforms into the inelastic range, as demonstrated in Fig. 20.3.7. Presented are the median values of story drift demands due to an ensemble of ground motions for beam-hinge models of five 9-story frames, designed for lateral force distribution specified in the 2009 International Building Code, and base shear given by Eq. (20.3.1), where A_y is chosen to correspond to the SDF-system ductility factor $\mu = 1, 1.5, 2, 4$, and 6; included

Figure 20.3.4   Large deformations in the first aboveground story of the Olive View Hospital building due to the San Fernando earthquake of February 9, 1971. (Courtesy of K. V. Steinbrugge Collection, Earthquake Engineering Research Center, University of California at Berkeley.)

Figure 20.3.5   Brittle failure of a tied corner column of the Olive View Hospital building. (From K. V. Steinbrugge Collection, courtesy of the Earthquake Engineering Research Center, University of California at Berkeley.)

Figure 20.3.6 Large permanent deformation of a spirally reinforced column of the Olive View Hospital building. (From K. V. Steinbrugge Collection, courtesy of the Earthquake Engineering Research Center, University of California at Berkeley.)

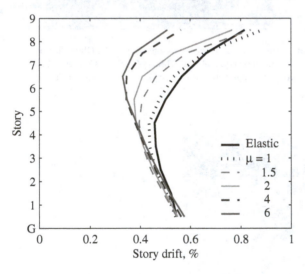

Figure 20.3.7 Variation of story drift demands in 9-story frames designed for different values of the SDF-system ductility factor μ.

are story drifts for the frame assumed to be linearly elastic. The story drift demands and their heightwise variation for inelastic systems differ from those of elastic systems and depend significantly on the ductility factor, a measure of the degree of inelastic behavior. The story drifts increase at the upper stories of the elastic frame, where the response contributions from higher vibration modes are known to be significant (see Chapter 19). As the ductility factor, μ, increases (i.e., the strength of the frame decreases, implying a larger degree of inelastic action), the drifts in upper stories decrease, and the largest drift occurs near the base of the structure. This trend can also be observed in Fig. 20.3.2.

20.4 STRENGTH DEMANDS FOR SDF AND MDF SYSTEMS

What base-shear yield strength is required in a multistory building to keep the earthquake-induced ductility demand in every story below a selected value? To address this question we examine the story ductility demands for a building with its base shear yield strength determined from Eq. (20.3.1) for the corresponding SDF system. By defining the yield base shear of the multistory building as the same as that of the corresponding SDF system, the ductility demands computed will permit direct comparison between the two systems and with the SDF-system ductility factor μ selected to determine A_y in Eq. (20.3.1). Before presenting the ductility demands for the multistory building, we note that the mean ductility demand imposed by the ensemble of ground motions on the corresponding SDF system will be identical to the μ selected (Chapter 7).

For multistory buildings, however, the ductility demands differ from the SDF-system ductility factor μ selected, and vary over the height. Figure 20.3.2 shows the mean values (over an ensemble of 15 ground motions) of story ductility factors for two 20-story building frames (beam-hinge model and column-hinge model) with the base shear strength defined according to Eq. (20.3.1) for $\mu = 8$. It is clear that story ductility demands differ from $\mu = 8$ and are not constant over height because of the more complex dynamics of MDF systems.

The first-story ductility demand (often, the largest among all stories) increases with the fundamental period, T_1 (or the number of stories) and may exceed the SDF-system ductility factor. It is also affected by the plastic hinge mechanism, as distinguished by column-hinge versus beam-hinge models. These trends are shown in Fig. 20.4.1, where the mean ductility demand in the first story of buildings 2, 5, 10, 20, 30, and 40 stories high is presented as a function of fundamental period T_1 for $\mu = 2$ and 8.

It is evident from the preceding results (Fig. 20.4.1) and related observations that unlike SDF systems, the base shear yield strength determined from Eq. (20.3.1) is not sufficient to limit the story ductility demands in a multistory building below the SDF-system

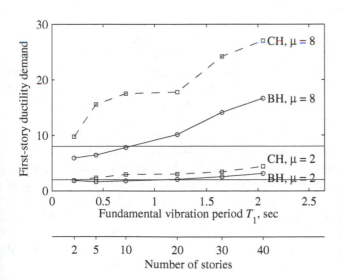

Figure 20.4.1 Mean first-story ductility demands for beam-hinge and column-hinge models of 2-, 5-, 10-, 20-, 30-, and 40-story frames designed for SDF-system ductility factors $\mu = 2$ and 8. (Data from Nassar and Krawinkler, 1991.)

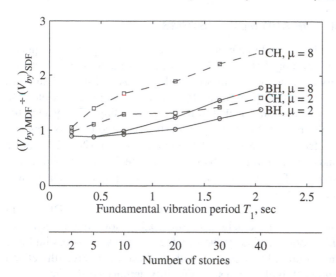

Figure 20.4.2 Modification factor to obtain the base shear yield strength of multistory frames from the base-shear yield strength of the corresponding SDF system for beam-hinge and column-hinge models. (Data from Nassar and Krawinkler, 1991.)

ductility factor μ. To achieve this design objective, the base shear yield strength V_{by} for SDF systems needs to be increased for MDF systems. The modification factor $(V_{by})_{\mathrm{MDF}} \div (V_{by})_{\mathrm{SDF}}$, where $(V_{by})_{\mathrm{MDF}}$ and $(V_{by})_{\mathrm{SDF}}$ are the base shear yield strengths of MDF and SDF systems, respectively, varies between 1 and 2.5 for the examples considered, increases with T_1 (or number of stories) and the μ value, and is also influenced by the plastic hinge mechanism, being larger for the column-hinge model than for the beam-hinge model (Fig. 20.4.2).

PART B: APPROXIMATE ANALYSIS PROCEDURES

20.5 MOTIVATION AND BASIC CONCEPT

At the present time, nonlinear RHA is an onerous task, for several reasons. First, an ensemble of ground motions compatible with the seismic design spectrum for the site must be selected. Second, despite increasing computing power, inelastic modeling remains challenging and nonlinear RHA remains computationally demanding, especially for unsymmetric-plan buildings—which require three-dimensional analysis to account for coupling between lateral and torsional motions—subjected to two horizontal components of ground motion. Third, such analyses must be repeated for several excitations because of the wide variability in demand due to plausible ground motions, and the record-to-record variability of response must be considered (see Section 20.2.3). Fourth, the structural model must be sophisticated enough to represent a building realistically, especially deterioration in its strength at large displacements (see Sections 20.2.1 and 20.2.2). Fifth, structural modeling and commercial computer programs analyzing the same structure should be robust enough to produce essentially identical response results. With additional research and software development, most of the preceding issues should eventually be resolved, and nonlinear RHA may then become common in structural engineering practice.

However, it may be unreasonable to require this onerous procedure for every building—no matter how simple—and of every structural engineering office—no matter how small. Therefore, we are interested in developing simplified methods that are approximate but are rooted in structural dynamics theory as an alternative to rigorous nonlinear RHA. For this purpose, we will utilize the notion of effective earthquake forces and their modal components introduced in Chapters 12 and 13.

The effective earthquake forces [Eq. (13.1.2)], repeated here for convenience, are

$$\mathbf{p}_{\text{eff}}(t) = -\mathbf{m}\iota\ddot{u}_g(t) \tag{20.5.1}$$

The spatial distribution of these forces over the structure is defined by the vector $\mathbf{s} = \mathbf{m}\iota$. This force distribution can be expanded as a summation of modal inertia force distributions $\mathbf{s}_n$ (Section 13.1.2), repeated here for convenience:

$$\mathbf{m}\iota = \sum_{n=1}^{N} \mathbf{s}_n = \sum_{n=1}^{N} \Gamma_n \mathbf{m}\phi_n \tag{20.5.2}$$

where

$$\Gamma_n = \frac{L_n}{M_n} \qquad L_n = \phi_n^T \mathbf{m}\iota \qquad M_n = \phi_n^T \mathbf{m}\phi_n \tag{20.5.3}$$

The effective earthquake forces can then be expressed as

$$\mathbf{p}_{\text{eff}}(t) = \sum_{n=1}^{N} \mathbf{p}_{\text{eff},n}(t) = \sum_{n=1}^{N} -\mathbf{s}_n \ddot{u}_g(t) \tag{20.5.4}$$

The contributions of the nth mode to $\mathbf{p}_{\text{eff}}(t)$ and $\mathbf{s}$ are

$$\mathbf{p}_{\text{eff},n}(t) = -\mathbf{s}_n \ddot{u}_g(t) \qquad \mathbf{s}_n = \Gamma_n \mathbf{m}\phi_n \tag{20.5.5}$$

The modal expansion of the force distribution $\mathbf{s}$ is illustrated for a perimeter frame of the SAC–Los Angeles 9-story building. The first three periods and modes of the building vibrating along an axis of plan symmetry are shown in Fig. 20.5.1, where we note that the

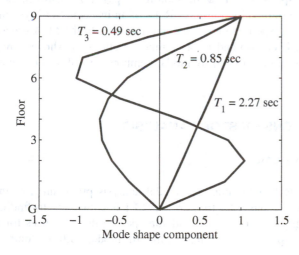

Figure 20.5.1   First three natural-vibration periods and modes of the SAC–Los Angeles 9-story building. (From Goel and Chopra, 2004.)

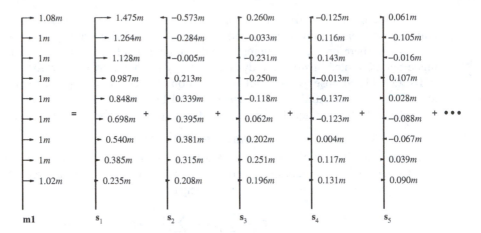

Figure 20.5.2 Modal expansion of the distribution $\mathbf{s} = \mathbf{m}\iota$ of effective earthquake forces.

floor displacements in the first (or fundamental) mode are all in the same direction, but they reverse direction in higher modes as one moves up the building. The modal expansion of the distribution $\mathbf{s} = \mathbf{m}\iota$ of the effective earthquake forces is shown in Fig. 20.5.2. As noted in Section 13.2, the direction of force s_{jn} at the jth floor level is controlled by the algebraic sign of ϕ_{jn}, the jth-floor displacement in mode ϕ_n. Hence, these forces for the first mode all act in the same direction, but for the second and higher modes they change direction as one moves up the building. The contribution of the first mode to the force distribution $\mathbf{s}$ is largest, and the modal contributions decrease progressively for higher modes.

Utilizing the modal expansion of $\mathbf{p}_{\text{eff}}(t)$ and $\mathbf{s}$, two procedures for approximate analysis of inelastic buildings will be described next: uncoupled modal response history analysis (UMRHA) and modal pushover analysis (MPA). Not intended for practical application, the UMRHA procedure is developed only to provide a rationale for the MPA procedure. In the UMRHA procedure, the response history of the building to $\mathbf{p}_{\text{eff},n}(t)$, the nth-mode component of the excitation, is determined by nonlinear RHA of an inelastic SDF system, and superposition of these "modal" responses gives the total response. In the MPA procedure, the peak response to $\mathbf{p}_{\text{eff},n}(t)$ is determined by a nonlinear static, or pushover, analysis, and the peak modal responses are combined by modal combination rules to estimate the total response.

20.6 UNCOUPLED MODAL RESPONSE HISTORY ANALYSIS

20.6.1 Linearly Elastic Systems

In this section we demonstrate that the classical modal analysis procedure for linearly elastic systems developed in Sections 12.4 to 12.6 and 13.1 is equivalent to finding the response of the structure to $\mathbf{p}_{\text{eff},n}(t)$ for each n and superposing the responses for all n. The response of the system to $\mathbf{p}_{\text{eff},n}(t)$ is entirely in the nth mode, with no contribution from other modes, which implies that the modes are uncoupled. Recall that this important property of the modal expansion of Eq. (20.5.2) was proven analytically in Section 12.8.

The equations governing the response of the linearly elastic MDF system to $\mathbf{p}_{\text{eff},n}(t)$, defined by Eq. (20.5.5a), are

$$\mathbf{m}\ddot{\mathbf{u}} + \mathbf{c}\dot{\mathbf{u}} + \mathbf{k}\mathbf{u} = -\mathbf{s}_n \ddot{u}_g(t) \tag{20.6.1}$$

and the resulting floor displacements are given by

$$\mathbf{u}_n(t) = \boldsymbol{\phi}_n q_n(t) \tag{20.6.2}$$

Substituting Eq. (20.6.2) in Eq. (20.6.1) and premultiplying the latter by $\boldsymbol{\phi}_n^T$ leads to the equation governing the modal coordinate q_n:

$$\ddot{q}_n + 2\zeta_n \omega_n \dot{q}_n + \omega_n^2 q_n = -\Gamma_n \ddot{u}_g(t) \tag{20.6.3}$$

in which ω_n is the natural frequency and ζ_n is the damping ratio, both for the nth mode, and Γ_n was defined by Eq. (20.5.3). As demonstrated in Section 13.1, the solution $q_n(t)$ of Eq. (20.6.3) is given by

$$q_n(t) = \Gamma_n D_n(t) \tag{20.6.4}$$

where $D_n(t)$ is the deformation response of the nth-mode linearly elastic SDF system, an SDF system with vibration properties—natural frequency ω_n (natural period $T_n = 2\pi/\omega_n$) and damping ratio ζ_n—of the nth mode of the MDF system, subjected to $\ddot{u}_g(t)$. It is governed by

$$\ddot{D}_n + 2\zeta_n \omega_n \dot{D}_n + \omega_n^2 D_n = -\ddot{u}_g(t) \tag{20.6.5}$$

Substituting Eq. (20.6.4) into Eq. (20.6.2) gives the lateral displacements of the floors:

$$\mathbf{u}_n(t) = \Gamma_n \boldsymbol{\phi}_n D_n(t) \tag{20.6.6}$$

and the story drift in the jth story is the difference between displacements of the jth and $(j-1)$th floors:

$$\Delta_{jn}(t) = \Gamma_n(\phi_{jn} - \phi_{j-1,n})D_n(t) \tag{20.6.7}$$

Equations (20.6.6) and (20.6.7) represent the response of the MDF system to $\mathbf{p}_{\text{eff},n}(t)$, and superposing the responses for all n gives the response of the system due to total excitation $\mathbf{p}_{\text{eff}}(t)$:

$$r(t) = \sum_{n=1}^{N} r_n(t) \tag{20.6.8}$$

This is the UMRHA procedure for exact analysis of linearly elastic systems, which is identical to the classical modal RHA. Equation (20.6.3) is the standard equation governing the modal coordinate $q_n(t)$; it is identical to Eq. (13.1.7). Equations (20.6.6) and (20.6.7) define the contribution of the nth mode to the response; they are identical to Eqs. (13.1.10) and (13.2.6). Equation (20.6.8) combines the response contributions due to all n excitation terms in the modal expansion of Eq. (20.5.4); it is identical to Eqs. (13.1.15) and (13.1.16).

However, these standard equations have now been derived in an unconventional way. In contrast to the classical derivation presented in Sections 12.4 and 13.1.3 to 13.1.5, we have used the modal expansion of the spatial distribution of the effective earthquake forces. This interpretation of modal analysis will provide a rational basis for the modal pushover analysis procedure developed later for inelastic systems.

20.6.2 Inelastic Systems

Although modal analysis is not valid for an inelastic system, its dynamic response can usefully be discussed in terms of the natural vibration modes of the corresponding linear system. Each structural element of this linear system is defined to have the same stiffness as its initial stiffness in the inelastic system; both systems have the same mass and damping. Therefore, the natural vibration periods and modes of the corresponding linear system are the same as the vibration properties of the inelastic system undergoing small oscillation, which are referred to as "*periods*" and "*modes*" of the inelastic system.[†] Thus Eqs. (20.5.2) to (20.5.5) are also valid for inelastic systems, where ϕ_n now represents the modes of the corresponding linear system.

The equations governing the response of the inelastic MDF system to $\mathbf{p}_{\mathrm{eff},n}(t)$ defined by Eq. (20.5.5a) are

$$\mathbf{m}\ddot{\mathbf{u}} + \mathbf{c}\dot{\mathbf{u}} + \mathbf{f}_s(\mathbf{u}) = -\mathbf{s}_n \ddot{u}_g(t) \tag{20.6.9}$$

The solution of Eq. (20.6.9) will no longer be described by Eq. (20.6.2) because modes other than the nth mode will also contribute to the system response, implying that the vibration modes are now coupled; thus the floor displacements are given by the first part of Eq. (20.6.10):

$$\mathbf{u}_n(t) = \sum_{r=1}^{N} \phi_r q_r(t) \simeq \phi_n q_n(t) \tag{20.6.10}$$

However, because for linear systems $q_r(t) = 0$ for all modes other than the nth mode, it is reasonable to expect that $q_r(t)$ may be small for inelastic systems, implying that the elastic modes are, at most, weakly coupled.

The above-mentioned expectation is confirmed numerically by the response of the SAC–Los Angeles 9-story building to ground motion intense enough to cause significant yielding of the structure. Its response to force vector $\mathbf{p}_{\mathrm{eff},n}(t)$ was determined by nonlinear RHA, solving Eq. (20.6.9) by the numerical methods described in Chapter 16, and the resulting floor displacements were decomposed into their modal components using the procedure in Section 10.7, applied at each time instant.

Figure 20.6.1 shows that the roof displacement due to the force vector $\mathbf{p}_{\mathrm{eff},n}(t)$ is due primarily to the nth mode but that other modes contribute to the response. The second, third, and fourth modes start responding to excitation $\mathbf{p}_{\mathrm{eff},1}(t)$ the instant the structure

---

[†]The quotation marks are included to emphasize that the concept of natural vibration periods and modes is strictly not valid for inelastic systems. Subsequently, however, the quotation marks are dropped, but are always implied in the context of inelastic systems.

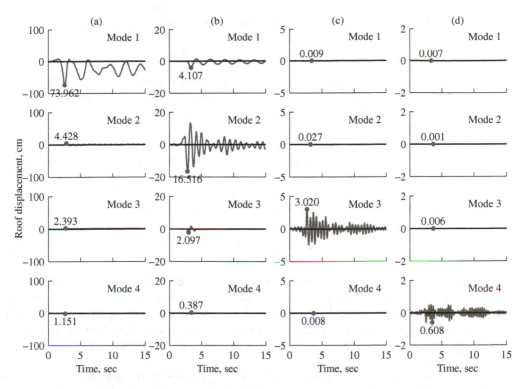

Figure 20.6.1 Modal decomposition of roof displacement due to $\mathbf{p}_{\text{eff},n}(t) = -\mathbf{s}_n \ddot{u}_g(t)$, $n = 1, 2, 3$, and 4, where $\ddot{u}_g(t) = $ LA25 ground motion: (a) $\mathbf{p}_{\text{eff},1} = -\mathbf{s}_1 \times$ LA25; (b) $\mathbf{p}_{\text{eff},2} = -\mathbf{s}_2 \times$ LA25; (c) $\mathbf{p}_{\text{eff},3} = -\mathbf{s}_3 \times$ LA25; (d) $\mathbf{p}_{\text{eff},4} = -\mathbf{s}_4 \times$ LA25.

first yields (Fig. 20.6.1a). Similarly, the first, third, and fourth modes start responding to excitation $\mathbf{p}_{\text{eff},2}(t)$ the instant the structure first yields (Fig. 20.6.1b).

Although the natural vibration modes are no longer uncoupled if the system responds in the inelastic range, modal coupling is weak. In the structural response due to $\mathbf{p}_{\text{eff},1}(t)$, the contributions to roof displacement of the second, third, and fourth modes are only 6, 3, and 2%, respectively, of the first-mode response (Fig. 20.6.1a). In the structural response to $\mathbf{p}_{\text{eff},2}(t)$, the contributions to roof displacement of the first, third, and fourth modes are 25, 13, and 2%, respectively, of the second-mode response (Fig. 20.6.1b). In the structural response to $\mathbf{p}_{\text{eff},3}(t)$, the contributions to the roof displacement of each of the first, second, and fourth modes are less than 1% of the third-mode response (Fig. 20.6.1c). In the structural response to $\mathbf{p}_{\text{eff},4}(t)$, the contributions to roof displacement of each of the first, second, and third modes are less than about 1% of the fourth-mode response (Fig. 20.6.1d).

This weak coupling of modes implies that the structural response due to excitation $\mathbf{p}_{\text{eff},n}(t)$ may be approximated by the second half of Eq. (20.6.10). Substituting this approximation into Eq. (20.6.9) and premultiplying by $\boldsymbol{\phi}_n^T$ gives

$$\ddot{q}_n + 2\zeta_n\omega_n\dot{q}_n + \frac{F_{sn}}{M_n} = -\Gamma_n\ddot{u}_g(t) \qquad (20.6.11)$$

where F_{sn} is a nonlinear hysteretic function of the nth modal coordinate q_n:

$$F_{sn} = F_{sn}(q_n) = \phi_n^T \mathbf{f}_s(q_n) \tag{20.6.12}$$

If the smaller contributions of other modes had not been neglected, F_{sn} would depend on all modal coordinates, and the set of equations defined by Eq. (20.6.11) for $n = 1, 2, \ldots, N$ would be coupled because of yielding of the structure and hence offer no advantage over Eq. (20.6.9).

What is a good way to express the solution of Eq. (20.6.11) for inelastic systems? To answer this question, we note the similarity between Eq. (20.6.11) and its counterpart, Eq. (20.6.3) for linearly elastic systems, and that the solution for the latter was related by Eq. (20.6.4) to the response $D_n(t)$ of the nth-mode elastic SDF system. Similarly, the solution of the former can be expressed as Eq. (20.6.4), where $D_n(t)$ is now governed by

$$\ddot{D}_n + 2\zeta_n \omega_n \dot{D}_n + \frac{F_{sn}}{L_n} = -\ddot{u}_g(t) \tag{20.6.13}$$

$D_n(t)$ may be interpreted as the deformation response of the nth-mode inelastic SDF system, an SDF defined by (1) small-oscillation vibration properties—natural frequency ω_n (natural period T_n) and damping ratio ζ_n—of the nth mode of the MDF system; and (2) the force–deformation ($F_{sn}/L_n - D_n$) relation. Introducing the nth-mode inelastic SDF system permitted extension to inelastic systems of the well-established concepts for elastic systems.

Solution of the nonlinear Eq. (20.6.13) provides $D_n(t)$, which is substituted into Eqs. (20.6.6) and (20.6.7) to obtain floor displacements and story drifts. They approximate the response of the inelastic MDF system to $\mathbf{p}_{\text{eff},n}(t)$, the nth-mode contribution to $\mathbf{p}_{\text{eff}}(t)$. Superposition of responses to $\mathbf{p}_{\text{eff},n}(t)$—$n = 1, 2, \ldots, N$—according to Eq. (20.6.8) provides the total response to $\mathbf{p}_{\text{eff}}(t)$. This is the UMRHA procedure for approximate analysis of inelastic systems. When specialized for linearly elastic systems, as mentioned in Section 20.6.1, it becomes identical to the classical modal RHA, an exact analysis procedure.

The UMRHA procedure for inelastic systems is based on two approximations, which can be identified by comparing the key equations in UMRHA for elastic and inelastic structural systems. Equations (20.6.4), (20.6.6), and (20.6.7) apply to both systems; Eqs. (20.6.3) and (20.6.5) differ from Eqs. (20.6.11) and (20.6.13) only in the resisting force; Eqs. (20.6.2) and (20.6.8) are exact for elastic systems but only approximate for inelastic systems. Commenting first on Eq. (20.6.8), the superposition of responses implied by this equation is strictly valid only for linearly elastic systems; however, it has been shown to be approximately valid for inelastic systems, a result that we do not demonstrate here because our intent in developing UMRHA is limited to justifying the modal uncoupling approximation needed for MPA. As is evident from Eq. (20.6.10), the second approximation comes from neglecting the coupling of modal coordinates, which permitted computing the response of the inelastic MDF system to $\mathbf{p}_{\text{eff},n}(t)$ from that of an SDF system. Supported by the numerical results of Fig. 20.6.1, this approximation is reasonable only because the excitation is $\mathbf{p}_{\text{eff},n}(t)$, the nth-mode contribution to the total excitation $\mathbf{p}_{\text{eff}}(t)$. It would not be valid for an excitation with lateral force distribution different from

$\mathbf{s}_n$ [e.g., the total excitation $\mathbf{p}_{\text{eff}}(t)$], pointing out that the modal expansion of Eq. (20.5.4) is a key concept underlying UMRHA.

To test the modal uncoupling approximation in UMRHA, the response of the SAC–Los Angeles 9-story building to $\mathbf{p}_{\text{eff},n}(t) = -\mathbf{s}_n \ddot{u}_g(t)$, where $\ddot{u}_g(t)$ is the same ground motion as the one selected earlier (Fig. 20.6.1), was determined by two methods and compared: (1) rigorous nonlinear RHA by solving the governing coupled equations [Eq. (20.6.9)]; and (2) approximate UMRHA procedure. Such a comparison for roof-displacement and top-story drift is presented in Figs. 20.6.2 and 20.6.3, respectively. The errors in UMRHA results are larger in drift than in displacement, but the errors in either

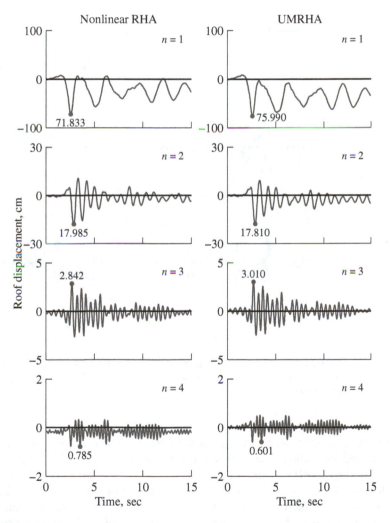

Figure 20.6.2 Comparison of approximate roof displacement from UMRHA and exact result from nonlinear RHA due to $\mathbf{p}_{\text{eff},n}(t) = -\mathbf{s}_n \ddot{u}_g(t)$, $n = 1, 2, 3$, and 4, where $\ddot{u}_g(t) = $ LA25 ground motion.

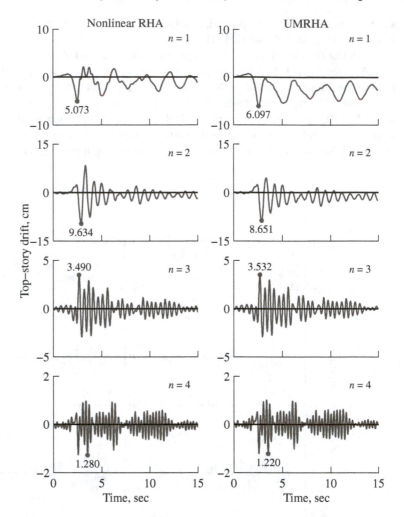

Figure 20.6.3   Comparison of approximate top-story drift from UMRHA and exact result from nonlinear RHA due to $\mathbf{p}_{\text{eff},n}(t) = -\mathbf{s}_n \ddot{u}_g(t)$, $n = 1, 2, 3$, and 4, where $\ddot{u}_g(t) = $ LA25 ground motion.

response quantity seem small enough to use the modal uncoupling approximation in developing approximate methods to estimate seismic demands for buildings.

The UMRHA procedure is based on the second half of Eq. (20.6.10), restricting the floor displacements due to $\mathbf{p}_{\text{eff},n}(t)$ to be proportional to the nth mode, which as stated earlier is an approximation for inelastic systems. This approximation is avoided in the MPA procedure, which is presented next, but a modal combination approximation must be introduced, as will be seen later. To provide a proper context, MPA is first presented for linearly elastic systems.

20.7 MODAL PUSHOVER ANALYSIS

20.7.1 Linearly Elastic Systems

The response spectrum analysis (RSA) procedure (Sections 13.7 and 13.8), which is a dynamic analysis procedure, can be interpreted in two ways: as static analysis or as pushover analysis. As demonstrated in Section 13.8.1, static analysis of the building subjected to lateral forces

$$\mathbf{f}_n = \mathbf{s}_n A_n = \Gamma_n \mathbf{m} \phi_n A_n \tag{20.7.1}$$

will provide the same value of r_n, the peak value of the nth-mode response $r_n(t)$, as in Eq. (13.7.1), where $A_n = A(T_n, \zeta_n)$, the pseudo-acceleration spectrum ordinate corresponding to the natural vibration period T_n and damping ratio ζ_n of the nth mode.

Alternatively, this peak modal response can be obtained by linear static analysis of the structure subjected to monotonically increasing lateral forces with an invariant height-wise distribution:

$$\mathbf{s}_n^* = \mathbf{m} \phi_n \tag{20.7.2}$$

pushing the structure up to the roof displacement, u_{rn} (the subscript r denotes "roof"), the peak value of the roof displacement due to the nth mode, which from Eq. (20.6.6) is

$$u_{rn} = \Gamma_n \phi_{rn} D_n \tag{20.7.3}$$

where $D_n \equiv D(T_n, \zeta_n)$ is the ordinate of the deformation response spectrum corresponding to the period T_n and damping ratio ζ_n of the nth mode. Figure 20.7.1 shows the lateral force distribution $\mathbf{s}_n^*$ for the first three modes of the SAC–Los Angeles 9-story building. But for scaling factors, Γ_n, these distributions are identical to those in Fig. 20.5.2.

The peak modal responses, r_n, each determined by one pushover analysis, can be combined according to the modal combination rules of Eq. (13.7.3) or (13.7.4), as

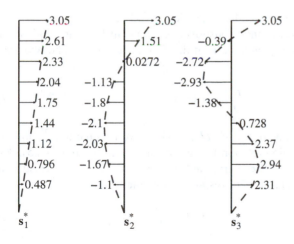

Figure 20.7.1   Lateral force distributions $\mathbf{s}_n^* = \mathbf{m}\phi_n$, $n = 1$, 2, and 3, for the first three modes of the SAC–Los Angeles 9-story building. (From Goel and Chopra, 2004.)

appropriate, to obtain an estimate of the peak value r of the total response. Equivalent to the standard RSA procedure described in Sections 13.7 and 13.8, the MPA procedure offers no advantage for linearly elastic systems, but this interpretation of RSA permits extension of MPA to approximate analysis of inelastic systems. Before doing so, note that r_n determined by pushover analysis can also be interpreted as the peak response of the linearly elastic system to $\mathbf{p}_{\text{eff},n}(t)$, the nth-mode component of the effective earthquake forces. This interpretation is valid because, as demonstrated in Section 12.8, the system responds only in its nth mode when subjected to this excitation.

20.7.2 Inelastic Systems

The peak response r_n of the inelastic system to $\mathbf{p}_{\text{eff},n}(t)$ is also determined by a pushover analysis, which is now a nonlinear static analysis instead of a linear static analysis, of the structure subjected to lateral forces distributed over the building height according to $\mathbf{s}_n^*$ [Eq. (20.7.2)] with the forces increased to push the structure up to roof displacement u_{rn}. This value of the roof displacement is also determined from Eq. (20.7.3), but D_n is now the peak deformation of the nth-mode inelastic SDF system (instead of the nth-mode elastic SDF system), determined by solving Eq. (20.6.13) for $D_n(t)$. At this roof displacement, the results of nonlinear static analysis provide an estimate of the peak value r_n of the response quantity $r_n(t)$: floor displacements, story drifts, and other deformation quantities.

Nonlinear static analysis using force distribution $\mathbf{s}_n^*$ leads to the nth-mode pushover curve, a plot of base shear V_{bn} versus roof displacement u_{rn}. From the nth-mode pushover curve is obtained the force–deformation $(F_{sn}/L_n - D_n)$ curve for the nth-mode inelastic SDF system, which is required in Eq. (20.6.13). The forces and displacements in the two sets of curves are related as follows (see Derivation 20.1):

$$\frac{F_{sn}}{L_n} = \frac{V_{bn}}{M_n^*} \qquad D_n = \frac{u_{rn}}{\Gamma_n \phi_{rn}} \tag{20.7.4}$$

where $M_n^* = L_n \Gamma_n$ is the effective modal mass (Section 13.2.5).

Figure 20.7.2 shows the nth-mode pushover curve and its bilinear idealization; at the yield point the base shear is V_{bny} and the roof displacement is u_{rny}. The two are related through (Derivation 20.2)

$$\frac{F_{sny}}{L_n} = \omega_n^2 D_{ny} \tag{20.7.5}$$

As it should be, the initial slope of the $F_{sn}/L_n - D_n$ curve is equal to ω_n^2, implying that it matches the force–deformation relation for the linear system in Eq. (20.6.5). Knowing F_{sny}/L_n and D_{ny} from the pushover curve and Eq. (20.7.4), the initial elastic vibration period T_n of the nth-mode inelastic SDF system is computed from

$$T_n = 2\pi \left(\frac{L_n D_{ny}}{F_{sny}} \right)^{1/2} \tag{20.7.6}$$

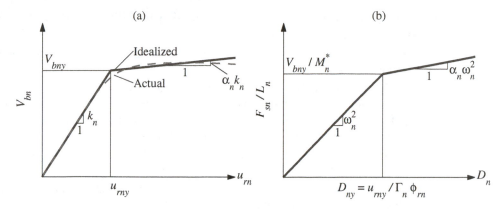

Figure 20.7.2   (a) An nth-mode pushover curve and its bilinear idealization; (b) force–deformation relation for the nth-mode inelastic SDF system.

This value of T_n, which may differ from the period of the corresponding linear system determined by solving the eigenvalue problem (Section 10.2), should be used in Eq. (20.6.13).

The response value r_n determined by pushover analysis is an estimate of the peak value of the response $r_n(t)$ of the inelastic structure to $\mathbf{p}_{\text{eff},n}(t)$, but it is not identical to another estimate determined by UMRHA. As mentioned earlier, r_n determined by pushover analysis of a linearly elastic system is the exact peak value of $r_n(t)$, the nth-mode contribution to response $r(t)$. Thus we refer to r_n as the peak modal response even in the case of inelastic systems. However, for inelastic systems the two—UMRHA and MPA—estimates of the peak modal response are both approximate and different from each other; the only exception is the roof displacement because it is deliberately matched in the two analyses. The two estimates differ because the underlying analyses involve different assumptions. UMRHA is based on the approximation contained in the second half of Eq. (20.6.10), which is avoided in MPA because the floor displacements, story drifts, and other deformation quantities are determined by nonlinear static analysis using force distribution $\mathbf{s}_n^*$. As a result, the floor displacements of the inelastic system are no longer proportional to the nth-mode shape, in contrast to the second half of Eq. (20.6.10). In this sense, the MPA procedure represents the nonlinear behavior of the structure better than UMRHA.

However, the MPA procedure contains a different source of approximation, which does not exist in UMRHA. The peak modal responses r_n, each determined by one nonlinear static analysis, are combined by a modal combination rule, just as in RSA of linearly elastic systems. This application of modal combination rules to inelastic systems lacks a rigorous theoretical basis, but seems reasonable because the modes are only weakly coupled.

20.7.3 Summary

The seismic deformation demands—floor displacements, story drifts, and plastic hinge rotations—for a symmetric-plan multistory building subjected to earthquake ground motion along an axis of symmetry can be estimated by the MPA procedure, which is summarized next in step-by-step form:

1. Compute the natural frequencies, ω_n, and modes, ϕ_n, for linearly elastic vibration of the building (Fig. 20.5.1).

2. For the nth mode, develop the base shear–roof displacement, V_{bn}–u_{rn}, pushover curve by nonlinear static analysis of the building using the lateral force distribution, s_n^* [Eq. (20.7.2) and Fig. 20.7.1]. Initial gravity (dead and live) loads are applied before the lateral forces, causing roof lateral displacement u_{rg}.

3. Convert the V_{bn}–u_{rn} pushover curve to the force–deformation, F_{sn}/L_n–D_n, relation for the nth-mode inelastic SDF system by utilizing Eq. (20.7.4).

4. Idealize the force–deformation relation for the nth-mode SDF system as a bilinear or trilinear curve, as appropriate, or by more sophisticated idealizations. Starting with this initial loading curve, define the unloading and reloading branches appropriate for the structural system and material.

5. Compute the peak deformation D_n of the nth-mode inelastic SDF system defined by the hysteretic force–deformation relation developed in step 4 and the damping ratio ζ_n. Compute the initial elastic vibration period [Eq. (20.7.6)] and estimate the damping ratio (Chapter 11). For this SDF system, D_n is determined by nonlinear RHA [i.e., by solving Eq. (20.6.13)].

6. Calculate the peak roof displacement u_{rn} associated with the nth-mode inelastic SDF system from Eq. (20.7.3).

7. From the pushover database (step 2), extract values of desired responses r_{n+g} due to the combined effects of gravity and lateral loads at roof displacement equal to $u_{rn} + u_{rg}$.

8. Repeat steps 3 to 7 for as many modes as required for sufficient accuracy.

9. Compute the dynamic response due to the nth mode: $r_n = r_{n+g} - r_g$, where r_g is the contribution of gravity loads alone.

10. Determine the total dynamic response r_d by combining the peak modal responses using an appropriate modal combination rule (Section 13.8).

11. Determine the total seismic demand by combining the initial response due to gravity loads and the peak dynamic response:

$$r \simeq \max(r_g \pm r_d) \tag{20.7.7}$$

Plastic hinge rotations and member forces.   The total floor displacements and story drifts are estimated by combining values obtained from gravity load analysis and modal pushover analyses (steps 10 and 11). This procedure may also be used to determine other deformation quantities, such as plastic hinge rotations. Alternatively, an improved estimate can be obtained by computing plastic hinge rotations from the total story drifts by a published procedure (Gupta and Krawinkler, 1999).

As summarized above, the MPA procedure can also be used to estimate internal forces in those structural members that remain within their linearly elastic range, but not in those that deform into the inelastic range. In the latter case, the member forces are estimated from the total member deformations—determined by step 11 of the MPA procedure. Researchers have developed such procedures to compute member forces but these are not included here.

Extension of MPA.   Restricted in the preceding sections to symmetric-plan buildings, the MPA procedure has been extended to unsymmetric-plan buildings, which respond in coupled lateral–torsional motions during earthquakes. This extension draws on the earlier development of modal RHA and RSA procedures for linear analysis of unsymmetric-plan buildings (Sections 13.3 and 13.9). The force distribution $\mathbf{s}_n^*$ used in the pushover analysis for each "mode" now includes two lateral forces and torque at each floor, and the modal demands are combined by the CQC rule, instead of the SRSS rule, to obtain an estimate of the total seismic demand.

Derivation 20.1

Equation (20.7.4b), which relates roof displacement u_{rn} of the MDF system in the modal pushover curve to deformation D_n of the SDF system, is obvious from Eq. (20.7.3), whereas Eq. (20.7.4a), relating forces in the two systems, may be derived as follows: At any stage of the nonlinear static procedure, the lateral forces are given by Eq. (20.7.2) times a scale factor, say, α: $\mathbf{f}_{sn} = \alpha \mathbf{m} \phi_n$. Substituting this $\mathbf{f}_{sn}$ into Eq. (20.6.12) and into the equation for base shear, $V_{bn} = \mathbf{1}^T \mathbf{f}_{sn}$, where $\mathbf{1}$ is a vector with all elements equal to unity, and utilizing Eq. (20.5.3b and c) leads to

$$F_{sn} = \alpha M_n \qquad V_{bn} = \alpha L_n \tag{a}$$

Thus
$$\frac{F_{sn}}{M_n} = \frac{V_{bn}}{L_n} \tag{b}$$

Dividing both sides of Eq. (b) by Γ_n, defined in Eq. (20.5.3a), gives Eq. (20.7.4a).

Derivation 20.2

Consider the lateral forces $\mathbf{f}_{sny} = \alpha_y \mathbf{m} \phi_n$ that cause base shear equal to its yield value V_{bny}. Corresponding to these lateral forces, Eq. (20.6.12) gives

$$F_{sny} = \alpha_y M_n \tag{a}$$

The resulting static displacements $\mathbf{u}_{ny}^{st}$ satisfy

$$\mathbf{k}\mathbf{u}_{ny}^{st} = \alpha_y \mathbf{m} \phi_n \tag{b}$$

Solving these equations and using Eq. (10.2.4) gives

$$\mathbf{u}_{ny} = \mathbf{k}^{-1}(\alpha_y \mathbf{m} \phi_n) = \frac{\alpha_y}{\omega_n^2} \phi_n \tag{c}$$

Equating two expressions for the roof displacement, one from Eq. (20.7.3) and the other from Eq. (c), gives

$$\Gamma_n \phi_{rn} D_{ny} = \frac{\alpha_y}{\omega_n^2} \phi_{rn} \tag{d}$$

Equating two expressions for α_y, one from Eq. (d) and the other from Eq. (a), and utilizing Eq. (20.5.3a) leads to Eq. (20.7.5).

20.8 EVALUATION OF MODAL PUSHOVER ANALYSIS

The dynamic response of each structural system to each of the 20 ground motions was determined by two procedures: nonlinear RHA and MPA. The "exact" peak value of structural response or demand r, determined by nonlinear RHA, is denoted by $r_{\text{NL-RHA}}$ and the approximate value from MPA by r_{MPA}. The response of each building was also computed under the assumption that the structure is strong enough to remain elastic. For elastic systems, nonlinear RHA specializes to linear RHA and MPA reduces to RSA; thus, these responses are denoted as r_{RHA} and r_{RSA}. Presented in this section are median values of seismic responses or demands for the 9- and 20-story buildings. The RSA and MPA procedures were implemented, including a variable number of modes: one, two, or three modes for the 9-story buildings; one, three, or five modes for the 20-story buildings.

20.8.1 Modal Pushover Curves and Roof Displacements

Figures 20.8.1 to 20.8.3 show pushover curves for the first, second, and third modes, respectively, and identify the modal roof displacements due to each of the 20 ground motions and their median value; these roof displacements were determined by the MPA procedure (see steps 5 and 6 in the MPA summary presented in Section 20.7.3). Excluded from the first-mode plot are roof displacements due to those ground motions that caused collapse of the SDF system: one, three, and six excitations in the case of the Seattle 9-story and Los

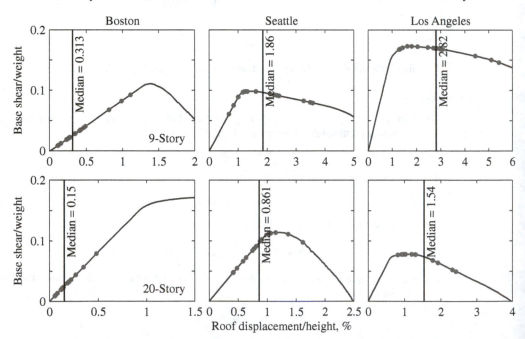

Figure 20.8.1   First-mode pushover curves for six SAC buildings; the roof displacement due to each of 20 ground motions is identified and the median value is also noted.

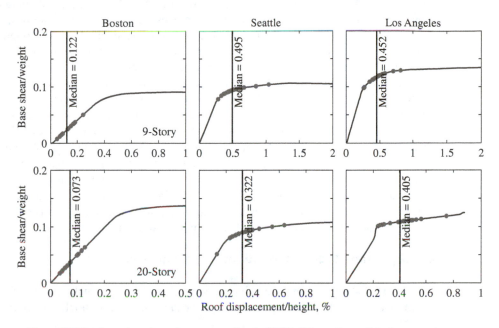

Figure 20.8.2   Second-mode pushover curves for six SAC buildings; the roof displacement due to each of 20 ground motions is identified and the median value is also noted.

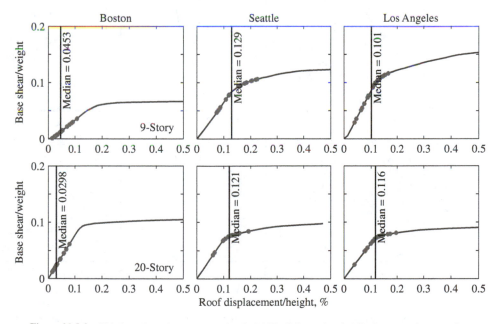

Figure 20.8.3   Third-mode pushover curves for six SAC buildings; the roof displacement due to each of 20 ground motions is identified and the median value is also noted.

Angeles 9- and 20-story buildings. Figures 20.8.1 to 20.8.3 permit the following observations: Boston buildings remain elastic for all modes during all ground motions, and their median roof displacement is well below the yield displacement. Several ground motions drive the Seattle 9-story building well beyond the elastic limit in the first two modes but not in the third mode. The median displacement is well beyond the yield displacement for the first mode but only slightly beyond for the second mode. Several ground motions drive the Seattle 20-story building well beyond the yield displacement in the first three modes; however, the median displacement exceeds the yield displacement significantly only for the second mode. The very intense Los Angeles motions, which include several near-fault ground motions, drive the Los Angeles buildings well beyond the yield displacement in the first two modes; even the median displacement exceeds the yield displacement, although more so in the first mode than in the second. The overall impression is that some excitations deform the Seattle and Los Angeles buildings into the inelastic range in the first three modes, but the median displacement in modes higher than the first is either close to or exceeds the yield displacement only by a modest amount.

20.8.2 Higher-Mode Contributions in Seismic Demands

Figure 20.8.4 shows the median values of story drift demands, including a variable number of modes in MPA superimposed with the "exact" result from nonlinear RHA. The first

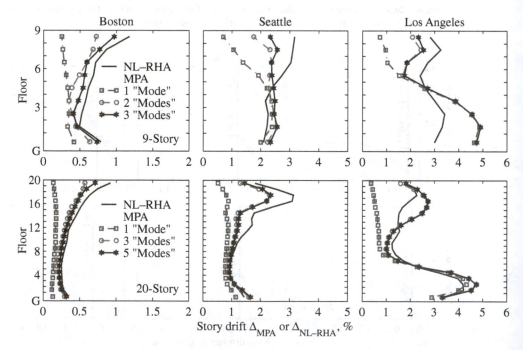

Figure 20.8.4 Median story drifts for six SAC buildings determined by nonlinear RHA and MPA, with a variable number of modes. (Adapted from Goel and Chopra, 2004.)

mode alone is inadequate in estimating story drifts, but with a few modes included, story drifts estimated by MPA are much better and resemble nonlinear RHA results; however, significant discrepancies are noted for the Los Angeles buildings. We return to these discrepancies later.

20.8.3 Accuracy of Modal Pushover Analysis

The MPA procedure for inelastic systems is based on two principal approximations: (1) neglecting the weak coupling of modes in computing the peak modal response r_n to $\mathbf{p}_{\text{eff},n}(t)$; and (2) combining the r_n by modal combination rules, known to be approximate in estimating the peak value of the total response. Because the latter is the only source of approximation in the widely used RSA procedure for linearly elastic system (Sections 13.7 and 13.8), the resulting error in the response of these systems serves as a baseline for evaluating the additional approximation in MPA for inelastic systems.

Figures 20.8.5 and 20.8.6 compare the accuracy of RSA in estimating the response of elastic systems with that of MPA in estimating the response of inelastic systems. For each of the six SAC buildings, the results are organized in two parts: (a) story drift demands for

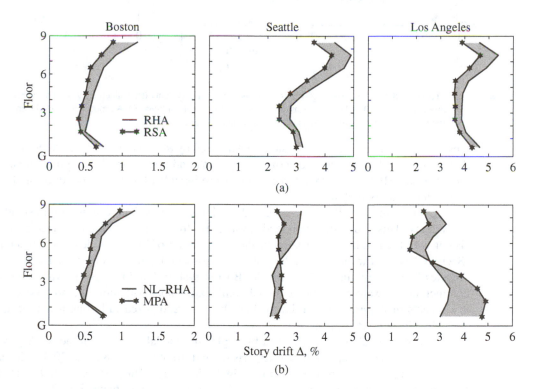

Figure 20.8.5 Median story drifts for (a) linearly elastic systems determined by RSA and RHA procedures, and (b) inelastic systems determined by MPA and nonlinear RHA procedures. Results are for SAC 9-story buildings. (Adapted from Goel and Chopra, 2004.)

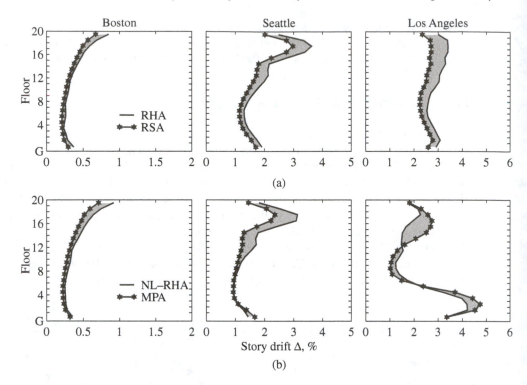

Figure 20.8.6 Median story drifts for (a) linearly elastic systems determined by RSA and RHA procedures, and (b) inelastic systems determined by MPA and nonlinear RHA procedures. Results are for SAC 20-story buildings. (Adapted from Goel and Chopra, 2004.)

these buildings treated as elastic systems determined by RSA and RHA procedures, and (b) demands for inelastic systems determined by MPA and nonlinear RHA. In implementing the RSA and MPA procedures, three modes were included for 9-story buildings and five modes for 20-story buildings.

Observe that the RSA procedure underestimates the median response of all six elastic systems. This underestimation tends to increase from bottom to top of buildings, consistent with the heightwise variation of the contribution of higher modes to the response (Section 19.6). The heightwise largest underestimation ranges from 15% for the Los Angeles 9-story building to 28% for the Boston 9-story building. By pervasive use of commercial software based on the modal combination approximation, the profession tacitly accepts this approximation, but perhaps has not recognized fully that it may lead to such significant underestimation of response.

The additional errors introduced by neglecting modal coupling in the MPA procedure, which are apparent by comparing parts (a) and (b) of Figs. 20.8.5 and 20.8.6, depend on how far the building responds into the inelastic range. This can be judged from the first-mode pushover curves and the peak values of roof displacement (Fig. 20.8.1). The additional errors in MPA (compared to those in RSA) are small for both Boston buildings because they remain essentially elastic; however, these errors increase slightly for

Seattle buildings because they are deformed moderately into the inelastic range; furthermore, they increase significantly for Los Angeles buildings, especially for the Los Angeles 9-story building, because they are deformed into the region of negative postyield stiffness and concomitant deterioration of lateral capacity, leading to collapse of its first-mode SDF system during several excitations.

20.9 SIMPLIFIED MODAL PUSHOVER ANALYSIS FOR PRACTICAL APPLICATION

For evaluating existing buildings or proposed designs of new buildings, the MPA procedure summarized in Section 20.7.3 can be simplified in two ways: The first simplification comes in determining the peak deformation D_n of the nth-mode inelastic SDF system that is needed in Eq. (20.7.3) to estimate the roof displacement u_{rn}, at which the nth-mode response r_n is determined by nonlinear static analysis of the structure (steps 7 to 9 in Section 20.7.3). In the results presented in Section 20.8, D_n was determined as the peak value of $D_n(t)$ obtained by nonlinear RHA of the SDF system for a given $\ddot{u}_g(t)$. Although implementation of such a numerical solution of Eq. (20.6.13) is straightforward using methods presented in Chapter 5, such computation can be avoided in practical applications of MPA.

One convenient approach is to estimate D_n directly from the earthquake design spectrum (Section 6.9) that defines the seismic hazard for the site, using the method presented in Section 7.12.2. Alternatively, D_n for an inelastic system can be estimated as the peak deformation of the corresponding linear system, which is read off the design spectrum, multiplied by the inelastic deformation ratio. Empirical equations for this ratio, defined as the ratio of peak deformations of inelastic and corresponding linear SDF systems, have been developed by several researchers.

The second simplification comes in computing the response contributions of modes higher than the first. The results of Fig. 20.8.1 to 20.8.3 and their interpretation suggested that consideration of inelastic behavior of the structure is essential in the first-mode pushover analysis, but may not be as important for the higher-mode analyses. The errors introduced in higher-mode demands by ignoring inelastic behavior of the structure are expected to be less significant in estimating the total demand, which contains important contributions of the first mode that are computed without introducing such an approximation.

Treating the building as linearly elastic in estimating the higher-mode contributions to seismic demands, a modified MPA procedure has been developed. It requires less computational effort because pushover analysis is required only for the first mode. Such a modified MPA is an attractive alternative for practical application because it leads to a larger estimate of seismic demand—although not necessarily more accurate estimate—than MPA, thus reducing the unconservatism of MPA results (relative to nonlinear RHA) in some cases and increasing their conservatism in others. Although this increase in demand is modest and acceptable for systems with moderate damping, at least 5%, it is unacceptably large for lightly damped systems.

FURTHER READING

Bobadilla, H., and Chopra, A. K., "Evaluation of the MPA Procedure for Estimating Seismic Demands: RC-SMRF Buildings," *Earthquake Spectra*, 24, 2008, pp. 827–845.

Chopra, A. K., and Chintanapakdee, C., "Inelastic Deformation Ratios for Design and Evaluation of Structures: Single-Degree-of-Freedom Bilinear Systems," *Journal of Structural Engineering, ASCE*, **130**, 2004, pp. 309–1319.

Chopra, A. K., and Goel, R. K., "A Modal Pushover Analysis Procedure for Estimating Seismic Demands for Buildings," *Earthquake Engineering and Structural Dynamics*, **31**, 2002, pp. 561–582.

Chopra, A. K., and Goel, R. K., "A Modal Pushover Analysis Procedure to Estimate Seismic Demands for Unsymmetric-Plan Buildings," *Earthquake Engineering and Structural Dynamics*, **33**, 2004, pp. 903–927.

Chopra, A. K., Goel, R. K., and Chintanapakdee, C., "Evaluation of a Modified MPA Procedure Assuming Higher Modes as Elastic to Estimate Seismic Demands," *Earthquake Spectra*, **20**, 2004, pp. 757–778.

Filippou, F. C., and Fenves, G. L., "Methods of Analysis for Earthquake-Resistant Strcutures," Chapter 6 in *Earthquake Engineering* (ed. Y. Bozorgnia and V. V. Bertero), CRC Press, Boca Raton, Fla., 2004.

Goel, R. K., and Chopra, A. K., "Evaluation of Modal and FEMA Pushover Analyses: SAC Buildings," *Earthquake Spectra*, **20**, 2004, pp. 225–254.

Goel, R. K., and Chopra, A. K., "Extension of Modal Pushover Analysis to Compute Member Forces," *Earthquake Spectra*, **21**, 2005, pp. 125–140.

Gupta, A., and Krawinkler, H., "Seismic Demands for Performance Evaluation of Steel Moment Resisting Frame Structures (SAC Task 5.4.3), *Report No. 132*, John A. Blume Earthquake Engineering Center, Stanford University, Stanford, Calif., 1999.

Gupta, A., and Krawinkler, H., "Behavior of Ductile SMRFs at Various Seismic Hazard Levels," *Journal of Structural Engineering, ASCE*, **126**, 2000a, pp. 98–107.

Gupta, A., and Krawinkler, H., "Dynamic *P*–Delta Effects for Flexible Inelastic Steel Structures," *Journal of Structural Engineering, ASCE*, **126**, 2000b, pp. 145–154.

Krawinkler, H., and Nassar, A. A., "Seismic Design Based on Ductility and Cumulative Damage Demands and Capacities," in *Nonlinear Seismic Analysis and Design of Reinforced Concrete Buildings* (ed. P. Fajfar and H. Krawinkler), Elsevier Applied Science, New York, 1992.

Nassar, A. A., and Krawinkler, H., "Seismic Demands for SDOF and MDF Systems," *Report No. 95*, John A. Blume Earthquake Engineering Center, Stanford University, Stanford, Calif., 1991.

21

Earthquake Dynamics of
Base-Isolated Buildings

PREVIEW

The concept of protecting a building from the damaging effects of an earthquake by introducing some type of support that isolates it from the shaking ground is an attractive one, and many mechanisms to achieve this result have been proposed. Although the early proposals go back 100 years, it is only in recent years that *base isolation* has become a practical strategy for earthquake-resistant design. In this chapter we study the dynamic behavior of buildings supported on base isolation systems with the limited objective of understanding why and under what conditions isolation is effective in reducing the earthquake-induced forces in a structure. Base isolation is currently an active and expanding subject, however, and a large body of literature exists on various aspects of base isolation: testing and mechanics of hardware in isolation systems, nonlinear dynamic analysis, shaking table tests, design projects, field installation, and field performance.

21.1 ISOLATION SYSTEMS

Despite wide variation in detail, base isolation techniques follow two basic approaches with certain common features. In the first approach the isolation system introduces a layer of low lateral stiffness between the structure and the foundation. With this isolation layer the structure has a natural period that is much longer than its fixed-base natural period. As shown by the elastic design spectrum of Fig. 21.1.1, this lengthening of period can reduce the pseudo-acceleration and hence the earthquake-induced forces in the structure,

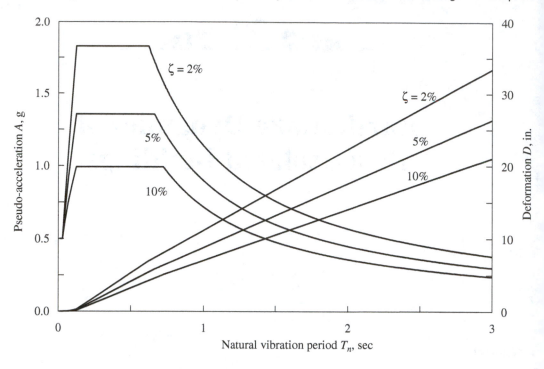

Figure 21.1.1 Elastic design spectrum.

but the deformation is increased; this deformation is concentrated in the isolation system, however, accompanied by only small deformations in the structure. This type of isolation system is effective even if the system is linear and undamped. Damping is beneficial, however, in further reducing the forces in the structure and the deformation in the isolation system.

The most common system of this type uses short, cylindrical bearings with one or more holes and alternating layers of steel plates and hard rubber (Fig. 21.1.2). Interposed between the base of the structure and the foundation, these laminated bearings are strong and stiff under vertical loads, yet very flexible under lateral forces (Fig. 21.1.3). Because the natural damping of the rubber is low, additional damping is usually provided by some form of mechanical damper. These have included lead plugs inserted into the holes, hydraulic dampers, steel bars, or steel coils. Metallic dampers provide energy dissipation through yielding, thus introducing nonlinearity in the system.

The second most common type of isolation system uses sliding elements between the foundation and the base of the structure. The shear force transmitted to the structure across the isolation interface is limited by keeping the coefficient of friction as low as practical. However, the friction must be sufficiently high to sustain strong winds and small earthquakes without sliding, a requirement that reduces the isolation effect. In this type of isolation system, the sliding displacements are controlled by high-tension springs or laminated rubber bearings, or by making the sliding surface curved; these

Figure 21.1.2   Section of a laminated rubber bearing. (Courtesy of I. D. Aiken.)

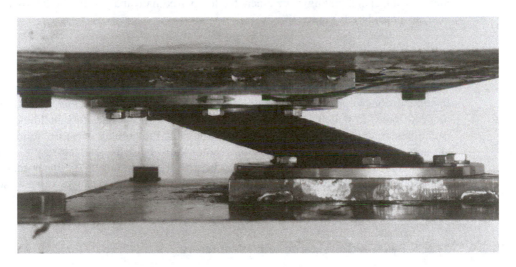

Figure 21.1.3   Deformed laminated rubber bearing. (Courtesy of I. D. Aiken.)

mechanisms provide a restoring force, otherwise unavailable in this type of system, to return the structure to its equilibrium position. The friction pendulum system (FPS) is a sliding isolation system wherein the weight of the structure is supported on spherical sliding surfaces that slide relative to each other when the ground motion exceeds a threshold level (Fig. 21.1.4). The restoring action is caused by raising the building slightly when sliding occurs on the spherical surface. The dynamics of structures on

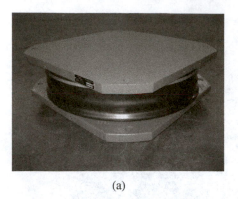
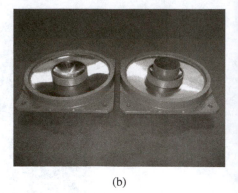

(a)   (b)

Figure 21.1.4   (a) Friction pendulum sliding bearing; (b) internal components. (Courtesy of Earth-quake Protection Systems.)

slider type of isolation systems is complicated because the slip process is intrinsically nonlinear.

   To avoid this complication, this introductory presentation is limited to understanding the dynamic behavior of structures using the isolation system with laminated rubber bearings. Such isolated buildings are amenable to approximate analysis by the familiar modal analysis procedure (Chapter 13).

21.2 BASE-ISOLATED ONE-STORY BUILDINGS

In this section we identify why base isolation is effective in reducing the earthquake-induced forces in buildings. For this purpose we consider a one-story building with an isolation system between the base of the building and the ground. Most isolation systems are nonlinear in their force–deformation relationships, but it is not necessary to consider these nonlinear effects in this introductory treatment of the subject. A linear analysis of the system would serve our purpose of gaining insight into the dynamics of base-isolated buildings. Nonlinearity in the force–deformation relation should be considered for final design, however.

21.2.1 System Considered and Parameters

The one-story building to be isolated is shown idealized in Fig. 21.2.1a together with its properties: lumped mass m, lateral stiffness k, and damping coefficient c. This is the familiar SDF system with natural frequency ω_n, natural period T_n, and damping ratio ζ. Here we use the subscript f instead of n to emphasize that these are properties of the structure on a fixed base (i.e., without any isolation system); thus

$$\omega_f = \sqrt{\frac{k}{m}} \qquad T_f = \frac{2\pi}{\omega_f} \qquad \zeta_f = \frac{c}{2m\omega_f} \tag{21.2.1}$$

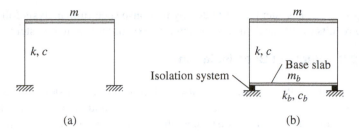

Figure 21.2.1   (a) Fixed-base structure; (b) isolated structure.

As shown in Fig. 21.2.1b, this one-story building is mounted on a base slab of mass m_b that in turn is supported on a base isolation system with lateral stiffness k_b and linear viscous damping c_b. Two parameters, T_b and ζ_b, are introduced to characterize the isolation system:

$$T_b = \frac{2\pi}{\omega_b} \quad \text{where} \quad \omega_b = \sqrt{\frac{k_b}{m + m_b}} \tag{21.2.2a}$$

$$\zeta_b = \frac{c_b}{2(m + m_b)\omega_b} \tag{21.2.2b}$$

We may interpret T_b as the natural vibration period, and ζ_b as the damping ratio, of the isolation system (with the building assumed to be rigid). For base isolation to be effective in reducing the forces in the building, T_b must be much longer than T_f, as we shall see later. The one-story building on a base isolation system (Fig. 21.2.1b) is a two-DOF system with mass, stiffness, and damping matrices denoted by **m**, **k**, and **c**, respectively. The disparity between the high damping in rubber bearings and the low damping of the building means that damping in the combined system is nonclassical.

21.2.2 Analysis Procedure

The response history of nonclassically damped systems can be determind by the extended modal analysis procedure (Chapter 14) or by numerical solution of the coupled equations of motion (Chapter 16). However, these approaches are not convenient for our objective to understand the dynamics of base-isolated buildings. Although, strictly speaking, classical modal analysis is not applicable to nonclassically damped systems, it can provide approximate results that suffice for our limited objective. This is the approach adopted here to determine the peak response of base-isolated structures to ground motion characterized by a design spectrum.

The two-DOF system that defines the one-story building on an isolation system is analyzed by the methods presented earlier in this book. With **m**, **k**, and **c** appropriately defined, Eq. (9.4.4) gives the equations of motion for the system; the natural vibration periods and modes of the system are determined following Example 10.4, and the earthquake response of the system is estimated by response spectrum analysis following Section 13.8. The results of this analysis are presented next for an example system.

21.2.3 Effects of Base Isolation

To understand the dynamics of base isolation, let us consider a specific system: $m_b = 2m/3$, $T_f = 0.4$ sec, $T_b = 2.0$ sec, $\zeta_f = 2\%$, and $\zeta_b = 10\%$. The base shear V_b in the building and the base displacement u_b are to be estimated using the elastic design spectrum of Fig. 21.1.1, shown for damping ratios 2, 5, and 10%. For 5% damping this design spectrum is the same as in Fig. 6.9.5 scaled to peak ground acceleration of 0.5g. The other two spectra were constructed similarly using appropriate amplification factors from Table 6.9.1; these were shown earlier in Figs 6.9.9 and 6.9.10.

Observe that we have chosen damping in the structure as 2% of critical damping, lower than the 5% typically assumed in earthquake analysis and design of structures. As mentioned in Chapter 11, the higher damping value accounts for the additional energy dissipation through nonstructural damage expected in conventional structures at the larger motion during earthquakes. The aim of base isolation is to reduce the forces imparted to the structure to such a level that no damage to the structure or nonstructural elements occur and thus a lower value of damping is appropriate.

Vibration properties. The natural vibration periods T_n and modes ϕ_n of the one-story building on an isolation system are shown in Fig. 21.2.2. In the first mode the isolator undergoes deformation but the structure behaves as essentially rigid; this mode is therefore called the *isolation mode*. The natural period of this mode, $T_1 = 2.024$ sec, indicates that the isolation system period, $T_b = 2.0$ sec, is changed only slightly by flexibility of the structure. The second mode involves deformation of the structure as well as in the isolation system, and the structural deformation is larger. Therefore, this is called the *structural mode*, although as we shall see later, this mode contributes little to the earthquake-induced forces in the structure. The natural period of this mode, $T_2 = 0.25$ sec, is significantly shorter than the fixed-base period, $T_f = 0.4$ sec, of the structure. The natural periods of the combined system are more separate than the isolation system period T_b and fixed-base period T_f of the structure.

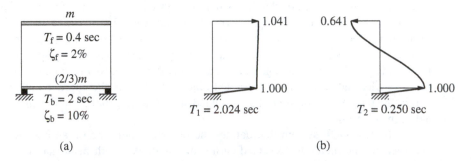

(a)   (b)

Figure 21.2.2 (a) One-story building on isolation system; (b) natural vibration modes and periods.

Modal static responses.   Introduced in Section 13.2.1, the modal expansion of the effective earthquake force distribution, $s = m1$, for the system of Fig. 21.2.2a is shown in Fig. 21.2.3. These striking results indicate that the first-mode forces s_1 are essentially the same as the total forces s, and the second-mode forces s_2 are very small. Static analysis of the system for forces s_n gives the modal static responses r_n^{st} for response quantity $r(t)$; see Table 13.2.1. In particular, for the base shear $V_b(t)$ in the structure and displacement $u_b(t)$ at the base, which is also the deformation of the isolation system, the modal static responses in the two modes are (see Fig. 21.2.3)

$$V_{b1}^{st} = 1.015m \qquad V_{b2}^{st} = -0.015m \qquad (21.2.3a)$$

$$\omega_1^2 u_{b1}^{st} = 0.976 \qquad \omega_2^2 u_{b2}^{st} = 0.024 \qquad (21.2.3b)$$

It is clear that the modal static responses for the second mode are negligible compared to the first mode. The second mode is therefore expected to contribute little to the earthquake response of the structure.

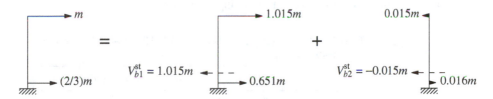

Figure 21.2.3   Modal expansion of effective earthquake forces and modal static responses for base shear.

Modal damping ratios.   The modal damping ratios are determined by Eq. (10.9.11), repeated here for convenience:

$$\zeta_n = \frac{C_n}{2M_n \omega_n} \qquad (21.2.4a)$$

where

$$M_n = \phi_n^T m \phi_n \quad \text{and} \quad C_n = \phi_n^T c \phi_n \qquad (21.2.4b)$$

For the system chosen, these equations give

$$\zeta_1 = 9.65\% \qquad \zeta_2 = 5.06\% \qquad (21.2.5)$$

Observe that the 9.65% damping in the first mode, the isolation mode, is very similar to the isolation-system damping, $\zeta_b = 10\%$; damping in the structure has little influence on modal damping in the isolation mode because the structure behaves as a rigid body for this mode. In contrast, high damping of the isolation system has increased the damping in the structural mode from 2% to 5.06%. Note that the coupling terms $C_{21} = C_{12} = \phi_1^T c \phi_2$ are nonzero, thus the modal equations are coupled (see Sections 10.9 and 12.4). It is this coupling we are neglecting in using classical modal analysis for estimating the peak response of this nonclassically damped system.

Peak modal and total responses.   The peak value of the nth-mode contribution $r_n(t)$ to response $r(t)$ is given by Eq. (13.7.1), repeated here for convenience:

$$r_n = r_n^{st} A_n$$

where $A_n \equiv A(T_n, \zeta_n)$ is the ordinate of the pseudo-acceleration response (or design) spectrum at period T_n for damping ratio ζ_n. Specializing this equation for the two response quantities of interest, base shear V_b in the structure and isolator deformation u_b, gives

$$V_{bn} = V_{bn}^{st} A_n \qquad u_{bn} = (\omega_n^2 u_{bn}^{st}) D_n \tag{21.2.6}$$

where $D_n = A_n/\omega_n^2$ is the deformation spectrum ordinate. These calculations are summarized in Table 21.2.1 using the A_n values noted in Fig. 21.2.4; these were obtained for the actual damping values, Eq. (21.2.5), by using Table 6.9.2 instead of interpolating between the spectrum curves for 2, 5, and 10% damping. Observe that the response due to the second mode, the structural mode, is negligible although the pseudo-acceleration is large—because the modal static response is small. Obtained by combining modal responses by the SRSS combination rule, the deformation in the isolator is 14.042 in. and the base shear is 36.5% of the building weight excluding the base slab.

TABLE 21.2.1   CALCULATION OF BASE SHEAR AND ISOLATOR DEFORMATION

| Mode | Base Shear | | | Isolator Deformation | | |
	A_n/g	V_{bn}^{st}/m	V_{bn}/w	D_n (in.)	$\omega_n^2 u_{bn}^{st}$	u_{bn} (in.)
1	0.359	1.015	0.365	14.390	0.976	14.042
2	1.347	−0.015	−0.021	0.823	0.024	0.020
SRSS			0.365			14.042

Reduction in base shear.   The base shear is much larger if the structure is not isolated. This fixed-base structure has a natural vibration period $T_f = 0.4$ sec and damping ratio $\zeta_f = 2\%$. For these parameters the design spectrum of Fig. 21.2.4 gives $A(T_f, \zeta_f) = 1.830g$. Thus the base shear in the fixed-base structure is

$$V_b = m A(T_f, \zeta_f) = m(1.830g) \quad \text{or} \quad \frac{V_b}{w} = 1.830 \tag{21.2.7}$$

that is, 183% of the weight w of the building excluding the base slab, about five times the base shear in the isolated building.

The isolation system reduces the base shear primarily because the natural period of the first mode, the isolation mode, providing most of the response, is much longer than the fixed-base period of the structure, leading to a smaller spectral ordinate, as seen in Fig. 21.2.4. This becomes clear by reexamining the terms entering into the base shear due to the first mode:

$$V_{b1} = V_{b1}^{st} A_1 = (1.015m)(0.359g) \tag{21.2.8}$$

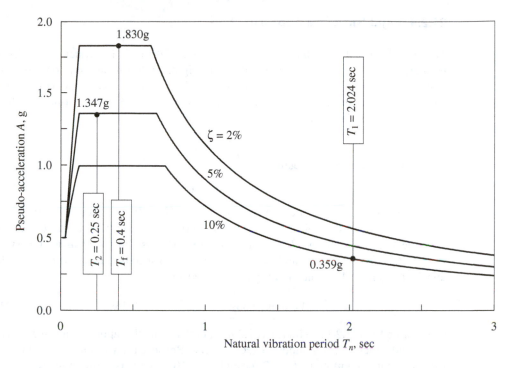

Figure 21.2.4   Design spectrum and spectral ordinates for fixed-base and isolated buildings.

Comparing Eq. (21.2.8) with (21.2.7a), it is apparent that because of the isolation system, the pseudo-acceleration is reduced from 1.830g to 0.359g, whereas the effective modal mass is essentially the same as the mass of the fixed-base building.

Why is base isolation effective?   Base isolation lengthens the fundamental vibration period of the structure, and thus reduces the pseudo-acceleration for this mode (for the design spectrum considered) and hence the earthquake-induced forces in the structure. The second mode that produces deformation in the structure is essentially not excited by the ground motion, although its pseudo-acceleration is large. This can be explained as follows: The first vibration mode of the base-isolated structure involves deformation only in the isolation system, the structure above being essentially rigid. Thus the first-mode component s_1 of the effective earthquake force distribution $s = m1$ is essentially the same as s, and the second-mode component s_2 is very small, causing very small modal static response in the second mode.

The primary reason for effectiveness of base isolation in reducing earthquake-induced forces in a building is the above-mentioned lengthening of the first-mode period. The damping in the isolation system and associated energy dissipation is only a secondary factor in reducing structural response.

21.2.4 Rigid-Structure Approximation

The base shear in the building and the deformation of the isolation system can be estimated by a simpler analysis treating the building as rigid. With this assumption the combined system has only one DOF. For this SDF system with natural period T_b and damping ratio ζ_b, the design spectrum gives the pseudo-acceleration $A(T_b, \zeta_b)$ and deformation $D(T_b, \zeta_b)$. Thus the isolator deformation is

$$u_b = D(T_b, \zeta_b) \qquad (21.2.9)$$

and the base shear in the structure is

$$V_b = m A(T_b, \zeta_b) \qquad (21.2.10)$$

The approximate results of Eqs. (21.2.9) and (21.2.10) are accurate for base-isolated systems if the period T_b of the isolation system (assuming rigid structure) is much longer than the fixed-base period T_f of the structure. This is illustrated using the system of Fig. 21.2.2: For $T_b = 2$ sec and $\zeta_b = 10\%$, the design spectrum gives $A(T_b, \zeta_b) = 0.359g$ and $D(T_b, \zeta_b) = 14.036$ in., and Eq. (21.2.10) gives

$$V_b = m(0.359g) \quad \text{or} \quad \frac{V_b}{w} = 0.359 \qquad (21.2.11)$$

Comparing Eq. (21.2.11) with Eq. (21.2.8), it is clear why this approximate analysis assuming a rigid structure gives almost "exact" results. Because the vibration properties with the rigid-structure assumption, $T_b = 2$ sec and $\zeta_b = 10\%$, are very close to the first-mode values, $T_1 = 2.024$ sec and $\zeta_1 = 9.65\%$, the spectral accelerations $A(T_b, \zeta_b)$ and $A(T_1, \zeta_1)$ are identical to three digits. Furthermore, the effective masses that enter into Eqs. (21.2.8) and (21.2.11) are essentially identical. Similarly, the isolator deformation from Eq. (21.2.9),

$$u_b = 14.036 \text{ in.} \qquad (21.2.12)$$

is essentially identical to the first-mode response in Table 21.2.1.

Because of its accuracy, the rigid-structure approximation provides an expedient means to estimate the effectiveness of a base isolation system and to estimate the isolator deformation. First, the ratio $A(T_b, \zeta_b)/A(T_f, \zeta_f)$ of two spectral ordinates gives the base shear in the isolated system as a fraction of the base shear in the fixed-base structure. Second, the deformation spectrum ordinate $D(T_b, \zeta_b)$ is the isolator deformation.

21.3 EFFECTIVENESS OF BASE ISOLATION

It is clear that the effectiveness of base isolation in reducing structural forces is closely tied to the lengthening of the natural period of the structure, and for this purpose the period ratio T_b/T_f should be as large as practical. In the example of the preceding section, the natural period of the fixed-base structure located the structure at the peak of the selected design spectrum. With base isolation, the natural period (of the isolation mode contributing almost all of the response) was shifted to the velocity-sensitive region of the spectrum with much smaller pseudo-acceleration. As a result, the base shear is reduced from 183% of the structural weight (excluding the base slab) to 36.5%. Whether the forces in the structure are reduced because of this period shift depends on the natural period of the fixed-base structure and on the shape of the earthquake design spectrum, among other factors. We illustrate these concepts next.

First, consider the same one-story building and base isolation system as in the preceding section to be located in Mexico City at the site where ground motions recorded during the 1985 earthquake produced the response spectrum shown in Fig. 21.3.1. Noted on this spectrum are the pseudo-acceleration values $A(T_f, \zeta_f) = 0.25g$ corresponding to $T_f = 0.4$ sec and $\zeta_f = 2\%$ for the fixed-base structure and $A(T_b, \zeta_b) = 0.63g$ associated with $T_b = 2.0$ sec and $\zeta_b = 10\%$ for the isolated structure (with the building assumed to be rigid). The ratio $A(T_b, \zeta_b)/A(T_f, \zeta_f) = 0.63g/0.25g = 2.52$ implies that the base shear in the isolated structure is approximately 2.52 times the base shear in the fixed-base structure. In this case base isolation is not helpful; in fact, it is harmful.

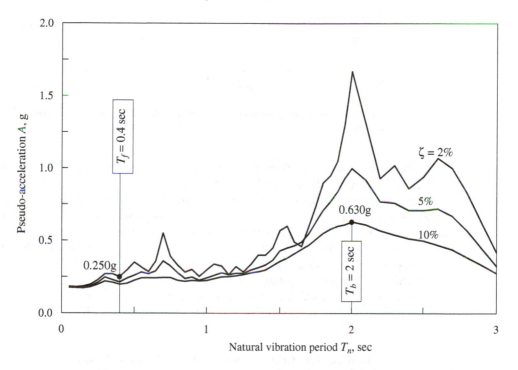

Figure 21.3.1   Response spectrum for ground motion recorded on September 19, 1985, at SCT site in Mexico City and spectral ordinates for fixed-base and isolated buildings.

Next, consider a structure with a relatively long fixed-base period and the ground motion characterized by the original design spectrum (Fig. 21.1.1). In this case we shall see that base isolation is only slightly beneficial—much less than when the fixed-base period was relatively short. To illustrate these results, consider a structure with a fixed-base period of 2 sec with other parameters for the structure and isolation system as before. Thus the system parameters are $T_f = 2$ sec, $\zeta_f = 2\%$, $m_b = \frac{2}{3}m$, $T_b = 2$ sec, and $\zeta_b = 10\%$.

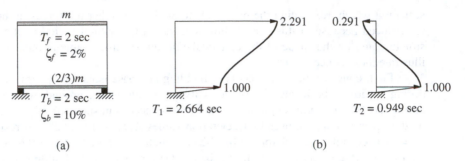

Figure 21.3.2 (a) One-story building on isolation system; (b) natural vibration modes and periods.

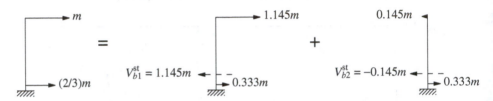

Figure 21.3.3 Modal expansion of effective earthquake forces and modal static responses for base shear.

Analysis of this system with $T_b = T_f$ by the procedures used for the example of Section 21.2 gives the natural vibration periods and modes (Fig. 21.3.2), the modal expansion of the effective earthquake force distribution: $\mathbf{s} = \mathbf{m1}$ (Fig. 21.3.3), and the modal damping ratios: 4.50% and 12.64%. In contrast to the previous system with $T_b \gg T_f$: (1) the structure does not behave as rigidly in the first mode, and this natural period is significantly affected by the flexibility of the structure; (2) the second-mode contribution to the effective earthquake forces is no longer negligible; and (3) the first-mode damping of 4.5% is no longer close to the isolation system damping of 10%.

A summary of the calculations to obtain the base shear and isolator deformation is presented in Table 21.3.1, with the spectral values identified in Fig. 21.3.4. In contrast

TABLE 21.3.1   CALCULATION OF BASE SHEAR AND ISOLATOR DEFORMATION

Mode	A_n/g	Base Shear		Isolator Deformation		
		V_{bn}^{st}/m	V_{bn}/w	D_n (in.)	$\omega_n^2 u_{bn}^{\text{st}}$	u_{bn} (in.)
1	0.348	1.145	0.398	24.136	0.500	12.068
2	0.691	−0.145	−0.101	6.095	0.500	3.047
SRSS			0.411			12.447

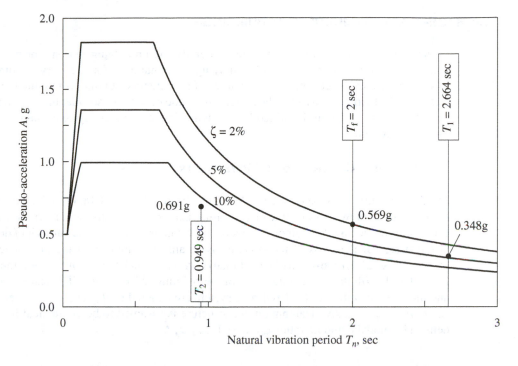

Figure 21.3.4   Design spectrum and spectral ordinates for fixed-base and isolated buildings.

to the previous system with $T_b \gg T_f$, the response of the second mode is significant, although it does not contribute much when it is combined—using the SRSS rule—with the first-mode response. We find the isolator deformation to be 12.447 in., and the base shear is 41.1% of the structural weight, excluding that of the base slab.

The fixed-base structure has a natural period $T_f = 2$ sec and damping ratio $\zeta_f = 2\%$, and the corresponding spectral ordinate is $A(T_f, \zeta_f) = 0.569g$ (Fig. 21.3.4). Thus if the structure were not isolated, the base shear is

$$V_b = m A(T_f, \zeta_f) \quad \text{or} \quad \frac{V_b}{w} = 0.569 \tag{21.3.1}$$

(i.e., 56.9% of the building weight). It is clear that some benefit is obtained by base isolation, although it is much less than if the vibration period of the structure had been shorter, as in the original example. It is for this reason that base isolation is rarely used for structures with natural period well into the velocity-sensitive region of the spectrum.

In passing we also note that the approximate analysis based on the assumption of a rigid structure is not accurate for a structure with a relatively long natural period. The approximate analysis gives a base shear coefficient of 35.9%, Eq. (21.2.11), compared to 41.1% from the complete analysis. The isolator deformation from the approximate analysis is 14.036 in., Eq. (21.2.12), compared to 12.447 in. from the complete analysis.

21.4 BASE-ISOLATED MULTISTORY BUILDINGS

In the preceding sections we were able to identify the underlying reasons for the effectiveness of a base isolation system by studying the dynamics of a one-story building. In this section we investigate how the dynamics of a multistory building is modified by base isolation. As before, we assume the system to be linear. We will see that the key concepts underlying base isolation, identified by the dynamics of one-story systems, carry over to multistory systems.

21.4.1 System Considered and Parameters

The N-story building to be isolated is shown idealized in Fig. 21.4.1a. On a fixed base, this system is defined by mass matrix $\mathbf{m}_f$, damping matrix $\mathbf{c}_f$, and stiffness matrix $\mathbf{k}_f$, which can be constructed by the methods developed in Chapters 9 and 11; the subscript f denotes "fixed base." If the mass of the structure is idealized as lumped at the floor levels, as shown in Fig. 21.4.1a, $\mathbf{m}_f$ is a diagonal matrix with diagonal element $m_{jj} = m_j$, the mass lumped at the jth floor. The total mass of the building, $M = \sum m_j$. The natural periods and modes of vibration of the fixed-base system are denoted by T_{nf} and ϕ_{nf}, respectively, where $n = 1, 2, \ldots, N$. Damping in the structure is assumed to be of classical form and defined by modal damping ratios $\zeta_{nf}, n = 1, 2, \ldots, N$.

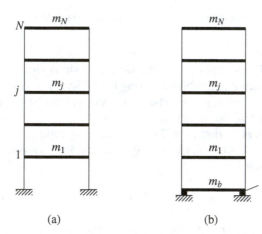

Figure 21.4.1 (a) Fixed-base N-story building; (b) isolated N-story building.

As shown in Fig. 21.4.1b, this N-story building is mounted on a base slab of mass m_b, supported in turn on a base isolation system with lateral stiffness k_b and linear viscous damping c_b. As in Section 21.2, two parameters, T_b and ζ_b, are introduced to characterize the isolation system:

$$T_b = \frac{2\pi}{\omega_b} \qquad \text{where} \quad \omega_b = \sqrt{\frac{k_b}{M + m_b}} \qquad (21.4.1a)$$

$$\zeta_b = \frac{c_b}{2(M + m_b)\omega_b} \qquad (21.4.1b)$$

As before, we may interpret T_b as the natural vibration period and ζ_b the damping ratio of the isolation system (with the building assumed to be rigid). For base isolation to be effective in reducing the earthquake-induced forces in the building, T_b must be much longer than T_{1f}, the fundamental period of the fixed-base building.

The N-story building on a base isolation system is an $(N + 1)$-DOF system with nonclassical damping because damping in the isolation system is typically much more than in the building. The mass, stiffness, and damping matrices of order $N + 1$ for the combined system are denoted by $\mathbf{m}$, $\mathbf{k}$, and $\mathbf{c}$, respectively.

21.4.2 Analysis Procedure

With ground motion characterized by a design spectrum, the RSA procedure of Chapter 13, Part B, will be used to analyze two systems: (1) a building on a fixed base, and (2) the same structure supported on an isolation system. In applying the RSA procedure to the isolated structure we are ignoring the coupling of modal equations due to nonclassical damping, typical of structures on isolation systems. The modal damping ratios of the isolated structure are given by Eq. (21.2.4). We focus on two response quantities: base shear in the building and the base displacement (or isolator deformation). The peak responses due to the nth mode of vibration are determined using Eq. (21.2.6), and these peak modal responses are combined by the SRSS rule. The results of such analyses are presented next for an example system.

21.4.3 Effects of Base Isolation

To understand how base isolation affects the dynamics of buildings, we consider a specific system. The fixed-base structure is a five-story shear frame (i.e., beam-to-column stiffness ratio $\rho = \infty$) with mass and stiffness properties uniform over its height: lumped mass $m = 100$ kips/g at each floor, and stiffnesses k for each story; k is chosen so that the fundamental natural vibration period $T_{1f} = 0.4$ sec. The classical damping matrix $\mathbf{c}_f = a_1 \mathbf{k}_f$ with a_1 chosen to obtain 2% damping in the fundamental mode. The base slab mass $m_b = m$ and the stiffness and damping of the isolation system are such that $T_b = 2.0$ sec and $\zeta_b = 10\%$ [Eq. (21.4.1)]. In this section we examine the vibration properties— natural periods and natural modes—modal damping ratios, and the earthquake response of two systems: (1) this five-story building on a fixed base, and (2) the same five-story building supported on the isolation system described above. The earthquake excitation is characterized by the design spectrum of Fig. 21.1.1.

Vibration properties.   The natural vibration periods and modes of both systems are presented in Fig. 21.4.2 and Table 21.4.1. The fixed-base structure has the familiar mode shapes and ratios of natural periods. In the first mode of the isolated building, the isolator undergoes deformation but the building behaves as essentially rigid; this mode is therefore called the isolation mode. The natural period of this mode, $T_1 = 2.030$ sec, indicates that the isolation-system period, $T_b = 2.0$ sec, is changed only slightly by the flexibility of the structure. The other modes involve deformation in the structure as well as in the isolation system. We refer to these modes as structural modes, although as we shall

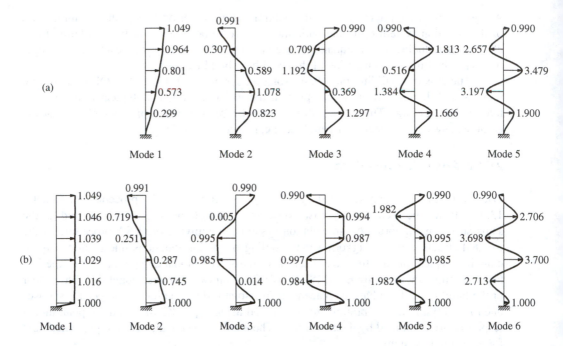

Figure 21.4.2   Natural vibration modes: (a) fixed-base building; (b) isolated building.

TABLE 21.4.1   NATURAL PERIODS AND MODAL DAMPING RATIOS

	Fixed-Base Building			Isolated Building	
Mode	T_{nf} (sec)	ζ_{nf} (%)	Mode	T_n (sec)	ζ_n (%)
			1	2.030	9.58
1	0.400	2.00	2	0.217	5.64
2	0.137	5.84	3	0.114	7.87
3	0.087	9.20	4	0.080	10.3
4	0.068	11.8	5	0.066	12.3
5	0.059	13.5	6	0.059	13.6

see later, these modes contribute little to the earthquake-induced forces in the structure. It is clear that the isolation system has a large effect on the natural period of the first structural mode but a decreasing effect on the higher-mode periods. In these higher modes the motion of the base mass decreases relative to the structural motions, and the base mass is acting essentially as a fixed base.

Modal damping ratios.   The modal damping ratios for both systems are presented in Table 21.4.1. The modal damping ratios for the fixed-base structure decrease linearly with natural period (i.e., increase linearly with natural frequency) because the damping is stiffness proportional (Section 11.4.1). The damping of 9.58% in the first mode of the isolated building, the isolation mode, is very similar to the isolation-system damping, $\zeta_b = 10\%$; damping in the structure has little influence on modal damping because the structure remains essentially rigid in this mode. The high damping of the isolation system has increased the damping in the first structural mode from 2.0% to 5.64%, but to a smaller degree in the higher modes.

Modal static responses.   We now compare the modal static response in the natural modes of both systems, the fixed-base and isolated buildings. The modal components $\mathbf{s}_n$ of the effective earthquake force distribution, $\mathbf{s} = \mathbf{m1}$, are shown in Fig. 13.2.4 for the fixed-base structure and in Fig. 21.4.3 for the base-isolated structure.[†] In the latter case, forces in the first mode, the isolation mode, are essentially the same as the total forces, and the forces associated with all the structural modes are very small. Static analysis of both systems for their respective modal forces gives the modal static shears V_{bn}^{st} at the base of the structure and modal static base displacements or isolator deformations u_{bn}^{st}; see Table 13.2.1. The results are given in Tables 21.4.2 and 21.4.3. It is clear that V_{bn}^{st} has

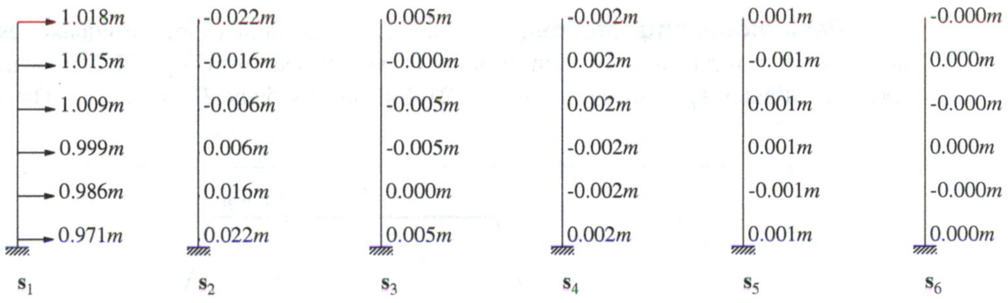

1.018m	-0.022m	0.005m	-0.002m	0.001m	-0.000m
1.015m	-0.016m	-0.000m	0.002m	-0.001m	0.000m
1.009m	-0.006m	-0.005m	0.002m	0.001m	-0.000m
0.999m	0.006m	-0.005m	-0.002m	0.001m	0.000m
0.986m	0.016m	0.000m	-0.002m	-0.001m	-0.000m
0.971m	0.022m	0.005m	0.002m	0.001m	0.000m
$\mathbf{s}_1$	$\mathbf{s}_2$	$\mathbf{s}_3$	$\mathbf{s}_4$	$\mathbf{s}_5$	$\mathbf{s}_6$

Figure 21.4.3   Modal components of effective earthquake forces for a five-story building on isolation system.

TABLE 21.4.2 CALCULATION OF BASE SHEAR IN FIXED-BASE AND ISOLATED BUILDINGS

Fixed-Base Building				Isolated Building			
Mode	A_n/g	V_{bn}^{st}/m	V_b/W	Mode	A_n/g	V_{bn}^{st}/m	V_b/W
				1	0.359	5.028	0.361
1	1.830	4.398	1.609	2	1.291	-0.021	-0.005
2	1.272	0.436	0.111	3	1.058	-0.005	-0.001
3	0.859	0.121	0.021	4	0.792	-0.002	-0.000
4	0.700	0.038	0.005	5	0.682	-0.0005	-0.000
5	0.638	0.008	0.001	6	0.635	-0.0001	-0.000
SRSS			1.613	SRSS			0.361

[†]Figure 13.2.4 is valid for this uniform five-story shear frame because its vibration modes are the same as those of the system considered in Section 13.2.6, although the story stiffness (and natural vibration periods) of the two systems are different.

TABLE 21.4.3 CALCULATION OF
ISOLATOR DEFORMATION

Mode	D_n	$\omega_n^2 u_{bn}^{st}$	u_{bn} (in.)
1	14.470	0.971	14.045
2	0.597	0.022	0.013
3	0.133	0.005	0.001
4	0.050	0.002	0.000
5	0.029	0.001	0.000
6	0.022	0.0001	0.000
SRSS			14.045

significant values in the first two modes of the fixed-base structure. However, for the isolated building, V_{bn}^{st} is small in all the structural modes, and the response in these modes is expected to be negligible. The isolation mode provides the dominant value V_{b1}^{st} and therefore will provide most of the response.

Peak modal and total responses.   The peak value of the earthquake response due to each natural mode of both systems is determined from Eq. (21.2.6), where the spectral ordinates A_n/g are shown in Figs. 21.4.4 and 21.4.5 and $D_n = A_n/\omega_n^2$. These peak

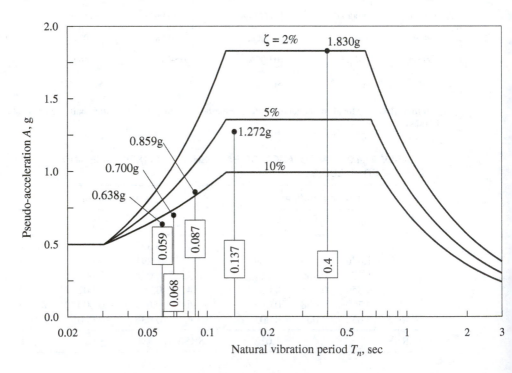

Figure 21.4.4   Design spectrum and spectral ordinates for fixed-base five-story building.

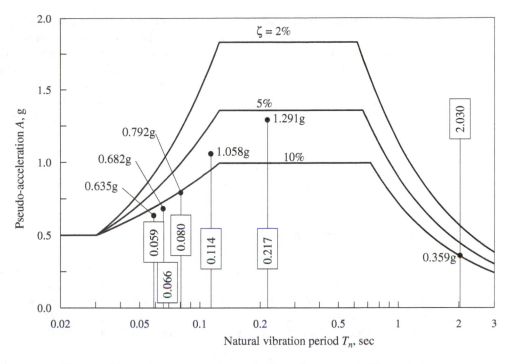

Figure 21.4.5   Design spectrum and spectral ordinates for five-story building on isolation system.

modal responses are presented in Tables 21.4.2 and 21.4.3 together with their combined value determined by the SRSS rule. Observe that as predicted from the modal static responses, the dynamic response of the isolated building due to all its structural modes is negligible although their pseudo-accelerations are large. The isolation mode alone produces essentially the entire response: isolation system deformation of 14.045 in. and base shear equal to 36.1% of W, the 500-kip weight of the building excluding that of the base slab. The response in the first two modes of the fixed-base building is significant; however, the second mode contributes little to the SRSS-combined value of 161.3% of W.

Reduction in base shear.   To understand the underlying reasons for this drastic reduction in base shear, we examine the peak modal responses in both fixed-base and isolated systems. Each peak modal response is the product of two parts: the modal static response V_{bn}^{st} and the pseudo-acceleration A_n. Each part is examined for the first mode of the base-isolated building and of the fixed-base building; this is the mode that provides most of the response in each case. Observe that $V_{b1}^{st} = 5.028m$ for the isolated building, which is slightly larger than $V_{b1}^{st} = 4.398m$ for the fixed-base building. However, $A_1 = 0.359g$ for the isolated building (Fig. 21.4.5) is only one-fifth of $A_1 = 1.830g$ for the fixed-base building (Fig. 21.4.4); as a result, the first-mode base shear coefficient of 36.1% in the base-isolated building is much smaller than 160.9% for the fixed-base building.

Why is base isolation effective?   The isolation system reduces the base shear primarily because the natural vibration period of the isolation mode, providing most of the response, is much longer than the fundamental period of the fixed-base structure, leading to a much smaller spectral ordinate. This is typical of design spectra on firm ground and fixed-base structures with fundamental period in the flat portion of the acceleration-sensitive region of the spectrum. The higher modes are essentially not excited by the ground motion—although their pseudo-accelerations are large—because their modal static responses are very small.

The primary reason for effectiveness of base isolation in reducing earthquake-induced forces in a building is the above-mentioned lengthening of the first mode period. The damping in the isolation system and associated energy dissipation is only a secondary factor in reducing structural response.

21.4.4 Rigid-Structure Approximation

The base shear in the isolated building and the deformation of the isolation system can be estimated by a simpler analysis, treating the building as rigid. The natural period of the resulting SDF system is T_b and its damping ratio is ζ_b [Eq. (21.4.1)]; the associated design spectrum ordinates are $A(T_b, \zeta_b)$ for the pseudo-acceleration and $D(T_b, \zeta_b)$ for the deformation. Thus the base shear in the structure and the isolator deformation are

$$V_b = M A(T_b, \zeta_b) \qquad u_b = D(T_b, \zeta_b) \qquad (21.4.2)$$

This approximate procedure will provide excellent results if the isolation-system period T_b is much longer than the fundamental period T_{1f} of the fixed-base structure. This is illustrated using the system of Fig. 21.4.1b, analyzed earlier. For this system, $T_b = 2.0$ sec and $\zeta_b = 10\%$ and the spectral values are $A(T_b, \zeta_b) = 0.359g$ and $D(T_b, \zeta_b) = 14.036$ in., as noted in Section 21.2.4. Substituting these values in Eq. (21.4.2) gives

$$V_b = 0.359W \qquad u_b = 14.036 \text{ in.} \qquad (21.4.3)$$

which are essentially identical to the responses due to the isolation mode (and to the total response) presented in Tables 21.4.2 and 21.4.3.

21.5 APPLICATIONS OF BASE ISOLATION

Base isolation provides an alternative to the conventional, fixed-base design of structures and may be cost-effective for some new buildings in locations where very strong ground shaking is likely. It is an attractive alternative for buildings that must remain functional after a major earthquake (e.g., hospitals, emergency communications centers, computer processing centers). Several new buildings have been isolated using rubber (or elastomeric) bearings or FPS isolators; these examples are described in references at the end of this chapter.

Figure 21.5.1   San Francisco City Hall. (Courtesy of S. Nasseh.)

Both types of isolation systems have also been used for retrofit of existing build-ings that are brittle and weak: for example, unreinforced masonry buildings or reinforced-concrete buildings of early design, not including the type of detailing of the reinforcement necessary for ductile performance. Conventional seismic strengthening designs require adding new structural members, such as shear walls, frames, and bracing. Base isolation minimizes the need for such strengthening measures by reducing the earthquake forces imparted to the building. It is therefore an attractive retrofit approach for buildings of his-torical or architectural merit whose appearance and character must be preserved. Many ex-amples of retrofitting existing buildings are described in end-of-chapter references. How-ever, it is difficult and expensive to construct a new foundation system for the isolators, to modify the base of the building so that it can be supported on isolators, and to shore up the building during construction of the isolation and foundation systems.

A good example of a retrofit application of laminated-rubber-bearing-isolation sys-tems is the San Francisco City Hall in San Francisco, California. Constructed in 1915 to replace the previous structure that was destroyed in the 1906 San Francisco earthquake, this building is an outstanding example of classical architecture and is listed in the National Register of Historic Places (Fig. 21.5.1). The five-story building with its dome rising 300 ft above the ground floor is 309 ft by 408 ft in plan, occupying two city blocks. The structural system is a steel frame and concrete slabs with unreinforced brick masonry integral with the granite cladding, hollow clay tile infill walls, and limestone or marble panels lining many of the interior spaces.

Substantial damage sustained from the 1989 Loma Prieta earthquake, centered about 60 miles away, necessitated repairs and strengthening. The fixed-base fundamental period of vibration of the building is approximately 0.9 sec, implying that large ductility demands can be imposed on the structure by strong shaking expected at the building site from an earthquake centered on a nearby segment of the San Andreas fault. To improve the earthquake resistance of this structure, base isolation was adopted especially because it preserved the historic fabric of this building. In addition, the superstructure was strengthened by new shear walls in the interior of the building. This retrofit project was completed in 1998.

The isolation system consisted of 530 isolators, each a laminated rubber bearing with lead plugs, located at the base of each column and at the base of the shear walls (Fig. 21.5.2). The 21-in.-high bearings varied from 31 to 36 in. in diameter. The columns are supported on one or more isolators under a cruciform-shaped steel structure; multiple isolators were provided for the heavily loaded columns. Installation of the isolators proved to be very complicated and required shoring up the columns, cutting the columns, and transferring the column loads to temporary supports. The plane of isolation is just above the existing foundation.

(a)   (b)

Figure 21.5.2   San Francisco City Hall: laminated rubber bearings at the base of (a) columns and (b) shear walls. [(a) Courtesy of J. M. Kelly; (b) Robert Canfield photo, courtesy of S. Nasseh.]

The isolated building is estimated to move 18 to 26 in. at an isolation period of 2.5 sec for a design earthquake with peak ground acceleration of 0.4g. To permit this motion, a moat was constructed around the building to provide a minimum seismic gap of 28 in. Flexible joints were provided for utilities—plumbing, electrical, and telephone lines—crossing this moat space to accommodate movement across the isolation system.

A good example of new construction using base isolation is the new (completed in 2000) International Terminal at the San Francisco Airport (Figs. 21.5.3 and 21.5.4), which was designed to remain operational after a Magnitude 8 earthquake on the San Andreas Fault approximately 3 miles away. To achieve this performance goal it was

Figure 21.5.3 International Terminal at San Francisco Airport. (T. Hursley, photo, courtesy of Skidmore, Owings, & Merrill LLP.)

Figure 21.5.4 International Terminal at San Francisco Airport. (T. Hursley, photo, courtesy of Skidmore, Owings, & Merrill LLP.)

Figure 21.5.5 International Terminal at San Francisco Airport: FPS bearing at base of column. (P. Lee, photo, courtesy of Skidmore, Owings, & Merrill LLP.)

decided to isolate the superstructure, which consists of steel concentric and eccentric braced frames with fully welded moment connections.

The isolation system consists of 267 isolators, one at the base of each column (Fig. 21.5.5). Each isolator is a friction pendulum sliding bearing. The cast steel bearing consists of a stainless steel spherical surface and articulated slider, which allows a lateral displacement up to 20 in. and provides an isolation period of 3 sec.

Base isolation had the effect of reducing the earthquake force demands on the superstructure to 30% of the demands for a fixed-base structure. With this force reduction it was feasible to design the superstructure to remain essentially elastic and hence undamaged under the selected design earthquake with peak ground acceleration of 0.6g.

FURTHER READING

Kelly, J. M., *Earthquake Resistant Design with Rubber*, 2nd ed., Springer-Verlag, London, 1996, Chapters 1, 3, and 4.

Naeim, F., and Kelly, J. M., *Design of Seismic Isolated Structures: From Theory to Practice*, Wiley, Chichester, U.K., 1999.

Seismic Isolation: From Idea to Reality, theme issue of *Earthquake Spectra*, **6**, 1990, pp. 161–432.

Skinner, R. I., Robinson, W. H., and McVerry, G. H., *An Introduction to Seismic Isolation*, Wiley, Chichester, U.K., 1993.

Warburton, G. B., *Reduction of Vibrations*, The 3rd Mallet–Milne Lecture, Wiley, Chichester, U.K., 1992, pp. 21–46.

22

Structural Dynamics in Building Codes

PREVIEW

Most seismic building codes permit the use of a static equivalent lateral force (ELF) procedure for many regular structures with relatively short periods. For other structures, dynamic analysis procedures are required. According to the ELF procedure, structures are designed to resist specified *static* lateral forces related to the properties of the structure and the seismicity of the region. Based on an estimate of the fundamental natural vibration period of the structure, formulas are specified for the base shear and the distribution of lateral forces over the height of the building. Static analysis of the building for these forces provides the design forces, including shears and overturning moments for the various stories, with some codes permitting reductions in the statically computed overturning moments. These seismic design provisions in four building codes—*International Building Code* (United States),[†] *National Building Code of Canada, Mexico Federal District Code,* and *Eurocode 8*—are presented in Part A of this chapter together with their relationship to the theory of structural dynamics developed in Chapters 6, 7, 8, and 13. The code provisions presented are not complete; those provisions that we are unprepared to evaluate based on this book have been excluded or only mentioned: effects of local soil conditions, torsional moments about a vertical axis, combination of earthquake forces due to the simultaneous action of ground motion components, and the requirements for detailing structures to ensure ductile behavior, among others. In Part B of the chapter the code provisions are evaluated in light of the results of the dynamic analysis of buildings, discussed in Chapters 19 and 20.

[†]In the 2009 IBC, most of the technical requirements are adopted by reference to the ASCE 7-05 document.

Most codes permit both response spectrum analysis (RSA) and response history analysis (RHA) procedures for dynamic analysis of structures. The code versions of these procedures are not presented because they are essentially equivalent to those that were developed in Chapter 13.

PART A: BUILDING CODES AND STRUCTURAL DYNAMICS[†]

22.1 *INTERNATIONAL BUILDING CODE* (UNITED STATES), 2009

22.1.1 Base Shear

The 2009 edition of the *International Building Code* (IBC) specifies the base shear as

$$V_b = C_s W \tag{22.1.1}$$

where W is the total dead load and applicable portions of other loads, and the seismic coefficient,

$$C_s = \frac{C_e}{R} \tag{22.1.2}$$

This coefficient corresponding to $R = 1$ is called the *elastic seismic coefficient:*

$$C_e = IC \tag{22.1.3}$$

where the importance factor $I = 1.0$, 1.25, or 1.5; $I = 1$ for most structures; $I = 1.25$ for structures that have a "substantial public hazard due to occupancy or use"; and $I = 1.5$ for "essential facilities" that are required in post-earthquake recovery and for facilities containing hazardous substances.

The period-dependent coefficient C depends on the location of the structure and the site class. The variation of C with site classes A, B, C, D, E, and F defined in the code accounts for local soil effects on earthquake ground motion, a topic not covered in the book. C is related to ordinates of the pseudo-acceleration design spectrum: $A(T_{n,\text{short}})$, the pseudo-acceleration at short periods; and $A(T_n = 1.0\,\text{sec})$, the pseudo-acceleration at a 1.0-sec period. Maps of the United States show these two A-values for ground motion due to the *maximum considered earthquake* (MCE). These A-values are multiplied by $\frac{2}{3}$ to obtain A-values for the *design basis earthquake* (DBE).

Consider a location representative of coastal California regions not in the near field of known active faults. For site class B the maps provide $A(T_{n,\text{short}}) \div g = 1.5$ and $A(T_n = 1.0\,\text{sec}) \div g = 0.6$, which are multiplied by $\frac{2}{3}$ to obtain 1.0 and 0.4, respectively.

[†]The notation employed in building codes is not used in presenting the code equations. Instead, equations from all four codes are presented in a common (to the extent possible) notation, consistent with preceding chapters. Supplementary equations are included to facilitate interpretation of code equations.

The numerical coefficient C is then given as

$$C = \begin{cases} 1.0 & T_1 \leq 0.4 \\ 0.4/T_1 & 0.4 \leq T_1 \leq T_L \\ 0.4T_L/T_1^2 & T_1 \geq T_L \end{cases} \qquad (22.1.4)$$

where T_1 is the fundamental natural vibration period of the structure in seconds, and $T_L = 12$ or 8 sec in coastal regions of northern California and southern California, respectively. For highly seismic areas, the value of C obtained from Eq. (22.1.4) for longer-period buildings is not permitted to be below a specified minimum value; it is 0.3 for the example location noted above.

The code permits computation of the fundamental natural vibration period by procedures presented in Chapters 8 and 10. Alternatively, the code gives empirical formulas for T_1 that depend on building material (steel, reinforced concrete, etc.), building type (frame, shear wall, etc.), and overall dimensions.

Figure 22.1.1 shows the elastic seismic coefficient for the foregoing example location and site class, and $I = 1$ is valid for most structures. The code also specifies (for use in dynamic analysis) the elastic pseudo-acceleration design spectrum that is the basis for defining C; shown in Fig. 22.1.1, this spectrum for the example location considered is

$$A/g = \begin{cases} 0.4 + 7.5T_n & 0 \leq T_n \leq 0.08 \\ 1.0 & 0.08 < T_n \leq 0.4 \\ 0.4/T_n & T_n > 0.4 \end{cases} \qquad (22.1.5)$$

where T_n is the natural vibration period (in seconds) of an SDF system.

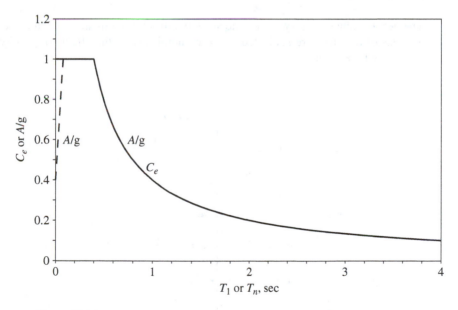

Figure 22.1.1 IBC (2009): elastic coefficient C_e and pseudo-acceleration A/g for a location in California coastal regions not in the near field of known active faults.

The response modification factor R in Eq. (22.1.2) depends on several factors, including the ductility capacity and inelastic performance of structural materials and systems during past earthquakes. Specified values of R vary between 1.5 (for certain bearing wall systems) and 8 (for moment-resisting frames with special detailing for ductile behavior). It is important to note that R is not equal to the ductility capacity required.

22.1.2 Lateral Forces

The distribution of lateral forces over the height of the building is determined from the base shear in accordance with the formula for the lateral (or horizontal) force at the jth floor (Fig. 22.1.2):

$$F_j = V_b \frac{w_j h_j^k}{\sum_{i=1}^{N} w_i h_i^k} \tag{22.1.6}$$

where w_i is the weight at the ith floor at height h_i above the base and k is a coefficient related to the vibration period T_1 as follows:

$$k = \begin{cases} 1 & T_1 \leq 0.5 \\ (T_1 + 1.5)/2 & 0.5 \leq T_1 \leq 2.5 \\ 2 & T_1 \geq 2.5 \end{cases} \tag{22.1.7}$$

22.1.3 Story Forces

The design values of story shears and story overturning moments are determined by static analysis of the structure subjected to these lateral forces; the effects of gravity and other loads should be included.

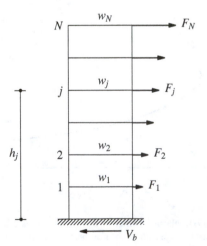

Figure 22.1.2   IBC lateral forces.

22.2 *NATIONAL BUILDING CODE OF CANADA*, 2010

22.2.1 Base Shear

The base shear formula in the 2010 edition of the *National Building Code of Canada* (NBCC) may also be expressed as Eq. (22.1.1), where the seismic coefficient is

$$C_s = \frac{C_e}{R_d R_O} \tag{22.2.1}$$

In this equation, R_O is the overstrength-related force modification factor that accounts for the reserve (or unaccounted for) strength in a code-designed structure. It varies from 1.0 for unreinforced masonry structures to 1.7 for ductile coupled-wall concrete structures.

The elastic seismic coefficient

$$C_e = I A_1 M_v \tag{22.2.2}$$

is the product of three factors:

1. The importance factor I varies between 0.8 and 1.5; it is 1.0 for "normal" buildings and 1.5 for buildings in the "post-disaster category."
2. The second factor is the pseudo-acceleration design spectrum value A_1 at the fundamental vibration period T_1 of the building. Tables are provided for the peak ground acceleration and the pseudo-acceleration spectral values (for 5% damping) for site class C (very dense soil and soft rock) for several period values. Two factors are specified to obtain spectral values for other site classes: one for the acceleration-sensitive region and the other for the velocity-sensitive region of the spectrum,
3. The factor M_v accounts for higher-mode contributions to the base shear. The following values of M_v are applicable for western Canada. For moment-resisting frames, braced frames, and coupled walls, $M_v = 1.0$ for all values of T_1. For walls and wall-frame systems, M_v is 1.0 for $T_1 \leq 1.0$ sec, 1.2 for $T_1 = 2.0$ sec, and 1.6 for $T_1 \geq 4$ sec. For other systems, $M_v = 1$ for $T_1 \leq 1.0$ sec and 1.2 for $T_1 \geq 2.0$ sec. For intermediate values of T_1, the product $A_1 M_v$ is obtained by interpolation.

For all structures other than walls, coupled walls, and wall-frame systems, the factor C_e is constant for $T_1 > 2.0$ sec and equal to its value at $T_1 = 2.0$ sec. For walls, coupled walls, and wall-frame systems, C_e at $T_1 = 4.0$ sec is half its value at $T_1 = 2.0$ sec, and is constant at the latter value for periods longer than 4.0 sec.

Although the design base shear is, in general, obtained from the seismic coefficient defined in Eq. (22.2.1), the code limits the value of C_e to two-thirds of $A(0.2$ sec$)$ if $R_a \geq 1.5$ and if the site does not belong to site class F (liquefiable soils, and sensitive, organic, and highly plastic clays).

In addition to empirical formulas for estimating the fundamental period, the NBCC gives the formula

$$T_1 = 2\pi \left[\frac{\sum_{i=1}^{N} w_i u_i^2}{g \sum_{i=1}^{N} w_i u_i} \right]^{1/2} \tag{22.2.3}$$

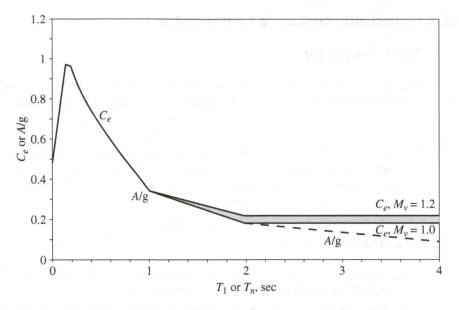

Figure 22.2.1   NBCC (2010): pseudo-acceleration A/g and elastic ($R_d R_O = 1$) seismic coefficient C_e for $I = 1$ and two values of M_v: 1.0 and 1.2; valid for all structures other than walls, coupled walls, and wall-frame systems.

where u_i are the floor displacements due to static application of a set of lateral forces w_i at floor levels in an N-story building.

Consider a location in Vancouver, B.C. The peak ground acceleration $\ddot{u}_{go}/g$ specified for site class C is 0.48; and the pseudo-acceleration spectral values $A/g = 0.83, 0.97, 0.96,$ 0.84, 0.74, 0.66, 0.34, 0.18, and 0.09 at $T_1 = 0.1, 0.15, 0.2, 0.3, 0.4, 0.5, 1.0, 2.0,$ and 4.0 sec, respectively. Figure 22.2.1 shows this design spectrum and C_e for the two values $M_v = 1.0$ and 1.2, and for buildings in the normal importance category ($I = 1$); C_e and A/g are identical up to a period of 2 sec for $M_v = 1.0$.

The force modification factor R_d in Eq. (22.2.1) is based on the ductility capacity of the structural system and materials. Specified values of R_d range from 1.0 for brittle structures such as unreinforced masonry to 5.0 for ductile moment-resisting steel frames.

22.2.2 Lateral Forces

The distribution of lateral forces over the height of the building is to be determined from the base shear in accordance with the formula for the lateral (or horizontal) force at the jth floor:

$$F_j = (V_b - F_t)\frac{w_j h_j}{\sum_{i=1}^{N} w_i h_i} \tag{22.2.4}$$

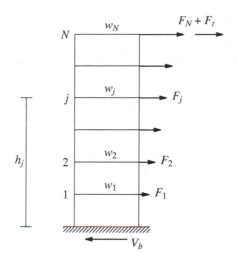

Figure 22.2.2   NBCC lateral forces.

with the exception that the force at the top floor (or roof) computed from Eq. (22.2.4) is increased by an additional force, the top force:

$$F_t = \begin{cases} 0 & T_1 \leq 0.7 \\ 0.07T_1V_b & 0.7 < T_1 < 3.6 \\ 0.25V_b & T_1 \geq 3.6 \end{cases} \qquad (22.2.5)$$

where h_j is the height of the jth floor above the base. These lateral forces are shown in Fig. 22.2.2.

22.2.3 Story Forces

The design values of story shears are determined by static analysis of the structure subjected to these lateral forces. Similarly determined overturning moments are multiplied by reduction factors J and J_i at the base of the structure and at the ith floor level, respectively. The following values of J are applicable for western Canada: for all structures $J = 1$ for $T_1 \leq 0.5$ sec; for moment-resisting frames $J = 0.9$ for $T_1 \geq 2.0$ sec; for coupled walls $J = 0.9$ for $T_1 = 2.0$ sec and $J = 0.8$ for $T_1 \geq 4.0$ sec; for braced frames $J = 0.8$ for $T_1 \geq 2.0$ sec; for walls and wall-frame systems $J = 0.6$ for $T_1 = 2.0$ sec and $J = 0.5$ for $T_1 \geq 4.0$ sec. Values of J at intermediate periods are determined by interpolation. The reduction factor $J_i = 1.0$ over the top 40% of the building height, and varies linearly from this value to the J value at the base.

22.3 *MEXICO FEDERAL DISTRICT CODE,* 2004

22.3.1 Base Shear

The base shear formula in the 2004 edition of the *Mexico Federal District Code* (MFDC) may also be expressed as Eq. (22.1.1) with the seismic coefficient

$$C_s = \frac{C_e}{Q'} \qquad (22.3.1)$$

The elastic seismic coefficient

$$C_e = \begin{cases} A/g & T_1 \le T_c \\ A/g\{1 + 0.5r[1 - (T_c/T_1)^r]\} & T_1 > T_c \end{cases} \tag{22.3.2}$$

where the pseudo-acceleration design spectrum A/g is given by

$$\frac{A}{g} = \begin{cases} a_o + (A_m - a_o)T_n/T_b & T_n < T_b \\ A_m & T_b \le T_n \le T_c \\ A_m(T_c/T_n)^r & T_n > T_c \end{cases} \tag{22.3.3}$$

and T_1 is the fundamental period; T_b and T_c denote the periods at the beginning and end of the constant pseudo-acceleration region of the design spectrum (Fig. 22.3.1). The coefficients a_o, A_m, and r and period values T_b and T_c are given in Table 22.3.1 for the seven

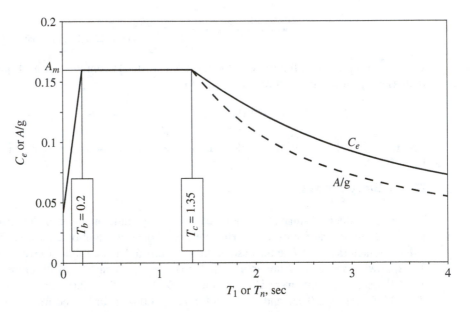

Figure 22.3.1 MFDC (2004): elastic $(Q' = 1)$ seismic coefficient C_e for zone I, and pseudo-acceleration A/g.

TABLE 22.3.1 PSEUDO-ACCELERATION DESIGN SPECTRUM PARAMETERS

Zone	a_o	A_m	T_b (sec)	T_c (sec)	r
I: Hard ground	0.04	0.16	0.20	1.35	1
II: Transition	0.08	0.32	0.35	1.35	$\frac{4}{3}$
IIIa: Soft soil a	0.10	0.40	0.53	1.80	2
IIIb: Soft soil b	0.11	0.45	0.85	3.00	2
IIIc: Soft soil c	0.10	0.40	1.25	4.20	2
IIId: Soft soil c	0.10	0.30	0.85	4.20	2

zones of the Mexico Federal District, depending primarily on the local soil conditions. Figure 22.3.1 shows A/g and C_e for zone I.

For the fundamental natural vibration period, the MFDC provides the formula

$$T_1 = 2\pi \left[\frac{\sum_{i=1}^{N} w_i u_i^2}{g \sum_{i=1}^{N} F_i u_i} \right]^{1/2} \tag{22.3.4}$$

where u_i are the floor displacements due to static application of the set of lateral forces F_i defined by Eq. (22.3.6).

The elastic seismic coefficient is divided by the seismic reduction factor:

$$Q' = \begin{cases} 1 + (T_1/T_b)(Q-1) & T_1 < T_b \\ Q & T_1 \geq T_b \end{cases} \tag{22.3.5}$$

where the seismic behavior factor Q varies between 1 and 4, depending on several factors, including the structural material, structural system, and level of detailing.

22.3.2 Lateral Forces

The formula for the lateral force F_j at the jth floor depends on whether $T_1 \leq T_c$ or $T_1 > T_c$:

$$F_j = V_b \frac{w_j h_j}{\sum_{i=1}^{N} w_i h_i} \qquad T_1 \leq T_c \tag{22.3.6a}$$

and

$$F_j = V_b^{(1)} \frac{w_j h_j}{\sum_{i=1}^{N} w_i h_i} + V_b^{(2)} \frac{w_j h_j^2}{\sum_{i=1}^{N} w_i h_i^2} \qquad T_1 > T_c \tag{22.3.6b}$$

where the base shear V_b has been separated into two parts, $V_b^{(1)}$ and $V_b^{(2)}$:

$$V_b^{(1)} = \frac{W(A/g)}{Q'} \left\{ 1 - r \left[1 - \left(\frac{T_c}{T_1} \right)^r \right] \right\} \tag{22.3.7a}$$

$$V_b^{(2)} = \frac{W(A/g)}{Q'} \left\{ 1.5r \left[1 - \left(\frac{T_c}{T_1} \right)^r \right] \right\} \tag{22.3.7b}$$

22.3.3 Story Forces

The design values of story shears are determined by static analysis of the structure subjected to the lateral forces defined by the foregoing equations. Similarly determined overturning moments are multiplied by a reduction factor that varies linearly from 1.0 at the top of the building to 0.8 at its base to obtain the design values. There is an additional requirement, however, that the reduced moments not be less than the product of the story shear at that elevation and the distance to the center of gravity of the portion of the building above the elevation being considered.

22.4 *EUROCODE 8*, 2004

22.4.1 Base Shear

The base shear formula in *Eurocode 8* (EC) may also be expressed as Eq. (22.1.1) with the seismic coefficient

$$C_s = \frac{C_e}{(OS)\, q'} \tag{22.4.1}$$

where $OS = 1.5$ is the overstrength factor, intended to account for the difference between the design and actual values of the structural strength. Applicable to selected structural systems is an additional overstrength factor that is included in q' and is intended to account for the difference between resistance values at first yield and at formation of a plastic mechanism.

The elastic seismic coefficient

$$C_e = \frac{A}{g} \tag{22.4.2}$$

except that for buildings taller than two stories and fundamental vibration period $T_1 < 2T_c$, the seismic coefficient is multiplied by 0.85. The pseudo-acceleration design spectrum A, normalized by the design peak ground acceleration $\ddot{u}_{go}$, is given by

$$\frac{A}{\ddot{u}_{go}} = \begin{cases} 1 + 1.5 T_n/T_b & 0 \le T_n \le T_b \\ 2.5 & T_b \le T_n \le T_c \\ 2.5(T_c/T_n) & T_c \le T_n \le T_d \\ 2.5 T_c T_d/T_n^2 & T_n \ge T_d \end{cases} \tag{22.4.3}$$

where T_n is the natural vibration period of an SDF system; T_b, T_c, and T_d denote the periods at the beginning of the constant-pseudo-acceleration, constant-pseudo-velocity, and constant-deformation regions of the design spectrum, respectively (Fig. 22.4.1). Implicit in Eq. (22.4.3) are importance factor $= 1$ and damping ratio $= 5\%$. Equation (22.4.3) is valid for all five ground types: A (rock), B (very stiff soils), C (medium-stiff soils), D (soft soils), and E (thin stratum of medium-stiff or soft soils over rock). However, the parameter values in Eq. (22.4.3) vary with ground type, as shown in Table 22.4.1. Included for each ground type are values for periods T_b, T_c, and T_d, and a value for S, a multiplier for the $\ddot{u}_{go}$ value specified for rock to obtain the $\ddot{u}_{go}$ value for another ground type. Figure 22.4.1 shows $A/\ddot{u}_{go}$ and $C_e/\ddot{u}_{go}$ for soil classification C. In addition to empirical formulas for estimating the fundamental period, the EC allows use of Eq. (22.3.4) for calculating T_1.

The elastic seismic coefficient is divided by a seismic reduction factor, which as implied by the other equations in the code is close to

$$q' = \begin{cases} 1 + (T_1/T_b)\,(q_y - 1) & T_1 < T_b \\ q_y & T_1 \ge T_b \end{cases} \tag{22.4.4}$$

where $q_y = q/1.5$, q being the seismic behavior factor, which varies between 1.5 and 8 depending on several factors, including the structural material and structural system. Although not included explicitly in the *Eurocode*, this decomposition of q into 1.5 and q_y facilitates this presentation and its interpretation later.

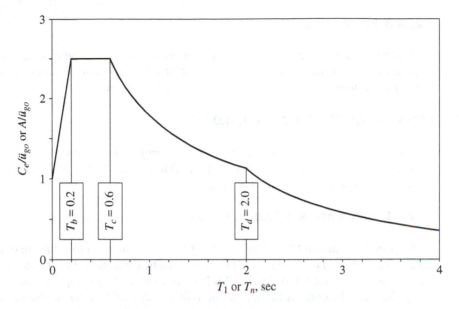

Figure 22.4.1   *Eurocode 8* (2004): elastic [$(OS)q' = 1$] seismic coefficient C_e and pseudo-acceleration, both normalized by $\ddot{u}_{go}$ for soil classification C.

TABLE 22.4.1   PSEUDO-ACCELERATION DESIGN
SPECTRUM PARAMETERS

Ground Type	S	Spectrum Type 1		
		T_b (sec)	T_c (sec)	T_d (sec)
A	1.00	0.15	0.4	2.0
B	1.20	0.15	0.5	2.0
C	1.15	0.20	0.6	2.0
D	1.35	0.20	0.8	2.0
E	1.40	0.15	0.5	2.0

22.4.2 Lateral Forces

The formula for the lateral force F_j at the jth floor is

$$F_j = V_b \frac{w_j \phi_{j1}}{\sum_{i=1}^{N} w_i \phi_{i1}} \tag{22.4.5}$$

where ϕ_{j1} is the displacement of the jth floor in the fundamental mode of vibration. The code permits linear approximation of this mode, in which case Eq. (22.4.5) becomes

$$F_j = V_b \frac{w_j h_j}{\sum_{i=1}^{N} w_i h_i} \tag{22.4.6}$$

22.4.3 Story Forces

The design values of story shears, story overturning moments, and element forces are determined by static analysis of the building subjected to these lateral forces; the overturning moments computed are not multiplied by a reduction factor.

22.5 STRUCTURAL DYNAMICS IN BUILDING CODES

Most of the seismic provisions contained in building codes are either derived from or related to the theory of structural dynamics. In this section we discuss this interrelationship for several aspects of the code provisions.

22.5.1 Fundamental Vibration Period

The period formula in MFDC and EC, Eq. (22.3.4), is identical to the result obtained from Rayleigh's method using the shape function given by the static deflections u_i due to a set of lateral forces F_i at the floor levels. This becomes apparent by comparing Eq. (22.3.4) with Eq. (8.6.4b). The period formula in the NBCC, Eq. (22.2.3), has the same basis except that the lateral forces used to determine the static deflections are assumed equal to the lumped weights at the floor levels. This becomes apparent by comparing Eq. (22.2.3) with Eq. (8.6.4c).

22.5.2 Elastic Seismic Coefficient

The elastic seismic coefficient C_e is related to the pseudo-acceleration spectrum for linearly elastic systems. In a linear SDF (one-story) system of weight w, the peak base shear is (see Section 6.6.3)

$$V_b = \frac{A}{g}w \qquad (22.5.1)$$

and $C_e = A/g$. For buildings the four codes—IBC, NBCC, MFDC, and EC—give the design base shear as

$$V_b = \frac{1}{R}C_e W \qquad V_b = \frac{1}{R_d R_O}C_e W \qquad V_b = \frac{1}{Q'}C_e W \qquad V_b = \frac{1}{(OS)q'}C_e W \quad (22.5.2)$$

respectively. By taking $R = R_d R_O = Q' = (OS)q' = 1$, it is clear that C_e in building codes corresponds to A/g, the pseudo-acceleration for linearly elastic systems normalized with respect to gravitational acceleration. The two, C_e and A/g, as specified in codes, are not necessarily identical, however, as seen in Figs. 22.1.1, 22.2.1, 22.3.1, and 22.4.1. The ratio $C_e \div A/g$ is plotted as a function of period in Fig. 22.5.1, where we note that for NBCC and MFDC it exceeds unity for most periods and increases with T_1. However, for the IBC and EC, this ratio is equal to unity for all periods (except very short periods in the case of IBC).

The seismic coefficient C_e is specified larger than A/g to account for the more complex dynamics of multistory buildings responding in several natural modes of vibration

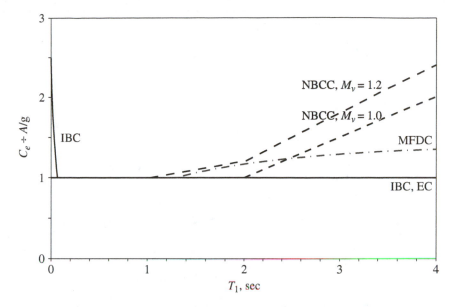

Figure 22.5.1   Ratio of elastic seismic coefficient C_e and pseudo-acceleration A/g for four building codes; NBCC plots are valid for all structures other than walls, coupled walls, and wall-frame systems.

and to recognize uncertainties in a calculated value of the fundamental vibration period. In Section 13.8 we demonstrated that the peak value of the base shear due to the nth mode is

$$V_{bn} = \frac{A_n}{g} W_n^* \tag{22.5.3}$$

where W_n^* is the effective weight and A_n/g is the normalized pseudo-acceleration, both for the nth mode. The peak value of the base shear considering several modes is generally estimated by the SRSS formula, Eq. (13.7.3). For the present purpose, however, we use the upper bound result of Eq. (13.7.2), specialized for the base shear:

$$V_b \leq \sum_{n=1}^{N} |V_{bn}| = \sum_{n=1}^{N} \frac{A_n}{g} W_n^* \tag{22.5.4}$$

If all the A_n values were equal to A_1, which they are not, Eq. (22.5.4) reduces to

$$V_b \leq \frac{A_1}{g} \sum_{n=1}^{N} W_n^* = \frac{A_1}{g} W \tag{22.5.5}$$

where the second half of this equation is obtained after using Eq. (13.2.14). Thus for an MDF system, C_e and A_1/g have similar but by no means identical meaning. In Part B we discuss these conceptual differences between C_e and A_1/g further, as well as their numerical differences, shown in Fig. 22.5.1.

It is of interest to compare the pseudo-acceleration design spectrum specified in the four codes with two levels—50th and 84.1th percentile—of the design spectra for firm

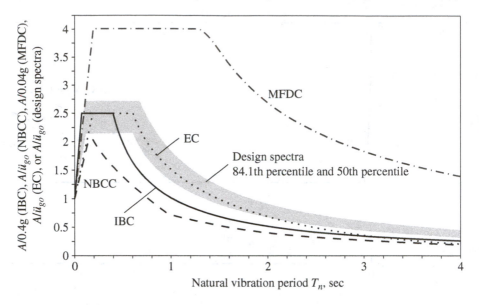

Figure 22.5.2 Comparison of pseudo-acceleration design spectrum in building codes
with the design spectra developed in Chapter 6; the latter are 84.1th and 50th percentile
spectra for 5% damping.

ground sites developed by the procedures of Fig. 6.9.3 (see also Figs. 6.9.4 and 6.9.5). All
five of these spectra are presented in Fig. 22.5.2, where the pseudo-acceleration is normalized relative to its value at zero period; such normalizing with respect to peak ground
acceleration is widely used but is not the best option. This normalization removes any
differences in the peak ground accelerations implied in the five spectra and provides a
comparison of the spectral shapes. The code spectra are generally quite different from the
design spectrum of Fig. 6.9.5 because the two are developed by different methods. A code
spectrum is "semi-custom-developed" for a site in the sense that it is based on probabilistic seismic hazard analysis considering all seismic sources relevant to the site. In contrast,
Fig. 6.9.5 is a "generic" spectrum based on the statistics of several ground motions recorded
in the western United States.

22.5.3 Design Force Reduction

Most codes specify the design base shear to be smaller than the elastic base shear (determined using the elastic seismic coefficient C_e). For the four codes described earlier, the
reduction factors are R, R_d, Q', and q' in Eq. (22.5.2). In this section we examine how
the reduction in design force specified in codes relates to the results obtained in Chapter 7
from dynamic response analysis of yielding SDF systems.

The reduction factors R, R_d, Q', and q' specified in the four codes are compared
with the yield-strength reduction factor R_y of elastoplastic systems. The code reduction
factors are plotted in Fig. 22.5.3 as a function of the fundamental vibration period T_1 for

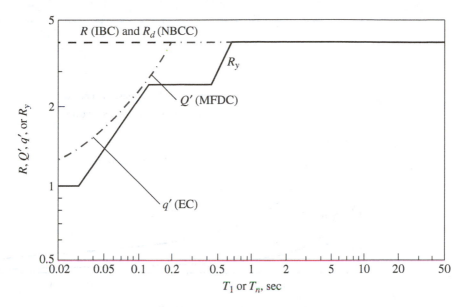

Figure 22.5.3 Comparison of yield-strength reduction factors—R, R_d, Q', and q'—in four building codes and yield-strength reduction factor R_y for an elastoplastic SDF system; $R = R_d = Q' = q_y = \mu = 4$.

$R = R_d = Q = q_y = 4$. They are independent of T_1 in IBC and NBCC, but their period dependence in MFDC and EC is defined by Eqs. (22.3.5) and (22.4.4), respectively. Determined from dynamic analysis of SDF systems, the reduction factor R_y for yield strength of elastoplastic systems corresponding to a ductility factor of 4 (Fig. 7.11.2) is also shown. It is clear from this comparison that the MFDC and EC seismic reduction factors vary with vibration period in a manner consistent with structural dynamics theory. However, the period independence of factor R in the IBC and R_d in the NBCC contradicts dynamic response results for structures with fundamental period in the acceleration-sensitive region of the design spectrum. The resulting discrepancy in the design spectra is seen in Fig. 22.5.4 for two values of the ductility factor μ. The inelastic design spectra shown are from Fig. 7.11.6 scaled by 0.4 so that they correspond to peak ground acceleration $\ddot{u}_{go} = 0.4g$. The elastic design spectrum reduced by the period-independent factor μ is lower in the acceleration-sensitive period region, as shown in Fig. 22.5.4. Thus, by ignoring the period dependence of the yield-strength reduction factor, the code may give excessively small design forces for structures in this period region.

This may imply that the code provisions are unconservative in certain situations, but we will not get into this issue because of several practical considerations, which are beyond the scope of this book. One of these is worth mentioning, however. The actual strength of a building exceeds its design strength, especially for short-period systems. Overstrength can come from a variety of sources. Examples are the difference between the design strength and the theoretical strength of structural elements because of the difference between allowable and yield stresses, the effects of gravity loads on element strengths, element

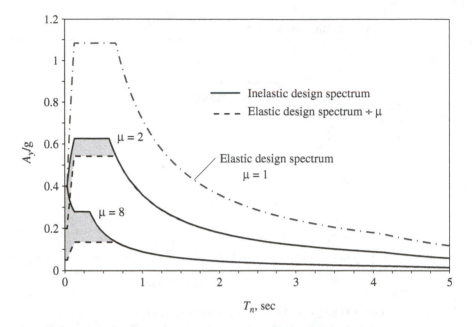

Figure 22.5.4 Comparison between inelastic design spectrum and elastic design spectrum reduced by the period-independent factor μ; results are presented for $\mu = 2$ and 8.

overstrength due to discrete choices of member sizes, element overstrength due to stiffness (drift) requirements, increase in the structural strength due to redistribution of element forces in the inelastic range, and the contributions of all structural and nonstructural elements that in the design process are not considered as part of the lateral force-resisting system. This overstrength of a building is recognized explicitly in some building codes (e.g., NBCC and EC).

22.5.4 Lateral Force Distribution

Structural dynamics gives the base shear and equivalent static lateral force at floor level j for mode n of a multistory building (Section 13.8.1):

$$V_{bn} = M_n^* A_n \qquad f_{jn} = \Gamma_n m_j \phi_{jn} A_n$$

Using the definitions for M_n^* and Γ_n, Eqs. (13.2.9a) and (13.2.3), f_{jn} can be expressed in terms of V_{bn}:

$$f_{jn} = V_{bn} \frac{w_j \phi_{jn}}{\sum_{i=1}^{N} w_i \phi_{in}} \qquad (22.5.6)$$

We now compare this force distribution from structural dynamics with code specifications. The IBC with $k = 1$ [Eq. (22.1.6)], the NBCC with the top force $F_t = 0$ [Eq. (22.2.4)] and the EC [Eq. (22.4.6)] all give

$$F_j = V_b \frac{w_j h_j}{\sum_{i=1}^{N} w_i h_i} \qquad (22.5.7)$$

This force distribution agrees with Eq. (22.5.6) if ϕ_{jn} is proportional to h_j, that is, if the mode shape is linear. The linear shape is a reasonable approximation to the fundamental mode of many buildings; as shown in Fig. 19.1.5, it is in between the fundamental mode shapes for the two extreme values, 0 and ∞, of the beam-to-column stiffness ratio ρ.

In the IBC the heightwise distribution of lateral forces is given by Eq. (22.1.6), based on the assumption that the lateral displacements are proportional to h_j for $T_1 \leq 0.5$ sec, to h_j^2 for $T_1 \geq 2.5$ sec, and to an intermediate power of h_j for intermediate values of T_1. These force distributions are intended to recognize the changing fundamental mode and increasing higher-mode contributions to structural response with increasing T_1.

Assignment of the additional force F_t at the top of the building—in addition to the forces from Eq. (22.2.4)—is intended by the NBCC to consider approximately and simplistically the influence of the higher vibration modes on the force distribution. The force F_t increases the shear force in the upper stories relative to the base shear. This is consistent with the predictions of structural dynamics that the higher modes affect the forces in the upper stories more than in lower stories (Section 19.6). Equation (22.2.5) gives F_t values that range from zero for short-period buildings to $0.25V_b$ for long-period buildings, for which structural dynamics theory demonstrates that higher-mode responses are more significant (Section 19.4).

In the MFDC, if $T_1 \leq T_c$, the heightwise distribution of lateral forces is also given by Eq. (22.5.7), which considers only the response in the fundamental vibration mode, assumed to have a linear shape. For $T_1 \geq T_c$, the MFDC specifies Eq. (22.3.6b), based on specifying floor displacements proportional to h_j at $T_1 = T_c$, to h_j^2 at T_1 much longer than T_c, and intermediate between the linear and parabolic shapes at intermediate values of T_1. This variation in deflected shape and hence force distribution is intended to recognize the changing shape of the fundamental vibration mode and increasing higher-mode responses with increasing fundamental period.

Equations (22.4.5) and (22.4.6) both appear in the EC, implying that the distribution of lateral forces is based entirely on the fundamental mode of vibration without considering the increasing higher-mode contributions to response with longer T_1.

22.5.5 Overturning Moments

Some building codes, including NBCC and MFDC, allow reduction of overturning moments relative to the values computed from lateral forces F_j by statics, because the response contributions of higher modes are more significant for story shears than for overturning moments (Chapter 19). In particular, if the first mode were linear, the higher modes would provide no contribution to the overturning moment at the base (Example 13.6), although they would affect overturning moments at higher levels and story shears at all levels. Thus the overturning moments computed from the code forces, supposedly calibrated against dynamic response results to provide the correct story shears, would exceed the values predicted by dynamic analysis and could therefore be reduced. The reduction factor at the building base is 0.8 in the MFDC, independent of T_1 or building height. In the NBCC the reduction factor at the building base ranges from 1.0 for buildings with $T_1 \leq 0.5$ sec to 0.7, 0.8, or 1.0 for buildings with $T_1 \geq 2$ sec in western Canada, depending on the structural system. The IBC and EC permit no reduction of overturning moments relative to their values computed from lateral forces by statics.

PART B: EVALUATION OF BUILDING CODES

In Part B we evaluate how well the seismic forces specified in building codes agree with the results of dynamic analysis presented in Chapters 19 and 20.

22.6 BASE SHEAR

The significance of the response contributions of the higher vibration modes in the dynamic response of buildings (Chapter 19) plays a central role in evaluating the code forces. For this reason we first recall that the combined responses of the second and higher modes depend mainly on two parameters: fundamental period T_1 and beam-to-column stiffness ratio ρ. With reference to Fig. 19.4.2, we had concluded that the base shear for buildings with T_1 within the acceleration-sensitive region of the spectrum is essentially all due to the first mode. However, for buildings with T_1 in the velocity- or displacement-sensitive regions of the spectrum, the higher-mode responses can be significant, increasing with increasing T_1 and with decreasing ρ, for reasons discussed in Chapter 19.

For buildings with T_1 in the acceleration-sensitive region of the spectrum, these results and Eq. (22.5.3) indicate that the code formula, Eq. (22.1.1), would accurately predict the base shear for elastic buildings if the seismic coefficient C_e were defined as A_1/g and the total weight W were replaced by the first-mode effective weight W_1^*. If W is used instead of W_1^*, as in building codes, the base shear is overestimated. This becomes obvious by renormalizing the base shear data of Fig. 19.4.2 with respect to the total weight, as shown in Fig. 22.6.1; recall that $W_1^* = 0.679W$, $0.796W$, and $0.880W$ for $\rho = 0$, $\frac{1}{8}$, and ∞, respectively (Table 19.1.1). Therefore, the overestimation varies with ρ, being least for shear buildings ($\rho = \infty$), largest for flexural buildings ($\rho = 0$), and in between for frame buildings with intermediate values of ρ.

For buildings with T_1 in the velocity- or displacement-sensitive spectral regions, however, the increase in base shear by using the total weight of the building may not be sufficient to compensate for the higher-mode response. This is clear from Fig. 22.6.1, where we observe that V_b/W exceeds A_1/g for longer T_1 and smaller ρ, conditions that produce increasingly significant higher-mode response. For this range of parameters, therefore, the seismic coefficient C_e should be larger than A_1/g.

The dynamic response (RSA) results of Fig. 19.4.2 provide insight into how the A/g spectrum should be modified to obtain the seismic coefficient C_e. For this purpose, curves of the form $\alpha T_1^{-\beta}$ are fitted to the base shear versus period curve from dynamic analysis, as shown in Fig. 22.6.2. The parameters α and β for each of the velocity- and displacement-sensitive regions of the spectrum are evaluated by a least-squared error fit to the V_b–T_1 curve. The curve-fitting procedure minimizes the error, defined as the integral

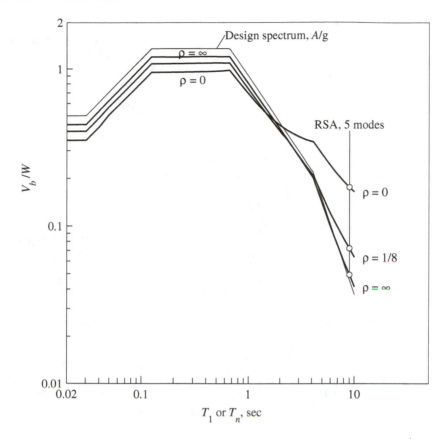

Figure 22.6.1   Base shear V_b (normalized by total weight W) in buildings with $\rho = 0$, $\frac{1}{8}$, or ∞, computed for the design spectrum shown.

over the period range considered of the squares of the differences between the logarithm of the ordinates of the "exact" and fitted curves. This curve-fitting procedure is designed to satisfy the following constraints. First, the ordinate of the fitted curve at $T_1 = T_c$ is equal to the ordinate of the flat portion of the A/g spectrum. Second, the curves fitted to the velocity- and displacement-sensitive regions of the spectrum have the same ordinates at $T_1 = T_d$. Third, the exponent β for the displacement-sensitive region should not be smaller than its value in the velocity-sensitive region. The resulting functions $\alpha T_1^{-\beta}$ are shown in Fig. 22.6.2 together with the V_b–T_1 curves from dynamic analyses.

This comparison suggests that to estimate V_b without dynamic analysis (RSA), the code seismic coefficient C_e should be defined by decreasing the coefficient β in $\alpha T_1^{-\beta}$ and thus raising the pseudo-acceleration design spectrum to account for the higher-mode response. For this purpose, building codes should define C_e in terms of the design spectrum A/g, which in turn should be specified explicitly. Once this format is adopted, C_e can be defined by raising the design spectrum in its velocity- and displacement-sensitive regions,

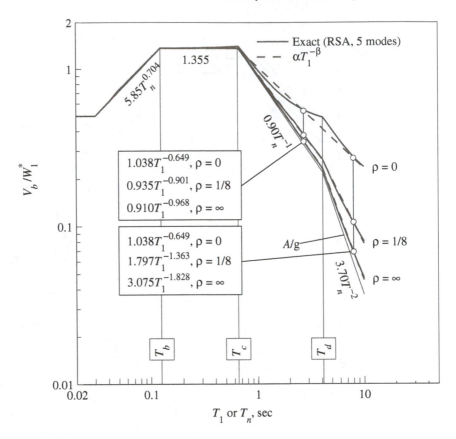

Figure 22.6.2   Comparison of functions $\alpha T_1^{-\beta}$ and "exact" base shear versus T_1 curves.

based on dynamic response results of the type presented here. The degree to which the spectrum should be raised depends on the beam-to-column stiffness ratio ρ; the spectrum needs to be raised very little for shear buildings ($\rho = \infty$) but to an increasing degree with increasing frame action (i.e., decreasing ρ). The spectral modifications presented also depend on the heightwise distribution of mass and stiffness, parameters that have not been varied here.

Having utilized dynamic response results to determine how the design spectrum A/g should be modified for higher-mode response, we now compare these results with building code provisions. For this purpose we return to Fig. 22.5.1, where the $C_e \div A/g$ in four codes was presented as a function of T_1. These results are compared in Fig. 22.6.3 with $V_b/W_1^* \div A/g$, the ratio of two values of base shear, the first including responses due to all modes and the other considering only the first mode (Fig. 22.6.2). It is clear that two of the four codes considered recognize the dependence of higher-mode response on the fundamental period T_1, and the NBCC also recognizes the influence of the stiffness ratio ρ. The NBCC appears to overcompensate for the higher-mode response over

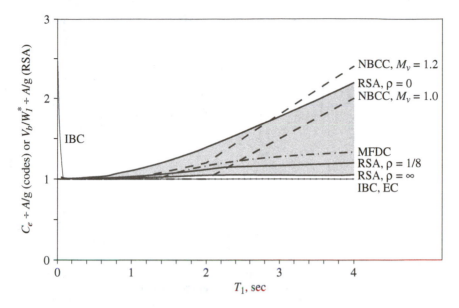

Figure 22.6.3   Comparison of $C_e \div A/g$ in four building codes and $V_b/W_1^* \div A/g$ from RSA for three values of ρ. Note that IBC, NBCC, and EC do not permit the use of a static ELF procedure for buildings with T_1 exceeding $3.5T_c$, 2.0 sec, and 2.0 sec, respectively.

a wide range of T_1 and ρ—not because of the M_v factor but because C_e is constant for $T_1 > 2$ sec—whereas MFDC seems to be reasonable except for buildings with very small ρ. Unfortunately, the IBC and EC ignore the higher-mode response and specify $C_e = A/g$.

The IBC and EC deal with the higher-mode contribution to base shear in a different way. Recognizing the results of Fig. 22.6.1, EC and IBC do not permit use of the ELF analysis procedure for buildings with T_1 exceeding 2.0 sec and 3.5 T_c, respectively, where T_c is the period separating the acceleration- and velocity-sensitive regions of the spectrum; $T_c = 0.66$ sec for the design spectrum of Fig. 6.9.5. For certain buildings, Eq. (22.4.2) is multiplied by 0.85 in EC to avoid overestimation of the base shear by using the total weight W in Eq. (22.1.1); see Fig. 22.6.1 and the related discussion.

Inelastic response behavior in multistory buildings and its differences relative to SDF systems are factors that should be considered in specifying the seismic coefficient in building codes. This important concept is illustrated by returning to Fig. 20.3.2 from nonlinear dynamic analysis, showing the ratio of base shear yield strengths in multistory buildings and SDF systems necessary to limit the ductility demand to the same allowable value, $\mu = 2$ or 8. Superimposed on these results in Fig. 22.6.4 is the ratio $C_e \div A/g$ for four building codes from Fig. 22.5.1; these code ratios also apply to yielding systems irrespective of the allowable reduction factors R, R_d, Q', or q'. In contrast, dynamic response results indicate that the strength increase required to account for MDF effects depends significantly on the ductility factor and on the plastic hinge mechanism.

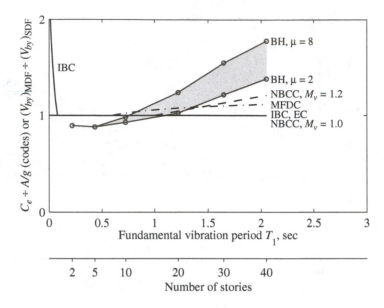

Figure 22.6.4   Comparison of $C_e \div A/g$ in four building codes and $(V_{by})_{\text{MDF}}/(V_{by})_{\text{SDF}}$ from nonlinear dynamic analysis of beam- and column-hinge models of frames for two values of the SDF-system ductility factor.   (Dynamic analysis data from Nassar and Krawinkler, 1991.)

22.7 STORY SHEARS AND EQUIVALENT STATIC FORCES

Having compared the code-specified base shear with the predictions of dynamic analysis, we now evaluate the heightwise distribution of story shears and lateral forces. As mentioned in Sections 22.1 to 22.4, the codes specify lateral forces in terms of the base shear, and static analysis of the structure for these forces provides the story shears. The story shears and lateral forces determined from four codes are divided by the base shear V_b and presented in Figs. 22.7.1 and 22.7.2; included in each case are two values of $T_1 = 0.5$ and 3 sec, chosen to be representative of the acceleration and velocity-sensitive spectral regions. Note in Fig. 22.7.2 that the numerical values of the equivalent static forces that are concentrated at the floor levels have been joined by straight lines between floors for easier visualization. Also included in Fig. 22.7.1 are the story shears of Fig. 19.6.1 from dynamic analysis (RSA), including response contributions of all modes; these have been normalized by the corresponding base shear. Similarly, Fig. 22.7.2 includes the equivalent static forces computed from the story shears of Fig. 19.6.1 as the differences between the shears in consecutive stories (equal to the discontinuity in shears at the floor levels).

Figures 22.7.1 and 22.7.2 permit the following observations: For buildings with T_1 in the acceleration-sensitive region of the spectrum, the heightwise distributions of lateral forces and story shears specified by the four codes are essentially identical to each other and fall between the dynamic response curves for $\rho = 0$ and ∞. With increasing T_1, the

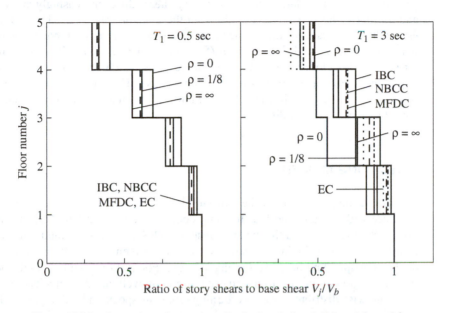

Figure 22.7.1 Comparison of story shear distributions in four building codes and from RSA for three values of ρ. Note that IBC, NBCC, and EC do not permit the use of a static ELF procedure for buildings with T_1 exceeding $3.5T_c$, 2.0 sec, and 2.0 sec, respectively.

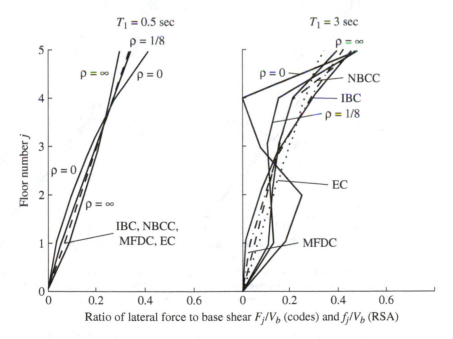

Figure 22.7.2 Comparison of static equivalent lateral force distributions in four building codes and from RSA for three values of ρ. Note that IBC, NBCC, and EC do not permit the use of a static ELF procedure for buildings with T_1 exceeding $3.5T_c$, 2.0 sec, and 2.0 sec, respectively.

code distributions for lateral forces and story shears differ increasingly among the codes, and all four codes differ increasingly from the dynamic response. These differences are especially significant for the smaller values of ρ because higher-mode response increases with increasing T_1 and decreasing ρ (Sections 19.4 and 19.5). It is clear that the code formulas do not closely follow the dynamic response results or recognize the effects of the important building parameters on dynamic response. These discrepancies are accentuated when we consider the influence of the heightwise distribution of stiffness and strength on inelastic response of buildings.

22.8 OVERTURNING MOMENTS

The overturning moments in the two buildings, $T_1 = 0.5$ and 3 sec, determined in accordance with three of the four building codes[†] (Sections 22.1 to 22.4) are presented in Fig. 22.8.1 together with the dynamic response (RSA) results, including responses due to all modes (Fig. 19.6.2); in each case, overturning moments at all elevations are normalized by the corresponding base overturning moment. For buildings with T_1 in the acceleration-sensitive region of the spectrum and even extending well into the velocity-sensitive region, the heightwise distributions of overturning moments specified by the three codes are close

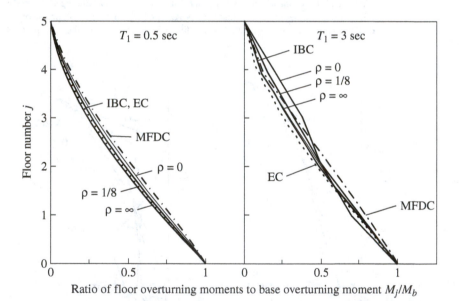

Figure 22.8.1 Comparison of floor overturning moment distributions in three building codes and from RSA for three values of ρ. Note that IBC and EC do not permit the use of a static ELF procedure for buildings with T_1 exceeding $3.5T_c$ and 2.0 sec, respectively.

[†]The format of Figs. 22.8.1 and 22.8.3 does not lend itself to plots for NBCC.

to each other and to the dynamic response. The discrepancy in code values relative to the dynamic response increases with increasing T_1, especially for buildings with smaller values of ρ, as the higher-mode response becomes increasingly significant (Sections 19.4 and 19.5). However, this discrepancy in overturning moments is much smaller than was noted in Section 22.7 for story shears, because the higher-mode response contributions to the overturning moments are less significant (Chapter 19).

We next compare the overturning moments computed by two methods, both based on dynamic analysis: (1) response spectrum analysis (RSA) considering all modes (Fig. 22.8.1), and (2) static analysis of the building subjected to the lateral forces of Fig. 22.7.2 determined, as described in Section 22.7, from RSA predictions of story shears. The latter method will provide the same story shears as the first method because the lateral forces were determined by statics from the story shears predicted by RSA, but not the correct overturning moments. This fact is demonstrated by presenting the ratio of overturning moments computed by the two methods. Akin to the reduction factor J specified in some building codes, this ratio is presented in Fig. 22.8.2 for the base overturning moment as a function of T_1 for three values of ρ; and in Fig. 22.8.3 for overturning moments at all floors of the building for two values of T_1. This reduction factor never exceeds unity, implying that the approximate value of the overturning moment obtained from the lateral forces (second method) always exceeds the "exact" value obtained from RSA (first method). The two values are identical if the response contribution of only the fundamental vibration mode

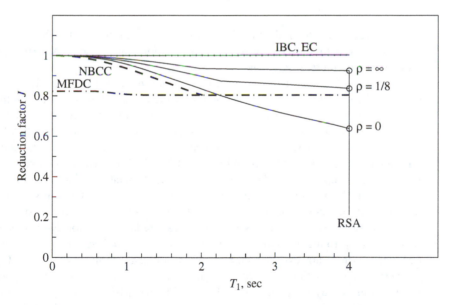

Figure 22.8.2 Comparison of reduction factors for base overturning moment in four building codes and from RSA for three values of ρ. Note that IBC, NBCC, and EC do not permit the use of static ELF procedure for buildings with T_1 exceeding $3.5T_c$, 2.0 sec, and 2.0 sec, respectively. NBCC plot valid for all structures other than walls, coupled walls, and wall-frame systems.

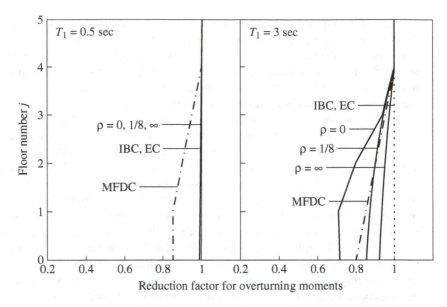

Figure 22.8.3 Comparison of reduction factors for overturning moments in three building codes and from RSA for three values of ρ. Note that IBC and EC do not permit the use of static ELF procedure for buildings with T_1 exceeding $3.5T_c$ and 2.0 sec, respectively.

is considered. Thus the reduction factor accounts for the fact that higher vibration modes contribute more to shears than to overturning moments; it decreases (implying greater reduction) with increasing T_1 and decreasing ρ.

Because the lateral forces specified in building codes are intended to provide the dynamically computed story shears, the preceding observations indicate that the overturning moments will be overestimated if they were also computed from the lateral forces by statics. Thus some building codes specify reduction factors by which statically computed overturning moments should be multiplied. These reduction factors, defined earlier for building codes, are also included in Figs. 22.8.2 and 22.8.3. The reduction factor specified in MFDC is independent of T_1, except for the slight variation arising from an equilibrium requirement. As a result, the reduction is excessive in the acceleration-sensitive region of the spectrum, but not enough in parts of the velocity- and displacement-sensitive regions, depending on ρ. However, the NBCC specifies a reduction factor that varies with T_1 in a manner similar to the dynamic response up to about $T_1 = 2$ sec; it does not recognize the dependence on ρ, however, which becomes significant at longer T_1. The IBC and EC do not permit any reduction in overturning moments, a provision that is not supported by the results of elastic dynamic analysis. The NBCC and MFDC specify no reduction in the top story and increasing reduction as one moves down to lower levels of the building. Although consistent in this regard with the results of dynamic analysis (Fig. 22.8.3), the code specifications do not fully recognize the dependence of the reduction factor on the building parameters T_1 and ρ.

22.9 CONCLUDING REMARKS

We have demonstrated that some of the concepts developed in this book about earthquake analysis, response, and design of structures are reflected in building codes, but they are not always stated explicitly or applied in accordance with structural dynamics results. Building codes should adopt a different approach, explicitly stating the underlying basis for each provision so that it can be improved as we develop a better understanding of structural dynamics and earthquake performance of structures. The seismic design approaches must also consider, much more realistically than has been done in the past, the demands imposed by earthquakes on structures and the structural capacity to meet these demands.

Building codes represent a consensus of the structural engineering profession on the seismic design of ordinary buildings where special earthquake considerations are not cost-effective. There can be major design deficiencies if the building code is applied to structures whose dynamic properties differ significantly from those of ordinary buildings. This is suggested by the collapse or irreparable damage of some buildings during major earthquakes. Similarly, building codes should not be applied to special structures, for they require special consideration: because of cost, potential hazard, or the need to maintain function. For critical projects such as high-rise buildings, dams, nuclear power plants, offshore oil-drilling platforms, long-span bridges, major industrial facilities, and so on, special earthquake considerations are necessary.

FURTHER READING

Canadian Commission on Building and Fire Codes, *The National Building Code of Canada, 2010*, National Research Council, Ottawa, 2010.

Committee European de Normalisation, European Standard EN 1998-1:2004, *Eurocode 8: Design of Structures for Earthquake Resistance,* Part 1, *General Rules, Seismic Actions and Rules for Buildings*, CEN, Brussels, Belgium, 2004.

Chopra, A. K., and Cruz, E. F. "Evaluation of Building Code Formulas for Earthquake Forces," *Journal of Structural Engineering, ASCE,* **112,** 1986, pp. 1881–1899.

Chopra, A. K., and Newmark, N. M., "Analysis," Chapter 2 in *Design of Earthquake Resistant Structures* (ed. E. Rosenblueth), Pentech Press, London, 1980.

Cruz, E. F., and Chopra, A. K., "Improved Code-Type Earthquake Analysis Procedure for Buildings," *Journal of Structural Engineering, ASCE,* **116,** 1990, pp. 679–700.

Government of the Federal District, "Complementary Technical Norms for Seismic Design," Official Gazette of the Federal District, October 6, 2004, Mexico.

International Code Council, *International Building Code 2009,* Washington, D.C., 2009.

Nassar, A. A., and Krawinkler, H., "Seismic Demands for SDOF and MDF Systems." *Report No. 95,* John A. Blume Earthquake Engineering Center, Stanford University, Stanford Calif., 1991.

Rosenblueth, E., "Seismic Design Requirements in a Mexican 1976 Code," *Earthquake Engineering and Structural Dynamics,* **7,** 1979, pp. 49–61.

23

Structural Dynamics in Building Evaluation Guidelines

PREVIEW

A major challenge for performance-based seismic engineering is to develop simple, yet sufficiently accurate methods for evaluating existing buildings to meet selected performance objectives. As reflected in guidelines and standards for evaluating existing buildings, such as the FEMA-273 (1997), its successors FEMA 356 (2000) and ASCE 41-06 (2007)—henceforth abbreviated as ASCE 41— and ATC-40 (1996) documents, the profession has shifted away from the traditional elastic analysis of structures subjected to seismic forces reduced to recognize indirectly inelastic response, as in current building codes (Chapter 22); instead, inelastic behavior of structures is considered explicitly in estimating seismic demands at low performance levels, such as life safety and collapse prevention. In this chapter, selected aspects of the aforementioned guidelines for computing seismic demands are discussed in light of structural dynamics theory presented in Chapters 6, 7, 13, and 20. For estimating seismic demands at the operational performance level, these guideline documents include linear dynamic analysis procedures that are generally consistent with those described in Chapter 13 and therefore not discussed here.

Currently, building evaluation guidelines permit use of two nonlinear analysis methods to estimate seismic demands at low performance levels: the nonlinear static procedure (NSP) and the nonlinear dynamic procedure (NDP). According to the NSP, seismic demands are computed by nonlinear static analysis of a structure subjected to monotonically increasing lateral forces (known as *pushover analysis*) with a specified, usually invariant heightwise distribution until a predetermined target displacement is reached; supplementary elastic analyses with relaxed acceptance criteria are required for structures with significant higher-mode responses, but the NSP is still permitted in ASCE 41. The target

863

displacement is estimated from the deformation of an inelastic SDF system derived from the pushover curve. In light of structural dynamics theory, issues related to NDP are mentioned in Section 23.1, issues related to estimating target displacement are examined in Sections 23.2 and 23.3, and issues related to pushover analysis are discussed in Section 23.4.

23.1 NONLINEAR DYNAMIC PROCEDURE: CURRENT PRACTICE

The nonlinear dynamic procedure (NDP) is essentially equivalent to nonlinear response history analysis (RHA), presented in Chapter 20. Therefore, we will not comment on the NDP except to point out the need to improve current guidelines for its implementation. The ASCE 41 specifications for the NDP, which are essentially the same as those in the International Building Code and ASCE7-05, state that the seismic demand may be estimated as (1) the maximum of demands due to three scaled ground motion records, or (2) the mean value of demands due to seven scaled records; a procedure for selecting and scaling ground motion records is also specified.

We demonstrate that these estimates can vary significantly. For this purpose, results are presented for a 9-story steel building, symmetric in plan, located in Aliso Viejo, California, subjected to an ensemble of 28 two-component ground motions recorded from earthquakes with moment magnitude varying from 6.5 to 7.6 at distances ranging from 7 to 28 km. The accuracy of the ASCE 7-05 procedure for scaling records (which is very similar to the ASCE 41-06 procedure) was evaluated as follows: The maximum value of the story drifts due to a set of three scaled records and the average of these demands due to a set of seven scaled records, both, were compared against the benchmark value defined as the median (or geometric mean) value of the story drifts due to 28 unscaled records.

The results, shown in Fig. 23.1.1, demonstrate a significant variation in the first-story drift (presented as percent of story height) estimated by three implementations (each selecting different sets of excitations from among the 28 records) of both versions of the

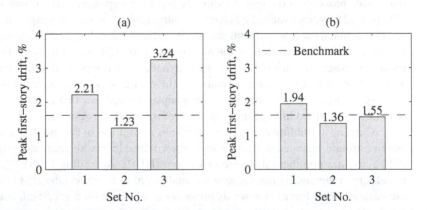

Figure 23.1.1 First-story drift, x-component: (a) maximum of demands due to three excitations; (b) average of demands due to seven excitations. The excitations were selected randomly three times from an ensemble of 28 records. Also shown is the benchmark value.

ASCE 7-05 procedure. Compared to the benchmark value, the estimate obtained as the maximum of drifts due to three scaled records errs by -23 to 101%, and the estimate obtained as the average of drifts due to seven scaled records errs by -4 to 21%. Such errors and variability obviously imply that different engineers following the same criteria could arrive at contradictory conclusions about seismic safety and rehabilitation requirements for an existing building. This observation points to the need for better criteria to select and scale ground motions, a topic of much ongoing research activity since 2005.

23.2 SDF-SYSTEM ESTIMATE OF ROOF DISPLACEMENT

How well can the roof displacement u_r of a multistory building be determined from the deformation of an SDF system, which is the concept underlying current NSP? To address this question, we compare the values of roof displacement determined by two methods: the "exact" value $(u_r)_{MDF}$, determined by nonlinear RHA of the multistory building treated as an MDF system (Section 20.1); and the SDF-system estimate: $(u_r)_{SDF} = \Gamma_1 \phi_{r1} D_1$ [see Eq. (20.7.3)], where Γ_1 was defined in Eq. (20.5.3a), ϕ_{r1} is the value at the roof in the first mode ϕ_1, and D_1 is the peak deformation of the inelastic SDF system with its force–deformation relation determined from the pushover curve, a plot of base shear versus roof displacement, obtained by nonlinear static analysis of the building using the first-mode force distribution: $\mathbf{s}_1^* = \mathbf{m}\phi_1$ [see Eq. (20.7.2)]; D_1 is determined by nonlinear RHA of the SDF system. Note that this procedure is equivalent to UMRHA (Section 20.6.2) considering only the first mode of vibration, and the resulting $(u_r)_{SDF}$ will be identical to u_{r1} for the first mode from step 6 in the MPA summary (Section 20.7.3). The response of seven of the SAC buildings to each of 20 SAC ground motions is computed and the displacement ratio is determined: $\left(u_r^*\right)_{SDF} = (u_r)_{SDF} \div (u_r)_{MDF}$. The difference between the median of this displacement ratio and unity indicates the bias in the SDF-system estimate of median roof displacement.

The SDF system estimates the median roof displacement of multistory buildings to a useful degree of accuracy. Figures 23.2.1 and 23.2.2 show histograms of the 20 values of the displacement ratio, together with its range of values and the median value for each

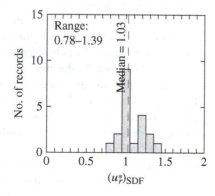

Figure 23.2.1   Histograms of ratio $\left(u_r^*\right)_{SDF}$ for the SAC–Los Angeles 3-story building. The range of values and the median value of this ratio are noted.

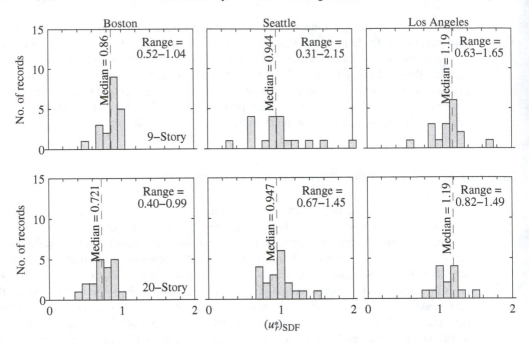

Figure 23.2.2 Histograms of ratio $\left(u_r^*\right)_{\text{SDF}}$ for SAC buildings: Boston, Seattle, and Los Angeles 9- and 20-story. The range of values and the median value of this ratio are noted.

of seven SAC buildings. Examining first the results for the three Los Angeles buildings, the SDF system estimates the median roof displacement accurately—within 3% of the "exact" value—for the 3-story building, but overestimates it by 19% for the 9- and 20-story buildings. The median roof displacement of taller buildings is not always overestimated by the SDF system; for example, it is underestimated by 14 and 18% for the Boston 9- and 20-story buildings, respectively, and by 5% for the Seattle 9- and 20-story buildings. The bias in the SDF-system estimate of median roof displacement depends on the vibration properties of the building and how far it is deformed into the inelastic range. For short-period buildings (e.g., the Los Angeles 3-story building), the bias is small because their response is dominated by the first mode. This bias is larger for long-period buildings (e.g., the SAC 9- and 20-story buildings) because they respond significantly in higher modes of vibration. Correction factors should be developed to eliminate the larger bias in taller buildings, thus improving the estimate of their roof displacement.

The SDF system may not estimate the roof displacement of multistory buildings due to individual excitations to a useful degree of accuracy. For the 3-story building, this SDF-system estimate varies (among 20 ground motions) from 78 to 139% of the exact value, perhaps a surprisingly large discrepancy for a first-mode-dominated structure. The SDF-system estimate can be alarmingly small (as low as 31 to 82% of the exact value among the six taller buildings) or unexpectedly large (as large as 145 to 215% of the exact value among the six taller buildings). The errors are actually worse than indicated by Fig. 23.2.2

because it does not include those cases where nonlinear RHA predicted collapse of the first-mode SDF system but not of the building as a whole.[†]

This large discrepancy arises because for individual ground motions the SDF system may significantly underestimate or overestimate the yielding-induced permanent drift in the response of the building. To demonstrate this assertion, the response history of the first-mode contribution determined by UMRHA (Section 20.6), and the "exact" response from nonlinear RHA of the MDF system are presented for the Los Angeles 9-story building due to three of the 20 ground motions in Fig. 23.2.3. The first-mode contribution was verified to be the dominant response for each of the three excitations. In the first case, the peak response occurs at the end of the first large inelastic excursion before the yielding-induced permanent drift away from the zero-displacement position takes place, and the SDF-system estimate is highly accurate (Fig. 23.2.3a). In the second case, the permanent drift in the first-mode response is much smaller than in the exact response of the MDF system, and the SDF-system method underestimates the roof displacement by 37% (Fig. 23.2.3b). In the third case, the permanent drift in the first-mode response away from the initial position is much larger than in the exact response of the MDF system, and the SDF-system method overestimates the roof displacement by 65% (Fig. 23.2.3c).

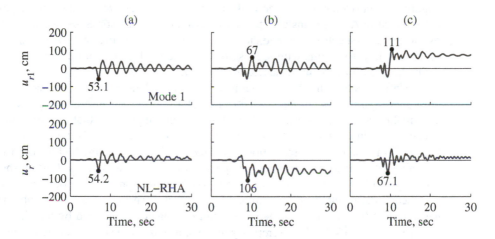

Figure 23.2.3   Response histories of roof displacement of the Los Angeles 9-story building due to three ground motions determined by two methods, first-mode contribution determined by UMRHA and exact response from nonlinear RHA: (a) record 39; (b) record 22; (c) record 21. Peak values are noted.

[†]Data for excitations that caused collapse of the SDF system are excluded, reducing the number of data to 19 for the Seattle 9-story building, 17 for the Los Angeles 9-story building, and 14 for the Los Angeles 20-story building; the median values for these buildings are computed by the counting method. The data values corresponding to 20 ground motions were sorted in ascending order, and the median was estimated as the average of the tenth and eleventh values, starting from the lowest value.

23.3 ESTIMATING DEFORMATION OF INELASTIC SDF SYSTEMS

As mentioned earlier, seismic demands for buildings are estimated in current engineering practice by pushover analysis up to a target displacement of the roof, determined from the deformation D of an inelastic SDF system. Methods to determine D are described in the ATC-40 guidelines and ASCE 41-06 standard.

23.3.1 ATC-40 Method

The approach adopted in ATC-40 was to estimate the earthquake response of inelastic SDF systems by approximate analytical methods in which the nonlinear system is replaced by an "equivalent" linear system. The natural period of vibration of the equivalent linear system is longer, and its damping ratio larger, than the corresponding properties of the inelastic system vibrating within its linearly elastic range. Developed in the early 1960s, equivalent linear systems were of much interest to researchers at that time, when analysis of inelastic systems and understanding of their earthquake response was in its infancy. Since then, the earthquake analysis and response of inelastic SDF systems has developed into a mature subject, which we studied in depth in Chapter 7. However, these procedures were not selected in ATC-40. Instead, the deformation of an inelastic SDF system is estimated by the capacity-spectrum method, an iterative method requiring analysis of a sequence of equivalent linear systems with successively updated values of period and damping ratio; the method is typically implemented graphically.

Unfortunately, the ATC-40 iterative procedure does not always converge; when it does converge, it does not lead to the exact deformation. Because convergence traditionally implies accuracy, the user could be left with the impression that the deformation calculated is accurate, but the ATC-40 estimate errs considerably. This is demonstrated in Fig. 23.3.1a, where the deformation estimated by the ATC-40 method is compared with the value determined from inelastic design spectrum theory and the well-established equation [Eq. (7.12.3)] using the $R_y-\mu-T_n$ equation from Section 7.11.1. Both the approximate and theoretical results were determined for systems covering a wide range of period values and ductility factors subjected to ground motions characterized by the elastic design spectrum of Fig. 6.9.5. The discrepancy in the approximate result presented in Fig. 23.3.1b shows that the ATC-40 method underestimates the deformation by 40 to 50% over a wide range of periods.

The two flaws in the ATC-40 capacity spectrum method—lack of convergence in some cases and large errors in many cases—appear to have been rectified in ASCE 41. It derives the optimal vibration period and damping ratio parameters for the equivalent linear system by minimizing the differences between its response and that of the actual inelastic system. Such an equivalent linear method would obviously give close to the correct deformation. However, the benefit in making the equivalent linearization detour is questionable when the deformation of an inelastic system can be determined readily using the inelastic design spectrum (Section 7.12.2) or by using available equations for the inelastic deformation ratio, a topic addressed in Section 23.3.3.

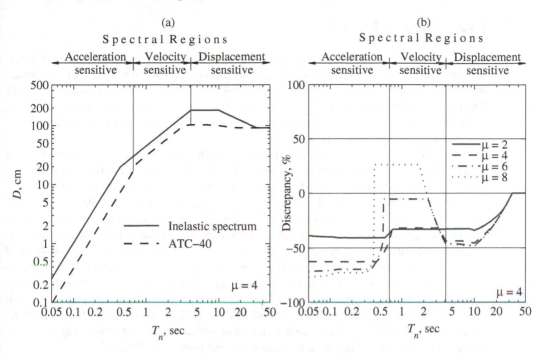

Figure 23.3.1 Deformation of SDF system computed by the ATC-40 method and from an inelastic design spectrum: (a) comparison of deformations; (b) discrepancy in the ATC-40 method. (Adapted from Chopra and Goel, 2000.)

23.3.2 ASCE 41-06 Method

The deformation of an inelastic SDF system is estimated by

$$D = C_1 C_2 \left(\frac{T_n}{2\pi} \right)^2 A \qquad (23.3.1)$$

Multiplying the deformation of the corresponding linear system [given by Eq. (6.7.1)] are two coefficients, C_1 and C_2. The coefficient C_1 represents the inelastic deformation ratio, u_m/u_o [see Eq. (7.2.4), Fig. 7.4.4b, and Section 23.3.3], for inelastic systems with stable, bilinear hysteresis loops (i.e., without pinching, cyclic stiffness degradation, or strength deterioration of the hysteresis loop). The coefficient C_2 accounts for the increase in deformation of the inelastic system due to these features not considered in C_1.

Equations and numerical values for these coefficients specified in ASCE 41 are based on research results and on judgment. However, some of the numerical values are not supported by research results; for example, C_1 is limited to values much smaller than the inelastic deformation ratio determined from dynamic response analyses for systems in the acceleration-sensitive region of the spectrum (Figs. 7.11.8 and 23.3.2 to 23.3.5); however, the value of $C_1 = 1.0$ at longer periods is theoretically correct (Section 7.11.3).

23.3.3 Improved Methods

Since the early 1990s there has been increasing emphasis on estimating structural defor-mations and on displacement-based design, which has been advocated as a more relevant and rational approach than traditional strength-based seismic design of structures. This has led to renewed interest in the relationship between peak deformations u_m and u_o of inelastic and corresponding linear SDF systems, respectively, a problem we studied first in Chapter 7, resulting in many research publications.

If expressed as a function of the initial elastic vibration period T_n and ductility factor μ, the inelastic deformation ratio can be used to determine the inelastic deformation of a new or rehabilitated structure where global ductility capacity can be estimated because deformation of the corresponding linear system is readily known from the elastic design spectrum [Eq. (6.7.1)]. If expressed as a function of T_n and yield-strength reduction factor R_y, the inelastic deformation ratio can be used to determine the deformation of an exist-ing structure with known lateral strength. The inelastic deformation ratio will be denoted by $C_\mu = u_m/u_o$ or $C_R = u_m/u_o$, where the subscripts μ and R represent systems with known ductility capacity μ or known yield strength defined by the reduction factor R_y, respectively; the peak deformations u_m and u_o are determined by numerical solution of Eqs. (7.3.2) and (6.2.1), respectively. Based on a comprehensive set of results, some of the observations in this section are similar to, and others are important refinements of, those based on limited data of response to a single ground motion in Chapter 7.

Figure 23.3.2a and b present the median values of C_μ and C_R, respectively, as a function of T_n for elastoplastic systems subjected to the LMSR[†] ensemble of 20 ground motions; the spectral regions are noted in the plots. In the acceleration-sensitive region, C_μ and $C_R \simeq 1$ at $T_n = T_c$ for smaller μ and R_y, but they exceed unity increasingly for shorter periods and larger μ or R_y, indicating greater inelastic action. For these short-period systems, C_μ and C_R are very sensitive to yield strength, increasing as the yield strength is reduced. For very short-period systems ($T_n < T_a$), even if their strength is only slightly smaller than the minimum strength required for the structure to remain elastic (e.g., $R_y = 1.5$), C_R is much larger than unity. At $T_n = 0$, for elastoplastic systems $C_\mu = \mu$ (see Section 7.11.3) and $C_R = \infty$, values that are a special case of those derived by researchers for bilinear systems. In the velocity-sensitive region, C_μ and $C_R \simeq 1$ and are essentially independent of ductility factor or yield strength. In the displacement-sensitive region, C_μ and $C_R < 1$ for systems in the period range T_d to T_f, where these ratios decrease as the ductility factor is increased or the strength is reduced; however, for systems with periods longer than T_f, C_μ and $C_R \simeq 1$ are essentially independent of ductility factor and strength, and both C_μ and $C_R = 1$ for very long-period systems for reasons mentioned in Section 7.4.2. Such results are the basis for the widely used equal-deformation rule (i.e., $u_m = u_o$), which is reasonable for systems in the velocity- and displacement-sensitive regions of the spectrum but not in the acceleration-sensitive region.

[†]Mentioned herein are seven ensembles of far-fault ground motions, each with 20 records; these ground motions and their parameters are listed in Chopra and Chintanapakdee (2004). The first group of ensembles, denoted by LMSR, LMLR, SMSR, and SMLR, represent four combinations of large ($M = 6.6$ to 6.9) or small ($M = 5.8$ to 6.5) magnitude and short ($R = 13$ to 30 km) or long ($R = 30$ to 60 km) distance.

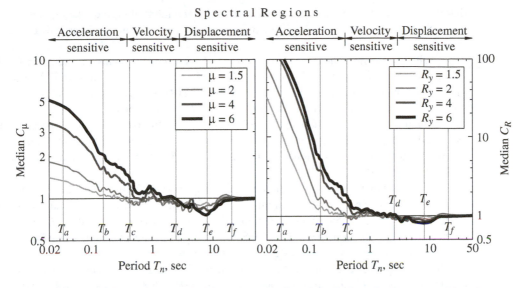

Figure 23.3.2   Median inelastic deformation ratios C_μ and $\bar{C}_R$ for elastoplastic systems subjected to the LMSR ensemble of ground motions.

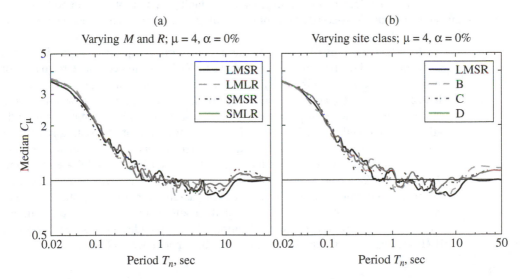

Figure 23.3.3   Comparison of C_μ for elastoplastic systems with $\mu = 4$ subjected to (a) LMSR, LMLR, SMSR, and SMLR ground motion ensembles; and (b) site class B, C, and D ensembles.

What is the influence of earthquake magnitude and distance on the inelastic deformation ratio? To answer this question, the median C_μ is plotted against T_n in Fig. 23.3.3a for the LMSR, LMLR, SMSR, and SMLR ground motion ensembles. These results indicate that the inelastic deformation ratio is essentially independent of earthquake magnitude and distance; however, it would be different for near-fault ground motions, as will be shown later.

What is the influence of soil conditions at the recording sites on the inelastic deformation ratio? To answer this question, the median C_μ is presented in Fig. 23.3.3b for three ensembles of ground motions recorded on NEHRP site classes B, C, and D,[†] all of which are firm soil sites. The median C_μ versus T_n curves for the three site classes are very similar to each other and to the LMSR result. Thus the inelastic deformation ratio is essentially independent of local soil conditions as long as they are firm soil sites, but it may be affected by soft soil conditions, such as in parts of Mexico City and around the margins of San Francisco Bay.

The median inelastic deformation ratios C_μ (and C_R) for near-fault ground motions described in Section 6.8 are significantly different from those for far-fault motions[‡] (Figs. 23.3.4a and 23.3.5a). This systematic difference between the values of C_μ (and C_R) for near-fault and far-fault ground motions, especially in the acceleration-sensitive spectral region, is due primarily to the differences between the spectral shapes and values of T_c for the two types of excitations (Section 6.8); recall that T_c is the period separating the acceleration- and velocity-sensitive spectral regions (Fig. 6.8.3). This assertion is demonstrated by plotting the ensemble median of the individual ground motion data for C_μ (and C_R) as a function of the normalized vibration period T_n/T_c (Figs. 23.3.4b and 23.3.5b). Now the inelastic deformation ratio plots for far-fault ground motions and both—fault normal and fault parallel—components of near-fault ground motions have become very similar in all spectral regions.

Simplified equations for inelastic deformation ratios C_μ and C_R provide the most direct estimation of the deformation of an inelastic SDF system because deformation of the corresponding linear system is readily known from the elastic design spectrum [Eq. (6.7.1)]. Such an equation for C_μ could be used to determine deformation of a new or rehabilitated structure where the global displacement ductility capacity can be estimated. Similarly, an equation for C_R could be used to determine the deformation of an existing structure with known lateral strength. Based on the data presented in this section, which is a part of a much larger data set, two equations have been developed for C_μ and C_R as functions of T_n/T_c, and μ or R_y, respectively. Because these equations are developed as a function of the normalized vibration period T_n/T_c, instead of the vibration period T_n, the same equation is valid for near-fault ground motions recorded on soil or rock, as well as far-fault ground motions associated with a wide range of earthquake magnitudes and distances recorded on NEHRP site classes B, C, and D (Chopra and Chintanapakdee, 2004).

[†]The second group of three ensembles are categorized by NEHRP site classes B, C, or D. These ground motions were recorded during earthquakes with magnitudes ranging from 6.0 to 7.4 at distances ranging from 11 to 118 km.

[‡]The two ensembles of near-fault (NF) ground motions, denoted by NF–FN and NF–FP, are the two horizontal components—fault normal (FN) and fault parallel (FP)—of 15 ground motions, recorded during earthquakes of magnitudes ranging from 6.2 to 6.9 at distances ranging from 0 to 9 km. These ground motions and their parameters are listed in Chopra and Chintanapakdee (2004).

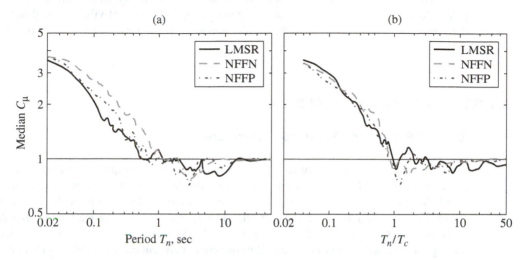

Figure 23.3.4 Comparison of C_μ for far-fault (LMSR) and two near-fault (NF) ground motion ensembles—fault normal (FN) components and fault parallel (FP) components—plotted versus (a) initial elastic vibration period T_n and (b) normalized period T_n/T_c; both plots are for elastoplastic systems with $\mu = 4$.

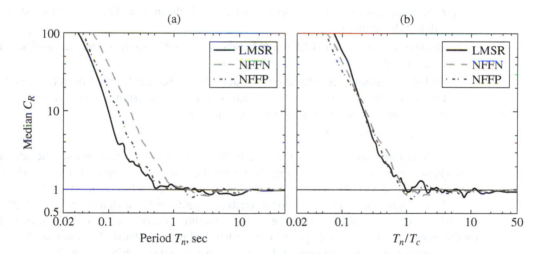

Figure 23.3.5 Comparison of C_R for far-fault (LMSR) and near-fault ground motion ensembles plotted versus (a) initial elastic vibration period T_n and (b) normalized period, T_n/T_c; both plots are for elastoplastic systems with $R_y = 4$.

These equations were developed for bilinear hysteretic systems with nonnegative postyield stiffness that unload and reload during repeated cycles of vibration without any deterioration of stiffness or strength. Similar equations for other hysteretic systems can be developed.

23.4 NONLINEAR STATIC PROCEDURES

23.4.1 ASCE 41 and FEMA 356 Procedure

The nonlinear static procedure in ASCE 41 and FEMA 356 requires development of a pushover curve, a plot of base shear versus roof displacement, by nonlinear static analysis of the structure subjected first to gravity loads, followed by monotonically increasing lateral forces with a specified invariant heightwise distribution. At least two force distributions must be considered according to FEMA 356. The first is to be selected from among the following: first-mode distribution, equivalent lateral force (ELF) distribution, and SRSS distribution. The second distribution is either the "uniform" distribution or an adaptive distribution; several options are mentioned for the latter, which varies with change in deflected shape of the structure as it yields. The other four force distributions mentioned above are defined as follows:

1. *First-mode distribution:* $s_j^* = m_j \phi_{j1}$, where m_j is the mass and ϕ_{j1} is the mode shape value at the jth floor.

2. *Equivalent lateral force* (ELF) *distribution:* $s_j^* = m_j h_j^k$, where h_j and k are as defined in Section 22.1.

3. *RSA distribution:* s^* is defined by the lateral forces back-calculated from (as the difference of) the story shears determined by response spectrum analysis of the structure, assumed to be linearly elastic (Section 13.8).

4. *Uniform distribution:* $s_j^* = m_j$.

Each of these force distributions pushes the building in the same direction over the height of the building, as demonstrated in Fig. 23.4.1 for the SAC–Los Angeles 9-story building. Subsequently, in ASCE 41 only the first mode force distribution has been retained.

If the higher modes of vibration contribute significantly, as defined in FEMA 356 and ASCE 41, to the elastic response of the structure, the NSP must be supplemented by the linear dynamic analysis procedure (LDP), and seismic demands computed by the two procedures are evaluated against their respective acceptance criteria. The SAC 9- and 20-story buildings fall into this category.

The potential and limitations of the FEMA 356 and ASCE 41 force distributions are demonstrated in Figs. 23.4.2 to 23.4.4, where the resulting estimates of the median story drift and plastic hinge rotation demands imposed on seven of the SAC buildings by the ensemble of 20 SAC ground motions are compared with the "exact" median value determined by nonlinear RHA. The target displacement for FEMA 356 and ASCE 41 analysis was not determined using Eq. (23.3.1) but was calculated accurately to ensure a meaningful

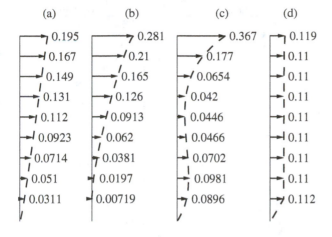

(a)   (b)   (c)   (d)

(a)	(b)	(c)	(d)
0.195	0.281	0.367	0.119
0.167	0.21	0.177	0.11
0.149	0.165	0.0654	0.11
0.131	0.126	0.042	0.11
0.112	0.0913	0.0446	0.11
0.0923	0.062	0.0466	0.11
0.0714	0.0381	0.0702	0.11
0.051	0.0197	0.0981	0.11
0.0311	0.00719	0.0896	0.112

Figure 23.4.1 FEMA 356 force distributions for the Los Angeles 9-story building: (a) first mode; (b) ELF; (c) RSA; (d) uniform. ASCE 41 specifies only the first-mode force distribution.

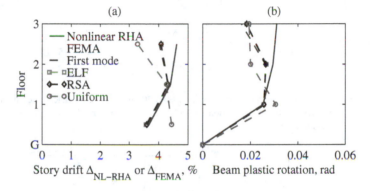

Figure 23.4.2 Median seismic demands for the Los Angeles 3-story building determined by nonlinear RHA and four FEMA 356 force distributions, which include the ASCE 41 force distribution: first mode, ELF, RSA, and uniform: (a) story drift; (b) plastic rotations.

comparison of the two sets of results. The FEMA 356 and ASCE 41 lateral force distributions provide an acceptable estimate of story drifts for the 3-story building (Fig. 23.4.2a). However, consistent with earlier results (Figs. 19.4.3 and 19.7.1) for elastic buildings, the first-mode force distribution grossly underestimates the story drifts, especially in the upper stories of the 9- and 20-story buildings, showing that higher-mode contributions are especially significant in the seismic demands for upper stories (Fig. 23.4.3). Although the ELF and RSA force distributions are intended to account for higher-mode responses, they do not provide satisfactory estimates of seismic demands for buildings that remain essentially elastic (Boston buildings), for buildings that are deformed moderately into the inelastic range (Seattle buildings), or for buildings that are deformed far into the inelastic range (Los Angeles buildings). The "uniform" force distribution seems unnecessary because it grossly underestimates drifts in upper stories and grossly overestimates them in lower stories (Fig. 23.4.3). Because FEMA 356 requires that seismic demands be estimated as the

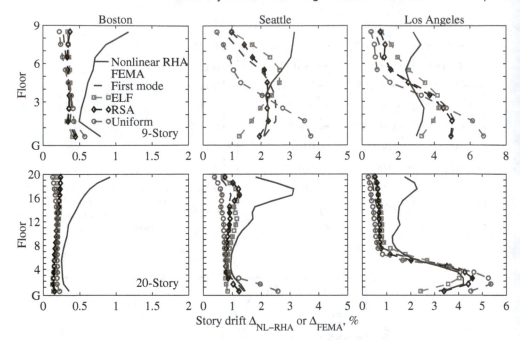

Figure 23.4.3 Median story drifts for 9- and 20-story buildings determined by nonlinear RHA and four FEMA 356 force distributions, which include the ASCE 41 force distribution: first mode, ELF, RSA, and uniform.

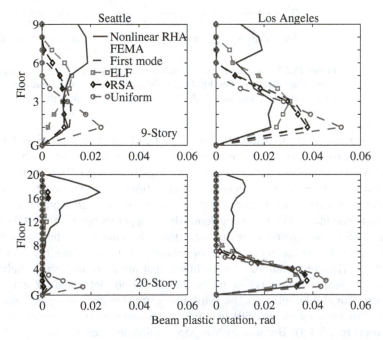

Figure 23.4.4 Median plastic rotations in interior beams of 9- and 20-story buildings determined by nonlinear RHA and four FEMA 356 force distributions, which include the ASCE 41 force distribution: first mode, ELF, RSA, and uniform. Boston buildings are excluded because they remained essentially elastic.

larger of the results from at least two lateral force distributions, it is useful to examine the upper bound of results from the four force distributions considered. This upper bound also significantly underestimates drifts in upper stories of taller buildings but overestimates them in lower stories (Fig. 23.4.3).

The potential and limitation of the FEMA 356 and ASCE 41 lateral force distributions in estimating plastic hinge rotations are also described by the preceding observations from results for story drifts. These lateral force distributions provide an acceptable estimate of plastic hinge rotations for the 3-story building (Fig. 23.4.2b), but either fail to identify or significantly underestimate plastic hinge rotations in beams at the upper floors of 9- and 20-story buildings (Fig. 23.4.4). Many discussions of pushover analysis and its potential and limitations are available in the literature.

23.4.2 Improved Nonlinear Static Procedures

It is clear from the preceding discussion that the seismic demand estimated by NSP using the first-mode force distribution in ASCE 41 (or others in FEMA 356) should be improved. One approach to reducing the discrepancy in this approximate procedure relative to nonlinear RHA is to include the contributions of higher modes of vibration to seismic demands. Recall that when higher-mode responses were included in the response spectrum analysis (RSA) procedure, improved results were obtained for linearly elastic systems (Fig. 19.7.1). Although modal analysis theory is strictly not valid for inelastic systems, earlier we demonstrated that the natural vibration modes of the corresponding linear system are coupled only weakly in the response of inelastic systems (Fig. 20.6.1), a property that permitted development of the modal pushover analysis procedure (MPA) for inelastic systems (Section 20.8).

The MPA procedure estimates seismic demands much better than do FEMA 356 or ASCE 41 force distributions for the SAC 9- and 20-story buildings (Figs. 23.4.5 to 23.4.8). For each building, the results are organized into two parts. In the upper part, the FEMA 356 (and ASCE 41) estimates of median (over 20 SAC ground motions) seismic demands are compared with the "exact" median value determined by nonlinear RHA, all taken from Fig. 23.4.3. In the lower part, the MPA estimate (including all significant modes) of median seismic demands is compared with the exact value, both taken from Fig. 20.8.4. It is obvious by comparing the two parts of the figure for each building that MPA provides much superior results for the entire range of buildings: from Boston buildings that remain essentially elastic, to Seattle buildings that are deformed moderately into the inelastic range, to Los Angeles buildings that are deformed far into the inelastic range. The MPA provides much-improved estimates of story drifts in the upper stories, where higher-mode effects are especially significant (Figs. 23.4.5 and 23.4.6). In contrast to the FEMA 356 and ASCE 41 force distributions, MPA is able to identify plastic hinging in upper stories and to provide good estimates of plastic hinge rotations throughout the height of the building (Figs. 23.4.7 and 23.4.8).

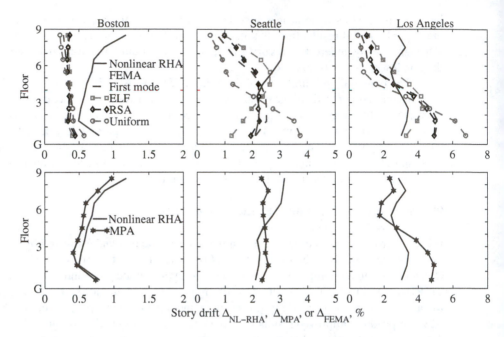

Figure 23.4.5 Median story drifts for 9-story buildings determined by three procedures: (1) nonlinear RHA, (2) four FEMA 356 force distributions, which include the ASCE 41 force distribution (upper boxes), and (3) MPA (lower boxes).

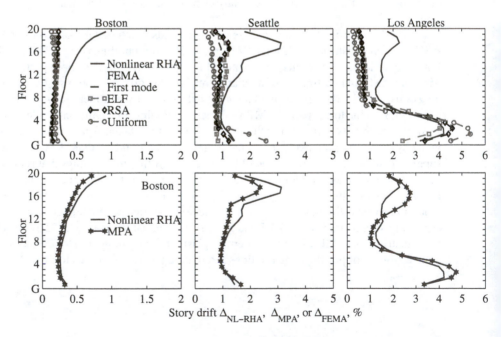

Figure 23.4.6 Median story drifts for 20-story buildings determined by three procedures: (1) nonlinear RHA, (2) four FEMA 356 force distributions, which include the ASCE 41 force distribution (upper boxes), and (3) MPA (lower boxes).

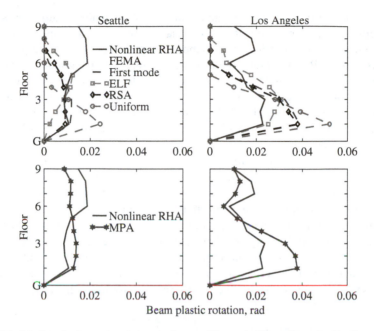

Figure 23.4.7   Median plastic rotations in interior beams of 9-story buildings determined by three procedures: (1) nonlinear RHA, (2) four FEMA 356 force distributions, which include the ASCE 41 force distribution (upper boxes), and (3) MPA (lower boxes). Boston building is excluded because it remained essentially elastic.

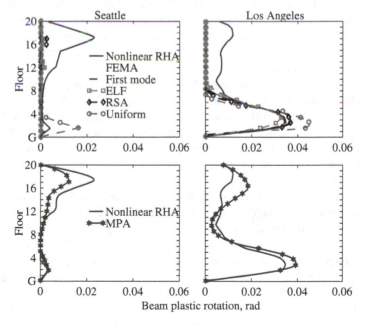

Figure 23.4.8   Median plastic rotations in interior beams of 20-story buildings determined by three procedures: (1) nonlinear RHA, (2) four FEMA 356 force distributions, which include the ASCE 41 force distribution (upper boxes), and (3) MPA (lower boxes). Boston building is excluded because it remained essentially elastic.

Based on structural dynamics theory, MPA achieves superior estimates of seismic demands while retaining the conceptual simplicity and computational attractiveness of nonlinear static procedures now standard in structural engineering practice (e.g., ASCE 41 procedure). Because higher-mode pushover analyses are similar to the first-mode pushover analyses included in ASCE 41, MPA is conceptually no more difficult than standard procedures. Because pushover analyses for the first few (two or three) modal force distributions are typically sufficient in MPA, it entails computational effort that is only slightly more than that required in the ASCE 41 procedure, which requires pushover analyses for one force distribution.

Although MPA is sufficiently accurate to be useful in seismic evaluation of many buildings for many ground motions—and is much more accurate than FEMA 356 or ASCE 41 procedures—it may not be highly accurate for buildings subjected to very intense ground motions that deform the structure far into the region of negative postyield stiffness, with significant deterioration of structural strength: for example, the SAC–Los Angeles 9- and 20-story buildings subjected to the very intense (2% probability of exceedance in 50 years or return period of 2475 years) SAC ensemble of ground motions, which included several near-fault ground motions. For such cases, nonlinear static or pushover procedures should be abandoned in favor of nonlinear RHA.

23.5 CONCLUDING REMARKS

The profession has come a long way in estimating seismic demands for buildings by abandoning traditional building code-based elastic analysis of the structure for reduced forces, and instead, developing procedures that explicitly consider inelastic behavior of the structure. However, these more recent methods, now standard in structural engineering practice to evaluate existing buildings, do not estimate seismic demands accurately, as was demonstrated in this chapter.

Seismic demands computed by nonlinear analysis procedures may depend significantly on assumptions in preparing an inelastic model of the building, as demonstrated in Chapter 20, and on software used in implementing the computation. Such variability implies that requirements for inelastic modeling valid at large structural deformations should be established and robust computer programs should be developed. With these developments, nonlinear analysis procedures would provide more realistic and reliable estimates of seismic demands.

A rigorous approach would require nonlinear RHA of a realistic three-dimensional idealization of the building subjected to an ensemble of multicomponent ground motions compatible with the site-specific earthquake design spectrum. However, such a requirement may be unreasonable for every building—no matter how simple—and of every structural engineering office—no matter how small. Therefore, nonlinear static (or pushover) analysis procedures may continue to be used in structural engineering practice; they are attractive because they are simpler and can work directly with the seismic design spectrum. Such approximate methods must be rooted in structural dynamics theory and their underlying assumptions and range of applicability identified. Nonlinear RHA may be employed for final evaluation of those combinations of buildings and ground motions where an approximate procedure begins to lose its accuracy.

Inelastic modeling of buildings, nonlinear response analysis procedures, and commercial software should be robust enough for structural engineers to predict with confidence the damage to a building during future seismic events. This requires a shift away from current building design codes and evaluation guidelines to methods capable of realistic prediction of structural performance—because in the words of Nathan M. Newmark and Emilio Rosenblueth, "earthquake effects on structures systematically bring out the mistakes made in analysis, design, and construction, even the minutest mistakes."

FURTHER READING

American Society of Civil Engineers, "Pre-Standard and Commentary for the Seismic Rehabilitation of Buildings," *FEMA 356*, Federal Emergency Management Agency, Washington, D.C., 2000.

American Society of Civil Engineers, "Minimum Design Loads for Building and Other Structures," *ASCE Standard 7-02*, ASCE, Reston, Va., 2002.

American Society of Civil Engineers, "Seismic Rehabilitation of Existing Buildings," *ASCE Standard ASCE/SEI 41-06*, ASCE, Reston, VA, 2007.

Applied Technology Council, "Seismic Evaluation and Retrofit of Concrete Buildings," *Report ATC-40*, ATC, Redwood City, Calif., 1996.

Applied Technology Council, "Improvement of Inelastic Seismic Analysis Procedures," *Report FEMA-440*, Federal Emergency Management Agency, Washington, D.C., 2005.

Building Seismic Safety Council, "NEHRP Guidelines for the Seismic Rehabilitation of Buildings," *FEMA-273*, Federal Emergency Management Agency, Washington, D.C., 1997.

Chopra, A. K., and Chintanapakdee, C., "Inelastic Deformation Ratios for Design and Evaluation of Structures: Single-Degree-of-Freedom Bilinear Systems," *Journal of Structural Engineering, ASCE*, **130**, 2004, pp. 1309–1319.

Chopra, A. K., and Goel, R. K., "Evaluation of NSP to Estimate Seismic Deformation: SDF Systems," *Journal of Structural Engineering, ASCE*, **26**, 2000, pp. 482–490.

Chopra, A. K., Goel, R. K., and Chintanapakdee, C., "Statistics of Single-Degree-of-Freedom Estimate of Displacement for Pushover Analysis of Buildings," *Journal of Structural Engineering, ASCE*, **119**, 2003, pp. 459–469.

Fajfar, P., "Structural Analysis in Earthquake Engineering: A Breakthrough of Simplified Nonlinear Methods," Paper 843, *Proceedings of the Twelfth European Conference on Earthquake Engineering*, London, 2002.

Fajfar, P., and Krawinkler, H. (eds.), *Seismic Design Methodologies for the Next Generation of Codes*, A.A. Balkema, Publishers, Lisse, The Netherlands, 1997.

Fajfar, P., and Krawinkler, H. (eds.), *Performance-Based Seismic Design Concepts and Implementation: Proceedings of an International Workshop*, Pacific Earthquake Engineering Research Center, University of California, Richmond, Calif., 2004.

Goel, R. K., and Chopra, A. K., "Evaluation of Modal and FEMA Pushover Analyses: SAC Buildings," *Earthquake Spectra*, **20**, 2004, pp. 225–254.

Krawinkler, H., and Miranda, E., "Performance-Based Earthquake Engineering," Chapter 9 in *Earthquake Engineering* (ed. Y. Bozorgnia and V. V. Bertero), CRC Press, Boca Raton, Fla., 2004.

Frequency-Domain Method of Response Analysis

PREVIEW

Presented in this appendix is the frequency-domain method for analysis of response of linear systems to excitations varying arbitrarily with time—an alternative to the time-domain method symbolized by Duhamel's integral (Section 4.2). We start by defining the *complex frequency-response function*, which is shown to contain the steady-state responses to sinusoidal and cosine forces derived in Section 3.2 by classical methods. This function in conjunction with the complex form of the Fourier series provides an alternative approach—relative to Section 3.13—to determine the response to periodic excitation. When the excitation is not periodic, it is represented by the *Fourier integral* that involves the *Fourier transform* of the excitation. The product of this transform and the complex frequency response function gives the direct Fourier transform of the response; the *inverse Fourier transform* then gives the response as a function of time. This is known as the frequency-domain method for analysis of dynamic response.

The direct and inverse Fourier transforms must be evaluated numerically for practical problems involving excitations varying arbitrarily with time. This numerical approach leads to the *discrete Fourier transform* (DFT) method, which is the subject of the rest of the appendix. After defining the direct and inverse discrete Fourier transforms, a general method is developed for numerical evaluation of response, a method that became a practical reality only with the publication of the Cooley–Tukey algorithm for the *fast Fourier transform* in 1965. The errors in the DFT solution—which represents the steady-state response to a periodic extension of the arbitrary excitation—are then examined with the objective of understanding the requirements for the solution to be accurate. Finally, an improved DFT solution is developed to determine the "exact" response from the steady-state response by the superposition of a corrective solution.

A.1 COMPLEX FREQUENCY-RESPONSE FUNCTION

A.1.1 SDF System with Viscous Damping

Consider a viscously damped SDF system subjected to external force $p(t)$. The equation of motion for the system is

$$m\ddot{u} + c\dot{u} + ku = p(t) \tag{A.1.1}$$

The particular solution of this differential equation for harmonic force was presented in Eqs. (3.2.3), (3.2.4), and (3.2.26). Known as the steady-state response, this solution is repeated here for convenience. The displacement (or deformation) $u(t)$ due to external force $p(t) = p_o \sin \omega t$ is

$$u(t) = \frac{p_o}{k} \frac{\left[1 - (\omega/\omega_n)^2\right] \sin \omega t - [2\zeta(\omega/\omega_n)] \cos \omega t}{\left[1 - (\omega/\omega_n)^2\right]^2 + [2\zeta(\omega/\omega_n)]^2} \tag{A.1.2}$$

and that due to $p(t) = p_o \cos \omega t$ is

$$u(t) = \frac{p_o}{k} \frac{\left[1 - (\omega/\omega_n)^2\right] \cos \omega t + [2\zeta(\omega/\omega_n)] \sin \omega t}{\left[1 - (\omega/\omega_n)^2\right]^2 + [2\zeta(\omega/\omega_n)]^2} \tag{A.1.3}$$

Now consider the external force:

$$p(t) = 1e^{i\omega t} \qquad \text{or} \qquad p(t) = 1(\cos \omega t + i \sin \omega t) \tag{A.1.4}$$

where $i = \sqrt{-1}$. Equation (A.1.4) is a compact representation of sinusoidal and cosine forces, together. The steady-state response of the system will be harmonic motion at the forcing frequency, ω, which can be expressed as

$$u(t) = H_u(\omega)e^{i\omega t} \tag{A.1.5}$$

where $H_u(\omega)$ remains to be determined. To do so, we differentiate Eq. (A.1.5) to obtain

$$\dot{u}(t) = i\omega H_u(\omega)e^{i\omega t} \qquad \ddot{u}(t) = -\omega^2 H_u(\omega)e^{i\omega t} \tag{A.1.6}$$

and substitute Eqs. (A.1.5) and (A.1.6) in Eq. (A.1.1):

$$H_u(\omega)e^{i\omega t}(-\omega^2 m + i\omega c + k) = e^{i\omega t}$$

Canceling the $e^{i\omega t}$ term from both sides of this equation gives

$$H_u(\omega) = \frac{1}{-\omega^2 m + i\omega c + k}$$

which can be expressed as

$$H_u(\omega) = \frac{1}{k} \frac{1}{\left[1 - (\omega/\omega_n)^2\right] + i\left[2\zeta(\omega/\omega_n)\right]} \tag{A.1.7}$$

where $\omega_n = \sqrt{k/m}$ and $\zeta = c/2m\omega_n$. Recall from Chapter 2 that ω_n is the natural frequency of vibration and ζ the damping ratio of the system.

Equation (A.1.7) contains the steady-state responses to both harmonic forces $p(t) = p_o \sin \omega t$ and $p(t) = p_o \cos \omega t$ defined by Eqs. (A.1.2) and (A.1.3). To demonstrate this fact, Eq. (A.1.7) is substituted in Eq. (A.1.5), which is then manipulated to obtain (see Derivation A.1)

$$u(t) = u^c(t) + i u^s(t) \tag{A.1.8}$$

where

$$u^c(t) = \frac{1}{k} \frac{\left[1 - (\omega/\omega_n)^2\right] \cos \omega t + [2\zeta(\omega/\omega_n)] \sin \omega t}{\left[1 - (\omega/\omega_n)^2\right]^2 + [2\zeta(\omega/\omega_n)]^2} \tag{A.1.9}$$

$$u^s(t) = \frac{1}{k} \frac{\left[1 - (\omega/\omega_n)^2\right] \sin \omega t - [2\zeta(\omega/\omega_n)] \cos \omega t}{\left[1 - (\omega/\omega_n)^2\right]^2 + [2\zeta(\omega/\omega_n)]^2} \tag{A.1.10}$$

Observe that Eqs. (A.1.9) and (A.1.10) are identical to Eqs. (A.1.3) and (A.1.2), respectively, specialized for $p_o = 1$. This implies that (1) the real part of Eq. (A.1.5) is the response to $p(t) = 1 \cos \omega t$, the real part of the force $p(t) = 1e^{i\omega t}$; and (2) the imaginary part of Eq. (A.1.5) is the response to $p(t) = 1 \sin \omega t$, the imaginary part of the force $p(t) = 1e^{i\omega t}$. This proves the assertion at the beginning of this paragraph.

Observe that Eqs. (A.1.5) and (A.1.7) are a more compact presentation of the response to harmonic excitation, relative to Eqs. (A.1.2) and (A.1.3). Also note that the derivation of Eq. (A.1.7) presented earlier is simpler than that for Eqs. (A.1.2) and (A.1.3), presented in Section 3.2. However, complex algebra is necessary to derive the complex frequency response function, whereas classical methods provided the particular solutions [Eqs. (A.1.2) and (A.1.3)] of the differential equation (A.1.1).

The function $H_u(\omega)$ is known as the *complex frequency-response function*. It describes the steady-state response of the system to the force defined by Eq. (A.1.4a), a harmonic force of unit amplitude (i.e., $p_o = 1$). Defined by Eq. (A.1.7), $H_u(\omega)$ is a complex-valued function of the forcing frequency and system parameters k, ω_n, and ζ. The absolute value of this complex-valued function is

$$\frac{|H_u(\omega)|}{(u_{st})_o} = \frac{1}{\sqrt{\left[1 - (\omega/\omega_n)^2\right]^2 + [2\zeta(\omega/\omega_n)]^2}} \tag{A.1.11}$$

where $(u_{st})_o \equiv p_o/k = 1/k$. Equation (A.1.11) is equivalent to Eq. (3.2.11) for the amplitude of the steady-state response of the system to harmonic excitation, which was plotted in Fig. 3.2.6. The real and imaginary parts of $H_u(\omega)$, denoted by Re(·) and Im(·), respectively, are related as follows:

$$\frac{-\text{Im}[H_u(\omega)]}{\text{Re}[H_u(\omega)]} = \frac{2\zeta(\omega/\omega_n)}{1 - (\omega/\omega_n)^2} \tag{A.1.12}$$

This equation is equivalent to Eq. (3.2.12) for the phase angle or phase lag of the response, plotted in Fig. 3.2.6. Thus, it is clear that the complex frequency response function defines the amplitude and phase angle of the response.

The subscript u in $H_u(\omega)$, defined by Eq. (A.1.7), denotes that this function describes the deformation u; complex frequency-response functions can be similarly derived for other response quantities—velocity $\dot{u}$, acceleration $\ddot{u}$, elastic resisting force $f_S = ku$, etc. Later, the subscript u will be dropped for convenience in notation.

Derivation A.1

Substituting Eq. (A.1.7) in Eq. (A.1.5) gives

$$u(t) = \frac{1}{k}\frac{1}{\left[1 - (\omega/\omega_n)^2\right] + i\,[2\zeta\,(\omega/\omega_n)]}e^{i\omega t} \tag{a}$$

Multiplying the numerator and denominator by $\left[1 - (\omega/\omega_n)^2\right] - i\,[2\zeta\,(\omega/\omega_n)]$, the complex conjugate of the denominator, and using Eq. (A.1.4) gives

$$u(t) = \frac{1}{k}\frac{\left[1 - (\omega/\omega_n)^2\right] - i\,[2\zeta\,(\omega/\omega_n)]}{\left[1 - (\omega/\omega_n)^2\right]^2 + [2\zeta\,(\omega/\omega_n)]^2}(\cos\omega t + i\sin\omega t) \tag{b}$$

Multiplying the two parts of the numerator and collecting real and imaginary terms separately leads to Eqs. (A.1.8) to (A.1.10).

A.1.2 SDF System with Rate-Independent Damping

The equation governing harmonic motion (at frequency ω) of an SDF system with rate-independent linear damping, first presented as Eq. (3.10.3), is

$$m\ddot{u} + \frac{\eta k}{\omega}\dot{u} + ku = p(t) \tag{A.1.13}$$

The steady-state response of the system to harmonic forcing function $p(t) = 1e^{i\omega t}$ is also given by Eq. (A.1.5). Substituting Eqs. (A.1.5) and (A.1.6a) in Eq. (A.1.13) leads to

$$m\ddot{u} + k(1 + i\eta)u = p(t) \tag{A.1.14}$$

The complex term $k(1+i\eta)u$ represents the elastic and damping forces together; $k(1+i\eta)$ is often referred to as the *complex stiffness* of the system.

Substituting Eqs. (A.1.5) and (A.1.6b) in Eq. (A.1.14) gives

$$H_u(\omega)e^{i\omega t}\left[-\omega^2 m + k(1 + i\eta)\right] = e^{i\omega t}$$

Canceling the $e^{i\omega t}$ term from both sides of this equation gives

$$H_u(\omega) = \frac{1}{-\omega^2 m + k(1 + i\eta)}$$

which can be expressed as

$$H_u(\omega) = \frac{1}{k}\frac{1}{\left[1 - (\omega/\omega_n)^2\right] + i\eta} \tag{A.1.15}$$

A.2 RESPONSE TO PERIODIC EXCITATION

In Chapter 3 a procedure was developed to determine the steady-state response of an SDF system to a periodic force. The excitation was separated into its harmonic (sine and cosine) components using the *Fourier series* (Section 3.12). Then the response to each force component was written by adapting Eqs. (A.1.2) and (A.1.3). Finally, these responses to individual terms in the Fourier series were combined to determine the response of a linear system to periodic excitation (Section 3.13). The complex frequency-response function provides an alternative approach to determine the response to periodic excitation. To develop this method we first develop an alternative form for the Fourier series.

A.2.1 Complex Fourier Series

An excitation $p(t)$ that is periodic with period T_0 can be separated into its harmonic components using the *complex Fourier series:*

$$p(t) = \sum_{j=-\infty}^{\infty} P_j e^{i(j\omega_0 t)} \tag{A.2.1}$$

where the fundamental or first harmonic in the excitation has the frequency

$$\omega_0 = \frac{2\pi}{T_0} \tag{A.2.2}$$

and $\omega_j \equiv j\omega_0$ is the circular frequency of the jth harmonic. The Fourier coefficients P_j can be expressed in terms of $p(t)$ because the exponential functions are orthogonal (see Derivation A.2):

$$P_j = \frac{1}{T_0} \int_0^{T_0} p(t) e^{-i(j\omega_0 t)} \, dt \qquad j = 0, \pm 1, \pm 2, \ldots \tag{A.2.3}$$

The complex-valued coefficient P_j defines the amplitude and phase of the jth harmonic. Observe that the complex Fourier series, Eqs. (A.2.1) and (A.2.3), is compact compared to the traditional form of the Fourier series, Eqs. (3.12.1) to (3.12.5). Also note that the time function is denoted by a lowercase letter, and the Fourier coefficients for the function by the same letter in uppercase.

Equation (A.2.3) indicates that

$$P_{-j} = \overline{P}_j \tag{A.2.4}$$

where the overbar denotes the complex conjugate, and

$$P_0 = \frac{1}{T_0} \int_0^{T_0} p(t) \, dt \tag{A.2.5}$$

In other words, P_0 is the average value of $p(t)$.

Although the applied force $p(t)$ is real valued, each term on the right side of Eq. (A.2.1) is a product of a complex-valued coefficient and a complex-valued exponential function. However, it can be shown that (1) the sum of each pair of jth and $-j$th terms is real valued because of Eq. (A.2.4); and (2) the $j = 0$ term simplifies to P_0, which is real valued [Eq. (A.2.5)]. Thus, the sum of all terms is real valued, as it should be for real-valued $p(t)$.

Alternatively, Eqs. (A.2.1) and (A.2.3) can be derived starting from the conventional form of the Fourier series, as represented by Eqs. (3.12.1) to (3.12.5). This is achieved by using De Moivre's theorem, which relates the sine and cosine functions to exponential functions with complex exponent:

$$\sin x = \frac{1}{2i}(e^{ix} - e^{-ix}) \qquad \cos x = \frac{1}{2}(e^{ix} + e^{-ix}) \qquad (A.2.6)$$

where $x \equiv j\omega_0 t$. Substituting Eq. (A.2.6) into the jth term of the sine series and the jth term of the cosine series in Eq. (3.12.1), the sum of the two terms can be expressed as $P_j e^{i(j\omega_0 t)} + P_{-j} e^{-i(j\omega_0 t)}$. These are two terms in the series of Eq. (A.2.1), indicating that it is equivalent to Eq. (3.12.1).

A.2.2 Steady-State Response

The response of a linear sytem to a periodic force can be determined by combining the responses to individual excitation terms in the Fourier series of Eq. (A.2.1). To determine these individual responses, we recall that the response to $p(t) = 1e^{i\omega t}$ is given by Eq. (A.1.5), where $H_u(\omega)$ is defined by Eq. (A.1.7). Therefore, the response $u_j(t)$ of the system to an applied force equal to the jth term in the Fourier series—$p_j(t) = P_j e^{i(j\omega_0 t)}$— is obtained by replacing ω by $j\omega_0$ in Eqs. (A.1.5) and (A.1.7) and multiplying Eq. (A.1.5) by P_j, leading to

$$u_j(t) = U_j e^{i(j\omega_0 t)} \qquad (A.2.7)$$

where

$$U_j = H(j\omega_0)P_j \qquad (A.2.8)$$

Adding such responses due to all excitation terms in Eq. (A.2.1) gives the total response:

$$u(t) = \sum_{j=-\infty}^{\infty} H(j\omega_0)P_j e^{i(j\omega_0 t)} \qquad (A.2.9)$$

where the Fourier coefficients P_j are defined by Eq. (A.2.3) and the complex frequency-response function by Eq. (A.1.7) or (A.1.15). Observe that Eq. (A.2.9) is a more compact presentation of the response to periodic excitation compared to the traditional form, Eq. (3.13.6).

The procedure symbolized by Eq. (A.2.9) is known as the frequency-domain method for analysis of structural response to periodic excitation; it is shown schematically in Fig. A.2.1. The excitation $p(t)$ is transformed from the time domain to the frequency domain, where it is described by the Fourier coefficients P_j [Eq. (A.2.3)]. The response to the jth harmonic is defined by Eq. (A.2.8) in the frequency domain. Adding the responses to all harmonic excitations gives response $u(t)$ in the time domain [Eq. (A.2.9)].

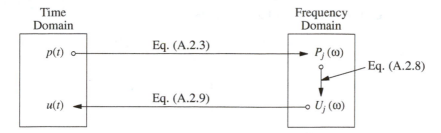

Figure A.2.1 Solution of response to periodic excitation by transformation to frequency domain.

Derivation A.2

Multiplying both sides of Eq. (A.2.1) by $e^{-i(n\omega_0 t)}$ and integrating over a period, 0 to T_0 gives

$$\int_0^{T_0} p(t)e^{-i(n\omega_0 t)}\, dt = \sum_{j=-\infty}^{\infty} P_j \int_0^{T_0} e^{-i(n\omega_0 t)} e^{i(j\omega_0 t)}\, dt \tag{a}$$

To evaluate P_j we note that

$$\int_0^{T_0} e^{-i(n\omega_0 t)} e^{i(j\omega_0 t)}\, dt = \begin{cases} 0 & j \neq n \\ T_0 & j = n \end{cases} \tag{b}$$

Thus all terms on the right side of Eq. (a) are equal to zero except the nth term, leading to

$$P_n = \frac{1}{T_0}\int_0^{T_0} p(t)e^{-i(n\omega_0 t)}\, dt \tag{c}$$

which is identical to Eq. (A.2.3) except for a different index—n instead of j.

Example A.1

Solve Example 3.8 by the frequency-domain method.

Solution   **a.** *Determine Fourier coefficients.*

$$P_j = \frac{1}{T_0}\int_0^{T_0} p(t)e^{-i(j\omega_0 t)}\, dt$$

$$= \frac{1}{T_0}\left[p_o \int_0^{T_0/2} e^{-i(j\omega_0 t)}\, dt + (-p_o)\int_{T_0/2}^{T_0} e^{-i(j\omega_0 t)}\, dt \right]$$

$$= \frac{-p_o}{ij\omega_0 T_0}\left[e^{-i(j\omega_0 t)}\Big|_0^{T_0/2} - e^{-i(j\omega_0 t)}\Big|_{T_0/2}^{T_0} \right] \tag{a}$$

To evaluate the exponential terms, we note that $\omega_0 T_0 = 2\pi$ from Eq. (A.2.2). Therefore,

$$e^{-i(j\omega_0 T_0/2)} = e^{-i(j\pi)} = \begin{cases} +1 & j \text{ even} \\ -1 & j \text{ odd} \end{cases} \tag{b}$$

$$e^{-ij\omega_0 T_0} = e^{-i(2j\pi)} = 1 \tag{c}$$

Using Eqs. (b) and (c), Eq. (a) becomes

$$P_j = \frac{ip_o}{2\pi j}\left[2e^{-i(j\pi)} - 1 - e^{-i(2j\pi)}\right]$$

or

$$P_j = \begin{cases} 0 & j \text{ even} \\ -\dfrac{2p_o i}{j\pi} & j \text{ odd} \end{cases} \tag{d}$$

b. *Determine response.*

$$U_j = H(j\omega_0)P_j \tag{e}$$

where P_j is given by Eq. (d) and $H(j\omega_0)$ for an undamped system is given by Eq. (A.1.7), specialized for $\zeta = 0$ and $\omega = j\omega_0$:

$$H(j\omega_0) = \frac{1}{k}\frac{1}{1 - \beta_j^2} \tag{f}$$

where $\beta_j = j\omega_0/\omega_n$. Substituting Eqs. (d) and (f) in Eq. (e) gives

$$U_j = -\frac{2p_o i}{\pi k}\frac{1}{j}\frac{1}{1 - \beta_j^2} \tag{g}$$

for odd values of j; $U_j = 0$ for even values of j. Substituting Eq. (g) in Eq. (A.2.7) gives the response in the time domain:

$$u(t) = (u_{\text{st}})_o\frac{2}{\pi}\sum_{\substack{j=-\infty \\ j \text{ odd}}}^{\infty} -i\frac{1}{j}\frac{1}{1 - \beta_j^2}e^{i(j\omega_0 t)} \tag{h}$$

It is of interest to compare this solution with that obtained earlier by the classical Fourier series. Specializing Eq. (f) of Example 3.8 for undamped systems gives

$$u(t) = (u_{\text{st}})_o\frac{4}{\pi}\sum_{j=1,3,5}^{\infty}\frac{1}{j}\frac{1}{1 - \beta_j^2}\sin j\omega_0 t \tag{i}$$

We note that the complex-valued coefficients in Eq. (h) have an amplitude that is half of the real-valued coefficients in Eq. (i). The contribution of the $-j$ terms in the complex Fourier series accounts for the difference.

A.3 RESPONSE TO ARBITRARY EXCITATION

A.3.1 Fourier Integral

In the preceding section we have seen that a periodic excitation can be represented by a Fourier series as in Eqs. (3.12.1) and (A.2.1). When the excitation $p(t)$ is not periodic, it can be represented by the *Fourier integral:*

$$p(t) = \frac{1}{2\pi}\int_{-\infty}^{\infty} P(\omega)e^{i\omega t}\,d\omega \tag{A.3.1}$$

where

$$P(\omega) = \int_{-\infty}^{\infty} p(t)e^{-i\omega t}\, dt \qquad\qquad (A.3.2)$$

Equation (A.3.2) represents the *Fourier transform* (also known as the *direct Fourier transform*) of the time function $p(t)$, and Eq. (A.3.1) is the *inverse Fourier transform* of the frequency function $P(\omega)$. The two equations together are called a *Fourer transform pair.* Note that the time function is denoted by a lowercase letter, and the Fourier transform of the function by the same letter in uppercase.

In Eq. (A.3.1) $p(t)$ has been expressed as the superposition of harmonic functions $[P(\omega)/2\pi]e^{i\omega t}$, where the complex-valued coefficient $P(\omega)$ for a given $p(t)$ is to be determined from Eq. (A.3.2). Included in the superposition is an infinite number of harmonic functions with their frequencies varying continuously. In contrast, a periodic function was represented as the superposition of an infinite number of harmonic functions with discrete frequencies $j\omega_0$, $j = 0,\ \pm 1,\ \pm 2,\ \dots$ Equations (A.3.1) and (A.3.2) can be derived starting from the Fourier series equations (A.2.1) and (A.2.3) and letting the period T_0 approach infinity.

A.3.2 Response to Arbitrary Excitation

The response of a linear system to excitation $p(t)$ can be determined by combining the responses to individual harmonic excitation terms in the Fourier integral of Eq. (A.3.1). The response of the system to the excitation $P(\omega)e^{i\omega t}$ is given by $H(\omega)P(\omega)e^{i\omega t}$. Superposing the responses to all harmonic terms in Eq. (A.3.1) gives the total response:

$$u(t) = \frac{1}{2\pi} \int_{-\infty}^{\infty} U(\omega)e^{i\omega t}\, d\omega \qquad\qquad (A.3.3)$$

where

$$U(\omega) = H(\omega)P(\omega) \qquad\qquad (A.3.4)$$

This is known as the frequency-domain method for analysis of structural response to arbitrary excitation. Equation (A.3.3) is the inverse Fourier transform of $U(\omega)$, the product of the complex frequency-response function and the Fourier transform of the excitation.

From Eq. (A.3.2) it is clear that straightforward integration is adequate to determine the direct Fourier transform. In contrast, contour integration in the complex plane is necessary to evaluate the inverse Fourier transform of Eq. (A.3.3). This integration procedure is not described here because rarely is it feasible analytically for structural dynamics problems arising in engineering practice.

A.4 RELATIONSHIP BETWEEN COMPLEX FREQUENCY RESPONSE AND UNIT IMPULSE RESPONSE

We digress briefly to develop the relationship between the complex frequency-response function $H(\omega)$, introduced in the preceding sections, and the unit impulse-response function $h(t)$, defined in Chapter 4. $H(\omega)$ describes the system response in the frequency domain to unit harmonic excitation. $h(t)$ describes the system response in the time domain to a unit impulse excitation, $p(t) = \delta(t)$. For example, for a viscously damped SDF system $H(\omega)$ is given by Eq. (A.1.7) and $h(t)$ by Eq. (4.1.7), which is specialized for $\tau = 0$ and repeated here for convenience:

$$h(t) = \frac{1}{m\omega_D} e^{-\zeta\omega_n t} \sin \omega_D t \qquad (A.4.1)$$

We will demonstrate that $H(\omega)$ and $h(t)$ are a Fourier transform pair. For this purpose we will use the frequency-domain analysis procedure (Section A.3) to determine the response to a unit impulse excitation $p(t) = \delta(t)$. Substituting this $p(t)$ in Eq. (A.3.2) gives the Fourier transform of the unit impulse:

$$P(\omega) = \int_{-\infty}^{\infty} \delta(t)e^{-i\omega t}\, dt = 1 \qquad (A.4.2)$$

Substituting $P(\omega) = 1$ in Eqs. (A.3.4) and (A.3.3) gives

$$h(t) = \frac{1}{2\pi} \int_{-\infty}^{\infty} H(\omega)e^{i\omega t}\, d\omega \qquad (A.4.3)$$

Comparing this result with the definitions of Fourier transform, Eq. (A.3.2), and inverse Fourier transform, Eq. (A.3.1), it is clear that $h(t)$ is the inverse Fourier transform of $H(\omega)$ and that $H(\omega)$ is the Fourier transform of $h(t)$:

$$H(\omega) = \int_{-\infty}^{\infty} h(t)e^{-i\omega t}\, dt \qquad (A.4.4)$$

Observe that the choice of symbol h to represent the unit impulse response and H to denote the complex frequency response conforms to our selected notations for a Fourier transform pair.

A.5 DISCRETE FOURIER TRANSFORM METHODS

The frequency-domain analysis of the dynamic response of structures developed in Section A.3 requires that the Fourier transform of $p(t)$, Eq. (A.3.2), and the inverse Fourier transform of $U(\omega)$, Eq. (A.3.3), both be determined. Analytical evaluation of these direct and inverse Fourier transforms is not possible except for excitations described by simple functions applied to simple structural systems. These integrals must be evaluated numerically for excitations varying arbitrarily with time, complex vibratory systems, or situations where the complex frequency response (or unit impulse response) is described numerically. Numerical evaluation requires truncating these integrals over infinite range to a finite range, and becomes equivalent to approximating the arbitrarily time-varying excitation $p(t)$ by a periodic function. We develop these ideas next.

A.5.1 Discretization of Excitation

The system is excited by a force $p(t)$ of duration t_d, as shown in Fig. A.5.1. Our objective is to determine the resulting displacement $u(t)$ of the system, which is presumed to be initially at rest. Since the peak (or absolute maximum) response of the system may be attained after the excitation has ended, the analysis should be carried out over a time duration T_0 that is longer than t_d. If the peak occurs after the excitation has ended, it will be attained in the first half-cycle of free vibration because the motion will decay in subsequent cycles due to damping. Therefore, we should choose

$$T_0 \geq t_d + \frac{T_n}{2} \tag{A.5.1}$$

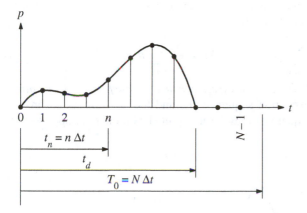

Figure A.5.1 Excitation $p(t)$ and its discretized version.

The forcing function $p(t)$ over the time duration T_0 is sampled at N equally spaced time instants, numbered from 0 to $N - 1$ (Fig. A.5.1). The sampling interval is denoted by Δt; thus

$$T_0 = N\, \Delta t \tag{A.5.2}$$

The forcing function $p(t)$ is then defined by a set of discrete values $p_n \equiv p(t_n) \equiv p(n\, \Delta t)$, shown as the series of dots in Fig. A.5.1.

The sampling interval Δt should be short enough compared both to the periods of significant harmonics in the excitation and to the natural period T_n of the system. The first requirement ensures accurate representation of the excitation and of the forced vibrational component of the response, and the second requirement ensures accurate representation of the free vibrational response component. The latter requirement also ensures accurate representation of the response of lightly damped SDF systems to broad frequency-band excitations such as most ground motions recorded during earthquakes; recall that the dominant period of such response is T_n (Fig. 6.4.1).

A.5.2 Fourier Series Representation of Excitation

Consider a periodic extension of the excitation $p(t)$, with its period defined as T_0 [Eq. (A.5.1)], shown schematically in Fig. A.5.2, and replace $p(t)$ by the array p_n describing the discretized forcing function. Starting with the complex Fourier series for the function $p(t)$ [Eq. (A.2.1)], the array p_n can be expressed (see Derivation A.3) as a superposition of N harmonic functions:

$$p_n = \sum_{j=0}^{N-1} P_j e^{i(j\omega_0 t_n)} = \sum_{j=0}^{N-1} P_j e^{i(2\pi nj/N)} \tag{A.5.3}$$

in which $\omega_0 = 2\pi/T_0$, the frequency of the fundamental or first harmonic in the periodic extension of $p(t)$; $\omega_j = j\omega_0$ is the circular frequency of the jth harmonic; and P_j is a complex-valued coefficient that defines the amplitude and phase of the jth harmonic. Starting with Eq. (A.2.3), which defines P_j for the function $p(t)$, P_j associated with the array p_n can be expressed (see Derivation A.3) as

$$P_j = \frac{1}{T_0} \sum_{n=0}^{N-1} p_n e^{-i(j\omega_0 t_n)} \Delta t = \frac{1}{N} \sum_{n=0}^{N-1} p_n e^{-i(2\pi nj/N)} \tag{A.5.4}$$

Equations (A.5.3) and (A.5.4) define a *discrete Fourier transform (DFT) pair;* the array P_j is the DFT of the excitation sequence p_n, and the array p_n is the inverse DFT of the sequence P_j. These equations may be interpreted as numerical approximations of Eqs. (A.2.1) and (A.2.3).

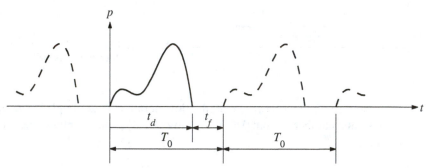

Figure A.5.2 Periodic extension of $p(t)$.

The continuous and discrete Fourier transforms differ in an important way. Whereas the continuous transform [Eq. (A.3.1)] is a true representation of the excitation function, the discrete transform [Eq. (A.5.3)] represents only a periodic version of the function. The implication of this distinction will be discussed later.

Observe that only positive frequencies are considered in Eqs. (A.5.3); therefore, we call this a one-sided Fourier expansion. In contrast, the original Eq. (A.2.1) is a two-sided Fourier expansion containing positive and negative frequencies. Just as the negative frequencies have no physical significance in the two-sided expansion, the frequencies corresponding to $N/2 < j \le N-1$ have no physical significance; they are the counterparts of the negative frequencies. If the summation in Eq. (A.2.1) were truncated to go from $j = -N/2$ to $N/2$, ω would extend from $-\omega_{N/2}$ to $\omega_{N/2}$. Thus, $\omega_{N/2}$ also defines the frequency of the highest harmonic included in Eq. (A.5.3). Denoted also by ω_{max}, this frequency is known as the *Nyquist frequency* or *folding frequency* and is given by

$$\omega_{max} = \frac{N}{2}\omega_0 = \frac{\pi}{\Delta t} \tag{A.5.5}$$

where the frequency ω_0 of the fundamental or first harmonic is defined in Eq. (A.2.2), which, together with Eq. (A.5.2), gives the second half of Eq. (A.5.5). The shortest and longest periods of the harmonics included in the Fourier expansion are determined from Eqs. (A.5.5) and (A.2.2) to be $2\Delta t$ and T_0, respectively.

Recall that in the two-sided Fourier expansion P_j and P_{-j} were complex conjugates of each other [Eq. (A.2.4)]. Consequently, in the one-sided expansion the values of P_j on either side of $\omega_{N/2}$ are complex conjugates of each other:

$$P_j = \overline{P}_{N-j} \qquad \frac{N}{2} < j \leq N - 1 \tag{A.5.6}$$

A.5.3 Complex Frequency-Response Function

This function $H(\omega)$ is computed from Eq. (A.1.7) or (A.1.15) for each $\omega = \omega_j$, and this value is denoted by H_j. A two-sided Fourier expansion includes both positive and negative frequencies ω_j and $-\omega_j$ [Eq. (A.2.1)] and H_{-j} is the complex conjugate of H_j; this assertion can be easily proven starting from Eq. (A.1.7). In a one-sided Fourier expansion only positive frequencies are included [Eq. (A.5.3)]; the frequencies corresponding to $N/2 < j \leq N - 1$ are the counterparts of the negative frequencies. Thus, the values of H_j on either side of $j = N/2$ must be complex conjugates of each other. The H_j values may be determined from Eq. (A.1.7) with the following interpretation of ω_j:

$$\omega_j = \begin{cases} j\omega_0 & 0 \leq j \leq N/2 \\ -(N-j)\omega_0 & N/2 < j \leq N - 1 \end{cases} \tag{A.5.7}$$

A.5.4 Computation of Response

In the DFT approach we first compute the response to each harmonic component of the excitation in the frequency domain. As seen in Eq. (A.2.8), this requires computation of the products

$$U_j = H_j P_j \qquad 0 \leq j \leq N - 1 \tag{A.5.8}$$

Then, the response $u_n \equiv u(t_n)$ at discrete time instants $t_n \equiv n\,\Delta t$ is computed from a truncated version of Eq. (A.2.9):

$$u_n = \sum_{j=0}^{N-1} U_j e^{i(j\omega_0 t_n)} = \sum_{j=0}^{N-1} U_j e^{i(2\pi nj/N)} \tag{A.5.9}$$

Equation (A.5.9) corresponds to Eq. (A.5.3), indicating that the sequence u_n represents the inverse DFT of the sequence U_j. This will be called the classical DFT solution.

A.5.5 Fast Fourier Transform

The DFT method to determine the dynamic response of a system requires computation of the direct DFT of the sequence p_n [Eq. (A.5.4)] and the inverse DFT of the sequence U_j [Eq. (A.5.9)]. These computations became a practical reality only with the publication of the Cooley–Tukey algorithm for the *fast Fourier transform* (FFT) in 1965. This is not a new type of transform, but is a highly efficient and accurate algorithm for computing the DFT and the inverse DFT. The original algorithm required that the number of points, N, be an integer power of 2, but it has been generalized to permit consideration of an arbitrary value of N.

It is important to recognize that the computational effort required is drastically reduced by use of the FFT algorithm. A measure of the amount of computation involved in Eq. (A.5.4) or (A.5.9) is the number of products of complex-valued quantities. It is clear that there are N sums, each of which requires N complex products, or there are N^2 products required for computing all of the P_j's or all of the u_n's. The number of complex products for the original FFT algorithm is given by $(N/2) \log_2 N$. For example, if $N = 2^{10} = 1024$, the FFT algorithm requires 0.5% of the computational effort necessary in standard computation.

A.5.6 Summary

The classical DFT procedure for response analysis of an SDF system [governed by Eq. (A.1.1)] can be summarized as a sequence of steps:

1. Define a periodic extension of the excitation $p(t)$ with its period defined as T_0 [Eq. (A.5.1)] and discretize $p(t)$ by an array $p_n \equiv p(t_n) \equiv p(n\Delta t)$, where $n = 0, 1, 2, \ldots, N - 1$.
2. Compute P_j, the DFT of p_n, according to Eq. (A.5.4); $j = 0, 1, 2, \ldots, N - 1$.
3. Determine the frequency response function $H_j \equiv H(\omega_j)$, where $H(\omega)$ is defined by Eq. (A.1.7) and ω_j by Eq. (A.5.7).
4. Compute U_j as the product defined by Eq. (A.5.8).
5. Compute the inverse DFT of the array U_j from Eq. (A.5.9) to obtain the response $u_n \equiv u(t_n)$ at discrete time instants $t_n \equiv n\Delta t$.

Derivation A.3

The periodic extension of $p(t)$ with period T_0 is represented by the Fourier series of Eq. (A.2.1), with the Fourier coefficients defined by Eq. (A.2.3). Truncating the series to include only a finite number of harmonic functions gives

$$p(t) = \sum_{j=-M}^{M} P_j e^{i(j\omega_0 t)} \tag{a}$$

The frequency of the highest harmonic included in Eq. (a) is $M\omega_0$.

The integral of Eq. (A.2.3) is evaluated numerically by the trapezoidal rule applied to the values of the integrand at discrete time instants $t_n = n\,\Delta t$, where $n = 0, 1, 2, \ldots, N$:

$$P_j = \frac{\Delta t}{T_0}\left[\frac{1}{2}p_0 e^{-i(j\omega_0 t_0)} + \sum_{n=1}^{N-1} p_n e^{-i(j\omega_0 t_n)} + \frac{1}{2}p_N e^{-i(j\omega_0 t_N)}\right] \tag{b}$$

where $t_0 = 0\,\Delta t = 0$, $t_n = n\,\Delta t$, and $t_N = N\,\Delta t$. The first term reduces to $p_0/2$ and the last term to $p_N/2$ because both exponentials can be shown to equal unity. Because the sequence p_n is periodic with period N, $p_0 = p_N$, and recognizing that $T_0 = N\,\Delta t$, Eq. (b) can be rewritten as

$$P_j = \frac{1}{N}\sum_{n=0}^{N-1} p_n e^{-i(j\omega_0 t_n)} \tag{c}$$

Now the exponential terms in Eqs. (a) and (c) are rewritten by recognizing that $\omega_0 = 2\pi/T_0$, $T_0 = N\,\Delta t$, and $t_n = n\,\Delta t$; thus

$$j\omega_0 t_n = j\frac{2\pi}{N\,\Delta t}n\,\Delta t = \frac{2\pi nj}{N} \tag{d}$$

Recall that j is the frequency number of the harmonic and n is the time step number. Substituting Eq. (d) in Eqs. (a) and (c) gives

$$p_n = \sum_{j=-M}^{M} P_j e^{i(2\pi nj/N)} \tag{e}$$

$$P_j = \frac{1}{N}\sum_{n=0}^{N-1} p_n e^{-i(2\pi nj/N)} \tag{f}$$

Suppose that we select the large positive integers M and N such that $2M + 1 = N$ (i.e., the number of frequencies is equal to the number of time steps). Then the sequence P_j is also periodic with period N and the summation in Eq. (e) over the range $j = -M$ to $j = M$ can be rewritten as a summation over the range $j = 0$ to $N - 1$, which is presented without proof:

$$p_n = \sum_{j=0}^{N-1} P_j e^{i(2\pi nj/N)} \tag{g}$$

This completes the derivation of Eqs. (A.5.3) and (A.5.4).

A.6 POSSIBLE ERRORS IN CLASSICAL DFT SOLUTION

It should be clear that the classical DFT solution given by Eq. (A.5.9) does not generally represent the desired response of the system to the excitation shown in Fig. A.5.1. Instead, it represents the steady-state response of the system to a periodic extension of the excitation

(Fig. A.5.2). In this section we examine the errors in the classical DFT solution, with the objective of understanding the requirements for the solution to be accurate. The classical DFT solution will become increasingly accurate as the duration t_f of free vibration, shown in Fig. A.5.2, becomes longer. This assertion should be obvious because longer t_f implies longer period T_0 of the periodic extension of the excitation, which is better because an arbitrary (nonperiodic) excitation can be interpreted as a periodic excitation with infinitely long period. However, to identify the factors that influence the t_f necessary for an accurate solution, we present numerical results.

It is desired to determine the dynamic response of a viscously damped SDF system, starting from at-rest conditions, to one full cycle of a sinusoidal force, $p(t) = p_o \sin \omega t$, shown in Fig. A.6.1. As mentioned in Section A.5.1, the dynamic response of the system should be determined over the time duration $T_0 = t_d + T_n/2$ or longer—thus, the shortest $t_f = T_n/2$. However, DFT solutions with longer t_f will also be presented to demonstrate the sensitivity of these solutions to the choice of t_f. With a selected t_f, the

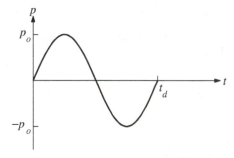

Figure A.6.1

periodic extension of the excitation is sampled over one period at intervals $\Delta t = t_d/40$, and the response is evaluated at the same intervals. Thus, the circular frequency of the highest harmonic in the Fourier series representation of the forcing function [Eq. (A.5.3)] will be $\omega_{\max} = 40\pi/t_d$ [from Eq. (A.5.5)] and the associated period is $t_d/20$. All DFTs in these solutions that will be presented were computed by the FFT routines in MATLAB.

In Figs. A.6.2 and A.6.3 are shown the periodic extension of the excitation and the displacement response of the system computed by the classical DFT method using two different values of t_f. The duration t_d of the force and the natural vibration period T_n of the system are chosen such that $t_d/T_n = 0.5$; the damping ratio ζ of the system is 5%. The time scale on the response history plots is normalized with respect to t_d, and the displacement $u(t)$ is normalized with respect to $(u_{\mathrm{st}})_o \equiv p_o/k$, the static displacement due to the peak value of the applied force. The results demonstrate that the DFT solution depends on the duration t_f of free vibration. It is clear from Fig. A.6.2 that $t_f = 4.75T_n$, which implies $T_0 = 10.5t_d$, does not provide a large enough number of cycles for free vibration (during $t > t_d$) of the system to damp out, leaving significant displacement and velocity at the end of the period T_0, thus violating the at-rest initial conditions. Therefore, the DFT solution cannot be expected to be accurate. Figure A.6.3 demonstrates that $t_f = 9.75T_n$,

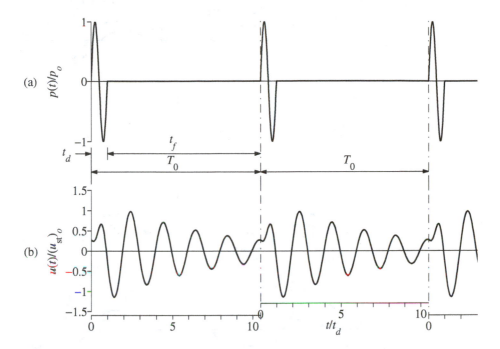

Figure A.6.2   (a) Periodic extension of $p(t)$ with $t_f = 4.75T_n$ (i.e., $T_0 = 10.5t_d$); (b) response determined by classical DFT method; $t_d/T_n = 0.5$; $\zeta = 5\%$.

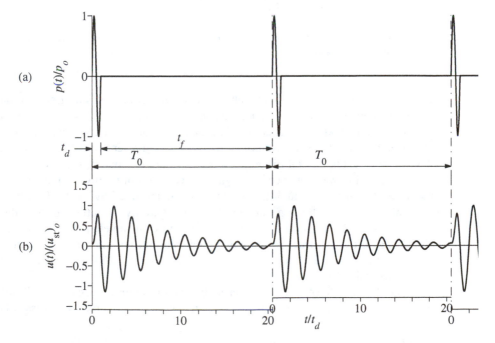

Figure A.6.3   (a) Periodic extension of $p(t)$ with $t_f = 9.75T_n$ (i.e., $T_0 = 20.5t_d$); (b) response determined by classical DFT method; $t_d/T_n = 0.5$; $\zeta = 5\%$.

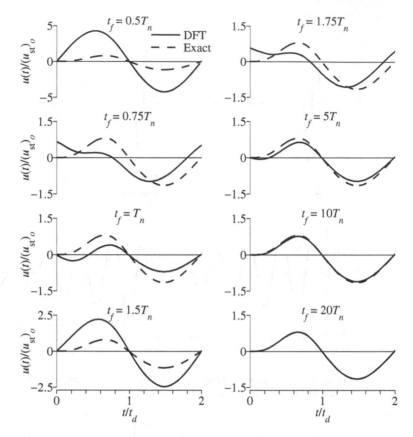

Figure A.6.4 Comparison of DFT solutions using different values of t_f with the exact response; $t_d/T_n = 0.5$; $\zeta = 5\%$.

which implies $T_0 = 20.5t_d$, provides an adequate number of cycles for free vibration of the system to damp out to small motion at the end of the period T_0, thus essentially satisfying the at-rest initial conditions. Therefore, the DFT solution is expected to be accurate.

These expectations are confirmed by the results presented in Fig. A.6.4, where the DFT solutions using several different values of t_f are compared with the exact solution. The exact solution was obtained by solving the equation of motion by the methods developed in Section 4.8. Although the response was determined by the DFT method for the period of the extended forcing function $T_0 = t_d + t_f$, only its initial part over the desired duration $t_d + T_n/2$ is plotted. The results clearly show that unless t_f is quite long, the DFT solution may differ significantly from the exact solution. For the example considered with $t_d/T_n = 0.5$ and $\zeta = 5\%$, the errors are noticeable in Fig. A.6.4 even for $t_f = 10T_n$, but become negligible for $t_f = 20T_n$.

The duration t_f of free vibration necessary to obtain an accurate DFT solution is controlled by the number of cycles necessary for free vibration to decay to essentially zero

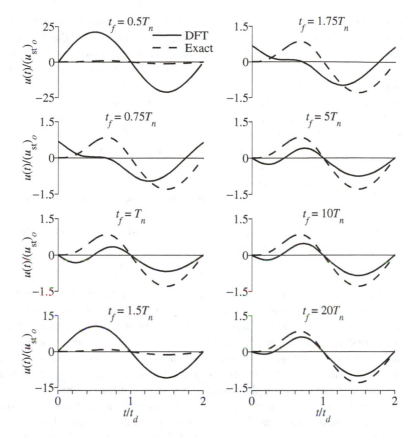

Figure A.6.5 Comparison of DFT solutions using different values of t_f with the exact response; $t_d/T_n = 0.5$; $\zeta = 1\%$.

displacement and velocity. More cycles are necessary for the motion of lightly damped systems to decay sufficiently (see Fig. 2.2.4). This implies that if t_f is chosen as a fixed multiple of T_n, the error in the DFT solution is expected to be larger for lightly damped systems. The DFT and exact solutions for systems with $\zeta = 1\%$ are plotted in Fig. A.6.5 for several different choices of t_f. Comparison of Fig. A.6.5 with Fig. A.6.4 demonstrates that for each t_f/T_n value, the DFT solution is less accurate for systems with 1% damping compared to systems with 5% damping. Thus, to keep the errors in the DFT solution below some selected tolerance limit, longer t_f would be necessary for systems with less damping.

A.7 IMPROVED DFT SOLUTION

We have demonstrated in Section A.6 that the DFT solution may err significantly unless the duration t_f of free vibration included in the periodic extension of the excitation is sufficiently long. t_f should be selected as a multiple of the natural vibration period T_n of the system. This multiple depends on the damping ratio of the system and the accuracy desired in the DFT solution. Thus, the DFT computations have to be implemented for a periodic extension of the excitation with period T_0 longer than the duration $t_d + T_n/2$ over which the response is desired; T_0 must be much longer than $t_d + T_n/2$ for lightly damped systems, especially systems with long period of vibration. Consider, for example, a system with $T_n = 10$ sec and $\zeta = 5\%$, for which $t_f = 20T_n$ is necessary to achieve sufficient accuracy in the classical DFT solution. If the duration of the excitation, t_d, is 30 sec, the response should be computed for $T_0 = t_d + 20T_n = 30 + 20(10) = 230$ sec, whereas we really need the response only for $t_d + T_n/2 = 35$ sec. Improved procedures have been developed to avoid the additional, seemingly unnecessary computational effort required in the classical DFT solution.

In the improved method the period of the periodic extension of the excitation is equal to the duration over which the response of the system is actually desired (i.e., $T_0 = t_d + T_n/2$); the steady-state response $\tilde{u}(t)$ over this period is computed by the classical DFT method;[†] and the "exact" response $u(t)$ is obtained from the steady-state response by the superposition of a corrective solution $v(t)$:

$$u(t) = \tilde{u}(t) + v(t) \tag{A.7.1}$$

Suppose that the response $u(t)$ to a given forcing function $p(t)$ is to be determined for a system starting with $u(0) = 0$ and $\dot{u}(0) = 0$. However, as shown in Fig. A.6.2, the classical DFT solution does not satisfy those initial conditions. Since the excitation over the period T_0 for both the DFT and exact responses is the same, the difference in the two solutions shown in Fig. A.6.4 must stem from differences in the initial states of the two motions. Therefore, the corrective solution is simply the free vibrational solution, which ensures that the initial displacement and velocity of the desired motion conforms to the prescribed initial conditions. If the initial displacement and velocity associated with the DFT solution are $\tilde{u}(0)$ and $\dot{\tilde{u}}(0)$, which are generally nonzero, the corrective solution is the free vibrational response due to initial displacement $-\tilde{u}(0)$ and initial velocity $-\dot{\tilde{u}}(0)$. This corrective solution for systems with viscous damping is then given by Eq. (2.2.4) with appropriate changes in notation:

$$v(t) = e^{-\zeta\omega_n t}\left[-\tilde{u}(0)\cos\omega_D t + \left(\frac{-\dot{\tilde{u}}(0) - \zeta\omega_n\tilde{u}(0)}{\omega_D}\right)\sin\omega_D t\right] \tag{A.7.2}$$

where

$$\omega_D = \omega_n\sqrt{1 - \zeta^2} \tag{A.7.3}$$

This improved procedure is illustrated by applying it to determine the response of an SDF system with $\zeta = 5\%$ subjected to the one cycle of sinusoidal force of duration t_d considered previously (Fig. A.6.1). The system is presumed to be initially at rest and its natural period $T_n = t_d/1.4$. The response is evaluated for a duration $T_0 = t_d + T_n/2 = 1.4T_n + 0.5T_n = 1.9T_n$. The sampling interval is chosen as before, $\Delta t = t_d/40$. Obtained by the classical DFT procedure, the steady-state displacement $\tilde{u}(t)$ is shown in Fig. A.7.1.

[†]From this point on, the classical DFT solution is denoted by $\tilde{u}(t)$ to distinguish it from the "exact" solution $u(t)$.

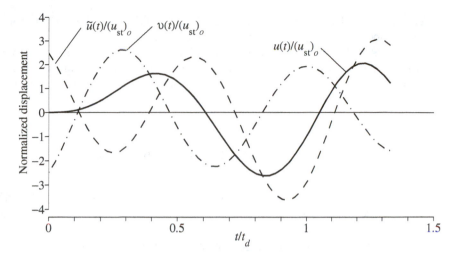

Figure A.7.1   Steady-state response $\tilde{u}(t)$, corrective solution $v(t)$, and "exact" response $u(t)$; $t_d/T_n = 1.4$ (i.e., $T_0 = 1.9T_n$); and $\zeta = 5\%$. [Adapted from Veletsos and Ventura (1985).]

The initial value of this displacement is $\tilde{u}(0) = 2.486(u_{\text{st}})_o$, and the initial velocity[†] is $\dot{\tilde{u}}(0) = -1.761\omega_n(u_{\text{st}})_o$. Substituting these values of $\tilde{u}(0)$ and $\dot{\tilde{u}}(0)$ in Eq. (A.7.2) gives the corrective solution $v(t)$ shown in Fig. A.7.1. Determined by combining $\tilde{u}(t)$ and $v(t)$ according to Eq. (A.7.1), the desired "exact" response is also shown in Fig. A.7.1. It is essentially identical to the analytical solution obtained by solving the differential equation of motion by the methods of Section 4.8.

A.8 MULTI-DEGREE-OF-FREEDOM SYSTEMS

The improved DFT procedure can readily be extended to determine the response of MDF systems for which the classical mode superposition method is applicable. As shown in Chapters 12 and 13, the equations of motion for classically damped systems can be transformed to a set of uncoupled equations in modal coordinates, as many equations as the number of DOFs in the system. Each modal equation is of the same form as the equation governing the motion of an SDF system. Therefore, each modal equation can be solved by the classical DFT method and combined with the corrective solution described in Section A.7 to determine the modal responses accurately. Thus, the frequency-domain method

[†]The initial velocity is given by

$$\dot{u} = -\frac{4\pi}{T_0} \sum_{j=0}^{N/2} j \, \text{Im}(U_j) \tag{A.7.4}$$

in which $\text{Im}(U_j)$ denotes the imaginary part of U_j. Equation (A.7.4) is obtained by differentiating Eq. (A.5.9), making use of the facts that $j\omega_0 = 2\pi j/T_0$ and the values of U_j for $j = N/2 + 1, \ N/2 + 2, \ \dots, \ N - 1$ are the complex conjugates of those for $j = N/2 - 1, \ N/2 - 2, \ \dots, \ 1$, respectively.

can be used to determine modal responses (step 3a of Section 12.5), but the remaining steps in the modal analysis procedure remain unchanged.

Researchers have also developed DFT procedures for analysis of nonclassically damped systems with either constant or frequency-dependent parameters. The latter situation arises in dynamic analysis of structures, including effects of soil–structure interaction or fluid–structure interaction.

FURTHER READING

Bergland, G. D., "A Guided Tour of the Fast Fourier Transform," *IEEE Spectrum*, **6**, 1969, pp. 41–52.

Blackwell, R., *The Fourier Transform and Its Applications*, McGraw-Hill, New York, 1978, pp. 232–236.

Brigham, E. O., *The Fast Fourier Transform*, Prentice Hall, Englewood Cliffs, N.J., 1974.

Clough, R. W., and Penzien, J., *Dynamics of Structures*, McGraw-Hill, New York, 1993, Chapters 4, 6, and 12.

Cooley, J. W., and Tukey, J. W., "An Algorithm for the Machine Calculation of Complex Fourier Series," *Mathematics of Computation*, **19**, 1965, pp. 297–301.

Humar, J. L., *Dynamics of Structures*, CRC Press/Balkema, Leiden, The Netherlands, 2012, Chapter 9.

Veletsos, A. S., and Ventura, C. E., "Dynamic Analysis of Structures by the DFT Method," *Journal of Structural Engineering, ASCE*, **111**, 1985, pp. 2625–2642.

B

Notation

All symbols used in this book are defined where they first appear. For the reader's convenience, this appendix, arranged in four parts to follow the organization of the text, contains the principal meanings of the commonly used notations. The reader is cautioned that some symbols denote more than one quantity, but the meaning should be clear when read in context.

Abbreviations

CM	center of mass	NBCC	*National Building Code of Canada*
CQC	complete quadratic combination	NDP	nonlinear dynamic procedure
DFT	discrete Fourier transform	NSP	nonlinear static procedure
DOF	degree of freedom	RHA	response history analysis
EC	*Eurocode 8*	RSA	response spectrum analysis
FFT	fast Fourier transform	SDF	single-degree-of-freedom
IBC	*International Building Code*	SRSS	square root of the sum of squares
MDF	multi-degree-of-freedom	UBC	*Uniform Building Code*
MFDC	*Mexico's Federal District Code*	UMRHA	uncoupled modal response history
MPA	modal pushover analysis		analysis

Accents

$(\bar{\ })$	modal contribution factor for ( )	$\hat{\mathbf{k}}$	condensed $\mathbf{k}$
$(\check{\ })$	shifted value of ( )	$(\tilde{\ })$	approximation to ( )
$(\dot{\ })$	$\dfrac{d}{dt}$ ( )	$(\tilde{\ })$	generalized ( )

Prefixes

$\delta, \boldsymbol{\delta}$ increment over extended time step
$\delta(\cdot)$ virtual $(\cdot)$
$\Delta, \boldsymbol{\Delta}$ increment over time step

Subscripts

A	acceleration	I	inertia; input
b	base; beam; base-isolation system	K	kinetic
		m	peak value for inelastic systems; maximum
c	column; complementary solution	n	natural; mode number
cr	critical	o	peak value
d	duration	p	particular solution
D	damping; damped; displacement	sec	secant
		st	static
e	eccentric; element	S	spring (elastic or inelastic); strain
eff	effective		
eq	equivalent	T	tangent; transmitted
f	fixed-base system	V	velocity
F	friction	x, y, θ	directions or components
g	ground	y	yield
i	time step number; peak number	Y	yielding
i, j	floor number; story number; DOF; frame number		

Superscripts

s	quasi-static	t	total
st	static		

PART I: CHAPTERS 1–8

Roman Symbols

a	see Table 5.3.1	$A(t)$	pseudo-acceleration
a_1, a_2	arbitrary constants	A_y	$\omega_n^2 u_y$; f_y/m
a_1, a_2, a_3	see Table 5.4.2 or 5.7.1	b	see Table 5.3.1; complex-valued constant
a_j	Fourier cosine coefficients		
a_0	Fourier coefficient	b_j	Fourier sine coefficients
A	integration constant; arbitrary constant; coefficient in Eq. (5.2.5); pseudo-acceleration spectrum ordinate	B	integration constant; arbitrary constant; coefficient in Eq. (5.2.5)
		B'	coefficient in Eq. (5.2.5)
A'	coefficient in Eq. (5.2.5)	B_1, B_2	arbitrary constants

c	damping coefficient
$\tilde{c}$	generalized damping
c_{cr}	critical damping coefficient
C	arbitrary constant; coefficient in Eq. (5.2.5)
C'	coefficient in Eq. (5.2.5)
D	arbitrary constant; coefficient in Eq. (5.2.5); deformation spectrum ordinate
D'	coefficient in Eq. (5.2.5)
D_y	yield deformation spectrum ordinate
e	eccentricity of rotating mass
E	modulus of elasticity
E_D	energy dissipated by damping
E_F	energy dissipated by friction
E_I	input energy
E_K	kinetic energy
E_{Ko}	maximum kinetic energy
E_S	strain energy
E_{So}	maximum strain energy
E_Y	energy dissipated by yielding
$EI(x)$	flexural rigidity
f	exciting or forcing frequency (Hz)
f_D	damping force
f_I	inertia force
$f_I(x, t)$	distributed inertia forces
f_{Ij}	inertia force in DOF j
f_{jo}	peak value of force at jth floor
f_n	natural frequency (undamped) (Hz)
$f_o(x)$	peak value of $f_S(x, t)$
f_S	elastic or inelastic resisting force; equivalent static force
$(f_s)^j$	f_s after j cycles of iteration
$(\hat{f}_s)_i$	see Eq. (5.7.10)
$(f_S)_i$	value of f_S at time i
$\tilde{f}_S(u)$	defined by Eq. (7.3.3)
f_{So}, f_o	peak value of $f_S(t)$
f_T	transmitted force
f_y	yield strength
$\bar{f}_y$	normalized yield strength
F	friction force
g	acceleration due to gravity
h	height of one-story frame; story height
$h(t)$	unit impulse response
$H(\omega)$	complex frequency response
i	time step number
I	second moment of area
$\mathcal{I}$	magnitude of impulse
I_b	I for a beam
I_c	I for a column
I_O	moment of inertia about O
k	stiffness or spring constant
$\mathbf{k}$	stiffness matrix
$\hat{k}$	see Eq. (5.3.5) or (5.4.11)
$\tilde{k}$	generalized stiffness
k_i	$(k_i)_T$
$\hat{k}_i$	defined in Eq. (5.7.6)
k_j	stiffness of jth story
k_T	tangent stiffness
$\hat{k}_T$	defined in Eq. (5.7.8)
$k_T^{(j)}$	tangent stiffness at $u^{(j)}$
L	width of frame; length of beam or tower
$\tilde{L}$	see Eqs. (8.3.12) and (8.4.12)
$\tilde{L}^\theta$	defined in Eq. (8.4.18)
$m(x)$	mass per unit length
m	mass
$\mathbf{m}$	mass matrix
$\tilde{m}$	generalized mass
m_e	eccentrically rotating mass
m_j	mass at jth DOF or jth floor
$\mathcal{M}(x, t)$	bending moments in a distributed-mass system
M_a, M_b	bending moments at nodes a and b
M_b	base overturning moment
M_{bo}	peak value of $M_b(t)$
M_{io}	peak value of ith-floor overturning moment
$\mathcal{M}_o(x)$	peak value of $\mathcal{M}(x, t)$
N	normal force between sliding surfaces; number of DOFs
p	external force
$P(\omega)$	Fourier transform of $p(t)$
$\tilde{p}(t)$	generalized external force
p_{eff}	effective earthquake force
$(p_{\text{eff}})_o$	peak value of $p_{\text{eff}}(t)$
p_i	value of $p(t)$ at time i

$\hat{p}_i$	see Eq. (5.3.6), (5.4.12) or (5.6.3)	$u_i^{(j)}$	u_i after j iterations
p_o	amplitude of $p(t)$	$\dot{u}_i$	velocity at time i
$r(t)$	any response quantity	$\ddot{u}_i$	acceleration at ith peak; acceleration at time i
r_o	$\max_t |r(t)|$, the peak response	u_j	relative displacement of jth floor
R_a	acceleration response factor	$u_j^c(t)$	response to $p(t) = p_0 \cos j\omega_0 t$
R_d	deformation (or displacement) response factor	$u_j^s(t)$	response to $p(t) = p_0 \sin j\omega_0 t$
$R^{(j)}$	residual force at the end of jth iteration cycle	u_j^t	total displacement of jth floor
$\hat{R}^{(j)}$	see Eq. (5.7.15)	u_{jo}	peak or maximum value of $u_j(t)$
R_v	velocity response factor	$\dot{u}_{jo}$	maximum value of $\dot{u}_j(t)$
R_y	yield reduction factor	u_m	$\max_t |u(t)|$ for an inelastic system
t	time		
t'	time variable		
t_d	duration of pulse force	u_m^-	$|\min_t [u(t)]|$
t_i	time at ith peak in free vibration; time at end of ith time step	u_m^+	$\max_t [u(t)]$
t_o	time when $u(t)$ is maximum	u_o	peak or maximum value of $u(t)$
t_r	rise time	$\dot{u}_o$	peak value of $\dot{u}(t)$
$T_a, T_b, T_c,$ T_d, T_e, T_f	periods that define spectral regions	$u_o(x)$	peak or maximum value of $u(x, t)$
T_D	natural period (damped)	$\dot{u}_o(x)$	maximum value of $\dot{u}_o(x, t)$
T_n	natural period (undamped)	u_o^t	peak value of $u^t(t)$
T_0	period of periodic excitation	$\ddot{u}_o^t$	peak value of $\ddot{u}^t(t)$
TR	transmissibility	u_p	particular solution; permanent deformation
u	displacement; deformation; displacement relative to ground	$u_{\text{st}}(t)$	static deformation due to $p(t)$
u^t	total displacement	$(u_{\text{st}})_o$	static deformation due to p_o
$\mathbf{u}$	vector of displacements u_j	u_x, u_y	x and y displacements
$\hat{u}(i\omega)$	Fourier transform of $u(t)$	u_y	yield deformation
$u(0)$	initial displacement	$u_0(t)$	response to $p(t) = a_0$
$\dot{u}(0)$	initial velocity	u_θ	rotation about a vertical axis
u_a, u_b	displacements of nodes a and b	v	velocity
u_c	complementary solution	V	pseudo-velocity spectrum ordinate
u_F	F/k	$\mathcal{V}(x, t)$	shearing forces in a distributed-mass system
u_g	ground (or support) displacement	V_a, V_b	shearing forces at nodes a and b
$\ddot{u}_g$	ground (or support) acceleration	V_b	base shear
u_{go}	peak ground displacement	V_{bo}	peak value of $V_b(t)$
$\dot{u}_{go}$	peak ground velocity	$\mathcal{V}_{bo}$	peak value of $\mathcal{V}_b(t)$
$\ddot{u}_{go}$	peak ground acceleration	V_j	shear in the jth story
$u_{g\theta}$	ground rotation about a vertical axis	V_{jo}	peak value of $V_j(t)$
u_i	displacement at ith peak; displacement at time i	$\mathcal{V}_o(x)$	peak value of $\mathcal{V}(x, t)$
$u^{(j)}$	u after j iterations		

V_y	$\omega_n u_y$	z	generalized displacement
w	weight	z_o	peak value of $z(t)$
x, y	Cartesian coordinates	**1**	vector of ones

Greek Symbols

$\alpha_A, \alpha_D, \alpha_V$	spectral amplification factors	ζ_{eq}	equivalent viscous damping ratio
β	parameter in Newmark's method	η	rate-independent damping coefficient
β_j	$j\omega_0/\omega_n$	$\ddot{\theta}$	rotational acceleration
γ	parameter in Newmark's method	θ_a, θ_b	rotations at nodes a and b
λ	constant in $e^{\lambda t}$	θ_g	ground rotation about a horizontal axis
$\tilde{\Gamma}$	$\tilde{L}/\tilde{m}$	$\kappa(x)$	curvature
δ	logarithmic decrement	μ	coefficient of friction; ductility factor
$\delta(\cdot)$	Dirac delta function		
δ_{st}	mg/k	ρ	beam-to-column stiffness ratio; coefficient in $\pm\rho e^{-\zeta\omega_n t}$
$\delta u(x)$	virtual displacements		
$\delta\mathbf{u}$	virtual displacement vector	σ	standard deviation
δu_j	virtual displacement u_j	τ	dummy time variable
δW_E	external virtual work	ϕ	phase angle
δW_I	internal virtual work	$\psi(x)$	shape function
δz	virtual displacement	ψ	shape vector
$\delta\kappa(x)$	virtual curvature	ψ_j	jth element of ψ
Δ_j	story drift in jth story	ω	exciting or forcing frequency (rad/sec)
Δt	time step		
Δt_i	time step i	ω_D	natural frequency (damped) (rad/sec)
$\Delta u^{(j)}$	change in u in iteration cycle		
ε	duration of an impulsive force	ω_n	natural frequency (undamped) (rad/sec)
$\varepsilon_R, \varepsilon_u, \varepsilon_w$	tolerance values in Eq. (5.7.8)		
ζ	damping ratio	ω_0	$2\pi/T_0$
$\underline{\zeta}$	numerical damping ratio		

PART II: CHAPTERS 9–18

Roman Symbols

$\mathbf{a}_e$	element transformation matrix	$\mathbf{a}_1, \mathbf{a}_2, \mathbf{a}_3$	see Table 16.2.2 or 16.3.3
		$\mathbf{a}_{xi}, \mathbf{a}_{yi}$	transformation matrices
a_{in}	defined by Eq. (14.4.11)	a_0, a_1	Rayleigh damping coefficients
a_l	coefficients in Caughey series		
a_x, a_y, a_z	x, y, and z components of ground acceleration	$\mathcal{A}$	direct assembly operator
		A_n	pseudo-acceleration spectrum ordinate $A(T_n, \zeta_n)$
a_1, a_2, a_3	principal components of ground acceleration		

$A_n(t)$	pseudo-acceleration of nth-mode SDF system	$\mathbf{f}_S$	elastic resisting forces
$A_{nl}(t)$	$A_n(t)$ due to $\ddot{u}_{gl}(t)$	$\mathbf{f}_S(\mathbf{u})$	inelastic resisting forces
$B_n, \bar{B}_n$	complex-conjugate pair of constants	f_{SA}	lateral force on frame A
		$\mathbf{f}_{Sg}, \mathbf{f}_{Sg}(t)$	equivalent static forces in support DOFs
B_n^g	see Eq. (14.3.2)	f_{Sj}	elastic or inelastic resisting force in DOF j
$\mathbf{c}$	damping matrix		
$\tilde{\mathbf{c}}$	defined by Eq. (15.3.4)	$\mathbf{f}_{yn}$	peak value of $\mathbf{f}_{yn}(t)$
c_{ij}	damping influence coefficient	$\mathbf{f}_{yn}(t)$	equivalent static lateral forces, mode n
c_j	jth-story damping coefficient	$\mathbf{f}_{\theta n}$	peak value of $\mathbf{f}_{\theta n}(t)$
		$\mathbf{f}_{\theta n}(t)$	equivalent static torques, mode n
$\mathbf{c}_n$	nth-mode damping matrix		
$\mathbf{C}$	$\mathbf{\Phi}^T \mathbf{c} \mathbf{\Phi}$, diagonal matrix of C_n	h	height of one-story frame; story height
C_n	generalized damping for nth mode	h_j	height of jth floor
C_{nr}	element of $\mathbf{C}$	h_n^*	effective modal height, mode n
$D_n(t)$	deformation of nth-mode SDF system	$h(t)$	unit impulse response function
$\dot{D}_n(t)$	relative velocity response, nth-mode SDF system	$\mathbf{h}(t)$	vector of unit impulse response functions
$D_{nl}(t)$	$D_n(t)$ due to $\ddot{u}_{gl}(t)$	$h_n(t)$	$h(t)$ for deformation of nth-mode SDF system
D_{no}	peak value of $D_n(t)$		
e_J	error in static response; error norm of Eq. (15.5.3)	I	second moment of area
		$\mathbf{I}$	identity matrix
E	modulus of elasticity	I_b	I for beam
EI	flexural rigidity	I_c	I for column
$\hat{\mathbf{f}}$	flexibility matrix	$\mathbf{I}_O$	diagonal matrix: $I_{jj} = I_{Oj}$
$\mathbf{f}_D, \mathbf{f}_D(t)$	damping forces	I_{Oj}	moment of inertia of jth floor about O
f_{Dj}	damping force in DOF j		
$\hat{f}_{ij}$	flexibility influence coefficient	I_{On}^*	defined by Eq. (13.3.10)
		J	number of Ritz vectors
$\mathbf{f}_I$	inertia forces	$\mathbf{k}$	stiffness matrix
f_{Ij}	inertia force in DOF j	$\check{\mathbf{k}}$	$\mathbf{k} - \mu \mathbf{m}$
$\mathbf{f}_{in}$	peak value of $\mathbf{f}_{in}(t)$	$\tilde{\mathbf{k}}$	defined by Eq. (15.3.4); matrix of $\tilde{k}_{ij}$ in Eq. (18.1.4)
$\mathbf{f}_{in}(t)$	equivalent static forces: frame i, mode n		
		$\mathbf{k}_A$	stiffness matrix of frame A in global DOF
f_{jn}	jth element of $\mathbf{f}_n$; peak value of $f_{jn}(t)$	$\mathbf{k}_e$	element stiffness matrix in global element DOFs
$f_{jn}(t)$	equivalent static force: DOF j, mode n	$\hat{\mathbf{k}}$	see Table 16.3.1, Eq. (16.3.5)
f_{jyn}	jth element of $\mathbf{f}_{yn}$	$\bar{\mathbf{k}}_e$	element stiffness matrix in local element coordinates
$f_{j\theta n}$	jth element of $\mathbf{f}_{\theta n}$		
f_n	nth natural frequency, Hz	$\mathbf{k}_{gg}^l$	lth column of $\mathbf{k}_{gg}$
$f_n(x, t)$	equivalent static forces, mode n	$\mathbf{k}_i$	stiffness matrix of frame i in global DOF
$\mathbf{f}_n$	peak value of $\mathbf{f}_n(t)$	$\hat{\mathbf{k}}_i$	see Tables 16.3.1 and 16.3.3
$\mathbf{f}_n(t)$	equivalent static forces, mode n	k_{ij}	stiffness influence coefficient
$f_{no}(x)$	peak value of $f_n(x, t)$	k_j	stiffness of jth story

$\hat{\mathbf{k}}_{tt}$	condensed stiffness matrix	p	external force
$\hat{\mathbf{k}}_T$	tangent stiffness matrix; see Table 16.3.3	$\mathbf{p}$	external forces
		$\tilde{\mathbf{p}}$	vector of $\tilde{p}_i$ in Eq. (18.2.6)
$k_{xi},\ k_{yi}$	lateral stiffness of frame i in x and y directions	$p(\lambda)$	polynomial in λ
$\mathbf{k}_{xi},\ \mathbf{k}_{yi}$	lateral stiffness matrix of frame i in x and y directions	$\mathbf{p}_e$	element force vector in global element DOFs
k_y	lateral stiffness of frame A	$\bar{\mathbf{p}}_e$	element force vector in local element coordinates
$\mathbf{k}_{xy}, \mathbf{k}_{x\theta}, \mathbf{k}_{\theta y}, \\ \mathbf{k}_{y\theta}, \mathbf{k}_{\theta\theta}$	submatrices of $\mathbf{k}$	p_{eff}	effective earthquake force
		$\mathbf{p}_{\text{eff}}$	effective earthquake force vector
$\mathbf{K}$	diagonal matrix of K_n		
$\hat{\mathbf{K}}$	see Tables 16.2.1 and 16.2.2	$\mathbf{p}_g$	support forces
K_n	generalized stiffness, mode n	$\mathbf{p}_g^s(t)$	quasi-static support forces
L	length of beam; length of finite element	p_j	external force at jth DOF or jth floor
L_n^h	see Eq. (13.2.3) or (17.6.2)	p_o	maximum value of $p(t)$
L_n^θ	see Eq. (13.2.9b) or (17.6.17)	$\mathbf{p}_t$	external forces in $\mathbf{u}_t$ DOF
L_{nl}	defined by Eq. (13.5.3)	$\mathbf{P}(t)$	$\boldsymbol{\Phi}^T \mathbf{p}(t)$
m	mass of an SDF system	$\mathbf{P}(t)$	vector of $P_n(t)$
$\mathbf{m}$	mass matrix	$\hat{\mathbf{P}}_i$	defined in Table 16.2.1
$\tilde{\mathbf{m}}$	defined by Eq. (15.3.4); matrix of $\tilde{m}_{ij}$ in Eq. (18.1.4)	$P_n(t)$	generalized force, mode n
		$\mathbf{q}$	modal coordinate vector
$m(x)$	mass per unit length	$\mathbf{q}_i$	modal coordinates at time i
$\mathbf{m}_e$	element mass matrix in global element DOFs	$q_n(t)$	nth modal coordinate
		r	radius of gyration
$\bar{\mathbf{m}}_e$	element mass matrix in local element coordinates	$r(t)$	any response quantity
		$r_a(t), r_b(t)$	response quantities
m_{ij}	mass influence coefficient	r_{ao}, r_{bo}	peak values of $r_a(t), r_b(t)$
m_j	mass at jth DOF or jth floor	$r_{an}^{\text{st}}, r_{bn}^{\text{st}}$	nth modal static responses r_a and r_b
$\mathbf{m}_{tt}$	mass matrix for $\mathbf{u}_t$ DOF		
$\mathbf{M}$	diagonal matrix of M_j	r_α	projection of response trajectory in direction α
$\mathcal{M}(x,t)$	bending moment in a distributed-mass system	r_{ano}, r_{bno}	peak values of $r_{an}(t), r_{bn}(t)$
$\mathcal{M}_b$	bending moment at the base	r_{abo}	cross term between responses r_a and r_b in Eq. (13.10.4)
$\mathcal{M}_{bn}(t)$	$\mathcal{M}_b(t)$ due to mode n		
$M_{bn}(t)$	$M_b(t)$ due to mode n	r_{as}, r_{bs}	responses r_a, r_b due to initial static loads
$\mathcal{M}_{bn}^{\text{st}}$	nth modal static response $\mathcal{M}_b$	r_{cr}	critical response; largest value of $r(\theta)$
M_{bn}^{st}	nth modal static response M_b		
M_i	ith-floor overturning moment	r_n	peak value of $r_n(t)$
M_{in}^{st}	nth modal static response M_i	$\bar{r}_n$	nth modal contribution factor
M_n	generalized mass, mode n	$r_n(t)$	$r(t)$ due to mode n
M_n^*	effective modal mass, mode n	r_{no}	peak value of $r_n(t)$
		r_o	peak value of $r(t)$
N	number of DOFs	r^{st}	static response to forces $\mathbf{s}$
N_d	number of modes responding dynamically	r_n^{st}	nth modal static response
N_e	number of finite elements	r_x, r_y, r_z	peak response r due to x, y, and z components of ground motion
N_g	number of ground (or support) displacements		
$\mathbf{O}$	null matrix		

r_{xy}	cross term between modal response contributions to r_x and r_y		$\mathbf{u}_i$	lateral displacements of frame i; displacements at time i
R_{dn}	dynamic response factor for nth-mode SDF system		$\mathbf{u}_{in}$	$\mathbf{u}_i$ due to mode n
$\mathbf{R}^{(j)}$	residual force vector after jth iteration cycle		u_j	peak value of $u_j(t)$
			$u_j(t)$	relative displacement at DOF j or floor j
$\mathbf{s}, \mathbf{s}_a, \mathbf{s}_b$	spatial distributions of $\mathbf{p}(t)$		u_j^s	quasi-static displacement at DOF j
s_{jn}	jth element of $\mathbf{s}_n$			
s_{jyn}	jth element of $\mathbf{s}_{yn}$		u_j^t	total displacement at DOF j or floor j
$s_{j\theta n}$	jth element of $\mathbf{s}_{\theta n}$			
$\mathbf{s}_n$	defined by Eq. (12.8.4) or (13.1.6)		u_{jn}	peak value of $u_{jn}(t)$
			u_{jn}^{st}	nth modal static response u_j
$\mathbf{s}_n(x)$	defined by Eq. (17.6.4)		$u_{jn}(t)$	$u_j(t)$ due to mode n
$\mathbf{s}_{yn}, \mathbf{s}_{\theta n}$	subvectors of $\mathbf{s}_n$		u_{jx}, u_{jy}	displacements of CM of floor j along x and y axes
t	time variable			
t_d	duration of pulse force		u_{jyn}	peak value of $u_{jyn}(t)$
T_b	base torque		$u_{j\theta}$	rotation of floor j about CM
T_{bn}	peak value of $T_{bn}(t)$		$u_{j\theta n}$	peak value of $u_{j\theta n}(t)$
T_{bn}^{st}	nth modal static response T_b		u_n	peak value of $\mathbf{u}_n(t)$
$T_{bn}(t)$	$T_b(t)$ due to mode n		$\mathbf{u}_n^{\text{st}}$	nth modal static response $\mathbf{u}$
T_i	ith-story torque		$u_n(x, t)$	$u(x, t)$ due to mode n
$T_{in}(t)$	$T_i(t)$ due to mode n		$u_n^{\text{st}}(x)$	nth modal static response $u(x)$
T_{in}^{st}	nth modal static response T_i			
T_n	nth natural period (undamped)		$\mathbf{u}_t$	dynamic DOF
			u_x, u_y	x and y displacements of CM
u	displacement or deformation		$\ddot{u}_x^t, \ddot{u}_y^t, \ddot{u}_\theta^t$	x, y, and θ components of total acceleration
$\mathbf{u}$	displacement vector			
$\mathbf{u}^{(j)}$	$\mathbf{u}$ after j cycles of iteration		$\mathbf{u}_y$	y-lateral displacements
$\dot{\mathbf{u}}_i$	velocities at time i		$\mathbf{u}_{yn}(t)$	$\mathbf{u}_y(t)$ due to mode n
$\ddot{\mathbf{u}}_i$	accelerations at time i		$\mathbf{u}_0$	DOF with zero mass
$\mathbf{u}_n(t)$	$\mathbf{u}(t)$ due to pair of modes $\psi_n, \bar{\psi}_n$; $\mathbf{u}(t)$ due to mode ϕ_n		$\bar{u}_{5n}$	nth modal contribution factor for u_5
			u_θ	rotation about CM
$\mathbf{u}^s$	quasi-static displacements		$\mathbf{u}_\theta$	floor rotations
$\mathbf{u}^t$	total displacements		$\mathbf{u}_{\theta n}(t)$	$\mathbf{u}_\theta(t)$ due to mode n
u_A	displacement at frame A		$\mathcal{V}(x, t)$	transverse shear force in a distributed-mass system
$\mathbf{u}_x$	x-lateral displacements			
$\mathbf{u}_{xn}(t)$	$\mathbf{u}_x(t)$ due to mode n		V_b	peak value of $V_b(t)$
$\mathbf{u}_e$	element displacements in global element DOFs		$\mathcal{V}_b$	base shear
			$V_b(t)$	base shear
u_g	ground (or support) displacement		V_b^{st}	V_b due to forces $\mathbf{s}$
			V_{bn}	peak value of $V_{bn}(t)$
$\mathbf{u}_g$	ground (or support) displacement vector		$\bar{V}_{bn}$	nth modal contribution factor for V_b
$\ddot{u}_g$	ground (or support) acceleration		$V_{bn}(t)$	$V_b(t)$ due to mode n
			$\mathcal{V}_{bn}(t)$	$\mathcal{V}_b(t)$ due to mode n
u_{gl}	lth support displacement		V_{bn}^{st}	nth modal static response V_b
$\ddot{u}_{gx}, \ddot{u}_{gy}, \ddot{u}_{g\theta}$	x, y, and θ components of ground acceleration		$\mathcal{V}_{bn}^{\text{st}}$	nth modal static response $\mathcal{V}_b$
u_i	displacement in DOF i			

V_{bo}	peak value of $V_b(t)$	$\mathbf{y}_n$	eigenvector of $\mathbf{A}$; defined by
V_i	peak value of $V_i(t)$		Eq. (14.4.10)
$V_i(t)$	ith-story shear	$\mathbf{z}$	generalized coordinate vector
V_{in}^{st}	nth modal static response V_i	z_j	generalized coordinates
x, y	Cartesian coordinates	$\mathbf{z}_n$	eigenvector
x_i, y_i	define location of frame i	$\mathbf{0}$	vector of zeros
$\mathbf{x}_i$	iteration vector	$\mathbf{1}$	vector of ones

Greek Symbols

α	arbitrary direction in response space; fraction used in Eq. (13.11.13)	$\varepsilon_R, \varepsilon_u, \varepsilon_w$	tolerances in Eq. (16.3.8)
		$\varepsilon'_R, \varepsilon'_u, \varepsilon'_w$	tolerances in Eq. (16.3.9)
		ϵ_{in}	defined by Eq. (13.7.7a)
α_n^g	defined in Eqs. (14.8.8) and (14.10.17)	ζ_n	damping ratio for nth mode
		ζ'_n	defined by Eq. (13.7.7b)
β	parameter in Newmark's method	θ	incident angle
β_{in}	ω_i/ω_n	θ_{cr}	critical incident angle
β_n	see Eqs. (14.2.7), (14.7.5), and (14.10.8)	θ_g	ground rotation about a horizontal axis
β_n^g	see Eqs. (14.3.4), (14.8.2), and (14.10.10)	ι	influence vector; influence matrix
γ	parameter in Newmark's method; ground motion intensity factor	ι_l	influence vector for u_{gl}
		κ	shear stress constant
		$\lambda(\chi)$	Rayleigh's quotient
γ_n	see Eqs. (14.2.7), (14.7.5), and (14.10.9)	$\lambda^{(j)}$	estimate of eigenvalue
		λ	eigenvalue
γ_n^g	see Eqs. (14.3.4), (14.8.2), and (14.10.12)	$\lambda_n, \bar{\lambda}_n$	complex conjugate pair of eigenvalues
Γ_n	see Eq. (12.8.3) or (13.2.3)	$\check{\lambda}$	$\lambda - \mu$
Γ_{nl}	defined by Eq. (13.5.3)	λ_n	nth eigenvalue
$\delta u(x)$	virtual displacement $u(x)$	μ	absorber mass ratio; shift of eigenvalue spectrum
δu_j	virtual displacement u_j		
δW_E	external virtual work	μ_{ij}	element of $\boldsymbol{\mu}$ defined in Eq. (13.11.1)
δW_I	internal virtual work		
$\delta(\cdot)$	Dirac delta function	$\boldsymbol{\mu}$	covariance matrix
Δ_j	peak value of $\Delta_j(t)$	ρ_{in}	cross-correlation coefficient for modes i and n
$\Delta_j(t)$	jth-story deformation or drift		
χ	generalized coordinate vector in Eq. (15.3.5)	ϕ_{jn}	jth element of $\boldsymbol{\phi}_n$
		$\phi_{jyn}, \phi_{j\theta n}$	jth elements of $\boldsymbol{\phi}_{yn}$ and $\boldsymbol{\phi}_{\theta n}$
χ_n	imaginary-valued part of ψ_n	$\phi_n(x)$	nth natural vibration mode
Δt	time step	$\tilde{\phi}_n(x)$	approximation to $\phi_n(x)$
$\Delta \mathbf{u}^{(j)}$	change in $\mathbf{u}$, jth iteration cycle	$\tilde{\boldsymbol{\phi}}_n$	approximation to $\boldsymbol{\phi}_n$
		$\boldsymbol{\phi}_n$	nth natural vibration mode; real-valued part of ψ_n
$\Delta_{jn}(t)$	$\Delta_j(t)$ due to mode n	$\boldsymbol{\phi}_{yn}, \boldsymbol{\phi}_{\theta n}$	subvectors of $\boldsymbol{\phi}_n$
Δ_{jn}	peak value of $\Delta_{jn}(t)$	$\boldsymbol{\Phi}$	modal matrix

ψ	complex-valued eigenvector; see Eq. (14.5.3)	ψ_j	shape vector or Ritz vector
ψ_n	nth (complex-valued) eigenvector	$\hat{\psi}_n$	defined by Eq. (14.4.12)
		ω	exciting or forcing frequency
$\boldsymbol{\Psi}$	$\langle \psi_1 \quad \psi_2 \quad \cdots \quad \psi_J \rangle$	ω_n	nth natural frequency (undamped) (rad/sec)
$\boldsymbol{\Psi}(x)$	$\langle \psi_1(x) \quad \psi_2(x) \quad \cdots \quad \psi_N(x) \rangle$	ω'_n	ω_n of a beam considering rotational inertia and shear effects
$\psi_i(x)$	trial, Ritz, or shape function; finite element interpolation function		
		$\tilde{\omega}_n$	approximation to ω_n
$\hat{\psi}_i(x)$	beam interpolation function	$\boldsymbol{\Omega}^2$	spectral matrix

PART III: CHAPTERS 19–23

Roman Symbols

A	pseudo-acceleration spectrum ordinate	k_j	stiffness of jth story
A_m	maximum A/g, MFDC	L_b	length of beams
A_y	A for yielding system	L_c	length of columns
c	damping coefficient of fixed-base system	m	lumped mass of fixed-based system
c_b	damping coefficient of isolation system	m_b	mass of base slab
$\mathbf{c}_f$	damping matrix of fixed-base system	$\mathbf{m}_f$	mass matrix of fixed-base system
		M_v	higher-mode factor, NBCC
C	period-dependent coefficient, IBC	OS	overstrength factor, EC
C_e	elastic seismic coefficient	q	seismic behavior factor, EC
C_s	seismic coefficient	q'	seismic reduction factor, EC
e_J	error in static response [Eq. (19.7.1)]	Q	seismic behavior factor, MFDC
		Q'	seismic reduction factor, MFDC
EI_b	flexural rigidity of beams	R	response modification factor, IBC
EI_c	flexural rigidity of columns	r	coefficient in C_e, MFDC
f_{jn}	lateral force: floor j, mode n	R_d	force modification factor, NBCC
f_y	design yield strength	R_O	overstrength-related force modification factor, NBCC
F_j	code lateral force at floor j		
F_{Sn}	defined by Eq. (20.6.12)	$\mathbf{s}_n^*$	defined by Eq. (20.7.2)
F_t	additional lateral force at the top floor, NBCC	T_b	isolation period
		T_b, T_c	periods defining the constant-A spectral region
h	story height		
h_j	height of jth floor	T_f	natural period of fixed-base system
I	importance factor, IBC and NBCC	T_n	natural period of SDF system; nth natural period of MDF system
J, J_i	reduction factors for overturning moments		
		T_{nf}	nth natural period of fixed-base system
k	lateral stiffness of fixed-base system	u_b	isolator deformation
k_b	lateral stiffness of isolation system	u_{bn}	u_b due to mode n
$\mathbf{k}_f$	stiffness matrix of fixed-base system	u_{bn}^{st}	nth modal static response u_b

u_i	ith-floor displacement due to forces F_j $(j = 1, 2, \ldots, N)$
u_{jm}	$\max_t |u_j(t)|$ for an inelastic system
u_{rn}	roof displacement, nth mode
u_y	yield displacement
V_b	base shear
$V_b^{(1)}$, $V_b^{(2)}$	two parts of V_b, MFDC
V_{by}	yield strength value of V_b

$\bar{V}_{by}$	normalized value of V_{by}
V_j	jth-story shear
V_{jy}	jth-story yield strength
w	weight of SDF system
w_i	weight at ith floor
W	total weight of building; total dead load and applicable portion of other loads

Greek Symbols

α, β	coefficients in least-square-error fit of V_b–T_1 curve
Δ_j	jth-story deformation or drift
Δ_{jm}	peak value of $\Delta_j(t)$ for an inelastic system
ζ	damping ratio
ζ_b	ζ of isolation system with rigid building
ζ_f	ζ for fixed-base system
ζ_{nf}	ζ for nth mode of fixed-base system

μ	ductility factor
ϕ_n	nth natural vibration mode of corresponding linear system
ϕ_{nf}	nth mode of fixed-base system
ϕ_{rn}	roof element of ϕ_n
ω_f	natural frequency of fixed-base system
ω_n	nth natural vibration frequency of corresponding linear system

APPENDIX A

Roman Symbols

c	damping coefficient
$h(t)$	unit impulse response
H_j	$H(\omega_j)$; see Eq. (A.5.7)
$H(\omega)$	complex frequency response
$H_u(\omega)$	complex frequency response for $u(t)$
k	stiffness
m	mass
M	number of harmonics in truncated series
N	number of equally spaced time instants
p	external force
p_n	$p(t_n) \equiv p(n\,\Delta t)$
p_o	amplitude of $p(t)$

$P(\omega)$	Fourier transform of $p(t)$
P_j	Fourier coefficient (complex valued) for $p(t)$
$\bar{P}_j$	complex conjugate of P_j
t_d	duration of excitation
t_f	duration of free vibration
T_n	natural period (undamped)
T_0	period of periodic extension of $p(t)$
u	displacement; deformation
$u(t)$	"exact" response
$\tilde{u}(t)$	steady-state response by DFT method
$(u_{st})_o$	static deformaton due to p_o
U_j	see Eq. (A.2.8) or (A.5.8)

Greek Symbols

β_j	$j\omega_0/\omega_n$	ω_D	natural frequency (damped) (rad/sec)
$\delta(\cdot)$	Dirac delta function		
Δt	sampling interval	ω_j	$j\omega_0$
ζ	damping ratio	ω_{max}	Nyquist frequency; folding frequency; see Eq. (A.5.5)
η	rate-independent damping coefficient	ω_n	natural frequency (undamped) (rad/sec)
$\upsilon(t)$	corrective solution		
ω	excitation or forcing frequency (rad/sec)	ω_0	$2\pi/T_0$

Answers to Selected Problems

Chapter 1

1.1 $k_e = k_1 + k_2;\ m\ddot{u} + k_e u = p(t)$

1.3 $k_e = \dfrac{(k_1 + k_2)k_3}{k_1 + k_2 + k_3};\ m\ddot{u} + k_e u = p(t)$

1.5 $\ddot{\theta} + \dfrac{3}{2}\dfrac{g}{L}\theta = 0;\ \omega_n = \sqrt{\dfrac{3}{2}\dfrac{g}{L}}$

1.8 $\dfrac{mR^2}{2}\ddot{\theta} + \dfrac{\pi d^4 G}{32L}\theta = 0$

1.10 $\dfrac{w}{g}\ddot{u} + \dfrac{3EI}{L^3}u = 0$

1.12 $\omega_n = \sqrt{\dfrac{k_e}{m}};\ k_e = \dfrac{k(48EI/L^3)}{k + 48EI/L^3}$

1.15 $m\ddot{u} + \left(\dfrac{120}{11}\dfrac{EI_c}{h^3}\right)u = p(t)$

1.16 $m\ddot{u} + \left(2\dfrac{EI_c}{h^3}\right)u = p(t)$

1.17 $m\ddot{u}_x + \left(\sqrt{2}\dfrac{AE}{h}\right)u_x = 0;\ m\ddot{u}_y + \left(\sqrt{2}\dfrac{AE}{h}\right)u_y = 0$

1.18 $\dfrac{mh^2}{6}\ddot{u}_\theta + \dfrac{AEh}{\sqrt{2}}u_\theta = 0$

Chapter 2

2.1 $w = 40$ lb; $k = 16.4$ lb/in.

2.2 $u(t) = 2\cos(9.82t)$

2.4 $u(t) = 0.565\sin(60.63t)$ in.

2.5 $u(t) = \dfrac{m_2 g}{k}(1 - \cos\omega_n t) + \dfrac{\sqrt{2gh}}{\omega_n}\dfrac{m_2}{m_1 + m_2}\sin\omega_n t$

2.7 $EI = 8827$ lb-ft^2

2.11 $j_{10\%} = 0.366/\zeta$

2.13 (a) $T_n = 0.353$ sec;   (b) $\zeta = 1.94\%$

2.14 (a) $c = 215.9$ lb-sec/in., $k = 1500$ lb/in.

   (b) $\zeta = 0.908$

   (c) $\omega_D = 5.28$ rad/sec

2.15 $k = 175.5$ lb/in.; $c = 0.107$ lb-sec/in.

2.16 $\omega_n = 21.98$ rad/sec; $\zeta = 0.163$; $\omega_D = 21.69$ rad/sec

2.17 $T_D = 0.235$ sec, $\zeta = 0.236\%$

2.19 1.449 in., 1.149 in.

2.20 0.536 in.; 8 cycles

Chapter 3

3.1 $m = 6.43/g$ lb/g; $k = 10.52$ lb/in.

3.2 $\zeta = 0.05$

3.3 $\zeta = 0.0576$

3.5 $u_o = 2.035 \times 10^{-3}$ in.; $\ddot{u}_o = 0.0052g$

3.10 $\zeta = 9.82\%$

3.11 $\zeta = 1.14\%$

3.12 (b) $\delta_{st} = 0.269$ in.

3.13 474.8 lb

3.15 11.6 kips/in.

3.17 Error $= 0, 0.9$, and 15% at $f = 10, 20$, and 30 Hz, respectively

3.19 $f \le 20.25$ Hz

3.21 $f \ge 2.575$ Hz

3.25 0.308 in.

3.26 (a) $p(t) = \dfrac{p_o}{2} + \dfrac{4p_o}{\pi^2} \displaystyle\sum_{j=1,3,5,\ldots}^{\infty} \dfrac{1}{j^2} \cos j\omega_0 t$

(b) $\dfrac{u(t)}{(u_{st})_o} = \dfrac{1}{2} + \dfrac{4}{\pi^2} \displaystyle\sum_{j=1,3,5,\ldots}^{\infty} \dfrac{1}{j^2(1-\beta_j^2)} \cos j\omega_0 t$, where $(u_{st})_o = p_o/k$,

$\beta_j = j\omega_0/\omega_n$, and $\beta_j \neq 1$

(c) Two terms are adequate.

Chapter 4

4.9 (a) $\dfrac{u(t)}{(u_{st})_o} = 1 - e^{-\zeta\omega_n t}\left(\cos \omega_D t + \dfrac{\zeta}{\sqrt{1-\zeta^2}} \sin \omega_D t\right)$

(b) $\dfrac{u_o}{(u_{st})_o} = 1 + \exp(-n\zeta/\sqrt{1-\zeta^2})$

4.12 $u_o \simeq 12.2$ in.; $u_0 \simeq 6.1$ in.

4.17 $u_o = 0.536$ in.; $\sigma = 18.9$ ksi

4.18 $u_o = 2.105$ in.; $\sigma = 37.2$ ksi

4.23 $u_o = \dfrac{p_o}{k} \dfrac{4}{\omega_n t_1} \sin \omega_n t_1 \sin \dfrac{\omega_n t_1}{2}$

4.24 $\dfrac{u_o}{(u_{st})_o} = \dfrac{4\pi}{3} \dfrac{t_d}{T_n}$; error $= 5.9\%$

4.26 (a) $V_b = 15.08$ kips, $M_b = 1206$ kip-ft; **(b)** increase in mass has the effect of reducing the dynamic response.

Chapter 5

5.2 Check numerical results against the theoretical solution in Tables E5.1a, b.

5.4 Check numerical results against the theoretical solution in Table E5.2.

5.8 Check numerical results against the theoretical solution in Table E5.3.

5.9 Check numerical results against the theoretical solution in Tables E5.1a, b; the numerical results are large in error, but the solution is stable.

5.11 Check numerical results against the theoretical solution in Table E5.4.

Chapter 6

6.4 $D = \dfrac{\ddot{u}_{go}}{2\pi} T_n \exp\left(-\dfrac{\zeta}{\sqrt{1-\zeta^2}} \tan^{-1} \dfrac{\sqrt{1-\zeta^2}}{\zeta}\right)$

$V = \dfrac{2\pi}{T_n} D \, ; \, A = \left(\dfrac{2\pi}{T_n}\right)^2 D$

6.10 $u_o = 1.91$ in.; $\sigma = 38.2$ ksi

6.11 (a) $u_o = 14.1$ in., $V_{bo} = 56.2$ kips

    (b) $u_o = 9.93$ in., $V_{bo} = 79.5$ kips

    (c) $u_o = 14.1$ in., $V_{bo} = 112.4$ kips

6.14 (a) $u_o = 0.674$ in., $M = 40.65$ kip-ft; **(b)** $u_o = 2.7$ in., $M = 81.30$ kip-ft

6.15 $u_o = 2.7$ in., $M = 81.30$ kip-ft at top of column

6.16 (a) $u_o = 0.316$ in., $M = 18.3$ kip-ft

    (b) $u_o = 0.101$ in., $p_{brace} = 3.53$ kips

6.17 $u_o = 1.3$ in., $M_{short} = 60.39$ kip-ft, $M_{long} = 15.1$ kip-ft

6.18 $u_o = 3.86$ in.; bending moments (kip-ft) in columns: 569 at base and 244 at top; bending moments (kip-ft) in beam: 244 at both ends

6.19 $u_o = 11.1$ in., bending moments (kip-ft) in columns: 0 at base and 428 at top; bending moments (kip-ft) in beam: 428 at both ends

6.20 $u_o = 0.76$ in.; stresses at base of column: bending stress due to earthquake = 19.06 ksi; axial stress due to earthquake and gravity forces = 3.28 ksi; total stress = 22.34 ksi

6.22 Corner displacements = 0.582 in., 0.388 in.; base torque = 58.2 kip-ft; bending moments at top and base of each column: $M_y = 5.24$ kip-ft, $M_x = 2.33$ kip-ft

6.23 (b) $A(T_n)/g = 0.5$ for $T_n \le \frac{1}{33}$ sec; $12.28 T_n^{0.916}$ for $\frac{1}{33} < T_n \le \frac{1}{8}$ sec; 1.83 for $\frac{1}{8} < T_n \le 0.623$ sec; $1.14 T_n^{-1}$ for $0.623 < T_n \le 3.91$ sec; $4.46 T_n^{-2}$ for $3.91 < T_n \le 10$ sec; $24.49 T_n^{-2.74}$ for $10 < T_n \le 33$ sec; and $1.84 T_n^{-2}$ for $T_n > 33$ sec

Chapter 7

7.1 (a) $T_n = 0.502$ sec, $\zeta = 2\%$

    (b) no

    (c) $T_n = 0.502$ sec, $\zeta = 2\%$

    (d) $\bar{f}_y = 0.323$, $R_y = 3.09$

7.5 $\mu = 1.44, 3.11$, and 7.36

7.6 $A_y(T_n)/g = 0.5$ for $T_n \le \frac{1}{33}$ sec; $1.68 T_n^{0.348}$ for $\frac{1}{33} < T_n \le \frac{1}{8}$ sec; 0.818 for $\frac{1}{8} < T_n \le 0.465$ sec; $0.380 T_n^{-1}$ for $0.465 < T_n \le 3.91$ sec; $1.487 T_n^{-2}$ for $3.91 < T_n \le 10$ sec; $8.16 T_n^{-2.74}$ for $10 < T_n \le 33$ sec; and $0.614 T_n^{-2}$ for $T_n > 33$ sec

7.7

		$T_n = 0.02$ sec	$T_n = 0.2$ sec		$T_n = 2$ sec	
μ	f_y/w	u_m (in.)	f_y/w	u_m (in.)	f_y/w	u_m (in.)
1	0.50	1.955×10^{-3}	1.355	0.530	0.448	17.57
2	0.50	3.910×10^{-3}	0.782	0.612	0.224	17.57
4	0.50	7.820×10^{-3}	0.512	0.801	0.112	17.57
8	0.50	15.640×10^{-3}	0.350	1.095	0.056	17.57

7.8 $u_m = 17.6$ in.

Chapter 8

8.2 **(a)** $\tilde{m}\ddot{\theta} + \tilde{c}\dot{\theta} + \tilde{k}\theta = \tilde{p}(t)$, $\tilde{m} = \dfrac{103mL^2}{64}$, $\tilde{c} = c$, $\tilde{k} = \dfrac{kL^2}{4}$, $\tilde{p}(t) = \dfrac{9L}{8}p(t)$

   (b) $\omega_n\sqrt{16k/103m}$, $\zeta = 8c/\sqrt{103kmL^4}$; **(c)** $u(x,t) = \dfrac{72x}{103mL\omega_D}e^{-\zeta\omega_n t}\sin\omega_D t$

8.4 **(a)** $\tilde{m}\ddot{\theta} + \tilde{c}\dot{\theta} + \tilde{k}\theta = \tilde{p}(t)$; $\tilde{m} = \dfrac{mL^3}{12}$; $\tilde{c} = \dfrac{cL^3}{12}$$\tilde{k} = \dfrac{kL^3}{12}$; $\tilde{p}(t) = \dfrac{L^2}{6}p(t)$

8.5 $z =$ vertical deflection of lower end of spring; $\tilde{m}\ddot{z} + \tilde{c}\dot{z} + \tilde{k}z = \tilde{p}(t)$; $\tilde{m} = \dfrac{m}{3}$;

   $\tilde{c} = \dfrac{c}{4}$; $\tilde{k} = \dfrac{k}{5}$; $\tilde{p}(t) = -\dfrac{2}{5}p(t)$

8.7 **(a)** $\mathcal{V}_o(L/2) = 1426$ kips, $\mathcal{M}_o(L/2) = 0.2399 \times 10^6$ kip-ft, $\mathcal{V}_{bo} = 1739$ kips, $\mathcal{M}_{bo} = 0.7368 \times 10^6$ kip-ft; **(b)** $u_o(L) = 25.1$ in.

8.9 **(a)** $\mathcal{V}_o(L/2) = 216$ kips, $\mathcal{M}_o(L/2) = 3.629 \times 10^4$ kip-ft, $\mathcal{V}_{bo} = 263$ kips, $\mathcal{M}_{bo} = 11.15 \times 10^4$ kip-ft

   (b) $u_o(L) = 3.79$ in.

8.10 $u_{1o} = 0.519$, $u_{2o} = 0.830$, $u_{3o} = 0.934$ in.; $V_{3o} = 41.2$, $V_{2o} = 114.4$, $V_{bo} = 160.2$ kips; $M_{2o} = 494$, $M_{1o} = 1867$, $M_{bo} = 3789$ kip-ft

8.12 $u_{1o} = 0.312$, $u_{2o} = 0.624$, $u_{3o} = 0.935$ in.; $V_{3o} = 48.2$, $V_{2o} = 112.4$, $V_{1o} = 144.5$ kips; $M_{2o} = 578$, $M_{1o} = 1927$, $M_{bo} = 3661$ kip-ft

8.15 Floor displacements (in.): 1.34, 2.39, 3.39, 3.98, 4.28
Story shears (kips): 38.5, 110, 171, 214, 238
Floor overturning moments (kip-ft): 462, 1782, 3833, 6399, 9255

8.17 $\omega_n = 0.726\sqrt{EI/mL^3}$

8.18 **(b)** $\omega_1 = \sqrt{2.536k/m}$, $\psi_1 = \langle 1 \quad 0.366 \rangle^T$; $\omega_2 = \sqrt{9.464k/m}$, $\psi_2 = \langle 1 \quad -1.366 \rangle^T$

8.19 $\omega_n^2 = \dfrac{\pi^4 EI/32L^3}{m\left[1 + (\pi^2/16)(R/L)^2\right]}$ if $\psi(x) = 1 - \cos\dfrac{\pi x}{2L}$

8.20 $\omega_n = \dfrac{1.657}{L^2}\sqrt{\dfrac{EI}{m}}$

8.21 $\omega_n = 20.612$ rad/sec

8.24 $\omega_n = 20.628$ rad/sec

8.25 $u(L/2, t) = \dfrac{2p_o}{\pi m}\dfrac{1}{\omega_n^2}\left[1 - \dfrac{\omega_n^2}{\omega_n^2 - (\pi v/L)^2}\cos\dfrac{\pi v}{L}t + \dfrac{(\pi v/L)^2}{\omega_n^2 - (\pi v/L)^2}\cos\omega_n t\right]$

$$t \le L/v$$

$$u(L/2, t) = \dfrac{2p_o}{\pi m}\dfrac{1}{\omega_n^2}\left\{2 - \left[1 - \dfrac{\omega_n^2}{\omega_n^2 - (\pi v/L)^2}\right]\cos\omega_n(t - t_d)\right.$$

$$\left. + \dfrac{(\pi v/L)^2}{\omega_n^2 - (\pi v/L)^2}\cos\omega_n t\right\} \qquad t \ge L/v$$

Chapter 9

9.2 $\dfrac{mL}{3}\begin{bmatrix} 1 & 0 \\ 0 & 1 \end{bmatrix}\begin{Bmatrix} \ddot{u}_1 \\ \ddot{u}_2 \end{Bmatrix} + \dfrac{162EI}{5L^3}\begin{bmatrix} 8 & -7 \\ -7 & 8 \end{bmatrix}\begin{Bmatrix} u_1 \\ u_2 \end{Bmatrix} = \begin{Bmatrix} p_1(t) \\ p_2(t) \end{Bmatrix}$

9.4 $\mathbf{m} = \dfrac{m}{6}\begin{bmatrix} 2 & 1 \\ 1 & 2 \end{bmatrix}, \mathbf{k} = \dfrac{EI}{L^3}\begin{bmatrix} 28 & -10 \\ -10 & 4 \end{bmatrix}$

9.6 $m\begin{bmatrix} 1 & \\ & 0.5 \end{bmatrix}\begin{Bmatrix} \ddot{u}_1 \\ \ddot{u}_2 \end{Bmatrix} + \dfrac{EI}{h^3}\begin{bmatrix} 37.15 & -15.12 \\ -15.12 & 10.19 \end{bmatrix}\begin{Bmatrix} u_1 \\ u_2 \end{Bmatrix} = \begin{Bmatrix} p_1(t) \\ p_2(t) \end{Bmatrix}$

9.9 $m\begin{bmatrix} 1 & & \\ & 1 & \\ & & 0.5 \end{bmatrix}\begin{Bmatrix} \ddot{u}_1 \\ \ddot{u}_2 \\ \ddot{u}_3 \end{Bmatrix} + \dfrac{EI}{h^3}\begin{bmatrix} 40.85 & -23.26 & 5.11 \\ -23.26 & 31.09 & -14.25 \\ 5.11 & -14.25 & 10.06 \end{bmatrix}\begin{Bmatrix} u_1 \\ u_2 \\ u_3 \end{Bmatrix} = \begin{Bmatrix} p_1(t) \\ p_2(t) \\ p_3(t) \end{Bmatrix}$

9.11 $m\begin{bmatrix} 1 & & \\ & 1 & \\ & & 0.5 \end{bmatrix}\begin{Bmatrix} \ddot{u}_1 \\ \ddot{u}_2 \\ \ddot{u}_3 \end{Bmatrix} + \dfrac{EI}{h^3}\begin{bmatrix} 32.67 & -14.58 & 2.17 \\ & 15.81 & -5.60 \\ (\text{sym}) & & 3.85 \end{bmatrix}\begin{Bmatrix} u_1 \\ u_2 \\ u_3 \end{Bmatrix} = \begin{Bmatrix} p_1(t) \\ p_2(t) \\ p_3(t) \end{Bmatrix}$

9.13 $\mathbf{u} = \langle u_1 \quad u_2 \quad u_3 \rangle$, where u_1 is the horizontal displacement of the masses and u_2 and u_3 are the vertical displacements of the right and left masses, respectively.

$$\mathbf{m}\ddot{\mathbf{u}} + \mathbf{k}\mathbf{u} = \mathbf{p}_{\text{eff}}(t)$$

$$\mathbf{m} = m\begin{bmatrix} 5 & & \\ & 1 & \\ & & 1 \end{bmatrix}, \mathbf{k} = \dfrac{3EI}{10L^3}\begin{bmatrix} 28 & 6 & -6 \\ 6 & 7 & 3 \\ -6 & 3 & 7 \end{bmatrix}; \mathbf{p}_{\text{eff}}(t) = -\mathbf{m}\boldsymbol{\iota}\ddot{u}_g(t)$$

$$\boldsymbol{\iota}^T = \begin{cases} \langle 1 \quad 0 \quad 0 \rangle & \text{for } \ddot{u}_g(t) = \ddot{u}_{gx}(t) \\ \langle 0 \quad 1 \quad 1 \rangle & \text{for } \ddot{u}_g(t) = \ddot{u}_{gy}(t) \\ \langle \frac{1}{\sqrt{2}} \quad \frac{1}{\sqrt{2}} \quad \frac{1}{\sqrt{2}} \rangle & \text{for } \ddot{u}_g(t) = \ddot{u}_{gbd}(t) \\ \langle -\frac{1}{\sqrt{2}} \quad \frac{1}{\sqrt{2}} \quad \frac{1}{\sqrt{2}} \rangle & \text{for } \ddot{u}_g(t) = \ddot{u}_{gbc}(t) \\ \langle -L \quad L \quad -L \rangle & \text{for } \ddot{u}_g(t) = \ddot{u}_{g\theta}(t) \end{cases}$$

9.15 $\mathbf{m}\ddot{\mathbf{u}} + \mathbf{k}\mathbf{u} = \mathbf{p}_{\text{eff}}(t)$

$$\mathbf{m} = m\begin{bmatrix} 2/3 & -1/6 & 1/2 \\ -1/6 & 2/3 & -1/2 \\ 1/2 & -1/2 & 1 \end{bmatrix}, \mathbf{k} = k\begin{bmatrix} 5 & -2 & 2 \\ -2 & 5 & -2 \\ 2 & -2 & 6 \end{bmatrix}, \mathbf{p}_{\text{eff}}(t) = -\mathbf{m}\boldsymbol{\iota}\ddot{u}_g(t)$$

$$[\mathbf{m}\boldsymbol{\iota}]^T = \begin{cases} m\langle 1/2 & 1/2 & 0 \rangle & \text{for ground motion in the } x \text{ direction} \\ m\langle 1/2 & -1/2 & 1 \rangle & \text{for ground motion in the } y \text{ direction} \\ m\langle 1/\sqrt{2} & 0 & 1/\sqrt{2} \rangle & \text{for ground motion in the } d\text{--}b \text{ direction} \end{cases}$$

9.16 $\mathbf{m}\ddot{\mathbf{u}} + \mathbf{k}\mathbf{u} = \mathbf{p}_{\text{eff}}(t)$

$$\mathbf{m} = m\begin{bmatrix} 1 & 0 & 0 \\ 0 & 1 & b/2 \\ 0 & b/2 & 5b^2/12 \end{bmatrix}, \mathbf{k} = k\begin{bmatrix} 6 & 0 & 0 \\ 0 & 6 & 2b \\ 0 & 2b & 7b^2/2 \end{bmatrix}, \mathbf{p}_{\text{eff}}(t) = -\mathbf{m}\boldsymbol{\iota}\ddot{u}_g(t)$$

$$[\mathbf{m}\boldsymbol{\iota}]^T = \begin{cases} m\langle 1 & 0 & 0 \rangle & \text{for ground motion in the } x \text{ direction} \\ m\langle 0 & 1 & b/2 \rangle & \text{for ground motion in the } y \text{ direction} \\ m\langle 1/\sqrt{2} & 1/\sqrt{2} & b/2\sqrt{2} \rangle & \text{for ground motion in the } d\text{--}b \\ & & \text{direction} \end{cases}$$

9.18 $\mathbf{m\ddot{u}} + \mathbf{ku} = -\mathbf{m}\boldsymbol{\iota}\ddot{u}_g(t)$

$$\mathbf{m} = m\begin{bmatrix} 1 & & \\ & 1 & \\ & & 1 \end{bmatrix},\ \mathbf{k} = \frac{EI}{L^3}\begin{bmatrix} 0.9283 & 0.9088 & 0.2345 \\ & 1.4294 & 0.2985 \\ \text{symm} & & 0.3234 \end{bmatrix}$$

$$\boldsymbol{\iota}^T = \begin{cases} \langle 1 & 0 & 0 \rangle & \text{for ground motion in the } x \text{ direction} \\ \langle 0 & 1 & 0 \rangle & \text{for ground motion in the } y \text{ direction} \\ \langle 0 & 0 & 1 \rangle & \text{for ground motion in the } z \text{ direction} \\ (1/\sqrt{3})\langle 1 & 1 & 1 \rangle & \text{for ground motion in the } a\text{–}d \text{ direction} \end{cases}$$

9.20 $m\ddot{u} + \dfrac{6EI}{L^3}u = -m\,\langle 1/2 \quad 1/2 \rangle \begin{Bmatrix} \ddot{u}_{g1}(t) \\ \ddot{u}_{g2}(t) \end{Bmatrix}$

9.21 $\mathbf{m\ddot{u}} + \mathbf{ku} = -\mathbf{m}\boldsymbol{\iota}\ddot{u}_g(t)$

$$\mathbf{u} = \begin{Bmatrix} u_1 \\ u_2 \end{Bmatrix},\ \mathbf{u}_g = \begin{Bmatrix} u_{g1} \\ u_{g2} \end{Bmatrix}$$

$$\mathbf{m} = m\begin{bmatrix} 1 & \\ & 1 \end{bmatrix},\ \mathbf{k} = \frac{6EI}{7L^3}\begin{bmatrix} 176 & -48 \\ -48 & 176 \end{bmatrix},\ \boldsymbol{\iota} = \frac{1}{448}\begin{bmatrix} 266 & 182 \\ 42 & -42 \end{bmatrix}$$

9.22 $\mathbf{m\ddot{u}} + \mathbf{ku} = -\mathbf{m}\boldsymbol{\iota}\ddot{u}_g(t)$

$$\mathbf{u} = \begin{Bmatrix} u_1 \\ u_2 \end{Bmatrix},\ \mathbf{u}_g = \begin{Bmatrix} u_{g1} \\ u_{g2} \end{Bmatrix}$$

$$\mathbf{m} = m\begin{bmatrix} 1 & \\ & 1/2 \end{bmatrix},\ \mathbf{k} = \frac{6EI}{7L^3}\begin{bmatrix} 128 & 0 \\ 0 & 140 \end{bmatrix},\ \boldsymbol{\iota} = \begin{bmatrix} 0,5 & 0.5 \\ 0.3 & -0.3 \end{bmatrix}$$

9.23 $\mathbf{m\ddot{u}} + \mathbf{ku} = -\mathbf{m}\boldsymbol{\iota}\ddot{u}_g(t)$

$$\mathbf{u}^T = \langle u_x \quad u_y \quad u_\theta \rangle \qquad \mathbf{u}_g^T = \langle u_{ga} \quad u_{gb} \quad u_{gc} \quad u_{gd} \rangle^T$$

$$\mathbf{m} = m\begin{bmatrix} 1 & & \\ & 1 & \\ & & b^2/6 \end{bmatrix} \qquad \mathbf{k} = k\begin{bmatrix} 4 & 0 & 0 \\ 0 & 4 & 0 \\ 0 & 0 & 2b^2 \end{bmatrix}$$

$$\boldsymbol{\iota} = \frac{1}{4}\begin{bmatrix} 1 & 1 & 1 & 1 \\ 0 & 0 & 0 & 0 \\ -1/b & -1/b & 1/b & 1/b \end{bmatrix}$$

9.25 $\mathbf{m\ddot{u}} + \mathbf{ku} = -\mathbf{m}\boldsymbol{\iota}\ddot{u}_g(t)$

$$\mathbf{u} = \begin{Bmatrix} u_1 \\ u_2 \end{Bmatrix},\ \mathbf{u}_g = \begin{Bmatrix} u_{g1} \\ u_{g2} \end{Bmatrix}$$

$$\mathbf{m} = \begin{bmatrix} 3.624 & \\ & 1.812 \end{bmatrix},\ \mathbf{k} = 10^4\begin{bmatrix} 0.9359 & 0.7701 \\ 0.7701 & 1.5088 \end{bmatrix},\ \boldsymbol{\iota} = \begin{bmatrix} 0.6035 & 0.3965 \\ -0.2143 & 1.2143 \end{bmatrix}$$

Chapter 10

10.3   **(a)** $u_1(t) = 0.5\cos\omega_1 t + 0.5\cos\omega_2 t;\ u_2(t) = 0.5\cos\omega_1 t - 0.5\cos\omega_2 t$

10.6   **(a)** $\omega_1 = 3.750\sqrt{EI/mh^3},\ \omega_2 = 9.052\sqrt{EI/mh^3},$
$\phi_1 = \langle 1 \quad \sqrt{2} \rangle^T,\ \phi_2 = \langle 1 \quad -\sqrt{2} \rangle^T$

10.7 $\omega_1 = 1.971\sqrt{EI/mh^3}, \omega_2 = 8.609\sqrt{EI/mh^3}, \phi_1 = \langle 0.919 \quad 1 \rangle^T,$
$\phi_2 = \langle -0.544 \quad 1 \rangle^T$

10.8 **(a)** $u_1(t) = 1.207\cos\omega_1 t - 0.207\cos\omega_2 t, u_2(t) = 1.707\cos\omega_1 t + 0.293\cos\omega_2 t$

10.10 $\omega_1 = 2.407\sqrt{EI/mh^3}, \omega_2 = 7.193\sqrt{EI/mh^3}$
$\phi_1 = \langle 0.482 \quad 1 \quad -0.490/h \quad -0.490/h \quad -0.304/h \quad -0.304/h \rangle^T$
$\phi_2 = \langle -1.037 \quad 1 \quad -0.241/h \quad -0.241/h \quad -1.677/h \quad -1.677/h \rangle^T$

10.12 $\omega_n = \alpha_n\sqrt{EI/mh^3}, \alpha_1 = 2.241, \alpha_2 = 4.899, \alpha_3 = 7.14; \phi_1 = \langle 0.314 \quad 0.686 \quad 1 \rangle$
$\phi_2 = \langle -0.5 \quad -0.5 \quad 1 \rangle^T, \phi_3 = \langle 3.186 \quad -2.186 \quad 1 \rangle^T$

10.14 $\omega_n = \alpha_n\sqrt{EI/mh^3}, \alpha_1 = 1.423, \alpha_2 = 4.257, \alpha_3 = 6.469; \phi_1 = \langle 0.7 \quad 0.873 \quad 1 \rangle^T,$
$\phi_2 = \langle -0.549 \quad -0.133 \quad 1 \rangle^T, \phi_3 = \langle 1.301 \quad -1.614 \quad 1 \rangle^T$

10.15 **(a)** $\begin{Bmatrix} u_1(t) \\ u_2(t) \\ u_3(t) \end{Bmatrix} = \begin{Bmatrix} 1.2440 \\ 2.1547 \\ 2.4880 \end{Bmatrix}\cos\omega_1 t + \begin{Bmatrix} -0.3333 \\ 0 \\ 0.3333 \end{Bmatrix}\cos\omega_2 t + \begin{Bmatrix} 0.0893 \\ -0.1547 \\ 0.1786 \end{Bmatrix}\cos\omega_3 t$

10.16 **(a)** $\begin{Bmatrix} u_1(t) \\ u_2(t) \\ u_3(t) \end{Bmatrix} = \begin{Bmatrix} 0.935 \\ 2.04 \\ 2.98 \end{Bmatrix}\cos\omega_1 t + \begin{Bmatrix} 0.065 \\ -0.0448 \\ 0.0205 \end{Bmatrix}\cos\omega_3 t$

10.23 **(a)** $\omega_n = \alpha_n\sqrt{EI/mL^3}; \alpha_n = 0.5259, 1.6135, 1.7321;$

$$\phi_1 = \begin{Bmatrix} 1 \\ -1.9492 \\ 1.9492 \end{Bmatrix}, \phi_2 = \begin{Bmatrix} 1 \\ 1.2826 \\ -1.2826 \end{Bmatrix}, \phi_3 = \begin{Bmatrix} 0 \\ 1 \\ 1 \end{Bmatrix}$$

(b) $\begin{Bmatrix} u_1(t) \\ u_2(t) \\ u_3(t) \end{Bmatrix} = \begin{Bmatrix} 0.3969 \\ -0.7736 \\ 0.7736 \end{Bmatrix}\cos\omega_1 t + \begin{Bmatrix} 0.6031 \\ 0.7736 \\ -0.7736 \end{Bmatrix}\cos\omega_2 t$

10.24 $\omega_n = 5.96, 6.21, 10.90$ rad/sec
$\phi_1 = \langle 0 \quad 2.0322 \quad 0.0033 \rangle^T, \phi_2 = \langle 2.071 \quad 0 \quad 0 \rangle^T,$
$\phi_3 = \langle 0 \quad -0.3988 \quad 0.0166 \rangle^T$

10.28 **(a)** $\omega_n = \alpha_n\sqrt{EI/mL^3}, \alpha_1 = 0.4834, \alpha_2 = 0.4990, \alpha_3 = 1.4827$

10.29–10.32 Compare against exact results.

Chapter 11

11.1 $\mathbf{c} = \begin{bmatrix} 0.824 & -0.257 & 0 \\ (\text{sym}) & 0.604 & -0.110 \\ & & 0.229 \end{bmatrix}, \zeta_2 = 0.0430$

11.2, 11.4 $\mathbf{c} = \begin{bmatrix} 0.848 & -0.234 & -0.023 \\ (\text{sym}) & 0.628 & -0.133 \\ & & 0.252 \end{bmatrix}$

Chapter 12

12.1 $u_1(t) = (p_o/k)(1.207C_1 - 0.207C_2)\sin \omega t$,

$u_2(t) = (p_o/k)(1.707C_1 + 0.293C_2)\sin \omega t$, where $C_n = \left[1 - (\omega/\omega_n)^2\right]^{-1}$

12.4 $u_1(t) = \dfrac{p_o}{2m}\left(\dfrac{\sin \omega_1 t}{\omega_1} + \dfrac{\sin \omega_2 t}{\omega_2}\right)$, $u_2(t) = \dfrac{0.707\,p_o}{m}\left(\dfrac{\sin \omega_1 t}{\omega_1} - \dfrac{\sin \omega_2 t}{\omega_2}\right)$

12.5 **(a)** $u_1(t) = \dfrac{p_o}{k}(1 - 0.853\cos \omega_1 t - 0.147\cos \omega_2 t)$, $u_2(t) =$

$\dfrac{p_o}{k}(1 - 1.207\cos \omega_1 t + 0.207\cos \omega_2 t)$

12.9 $u_{3o} = \omega^2 p_o\sqrt{\left(\dfrac{C_1}{K_1} + \dfrac{C_2}{K_2} + \dfrac{C_3}{K_3}\right)^2 + \left(\dfrac{D_1}{K_1} + \dfrac{D_2}{K_2} + \dfrac{D_3}{K_3}\right)^2}$, $\ddot{u}_{3o} = \omega^2 u_{3o}$,

where $p_o = 1.242$ lb; $K_1 = 131.16$, $K_2 = 978.97$, and

$K_3 = 1826.80$ kips/in.; and C_n and D_n are as defined in Example 12.5.

12.16

$\begin{Bmatrix} u_1(t) \\ u_2(t) \end{Bmatrix} = \begin{Bmatrix} 1 \\ 1 \end{Bmatrix} 1.736(1 - \cos 10t) + \begin{Bmatrix} 1 \\ -1 \end{Bmatrix}(-0.116)(1 - \cos 38.73t); \quad t \leq 0.3\ \text{sec}$

$\begin{Bmatrix} u_1(t) \\ u_2(t) \end{Bmatrix} = \begin{Bmatrix} 1 \\ 1 \end{Bmatrix} [3.455\cos 10(t - 0.3) + 0.245\sin 10(t - 0.3)]$

$+ \begin{Bmatrix} 1 \\ -1 \end{Bmatrix} [-0.0483\cos 38.73(t - 0.3) + 0.09419\sin 38.73(t - 0.3)]; \quad t \geq 0.3\ \text{sec}$

12.18 Combined values of shears (in kips) and bending moments (in kip-in.):

$$V_a = V_b = 52.08; \quad V_c = V_d = -58.16; \quad V_e = V_f = 6.12$$

$$M_a = 0; \quad M_b = M_c = -2604; \quad M_d = M_e = 306; \quad M_f = 0$$

12.19 $\mathbf{u}_o = \begin{Bmatrix} 0.133 \\ 0.529 \end{Bmatrix}$ in., $\ddot{\mathbf{u}}_o = \begin{Bmatrix} 83.125 \\ 330.63 \end{Bmatrix}$ in./sec^2

12.20 $M_{bo} = 948.8$ kip-in., $M_{do} = -1902$ kip-in.

12.24 **(a)** $M(t) = \sum M^{\text{st}}\bar{M}_n\left[\omega_n^2 D_n(t)\right]$, where

$$D_n(t) = \dfrac{p_o}{\omega_n^2}\dfrac{1}{1 - (T_n/2t_d)^2}\left[\sin\left(\pi\dfrac{t}{t_d}\right) - \dfrac{T_n}{2t_d}\sin\left(2\pi\dfrac{t}{T_n}\right)\right]; \quad t \leq t_d$$

$$D_n(t) = D_n(t_d)\cos \omega_n(t - t_d) + \dfrac{\dot{D}_n(t_d)}{\omega_n}\sin \omega_n(t - t_d); \quad t \geq t_d$$

$$M^{\text{st}} = -0.3125L; \quad \bar{M}_1 = 0.3414, \quad \bar{M}_2 = 0.6000, \quad \bar{M}_3 = 0.0586$$

(b) $M(t) = M^{\text{st}}\left\{p(t) + \bar{M}_1\left[\omega_1^2 D_1(t) - p(t)\right]\right\}$

Chapter 13

13.1 (a) $s_1 = m \langle 0.854 \quad 0.604 \rangle^T$, $s_2 = m \langle 0.146 \quad -0.104 \rangle^T$

(b) $u_1(t) = 0.854 D_1(t) + 0.146 D_2(t)$, $u_2(t) = 1.207 D_1(t) - 0.207 D_2(t)$

(c) $V_1(t) = 1.458 m A_1(t) + 0.042 m A_2(t)$, $V_2(t) = 0.604 m A_1(t) - 0.104 m A_2(t)$

(d) $M_b(t) = 2.062 mh A_1(t) - 0.062 mh A_2(t)$, $M_1(t) = 0.604 mh A_1(t) - 0.104 mh A_2(t)$

13.2 (c) Peak values of total responses: $u_1 = 0.679$ in., $u_2 = 0.964$ in., $V_b = 115.11$ kips, $V_2 = 49.56$ kips, $M_b = 1959.25$ kip-ft, $M_2 = 594.65$ kip-ft

13.4 (a) Floor displacements: $u_1(t) = 0.647 D_1(t) + 0.353 D_2(t)$,
$u_2 = 1.341 D_1(t) - 0.341 D_2(t)$
Joint rotations at first and second floors: $u_3(t) = (1/h)[-0.657 D_1(t) + 0.082 D_2(t)]$, $u_5(t) = (1/h)[-0.407 D_1(t) + 0.572 D_2(t)]$

(b) Bending moments in first-story column:
Top: $M_a = mh[0.216 A_1(t) + 0.0473 A_2(t)]$
Bottom: $M_b = mh[0.443 A_1(t) + 0.0441 A_2(t)]$
Bending moment at ends of second-floor beam:
$M_a = M_b = mh[-0.211 A_1(t) + 0.0332 A_2(t)]$

13.5 (a) $s_1 = m \langle 0.622 \quad 1.077 \quad 0.622 \rangle^T$, $s_2 = m \langle 0.333 \quad 0 \quad -0.167 \rangle^T$,
$s_3 = m \langle 0.045 \quad -0.077 \quad 0.045 \rangle^T$

(b) $u_1(t) = 0.622 D_1(t) + 0.333 D_2(t) + 0.045 D_3(t)$,
$u_2(t) = 1.077 D_1(t) - 0.077 D_3(t)$,
$u_3(t) = 1.244 D_1(t) - 0.333 D_2(t) + 0.089 D_3(t)$

(c) $V_1(t) = 2.321 m A_1(t) + 0.167 m A_2(t) + 0.012 m A_3(t)$,
$V_2(t) = 1.699 m A_1(t) - 0.167 m A_2(t) - 0.0326 m A_3(t)$,
$V_3(t) = 0.622 m A_1(t) - 0.167 m A_2(t) + 0.045 m A_3(t)$

(d) $M_b(t) = mh[4.642 A_1(t) - 0.167 A_2(t) + 0.024 A_3(t)]$

13.7 Peak values of total responses: $u_1 = 0.580$ in., $u_2 = 0.957$ in., $u_3 = 1.103$ in., $V_b = 189.3$ kips, $V_2 = 138.1$ kips, $V_3 = 52.2$ kips, $M_b = 4321$ kip-ft, $M_2 = 2268$ kip-ft, $M_1 = 627$ kip-ft

13.9 $M_1^* = 2.3213 m$, $M_2^* = 0.1667 m$, $M_3^* = 0.0121 m$
$h_1^* = 2h$, $h_2^* = -h$, $h_3^* = 2h$

13.11 (a) Floor displacements: $u_1(t) = 0.427 D_1(t) + 0.377 D_2(t) + 0.197 D_3(t)$,
$u_2(t) = 1.007 D_1(t) + 0.182 D_2(t) - 0.189 D_3(t)$,
$u_3(t) = 1.352 D_1(t) - 0.508 D_2(t) + 0.157 D_3(t)$

(b) Bending moments in first-story column:
$M_{top} = mh[0.301 A_1(t) + 0.068 A_2(t) + 0.023 A_3(t)]$,
$M_{base} = mh[0.753 A_1(t) + 0.084 A_2(t) + 0.020 A_3(t)]$
Bending moments at ends of second-floor beams:
$M_a = M_b = mh[-0.541 A_1(t) + 0.057 A_2(t) + 0.003 A_3(t)]$

13.15 (a) $s_1 = m \langle 0.849 \quad 0.594 \rangle^T$, $s_2 = m \langle -0.849 \quad 0.406 \rangle^T$

(b) $u_1(t) = 0.283D_1(t) - 0.283D_2(t)$, $u_2(t) = 0.594D_1(t) + 0.406D_2(t)$
(c) $M_b(t) = 1.443mLA_1(t) - 0.443mLA_2(t)$

13.17 (a) $s_1 = m\langle 1.985 \quad -0.774 \quad 0.774\rangle^T$, $s_2 = m\langle 3.015 \quad 0.774 \quad -0.774\rangle^T$,
$s_3 = \langle 0 \quad 0 \quad 0\rangle^T$
(b) $u_1(t) = 0.397D_1(t) + 0.603D_2(t)$, $u_2(t) = -0.774D_1(t) + 0.774D_2(t)$,
$u_3(t) = 0.774D_1(t) - 0.774D_2(t)$
(c) $M_b(t) = 3.533mLA_1(t) + 1.467mLA_2(t)$, $M_a(t) = -0.774mLA_1(t) +$
$0.774mLA_2(t)$

13.23 (c) Peak responses: $u_{3o} = 44.58$ in., $V_{ao} = 0.879$ kip, $V_{bo} = 159.83$ kips
(d) Seismic coefficients: 4.69 and 0.426 for appendage and for tower, respectively

13.25 (a) $s_1 = \langle 0 \quad 0.1588 \quad 3.816\rangle^T$, $s_2 = \langle 0.1649 \quad 0 \quad 0\rangle^T$,
$s_3 = \langle 0 \quad 0.0061 \quad -3.816\rangle^T$
(c) $u_x(t) = 0.7071D_2(t)$, $u_y(t) = 0.6809D_1(t) + 0.0262D_3(t)$, $u_\theta(t) =$
$0.0011D_1(t) - 0.0011D_3(t)$
(d) $V_{bx}(t) = 0.1649A_2(t)$, $V_{by}(t) = 0.1588A_1(t) + 0.0061A_3(t)$, $T_b(t) =$
$3.816A_1(t) - 3.816A_3(t)$

13.26 (c) Peak responses: $u_{yo} = 4.182$ in., $(b/2)u_{\theta o} = 1.22$ in., $V_{bo} = 35.6$ kips,
$T_{bo} = 1899$ kip-in.

13.27 (a) $s_1 = m\langle 0.603 \quad -0.382 \quad -0.305\rangle^T$,
$s_2 = m\langle 0.043 \quad -0.081 \quad 0.187\rangle^T$,
$s_3 = m\langle 0.353 \quad 0.463 \quad 0.118\rangle^T$
(b) $u_x(t) = 0.603D_1(t) + 0.043D_2(t) + 0.353D_3(t)$,
$u_y(t) = -0.382D_1(t) - 0.081D_2(t) + 0.463D_3(t)$,
$u_z(t) = -0.305D_1(t) + 0.187D_2(t) + 0.118D_3(t)$
(c) $M_x(t) = mL[-0.077A_1(t) - 0.268A_2(t) + 0.345A_3(t)]$,
$M_y(t) = mL[-0.908A_1(t) + 0.144A_2(t) - 0.235A_3(t)]$,
$T(t) = mL[0.986A_1(t) + 0.124A_2(t) - 0.110A_3(t)]$

13.32 (i) $u(t) = 0.8889u_{go}\sin 6.667t$, $u^t(t) = 1.3889u_{go}\sin 6.667t$,
$M = -(3EI/L^2)u(t)$; **(ii)** $u(t) = 1.7778u_{go}\sin 6.667t$,
$u^t(t) = 2.7778u_{go}\sin 6.667t$, $M = -(3EI/L^2)u(t)$

13.34 (a) $u_1^t(t) = 0.5u_g(t) + 0.5u_g(t - t') + 0.5D_1(t)$
$\qquad + 0.5D_1(t - t') + 0D_2(t) + 0D_2(t - t')$
$u_2^t(t) = 0.3u_g(t) - 0.3u_g(t - t') + 0D_1(t)$
$\qquad + 0D_1(t - t') + 0.30D_2(t) - 0.30D_2(t - t')$
$M_a(t) = (EI/L^2)[-7.2u_g(t) + 7.2u_g(t - t')]$
$\qquad + mL[0.0781A_1(t) + 0.0781A_1(t - t')$
$\qquad + 0.0075A_2(t) - 0.0075A_2(t - t')]$; etc.

13.36 (a) $u^t_y(t) = -0.088u_g(t) + 0.088u_g(t - t') - 0.126D_1(t)$
$\quad\quad\quad + 0.126D_1(t - t') + 0D_2(t) + 0D_2(t - t')$
$\quad\quad\quad + 0.038D_3(t) - 0.038D_3(t - t'); \text{ etc.}$
$\quad\quad V_{ax} = -0.69u_g(t) + 0.69u_g(t - t') + 0.0025A_1(t)$
$\quad\quad\quad + 0.0389A_2(t) + 0.0057A_3(t) - 0.0025A_1(t - t')$
$\quad\quad\quad + 0.0389A_2(t - t') - 0.0057A_3(t - t');$
$\quad\quad V_{ay} = 0.5454u_g(t) - 0.5454u_g(t - t') - 0.0081A_1(t)$
$\quad\quad\quad + 0.0066A_3(t) + 0.0081A_1(t - t') - 0.0066A_3(t - t'); \text{ etc.}$

13.38 (a) $u^t_2(t) = -0.2143u_g(t) + 1.2143u_g(t - t') - 0.3416D_1(t)$
$\quad\quad\quad -0.0150D_1(t - t') + 0.1273D_2(t) + 1.2294D_2(t - t')$
     (b) $u^t_2(t) = u_g(t) - 0.3567D_1(t) + 1.3566D_2(t)$

13.39 (a) $u_1 = 1.43$ in., $u_2 = 2.96$ in.
     (b) First-story column: $M_{top} = 240$, $M_{base} = 483$ kip-ft
        Second-floor beam: $M_{left} = M_{right} = 232$ kip-ft

13.40 $u_1 = 0.680$ in., $u_2 = 0.962$ in.
$\quad\quad V_b = 115.22$ kips, $V_2 = 48.24$ kips,
$\quad\quad M_b = 1955.43$ kip-ft, $M_1 = 578.87$ kip-ft

13.41 (a) $u_1 = 1.340$ in., $u_2 = 3.152$ in., $u_3 = 4.233$ in.
     (b) First-story column: $M_{top} = 336$ kip-ft, $M_{base} = 821$ kip-ft
        Second-floor beam: $M_{left} = M_{right} = 590$ kip-ft

13.47 $u_1 = 1.207$ in., $u_2 = 2.484$ in., $M_b = 188.2$ kip-in.
13.50 (a) $u_1 = 1.57$ in., $u_2 = 3.01$ in., $u_3 = 3.01$ in.
     (b) $M_b = 21.25$ kip-ft, $M_a = 7.38$ kip-ft

13.54

	u_3 (in.)	V_a (kips)	V_b (kips)
SRSS	165.0	3.158	144.3
CQC	77.27	1.465	182.3

13.58

	u_x (in.)	u_y (in.)	$(b/2)u_\theta$ (in.)	V_x (kips)	V_y (kips)	T (kip-in.)
CQC	6.286	6.309	1.662	56.6	52.4	2331.4
SRSS	6.286	6.306	1.678	56.6	52.3	2357.8

Bending moments (kip-in.):

	M_{ay}	M_{ax}	M_{by}	M_{bx}	M_{cy}	M_{cx}	M_{dy}	M_{dx}
CQC	1108	1055	554.0	845.5	824.2	845.5	1648	1055
SRSS	1405	1050	702.7	847.1	702.7	847.1	1405	1050

13.59

Displacement	SRSS Rule (in.)	CQC Rule (in.)
u_x	0.724	0.767
u_y	0.470	0.544
u_z	0.421	0.195

Response	SRSS Rule (kip-in.)	CQC Rule (kip-in.)
M_x	8.66	9.42
M_y	18.52	15.92
T	19.50	21.55

13.63 $M_\alpha = 11.21$ kip-in. (SRSS rule) and 13.52 kip-in. (CQC rule)

13.67 (a) $M_\alpha = (A \sin^2 \alpha + B \sin \alpha \cos \alpha + C \cos^2 \alpha)^{1/2}$
where $A = 343.1$, $B = -37.70$, $C = 75.00$, all in kip-in. (SRSS rule);
$A = 253.6$, $B = 122.3$, $C = 88.75$ (CQC rule)

(b) Maximum value of $M_\alpha = 18.56$ kip-in. at $\alpha = 86.0°$ (SRSS rule); $M_\alpha = 16.55$ kip-in. at $\alpha = 71.7°$ (CQC rule)

Chapter 14

14.2 $\omega_1 = 0.7709\sqrt{\dfrac{k}{m}}$ $\qquad\qquad$ $\omega_2 = 1.8346\sqrt{\dfrac{k}{m}}$

$\psi_1 = \begin{Bmatrix} 0.7132 - 0.0775\,i \\ 1.0000 \end{Bmatrix}$ $\qquad$ $\psi_2 = \begin{Bmatrix} -0.6732 - 0.1806\,i \\ 1.0000 \end{Bmatrix}$

$\zeta_1 = 0.1316$ $\qquad\qquad\qquad$ $\zeta_2 = 0.0537$

14.4 $\mathbf{u}(t) = e^{-0.1316\,\omega_1 t}\left[\begin{Bmatrix} -0.1324 \\ -0.2294 \end{Bmatrix} \cos \omega_{1D}t - \begin{Bmatrix} 0.3046 \\ 0.4022 \end{Bmatrix} \sin \omega_{1D}t\right]$

$+ e^{-0.0537\,\omega_2 t}\left[\begin{Bmatrix} -0.8676 \\ 1.2294 \end{Bmatrix} \cos \omega_{2D}t + \begin{Bmatrix} 0.0731 \\ 0.2212 \end{Bmatrix} \sin \omega_{2D}t\right]$

14.6 $\mathbf{u}(t) = \begin{Bmatrix} 0.8681 \\ 1.2287 \end{Bmatrix} h_1(t) + \begin{Bmatrix} -0.0714 \\ 0.0736 \end{Bmatrix}\sqrt{\dfrac{m}{k}}\,\dot{h}_1(t) + \begin{Bmatrix} 0.1315 \\ -0.2283 \end{Bmatrix} h_2(t)$

$+ \begin{Bmatrix} 0.0714 \\ -0.0736 \end{Bmatrix}\sqrt{\dfrac{m}{k}}\,\dot{h}_2(t)$

where $h_n(t)$ and $\dot{h}_n(t)$ are given by Eq. (14.8.4) and (14.8.5).

14.8 $\mathbf{u}(t) = \begin{Bmatrix} 0.8681 \\ 1.2287 \end{Bmatrix} D_1(t) + \begin{Bmatrix} -0.0714 \\ 0.0736 \end{Bmatrix}\sqrt{\dfrac{m}{k}}\,\dot{D}_1(t) + \begin{Bmatrix} 0.1315 \\ -0.2283 \end{Bmatrix} D_2(t)$

$+ \begin{Bmatrix} 0.0714 \\ -0.0736 \end{Bmatrix}\sqrt{\dfrac{m}{k}}\,\dot{D}_2(t)$

14.9 $\omega_1 = 0.7743\sqrt{\dfrac{k}{m}}$ $\omega_2 = 1.8265\sqrt{\dfrac{k}{m}}$

$\zeta_1 = 0.2553$ $\zeta_2 = 1.0415$

$\psi_1 = \left\{ \begin{array}{c} 0.7611 + 0.0873\,i \\ 1.0000 \end{array} \right\}$ $\psi_2 = \left\{ \begin{array}{c} -0.4568 \\ 1.0000 \end{array} \right\}$ $\psi_3 = \left\{ \begin{array}{c} -0.5421 \\ 1.0000 \end{array} \right\}$

14.10 $\mathbf{u}(t) = \mathbf{u}_1(t) + \mathbf{u}_2(t)$

$$\mathbf{u}_1(t) = e^{-0.2553\,\omega_1 t}\left[\left\{ \begin{array}{c} 1.1901 \\ 1.4919 \end{array} \right\} \cos\omega_{1D}t + \left\{ \begin{array}{c} 0.3454 \\ 0.6250 \end{array} \right\} \sin\omega_{1D}t \right]$$

$$\mathbf{u}_2(t) = e^{-1.0415\,\omega_2 t}\left[\left\{ \begin{array}{c} -0.1901 \\ 0.5081 \end{array} \right\} \cosh\omega_{2D}t - \left\{ \begin{array}{c} 0.7240 \\ -1.4928 \end{array} \right\} \sinh\omega_{2D}t \right]$$

14.12 $\mathbf{u}(t) = \left\{ \begin{array}{c} 0.9178 \\ 1.1607 \end{array} \right\} D_1(t) + \left\{ \begin{array}{c} 0.0544 \\ -0.1097 \end{array} \right\} \sqrt{\dfrac{m}{k}}\,\dot{D}_1(t) + \left\{ \begin{array}{c} -0.1034 \\ 0.2133 \end{array} \right\} D_2(t)$

$\qquad\qquad + \left\{ \begin{array}{c} -0.0544 \\ 0.1097 \end{array} \right\} \sqrt{\dfrac{m}{k}}\,\dot{D}_2(t)$

where $D_n(t)$ and $\dot{D}_n(t)$ are defined in Eqs. (14.9.3) and (14.9.4).

Chapter 15

15.3 $\tilde{\omega}_1 = 8.389$, $\tilde{\omega}_2 = 23.59$ rad/sec

$\tilde{\phi}_1 = \langle 0.2319\quad 0.4639\quad 0.6311\quad 0.7983\quad 0.9332 \rangle^T$

$\tilde{\phi}_2 = \langle -0.4366\quad -0.8732\quad -0.3396\quad 0.1940\quad 1.2126 \rangle^T$

15.4 $\tilde{\omega}_1 = 8.263$, $\tilde{\omega}_2 = 23.428$ rad/sec

$\tilde{\phi}_1 = \langle 0.2319\quad 0.4449\quad 0.6753\quad 0.8249\quad 0.9038 \rangle^T$

$\tilde{\phi}_2 = \langle -0.4366\quad -0.4200\quad -0.1042\quad 0.2095\quad 0.4108 \rangle^T$

15.9 Peak values of displacements using:
Two Ritz vectors: 0.0548, 0.213, 0.460, 0.648, 0.757 in.
Two modes: 0.0823, 0.205, 0.430, 0.646, 0.799 in.
Five modes: 0.0599, 0.202, 0.447, 0.6513, 0.778 in.

Chapter 16

16.1–16.4 Check numerical results against the theoretical solution.
16.5 Check numerical results against Table E16.1.
16.6 Check numerical results against Table E16.1 and the theoretical solution.
16.7 Check numerical results against Table E16.2.

16.8 Check numerical results against Table E16.2.

16.10 Check numerical results against Table E16.3.

16.11 Check numerical results against Table E16.4.

Chapter 17

17.2 $\omega_n = \alpha_n \sqrt{EI/mL^4}$; $\alpha_1 = 15.42$, $\alpha_2 = 49.97$, and $\alpha_3 = 104.2$

17.4 $u\left(\dfrac{L}{2}, t\right) = -\dfrac{2WL^3}{\pi^4 EI}\left(\cos\omega_1 t + \dfrac{\cos\omega_3 t}{81} + \dfrac{\cos\omega_5 t}{625} + \dfrac{\cos\omega_7 t}{2401} + \cdots\right)$

17.6 $u\left(\dfrac{L}{4}, t\right) = \dfrac{8pL^4}{\pi^5 EI}\left(\dfrac{1-\cos\omega_2 t}{32} - \dfrac{1-\cos\omega_6 t}{7776} + \dfrac{1-\cos\omega_{10} t}{100,000} - \cdots\right)$

17.7 $u(L/2,\ t) = q_1(t) - q_3(t) + q_5(t) - q_7(t) - q_9(t) - \cdots$

where $q_n(t)$ is given by (valid if $\omega_n \neq n\pi v/L$)

$$q_n(t) = \frac{2p_o}{n\pi m}\frac{1}{\omega_n^2}\left[1 - \frac{\omega_n^2}{\omega_n^2 - (n\pi v/L)^2}\cos\frac{n\pi v}{L}t + \frac{(n\pi v/L)^2}{\omega_n^2 - (n\pi v/L)^2}\cos\omega_n t\right]$$

$$t \leq L/v$$

$$q_n(t) = \frac{2p_o}{n\pi m}\frac{1}{\omega_n^2}\left\{2 - \left[1 + \frac{\omega_n^2}{\omega_n^2 - (n\pi v/L)^2}(-1)^n\right]\cos\omega_n(t-t_d)\right.$$

$$\left. + \frac{(n\pi v/L)^2}{\omega_n^2 - (n\pi v/L)^2}\cos\omega_n t\right\} \qquad t \geq L/v$$

17.11 $u_o(L) = 7.410$ in., $V_{bo} = 1365$ kips, $\mathcal{M}_{bo} = 186{,}503$ kip-ft

Chapter 18

18.2 $\tilde{\omega}_1 = \dfrac{11.765}{L^2}\sqrt{\dfrac{EI}{m_o}}$, $\tilde{\omega}_2 = \dfrac{130.467}{L^2}\sqrt{\dfrac{EI}{m_o}}$

$\tilde{\phi}_1(x) = 1.0\sin(\pi x/L) - 0.0036\sin(3\pi x/L)$
$\tilde{\phi}_2(x) = 0.2790\sin(\pi x/L) + 0.9603\sin(3\pi x/L)$

18.3 (a) $\omega_n = \alpha_n\sqrt{EI/mL^4}$; $\alpha_1 = 9.9086$, $\alpha_2 = 43.818$, $\alpha_3 = 110.14$, $\alpha_4 = 200.80$

(b) $\omega_1 = 9.798\sqrt{EI/mL^4}$

18.5 (a) $\omega_n = \alpha_n\sqrt{EI/mL^4}$; $\alpha_1 = 15.56$, $\alpha_2 = 58.41$, $\alpha_3 = 155.64$

(b) $\omega_1 = 14.81\sqrt{EI/mL^4}$

18.6 (a) $\omega_n = \alpha_n \sqrt{EI/mh^4}$; $\alpha_1 = 1.5354$, $\alpha_2 = 4.0365$, $\alpha_3 = 10.7471$

$\phi_1 = \langle\, 0.5440 \quad -0.5933/h \quad -0.5933/h \,\rangle^T$

$\phi_2 = \langle\, 0 \quad -1/\sqrt{2}h \quad 1/\sqrt{2}h \,\rangle^T$

$\phi_3 = \langle\, -0.0001 \quad 1/\sqrt{2}h \quad 1/\sqrt{2}h \,\rangle^T$

18.7 $\omega_1 = 1.477\sqrt{EI/mh^4}$

$\phi = \langle\, 0.544 \quad -0.594/h \quad -0.594/h \,\rangle^T$

Index